Classical and Quantum Information Theory

Discover the foundations of classical and quantum information theory in the digital age with this modern introductory textbook.

From foundational concepts to modern tools
Familiarise yourself with core topics such as uncertainty, correlation, and entanglement before exploring modern techniques and concepts including tensor networks, quantum circuits and quantum discord.

Deepen your understanding and extend your skills
Over 250 thought-provoking end-of-chapter problems, with solutions for instructors, accompanied by curated further reading, provide opportunities for further study.

Understand how abstract concepts connect to real-world scenarios
Over 400 examples, including numerical and conceptual illustrations, expand on the material in the main text, and emphasise practical applications.

Build confidence with flexible learning pathways
Progressing from foundational classical information measures to advanced quantum counterparts, with alternating chapters that compare the two domains, this textbook is ideal for senior undergraduate and graduate students in electrical engineering, computer science, and applied mathematics seeking a rigorous, self-contained path to the essentials of information theory.

Osvaldo Simeone is Professor of Information Engineering at Northeastern University London. He previously held academic positions at King's College London and the New Jersey Institute of Technology. He is the author of Machine Learning for Engineers (2022), along with several other monographs. Professor Simeone has received numerous best paper awards and is a Fellow of the IEEE, EPSRC, and IET.

"Professor Simeone presents a comprehensive synthesis of classical and quantum information theory, bridging foundational principles with modern perspectives. This book provides students and researchers with the essential foundations for exploring the central role of information theory in signal processing and communications, as well as its emerging quantum dimensions. The presentation is very well organized pedagogically, with clear learning objectives and exposition of needed background materials throughout."

Vincent Poor, Princeton University

"In this most timely book, Osvaldo Simeone masterfully unifies classical and quantum perspectives on information theory, offering clarity and depth that will inspire both students and researchers. This book addressing Classical and Quantum Information Theory, provides a rigorous yet highly accessible guide to a field that underpins engineering, computer science, and the sciences at large."

Shlomo Shamai Shitz, Technion Israel Institute of Technology

"I thought there was no longer any gap left in the literature of information theory for another fundamental book, but here it is, innovatively juxtaposing classical and quantum theory..."

Lajos Hanzo, University of Southampton

Classical and Quantum Information Theory

Uncertainty, Information, and Correlation

Osvaldo Simeone
Northeastern University London

CAMBRIDGE
UNIVERSITY PRESS

CAMBRIDGE
UNIVERSITY PRESS

Shaftesbury Road, Cambridge CB2 8EA, United Kingdom

One Liberty Plaza, 20th Floor, New York, NY 10006, USA

477 Williamstown Road, Port Melbourne, VIC 3207, Australia

314–321, 3rd Floor, Plot 3, Splendor Forum, Jasola District Centre,
New Delhi – 110025, India

103 Penang Road, #05–06/07, Visioncrest Commercial, Singapore 238467

Cambridge University Press is part of Cambridge University Press & Assessment,
a department of the University of Cambridge.

We share the University's mission to contribute to society through the pursuit of
education, learning and research at the highest international levels of excellence.

www.cambridge.org
Information on this title: www.cambridge.org/highereducation/isbn/9781009579520

DOI: 10.1017/9781009579551

When citing this work, please include a reference to the DOI 10.1017/9781009579551

First published 2026

A catalogue record for this publication is available from the British Library

A Cataloging-in-Publication data record for this book is available from the Library of Congress

ISBN 978-1-009-57952-0 Hardback

Additional resources for this publication at www.cambridge.org/simeone-IT

Contents

8 Quantum Relative Entropy

Preface

Uncertainty, Correlation, and Entanglement

Public information, insider information, background information, fake information, classified information, personal information, free information, misinformation, disinformation – information, like energy, is difficult to define yet easy to identify, in different forms, all around us.

Information theory provides a mathematical framework for understanding and quantifying uncertainty, as well as for exploring the relationship between uncertainty and information. In the classical world of macroscopic or virtual things, uncertainty reflects a lack of knowledge about the actual state of a system, and information serves as the means to resolve the uncertainty. We may be uncertain about whether or not our application was approved, about the future movements of the market, or about the existence of life elsewhere in the universe. Information is what resolves the state of uncertainty, and the greater the uncertainty, the more information is needed to resolve it.

The concept of uncertainty is closely related to the idea of disorder in thermodynamics. A more disordered system is one on whose state an observer must have a larger uncertainty. However, while thermodynamic disorder is concerned with the physical properties of a system, determining how much work can be extracted from it, uncertainty is better understood as a feature of the observer's relationship with the system.

Uncertainty can be classified as aleatoric when it refers to inherently unpredictable events, and as epistemic when it involves information that is knowable but not currently known. For instance, the outcome of a coin toss that has not yet occurred is an example of aleatoric uncertainty, while the result of a coin toss that has already happened but is hidden behind a closed door represents epistemic uncertainty. However, neatly distinguishing between aleatoric and epistemic uncertainties is not always possible. As an example, a scientist may be epistemically uncertain about the aleatoric uncertainty associated with some phenomenon under study.

There are various ways to quantify uncertainty, and thus information, depending on the objective of the analysis. If we aim to determine the number of servers needed for an online service, we might focus on the typical traffic load, accepting the possibility of outages during extreme peak periods. In contrast, if we are assessing the security of an authentication system, we must consider worst-case scenarios to prevent breaches, even against the most skilled and well-resourced adversaries. In the first case, uncertainty measures should capture typical outcomes, whereas in the latter, they must account for the most extreme situations.

Different observers can experience varying levels of uncertainty about the same object. For example, two weather forecasters might predict sunny weather, yet assign different levels of confidence to their forecasts. Furthermore, one forecaster may predict an extreme-weather event with low, but positive, probability, while the other may discard this possibility altogether. Comparing different states of uncertainty can help determine the extent to which the perspectives of different observers align or diverge.

While in the classical world uncertainty is tied to a lack of knowledge about an object's actual state, in the microscopic realm of quantum systems a different kind of uncertainty emerges. This new form of uncertainty does not refer to some state pre-existing the measurement; rather, it is introduced by the act of measurement itself. A quantum measurement does not uncover a previously unknown state; instead, it produces a non-deterministic outcome, whose uncertainty is shaped by the specific measurement setting. This way, the relationship between information and the state of an object is inherently ambiguous, as the quantum state encapsulates the uncertainty and the information related to any potential measurement setting.

When a classical composite system consists of multiple components – such as the pages of a book, lines of code, pixels in an image, or coordinates of a trajectory – the uncertainty about one part of the system can be influenced by our knowledge of other parts of the system. For instance, reading the first few chapters of a book or watching the first half of a movie reduces our uncertainty about the rest of the book or movie as compared to knowing only their titles. When information about one part of a system modifies the uncertainty about another part, the two parts are said to be correlated.

Correlations can increase or decrease uncertainty. In particular, observations of part of a system can either augment or reduce our uncertainty about other components of the system. Knowing that a movie is categorized as a comedy might make a happy ending more predictable. However, learning about a recent unexpected event, such as a volcanic eruption, could increase our uncertainty about future developments, such as our travel plans in the affected area.

Correlations in composite quantum systems extend beyond the shared information that characterizes correlations in classical systems. In quantum systems, correlations also include entanglement, a phenomenon with no classical equivalent. Entangled systems behave as if they share an instantaneous – "spooky" in Einstein's famous expression – communication channel. This channel enables classically impossible communication tasks, such as quantum teleportation. (In quantum teleportation, a quantum state is transmitted between two distant parties without the need to physically transfer any quantum particles, utilizing the pre-existing entangled states as a resource.) However, the channel effectively created by entanglement does not allow for faster-than-light communication, in accordance with Einstein's theory of relativity.

Why Read This Book?

Information theory originated at the end of the Second World War through the pioneering work of Claude Shannon, an American mathematician and engineer. Since then, it has driven major advances in digital technologies, including wireless connectivity and artificial intelligence. Beyond technology, information theory also plays a crucial role in the sciences, offering a valuable framework for understanding and explaining physical phenomena across scales – from the microscopic realm of quantum mechanics to the macroscopic domain of thermodynamics. As the language of information continues to gain relevance, especially for engineers and scientists utilizing machine learning and data analytics tools, the significance of information theory is expected to grow even further.

The traditional approach to introducing information theory, adopted in most classical textbooks like the standard reference by T. M. Cover and J. A. Thomas, emphasizes the operational significance of information measures for tasks such as compression and digital communication in the context of an asymptotic framework. Accordingly, these textbooks build on results like the asymptotic equipartition property, which apply to long sequences. This asymptotic perspective is also prevalent in quantum information theory texts, including the excellent books of M. Wilde and J. Watrous.

However, engineers and scientists typically engage with information theory in a different manner, focusing less on the asymptotic interpretation of information measures and more on their properties as quantifiers of uncertainty, information, correlation, and entanglement. Some books, such as the classical text by J. Aczél, develop information measures from axioms that formalize desirable properties. The axiomatic approach, which emphasizes the mathematical consequences of specific assumptions, often falls short in providing practical and intuitive insights for the use of information measures.

The aim of this book is to offer an intuitive yet rigorous introduction to information measures, covering both classical and quantum systems. The book's scope includes static classical and quantum information measures – namely classical and quantum Shannon, Rényi, and Tsallis entropy metrics and their associated relative entropies; the dynamic transformations of classical and quantum systems and their effects on information measures; correlations in classical systems analyzed in terms of shared information and residual uncertainties; classical and quantum correlations in quantum systems, including entanglement and discord; and multipartite classical and quantum correlations explored through the framework of tensor networks.

The text is structured around intuition and examples, progressively guiding the reader through increasingly complex scenarios: from static to dynamic, from single systems to bipartite and multipartite systems, from classical to quantum. The book alternates between sections focused on classical systems and those on quantum systems. Readers interested only in classical systems may concentrate on the chapters devoted to classical information theory, while those interested in quantum systems can proceed through the text sequentially. Each chapter concludes with problems for practice and deeper understanding.

To my knowledge, there are few books that offer a unified treatment of both classical and quantum information theory, and none appear to match the breadth and focus of this text. For instance, E. Desurvire's book covers foundational topics in classical information theory before transitioning to quantum computing topics like quantum error correction, primarily within communication contexts. The classic reference by M. Nielsen and I. Chuang stands in a league of its own, offering a comprehensive introduction to quantum computing, with select chapters addressing information measures and asymptotic results.

How To Use This Book

My intention is for this text to become a valuable resource for instructors aiming to revitalize undergraduate and graduate information theory courses – often challenged by declining enrolment – by incorporating topics with broader appeal and relevance. It is also intended for information theory experts who wish to explore the field of quantum information theory, as well as for electrical and computer engineers, computer scientists, and scientists seeking a comprehensive reference to expand their information theory toolkit.

In line with these intentions, this book does not cover coding theorems, which are well explained in many references, including the standard textbooks by T. M. Cover and J. A. Thomas for classical systems and M. Wilde for quantum systems.

This book is designed to support multiple courses within departments of engineering, computer science, and physics, including:

- An **undergraduate course on classical information theory**, focusing on static and dynamic information measures (covering selected sections of Part I and Chapter 9).
- An **undergraduate course on quantum information theory**, concentrating on static and dynamic quantum information measures (covering selected sections of Part II and Chapters 10 and 11, along with any necessary background review material from Part I).
- A **graduate course on classical information theory**, emphasizing bipartite and multipartite information (covering selected sections of Part IV, along with Chapters 17–19).
- A **graduate course on quantum information theory**, exploring entanglement and discord in both bipartite and multipartite systems (covering selected sections of Part V, along with Chapters 17, 20, and 21).

Throughout the book, definitions are highlighted in light-gray boxes for easy identification, while darker-gray boxes are used to present key results, such as lemmas, propositions, and theorems. Proofs for these results are either provided within the text or referenced in the Recommended Resources. Examples throughout the book include both numerical and conceptual illustrations to clarify and expand on the material discussed in the main text.

Each chapter concludes with a concise, itemized summary, a curated list of Recommended Resources with commentary, a set of problems for practice, and appendices containing proofs and supplementary material.

To assist readers who are new to the topic, sections featuring advanced material that can be skipped on a first read are clearly marked with the fast-forward symbol ⏭.

Organization

Part I: Classical Information

Chapter 1. Uncertainty, Information, and Entropy. This chapter introduces entropy as a fundamental measure of information. It begins by establishing the relationship between information and uncertainty, differentiating between subjective and objective informational states. The chapter provides an intuitive list of properties expected of an entropy measure, and discusses various informational tasks that motivate and justify the derivation of entropy metrics.

Chapter 2. Information Quantification by Asking, Compressing, or Sampling: Shannon Entropy. This chapter examines how uncertainty can be quantified using resources required for tasks such as asking questions, compressing data, and sampling. In each case, the same foundational concept – the Shannon entropy – emerges as a consistent measure of uncertainty, expressed in binary units or bits.

Chapter 3. Information Quantification by Predicting or Guessing: Tsallis Entropy and Rényi Entropy. This chapter introduces two alternative measures of uncertainty: the Tsallis entropy and the Rényi entropy. The Tsallis entropy quantifies prediction errors, representing a family of entropies parameterized by a scalar that modulates the sensitivity to prediction loss. In contrast, the Rényi entropy is derived from tasks involving guessing and alternative data compression strategies. The chapter concludes with a unified framework integrating Shannon, Tsallis, and Rényi entropies.

Chapter 4. Relative Entropy. Focusing on monotonicity, this chapter examines how entropy measures evolve as distributions approach uniformity. It introduces distance and fidelity metrics in the space of distributions, presenting Shannon, Tsallis, and Rényi relative entropies as solutions to detection problems involving samples from different distributions. Finally, it generalizes relative entropies using the f-divergence framework.

Part II: Quantum Information

Chapter 5. From Classical to Quantum Information. This chapter begins the study of quantum information by exploring its distinctions from classical information. It introduces quantum states and measurements, emphasizing uniquely quantum features such as contextuality and the indistinguishability of quantum states.

Chapter 6. Quantum Uncertainty: Measured Entropy, Coherence, and the Uncertainty Principle. The measured entropy is introduced as a tool to quantify the uncertainty in quantum measurement outcomes. This chapter explores how the measured entropy varies with measurement settings, leading to the concepts of measurement compatibility and to the uncertainty principle. The chapter also extends the relationship between coherence and measurement compatibility from pure states to mixed states, building on concepts from Chapter 5.

Chapter 7. Classical and Quantum Uncertainty: Mixed States and Quantum Entropy. Integrating classical and quantum uncertainties, this chapter introduces mixed states, which are represented by density matrices and ensembles of pure states. Quantum entropy is defined as a measure of uncertainty, with non-selective quantum measurements shown to generally increase entropy. The chapter extends coherence and compatibility concepts from pure to mixed states.

Chapter 8. Quantum Relative Entropy. This chapter parallels the treatment of classical relative entropy, introducing quantum analogs that quantify differences between density matrices. Tools such as quantum trace distance and fidelity are presented, with fidelity explored from multiple perspectives, including connections to Tsallis entropy and the fidelity of pure and mixed states. The concept of state purification is also introduced to provide deeper insights into fidelity measures.

Part III: Dynamic Information

Chapter 9. Dynamic Classical Information. This chapter transitions from static to dynamic systems, exploring how state evolution affects uncertainty. It introduces majorization, a partial ordering of probability distributions, to formalize changes in uncertainty. For more general transformations, the data processing inequality is introduced to capture the reduction in distinguishability between distributions.

Chapter 10. Quantum Dynamic Information in Closed Systems. This chapter examines transformations in closed quantum systems, where state evolution is governed by unitary matrices. Unitary transformations are described as rotations or through Hermitian matrices known as Hamiltonians. The chapter also introduces the phenomenon of quantum interference, and it analyzes how closed-system dynamics impact information and relative information.

Chapter 11. Quantum Dynamic Information in Open Systems. Focusing on open quantum systems interacting with their environment, this chapter introduces quantum channels and positive operator-valued measure (POVM) representations. It discusses entropy changes under quantum channels and the data processing inequality, paralleling concepts from classical systems.

Part IV: Bipartite Classical Information

Chapter 12. Bipartite Classical Information as Correlation. This chapter explores correlations in systems composed of two subsystems, defining the Shannon mutual information, as well as common information measures and the maximal correlation.

Common information is analyzed in terms of shared randomness needed for local generation of system states and of maximum distillable information via distributed operations.

Chapter 13. Bipartite Classical Information via Residual Uncertainty. Focusing on sequential observations, this chapter examines how observing one subsystem reduces uncertainty about the other. It introduces ensemble entropy and locally accessible information as measures of bipartite information. Conditional entropy is also presented as a complementary perspective on residual uncertainty.

Part V: Bipartite Quantum Information

Chapter 14. Bipartite Classical and Quantum Information for Pure States. This chapter introduces quantum entanglement as a unique form of correlation in bipartite pure states. It discusses the non-local properties of entangled states and their applications in informational tasks. Measures such as mutual information, entanglement entropy, and quantum mutual information are introduced to quantify correlations.

Chapter 15. Bipartite Dynamic Classical and Quantum Information for Pure States. In this chapter, the dynamic properties of entanglement under local operations and classical communication (LOCC) are explored, using majorization to analyze state transformations. Entanglement measures, monotones, and operations that generate entanglement are also covered, alongside fundamental constraints like the no-cloning and no-deletion theorems.

Chapter 16. Bipartite Quantum and Classical Information for Mixed States. This chapter extends the analysis of bipartite quantum systems to mixed states, distinguishing between classical and quantum correlations. Measures such as entanglement of formation, relative entropy of entanglement, and quantum discord are introduced, emphasizing their role in quantifying correlations beyond entanglement.

Part VI: Multipartite Classical and Quantum Information

Chapter 17. A Brief Introduction to Tensor Networks. This chapter introduces tensor networks as tools to describe multipartite systems, encoding correlations in large-dimensional systems efficiently. The focus is on the foundational aspects of tensor networks relevant to subsequent chapters.

Chapter 18. Multipartite Classical Information: Fragmentation, Scale, and Strength. This chapter examines the fragmentation, scale, and strength of correlations in multipartite classical systems. Fragmentation captures the independence of subsystems, scale addresses the largest correlated subsets, and strength quantifies deviations from mutual independence.

Chapter 19. Multipartite Classical Information: Structure. This chapter explores the structure of multipartite correlations using tensor networks, canonical decompositions, and classical circuits. These frameworks describe how global and local latent variables mediate correlations and how dynamic transformations generate desired joint distributions.

Chapter 20. Multipartite Quantum Information: Fragmentation, Scale, and Strength. This chapter generalizes entanglement concepts to multipartite systems, introducing entanglement classes and cut-based measures. It also discusses the robustness and scalability of entanglement in systems with three or more qudits.
Chapter 21. Multipartite Quantum Information: Structure. This chapter focuses on matrix product states and quantum circuits as frameworks for describing multipartite quantum systems. Tools like ZX calculus are seen to enable diagrammatic simplifications for analyzing state properties without resorting solely to numerical or analytical techniques.

Recommended Resources

The book [1] offers an engaging and non-technical introduction to the concept of classical uncertainty. The foundational textbook on classical information theory is [2], which comprehensively covers the core principles, while the axiomatic perspective is explored in [3]. For a more contemporary and thorough treatment of the subject, [4] serves as an excellent reference.

In the realm of quantum information theory, well-established textbooks include [5, 6], which provide rigorous and detailed coverage of the field. Additionally, the book [7] bridges classical information theory with quantum computing concepts. For a definitive introduction to quantum computing, the classical textbook is [8].

REFERENCES

[1] D. Spiegelhalter, *The Art of Uncertainty: How to Navigate Chance, Ignorance, Risk, and Luck*. London: Penguin Books, 2024.
[2] T. M. Cover and J. A. Thomas, *Elements of Information Theory*, 2nd ed. Hoboken, NJ: Wiley-Interscience, 2006.
[3] J. Aczél and Z. Daróczy, *On Measures of Information and Their Characterizations*. New York: Academic Press, 1975.
[4] Y. Polyanskiy and Y. Wu, *Information Theory: From Coding to Learning*. Cambridge: Cambridge University Press, 2024.
[5] M. M. Wilde, *Quantum Information Theory*. Cambridge: Cambridge University Press, 2013.
[6] J. Watrous, *The Theory of Quantum Information*. Cambridge: Cambridge University Press, 2018.
[7] E. Desurvire, *Classical and Quantum Information Theory: An Introduction for the Telecom Scientist*. Cambridge: Cambridge University Press, 2009.
[8] M. A. Nielsen and I. L. Chuang, *Quantum Computation and Quantum Information*. Cambridge: Cambridge University Press, 2010.

Acknowledgments

This book is the result of many years of studying, teaching, and applying the tools of information theory. Over these years, I have used information theory to address challenges such as designing wireless systems, exploring the learning capabilities of artificial intelligence, optimizing neuromorphic computing systems, and studying cortex–muscle communication. Along the way, my thinking has been profoundly shaped by numerous conversations with researchers from a wide range of fields, both within and beyond the core information theory community.

I am deeply grateful to my collaborators, colleagues, and friends who share my passion for information theory. Some names stand out immediately – Leonardo Banchi, Elza Erkip, Deniz Gündüz, Lajos Hanzo, Ubli Mitra, Vincent Poor, Petar Popovski, Bipin Rajendran, Onur Sahin, Shlomo Shamai, Aldo Scutari, Ravi Tandon, and Aylin Yener – though the list is far longer than I can include here. My thanks extend to all those who have contributed to my intellectual journey (not to mention the physical journeys across the United States and Europe).

I had the privilege of teaching material from this book during a year-long series of seminars at King's College London. I would like to express my appreciation to the students and researchers who participated in the lectures – Francesco Binucci, Jiechen Chen, Kfir Cohen, Amirmohammad Farzaneh, Qiushuo Hou, Jiayi Huang, Dongwon Kim, Sangwoo Park, Wen Shang, Yixing Shen, Houssem Sifaou, Seonghoon Yoo, Matteo Zecchin, Yunchuan Zhang, and Meiyi Zhu. Your insightful feedback has been invaluable. The careful annotations by Kfir Cohen on the first two parts of the book have been particularly appreciated.

At Cambridge University Press, I had the pleasure to work with Elizabeth Horne and Amy Jacobsen, whose expertise and professionalism contributed to the successful realization of the project.

Finally, I dedicate this book to my parents, Rita and Antonello, and to my family – Lena, Noah, and Lisa.

Notations

General

$[N] = \{0, 1, \ldots, N-1\}$	set including all integer numbers from 0 to $N-1$
$\{a_n\}_{n\in[N]}$	set $\{a_0, a_1, \ldots, a_{N-1}\}$
$\lvert\mathcal{S}\rvert$	cardinality of a set $\mathcal{S}$
$\bar{\mathcal{S}} = [S]\backslash\mathcal{S}$	complement of a subset $\mathcal{S} \subseteq [S]$
$\propto$	proportional to
$\sum_{x\in[N]}$	sum over all values for variable $x \in [N]$
$\sum_x$	sum over all values for variable x
1_2	$\log_2(e) \approx 1.44$
$A \Longrightarrow B$	A implies B
$A \Longleftrightarrow B$	A is equivalent to B
xy, with $y \in [M]$	bi-index $Mx + y$
$x^0x^1...x^{S-1}$, with $x^s \in [D^s]$	multi-index $\sum_{s\in[S]} D^{S-1-s}x^s$

Functions

δ_{xy}	copy function ($\delta_{xx} = 1$ and $\delta_{xy} = 0$ otherwise)
$\lceil\cdot\rceil$	ceiling function (closest larger integer)
$\text{step}(\cdot)$	step function: $\text{step}(x) = 1$ if $x > 0$ and $\text{step}(x) = 0$ if $x < 0$
$\min_x f(x)$	minimization problem for function $f(\cdot)$ over x
$\arg\min_x f(x)$	an element x in the set of minimizers of function $f(x)$
$\log_{2,\alpha}(q)$	α-logarithm
$\text{Re}(\cdot)$	real part of a complex number

Linear Algebra

$(\cdot)^*$	complex conjugate
a_x or $[a]_x$	xth element of vector a
$A_{x,y}$ or $[A]_{x,y}$	(x, y)th element of matrix A
a^T and A^T	transpose of vector a and matrix A
$a^\dagger$ and $A^\dagger$	Hermitian transpose of vector a and matrix A
$\lvert\lvert a\rvert\rvert_\alpha$	α-norm of vector a
I	identity matrix
$\mathcal{S}(\{\lvert v_x\rangle\}_{x\in[L]})$	span of the vectors $\{\lvert v_x\rangle\}_{x\in[L]}$
$\text{rank}(A)$	rank of matrix A
$\lvert\lvert A\rvert\rvert_\alpha$	α-Schatten norm of matrix A
$\text{tr}(A)$	trace of matrix A
$\text{diag}(a)$	diagonal matrix with diagonal elements given by vector a
$\otimes$	Kronecker product or tensor product
$\oplus$	direct sum
$\lvert\lvert a\rvert\rvert_{(k)}$	kth Ky Fan norm of vector a
$x^{[S]}$	tensor indices and multi-index

$x^{\mathcal{S}}$	(ordered) subset of indices and multi-index
$T_{x^{[S]}}$	$x^{[S]}$th entry of tensor T
$\mathrm{vec}(T)$	matrix or tensor vectorization
$\mathrm{mat}_{\mathcal{S}}(T)$	tensor matrization with row multi-index $x^{\mathcal{S}}$

Probability

$\mathrm{x} \sim p(x)$	x is distributed according to distribution $p(x)$
$\mathrm{x} = x$	random variable x takes value x
$\mathrm{E}_p[f(\mathrm{x})]$	expected value $\sum_x p_x f(x)$
$\mathrm{Var}_p(\mathrm{x})$	variance $\mathrm{E}_p[(\mathrm{x} - \mathrm{E}_p[\mathrm{x}])^2]$
$\delta_{\bar{x}}$	one-hot vector with a 1 in position $\bar{x}$
u	uniform distribution
$\mathrm{supp}(p)$	support of distribution p
$p \succ q$	p majorizes q
$p^{\downarrow}$	ordered probability distribution
$p_{x \to y}$	channel
p^{XY}	joint distribution for a bipartite classical system
$p_x^{Y\|X}$	conditional distribution of Y given X for $\mathrm{x} = x$
$\mathcal{E}^{Y\|X}$	ensemble of conditional distributions
$p^{[S]}$ and $p^{[S]}_{x^{[S]}}$	joint distribution over S subsystems and $x^{[S]}$th entry
$p^{\mathcal{S}}$ and $p^{\mathcal{S}}_{x^{\mathcal{S}}}$	joint distribution over group $\mathcal{S}$ and $x^{\mathcal{S}}$th entry

Classical Information Theory

$\mathrm{H}(p)$	Shannon entropy
$\mathrm{h}(p_1)$	binary Shannon entropy
$\mathrm{H}^{\mathrm{T}}_{\alpha}(p)$	Tsallis entropy of order α
$\mathrm{h}^{\mathrm{T}}_{\alpha}(p_1)$	binary Tsallis entropy of order α
$\mathrm{H}^{\mathrm{R}}_{\alpha}(p)$	Rényi entropy
$\mathrm{h}^{\mathrm{T}}_{\alpha}(p_1)$	binary Rényi entropy
$\mathrm{H}_{\alpha}(p)$	any (Shannon, Tsallis, or Rényi) entropy
$\mathrm{H}^{\mathrm{U}}_{h,\phi}(p)$	(h, ϕ)-entropy
$\mathrm{H}(p\|\|q)$	Shannon cross entropy
$\mathrm{D}(p\|\|q)$	Shannon relative entropy
$\mathrm{d}(p_1\|\|q_1)$	binary Shannon relative entropy
$\mathrm{D}^{\mathrm{T}}_{\alpha}(p\|\|q)$	Tsallis relative entropy
$\mathrm{d}^{\mathrm{T}}_{\alpha}(p_1\|\|q_1)$	binary Tsallis relative entropy
$\mathrm{D}^{\mathrm{R}}_{\alpha}(p\|\|q)$	Rényi relative entropy
$\mathrm{d}^{\mathrm{R}}_{\alpha}(p\|\|q)$	binary Rényi relative entropy
$\mathrm{D}^{\mathrm{tr}}(p, q)$	trace distance, aka total variation
$\mathrm{D}_f(p\|\|q)$	f-divergence
$\mathrm{H}_{\alpha}(p^{XY}) = \mathrm{H}_{\alpha}(X, Y)_{p^{XY}}$	joint entropy
$\mathrm{I}(X;Y)_{p^{XY}}$	Shannon mutual information
$\mathrm{I}_{\mathrm{D}_{\alpha}}(X;Y)_{p^{XY}}$	relative entropy of correlation
$\mathrm{I}^{\mathrm{T}}_{\alpha}(X;Y)_{p^{XY}}$	Tsallis mutual information
$\mathrm{I}^{\mathrm{R}}_{\alpha}(X;Y)_{p^{XY}}$	Rényi mutual information
$\mathrm{C}_{\alpha}(X;Y)_{p^{XY}}$	distributed-sampling common information
$\mathfrak{C}_{\alpha}(X;Y)_{p^{XY}}$	distillable common information

$\mathfrak{C}^{\max}(X;Y)_{p^{XY}}$	maximal correlation
$\mathrm{H}_\alpha(\mathcal{E}^{Y\|X})_{p^{XY}}$	ensemble entropy
$\mathrm{J}_\alpha(X \to Y)_{p^{XY}}$	locally accessible information
$\mathrm{H}_\alpha(Y\|X)_{p^{XY}}$	conditional entropy
$\mathrm{H}(X^{\mathcal{S}})_{p^{\mathcal{S}}}$	joint entropy of subset $\mathcal{S}$
$\mathrm{I}(X^{\mathcal{U}};X^{\mathcal{V}})_{p^{\mathcal{U},\mathcal{V}}}$	mutual information between subsets $\mathcal{U}$ and $\mathcal{V}$
$\mathrm{I}(X^{\mathcal{S}};X^{\bar{\mathcal{S}}})_{p^{[S]}}$	mutual information across cut $\mathcal{S}$
$\mathrm{I}_{\max}(X^0;X^1;\ldots;X^{S-1})_{p^{[S]}}$	maximum cut-based multipartite information
$\mathrm{I}_{\min}(X^0;X^1;\ldots;X^{S-1})_{p^{[S]}}$	minimum cut-based multipartite information
$\mathrm{I}_{\mathrm{D}}(X^0;X^1;\ldots;X^{S-1})_{p^{[S]}}$	total correlation
$\mathrm{I}(X;Y\|Z)_{p^{XYZ}}$	conditional mutual information

Quantum Mechanics

$p(\|\psi\rangle\|\mathcal{M})$ and $p_x(\|\psi\rangle\|\mathcal{M})$	measurement probability distribution and xth entry
$\mathrm{Pr}_{\|\psi\rangle}[\cdot\|\mathcal{M}]$	measurement probability under setting $\mathcal{M}$ for state $\|\psi\rangle$
$\langle O\rangle_{\|\psi\rangle}$	expected value of observable O for pure state $\|\psi\rangle$
$\mathrm{supp}(\rho)$	support of density matrix ρ
$\ker(\rho)$	kernel of density matrix ρ
$\|x\rangle$ with integer x	computational basis vector
$\|+\rangle, \|-\rangle$	Hadamard basis
P_0	2×2 identity matrix
P_1	Pauli X matrix
P_2	Pauli Y matrix
P_3	Pauli Z matrix
H	Hadamard matrix
$\mathcal{K}$	POVM setting
$\mathcal{K}(\cdot)$	quantum channel
$\|\psi^{XY}\rangle$	joint pure state for a bipartite quantum system
ρ^{XY}	joint density matrix for a bipartite quantum system
$\|\Phi_+^{XY}\rangle = \frac{1}{\sqrt{2}}(\|00\rangle + \|11\rangle)$	Bell state
$\|\Psi_+^{XY}\rangle = \frac{1}{\sqrt{2}}(\|01\rangle + \|10\rangle)$	Bell state
$\|\Phi_-^{XY}\rangle = \frac{1}{\sqrt{2}}(\|00\rangle - \|11\rangle)$	Bell state
$\|\Psi_-^{XY}\rangle = \frac{1}{\sqrt{2}}(\|01\rangle - \|10\rangle)$	Bell state
Ψ^{XY}	amplitude matrix
$\mathrm{tr}_X(\cdot)$	partial trace over subsystem X
$\|\psi^{XY}\rangle \underset{\mathrm{LOCC}}{\longrightarrow} \|\phi^{XY}\rangle$	LOCC protocol
$\|\psi^{XY}\rangle \underset{\mathrm{SLOCC}}{\longrightarrow} \{(\|\phi_z^{XY}\rangle, p_z)\}_{z\in[M]}$	SLOCC protocol
χ^{XY}	classical–classical state
$\overrightarrow{\omega}^{XY}$	classical–quantum state
$(\cdot)^{T^X}$	partial transpose over subsystem X

Quantum Information Theory

$\mathrm{F}(\|\psi\rangle, \|\phi\rangle)$	fidelity between pure states $\|\psi\rangle$ and $\|\phi\rangle$
$\mathrm{D}^{\mathrm{tr}}(\|\psi\rangle, \|\phi\rangle)$	trace distance between pure states $\|\psi\rangle$ and $\|\phi\rangle$
$\mathrm{H}(\|\psi\rangle\|\mathcal{M})$	Shannon measured entropy
$\mathrm{H}_\alpha^{\mathrm{T}}(\|\psi\rangle\|\mathcal{M})$	Tsallis measured entropy
$\mathrm{H}_\alpha^{\mathrm{R}}(\|\psi\rangle\|\mathcal{M})$	Rényi measured entropy

$\mathrm{H}_\alpha(\vert\psi\rangle\vert\mathcal{M})$	any (Shannon, Tsallis, or Rényi) measured entropy
$\mathrm{K}^{\mathrm{tr}}(\vert\psi\rangle\vert\mathcal{M})$	trace distance of coherence for pure states
$\mathrm{H}(\rho)$	Shannon quantum entropy
$\mathrm{H}^{\mathrm{T}}_\alpha(\rho)$	Tsallis quantum entropy of order α
$\mathrm{H}^{\mathrm{R}}_\alpha(\rho)$	Rényi quantum entropy of order α
$\mathrm{H}_\alpha(p)$	any (Shannon, Tsallis, or Rényi) quantum entropy
$\mathrm{D}(\rho\vert\vert\sigma)$	Shannon quantum relative entropy
$\mathrm{D}^{\mathrm{T}}_\alpha(\rho\vert\vert\sigma)$	Tsallis relative entropy
$\mathrm{D}^{\mathrm{R}}_\alpha(\rho\vert\vert\sigma)$	Rényi relative entropy
$\mathrm{D}^{\mathrm{tr}}(\rho,\sigma)$	trace distance
$\mathrm{K}(\rho\vert\mathcal{M})$	coherence measure for state ρ and basis $\mathcal{M}$
$\mathrm{F}^{\mathrm{UJ}}(\rho,\sigma)$	Uhlmann–Josza fidelity
$\mathrm{J}(X;Y)_{\rho^{XY}}$	maximum measured mutual information
$\mathrm{I}(X;Y)_{\rho^{XY}}$	quantum mutual information
$\mathrm{E}(X;Y)_{\vert\psi^{XY}\rangle}$	entanglement entropy
$\mathrm{E}(X;Y)_{\rho^{XY}}$	entanglement of formation
$\mathrm{H}(X\vert Y)_{\rho^{XY}}$ and $\mathrm{H}(Y\vert X)_{\rho^{XY}}$	quantum conditional entropy
$\mathcal{E}(X;Y)_{\rho^{XY}}$	entanglement measure
$\mathfrak{E}(X;Y)_{\rho^{XY}}$	entanglement monotone
$\mathrm{Q}(X;Y)_{\rho^{XY}}$	quantum symmetric discord
$\mathrm{Q}_{\mathrm{R}}(X;Y)_{\rho^{XY}}$	relative entropy of quantumness
$\mathfrak{E}_R(X;Y)_{\rho^{XY}}$	relative entropy of entanglement
$\mathrm{H}(\mathcal{E}^{Y\vert X})_{\rho^{XY}}$	quantum ensemble entropy
$\mathrm{J}(X\to Y)_{\rho^{XY}}$	quantum locally accessible information
$\mathrm{Q}(X\to Y)_{\rho^{XY}}$	quantum discord

Acronyms

AI	artificial intelligence
CPD	canonic polyadic decomposition
CPTP	completely positive trace-preserving
i.i.d.	independent and identically distributed
KL	Kullback–Leibler
LOCC	local operations and classical communication
MPS	matrix product state
PEPS	projected entangled-pair state
POVM	positive operator-valued measure
PPT	positive partial transpose
SLOCC	stochastic local operations and classical communication
SVD	singular value decomposition

Part I

Classical Information

1 Uncertainty, Information, and Entropy

1.1 Overview

This chapter introduces the concept of entropy as a measure of information and uncertainty. Starting with two examples, a distinction is first made between subjective and objective information. Then, the definitions of system, state of the system, and informational state of an observer of the system are formally introduced. Based on these definitions, the chapter provides an intuitive list of properties that an entropy measure is expected to satisfy to serve as a meaningful measure of uncertainty and information. Finally, examples are provided to illustrate the operational approach to the definition of an entropy measure.

1.1.1 Learning Objectives and Organization

By the end of this chapter, the reader should be able to:

- understand at an intuitive level the relation between information, uncertainty, and surprise;
- describe the informational state of an observer in terms of a probability distribution;
- distinguish between minimal and maximal states of uncertainty;
- list basic properties expected from an entropy to serve as an uncertainty measure;
- make an educated guess about whether a function of a probability vector is a valid entropy;
- and understand the axiomatic and operational approaches to the definition of entropy measures.

1.2 Uncertainty and Information

This section introduces the ideas of uncertainty and information, distinguishing between subjective and objective uncertainty.

1.2.1 Two Examples

In order to start thinking about uncertainty and information, it is useful to consider the following two examples.

Example 1.1

For a typical textbook example, consider a closed box that includes only black and gray balls. As shown in Figure 1.1, the black balls are identified as 1 and the gray balls as 0. Assume also that we know the composition of the box. Specifically, the box is known to contain a fraction p_1, with $0 \leq p_1 \leq 1$, of black balls, and a fraction $p_0 = 1 - p_1$ of gray balls. Therefore, our **informational state** as observers of the system can be described by the probability distribution vector

$$p = \begin{bmatrix} p_0 \\ p_1 \end{bmatrix}. \tag{1.1}$$

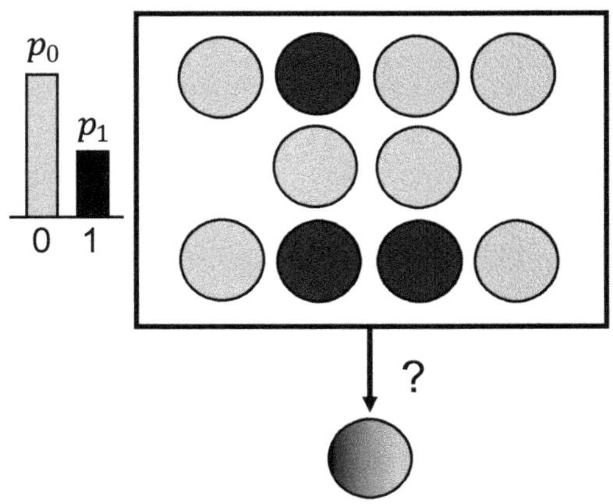

Figure 1.1 Given the known fraction of black and gray balls, how much uncertainty is there about the color of the selected ball *before* observing it? How much information is received *after* observing the color of the selected ball?

One ball is picked at random from the box. How much uncertainty do we have about the color of the selected ball *before* observing it? How much information do we receive *after* observing the color of the selected ball?

If $p_1 = 0$ or $p_1 = 1$, there is clearly no prior uncertainty on the color of the selected ball. In fact, the ball drawn from the box is always either black, if $p_1 = 1$, or gray, if $p_1 = 0$. Equivalently, we receive no new information upon observing the color of the ball. As a result, both the uncertainty before the observation and the information after the observation may be considered to equal zero in this case.

In all other cases, i.e., when $0 < p_1 < 1$, we have some uncertainty on the color of the selected ball before observing it, and we do receive new information upon observing the ball's color. The "amount" of prior uncertainty, as well as the "amount" of information gathered from the observation of the ball's color, intuitively depend on the value of the fraction p_1. For example, if the fraction of black balls is $p_1 = 0.1$, we may be relatively certain that the ball's color is gray before observing it, and hence our prior uncertainty is low. Sometimes we will see a black ball, but 90% of the time a gray ball will be observed.

If, instead, there is a fraction $p_1 = 0.4$ of black balls, we are less certain a priori about the outcome of a draw, and we receive more information by observing the color of the ball selected from the box. In the extreme case with $p_1 = 0.5$, there are as many black as gray balls in the box. Therefore, in this setting, we are maximally uncertain before the observation, and we receive a maximal amount of information after the observation.

Example 1.2

Consider the problem of predicting whether the next day will be sunny or rainy. Label sunny weather as 1 and rain as 0 (see Figure 1.2). Given the available information about past weather-related variables, a forecasting model produces a **prediction** of future weather conditions. For instance, a weather forecasting model may indicate a chance of rain for the next day of 60%, i.e., a probability $p_0 = 0.6$, while another model may put this chance at 20%, i.e., at probability $p_0 = 0.2$.

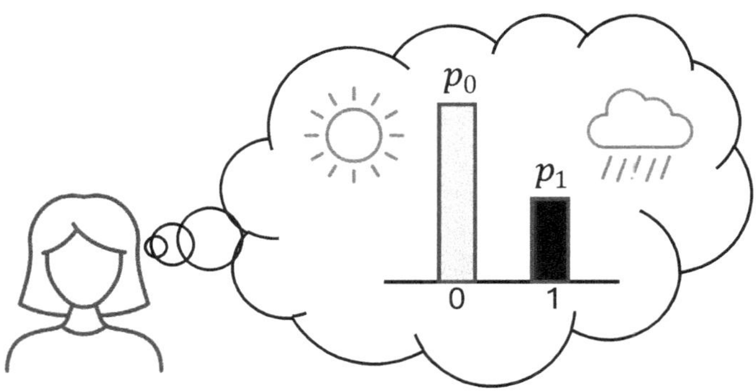

Figure 1.2 A weather predictor assigns a probability to all possible weather conditions (here only two: rain or sun).

Accordingly, the informational states of the two predictors are defined, respectively, by the probability vectors $p = [p_0 = 0.6, p_1 = 0.4]^T$ and $p = [p_0 = 0.2, p_1 = 0.8]^T$. (The notation $[\cdot]^T$ represents the **transpose** operation, which transforms a row vector into a column vector and vice versa.) Thus, the first model is more uncertain about its prediction than the second model. In fact, the informational state of the first model assigns more balanced probabilities between the two possible weather conditions, while the second model assigns a more skewed probability split.

The two examples above illustrate a number of important concepts that will be elaborated on in this part of the book. The first is the dual nature of information and uncertainty. Before observing the state of the system, we are generally uncertain about the outcome of the observation; meanwhile, after making the observation, we have acquired information, resolving the state of uncertainty. Therefore, defining and quantifying the information acquired after making an observation is tantamount to defining and quantifying the uncertainty before making the observation. In other words, if we can quantify how uncertain we are about something, we can also quantify how much information we will need to resolve this state of uncertainty.

The second concept illustrated by the examples is the distinction between objective and subjective informational states, which is discussed next.

1.2.2 Subjective vs. Objective Uncertainty

In some contexts, uncertainty represents an **objective property of a system**, while in others, it reflects a **subjective state of knowledge** held by the observer. In the first example above, the informational state provided an accurate and faithful description of the physical system, offering an objective measure of uncertainty. By contrast, in the second example, uncertainty was subjective, representing the forecaster's state of knowledge rather than the true probabilities governing potential weather outcomes.

That said, even in the first example, interpreting uncertainty as an objective property is not always tenable. Consider a scenario involving two observers. The first observer holds the ball drawn from a box and knows its color, say black, while the second observer, located elsewhere – perhaps in a different room – remains unaware of the ball's color.

From the perspective of the first observer, there is no uncertainty about the ball's color. The informational state of the first observer can thus be described by the vector $p = [0, 1]^T$, assigning probability 1 to the ball being black. In contrast, the second observer, lacking direct knowledge, considers the probability of the ball being black to be p_1, which equals the fraction of black balls in the box. For this observer, the informational state is defined by the probability vector $p = [p_0, p_1]^T$, where p_0 is the probability of the ball being gray.

A more nuanced case arises when the second observer, while still in a different room, overhears a comment suggesting the ball might be black. In this situation, her informational state might be better described by a probability $p'_1 \neq p_1$, reflecting an updated subjective belief that the ball is black, even if this belief exceeds the true fraction p_1.

Overall, while the objective interpretation of uncertainty may apply in specific circumstances, it is often more appropriate to view uncertainty as a subjective state of knowledge.

In keeping with this subjective perspective, this book will adopt the notion of an observer when discussing uncertainty and information, using the pronoun "we" to describe the observer's perspective. As the Lebanese-American author Nassim Taleb aptly noted, probability – and by extension, the quantification of uncertainty – ultimately concerns "*the acceptance of the lack of certainty in our knowledge and the development of methods for dealing with our ignorance.*"

1.3 System, States, and Informational State

1.3.1 System and States

In order to quantify uncertainty in classical settings, this book will focus on a **system** that can be in one of N possible **states**, with N being a finite integer. The set of N possible states is mapped into the set of integers

$$[N] = \{0, 1, \ldots, N-1\}. \tag{1.2}$$

Importantly, the specific way in which the actual states of a system are represented by integers in the set $[N]$ is arbitrary. For instance, when analyzing the outcome of a coin flip, we could assign 0 to heads and 1 to tails, or vice versa. Uncertainty and information do not depend on this arbitrary choice. In the two examples above, the system of interest had $N = 2$ possible states. For the first example, the color gray was mapped to integer 0 and the color black to 1. For the second example, rainy and sunny weather conditions were assigned to integers 0 and 1, respectively.

1.3.2 Informational State

From the perspective of an observer, the state of the system is generally unknown, and is thus described by a discrete **random variable** x. The distribution of the random variable x is described by a **probability vector**

$$p = \begin{bmatrix} p_0 \\ \vdots \\ p_{N-1} \end{bmatrix}, \tag{1.3}$$

where p_x is the probability that the system is observed to be in state $x \in [N]$, i.e.,

$$p_x = \Pr[\text{the system is in state } x]. \tag{1.4}$$

The fact that random variable x follows distribution p is denoted as

$$\mathrm{x} \sim p. \tag{1.5}$$

Notation-wise, Roman fonts, as in x, are reserved for random variables, and the corresponding regular fonts, such as x, are used for realizations of the random variable.

Probabilities are non-negative and they must sum to 1. That is, the inequalities $p_x \geq 0$ hold for all $x \in [N]$, along with the equality

$$\sum_{x \in [N]} p_x = 1. \tag{1.6}$$

Note that, throughout the book, entries of a vector, like p in (1.3), are indexed in an ascending way from top to bottom, with the first entry being indexed as $x = 0$.

To summarize, the state of knowledge of the observer about the state of the system is characterized by a probability distribution p. Accordingly, the probability distribution p is referred to as the informational state of the observer.

Definition 1.1 Informational state
For a system with N possible states, the informational state of the observer is described by an $N \times 1$ probability vector p as in (1.3). Accordingly, from the perspective of the observer, the state of the system is a random variable $\mathrm{x} \sim p$ with distribution p.

Following this definition, an informational state p describes the uncertainty of the observer about the state of the system. This uncertainty is resolved once the observer gains access to the system, revealing the true state of the system x.

Example 1.3

Editing programs, email clients, and chatbots all rely on the ability to predict the next word based on preceding words in a text. Specifically, given a sequence of previous words – either entered by the user or generated by the software – the system produces a probability vector p over a set of N possible next words. A vocabulary size of around 10,000 words is typically sufficient to cover most scenarios, making $N \approx 10{,}000$ a common choice for such systems.

1.3.3 Simple vs. Composite Systems

Finally, a distinction must be made between simple and composite systems. A **simple system** is an entity that is treated as indivisible. All the examples seen so far – a ball, a binary weather condition, the next word in a text – fall into this category. In the case of simple systems, the value of N is typically relatively modest.

Composite systems, in contrast, are treated as consisting of multiple, simple, subsystems. For instance, a text may be viewed as a composite system consisting of a sequence of letters, or of a sequence of words. In the former case, the constituent subsystems are letters, while in the latter case, the subsystems are words.

In a composite system, the individual subsystems may be physically and/or logically separated. As an example, the system under study may consist of different geographical locations, different individuals, or different elementary particles, entailing a physical separation. In physics, subsystems are also referred to as degrees of freedom of the system.

However, subsystems may also refer to logical quantities, such as the marks of a student throughout her graduate studies, the preferences of a user of a streaming service across different movie genres, or financial indicators. In all of these cases, the different subsystems – marks, ratings, indicators – are better thought of as being logically separated.

A composite system consists of S subsystems, with each subsystem being in one of D possible states. For instance, in a text made of S letters, each letter can be in one of $D = 27$ possible states in English, including space.

How large is the number N of possible states of the composite system? Since the state of a composite system is given by the collection of states for all the constituent subsystems, there are

$$N = D^S \tag{1.7}$$

possible states for the composite system. Thus, the number of states increases exponentially with the number of subsystems, S. Accordingly, the size of the probability vector p describing a state of uncertainty over the state of a composite system also grows exponentially with S.

The study of composite systems will be the subject of the later parts of this book. It is anticipated here that the characterization of uncertainty and information in composite systems is more complex than in simple systems, depending on how the states of the subsystems are related to each other. For example, if two coins always land on the same side, then knowing the outcome of one coin flip completely eliminates our uncertainty about the other. These relationships between different subsystems are referred to as **correlations**.

1.4 Uncertainty and Entropy

1.4.1 Maximal Uncertainty

When is the uncertainty associated with a state of knowledge described by a probability vector p **maximal**? Intuitively, if we have no information about the state of the system, we should assign an equal probability to all alternatives in set $[N]$. Accordingly, the following definition holds.

Definition 1.2 Informational state of maximal uncertainty
The informational state of maximal uncertainty is described by the **uniform distribution** u, which assigns the same probability

$$p_x = u_x = \frac{1}{N} \tag{1.8}$$

to all states $x \in [N]$. The uniform distribution is written in vector form as

$$u = \begin{bmatrix} u_0 \\ u_1 \\ \vdots \\ u_{N-1} \end{bmatrix} = \frac{1}{N} \begin{bmatrix} 1 \\ 1 \\ \vdots \\ 1 \end{bmatrix}. \tag{1.9}$$

Example 1.4

In Examples 1.1 and 1.2, since the number of states is $N = 2$, the informational state of maximal uncertainty corresponds to an equal probability $p_0 = p_1 = 1/2$ of observing either color or either weather condition. Given the duality of uncertainty and information, an observer characterized by an informational state of maximal uncertainty receives a maximum amount of information when making an observation.

1.4.2 Minimal Uncertainty

Conversely, the uncertainty associated with a state of knowledge p is **minimal** when the observer has no uncertainty about the state of the system, even before an observation is made. In this case, the observer knows that, with probability 1, the system is in a certain state $\bar{x} \in [N]$. Accordingly, the probability vector describing a state of knowledge with minimal uncertainty has entries given by $p_{\bar{x}} = 1$ for some state $\bar{x} \in [N]$ and $p_x = 0$ for all other states $x \neq \bar{x}$.

To elaborate, define the **copier function** as

$$\delta_{xy} = \begin{cases} 1 & \text{if } x = y, \\ 0 & \text{otherwise.} \end{cases} \tag{1.10}$$

In words, the copier function returns 1 if x and y are a "copy" of each other, i.e., if $x = y$; otherwise, it returns 0. Note that the copier function can also be referred to as the Kronecker delta function.

With this definition, an informational state of minimal uncertainty, p, has entries of the form

$$p_x = \delta_{x\bar{x}} = \begin{cases} 1 & \text{if } x = \bar{x}, \\ 0 & \text{otherwise.} \end{cases} \tag{1.11}$$

In vector form, the informational state p in (1.11) is a **"one-hot" vector** $\delta_{\bar{x}}$ that has all entries equal to zero except for a 1 ("hot") entry in position $\bar{x}$.

Definition 1.3 Informational state of minimal uncertainty
An informational state of minimal uncertainty is defined as the one-hot vector

$$p = \delta_{\bar{x}} \tag{1.12}$$

for some integer $\bar{x} \in [N]$.

1.4.3 Entropy

How can one quantify the uncertainty level associated with any informational state p? The answer to this question is given by the **entropy**, a function of probability vector p that satisfies the following basic three properties.

Definition 1.4 Non-negativity and minimum-value property of an entropy
An entropy $\mathrm{F}(p)$ must be non-negative, i.e.,

$$\mathrm{F}(p) \geq 0 \text{ for all probability vectors } p;$$

and it must equal zero if $p = \delta_{\bar{x}}$ for some integer $\bar{x} \in [N]$, i.e.,

$$p = \delta_{\bar{x}} \text{ for some } \bar{x} \in [N] \Longrightarrow \mathrm{F}(p) = 0. \tag{1.13}$$

When the entropy is zero, $\mathrm{F}(p) = 0$, one may say that there is **no uncertainty** on the state of the system according to the entropy function $\mathrm{F}(p)$.

Definition 1.5 Maximum-value property of an entropy
An entropy $\mathrm{F}(p)$ must be maximized when the distribution p is uniform, that is, the following inequality must hold:

$$\mathrm{F}(p) \leq \mathrm{F}(u) \text{ for all probability vectors } p. \tag{1.14}$$

When equality $\mathrm{F}(p) = \mathrm{F}(u)$ is satisfied, one may say there is maximal uncertainty on the state of the system according to the entropy function $\mathrm{F}(p)$.

Definition 1.6 Monotonicity property of an entropy
As illustrated in Figure 1.3, an entropy $\mathrm{F}(p)$ must increase as the distribution p becomes "closer" to the uniform distribution u, i.e., as the distance between distributions p and u decreases.

The first two are well-defined formal properties on the definition of an entropy function. In contrast, the monotonicity property is stated above rather informally, capturing the expectation that a "more uniform" distribution should have a larger

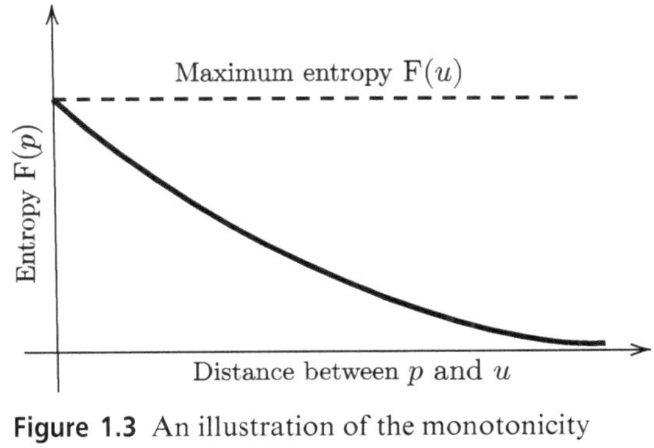

Figure 1.3 An illustration of the monotonicity property of an entropy measure: an entropy measure must increase as the distribution p becomes "closer" to the uniform distribution u.

entropy. A formalization of this property will require the definition of a distance measure in the space of probability vectors that quantifies the "closeness" of a vector p to the uniform distribution u, and it will be addressed in Chapter 4.

As will be discussed in Chapter 4, the specification of a distance between distributions effectively defines an entropy, and different distance measures yield entropies with distinct applications.

As a note about terminology, in this book terms such as "measure" or "metric" (used interchangeably) or "distance" will not be intended according to their strict mathematical definitions. Rather, they will be used to indicate functions satisfying some intuitive properties of measures, metrics, or distances.

1.4.4 Entropy vs. Moments

A final comment is in order about the difference between entropy measures and other measures of uncertainty based on **moments** of the distribution p, such as the **variance**. As discussed in this chapter, an entropy measure $\mathrm{F}(p)$ depends solely on the probabilities in vector p, and not on the specific significance and numerical values corresponding to each state.

Example 1.5

A temperature variable x that takes all values in the set $\{-30, -20, -10, 0, 10, 20, 30\}$ (in Celsius scale) with equal probability has the same entropy as one that takes all values in the set $\{15, 18, 21, 24, 27, 30, 33\}$ with equal probability. This is because, in both cases, the possible values of the temperature are mapped to the same set, namely [7]. Thus, the probability vector p is the same, i.e., the uniform distribution $p = u$. However, clearly, the two situations are quite different in terms of their semantics – one would likely choose to be uncertain about the second set of values rather than about the first!

In contrast to the entropy, there are alternative measures of uncertainty that do depend on the specific numerical values that may be taken by the state of the system. To elaborate, consider now a random variable x taking values in some set of numerical levels of interest as in Example 1.5. The **variance** of the random variable $\mathrm{x} \sim p$ is defined as

$$\mathrm{Var}_p(\mathrm{x}) = \mathrm{E}_p[(\mathrm{x} - \mathrm{E}_p[\mathrm{x}])^2], \tag{1.15}$$

where

$$\mathrm{E}_p[f(\mathrm{x})] = \sum_x p_x f(x) \tag{1.16}$$

represents the **expectation** of random variable $f(\mathrm{x})$ for some function $f(\cdot)$. The notation $\sum_x$ corresponds to the sum over all values of variable x.

The variance measures the average deviation of the observed value from the average value. Accordingly, it depends on the numerical values that can be assumed by the random variable x. Beyond the variance, one could consider other (central) moments of random variable x as measures of uncertainty. Such **moment-based** measures are distinct from **entropic** measures, which are only sensitive to the spread of probabilities in vector p.

By accounting for the numerical values of the observations, moment-based measures have the key disadvantage, as compared to entropic measures, of depending on the specific numerical labeling of the outputs. For example, unlike entropies, moment-based measures are sensitive to a simple **rescaling** of the output values. Consider what happens to the variance if the random variable x is multiplied by a constant. Furthermore, in the case of non-numerical states, such as colors, a meaningful numerical labeling may simply not be applicable.

1.5 Axiomatic vs. Operational Definitions of Entropy

1.5.1 Axiomatic Definitions of Entropy

Section 1.4 introduced the concept of entropy as a function $F(p)$ of an informational state p that satisfies three intuitive properties for a measure of uncertainty. This approach exemplifies at a high level the **axiomatic method** for the definition of entropy measures. When following an axiomatic approach, one first specifies a list of mathematical properties that must be met by a function $F(p)$ of vector p to be considered an entropy. Then, the class of functions that satisfy these properties is derived.

An alternative direction, favored by engineers, is to define entropy measures via the **operational approach**, which is discussed next.

1.5.2 Operational Definitions of Entropy

The operational approach to the definition of entropy measures formulates an **information processing task**, and then relates the optimal solution of the task to a measure of uncertainty. The resulting measures of uncertainty are often expressed in terms of the number of **binary information units**, or **bits**, needed to carry out the given information processing task in the most efficient way. The next examples illustrate four informational tasks, namely asking, compressing, sampling, and guessing, which will be further elaborated on in Chapters 2 and 3.

Example 1.6

Asking. Think of a band. I, the observer, wish to find out the identity of the band you have just selected. To start with, my informational state is defined by a probability distribution p, describing my state of knowledge about your musical taste. As illustrated in Figure 1.4, to resolve this state of uncertainty, I am only allowed to ask questions with a yes or no answer. How many questions will I need to identify your choice? An optimal questioning strategy yields the minimum average number of binary questions, or bits, providing a way to quantify my initial uncertainty on the identity of the band you selected.

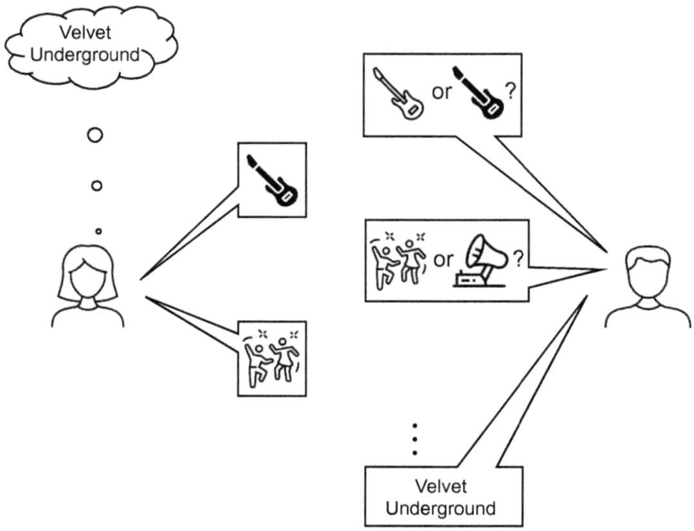

Figure 1.4 Illustration of the informational task of asking. I, the observer, wish to find out the identity of the band you have just selected. How many binary questions will I need to ask in order to identify your choice?

Example 1.7

Compressing. As shown in Figure 1.5, a satellite takes pictures of a remote site, which is currently unreachable by the observer. The observer may have some prior

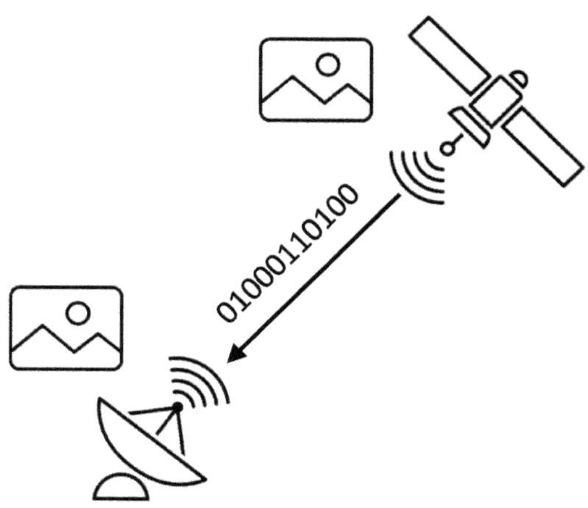

Figure 1.5 Illustration of the informational task of compressing. Given the prior state of knowledge of the observer, how many bits does the satellite need to communicate in order to describe a picture of a remote site?

expectations on the conditions on the ground, which can be described by a probability vector p. How many bits does the satellite need to send to the observer in order to inform her of the actual state of the remote site? In this case, the informational task is compressing the images taken by the satellite. An optimal compression strategy produces the **minimum average number of bits** to be transmitted so as to inform the observer, quantifying the initial uncertainty of the observer.

Example 1.8

Sampling. As seen in Figure 1.6, an artificial intelligence (AI) is designed to generate images x resembling famous actors. Valid images have a desired distribution p over the set of all possible images. How many bits of randomness should the program have access to in order to sample a valid image? This task can also be viewed as the problem of simulating an output $\mathrm{x} \sim p$ produced by an observer in informational state p.

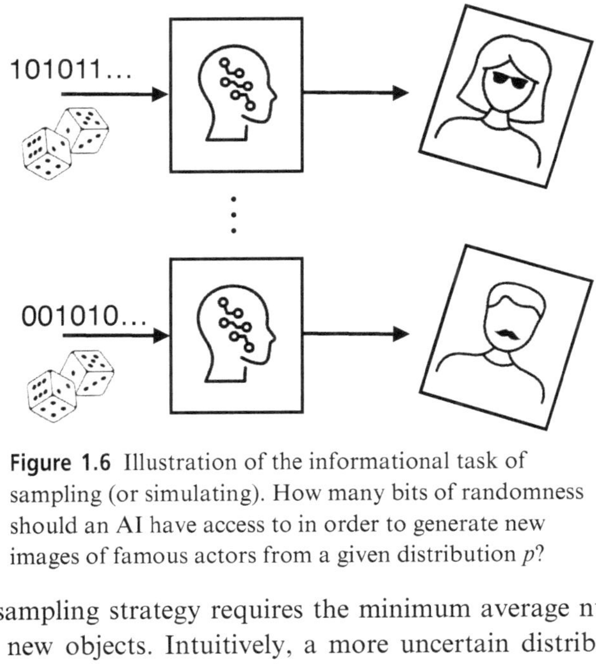

Figure 1.6 Illustration of the informational task of sampling (or simulating). How many bits of randomness should an AI have access to in order to generate new images of famous actors from a given distribution p?

The optimal sampling strategy requires the minimum average number of random bits to produce new objects. Intuitively, a more uncertain distribution p would be spread out among a larger number of images or texts, requiring a larger number of random bits. Accordingly, the resulting minimum average number of bits can be considered to be a measure of uncertainty associated with the original distribution p.

Example 1.9

Guessing. As illustrated in Figure 1.7, an attacker is trying to guess the password of a legitimate user. The operational task for the attacker is identifying the correct password via trial and error, starting from an initial state of knowledge described by a probability distribution p over passwords. At each trial, the attacker tries a password, and is told whether the guessed password is correct or not via a single bit of information. An optimal guessing strategy requires the minimum number of guesses, on average, to identify the password. The resulting minimum average number of guesses, or bits, provides a measure of the uncertainty associated with the initial informational state p.

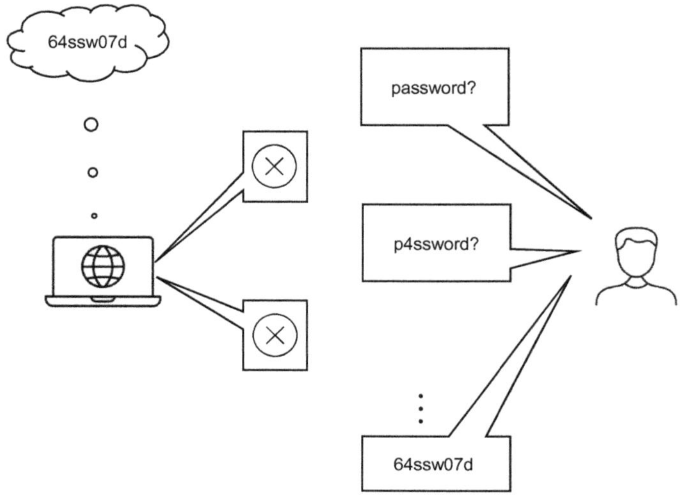

Figure 1.7 Illustration of the informational task of guessing. Given an initial state of knowledge about the password, an attacker guesses possible passwords. How many guesses are needed to identify the password?

The previous examples illustrate four informational tasks whose solutions entail the design of strategies that collect bits of information to resolve or represent a state of uncertainty. A conceptually different operational approach to the quantification of uncertainty is via the informational task of **prediction**, in which the observer wishes to predict the state of a system, x, based on knowledge of the distribution p. The two examples given at the beginning of this chapter (Examples 1.1 and 1.2) can be viewed as prediction tasks, whereby the observer wishes to predict the color of a ball to be extracted from a closed box or the future weather conditions. An optimal predictor obtains the minimum possible **average prediction error**, which serves as another way to quantify the initial uncertainty of the observer. Equivalently, a predictor's uncertainty is larger if the predictor is, on average, more surprised upon observing the actual state of the system.

1.5.3 Outlook

Chapters 2 to 4 will elaborate on entropy measures derived by following operational approaches based on asking, compressing, sampling, guessing, and predicting. The derived measures will also be verified to satisfy the desired properties listed above of non-negativity and minimal entropy, maximal entropy, and monotonicity.

Before concluding this chapter, two remarks are in order.

- **Choice of entropy:** Since different operational approaches yield distinct entropy measures, the choice of one entropy metric over another depends on many factors, and it should ideally be informed by the operational significance of the uncertainty measure for the given problem of interest. For instance, in the design of a communication system, entropies that relate to the task of compression are a natural choice; meanwhile, for the problem of weather forecasting, a prediction-based entropy may be preferable.
- **Bit and it:** In engineering and scientific contexts, bits may refer not only to the logical binary digits encoding information, but also to the physical embodiment of the information. Bits are practically transmitted and stored using physical resources, such as the electromagnetic waves emitted by a satellite toward a ground station and the memory elements in an AI chip. That said, while needing a physical substrate to be stored or communicated, information bits have a logical identity that is independent from the medium used to embody the underlying information. In other words, a bit is a bit, irrespective of the "it" that encodes the binary digit. (And, according to American physicist John Archibald Wheeler, the opposite – "it from bit" – relation may also hold true, with the material world being the result of an informational blueprint.)

1.6 Summary

- Uncertainty and information play a dual role: the observer is uncertain before making an observation; meanwhile, after the observation, the acquired information resolves the state of uncertainty.
- Uncertainty can in some cases be viewed as an objective property of a system, but it is more often understood as a subjective state of knowledge of the observer.
- Throughout this book, when discussing classical information, the focus will be on systems that can be in one of N possible states, with state set $[N] = \{0, 1, \ldots, N-1\}$. The mapping between state identifiers and integers in the set $[N]$ is arbitrary and it does not affect the definition of information measures.
- An informational state about the system is described by a probability vector $p = [p_0, \ldots, p_{N-1}]^T$.
- From the perspective of an observer having informational state p, the state of the system is a random variable $\mathrm{x} \sim p$.
- An informational state of maximal uncertainty corresponds to the uniform distribution $p = u$, while minimal uncertainty is attained when the probability vector is concentrated at any one specific state $\bar{x} \in [N]$, which is written as $p = \delta_{\bar{x}}$.
- The next chapters of this part of the book will elaborate on the informational tasks of asking, compressing, sampling, guessing, and predicting, whose solutions will lead to the definition of several valid entropy measures.

1.7 Recommended Resources

High-level discussions on the role of uncertainty across fields like economics and the sciences can be found in [1] and [2]. For a technical exploration of information measures, see [3] and [4]. Philosophical perspectives on how uncertainty and information shape our understanding of the nature of reality are presented in [5].

PROBLEMS

1.1 Consider an observer with informational state p. Define the function $f(p_x) = 1/p_x$, where $p_x \in [0,1]$ is the probability for state $x \in [N]$. (a) Plot this function in the interval $[0,1]$. (b) Can this function be a valid measure of the surprise experienced by the observer upon being informed that the state of the system is x? *Hint:* A larger value p_x implies that the observer is more confident that the state of the system is x. (c) Would function $f(p_x) = p_x$ be a valid measure of surprise? (d) Would function $f(p_x) = -\log_2(p_x)$ be a valid measure of surprise?

1.2 (a) Which one among the distributions $p = [0, 0.5, 0.5, 0]^T$, $p = [0.25, 0.25, 0.25, 0.25]^T$, and $p = [1, 0, 0, 0]^T$ has maximal entropy? (b) Which one has minimal entropy?

1.3 For each of the five informational tasks asking, compressing, sampling, guessing, and predicting, argue that the resulting entropy measures must be consistent with the monotonicity property. For instance, if the observer is trying to identify a state that is uniformly distributed within a set of N options, the average number of required yes/no questions cannot be smaller than if the state had a different distribution over the same number of options, N. *Hint:* A possible argument may be built around the fact that considering the average of all permutations of a distribution p, one obtains a uniform distribution. In fact, a random permutation should not make a task easier.

REFERENCES

[1] N. N. Taleb, *Fooled by Randomness: The Hidden Role of Chance in Life and in the Markets*. New York: Random House, 2005.

[2] E. Thompson, *Escape from Model Land: How Mathematical Models Can Lead Us Astray and What We Can Do About It*. London: Basic Books, 2022.

[3] J. Aczél and Z. Daróczy, *On Measures of Information and Their Characterizations*. New York: Academic Press, 1975.

[4] Y. Polyanskiy and Y. Wu, *Information Theory: From Coding to Learning*. Cambridge: Cambridge University Press, 2024.

[5] W. Egginton, *The Rigor of Angels: Borges, Heisenberg, Kant, and the Ultimate Nature of Reality*. New York: Pantheon, 2023.

2 Information Quantification by Asking, Compressing, or Sampling: Shannon Entropy

2.1 Overview

Chapter 1 showed that uncertainty can be quantified by formulating information processing tasks whose solutions require the use of a number of bits that grow with the uncertainty of the observer. This chapter elaborates on three fundamental information processing tasks, namely asking, compressing, and sampling. In all cases, uncertainty will be shown to be quantified by the same fundamental function of the informational state, the Shannon entropy. The Shannon entropy will also be seen to satisfy the properties of an entropy presented in Chapter 1.

2.1.1 Learning Objectives and Organization

By the end of this chapter, the reader should be able to:

- formulate the task of identifying a state via an optimal sequence of binary, i.e., yes/no, questions;
- understand and apply the Huffman algorithm;
- understand the properties of the Shannon surprise;
- understand the definition of the Shannon entropy, as well as its relation to the solution of the problem of determining the optimal sequence of binary questions;
- formulate the task of compressing the state of a system;
- understand the role of the Shannon entropy in the solution of the compression problem;
- formulate the task of sampling the state of a system – i.e., simulating a system – by using the minimum number of random bits;
- and understand the role of the Shannon entropy in the solution of the sampling problem.

2.2 Quantifying Information by Asking

This section focuses on the informational task of asking yes/ no questions, also referred to as **binary questions**, in order to identify the state of a system. Each yes/no answer can be thought of as 1 bit of information collected by the observer. This section formalizes the problem by presenting two equivalent ways of representing an asking strategy, namely as a **binary tree** and as a **binary codebook**.

2.2.1 The Informational Task of Asking

Consider an observer with an informational state regarding the true state of the system that is described by an N-dimensional probability vector $p = [p_0, \ldots, p_{N-1}]^T$. Accordingly, as far as the observer is concerned, each of the N states of the system, indexed via an integer $x \in [N]$, has probability p_x. In other words, from the viewpoint of the observer, the true state of the system is a random variable x with distribution p, which is denoted as x $\sim p$.

The key question of interest is: How many yes/no questions are sufficient, on average with respect to the probability vector p, to unambiguously identify the true state x? Intuitively, the more spread out the distribution p is, the more questions will be needed, signaling a larger uncertainty for the observer.

It is emphasized that the minimum average number of questions is evaluated with respect to the probability vector p, implicitly assuming that the "true" probability of state x being equal to any value $x \in [N]$ is indeed p_x. As a result, the setting here can be interpreted in one of two ways.

- **Objective uncertainty measure:** The probability vector p represents an **objective** description of the uncertainty about the system state x, and thus the average represents the true expected value of the number of questions.
- **Subjective uncertainty measure:** The probability vector p describes the **subjective** state of knowledge of the observer, and thus the average is the value expected by the observer based on the information it has available about the state variable x. The situation in which the subjective informational state p is mismatched as compared to the true, objective, distribution of the state will be discussed in Chapter 4.

As illustrated by the following example, each binary question divides the set $[N]$ of possible state values into two disjoint subsets. Specifically, the bit of information carried by the answer to a binary question distinguishes the subset of values in $[N]$ that is consistent with the answer from the subset of values that is not consistent and hence can be excluded. (The reader may be reminded of the "Guess Who?" board game.) Therefore, as detailed in the next section, the evolution of the state of knowledge of the observer can be described through a binary tree. In such a tree, the nodes report the current subset of states that are compatible with all the answers received so far.

Example 2.1

Consider a system with state set $[N] = \{0, 1, 2, 3\}$, i.e., with $N = 4$ possible states. Assume that all four states have equal probability, so that the distribution p is uniform, i.e., $p = u$. Figure 2.1 reports a possible strategy for the selection of binary questions, along with the corresponding binary tree describing the state of knowledge of the observer. As represented by the root of the tree, at the beginning all states are consistent with the available information. As a first question, we may ask "is x even?" The answer – yes or no – divides the set $[N]$ into two subsets: the subset of states that are consistent with the answer "yes" – the even numbers $\{0, 2\}$ – and the subset of states that are consistent with the answer "no" – the odd numbers $\{1, 3\}$.

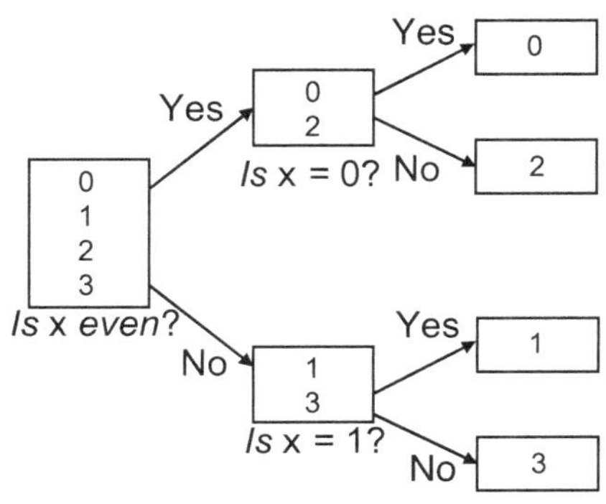

Figure 2.1 Binary tree representing an asking strategy to identify an integer in set [4].

One more question is sufficient to resolve the residual uncertainty within the sets of even and odd numbers. For example, according to the asking strategy in the figure, if we are at the node of the tree reporting odd numbers, we are only uncertain between the possibilities x = 1 and x = 3. Hence, we can ask the question "is x = 1?" Having received the answer "yes," the observer can conclude unambiguously that the state is x = 1, while the answer "no" allows us to determine that the state is x = 0. In a similar way, from the node containing even numbers, we can ask the question "is x = 0?" to identify the correct value of the state x. The strategy in the figure can thus identify any state $x \in [4]$ with two questions, i.e., with two bits.

The previous example demonstrates that two questions are sufficient to identify the true state x out of a set of $N = 4$ possible states. In particular, the asking strategy adopted in the example, illustrated in Figure 2.1, always ends up asking two questions, no matter what the true state x is. Could we reduce the number of questions by asking a single question? This is clearly not possible in general, since a binary answer can only distinguish between two options. The next example shows that it may be possible to ask fewer than two questions on average, as long as the distribution p is not uniform.

Example 2.2

An important limitation of the asking strategy in the previous example is that it ends up asking the *same* number of questions irrespective of the true state of the system. This approach is suboptimal if there are states that are more likely to occur than others. For example, consider the informational state $p = [0.9, 0.05, 0.05, 0]^T$ in which state $\mathrm{x} = 0$ is much more likely than any other state, and state $\mathrm{x} = 3$ cannot occur since it has zero probability, i.e., $p_3 = 0$. If we used the strategy in Example 2.1, the average number of questions would be equal to two, since two questions are required to identify each of the states.

This example studies a different asking strategy that uses knowledge of the distribution p to decrease the average number of questions. To this end, first, we can exclude the state $\mathrm{x} = 3$ from the initial set of possible states indicated in the root of the tree. In fact, this state is impossible under the distribution p. As illustrated in Figure 2.2, the resulting strategy starts by asking the question "is $\mathrm{x} = 0$?" This way, a "yes" answer would allow us to unambiguously identify the state as $\mathrm{x} = 0$ with a single question. Note that this happens with probability 0.9.

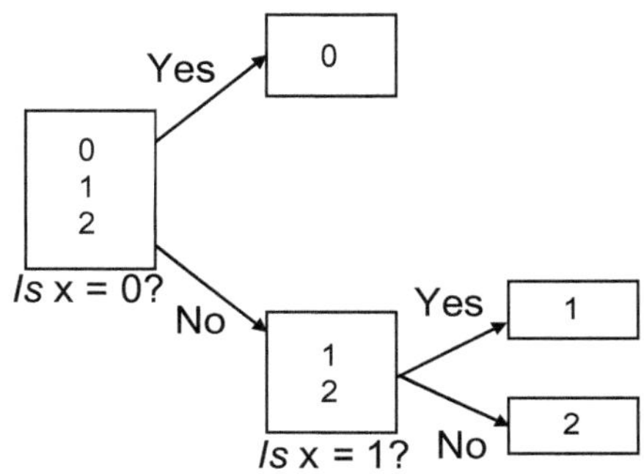

Figure 2.2 Binary tree representing an asking strategy for Example 2.2.

In the unlikely event of a "no" answer, which occurs with probability 0.1, the observer only needs to determine whether the state is $\mathrm{x} = 1$ or $\mathrm{x} = 2$. To this end, the strategy in Figure 2.2 asks "is $\mathrm{x} = 1$?"

With probability 0.9, this asking strategy requires a single question, while with probability 0.1 two questions are needed. The average number of questions required by this strategy is thus

$$0.9 \cdot 1 + 0.1 \cdot 2 = 1.1, \tag{2.1}$$

which is significantly smaller than the two questions required on average by the strategy in Figure 2.1.

The key property of the policy in Figure 2.2 is that it allocates a different number of questions to distinct values x of the state variable, with *more* questions asked for values x having a smaller probability p_x.

2.2.2 Asking Strategy as a Binary Tree

Building on the previous examples, this subsection formalizes the use of binary trees to represent asking strategies. As seen in the examples, an **asking strategy** determines a sequence of binary questions as a function of the past received binary answers. An asking strategy will also be referred to as a code for reasons that will become clear in Section 2.2.4.

The **root node** of the binary tree is labeled with the subset of values x in set $[N]$ that have a positive probability $p_x > 0$. Values x with a zero probability, i.e., $p_x = 0$, are impossible under distribution p, and hence they can be excluded from the initial subset of states. In the following, the probabilities p_x are assumed to be positive for all states $x \in [N]$. This comes without loss of generality, as one can simply redefine N to be the number of states with positive probabilities.

Starting from the root, the tree develops in layers, with each layer corresponding to one step of the asking process, i.e., to one question asked by the observer. Layer 0 includes only the root node that contains the entire set $[N]$. In going from any step $l > 0$ to step $l + 1$, each node at layer l that contains at least two states splits into two **child nodes** at layer $l + 1$. This way, the subset of states associated with any node at layer l is partitioned into two disjoint subsets, each of which is assigned to one of the two child nodes at layer $l + 1$.

Nodes that are labeled with a single state are known as the **leaves** of the tree, and are assigned a specific state $x \in [N]$. By this construction, there must be N leaves, one for each state value $x \in [N]$. All other, non-leaf, nodes of the tree are labeled with a binary question. The child nodes are labeled such that one represents the answer "yes" and the other represents the answer "no."

Definition 2.1 Asking strategy, aka code

An asking strategy, also known as a code, operates sequentially by asking the questions associated with the nodes of the tree that are visited as a function of the binary answers. Specifically, an asking strategy first asks the question associated with the root at layer 0; then the question associated with the children in layer 1 corresponding to the received answer; and so on. When reaching a leaf, the sequence of questions terminates with the identification of the corresponding state $\mathrm{x} = x$. The **depth of the leaf**, labeled as $x \in [N]$, is the number of steps needed to reach the leaf, and thus it quantifies the number of questions required by the asking strategy to identify the given state x. The **depth of the tree** equals the maximum depth across all the leaves of the tree.

Example 2.3

Figure 2.3 illustrates the following asking strategy or code.

- **Step 0:** Ask "is $\mathrm{x} = 1$?"
- **Step 1:** If the answer to the first question is "yes," determine the state as $\mathrm{x} = 1$; meanwhile, if the answer is "no," ask "is $\mathrm{x} = 0$?"
- **Step 2:** If the answer to the second question is "yes," determine the state as $\mathrm{x} = 0$; meanwhile, if the answer is "no," ask "is $\mathrm{x} = 2$?"

- **Step 3:** If the answer to the third question is "yes," determine the state as $\mathrm{x} = 2$; meanwhile, if the answer is "no," determine the state as $\mathrm{x} = 3$.

With this code, the number of questions depends on the actual value of the state x. In fact, the leaf corresponding to $x = 1$ is reached after one question, i.e., the depth of the leaf for state $x = 1$ equals 1. Hence, if the true state is indeed $\mathrm{x} = 1$, a single question is sufficient. In contrast, the depth of the leaf for $x = 0$ equals 2, and the depths of the leaves for $x = 2$ and $x = 3$ equal 3.

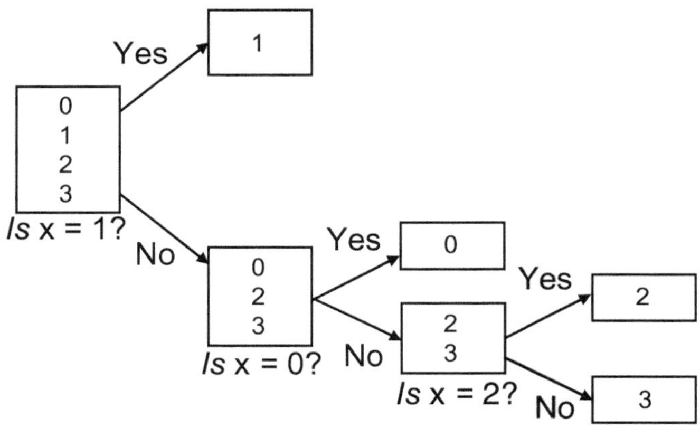

Figure 2.3 Binary tree representing the asking strategy, or code, in Example 2.3.

2.2.3 Average Number of Questions

For a given code, define as L_x the number of binary questions, or bits, required to identify the state when the true value of the state is $\mathrm{x} = x \in [N]$. From the discussion above, the integer L_x coincides with the depth of the leaf in the tree that is labeled with state x. Define also the $N \times 1$ vector

$$L = \begin{bmatrix} L_0 \\ L_1 \\ \vdots \\ L_{N-1} \end{bmatrix} \tag{2.2}$$

collecting the number of questions for all states.

Given an informational state defined by a distribution p, the goal is to find an asking strategy, i.e., a code, that minimizes the **average number of questions**

$$\begin{aligned} L_{\mathrm{avg}}(p) &= \mathrm{E}_p[L_{\mathrm{x}}] \\ &= \sum_{x \in [N]} p_x L_x. \end{aligned} \tag{2.3}$$

Recall that the notation $\mathrm{E}_p[\cdot]$ represents the expectation with respect to the distribution p. In particular, in the expectation $\mathrm{E}_p[L_{\mathrm{x}}]$, the average is taken over the random variable $\mathrm{x} \sim p$, which is emphasized also by the notation $L_{\text{avg}}(p)$. In contrast, the dependence of the quantity $L_{\text{avg}}(p)$ on the code is kept implicit in order to simplify the notation.

Definition 2.2 Optimal asking strategy, aka optimal code
The minimum average number of questions is defined as the solution of the problem

$$L^*_{\text{avg}}(p) = \min_{\text{code}} L_{\text{avg}}(p), \tag{2.4}$$

where the minimum is evaluated over all possible codes. A code that obtains the minimum average number of questions $L^*_{\text{avg}}(p)$ is referred to as an optimal asking strategy, or as an optimal code.

The minimum average number of questions, $L^*_{\text{avg}}(p)$, provides a measure of the uncertainty associated with distribution p. In fact, a more uncertain state $\mathrm{x} \sim p$ will require more questions, on average, to be identified.

The function $L^*_{\text{avg}}(p)$ is permutation-invariant in the sense that permuting the order of the states does not change the value of the minimum average number of questions. This property is aligned with the fact that uncertainty measures should not depend on the specific labeling of the states (see Section 1.4.4).

Example 2.4

Consider an informational state described by distribution $p = [1/4, 1/2, 1/8, 1/8]^T$. With the code in Figure 2.1, the average number of questions is $L_{\text{avg}}(p) = 2$, since all states $x \in [4]$ are assigned the same number of questions $L_x = 2$.

In contrast, the code in Figure 2.3 has leaves at depths defined by the vector

$$L = \begin{bmatrix} L_0 \\ L_1 \\ L_2 \\ L_3 \end{bmatrix} = \begin{bmatrix} 2 \\ 1 \\ 3 \\ 3 \end{bmatrix}. \tag{2.5}$$

This yields the average number of questions

$$\begin{aligned} L_{\text{avg}}(p) &= \mathrm{E}_p[L_{\mathrm{x}}] \\ &= \frac{1}{4}L_0 + \frac{1}{2}L_1 + \frac{1}{8}L_2 + \frac{1}{8}L_3 \\ &= \frac{1}{4}2 + \frac{1}{2}1 + \frac{1}{8}3 + \frac{1}{8}3 = \frac{7}{4} < 2. \end{aligned} \tag{2.6}$$

Therefore, this second code leads to a smaller average number of questions. It does so by asking fewer questions for the states that are most likely to occur. For instance, the most likely state, namely $x = 1$, is assigned a single question.

As will be seen in the next section, this is in fact the optimal code for the given distribution p, and hence the minimum average number of questions is $L^*_{\text{avg}}(p) = 7/4$.

2.2.4 Asking Strategy as a Codebook

So far, a code has been described in the form of a binary tree, with the number of leaves equal to the number of possible state values, N. As discussed next, a code can equivalently be described as a table of N **binary codewords**, where a binary codeword is a sequence of bits. Each of the N binary codewords is associated with a value $x \in [N]$ or, equivalently, with the corresponding leaf in the binary tree.

Definition 2.3 Asking strategy as a codebook of binary codewords
Each asking strategy can be represented by a codebook of N binary codewords. To construct the codebook, assign to each state $x \in [N]$ the sequence of answers leading to the leaf labeled as x in the binary tree. This is done by encoding a "yes" answer with the bit value 0 and a "no" answer with the bit value 1. The list of binary codewords, known as a **codebook**, is equivalent to the binary tree representation of the asking strategy and thus it provides an alternative formulation of the code.

Note that the mapping between yes/no answers and 0/1 is arbitrary, and it can be reversed without affecting the main conclusions.

Owing to the direct correspondence between bits and answers to binary questions, the number of binary questions, L_x, assigned to state $x \in [N]$ can also be interpreted as the **length** of the codeword assigned to state $x \in [N]$. Accordingly, the minimum average number of questions, $L^*_{\text{avg}}(p)$, is also the **minimum average codeword length**.

As detailed in Sections 2.4 and 2.6.4, not every table containing N different sequences of bits is a codebook corresponding to an asking strategy. In other words, unless the table has specific properties, there is generally no asking strategy producing the given table of bits. A codebook that describes an asking strategy will be said henceforth to be **valid**.

Example 2.5

Continuing Example 2.3, the table of codewords describing the sequence of questions for the tree in Figure 2.3 is given as in Table 2.1. For $x = 0$, two answers are needed to identify the state, and the corresponding sequence of binary answers is ("no", "yes"), yielding the binary codeword 10 of length $L_0 = 2$. In contrast, for $x = 1$, a single

Table 2.1 Codebook corresponding to the binary tree in Figure 2.3

State x	Probability p_x	Answers/codeword	L_x
0	1/4	10	2
1	1/2	0	1
2	1/8	110	3
3	1/8	111	3

"yes" answer is sufficient to identify the state, and hence the corresponding codeword is given by the single-bit codeword 0 with length $L_1 = 1$. The entries in the table for states $x = 2$ and $x = 3$ follow in a similar way.

2.3 Huffman Code

Given an informational state p, how can an optimal code be determined? What is the corresponding minimum average codeword length $L^*_{\text{avg}}(p)$? This section addresses these questions by first deriving a simple upper bound on the quantity $L^*_{\text{avg}}(p)$ that is based on the performance of the **uniform code**, which assigns the same number of questions/bits to all states. The optimal code, and hence also the minimum average number of questions $L^*_{\text{avg}}(p)$, is then obtained via the **Huffman algorithm**. The resulting optimal asking sequence is known as the **Huffman code**.

2.3.1 Uniform Code

Consider first the uniform code, which assigns the same number of questions $L_x = l$ to all states $x \in [N]$. The uniform code corresponds to a binary tree with all leaves having the same depth l. That is, all leaves of the tree are at the same distance of l steps from the root of the tree. A tree with all leaves at depth l has 2^l leaves, since at every step each node gives rise to two child nodes. At layer 1 there are 2 leaves; at layer 2 there are $2^2 = 4$ leaves; and so on.

Expressing a uniform code as a codebook, all the N codewords have the same length l bits. Thus, the number of distinct codewords, N, is limited to 2^l, since there are 2^l distinct sequences of l bits.

To uniquely identify all states, it suffices to have as many leaves – or equivalently, as many codewords – as the total number of states. Thus, the number of questions, l, must satisfy the inequality

$$2^l \geq N \text{ or } l \geq \log_2(N). \tag{2.7}$$

This ensures that the required number of questions is sufficient to identify all possible states.

Since l must be an integer, this demonstrates that a uniform code with

$$l = \lceil \log_2(N) \rceil \tag{2.8}$$

questions, where $\lceil a \rceil$ rounds up the real number a to the closest larger integer, is valid. Since the optimal code can be no worse than a uniform code, this in turn implies that the minimum average number of questions cannot exceed $\lceil \log_2(N) \rceil$, yielding the first inequality in

$$L^*_{\text{avg}}(p) \leq \lceil \log_2(N) \rceil < \log_2(N) + 1. \tag{2.9}$$

The second, strict, inequality is a simple consequence of the fact that rounding up can increase a number by no more than 1.

As a special case, if the number of states, N, is a power of 2, i.e., if $N = 2^l$ for some integer l, the bound (2.9) simplifies as

$$L^*_{\text{avg}}(p) \leq \left\lceil \log_2(2^l) \right\rceil = \log_2(2^l) = l. \tag{2.10}$$

2.3.2 Huffman Code

As demonstrated in Section 2.2, the uniform code is generally suboptimal, since an optimal code must generally assign shorter codewords to more likely states. Using this principle, an optimal code can be designed through the Huffman algorithm, which is described in this subsection.

The Huffman algorithm builds an optimal code backwards, defining the corresponding binary tree from the leaves to the root. To start, recall that the binary tree must have N leaves, each assigned to one of the states $x \in [N]$. Furthermore, in order for the code to be optimal, it is necessary for the states corresponding to states x with smaller probabilities p_x to have a larger depth L_x.

Accordingly, building the tree from the leaves at maximal depth, the Huffman code processes states in order of increasing probability, starting from the states with the lowest probability. Warning the reader that the Huffman algorithm is best understood via examples, which are provided next, a general description of the algorithm is provided in Table 2.2.

Throughout the process described in the table, ties between probabilities are broken arbitrarily, highlighting the fact that Huffman code is generally not unique. The optimality of the Huffman code can be proved by induction, as detailed in Appendix 2.A.

Table 2.2 Huffman coding algorithm

Huffman coding
input: distribution p
output: an optimal code in the form of a binary tree
add a node for each state $x \in [N]$ into a queue
list the states in the queue from the most to the least probable
while the queue contains more than one node:
remove the last two (i.e., lowest-probability) nodes from the queue
connect the two removed nodes to a newly created node
add the new node to the queue
reorder, if needed, the nodes in the queue from the most to the least probable
end
set the remaining node in the queue as the root
unroll the created connections between nodes starting from the root
return the resulting binary tree

Example 2.6

To illustrate the Huffman code, consider again the probability distribution $p = [1/4, 1/2, 1/8, 1/8]^T$. Figure 2.4 depicts the steps of the Huffman algorithm, with each step corresponding to one execution of the instructions within the while loop in Table 2.2. In the figure, the nodes are ordered from most likely to least likely state, reflecting the order in the queue.

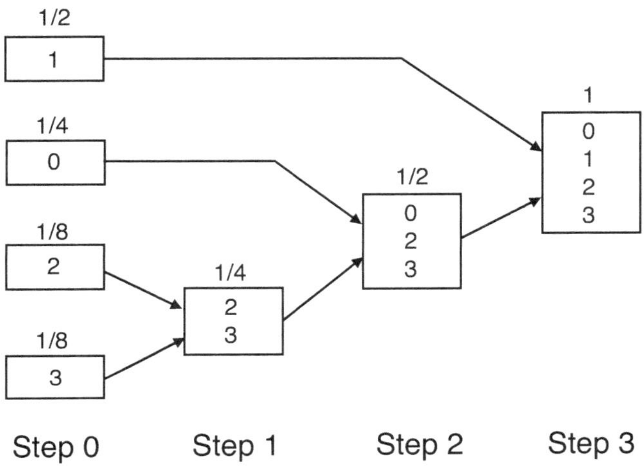

Figure 2.4 Huffman coding derivation (using Table 2.2) for distribution $p = [1/4, 1/2, 1/8, 1/8]^T$.

The algorithm proceeds as follows.

- **Step 0:** Start from the initial queue containing all elements in set $[N] = [4]$ listed in the order of decreasing probabilities as $\{1, 0, 2, 3\}$. The relative ordering of states 2 and 3 is arbitrary, since they have the same probability in vector p.
- **Step 1:** Remove the last two nodes, namely 2 and 3, creating a new node with probability equal to the sum $1/8 + 1/8 = 1/4$. The new node is connected to the nodes 2 and 3. The updated queue is given by states 1, 0, and by the new node $\{2, 3\}$, which has probability $1/4$. The new node is added at the bottom of the queue, since it has the smallest probability in the queue (tied with state 0). Note that no further reordering of the queue is required.
- **Steps 2 and 3:** The process is repeated in the same way for steps 2 and 3. At the end of step 3, the queue contains a single node and the process ends.
- **Unrolling the tree:** As shown in Figure 2.4, the tree that defines the optimal code is finally obtained by inverting the directions of the arrows starting from the root.

It can readily be checked that the binary tree produced by this approach corresponds to the code in Figure 2.3, demonstrating the code's optimality for the given distribution p. In this regard, it is recalled that the choice of the labeling of each of the two edges stemming out of a node as "yes" or as "no" – or as 0 or as 1 – is arbitrary.

As anticipated, the minimum average number of questions is given by

$$L^*_{\text{avg}}(p) = \frac{7}{4}. \tag{2.11}$$

For comparison, the uniform code would require $\lceil \log_2(N) \rceil = \log_2(4) = 2$ bits, which is larger than the minimum codeword length $L^*_{\text{avg}}(p) = 7/4$.

Example 2.7

Consider now an example with $N = 5$ states characterized by the probability vector

$$p = \begin{bmatrix} 0.25 \\ 0.25 \\ 0.2 \\ 0.15 \\ 0.15 \end{bmatrix}. \tag{2.12}$$

The Huffman code is obtained as shown in Figure 2.5 by following the steps of the Huffman algorithm in Table 2.2. An important aspect of this example is that, as illustrated in the figure, at each step, the queue is reordered in order to ensure that the nodes are ordered from largest to smaller probability.

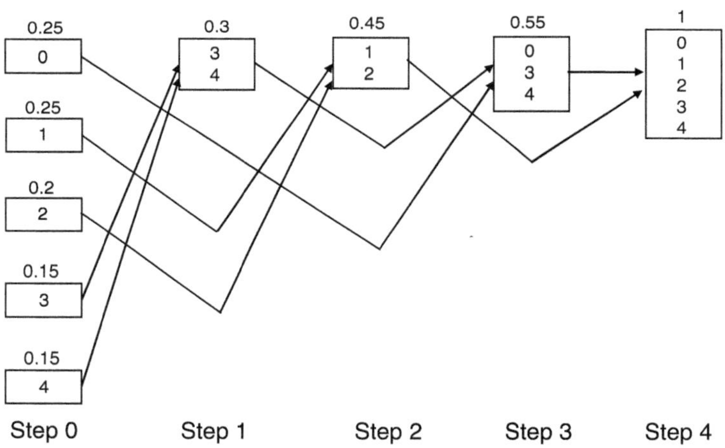

Figure 2.5 Huffman coding derivation (using Table 2.2) for distribution (2.12).

In particular, at step 1, a new node is created by combining the two nodes, namely 3 and 4, with the lowest probabilities. The resulting node {3, 4} is placed at the top of the queue, since it has the largest probability $0.15 + 0.15 = 0.3$. Correspondingly, the nodes 0, 1, and 2 are moved down by one position in the queue, as represented in the figure by the downward lines moving from step 0 to step 1.

Table 2.3 Codebook for the Huffman code in Figure 2.6

State x	Probability p_x	Answers/codewords	L_x
0	0.25	01	2
1	0.25	10	2
2	0.2	11	2
3	0.15	000	3
4	0.15	001	3

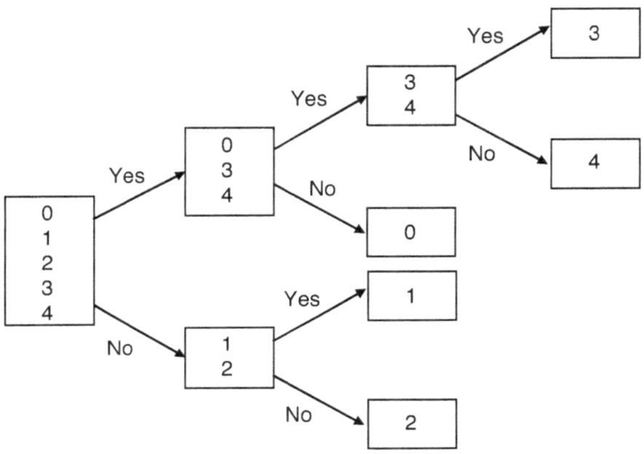

Figure 2.6 Binary tree representation of the Huffman code obtained from Figure 2.5.

Similar operations are applied for the following steps. The final unrolled tree, obtained by reversing the direction of the arrows, is illustrated in Figure 2.6. Furthermore, the codebook corresponding to the derived Huffman code is given in Table 2.3. Accordingly, the minimum average number of questions is given by

$$\begin{aligned} L^*_{\text{avg}}(p) &= 0.25 \cdot 2 + 0.25 \cdot 2 + 0.2 \cdot 2 + 0.15 \cdot 3 + 0.15 \cdot 3 \\ &= 2.3. \end{aligned} \tag{2.13}$$

2.3.3 Optimal Code for Dyadic Distributions

In general, evaluating the minimum average number of questions, $L^*_{\text{avg}}(p)$, requires running the Huffman algorithm for the given distribution vector p of interest. There is, however, a class of distributions, known as **dyadic distributions**, for which the quantity $L^*_{\text{avg}}(p)$ can be obtained in analytical form.

To start, consider the special case of a **uniform** dyadic distribution, that is, a distribution p that has all entries $x \in [N]$ equal to

$$p_x = 2^{-l} \tag{2.14}$$

for some integer l. For uniform dyadic distributions, the uniform code is optimal, and thus the minimum codeword length is given by

$$L^*_{\text{avg}}(p) = l. \tag{2.15}$$

This result, which can be verified via the Huffman algorithm, is also intuitively clear: since all codewords share the same probability (2.14), the average codeword length cannot be reduced by assigning different codeword lengths to different values of x.

Definition 2.4 Dyadic distributions
A distribution p is dyadic if each probability p_x can be written as

$$p_x = 2^{-l_x}, \tag{2.16}$$

where the integer l_x is allowed to depend on the state $x \in [N]$. Equivalently, a distribution p is dyadic if the negative logarithm of each probability p_x is an integer l_x, i.e.,

$$-\log_2(p_x) = l_x \tag{2.17}$$

for all $x \in [N]$.

Uniform dyadic distributions are clearly a special case of dyadic distributions with the equality $l_x = l$ holding for all $x \in [N]$.

Key Result 2.1 Optimal code for dyadic distributions
For dyadic distributions, an optimal code is characterized by the **optimal codeword lengths**

$$L_x^* = -\log_2(p_x) = l_x \tag{2.18}$$

for all $x \in [N]$. Thus, the minimum average codeword length is given by

$$\begin{aligned} L_{\text{avg}}^*(p) &= \sum_{x\in[N]} p_x l_x \\ &= -\sum_{x\in[N]} p_x \log_2(p_x). \end{aligned} \tag{2.19}$$

This result can be concluded by analyzing the performance of the Huffman code, and an alternative proof approach will be presented in the next section. As elaborated in the next two sections, the expression (2.18) plays a key role in information theory.

Example 2.8

Consider again the distribution $p = [1/4, 1/2, 1/8, 1/8]^T$ studied in the previous example. This is a dyadic distribution with integers $l_0 = 2$, $l_1 = 1$, $l_2 = 3$, and $l_3 = 3$. Therefore, the optimal codeword lengths (2.18) are given as in Table 2.1.

2.4 Ideal Codeword Lengths and Shannon Surprise

2.4.1 Optimizing the Codeword Lengths

In Section 2.3, optimal codes were derived via the Huffman algorithm. In this section, an alternative optimization-based approach is presented, in which the minimum codeword length $L_{\text{avg}}(p)$ is directly minimized with respect to the vector L of codeword lengths. Through this analysis, it will be concluded that an optimal vector L can be well approximated by setting the codeword lengths as $L_x = -\log_2(p_x)$ for all states $x \in [N]$. Note that, by (2.19), this solution is exactly optimal for dyadic distributions. The quantity $-\log_2(p_x)$ is central in information theory, and is known as the **Shannon surprise**. The Shannon surprise $-\log_2(p_x)$ will be interpreted as a measure of the information acquired by the observer when the observation of the system returns state x.

The problem of interest is the minimization of the average codeword length in (2.4). We define as valid a vector L of codeword lengths corresponding to an asking strategy. Accordingly, a vector L is valid if it describes the depths of the leaves of a binary tree. The following fundamental result, proved in Appendix 2.B, identifies a condition:

Key Result 2.2 Kraft–McMillan inequality

A vector $L = [L_0 \cdots L_{N-1}]^T$ of codeword lengths is valid, i.e., it is possible to construct a binary tree with leaves at depths specified by vector L, if and only if the inequality

$$\sum_{x\in[N]} 2^{-L_x} \leq 1 \tag{2.20}$$

is satisfied.

Intuitively, the Kraft–McMillan inequality (2.20) says that a binary tree cannot have too many leaves with short codewords, since a smaller value of codeword length L_x implies a larger value of the quantity 2^{-L_x}.

Using the Kraft–McMillan inequality (2.20), the problem of minimizing the average codeword length can now be formally stated as the minimization

$$\begin{aligned} &\underset{L}{\text{minimize}} \left\{ L_{\text{avg}}(p) = \sum_{x\in[N]} p_x L_x \right\} \\ &\text{subject to} \sum_{x\in[N]} 2^{-L_x} \leq 1 \end{aligned} \tag{2.21}$$

over the codeword length vector L. By Kraft–McMillan's inequality (2.20), the constraint in (2.21) enforces that the vector L obtained as the solution of problem (2.21) be valid.

Denoting as $L^* = [L_0^*, \ldots, L_{N-1}^*]^T$ an optimal solution for the problem (2.21), the corresponding optimal value of the objective, $\sum_{x\in[N]} p_x L_x^*$, equals the minimum average codeword length:

$$L^*_{\text{avg}}(p) = \sum_{x \in [N]} p_x L^*_x. \tag{2.22}$$

As discussed in Section 2.3, an exact solution to problem (2.21) can be found via the Huffman algorithm (Table 2.2). This section focuses on deriving and analyzing a simpler, but generally approximate, solution.

To this end, the problem (2.21) is now studied through a simplified formulation that is obtained by removing the constraint that the entries of vector L be integer numbers. As proved in Appendix 2.C, the solution of this problem can be computed explicitly as follows:

Key Result 2.3 Ideal codeword lengths
Consider the relaxed problem (2.21) in which the entries of vector L can take any non-negative values. The solution of this problem is given by the vector $\tilde{L}^* = [\tilde{L}^*_0, \ldots, \tilde{L}^*_{N-1}]^T$ with

$$\tilde{L}^*_x = -\log_2(p_x) \tag{2.23}$$

for all $x \in [N]$. The values $\{\tilde{L}^*_x = -\log_2(p_x)\}_{x \in [N]}$ are known as the ideal codeword lengths for distribution p.

In general, the ideal codeword lengths $\tilde{L}^* = [-\log_2(p_0), \ldots, -\log_2(p_{N-1})]^T$ are not valid codeword lengths, since the entries $\{\tilde{L}^*_x = -\log_2(p_x)\}_{x \in [N]}$ are not necessarily integers. However, being the solution of a relaxed minimization problem, the ideal codeword lengths yield an average codeword length that is no larger than the optimal average codeword length. This result is summarized next.

Key Result 2.4 Lower bound on the minimum average codeword length
The minimum average codeword length is lower bounded as

$$\begin{aligned} L^*_{\text{avg}}(p) &= \sum_{x \in [N]} p_x L^*_x \\ &\geq \sum_{x \in [N]} p_x \tilde{L}^*_x \\ &= -\sum_{x \in [N]} p_x \log_2(p_x). \end{aligned} \tag{2.24}$$

The next example shows that for dyadic distributions the ideal codeword lengths actually correspond to the optimal codeword lengths for the original problem (2.21). This provides an alternative proof for the result (2.18) derived in Section 2.3.

Example 2.9

In the special case of the dyadic distribution $p_x = 2^{-l_x}$ for $x \in [N]$, by definition (2.17), the ideal codeword lengths are integer valued, as the equality

$$\tilde{L}_x^* = -\log_2(p_x) = l_x \tag{2.25}$$

holds for all $x \in [N]$. Therefore, for any dyadic distribution, the ideal codeword lengths solve the original problem (2.21), and hence coincide with the optimal codeword lengths, i.e., $L_x^* = \tilde{L}_x^*$ for all $x \in [N]$. It follows that the minimum average codeword length is given by (2.19), as also proved in Section 2.3.

2.4.2 Shannon Surprise

The ideal codeword length $\tilde{L}_x^*$ is also known as the Shannon surprise associated with the state x. As discussed next, it can be interpreted as the amount of information accrued by the observer when observing state x.

Definition 2.5 Shannon surprise, aka Shannon self-information
Given a distribution p, the Shannon surprise, or Shannon self-information, associated with state x is defined as

$$-\log_2(p_x). \tag{2.26}$$

The interpretation of the Shannon surprise as the amount of information associated with state x is a consequence of the behavior illustrated in Figure 2.7. First,

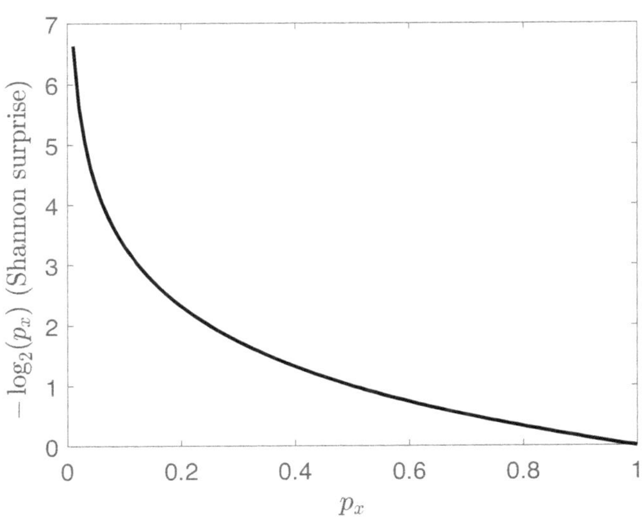

Figure 2.7 Shannon surprise $-\log_2(p_x)$ vs. probability p_x.

as seen in the figure, the Shannon surprise is non-negative, and it equals zero if and only if there is no uncertainty – and thus no surprise – associated with state x.

Key Result 2.5 Non-negativity and minimum-value property of the Shannon surprise
For any probability $p_x \in (0, 1]$, the Shannon surprise is positive

$$-\log_2(p_x) > 0, \tag{2.27}$$

with equality

$$-\log_2(p_x) = 0 \tag{2.28}$$

holding if and only if $p_x = 1$.

Furthermore, as also illustrated in Figure 1.7, the Shannon surprise decreases as the corresponding probability p_x gets smaller, making the event x more surprising. Indeed, the Shannon surprise tends to infinity as the probability p_x tends to zero.

Key Result 2.6 Monotonicity property of the Shannon surprise
The Shannon surprise $-\log_2(p_x)$ is a strictly decreasing function of probability p_x. Furthermore, in the limit $p_x \to 0$, the Shannon surprise tends to infinity, i.e.,

$$\lim_{p_x \to 0} -\log_2(p_x) = \infty. \tag{2.29}$$

Example 2.10

A letter of the English alphabet, excluding the space, is picked uniformly at random from a text. Recalling that there are 27 letters in the English alphabet, the resulting observation is an integer x $\in$ [26]. The probability distribution of random variable x depends on the **frequency** with which each letter, assigned to an integer x, appears in the text.

Figure 2.8(a) shows a typical distribution over letters for an English text, where letters are numbered in alphabetical order from 0 to 25. For example, one can see that the letter indexed as 4, namely e, has the largest probability. (The frequency of a letter can also be used to quantify the difficulty of writing a text without it, i.e., a **lipogram**.)

Figure 2.8(b) displays the corresponding Shannon surprise $-\log_2(p_x)$. Letters with a smaller probability are associated with a larger Shannon surprise, while more likely letters have a smaller Shannon surprise.

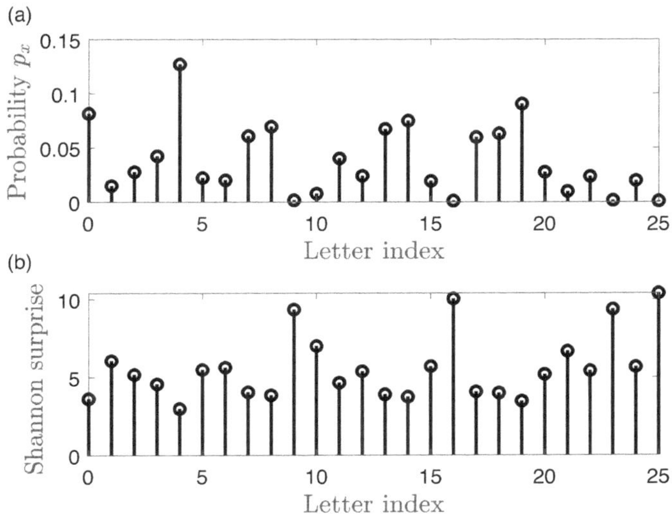

Figure 2.8 (a) Probabilities p_x and (b) corresponding Shannon surprises $-\log_2(p_x)$ for the letters of the English alphabet.

Example 2.11

Consider again the distribution (2.12), i.e., $p = [0.25, 0.25, 0.2, 0.15, 0.15]^T$, over $N = 5$ states. Table 2.4 reports the Shannon surprises, as well as the optimal codeword lengths derived in Example 2.7 via the Huffman algorithm. The distribution is not dyadic, and the Shannon surprises do not correspond to valid codeword lengths, as they are not all integers. Therefore, in this case, the simple relation $L_x^* = -\log_2(p_x)$ between Shannon surprise and optimal codeword lengths L_x^* for all $x \in [N]$ does not hold.

Table 2.4 Probabilities, Shannon surprises, and optimal codeword lengths for distribution (2.12)

State x	Probability p_x	Shannon surprise $-\log_2(p_x)$	L_x
0	0.25	2	2
1	0.25	2	2
2	0.2	2.32	2
3	0.15	2.73	3
4	0.15	2.73	3

2.5 Shannon Entropy

This section introduces the Shannon entropy, the most fundamental measure of uncertainty and information. It will be demonstrated that the Shannon entropy provides a close approximation of the minimum average number of questions, $L_{\text{avg}}^*(p)$, i.e., of the minimum average codeword length. Furthermore, the Shannon entropy will be shown in this section to satisfy the properties of non-negativity and minimal entropy, as well as of maximal entropy, described in Chapter 1. Chapter 4 will finally prove that the Shannon entropy also guarantees the third, monotonicity, property listed in Chapter 1, thus providing a valid entropy measure.

2.5.1 Defining the Shannon Entropy

Definition 2.6 Shannon entropy
Given a distribution vector p, the Shannon entropy $\mathrm{H}(p)$ is defined as the **average Shannon surprise**

$$\begin{aligned} \mathrm{H}(p) &= -\mathrm{E}_p[\log_2(p_{\mathrm{x}})] \\ &\sum_{x\in[N]} p_x(-\log_2(p_x)), \end{aligned} \tag{2.30}$$

where the average is taken with respect to the random variable $\mathrm{x} \sim p$.

The Shannon entropy is measured in bits. Furthermore, like $L^*_{\mathrm{avg}}(p)$, the Shannon entropy is **permutation-invariant**, in the sense that permuting the order of the states does not change the value of the Shannon entropy.

Since some of the probabilities p_x may equal zero, in evaluating the Shannon entropy (2.30), one applies the equality

$$0\log_2(0) = 0, \tag{2.31}$$

effectively discarding all values x with zero probability. One may thus say that **impossible** states x, i.e., states with probability $p_x = 0$, do not contribute to the Shannon entropy. The equality (2.31) is justified by the limit

$$\lim_{p_x \to 0} p_x \log_2(p_x) = 0. \tag{2.32}$$

Example 2.12

As discussed in Section 1.4, any distribution of the form $p = \delta_{\bar{x}}$ for some state $\bar{x} \in [N]$ corresponds to an informational state of minimal uncertainty, in which the observer knows the state to be $\bar{x}$. In a manner consistent with this intuition, the Shannon entropy for distribution $p = \delta_{\bar{x}}$ equals zero, i.e.,

$$\begin{aligned} \mathrm{H}(\delta_{\bar{x}}) &= -p_{\bar{x}}\log_2(p_{\bar{x}}) - \sum_{x\in[N]:\, x\neq\bar{x}} p_x \log_2(p_x) \\ &= -1\cdot\log_2(1) - (N-1)\cdot 0\cdot\log_2(0) \\ &= 0. \end{aligned} \tag{2.33}$$

Example 2.13

As also seen in Section 1.4, the uniform distribution $p = u$, with $p_x = u_x = 1/N$ for all states $x \in [N]$, corresponds to an informational state of maximal uncertainty. This yields the Shannon entropy

$$H(u) = N\frac{1}{N}\log_2(N) = \log_2(N), \tag{2.34}$$

which will be seen to be the maximum possible value for the Shannon entropy. Accordingly, the Shannon entropy grows at most logarithmically with the number of states, N.

For the special case of a uniform dyadic distribution (2.14), i.e., with $p_x = N^{-1} = 2^{-l}$ for an integer l, using (2.34), the Shannon entropy equals

$$H(u) = \log_2(N) = \log_2(2^l) = l, \tag{2.35}$$

which, as seen in Section 2.3.3, corresponds to the minimum average codeword length for this informational state.

Example 2.14

As another illustration of the result (2.34), take a composite system as described in Section 1.3.3. For a system with S subsystems, each with D possible states, the number of states is $N = D^S$, and thus a uniform distribution assigns probability $u_x = D^{-S}$ to all states $x \in [N]$. The corresponding Shannon entropy is given by

$$H(u) = \log_2(N) = S\log_2(D). \tag{2.36}$$

Therefore, under a uniform distribution, the uncertainty of a composite system, as measured by the Shannon entropy, grows linearly with the number of subsystems. For this reason, the Shannon entropy is said to be **extensive**, as it increases proportionally with the number of subsystems, S.

2.5.2 Shannon Entropy and Minimum Average Codeword Length

How is the Shannon entropy $H(p)$ related to the minimum average number of questions $L^*_{\text{avg}}(p)$? For a dyadic distribution, as concluded in (2.19), the two quantities coincide, i.e.,

$$L^*_{\text{avg}}(p) = H(p). \tag{2.37}$$

More generally, for any distribution vector p, the Shannon entropy will be seen next to provide an approximation of the minimum average number of questions that is correct with an error of at most one binary question, or bit.

Key Result 2.7 Shannon entropy and minimum average codeword length
For any probability distribution p, the Shannon entropy $\mathrm{H}(p)$ and the minimum average codeword length $L^*_{\mathrm{avg}}(p)$ are related as

$$\mathrm{H}(p) \leq L^*_{\mathrm{avg}}(p) < \mathrm{H}(p) + 1. \tag{2.38}$$

The **lower bound** $L^*_{\mathrm{avg}}(p) \geq \mathrm{H}(p)$ states that the minimum average number of questions cannot be smaller than the Shannon entropy. This result follows directly from the inequality (2.24), as the Shannon entropy coincides with the average of the ideal codeword lengths.

The **upper bound** $L^*_{\mathrm{avg}}(p) < \mathrm{H}(p) + 1$ states that the minimum average codeword length cannot exceed the Shannon entropy by more than 1 bit. This result is proved by showing the existence of a code with average codeword length $L_{\mathrm{avg}}(p)$ no larger than $\mathrm{H}(p) + 1$, i.e., $L_{\mathrm{avg}}(p) < \mathrm{H}(p) + 1$. In fact, an optimal code cannot have an average codeword length larger than that of any other code.

The existence of a code with average codeword length $L_{\mathrm{avg}}(p) \leq \mathrm{H}(p) + 1$ is a direct consequence of the Kraft–McMillan inequality. To see this, take the following code, which is also referred to as the **Shannon code**. Under the Shannon code, the codeword length for state $x \in [N]$ is given by the **rounded-up ideal codeword length**

$$L_x = \lceil -\log_2(p_x) \rceil < -\log_2(p_x) + 1. \tag{2.39}$$

As indicated in (2.39), the codeword length L_x cannot exceed the ideal codeword length $\tilde{L}^*_x = -\log_2(p_x)$ by more than 1 bit, since rounding up can increase the input number by no more than 1.

The assignment (2.39) corresponds to a valid code, since it satisfies the Kraft–MacMillan inequality (2.20). This can readily be verified as

$$\begin{aligned} \sum_{x\in[N]} 2^{-L_x} &< \sum_{x\in[N]} 2^{\log_2(p_x)-1} \\ &= \frac{1}{2}\sum_{x\in[N]} p_x \\ &= \frac{1}{2} < 1, \end{aligned} \tag{2.40}$$

concluding the proof of the upper bound in (2.38).

As a final remark, it is important to emphasize that, while the minimum average number of questions is well approximated by replacing the codeword length L_x with the Shannon surprise $-\log_2(p_x)$, this does not mean that the vector of Shannon surprises, $[-\log_2(p_0), \ldots, -\log_2(p_{N-1})]^T$, provides a good approximation of a vector L^* of optimal codeword lengths. This fact is illustrated by the following example.

Example 2.15

Consider a binary system ($N = 2$) with informational state $p = [2^{-l}, 1-2^{-l}]^T$ for some finite integer $l > 0$. The optimal codeword lengths are given by vector $L = [1, 1]^T$, since one question is clearly sufficient to identify a binary state. However, the vector of Shannon surprises is given by $[-\log_2(2^{-l}) = l, -\log_2(1 - 2^{-l})]^T$, and thus the Shannon surprise of the first state, $x = 0$, grows unbounded as l increases.

2.5.3 Properties of the Shannon Entropy

As illustrated by the next example, the Shannon entropy satisfies the three defining properties of an entropy introduced in Chapter 1. The first two properties are formalized and proved after the example, while monotonicity will be demonstrated in Chapter 4.

Example 2.16

Take again a binary system, this time with a general distribution $p = [p_0, p_1]^T$. The Shannon entropy can be expressed as

$$\begin{aligned} \mathrm{H}(p) &= \mathrm{h}(p_1) \\ &= -(1-p_1)\log_2(1-p_1) - p_1\log_2(p_1), \end{aligned} \tag{2.41}$$

where the function $\mathrm{h}(p_1)$ of probability $p_1 \in [0, 1]$ will be referred to henceforth as the **binary Shannon entropy**. Note that the equality

$$\mathrm{h}(p_1) = \mathrm{h}(p_0) \tag{2.42}$$

holds, reflecting the permutation invariance of the Shannon entropy.

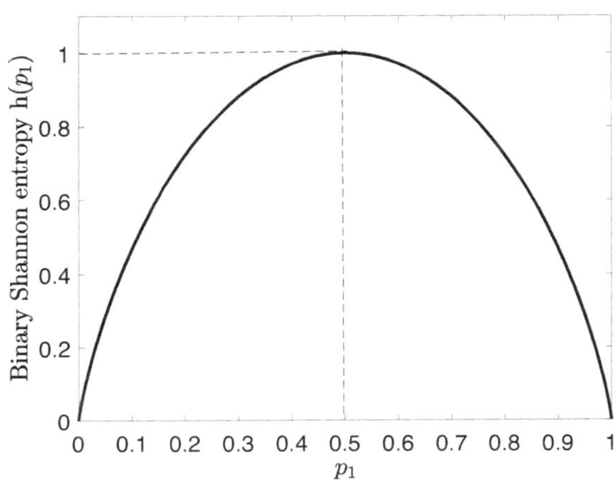

Figure 2.9 Binary Shannon entropy $\mathrm{h}(p_1) = \mathrm{h}(p_0)$, i.e., Shannon entropy of a bit with generic distribution $p = [p_0, p_1]^T$.

As illustrated in Figure 2.9, the binary Shannon entropy is non-negative and it achieves the minimum value of zero if and only if $p = \delta_0$ or $p = \delta_1$, i.e., if and only if $p_1 = 0$ or $p_1 = 1$, respectively. In either case, the observer is certain about the state of the system being either 0 or 1, and the binary Shannon entropy equals zero, $\mathrm{h}(0) = \mathrm{h}(1) = 0$.

Furthermore, the binary Shannon entropy is maximized when the distribution is uniform, i.e., when $p_0 = p_1 = 0.5$ and thus the observer considers both states to be equally likely, in which case it equals $\mathrm{H}(p) = \log_2(2) = 1$ bit.

Finally, the binary Shannon entropy is **monotonically increasing** as p_1 moves away from $p_1 = 0$ or $p_1 = 1$ toward the value $p_1 = 0.5$, which corresponds to a uniform distribution.

Generalizing the example above, the Shannon entropy satisfies the following properties, which were first introduced in Chapter 1 as general defining properties of entropy measures.

The first property states that the Shannon entropy is non-negative and that informational states of minimal uncertainty – as measured by the Shannon entropy – are exactly those that are fully concentrated at one state $\bar{x} \in [N]$.

Key Result 2.8 Non-negativity and minimum-value property of the Shannon entropy

The Shannon entropy is non-negative,

$$\mathrm{H}(p) \geq 0, \tag{2.43}$$

for all distributions p, and it equals zero if and only if the distribution is of the form $p = \delta_{\bar{x}}$ for some state $\bar{x} \in [N]$,

$$\mathrm{H}(\delta_{\bar{x}}) = 0. \tag{2.44}$$

This property follows directly from the non-negativity and minimum value property, (2.27) and (2.28), of the Shannon surprise.

The second property states that the Shannon entropy grows at most logarithmically with the size of the state space, N. Therefore, doubling the state space can increase the Shannon entropy by 1 bit.

Key Result 2.9 Maximum-value property of the Shannon entropy

The Shannon entropy is upper bounded as

$$\mathrm{H}(p) \leq \log_2(N), \tag{2.45}$$

achieving the upper bound if and only if the distribution is uniform, i.e.,

$$\mathrm{H}(u) = \log_2(N). \tag{2.46}$$

A proof of this property can be found in Appendix 2.E, and another proof will be provided in Chapter 3. The monotonicity of the Shannon entropy will be proved in Chapter 4.

Example 2.17

The maximum logarithmic growth of the Shannon entropy described by the upper bound (2.45) is the reason why, now and for the foreseeable future, all devices on the Internet – around $N = 15 \times 10^9$ at the time of writing – can be identified with IP addresses of only 128 bits. In fact, by (2.34), with 128 bits one can identify up to $N = 2^{128} \approx 3 \times 10^{38}$ devices.

2.5.4 Concavity of the Shannon Entropy

Another useful property of the Shannon entropy, which is not necessarily satisfied by all entropy measures, is its **concavity**. Concavity may be informally stated as the property that mixing increases uncertainty. To understand this interpretation, consider a situation in which the observer is uncertain between two possible informational states p' and p''. As an example, the observer may have access to two experts or two predictive models, each offering generally distinct confidence levels p' and p'' over the possible state values $x \in [N]$.

Not knowing which informational state to adopt, the observer assigns probability $0 \leq \gamma \leq 1$ to informational state p' and $1 - \gamma$ to informational state p''. So, the probability γ defines a higher-level form of uncertainty of the observer between the two informational states. By the **law of total probability**, the resulting informational state is the average

$$p = \gamma p' + (1 - \gamma)p''. \tag{2.47}$$

It is interesting now to compare the Shannon entropy of the average informational state (2.47), i.e.,

$$\mathrm{H}(p) = \mathrm{H}(\gamma p' + (1 - \gamma)p''), \tag{2.48}$$

with the Shannon entropies $\mathrm{H}(p')$ and $\mathrm{H}(p'')$ of the individual informational states p' and p''. Intuitively, mixing, as per (2.47), may be expected to increase uncertainty. After all, because of mixing, the observer is not even sure about which of the two informational states p' and p'' should be adopted, introducing a higher level of uncertainty.

As a reference, one accordingly takes the average Shannon entropy

$$\gamma \mathrm{H}(p') + (1 - \gamma)\mathrm{H}(p''), \tag{2.49}$$

which would be obtained on average for an observer that is either in informational state p' or p'' – and not in the mixed informational state (2.47). The concavity property of the Shannon entropy amounts to a relationship between the Shannon entropy $\mathrm{H}(p)$ of the mixed informational state in (2.48) and the average Shannon entropy in (2.49).

Key Result 2.10 Concavity of the Shannon entropy (mixing increases uncertainty)

The Shannon entropy is a **concave function** of the probability vector p. This means that, given any two distinct probability vectors p' and p'', and for any probability $\gamma \in [0, 1]$, the inequality

$$\mathrm{H}(\gamma p' + (1-\gamma)p'') \geq \gamma \mathrm{H}(p') + (1-\gamma)\mathrm{H}(p'') \quad (2.50)$$

is satisfied, with equality holding if and only if $\gamma = 0$ or $\gamma = 1$.

As anticipated, this result may be equivalently summarized by saying that mixing increases uncertainty when the latter is measured by the Shannon entropy.

Mathematically, the concavity property (2.50) indicates that the Shannon entropy has a downward-pointing curvature as a function of the probability vector p. Therefore, the result (2.50) can be proved by evaluating the second derivatives – more precisely, the Hessian matrix – of the Shannon entropy. Equivalently, taking a geometric perspective, the inequality (2.50) states that the Shannon entropy is never below the segment drawn between any two points on the function. This is illustrated by the next example.

Example 2.18

The binary Shannon entropy (2.41) is plotted again in Figure 2.10. From the figure, it can be observed that, by its concavity, the Shannon entropy has a downward-pointing,

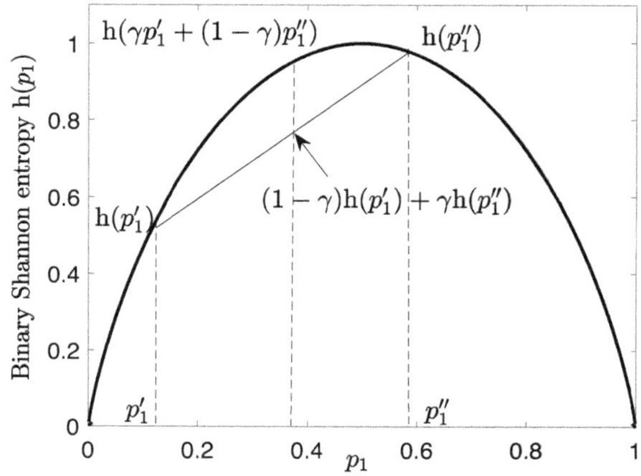

Figure 2.10 Illustration of the concavity of the Shannon entropy. Concavity indicates that the Shannon entropy is never below the segment drawn between any two points on the function.

i.e., "∩-shaped," curvature. As mentioned above, this can also be shown analytically by evaluating the second derivative of the binary Shannon entropy, and by showing that it is negative.

To provide an equivalent, geometric, definition of concavity, consider, as in the figure, two probabilities p_1' and p_1'', as well as any mixed probability $\gamma p_1' + (1-\gamma)p_1''$ in between the two, for some probability $0 \leq \gamma \leq 1$. The Shannon entropy (2.48) at this mixed probability equals $\mathrm{h}(\gamma p_1' + (1-\gamma)p_1'')$.

As shown in Figure 2.10, the binary Shannon entropy $\mathrm{h}(\gamma p_1' + (1-\gamma)p_1'')$ is larger than the average Shannon entropy (2.49), i.e., $\gamma\mathrm{h}(p_1') + (1-\gamma)\mathrm{h}(p_1'')$, which is obtained geometrically by connecting the two points $(p_1', \mathrm{h}(p_1'))$ and $(p_1'', \mathrm{h}(p_1''))$ with a segment. Accordingly, the concavity of the binary Shannon entropy can be expressed as the following special case of the inequality (2.50):

$$\mathrm{h}(\gamma p_1' + (1-\gamma)p_1'') \geq \gamma\mathrm{h}(p_1') + (1-\gamma)\mathrm{h}(p_1''), \tag{2.51}$$

which states that the points on the segment between any two points on the Shannon entropy curve are never above the Shannon entropy itself.

Appendix 2.D provides further discussion on concavity, and Appendix 2.E uses the concavity of the Shannon entropy to prove the upper bound (2.45). As a summary, the key idea behind the proof of the inequality (2.45) is that mixing all possible permutations of any probability vector p (with all positive entries) yields the uniform distribution u. Therefore, by concavity (2.50) and by the permutation invariance of the Shannon entropy, the Shannon entropy of the mixed, uniform, distribution is no smaller than the Shannon entropy of the distribution p.

2.5.5 Thermodynamic Information and Shannon Entropy

This section is concluded by making a connection with physics, returning to the distinction presented in Chapter 1 between objective and subjective uncertainty. In thermodynamics, one views a system characterized by the maximally uncertain distribution $p = u$ over the states as a maximally disordered state from which no work can be extracted. Conversely, any other informational state $p \neq u$ contains sufficient order that can be, loosely speaking, turned into work. Accordingly, the level of order can be thought of as a form of objective information about the system, given its physical significance in terms of work extraction.

Based on these considerations, when the distribution p describes the physical, thermodynamic, state of a system, it is common to take the quantity

$$\mathrm{I}(p) = \log_2(N) - \mathrm{H}(p) \tag{2.52}$$

as a measure of the objective, thermodynamic, information present in a state p. The quantity $\mathrm{I}(p)$, known as **negentropy**, is non-negative, i.e., $\mathrm{I}(p) \geq 0$, by the maximum-value property (2.45) of the Shannon entropy. In thermodynamics, the amount of work that can be extracted from a heat bath of temperature T using a system with distribution p is given by $kT \cdot \mathrm{I}(p)$, where k is the Boltzmann constant.

The negentropy $\mathrm{I}(p)$ in (2.52) has a dual relationship with the entropy $\mathrm{H}(p)$. In fact, a larger Shannon entropy $\mathrm{H}(p)$, and thus a larger uncertainty, reduces the negentropy $\mathrm{I}(p)$ associated with the state. Indeed, the sum of the Shannon entropy $\mathrm{H}(p)$ and of the negentropy $\mathrm{I}(p)$ must always equal the logarithm of the number of possible states of the system, i.e.,

$$\mathrm{I}(p) + \mathrm{H}(p) = \log_2(N). \tag{2.53}$$

2.6 Quantifying Information by Compressing

The previous sections have studied metrics that quantify uncertainty by counting the minimum average number of binary questions, $L^*_{\text{avg}}(p)$, required to identify the state of a system under a given informational state p. Notably, the Shannon entropy, $\mathrm{H}(p)$, was seen to offer a close, easily computable, approximation of $L^*_{\text{avg}}(p)$ that also satisfies the key desirable properties for an uncertainty measure mentioned in Chapter 1. This section moves on to the different informational task of **compressing** the state of the system.

2.6.1 Compression

As illustrated in Figure 2.11, consider an **information source** that produces a random output $\mathrm{x} \sim p$. The output of the information source may be, for example, an image, an audio signal, a video, and it is assumed to be selected within some repository of N contents indexed by the state variable $\mathrm{x} \in [N]$.

An **encoder** observes the actual state $\mathrm{x} \sim p$, and the goal is for the encoder to describe x to a **decoder** by communicating the minimum average number of bits. To this end, the message sent by encoder to decoder can have a different number of bits depending on the specific value x. In designing encoder and decoder, the probability vector p describing the informational state is assumed to be known at both ends, but only the encoder observes the realization $\mathrm{x} \sim p$. The decoder thus plays the role of the observer with prior knowledge represented by the distribution p.

Definition 2.7 Compression codebook, aka compression code
The encoder obtains a realization $\mathrm{x} = x$ of the state, and it assigns to it a message of L_x bits, which is referred to as the **codeword** c_x. The list of all codewords $\{c_x\}_{x\in[N]}$ is known as the compression codebook, which is shared between transmitter and receiver. Knowing the codebook, the decoder maps the received codeword c_x to the original observation x.

The only constraint imposed on the design of the compression codebook is that the codewords be distinct, so as to enable decoding. Accordingly, unlike for a codebook corresponding to an asking strategy, no binary tree structure is imposed on

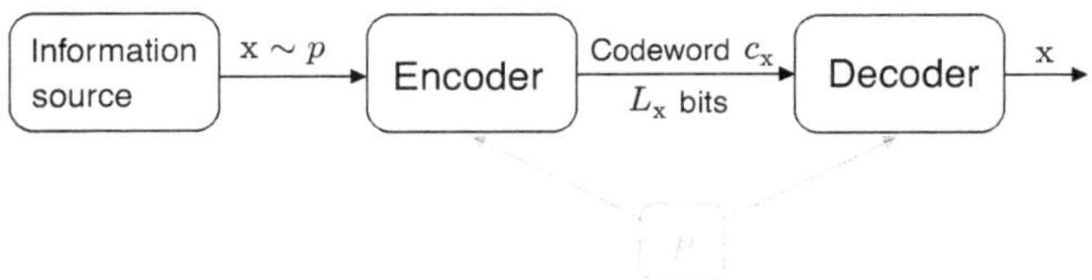

Figure 2.11 Illustration of the problem of compressing an information source $x \sim p$.

the compression codebook. This point will be discussed extensively in Section 2.6.4. Furthermore, it is emphasized that knowledge of the codebook is essential for decoding: for a decoder without knowledge of the codebook, the received bits are effectively **encrypted**, and decoding is not possible in general.

An optimal compression codebook is one that minimizes the average codeword length, obtaining, as in (2.4), the minimum average codeword length

$$L^*_{\text{avg}}(p) = \min_{\text{compression code}} L_{\text{avg}}(p), \tag{2.54}$$

with the minimum evaluated over all possible compression codebooks $\{c_x\}_{x \in [N]}$.

Example 2.19

The extreme case of **minimum-entropy distributions** lends itself to a straightforward solution of the problem (2.54). In fact, if the distribution is concentrated at some state $\bar{x} \in [N]$, i.e., if $p = \delta_{\bar{x}}$, the decoder already knows the true state $\bar{x}$ of the information source, and a message of 0 bits is sufficient, yielding the optimal average codeword length $L^*_{\text{avg}}(p) = 0$.

2.6.2 Uncertainty and Compression

Compression relies on the assignment of a smaller number of bits L_x to observations x with a larger probability p_x. This is indeed the principle underlying many practical compression schemes, from the Morse code and the Braille code to modern algorithms such as arithmetic coding or asymmetric numeral systems.

In this regard, as a **baseline** for the evaluation of the performance of compression schemes, one can consider a **uniform compression code** that assigns the same number of bits, L, to all states $x \in [N]$ (see Section 2.3.1). Since a distinct codeword must be assigned to each state in order to enable successful decoding, and given that there are 2^L distinct messages with L bits, a uniform compression code requires a number of bits L that satisfies the inequality

$$2^L \geq N. \tag{2.55}$$

The minimum such number is thus given by $L = \lceil \log_2(N) \rceil$ bits.

Table 2.5 Optimal compression algorithm

Optimal compression
input: distribution p
output: an optimal compression codebook
list all values x in order of decreasing probabilities p_x, with ties broken arbitrarily
for n from 1 to N:
assign to the nth most likely state the shortest as-of-yet unassigned bit string
end
return the resulting compression codebook

Since an optimal compression code cannot have an average codeword length larger than that of any code, the minimum average codeword length must satisfy the inequality

$$L^*_{\text{avg}}(p) \leq \lceil \log_2(N) \rceil, \tag{2.56}$$

and compression is attained when the average number of bits is smaller than the upper bound $\lceil \log_2(N) \rceil$ bits. Intuitively, compression must hence only be possible when some states x are more likely than others, allowing the encoder to assign shorter codewords to the more probable states.

2.6.3 Optimal Compression

Two properties must be satisfied by an optimal compression code: (i) it should contain distinct codewords; and (ii) it must assign shorter codewords to more likely states. Given that a compression code need not satisfy any tree structure, an optimal code can be designed by simply assigning progressively longer codewords to progressively less likely states. Specifically, as detailed in Table 2.5, first one lists the states in order of decreasing probabilities. Then, binary codewords are assigned starting from the most likely states by using the shortest as-of-yet unassigned bit string.

Note that the optimal compression code is not unique in general, given that in Table 2.5: (i) ties are broken arbitrarily; and (ii) there is some freedom in assigning binary codewords in order of length.

The procedure in Table 2.5 can conveniently be described via a **permutation function** $s(n)$ that returns the nth most probable value x. That is, $s(0)$ identifies the most probable state, and $s(N-1)$ the least probable state. The next example illustrates the optimal coding procedure.

Example 2.20

Consider again the dyadic distribution $p = [1/4, 1/2, 1/8, 1/8]^T$. In the first step of the algorithm in Table 2.5, the states are reordered according to the permutation function $s(0) = 1$, $s(1) = 0$, $s(2) = 3$, and $s(3) = 4$, where the tie between the last two states, both with probability 1/8, is broken arbitrarily. The algorithm returns the compression codebook in Table 2.6, as detailed next.

Table 2.6 Optimal compression codebook for the dyadic distribution $p = [1/4, 1/2, 1/8, 1/8]^T$

State x	Probability p_x	Codeword c_x	Length L_x
0	1/4	1	1
1	1/2	0	1
2	1/8	00	2
3	1/8	01	2

The algorithm assigns the shortest codeword to the first state $s(0) = 1$. For this purpose, one can choose either codeword 0 or 1, and Table 2.6 illustrates the choice 1. For the second state on the list, $s(1) = 0$, the algorithm now assigns the shortest as-of-yet unassigned codeword, namely 0.

Having exhausted all codewords of length 1, the algorithm assigns arbitrary, distinct, codewords of length 2 bits to the remaining two states, completing the codebook in Table 2.6.

Accordingly, the minimum average codeword length is

$$L^*_{\text{avg}}(p) = \frac{1}{4}1 + \frac{1}{2}1 + \frac{1}{8}2 + \frac{1}{8}2 = \frac{5}{4} \text{ bits.} \tag{2.57}$$

Note that compression has been achieved, in the sense that the average number of bits is smaller than the maximum value $\lceil \log_2(N) \rceil = \log_2(4) = 2$ bits in (2.56).

Generalizing the example, the following result holds.

Key Result 2.11 Minimum average codeword length for compression

For a distribution p, write as $s(n)$ the permutation function that returns the nth most probable value x. The minimum average codeword length for compression is given by

$$L^*_{\text{avg}}(p) = \sum_{n=0}^{N-1} p_{s(n)} \lfloor \log_2(n+2) \rfloor . \tag{2.58}$$

This result follows since the first two most likely states – with $n = 0$ and $n = 1$ – require 1 bit; the second four most likely states – with n ranging from 2 to 5 – require 2 bits; and so on. Note that the minimum average codeword length (2.58) scales at most logarithmically with the number of states N.

2.6.4 Compressing vs. Asking

So far, this chapter has introduced two ways to quantify uncertainty – in terms of the minimum average number of binary questions for identification of the state and in terms of the minimum number of bits for the compression of the state. How are these two uncertainty measures related? To address this question, the following example provides some useful insight.

Example 2.21

Take again the case of the dyadic distribution $p = [1/4, 1/2, 1/8, 1/8]^T$. As studied in the previous sections, the minimum average number of binary questions is 7/4 (see (2.6)), while the minimum average number of bits for compression is 5/4 (see (2.58)). Therefore, at least for the example, the minimum average number of bits required for compression is **smaller** than the minimum average number of questions.

To understand why this is the case, it is useful to compare the optimal compression codebook $\{c_x\}_{x\in[N]}$ with the optimal asking strategy obtained in the previous sections. This comparison raises the question: Could we use the optimal compression codebook $\{c_x\}_{x\in[N]}$ to specify a tree of binary questions? Suppose that the state is $\mathrm{x} = 1$. By the compression codebook in Table 2.6, the encoder would send the codeword $c_1 = 0$. In the asking problem, this would correspond to receiving the answer "yes" to the first question.

However, based on this answer, the observer could not conclude unambiguously that the state equals $\mathrm{x} = 1$. In fact, the compression codebook, when interpreted as a sequence of answers, assigns the sequence of answers (yes, yes) and (yes, no) to states $x = 2$ and $x = 3$. Therefore, the first "yes" answer may as well be, as far as the decoder is concerned, also the first answer of the pair (yes, yes) or of the pair (yes, no).

The key point here is that in the compression problem the reception of a codeword at the decoder implicitly brings information also about the length of the codeword. In contrast, the binary answer to a question does not by itself indicate whether that is the last question of the sequence or whether more questions are needed to identify the state.

As illustrated by the example, a generic compression codebook cannot be used an asking strategy (following the approach described in Section 2.2.4). In contrast, any asking strategy can be used as a compression codebook by replacing "yes" answers with 0 and "no" answers with 1 (or vice versa). It follows that the minimum average number of binary questions is always no smaller than the minimum average codeword length for compression.

Looking again at the example above, the key issue preventing the use of the compression codebook as an asking strategy is that codeword c_1 is the **prefix** of the codewords c_2 and c_3. In fact, this is the reason why, once we obtain answer "yes" (0) to the first question, we cannot determine whether we should stop, concluding that the state equals 1, or ask another question.

In order to be able to repurpose a compression code as a sequence of binary questions, one hence needs to ensure that the codebook is prefix-free.

Definition 2.8
A codebook, i.e., a collection of N binary sequences, is said to be **prefix-free** if no codeword is the prefix of any other codeword, that is, no codeword is given by the concatenation of another codeword followed by other bits.

With this definition, the following result follows:

Key Result 2.12 Equivalence between prefix-free codebooks and asking strategies
Any prefix-free codebook with codeword lengths given by vector $L = [L_0, \ldots, L_{N-1}]^T$ defines a binary tree with leaves at depths defined by vector L; and, vice versa, any binary tree with leaves at depths defined by vector L specifies a prefix-free codebook with codewords of lengths in vector L.

Given the equivalence between prefix-free codebooks and sequences of binary questions, an **optimal prefix-free compression codebook** can be obtained via the Huffman algorithm. Accordingly, by (2.38), not only does the Shannon entropy $\mathrm{H}(p)$ closely approximate the minimum average number of binary questions, but it also approximates the minimum average number of bits needed to compress an information source under the prefix-free constraint.

2.6.5 Practical Considerations on Huffman Coding for Compression

This section is concluded with two practical considerations on Huffman coding when used for compression. The first remark has to do with the problem of compressing a sequence of observations, while the second with the computational complexity of evaluating the Huffman code.

- **Compressing a sequence of states:** It is often important to compress a sequence of T observations $x_1, x_2, \ldots, x_T$, with each value x_t being selected from the set $[N]$. To fix the ideas, consider the case of a text, in which each value x_t represents a letter. An important advantage of a prefix-free codebook is that it can be used to compress a sequence of letters from the alphabet $[N]$ without requiring explicit **separators** between consecutive letters. This is because one can identify unambiguously the ending of a codeword as soon as it is received or read, as no codeword is the prefix for another codeword. For example, using the Huffman code in Table 2.1, the sequence of values $1, 3, 0, 2$ is encoded as the bit string 011110110. This bit string can be automatically divided into codewords as $0, 111, 10, 110$ by simply placing a comma at the end of each codeword. One can readily check that this is generally not the case for a codebook that does not satisfy the prefix-free constraint.
- **Complexity of the Huffman code:** Evaluating and storing the Huffman code becomes impractical when the number of values, N, is large. For such situations,

one can consider alternative compression algorithms that approach the average codeword length of Huffman coding, while also enabling the evaluation of the codeword c_x on the fly, based on the vector p for any input state x. Like Huffman coding, such schemes are referred to as **entropy coding** algorithms, since they produce an average codeword length that is approximately equal to the Shannon entropy $\mathrm{H}(p)$. Two of the main entropy coding algorithms are **arithmetic coding** and the **asymmetric numeral system**. As an example, arithmetic coding, also known as Shannon–Fano–Elias coding, requires $\lceil -\log_2(p_x) \rceil + 1$ bits, i.e., one more bit than the rounded-up Shannon surprise.

2.7 Quantifying Information as Randomness for Sampling ⏭

This section presents a different perspective on the quantification of uncertainty, which hinges on the informational task of sampling. Accordingly, uncertainty is measured in terms of the complexity of simulating a given system from the perspective of the observer. Intuitively, a more uncertain random variable – one for which the probability distribution over possible realizations is more spread out – requires a more powerful source of randomness to be simulated. The question of interest is: Given a distribution p, how much "randomness" is needed in order to generate random variable $\mathrm{x} \sim p$?

2.7.1 Sampling

To start, one needs to decide how to quantify the amount of randomness. Following the same principle adopted throughout this chapter, randomness can be measured in bits. More precisely, then, the goal is to determine the average number of independent **"fair" bits** needed to generate a realization of random variable $\mathrm{x} \sim p$. Here, "fair" bits are those with an equal probability of being 0 or 1. The average number of required random bits serves as a measure of the randomness, and thus the uncertainty, inherent in distribution p.

This section will demonstrate that this measure is again related to the Shannon entropy. For simplicity, the presentation here will focus on dyadic distributions, and the extension to non-dyadic distributions will be discussed at the end of the section.

To formulate the problem, assume access to a sequence of uniformly distributed bits $\mathrm{b}_0, \mathrm{b}_1, \mathrm{b}_2, \ldots$. The bits $\mathrm{b}_0, \mathrm{b}_1, \mathrm{b}_2, \ldots$ are independent, which means, as detailed in Part IV, that none of the bits provide any information – i.e., reduces the uncertainty – on any of the other bits. These random bits are used as **seed randomness** for sampling.

2.7.2 Sampling via a Complete Binary Tree

In a manner analogous to asking policies, a sampling policy can be described by a binary tree. In it, as exemplified in Figure 2.12, each random bit determines which

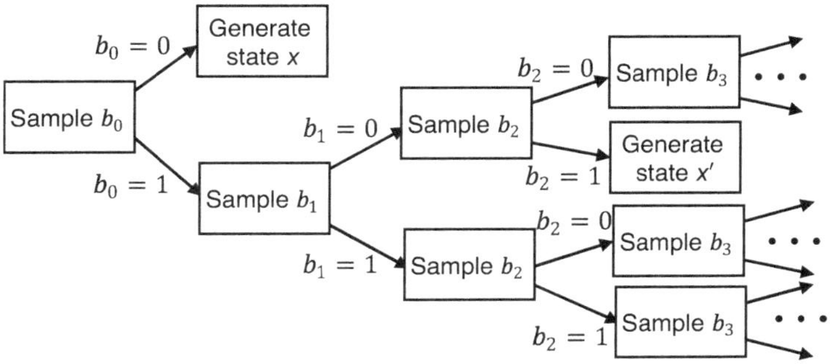

Figure 2.12 Illustration of a complete binary tree for the informational task of sampling.

edge is followed in the transition from one layer to the next, and each state $x \in [N]$ is associated with a leaf. When a leaf is reached, the corresponding state x is generated. The number of random bits L_x required to produce sample $\mathrm{x} = x$ equals the depth of the leaf associated with state x, and is denoted as L_x. Accordingly, a state x is produced when a specific sequence of L_x bits is generated.

A possible way to gain an intuition for the outlined sampling strategy is that each random bit decides on some feature of the sample x, with a specific collection of binary features $(b_0, b_1, \ldots, b_{L_x - 1})$ identifying each sample x.

In order for the sampling procedure to be successful, it is necessary for each state to be associated with a leaf, since the sampler must know when to stop and generate a sample. Furthermore, it is also necessary for the binary tree to be **complete**, that is, for each non-leaf node to have two children. In fact, otherwise, there would be at least one bit sequence $\mathrm{b}_0, \mathrm{b}_1, \mathrm{b}_2, \ldots$ that cannot be assigned to any state x, causing the sampling procedure not to stop when such a bit sequence is generated.

As shown in Appendix 2.B, a binary tree is complete if and only if the vector of leaves' depths L satisfies the Kraft–McMillan inequality (2.20) with an equality sign, i.e.,

$$\sum_{x \in [N]} 2^{-L_x} = 1. \tag{2.59}$$

Furthermore, as demonstrated in Appendix 2.A, the Huffman code satisfies condition (2.59), and thus it always returns a complete binary tree. The intuitive reason for this property is that any code that is not complete can be shortened, and thus it cannot be optimal (Problem 2.7).

Since sample $\mathrm{x} = x$ is produced as a result of the generation of a sequence of L_x fair bits, the probability q_x of generating state x equals the probability of producing this specific sequence of L_x bits. Given that each bit b_l is uniformly distributed and the bits $\{\mathrm{b}_l\}_{l \in [L_x]}$ are independent, the probability of generating sample x is

$$q_x = \left(\frac{1}{2}\right)^{L_x} = 2^{-L_x}. \tag{2.60}$$

In fact, as will be detailed in Part IV, the probability of independent events multiplies.

Importantly, by (2.59), the probabilities $\{q_x\}_{x\in[N]}$ in (2.60) define a valid probability vector $q = [q_0, \ldots, q_{N-1}]^T$. Furthermore, by (2.60), the probability distribution q is dyadic.

2.7.3 Sampling from Dyadic Distributions

The upshot of the discussion in the previous subsection is that a Huffman code can be used to generate samples from any dyadic source with probability distribution given by $p_x = 2^{-L_x}$ for $x \in [N]$, where vector L collects the codeword lengths of the Huffman code. With this strategy, using the results in Example 2.9, the number of random bits L_x consumed to produce the output $\mathrm{x} = x$ is related to the corresponding probability p_x via the Shannon surprise as $L_x = -\log_2(p_x)$. This way, the average number of bits required to generate a sample $x \sim p$ from a dyadic distribution p is

$$\begin{aligned} \sum_{x\in[N]} p_x L_x &= -\sum_{x\in[N]} p_x \log_2(p_x) \\ &= \mathrm{H}(p), \end{aligned} \tag{2.61}$$

which recovers again the Shannon entropy.

For dyadic distributions, this is also the minimum average number of bits, $L^*_{\text{avg}}(p)$, required for simulation of random variable $\mathrm{x} \sim p$, i.e., $L^*_{\text{avg}}(p) = \mathrm{H}(p)$. In fact, if one could find a complete binary tree yielding a smaller average number of bits than $\mathrm{H}(p)$, the same tree could be directly used as an asking strategy, or a prefix-free compression codebook, contradicting the lower bound in (2.38).

Example 2.22

For the dyadic distribution $p = [1/4, 1/2, 1/8, 1/8]^T$, the sampling strategy obtained from Huffman coding is shown in Figure 2.13. The corresponding optimal sampler first samples a random bit b_0. If this bit is equal to 0, the leaf corresponding to state 1

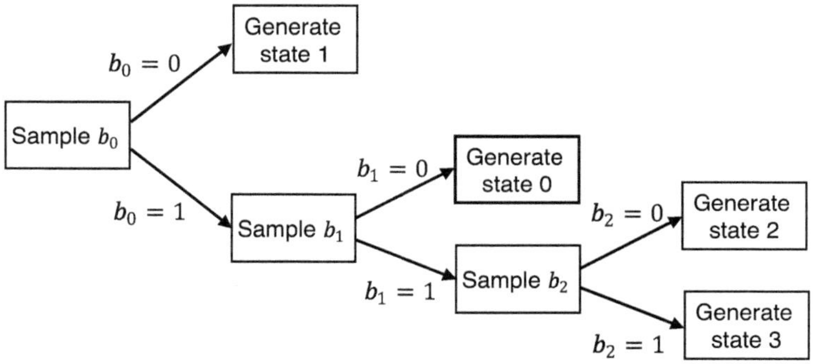

Figure 2.13 Huffman code for the dyadic distribution $p = [1/4, 1/2, 1/8, 1/8]^T$ repurposed as an optimal sampling strategy.

is reached, and thus state x $= 1$ is generated, requiring $L_0 = 1$ random bit. If, instead, the generated bit is $\mathrm{b}_0 = 1$, the sampler generates another random bit, b_1, and so on, as detailed in the figure.

2.7.4 Sampling from Non-Dyadic Distributions

Non-dyadic distributions cannot be represented in the form (2.60). Therefore, it is not possible to construct a sampling strategy by assigning every state x to a leaf of a complete binary tree. An approach to address this problem is to assign the same state x to more than one leaf of a complete binary tree. This method can be used to generate any state x that has a probability of the form

$$p_x = \sum_{y \in [M]} 2^{-l_{y|x}}, \tag{2.62}$$

where $\{l_{y|x}\}$ are integers equal to the depths of the M leaves assigned to state x.

But *any* probability p_x can be expressed as in (2.62), as long as one allows the sum to extend over an arbitrarily large, possibly infinite, number of terms, M. In fact, the expression (2.62) is equivalent to a binary expansion of the probability p_x that contains a 1 at all the positions indexed by the integers $\{l_{y|x}\}_{y \in [M]}$. For example, the probability $p_x = 1/3$ can be expressed using the infinite binary expansion 0.01010101 . . ., which corresponds to depths $\{l_{y|x}\} = \{2, 4, 6, 8, \ldots\}$.

Once one has represented all probabilities p_x in the target distribution p in the form (2.62), the problem is reduced to that of sampling from a dyadic distribution with probabilities $2^{-l_{y|x}}$ for all $x \in [N]$ and $y \in [M]$. Note that the probability vector of the resulting dyadic distribution may be infinitely long. Using the Huffman code on this dyadic distribution yields an optimal sampling strategy.

It can be proved that the minimum average number of fair bits required for generation of a sample $\mathrm{x} \sim p$ satisfies the inequalities

$$\mathrm{H}(p) \leq L^*_{\mathrm{avg}}(p) \leq \mathrm{H}(p) + 2. \tag{2.63}$$

2.8 Summary

- In this chapter, three distinct information processing tasks – asking, compressing, and sampling – were used as frameworks to quantify the uncertainty present in an informational state p, and thus in a random variable $\mathrm{x} \sim p$.
- It was concluded that the three tasks – under the prefix-free condition for compression – are informationally equivalent, in the sense that the Shannon entropy provides a characterization of the relevant uncertainty measures in all cases.
- Underlying all three problems is a binary tree representation of the policy used to solve the given informational task, namely the sequence of binary questions, the prefix-free compression codewords, and the sampling strategy.

- For any given distribution p, the Huffman code obtains an optimal binary tree that minimizes the average depth of the leaves, and hence also the average number of bits associated with each state $x \in [N]$.
- For all three problems, the minimum average number of bits is given by the Shannon entropy $\mathrm{H}(p)$ with an error of up to 1 bit.
- Furthermore, the Shannon entropy satisfies the three properties listed in Chapter 1 as desirable for an uncertainty measure. In fact, it is non-negative; it achieves the minimum value, zero, for an informational state concentrated at a single state; and it achieves the maximum value for a uniform distribution.
- Finally, the Shannon entropy is concave, implying that mixing increases uncertainty when uncertainty is measured by the Shannon entropy.

2.9 Recommended Resources

The standard textbook [1] offers detailed coverage on compression and sampling, including also extensions to asymptotic settings. Further insights into optimal codeword lengths for compression without the prefix-free condition are discussed in [2]. Connections between literature and mathematics, including lipograms, are explored in [3]. For an in-depth treatment of thermodynamic concepts related to information, see [4] and [5]. A comprehensive introduction to convexity and convex optimization is provided in [6].

PROBLEMS

2.1 (a) Show that when $N = 2^l$ for some integer l and the distribution p is uniform, i.e., when $p = u$, the Huffman code returns the minimum average number of questions $L^*_{\mathrm{avg}}(u) = l$. (b) Show that, with $p = u$, an optimal code is not necessarily uniform when the condition $N = 2^l$ is not satisfied (e.g., try with $N = 3$).

2.2 Show that the average number of questions, $L^*_{\mathrm{avg}}(p)$, is minimized when the distribution is concentrated at a single state $\bar{x}$, i.e., when $p = \delta_{\bar{x}}$ for some $\bar{x} \in [N]$.

2.3 (a) Consider a temperature variable x that takes all values in the set $\{-30, -20, -10, 0, 10, 20, 30\}$ (in Celsius scale) with equal probability. Compute the variance and Shannon entropy. (b) Compare your results with the variance and Shannon entropy for a temperature variable that takes all values in the set $\{15, 18, 21, 24, 27, 30, 33\}$ with equal probability.

2.4 (a) For the informational state $p = [0.125, 0.25, 0.125, 0.25, 0.25]^T$, derive the Huffman code and the corresponding minimum average codeword length under the prefix-free condition. (b) Compare your result in part (a) with the minimum average number of bits required to compress an information source with distribution p without imposing the prefix-free condition.

2.5 (a) For the informational state $p = [0.05, 0.2, 0.15, 0.3, 0.3]^T$, derive the Huffman code and the corresponding minimum average codeword length under the prefix-free condition. (b) Compare your result with the minimum average number of bits required to compress an information source with distribution p without imposing the prefix-free condition.

2.6 Using a programming language of your choosing, use the Huffman code for the distribution $p = [0.125, 0.25, 0.125, 0.25, 0.25]^T$ to implement a sampler, and verify that the sampler produces a random variable with the correct distribution p. *Hint:* You can estimate the probabilities in vector p by repeating the simulation a sufficiently large number of times.

2.7 Show that any code that minimizes the average codeword length must correspond to a complete binary tree. *Hint:* Argue that, for a tree that is not complete, it is possible to reduce the depth of some leaf.

2.8 Consider a source emitting symbols $\{A, B, C\}$ with probabilities $p = [0.5, 0.3, 0.2]$. (a) Construct the optimal binary Huffman code for this source and compute the average codeword length in bits. (b) Compute the Shannon entropy of the source and compare it with the average codeword length. (c) Suppose symbol A is replaced by two subsymbols, A_1 and A_2, with probabilities 0.3 and 0.2, respectively, while the probabilities of B and C remain unchanged. Construct the new binary Huffman code for the updated source and calculate the new average codeword length. (d) Compare the new average codeword length with the original. Discuss the impact of symbol splitting on compression efficiency.

2.9 In information theory, information can be quantified using bases other than binary (base-2). For this problem, consider using **ternary digits**, which take values in the set $[3] = \{0, 1, 2\}$, instead of binary digits (bits), to compress a source. Specifically, consider a source with probability distribution $p = [0.4, 0.2, 0.15, 0.1, 0.1, 0.05]$. Instead of using a binary Huffman tree, use a ternary Huffman tree to encode the source. (a) Construct the ternary tree and derive the codewords. (b) Compute the minimum average codeword length for this ternary encoding. (c) Compare the result in part (b) with the Shannon entropy of the source measured in bits. (d) How would you modify the definition of the Shannon entropy given in the text to match the different base used in this example?

2.10 Information, as intended in information theory, is distinct from semantics. To elaborate on this, consider a language model that, when prompted with the question "Who is the father of information theory?," produces the following answers: "Claude Shannon" with probability 0.6; "Shannon" with probability 0.1; "Claude Elwood Shannon" with probability 0.1; "Norbert Wiener" with probability 0.1; "Wiener" with probability 0.05; and "Warren Weaver" with probability 0.05. (a) Compute the Shannon surprises associated with each answer. (b) Compute the Shannon entropy of the answers. (c) The **semantic entropy** is obtained by first grouping all answers that have the same meaning into a single answer and computing the Shannon entropy of the resulting random variable. Compute the semantic entropy of the answers. (d) Prove that the semantic entropy is no larger than the Shannon entropy for any distribution p.

Appendix

2.A Proof of the Optimality of the Huffman Code

Order the states from most to least probable, breaking ties arbitrarily. The key observation underlying the proof of the optimality of the Huffman code is that, in order for a code to be optimal, the two last states, i.e., the two states with the lowest probabilities, must be assigned to the maximal-depth leaves of the tree.

In fact, the state with the smallest probability must be clearly assigned the leaf with the largest depth. Otherwise, another code could reduce, or at least not increase, the average length. This could be done by swapping the labels of two leaves so that the lowest-probability state has the longest codeword. But, once we have one leaf at the maximum depth, we also get "for free" another leaf at the same length, since leaves always come in pairs.

With this in mind, the proof is complete by **induction**. For $N = 2$, the Huffman code is clearly optimal since one cannot use less than 1 bit to represent two states. Assume now that the Huffman code is optimal for $N - 1$. The proof is completed by demonstrating that the optimality of the Huffman code for $N - 1$ implies its optimality also for N.

Without loss of optimality, by the argument above, one can combine the two states at the bottom of the queue in the first step of the construction of the tree. In fact, this ensures that these two states will end up being assigned to the leaves with the largest depth. This coincides with the first step of the Huffman algorithm.

But after this first step, one has effectively obtained a system with $N - 1$ states. The new system has the same $N - 2$ states on top of the queue as the original N-state system, with the addition of a state corresponding to the newly created node encompassing the two states with the lowest probabilities. By assumption, the Huffman code is optimal for this system. This concludes the proof of the optimality of the Huffman code.

Using the same argument, one can also show that the Huffman code always corresponds to a complete binary tree. In fact, this is clearly true for $N = 2$, and, by the induction-based reasoning above, it is also the case for any value of N.

2.B Proof of the Kraft–McMillan Inequality

In this appendix, the Kraft–McMillan inequality $\sum_{x\in[N]} 2^{-L_x} \leq 1$ is shown to be necessary and sufficient for a code with codeword length vector L to exist. To this end, consider a binary tree, i.e., a code, and denote as L_x the depth of the leaf assigned to each state $x \in [N]$. Define as $L_{\max} = \max_{x\in[N]} L_x$ the depth of the tree. The maximum number of leaves for the tree is hence $2^{L_{\max}}$, and this maximum number is obtained if all leaves are at the maximum depth $L_{\max}$.

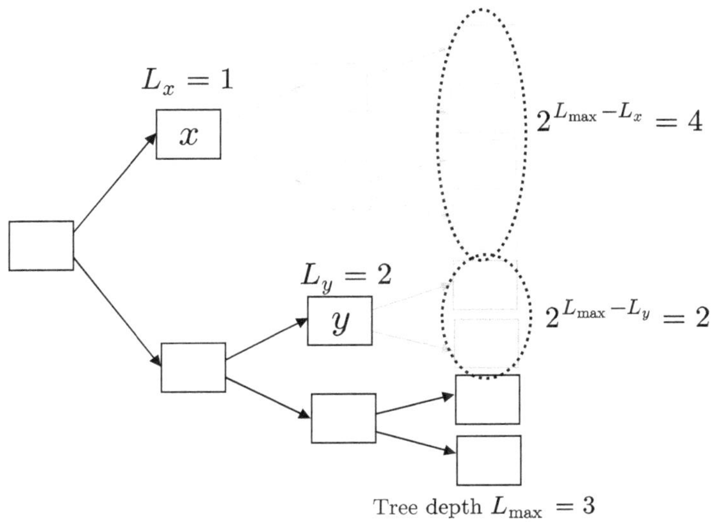

Figure 2.14 Binary tree used to illustrate the proof of the Kraft–McMillan inequality.

In contrast, if the leaf assigned to some state x is at depth $L_x \leq L_{\max}$, this implies that $2^{L_{\max}-L_x}$ leaves at depth L are no longer reachable by the code. This is because, as illustrated in Figure 2.14, all paths from this leaf are blocked by the leaf. Therefore, in order for a codeword length assignment L to be valid, it is necessary and sufficient to verify the inequality

$$\sum_{x\in[N]} 2^{L_{\max}-L_x} \leq 2^L, \tag{2.64}$$

which is equivalent to the Kraft–McMillan inequality. Furthermore, by the same argument, if each non-leaf node has two children, the relation (2.64) is satisfied with equality, proving the condition (2.59).

2.C Solving the Relaxed Problem (2.21)

In this appendix, the optimization problem (2.21) is addressed. Applying the change of variable $q_x = 2^{-L_x}$, the problem can equivalently be restated as

$$\begin{aligned} &\underset{L}{\text{maximize}} \sum_{x\in[N]} p_x \log_2(q_x) \\ &\text{subject to} \sum_{x\in[N]} q_x \leq 1. \end{aligned} \tag{2.65}$$

While L_x is constrained to be an integer, the problem (2.65) is now relaxed by allowing any value for the optimization variables $0 \leq q_x \leq 1$.

Observe that, at an optimum solution, the equality $\sum_{x\in[N]} q_x = 1$ must hold, as, otherwise, we could increase the objective due to monotonicity of function $\log_2(q_x)$. Referring to standard textbooks on convex optimization for the necessary background (see Recommended Resources and Appendix 2.D), since the relaxed problem is convex, Lagrangian duality can be used to solve the problem. Accordingly, by differentiating the Lagrangian function, one obtains the necessary and sufficient optimality conditions $p_x/q_x = \lambda$, where λ is the Lagrange multiplier. Finally, imposing the equality constraint yields the optimal solution $q = p$. This concludes the proof that the ideal codeword lengths $\{\tilde{L}_x^* = -\log_2(p_x)\}_{x\in[N]}$ are optimal for the relaxed problem (2.65).

2.D Convexity, Concavity, and Jensen's Inequality

Generalizing the discussion in the main text, the following definitions hold.

Definition 2.9 Convexity and concavity
A function $f(x)$ is said to be **convex** if, for any x' and x'' with $x' \neq x''$ in the domain of the function, the inequality

$$f(\gamma x' + (1-\gamma)x'') \leq \gamma f(x') + (1-\gamma)f(x'') \tag{2.66}$$

holds for any $0 \leq \gamma \leq 1$. A function $f(x)$ is **concave** if $-f(x)$ is convex.

Definition 2.10 Strict convexity and strict concavity
A function is said to be **strictly convex** if the strict inequality

$$f(\gamma x' + (1-\gamma)x'') < \gamma f(x') + (1-\gamma)f(x'') \tag{2.67}$$

holds for $x' \neq x''$ unless either of the equalities $\gamma = 0$ or $\gamma = 1$ holds. A function $f(x)$ is **strictly concave** if $-f(x)$ is strictly convex.

For the case of a scalar variable x, a function $f(x)$ is convex if its second derivative is non-negative and strictly convex if the second derivative is strictly positive over its domain. In the more general case of a vector x, a function $f(x)$ is convex if all the eigenvalues of the Hessian matrix are non-negative, while it is strictly convex if all eigenvalues are strictly positive over its domain. By using these conditions, the Shannon entropy $\mathrm{H}(p)$ can be shown to be strictly concave, and hence also concave, in the probability vector p.

Key Result 2.13 Jensen inequality
A convex function $f(x)$ satisfies the Jensen inequality

$$f(\mathrm{E}_p[\mathrm{x}]) \leq \mathrm{E}_p[f(\mathrm{x})], \tag{2.68}$$

for any random variable $\mathrm{x} \sim p$ taking values in its domain. A concave function satisfies the Jensen inequality

$$f(\mathrm{E}_p[\mathrm{x}]) \geq \mathrm{E}_p[f(\mathrm{x})]. \tag{2.69}$$

Furthermore, for a strictly convex function, one has the strict inequality

$$f(\mathrm{E}_p[\mathrm{x}]) < \mathrm{E}_p[f(\mathrm{x})] \tag{2.70}$$

unless distribution p is of the form $p = \delta_{\bar{x}}$ for some $\bar{x} \in [N]$. Analogously, for a strictly concave function, one has the strict inequality

$$f(\mathrm{E}_p[\mathrm{x}]) > \mathrm{E}_p[f(\mathrm{x})] \tag{2.71}$$

unless distribution p is of the form $p = \delta_{\bar{x}}$ for some $\bar{x} \in [N]$.

Note that the Jensen inequality generalizes the definition (2.65) of the convex function. Furthermore, by the Jensen inequality, the non-negative difference

$$\mathrm{E}_p[f(\mathrm{x})] - f(\mathrm{E}_p[\mathrm{x}]) \geq 0, \tag{2.72}$$

also known as the **Jensen gap**, measures the degree of convexity of the function.

2.E Proof of the Maximum-Value Property of the Shannon Entropy

To prove (2.45), consider any distribution p, as well as all the $N!$ distributions obtained by permuting the states in set $[N]$. Denote each such distribution as $\tilde{p}^y$, where index $y \in [N!]$ runs over all the possible permutations, with $\tilde{p}_0 = p$. Note that a permutation of the entries does not change the Shannon entropy, i.e., the equality $\mathrm{H}(\tilde{p}^y) = \mathrm{H}(p)$ holds for all $y \in [N!]$. Furthermore, the mixed distribution with equal mixing weights returns the uniform distribution

$$u = \frac{1}{N!} \sum_{y \in [N!]} \tilde{p}^y. \tag{2.73}$$

Therefore, by the concavity of the Shannon entropy and by the Jensen inequality (2.69), the following inequality is satisfied:

$$\begin{aligned} \mathrm{H}(u) &\geq \frac{1}{N!} \sum_{y \in [N!]} \mathrm{H}(\tilde{p}^y) \\ &= \mathrm{H}(p), \end{aligned} \tag{2.74}$$

completing the proof. Note that in going from (2.69) to (2.74), the function $f(\cdot)$ in (2.69) is replaced by the Shannon entropy $\mathrm{H}(\cdot)$ and the random variable x is replaced by the random variable $\tilde{p}^{\mathrm{y}}$ with variable y being uniformly distributed in the set $[N!]$.

REFERENCES

[1] T. M. Cover and J. A. Thomas, *Elements of Information Theory*, 2nd ed. Hoboken, NJ: Wiley-Interscience, 2006.

[2] I. Kontoyiannis and S. Verdú, "Optimal lossless data compression: Non-asymptotics and asymptotics," *IEEE Transactions on Information*

Theory, vol. 60, no. 2, pp. 777–795, 2013.
[3] S. Hart, *Once Upon a Prime: The Wondrous Connections Between Mathematics and Literature*. New York: Flatiron Books, 2023.
[4] J. Oppenheim, M. Horodecki, P. Horodecki, and R. Horodecki, "Thermodynamical approach to quantifying quantum correlations," *Physical Review Letters*, vol. 89, no. 18, art. 180402, 2002.
[5] M. Horodecki, P. Horodecki, R. Horodecki, J. Oppenheim, A. Sen, U. Sen, and B. Synak-Radtke, "Local versus nonlocal information in quantum-information theory: Formalism and phenomena," *Physical Review A*, vol. 71, no. 6, art. 062307, 2005.
[6] S. Boyd and L. Vandenberghe, *Convex Optimization*. Cambridge: Cambridge University Press, 2004.

3 Information Quantification by Predicting or Guessing: Tsallis Entropy and Rényi Entropy

3.1 Overview

Chapter 2 explored how the Shannon entropy arises naturally as the solution to three distinct information processing tasks, all of which require more bits when the underlying informational state reflects greater uncertainty. In this chapter, two additional informational tasks will be studied – predicting and guessing – to motivate the development of alternative measures of uncertainty.

First, the chapter introduces the Tsallis entropy, an uncertainty measure designed to quantify the prediction error associated with a given informational state. The underlying principle of this definition is that larger prediction errors correspond to greater uncertainty. Unlike the Shannon entropy, the Tsallis entropy constitutes a family of entropies parameterized by a scalar value called the order, which determines the "severity" of the prediction loss.

By varying the order, Tsallis entropy offers a spectrum of uncertainty measures with distinct properties. For example, while Shannon entropy grows at most logarithmically with the number of possible states N, the growth rate of the Tsallis entropy can vary depending on the chosen order. This flexibility proves crucial in characterizing the optimal solution to the informational task of guessing, which can be exponentially more demanding, in terms of bits required, than the tasks analyzed in Chapter 2.

Then, the chapter introduces another entropy function: the Rényi entropy. Unlike the Tsallis entropy, the Rényi entropy family shares the same maximal logarithmic scaling with the number of states as the Shannon entropy. The Rényi entropy emerges as the solution to a generalized version of the problem of compression, where the objective incorporates a broader measure of the average codeword length to avoid excessively long codewords.

Finally, a unification and generalization of the Shannon, Tsallis, and Rényi entropies is presented.

3.1.1 Learning Objectives and Organization

By the end of this chapter, the reader should be able to:

- understand the use and properties of the log-loss and of the generalized log-loss;
- understand the Shannon entropy as an average prediction loss;

- understand the definition and properties of the Tsallis entropy;
- demonstrate that a member of the family of Tsallis entropy emerges as the solution to the problem of optimal guessing;
- understand the definition and properties of the Rényi entropy;
- derive the Rényi entropy via the solution of a generalized version of the problem of optimal compression;
- and understand how to unify and generalize the Shannon, Tsallis, and Rényi entropies.

3.2 Information Quantification by Predicting

This section introduces a method for quantifying uncertainty based on the following principle: uncertainty can be measured by the degree of difficulty in predicting the outcome of an observation.

3.2.1 Prediction Loss

To begin, consider the random state x of a system, which the observer characterizes using a distribution vector p. As in previous chapters, the vector p is referred to as the observer's **informational state**, representing their uncertainty about the system's state x. Specifically, each entry p_x of the vector quantifies the observer's **confidence** that the random state $\mathrm{x} \sim p$ is equal to x.

A helpful example is weather prediction, where x represents possible weather conditions, and p_x expresses the observer's **belief** in the weather being in state x. For instance, p_x might represent the likelihood the observer assigns to rain, sunshine, or snow on a given day.

Now, suppose the true observation $\mathrm{x} = x$ is recorded. How well did the observer perform based on their assigned informational state p? If the observer assigned a high probability p_x to the actual outcome x, they can be said to have performed well, incurring a small **prediction loss**. Conversely, if p_x was small, the observer should be judged more harshly, incurring a larger prediction loss. For example, if a weather forecaster assigns a probability of $p_x = 0.1$ to rain and it rains, the prediction loss should be much greater than if they had predicted rain with a high probability of $p_x = 0.9$.

Following this intuition, the predictor's performance can be evaluated using a **loss function** that adheres to the following definition, which is illustrated in Figure 3.1.

Definition 3.1 Loss function
A loss function $\ell(p_x)$ is a function of the probability p_x that satisfies the following properties:

- **Non-negativity and minimum-value property:** The loss function $\ell(p_x)$ is non-negative, i.e., $\ell(p_x) \geq 0$, and equals zero when $p_x = 1$, i.e., $\ell(p_x) = 0$.

- **Monotonicity and maximum-value property:** The loss function $\ell(p_x)$ is non-increasing in the interval $0 \leq p_x \leq 1$, reaching its maximum value in the limit as $p_x \to 0$.

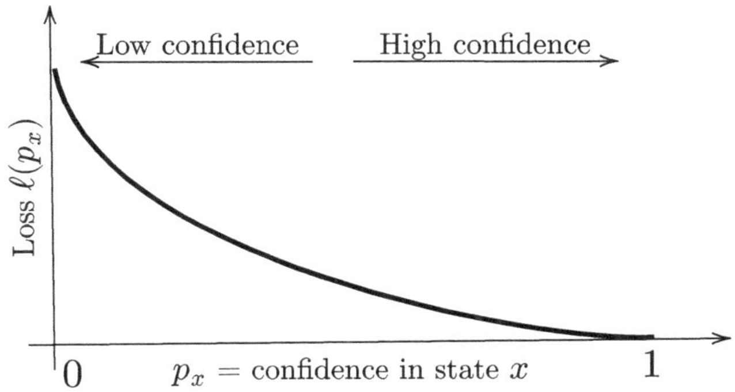

Figure 3.1 Illustration of the definition of a loss function $\ell(p_x)$ satisfying the non-negativity and minimum-value property as well as the monotonicity and maximum-value property.

The two defining properties of a loss function can thus be summarized as follows: (i) the loss is a non-negative quantity that vanishes for a "perfect" prediction (i.e., when 100% confidence is assigned to the correct outcome); (ii) the loss decreases as the observer's confidence in the correct outcome increases. It is worth noting that the limit $p_x \to 0$ is used instead of the equality $p_x = 0$ for technical reasons. Many common loss functions are in fact unbounded when $p_x = 0$, making the limit formulation more practical.

Given a specified loss function, the observer's state of uncertainty can be now evaluated as the **average prediction loss**:

$$\mathrm{H}_\ell(p) = \mathrm{E}_p[\ell(p_\mathrm{x})]. \tag{3.1}$$

Problem 3.1 elaborates on the conditions under which the average loss $\mathrm{H}_\ell(p)$ ensures the desired properties of entropy functions discussed in Chapter 1.

3.2.2 Average Prediction Loss and Shannon Entropy

To consolidate the notion of the average prediction loss $\mathrm{H}_\ell(p) = \mathrm{E}_p[\ell(p_\mathrm{x})]$ as an uncertainty measure, it is now observed that the average prediction loss recovers the Shannon entropy for a particular choice of the loss function.

Definition 3.2 Log-loss
The log-loss is defined as

$$\ell(p_x) = -\log_2(p_x). \tag{3.2}$$

The log-loss thus coincides with the Shannon surprise associated with state x under distribution p (see Section 2.4).

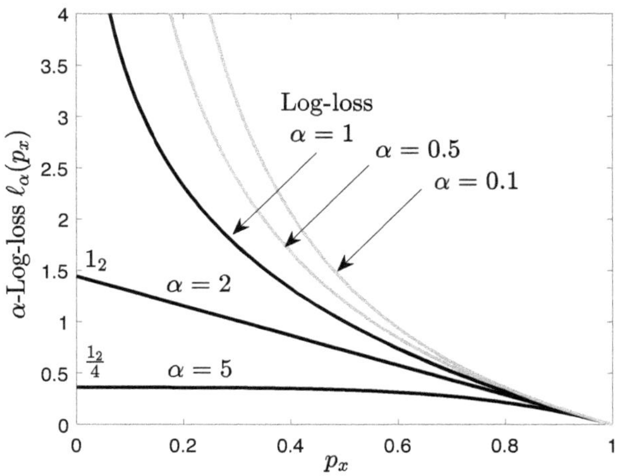

Figure 3.2 α-Log-loss $\ell_\alpha(p_x) = -\log_{2,\alpha}(p_x)$ as a function of the probability p_x for different values of the order α. With $\alpha = 1$, the α-log-loss recovers the log-loss $\ell(p_x) = -\log_2(p_x)$.

As illustrated in Figure 3.2, it can readily be verified that the log-loss satisfies the two defining properties for a loss function, namely the non-negativity and minimum-value property as well as the monotonicity and maximum-value property.

With the log-loss (3.2), the average prediction loss $H_\ell(p)$ equals the Shannon entropy:

$$\begin{aligned} H_\ell(p) &= E_p[\ell(p_x)] \\ &= -E_p[\log_2(p_x)] = H(p). \end{aligned} \tag{3.3}$$

This provides a fourth interpretation of the Shannon entropy: the Shannon entropy not only quantifies (approximately) the number of bits required for asking, compressing, and sampling, but it is also the average prediction loss when the log-loss is adopted as the loss metric.

3.3 α-Log-Loss

As illustrated in Figure 3.2, the log-loss penalizes harshly predictors that assign low probabilities to the outcome x. In fact, in the limit of small probabilities p_x, the log-loss attains an infinite value:

$$\lim_{p_x \to 0} (-\log_2(p_x)) = \infty. \tag{3.4}$$

As an example, if a weather forecaster assigns a zero probability to some weather condition x that actually occurs, the log-loss measure would mark this predictor as infinitely unreliable. (Events that were deemed impossible before being observed are also known as black swans – this is discussed further in Chapter 4.)

This section introduces a broader class of loss functions that allow a finer-grained control of the penalty accrued when a low-confidence event occurs. This family of loss functions is known as the **α-log-loss** and denoted as $\ell_\alpha(p_x)$. Plots of members of this family can be found in Figure 3.2. As studied in the next section, when averaged,

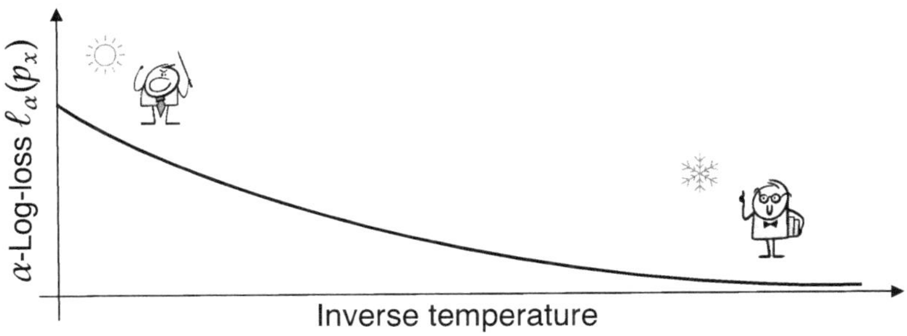

Figure 3.3 A cartoon illustration of the impact of the order α, interpreted as an inverse temperature, on the α-log-loss measure of prediction loss.

this family of loss functions yields a new entropy measure known as the **Tsallis entropy**.

To start, as in the cartoon illustration in Figure 3.3, it is useful to think of the order α as an **inverse temperature**. One can indeed interpret the order α as an inverse temperature in the following metaphorical sense: as the order α increases, the α-log-loss function $\ell_\alpha(p_x)$ "cools down" (becomes less "severe," that is, smaller), and, conversely, a smaller α indicates a "hotter" (more punishing) loss.

The mathematical definition of the α-log-loss requires the introduction of the α-logarithm, which is discussed next.

3.3.1 α-Logarithm

The α-loss function is defined as the negative of the α-logarithm, i.e., as $\ell_\alpha(p_x) = -\log_{2,\alpha}(p_x)$.

> **Definition 3.3 α-logarithm, or α-log**
> For any order $\alpha \geq 0$, the α-logarithm, or α-log, in base 2, is defined as
> $$\log_{2,\alpha}(q) = \begin{cases} \frac{\mathrm{l}_2}{\alpha-1}(q^{\alpha-1} - 1) & \text{for } \alpha \geq 0 \text{ with } \alpha \neq 1, \\ \log_2(q) & \text{for } \alpha = 1, \end{cases} \tag{3.5}$$
> with the notation
> $$\mathrm{l}_2 = \log_2(e) \approx 1.44. \tag{3.6}$$

The α-logarithm generalizes the conventional logarithm in base 2, which is obtained in the special case with $\alpha = 1$, i.e.,

$$\log_{2,1}(q) = \log_2(q). \tag{3.7}$$

Mathematically, the definition of the α-log for $\alpha = 1$ is justified via the limit

$$\lim_{\alpha \to 1} \frac{\mathrm{l}_2}{\alpha - 1}(q^{\alpha-1} - 1) = \log_2(q), \tag{3.8}$$

ensuring that the α-log is a continuous function of α. In this regard, the constant l_2 is introduced in the definition (3.5) in order to obtain the logarithm in base 2 when $\alpha \to 1$. Removing this term would lead to the natural logarithm in this limit.

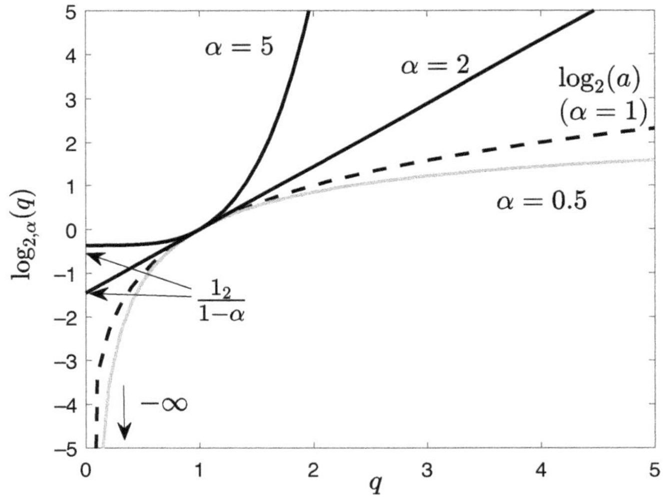

Figure 3.4
α-Logarithm $\log_{2,\alpha}(q)$ as a function of probability q for different values of the order α.

As illustrated in Figure 3.4, the α-logarithm $\log_{2,\alpha}(q)$ inherits some properties of the logarithm, such as being increasing in q. In this regard, note that the multiplication by $1_2/(\alpha - 1)$ in the definition (3.5) is critical for this property to hold. Furthermore, like the logarithm, the α-log with low inverse temperatures $0 \leq \alpha < 1$ is defined only for $q > 0$, as it tends to negative infinity as q tends to zero.

However, the α-logarithm fails to satisfy other properties of the logarithm. Notably, when $\alpha > 1$, the α-logarithm is defined for all values $q \geq 0$, including also $q = 0$, at which point it takes the **finite** value $\log_{2,\alpha}(0) = 1_2/(1-\alpha)$.

Another notable difference between logarithm and α-logarithm is that, for any $\alpha \neq 1$, the α-logarithm is non-additive.

Key Result 3.1 Non-additivity of the α-log

While the conventional logarithm satisfies the additive property

$$\log_2(pq) = \log_2(p) + \log_2(q) \tag{3.9}$$

for all probabilities $p > 0$ and $q > 0$, this is generally not the case for the α-logarithm. In other words, the inequality

$$\log_{2,\alpha}(pq) \neq \log_{2,\alpha}(p) + \log_{2,\alpha}(q) \tag{3.10}$$

generally holds for $\alpha \neq 1$.

Appendix 3.A provides further details on the α-logarithm.

3.3.2 α-Log-Loss

The α-log-loss generalizes the log-loss by replacing the logarithm with the α-logarithm as per the following definition.

Definition 3.4 α-log-loss
For any order $\alpha \geq 0$, the α-log-loss is defined as

$$\ell_\alpha(p_x) = -\log_{2,\alpha}(p_x)$$
$$= \begin{cases} \frac{l_2}{1-\alpha}(p_x^{\alpha-1} - 1) & \text{for } \alpha \geq 0 \text{ with } \alpha \neq 1, \\ -\log_2(p_x) & \text{for } \alpha = 1. \end{cases} \tag{3.11}$$

Accordingly, the log-loss is obtained as a special case of the α-log-loss for $\alpha = 1$, i.e.,

$$\ell_1(p_x) = -\log_2(p_x). \tag{3.12}$$

As formalized next, like the log-loss, the α-log-loss satisfies the required properties of a loss function. Readers primarily interested in the high-level features of the Tsallis entropy may move directly to Section 3.4.

3.3.3 Properties of the α-Log-Loss ⏭

Key Result 3.2 Non-negativity and minimum-value property of the α-log loss
The α-log-loss is non-negative, $\ell_\alpha(p_x) \geq 0$, and it equals 0 for $p_x = 1$, i.e.,

$$\ell_\alpha(1) = -\log_{2,\alpha}(1) = 0 \tag{3.13}$$

for all $\alpha \geq 0$.

Key Result 3.3 Monotonicity and maximum-value property of the α-log loss
The α-log-loss $\ell_\alpha(p_x)$ decreases as the probability $0 \leq p_x \leq 1$ increases, achieving the maximal value in the limit $p_x \to 0$ for $0 \leq \alpha \leq 1$ and at $p_x = 0$ for $\alpha > 1$. In particular, with $0 \leq \alpha \leq 1$, the α-log-loss tends to infinity as $p_x \to 0$:

$$\lim_{p_x \to 0} \ell_\alpha(p_x) = \infty \text{ for } 0 \leq \alpha \leq 1; \tag{3.14}$$

while for $\alpha > 1$, a finite loss value is attained even in the worst case of a zero predictive probability, $p_x = 0$, namely

$$\ell_\alpha(0) = \frac{l_2}{\alpha - 1} \text{ for } \alpha > 1. \tag{3.15}$$

By varying the order α, the α-log-loss yields different measures of the prediction error, with a more "lenient" loss $\ell_\alpha(p_x)$ obtained for any fixed probability p_x as the order α increases. This observation is formalized by the following property, which is also illustrated in Figure 3.2.

Key Result 3.4 The α-log-loss decreases with the inverse temperature α
For any fixed probability p_x, the α-log-loss $\ell_\alpha(p_x)$ is a decreasing function of the order $\alpha \geq 0$, i.e., the inequality

$$\ell_{\alpha'}(p_x) \leq \ell_\alpha(p_x) \tag{3.16}$$

holds for any two values α' and α with $\alpha' \geq \alpha$.

One may say that a smaller inverse temperature α – i.e., a "hotter" loss – magnifies the loss assigned to small predictive probabilities.

3.3.4 Scoring Rules ⏭

This chapter focuses on the α-log-loss for the important role it plays in information theory. However, there exist even more general ways to define prediction loss functions. In particular, the α-log-loss is an example of a **scoring rule**. A scoring rule is a function of the form $\ell(p, x)$, where p is the $N \times 1$ probability vector defining the probabilistic predictor and $x \in [N]$ is a state. The value $\ell(p, x)$ quantifies the loss of the predictor when the true state of the system is x. For the α-log-loss, the scoring rule is given by $\ell(p, x) = -\log_{2,\alpha}(p_x)$.

Other scoring rules include the **Brier score**

$$\ell(p, x) = -2p_x + ||p||_2^2, \tag{3.17}$$

where $||p||_2^2 = \sum_{x' \in [N]} p_{x'}^2$ is the 2-norm of vector p. The reader is referred to the Recommended Resources for further discussion.

3.4 Tsallis Entropy

This section defines and elaborates on the Tsallis entropy, which generalizes the Shannon entropy by evaluating the prediction loss via the α-log-loss.

3.4.1 Defining the Tsallis Entropy

Using the α-log-loss in (3.11), the Tsallis entropy is defined as follows.

Definition 3.5 Tsallis entropy
For any $\alpha \geq 0$, the Tsallis entropy of order α (also known as the Havrda–Charvat–Tsallis entropy) is defined as

$$\begin{aligned} \mathrm{H}_\alpha^{\mathrm{T}}(p) &= \mathrm{E}_p[\ell_\alpha(p_{\mathrm{x}})] \\ &= \begin{cases} \frac{1_2}{1-\alpha}\left(\sum_{x \in [N]} p_x^\alpha - 1\right) & \text{for } \alpha \geq 0 \text{ with } \alpha \neq 1, \\ \mathrm{H}(p) & \text{for } \alpha = 1, \end{cases} \end{aligned} \tag{3.18}$$

where one assumes the equality

$$0^0 = 0. \tag{3.19}$$

The condition (3.19) ensures that, as for the Shannon entropy, impossible states, with $p_x = 0$, do not affect the Tsallis entropy. Furthermore, by (3.18), the Tsallis entropy coincides with the Shannon entropy by setting $\alpha = 1$, i.e., $\mathrm{H}_1^{\mathrm{T}}(p) = \mathrm{H}(p)$. This ensures that the Tsallis entropy is a continuous function of the order $\alpha \geq 0$, i.e.,

$$\lim_{\alpha \to 1} \mathrm{H}_\alpha^{\mathrm{T}}(p) = \mathrm{H}_1^{\mathrm{T}}(p) = \mathrm{H}(p), \tag{3.20}$$

which follows from the limit (3.8). Note also that, like the Shannon entropy, the Tsallis entropy is **permutation-invariant**, as it does not depend on the ordering of the states.

The Tsallis entropy (3.18) provides a family of entropy functions, which are parameterized by the order $\alpha \geq 0$. In particular, when varying the order $\alpha \geq 0$, one obtains an uncertainty measure $\mathrm{H}_\alpha^{\mathrm{T}}(p)$ that weighs in different ways the contributions of the most uncertain states, which correspond to the largest prediction errors. Formally, by the monotonicity property (3.16) of the α-log-loss, the following result holds.

Key Result 3.5 The Tsallis entropy decreases with the inverse temperature α

For any fixed probability vector p, the Tsallis entropy $\mathrm{H}_\alpha^{\mathrm{T}}(p)$ is a decreasing function of the order $\alpha \geq 0$, i.e., the inequality

$$\mathrm{H}_{\alpha'}^{\mathrm{T}}(p) \leq \mathrm{H}_\alpha^{\mathrm{T}}(p) \tag{3.21}$$

holds for any two orders α' and α with $\alpha' \geq \alpha$.

So, a "cooler" order α, i.e., $\alpha > 1$, yields smaller Tsallis entropies, while a "hotter" α, i.e., $\alpha < 1$, produces larger Tsallis entropies.

3.4.2 Tsallis Sum

A useful way to understand the Tsallis entropy (3.18) is to note that, for any $\alpha \neq 1$, it depends on the distribution p only through the following fundamental quantity, whose properties will be studied in this subsection.

Definition 3.6 Tsallis sum, aka information generating function

For any order $\alpha \geq 0$, the Tsallis sum of order α is defined as

$$\text{Tsallis sum of order } \alpha = \sum_{x \in [N]} p_x^\alpha, \tag{3.22}$$

where one assumes the equality (3.19).

Table 3.1 Interpretation of the Tsallis sum as a function of the order α

Order	Tsallis sum α-norm	$1/(1-\alpha)$
$0 \leq \alpha < 1$	Uncertainty measure	Positive
$\alpha > 1$	Certainty measure	Negative

The Tsallis sum of order α is closely related to the α-norm of the vector p, which is defined as follows.

Definition 3.7 α-norm
For any $\alpha \geq 1$, the α-norm of vector p is defined as

$$||p||_\alpha = \left(\sum_{x \in [N]} p_x^\alpha \right)^{\frac{1}{\alpha}}. \tag{3.23}$$

Following this definition, the α-norm (3.23) satisfies all the criteria required to be a mathematical norm, including the triangle inequality, only for $\alpha \geq 1$. Nevertheless, for simplicity, the term α-norm may also be used in the text to refer to the quantity (3.23) for $0 \leq \alpha < 1$.

As illustrated in the next examples and summarized in Table 3.1, the Tsallis sum – and thus also the α-norm $||p||_\alpha$ – have different interpretations depending on whether α is in the range $0 \leq \alpha < 1$ or $\alpha > 1$. Specifically, when $0 \leq \alpha < 1$, the Tsallis sum of order α is a measure of uncertainty, increasing as the distribution p becomes more spread out. Conversely, when $\alpha > 1$, the Tsallis sum of order α is a measure of certainty, decreasing for more spread-out distributions.

To borrow a term from quantum information theory, for $\alpha > 1$, the Tsallis sum becomes a measure of **purity**, which increases as distributions become more concentrated at one state, i.e., more "pure" (see Chapter 7). A formal proof of this interpretation is left as an exercise in Problem 3.5.

Example 3.1

Take the two extreme situations in terms of uncertainty, namely the minimum-entropy probability vector $p = \delta_{\bar{x}}$ for some $\bar{x} \in [N]$ and the maximum-entropy probability vector $p = u$. With $p = \delta_{\bar{x}}$, the Tsallis sum of order α equals

$$\sum_{x \in [N]} p_x^\alpha = 1 \tag{3.24}$$

for any value of the order $\alpha \geq 0$. In contrast, with $p = u$, the Tsallis sum of order α is evaluated as

$$\sum_{x \in [N]} p_x^\alpha = \frac{1}{N^{\alpha-1}}, \tag{3.25}$$

which is larger than 1 for $0 \leq \alpha < 1$, and smaller than 1 for $\alpha > 1$.

Accordingly, with $0 \leq \alpha < 1$, the Tsallis sum of order α increases in going from $p = \delta_{\bar{x}}$ to $p = u$, confirming that the Tsallis sum of order α is a measure of uncertainty in this regime. In contrast, when $\alpha > 1$, the Tsallis sum of order α decreases in going from $p = \delta_{\bar{x}}$ to $p = u$, which is consistent with the interpretation of the Tsallis sum of order α as a measure of certainty in this regime.

Example 3.2

To further elaborate on the different behaviors of the α-Tsallis sum as a function of the order α, consider a bit with distribution $p = [1 - p_1, p_1]^T$. The Tsallis sum of order α, which can be computed as $p_1^\alpha + (1 - p_1)^\alpha$, is shown in Figure 3.5 as a function of probability p_1. The probability ranges from $p_1 = 0$, corresponding to no uncertainty, to $p_1 = 0.5$, corresponding to a uniform distribution.

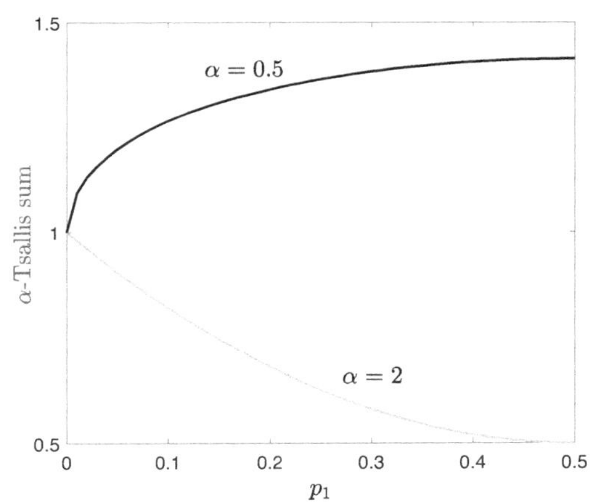

Figure 3.5 Tsallis sum of orders $\alpha = 0.5$ and $\alpha = 2$ as a function of the probability p_1 for a single-bit system ($N = 2$).

When $\alpha = 0.5$ – representing the interval $0 \leq \alpha < 1$ – the Tsallis sum increases with the uncertainty level, showcasing its interpretation as an uncertainty measure. In contrast, for $\alpha = 2$ – representing the range $\alpha > 1$ – the Tsallis sum of order α decreases with p_1, thus providing a measure of certainty.

In a manner consistent with the previous examples, the following bounds apply for the Tsallis sum.

Key Result 3.6 Bounds on the Tsallis sum

For $0 \leq \alpha < 1$, the Tsallis sum of order α satisfies the inequalities

$$\underbrace{1}_{\text{minimum entropy}} \leq \sum_{x \in [N]} p_x^\alpha \leq \underbrace{\frac{1}{N^{\alpha-1}}}_{\text{maximum entropy}}, \tag{3.26}$$

with the lower bound attained for a minimum-entropy distribution $p = \delta_{\bar{x}}$ for some $\bar{x} \in [N]$ and the upper bound obtained for the maximum-entropy distribution $p = u$.

For $\alpha > 1$, the Tsallis sum of order α satisfies the inequalities

$$\underbrace{\frac{1}{N^{\alpha-1}}}_{\text{maximum entropy}} \leq \sum_{x\in[N]} p_x^\alpha \leq \underbrace{1}_{\text{minimum entropy}}, \tag{3.27}$$

with the lower bound obtained for the maximum-entropy distribution $p = u$ and the upper bound attained for the minimum-entropy distribution $p = \delta_{\bar{x}}$ for some $\bar{x} \in [N]$.

3.4.3 Tsallis Sum and Tsallis Entropy

For all orders $\alpha \neq 1$, the Tsallis entropy (3.18) is obtained by transforming the Tsallis sum of order α through the function

$$\frac{1_2}{1-\alpha}(\cdot - 1). \tag{3.28}$$

As illustrated in Figure 3.6, the function (3.28) is increasing in the regime $0 \leq \alpha < 1$ for which the Tsallis sum of order α is a measure of uncertainty; while it is decreasing in the regime $\alpha > 1$ for which the Tsallis sum of order α is a measure of certainty.

3.4.4 Special Cases of the Tsallis Entropy

This section reviews some special cases of the Tsallis entropy, which are obtained by setting the order α to specific values.

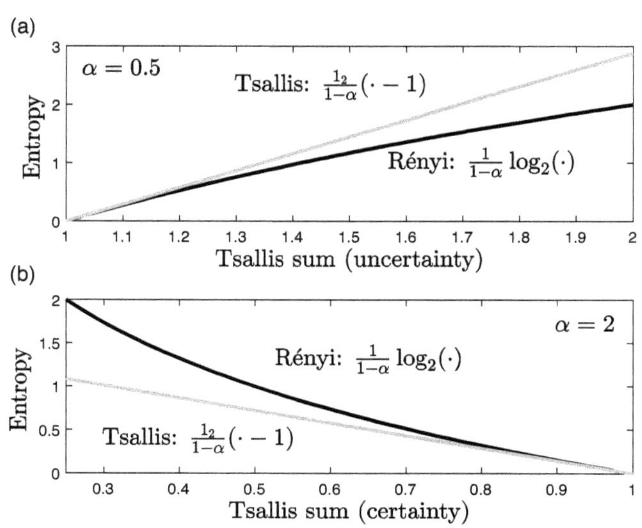

Figure 3.6 Plots of the function (3.28) and of the function (3.58), which relate the Tsallis sum respectively to the Tsallis entropy and the Rényi entropy, for (a) $\alpha = 0.5$ and (b) $\alpha = 2$.

Tsallis Entropy of Order $\alpha = 0$, aka Tsallis Max-Entropy. By (3.21), for any given probability vector p, the Tsallis entropy that returns the largest value is obtained by setting $\alpha = 0$ in (3.18). This is known as the Tsallis max-entropy.

To describe it, it is useful to define the support of a distribution.

Definition 3.8 Support of a distribution
The support of distribution p is the subset of states $x \in [N]$ for which the probability p_x is positive, i.e.,

$$\text{supp}(p) = \{x \in [N] \colon p_x > 0\}. \tag{3.29}$$

Note that the subset supp(p) "supports" the distribution p in the sense that the equality

$$\sum_{x \in \text{supp}(p)} p_x = 1 \tag{3.30}$$

holds.

For example, for the probability vector $p = [0.1, 0, 0.9, 0]^T$, the support set is $\text{supp}(p) = \{0, 2\}$ and the cardinality of the support set is $|\text{supp}(p)| = 2$.

The Tsallis sum with $\alpha = 0$ equals the cardinality of the support set, i.e.,

$$\sum_{x \in [N]} p_x^0 = |\text{supp}(p)|. \tag{3.31}$$

Therefore, using the function (3.28), the corresponding Tsallis entropy is given as follows.

Definition 3.9 Tsallis entropy of order $\alpha = 0$, aka Tsallis max-entropy
The Tsallis entropy of order $\alpha = 0$, also known as the Tsallis max-entropy, is evaluated as

$$\text{H}_0^{\text{T}}(p) = \text{l}_2 \cdot (|\text{supp}(p)| - 1). \tag{3.32}$$

Intuitively, a larger support set supp(p) indicates a distribution that is more spread out, and hence more uncertain. Reflecting this intuition, the Tsallis max-entropy grows linearly with the cardinality $|\text{supp}(p)|$ of the support.

Therefore, with $\alpha = 0$, all non-zero probabilities, no matter how small, have the same contribution to the Tsallis entropy, and the uncertainty measure $\text{H}_0^{\text{T}}(p)$ only depends on the size of the support of distribution p. Note that, unlike for other values of α, the Tsallis max-entropy is not continuous in p, as it "jumps" when a probability becomes strictly positive.

The name, Tsallis max-entropy, provides a reminder that the Tsallis entropy of order $\alpha = 0$, $\text{H}_0^{\text{T}}(p)$, is maximal among all Tsallis entropies for any distribution p.

Tsallis Entropy of Order $\alpha > 0$. Increasing the inverse temperature α causes the Tsallis entropy to assign a smaller weight to the contributions of the lower-probability states, yielding smaller entropy values. The following are two notable Tsallis entropies for $\alpha > 0$.

Definition 3.10 Tsallis entropy of order $\alpha = 0.5$
The Tsallis entropy of order $\alpha = 0.5$ is defined as

$$\mathrm{H}^{\mathrm{T}}_{1/2}(p) = 2 \cdot \mathrm{l}_2 \cdot \left(\sum_{x \in [N]} \sqrt{p_x} - 1 \right). \tag{3.33}$$

Definition 3.11 Tsallis entropy of order $\alpha = 2$, aka Tsallis collision entropy
The Tsallis entropy of order $\alpha = 2$, also known as the Tsallis collision entropy, is given by

$$\mathrm{H}^{\mathrm{T}}_{2}(p) = \mathrm{l}_2 \left(1 - \sum_{x \in [N]} p_x^2 \right). \tag{3.34}$$

The name "collision" entropy refers to the presence of the probability $\sum_{x \in [N]} p_x^2$, which can be interpreted as the probability that two independent realizations of the random variable $\mathrm{x} \sim p$ "collide," i.e., produce the same output $x \in [N]$. In fact, as detailed in Part IV, the probabilities of independent random variables multiply.

Finally, in the limit of an infinite α, the Tsallis entropy becomes vanishingly small, yielding the **Tsallis min-entropy** as

$$\mathrm{H}^{\mathrm{T}}_{\infty}(p) = 0. \tag{3.35}$$

Therefore, in the limit $\alpha \to \infty$, the average prediction error equals zero irrespective of the predictor p.

Example 3.3

This example studies the **scaling** of the Tsallis entropy as a function of the number of states N for different values of the order α. For reference, Chapter 2 showed that the Shannon entropy grows at most logarithmically with the number of states N, since the Shannon entropy of a uniform distribution is given by $\mathrm{H}(u) = \log_2(N)$. By the property (3.21), for the Tsallis entropy, one may expect a faster growth for low orders and a slower growth for larger orders.

For any order $\alpha \neq 1$, the Tsallis entropy evaluated for a uniform distribution is given as

$$\mathrm{H}^{\mathrm{T}}_{\alpha}(u) = \frac{\mathrm{l}_2}{1 - \alpha} \left(N^{1-\alpha} - 1 \right). \tag{3.36}$$

Figure 3.7 plots the Tsallis entropy $\mathrm{H}^{\mathrm{T}}_{\alpha}(u)$ for different values of the order α.

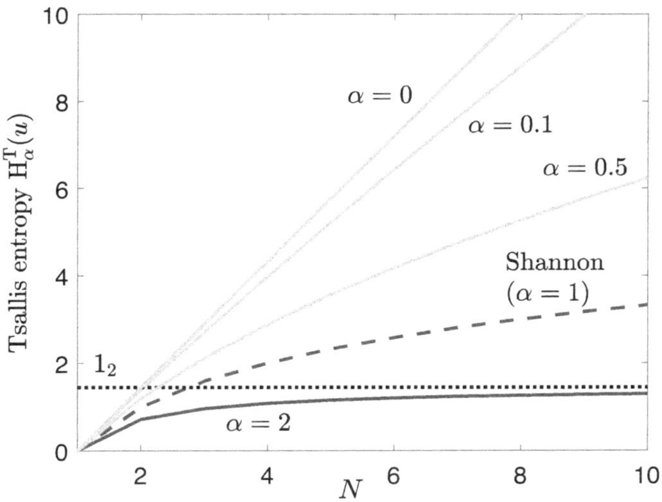

Figure 3.7 Tsallis entropy $\mathrm{H}^{\mathrm{T}}_{\alpha}(u)$ for a uniform distribution u versus the number of states, N, for different values of the order α.

To elaborate on the results in the figure, take first the case of smaller orders $\alpha < 1$. By (3.36), under a uniform distribution, the Tsallis max-entropy (3.32),

$$\mathrm{H}^{\mathrm{T}}_{0}(u) = \mathrm{l}_2 \cdot (\mathrm{supp}(p) - 1) = \mathrm{l}_2 \cdot (N - 1), \tag{3.37}$$

grows linearly and not logarithmically in N. As another example, with $\alpha = 1/2$, the Tsallis entropy (3.33) grows as the square root of N, i.e.,

$$\mathrm{H}^{\mathrm{T}}_{1/2}(u) = 2 \cdot \mathrm{l}_2 \cdot \left(\sqrt{N} - 1\right). \tag{3.38}$$

Accordingly, the Tsallis entropy with low orders $\alpha < 1$ can grow exponentially faster than the Shannon entropy as the number of states, N, increases.

In the opposite regime of larger inverse temperatures, $\alpha > 1$, the Tsallis entropy (3.36) remains bounded as N grows, satisfying the limit

$$\lim_{N \to \infty} \mathrm{H}^{\mathrm{T}}_{\alpha}(u) = \frac{\mathrm{l}_2}{\alpha - 1}. \tag{3.39}$$

Thus, with $\alpha > 1$, the Tsallis entropy does not even grow indefinitely as N increases.

Overall, by varying the order α, one can obtain a bounded growth (for $\alpha > 1$), a logarithmic growth (for $\alpha = 1$), or a faster growth for $\alpha \leq 1$, up to a linear growth with N (for $\alpha = 0$).

Example 3.4

Take now a composite system consisting of S subsystems, each having D states, so that the number of states grows exponentially with the number of subsystems as $N = D^S$ (see Section 1.3.3). With the maximum-entropy distribution u, the Shannon entropy grows proportionally to the number of subsystems S, i.e., as $\mathrm{H}(u) = \log_2(N) = S\log_2(D)$. The Shannon entropy is thus said to be **extensive**. In contrast, the Tsallis entropy with $\alpha \neq 1$ is not extensive. In fact, as detailed in Example 3.3, the Tsallis entropy $\mathrm{H}_\alpha^\mathrm{T}(u)$ for a uniform distribution u does not grow logarithmically with N, unless one sets $\alpha = 1$.

That said, in some physical systems, due to interactions among subsystems, the effective number of states – also known as microstates – may grow more slowly than exponentially with the number of subsystems S. For example, the effective number of states may follow a polynomial law of the form $N = S^a$ for some $a > 1$.

With this growth law, assuming a uniform distribution, the Shannon entropy would yield a logarithmic scaling with the number of subsystems, i.e., $\mathrm{H}(u) = a\log_2(S)$. In contrast, with the Tsallis entropy (3.43), one can recover a linear scaling in S by setting the order as $\alpha = 1 - 1/a < 1$ (see Problem 3.8). The reader is referred to the Recommended Resources for further information.

3.4.5 Properties of the Tsallis Entropy

The Tsallis entropy satisfies the three properties desired for an entropy measure, as listed in Chapter 1. Referring to Chapter 4 for a formal statement of the third property, the first two properties are as follows.

Key Result 3.7 Non-negativity and minimum-value property of the Tsallis entropy

The Tsallis entropy of any finite order $\alpha \geq 0$ is non-negative,

$$\mathrm{H}_\alpha^\mathrm{T}(p) \geq 0, \tag{3.40}$$

for all distributions p, and it equals zero if and only if the distribution is of the form $p = \delta_{\bar{x}}$ for some state $\bar{x} \in [N]$,

$$\mathrm{H}_\alpha^\mathrm{T}(\delta_{\bar{x}}) = 0. \tag{3.41}$$

The conditions above follow directly from the non-negativity and minimum-value property of the α-log-loss.

Key Result 3.8 Maximum-value property of the Tsallis entropy

The Tsallis entropy of any order $\alpha \geq 0$ is upper bounded as

$$\mathrm{H}_\alpha^\mathrm{T}(p) \leq \mathrm{H}_\alpha^\mathrm{T}(u), \tag{3.42}$$

with the maximum entropy given by

$$\begin{aligned}\mathrm{H}_\alpha^\mathrm{T}(u) &= -\log_{2,\alpha}\left(\frac{1}{N}\right)\\ &= \frac{\mathrm{l}_2}{1-\alpha}\left(N^{1-\alpha} - 1\right).\end{aligned} \tag{3.43}$$

Furthermore, for $\alpha > 0$, the upper bound is achieved if and only if the distribution p is uniform, i.e., $p = u$, while for $\alpha = 0$, the equality $\mathrm{H}^{\mathrm{T}}_{\alpha}(p) = \mathrm{H}^{\mathrm{T}}_{\alpha}(u)$ holds as long as the support of the distribution is $\mathrm{supp}(p) = [N]$.

Furthermore, like the Shannon entropy (see Section 2.5.4), the Tsallis entropy is concave.

Key Result 3.9 Concavity of the Tsallis entropy
The Tsallis entropy of any order $\alpha \geq 0$ is concave in the probability vector p.

As explained in Chapter 2, concavity implies that mixing increases uncertainty. Therefore, an observer that is unsure between two informational states, e.g., between the opinions of two experts, is no less uncertain, on average, than an observer that is in either informational state. The concavity of the Tsallis entropy, together with its permutation invariance, imply the upper bound (3.42) by following the same reasoning as in Appendix 2.E.

Example 3.5

Consider a single-bit system ($N = 2$). The corresponding **binary Tsallis entropy** of order α is evaluated from the general definition (3.18) as

$$\mathrm{h}^{\mathrm{T}}_{\alpha}(p_1) = \begin{cases} \frac{1_2}{1-\alpha}\left(p_1^{\alpha} + (1-p_1)^{\alpha} - 1\right) & \text{for } \alpha \geq 0 \text{ and } \alpha \neq 1 \\ \mathrm{h}(p_1) & \text{for } \alpha = 1, \end{cases} \tag{3.44}$$

where the binary Shannon entropy $\mathrm{h}(p_1)$ was introduced in Chapter 2. Note the equality $\mathrm{h}^{\mathrm{T}}_{\alpha}(p_1) = \mathrm{h}^{\mathrm{T}}_{\alpha}(p_0)$. Figure 3.8 plots the binary Tsallis entropy $\mathrm{h}^{\mathrm{T}}_{\alpha}(p_1)$ as a function of probability p_1 for different values of the order α.

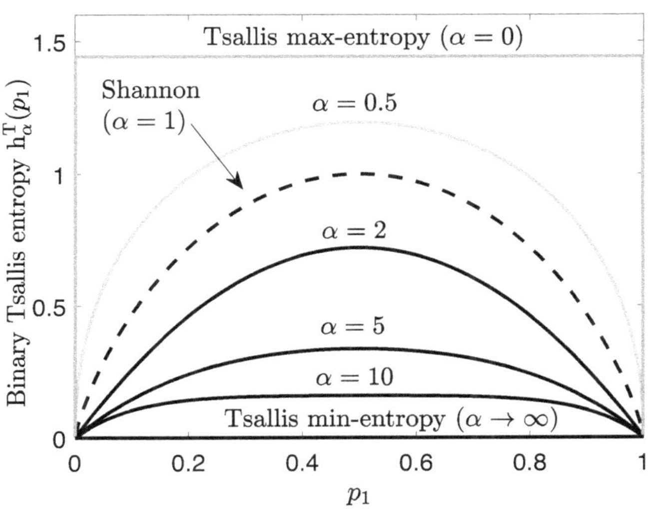

Figure 3.8 Binary Tsallis entropy $\mathrm{h}^{\mathrm{T}}_{\alpha}(p_1)$ as a function of probability p_1 for different values of the order α.

First, the figure illustrates how the Tsallis entropy is a decreasing function of the inverse temperature α (see the property (3.21)). It obtains smaller values than the Shannon entropy for $\alpha > 1$, and larger values than the Shannon entropy for $0 \leq \alpha < 1$. The maximum value of the entropy is given by the Tsallis max-entropy, which is not continuous in p_1, jumping from 0 to the maximum value 1_2 when moving away from the extreme values $p_1 = 0$ and $p_1 = 1$.

Second, the figure confirms that the Tsallis entropy, for any fixed order $\alpha \geq 0$, satisfies the basic characterizing properties of an entropy. Specifically, it is non-negative; it achieves the minimum value at minimum-uncertainty distributions $p_1 = 0$ and $p_1 = 1$; and it obtains the maximum value at the maximum-uncertainty, uniform, distribution $p_1 = 0.5$. In particular, as p_1 approaches the uniform-distribution setting $p_1 = 0.5$ starting from $p_1 = 0$ or $p_1 = 1$, the Tsallis entropy increases up to the maximum value $\mathrm{h}_\alpha^\mathrm{T}(0.5) = 1_2 \cdot (2^{1-\alpha} - 1)/(1 - \alpha)$ in (3.36).

Finally, like the Shannon entropy, the Tsallis entropy is also seen to be a concave function of the probability vector for any $\alpha \geq 0$.

3.5 Quantifying Information by Guessing

As detailed in the previous section, one of the key properties of the Tsallis entropy is that, with a low order α, it can grow more quickly than the Shannon entropy with respect to the number of states N (see Figure 3.7). This raises the question of whether there exist informational tasks in which the number of required bits may scale more quickly than the logarithmic maximal growth exhibited by the Shannon entropy.

This section studies one such informational task, namely that of identifying the state $\mathrm{x} \sim p$ of the system through a **trial-and-error** sequence of guesses. As a running example, the reader may think of the state x as a password that is known to lie in some set $[N]$ of possible options.

3.5.1 Guessing

The observer – the guesser here – has a state of knowledge concerning the true password x that is described by the probability vector p. Accordingly, the observer's belief in the password being equal to x is given by p_x.

To identify the state x, the observer proceeds sequentially via a number of guesses, or trials. For each trial, the observer guesses the state x to be equal to some value $x \in [N]$; it submits this guess to the system; and it receives a message from the system indicating whether the guess was correct, i.e., whether $\mathrm{x} = x$, or rather incorrect, i.e., $\mathrm{x} \neq x$. Therefore, a **guess** corresponds to submitting the question "is $\mathrm{x} = x$?" for some guess $x \in [N]$ to the system, and receiving a yes/no answer. This procedure is repeated for as many trials as needed until success, that is, until the system confirms the correctness of the observer's guess.

Intuitively, the more uncertain we are about the true password, the more rounds of trials we would need to guess the password before we can identify the correct one. Therefore, the average number of guesses can be considered as another measure of the uncertainty associated with the informational state p.

For a given guessing strategy, define as G_x the number of guesses required to identify the state x when it equals $x \in [N]$. The goal is to minimize the **average number of guesses**, also known as **guesswork**,

$$G_{\text{avg}}(p) = \mathrm{E}_p[G_{\mathrm{x}}] \tag{3.45}$$

by finding the best **guessing strategy**.

Example 3.6

According to some statistical analysis, the probabilities of the five most likely 4-digit PINs are given in Table 3.2. Therefore, more than 20% of all passwords can be guessed by trying out these five combinations one after another.

Table 3.2 Top five most likely 4-digit PINs

PIN	Probability
1234	0.107
1111	0.061
0000	0.018
1212	0.011
7777	0.007

3.5.2 Guessing vs. Asking

Before addressing the problem of finding the best guessing strategy, it is useful to elaborate on the relationship between the problem of guessing considered here and the problem of asking studied in Chapter 2.

In the asking problem, the observer is allowed to ask any question with a binary answer. In contrast, in the guessing problem, the observer is only permitted questions of the form "is $\mathrm{x} = x$?" This type of question is significantly more constrained as compared to the asking problem. In fact, a general binary question can inquire about x being in any subset of possible states, while the types of questions allowed by guessing can only rule out a single state value x.

3.5.3 Optimal Guessing

Minimizing the average number of guesses (3.45) requires guessing first the most probable values of the state under the informational state p, and then proceeding in order from most likely to least likely state, with ties broken arbitrarily. The strategy

Table 3.3 Optimal guessing strategy

Optimal guessing

input: distribution p
output: an optimal guessing strategy
list all values x in order of decreasing probabilities p_x, with ties broken arbitrarily
for n from 1 to N:
ask the question "is $x = x_n$?" for the nth state in the list
end
return the resulting guessing strategy

is similar to that used for optimal compression without the prefix-free condition studied in Chapter 2. However, there, one additional bit can "cover" two states, while here one guess can only address one state x. The optimal guessing procedure is summarized in Table 3.3.

To elaborate further on the optimal guessing strategy, as in Chapter 2, define a permutation function $s(n)$ that returns the nth most probable state, with $n = 0$ representing the most probable state, and $n = N - 1$ the least probable state. Ties are broken arbitrarily. Then, using the optimal procedure in Table 3.3, the **minimum average number of guesses** is given by

$$G^*_{\text{avg}}(p) = \sum_{n=0}^{N-1} p_{s(n)} \cdot (n+1). \tag{3.46}$$

The next examples illustrate the general fact that the minimum average number of guesses is no smaller than the minimum average number of binary questions. This highlights the added constraint that guesses must adhere to when evaluating potential state values.

Example 3.7

For the running example studied in Chapter 2 with the dyadic probability vector $p = [1/4, 1/2, 1/8, 1/8]^T$, an optimal guessing order is $s(0) = 1$, $s(1) = 0$, $s(2) = 2$, and $s(3) = 3$. Note that the optimal sequence is not unique, since the permutation $s(0) = 1$, $s(1) = 0$, $s(2) = 3$, and $s(3) = 2$ would also be optimal. The corresponding minimum average number of guesses is

$$G^*_{\text{avg}}(p) = \frac{1}{2}1 + \frac{1}{4}2 + +\frac{1}{8}3 + \frac{1}{8}4 = \frac{15}{8}. \tag{3.47}$$

The minimum average number of guesses, 15/8, is thus larger than the average number of binary questions in the asking problem, which was found in Chapter 2 to be equal to $7/4 = 14/8$.

Example 3.8

This example studies a uniform dyadic informational state p, that is, a distribution of the form $p_x = 1/N = 2^{-l}$, given an integer l, for all $x \in [N]$. As seen in Chapter 2, in this case, the minimum average number of binary questions is given by the Shannon entropy $L^*_{\text{avg}}(u) = \mathrm{H}(u) = \log_2(N)$. Therefore, the uncertainty, measured in terms of questions/ bits, grows logarithmically in the size, N, of the set $[N]$.

The minimum average number of guesses is obtained via any guessing order, since all states have the same probability. The resulting minimum average number of guesses is given by (3.46), which is evaluated as

$$G^*_{\text{avg}}(u) = \frac{1}{N}\sum_{n=1}^{N} n = \frac{N+1}{2}. \tag{3.48}$$

By (3.48), the minimum average number of guesses grows linearly with N, and thus it is exponentially larger than the number of questions/bits. Explicitly, the minimum average number of guesses, $G^*_{\text{avg}}(u)$, can be written as

$$G^*_{\text{avg}}(u) = \frac{2^{\mathrm{H}(u)}+1}{2} = \frac{2^{L^*_{\text{avg}}(u)}+1}{2}. \tag{3.49}$$

So, guessing can be exponentially harder than asking.

As a special case, for the case of a composite system with S subsystems and $N = D^S$ states, the optimal number of guesses, $G^*_{\text{avg}}(u)$, grows exponentially with the number of subsystems S, since (3.48) implies the equality

$$G^*_{\text{avg}}(u) = \frac{N+1}{2} = \frac{D^S+1}{2}. \tag{3.50}$$

This is unlike the number of binary questions $L^*_{\text{avg}}(u)$, which grows linearly with S, since the Shannon entropy is given by $\mathrm{H}(u) = S\log_2(D)$.

3.5.4 Optimal Guessing and Entropy ⏭

In general, for any distribution p, the minimum average number of guesses can be lower bounded in terms of the Shannon entropy as follows (see Appendix 3.D for a proof).

Key Result 3.10 Lower bound on the minimum average number of guesses
The minimum average number of guesses satisfies the inequality

$$G^*_{\text{avg}}(p) \geq \frac{2^{\mathrm{H}(p)}}{1+\ln(N)}, \tag{3.51}$$

where $\ln(\cdot)$ represents the natural logarithm.

This inequality says that the number of guesses is at least as large as $2^{\mathrm{H}(p)}$ times a number dependent on the number of states, N. The dependence of the bound (3.51) on the Shannon entropy $\mathrm{H}(p)$ is thus exponential. This stands in stark contrast to the minimum average number of questions, which is linear in the Shannon entropy $\mathrm{H}(p)$.

Unlike the Shannon entropy, the Tsallis entropy can reproduce the faster scaling (3.48) with N that was observed in Example 3.8. As a result, the Tsallis entropy offers a natural characterization of the minimum average number of guesses. Specifically, the minimum average number of guesses can be upper bounded in terms of the Tsallis max-entropy (3.32) as follows.

Key Result 3.11 Upper bound on the minimum average number of guesses

The minimum average number of guesses satisfies the inequality

$$\begin{aligned} G^*_{\mathrm{avg}}(p) &\leq \frac{1}{2 \cdot 1_2} \mathrm{H}^{\mathrm{T}}_0(p) + 1 \\ &= \frac{|\mathrm{supp}(p)| + 1}{2}. \end{aligned} \tag{3.52}$$

In fact, for the worst case of a uniform distribution with non-zero probabilities only inside the support set $\mathrm{supp}(p)$, by (3.48), the number of guesses equals (3.52). However, the bound is generally not tight. Indeed, it is not even sensitive to changes in the distribution vector p as long as the cardinality of the support, $|\mathrm{supp}(p)|$, remains constant.

Upper and lower bounds that depend directly on the numerical values of the entries in vector p can be obtained by using the Tsallis entropy of order $1/2$ (see Appendix 3.D).

Key Result 3.12 Upper and lower bounds on the minimum average number of guesses

The minimum average number of guesses satisfies the inequalities

$$\frac{\left(\frac{1}{2 \cdot 1_2} \mathrm{H}^{\mathrm{T}}_{1/2}(p) + 1\right)^2}{1 + \ln(N)} \leq G^*_{\mathrm{avg}}(p) \leq \left(\frac{1}{2 \cdot 1_2} \mathrm{H}^{\mathrm{T}}_{1/2}(p) + 1\right)^2, \tag{3.53}$$

or equivalently

$$\frac{\left(\sum_{x \in [N]} \sqrt{p_x}\right)^2}{1 + \ln(N)} \leq G^*_{\mathrm{avg}}(p) \leq \left(\sum_{x \in [N]} \sqrt{p_x}\right)^2. \tag{3.54}$$

Therefore, apart from a multiplicative constant, the minimum average number of guesses $G^*_{\mathrm{avg}}(p)$ can be approximated with a quadratic function of the Tsallis entropy of order $1/2$, or of the corresponding Tsallis sum of order $1/2$, i.e., $\sum_{x \in [N]} \sqrt{p_x}$.

Example 3.9

For this example, distribution vectors p with dimension $N = 10$ are generated at random by drawing each entry from a continuous-valued uniform distribution in the interval $[0, 1]$ and then normalizing the resulting vector. Note, for such random distributions p, the support equals N with probability 1. For each randomly drawn distribution p, the minimum average number of guesses $G^*_{\text{avg}}(p)$ in (3.46) was evaluated, along with the bound in (3.52) based on the Tsallis entropy of order 0 and the bounds in (3.54) based on the Tsallis entropy of order $1/2$.

Figure 3.9 shows that the lower bound in (3.54) provides a reasonable approximation to the minimum average number of guesses. Furthermore, while the upper bound (3.52) based on the Tsallis entropy of order 0 is tighter than the upper bound (3.54), it does not capture the dependence of the minimum average number of guesses on the entries of vector p beyond the support $|\text{supp}(p)| = N = 10$.

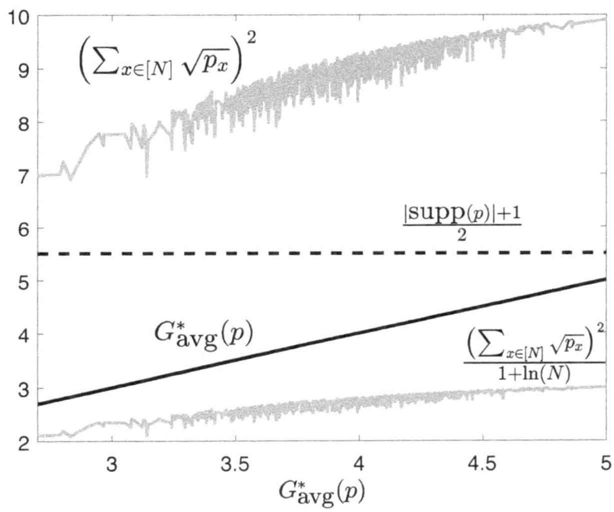

Figure 3.9 Comparison between the true minimum average number of guesses, $G^*_{\text{avg}}(p)$, and the upper bound in (3.52), as well as the upper and lower bounds in (3.54). The figure was obtained by drawing a number of distributions p at random as explained in the text.

3.6 Rényi Entropy

As discussed in the previous sections, the Tsallis entropy provides a family of uncertainty measures that is controlled by the order α. In particular, it was observed that the Tsallis entropy supports distinct scaling laws as a function of the number of states, N, from linear to bounded. While it is useful to have the freedom to choose among such a range of uncertainty measures for many applications, including the problem of guessing, in other settings it may be preferable to have a family of entropy measures that share the maximum logarithmic scaling of the Shannon entropy. In fact, as noted in Chapter 2, in many informational tasks, uncertainty and information are quantified by a number of bits that grows no faster than logarithmically with the number of possible states.

This section introduces one such family of entropies, which will be shown in the next section to relate to the solution of a modified version of the informational task of compression.

Like the Tsallis entropy, the Rényi entropy depends on an order parameter $\alpha \geq 0$, with the properties that a larger α yields a smaller entropy and that for $\alpha = 1$ one recovers the Shannon entropy. However, the Rényi entropy is upper bounded by $\log_2(N)$, even for small inverse temperatures $0 \leq \alpha < 1$.

The Rényi entropy has other important properties that separate it from the Tsallis entropy. The most consequential of these properties, additivity, stipulates that the entropy of two independent subsystems is the sum of the per-subsystem entropies. This intuitive property implies extensivity, which, as seen in Example 3.3, is not satisfied by the Tsallis entropy. Detailed discussions about additivity can be found in Chapter 12.

Another distinguishing property is that, while the Tsallis entropy is concave like the Shannon entropy, this is not necessarily the case for the Rényi entropy. So, mixing may not always increase the Rényi entropy.

3.6.1 Defining the Rényi Entropy

The Rényi entropy is defined as follows.

Definition 3.12 Rényi entropy

For any $\alpha \geq 0$, the Rényi entropy of order α is defined as

$$\mathrm{H}_{\alpha}^{\mathrm{R}}(p) = \begin{cases} \frac{1}{1-\alpha} \log_2 \left(\sum_{x \in [N]} p_x^{\alpha} \right) & \text{for } \alpha \geq 0 \text{ with } \alpha \neq 1 \\ \mathrm{H}(p) & \text{for } \alpha = 1. \end{cases} \tag{3.55}$$

By the definition (3.55), like the Tsallis entropy, the Rényi entropy coincides with the Shannon entropy with the choice $\alpha = 1$, and **continuity** is guaranteed by the limit

$$\lim_{\alpha \to 1} \mathrm{H}_{\alpha}^{\mathrm{R}}(p) = \mathrm{H}_{1}^{\mathrm{R}}(p) = \mathrm{H}(p). \tag{3.56}$$

Furthermore, the Rényi entropy is also permutation-invariant with respect to the probability vector p.

Like the Tsallis entropy, the Rényi entropy provides a family of decreasing uncertainty measures with the order α.

Key Result 3.13 The Rényi entropy decreases with the inverse temperature α
For any fixed probability vector p, the Rényi entropy $\mathrm{H}^{\mathrm{R}}_{\alpha}(p)$ is a decreasing function of the order $\alpha \geq 0$, i.e., the inequality

$$\mathrm{H}^{\mathrm{R}}_{\alpha'}(p) \leq \mathrm{H}^{\mathrm{R}}_{\alpha}(p) \tag{3.57}$$

holds for any two orders α' and α with $\alpha' \geq \alpha$.

3.6.2 Rényi Entropy and Tsallis Sum

An intuitive way to understand the Rényi entropy (3.55) is as a different transformation of the Tsallis sum of order α in (3.22) or of the α-norm in (3.23). In particular, as shown in Figure 3.6, while the Tsallis entropy applies the linear function in (3.28) to the Tsallis sum, the Rényi entropy applies the logarithmic function

$$\frac{1}{(1-\alpha)} \log_2(\cdot). \tag{3.58}$$

This observation makes it possible to compare the Rényi and Tsallis entropies by using the simple inequality $\log_2(x) \leq 1_2 \cdot (x-1)$, which holds for all $x > 0$. Specifically, the following property follows (Problem 3.6).

Key Result 3.14 Rényi entropy vs. Tsallis entropy
When $0 \leq \alpha < 1$, the Rényi entropy is no larger than the Tsallis entropy

$$\mathrm{H}^{\mathrm{R}}_{\alpha}(p) \leq \mathrm{H}^{\mathrm{T}}_{\alpha}(p); \tag{3.59}$$

with $\alpha = 1$, the two entropies are equal and they coincide with the Shannon entropy

$$\mathrm{H}^{\mathrm{R}}_{1}(p) = \mathrm{H}^{\mathrm{T}}_{1}(p) = \mathrm{H}(p); \tag{3.60}$$

while for $\alpha > 1$, the following inequality holds

$$\mathrm{H}^{\mathrm{R}}_{\alpha}(p) \geq \mathrm{H}^{\mathrm{T}}_{\alpha}(p). \tag{3.61}$$

A numerical example comparing Rényi and Tsallis entropies will be provided later in this section (see Figure 3.12).

3.6.3 Special Cases of the Rényi Entropy

This subsection examines specific cases of the Rényi entropy, presented in descending order of their values.

Rényi Entropy of order $\alpha = 0$, aka Rényi max-entropy. Like the Tsallis entropy, when $\alpha = 0$, the Rényi entropy depends only on the cardinality $|\mathrm{supp}(p)|$ of the support of the distribution p, yielding the following.

Definition 3.13 Rényi entropy of order $\alpha = 0$, aka Rényi max-entropy, or Hartley entropy

The Rényi entropy of order $\alpha = 0$, also known as the Hartley entropy, is given by

$$\mathrm{H}_0^{\mathrm{R}}(p) = \log_2(|\mathrm{supp}(p)|). \tag{3.62}$$

As suggested by the name, the Rényi max-entropy (3.62) is maximal among all Rényi entropies $\mathrm{H}_\alpha^{\mathrm{R}}(p)$ for any probability vector p. However, unlike the Tsallis max-entropy (3.32), the dependence of the Rényi max-entropy (3.62) on the support size $|\mathrm{supp}(p)|$ is logarithmic, and not linear. Note also that the Rényi max-entropy is not continuous in p.

Operationally, a variant of the Rényi max-entropy is obtained when studying the problem of fixed-length compression, which is detailed in Appendix 3.F.

Rényi Entropy of order $\alpha > 0$. Larger values of the order α yield smaller values for the Rényi entropy, and the following special cases are notable examples.

Definition 3.14 Rényi entropy of order $\alpha = 0.5$

The Rényi entropy of order $\alpha = 0.5$ is given by

$$\mathrm{H}_{1/2}^{\mathrm{R}}(p) = 2\log_2\left(\sum_{x\in[N]} \sqrt{p_x}\right). \tag{3.63}$$

Definition 3.15 Rényi entropy of order $\alpha = 2$, aka Rényi collision entropy

The Rényi entropy of order $\alpha = 2$, also known as the Rényi collision entropy, is given by

$$\mathrm{H}_2^{\mathrm{R}}(p) = -\log_2\left(\sum_{x\in[N]} p_x^2\right). \tag{3.64}$$

For an application of the Rényi collision entropy to the problem of randomness extraction, the reader is referred to the Recommended Resources.

In the limit $\alpha \to \infty$, unlike the Tsallis min-entropy (3.35), the Rényi min-entropy is generally positive. As derived in Appendix 3.B, it is given as follows.

Definition 3.16 Rényi entropy of order $\alpha = \infty$, aka Rényi min-entropy

The Rényi entropy of order $\alpha = \infty$, also known as the Rényi min-entropy, is given by

$$\mathrm{H}_\infty^{\mathrm{R}}(p) = -\log_2\left(\max_{x\in[N]} p_x\right). \tag{3.65}$$

To understand the definition (3.65) of the Rényi min-entropy, take again the perspective of an observer trying to predict the state x $\sim p$. As seen in the previous section, the best single guess of an observer in informational state p is to select a state x with maximum probability p_x, i.e., with probability $\max_{x\in[N]} p_x$. Therefore, the probability $\max_{x\in[N]} p_x$, which is known as the **guessing probability**, is larger for informational states p corresponding to easier-to-predict systems. Conversely, a highly uncertain system would be characterized by a small value of the guessing probability $\max_{x\in[N]} p_x$. In a manner consistent with this intuition, the Rényi min-entropy $\mathrm{H}^{\mathrm{R}}_{\infty}(p)$ in (3.65) yields an uncertainty value that increases as the guessing probability $\max_{x\in[N]} p_x$ decreases.

Operationally, the Rényi min-entropy can be related to the task of extracting fair random bits from a random variable x $\sim p$. Furthermore, it can be proved that for all $\alpha > 1$, the Rényi min-entropy is bounded as

$$\frac{\alpha-1}{\alpha}\mathrm{H}^{\mathrm{R}}_{\alpha}(p) \leq \mathrm{H}^{\mathrm{R}}_{\infty}(p) \leq \mathrm{H}^{\mathrm{R}}_{\alpha}(p). \tag{3.66}$$

For further details on the Rényi min-entropy, the reader is referred to the Recommended Resources.

Example 3.10

For the probability vector $p = [1/4, 1/2, 1/8, 1/8]^T$, the Rényi entropy is plotted in Figure 3.10 as a function of α. The figure illustrates the decrease of the Rényi entropy as a function of the order α, ranging from the Rényi max-entropy at $\alpha = 0$ to the Rényi min-entropy in the limit $\alpha \to \infty$.

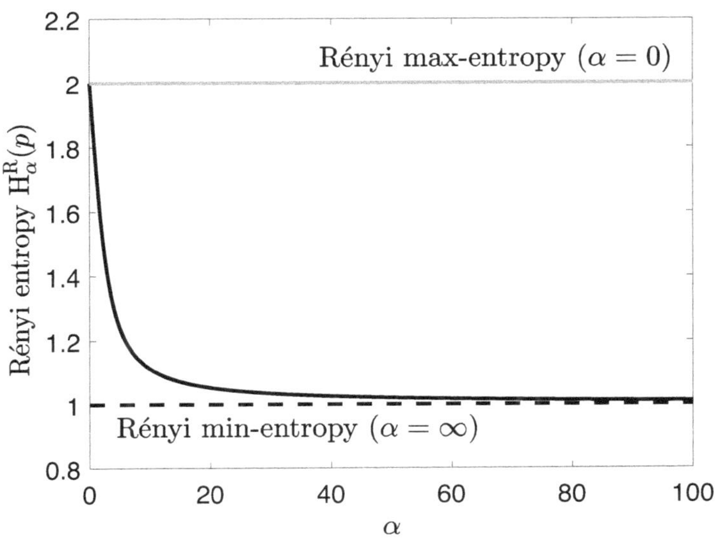

Figure 3.10 Rényi entropy for the probability vector $p = [1/4, 1/2, 1/8, 1/8]^T$ as a function of the order α.

Example 3.11

An ideal **cryptographic key** $\mathrm{x} \sim p$ is uniformly distributed, i.e., $p = u$, thus having maximum entropy. Consider a non-ideal cryptographic key $\mathrm{x} \sim p$ with distribution

$$p = \begin{cases} \frac{1}{2} & \text{for } x = 0 \\ \frac{1}{2(N-1)} & \text{otherwise.} \end{cases} \tag{3.67}$$

How good is the random variable $\mathrm{x} \sim p$ as a secret key? Intuitively, the more "random" x is, the better it serves as a secret key.

Considering the Shannon entropy as a measure of randomness would yield $\mathrm{H}(p) = 1 + 0.5\log_2(N-1)$, which is close to the maximum entropy $\mathrm{H}(u) = \log_2(N)$ attained by an ideal key. However, the key $\mathrm{x} \sim p$ takes value $x = 0$ with probability 1/2, and thus an eavesdropper can guess it correctly with probability 0.5.

The Rényi min-entropy provides a more relevant measure of the usefulness of the key $\mathrm{x} \sim p$, as it captures the difficulty in guessing the random variable x. For this example, the Rényi min-entropy is indeed equal to $\mathrm{H}^{\mathrm{R}}_{\infty}(p) = 1$, which does not grow with N, and is thus far from the maximum entropy $\mathrm{H}^{\mathrm{R}}_{\infty}(u) = \log_2(N)$ when N is large.

3.6.4 Properties of the Rényi Entropy

The Rényi entropy satisfies the following two properties, which follow directly from the analogous property of the Tsallis entropy given that there exists a **one-to-one mapping** between the two entropies (see Appendix 3.C). Chapter 4 states and proves also the monotonicity property described in Chapter 1.

Key Result 3.15 Non-negativity and minimum-value property of the Rényi entropy

The Rényi entropy for any order $\alpha \geq 0$ is non-negative,

$$\mathrm{H}^{\mathrm{R}}_{\alpha}(p) \geq 0, \tag{3.68}$$

for all distributions p, and it equals zero if and only if the distribution is of the form $p = \delta_{\bar{x}}$ for some state $\bar{x} \in [N]$,

$$\mathrm{H}^{\mathrm{R}}_{\alpha}(\delta_{\bar{x}}) = 0. \tag{3.69}$$

Key Result 3.16 Maximum-value property of the Rényi entropy

The Rényi entropy for any order $\alpha \geq 0$ is upper bounded as

$$\mathrm{H}^{\mathrm{R}}_{\alpha}(p) \leq \mathrm{H}^{\mathrm{R}}_{\alpha}(u) = \log_2(N). \tag{3.70}$$

Furthermore, for $\alpha > 0$, the upper bound is achieved if and only if the distribution p is uniform, i.e., $p = u$, while for $\alpha = 0$, the equality $\mathrm{H}_\alpha^\mathrm{R}(p) = \log_2(N)$ holds as long as the support of the distribution equals $\mathrm{supp}(p) = [N]$.

The following property provides an important distinction between the Rényi entropy and the Shannon and Tsallis entropies.

Key Result 3.17 Concavity of the Rényi entropy

The Rényi entropy $\mathrm{H}_\alpha^\mathrm{R}(p)$ is a concave function of the probability vector p for $0 < \alpha \leq 1$, but it is *not* guaranteed to be concave for $\alpha > 1$.

In fact, it can be shown that there exists a maximum value of α, say $\alpha_{\max}(N) \geq 1$, generally dependent on N, such that the Rényi entropy $\mathrm{H}_\alpha^\mathrm{R}(p)$ remains concave in the range $\alpha \in [0, \alpha_{\max}(N)]$ (see Recommended Resources). It follows that, when uncertainty is measured by a Rényi entropy of order $\alpha > 1$, mixing may not increase uncertainty.

Example 3.12

For a single-bit system, the Rényi entropy can be written as the **binary Rényi entropy**

$$\mathrm{h}_\alpha^\mathrm{R}(p_1) = \begin{cases} \frac{1}{1-\alpha}\log_2\left(p_1^\alpha + (1-p_1)^\alpha\right) & \text{for } \alpha \geq 0 \text{ with } \alpha \neq 1, \\ \mathrm{h}(p_1) & \text{for } \alpha = 1. \end{cases} \quad (3.71)$$

Note the equality $\mathrm{h}_\alpha^\mathrm{R}(p_1) = \mathrm{h}_\alpha^R(p_0)$.

Figure 3.11 plots the binary Rényi entropy as a function of probability p_1. A first observation is that, like the Tsallis and Shannon entropies, the Rényi entropy is

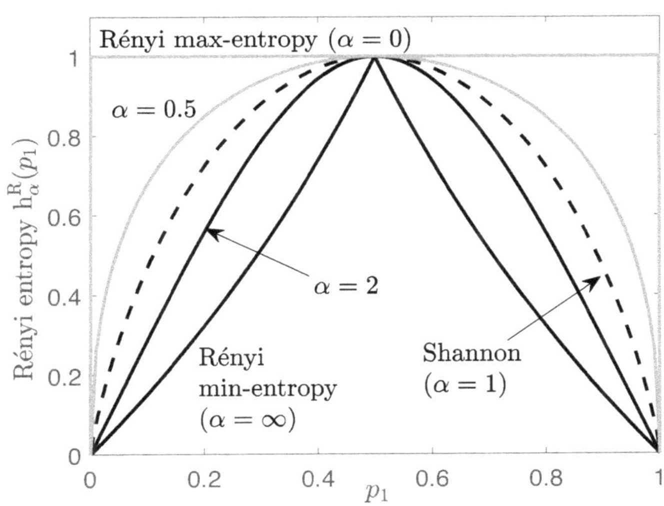

Figure 3.11 Binary Rényi entropy (3.71) as a function of the probability p_1 for different values of the order α.

minimal for the zero-uncertainty informational states $p_1 = 0$ and $p_1 = 1$, and maximal for the maximum-uncertainty setting $p_1 = 0.5$ corresponding to a uniform distribution. Furthermore, as the probability p_1 gets closer to $p_1 = 0.5$ starting from either $p_1 = 0$ or $p_1 = 1$, the Rényi entropy increases to the maximum value of 1 bit.

Another property in common with the Tsallis entropy is the monotonically decreasing trend of the Rényi entropy as a function of α, with the max-entropy $\mathrm{h}_0^{\mathrm{R}}(p_1)$ yielding the maximum value, and the Rényi min-entropy

$$\mathrm{h}_\infty^{\mathrm{R}}(p_1) = -\log_2(\max(p_1, 1 - p_1)) \tag{3.72}$$

providing the minimum value.

Figure 3.11, however, also illustrates an important difference between the Tsallis and the Rényi entropies. In fact, as can be verified using (3.72), the Rényi min-entropy $\mathrm{h}_\infty^{\mathrm{R}}(p_1)$ is not a concave function of p_1.

Example 3.13

Returning to the single-bit case studied in the previous example, Figure 3.12 compares the Tsallis and the Rényi entropies for different values of the order α as a function of the probability p_1. The figure confirms that, while for $\alpha = 1$ both entropies coincide and they are equal to the Shannon entropy, for $\alpha = 1/2$ the Tsallis entropy is larger than the Rényi entropy, while the opposite is true for $\alpha = 2$.

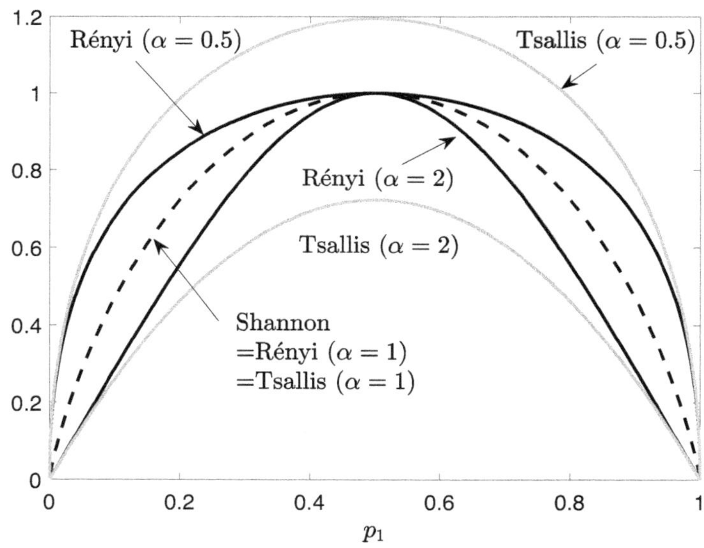

Figure 3.12 Comparison between binary Tsallis entropy (3.44) and binary Rényi entropy (3.71) as a function of the probability p_1 for different values of the order α.

3.7 Quantifying Information via Compression Beyond Average Codeword Length ⏭

In Chapter 2, the problem of designing an optimal compression scheme that minimizes the average codeword length $L_{\text{avg}}(p) = \mathrm{E}_p[L_{\mathrm{x}}]$ was studied with and without the prefix-free constraint. When imposing the prefix-free constraint, it was observed that a code with codeword lengths given by the **rounded-up Shannon surprises**

$$L_x = \lceil -\log_2(p_x) \rceil < -\log_2(p_x) + 1 \tag{3.73}$$

is almost optimal. This is in the sense that the corresponding average codeword length $\mathrm{E}_p[L_{\mathrm{x}}]$ exceeds the minimum average codeword length $L^*_{\text{avg}}(p)$ by at most 1 bit.

While approximately minimizing the average codeword length $L^*_{\text{avg}}(p)$, a code with codeword lengths (3.73) may be undesirable for some applications. To see why, consider a setting in which it is not possible – or very costly – for the decoder to process long codewords. This may be the case in applications characterized by delay or complexity constraints, in which the decoder does not have time to parse or store long codewords.

With the codeword lengths in (3.73), if the state x happens to have a low probability p_x, the associated codeword length L_x may be excessively large, violating the mentioned constraints at the decoder. To address this problem, one may modify the objective function – from the minimization of the average codeword length to the minimization of an alternative criterion that penalizes codewords that are excessively long. As will be detailed next, a particular choice of such a design criterion yields the Rényi entropy as a solution when imposing the prefix-free constraint.

3.7.1 Compression Design Problem

Consider the following function of the codeword length vector L and of the probability distribution p:

$$L_\beta(p) = \frac{1}{\beta}\log_2\left(\mathrm{E}_p\left[2^{\beta L_{\mathrm{x}}}\right]\right), \tag{3.74}$$

where $\beta > 0$ is fixed parameter. The criterion $L_\beta(p)$ in (3.74) corresponds to a form of **generalized average** of the codeword length L_{x} with respect to the state variable $\mathrm{x} \sim p$ that gives more and more weight to larger values of the codeword length as β increases.

To see this, take first the case of a small value of β. Using the first-order Taylor expansions $2^x \approx 1 + x/1_2$ and $\log_2(1+x) \approx 1_2 \cdot x$, for small values of parameter β, the criterion $L_\beta(p)$ in (3.74) is approximately equal to the conventional average codeword length $L_{\text{avg}}(p)$, i.e.,

$$\frac{1}{\beta}\log_2\left(\mathrm{E}_p\left[2^{\beta L_{\mathrm{x}}}\right]\right) \approx \mathrm{E}_p\left[L_{\mathrm{x}}\right] = L_{\text{avg}}(p). \tag{3.75}$$

In contrast, with larger values of $\beta > 0$, the objective in (3.76) assigns an exponentially increasing cost $2^{\beta L_x}$ to codeword length L_x, thus penalizing long codewords. The reader is referred to the Recommended Resources for further discussion on generalized averages.

Using this generalized design criterion, one can now pose the following extension of the code design problem (2.21) studied in Chapter 2:

$$\begin{aligned} &\underset{L}{\text{minimize}}\ L_\beta(p) \\ &\text{subject to} \sum_{x\in[N]} 2^{-L_x} \leq 1, \end{aligned} \tag{3.76}$$

where minimization is over all integer codeword lengths L. It is recalled from Chapter 2 that the constraint in (3.76) ensures the existence of a prefix-free code by the Kraft–McMillan inequality.

3.7.2 Escort Distribution

To make progress on problem (3.76) for any $\beta > 0$, it is useful to introduce the escort distribution.

Definition 3.17 Escort distribution
The escort distribution of order $\alpha \geq 0$ for distribution p is defined as the vector $\tilde{p}_\alpha = [\tilde{p}_{\alpha,0}, \ldots, \tilde{p}_{\alpha,N-1}]^T$ with entries

$$\tilde{p}_{\alpha,x} = \frac{p_x^\alpha}{\sum_{x'\in[N]} p_{x'}^\alpha} \tag{3.77}$$

for $x \in [N]$.

By (3.77), the escort distribution $\tilde{p}_\alpha$ is obtained by applying the exponent α to each entry p_x of vector p, and then scaling the resulting entries so as to ensure the normalization condition $\sum_{x\in[N]} \tilde{p}_{\alpha,x} = 1$.

Accordingly, since the denominator in (3.77) is uniquely determined by the need to guarantee normalization, one may also identify the escort distribution $\tilde{p}_\alpha$ using the **proportionality** sign $\propto$ as

$$\tilde{p}_{\alpha,x} \propto p_x^\alpha. \tag{3.78}$$

As the order α tends to 0, all the non-zero entries of the escort distribution (3.77) tend to become equal. In particular, in the limit $\alpha \to 0$, if all probabilities in vector p are positive, the escort distribution becomes closer and closer to a uniform distribution, with

$$\tilde{p}_0 = u. \tag{3.79}$$

So, thinking of the order α as the inverse temperature, increasing the temperature produces a "hot" system in which all states are equally likely.

Conversely, as the inverse temperature α grows large, the largest probability $\max_{x\in[N]} p_x$ ends up being dominant, and the escort distribution tends to a distribution concentrated at the states x that attain the maximal probability $\max_{x\in[N]} p_x$. For instance, if there is a single value $\bar{x}$ for which the probability $p_{\bar{x}}$ is maximal in vector p, i.e., for which the equality $p_{\bar{x}} = \max_{x\in[N]} p_x$ holds, the limit

$$\lim_{\alpha\to\infty} \tilde{p}_\alpha = \delta_{\bar{x}} \tag{3.80}$$

is satisfied. Thus, in "cooler" systems, only the largest-probability states matter.

Example 3.14

For the distribution $p = [0.1, 0.2, 0.05, 0.3, 0.05, 0.2, 0.1]^T$, consider the two escort distributions corresponding to orders $\alpha = 0.1$ and $\alpha = 5$. The distribution p is depicted in the central panel of Figure 3.13, while the top panel reports the escort distribution with the lower inverse temperature $\alpha = 0.1$ and the bottom panel shows the escort distribution with the higher inverse temperature $\alpha = 5$.

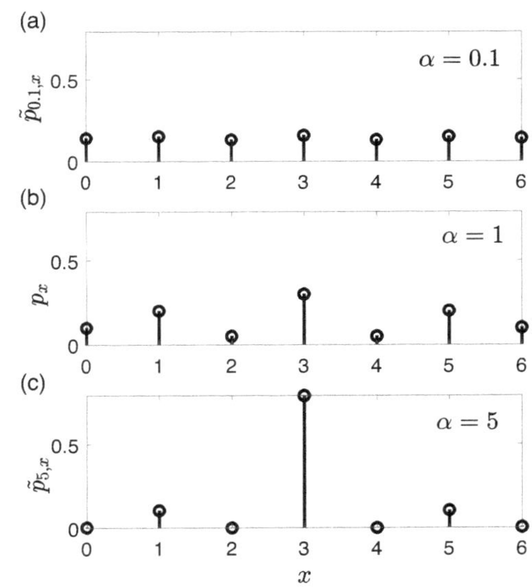

Figure 3.13 (b) Distribution $p = [0.1, 0.2, 0.05, 0.3, 0.05, 0.2, 0.1]^T$, together with the corresponding escort distributions (3.77) for orders (inverse temperatures) (a) $\alpha = 0.1$ and (b) $\alpha = 5$.

The figure confirms that a lower order equalizes the probabilities, returning a close-to-uniform distribution; while a larger order produces a more concentrated distribution at the state $x = 3$, which has the largest probability in vector p.

3.7.3 Optimal Code

In a manner similar to Section 2.4, consider a relaxed version of problem (3.76) in which the codeword lengths are allowed to be non-integer numbers. Recall that with β small, i.e., when minimizing the average codeword length, the resulting ideal codeword lengths are given by the Shannon surprises as $\{\tilde{L}_x^* = -\log_2(p_x)\}_{x\in[N]}$.

More generally, as shown in Appendix 3.E, the ideal codeword lengths for the problem (3.76) are still given by the Shannon surprises, but not with respect to distribution p. Rather, the Shannon surprises are to be evaluated with respect to the escort distribution (3.77) with order

$$\alpha = \frac{1}{1+\beta}. \tag{3.81}$$

Accordingly, the optimal solution to the relaxed problem (3.76) is given by the ideal codeword lengths

$$\tilde{L}_x^* = -\log_2(\tilde{p}_{\alpha,x}). \tag{3.82}$$

For $\beta = 0$, and thus $\alpha = 1$, the resulting escort distribution $\tilde{p}_{1,x}$ coincides with the original distribution p, confirming that the ideal codeword lengths recover the Shannon surprises $\{\tilde{L}_x^* = -\log_2(p_x)\}_{x\in[N]}$. In contrast, as β increases, the inverse temperature α decreases, and the escort distribution $\tilde{p}_{\alpha,x}$ tends to the uniform distribution as in (3.79). Accordingly, all probabilities in the escort distribution $\tilde{p}_{\alpha,x}$ tend to be equalized, avoiding excessively long codewords for states x with smaller probabilities.

The ideal codeword lengths (3.82) provide a solution to the original problem (3.76) if they are all integer numbers, which happens if the escort distribution of order α is dyadic. In this case, the optimal value for the problem (3.76) can be evaluated as (see Appendix 3.E)

$$L_\beta^*(p) = \mathrm{H}_\alpha^{\mathrm{R}}(p) = \mathrm{H}_{\frac{1}{1+\beta}}^{\mathrm{R}}(p), \tag{3.83}$$

with parameters α and β related as in (3.81).

Therefore, generalizing the case $\beta \to 0$, which recovers the Shannon entropy, the Rényi entropy results from the optimal solution of problem (3.76). In particular, in order to penalize long codewords, corresponding to larger values of β, one must select smaller inverse temperatures $\alpha = 1/(1+\beta)$ or larger temperatures $1+\beta$, as per (3.81), making the system "hotter."

Beyond dyadic distributions, a valid code satisfying Kraft–McMillan's inequality can be obtained by rounding up the ideal codeword lengths as

$$L_x = \lceil -\log_2(\tilde{p}_{\alpha,x}) \rceil \leq -\log_2(\tilde{p}_{\alpha,x}) + 1. \tag{3.84}$$

Example 3.15

For the distribution $p = [0.001, 0.299, 0.05, 0.3, 0.05, 0.299, 0.001]^T$, Figure 3.14 shows the ideal codeword lengths (3.82) for different values of parameter β. Recall that a larger value of β targets codes that avoid long codewords. The top panel of the figure shows the result for $\beta \to 0$, which corresponds to the average codeword length criterion. In this case, the ideal codeword lengths are given by the Shannon surprises. Accordingly, the codeword lengths are large for the states $x = 0$ and $x = 6$, which are characterized the lowest probabilities $p_0 = p_6 = 0.1$.

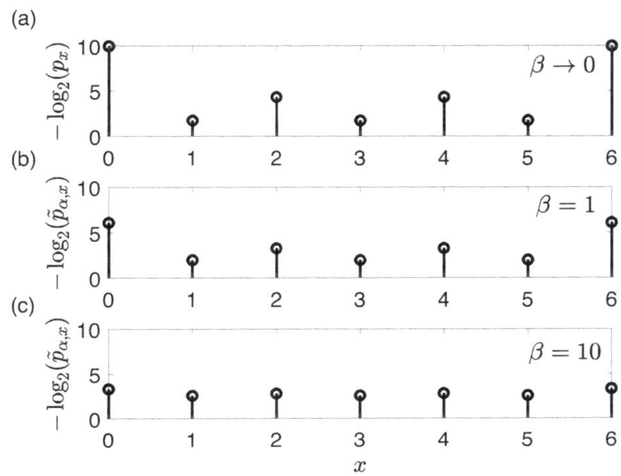

Figure 3.14 Ideal codeword lengths (3.82) for the distribution $p=[0.001, 0.299, 0.05, 0.3, 0.05, 0.299, 0.001]^T$ for: (a) $\beta \to 0$, i.e., Shannon surprises; (b) $\beta = 1$; and (c) $\beta = 10$.

The central and lower panels display the ideal codeword lengths obtained for $\beta = 1$ and $\beta = 10$, respectively. By comparison with the first panel, it is observed that increasing the value of β – and thus the temperature $1 + \beta$ – reduces the length of the longest codewords – for states $x = 0$ and $x = 6$ – increasingly equalizing all codeword lengths.

3.8 Unifying Entropies ⏭

This last section shows that the Shannon, Tsallis, and Rényi entropies can be unified into a single generalized formulation. This formulation hinges on the following definition.

Definition 3.18 (h,ϕ)-entropy
Fix any two functions $h(\cdot)$ and $\phi(\cdot)$ satisfying *either* of the following two properties:

- function $\phi(\cdot)$ is concave and $h(\cdot)$ is increasing;
- *or* function $\phi(\cdot)$ is convex and $h(\cdot)$ is decreasing.

The (h, ϕ)-entropy for probability vector p is defined as

$$\mathrm{H}^{\mathrm{U}}_{h,\phi}(p) = h\left(\sum_{x\in[N]} \phi(p_x)\right). \tag{3.85}$$

Table 3.4 Selections of functions $\phi(\cdot)$ and $h(\cdot)$ recovering the three entropies from the definition (3.85) of the (h, ϕ)-entropy

Entropy	$\phi(x)$	$h(x)$
Shannon	$-x\log_2(x)$	x
Tsallis	x^α	$\frac{1_2}{1-\alpha}(x-1)$
Rényi	x^α	$\frac{1}{1-\alpha}\log_2(x)$

As summarized in Table 3.4, the formulation (3.85) can readily be seen to recover the Shannon, Tsallis, and Rényi entropies as special cases:

- The Shannon entropy is obtained by setting $\phi(x) = -x\log_2(x)$, which is a concave function, and $h(x) = x$, which is increasing.
- For the Tsallis entropy, one sets $\phi(x) = x^\alpha$ and $h(x) = 1_2(x-1)/(1-\alpha)$. Accordingly, for $0 \leq \alpha < 1$, the function $\phi(x)$ is concave, and $h(x)$ is increasing; while for $\alpha > 1$ function $\phi(x) = x^\alpha$ is convex, and $h(x)$ is decreasing.
- For the Rényi entropy, the two functions are to be selected as $\phi(x) = x^\alpha$ and $h(x) = \log_2(x)/(1-\alpha)$. In a manner similar to Tsallis entropy, for $0 \leq \alpha < 1$, the function $\phi(x)$ is concave, and $h(x)$ is increasing; while for $\alpha > 1$ function $\phi(x) = x^\alpha$ is convex, and $h(x)$ is decreasing.

Any (h, ϕ)-entropy can be shown to be a concave function of the distribution vector p in the special cases in which function $\phi(\cdot)$ is concave and $h(\cdot)$ is concave and increasing; or else if function $\phi(\cdot)$ is convex and $h(\cdot)$ is concave and decreasing. These properties are satisfied by the Shannon entropy, by the Tsallis entropy for any $\alpha \geq 0$, and by the Rényi entropy for $0 \leq \alpha \leq 1$. Furthermore, it can be shown that the (h, ϕ)-entropy satisfies a form of monotonicity based on the theory of majorization (see Part III of the book).

3.9 Summary

- In this chapter, two strongly related families of entropies, the Tsallis entropy and the Rényi entropy, were introduced.
- The Tsallis entropy $\mathrm{H}^{\mathrm{T}}_{\alpha}(p)$ was presented as the average prediction loss when the prediction error is evaluated by using the α-log-loss.
- With $\alpha = 1$, the α-log-loss reduces to the conventional log-loss, and thus the Tsallis entropy recovers the Shannon entropy, i.e., $\mathrm{H}^{\mathrm{T}}_{1}(p) = \mathrm{H}(p)$.

- The α-log-loss decreases with the inverse temperature parameter $\alpha \geq 0$, and so does the Tsallis entropy $H_\alpha^T(p)$. Therefore, a larger order α entails a more lenient measure of average prediction loss, or uncertainty, for an observer in the informational state p.
- By varying the order α, the Tsallis entropy can exhibit different maximum-growth trends as a function of the number of states, N, from bounded to linear, generalizing the logarithmic trend of the Shannon entropy.
- Thanks to this flexibility, the Tsallis entropy is related to the solution to the informational task of guessing, which can be exponentially harder than the tasks studied in Chapter 2.
- The Rényi entropy $H_\alpha^R(p)$ also decreases with the order α, and it recovers the Shannon entropy for $\alpha = 1$, i.e., $H_\alpha^R(p) = H(p)$. However, unlike the Tsallis entropy, the Rényi entropy $H_\alpha^R(p)$ is upper bounded by $\log_2(N)$ for all values of the order $\alpha \geq 0$.
- In terms of operational interpretations, the Rényi entropy was proved to describe the optimal cost for a compression problem that penalizes long codewords.
- The Tsallis and Rényi entropies were seen to satisfy the non-negativity and minimum-value, as well as the maximum-value, properties of an entropy function, and they will also be demonstrated in Chapter 4 to satisfy the monotonicity property.
- Moreover, concavity – i.e., "mixing increases entropy" – holds for the Tsallis entropy, but only for a subset of Rényi entropies.
- While this chapter and Chapter 2 present the Shannon, Tsallis, and Rényi entropies as the solutions to a number of informational tasks, which satisfy some desirable properties, an axiomatic approach, whereby entropy measures are introduced as all functions that satisfy a number of mathematical conditions (see Chapter 1), is briefly elaborated on in Appendix 3.G.

3.10 Recommended Resources

Reference [1] offers a useful discussion on scoring rules for prediction. Extensive treatments of information measures and their applications in physics are available in [2] and [3]. The Tsallis sum, referred to as information generating function, was studied in [4]. An algebraic framework addressing the non-additivity of the α-logarithm is presented in [5]. Operational uses of Rényi collision entropy, Rényi min-entropy, and ϵ-smooth max-entropy in cryptography are discussed in [6]. For a proof of (3.66), see [7], and further insights into the concavity of Rényi entropy are provided in [8].

A comprehensive resource on convex functions, convex sets, and convex optimization is the textbook [9], which also covers composition rules ensuring concavity for defining (h, ϕ)-entropy. Extensive coverage of (h, ϕ)-entropy can be found in [10].

The generalized average in (3.74) exemplifies a **quasi-arithmetic mean**, which is generally expressed as $g^{-1}(\mathrm{E}_p[g(\mathrm{x})])$, where $g(\cdot)$, an invertible function, serves as the mean's generator. Specifically, in (3.74), the generator function is $g(x) = 2^{\beta x}$ with inverse $g^{-1}(x) = \log_2(x)/\beta$. Further discussion on quasi-arithmetic means is provided in [11].

Proofs relevant to the guessing problem, including the relation in (3.104), are detailed in [12]. The compression problem discussed in Section 3.7 is analyzed in [13], and an axiomatic approach to defining entropy measures is presented in [14].

PROBLEMS

3.1 (a) Using the two defining properties of a loss function $\ell(\cdot)$, namely non-negativity and minimum-value and monotonicity and maximum-value, show that the average loss $\mathrm{H}_\ell(p)$ in (3.1) satisfies the non-negativity and minimum-value property of an entropy measure introduced in Chapter 1. (b) For the maximum-value property to be verified, which other condition should the loss function $\ell(\cdot)$ satisfy?

3.2 Consider the $N \times 1$ probability vector p defined as

$$p_x = \begin{cases} \frac{1}{2} & \text{if } x = 0, \\ \frac{1}{2(N-1)} & \text{otherwise.} \end{cases} \tag{3.86}$$

(a) Evaluate the Shannon and Tsallis entropies. (b) How do the Shannon and Tsallis entropies scale as N increases? (c) Repeat Problems (a) and (b) for the Rényi entropy.

3.3 Show that the Tsallis entropy satisfies the inequality $1 + \frac{(1-\alpha)}{1_2}\mathrm{H}_\alpha^{\mathrm{T}}(p) > 0$ for all values $\alpha \geq 0$.

3.4 Prove the existence of a prefix-free code with codeword lengths $L_x = \lceil -\log_2(\tilde{p}_{\alpha,x}) \rceil \leq -\log_2(\tilde{p}_{\alpha,x}) + 1$, where $\tilde{p}_\alpha$ is the escort distribution in (3.77).

3.5 (a) Show that the Tsallis sum of order α is a concave function of distribution p when $0 \leq \alpha < 1$, while it is a convex function for $\alpha > 1$. (b) Use this result to argue that mixing increases the Tsallis sum of order α for $0 \leq \alpha < 1$, while it decreases it for $\alpha > 1$. (c) Conclude that the Tsallis sum of order α is an uncertainty measure for $0 \leq \alpha < 1$, while it is a certainty measure for $\alpha > 1$.

3.6 Using the inequality $\log_2(x) \leq 1_2 \cdot (x - 1)$, show the two inequalities (3.59) and (3.61).

3.7 Prove the upper bound in (3.52), namely $G^*_{\mathrm{avg}}(p) \leq (|\mathrm{supp}(p)| + 1)/2$, by arguing that, given a distribution p, the minimum average number of guesses, $G^*_{\mathrm{avg}}(p)$, is no smaller than that of a distribution that assigns equal probability to all states x with $p_x > 0$ and assigns zero probability to all other states. *Hint:* Use (3.46).

3.8 Assume that for a composite system the number of states grows as $N = S^a$ for some $a > 1$ with respect to the number of subsystems S. Prove that the Tsallis entropy $\mathrm{H}_{1-1/a}(u)$ grows linearly with S.

3.9 Consider a source emitting symbols $\{A, B, C, D\}$ with the probability distribution $p = [0.7, 0.2, 0.08, 0.02]$. (a) Compute the Shannon entropy $\mathrm{H}(p)$. (b) Compute the Tsallis entropy $\mathrm{H}^{\mathrm{T}}_{\alpha}(p)$ for $\alpha = 2$. (c) Compute the Rényi entropy $\mathrm{H}^{\mathrm{R}}_{\alpha}(p)$ for $\alpha = 2$. (d) Interpret the differences between these entropies and explain why they differ for this skewed distribution.

3.10 In cryptography, the security of a key depends not only on its Shannon entropy but also on how predictable the key is under certain attacks. Consider a system that generates a key k with the following distribution:

$$p = \begin{cases} 0.5 & \text{if k} = k_0, \\ 0.25 & \text{if k} = k_1, \\ 0.25 & \text{if k} = k_2, \end{cases} \tag{3.87}$$

where $\{k_0, k_1, k_2\}$ are the three possible values for the key. (a) Compute the Shannon entropy $\mathrm{H}(p)$ of the key distribution. (b) Compute the Rényi entropy $\mathrm{H}^{\mathrm{R}}_{\alpha}(p)$ for $\alpha = 2$ and $\alpha = \infty$, and explain their significance in the context of cryptographic attacks. (c) Discuss why Rényi entropy is a better metric than Shannon entropy for assessing the security of cryptographic keys.

3.11 Consider an ecosystem with species $\{A, B, C, D\}$, where the relative abundances of the species are given by $p = [0.4, 0.3, 0.2, 0.1]$, respectively. Different entropy measures are used to quantify species diversity. (a) Compute the Shannon entropy. (b) Compute the Rényi entropy with order $\alpha = 2$ and $\alpha = \infty$. Discuss the impact of increasing α, and how this choice reflects the relative importance of less abundant species. Explain the implications of using each entropy measure in the context of biodiversity assessment.

3.12 Consider a physical system in thermal equilibrium with three possible energy states $\{E_0, E_1, E_2\}$. The probability distribution over these states is given by the Boltzmann distribution:

$$p_x = \frac{\exp(-\beta E_x)}{Z}, \quad Z = \sum_{x \in [3]} \exp(-\beta E_x), \tag{3.88}$$

where $\beta = \frac{1}{k_B T}$ is the inverse temperature, k_B is the Boltzmann constant, and T is the temperature. Suppose that the energy levels are $E_0 = 0$, $E_1 = 1$, and $E_2 = 2$ in units where $k_B = 1$. (a) Compute the Shannon entropy $\mathrm{H}(p)$ for the system at temperature $T = 1$. (b) Compute the Rényi entropy $\mathrm{H}^{\mathrm{R}}_{\alpha}(p)$ for $\alpha = 2$ and discuss its significance in the context of quantifying the system's disorder.

Appendix

3.A α-Log Function

This appendix lists some key properties of the α-log function $\log_{2,\alpha}(q)$ with order $\alpha \geq 0$. The properties are best understood with reference to the plots of the α-logarithm in Figure 3.4.

- **Domain:** The domain of the function is given by all positive values q, i.e., $q > 0$, when $\alpha \leq 1$; and by all non-negative values $q \geq 0$ when $\alpha > 1$.
- **Special cases:** For $\alpha = 1$, the α-log coincides with the standard logarithm, and for $\alpha = 2$ it is a linear function of q, namely $\log_{2,2}(q) = \mathrm{l}_2(q-1)$.
- **Convexity and concavity:** For high inverse temperatures, i.e., for $\alpha > 1$, the α-log is a convex function of q; while for lower inverse temperatures, i.e., for $\alpha \leq 1$, it is a concave function of q.
- **Dependence on the input** q**:** Like the logarithm, the α-log is an increasing function of q, crossing the horizontal axis at $q = 1$, i.e., $\log_{2,\alpha}(1) = 0$. In the limit $q \to 0$, the α-log function $\log_{2,\alpha}(q)$ tends to $-\infty$ for $\alpha \leq 1$; while, for $\alpha > 1$, the inequality

$$\log_{2,\alpha}(q) \geq \frac{\mathrm{l}_2}{\alpha - 1} \tag{3.89}$$

is satisfied, with the minimum value $\mathrm{l}_2/(\alpha-1)$ attained for $q = 0$. In the complementary regime given by the limit $q \to \infty$, the α-logarithm tends to $\mathrm{l}_2/(1-\alpha)$ for $\alpha < 1$, and to infinity for $\alpha \geq 1$, i.e.,

$$\lim_{q\to\infty} \log_{2,\alpha}(q) = \begin{cases} \frac{\mathrm{l}_2}{1-\alpha} & \text{for } \alpha < 1, \\ \infty & \text{for } \alpha \geq 1. \end{cases} \tag{3.90}$$

- **Dependence on the order** α**:** For a fixed value of $q > 0$, the α-log is an increasing function of $\alpha \geq 0$, ranging from $\mathrm{l}_2 \cdot (1 - 1/q)$ when $\alpha = 0$ to either 0 or ∞ in the limit $\alpha \to \infty$ depending on whether q satisfies the inequalities $0 < q \leq 1$ or $q > 1$, respectively.
- **α-Additivity:** The α-log, and hence also the α-log-loss, do not satisfy the additivity property of the logarithm. In fact, for any q and r in the domain of the function, the following equality holds:

$$\log_{2,\alpha}(qr) = \log_{2,\alpha}(q) + \log_{2,\alpha}(r) + \frac{(\alpha-1)}{\mathrm{l}_2}\log_{2,\alpha}(q)\log_{2,\alpha}(r). \tag{3.91}$$

This property, known as α-additivity, recovers the usual additivity property of the logarithm for $\alpha = 1$.
- **Consequences of α-additivity:** For $r > 0$ and q in the domain of the function, the equality

$$\log_{2,\alpha}\left(\frac{q}{r}\right) = r^{1-\alpha}(\log_{2,\alpha}(q) - \log_{2,\alpha}(r)) \tag{3.92}$$

holds, which recovers the usual counterpart result for the logarithm when $\alpha = 1$. As a special case of the relation (3.92), the following equality holds:

$$\log_{2,\alpha}\left(\frac{1}{r}\right) = -r^{1-\alpha}\log_{2,\alpha}(r). \tag{3.93}$$

Furthermore, more generally, one can relate the α-logarithm of a ratio and of its reciprocal as

$$\log_{2,\alpha}\left(\frac{q}{r}\right) = -\left(\frac{r}{q}\right)^{1-\alpha}\log_{2,\alpha}\left(\frac{r}{q}\right). \tag{3.94}$$

As a final note, in physics, a specific algebra has been introduced that defines multiplications and divisions so as to facilitate the use of the α-logarithm, and of its inverse function, known as the **α-exponential function** (see Recommended Resources).

3.B Derivation of the Rényi Min-Entropy

In this appendix, the min-entropy (3.65) is derived as a special case of the Rényi entropy (3.55). To this end, it is useful to leverage the inequalities

$$\max_{x\in[N]} p_x \leq \sum_{x\in[N]} p_x \leq N \max_{x\in[N]} p_x, \tag{3.95}$$

as well as the equality $(\max_{x\in[N]} p_x)^\alpha = \max_{x\in[N]} p_x^\alpha$. In fact, these relations yield the inequalities

$$\begin{aligned}\frac{\alpha}{1-\alpha}\log_2\left(\max_{x\in[N]} p_x\right) &\leq \frac{1}{1-\alpha}\log_2\left(\sum_{x\in[N]} p_x^\alpha\right)\\ &\leq \frac{\log_2(N)}{1-\alpha} + \frac{\alpha}{1-\alpha}\log_2\left(\max_{x\in[N]} p_x\right).\end{aligned} \tag{3.96}$$

Therefore, as $\alpha \to \infty$, the desired result is obtained as the limit

$$\begin{aligned}\lim_{\alpha\to\infty}\frac{1}{1-\alpha}\log_2\left(\sum_{x\in[N]} p_x^\alpha\right) &= -\log_2\left(\max_{x\in[N]} p_x\right)\\ &= \mathrm{H}^{\mathrm{R}}_\infty(p).\end{aligned} \tag{3.97}$$

3.C Rényi Entropy and Tsallis Entropy

Given the dependence of both Tsallis and Rényi entropies on the Tsallis sum of order α, $\sum_{x\in[N]} p_x^\alpha$, it can be verified that there exists a one-to-one mapping between Rényi and Tsallis entropies. Specifically, the Rényi entropy can be expressed as a function of the Tsallis entropy as

$$\mathrm{H}^{\mathrm{R}}_\alpha(p) = \mathrm{F}^{\mathrm{T}\to\mathrm{R}}_\alpha(\mathrm{H}^{\mathrm{T}}_\alpha(p)) \tag{3.98}$$

through the function

$$\mathrm{F}^{\mathrm{T}\to\mathrm{R}}_\alpha(h) = \frac{1}{1-\alpha}\log_2\left(1 + \frac{(1-\alpha)}{1_2}h\right). \tag{3.99}$$

Note that the function $\mathrm{F}^{\mathrm{T}\to\mathrm{R}}_\alpha(h)$ is defined for all values of non-negative values of h such that the inequality $1 + (1-\alpha)h/1_2 > 0$ holds, which is always the case when $h = \mathrm{H}^{\mathrm{R}}_\alpha(p)$ (see Problem 3.3).

Furthermore, the Tsallis entropy can be expressed as a function of the Rényi entropy as

$$\mathrm{H}^{\mathrm{T}}_\alpha(p) = \mathrm{F}^{\mathrm{R}\to\mathrm{T}}_\alpha(\mathrm{H}^{\mathrm{R}}_\alpha(p)), \tag{3.100}$$

with function

$$\mathrm{F}_{\alpha}^{\mathrm{R}\to\mathrm{T}}(h) = \frac{1_2}{1-\alpha}\left(2^{(1-\alpha)h} - 1\right), \tag{3.101}$$

which can be obtained as the inverse of function $\mathrm{F}_{\alpha}^{\mathrm{T}\to\mathrm{R}}(h)$.

The functions $\mathrm{F}_{\alpha}^{\mathrm{T}\to\mathrm{R}}(h)$ and $\mathrm{F}_{\alpha}^{\mathrm{R}\to\mathrm{T}}(h)$ can easily be checked to be monotonically increasing in h for any fixed α. This indicates that, for any fixed order α, as one of the two entropies increases, so does the other; and, vice versa, when one entropy decreases, the other decreases as well.

3.D Proofs for the Guessing Problem

In this appendix, the optimal average number of guesses, $G^*_{\text{avg}}(p)$, is investigated for any distribution vector p. Denoting as $\{G^*_x\}_{x\in[N]}$ an optimal sequence of number of guesses, the optimal minimum average number of guesses can be written as

$$G^*_{\text{avg}}(p) = \mathrm{E}_p[G^*_{\mathrm{x}}] = \sum_{x\in[N]} p_x G^*_x. \tag{3.102}$$

Without loss of generality, assume that the states are listed in order of decreasing probability so that the inequalities $p_0 \geq p_1 \geq \cdots \geq p_{N-1}$ hold. Therefore, by the optimal guessing algorithm described in Section 3.5, the optimal number of guesses can be set as $G^*_x = x+1$ for $x \in [N]$. In fact, state $x = 0$ is identified after one guess, state $x = 1$ after two guesses, and so on. It follows that the equality

$$\sum_{x\in[N]} \frac{1}{G^*_x} = \sum_{x\in[N]} \frac{1}{x+1} \tag{3.103}$$

holds, which can be bounded as

$$\sum_{x\in[N]} \frac{1}{x+1} \leq 1 + \ln(N). \tag{3.104}$$

This result is now used to relate the minimum average number of guesses to the Shannon entropy, proving inequality (3.51).

Using the definition of the logarithm for the first equality and the Jensen inequality, one obtains

$$\begin{aligned} G^*_{\text{avg}}(p) &= \sum_{x\in[N]} p_x 2^{-\log_2\left(\frac{1}{G^*_x}\right)} \\ &\geq 2^{-\sum_{x\in[N]} p_x \log_2\left(\frac{1}{G^*_x}\right)}, \end{aligned} \tag{3.105}$$

where the Jensen inequality leverages the convexity of the function $f(x) = 2^{-x}$ (see Appendix 2.D).

Multiplying and dividing by p_x, the term in (3.105) can be rewritten as

$$\begin{aligned}2^{-\sum_{x\in[N]} p_x \log_2\left(\frac{p_x}{G_x^* p_x}\right)} &= 2^{\mathrm{H}(p)-\sum_{x\in[N]} p_x \log_2\left(\frac{1}{G_x^* p_x}\right)}\\ &\geq 2^{\mathrm{H}(p)-\log_2\left(\sum_{x\in[N]} p_x \frac{1}{G_x^* p_x}\right)}\\ &= \frac{2^{\mathrm{H}(p)}}{\sum_{x\in[N]} \frac{1}{G_x^*}}\\ &\geq \frac{2^{\mathrm{H}(p)}}{1+\ln(N)},\end{aligned} \tag{3.106}$$

where the first inequality is from Jensen's inequality given the convexity of the function $f(x) = -\log_2(x)$ and the last inequality follows from the inequality (3.104). This concludes the derivation of (3.51).

Next, the inequalities (3.53) are derived, focusing first on the lower bound. To start, the inequalities (3.53) are rewritten in terms of the Rényi entropy of order 1/2 as

$$\frac{2^{\mathrm{H}^{\mathrm{R}}_{1/2}(p)}}{1+\ln(N)} \leq G^*_{\mathrm{avg}}(p) \leq 2^{\mathrm{H}^{\mathrm{R}}_{1/2}(p)} \tag{3.107}$$

by using the relations between Tsallis and Rényi entropy.

To simplify the notation in what follows, it is useful to introduce the **Shannon relative entropy**

$$\mathrm{D}(p||q) = \mathrm{E}_p\left[\log_2\left(\frac{p_{\mathrm{x}}}{q_{\mathrm{x}}}\right)\right] = \sum_{x\in[N]} p_x \log_2\left(\frac{p_x}{q_x}\right), \tag{3.108}$$

which will be extensively discussed in Chapter 4.

Following the same steps in (3.105) and (3.106), but multiplying and dividing by q_x, rather than p_x, the following steps can be verified:

$$\begin{aligned}G^*_{\mathrm{avg}}(p) &= \sum_{x\in[N]} p_x \frac{q_x}{q_x} G_x^*\\ &= \sum_{x\in[N]} q_x 2^{-\log_2\left(\frac{q_x}{G_x^* p_x}\right)}\\ &\geq 2^{-\sum_{x\in[N]} q_x \log_2\left(\frac{q_x}{G_x^* p_x}\right)}.\end{aligned} \tag{3.109}$$

Now, using the definition of the Shannon relative entropy (3.108), the lower bound in (3.109) can be rewritten as

$$\begin{aligned}2^{-\sum_{x\in[N]} q_x \log_2\left(\frac{q_x}{G_x^* p_x}\right)} &= 2^{-\mathrm{D}(q||p)-\sum_{x\in[N]} q_x \log_2\left(\frac{1}{G_x^*}\right)}\\ &= 2^{-\mathrm{D}(q||p)-\sum_{x\in[N]} q_x \log_2\left(\frac{q_x}{G_x^* q_x}\right)}\\ &= 2^{-\mathrm{D}(q||p)+\mathrm{H}(q)-\sum_{x\in[N]} q_x \log_2\left(\sum_{x\in[N]} \frac{1}{G_x^* q_x}\right)}\\ &= 2^{-\mathrm{D}(q||p)+\mathrm{H}(q)-\sum_{x\in[N]} q_x \log_2\left(\sum_{x\in[N]} \frac{1}{G_x^* q_x}\right)}.\end{aligned} \tag{3.110}$$

Since the lower bound (3.110) applies for any distribution q, the tightest, i.e., largest, lower bound can be obtained by minimizing the difference $\mathrm{D}(q||p) - \mathrm{H}(q)$ over distribution q. Through standard Lagrange duality from convex optimization, imposing the constraint $\sum_{x\in[N]} q_x = 1$, one obtains the optimality condition $q_x^2/p_x = c$, where c is a constant chosen to satisfy the constraint (see Recommended Resources for convex optimization). This is done by imposing that the derivative of the Lagrangian function be equal to zero. As a result, the solution q^* is given by the escort distribution of order 1/2, i.e.,

$$q_x^* = \tilde{p}_{1/2,x} = \frac{\sqrt{p_x}}{\sum_{x'\in[N]}\sqrt{p_{x'}}}. \tag{3.111}$$

Plugging this distribution into the bound (3.110) returns the desired inequality in (3.107):

$$G_{\text{avg}}^*(p) \geq \frac{2^{-(\mathrm{D}(q^*||p)-\mathrm{H}(q^*))}}{1+\ln(N)} = \frac{2^{\mathrm{H}_{1/2}^{\mathrm{R}}(p)}}{1+\ln(N)}. \tag{3.112}$$

Finally, consider the upper bound in (3.107). By the optimal guessing strategy, one must have the equality

$$G_x^* \leq \sum_{x'\in[N]:p_{x'}\geq p_x} 1, \tag{3.113}$$

since the right-hand side simply counts the number of states x' with probability no smaller than x. Noting that the inequality $p_{x'}/p_x \geq 1$ holds for the state values x' with $p_{x'} \geq p_x$, and that the inequality $y^{1/2} \geq 1$ holds for $y \geq 1$, one obtains

$$\begin{aligned} G_x^* &\leq \sum_{x'\in[N]:p_{x'}\geq p_x} \left(\frac{p_{x'}}{p_x}\right)^{1/2} \\ &\leq \sum_{x'\in[N]} \left(\frac{p_{x'}}{p_x}\right)^{1/2}, \end{aligned} \tag{3.114}$$

where the second inequality follows by extending the sum to all values of the state.

Therefore, taking the average over x $\sim p$, the desired upper bound is recovered as

$$\begin{aligned} G_{\text{avg}}^*(p) &\leq \sum_{x\in[N]} p_x \sum_{x'\in[N]} \left(\frac{p_{x'}}{p_x}\right)^{1/2} \\ &= \left(\sum_{x\in[N]}\sqrt{p_x}\right)\left(\sum_{x'\in[N]}\sqrt{p_{x'}}\right) \\ &= \left(\sum_{x\in[N]}\sqrt{p_x}\right)^2 \\ &= 2^{\mathrm{H}_{1/2}^{\mathrm{R}}(p)}. \end{aligned} \tag{3.115}$$

3.E Optimal Codeword Lengths for Problem (3.76)

In this appendix, problem (3.76) is tackled by relaxing the variables in vector L to take any non-negative real value. Applying the change of variable $q_x = 2^{-L_x}$, i.e., $L_x = -\log_2(q_x)$, problem (3.76) can equivalently be stated as

$$\begin{aligned} \min_{L} \; & \sum_{x\in[N]} p_x q_x^{-\beta} \\ \text{subject to} \; & \sum_{x\in[N]} q_x \leq 1, \end{aligned} \tag{3.116}$$

which is an optimization over vector q. While the lengths $\{L_x\}_{x\in[N]}$ are constrained to be integer numbers, the problem is now relaxed by allowing any value for $0 \leq q_x \leq 1$, and hence any non-negative real value for the variables in vector L. Note that, if the solution to the relaxed problem yields integer values for the variables $\{L_x = -\log_2(q_x)\}_{x\in[N]}$, then the obtained integer vector must also be a solution of the original, integer-constrained, problem.

To start, it is observed that an optimal solution for problem (3.116) must satisfy the equality $\sum_{x\in[N]} q_x = 1$. In fact, otherwise, one could decrease the objective due to monotonicity of function $q_x^{-\beta}$. Moreover, since the relaxed problem is convex, one can differentiate the Lagrangian function to obtain the necessary and sufficient optimality conditions $p_x/q_x^{(1+\beta)} = c$, where c is a constant. As done in Appendix 3.D, imposing the normalization condition yields the escort distribution (3.77) with order $\alpha = 1/(1+\beta)$ as the optimal solution.

Finally, plugging the ideal optimal codewords (3.82) into the objective (3.74) and recalling the relation $\beta = (1-\alpha)/\alpha$ yields the minimum value of the objective as

$$\begin{aligned} L_\beta^*(p) &= \frac{1}{\beta}\log_2\left(\mathrm{E}_p\left[2^{\left(1-\frac{1}{\alpha}\right)\log_2\left(\frac{p_x^\alpha}{\sum_{x'\in[N]} p_{x'}^\alpha}\right)}\right]\right) \\ &= \frac{1}{\beta}\log_2\left(\mathrm{E}_p\left[\left(\frac{p_x^\alpha}{\sum_{x'\in[N]} p_{x'}^\alpha}\right)^{\frac{\alpha-1}{\alpha}}\right]\right) \\ &= \frac{1}{\beta}\log_2\left(\sum_{x\in[N]} p_x \frac{p_x^{\alpha-1}}{(\sum_{x'\in[N]} p_{x'}^\alpha)^{\frac{\alpha-1}{\alpha}}}\right) \\ &= \frac{1}{\beta}\log_2\left(\left(\sum_{x\in[N]} p_x^\alpha\right)^{\frac{1}{\alpha}}\right) = \mathrm{H}^{\mathrm{R}}(p). \end{aligned} \tag{3.117}$$

3.F Rényi Entropy and Fixed-Length Compression

This appendix explores another connection between the Rényi entropy and the task of compression. So far, the text has focused on *variable-length* compression, whereby each state x is generally assigned a codeword of different length L_x. In contrast, this appendix considers *fixed-length* compression, in which every state x is assigned a binary string of the same number of bits.

With fixed-length compression, one must use codewords of length $\lceil \log_2(N) \rceil$ bits in order to ensure that the decoder can unambiguously recover the state x. It is, however, possible to use fewer than $\lceil \log_2(N) \rceil$ bits if one allows for decoding errors.

To elaborate, define as $\epsilon > 0$ the allowed probability of decoding error. Under the constraint dictated by the probability of decoding error ϵ, the optimal compression strategy assigns distinct codewords only to the ϵ-support of distribution p. This corresponds to the smallest subset of states x for which the sum of the probabilities exceeds $1 - \epsilon$. Mathematically, it is defined as

$$\mathrm{supp}_\epsilon(p) = \arg\min_{\mathcal{X} \subseteq [N]:\ \sum_{x \in \mathcal{X}} p_x \geq 1-\epsilon} |\mathcal{X}|, \tag{3.118}$$

where $\mathcal{X} \subseteq [N]$ is a subset of the state set $[N]$.

An optimal code assigns a distinct codeword to each state in subset $\mathrm{supp}_\epsilon(p)$. Instead, to all other values $x \notin \mathrm{supp}_\epsilon(p)$, the same codeword can be assigned as for an arbitrary value x in subset $\mathrm{supp}_\epsilon(p)$. This way, by construction, the probability of error is no larger than ϵ.

The minimum number of bits required by this code is given as $\left\lceil \mathrm{H}^R_{0,\epsilon}(p) \right\rceil$, where

$$\mathrm{H}^{\mathrm{R}}_{0,\epsilon}(p) = \log_2(\mathrm{supp}_\epsilon(p)) \tag{3.119}$$

is known as the **ϵ-smooth max-entropy**. Note that the ϵ-smooth max-entropy reduces to the Rényi max-entropy (3.62) for $\epsilon = 0$, i.e., $\mathrm{H}^{\mathrm{R}}_{0,0}(p) = \mathrm{H}^{\mathrm{R}}_0(p)$.

3.G Axiomatic Definition of Entropy

In this appendix, the entropy measures introduced in this chapter and Chapter 2 are reviewed through the lens of the axiomatic approach to the definition of entropy measures. Following this approach, mathematical axioms are first defined that formalize intuitions about desired properties of uncertainty measures. Then, entropy measures are derived by investigating functions of the informational state that satisfy the mathematical conditions expressed by the axioms.

To start, consider a function $f(p_x)$ of probability p_x whose goal is to quantify the information brought by an observation x to an observer that assigns to it probability p_x. The following axioms formalize reasonable properties that such a function is expected to satisfy.

- **Non-negativity axiom:** $f(p_x) \geq 0$ for all probabilities $0 \leq p_x \leq 1$.
- **Additivity axiom:** $f(p_x p_y) = f(p_x) + f(p_y)$.
- **Normalization axiom:** $f(0.5) = 1$.

The non-negativity axiom merely formalizes the fact that the degree of uncertainty should be no smaller than zero.

The additivity axiom states that, if we make two *independent* observations, one with probability p_x and the other with probability p_y, the overall uncertainty should be the sum of the two respective uncertainty levels. Referring to Chapter 12 for a formal definition of independent random variables, it is sufficient to note here that the probability that independent observations produce outputs x and y is given by the product $p_x p_y$.

Finally, the normalization axiom is introduced to regard an observation that has a 50% chance of occurring as carrying 1 bit of uncertainty. It should be suitably replaced if one wishes to measure information in units other than the bit.

It can be proved that only functions of the form

$$f(p_x) = -c \log_2(p_x) \tag{3.120}$$

for some constant $c > 0$ satisfy the first two axioms, and that the normalization axiom imposes the choice $c = 1$ (see Recommended Resources). Therefore, only the Shannon surprise $f(p_x) = -\log_2(p_x)$ satisfies all three axioms.

So far, in this appendix, the focus has been on a function $f(p_x)$ of a single probability p_x. It is now of interest to define information measures that apply to a probability vector p. According to the axiomatic framework, a function $\mathrm{F}(p)$ of a distribution p is said to be an **entropy measure** if it satisfies the following three axioms.

- **Continuity axiom:** The function $\mathrm{F}(p)$ depends continuously on all variables in vector p.
- **Maximum-value axiom:** The function $\mathrm{F}(p)$ is maximized when p is the uniform distribution, i.e., when $p = u$; that is, the inequality $\mathrm{F}(p) \leq \mathrm{F}(p)$ holds for every distribution vector p.
- **Expansibility axiom:** Adding values in the set $[N]$ of possible observation outcomes does not change function $\mathrm{F}(p)$ if the added values have zero probability; that is, the equality $F(p') = F(p)$ holds if the $N' \times 1$ probability vector p', with $N' > N$, has the same non-zero elements of vector p with all other, $N' - N$, elements being equal to 0.

The continuity axiom is intuitively important in order to avoid abrupt changes in the uncertainty measure as the entries of vector p are varied by a small amount. The maximum-value axiom corresponds to the basic condition discussed in Chapter 1 that uncertainty should be maximized for uniform distributions. Finally, the expansibility axiom expresses the fact that zero-probability outcomes cannot change the uncertainty associated with a probability vector.

All entropy measures, Shannon, Tsallis, and Rényi, satisfy all these axioms for $\alpha > 0$. However, the Shannon entropy is unique in that it also satisfies a fourth axiom, known as strong additivity, which will be discussed in Appendix 13.D.

REFERENCES

[1] R. Tibshirani, "Statistical learning theory." https://www.stat.berkeley.edu/ryantibs/statlearn-s23/, 2023, University of California, Berkeley, Department of Statistics.

[2] H. Touchette, "When is a quantity additive, and when is it extensive?" *Physica A: Statistical Mechanics and its*

Applications, vol. 305, no. 1–2, pp. 84–88, 2002.

[3] H. J. Jensen, *Complexity Science: The Study of Emergence*. Cambridge: Cambridge University Press, 2022.

[4] S. Golomb, "The information generating function of a probability distribution (corresp.)," *IEEE Transactions on Information Theory*, vol. 12, no. 1, pp. 75–77, 1966.

[5] F. Futami, I. Sato, and M. Sugiyama, "Expectation propagation for t-exponential family using q-algebra," *Advances in Neural Information Processing Systems*, vol. 30, 2017.

[6] H. Tyagi and S. Watanabe, *Information-Theoretic Cryptography*. Cambridge: Cambridge University Press, 2023.

[7] H. Wilming, M. Goihl, I. Roth, and J. Eisert, "Entanglement-ergodic quantum systems equilibrate exponentially well," *Physical Review Letters*, vol. 123, no. 20, art. 200604, 2019.

[8] I. Bengtsson and K. Życzkowski, *Geometry of Quantum States: An Introduction to Quantum Entanglement*. Cambridge: Cambridge University Press, 2017.

[9] S. Boyd and L. Vandenberghe, *Convex Optimization*. Cambridge: Cambridge University Press, 2004.

[10] J. M. Amigó, S. G. Balogh, and S. Hernández, "A brief review of generalized entropies," *Entropy*, vol. 20, no. 11, p. 813, 2018.

[11] F. Nielsen, "What is... an inductive mean?" *Notices of the American Mathematical Society*, vol. 70, no. 11, p. 1, 2023. [Online]. Available at: http://dx.doi.org/10.1090/noti2832

[12] E. Arıkan, "Guessing and cryptography 1," *Aspects of Network and Information Security*, vol. 17, p. 211, 2008.

[13] L. L. Campbell, "A coding theorem and Rényi's entropy," *Information and Control*, vol. 8, no. 4, pp. 423–429, 1965.

[14] J. Aczél and Z. Daróczy, *On Measures of Information and Their Characterizations*. New York: Academic Press, 1975.

4 Relative Entropy

4.1 Overview

In the task of prediction studied in Chapter 3, the observer assigns a probability to every possible state of the system. The prediction error is then evaluated as an average over the same probability distribution assumed by the predictor. This approach, however, has a significant limitation: If the true distribution of the system differs from the assumed distribution, then the resulting prediction error reflects the observer's perspective rather than providing a realistic assessment of the expected error. A truly meaningful prediction error metric should account for the distinction between a predictor that is well aligned with the true distribution and one that is based on a mismatched distribution.

This chapter addresses this limitation by focusing on ways to quantify the "distance" between two distributions – such as the assumed and true predictive distributions – through measures known as relative entropies.

Relative entropies not only support the study of informational tasks involving a mismatch between true and assumed distributions, but also provide a formal framework for the monotonicity property of entropy measures, including Shannon, Tsallis, and Rényi entropies. As introduced in Chapter 1, the monotonicity property asserts that an entropy measure should increase as the informational state approaches the maximum-entropy (uniform) distribution. When formalizing this property, the notion of "distance" is measured differently depending on the entropy measure, giving rise to the Shannon, Tsallis, and Rényi relative entropies. Each entropy measure, therefore, represents a distinct perspective on how to quantify the "closeness" of a given informational state to the state of maximum entropy.

Furthermore, relative entropies are closely linked to the informational task of detecting which of two distributions generated a given sample. From this perspective, the chapter also introduces the trace distance, an alternative distance metric that has significant applications throughout the remainder of the book.

Finally, this chapter demonstrates how the trace distance, alongside the Shannon and Tsallis relative entropies, can be unified and generalized under the framework of f-divergences.

4.1.1 Learning Objectives and Organization

By the end of this chapter, the reader should be able to:

- understand the definition and properties of the Shannon, Tsallis, and Rényi relative entropies;
- formalize the property of monotonicity of the Shannon, Tsallis, and Rényi entropies in terms of the corresponding relative entropies;
- understand the dependence of the Tsallis and Rényi relative entropies on the order α;
- derive the optimal detection probability for the problem of distinguishing two distributions on the basis of an observation from either distribution;
- understand the definition and properties of the trace distance, as well as its connection to the optimal detection probability;
- bound the optimal detection probabilities in terms of the relative entropies;
- and leverage the framework of f-divergences to generalize the trace distance, as well as the Shannon and Tsallis relative entropies.

4.2 Shannon Relative Entropy

4.2.1 Mismatched Prediction

In order to introduce the Shannon relative entropy, this section starts by studying the **mismatched prediction** setting anticipated in the Overview. The formulation of this problem encompasses two distributions of the state $\mathrm{x} \in [N]$ of the system.

- **Reference distribution:** The reference distribution p is the informational state of some **reference observer**. The reference observer may be "Nature," making it useful to think of the distribution p as the *true* distribution of the state.
- **Informational state:** The observer has an informational state defined by a distribution q, which is generally mismatched with respect to the reference distribution p.

Informally, this means that the observer "thinks" that a state $x \in [N]$ will occur with probability q_x, while actually – or at least from the perspective of the reference observer – the event $\mathrm{x} = x$ occurs with probability p_x. Within this interpretation, the observer has a confidence q_x in the state being equal to $\mathrm{x} = x$, while the true probability of this event is p_x.

4.2.2 Impossible States, Black Swans, and Hallucinations

To quantify the loss obtained by a predictor in informational state q under the reference distribution p, it is useful to identify the following special types of states, which are summarized in Table 4.1.

Definition 4.1 Impossible states, black swans, and hallucinations

For reference distribution p and informational state q, one can distinguish the following special types of states.

- **Impossible states:** A state $x \in [N]$ is impossible if it has zero probability under the reference distribution, i.e.,

$$p_x = 0. \tag{4.1}$$

- **Black swans:** A state $x \in [N]$ is a black swan if the reference probability is positive, $p_x > 0$, and thus the state is possible, but the predicted probability is zero, i.e., $q_x = 0$:

$$p_x > 0 \text{ and } q_x = 0. \tag{4.2}$$

  A black swan is thus an occurrence that is deemed impossible by the observer, until it is actually observed.
- **Hallucinations:** A state $x \in [N]$ is a hallucination if it is impossible, i.e., $p_x = 0$, but the predicted probability is positive, i.e., $q_x > 0$:

$$q_x > 0 \text{ and } p_x = 0. \tag{4.3}$$

  Thus, the observer "thinks" that the state x is possible, while it is actually not. The term "hallucination" is borrowed from the analysis of large language models in AI (see Recommended Resources).

Table 4.1 Special states arising from the comparison of two distributions *p* and *q*, where *p* represents the reference, or true, distribution and *q* the informational state

True distribution p	Informational state q	State type
$p_x = 0$	$q_x \geq 0$	Impossible
$p_x > 0$	$q_x = 0$	Black swan
$p_x = 0$	$q_x > 0$	Hallucination

Understanding whether a state x is possible or not requires analyzing the probability p_x. For a possible state, with $p_x > 0$, determining whether the state is a black swan demands the consideration also of the probability q_x. In a similar way, for an impossible state, with $p_x = 0$, finding out whether it is a hallucination calls for an analysis of the probability q_x. Specifically, as summarized next, the presence or absence of black swans and hallucinations can be ascertained by comparing the supports (3.29) of the distributions p and q.

Key Result 4.1 Absence of black swans

For reference distribution p and informational state q, there are no black swans if and only if the support of the distribution q includes the support of the reference distribution p, i.e.,

$$\text{no black swans} \iff \text{supp}(p) \subseteq \text{supp}(q). \tag{4.4}$$

So, in order to avoid black swans, the observer must assign a positive probability q_x to all the events x that may occur, i.e., that have $p_x > 0$. In words, the observer must "believe" in everything that is possible. In measure-theoretic terms, this property is expressed by saying that distribution p is **absolutely continuous** with respect to distribution q, which is also indicated with the notation

$$p \ll q. \tag{4.5}$$

Example 4.1

Take the two distributions p and q with $N = 10$ as illustrated in Figure 4.1, where $q = 1/4 \cdot [0, 0, 0, 1, 1, 1, 1, 0, 0, 0]^T$ and $p = [0.18, 0.19, 0.18, 0.15, 0.11, 0.07, 0.05, 0.03, 0.01, 0.03]^T$. The supports of the two distributions are given by $\text{supp}(q) = \{3, 4, 5, 6\}$ and $\text{supp}(p) = [10]$. Thus, the support of distribution q does not cover that of distribution p. This indicates that there is at least one black swan. For example, the state $x = 0$ is a black swan, since the corresponding probabilities are given by $p_x = 0.18 > 0$ and $q_x = 0$.

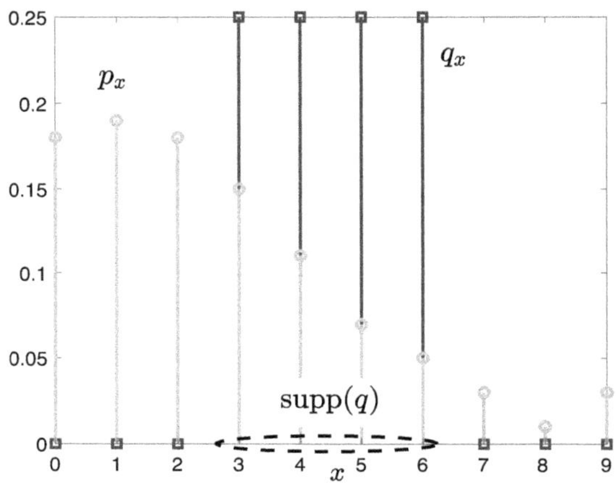

Figure 4.1 Given the reference distribution p and informational state q in the figure, the condition (4.4) does not hold and thus there is at least one black swan. In fact, states $\{0, 1, 2, 7, 8, 9\}$ are all black swans. Conversely, there are no hallucinations, since the inclusion (4.6) is verified.

Note that, if the roles of the two distributions were swapped, so that p played the role of the informational state and q that of the reference distribution, there would be no black swans, since the inclusion condition $\text{supp}(q) \subseteq \text{supp}(p)$ holds.

Key Result 4.2 Absence of hallucinations

For reference distribution p and informational state q, there are no hallucinations if and only if the support of the distribution p includes the support of the reference distribution q, i.e.,

$$\text{no hallucinations} \iff \text{supp}(q) \subseteq \text{supp}(p). \tag{4.6}$$

This condition may be paraphrased as stating that everything that is "believed" by the observer is possible. This requirement is dual to that in (4.4) for the absence of black swans. In fact, there are no black swans when the reference distribution is p and the informational state is q if and only if there are no hallucinations when the reference distribution is q and the informational state is p.

Example 4.2

For the two distributions in Figure 4.1, which were studied in Example 4.1, there are no hallucinations, since the condition $\text{supp}(q) \subseteq \text{supp}(p)$ is satisfied.

A final important setting to highlight is one in which the supports of the two distributions are **disjoint**. The following result concerning this situation follows directly from the definition of support of a distribution.

Key Result 4.3 Disjoint supports
For reference distribution p and informational state q, if the two supports are disjoint, i.e.,

$$\text{supp}(p) \cap \text{supp}(q) = \emptyset, \tag{4.7}$$

all possible states are black swans and all "believed" states are hallucinations, i.e.,

$$\begin{aligned} p_x > 0 &\Rightarrow q_x = 0 \\ \text{and } \; q_x > 0 &\Rightarrow p_x = 0. \end{aligned} \tag{4.8}$$

An implication of this result is that, when the supports are disjoint, swapping the roles of the two distributions still yields a situation in which all possible states are black swans and all "believed" states are hallucinations. In this sense, one may consider distributions with disjoint supports as being **maximally dissimilar**.

4.2.3 Shannon Cross Entropy

Having assigned a predictive probability q_x to state x, if the actual state x turns out to be equal to x, the observer experiences a loss that can be measured by the log-loss $-\log_2(q_x)$ (see Section 3.2.2). Accordingly, if the observer assigns a lower probability q_x, a larger loss is incurred. The **average prediction log-loss** obtained by the observer is given by the Shannon cross entropy, which is defined as follows.

Definition 4.2 Shannon cross entropy
Given two $N \times 1$ distributions p and q, the Shannon cross entropy is defined as the average log-loss

$$\begin{aligned} \text{H}(p||q) &= -\text{E}_p[\log_2(q_\text{x})] \\ &= -\sum_{x \in [N]} p_x \log_2(q_x). \end{aligned} \tag{4.9}$$

In evaluating (4.9), the following equalities are assumed.

- **Impossible states:** For an impossible state, i.e., for $p_x = 0$, one sets

$$0 \log_2(q_x) = 0 \tag{4.10}$$

for all probabilities $0 \leq q_x \leq 1$.
- **Black swans:** For a black swan, i.e., for $p_x > 0$ and $q_x = 0$, one sets

$$-p_x \log_2(0) = \infty. \tag{4.11}$$

By the conditions (4.9) and (4.10), impossible states do not contribute to the Shannon cross entropy, while black swans cause the Shannon cross entropy to be infinite (see Table 4.2). Furthermore, as a special case of the definition (4.9), if the observer's distribution q coincides with the reference distribution p, the Shannon cross entropy equals the Shannon entropy, i.e., the equality

$$\mathrm{H}(p||p) = \mathrm{H}(p) \tag{4.12}$$

holds.

The Shannon cross entropy is not symmetric in p and q, in the sense that the inequality

$$\mathrm{H}(p||q) \neq \mathrm{H}(q||p) \tag{4.13}$$

is generally verified. This implies that the average prediction error, as measured by the Shannon cross entropy, is different when the reference distribution is p and the observer's informational state is q from the case where the reference distribution is q and the observer's informational state is p. Throughout this book, following the standard convention in information theory, the "double bar" notation, as in $\mathrm{H}(p||q)$, will be dedicated for quantities that are not symmetric with respect to the two input vectors.

4.2.4 Other Mismatched Informational Tasks

As discussed in Section 4.2.3, if the predictor assumes a distribution q and the true distribution is p, the average prediction log-loss equals the Shannon cross

Table 4.2 Contribution of black swan states ($p_x > 0$ and $q_x = 0$) to different distance measures

Distance measure	Contribution
Shannon cross entropy and Shannon relative entropy	Infinite
Tsallis relative entropy of order $0 \leq \alpha < 1$	$p_x 1_2/(1-\alpha)$
Tsallis relative entropy of order $\alpha > 1$	Infinite
Rényi relative entropy of order $0 \leq \alpha < 1$	Finite*
Rényi relative entropy of order $\alpha > 1$	Infinite

*The contribution is infinite if *all* possible states are black swans.

entropy $\mathrm{H}(p||q)$. Therefore, the Shannon cross entropy quantifies the performance of a mismatched predictor. In a similar way, the Shannon cross entropy can also be introduced through the informational tasks of asking, compressing, and sampling, whenever the observer assumes a distribution q that is mismatched with the reference distribution p.

As discussed in Chapter 2, in all these tasks, with limited or no loss of optimality, an observer in informational state q can associate a number of bits to a state x that is given by the Shannon surprise $-\log_2(q_x)$. Therefore, when the true distribution is p, the average number of bits is given by

$$-\mathrm{E}_p[\log_2(q_{\mathrm{x}})] = \mathrm{H}(p||q), \tag{4.14}$$

recovering the Shannon cross entropy.

4.2.5 Shannon Relative Entropy

Given that it evaluates the performance of an observer under a mismatched informational state, the Shannon cross entropy should be no smaller than the Shannon entropy $\mathrm{H}(p) = \mathrm{H}(p||p)$. This motivates the introduction of the Shannon relative entropy as the difference between the Shannon cross entropy and the Shannon entropy, i.e.,

$$\begin{aligned}
\mathrm{D}(p||q) &= \mathrm{H}(p||q) - \mathrm{H}(p) \\
&= -\mathrm{E}_p[\log_2(q_{\mathrm{x}})] + \mathrm{E}_p[\log_2(p_{\mathrm{x}})] \\
&= \mathrm{E}_p\left[\log_2\left(\frac{p_{\mathrm{x}}}{q_{\mathrm{x}}}\right)\right] \\
&= \sum_{x\in[N]} p_x \log_2\left(\frac{p_x}{q_x}\right).
\end{aligned} \tag{4.15}$$

By its definition (4.15), the Shannon relative entropy $\mathrm{D}(p||q)$ measures the additional average prediction error – or the additional number of bits – caused by a mismatch between the informational state q and the true distribution p. The Shannon relative entropy $\mathrm{D}(p||q)$ may thus be interpreted as quantifying the regret of the observer for having assumed the informational state q when the true distribution is p. Accordingly, the Shannon relative entropy $\mathrm{D}(p||q)$ is a meaningful measure of the distance between distributions p and q.

Mathematically, the derivation (4.15) shows that the Shannon relative entropy compares the distributions p and q by evaluating the **log-probability ratio** $\log_2(p_x/q_x)$ for all states $x \in [N]$, and then computing the average of this ratio with respect to distribution p.

Considering that the Shannon cross entropy is infinite in the presence of black swans, and recalling the equivalent characterization (4.4) of the absence of black swans, yields the following formal definition.

Definition 4.3 Shannon relative entropy, aka Kullback–Leibler (KL) divergence Given two $N \times 1$ distributions p and q, the Shannon cross entropy is defined as

$$\mathrm{D}(p||q) = \begin{cases} \mathrm{E}_p\left[\log_2\left(\frac{p_\mathrm{x}}{q_\mathrm{x}}\right)\right] & \text{if } \mathrm{supp}(p) \subseteq \mathrm{supp}(q), \\ \infty & \text{otherwise.} \end{cases} \tag{4.16}$$

Accordingly, like the Shannon cross entropy, as summarized in Table 4.2, the Shannon relative entropy is finite if and only if the support of distribution q includes that of distribution p.

As anticipated by the notation, the Shannon relative entropy is *not symmetric*, in the sense that the inequality

$$\mathrm{D}(p||q) \neq \mathrm{D}(q||p) \tag{4.17}$$

generally holds. As for the Shannon cross entropy, this inequality reflects the different roles played by the two vectors p and q – one as the true distribution and the other as the distribution assumed by the observer.

Furthermore, both Shannon cross entropy and Shannon relative entropy are *permutation-invariant*; that is, their values do not change if the entries of both distributions p and q are permuted in the same way.

Example 4.3

Consider a single-bit system, i.e., with $N = 2$, for which the reference distribution $p = [1 - p_1, p_1]^T$ and the informational state $q = [1 - q_1, q_1]^T$ are defined by some probabilities $0 \leq p_1 \leq 1$ and $0 \leq q_1 \leq 1$. For this special case, the Shannon relative entropy, referred to as the **binary Shannon relative entropy**, is evaluated as

$$\begin{aligned} \mathrm{D}(p||q) &= \mathrm{d}(p_1||q_1) \\ &= p_1 \log_2\left(\frac{p_1}{q_1}\right) + (1 - p_1)\log_2\left(\frac{1 - p_1}{1 - q_1}\right). \end{aligned} \tag{4.18}$$

Note the equality $\mathrm{d}(p_1||q_1) = \mathrm{d}(p_0||q_0)$, with $p_0 = 1 - p_1$ and $q_0 = 1 - q_1$, which reflects the permutation invariance of the Shannon relative entropy.

To start, it is instructive to identify conditions under which the binary Shannon relative entropy is finite. The binary Shannon relative entropy takes an infinite value, $\mathrm{d}(p_1||q_1) = \infty$, if the support of q does not include the support of p. This happens in one of two cases: (i) $q_0 = 0$, so that the support of q is $\mathrm{supp}(q) = \{1\}$, while $p_0 > 0$, so that the support of p includes the black swan $x = 0$; or (ii) $q_1 = 0$, so that the support of q is $\mathrm{supp}(q) = \{0\}$, while $p_1 > 0$, so that the support of p includes the black swan $x = 1$.

To illustrate this point, Figure 4.2 shows the binary Shannon relative entropies $\mathrm{d}(p_1||q_1)$ and $\mathrm{d}(q_1||p_1)$ for a fixed value $q_1 = 0.3$ as a function of p_1, i.e., $\mathrm{d}(p_1||0.3)$

and $\mathrm{d}(0.3||p_1)$. It is observed that $\mathrm{d}(0.3||p_1)$ tends to infinity when p_1 tends to 0 or 1 in accordance with the discussion above. In contrast, $\mathrm{d}(p_1||0.3)$ is always finite since the support of distribution q encompasses the entire set $[2] = \{0, 1\}$. Note, in particular, that for $p_0 = 0$ and $p_1 = 1$, the binary Shannon relative entropy $\mathrm{d}(p_1||q_1)$ returns

$$\mathrm{d}(0||q_1) = -\log_2(1-q_1) \text{ and } \mathrm{d}(1||p_1) = -\log_2(q_1), \tag{4.19}$$

as seen in the figure.

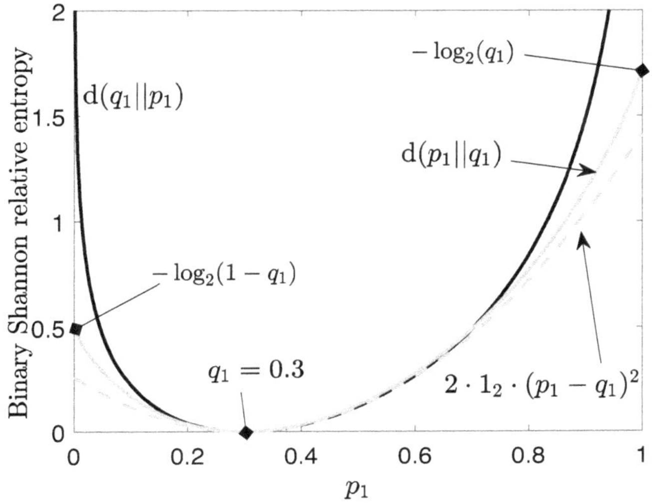

Figure 4.2 Binary Shannon relative entropies $\mathrm{d}(p_1||q_1)$ and $\mathrm{d}(q_1||p_1)$ with $q_1 = 0.3$ as a function of probability p_1, along with the lower bound (4.19).

Both binary Shannon relative entropies $\mathrm{d}(p_1||0.3)$ and $\mathrm{d}(0.3||p_1)$ increase from zero as p_1 deviates from q_1 in either direction, with $\mathrm{d}(q_1||p_1)$ tending to infinity as p_1 tends to 0 or 1. This trend can be approximately described by a quadratic function of the difference $p_1 - q_1 = q_0 - p_0$ via the lower bounds

$$\mathrm{d}(p_1||q_1) \geq 2 \cdot 1_2 \cdot (p_1 - q_1)^2 \tag{4.20}$$

and thus also $\mathrm{d}(q_1||p_1) \geq 2 \cdot 1_2 \cdot (p_1 - q_1)^2$. As seen in the figure, this bound is tight when the difference $p_1 - q_1 = q_0 - p_0$ is small, in which regime the two relative entropy measures take similar values.

Example 4.4

For $N = 4$, take the uniform distribution $p = u = [1/4, 1/4, 1/4, 1/4]^T$ and the distribution $q = [1/2, 1/2, 0, 0]^T$. The Shannon relative entropy $\mathrm{D}(u||q)$ takes an infinite value, since the support of distribution q does not include the support of the distribution u. In contrast, when the uniform distribution u plays the role of the observer's distribution, the Shannon relative entropy is finite and given by

$$\mathrm{D}(q||u) = \frac{1}{2}\log_2\left(\frac{4}{2}\right) + \frac{1}{2}\log_2\left(\frac{4}{2}\right)$$
$$= 1. \tag{4.21}$$

Example 4.5

Consider two distributions with $N = 4$ given by $p = [0, 0, 1/2, 1/2]^T$ and $q = [1/2, 1/2, 0, 0]^T$. In this case, the two distributions have disjoint supports, and thus all possible states are black swans for both settings in which either distribution is the true distribution or the informational state. As a result, *both* Shannon relative entropies $\mathrm{D}(p||q)$ and $\mathrm{D}(q||p)$ take an infinite value.

Example 4.6

Still for $N = 4$, take distributions $p = [0, 1/3, 1/3, 1/3]^T$ and $q = [1/2, 1/2, 0, 0]^T$. For these distributions, the support sets are not disjoint, but neither support includes the other. However, intuitively, these distributions are "closer" to each other than the distributions studied in the previous example. In fact, unlike the previous example, the supports of these two distributions overlap at one state. That said, here too, *both* Shannon relative entropies $\mathrm{D}(p||q)$ and $\mathrm{D}(q||p)$ take an infinite value. One can thus conclude that the Shannon cross entropy is not able to distinguish between the case in which the two distributions have disjoint supports from the case in which the supports are overlapped but neither support includes the other (causing the existence of black swans).

Example 4.7

Moving beyond the task of prediction, this example describes a practical situation in which it is important to assign an infinite penalty in the presence of black swans, i.e., when $\mathrm{supp}(p) \not\subseteq \mathrm{supp}(q)$. Take the case of an autonomous vehicle whose **control policy** applies steering actions in response to state information x obtained via the vehicle's sensors. Each action is associated with some numerical value, referred to as a reward, indicating how valuable the action is in the given state. The reward may be assigned by human annotators.

In order to design the control policy, data about actions and rewards are collected by exploring each state x with some probability q_x. The probability q_x may be affected by factors such as the amount of time available to drive a car around to collect data, or the availability of an accurate driving simulator. When testing the system, however, the actual probability of encountering a given state is p_x.

A black swan x, for which $q_x = 0$ and $p_x > 0$, is a state that cannot be represented in the data set, although it can be encountered by the vehicle when tested. This is a situation that is important to avoid, as it may lead the autonomous vehicle to face completely new conditions that may affect the safety of the passengers.

4.2.6 Properties of the Shannon Relative Entropy

The Shannon relative entropy was introduced in (4.15) as the difference of the Shannon cross entropy and the Shannon entropy. This definition was motivated by the intuition that the difference in (4.15) should be non-negative, as the Shannon cross entropy represents the performance of a mismatched observer, while the Shannon entropy accounts for a matched predictor. This intuition is verified by the following property.

Key Result 4.4 Non-negativity and minimum-value property of the Shannon relative entropy
The Shannon relative entropy is non-negative,

$$\mathrm{D}(p||q) \geq 0, \tag{4.22}$$

and it equals zero if and only if the two distributions are equal, i.e., if and only if $p = q$.

This property follows directly from the convexity of the function $-\log_2(\cdot)$ (see Problem 4.1). The next example illustrates another property that will be useful in Section 4.3.

Example 4.8

In this example, the observer adopts as informational state the uniform distribution u, thus being maximally uncertain across all possible N outcomes. What is the true distribution p that causes the Shannon relative entropy $\mathrm{D}(p||u)$ to be maximal? In other words, under which true distribution would such an observer suffer a maximal regret?

Using the definition (4.15), the Shannon relative entropy $\mathrm{D}(p||u)$ is given by

$$\begin{aligned} \mathrm{D}(p||u) &= \mathrm{H}(p||u) - \mathrm{H}(p) \\ &= \mathrm{E}_p[\log_2(N)] - \mathrm{H}(p) \\ &= \log_2(N) - \mathrm{H}(p). \end{aligned} \tag{4.23}$$

By the non-negativity and minimum-value property of the Shannon entropy $\mathrm{H}(p)$, the inequality

$$\mathrm{D}(p||u) \leq \log_2(N) \tag{4.24}$$

is satisfied, with equality holding if and only if $p = \delta_{\bar{x}}$ for some state $\bar{x} \in [N]$. It is concluded that the regret of a maximally uncertain observer takes the maximum value when the true distribution describes a situation of no uncertainty.

4.3 Monotonicity of the Shannon Entropy

This section focuses on formalizing and proving the property of monotonicity for the Shannon entropy. As introduced in Chapter 1, this property states that the Shannon entropy must increase as the distribution p becomes "closer" to the uniform distribution u, which corresponds to the case of maximum entropy. To describe and demonstrate the monotonicity property, the key idea is to leverage the Shannon relative entropy as a measure of distance between p and u, as detailed next.

Key Result 4.5 Monotonicity property of the Shannon entropy
For any distribution p, the Shannon entropy $\mathrm{H}(p)$ can be written as

$$\begin{aligned}\mathrm{H}(p) &= \mathrm{H}(u) - \mathrm{D}(p||u) \\ &= \log_2(N) - \mathrm{D}(p||u). \end{aligned} \tag{4.25}$$

Therefore, as the Shannon relative entropy $\mathrm{D}(p||u)$ between the distribution p and the uniform distribution u decreases to its minimum value of zero, the Shannon entropy approaches the maximum value $\mathrm{H}(u) = \log_2(N)$ from below.

The equality (4.25) can easily be verified by direct computation of the Shannon relative entropy $\mathrm{D}(p||u)$ as

$$\begin{aligned}\mathrm{D}(p||u) &= \mathrm{E}_p\left[\log_2\left(\frac{p_{\mathrm{x}}}{1/N}\right)\right] \\ &= \mathrm{E}_p\left[\log_2(p_{\mathrm{x}}) + \log_2(N)\right] \\ &= -\mathrm{H}(p) + \log_2(N). \end{aligned} \tag{4.26}$$

Example 4.9

The distribution

$$p_x = \frac{\exp\left(-\frac{x}{\gamma}\right)}{\sum_{x' \in [N]} \exp\left(-\frac{x'}{\gamma}\right)} \tag{4.27}$$

over a state variable $x \in [N]$ ranges from the minimum-entropy distribution to the maximum-entropy distribution by varying the parameter $\gamma > 0$. In fact, it tends to the minimum-entropy distribution δ_0 when $\gamma \to 0$, while it tends to the maximum-entropy distribution u when $\gamma \to \infty$.

Figure 4.3 plots the Shannon entropy $\mathrm{H}(p)$ and the Shannon relative entropy $\mathrm{D}(p||u)$ as a function of parameter γ for $N = 10$. As γ grows larger, the Shannon entropy increases from the minimum value of 0 to the maximum value

$\log_2(N) = \log_2(10) = 3.3219$, while the Shannon relative entropy $D(p||u)$ decreases from the maximum value $\log_2(N)$ to the minimum value 0.

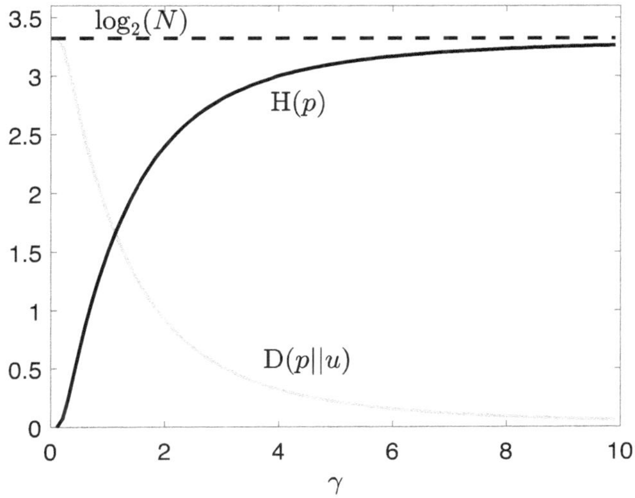

Figure 4.3 Shannon entropy $H(p)$ and Shannon relative entropy $D(p||u)$ for distribution (4.27) with $N = 10$ as a function of parameter γ.

4.4 Tsallis Relative Entropy and Monotonicity of the Tsallis Entropy

Sections 4.2 and 4.3 have introduced the Shannon relative entropy as a measure of the distance between two distributions, noting also that the Shannon relative entropy allows the formalization of the property of monotonicity for the Shannon entropy. This section generalizes the results by introducing the **Tsallis relative entropy** and connecting it, in a manner similar to the decomposition (4.25), to the Tsallis entropy.

4.4.1 Defining the Tsallis Relative Entropy

As introduced in Chapter 3, the Tsallis entropy $H_\alpha^T(p)$ is obtained from the Shannon entropy by replacing the logarithm with the α-logarithm $\log_{2,\alpha}(\cdot)$. The key advantage of this generalization is that it offers a spectrum of measures of the uncertainty associated with a distribution p that exhibit different scalings of the entropy with the number of states, N. In a similar way, the Tsallis relative entropy is obtained from the Shannon relative entropy (4.15) by replacing the logarithm with the α-logarithm $\log_{2,\alpha}(\cdot)$. This yields a family of distance measures, which include the Shannon relative entropy as a special case.

Definition 4.4 Tsallis relative entropy
Given two $N \times 1$ distributions p and q, the Tsallis relative entropy of order $\alpha \geq 0$ is defined as

$$\begin{aligned} \mathrm{D}_\alpha^{\mathrm{T}}(p||q) &= \mathrm{E}_p\left[\log_{2,\alpha}\left(\frac{p_{\mathrm{x}}}{q_{\mathrm{x}}}\right)\right] \\ &= \sum_{x\in[N]} p_x \log_{2,\alpha}\left(\frac{p_x}{q_x}\right). \end{aligned} \tag{4.28}$$

In evaluating (4.28), the following equalities are assumed.

- **Impossible states:** For an impossible state, i.e., for $p_x = 0$, one sets

$$0\log_{2,\alpha}\left(\frac{0}{q_x}\right) = 0 \tag{4.29}$$

for all probabilities $0 \leq q_x \leq 1$ and all orders $\alpha \geq 0$.
- **Black swans:** For a black swan, i.e., for $p_x > 0$ and $q_x = 0$, one sets

$$p_x \log_{2,\alpha}\left(\frac{p_x}{0}\right) = \begin{cases} p_x\left(\frac{1_2}{1-\alpha}\right) & \text{if } 0 \leq \alpha < 1, \\ \infty & \text{if } \alpha \geq 1. \end{cases} \tag{4.30}$$

Given that the α-logarithm recovers the logarithm for $\alpha = 1$, the Tsallis relative entropy reduces to the Shannon relative entropy for $\alpha = 1$, i.e.,

$$\mathrm{D}_1^{\mathrm{T}}(p||q) = \mathrm{D}(p||q). \tag{4.31}$$

Furthermore, for the Shannon relative entropy, condition (4.29) indicates that impossible states do not affect the Tsallis relative entropy for any order $\alpha \geq 0$. In contrast, condition (4.30) indicates that a black swan x causes an infinite Tsallis relative entropy only for orders $\alpha \geq 1$. For lower orders, the contribution of a black swan x is finite and given by $p_x 1_2/(1-\alpha)$ (see Table 4.2).

Accordingly, using the characterization (4.4), a complete definition of the Tsallis relative entropy of order $\alpha \geq 0$ is as follows.

Definition 4.5 Tsallis relative entropy (complete definition)
Given two $N \times 1$ distributions p and q, the Tsallis relative entropy of order $\alpha \geq 0$ is defined as

$$\mathrm{D}_\alpha^{\mathrm{T}}(p||q) = \begin{cases} \mathrm{E}_p\left[\log_{2,\alpha}\left(\frac{p_{\mathrm{x}}}{q_{\mathrm{x}}}\right)\right] & \text{if } 0 \leq \alpha < 1, \\ & \text{or if } \alpha \geq 1 \text{ and } \mathrm{supp}(p) \subseteq \mathrm{supp}(q), \\ \infty & \text{otherwise.} \end{cases} \tag{4.32}$$

As a final note on the definition (4.32), unlike the Shannon relative entropy, the Tsallis relative entropy with order $\alpha \neq 1$ does *not* have the interpretation of regret in a mismatched prediction problem. In fact, for any non-negative value $\alpha \neq 1$, in a manner analogous to (4.15), using the α-log-loss as the prediction loss measure would yield the regret

$$-\mathrm{E}_p[\log_{2,\alpha}(q_\mathrm{x})]-\mathrm{H}_\alpha^\mathrm{T}(p) = \frac{\mathrm{l}_2}{1-\alpha}\left(\sum_{x\in[N]} p_x(q_x^{\alpha-1} - p_x^{\alpha-1})\right), \tag{4.33}$$

which cannot be expressed as in (4.28).

4.4.2 Equivalent Definition of the Tsallis Relative Entropy

As discussed in Section 4.4.1, the Tsallis relative entropy of order $\alpha = 1$ coincides with the Shannon relative entropy, while offering a larger class of distance measures for a non-negative order $\alpha \neq 1$. Using the definition of the α-logarithm, the Tsallis relative entropy (4.28) can be written more explicitly in terms of the two distributions in one of the following ways.

Key Result 4.6 Equivalent expressions of the Tsallis relative entropy
For $\alpha \neq 1$, assuming the condition $\mathrm{supp}(p) \subseteq \mathrm{supp}(q)$, i.e., the absence of black swans, for $\alpha \geq 1$, the Tsallis relative entropy of order α can be expressed as

$$\mathrm{D}_\alpha^\mathrm{T}(p||q) = \frac{\mathrm{l}_2}{\alpha-1}\left(\mathrm{E}_p\left[\left(\frac{p_\mathrm{x}}{q_\mathrm{x}}\right)^{\alpha-1}\right]-1\right) \tag{4.34}$$

$$= \frac{\mathrm{l}_2}{\alpha-1}\left(\mathrm{E}_q\left[\left(\frac{p_\mathrm{x}}{q_\mathrm{x}}\right)^{\alpha}\right]-1\right) \tag{4.35}$$

$$= \frac{\mathrm{l}_2}{\alpha-1}\left(\sum_{x\in[N]} p_x^\alpha q_x^{1-\alpha} - 1\right). \tag{4.36}$$

The first expression (4.34) is in terms of an average with respect to distribution p, and it is a direct application of the definition of the α-logarithm; while in the second expression (4.35) the average is with respect to distribution q. Furthermore, the first two expressions (4.34) and (4.35), like the Shannon relative entropy, depend on the probability ratio p_x/q_x.

The third expression, namely (4.36), provides the most common way to write the Tsallis relative entropy. This expression makes it particularly clear that, if $0 \leq \alpha < 1$, a black swan x – which has $p_x > 0$ and $q_x = 0$ – does not cause the Tsallis relative entropy to take an infinite value. Thus, the Tsallis relative entropy with a small order provides a more useful distance metric in settings for which black swans are acceptable.

The third expression (4.36) will be further elaborated on in Section 4.4.4.

4.4.3 Properties of the Tsallis Relative Entropy

By the definition and properties of the α-logarithm reviewed in Chapter 3, the Tsallis relative entropy of order α satisfies the following properties.

Key Result 4.7 Non-negativity and minimum-value property of the Tsallis relative entropy
The Tsallis relative entropy is non-negative,

$$D_\alpha^T(p||q) \geq 0, \tag{4.37}$$

for all orders $\alpha \geq 0$. Furthermore, for all strictly positive orders $\alpha > 0$, the Tsallis relative entropy equals zero if and only if the two distributions are equal, i.e., if and only if the equality $p = q$ holds.

The special case $\alpha = 0$ will be discussed further in Section 4.4.7.

Key Result 4.8 The Tsallis relative entropy increases with the order α
For any fixed probability vectors p and q, the Tsallis relative entropy $D_\alpha^T(p||q)$ is an increasing function of the order $\alpha \geq 0$, i.e., the inequality

$$D_{\alpha'}^T(p||q) \geq D_\alpha^T(p||q) \tag{4.38}$$

holds for any two orders α' and α with $\alpha' \geq \alpha$.

In this regard, it may be useful to emphasize that the dependence on the order α follows an *opposite trend* as compared to the Tsallis entropy due to the different sign of the α-log in the two definitions. Specifically, while the Tsallis entropy *decreases* with α, thus providing less severe measures of the uncertainty associated with an informational state, the Tsallis relative entropy *increases* with α, magnifying the differences between two distributions.

Example 4.10

Consider the two distributions with $N = 4$ given by $p = [0, 0, 1/2, 1/2]^T$ and $q = [1/2, 1/2, 0, 0]^T$, which have *disjoint* supports (see Example 4.5). For all $\alpha \geq 1$, both Tsallis relative entropies $D_\alpha^T(p||q)$ and $D_\alpha^T(q||p)$ take infinite values, as does the Shannon relative entropy, which is obtained in the special case $\alpha = 1$. In contrast, for $0 \leq \alpha < 1$, the Tsallis relative entropy can be evaluated using (4.29) and (4.30) as $D_\alpha^T(p||q) = D_\alpha^T(q||p) = 1_2/(1-\alpha)$. In fact, all possible states are black swans in this case.

Example 4.11

As in Example 4.10, the distributions $p = [0, 1/3, 1/3, 1/3]^T$ and $q = [1/2, 1/2, 0, 0]^T$ are such that the support of neither distribution covers the support of the other distribution (see also Example 4.6). Therefore, again, for all $\alpha \geq 1$, both Tsallis relative entropies $\mathrm{D}^{\mathrm{T}}_{\alpha}(p||q)$ and $\mathrm{D}^{\mathrm{T}}_{\alpha}(q||p)$ take infinite values. However, the two distributions have partially overlapping supports, and setting $0 \leq \alpha < 1$ allows the Tsallis relative entropy to distinguish this case from Example 4.10.

In fact, using again (4.29) and (4.30), the Tsallis relative entropy with $0 \leq \alpha < 1$ is computed as

$$\mathrm{D}^{\mathrm{T}}_{\alpha}(p||q) = \frac{1}{3}\log_{2,\alpha}\left(\frac{2}{3}\right) + \frac{2}{3}\frac{1_2}{1-\alpha} = \frac{1}{3}\frac{1_2}{1-\alpha}\left(\left(\frac{2}{3}\right)^{\alpha-1} + 1\right), \tag{4.39}$$

which is smaller than the value $1_2/(1-\alpha)$ attained for disjoint supports. In a similar way, one obtains

$$\mathrm{D}^{\mathrm{T}}_{\alpha}(q||p) = \frac{1}{2}\frac{1_2}{1-\alpha}\left(\frac{3}{2}\right)^{\alpha-1}, \tag{4.40}$$

which is also smaller than the value $1_2/(1-\alpha)$ attained for disjoint supports. The resulting Tsallis relative entropies are plotted in Figure 4.4.

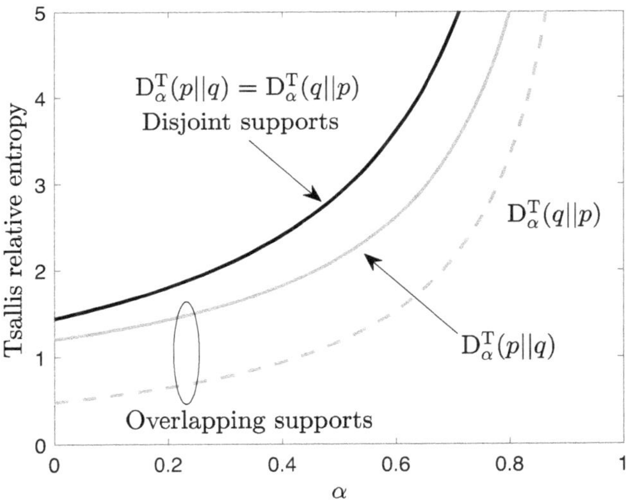

Figure 4.4 Tsallis relative entropies $\mathrm{D}^{\mathrm{T}}_{\alpha}(p||q)$ and $\mathrm{D}^{\mathrm{T}}_{\alpha}(q||p)$ for the distributions $p = [0, 0, 1/2, 1/2]^T$ and $q = [1/2, 1/2, 0, 0]^T$ (marked as "Disjoint supports"), as well as for the distributions $p = [0, 1/3, 1/3, 1/3]^T$ and $q = [1/2, 1/2, 0, 0]^T$ (marked as "Overlapping supports"), as a function of the order α.

Generalizing the two last examples, the following property holds.

> **Key Result 4.9 Upper bound on the Tsallis relative entropy for $0 \leq \alpha < 1$**
> When $0 \leq \alpha < 1$, the Tsallis relative entropy $\mathrm{D}^{\mathrm{T}}_{\alpha}(p||q)$ satisfies the upper bound
> $$\mathrm{D}^{\mathrm{T}}_{\alpha}(p||q) \leq \frac{1_2}{1-\alpha}, \tag{4.41}$$
> with equality if and only if the supports of the two distributions p and q are disjoint.

Thus, while for $\alpha \geq 1$ one obtains an infinite Tsallis relative entropy as soon as there is *at least one* black swan, the Tsallis relative entropy with $0 \leq \alpha < 1$ attains the maximum, finite, value in (4.41) if *all* possible states are black swans. This result substantiates the claim that distributions with disjoint supports are maximally distinct.

Example 4.12

To further illustrate the dependence of the Tsallis relative entropy on the order α, this example studies the case of a single-bit system. Given any two distributions $p = [p_0, p_1]^T$ and $q = [q_0, q_1]^T$, the Tsallis relative entropy, referred to as the **binary Tsallis relative entropy**, is written as

$$\begin{aligned}\mathrm{D}^{\mathrm{T}}_{\alpha}(p||q) &= \mathrm{d}^{\mathrm{T}}_{\alpha}(p_1||q_1) = \mathrm{d}^{\mathrm{T}}_{\alpha}(p_0||q_0) \\ &= \frac{1_2}{1-\alpha}\left(1 - p_1^{\alpha} q_1^{1-\alpha} - (1-p_1)^{\alpha}(1-q_1)^{1-\alpha}\right),\end{aligned} \tag{4.42}$$

where the last equality follows by applying the definition of the α-logarithm.

Figure 4.5 plots the binary Tsallis relative entropy $\mathrm{d}^{\mathrm{T}}_{\alpha}(p_1||q_1)$ as a function of q_1 for $p_1 = 0.3$. First, it is observed that for $\alpha = 1$, the binary Tsallis relative entropy coincides with the binary Shannon relative entropy, i.e., $\mathrm{d}^{\mathrm{T}}_{1}(p_1||q_1) = \mathrm{d}(p_1||q_1)$. Second,

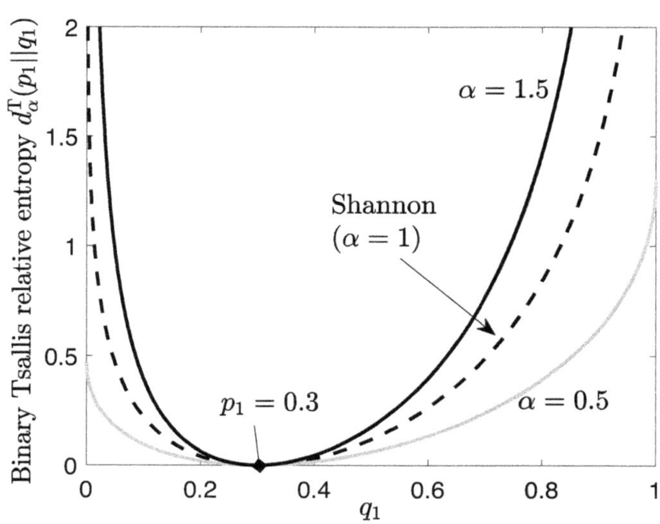

Figure 4.5 Binary Tsallis relative entropy $\mathrm{d}^{\mathrm{T}}_{\alpha}(p_1||q_1)$ as a function of q_1 for $p_1 = 0.3$.

the figure confirms that, as the order α increases, the Tsallis relative entropy increases. In particular, with $0 \le \alpha < 1$, the binary Tsallis relative entropy remains bounded even when the support of distribution q does not include that of distribution p, which occurs when $q_1 = 0$ and $q_1 = 1$. In contrast, for $\alpha \ge 1$, the Tsallis relative entropy tends to infinity as q_1 approaches $q_1 = 0$ or $q_1 = 1$.

Example 4.13

One of the applications of the Tsallis relative entropy is as a training criterion in **machine learning** applications. In such settings, one wishes to optimize the distribution q, representing a machine learning model, to match the true data-generating distribution p, which is typically estimated from data. To this end, the learner may seek a distribution q that minimizes the Tsallis relative entropy by addressing the problem

$$\underset{q}{\text{minimize}}\ \mathrm{D}^{\mathrm{T}}_{\alpha}(p||q). \tag{4.43}$$

The Tsallis relative entropy $\mathrm{D}^{\mathrm{T}}_{\alpha}(p||q)$ serves as a flexible learning criterion. In fact, a solution to the learning problem (4.43) would generally have distinct features when adopting different values for the order α. With a large order $\alpha \ge 1$, a solution q would tend to be *inclusive*, in the sense that it would have a support supp(q) that includes that of the target distribution p. This is, in fact, necessary to avoid large values of the objective $\mathrm{D}^{\mathrm{T}}_{\alpha}(p||q)$. This type of solution may be desirable if one wishes to cover the entire true distribution p, as may be the case for generative models trained on high-quality data.

However, when data are noisy, distribution p may also describe the presence of outliers, resulting in parts of the distribution that should ideally not be covered by the learned model q. In this case, choosing $\alpha \ge 1$ may yield suboptimal solutions. In contrast, with a lower order in the range $0 \le \alpha \le 1$, a solution of problem (4.43) may be *exclusive*, in the sense that the distribution q could focus on the states x with the largest probability p_x. Concentrating on higher-probability states may be useful to filter out outliers and noise in the distribution p.

4.4.4 Tsallis Relative Sum

In Section 3.4, the Tsallis entropy was shown to be a function of the Tsallis sum, which provides an uncertainty measure or certainty measure depending on the value of the order α. It is now observed that the Tsallis relative entropy can be expressed in terms of a related fundamental quantity referred to as the Tsallis relative sum.

Definition 4.6 Tsallis relative sum, aka relative information generating function Given two $N \times 1$ distributions p and q, the Tsallis relative sum of order $\alpha \ge 0$ with $\alpha \ne 1$ is defined as

$$\text{Tsallis relative sum of order } \alpha = \sum_{x\in[N]} p_x^{\alpha} q_x^{1-\alpha}. \tag{4.44}$$

In evaluating (4.44), the following equalities are assumed.

- **Impossible states:** For an impossible state, i.e., for $p_x = 0$, one sets

$$0^\alpha q_x^{1-\alpha} = 0 \tag{4.45}$$

for any probability $0 \leq q_x \leq 1$ and any order $\alpha \geq 0$.
- **Black swans:** For a black swan, i.e., for $p_x > 0$ and $q_x = 0$, one sets

$$p_x^\alpha 0^{1-\alpha} = \begin{cases} 0 & \text{if } 0 \leq \alpha < 1, \\ \infty & \text{if } \alpha > 1. \end{cases} \tag{4.46}$$

Table 4.3 Interpretation of the Tsallis relative sum as a function of the order α

Order	Tsallis relative sum	$1/(\alpha - 1)$
$0 \leq \alpha < 1$	Similarity measure	Negative
$\alpha > 1$	Dissimilarity measure	Positive

The conditions (4.46) and (4.45) are consistent with (4.29) and (4.30), respectively, and they ensure that impossible states under the reference distribution p do not affect the Tsallis relative sum and that black swans cause an infinite Tsallis relative sum for orders $\alpha > 1$.

In a manner analogous to the Tsallis sum, the Tsallis relative sum has a different interpretation depending on the value of α. In particular, as summarized in Table 4.3, the Tsallis relative sum will be shown next to serve as a measure of the *similarity* of the probability vectors p and q for $0 \leq \alpha < 1$; while, for $\alpha > 1$, it is a measure of *dissimilarity*.

To see this, it is useful to interpret the Tsallis relative sum as follows.

- **Tsallis relative sum as an inner product:** For $0 \leq \alpha < 1$, the Tsallis relative sum can be interpreted as the inner product between vectors $p^\alpha = [p_0^\alpha, p_1^\alpha, \ldots, p_{N-1}^\alpha]^T$ and $q^{1-\alpha} = [q_0^{1-\alpha}, q_1^{1-\alpha}, \ldots, q_{N-1}^{1-\alpha}]^T$, which is a measure of similarity of the probability distributions p and q (see also Section 5.3.1).
- **Tsallis relative sum as a probability ratio:** For $\alpha > 1$, the α-Tsallis relative sum can be written as the probability ratio

$$\sum_{x \in [N]} \frac{p_x^\alpha}{q_x^{\alpha-1}}, \tag{4.47}$$

representing a measure of dissimilarity.

Example 4.14

To illustrate the different roles played by the Tsallis relative sum as a function of the order α, consider two extreme situations. In the first, the two distributions p and q

are maximally dissimilar in the sense that they have disjoint support sets; while in the second they are maximally similar, i.e., $p = q$.

When $0 \leq \alpha < 1$, the Tsallis relative sum of order α equals $\sum_{x\in[N]} p_x^\alpha q_x^{1-\alpha} = 0$ for distributions with disjoint supports; while it equals $\sum_{x\in[N]} p_x^\alpha q_x^{1-\alpha} = \sum_{x\in[N]} p_x = 1$ when the distributions are identical. Therefore, the Tsallis sum of order $0 \leq \alpha < 1$ *increases* with the similarity between the two distributions.

In contrast, when $\alpha > 1$, by (4.46), the Tsallis relative sum takes an infinite value for distributions with disjoint supports; while it equals $\sum_{x\in[N]} p_x^\alpha q_x^{1-\alpha} = 1$ for identical distributions. Accordingly, when $\alpha > 1$, the Tsallis relative sum *decreases* with the similarity between the two distributions.

Example 4.15

To further elaborate on the different behavior of the Tsallis relative sum in different regimes of the order α, consider a single-bit system with generic distributions $p = [1 - p_1, p_1]^T$ and $q = [1 - q_1, q_1]^T$. The Tsallis relative sum of order α, which can be computed as

$$p_1^\alpha q_1^{1-\alpha} + (1 - p_1)^\alpha (1 - q_1)^{1-\alpha}, \tag{4.48}$$

is shown in Figure 4.6 as a function of probability q_1 for $p_1 = 0.3$ and for different values of the order α.

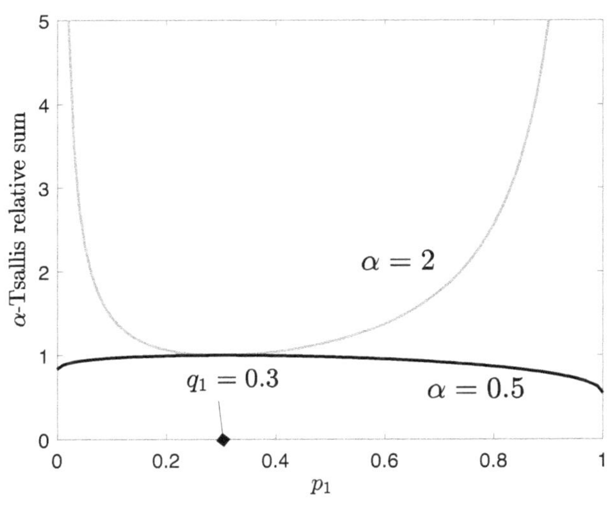

Figure 4.6 Tsallis relative sum of order α in (4.48) for a single-bit system as a function of the probability q_1 for $p_1 = 0.3$ and for different values of the order α.

When $\alpha = 1/2$ – representing the interval $0 \leq \alpha < 1$ – the Tsallis relative sum decreases as p_1 moves away from $q_1 = 0.3$. In contrast, the opposite is true for the order $\alpha = 2$, which represents the range $\alpha > 1$.

4.4.5 Properties of the Tsallis Relative Sum ⏭

Consider first an order $\alpha > 0$, thus temporarily excluding the case $\alpha = 0$. In a manner consistent with the interpretations outlined in Table 4.3, the following bounds apply to the Tsallis relative sum.

Key Result 4.10 Bounds on the Tsallis relative sum

For $0 < \alpha < 1$, the Tsallis relative sum satisfies the inequalities

$$\underbrace{0}_{\mathrm{supp}(p)\cap\mathrm{supp}(q)=\emptyset} \leq \sum_{x\in[N]} p_x^\alpha q_x^{1-\alpha} \leq \underbrace{1}_{p=q}, \tag{4.49}$$

with the lower bound attained for distributions with disjoint supports ($\mathrm{supp}(p) \cap \mathrm{supp}(q) = \emptyset$) and the upper bound attained with identical distributions ($p = q$).

For $\alpha > 1$, the Tsallis relative sum satisfies the inequalities

$$\underbrace{1}_{p=q} \leq \sum_{x\in[N]} p_x^\alpha q_x^{1-\alpha} \leq \underbrace{\infty}_{\mathrm{supp}(p)\not\subseteq\mathrm{supp}(q)}, \tag{4.50}$$

with the lower bound attained with identical distributions ($p = q$), and the upper bound achieved when the support of distribution q does not cover the support of distribution p ($\mathrm{supp}(p) \not\subseteq \mathrm{supp}(q)$).

In the special case of order $\alpha = 0$, using (4.45), the Tsallis relative sum equals

$$\sum_{x\in[N]} p_x^0 q_x = \sum_{x\in[N]:\, p_x>0} q_x = \sum_{x\in\mathrm{supp}(p)} q_x. \tag{4.51}$$

Accordingly, it measures the fraction of the "mass" in distribution q that is covered within the support of distribution p, providing a measure of similarity between the two distributions. This is consistent with the case $0 < \alpha < 1$ covered by the bound (4.49). However, the Tsallis relative sum of order $\alpha = 0$ takes the maximum value of 1 under more relaxed conditions as compared to the Tsallis relative sum of order $0 < \alpha < 1$.

Key Result 4.11 Bounds on the Tsallis relative sum of order $\alpha = 0$

The Tsallis relative sum of order $\alpha = 0$ satisfies the inequalities

$$\underbrace{0}_{\mathrm{supp}(p)\cap\mathrm{supp}(q)=\emptyset} \leq \sum_{x\in\mathrm{supp}(p)} q_x \leq \underbrace{1}_{\mathrm{supp}(q)\subseteq\mathrm{supp}(p)}, \tag{4.52}$$

where the lower bound is attained when the supports are disjoint and the upper bound is obtained when the support of distribution q is included in the support of distribution p.

By (4.52), the Tsallis relative sum of order $\alpha = 0$ equals 1 as long as the support of distribution q is included in that of distribution p, and thus there are no hallucinations. Thus, for $\alpha = 0$, unlike the case $0 < \alpha < 1$ in (4.49), there is no need to have identical distributions for the Tsallis relative sum to be maximal.

4.4.6 Tsallis Relative Entropy and Tsallis Relative Sum ⏭

Using the expression in (4.36), the Tsallis relative entropy of order $\alpha \neq 1$ can be obtained by applying the function

$$\frac{1_2}{(\alpha - 1)}(\cdot - 1) \tag{4.53}$$

to the Tsallis relative sum.

The function (4.53) relating the Tsallis relative entropy to the Tsallis relative sum of order α is the negative of the function (3.28) that connects the Tsallis entropy to the α-Tsallis sum. Therefore, the Tsallis relative sum of order α plays an analogous role to the α-Tsallis sum in determining the Tsallis relative entropy.

The function (4.53) is plotted in Figure 4.7. As illustrated, the function is decreasing in the regime $0 \leq \alpha < 1$ for which the Tsallis relative sum of order α is a measure of similarity; while it is increasing in the regime $\alpha > 1$ for which the Tsallis relative sum is a measure of dissimilarity. This ensures that the Tsallis relative entropy is always a measure of dissimilarity, or distance, between the two distributions.

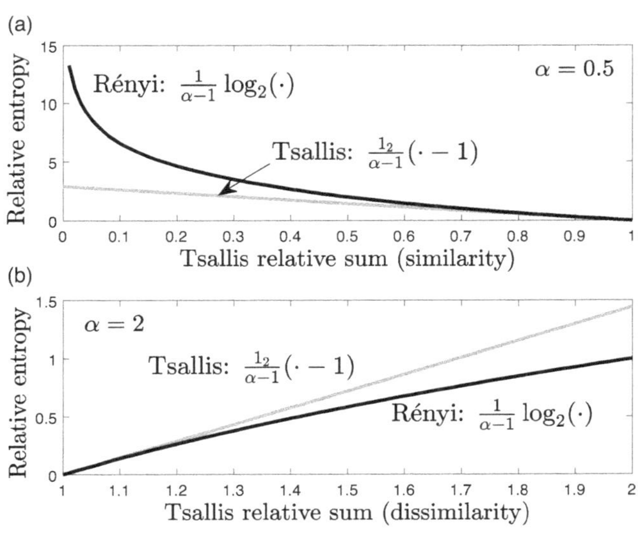

Figure 4.7 Function (4.53) relating the Tsallis relative entropy to the Tsallis relative sum and function (4.72) relating the Rényi relative entropy to the Tsallis relative sum for (a) $\alpha = 0.5$ and (b) $\alpha = 2$.

4.4.7 Special Cases of the Tsallis Relative Entropy ⏭

By varying the order α, one obtains the following notable special cases of the Tsallis relative entropy, which are presented in increasing order. Throughout this discussion, the conditions for a finite value of the Tsallis relative entropy are as in (4.32), and will not be indicated explicitly.

Definition 4.7 Tsallis entropy of order $\alpha = 0$, aka Tsallis min-relative entropy
The Tsallis relative entropy of order $\alpha = 0$, also known as the Tsallis min-relative entropy, can be evaluated from the definition (4.36) using (4.51) as

$$\mathrm{D}_0^{\mathrm{T}}(p||q) = 1_2 \cdot \left(1 - \sum_{x \in \mathrm{supp}(p)} q_x\right). \tag{4.54}$$

The name, min-relative entropy, serves as a reminder that, by (4.38), the Tsallis relative entropy of order 0 yields the minimum value among all possible orders $\alpha \geq 0$ for any fixed pair of distributions p and q.

By the bounds (4.52), the Tsallis min-relative entropy $\mathrm{D}_0^{\mathrm{T}}(p||q)$ is finite for all choices of distributions p and q, obtaining the maximum value $\mathrm{D}_0^{\mathrm{T}}(p||q) = 1_2$ in (4.41) when the two supports are disjoint. Moreover, the bounds (4.52) also imply that the Tsallis min-relative entropy $\mathrm{D}_0^{\mathrm{T}}(p||q)$ equals zero if and only if the support of distribution p covers that of distribution q, i.e.,

$$\mathrm{D}_0^{\mathrm{T}}(p||q) = 0 \Longleftrightarrow \mathrm{supp}(q) \subseteq \mathrm{supp}(p). \tag{4.55}$$

That is, the equality $\mathrm{D}_0^{\mathrm{T}}(p||q) = 0$ holds if there are no hallucinations: all states x that are thought to be possible, i.e., $q_x > 0$, are actually possible, i.e., $p_x > 0$. Note that this holds irrespective of whether there are any black swan.

Example 4.16

Suppose that a machine learning algorithm addresses the learning problem (4.43) with $\alpha = 0$. In this case, any distribution q whose support is fully contained in the support of the true distribution p, satisfying the condition $\sum_{x \in \mathrm{supp}(p)} q_x = 1$, would be a valid optimal solution. Therefore, even a distribution $q = \delta_{\bar{x}}$ that is fully concentrated at any state $\bar{x}$ with $p_{\bar{x}} > 0$ could be obtained as an output of the learning process. This type of solution may be **mode-seeking**, focusing solely on one part of the true distribution p and failing to properly cover the support of distribution p. In practice, this is an important problem for applications such as generative models q that aim at providing a diverse range of samples $\mathrm{x} \sim q$.

Definition 4.8 Tsallis relative entropy of order $\alpha = 1/2$, aka square Hellinger relative entropy
The Tsallis relative entropy of order $\alpha = 1/2$, also known as the square Hellinger relative entropy, is evaluated as

$$\mathrm{D}_{1/2}^{\mathrm{T}}(p, q) = 1_2 \cdot 2 \left(1 - \sum_{x \in [N]} \sqrt{p_x q_x}\right). \tag{4.56}$$

An interesting aspect of the square Hellinger distance is that it is symmetric with respect to the input vectors p and q, as it depends on the two probabilities only through the 1/2-Tsallis relative sum $\sum_{x\in[N]}\sqrt{p_x q_x}$. The symmetry of the square Hellinger relative entropy (4.56) is highlighted by the notation $\mathrm{D}^{\mathrm{T}}_{1/2}(p,q)$, which replaces the double bar between the two distributions with a comma. The special case $\sum_{x\in[N]}\sqrt{p_x q_x}$ of the Tsallis relative sum, also known as the **fidelity** between the two vectors p and q, will be further studied later in Section 4.5.

Definition 4.9 Tsallis relative entropy of order $\alpha = 2$, aka chi-square relative entropy

The Tsallis relative entropy of order $\alpha = 2$, also known as the chi-square relative entropy or Pearson divergence, is given by

$$\mathrm{D}^{\mathrm{T}}_2(p||q) = 1_2 \cdot \left(\sum_{x\in[N]} \frac{p_x^2}{q_x} - 1 \right). \tag{4.57}$$

As briefly discussed in Section 4.8, the chi-square divergence plays a special role as a **local approximation** of other relative entropies.

Definition 4.10 Tsallis relative entropy of order $\alpha = \infty$, aka Tsallis max-relative entropy

The Tsallis relative entropy of order $\alpha = \infty$, also known as the Tsallis max-relative entropy, is given by

$$\mathrm{D}^{\mathrm{T}}_\infty(p||q) = \begin{cases} 0 & \text{if } p = q, \\ \infty & \text{otherwise.} \end{cases} \tag{4.58}$$

Thus, trivially, the Tsallis max-relative entropy produces an infinite value for any two non-identical distributions.

As a last note on the impact of the temperature on the Tsallis entropy, consider two complementary values of the order, namely α and $1-\alpha$, within the range $0 \leq \alpha < 1$. By the definition of the Tsallis relative entropy, the equality

$$\mathrm{D}^{\mathrm{T}}_\alpha(p||q) = \frac{\alpha}{1-\alpha}\mathrm{D}^{\mathrm{T}}_{1-\alpha}(q||p) \tag{4.59}$$

holds. This shows that, in the regime $0 \leq \alpha < 1$, switching the roles of p and q amounts essentially to a change in the inverse temperature used to evaluate the Tsallis relative entropy.

4.4.8 α-Relative Entropy

Finally, a quantity closely related to the Tsallis relative entropy is the following.

Definition 4.11 α-relative entropy

Given two $N \times 1$ distributions p and q, the α-relative entropy is defined as

$$\frac{1}{\alpha}\mathrm{D}_{\alpha}^{\mathrm{T}}(p||q). \tag{4.60}$$

Most of the properties of the α-relative entropy are inherited from the Tsallis relative entropy. That said, an interesting advantage of the division by α in (4.60) is that, with it, the α-relative entropy not only recovers the Shannon entropy $\mathrm{D}(p||q)$ for $\alpha = 1$, but it also recovers the Shannon entropy $\mathrm{D}(q||p)$ in the limit $\alpha \to 0$, i.e.,

$$\lim_{\alpha \to 0} \frac{1}{\alpha}\mathrm{D}_{\alpha}^{\mathrm{T}}(p||q) = \mathrm{D}(q||p). \tag{4.61}$$

4.4.9 Monotonicity of the Tsallis Entropy

As concluded in Section 4.3, the Shannon relative entropy supports the decomposition (4.25) of the Shannon entropy $\mathrm{H}(p)$ as the sum of the maximum entropy value $\mathrm{H}(u) = \log_2(N)$, attained for a uniform distribution u, and of the distance $\mathrm{D}(p||u)$ between the given distribution p and u. The decomposition (4.25) provides a formalization of the monotonicity property of the Shannon entropy. The Tsallis entropy affords a similar decomposition with the Tsallis relative entropy in lieu of the Shannon entropy.

Key Result 4.12 Monotonicity property of the Tsallis entropy

For any distribution p, the Tsallis entropy $\mathrm{H}_{\alpha}^{\mathrm{T}}(p)$ can be written as

$$\begin{aligned}\mathrm{H}_{\alpha}^{\mathrm{T}}(p) &= \mathrm{H}_{\alpha}^{\mathrm{T}}(u) - N^{1-\alpha}\mathrm{D}_{\alpha}^{\mathrm{T}}(p||u) \\ &= -\log_{2,\alpha}\left(\frac{1}{N}\right) - N^{1-\alpha}\mathrm{D}_{\alpha}^{\mathrm{T}}(p||u).\end{aligned} \tag{4.62}$$

The decomposition (4.62) of the Tsallis entropy bears the same interpretation as the decomposition (4.25) of the Shannon entropy. Specifically, as the Tsallis relative entropy $\mathrm{D}_{\alpha}^{\mathrm{T}}(p||u)$ decreases to its minimum value of zero, and thus as distribution p becomes closer to the uniform distribution u, the Tsallis entropy approaches the maximum value $\mathrm{H}(u) = -\log_{2,\alpha}(1/N)$. That said, comparing (4.62) and (4.25), one may note in (4.62) the multiplication of the relative entropy term $\mathrm{D}_{\alpha}^{\mathrm{T}}(p||u)$ by the constant $N^{1-\alpha}$, which does not appear in (4.25).

The decomposition (4.62) follows from the properties of the α-logarithm (see Appendix 3.A), which imply the equalities

$$\begin{aligned}\mathrm{D}_{\alpha}^{\mathrm{T}}(p||u) &= \mathrm{E}_p\left[\log_{2,\alpha}\left(\frac{p_{\mathrm{x}}}{1/N}\right)\right] \\ &= \mathrm{E}_p\left[N^{\alpha-1}\left(\log_{2,\alpha}(p_{\mathrm{x}}) - \log_{2,\alpha}\left(\frac{1}{N}\right)\right)\right] \\ &= N^{\alpha-1}\left(\mathrm{H}_{\alpha}^{\mathrm{T}}(p) - \log_{2,\alpha}\left(\frac{1}{N}\right)\right).\end{aligned} \tag{4.63}$$

Example 4.17

Consider again the distribution (4.27), which is recalled to tend to the minimum-entropy distribution δ_0 when $\gamma \to 0$, while recovering the uniform distribution u when $\gamma \to \infty$. Figure 4.8 plots the Tsallis entropy $\mathrm{H}^{\mathrm{T}}_{\alpha}(p)$ and the Tsallis relative entropy $\mathrm{D}^{\mathrm{T}}_{\alpha}(p||u)$ for $N = 10$ and for two values of the order α as representatives of the two intervals $0 \leq \alpha < 1$ and $\alpha \geq 1$. In both figure panels, the Tsallis entropy is observed to increase from the minimum value of 0 to the maximum value $-\log_{2,\alpha}(1/N)$ as γ grows larger. At the same time, by the decomposition (4.62), the Tsallis relative entropy $\mathrm{D}^{\mathrm{T}}_{\alpha}(p||u)$ decreases from the maximum value $-N^{\alpha-1}\log_{2,\alpha}(1/N)$ to the minimum value 0.

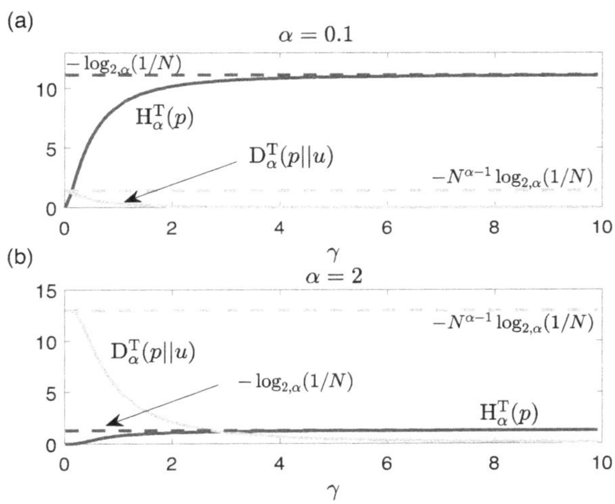

Figure 4.8 Tsallis entropy $\mathrm{H}^{\mathrm{T}}_{\alpha}(p)$ and Tsallis relative entropy $\mathrm{D}^{\mathrm{T}}_{\alpha}(p||u)$ for distribution (4.27) with $N = 10$ with (a) $\alpha = 0.1$ and (b) $\alpha = 2$.

As a final comment on this figure, while the Shannon entropy $\mathrm{H}(p)$ and the Shannon relative entropy $\mathrm{D}(p||u)$ take the same maximum values $\log_2(N)$, a look at the two figure panels demonstrates that this is not the case for the Tsallis entropy and the Tsallis relative entropy. In particular, the maximum value $-\log_{2,\alpha}(1/N)$ of the Tsallis entropy is larger than the maximum value $-N^{\alpha-1}\log_{2,\alpha}(1/N)$ for the Tsallis relative entropy when $0 \leq \alpha < 1$, while the opposite is true for $\alpha > 1$.

4.5 Fidelity ⏭

As discussed in Section 4.4, the Tsallis relative sum $\sum_{x\in[N]} p_x^{\alpha} q_x^{1-\alpha}$ in (4.44) is a measure of the similarity of the probability vectors p and q for $0 \leq \alpha < 1$, while it is a measure of dissimilarity for $\alpha > 1$. This section briefly elaborates on the Tsallis relative sum as a similarity metric by focusing on two distinct choices of the order α in the given range $0 \leq \alpha < 1$ that yield the **fidelity** and the **Chernoff fidelity**.

4.5.1 Defining the Fidelity

With $\alpha = 1/2$, the Tsallis relative sum (4.44) yields the fidelity between distributions p and q.

Definition 4.12 Fidelity
The fidelity between distributions p and q is defined as

$$\mathrm{F}(p, q) = \sum_{x \in [N]} \sqrt{p_x q_x}. \tag{4.64}$$

The fidelity can be interpreted as the inner product of the "square-root-probability" vectors $\sqrt{p} = [\sqrt{p_0}, \ldots, \sqrt{p_{N-1}}]^T$ and $\sqrt{q} = [\sqrt{q_0}, \ldots, \sqrt{q_{N-1}}]^T$, i.e.,

$$\mathrm{F}(p, q) = (\sqrt{p})^T \sqrt{q} = (\sqrt{q})^T \sqrt{p}. \tag{4.65}$$

Geometrically, vectors $\sqrt{p}$ and $\sqrt{q}$ lie on the surface of a hypersphere in the N-dimensional space with radius $||\sqrt{p}||_2^2 = ||\sqrt{q}||_2^2 = 1$. By its definition, the fidelity thus measures the **alignment** between the two vectors: the more aligned these two vectors are, the larger the fidelity $\mathrm{F}(p, q)$ is (see also Section 5.3.1).

Note that the fidelity $\mathrm{F}(p, q)$ is symmetric, and is related to the square Hellinger relative entropy (4.56) by the equality

$$\mathrm{D}^{\mathrm{T}}_{1/2}(p, q) = 1_2 \cdot 2(1 - \mathrm{F}(p, q)). \tag{4.66}$$

Therefore, a larger fidelity $\mathrm{F}(p, q)$ is equivalent to a smaller distance between the two distributions as measured by the Tsallis relative entropy $\mathrm{D}^{\mathrm{T}}_{1/2}(p, q)$. Specializing (4.49), the following property holds.

Key Result 4.13 Bounds on the fidelity
The fidelity satisfies the inequalities

$$\underbrace{0}_{\mathrm{supp}(p) \cap \mathrm{supp}(q) = \emptyset} \leq \mathrm{F}(p, q) \leq \underbrace{1}_{p=q}, \tag{4.67}$$

with equality $\mathrm{F}(p, q) = 0$ if p and q have disjoint supports and with equality $\mathrm{F}(p, q) = 1$ if and only if the distributions are identical.

4.5.2 Chernoff Fidelity

Another useful measure of similarity is obtained via the Tsallis relative sum by minimizing over the value of α in the range $0 \leq \alpha \leq 1$.

Definition 4.13 Chernoff fidelity
The Chernoff fidelity is defined as

$$\mathrm{F_C}(p,q) = \min_{\alpha:\, 0\leq\alpha\leq 1} \sum_{x\in[N]} p_x^{\alpha} q_x^{1-\alpha}. \tag{4.68}$$

By the definition (4.68), the metric $\mathrm{F_C}(p,q)$ finds the minimum similarity between the two probability vectors as measured by the Tsallis relative sum. In this regard, note that, unlike the Tsallis relative entropy, the Tsallis relative sum is not monotonic in α in the range $0 \leq \alpha \leq 1$. Therefore, the minimization in (4.68) is generally non-trivial.

By (4.49), the Chernoff fidelity satisfies the same property (4.67) as the fidelity. However, definition (4.68) implies the inequality

$$\mathrm{F_C}(p,q) \leq \mathrm{F}(p,q), \tag{4.69}$$

indicating that the Chernoff fidelity $\mathrm{F_C}(p,q)$ is generally a stricter measure of similarity as compared to the fidelity $\mathrm{F}(p,q)$.

The negative logarithm $-\log_2(\mathrm{F_C}(p,q))$ is known in information theory as the **Chernoff information**. The Chernoff information characterizes the asymptotic error incurred in discriminating between samples drawn from either one of the two distributions p and q. The problem of distinguishing between a single sample from either distribution will be covered in Section 4.7.

4.6 Rényi Relative Entropy and Monotonicity of the Rényi Entropy

In this section, the **Rényi relative entropy** is introduced as an alternative class of distance measures between two probability distributions that has properties similar to the Tsallis relative entropy. Specifically, like the Tsallis relative entropy, the Rényi relative entropy recovers the Shannon relative entropy as a special case, and is an increasing function of the order α. Furthermore, the Rényi relative entropy supports a decomposition of the Rényi entropy that formalizes the monotonicity property of the Rényi entropy.

4.6.1 Defining the Rényi Relative Entropy

The Rényi relative entropy is defined as follows.

Definition 4.14 Rényi relative entropy
The Rényi relative entropy of order $\alpha \geq 0$ with $\alpha \neq 1$ is defined as

$$\mathrm{D}_\alpha^{\mathrm{R}}(p||q) = \begin{cases} \frac{1}{\alpha-1}\log_2\left(\sum_{x\in[N]} p_x^{\alpha} q_x^{1-\alpha}\right) & \text{if } 0\leq\alpha<1 \text{ and } \mathrm{supp}(p)\bigcap\mathrm{supp}(p)\neq\emptyset, \\ & \text{or } \alpha>1 \text{ and } \mathrm{supp}(p)\subseteq\mathrm{supp}(q), \\ \infty & \text{otherwise,} \end{cases} \tag{4.70}$$

and, by continuity, with $\alpha = 1$, the Rényi relative entropy equals the Shannon relative entropy

$$\mathrm{D}_1^{\mathrm{R}}(p||q) = \mathrm{D}(p||q). \tag{4.71}$$

As described by (4.70), the Rényi relative entropy is finite under the same conditions as for the Tsallis relative entropy, with the only caveat that, when $0 \leq \alpha < 1$, one should also check that the support sets of the distributions are not disjoint. That is, even for low orders $0 \leq \alpha < 1$, the Rényi entropy takes an infinite value in the extreme case in which the two supports are disjoint.

4.6.2 Rényi Relative Entropy and Tsallis Relative Sum ⏭

As seen in Section 4.4.6, the Tsallis relative entropy can be written as a function of the Tsallis relative sum. In a similar manner, the Rényi relative entropy can also be introduced as a transformation of the Tsallis relative sum. Specifically, the Rényi relative entropy of order $\alpha \geq 0$, with $\alpha \neq 1$, applies the function

$$\frac{1}{(\alpha - 1)} \log_2(\cdot) \tag{4.72}$$

to the Tsallis relative sum of order α. The function (4.72) is the negative of the function (3.58) relating the Rényi entropy to the Tsallis sum (see Section 3.6.2), and is illustrated in Figure 4.7.

The figure can be used to explain why, for $0 \leq \alpha < 1$, the Rényi entropy takes an infinite value when the two supports are disjoint. In fact, the function (3.58) tends to infinity when $0 \leq \alpha < 1$ and the input tends to zero.

4.6.3 Properties of the Rényi Relative Entropy ⏭

The Rényi relative entropy satisfies the following properties.

Key Result 4.14 Non-negativity and minimum-value property of the Rényi relative entropy

The Rényi relative entropy is non-negative,

$$\mathrm{D}_\alpha^{\mathrm{R}}(p||q) \geq 0, \tag{4.73}$$

for all orders $\alpha \geq 0$. Furthermore, for all orders $\alpha > 0$, the Tsallis relative entropy equals zero if and only if the two distributions are equal, i.e., $p = q$.

The inequality (4.73) can be proved using the definition (4.70) with (4.49) and (4.50).

Furthermore, as one varies the order α, the Rényi relative entropy increases, as does the Tsallis relative entropy.

Key Result 4.15 The Rényi relative entropy increases with the order α
For any fixed probability vectors p and q, the Rényi relative entropy $\mathrm{D}_\alpha^{\mathrm{R}}(p||q)$ is an increasing function of the order $\alpha \geq 0$, i.e., the inequality

$$\mathrm{D}_{\alpha'}^{\mathrm{R}}(p||q) \geq \mathrm{D}_\alpha^{\mathrm{R}}(p||q) \tag{4.74}$$

holds for any two orders α' and α with $\alpha' \geq \alpha$.

Finally, the relation between the functions (4.53) and (4.72) illustrated in Figure 4.7 implies the following result.

Key Result 4.16 Rényi relative entropy vs. Tsallis relative entropy
When $0 \leq \alpha < 1$, the Rényi relative entropy is no smaller than the Tsallis relative entropy

$$\mathrm{D}_\alpha^{\mathrm{R}}(p||q) \geq \mathrm{D}_\alpha^{\mathrm{T}}(p||q); \tag{4.75}$$

with $\alpha = 1$, the two relative entropies are equal and they coincide with the Shannon relative entropy

$$\mathrm{D}_\alpha^{\mathrm{R}}(p||q) = \mathrm{D}_\alpha^{\mathrm{T}}(p||q) = \mathrm{D}(p||q); \tag{4.76}$$

while for $\alpha > 1$, the following inequality holds

$$\mathrm{D}_\alpha^{\mathrm{R}}(p||q) \leq \mathrm{D}_\alpha^{\mathrm{T}}(p||q). \tag{4.77}$$

Example 4.18

To illustrate the properties of the Rényi relative entropy, take again the case of a bit with generic distributions $p = [p_0, p_1]^T$ and $q = [q_0, q_1]^T$. For $\alpha \neq 1$, the Rényi relative entropy of order α, referred to as the **binary Rényi relative entropy**, is evaluated as

$$\begin{aligned} \mathrm{D}_\alpha^{\mathrm{R}}(p||q) &= \mathrm{d}_\alpha^{\mathrm{R}}(p_1||q_1) = \mathrm{d}_\alpha^{\mathrm{R}}(p_0||q_0) \\ &= \frac{1}{\alpha - 1} \log_2 \left(p_1^\alpha q_1^{1-\alpha} + (1 - p_1)^\alpha (1 - q_1)^{1-\alpha} \right). \end{aligned} \tag{4.78}$$

As shown in Figure 4.9 for $p_1 = 0.3$, the behaviors of the Tsallis and Rényi relative entropies in terms of their dependence on the probabilities and on the order α are quite similar. In particular, both relative entropies increase as q_1 moves away from $p_1 = 0.3$, remaining bounded for $0 \leq \alpha < 1$ and tending to infinity for $\alpha > 1$. Furthermore, the figure also illustrates how the Rényi relative entropy is larger than the Tsallis relative entropy for $0 \leq \alpha < 1$, while the opposite is true for $\alpha > 1$.

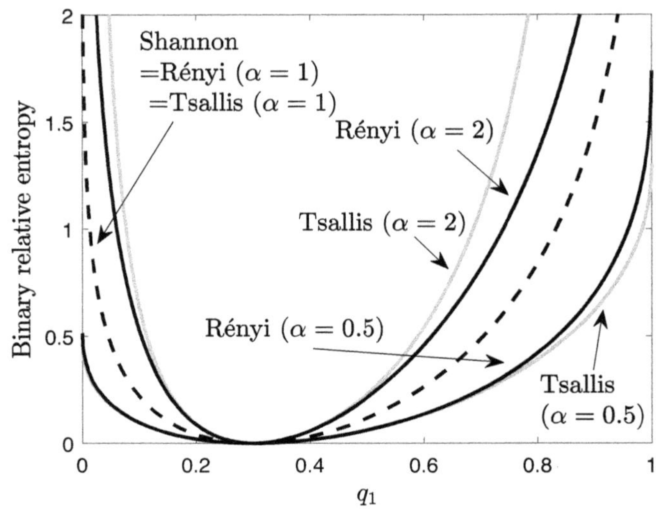

Figure 4.9 Binary Tsallis and Rényi relative entropies $d_\alpha^T(p_1||q_1)$ and $d_\alpha^R(p_1||q_1)$ as a function of q_1 for $p_1 = 0.3$.

4.6.4 Special Cases of the Rényi Relative Entropy ⏭

The following special cases of the Rényi relative entropy are worth highlighting, proceeding in order from the smallest to the largest. Note that the conditions for a finite value of the Rényi relative entropy are as in (4.70), and will not be made explicit in the following.

Definition 4.15 Rényi relative entropy of order $\alpha = 0$, aka Rényi min-relative entropy

The Rényi relative entropy of order $\alpha = 0$, also known as the Rényi min-relative entropy, is given by

$$D_0^R(p||q) = -\log_2\left(\sum_{x \in \text{supp}(p)} q_x\right). \tag{4.79}$$

In a manner analogous to the Tsallis min-relative entropy (4.54), the Rényi min-relative entropy equals zero as long as there are no hallucinations, i.e., as long as the support of distribution p covers that of distribution q, i.e., when $\text{supp}(q) \subseteq \text{supp}(p)$. However, unlike the Tsallis min-relative entropy, which is always finite, the Rényi min-relative entropy takes an infinite value when the supports of distributions p and q are disjoint.

Definition 4.16 Rényi relative entropy of order $\alpha = 0.5$
The Rényi relative entropy of order $\alpha = 0.5$ can be written as

$$\mathrm{D}^{\mathrm{R}}_{0.5}(p, q) = -2\log_2(\mathrm{F}(p, q)) \tag{4.80}$$

and is hence a decreasing function of the fidelity $\mathrm{F}(p, q)$ in (4.64).

Like the Tsallis relative entropy of order 1/2 in (4.56), the Rényi relative entropy is a *symmetric* function of the distribution p and q.

Definition 4.17 Rényi relative entropy of order $\alpha = 2$
The Rényi relative entropy of order $\alpha = 2$ is given by

$$\mathrm{D}^{\mathrm{R}}_{2}(p||q)=\log_2\left(\sum_{x\in[N]}\frac{p_x^2}{q_x}\right). \tag{4.81}$$

Definition 4.18 Rényi relative entropy of order $\alpha = \infty$, aka Rényi max-relative entropy
With $\alpha \to \infty$, the Rényi relative entropy, known as the Rényi max-relative entropy, is obtained as

$$\begin{aligned}\mathrm{D}^{\mathrm{R}}_{\infty}(p||q) &= \log_2\left(\max_{x\in[N]}\frac{p_x}{q_x}\right)\\ &= \max_{x\in[N]}\log_2\left(\frac{p_x}{q_x}\right).\end{aligned} \tag{4.82}$$

Accordingly, the Rényi max-relative entropy is the maximum value of the log-likelihood ratio $\log_2(p_x/q_x)$. By the monotonicity of the Rényi relative entropy with respect to the order α, it provides the maximum value across all Rényi relative entropies for a given pair of distributions p and q.

4.6.5 Monotonicity of the Rényi Entropy

Finally, the Rényi relative entropy is leveraged to formalize the monotonicity property of the Rényi entropy.

Key Result 4.17 Monotonicity property of the Rényi entropy
For any distribution p, the Rényi entropy $\mathrm{H}^{\mathrm{R}}_{\alpha}(p)$ can be written as

$$\mathrm{H}^{\mathrm{R}}_{\alpha}(p) = \log_2(N) - \mathrm{D}^{\mathrm{R}}_{\alpha}(p||u). \tag{4.83}$$

Generalizing the relation (4.25) for the Shannon entropy, the decomposition (4.83) indicates that the Rényi entropy increases toward its maximum value $\log_2(N)$ as the distribution p approaches the uniform distribution u, with distances measured via the Rényi relative entropy.

To prove the decomposition (4.83), it is sufficient to evaluate the Rényi relative entropy between distribution p and u as

$$\begin{aligned} \mathrm{D}_\alpha^{\mathrm{R}}(p||u) &= \frac{1}{\alpha-1}\log_2\left(N^{\alpha-1}\sum_{x\in[N]} p_x^\alpha\right) \\ &= \log_2(N) - \mathrm{H}_\alpha^{\mathrm{R}}(p). \end{aligned} \tag{4.84}$$

4.7 Optimal Detection, Relative Entropy, and Trace Distance

At the beginning of this chapter, the Shannon relative entropy was introduced by studying the informational tasks of mismatched prediction, compression, asking, or sampling. This section returns to the issue of interpreting relative entropies through the lens of informational tasks by studying the problem of detecting a sample as being generated from either one of two distributions. In the process of addressing this problem, another important distance measure between distributions, known as the trace distance or the total variation distance, will be introduced.

4.7.1 Optimal Detection

As illustrated in Figure 4.10, consider a detector that obtains a sample $\mathrm{x} \in [N]$, which may have been generated from distribution p, i.e., as $\mathrm{x} \sim p$, or from distribution q, i.e., as $\mathrm{x} \sim q$. The detector wishes to find out if the sampling distribution of x is p or q. Accordingly, for each input $\mathrm{x} = x$, the detector outputs a binary variable $d_x \in \{0, 1\}$, where

$$d_x = \begin{cases} 0 & \text{if the detector chooses distribution } q, \\ 1 & \text{if the detector chooses distribution } p. \end{cases} \tag{4.85}$$

A detector is thus specified by the $N \times 1$ vector $d = [d_0, d_1, \ldots, d_{N-1}]^T$, which describes the decisions of the detector for every input $x \in [N]$.

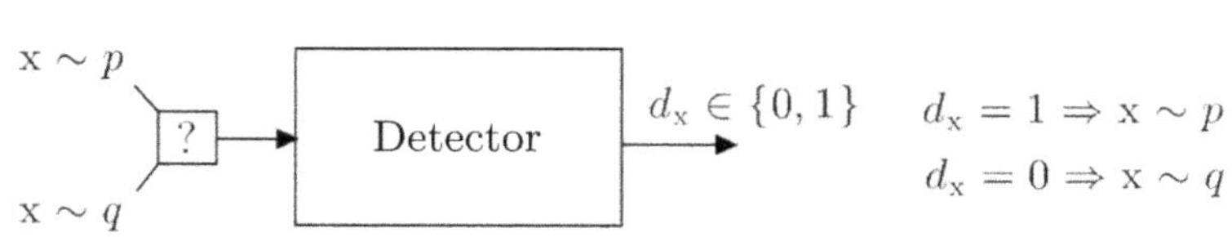

Figure 4.10 Illustration of the problem of optimal detection.

Example 4.19

Take first the simple case in which the supports of distributions p and q are disjoint. In this case, one can partition the set $[N]$ of possible observations into three distinct subsets:

- states x certainly generated from q, for which $p_x = 0$ and $q_x > 0$;
- states x certainly generated from p, for which $q_x = 0$ and $p_x > 0$; and
- impossible states, for which $p_x = q_x = 0$.

Note that the subset of impossible states may be empty.

A perfect detector d^* can hence be designed by simply setting $d_x^* = 0$ for the first subset of states; $d_x^* = 1$ for the second subset of states; and any arbitrary value $d_x^* \in \{0, 1\}$ for the third subset of states. Optimality follows since this detector *always* outputs the correct decision, irrespective of the observation x.

As an opposite situation, assume that the two distributions are equal, $p = q$. In this case, upon observing sample x, the detector does not acquire any useful information about the generating distribution, since any output x has the same probability of being generated under both distributions.

To optimize the detector, there are two, generally conflicting, performance criteria to consider. To elaborate, write as $\Pr_p[\cdot]$ and $\Pr_q[\cdot]$ the probabilities evaluated under distributions p and q, respectively. The first criterion is the **detection probability when the true distribution is** p, which is given by

$$\Pr_p[d_\mathrm{x} = 1] = \sum_{x \in [N]: d_x = 1} p_x = \sum_{x \in [N]} d_x p_x. \tag{4.86}$$

Maximizing this probability would call for a detection vector d with all entries equal to 1.

The second is the **detection probability when the true distribution is** q, i.e.,

$$\Pr_q[d_\mathrm{x} = 0] = \sum_{x \in [N]: d_x = 0} q_x = 1 - \sum_{x \in [N]} d_x q_x, \tag{4.87}$$

which is maximized by a detection vector d with all entries equal to 0.

Zero Detection Error Probability Under One Distribution. Suppose that, for the given application, it is very important not to miss any case in which the true distribution is p. Perhaps distribution p describes observations of critical phenomena such as a fire or an earthquake. To capture this requirement, one may formulate the design problem by imposing the constraint that the detection probability be equal to 1 when the true distribution is p, i.e., $\Pr_p[d_\mathrm{x} = 1] = 1$.

Under this constraint, one can then maximize the detection probability $\Pr_q[d_{\mathrm{x}} = 0]$ under the true distribution q as in the problem

$$\begin{aligned}&\underset{d}{\text{maximize}}\ \Pr_q[d_{\mathrm{x}} = 0]\\ &\text{subject to}\ \Pr_p[d_{\mathrm{x}} = 1] = 1.\end{aligned} \tag{4.88}$$

A solution to this problem is given by the detector

$$d_x^* = \begin{cases} 0 & \text{if } x \notin \text{supp}(p),\\ 1 & \text{if } x \in \text{supp}(p),\end{cases} \tag{4.89}$$

which chooses distribution p, by setting $d_x^* = 1$, for all states x that are possible under distribution p. For all other states x, in an attempt to maximize the detection probability $\Pr_q[d_{\mathrm{x}} = 0]$, the detector sets $d_x^* = 0$. Note that, accordingly, the optimal detector (4.89) depends only on distribution p, highlighting the asymmetry of the problem formulation (4.88) with respect to the roles of p and q.

Using (4.87), this yields the optimal value for the objective in problem (4.88) given by

$$\begin{aligned}\Pr_q[d_{\mathrm{x}}^* = 0] &= 1 - \sum_{x \in \text{supp}(p)} q_x\\ &= \frac{1}{1_2}\mathrm{D}_0^{\mathrm{T}}(p||q)\\ &= 1 - 2^{-\mathrm{D}_0^{\mathrm{R}}(p||q)}.\end{aligned} \tag{4.90}$$

This shows that the optimal detection probability under distribution q as per problem (4.88) increases with the Tsallis or Rényi min-relative entropies.

Note that the dependence of the optimal detection probability is *linear* with respect to the Tsallis min-relative entropy, and *exponential* with respect to the Rényi min-relative entropy. Furthermore, the asymmetry in the relative entropies reflects the different role played by the distribution p, which determines the detection probability constrained to be equal to 1.

By (4.90), the more "distant" the two distributions p and q are, the simpler it is to detect a sample drawn from either distribution. In particular, by (4.55), the optimal probability $\Pr_q[d_{\mathrm{x}}^* = 0]$ takes the minimum value of 0 whenever the support of distribution p includes that of distribution q. In fact, in this case, for all states x with $q_x > 0$, the optimal detector must set $d_x^* = 1$. Moreover, the maximum value $\Pr_q[d_{\mathrm{x}}^* = 0] = 1$ is attained when the two distributions have disjoint supports.

Average Detection Probability. For a different formulation of the optimal detection problem, assume that, a priori, the two distributions are equally likely to have generated x. The performance of a detector d is now measured by the probability that the detector correctly identifies the distribution of the input sample, on average with respect to the choice of the true distribution. The resulting detection probability of any detector d can be expressed as the average

$$\mathrm{P_S}(d) = \frac{1}{2}\mathrm{Pr}_p[d_{\mathrm{x}} = 1] + \frac{1}{2}\mathrm{Pr}_q[d_{\mathrm{x}} = 0] \tag{4.91}$$

of the detection probability $\mathrm{Pr}_p[d_{\mathrm{x}} = 1]$ obtained when the true distribution is p and of the detection probability $\mathrm{Pr}_q[d_{\mathrm{x}} = 0]$ obtained when the true distribution is q.

Accordingly, the design goal is to optimize the detector d by maximizing the detection probability as per the problem

$$\underset{d}{\text{maximize}}\ \mathrm{P_S}(d), \tag{4.92}$$

where the optimization is over all $N \times 1$ binary vectors d. The subscript in $\mathrm{P_S}(d)$ is a reminder that $\mathrm{P_S}(d)$ is the probability of a *successful* detection.

Plugging (4.86) and (4.87) into (4.91), the detection probability for any detector d can be expressed as

$$\mathrm{P_S}(d) = \frac{1}{2}\left(1 + \sum_{x\in[N]} d_x(p_x - q_x)\right). \tag{4.93}$$

Therefore, the design problem (4.92) can equivalently be written as

$$\underset{d}{\text{maximize}} \left\{ \mathrm{P_S}(d) = \frac{1}{2}\left(1 + \sum_{x\in[N]} d_x(p_x - q_x)\right)\right\}, \tag{4.94}$$

where maximization is over the space of $N \times 1$ binary vectors d.

By (4.94), an optimal detector satisfies the equalities (see Problem 4.8)

$$d_x = \begin{cases} 1 & \text{if } p_x > q_x, \\ 0 & \text{if } p_x < q_x. \end{cases} \tag{4.95}$$

When $p_x = q_x$, an optimal predictor can set either $d_x = 0$ or $d_x = 1$, since this choice does not affect the detection probability in (4.94). Therefore, without loss of generality, an optimal detector d^* can be defined as

$$d^* = \mathrm{step}(p - q), \tag{4.96}$$

where function $\mathrm{step}(z)$ returns 1 if $z \geq 0$ and 0 otherwise, and the function is applied element-wise to the input vector z.

4.7.2 Trace Distance

In Appendix 4.A, the expression (4.96) for the optimal detector d^* is leveraged to calculate the resulting maximal probability of detection $\mathrm{P_S}(d^*)$ in (4.93) as

$$\mathrm{P_S}(d^*) = \frac{1}{2}\left(1 + \mathrm{D^{tr}}(p, q)\right), \tag{4.97}$$

where the **trace distance**, or **total variation**, between distributions p and q is defined as follows.

Definition 4.19 Trace distance, aka total variation
For two distributions p and q, the trace distance, also known as the total variation, is defined as

$$\mathrm{D}^{\mathrm{tr}}(p,q) = \frac{1}{2}||p-q||_1$$
$$= \frac{1}{2}\sum_{x\in[N]} |p_x - q_x|, \quad (4.98)$$

where the 1-**norm** $||a||_1$ for a vector $a = [a_0, \ldots, a_{N-1}]^T$ is defined as $||a||_1 = \sum_{x\in[N]} |a_x|$.

4.7.3 Properties of the Trace Distance

As reflected by the notation $\mathrm{D}^{\mathrm{tr}}(p,q)$, the trace distance (4.98) is symmetric with respect to the two distributions p and q. Furthermore, it satisfies the following property.

Key Result 4.18 Bounds on the trace distance
The trace distance satisfies the inequalities

$$\underbrace{0}_{p=q} \leq \mathrm{D}^{\mathrm{tr}}(p,q) \leq \underbrace{1}_{\mathrm{supp}(p)\cap\mathrm{supp}(q)=\emptyset}, \quad (4.99)$$

where the lower bound, $\mathrm{D}^{\mathrm{tr}}(p,q) = 0$, is attained if and only if the distributions p and q are equal; while the upper bound, $\mathrm{D}^{\mathrm{tr}}(p,q) = 1$, is obtained if and only if p and q have disjoint supports.

Note that the boundedness of the trace distance is a property shared with the Tsallis relative entropy with orders $0 \leq \alpha < 1$ (see (4.41)). Furthermore, the bounds (4.99) can easily be interpreted by using the expression (4.97) of the optimal detection probability. In fact, as the trace distance $\mathrm{D}^{\mathrm{tr}}(p,q)$ increases from the minimum value 0 to the maximum value 1, the maximum detection probability $\mathrm{P_S}(d^*)$ decreases from the maximum value 1 to the minimum value 1/2. Therefore, the *maximum* detection probability $\mathrm{P_S}(d^*) = 1$ is attained when the two distributions have disjoint supports, and hence they can be perfectly distinguished based on the observation. In contrast, the *minimum* value of the detection probability, $\mathrm{P_S}(d^*) = 1/2$, is obtained when the two distributions are identical.

The trace distance is related to the Shannon relative entropy as follows.

> **Key Result 4.19 Pinsker inequality**
> The trace distance is related to the Shannon relative entropy between distributions p and q through the inequality
>
> $$\mathrm{D}^{\mathrm{tr}}(p,q) \leq \sqrt{\frac{1}{2\cdot 1_2}\mathrm{D}(p||q)}. \tag{4.100}$$

Pinsker's inequality (4.100) demonstrates that the Shannon relative entropy is a stricter, i.e., larger, measure of distance as compared to the squared trace distance, apart from a multiplicative factor. Please see the Recommended Resources for a proof.

By Pinsker's inequality (4.100), the maximum detection probability (4.97) can be upper bounded as

$$\mathrm{P_S}(d^*) \leq \frac{1}{2}\left(1+\sqrt{\frac{1}{2\cdot 1_2}\mathrm{D}(p||q)}\right). \tag{4.101}$$

Therefore, small values of the Shannon relative entropy $\mathrm{D}(p||q)$ entail that it is impossible to distinguish two distributions p and q with a large success probability. At values of the inverse temperatures in the range $0 \leq \alpha \leq 1$, bounds similar to the Pinsker inequality apply also to the Tsallis and Rényi relative entropies (see Recommended Resources).

Appendix 4.B develops a class of distances, known as **integral probability metrics**, that generalize the trace distance.

4.8 Unifying Relative Entropies via *f*-Divergences ⏭

This last section describes a unification and generalization of some of the distance measures between two distributions presented in this chapter, namely the Shannon relative entropy, the Tsallis relative entropy, and the trace distance. The unified framework is provided by ***f*-divergences**.

An f-divergence is defined as follows.

> **Definition 4.20 *f*-divergence**
> Fix any convex function $f(t)$ with the following properties: (i) $f(1) = 0$; and (ii) $0 \cdot f(0/0) = 0$. For any two distributions p and q, the f-divergence $\mathrm{D}_f(p||q)$ is defined as the expectation
>
> $$\mathrm{D}_f(p||q) = \mathrm{E}_q\left[f\left(\frac{p_{\mathrm{x}}}{q_{\mathrm{x}}}\right)\right]. \tag{4.102}$$

Table 4.4 Functions $f(t)$ yielding different distance measures as special cases of the f-divergence

Measure	Function
Shannon relative entropy	$f(t) = t\log_2(t)$
Tsallis relative entropy	$f(t) = \frac{1_2}{\alpha-1}(t^\alpha - 1)$
Trace distance	$f(t) = \frac{1}{2}\lvert t-1\rvert$

One can check that the Shannon relative entropy, the Tsallis divergence, and the trace distance are all f-divergences with the convex functions indicated in Table 4.4. The Rényi relative entropy is the only measure described in this chapter that is not an f-divergence, although it can be written as a monotonically increasing transformation of one, namely of the Tsallis relative entropy. This is elaborated on in Appendix 4.C, while Appendix 4.D describes another notable f-divergence, namely the Jensen-Shannon relative entropy.

The main advantage of the f-divergence formalism is that a number of properties can be stated for the general class of f-divergences without requiring a separate proof for each of the specific divergence measures. For instance, every f-divergence can be proved to be non-negative,

$$\mathrm{D}_f(p||q) \geq 0, \tag{4.103}$$

and to equal zero if the two distributions are equal (see Problem 4.11).

As another interesting general conclusion, an f-divergence satisfies the **joint convexity** property. This states that given any $N \times 1$ distributions p', p'', q', and q'', the inequality

$$\mathrm{D}_f(\gamma p' + (1-\gamma)p''||\gamma q' + (1-\gamma)q'') \leq \gamma \mathrm{D}_f(p'||q') + (1-\gamma)\mathrm{D}_f(p''||q'')$$

holds for any probability $0 \leq \gamma \leq 1$. This property implies the convexity of entropy measures derived from f-divergences using decompositions such as (4.25) and (4.62).

All f-divergences are well approximated by the chi-square relative entropy $\mathrm{D}_2^{\mathrm{T}}(p||q)$ when the two distributions p and q are close (see Recommended Resources for details).

Finally, while all f-divergences share basic properties such as non-negativity and joint convexity, they also have distinct characteristics. As an example, the trace distance satisfies the **triangular inequality**

$$\mathrm{D}^{\mathrm{tr}}(p,q) \leq \mathrm{D}^{\mathrm{tr}}(p,r) + \mathrm{D}^{\mathrm{tr}}(r,q) \tag{4.104}$$

for any three $N \times 1$ probability distributions p, q, and r, while this is not the case for the Shannon and Tsallis relative entropies.

4.9 Summary

- This chapter has introduced four main measures of distance between two probability distributions. All introduced measures are non-negative, and are equal to zero when the two distributions are equal.
- The Shannon relative entropy $\mathrm{D}(p||q)$ was defined as a measure of the regret for a predictor that uses distribution q as informational state when the true distribution is p.
- The Tsallis relative entropy $\mathrm{D}_\alpha^{\mathrm{T}}(p||q)$ and the Rényi relative entropy $\mathrm{D}_\alpha^{\mathrm{R}}(p||q)$ were seen to include as a special case the Shannon relative entropy, while providing more flexibility via the choice of the order $\alpha \geq 0$.
- In a way that is dual to the dependence of the entropy on the order α, the Tsallis and Rényi relative entropies tend to become more "severe," i.e., larger, as α increases. In particular, for $0 \leq \alpha < 1$, they return finite values even when the support of distribution q does not cover that of distribution p, which is not the case for the Shannon entropy.
- The Tsallis and Rényi relative entropies are both obtained as functions of the Tsallis relative sum $\sum_{x \in [N]} p_x^\alpha q_x^{1-\alpha}$, which bears the interpretation of similarity or dissimilarity between the two distributions depending on whether the order is in the range $0 \leq \alpha \leq 1$ or in the range $\alpha > 1$, respectively.
- Notably, for $\alpha = 1/2$, the Tsallis relative sum returns the fidelity $\mathrm{F}(p, q)$ between distributions p and q.
- Each of the three relative entropies was seen to support a formalization of the monotonicity property of the corresponding entropy, whereby a larger entropy is obtained as the distribution at hand becomes closer to the uniform distribution.
- The relative entropies are also related to the problem of optimal detection between two distributions.
- The solution of the optimal detection problem depends on the trace distance $\mathrm{D}^{\mathrm{tr}}(p, q)$ between two distributions, and hence, via the Pinsker inequality, also on the Shannon relative entropy.
- Shannon and Tsallis relative entropies, as well as the trace distance, can be recovered as special cases of the broader family of f-divergences.

4.10 Recommended Resources

Applications of information-theoretic measures in machine learning are explored in [1], with applications to cryptography discussed in [2]. Hallucinations in language models are analyzed in [3]. For details on α-relative entropy, see [4], and for coverage of Chernoff information, refer to [5, 6]. The Tsallis relative sum, referred to as the relative information generating function, is analyzed in [7]. Further insights into the properties of Rényi relative entropy can be found in [8], while extensive discussions on relative entropies and f-divergences are provided in [9]. The Pinsker inequality for Tsallis and Rényi relative entropies is presented in [10, Corollary 6].

Distance measures between distributions that, unlike the metrics discussed in this chapter, account for the specific mapping between states and the numerical values assigned to them are explored in the rich field of optimal transport [11]. A prominent example within this field is the family of Wasserstein distances.

PROBLEMS

4.1 Prove that the Shannon relative entropy is non-negative by using Jensen's inequality (see Appendix 2.D).

4.2 Argue that the equality (4.25) recovers the non-negativity and minimum-value property, as well as the maximum-value property, of the Shannon entropy.

4.3 (a) Prove the inequality $\mathrm{D}_\alpha^\mathrm{T}(p||q) \leq 1_2/(1-\alpha)$ for $0 \leq \alpha < 1$ (see (4.41)). *Hint:* Use the properties of the α-log reviewed in Appendix 3.A. (b) Prove that, for $0 \leq \alpha < 1$, the upper bound $\mathrm{D}_\alpha^\mathrm{T}(p||q) = 1_2/(1-\alpha)$ is attained if and only if the two distributions have disjoint support sets. (c) Show that, when $\alpha \geq 1$, the Tsallis relative entropy $\mathrm{D}_\alpha^\mathrm{T}(p||q)$ can diverge to infinity.

4.4 Prove the equivalence of the three expressions for the Tsallis relative entropy in (4.34)–(4.36).

4.5 (a) For $\alpha > 0$, prove that the Tsallis relative entropy is non-negative. *Hint:* Use the strict convexity of the function $(\alpha - 1)^{-1}x^\alpha$ as a function of variable $x \geq 0$ and Jensen's inequality. (b) For $\alpha > 0$, show also that the Tsallis relative entropy $\mathrm{D}_\alpha^\mathrm{T}(p||q)$ equals zero if and only if $p = q$. (c) For the special case $\alpha = 0$, use the expression (4.54) to show that the Tsallis relative entropy $\mathrm{D}_0^\mathrm{T}(p||q)$ is non-negative, and identify conditions under which it equals zero.

4.6 (a) For the distributions $p = [0, 1/3, 1/3, 1/3]^T$ and $q = [1/2, 1/2, 0, 0]^T$, evaluate the Tsallis relative entropy for $\alpha = 0.1$, $\alpha = 1$, and $\alpha = 10$. Explain your results in terms of the supports of the distributions. (b) Repeat for the distributions $p = [0, 1/3, 1/3, 1/3, 0]^T$ and $q = [1/2, 0, 0, 0, 1/2]^T$.

4.7 (a) For the distribution $p_x = \exp(-x/\gamma)/\sum_{x' \in [N]} \exp(-x'/\gamma)$ for $x \in [N]$ and $N = 10$, plot the Rényi entropy $\mathrm{H}_\alpha^\mathrm{R}(p)$ and the Rényi relative entropy $\mathrm{D}_\alpha^\mathrm{R}(p||u)$ for $\alpha = 0.1$ and for $\alpha = 2$ as a function of γ. (b) What are the values of the Rényi entropy $\mathrm{H}_\alpha^\mathrm{R}(p)$ and of the Rényi relative entropy $\mathrm{D}_\alpha^\mathrm{R}(p||u)$ when γ tends to zero and when γ goes to infinity?

4.8 Using the expression for the detection probability, $\mathrm{P_S}(d)$, in (4.94), prove that an optimal predictor is such that vector d has entries $d_x = 1$ for values of x with $p_x > q_x$ and $d_x = 0$ for values of x with $p_x < q_x$.

4.9 Show that the trace distance takes the maximum value 1 if the two distributions have disjoint supports.

4.10 (a) A sample x is generated from distribution p with probability γ and from probability q with probability $1 - \gamma$. Following the derivation in Appendix 4.A, show that the maximal probability of success can be expressed as

$$\mathrm{P_S}(d^*) = \frac{1}{2}(1 + ||\gamma p - (1-\gamma)q||_1). \tag{4.105}$$

(b) Show that the term $||p - \eta q||_1$ can be expressed in terms of the quantity $\sum_{x\in[N]}(p_x - \eta q_x)^+ - (1-\eta)^+$, which is known as the E_η**-divergence** for $\eta \geq 0$. (c) Show that the E_η-divergence is an f-divergence.

4.11 Prove that any f-divergence $\mathrm{D}_f(p||q)$ is non-negative. Furthermore, if function $f(\cdot)$ is strictly convex, the f-divergence $\mathrm{D}_f(p||q)$ is equal to zero if and only if the two distributions p and q are equal.

4.12 (a) Prove that, for $N \times 1$ probability vectors p and q with $\mathrm{supp}(p) \subseteq \mathrm{supp}(q)$, the Shannon relative entropy can be obtained as a result of the optimization

$$\mathrm{D}(p||q) = \max_r \sum_{x\in[N]} p_x \log_2\left(\frac{r_x}{q_x}\right), \tag{4.106}$$

where the maximization is over the space of $N \times 1$ probability distribution vectors r. *Hint:* Show the equality $\mathrm{D}(p||q) = \sum_{x\in[N]} p_x \log_2(r_x/q_x) + \mathrm{D}(p||r)$. (b) Using this result, prove the **Donsker–Varadhan variational representation** of the Shannon relative entropy

$$\mathrm{D}(p||q) = \max_g \mathrm{E}_p[g_\mathrm{x}] - \log_2\left(\mathrm{E}_q\left[2^{g_\mathrm{x}}\right]\right), \tag{4.107}$$

where the maximization is over all $N \times 1$ vectors g. *Hint:* Reparameterize each xth entry of the probability vector r in terms of vector q as $r_x = Z^{-1}q_x \cdot 2^{g_x}$, where $Z > 0$ is a suitably designed constant that ensures that vector r represents a probability distribution.

4.13 Show that an equivalent way to express the Rényi relative entropy of order ∞ is as $\mathrm{D}_\infty^\mathrm{R}(p||q) = \log_2(r^*)$, where

$$r^* = \min\{r > 0\colon r \cdot q_x - p_x \geq 0 \text{ for all } x \in [N]\}. \tag{4.108}$$

4.14 Prove that, given any two $N \times 1$ distributions p and q, the difference in expected values $|\mathrm{E}_p[f(\mathrm{x})] - \mathrm{E}_q[f(\mathrm{x})]|$ can be upper bounded by $2(\max_{x\in[N]} f(x))\mathrm{D}^{\mathrm{tr}}(p,q)$. Therefore, if the two distributions are close in terms of trace distance, all expectations under the two distributions are also approximately equal.

4.15 (a) Show that, for two different $N \times 1$ distribution vectors p and q, it is always possible to find a function o_x that makes the difference $\mathrm{E}_p[o_\mathrm{x}] - \mathrm{E}_q[o_\mathrm{x}]$ arbitrarily large. (b) Prove that, when the entries of vector $o = [o_0, \ldots, o_{N-1}]^T$ are limited in the interval $[-1, 1]$, the optimal solution is given by $o^* = \mathrm{sign}(p-q)$, where function $\mathrm{sign}(z)$ returns $+1$ if $z \geq 0$ and -1 otherwise, and the function is applied element-wise.

4.16 Using the inequality $\log_2(x) \leq 1_2 \cdot (x-1)$, show the inequality (4.75) for $0 \leq \alpha < 1$, as well as inequality (4.77) for $\alpha > 1$. *Hint:* Use the definitions of the two relative entropies.

Appendix

4.A Derivation of the Optimal Detection Probability

By the derived optimal detector (4.90), the maximal detection probability can be written, using (4.93), as

$$\begin{aligned} \mathrm{P_S}(d^*) &= \frac{1}{2}\left(1 + \sum_{x\in[N]} d^*_x(p_x - q_x)\right) \\ &= \frac{1}{2}\left(1 + \sum_{x:p_x>q_x} (p_x - q_x)\right) \\ &= \frac{1}{2}\left(1 + \sum_{x\in[N]} (p_x - q_x)^+\right), \end{aligned} \tag{4.109}$$

where $(a)^+$ equals the input a when $a > 0$, while we have $(a)^+ = 0$ otherwise.

Define the quantities $\Delta_x^+ = (p_x - q_x)^+$ that appear in (4.109), along with the related quantities $\Delta_x^- = (q_x - p_x)^+$, for $x \in [N]$. As can be directly verified, these quantities can be related to the probability vectors via the equalities

$$0 = \sum_{x\in[N]} (p_x - q_x) = \sum_{x\in[N]} (\Delta_x^+ - \Delta_x^-) \tag{4.110}$$

and

$$\sum_{x\in[N]} |p_x - q_x| = \sum_{x\in[N]} (\Delta_x^+ + \Delta_x^-). \tag{4.111}$$

The first result (4.110) implies the equality $\sum_{x\in[N]} \Delta_x^+ = \sum_{x\in[N]} \Delta_x^-$, which, in turn, when plugged into (4.111), yields

$$\sum_{x\in[N]} |p_x - q_x| = 2\sum_{x\in[N]} \Delta_x^+. \tag{4.112}$$

Hence, the optimal detection probability (4.109) is given as

$$\begin{aligned} \mathrm{P_S}(d^*) &= \frac{1}{2}\left(1 + \sum_{x\in[N]} \Delta_x^+\right) \\ &= \frac{1}{2}\left(1 + \frac{1}{2}\sum_{x\in[N]} |p_x - q_x|\right), \end{aligned} \tag{4.113}$$

concluding the proof.

4.B Integral Probability Metrics

Returning to the detection problem, suppose that, rather than making a single observation $\mathrm{x} \sim p$ or $\mathrm{x} \sim q$, the detector is allowed to make more observations, in a number large enough to enable the estimate of the expectation of some scalar function o_x of the observation x. Note that, for the problem to be practically interesting, one should also assume that the number of observations is not so large as to allow good estimates of the entire probability vector to be obtained. This may be indeed the case when the dimension N of the probability vector is large, since estimating a single scalar o_x can be done with fewer observations than estimating the entire distribution.

In this setting, the detector obtains the expectation $\mathrm{E}_p[o_\mathrm{x}]$ if the true distribution is p, and the expectation $\mathrm{E}_q[o_\mathrm{x}]$ if the true distribution is q. Unlike the single-observation problem studied in the text, in this case, unless the two distributions are identical, it is always possible to make sure that the two outputs $\mathrm{E}_p[o_\mathrm{x}]$ and $\mathrm{E}_q[o_\mathrm{x}]$ are distinct through a suitable choice of function o_x. For instance, the detector can ensure that $\mathrm{E}_p[o_\mathrm{x}]$ is positive, while $\mathrm{E}_q[o_\mathrm{x}]$ is negative.

In fact, one can even choose a function that makes the difference $\mathrm{E}_p[o_\mathrm{x}] - \mathrm{E}_q[o_\mathrm{x}]$ arbitrarily large (Problem 4.15). Imposing suitable constraints on function o, one may then be interested in the solution of the problem

$$\max_{o} \left\{ \mathrm{E}_p[o_\mathrm{x}] - \mathrm{E}_q[o_\mathrm{x}] = \sum_{x \in [N]} o_x (p_x - q_x) \right\} \tag{4.114}$$

as a measure of the distance between the two distributions. By varying the set of possible constraints on vector o, problem (4.114) defines a class of divergence measures known as **integral probability metrics**.

For instance, imposing the constraint that each entry of vector o be in the interval $[0, 1]$, following the same arguments put forth in the text, the optimal solution of the problem can be seen to be exactly $o^* = d^*$ in (4.96), and the optimal value of problem (4.114) equals the trace distance $\mathrm{D}^{\mathrm{tr}}(p, q)$.

4.C Rényi Relative Entropy and Tsallis Relative Entropy

For any fixed order $\alpha \geq 0$, the two relative entropies are related to one another through the same monotonically increasing functions $\mathrm{F}^{\mathrm{T}\to\mathrm{R}}_{2-\alpha}(\cdot)$ and $\mathrm{F}^{\mathrm{R}\to\mathrm{T}}_{2-\alpha}(\cdot)$ that were introduced in Appendix 3.C to relate the Rényi entropy and the Tsallis entropy. Specifically, the equalities

$$\mathrm{D}^{\mathrm{R}}_{\alpha}(p||q) = \mathrm{F}^{\mathrm{T}\to\mathrm{R}}_{2-\alpha}(\mathrm{D}^{\mathrm{T}}_{\alpha}(p||q)) \tag{4.115}$$

and

$$\mathrm{D}^{\mathrm{T}}_{\alpha}(p||q) = \mathrm{F}^{\mathrm{R}\to\mathrm{T}}_{2-\alpha}(\mathrm{D}^{\mathrm{R}}_{\alpha}(p||q)) \tag{4.116}$$

can be verified to hold.

4.D Jensen-Shannon Relative Entropy

Among other f-divergences, it is worth highlighting the **Jensen–Shannon relative entropy**, which may be viewed as a symmetric version of the Shannon relative entropy and is defined as

$$\mathrm{D}^{\mathrm{JS}}(p,q) = \frac{1}{2}\mathrm{D}\left(p \,\middle\|\, \frac{p+q}{2}\right) + \frac{1}{2}\mathrm{D}\left(q \,\middle\|\, \frac{p+q}{2}\right). \tag{4.117}$$

The Jensen–Shannon relative entropy (4.117) compares each distribution to an equal mixture of the two distributions, and then averages the result. The corresponding convex function in (4.102) is $f(t) = 0.5t\log_2(2t/(t+1)) + 0.5\log_2(2/(t+1))$.

A useful property of the Jensen–Shannon relative entropy is that, unlike the Shannon relative entropy, it is bounded, even when the support of distribution q does not cover that of distribution p. In fact, the Jensen–Shannon divergence is upper bounded by 1, i.e., $\mathrm{D}^{\mathrm{JS}}(p,q) \leq 1$, with the upper bound $\mathrm{D}^{\mathrm{JS}}(p,q) = 1$ being attained when the two distributions have disjoint support sets.

REFERENCES

[1] O. Simeone, *Machine Learning for Engineers*. Cambridge: Cambridge University Press, 2022.

[2] H. Tyagi and S. Watanabe, *Information-Theoretic Cryptography*. Cambridge: Cambridge University Press, 2023.

[3] S. Farquhar, J. Kossen, L. Kuhn, and Y. Gal, "Detecting hallucinations in large language models using semantic entropy," *Nature*, vol. 630, no. 8017, pp. 625–630, 2024.

[4] A. Cichocki and S.-i. Amari, "Families of alpha- beta- and gamma-divergences: Flexible and robust measures of similarities," *Entropy*, vol. 12, no. 6, pp. 1532–1568, 2010.

[5] T. M. Cover and J. A. Thomas, *Elements of Information Theory*. New York: Wiley, 2006.

[6] S. M. Kay, *Information-Theoretic Signal Processing and its Applications*. Independently published, 2020.

[7] S. Saha and S. Kayal, "General weighted information and relative information generating functions with properties," *IEEE Transactions on Information Theory*, vol. 70, no. 8, pp. 5498–5515, 2024.

[8] T. Van Erven and P. Harremos, "Rényi divergence and Kullback–Leibler divergence," *IEEE Transactions on Information Theory*, vol. 60, no. 7, pp. 3797–3820, 2014.

[9] Y. Polyanskiy and Y. Wu, *Information Theory: From Coding to Learning*. Cambridge: Cambridge University Press, 2024.

[10] G. L. Gilardoni, "On Pinsker's and Vajda's type inequalities for Csiszár's *f*-divergences," *IEEE Transactions on Information Theory*, vol. 56, no. 11, pp. 5377–5386, 2010.

[11] S. Chewi, J. Niles-Weed, and P. Rigollet, "Statistical optimal transport," *arXiv*, 2024. [Online]. Available at: https://arxiv.org/abs/2407.18163

Part II

Quantum Information

5 From Classical to Quantum Information

5.1 Overview

Part I of this book has explored the classical notions of information. At its foundation, classical information concerns the properties of probability distributions over the possible states of a system. These distributions may represent either the subjective knowledge of an observer about the system's state or the objective properties of the system itself, such as its level of disorder or its capacity to perform work. In both cases, classical information is inherently *physical*.

Regardless of whether the system is logical or physical, information about the system must be represented, stored, and communicated using physical resources – be it neural spikes in a biological brain, transistors in a memory chip, or electromagnetic waves in a communication link. Quantum information theory examines the implications of the physical embodiment of information in the microscopic world. For instance, information can be encoded in the quantum properties of elementary particles, such as the spin of an electron.

At microscopic scales, our current understanding of nature is governed by the laws of quantum physics. Accordingly, quantum information theory examines how these laws influence the representation, storage, and communication of information.

This initial chapter on quantum information elaborates on the differences between classical and quantum information by covering the formalism of quantum states and quantum measurements. A key takeaway is that quantum information introduces a fundamental form of uncertainty, absent in classical systems, which is referred to as quantum uncertainty. Quantum uncertainty arises from the distinctive properties of quantum states and quantum measurements, which differ profoundly from their classical counterparts.

To facilitate a clear comparison between classical and quantum settings, this chapter begins by reviewing classical uncertainty, adopting a notation that eases the transition to quantum concepts. It then introduces quantum states and measurements, highlighting uniquely quantum phenomena, such as contextuality. Finally, it addresses the challenge of distinguishing quantum states, demonstrating that, unlike classical systems, it is generally impossible to perform a measurement that unequivocally identifies the true state of a quantum system.

5.1.1 Learning Objectives and Organization

By the end of this chapter, the reader should be able to:

- list and appreciate the main differences between classical and quantum systems in terms of measurements and uncertainty;
- use the formalism of quantum states;
- define complete quantum measurements;
- apply the Born rule to evaluate the probability of the outcome of a complete quantum measurement;
- compute the post-measurement state corresponding to a given measurement output;
- and evaluate the performance of a complete quantum measurement aimed at distinguishing one state against another.

5.2 Classical vs. Quantum Information

This first section provides an introduction to the main differences between classical and quantum information. This discussion will serve as a reference for the rest of this part of the book, which will elaborate on each of the outlined distinctions. Throughout this discussion, a number of terms will be placed between quotation marks. These terms are part of the language of quantum physics, and their significance will become clearer in the following chapters.

5.2.1 Classical Information

To start, some key properties of classical information are reviewed from Part I. The reader is warned that all of the outlined classical characteristics appear at first glance to be self-evident, but they should be considered carefully, since quantum information, as will be seen shortly, violates all of them.

Property 1. No uncertainty for classical "pure" states
Consider a classical system that is known to be in a given state x within the set $[N]$ of possible states. One may describe this situation by saying that the system is in the "pure" state $x \in [N]$. In this case, the observer has no uncertainty on the system's state, and the corresponding informational state is described by a distribution $p = \delta_x$ concentrated at the value x. This indicates that the observer has confidence equal to 1 that the state of the system is x.

Example 5.1

A (trivial) example of a classical pure state is a box containing only gray balls. In this case, the ball drawn from the box is known to be gray, a state identified as $x = 0$. Therefore, the corresponding distribution vector is $p = \delta_0$.

Property 2. Uniqueness of classical "complete" measurements
When we observe a classical system, we merely retrieve information about its true state x. Following the terminology in quantum physics, one may refer to such conventional measurement of a system as being "complete," as it returns as much information as it could possibly produce about the system. In classical systems, there is clearly a single way to define a "complete" measurement, namely the measurement that returns the state x. In contrast, one may define multiple "partial" measurements that only retrieve some information about state x.

Example 5.2

An example of a partial measurement is the answer to a binary, yes/no, question about the state, such as "is x an even number?"

Property 3. Classical state invariance to "complete" measurements
If a classical system is in state $x \in [N]$, making an observation of the system retrieves the value x without changing it. For instance, evidently, if a ball is black, the act of observing it does not change its color to gray.

Property 4. Distinguishability of different classical states
If we are unsure about whether a classical system is in state $x' \in [N]$ or in state $x'' \in [N]$, with $x' \neq x''$, making a "complete" measurement of the system can always determine the correct state with no ambiguity. For instance, if we are unsure about whether a ball drawn from a box is black ($x' = 1$) or gray ($x'' = 0$), we can distinguish these two states by simply making a measurement of the ball's color.

5.2.2 Quantum Information

While self-evident for classical systems, the facts presented in Section 5.2.1 no longer hold for quantum systems. As this chapter will show, this implies that quantum systems are characterized by a new form of uncertainty, and, correspondingly, of information: **quantum uncertainty**. An informal statement of these properties is provided as follows. Note that the quotation marks are removed from the terms introduced in Section 5.2.1 given that their use is standard in quantum information theory.

Property 1. Quantum uncertainty for quantum pure states
Even when the state of a quantum system is known – i.e., even when the state of the quantum system is pure – the outcome of a measurement of the system is inherently uncertain.

Property 2. Non-uniqueness of complete quantum measurements
There are many distinct ways to implement complete measurements of a quantum system. For example, the spin of an electron, a binary quantity, can be measured along different axes.

Property 3. Quantum state transformation due to measurements
A measurement generally changes the state of a quantum system. One may say that a measurement collapses, or disturbs, the state of a quantum system. For instance, measuring the spin of an electron along some axis modifies the electron in an irreversible way.

Property 4. Indistinguishability of different quantum states
Distinct quantum states are generally not perfectly distinguishable. That is, given two arbitrary quantum states, there is no measurement that can perfectly distinguish the two states.

It is important to stress that these properties hold as long as one assumes the validity of quantum physics. Early critics of quantum physics, like Albert Einstein, believed that quantum uncertainty was not fundamentally different from classical uncertainty. In particular, they claimed that the properties listed above could be explained away by positing the presence of some unobserved, or latent, variables, dictating the behavior of the system. This way, quantum uncertainty would be understood as classical uncertainty caused by ignorance on the part of the observer about the hidden variables. Extensive experimental evidence based on the characterization of the possible behaviors allowed by the presence of hidden variables has, however, ruled out this hypothesis. Therefore, quantum physics remains the dominant physical theory for the explanation of natural phenomena at the microscopic scale.

5.2.3 Quantum Pure States vs. Quantum Mixed States

This chapter and the next study the case in which the state of the quantum system is known – i.e., the quantum state is pure. As anticipated by Property 1 above, in a classical system a pure state would trivially result in zero entropy measures, while this is not the case for quantum systems, which are characterized by a new form of uncertainty. Chapter 7 will investigate more general settings in which there is uncertainty on the preparation of the state of the system. In the latter case, the state is said to be mixed.

An illustration of the differences between classical and quantum states, as well as between pure and mixed states, is provided in Figure 5.1. As detailed in Section 5.4, like a classical informational state, a quantum pure state is described by an N-dimensional vector. This is shown in Figure 5.1 as an arrow within a space of $N = 2$ dimensions. Classical pure states are aligned with the horizontal and vertical axes of the space, while quantum pure states may follow an arbitrary direction.

As seen in the first row of the figure, for quantum pure states, the state of the system is known, and it corresponds to a single vector. Thus, *classical uncertainty* on the preparation of the state equals zero. However, *quantum uncertainty*, resulting from the inherent randomness of quantum measurements, is generally positive.

In contrast, as sketched in the lower part of Figure 5.1, a quantum mixed state is described by an ensemble of pure states, each with an assigned probability of being the true state. Therefore, quantum mixed states are characterized by both quantum uncertainty and classical uncertainty. Classical uncertainty describes the uncertainty about the identity of the true pure state within the ensemble. As will be seen in Chapter 7, given the indistinguishability of quantum states (Property 4), the contributions of classical and quantum uncertainties to the overall uncertainty associated with a mixed state cannot be isolated in general.

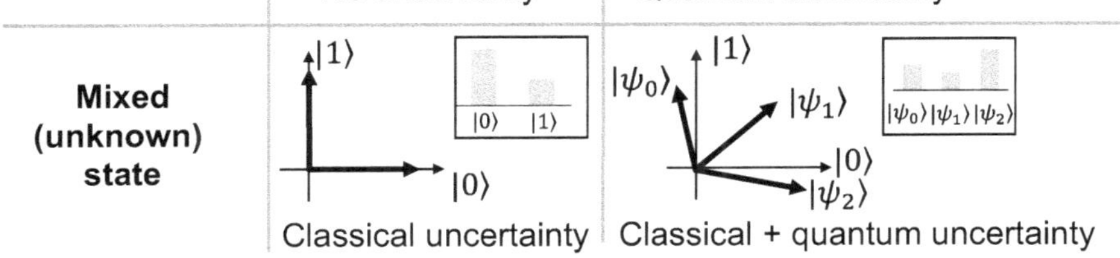

Figure 5.1 Illustration of the differences between classical and quantum states, as well as between pure and mixed states.

5.2.4 What Is a Quantum State?

As a final note, it is worth reflecting on the significance of a quantum "state." As discussed in Chapter 1, classical systems allow for a clear distinction between the true state of a system and the informational state of an observer. However, such a distinction does not apply to quantum pure states. In quantum physics, a pure state serves a dual role: it provides the only valid representation of the system's true state, while simultaneously acting as an informational state that enables an observer to determine the probability distribution of outcomes for any measurement.

The real nature of quantum information, however, remains a topic of active debate, engaging generations of physicists and philosophers alike. This is also known as the **reality problem**. Central to this discussion is the interpretation of quantum pure states, whether as subjective, objective, or even relational entities (see Recommended Resources). This book does not address the interpretation problem of quantum states, as it lies beyond the scope of quantum information theory.

5.3 Reviewing Classical Uncertainty through Dirac's Notation

This section revisits the description of a classical system with the main goal of introducing **Dirac's notation**, the standard mathematical language of quantum physics. To start, some basic definitions from linear algebra are introduced.

5.3.1 Transposition, Inner Product, and Norm

Throughout this book, the set of all $N \times 1$ vectors with **complex** entries will be generically referred to as the N-dimensional vector space, or more simply as the **N-dimensional space**.

Given N-dimensional vectors $v = [v_0, \ldots, v_{N-1}]^T$ and $u = [u_0, \ldots, u_{N-1}]^T$ in the N-dimensional space, the **inner product** between v and u is defined as

$$v^\dagger u = \sum_{x \in [N]} v_x^* u_x, \tag{5.1}$$

where $(\cdot)^*$ represents the complex conjugate operation and the $1 \times N$ vector

$$v^\dagger = [v_0^*, \ldots, v_{N-1}^*] \tag{5.2}$$

is the **Hermitian transpose** of vector v. The Hermitian transpose (5.2) can also be interpreted as the concatenation of conventional transposition $(\cdot)^T$ and complex conjugation $(\cdot)^*$, i.e.,

$$v^\dagger = (v^*)^T = (v^T)^*, \tag{5.3}$$

where the complex conjugate is applied entry-wise. Note that, by (5.3), one can apply conjugation and transposition in any order.

Geometrically, the inner product evaluates the degree of **alignment** of the two vectors. In particular, if

$$v^\dagger u = 0, \tag{5.4}$$

which is also equivalent to the equality $u^\dagger v = 0$, the vectors v and u are said to be **orthogonal**. The condition (5.4) indicates that the angle between the two vectors v and u equals $\pi/2$ radians.

The 2-**norm** – sometimes referred to just as **norm** for short – of an $N \times 1$ vector v is a special case of the α-norm introduced in Chapter 3 (see (3.23)), and is defined as

$$||v||_2 = \sqrt{v^\dagger v} = \sqrt{\sum_{x \in [N]} |v_x|^2}. \tag{5.5}$$

The 2-norm represents the length of vector v.

A vector v is said to be **normalized** if its norm is equal to 1, i.e., $||v||_2 = 1$. A normalized vector v lies on the surface of a **hypersphere** of radius 1 in the N-dimensional space.

5.3.2 Transposition for Matrices

For future reference, the **Hermitian transpose** operation is generalized to any $M \times N$ matrix A as follows. Denote as $A_{x,y}$ the (x, y)**th entry** of matrix A, with $x = 0$ indicating the first row, $x = 1$ the second row, and so on; while index $y = 0$ indicates the first column, $y = 1$ the second column, and so on:

$$A = \begin{bmatrix} A_{0,0} & A_{0,1} & \cdots & A_{0,N-1} \\ A_{1,0} & A_{1,1} & \cdots & A_{1,N-1} \\ \vdots & \vdots & \ddots & \vdots \\ A_{M-1,0} & A_{M-1,1} & \cdots & A_{M-1,N-1} \end{bmatrix}. \tag{5.6}$$

The transpose of an $M \times N$ matrix A is the $N \times M$ matrix A^T with entries $[A^T]_{x,y} = A_{y,x}$. Furthermore, the Hermitian transpose of an $M \times N$ matrix A is the $N \times M$ matrix

$$A^\dagger = (A^*)^T = (A^T)^*, \tag{5.7}$$

where the complex conjugate operation is again applied entry-wise.

The following properties of the Hermitian transposition are often useful.

Key Result 5.1 Hermitian transposition of a matrix product

For any two matrices B and C for which the product BC is well defined, the equality

$$(BC)^\dagger = C^\dagger B^\dagger \tag{5.8}$$

holds.

Key Result 5.2 Linearity of the Hermitian transposition
The Hermitian transposition is linear, in the sense that the equality

$$\left(\sum_x A_x\right)^\dagger = \sum_x A_x^\dagger \tag{5.9}$$

holds for any collection of $N \times N$ matrices $\{A_x\}$.

5.3.3 Basis

Definition 5.1 Basis, amplitudes, and superposition
A set of N vectors $\{u_x\}_{x\in[N]}$, all of dimension $N \times 1$, is said to be a **basis** for the N-dimensional space if every $N \times 1$ vector v can be expressed as a linear combination of the vectors $\{u_x\}_{x\in[N]}$, i.e., if the equality

$$v = \sum_{x\in[N]} \gamma_x u_x \tag{5.10}$$

holds for some, generally complex, numbers $\{\gamma_x\}_{x\in[N]}$. The number γ_x is also known as the **amplitude** of vector x with respect to basis vector u_x. Given a decomposition for the form (5.10), one may say that vector v is a **superposition** of all vectors $\{u_x\}_{x\in[N]}$ for which the amplitude γ_x is not equal to zero.

Definition 5.2 Orthogonal and orthonormal bases
Any set of N vectors $\{u_x\}_{x\in[N]}$ of dimension $N \times 1$ that are mutually orthogonal, i.e.,

$$u_x^\dagger u_y = 0 \text{ for all } x \neq y \text{ with } x, y \in [N], \tag{5.11}$$

form a basis, which is referred to as an **orthogonal basis**. Furthermore, if the vectors $\{u_x\}_{x\in[N]}$ are also normalized, i.e.,

$$||u_x||_2 = 1 \text{ for all } x \in [N], \tag{5.12}$$

the basis is said to be **orthonormal**.

Given an orthonormal basis, the amplitudes $\{\gamma_x\}_{x\in[N]}$ in the decomposition (5.10) can be evaluated by computing the inner product between the corresponding basis vector u_x and v, i.e.,

$$\gamma_x = u_x^\dagger v. \tag{5.13}$$

In fact, given the decomposition (5.10), one obtains

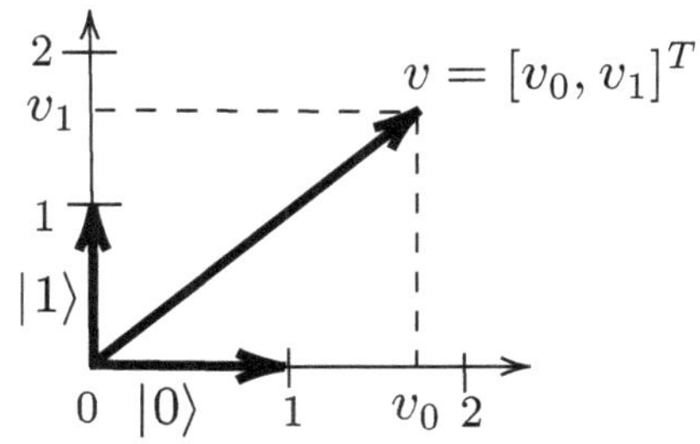

Figure 5.2 Illustration of the computational basis $\{|0\rangle, |1\rangle\}$ for the two-dimensional space. Any vector v in the space can be expressed as the superposition (5.21) with the two amplitudes v_0 and v_1 obtained by projecting vector v onto the directions specified by the basis vectors $|0\rangle$ and $|1\rangle$, respectively.

$$u_x^\dagger v = u_x^\dagger \left(\sum_{y \in [N]} \gamma_y u_y \right) = \sum_{y \in [N]} \gamma_y u_x^\dagger u_y = \gamma_x, \tag{5.14}$$

where the last equality follows from the normalization and mutual orthogonality of the vectors $\{u_x\}_{x \in [N]}$. The operation (5.13) can be interpreted geometrically as the orthogonal projection of vector v into the direction defined by the basis vector u_x.

5.3.4 Ket Notation

The central element of the Dirac notation is the **ket**.

Definition 5.3 Ket
A ket vector, or ket, is an $N \times 1$ vector with generally complex-valued entries that is normalized. A ket is denoted as

$$|\text{descriptor}\rangle, \tag{5.15}$$

with the descriptor of the state surrounded by a straight line on the left and a triangular parenthesis on the right. Being normalized, a ket satisfies the requirement

$$|||\text{descriptor}\rangle||_2 = 1. \tag{5.16}$$

As discussed next, kets can be used to describe classical pure states. The playful reason for the name "ket" will be made clear later.

5.3.5 Classical Pure States as Kets

As seen in Part I, a classical system, such as a gray or black ball, can be in a given number, N, of possible distinguishable states, collected in the set $[N] = \{0, 1, \ldots, N-1\}$. The states are distinguishable in the sense that a measurement of the system returns the true state $x \in [N]$ of the system with no ambiguity. When the state of the system $x \in [N]$ is known, the corresponding informational state, which is said to be pure, is described by the **zero-entropy** distribution $p = \delta_x$.

Zero-entropy probability distributions δ_x for any $x \in [N]$ can be viewed as kets, since they are normalized. They are one-hot vectors with all zero entries, except for a 1 in position x. In Dirac's notation, the $N \times 1$ vector δ_x is denoted as

$$\delta_x = |x\rangle, \tag{5.17}$$

where the descriptor is given by the corresponding known state x, and thus by the position of the entry equal to 1 in the vector.

Example 5.3

For $N = 2$, the two possible pure states in set $[N] = \{0, 1\}$ are represented via the 2×1 kets listed in Table 5.1.

Table 5.1 The two possible classical pure states for a two-dimensional classical system ($N = 2$) represented via 2 x 1 kets

State	0	1
State vector	$\|0\rangle = \begin{bmatrix} 1 \\ 0 \end{bmatrix}$	$\|1\rangle = \begin{bmatrix} 0 \\ 1 \end{bmatrix}$

Generalizing the example, the N states in set $[N] = \{0, 1, \ldots, N-1\}$ are represented via the kets in Table 5.2.

5.3.6 Computational Basis

The set $\{|x\rangle\}_{x\in[N]}$ of kets corresponding to all classical pure states $x \in [N]$ forms an orthonormal basis for the N-dimensional vector space in which they reside. This set consists of N vectors of dimension $N \times 1$ that are normalized, i.e.,

Table 5.2 The N possible classical pure states for an N-dimensional classical system represented via N x 1 kets

State	0	1	...	$N-1$
State vector	$\|0\rangle = \begin{bmatrix} 1 \\ 0 \\ \vdots \\ 0 \end{bmatrix}$	$\|1\rangle = \begin{bmatrix} 0 \\ 1 \\ \vdots \\ 0 \end{bmatrix}$	...	$\|N-1\rangle = \begin{bmatrix} 0 \\ 0 \\ \vdots \\ 1 \end{bmatrix}$

$$|||x\rangle||_2^2 = |x\rangle^\dagger |x\rangle = 1, \tag{5.18}$$

and mutually orthogonal, i.e.,

$$|x\rangle^\dagger |y\rangle = |y\rangle^\dagger |x\rangle = 0, \tag{5.19}$$

for all $x \neq y$ and $x, y \in [N]$. This yields the following definition.

Definition 5.4 Computational basis
The set of ket vectors

$$\{|x\rangle\}_{x\in[N]} \tag{5.20}$$

in Table 5.2 forms an orthonormal basis for the N-dimensional space, which is known as the computational basis.

The name "computational" is common in quantum computing, and it relates to the role of the basis (5.20) in defining the reference coordinates for quantum algorithms (see Recommended Resources).

Example 5.4

Consider the $N = 2$-dimensional vector space, for which Figure 5.2 shows the vectors $|0\rangle$ and $|1\rangle$ in the computational basis. These vectors have length equal to 1, and hence they are normalized. Furthermore, they are aligned with the two axes of the plane, and thus they are orthogonal, with the angle between them equal to $\pi/2$. Therefore, they form an orthonormal basis.

By the definition of a basis, any vector $v = [v_0, v_1]^T$ on the plane can be expressed via the two amplitudes v_0 and v_1 obtained by projecting vector v onto the two directions specified by vectors $|0\rangle$ and $|1\rangle$, i.e.,

$$v = v_0|0\rangle + v_1|1\rangle \tag{5.21}$$

with $v_0 = |0\rangle^\dagger v$ and $v_1 = |1\rangle^\dagger v$.

Generalizing the example above, any N-dimensional vector v can be expressed as a linear combination, or superposition, of the vectors in the computational basis as

$$\begin{aligned} v &= \begin{bmatrix} v_0 \\ v_1 \\ \vdots \\ v_{N-1} \end{bmatrix} \\ &= v_0 \begin{bmatrix} 1 \\ 0 \\ \vdots \\ 0 \end{bmatrix} + v_1 \begin{bmatrix} 0 \\ 1 \\ \vdots \\ 0 \end{bmatrix} + \cdots + v_{N-1} \begin{bmatrix} 0 \\ 0 \\ \vdots \\ 1 \end{bmatrix} \\ &= v_0|0\rangle + v_1|1\rangle + \cdots + v_{N-1}|N-1\rangle \\ &= \sum_{x\in[N]} v_x|x\rangle, \end{aligned} \tag{5.22}$$

where each xth amplitude is obtained via the inner product (5.14) as

$$v_x = |x\rangle^\dagger v. \tag{5.23}$$

5.4 Quantum Pure States

As seen in Section 5.3, a classical pure state corresponds to a zero-entropy distribution $p = \delta_x$, which can be written in Dirac's notation as $p = \delta_x = |x\rangle$. This section will introduce the formalism of quantum pure states, which generalize classical pure states by encompassing all kets, i.e., all complex-valued normalized vectors in the N-dimensional space.

5.4.1 Quantum Pure States

In a similar way to classical systems, complete measurements on a quantum system return one out of N possible values for some system-specific integer N. The integer N is known as the **dimension** of the quantum system, and it depends on the specific features of the physical system under study.

For individual quantum systems, such as the spin of an electron or the polarization of a photon, the value of N can be modest, e.g., $N = 2$ in the case of the electron spin. In contrast, for composite quantum systems, the dimension N grows exponentially with the number of subsystems. In particular, for a system consisting of S subsystems, in which a complete measurement of one subsystem returns one of D possible values, the dimension of the overall system equals

$$N = D^S. \tag{5.24}$$

For instance, for a system of S interacting electrons, the state vector of the spins is of dimension 2^S (see also Section 1.3.3).

When $N = 2$, the quantum system is also referred to as a **qubit**; with $N = 3$ as a **qutrit**; and with $N = 4$ as a **ququart**. More generally, with any N, a quantum state is also known as a **qudit**.

Formally, for an N-dimensional quantum system, a quantum pure state is described by an $N \times 1$ normalized complex-valued vector.

Definition 5.5 Quantum pure state

For an N-dimensional quantum system, a quantum pure state is described by an N-dimensional complex-valued vector

$$|\psi\rangle = \begin{bmatrix} \psi_0 \\ \psi_1 \\ \vdots \\ \psi_{N-1} \end{bmatrix} \tag{5.25}$$

that is normalized, i.e.,

$$|||\psi\rangle||_2^2 = \sum_{x\in[N]} |\psi_x|^2 = 1. \tag{5.26}$$

The state vector $|\psi\rangle$ is also known as the **waveform** of the system.

By this definition, all kets are valid quantum pure states.

Example 5.5

The vectors in the computational basis $\{|x\rangle\}_{x\in[N]}$ represent valid quantum pure states, as they are kets and thus they are normalized.

Any quantum pure state (5.27) can be written as the superposition of states in the computational basis

$$|\psi\rangle = \sum_{x\in[N]} \psi_x |x\rangle. \tag{5.27}$$

It is useful to pause to emphasize the physical significance of quantum pure states. In classical physics, a system in state x is described by a scalar x. In contrast, in quantum physics, despite the fact that the measurement output is still an integer $x \in [N]$, the state of the system *must* be described via an N-dimensional complex-valued vector. The idea that a system producing scalar output requires a vector space to be modeled is a significant conceptual leap. In this regard, the additional dimensions required by quantum physics are common to other physical theories such as general relativity or string theory (see Recommended Resources).

5.4.2 Bra and Inner Product

As discussed in Section 5.3, the inner product plays a central element in vector spaces. This subsection shows how to express the inner product using Dirac's notation.

Definition 5.6 Bra
The Hermitian transpose $|\psi\rangle^\dagger$ of a quantum pure state $|\psi\rangle$ is denoted as

$$\langle\psi| = |\psi\rangle^\dagger. \tag{5.28}$$

The $1 \times N$ vector $\langle\psi|$ in (5.28) is known as the bra vector, or bra, corresponding to the ket $|\psi\rangle$.

With this notation, the inner product (5.1) between two ket vectors $|\psi\rangle$ and $|\phi\rangle$ can be expressed as

$$|\psi\rangle^\dagger|\phi\rangle = \langle\psi|\cdot|\phi\rangle = \langle\psi|\phi\rangle, \tag{5.29}$$

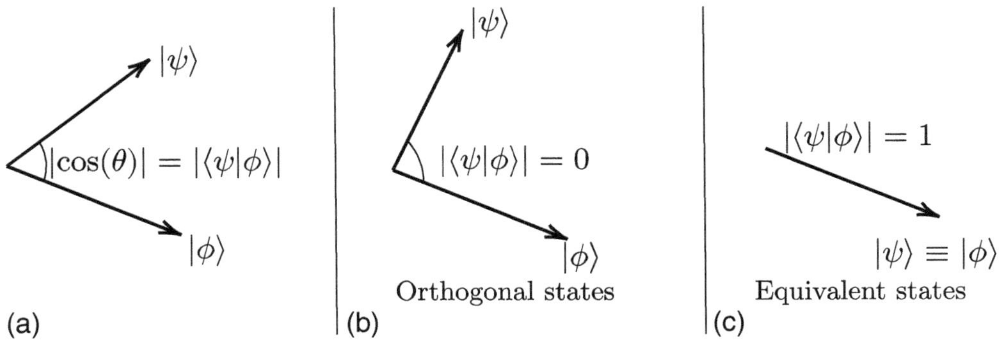

Figure 5.3 Illustration of the geometric significance of the absolute value $|\langle\phi|\psi\rangle| = |\langle\psi|\phi\rangle|$ of the inner product between two quantum pure states $|\psi\rangle$ and $|\phi\rangle$ as the cosine of the angle between the two vectors (a). Orthogonal states have an inner product equal to zero, $\langle\phi|\psi\rangle = 0$ (b). Equivalent states, satisfying the condition (5.32), are perfectly aligned, $|\langle\phi|\psi\rangle| = 1$ (c).

where the latter expression, which drops the second vertical bar for brevity, is known as **Dirac's bra-ket notation** for the inner product (completing the pun).

The space of N-dimensional vectors, when coupled with the inner product, is referred to in quantum physics as the **N-dimensional Hilbert space**. Accordingly, physicists may say that a quantum pure state belongs to an N-dimensional Hilbert space. A key conceptual leap in quantum physics is that from reasoning in the so-called **phase space** of the possible state values to reasoning in the Hilbert space of N-dimensional vectors.

Geometrically, as illustrated in Figure 5.3, the quantity $|\langle\phi|\psi\rangle| = |\langle\psi|\phi\rangle|$ of the inner product can be interpreted as the absolute value of the cosine of the angle θ between the states $|\psi\rangle$ and $|\phi\rangle$, i.e.,

$$|\langle\phi|\psi\rangle| = |\cos(\theta)|. \tag{5.30}$$

The quantity $|\langle\phi|\psi\rangle|$ is thus the magnitude of the orthogonal projection of vector $|\psi\rangle$ into the direction defined by vector $|\phi\rangle$; or, equivalently, the magnitude of the orthogonal projection of vector $|\phi\rangle$ into the direction defined by vector $|\psi\rangle$.

As will be discussed later in this chapter, the squared absolute value $|\langle\phi|\psi\rangle|^2$ is a common measure of similarity of two states, which is known as fidelity. In line with the geometric interpretation (5.30), the following bounds hold.

Key Result 5.3 Bounds on the absolute value of the inner product
The magnitude of the inner product $|\langle\phi|\psi\rangle|$ satisfies the inequality

$$\underbrace{0}_{\text{orthogonal}} \leq |\langle\phi|\psi\rangle| \leq \underbrace{1}_{\text{equivalent}}, \tag{5.31}$$

with equality $|\langle\phi|\psi\rangle| = \langle\phi|\psi\rangle = 0$ if and only if states $|\psi\rangle$ and $|\phi\rangle$ are orthogonal, and equality $|\langle\phi|\psi\rangle| = 1$ if and only if the equality

$$|\psi\rangle = \exp(i\gamma)|\phi\rangle \tag{5.32}$$

holds for some real number γ. The number γ is known as a **global phase** as it is applied to all the entries of vector $|\psi\rangle$. When condition (5.32) is satisfied, the states $|\psi\rangle$ and $|\phi\rangle$ are said to be **equivalent**.

Figure 5.3 illustrates the two extreme conditions identified by the bounds (5.31). When the absolute value $|\langle\phi|\psi\rangle|$ equals zero, the angle between the two states equals $\pi/2$ and the two states are orthogonal. In contrast, when the absolute value $|\langle\phi|\psi\rangle|$ takes the maximum value of 1, the two states are perfectly aligned. In this case, as per (5.32), the two states differ only for a global phase factor $\exp(i\gamma)$. In this regard, note that the alignment condition must preserve the normalization of the states, and thus the proportionality constant $\exp(i\gamma)$ has absolute value equal to 1, i.e., $|\exp(i\gamma)| = 1$.

The inner product is not symmetric, but it is **Hermitian symmetric**, in the sense that the inner product $\langle\psi|\phi\rangle$ equals the complex conjugate of the inner product $\langle\phi|\psi\rangle$ with the roles of the two vectors reversed, i.e.,

$$(\langle\psi|\phi\rangle)^* = \langle\phi|\psi\rangle. \tag{5.33}$$

Finally, using Dirac's notation, by (5.23), the xth amplitude ψ_x of a quantum pure state (5.25) can be expressed as

$$\psi_x = \langle x|\psi\rangle. \tag{5.34}$$

5.4.3 Beyond the Computational Basis

In Section 5.4.1, quantum pure states were introduced as the superposition (5.25) of vectors in the computational basis. This section elaborates on the fact that, since a quantum pure state $|\psi\rangle$ is an $N \times 1$ vector, it can also be written as a linear combination of vectors of any other orthonormal basis for the N-dimensional space. As discussed in Section 5.5, the freedom of selecting a basis has important implications in the definition and study of quantum measurements.

Example 5.6

Take the case of a qubit, that is, of a quantum system with two-dimensional states, i.e., with $N = 2$. Beyond the computational basis $\{|0\rangle, |1\rangle\}$, a useful orthonormal basis is given by the Hadamard basis.

Definition 5.7 Hadamard basis for qubits

The two-dimensional vectors

$$|+\rangle = \frac{1}{\sqrt{2}}(|0\rangle + |1\rangle) = \frac{1}{\sqrt{2}}\begin{bmatrix} 1 \\ 1 \end{bmatrix} \tag{5.35}$$

and

$$|-\rangle = \frac{1}{\sqrt{2}}(|0\rangle - |1\rangle) = \frac{1}{\sqrt{2}}\begin{bmatrix} 1 \\ -1 \end{bmatrix} \tag{5.36}$$

form an orthonormal basis that is known as the Hadamard basis.

With the Hadamard basis $\{|+\rangle, |-\rangle\}$, any qubit state $|\psi\rangle$ can be expressed as the superposition

$$|\psi\rangle = \begin{bmatrix} \psi_0 \\ \psi_1 \end{bmatrix} = \psi_0|0\rangle + \psi_1|1\rangle$$
$$= \frac{1}{\sqrt{2}}(\psi_0 + \psi_1)|+\rangle + \frac{1}{\sqrt{2}}(\psi_0 - \psi_1)|-\rangle. \quad (5.37)$$

This can easily be checked by computing the amplitudes (5.14) via the inner products

$$\langle +|\psi\rangle = \frac{1}{\sqrt{2}}(\psi_0 + \psi_1) \quad (5.38)$$

and

$$\langle -|\psi\rangle = \frac{1}{\sqrt{2}}(\psi_0 - \psi_1). \quad (5.39)$$

Figure 5.4 illustrates the relation between the amplitudes $\langle 0|\psi\rangle$ and $\langle 1|\psi\rangle$ calculated with respect to the computational basis and the amplitudes $\langle +|\psi\rangle$ and $\langle -|\psi\rangle$ evaluated using the Hadamard basis.

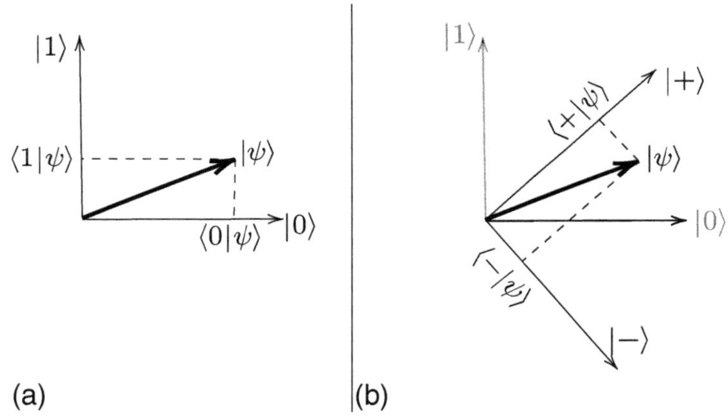

Figure 5.4 A qubit state $|\psi\rangle$ can be expressed as the superposition of states in any orthonormal basis of the two-dimensional space. (a) The example of the computational basis $\{|0\rangle, |1\rangle\}$, and (b) the Hadamard basis $\{|+\rangle, |-\rangle\}$.

Generalizing Example 5.6, given an orthonormal basis $\{|v_x\rangle\}_{x\in[N]}$, any quantum state $|\psi\rangle$ can be expressed as the superposition of state vectors $\{|v_x\rangle\}_{x\in[N]}$ as

$$|\psi\rangle = \sum_{x\in[N]} \alpha_x|v_x\rangle, \quad (5.40)$$

where each, generally complex, amplitude α_x is obtained by (5.14) via the inner product

$$\alpha_x = \langle v_x|\psi\rangle \quad (5.41)$$

for all $x \in [N]$.

The normalization condition (5.26) is equivalent to the equality

$$|||\psi\rangle||_2^2 = \sum_{x\in[N]} |\alpha_x|^2 = 1 \tag{5.42}$$

and thus the vector of amplitudes, $\alpha = [\alpha_0, \ldots, \alpha_{N-1}]^T$, must be normalized for any orthonormal basis, i.e., $||\alpha||_2 = 1$.

5.4.4 Unitary Matrices and Orthonormal Bases

Orthonormal bases can be constructed as the columns of a specific type of matrix.

Definition 5.8 Unitary matrix
An $N \times N$ matrix V is said to be unitary if it satisfies the equalities

$$V^\dagger V = VV^\dagger = I, \tag{5.43}$$

where I is the $N \times N$ identity matrix. This condition equivalently states that matrix V is invertible, with inverse matrix V^{-1} given by its Hermitian transpose, i.e.,

$$V^{-1} = V^\dagger. \tag{5.44}$$

The main reason for introducing unitary matrices here is the following.

Key Result 5.4 Orthonormal basis from unitary matrix
Given any $N \times N$ unitary matrix

$$V = [|v_0\rangle, |v_1\rangle, \ldots, |v_{N-1}\rangle], \tag{5.45}$$

the columns $\{|v_x\rangle\}_{x\in[N]}$ form an orthonormal basis.

This can be verified directly from the definition (5.43), which implies that the columns of matrix V are normalized and mutually orthogonal.

Example 5.7

The Hadamard-basis vectors (5.35) and (5.36) are the columns of the 2×2 **Hadamard matrix**, which is defined as

$$H = \frac{1}{\sqrt{2}}\begin{bmatrix} 1 & 1 \\ 1 & -1 \end{bmatrix}. \tag{5.46}$$

The Hadamard matrix satisfies the condition (5.43) as

$$H^\dagger H = HH^\dagger = I, \tag{5.47}$$

where I is the 2×2 identity matrix.

Example 5.8

An important class of 2×2 unitary matrices is given by the so-called Pauli matrices, which are defined as follows.

Definition 5.9 Pauli matrices
The **Pauli X matrix** is defined as

$$P_1 = \begin{bmatrix} 0 & 1 \\ 1 & 0 \end{bmatrix}; \tag{5.48}$$

the **Pauli Y matrix** is defined as

$$P_2 = \begin{bmatrix} 0 & -i \\ i & 0 \end{bmatrix}; \tag{5.49}$$

and the **Pauli Z matrix** is defined as

$$P_3 = \begin{bmatrix} 1 & 0 \\ 0 & -1 \end{bmatrix}. \tag{5.50}$$

It is common to include in the set of Pauli matrices also the 2×2 identity matrix, which is accordingly also written as

$$P_0 = I = \begin{bmatrix} 1 & 0 \\ 0 & 1 \end{bmatrix}. \tag{5.51}$$

Overall, the set of Pauli matrices is given by

$$\{P_0, P_1, P_2, P_3\}. \tag{5.52}$$

One can readily verify that all four Pauli matrices are indeed unitary, as they satisfy condition (5.43). For all four matrices, the columns yield basis vectors equivalent to the computational basis.

Example 5.9

The Hadamard basis $\{|+\rangle, |-\rangle\}$ in (5.35) and (5.36) can be generalized from qubits to qudits, i.e., from $N = 2$ to any $N > 2$. This can be done by defining a class of $N \times N$ unitary matrices that reduce to the 2×2 Hadamard matrix when $N = 2$.

An $N \times N$ Hadamard matrix is any matrix H with entries having magnitude equal to $1/\sqrt{N}$ and arbitrary phases, i.e.,

$$H_{x,y} = \frac{1}{\sqrt{N}} \exp(i\theta_{x,y}), \tag{5.53}$$

for phases $\{\theta_{x,y}\}_{x,y \in [N]}$ selected such that the unitarity condition (5.43), i.e., $H^\dagger H = HH^\dagger = I$, is satisfied. One can check that the 2×2 Hadamard matrix H in (5.46) is indeed a Hadamard matrix according to the more general definition (5.53). The columns of a Hadamard matrix form an orthonormal basis that is referred to as a Hadamard basis.

Two specific classes of Hadamard matrices, the Walsh matrices and Fourier matrices, both of which reduce to the 2×2 Hadamard matrix (5.46) for $N = 2$, are reviewed next.

The **Walsh matrix** is a Hadamard matrix with phases in (5.53) equal to either $\theta_{x,y} = 0$ or $\theta_{x,y} = \pi$. Specifically, the $N \times N$ Walsh matrix W_N is defined recursively as $W_1 = 1$ and

$$W_N = \frac{1}{\sqrt{2}} \begin{bmatrix} W_{N-1} & W_{N-1} \\ W_{N-1} & -W_{N-1} \end{bmatrix}. \tag{5.54}$$

Note that the construction (5.54) only applies to dimensions of the form $N = 2^n$ for some integer n. For example, setting $n = 2$ and $N = 2^2 = 4$, the 4×4 Walsh matrix is given by

$$W_4 = \frac{1}{2} \begin{bmatrix} 1 & 1 & 1 & 1 \\ 1 & -1 & 1 & -1 \\ 1 & 1 & -1 & -1 \\ 1 & -1 & -1 & 1 \end{bmatrix}. \tag{5.55}$$

Another class of Hadamard matrices is given by the **Fourier matrix**, which is a Hadamard matrix (5.53) with phases $\theta_{x,y} = 2\pi xy/N$. In particular, an $N \times N$ Fourier matrix is defined as

$$F_N = [|f_0\rangle |f_1\rangle \cdots |f_{N-1}\rangle], \tag{5.56}$$

with $N \times 1$ columns $\{|f_x\rangle\}_{x \in [N]}$, forming the **Fourier basis**, given as

$$\begin{aligned} |f_x\rangle &= \frac{1}{\sqrt{N}} \sum_{y \in [N]} |y\rangle \exp\left(i\frac{2\pi}{N}xy\right) \\ &= \frac{1}{\sqrt{N}} \begin{bmatrix} 1 \\ \exp\left(i\frac{2\pi}{N}x\right) \\ \vdots \\ \exp\left(i\frac{2\pi}{N}x(N-1)\right) \end{bmatrix}. \end{aligned} \tag{5.57}$$

5.4.5 Resolution of the Identity

The decomposition (5.40) and (5.41) can be expressed more succinctly as

$$|\psi\rangle = \sum_{x \in [N]} \langle v_x|\psi\rangle |v_x\rangle. \tag{5.58}$$

Equivalently, moving the scalar $\langle v_x|\psi\rangle$ to the right of the ket $|v_x\rangle$, the linear combination (5.58) can be written as

$$|\psi\rangle = \sum_{x\in[N]} |v_x\rangle\langle v_x|\psi\rangle$$
$$= \left(\sum_{x\in[N]} |v_x\rangle\langle v_x|\right)|\psi\rangle. \quad (5.59)$$

This in turn implies the following result.

Key Result 5.5 Resolution-of-the-identity equality
Given any orthonormal basis $\{|v_x\rangle\}_{x\in[N]}$, the following equality holds

$$\sum_{x\in[N]} |v_x\rangle\langle v_x| = I, \quad (5.60)$$

where I is the $N \times N$ identity matrix, which is referred to as the resolution-of-the-identity equality.

5.5 Complete Quantum Measurements

Building on the background on quantum states presented in the previous sections, this section introduces complete quantum measurements. As will be discussed, unlike classical systems, a quantum measurement has the following properties.

- **Quantum uncertainty:** The outcome of a complete quantum measurement is inherently uncertain, even when the system is in a pure, and hence known, state (see Property 1 in Section 5.2.2).
- **Non-uniqueness of complete quantum measurements:** There is an infinite number of ways to carry out a complete quantum measurement (see Property 2 in Section 5.2.2).
- **Disturbance to the quantum state:** Measuring quantum systems generally changes the state of the system (see Property 3 in Section 5.2.2).

5.5.1 Measurement Setting

As illustrated in Figure 5.5, a complete measurement of an N-dimensional quantum system returns a classical output, described by a random variable $\mathrm{x} \in [N]$, while leaving the quantum system in a, generally different, post-measurement state. This high-level description of a complete measurement has the following two important implications.

- **Classical information as output:** The information extracted by an observer via a measurement is classical, amounting to the scalar $x \in [N]$.
- **Limited information about the state:** The observer can only make *one* measurement on a given state $|\psi\rangle$. In fact, the post-measurement state is generally distinct from the original state $|\psi\rangle$, which is no longer accessible after the measurement. This

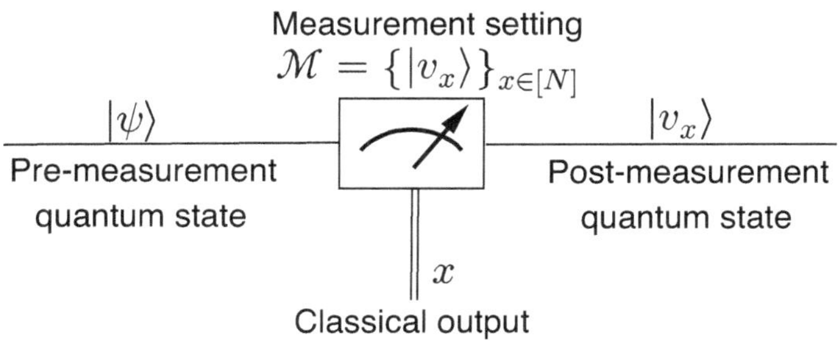

Figure 5.5 A complete measurement of an N-dimensional quantum system returns a classical output, described by a random variable $x \in [N]$, while leaving the quantum system in a, generally different, post-measurement state. The measurement is fully specified by the measurement basis $\mathcal{M}$, which defines the measurement setting. The double line in the figure represents a classical variable.

limits the amount of information that can be extracted by an observer about a quantum state $|\psi\rangle$.

In order to specify a complete measurement, it is necessary to define a frame of reference along which to probe the pre-measurement state $|\psi\rangle$ of the system.

Definition 5.10 Measurement setting
A measurement setting is defined by an orthonormal basis for the N-dimensional vector space

$$\mathcal{M} = \{|v_x\rangle\}_{x\in[N]}, \tag{5.61}$$

which is also known as the **measurement basis**.

An orthonormal basis thus fully specifies a complete measurement, with different bases generally corresponding to distinct measurements.

Example 5.10

The most typical measurement setting is given by the computational basis, which corresponds to the choice of the measurement basis $\mathcal{M} = \{|x\rangle\}_{x\in[N]}$. This is also known as a **standard measurement**.

5.5.2 Classical Output

Given a measurement setting $\mathcal{M}$ and a pre-measurement pure state $|\psi\rangle$, the **classical output** of a complete measurement is a random variable x that returns value $x \in [N]$ with probability denoted as

$$p_x(|\psi\rangle|\mathcal{M}) = \Pr_{|\psi\rangle}[\mathrm{x} = x|\mathcal{M}]. \tag{5.62}$$

The notation $\Pr_{|\psi\rangle}[\cdot|\mathcal{M}]$ in (5.62) indicates that the system being measured is in state $|\psi\rangle$ and that the measurement is made in setting $\mathcal{M}$. Accordingly, the classical output is the random variable

$$\mathrm{x} \sim p(|\psi\rangle|\mathcal{M}) \tag{5.63}$$

with $N \times 1$ **measurement probability distribution** defined as

$$p(|\psi\rangle|\mathcal{M}) = \begin{bmatrix} p_0(|\psi\rangle|\mathcal{M}) \\ p_1(|\psi\rangle|\mathcal{M}) \\ \vdots \\ p_{N-1}(|\psi\rangle|\mathcal{M}) \end{bmatrix}. \tag{5.64}$$

One of the key postulates in quantum mechanics is that the measurement probabilities $p_x(|\psi\rangle|\mathcal{M})$ are given by the Born rule.

Definition 5.11 Born rule
Given a measurement setting $\mathcal{M}$ and the pre-measurement state $|\psi\rangle$, the measurement probability is obtained as

$$p_x(|\psi\rangle|\mathcal{M}) = |\langle v_x|\psi\rangle|^2 \tag{5.65}$$

for all possible classical outputs $x \in [N]$.

According to Born's rule (5.65), the measurement probability $p_x(|\psi\rangle|\mathcal{M})$ equals the squared magnitude of the amplitude (5.41) of vector $|\psi\rangle$ with respect to the basis vector $|v_x\rangle$ corresponding to the output $x \in [N]$.

To see that the Born rule (5.65) defines a valid probability distribution, observe that the sum

$$\begin{aligned} \sum_{x\in[N]} p_x(|\psi\rangle|\mathcal{M}) &= \sum_{x\in[N]} \langle\psi|v_x\rangle\langle v_x|\psi\rangle \\ &= \langle\psi|\left(\sum_{x\in[N]} |v_x\rangle\langle v_x|\right)|\psi\rangle \\ &= \langle\psi|\psi\rangle = 1 \end{aligned} \tag{5.66}$$

holds, where the first equality follows from (5.33) and the third from the resolution-of-the-identity equality (5.60).

Example 5.11

The two rows of Figure 5.6 depict the same pre-measurement qubit state $|\psi\rangle$, along with two different measurement bases $\mathcal{M} = \{|v_0\rangle, |v_1\rangle\}$. As discussed, changing the measurement basis modifies the measurement setting, yielding different measurement probability distributions (5.65).

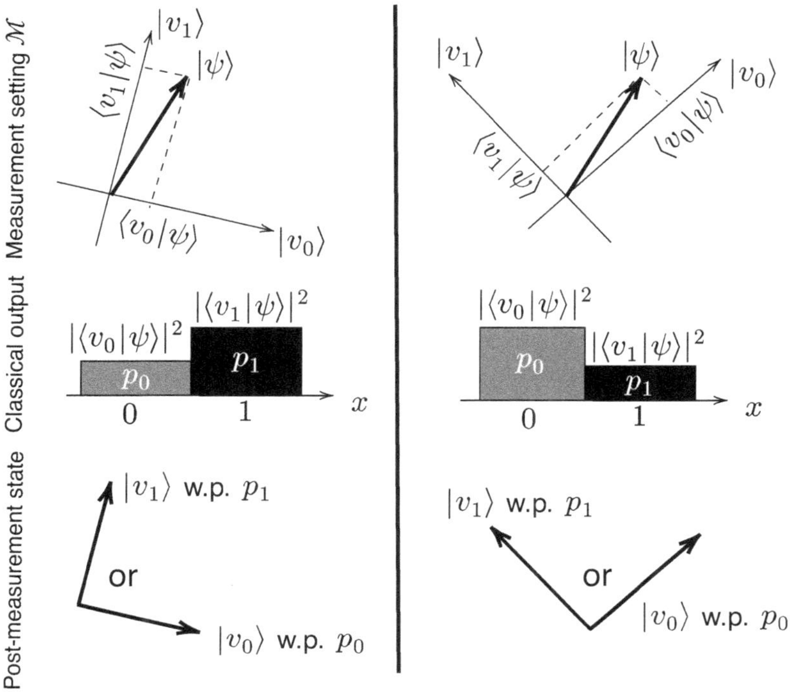

Figure 5.6 The two columns illustrate distinct measurement bases, along with a pre-measurement qubit state $|\psi\rangle$. The top row shows the measurement bases $\mathcal{M} = \{|v_0\rangle, |v_1\rangle\}$, while the middle and bottom rows depict the measurement probabilities $p(|\psi\rangle|\mathcal{M})$ and the post-measurement states, respectively ("w.p." stands for "with probability").

Specifically, by the Born rule (5.65), the probability $p_x(|\psi\rangle|\mathcal{M})$ of producing the classical output $\mathrm{x} = x$ depends on how aligned the state $|\psi\rangle$ is to the measurement basis vector $|v_x\rangle$. As shown in the second row, for the measurement basis illustrated in the first column, the state $|\psi\rangle$ is more aligned with basis vector v_1, and thus the measurement probability $p_1(|\psi\rangle|\mathcal{M})$ is larger than $p_0(|\psi\rangle|\mathcal{M})$. The opposite is true for the measurement basis in the second column, in which the probability $p_0(|\psi\rangle|\mathcal{M})$ is larger, since the cosine of the angle between state $|\psi\rangle$ and basis vector v_0 is larger. The last row will be discussed in Section 5.5.3.

The following remarks on the Born rule are in order.

- **Informational state or true state:** In classical systems, a probability vector p represents the informational state of a system that exists in a true but typically unknown state $x \in [N]$. This true state is revealed through a complete measurement. In quantum systems, a quantum pure state $|\psi\rangle$ describes the known state of the system, analogous to a classical pure state $|x\rangle$. However, unlike in classical systems, a complete measurement of a quantum system does not directly reveal the state $|\psi\rangle$. Instead, the observer obtains a single outcome $x \in [N]$, sampled from the probability distribution $p(|\psi\rangle|\mathcal{M})$, which depends on the state $|\psi\rangle$ and the chosen measurement $\mathcal{M}$.

- **Contextuality:** The measurement of a quantum system cannot be considered independently of the context in which the measurement is performed, i.e., of the measurement setting. In fact, by the Born rule, the measurement probability distribution is dictated by the measurement basis $\mathcal{M}$. This property is known as **contextuality**, and is a key feature of quantum information. When viewed as a function of the measurement setting $\mathcal{M}$, the distribution $p(|\psi\rangle|\mathcal{M})$ is referred to as the **behavior** of state $|\psi\rangle$.
- **Equivalent states and global phase:** By the Born rule, equivalent states – defined as in (5.32) – yield the same measurement probabilities. In fact, the measurement probabilities (5.65) do not depend on the global phase γ in (5.32), as one has the equality

$$p_x(\exp(i\gamma)|\psi\rangle|\mathcal{M}) = |\langle v_x|\psi\rangle|^2 = p_x(|\psi\rangle|\mathcal{M}) \tag{5.67}$$

for all measurement settings $\mathcal{M} = \{|v_x\rangle\}_{x\in[N]}$, irrespective of the value of the global phase γ. Equivalent states produce the same behavior, and are thus impossible to distinguish via any measurement. For this reason, throughout this book, equivalent states will be denoted with an equality sign, i.e.,

$$|\psi\rangle = |\psi\rangle \exp(i\gamma). \tag{5.68}$$

For instance, one may write $|0\rangle = -|0\rangle$ for a global phase $\gamma = \pi$.
- **Quantum uncertainty:** The measurement of a quantum pure state does not provide any new information about the quantum state, since the quantum state is known to the observer prior to the measurement (at least conceptually – see next bullet point). Accordingly, the uncertainty associated with the measurement of a quantum system does not reflect the ignorance of the observer about the *present*, pre-measurement, state of the system, but only about the *future* classical output of the measurement and about the corresponding post-measurement state of the system. In this sense, quantum uncertainty may be viewed as a "*new sort of ignorance about something that doesn't yet have a truth value, something that simply isn't one way or the other before we measure, something that requires us to act and do something that we call a measurement before nature supplies a truth value*" (see Recommended Resources for the source of this quote).
- **Informational vs. computational uncertainty:** It is important to distinguish the *informational perspective* adopted in this book from the *computational framework* within which quantum algorithms are designed in quantum computing. To illustrate, consider a quantum computer operating on a quantum system of S qubits, which transforms an initial state into a final state. From an informational viewpoint, the transformed state of the qubit system is, in principle, knowable. This is because, as detailed in Part III, the transformations applied by an ideal quantum computer are deterministic and free from random disturbances. However, simulating the operation of a quantum computer using classical resources becomes

computationally prohibitive as the number of qubits S increases. For example, with $S = 50$, the size of the state vector is $N = 2^{50} \approx 10^{15}$. Storing this state vector would require approximately 1 petabyte of memory – assuming one byte per complex amplitude – without accounting for the additional memory and computational resources needed to simulate the transformations of the quantum state. Thus, while the transformed state is, in principle, known from an informational perspective, it is practically incomputable using classical methods. This computational limitation implies that measurements of the pure state at the output of a quantum computer can reveal new information to a computationally constrained observer, even though the state itself is theoretically pure and fully specified (see also Appendix 5.A).

- **Geometry vs. logic:** According to Born's rule, the behavior of a quantum state $|\psi\rangle$ is governed by a geometrical framework that is fundamentally distinct from the traditional logical or set-theoretic framework used to formulate classical probability. This distinctive nature of "quantum probability" has inspired the application of the underlying geometrical principles to model phenomena – such as human behavior and natural language – that often deviate from classical probabilistic rules (see Recommended Resources).

5.5.3 Post-Measurement State

The fact that a measurement of a quantum state produces only a classical random output x may seem of little consequence. In fact, if the pre-measurement state $|\psi\rangle$ was still available after a measurement, even a computationally constrained observer could make multiple measurements on the state to extract as much information as required (see Appendix 5.A). However, as illustrated in Figure 5.5, after the measurement, the state of the system is transformed, and, except in special cases, it no longer equals the original state $|\psi\rangle$. Therefore, the classical observation x is the only observation that is obtained from the state before the quantum measurement irreparably disturbs it.

Definition 5.12 Post-measurement state
Given a measurement basis $\mathcal{M} = \{|v_x\rangle\}_{x\in[N]}$, if the classical output is $x \in [N]$, the post-measurement state of the quantum system is $|v_x\rangle$. Therefore, the post-measurement state is given by

$$|v_x\rangle \text{ with probability } p_x(|\psi\rangle|\mathcal{M}). \tag{5.69}$$

According to (5.69), a complete quantum measurement transforms the state $|\psi\rangle$ into vector $|v_x\rangle$ within the measurement basis $\mathcal{M}$ with probability $p_x(|\psi\rangle|\mathcal{M})$.

Example 5.12

Returning to Example 5.11, the last row of Figure 5.6 shows how the post-measurement state is fully determined by the measurement basis $\mathcal{M} = \{|v_0\rangle, |v_1\rangle\}$ and by the classical output x of the measurement. In particular, in the figure, given the classical output $x \in [2]$, the post-measurement state is $|v_x\rangle$.

The scientific and philosophical interpretation of the disturbance caused by the measurement on the quantum state is at the core of the debate on the true nature of the quantum state $|\psi\rangle$. Apart from the classical **Copenhagen interpretation**, which refrains from making statements that go beyond the observable behavior of a system, there are many theories that attempt to explain the collapse of the waveform. Famously, these include the **many-world hypothesis** that any measurement creates new universes, each corresponding to one of the possible post-measurement states $|v_x\rangle$.

5.5.4 Summary and Examples

Before considering a number of examples, it is worth summarizing the operation of a complete quantum measurement.

Definition 5.13 Complete quantum measurement of a pure state
Given a measurement basis $\mathcal{M} = \{|v_x\rangle\}_{x\in[N]}$, a complete quantum measurement applied to a pre-measurement state $|\psi\rangle$ results in the following.

- **Classical output:** The measurement yields a classical output $\mathrm{x} \sim p(|\psi\rangle|\mathcal{M})$, where the probability distribution depends on the pre-measurement state $|\psi\rangle$ and the chosen basis $\mathcal{M}$.
- **Post-measurement state:** The quantum system is left in the corresponding post-measurement state $|v_\mathrm{x}\rangle$ associated with the observed outcome x.

Example 5.13

As illustrated in Figure 5.7, consider a general qubit state $|\psi\rangle = \psi_0|0\rangle + \psi_1|1\rangle$ undergoing a measurement in the computational basis $\mathcal{M} = \{|0\rangle, |1\rangle\}$. Note that, when the measurement block does not report the setting, as in the figure, it is implicitly assumed to be in the computational basis. By Born's rule, a measurement in the computational basis returns the classical output $x = 0$ and the post-measurement state $|0\rangle$ with probability $p_0(|\psi\rangle|\mathcal{M}) = |\psi_0|^2$; while the classical output $x = 1$ and post-measurement state $|1\rangle$ are produced with probability $p_1(|\psi\rangle|\mathcal{M}) = |\psi_1|^2$.

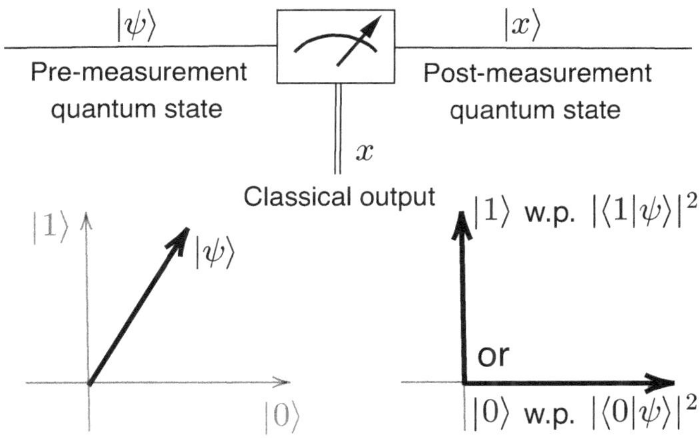

Figure 5.7 Illustration of a measurement in the computational basis, i.e., of a standard measurement, for a qubit system ("w.p." stands for "with probability").

In particular, if the state $|\psi\rangle$ coincides with one of the two vectors in the measurement basis, namely $|\psi\rangle = |0\rangle$ or $|\psi\rangle = |1\rangle$, the post-measurement state is unchanged with respect to the original state $|\psi\rangle$. Otherwise, the pre-measurement state $|\psi\rangle$ is disturbed by the measurement, collapsing to either state $|0\rangle$ or state $|1\rangle$.

Note that these conclusions hold also for any state that is equivalent to state $|\psi\rangle$, i.e., for any state of the form $|\psi\rangle\exp(i\gamma)$ in (5.67) for any global phase γ. In fact, even for such a state, the measurement probabilities equal

$$p_0(|\psi\rangle|\mathcal{M}) = |\psi_0\exp(i\gamma)|^2 = |\psi_0|^2 \tag{5.70}$$

and

$$p_1(|\psi\rangle|\mathcal{M}) = |\psi_1\exp(i\gamma)|^2 = |\psi_1|^2.$$

Example 5.14

This example illustrates the important phenomenon of interference between the amplitudes of a quantum state. To this end, take a qubit state $|\psi\rangle = \psi_0|0\rangle + \psi_1|1\rangle$ measured in the Hadamard basis $\mathcal{M} = \{|+\rangle, |-\rangle\}$. By Born's rule, using (5.38) and (5.39), this measurement returns the classical output $x = 0$ and the post-measurement state $|+\rangle$ with probability

$$p_0(|\psi\rangle|\mathcal{M}) = |\langle +|\psi\rangle|^2 = \frac{1}{2}|\psi_0 + \psi_1|^2; \tag{5.71}$$

while the classical output $x = 1$ and post-measurement state $|-\rangle$ are produced with probability

$$p_1(|\psi\rangle|\mathcal{M}) = 1 - p_0(|\psi\rangle|\mathcal{M}) = \frac{1}{2}|\psi_0 - \psi_1|^2. \tag{5.72}$$

Hence, with the Hadamard basis, the measurement probabilities are obtained through the sum and difference of the two complex amplitudes ψ_0 and ψ_1. These combinations may increase or decrease the measurement probabilities depending on the relative phase of the amplitudes ψ_0 and ψ_1, causing the constructive or the destructive interference between the two amplitudes.

Assume that the amplitude ψ_0 is real and positive, so that it equals its absolute value, $\psi_0 = |\psi_0|$, while the other amplitude is expressed as $\psi_1 = |\psi_1| \exp(i\gamma_1)$, where γ_1 is some phase. Using (5.71), the measurement probability for outcome $x = 0$ can be expressed as

$$\begin{aligned} p_0(|\psi\rangle|\mathcal{M}) &= \frac{1}{2}\left(|\psi_0| + |\psi_1| \exp(i\gamma_1)\right)^2 \\ &= \frac{1}{2}\left((|\psi_0| + |\psi_1| \cos(\gamma_1))^2 + (|\psi_1| \sin(\gamma_1))^2\right) \\ &= \frac{1}{2}\left(1 + 2|\psi_0||\psi_1| \cos(\gamma_1)\right). \end{aligned} \tag{5.73}$$

Taking as an example the case with equal magnitudes $|\psi_0| = |\psi_1| = 1/\sqrt{2}$, the measurement probability $p_0(|\psi\rangle|\mathcal{M}) = (1 + \cos(\gamma_1))/2$ in (5.73) ranges from $p_0(|\psi\rangle|\mathcal{M}) = 1$ for relative phase $\gamma_1 = 0$ to $p_0(|\psi\rangle|\mathcal{M}) = 0$ for relative phase $\gamma_1 = \pi$. Thus, when the relative phase is zero, $\gamma_1 = 0$, and thus the two amplitudes have the same phase, the two amplitudes ψ_0 and ψ_1 **interfere constructively** in the probability (5.73). In contrast, they **interfere destructively** when the relative phase is $\gamma_1 = \pi$, and thus the two amplitudes have a different sign.

This example highlights the difference between global phases, affecting all amplitudes, which do not change the behavior of a state, and relative phases, affecting only a subset of amplitudes, which can cause interference between the amplitudes. The phenomenon of interference is related to the traditional interpretation of a quantum state as a *wave*, whose amplitudes can interfere, boosting or depressing the wave in different parts of the state space.

As a general conclusion, interference illustrates that in quantum mechanics the outcome of an event may depend on the amplitudes of other possible events that could have occurred. Indeed, the complex amplitude for a given event can be altered – and even reduced to zero – by the amplitudes associated with other potential events.

5.5.5 Observables

As detailed in this section, a complete quantum measurement returns an integer $x \in [N]$, which identifies the post-measurement state $|v_x\rangle$ within the measurement basis $\mathcal{M} = \{|v_x\rangle\}_{x\in[N]}$. In practice, the integer x may be associated with a numerical value O_x. For instance, when measuring the state of a particle, the numerical value O_x may represent a position, a momentum, or an energy level, depending on the measurement setting. (Note that two values of the integer $x \in [N]$ may be assigned the same numerical value O_x.) Numerical outputs associated with a quantum measurement are known as observables.

Definition 5.14 Observable
An observable O is defined by a measurement basis $\mathcal{M}$ and by a set of scalars $\{O_x\}_{x\in[N]}$, where the numerical value O_x is associated with the measurement output $x \in [N]$. When the measurement in setting $\mathcal{M}$ returns classical output x, the observable O returns the value O_x. Therefore, an observable is described by a random variable O_x taking value O_x with probability $p_x(|\psi\rangle|\mathcal{M})$:

$$O_\mathrm{x} = O_x \text{ with probability } p_x(|\psi\rangle|\mathcal{M}). \tag{5.74}$$

The average value of the observable is defined as follows.

Definition 5.15 Expected value of an observable
Given a quantum state $|\psi\rangle$, the expected value of observable O, denoted as $\langle O\rangle_{|\psi\rangle}$, is evaluated as

$$\begin{aligned}\langle O\rangle_{|\psi\rangle} &= \mathrm{E}_{p(|\psi\rangle|\mathcal{M})}[O_\mathrm{x}] \\ &= \sum_{x\in[N]} O_x p_x(|\psi\rangle|\mathcal{M}).\end{aligned} \tag{5.75}$$

Note that the expected value $\langle O\rangle_{|\psi\rangle}$ can be estimated if one has access to multiple copies of state $|\psi\rangle$. In fact, in this case, it is possible to make multiple measurements of the observable O, one on each copy of the state. The measurements can be averaged to estimate the expected value of the observable.

Example 5.15

Consider Figure 5.8, in which the index x identifies the position of an object on the plane. The positions are quantized into N squares, and thus x can take N values. A particle has a state $|\psi\rangle$ assigning a complex amplitude ψ_x to each discrete position $x \in [N]$. The observable of interest here is the temperature O_x measured at the given position x. As shown in the figure, multiple positions x may share the same temperature value O_x. It is emphasized that this example is not meant to represent any physically realistic situation, but just to illustrate the concept of observable.

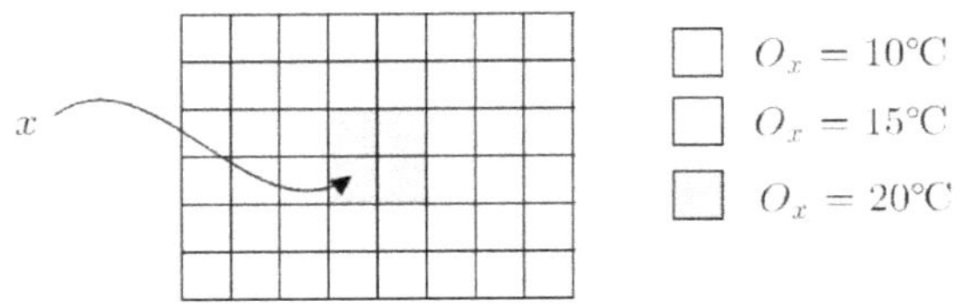

Figure 5.8 A cartoon example of an observable O. The position of a particle is quantized into N squares $x \in [N]$. Each position is assigned a temperature O_x. When the particle is observed to be in position x, the temperature value O_x is returned by the observable.

Example 5.16

Consider the qubit state

$$|\theta\rangle = \begin{bmatrix} \cos(\theta) \\ \sin(\theta) \end{bmatrix} \tag{5.76}$$

for some angle $0 \leq \theta < \pi$. Note that the state (5.76) recovers as special cases state $|\theta\rangle = |0\rangle$ for $\theta = 0$; $|\theta\rangle = |+\rangle$ for $\theta = \pi/4$; $|\theta\rangle = |1\rangle$ for $\theta = \pi/2$; and $|\theta\rangle = |-\rangle$, apart from a global phase, for $\theta = 3\pi/4$. Furthermore, the states $|\theta\rangle$ with θ outside the interval $0 \leq \theta < \pi$ are equivalent, as per definition (5.32), to states $|\theta\rangle$ with angle θ in the range $0 \leq \theta < \pi$. For instance, $|\pi\rangle = -|0\rangle$ is equivalent to state $|0\rangle$.

Suppose that the measurement implements an observable O in the computational basis with numerical values $O_0 = -1$ and $O_1 = 1$. Using (5.75), the expected value of the observable is given by

$$\begin{aligned} \langle O\rangle_{|\theta\rangle} &= O_0 p_0(|\theta\rangle|\mathcal{M}) + O_1 p_1(|\theta\rangle|\mathcal{M}) \\ &= -p_0(|\theta\rangle|\mathcal{M}) + p_1(|\theta\rangle|\mathcal{M}) \\ &= -(\cos(\theta))^2 + (\sin(\theta))^2. \end{aligned} \tag{5.77}$$

For example, for $\theta = 0$, the expected value is $\langle O\rangle_{|0\rangle} = -1$; for $\theta = \pi/4$, one obtains $\langle O\rangle_{|+\rangle} = 0$; $\theta = \pi/2$ yields the expected value $\langle O\rangle_{|1\rangle} = 1$; and with $\theta = 3\pi/4$ one gets $\langle O\rangle_{|-\rangle} = 0$.

5.6 Distinguishing Quantum States

For a classical system, recognizing whether the system is in either one of two distinct states would be simply a matter of observing the system to find out what the true state is. This section will demonstrate that this is not the case for quantum systems. Apart from special cases, any two quantum states, while distinct, are not perfectly distinguishable (see Property 4 in Section 5.2.2). Specifically, it will be concluded that, unless states $|\psi\rangle$ and $|\phi\rangle$ are orthogonal, there is no measurement that allows the two states to be perfectly distinguished. In the process, this section will introduce a measure of similarity – the **fidelity** – and one of dissimilarity – the **trace distance** – between quantum states that play an important role in quantum information theory.

5.6.1 Distinguishing Quantum States

As shown in Figure 5.9, consider a quantum system that, unbeknownst to the observer, may be either in state $|\psi\rangle$ or in state $|\phi\rangle$. The observer wishes to identify the state via a measurement.

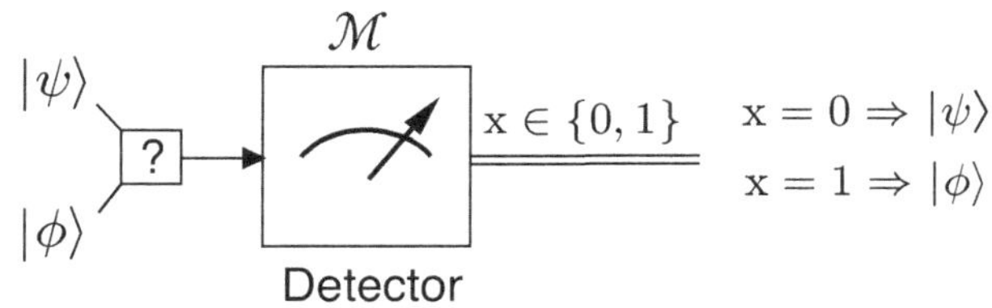

Figure 5.9 Illustration of the problem of distinguishing quantum pure states.

Example 5.17

Take first the special case in which states $|\psi\rangle$ and $|\phi\rangle$ are orthogonal. For this situation, consider any measurement setting $\mathcal{M} = \{|v_x\rangle\}_{x\in[N]}$ whose measurement basis includes the two states of interest by choosing

$$|v_0\rangle = |\psi\rangle \text{ and } |v_1\rangle = |\phi\rangle. \tag{5.78}$$

As shown in Figure 5.9, the detector selects state $|\psi\rangle$ if the classical output is $\mathrm{x} = 0$ and $|\phi\rangle$ if the classical output is $\mathrm{x} = 1$.

For this detector, if the true state is $|\psi\rangle$, the detection probability is

$$\begin{aligned}\Pr_{|\psi\rangle}[\mathrm{x} = 0] &= p_0(|\psi\rangle|\mathcal{M}) \\ &= |\langle v_0|\psi\rangle|^2 = |\langle\psi|\psi\rangle|^2 = 1;\end{aligned} \tag{5.79}$$

while, if the true state is $|\phi\rangle$, the detection probability is

$$\begin{aligned}\Pr_{|\phi\rangle}[\mathrm{x} = 1] &= p_1(|\phi\rangle|\mathcal{M}) \\ &= |\langle v_1|\phi\rangle|^2 = |\langle\phi|\phi\rangle|^2 = 1.\end{aligned} \tag{5.80}$$

Therefore, the detector outputs the correct decision irrespective of whether the true state is $|\psi\rangle$ or $|\phi\rangle$.

Example 5.17 has shown that for orthogonal states $|\psi\rangle$ and $|\phi\rangle$ it is possible to ensure that the detection probabilities for *both* states are equal to 1. This is not the case for non-orthogonal states. In fact, it is impossible to design a measurement setting whose measurement basis includes non-orthogonal vectors $|\psi\rangle$ and $|\phi\rangle$.

To elaborate on this point, consider a measurement setting $\mathcal{M} = \{|v_x\rangle\}_{x\in[N]}$ with $|v_0\rangle = |\psi\rangle$ equal to the first state, and any set of $N-1$ orthonormal vectors $\{|v_x\rangle\}_{x=1}^{N-1}$ that are orthogonal to $|v_0\rangle = |\psi\rangle$. The detector chooses state $|\psi\rangle$ if the measurement outcome is $\mathrm{x} = 0$, while it selects state $|\phi\rangle$ otherwise, i.e., when the classical output equals $\mathrm{x} \neq 0$. By (5.79), if the true state is $|\psi\rangle$, the detection probability $\Pr_{|\psi\rangle}[\mathrm{x} = 0]$ equals 1.

In contrast, if the true state is $|\phi\rangle$, the detection probability is given by

$$\begin{aligned}\Pr_{|\phi\rangle}[\mathrm{x} \neq 0|\mathcal{M}] &= 1 - p_0(|\phi\rangle|\mathcal{M}) \\ &= 1 - |\langle v_0|\phi\rangle|^2 \\ &= 1 - |\langle\psi|\phi\rangle|^2 \\ &= 1 - (\cos(\theta))^2,\end{aligned} \tag{5.81}$$

using the definition (5.30) for the cosine of the angle between the states $|\psi\rangle$ and $|\phi\rangle$.

Since the two states $|\psi\rangle$ and $|\phi\rangle$ are not orthogonal, the squared cosine $(\cos(\theta))^2$ is larger than zero, implying that there is a non-zero chance (5.81) that the detector chooses state $|\psi\rangle$ when the correct state is actually $|\phi\rangle$. Furthermore, the detection probability (5.81) gets smaller as the states $|\psi\rangle$ and $|\phi\rangle$ become more aligned, i.e., as angle θ becomes smaller.

5.6.2 Fidelity

The squared cosine $|\langle\psi|\phi\rangle|^2 = (\cos(\theta))^2$ was identified in (5.81) as the key quantity dictating whether two states $|\psi\rangle$ and $|\phi\rangle$ can be distinguished via a measurement. Accordingly, this quantity is widely adopted in quantum information theory as a measure of similarity of two states.

Definition 5.16 Fidelity, aka overlap, between pure states
The fidelity, or overlap, between two $N \times 1$ quantum pure states $|\psi\rangle$ and $|\phi\rangle$ is defined as

$$\mathrm{F}(|\psi\rangle, |\phi\rangle) = |\langle\psi|\phi\rangle|^2. \tag{5.82}$$

Using the corresponding bounds (5.31) for the inner product, the fidelity satisfies the following bounds.

Key Result 5.6 Bounds on the fidelity between quantum pure states
The fidelity between two quantum pure states $|\psi\rangle$ and $|\phi\rangle$ satisfies the inequalities

$$\underbrace{0}_{\text{orthogonal}} \leq \mathrm{F}(|\psi\rangle, |\phi\rangle) \leq \underbrace{1}_{\text{equivalent}}. \tag{5.83}$$

A zero fidelity $\mathrm{F}(|\psi\rangle, |\phi\rangle) = 0$ indicates orthogonal, and thus perfectly distinguishable, states, while a maximum fidelity $\mathrm{F}(|\psi\rangle, |\phi\rangle) = 1$ indicates equivalent, and thus indistinguishable, states.

Example 5.18

Consider the state $|0\rangle$ and the state $|\theta\rangle = [\cos(\theta), \sin(\theta)]^T$ in (5.76) with $0 \leq \theta < \pi$. The fidelity between the two states is

$$\mathrm{F}(|0\rangle, |\theta\rangle) = |\langle 0|\theta\rangle|^2 = (\cos(\theta))^2, \tag{5.84}$$

which is plotted in Figure 5.10 as a function of the angle θ in the range $0 \leq \theta < \pi$. As shown in the figure, the fidelity is maximal when the two states are equivalent, i.e., when $\theta = 0$ or $\theta = \pi$, while it is minimal when the two states are orthogonal, i.e., for $\theta = \pi/2$.

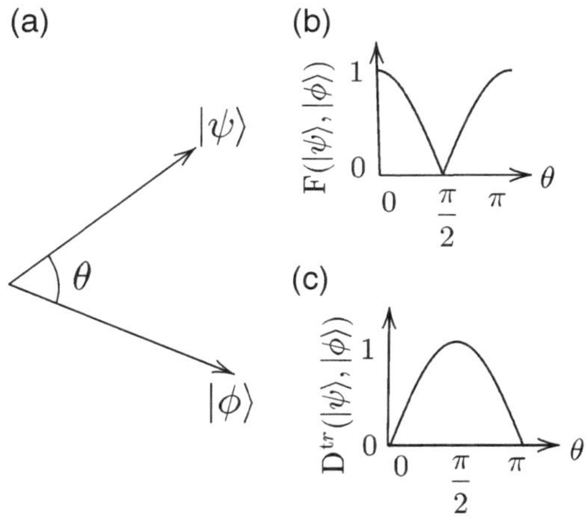

Figure 5.10 (a) Two states $|\psi\rangle = |0\rangle$ and $|\phi\rangle = |\theta\rangle$, along with the corresponding (b) fidelity $F(|\psi\rangle, |\phi\rangle)$ and (c) trace distance $D^{tr}(|0\rangle, |\theta\rangle)$.

Example 5.19

The Born rule (5.65) can be expressed in terms of the fidelity between the measurement basis state $|v_x\rangle$ and the pre-measurement state $|\psi\rangle$ as

$$p_x(|\psi\rangle|\mathcal{M}) = |\langle v_x|\psi\rangle|^2 = F(|v_x\rangle, |\psi\rangle). \tag{5.85}$$

Therefore, as the fidelity between the pre-measurement state $|\psi\rangle$ and the measurement basis state $|v_x\rangle$ increases, so does also the probability of obtaining output $\mathrm{x} = x$ from a measurement of state $|\psi\rangle$ in setting $\mathcal{M} = \{|v_x\rangle\}_{x\in[N]}$.

5.6.3 Trace Distance

While the fidelity is a measure of similarity, a closely related measure of dissimilarity between quantum states can be defined from the fidelity by considering the complementary quantity $1 - F(|\psi\rangle, |\phi\rangle)$.

Definition 5.17 Trace distance between pure states
The trace distance between two $N \times 1$ quantum pure states $|\psi\rangle$ and $|\phi\rangle$ is defined as

$$\begin{aligned} D^{tr}(|\psi\rangle, |\phi\rangle) &= \sqrt{1 - F(|\psi\rangle, |\phi\rangle)} \\ &= \sqrt{1 - |\langle\psi|\phi\rangle|^2}. \end{aligned} \tag{5.86}$$

Geometrically, recalling the trigonometric relation $\cos(\theta)^2 + \sin(\theta)^2 = 1$, the trace distance (5.86) can be interpreted as the absolute value of the sine of the angle θ between the states $|\psi\rangle$ and $|\phi\rangle$, i.e.,

$$D^{tr}(|\psi\rangle, |\phi\rangle) = |\sin(\theta)|. \tag{5.87}$$

Therefore, while the fidelity decreases as the states $|\psi\rangle$ and $|\phi\rangle$ become more orthogonal, the trace distance increases, providing a measure of how distinguishable the two vectors are. The following bounds follow immediately from the inequalities (5.87).

Key Result 5.7 Bounds on the trace distance between quantum pure states
The trace distance satisfies the inequalities

$$\underbrace{0}_{\text{equivalent}} \leq D^{tr}(|\psi\rangle, |\phi\rangle) \leq \underbrace{1}_{\text{orthogonal}}. \tag{5.88}$$

Another related measure of dissimilarity is the angle $\theta \in [0, \pi/2]$, which is known as the **Bhattacharyya–Wooters distance** (see Recommended Resources).

Example 5.20

Continuing Example 5.18, for the states $|\psi\rangle = |0\rangle$ and $|\phi\rangle = [\cos(\theta), \sin(\theta)]^T$, the trace distance is given by

$$D^{tr}(|\psi\rangle, |\phi\rangle) = |\sin(\theta)|. \tag{5.89}$$

As illustrated in Figure 5.10, the trace distance exhibits an opposite trend as compared to the fidelity, being minimal when the two states are equivalent, i.e., when $\theta = 0$ or $\theta = \pi$, while being maximal when the two states are orthogonal, i.e., for $\theta = \pi/2$.

5.7 Summary

- Quantum information theory – a field at the intersection of information theory and physics – studies systems that encode information in microscopic degrees of freedom that are subject to the rules of quantum mechanics.
- This chapter has introduced the key concepts of quantum pure states and complete measurements for quantum information.
- Quantum systems are characterized by a new form of uncertainty – quantum uncertainty – that arises due to the inherent randomness of the output of quantum measurements.
- Quantum uncertainty is fundamentally different from classical uncertainty, which arises due to ignorance about the true state of the system, given that it applies also to systems in a known, i.e., pure, quantum state.
- Unlike classical systems, measurement outputs are generally uncertain, and the probability distribution over the possible measurement outputs is contextual,

in the sense that it depends on the specific measurement setting selected by the observer.

- A complete quantum measurement is defined by a measurement setting $\mathcal{M} = \{|v_x\rangle\}_{x\in[N]}$, consisting of an orthonormal basis $\{|v_x\rangle\}_{x\in[N]}$.
- Given a pre-measurement state $|\psi\rangle$, a complete quantum measurement has the following effects: (i) it produces a classical output $\mathrm{x} \sim p(|\psi\rangle|\mathcal{M})$; and (ii) it leaves the system in the post-measurement state $|v_{\mathrm{x}}\rangle$.
- The probability distribution $p(|\psi\rangle|\mathcal{M})$, when viewed as a function of the measurement setting $\mathcal{M}$, is known as the behavior of the quantum state $|\psi\rangle$.
- By Born's rule, the probability $p_x(|\psi\rangle|\mathcal{M})$ of obtaining any classical output $x \in [N]$ is given by the squared absolute value of the inner product between states $|v_x\rangle$ and $|\psi\rangle$, i.e., $p_x(|\psi\rangle|\mathcal{M}) = |\langle v_x|\psi\rangle|^2$.
- The quantity $|\langle\phi|\psi\rangle|$ is the absolute value of the cosine of the angle between states $|\psi\rangle$ and $|\phi\rangle$, while the squared absolute value $\mathrm{F}(|\psi\rangle, |\phi\rangle) = |\langle\phi|\psi\rangle|^2$ is known as the fidelity between the two states.
- Only orthogonal states, i.e., states with zero fidelity, can be perfectly distinguished, while for non-orthogonal quantum states, no measurement can unambiguously determine the true quantum state.
- While the fidelity provides a measure of similarity between states, the trace distance $\mathrm{D}^{\mathrm{tr}}(|\psi\rangle, |\phi\rangle) = \sqrt{1 - \mathrm{F}(|\psi\rangle, |\phi\rangle)}$, which can be interpreted as the absolute value of the sine of the angle between the two states, is a measure of dissimilarity.

5.8 Recommended Resources

For high-level discussions on the use and interpretation of quantum states, the reader is referred to [1] and [2]. Standard references for quantum computing include [3] and [4]. A popular account of physical theories that leverage additional dimensions can be found in [5]. For a catalog of Hadamard matrices, please see [6]. A comprehensive discussion of how the geometric properties of quantum probabilities can be used to model cognition and decision processes in humans is presented in [7]. The quote used in Section 5.5.2 comes from [8]. Interesting historical and mathematical aspects of Pauli matrices and quantum mechanics are offered in [9].

PROBLEMS

5.1 (a) Show that, given an $N \times N$ matrix V, the condition $V^\dagger V = I$ implies that the columns of matrix V are normalized and mutually orthogonal. *Hint:* Plug (5.45) into the equality $V^\dagger V = I$. (b) Show that for a unitary matrix V the rows of matrix V also form an orthonormal basis. *Hint:* Consider the columns of the Hermitian transpose matrix $V^\dagger$.

5.2 Verify that the Pauli matrices are unitary.

5.3 Prove that if matrix V is unitary, then so is the matrix V^*, whose entries are the complex conjugate of matrix V, and the matrix V^T.

5.4 Prove that, for any orthonormal basis $\{|v_x\rangle\}_{x\in[N]}$, the equality $\sum_{x\in[N]}|\alpha_x|^2 = 1$ holds for the vector of amplitudes $\alpha_x = \langle v_x|\psi\rangle$ with $x \in [N]$. *Hint:* Express the equality by using the identity $\langle\psi|v_x\rangle\langle v_x|\psi\rangle = |\langle v_x|\psi\rangle|^2$ and use the resolution-of-the-identity equality $\sum_{x\in[N]}|v_x\rangle\langle v_x| = I$.

5.5 Derive the resolution-of-the-identity equality $\sum_{x\in[N]}|v_x\rangle\langle v_x| = I$ for any orthonormal basis $\{|v_x\rangle\}_{x\in[N]}$ from the equality $VV^\dagger = I$, where the matrix V has the vectors $\{|v_x\rangle\}_{x\in[N]}$ as its columns. *Hint:* Note that matrix V can be written as $V = \sum_{x\in[N]}|v_x\rangle\langle x|$, and then use the fact that vectors $\{|x\rangle\}_{x\in[N]}$ form an orthonormal basis.

5.6 For a general qubit state $|\psi\rangle = \psi_0|0\rangle+\psi_1|1\rangle$, calculate the measurement probability vector and post-measurement states when the following measurement bases are used: (a) the Hadamard basis; and (b) the orthonormal basis with vectors $|+i\rangle = (|0\rangle + i|1\rangle)/\sqrt{2}$ and $|-i\rangle = (|0\rangle - i|1\rangle)/\sqrt{2}$.

5.7 (a) Evaluate fidelity and trace distance for the qutrit states $|\psi\rangle = 1/\sqrt{3}(|0\rangle + |1\rangle + |2\rangle)$ and $|\phi\rangle = 1/\sqrt{2}(|0\rangle + |1\rangle)$. (b) Evaluate fidelity and trace distance for the qutrit states $|\psi\rangle = 1/\sqrt{2}(|0\rangle + |1\rangle)$ and $|\phi\rangle = |2\rangle$.

5.8 Consider a qutrit system described by the state

$$|\psi\rangle = \frac{1}{\sqrt{6}}|0\rangle + \frac{1}{\sqrt{6}}|1\rangle + \sqrt{\frac{2}{3}}|2\rangle. \tag{5.90}$$

This problem studies measurements in the computational basis $\{|0\rangle, |1\rangle, |2\rangle\}$ and in the Hadamard basis $\{|v_0\rangle, |v_1\rangle, |v_2\rangle\}$, where

$$|v_0\rangle = \frac{1}{\sqrt{3}}(|0\rangle + |1\rangle + |2\rangle), \quad |v_1\rangle = \frac{1}{\sqrt{3}}(|0\rangle + \omega|1\rangle + \omega^2|2\rangle), \tag{5.91}$$

and

$$|v_2\rangle = \frac{1}{\sqrt{3}}(|0\rangle + \omega^2|1\rangle + \omega|2\rangle), \quad \omega = \exp(2\pi i/3). \tag{5.92}$$

(a) Compute the probability distribution over classical outcomes in both the computational basis and the Hadamard basis. (b) Discuss how the measurement basis affects the type of information obtained about state $|\psi\rangle$, explaining why this measurement dependence highlights contextuality in quantum systems.

Appendix

5.A Estimating the Measurement Distribution

When the state $|\psi\rangle$ is computationally unknown, it is often important to estimate the measurement probability vector $p(|\psi\rangle|\mathcal{M})$ from a number of classical measurement outputs. To describe this situation, assume that M independent measurement outputs are available, with each measurement output being a realization of random variable $\mathrm{x} \sim p(|\psi\rangle|\mathcal{M})$. The notion of independent random variables will be formally defined in Part IV. The measurement probability vector $p(|\psi\rangle|\mathcal{M})$ can be estimated by counting the number of times, say m_x, that each outcome x is measured, out of the total number M of measurements.

An estimate of the probability vector $p(|\psi\rangle|\mathcal{M})$ can thus be obtained by evaluating the M fractions

$$\hat{\mathrm{p}}(|\psi\rangle|\mathcal{M}) = \frac{1}{M}\begin{bmatrix} \mathrm{m}_0 \\ \vdots \\ \mathrm{m}_{N-1} \end{bmatrix}. \tag{5.93}$$

The distribution of the random vector $\hat{\mathrm{p}}(|\psi\rangle|\mathcal{M})$ is known as **multinomial**. Using this distribution, it can be proved that the maximum error $|p_x(|\psi\rangle|\mathcal{M}) - \hat{\mathrm{p}}_x(|\psi\rangle|\mathcal{M})|$ across all values $x \in [N]$ decreases as $1/\sqrt{M}$ as one increases the number of observations M.

REFERENCES

[1] S. Carroll, *The Biggest Ideas in the Universe 2: Quanta and Fields*. London: Oneworld Publications, 2024.

[2] C. Rovelli, *Helgoland: The Strange and Beautiful Story of Quantum Physics*. Harmondsworth: Penguin Books, 2022.

[3] M. A. Nielsen and I. L. Chuang, *Quantum Computation and Quantum Information*. Cambridge: Cambridge University Press, 2010.

[4] N. D. Mermin, *Quantum Computer Science: An Introduction*. Cambridge: Cambridge University Press, 2007.

[5] P. Halpern, *The Allure of the Multiverse: Extra Dimensions, Other Worlds, and Parallel Universes*. New York: Basic Books, 2024.

[6] K. Życzkowski, "Hadamard matrices." https://chaos.if.uj.edu.pl/ karol/ hadamard/, 2024 [accessed on July 28, 2024].

[7] J. R. Busemeyer and P. D. Bruza, *Quantum Models of Cognition and Decision*. Cambridge: Cambridge University Press, 2012.

[8] J. Bub, "Foreword to *Understanding Quantum Raffles* by Michael Janas, Michael E. Cuffaro, Michel Janssen," 2023. [Online]. Available at: https://arxiv.org/abs/2301.12266

[9] R. Arianrhod, *Vector: A Surprising Story of Space, Time, and Mathematical Transformation*. Chicago, IL: The University of Chicago Press, 2024.

6 Quantum Uncertainty: Measured Entropy, Coherence, and the Uncertainty Principle

6.1 Overview

Chapter 5 demonstrated that measurements of a quantum system are inherently uncertain. This chapter elaborates on the resulting concept of quantum uncertainty from a quantitative, information-theoretic perspective by analyzing the behavior of a quantum state through the entropy of the corresponding classical measurement outcomes.

To begin, the notion of measured entropy is introduced as a metric of uncertainty for a given measurement setting. The measured entropy is shown to depend on the coherence of the quantum state with respect to the chosen measurement basis, where higher coherence corresponds to a greater measured entropy. Since a quantum state typically exhibits varying degrees of coherence depending on the measurement basis, adjusting the measurement setting can yield classical outputs with differing levels of uncertainty.

The relationship between measurement settings is further explored through the concept of compatibility, which is influenced by the coherence of states in one measurement basis relative to another. Greater coherence between states in two measurement bases implies lower compatibility. For two incompatible measurement settings, a quantum state producing low measured entropy under one setting must yield high measured entropy under the other. In other words, a state with low coherence in one basis will exhibit high coherence in the complementary basis, resulting in a larger measured entropy. This relationship is an expression of the uncertainty principle, which will be formulated in this chapter using an information-theoretic framework.

Finally, the concept of projective measurements is introduced as a generalization of complete measurements. Projective measurements enable the partial observation of a quantum system, balancing the trade-off between the amount of information extracted from the quantum state and the disturbance caused to the system. This property is formalized by showing that the measured entropy of a projective measurement is always smaller than or equal to the measured entropy of the corresponding complete measurement.

6.1.1 Learning Objectives and Organization

By the end of this chapter, the reader should be able to:

- understand and use the definition of measured entropy;
- evaluate the coherence of a quantum state with respect to a measurement basis;
- evaluate the compatibility of two measurement settings;
- compare the uncertainty levels corresponding to two quantum measurement settings by using the uncertainty principle;
- describe and apply projective measurements;
- and evaluate the measured entropy of projective measurements.

6.2 Quantum Uncertainty

6.2.1 Measured Entropy

As seen in Chapter 5, for any $N \times 1$ quantum pure state $|\psi\rangle$, the uncertainty on the classical measurement output depends on the measurement setting $\mathcal{M}$ via the Born rule. To review, fix a measurement setting $\mathcal{M} = \{|v_x\rangle\}_{x\in[N]}$, specified by an orthonormal basis $\{|v_x\rangle\}_{x\in[N]}$ of the N-dimensional complex vector space. For a quantum state $|\psi\rangle$, the classical output produced by a complete measurement in setting $\mathcal{M}$ is a random variable $\mathrm{x} \sim p(|\psi\rangle|\mathcal{M})$, with $N \times 1$ probability vector having entries

$$p_x(|\psi\rangle|\mathcal{M}) = |\langle v_x|\psi\rangle|^2 \tag{6.1}$$

for $x \in [N]$. The probability distribution $p(|\psi\rangle|\mathcal{M})$ viewed as a function of the basis $\mathcal{M}$ determines the behavior of the quantum state $|\psi\rangle$.

The uncertainty of the measurement outcome x can thus be quantified by applying any of the entropy measures described in the previous chapters to the measurement probability distribution $p(|\psi\rangle|\mathcal{M})$.

Definition 6.1 Shannon, Tsallis, and Rényi measured entropies
The Shannon, Tsallis, and Rényi measured entropies of quantum state $|\psi\rangle$ under measurement setting $\mathcal{M}$ are defined as

$$\mathrm{H}(|\psi\rangle|\mathcal{M}) = \mathrm{H}(p(|\psi\rangle|\mathcal{M})), \tag{6.2}$$

$$\mathrm{H}^{\mathrm{T}}_{\alpha}(|\psi\rangle|\mathcal{M}) = \mathrm{H}^{\mathrm{T}}_{\alpha}(p(|\psi\rangle|\mathcal{M})), \tag{6.3}$$

and

$$\mathrm{H}^{\mathrm{R}}_{\alpha}(|\psi\rangle|\mathcal{M}) = \mathrm{H}^{\mathrm{R}}_{\alpha}(p(|\psi\rangle|\mathcal{M})), \tag{6.4}$$

respectively.

In the following, when referring generically to any measured entropy, the notation $\mathrm{H}_\alpha(|\psi\rangle|\mathcal{M})$ will be used; and, similarly, the notation $\mathrm{H}_\alpha(p)$ will be adopted to refer to the Shannon, Tsallis, and Rényi entropies for a probability vector p. Accordingly, the measured entropy will be generically written as

$$\mathrm{H}_\alpha(|\psi\rangle|\mathcal{M}) = \mathrm{H}_\alpha(p(|\psi\rangle|\mathcal{M})). \tag{6.5}$$

The measured entropies can be written explicitly in terms of the measurement probabilities by using the definitions of the corresponding classical entropies. For instance, the **Shannon measured entropy** can be written explicitly as a function of the state $|\psi\rangle$ and of the measurement basis $\mathcal{M}$ as

$$\begin{aligned}\mathrm{H}(|\psi\rangle|\mathcal{M}) &= -\sum_{x\in[N]} p_x(|\psi\rangle|\mathcal{M}) \log_2(p_x(|\psi\rangle|\mathcal{M})) \\ &= -\sum_{x\in[N]} |\langle v_x|\psi\rangle|^2 \log_2(|\langle v_x|\psi\rangle|^2).\end{aligned} \tag{6.6}$$

By its definition, operationally, the measured entropy $\mathrm{H}_\alpha(|\psi\rangle|\mathcal{M})$ can be interpreted through the lens of the informational tasks described in Part I for classical systems. In particular, the measured entropy relates to the minimum average number of questions, bits, guesses, and random bits, or to the minimum average prediction loss, for the random variable $\mathrm{x} \sim p(|\psi\rangle|\mathcal{M})$ describing the classical output produced by measurement $\mathcal{M}$.

6.2.2 Optimizing the Measured Entropy

Due to the contextuality of quantum measurements, the measured entropy $\mathrm{H}_\alpha(|\psi\rangle|\mathcal{M})$ depends on the measurement setting $\mathcal{M}$. Given a state $|\psi\rangle$, one may thus optimize the measurements setting $\mathcal{M}$ so as to minimize or maximize the measured entropy $\mathrm{H}_\alpha(|\psi\rangle|\mathcal{M})$.

To start, recall from Part I that the entropy $\mathrm{H}_\alpha(p)$ ranges from the minimal value 0, attained for a distribution concentrated at a single state, to the maximal value $\mathrm{H}_\alpha(u)$, which is obtained for a uniform distribution. As will be demonstrated next, for any fixed pure state $|\psi\rangle$, it is always possible to find measurement settings that yield both a *minimal* measured entropy equal to zero and a *maximal* measured entropy equal to $\mathrm{H}_\alpha(u)$.

Consider first the problem of minimizing the measured entropy $\mathrm{H}_\alpha(|\psi\rangle|\mathcal{M})$ over all possible complete measurement settings $\mathcal{M}$, i.e., over all possible orthonormal bases of the N-dimensional space. The stated problem amounts to the minimization

$$\underset{\mathcal{M}}{\text{minimize}}\ \mathrm{H}_\alpha(|\psi\rangle|\mathcal{M}). \tag{6.7}$$

A solution to this problem is readily seen to be given by any measurement basis $\mathcal{M}$ that includes state $|\psi\rangle$.

In fact, with any such basis, which is denoted as $\mathcal{M}_*^{\min}$, the measurement distribution $p(|\psi\rangle|\mathcal{M}_*^{\min})$ is concentrated at a single output value (see Problem 6.1). Therefore, the resulting value of the measured entropy is zero, i.e., $\mathrm{H}_\alpha(|\psi\rangle|\mathcal{M}_*^{\min}) = 0$, yielding the minimum value for problem (6.5). As detailed in Section 6.3, state $|\psi\rangle$ is said to be **incoherent** with respect to any entropy-minimizing measurement basis $\mathcal{M}_*^{\min}$.

Example 6.1

For a qubit system in state $|\psi\rangle$, an entropy-minimizing measurement setting is given by

$$\mathcal{M}_*^{\min} = \{|\psi\rangle, |\psi^\perp\rangle\}, \tag{6.8}$$

where $|\psi^\perp\rangle$ is any state that is orthogonal to $|\psi\rangle$, i.e.,

$$\langle\psi|\psi^\perp\rangle = 0. \tag{6.9}$$

Numbering the states in the measurement basis $\mathcal{M}_*^{\min} = \{|v_x\rangle\}_{x\in[2]}$ as $|v_0\rangle = |\psi\rangle$ and $|v_1\rangle = |\psi^\perp\rangle$, the measurement probability is

$$p_0(|\psi\rangle|\mathcal{M}_*^{\min}) = |\langle\psi|\psi\rangle|^2 = 1$$

for output $x = 0$. Therefore, the measured entropy under setting $\mathcal{M}_*^{\min}$ is minimal, i.e., $\mathrm{H}_\alpha(|\psi\rangle|\mathcal{M}_*^{\min}) = 0$.

Take now the problem of maximizing the measured entropy with respect to the measurement setting $\mathcal{M}$ for a given state $|\psi\rangle$, which can be stated as the maximization

$$\underset{\mathcal{M}}{\text{maximize}}\ \mathrm{H}_\alpha(|\psi\rangle|\mathcal{M}). \tag{6.10}$$

If a measurement basis $\mathcal{M} = \{|v_x\rangle\}_{x\in[N]}$ exists for which the measurement probability vector $p(|\psi\rangle|\mathcal{M})$ is uniform, i.e., for which $p(|\psi\rangle|\mathcal{M}) = u$, this choice would maximize the measured entropy, achieving the maximum value $\mathrm{H}_\alpha(u)$.

To find one such solution, it is sufficient to identify an orthonormal basis $\mathcal{M}_*^{\max}$ for which the state $|\psi\rangle$ can be written as the **equal superposition**

$$|\psi\rangle = \frac{1}{\sqrt{N}} \sum_{x\in[N]} |v_x\rangle \tag{6.11}$$

of states $\{|v_x\rangle\}_{x\in[N]}$ in the basis. In fact, with any such orthonormal basis $\mathcal{M}_*^{\max}$, the measurement probabilities are given by

$$p_x(|\psi\rangle|\mathcal{M}_*^{\max}) = |\langle v_x|\psi\rangle|^2 = \frac{1}{N} \tag{6.12}$$

for all $x \in [N]$, yielding the desired uniform measurement probability vector $p(|\psi\rangle|\mathcal{M}_*^{\max}) = u$. As proved in Appendix 6.A, it is always possible to find such a basis.

As detailed in Section 6.3, a state $|\psi\rangle$ satisfying condition (6.11) is said to be **maximally coherent** with respect to the entropy-maximizing basis $\mathcal{M}_*^{\max}$.

Example 6.2

For an arbitrary qubit state $|\psi\rangle$, the measurement basis

$$\mathcal{M}_*^{\max} = \left\{ \frac{1}{\sqrt{2}}(|\psi\rangle + |\psi^\perp\rangle), \frac{1}{\sqrt{2}}(|\psi\rangle - |\psi^\perp\rangle) \right\}, \tag{6.13}$$

where $|\psi^\perp\rangle$ is a state orthogonal to $|\psi\rangle$, satisfies the condition (6.11). This can be directly verified by evaluating the sum (6.11) as

$$\frac{1}{\sqrt{2}} \left(\frac{1}{\sqrt{2}}(|\psi\rangle + |\psi^\perp\rangle) + \frac{1}{\sqrt{2}}(|\psi\rangle - |\psi^\perp\rangle) \right) = |\psi\rangle. \tag{6.14}$$

Therefore, the state $|\psi\rangle$ is maximally coherent with respect to the measurement basis $\mathcal{M}_*^{\max}$, and a measurement in setting $\mathcal{M}_*^{\max}$ yields the maximum value $\mathrm{H}_\alpha(u)$ for the measured entropy.

6.3 Coherence

As seen in Section 6.2, given a measurement setting $\mathcal{M}$, if the state of the quantum system corresponds to one of the vectors in the measurement basis, the measurement uncertainty is zero; while, if the state is a proper superposition of at least two vectors in the measurement basis, the measured entropy is positive. Indeed, the measured entropy is maximized for states corresponding to the equal superposition (6.11) of all measurement basis states. This section formalizes the concept of coherence, which quantifies the level of superposition of a quantum state $|\psi\rangle$ with respect to a given measurement basis $\mathcal{M}$.

6.3.1 Defining Coherence

To start thinking about coherence, it is useful to formalize situations characterized by minimal and maximal coherence for a state $|\psi\rangle$ with respect to a measurement setting $\mathcal{M} = \{|v_x\rangle\}_{x\in[N]}$.

Definition 6.2 Minimal coherence, or incoherence
When the state $|\psi\rangle$ is equivalent to one of the basis vectors, i.e., when

$$|\psi\rangle = \exp(i\gamma)|v_x\rangle \tag{6.15}$$

for some integer $x \in [N]$ and some global phase γ, the state $|\psi\rangle$ is said to be **incoherent** with respect to the orthonormal basis $\mathcal{M}$. When this condition is not met, the state $|\psi\rangle$ is said to be **coherent** with respect to the basis $\mathcal{M}$.

As seen in Section 6.2, a measurement in setting $\mathcal{M}$ yields a zero measured entropy for any state that is incoherent with respect to basis $\mathcal{M}$.

Definition 6.3 Maximal coherence
When the state $|\psi\rangle$ can be expressed as

$$|\psi\rangle = \frac{1}{\sqrt{N}} \sum_{x \in [N]} \exp(i\theta_x)|v_x\rangle \tag{6.16}$$

for some phases $\{\theta_x\}_{x\in[N]}$, the state $|\psi\rangle$ is said to be maximally coherent with respect to the orthonormal basis $\mathcal{M}$.

Condition (6.16) generalizes the equal superposition (6.11), which is obtained by setting all phases $\{\theta_x\}_{x\in[N]}$ to zero. In this situation, following the reasoning in Section 6.2, the measured entropy is maximal, attaining the value $\mathrm{H}_\alpha(u)$.

According to the definitions given above, coherence measures must range from a minimum value in situations of minimal coherence to a maximal value in situations of maximal coherence. Section 6.3.2 introduces a notable such measure.

6.3.2 Trace Distance of Coherence ⏭

Since minimal coherence is attained for states corresponding to vectors in the measurement basis $\mathcal{M}$, one may measure coherence by evaluating the distance between the state $|\psi\rangle$ and the closest state in basis $\mathcal{M}$. Using the trace distance introduced in Chapter 5 to quantify distances between states, one obtains the following definition.

Definition 6.4 Trace distance of coherence
The trace distance of coherence of state $|\psi\rangle$ with respect to orthonormal basis $\mathcal{M}$ is defined as

$$\begin{aligned}\mathrm{K}^{\mathrm{tr}}(|\psi\rangle|\mathcal{M}) &= \min_{|v\rangle\in\mathcal{M}} \mathrm{D}^{\mathrm{tr}}(|\psi\rangle, |v\rangle) \\ &= \min_{x\in[N]} \mathrm{D}^{\mathrm{tr}}(|\psi\rangle, |v_x\rangle) \\ &= \sqrt{1 - \max_{x\in[N]} \mathrm{F}(|\psi\rangle, |v_x\rangle)}.\end{aligned} \tag{6.17}$$

The significance of this definition is made clear by noting the following bounds.

Key Result 6.1 Bounds on the trace distance of coherence
The trace distance of coherence (6.17) satisfies the inequalities

$$\underbrace{0}_{\text{incoherent}} \leq \mathrm{K}^{\mathrm{tr}}(|\psi\rangle|\mathcal{M}) \leq \underbrace{\sqrt{1 - \frac{1}{N}}}_{\text{maximally coherent}}. \tag{6.18}$$

The lower bound holds if and only if the state $|\psi\rangle$ is incoherent with respect to basis $\mathcal{M}$, while the upper bound is attained if and only if the state $|\psi\rangle$ is maximally coherent with respect to basis $\mathcal{M}$.

In accordance with these bounds, any state $|\psi\rangle$ with a positive trace distance of coherence $\mathrm{K}^{\mathrm{tr}}(|\psi\rangle|\mathcal{M}) > 0$ is coherent with respect to the orthonormal basis $\mathcal{M}$.

The lower bound in (6.18) follows directly from the definition of trace distance of coherence, while the upper bound deserves further discussion. To elaborate on this, using the definition of trace distance and recalling the equality $\mathrm{F}(|\psi\rangle, |v_x\rangle) = p_x(|\psi\rangle|\mathcal{M})$ between fidelity and measurement probability (see Section 5.6), the trace distance of coherence can equivalently be written as follows.

Key Result 6.2 Trace distance of coherence and entropy
The trace distance of coherence can be expressed as

$$\mathrm{K}^{\mathrm{tr}}(|\psi\rangle|\mathcal{M}) = \sqrt{1 - 2^{-\mathrm{H}^{\mathrm{R}}_{\infty}(|\psi\rangle|\mathcal{M})}}, \tag{6.19}$$

where the last equality uses the definition of the Rényi min-entropy $\mathrm{H}^{\mathrm{R}}_{\infty}(|\psi\rangle|\mathcal{M})$ introduced in Section 3.6:

$$\mathrm{H}^{\mathrm{R}}_{\infty}(|\psi\rangle|\mathcal{M}) = -\log_2\left(\max_{x\in[N]} p_x(|\psi\rangle|\mathcal{M})\right). \tag{6.20}$$

The expression (6.19) demonstrates that the trace distance of coherence increases with the uncertainty on the classical measurement output, when the latter is measured by the Rényi min-entropy $\mathrm{H}^{\mathrm{R}}_{\infty}(|\psi\rangle|\mathcal{M})$. Furthermore, since the Rényi min-entropy is upper bounded as $\mathrm{H}^{\mathrm{R}}_{\infty}(|\psi\rangle|\mathcal{M}) \leq \log_2(N)$, with equality if and only if the measurement probability distribution $p(|\psi\rangle|\mathcal{M})$ is uniform (see Section 3.6.4), the upper bound (6.18) is verified.

Finally, note that the bounds (6.18) also imply that the maximum fidelity between a state $|\psi\rangle$ and the states in a measurement basis $\mathcal{M} = \{|v_x\rangle\}_{x\in[N]}$ satisfies the inequalities

$$\underbrace{\frac{1}{N}}_{\text{maximally coherent}} \leq \max_{|v\rangle\in\mathcal{M}} \mathrm{F}(|\psi\rangle, |v\rangle) \leq \underbrace{1}_{\text{incoherent}}. \tag{6.21}$$

Example 6.3

For the qubit state $|\theta\rangle = [\cos(\theta), \sin(\theta)]^T$, with angle $0 \leq \theta < \pi$, using (6.17), the trace distance of coherence with respect to the computational basis $\mathcal{M} = \{|0\rangle, |1\rangle\}$ is given by

$$\begin{aligned}\mathrm{K}^{\mathrm{tr}}(|\psi\rangle|\mathcal{M}) &= \sqrt{1 - \max((\cos(\theta))^2, (\sin(\theta))^2)} \\ &= \min(|\cos(\theta)|, |\sin(\theta)|).\end{aligned} \tag{6.22}$$

In contrast, the coherence with respect to the Hadamard basis $\mathcal{M} = \{|+\rangle, |-\rangle\}$ is

$$\mathrm{K}^{\mathrm{tr}}(|\psi\rangle|\mathcal{M}) = \sqrt{1 - \frac{1}{2}\max((\cos(\theta)+\sin(\theta))^2, (\cos(\theta)-\sin(\theta))^2)}$$
$$= \sqrt{\frac{1}{2}(1 - |\sin(\theta)|)}. \quad (6.23)$$

The trace distance of coherence with respect to the two bases is plotted in Figure 6.1 as a function of the angle θ. It is observed that when the state is maximally coherent with respect to one basis, it is minimally coherent with respect to the other basis. Indeed, as the coherence of state $|\psi\rangle$ with respect to the computational basis increases, the coherence of state $|\psi\rangle$ with respect to the Hadamard basis decreases, and vice versa.

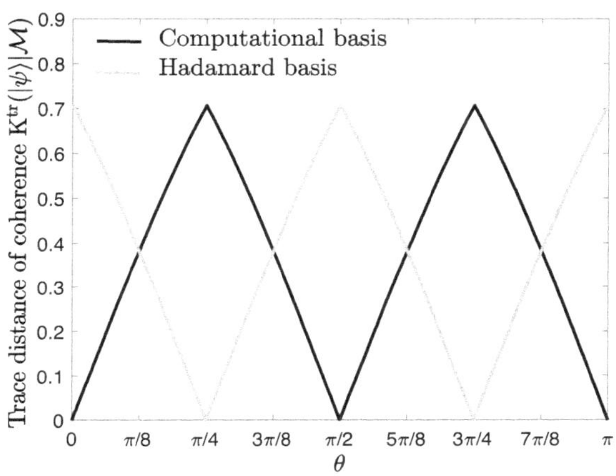

Figure 6.1 Trace distance of coherence for the qubit state $|\theta\rangle$ with respect to the computational basis and to the Hadamard basis as a function of the angle θ.

As will be discussed in Section 6.4, the computational and Hadamard bases provide an example of incompatible measurement settings.

6.3.3 Coherence and Disturbance

So far, coherence has been presented as a measure of the amount of uncertainty associated with the classical output produced by a measurement setting $\mathcal{M}$ for a state $|\psi\rangle$. As detailed next, one can equivalently view coherence in relation to the **disturbance** caused by measurement $\mathcal{M}$ to a system in state $|\psi\rangle$.

Key Result 6.3 Coherence and disturbance
A state $|\psi\rangle$ is incoherent with respect to basis $\mathcal{M}$ if and only if a measurement in setting $\mathcal{M}$ causes *no disturbance* to the state. Conversely, if a state $|\psi\rangle$ is coherent with respect to basis $\mathcal{M}$, a measurement in basis $\mathcal{M}$ causes the post-measurement state to differ from the pre-measurement state $|\psi\rangle$.

This equivalent characterization of coherence follows directly from the definition (6.15) of an incoherent state. In fact, an incoherent state (6.15) satisfies the equality $|\psi\rangle = |v_x\rangle$ for some $x \in [N]$.

According to this perspective, one may define a **coherence measure** for state $|\psi\rangle$ with respect to an orthonormal basis $\mathcal{M} = \{|v_x\rangle\}_{x\in[N]}$ as a function of the form

$$\begin{aligned} \mathrm{K}(|\psi\rangle|\mathcal{M}) &= \min_{|v\rangle\in\mathcal{M}} \mathrm{D}(|\psi\rangle, |v\rangle) \\ &= \min_{x\in[N]} \mathrm{D}(|\psi\rangle, |v_x\rangle), \end{aligned} \tag{6.24}$$

where $\mathrm{D}(\cdot,\cdot)$ is any distance measure between two states. The quantity (6.24) measures the minimum distance between the pre-measurement state $|\psi\rangle$ and a post-measurement state. This way, a coherence measure $\mathrm{K}(|\psi\rangle|\mathcal{M})$ of the form (6.24) gauges the disturbance caused by measurement $\mathcal{M}$ to state $|\psi\rangle$. The trace distance of coherence in (6.17) provides an example of such a coherence measure.

6.3.4 Coherence as a Resource ⏭

Coherence is a key property in quantum computing, as quantum states in genuine superpositions of basis states constitute a critical **resource** for quantum information processing tasks. For instance, superposition enables quantum computers to simultaneously store and process information across multiple computational branches. This phenomenon, known as **quantum parallelism**, is fundamental to several quantum algorithms, such as Grover's search, which demonstrate provable advantages over classical solutions.

Adopting this perspective, coherence can be quantitatively studied through the framework of **resource theory**. Informally, a resource is defined as a property of a quantum state that is valuable for specific information processing tasks. This concept is formalized by specifying a set of **free states**, which lack the property of interest and are therefore not useful for the task.

As illustrated in Figure 6.2, the degree to which a state $|\psi\rangle$ possesses the desired property is quantified by a measure of the distance between the state and the set of free states. This approach provides a rigorous means to evaluate the utility of a quantum state as a resource in quantum computation and other applications.

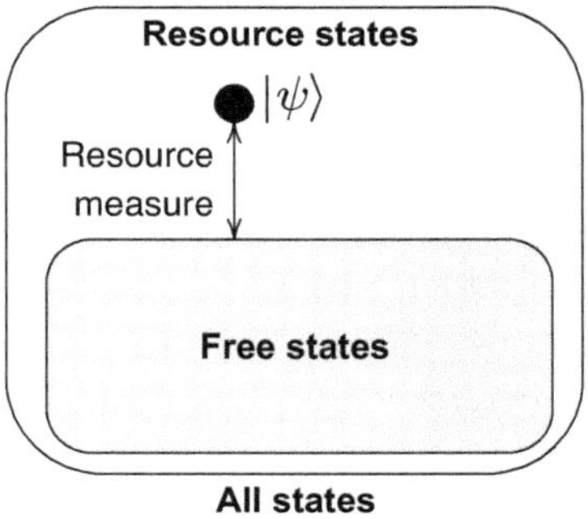

Figure 6.2 In resource theory, the notion of a resource is formalized by specifying a set of free states that do not possess the property of interest. This way, the level of this resource present in a state $|\psi\rangle$ is measured by the distance between the state and the set of free states.

In the case of coherence as a resource with respect to a given measurement basis $\mathcal{M} = \{|v_x\rangle\}_{x\in[N]}$, the set of free states consists of the states $\{|v_x\rangle\}_{x\in[N]}$ within the measurement basis, which are incoherent with respect to $\mathcal{M}$. These basis states in $\mathcal{M}$ are therefore referred to as **coherence-free states** with respect to the measurement basis $\mathcal{M}$.

Accordingly, the trace distance of coherence (6.17) and the more general coherence measure (6.24) function as resource measures. They quantify the distance between the state $|\psi\rangle$ and the set of coherence-free states, thus assessing the degree of coherence present in $|\psi\rangle$.

A more comprehensive discussion of resource theory, including the concept of free operations, will be presented in Chapter 12.

6.4 Compatibility of Measurements and Uncertainty Inequality ⏭

6.4.1 Uncertainty Principle

Consider a scientist studying two distinct physical properties of a quantum system. A standard textbook example involves measuring the position and momentum of a particle. For a given quantum state $|\psi\rangle$, the scientist is interested in the classical outputs produced under two different measurement settings, $\mathcal{M}_0$ and $\mathcal{M}_1$. A key question arises: How are the classical outputs from these two measurement settings related? Furthermore, can both measurements be applied simultaneously on the same copy of the quantum state?

To explore these questions, first consider a simple scenario. Suppose that for each state in one measurement basis, there exists an equivalent, state in the other basis. In this case, performing one measurement effectively also performs the other. Measurement bases with this property are said to be compatible. Accordingly, two compatible bases are such that each state in one basis is incoherent with respect to the other basis.

In contrast, if at least one state in one basis is mutually *coherent* with respect to the other basis, the two bases are classified as incompatible. This section demonstrates that when the states corresponding to the measurement settings $\mathcal{M}_0$ and $\mathcal{M}_1$ are sufficiently incompatible, a low measured entropy for one setting must imply a high measured entropy for the other. This relationship is an expression of the well-known **uncertainty principle**: For any given quantum state $|\psi\rangle$, if one measurement setting $\mathcal{M}_0$ produces a low-uncertainty – or sharp – observation, then a sufficiently incompatible measurement setting $\mathcal{M}_1$ must yield a highly uncertain output.

Historically, the uncertainty principle was first formulated by Werner Heisenberg in terms of observables, such as the position and momentum of a particle. In this formulation, the principle states that if two observables have sufficiently incompatible measurement bases, then achieving precise measurement of one observable necessarily leaves the other uncertain.

In this chapter, the uncertainty principle is explored using the framework of measurement settings, while Chapter 7 will revisit the traditional formalization in terms of observables.

6.4.2 Compatibility of Measurement Settings

To formalize the uncertainty principle, it is useful to first define compatible and incompatible measurement bases as follows.

Definition 6.5 Compatible and incompatible measurement settings
Two measurement settings $\mathcal{M}_0 = \{|v_x\rangle\}_{x\in[N]}$ and $\mathcal{M}_1 = \{|u_x\rangle\}_{x\in[N]}$ are said to be *compatible* if for each state $|v_x\rangle$ in basis $\mathcal{M}_0$ there is an equivalent state $|u_y\rangle$ in basis $\mathcal{M}_1$. The two measurement settings $\mathcal{M}_0$ and $\mathcal{M}_1$ are said to be *incompatible* otherwise.

This definition can equivalently be expressed using the notion of coherence. In particular, the two bases are compatible if each state in one measurement basis is incoherent with respect to the other measurement basis. That is, for all $x \in [N]$, there exists a $y \in [N]$ such that the fidelity between the states $|v_x\rangle$ and $|u_y\rangle$ is maximal, i.e.,

$$\mathrm{F}(|v_x\rangle, |u_y\rangle) = |\langle v_x|u_y\rangle|^2 = 1. \tag{6.25}$$

In contrast, if at least one state in one basis is coherent with the other basis, the two measurement bases are incompatible.

Definition 6.6 Maximally incompatible measurement settings
Measurement settings $\mathcal{M}_0$ and $\mathcal{M}_1$ are said to be maximally incompatible if every state in one basis is maximally coherent with respect to the other basis. That is, for all pairs $x \in [N]$ and $y \in [N]$, the fidelity between states $|v_x\rangle$ and $|u_y\rangle$ satisfies the equality (6.21), i.e.,

$$\mathrm{F}(|v_x\rangle, |u_y\rangle) = \frac{1}{N}. \tag{6.26}$$

Maximally incompatible bases are also known as **mutually unbiased** or **complementary**.

In order to study the degree of compatibility of two measurement settings $\mathcal{M}_0$ and $\mathcal{M}_1$, the following matrix plays a key role.

Definition 6.7 Compatibility matrix
Given two measurement bases $\mathcal{M}_0$ and $\mathcal{M}_1$, the $N \times N$ compatibility matrix

$$C(\mathcal{M}_0, \mathcal{M}_1) = \begin{bmatrix} F(|v_0\rangle, |u_0\rangle) & F(|v_0\rangle, |u_1\rangle) & \cdots & F(|v_0\rangle, |u_{N-1}\rangle) \\ F(|v_1\rangle, |u_0\rangle) & F(|v_1\rangle, |u_1\rangle) & \cdots & F(|v_1\rangle, |u_{N-1}\rangle) \\ \vdots & \vdots & \ddots & \vdots \\ F(|v_{N-1}\rangle, |u_0\rangle) & F(|v_{N-1}\rangle, |u_1\rangle) & \cdots & F(|v_{N-1}\rangle, |u_{N-1}\rangle) \end{bmatrix} \tag{6.27}$$

reports the fidelity values for every pair of states from the two bases.

Since the fidelity is a symmetric function of its inputs, the compatibility matrix satisfies the equality (Problem 6.3)

$$C(\mathcal{M}_0, \mathcal{M}_1)^T = C(\mathcal{M}_1, \mathcal{M}_0). \tag{6.28}$$

Furthermore, it belongs to a special class of matrices defined next.

Definition 6.8 Doubly stochastic matrix
An $N \times N$ matrix A is said to be doubly stochastic if it has non-negative entries $A_{x,y} \geq 0$, and the sums of the elements over any column and over any row are equal to one, i.e.,

$$\sum_{x \in [N]} A_{x,y} = 1 \text{ for all } y \in [N] \tag{6.29}$$

and

$$\sum_{y \in [N]} A_{x,y} = 1 \text{ for all } x \in [N]. \tag{6.30}$$

The following are two extreme examples of doubly stochastic matrices.

Example 6.4

An $N \times N$ matrix A is said to be a **permutation matrix** if every row has exactly one entry equal to 1 with all other entries equal to 0, and every column has exactly one entry equal to 1 with all other entries equal to 0. Permutation matrices are doubly stochastic.

Example 6.5

A matrix with all entries equal to $1/N$ is doubly stochastic.

Key Result 6.4 Compatibility of measurement settings and compatibility matrix Any compatibility matrix $C(\mathcal{M}_0, \mathcal{M}_1)$ is doubly stochastic. Furthermore, two measurement settings $\mathcal{M}_0$ and $\mathcal{M}_1$ are *compatible* if and only if the compatibility matrix $C(\mathcal{M}_0, \mathcal{M}_1)$ is a permutation matrix; while they are *maximally incompatible* if and only if all the entries of matrix $C(\mathcal{M}_0, \mathcal{M}_1)$ are equal to $1/N$.

To prove that any compatibility matrix is doubly stochastic, sum the entries over any xth row to obtain

$$\sum_{y\in[N]} F(|v_x\rangle, |u_y\rangle) = \sum_{y\in[N]} |\langle v_x|u_y\rangle|^2 = \sum_{y\in[N]} p_y(|v_x\rangle|\mathcal{M}_1) = 1; \tag{6.31}$$

and a similar calculation applies for the columns (Problem 6.3):

$$\sum_{x\in[N]} F(|v_x\rangle, |u_y\rangle) = 1. \tag{6.32}$$

Example 6.6

Consider the measurement settings $\mathcal{M}_0 = \{|0\rangle, |1\rangle\}$, corresponding to the computational basis, and the measurement setting $\mathcal{M}_1 = \{|\theta\rangle, |\theta^{\perp}\rangle\}$, where $|\theta\rangle = [\cos(\theta), \sin(\theta)]^T$ and $|\theta^{\perp}\rangle = [-\sin(\theta), \cos(\theta)]^T$ for some angle $0 \leq \theta < \pi$. Note that the measurement basis $\mathcal{M}_1$ coincides with the Hadamard basis when θ equals $\pi/4$ or $3/4\pi$. The compatibility matrix between the two measurement settings is

$$C(\mathcal{M}_0, \mathcal{M}_1) = \begin{bmatrix} (\cos(\theta))^2 & (\sin(\theta))^2 \\ (\sin(\theta))^2 & (\cos(\theta))^2 \end{bmatrix}. \tag{6.33}$$

For any angle θ, the matrix $C(\mathcal{M}_0, \mathcal{M}_1)$ is doubly stochastic, as the sums over rows and columns all equal to $(\cos(\theta))^2 + (\sin(\theta))^2 = 1$.

The two settings are compatible if θ equals 0 or $\pi/2$, in which case the compatibility matrix has a single 1 in each row and column, and thus is a permutation matrix. Conversely, the settings are complementary if θ equals $\pi/4$ or $3/4\pi$, as in this case all entries of the compatibility are equal, i.e.,

$$C(\mathcal{M}_0, \mathcal{M}_1) = \frac{1}{2}\begin{bmatrix} 1 & 1 \\ 1 & 1 \end{bmatrix}. \tag{6.34}$$

6.4.3 Compatibility Measure

The compatibility matrix provides a detailed description of the compatibility of two measurement settings. It is often preferable, however, to have a scalar compatibility metric, which describes the compatibility between two settings using a single number. One such measure is described next.

Definition 6.9 Compatibility measure

The compatibility between two measurement settings $\mathcal{M}_0$ and $\mathcal{M}_1$ is given by the maximum value of the entries of the compatibility matrix, i.e.,

$$\mathrm{c}(\mathcal{M}_0, \mathcal{M}_1) = \max_{x,y\in[N]} \mathrm{F}(|v_x\rangle, |u_y\rangle) = \max_{x,y\in[N]} |\langle v_x|u_y\rangle|^2. \tag{6.35}$$

This corresponds to the maximum fidelity between states from the two measurement bases.

Example 6.7

For $N = 2$, i.e., for single qubits, the compatibility $\mathrm{c}(\mathcal{M}_0, \mathcal{M}_1)$ provides a full description of the compatibility matrix $\mathrm{C}(\mathcal{M}_0, \mathcal{M}_1)$ In fact, apart from the ordering of the rows and columns, which is not relevant, knowing one entry of the matrix allows the reconstruction of all the other entries via the conditions (6.31) and (6.32) (see also Problem 6.7).

More generally, for $N > 2$, the compatibility $\mathrm{c}(\mathcal{M}_0, \mathcal{M}_1)$ offers incomplete information about the compatibility matrix $\mathrm{C}(\mathcal{M}_0, \mathcal{M}_1)$. Specifically, the following property follows directly from the fact that the compatibility matrix is doubly stochastic (see Problem 6.6).

Key Result 6.5 Bounds on the compatibility

The compatibility measure $\mathrm{c}(\mathcal{M}_0, \mathcal{M}_1)$ satisfies the inequalities

$$\underbrace{\frac{1}{N}}_{\text{maximally incompatible}} \leq \mathrm{c}(\mathcal{M}_0, \mathcal{M}_1) \leq \underbrace{1}_{\text{may be compatible}}. \tag{6.36}$$

The lower bound $\mathrm{c}(\mathcal{M}_0, \mathcal{M}_1) = 1/N$ is obtained if and only if the measurement bases $\mathcal{M}_0$ and $\mathcal{M}_1$ are maximally incompatible, while the upper bound $\mathrm{c}(\mathcal{M}_0, \mathcal{M}_1) = 1$ *must* be obtained if the measurement bases $\mathcal{M}_0$ and $\mathcal{M}_1$ are compatible. Thus, the condition $\mathrm{c}(\mathcal{M}_0, \mathcal{M}_1) = 1$ is *necessary*, but not sufficient, for the two measurement bases to be compatible. This result can be stated as the implication

$$\mathrm{c}(\mathcal{M}_0, \mathcal{M}_1) < 1 \Longrightarrow \mathcal{M}_0 \text{ and } \mathcal{M}_1 \text{ are incompatible.} \tag{6.37}$$

Example 6.8

For a qubit system, consider the computational basis $\mathcal{M}_0 = \{|0\rangle, |1\rangle\}$ and the Hadamard basis $\mathcal{M}_1 = \{|+\rangle, |-\rangle\}$. From (6.34), the compatibility (6.35) is evaluated as

$$c(\mathcal{M}_0, \mathcal{M}_1) = \frac{1}{2}. \tag{6.38}$$

In this case, the compatibility $c(\mathcal{M}_0, \mathcal{M}_1)$ takes its minimal value in (6.36), which implies that the measurement bases are maximally incompatible.

Example 6.9

For a system with $N = 4$ dimensions, consider the computational basis $\mathcal{M}_0 = \{|0\rangle, |1\rangle, |2\rangle, |3\rangle\}$ and the basis

$$\mathcal{M}_1 = \left\{\frac{1}{2}(|0\rangle + |1\rangle + |2\rangle + |3\rangle), \frac{1}{\sqrt{2}}(|0\rangle - |1\rangle), \frac{1}{\sqrt{2}}(|2\rangle - |3\rangle), \frac{1}{2}(|0\rangle + |1\rangle - |2\rangle - |3\rangle)\right\}. \tag{6.39}$$

The compatibility matrix is evaluated as

$$C(\mathcal{M}_0, \mathcal{M}_1) = \frac{1}{2}\begin{bmatrix} 0.5 & 1 & 0 & 0.5 \\ 0.5 & 1 & 0 & 0.5 \\ 0.5 & 0 & 1 & 0.5 \\ 0.5 & 0 & 1 & 0.5 \end{bmatrix}. \tag{6.40}$$

Accordingly, the compatibility is given by $c(\mathcal{M}_0, \mathcal{M}_1) = 1/2$, which is in the middle of the range (6.36) of possible values $1/4 \leq c(\mathcal{M}_0, \mathcal{M}_1) \leq 1$. Therefore, by (6.37), the two bases are incompatible, although they are not maximally incompatible.

Example 6.10

For a system with $N = 4$ dimensions, consider again the computational basis $\mathcal{M}_0 = \{|0\rangle, |1\rangle, |2\rangle, |3\rangle\}$, along with the basis

$$\mathcal{M}_1 = \left\{|0\rangle, |1\rangle, \frac{1}{\sqrt{2}}(|2\rangle - |3\rangle), \frac{1}{\sqrt{2}}(|2\rangle + |3\rangle)\right\}. \tag{6.41}$$

The compatibility matrix is

$$C(\mathcal{M}_0, \mathcal{M}_1) = \begin{bmatrix} 1 & 0 & 0 & 0 \\ 0 & 1 & 0 & 0 \\ 0 & 0 & 0.5 & 0.5 \\ 0 & 0 & 0.5 & 0.5 \end{bmatrix}, \tag{6.42}$$

and thus the compatibility is $c(\mathcal{M}_0, \mathcal{M}_1) = 1$. While the compatibility $c(\mathcal{M}_0, \mathcal{M}_1)$ attains the maximum value in (6.36), the two bases are not compatible.

As illustrated by the previous example, in general, the condition $c(\mathcal{M}_0, \mathcal{M}_1) = 1$ is only necessary for the compatibility of two bases. However, in the special case $N = 2$, this condition is both necessary and sufficient (see Example 6.7).

Example 6.11

The computational basis $\mathcal{M}_0$ is maximally incompatible with any Hadamard basis $\mathcal{M}_1$ for any dimension N (see Section 5.4 for the definition of Hadamard bases). In fact, in this case the compatibility matrix $\mathrm{C}(\mathcal{M}_0, \mathcal{M}_1)$ has all entries equal to $1/N$, and thus the compatibility $\mathrm{c}(\mathcal{M}_0, \mathcal{M}_1)$ also equals $1/N$.

As an example, take the computational basis $\mathcal{M}_0$ and the Fourier basis $\mathcal{M}_1$, which, as seen in Section 5.4, is a Hadamard basis. When the states in the computational basis represent spatial coordinates along a line or along the time axis, the states in the Fourier basis represent the corresponding frequency components in space or time. Accordingly, measurements of position or time are incompatible with the corresponding measurements in the frequency domain.

6.4.4 Uncertainty Inequality

The definition of the compatibility between two measurement settings allows the formalization of the uncertainty principle via the **uncertainty inequality**. The uncertainty inequality states that it is not possible for incompatible measurement settings to both yield low-entropy outputs when applied to the same state. Formally, using the Shannon entropy as a measure of uncertainty, the following key result, proved in Appendix 11.A, holds.

Key Result 6.6 Uncertainty inequality
For any quantum state $|\psi\rangle$, given two measurement settings $\mathcal{M}_0$ and $\mathcal{M}_1$, the corresponding Shannon measured entropies satisfy the inequality

$$\mathrm{H}(|\psi\rangle|\mathcal{M}_0) + \mathrm{H}(|\psi\rangle|\mathcal{M}_1) \geq \log_2\left(\frac{1}{\mathrm{c}(\mathcal{M}_0, \mathcal{M}_1)}\right). \tag{6.43}$$

In words, the uncertainty inequality (6.43) says that, if the inequality $\mathrm{c}(\mathcal{M}_0, \mathcal{M}_1) < 1$ holds – and hence by (6.37) the two measurements are incompatible – it is not possible for both measurements to yield zero-entropy outputs:

$$\mathrm{H}(|\psi\rangle|\mathcal{M}_0) + \mathrm{H}(|\psi\rangle|\mathcal{M}_1) > 0 \text{ for } \mathrm{c}(\mathcal{M}_0, \mathcal{M}_1) < 1. \tag{6.44}$$

In fact, by (6.44), if one of the two measurement entropies is zero, say $\mathrm{H}(|\psi\rangle|\mathcal{M}_0) = 0$, then the other measured entropy must be positive, $\mathrm{H}(|\psi\rangle|\mathcal{M}_1) > 0$.

Example 6.12

For the case of a qubit system, take again the maximally incompatible measurement settings given by the computational basis $\mathcal{M}_0$ and by the Hadamard basis $\mathcal{M}_1$. As seen in Example 6.11, for these bases, the compatibility is minimal, i.e., $\mathrm{c}(\mathcal{M}_0, \mathcal{M}_1) = 1/2$.

Therefore, the uncertainty inequality (6.43) can be evaluated as

$$H(|\psi\rangle|\mathcal{M}_0) + H(|\psi\rangle|\mathcal{M}_1) \geq 1. \tag{6.45}$$

To see the inequality in action, consider the state $|\psi\rangle = |0\rangle$. The computational-basis setting yields the measured entropy $H(|0\rangle|\mathcal{M}_0) = 0$, since the state $|0\rangle$ is incoherent with respect to the computational basis $\mathcal{M}_0$. By the uncertainty inequality (6.45), this implies that the measured entropy must be maximal for the Hadamard basis, i.e., $H(|0\rangle|\mathcal{M}_1) = \log_2(2) = 1$, as can readily be verified.

Taking the computational basis vectors to represent coordinates in space or time instants, the state $|0\rangle$ corresponds to a system in a given position in space or time. Accordingly, since the Hadamard basis is the Fourier basis for $N = 2$, a state like $|0\rangle$, which is perfectly certain in terms of its location in space or time, is maximally uncertain when measured in the frequency domain.

Example 6.13

Generalizing the previous example, consider any two maximally incompatible measurements. By (6.36), maximally incompatible measurement settings have minimal compatibility metric $c(\mathcal{M}_0, \mathcal{M}_1) = 1/N$, and thus the uncertainty inequality (6.43) imposes the condition

$$H(|\psi\rangle|\mathcal{M}_0) + H(|\psi\rangle|\mathcal{M}_1) \geq \log_2(N). \tag{6.46}$$

Since the maximum value of the Shannon entropy is $\log_2(N)$, this condition in turn implies that, if the measured entropy is zero under one measurement setting, then the measured entropy under the other setting must be maximal, i.e., equal to $\log_2(N)$.

For the case $N = 4$ with the computational basis $\mathcal{M}_0$ and the basis $\mathcal{M}_1$ in (6.41), the compatibility equals $c(\mathcal{M}_0, \mathcal{M}_1) = 1$, yielding the trivial lower bound $-\log_2(c(\mathcal{M}_0, \mathcal{M}_1)) = 0$. In this case, while the two measurements are incompatible, the uncertainty inequality is not informative, as it yields the evident inequality $H(|\psi\rangle|\mathcal{M}_0) + H(|\psi\rangle|\mathcal{M}_1) \geq 0$.

6.4.5 Uncertainty Principle and Simultaneous Measurements

The uncertainty principle is often stated as implying that two compatible measurement settings $\mathcal{M}_0$ and $\mathcal{M}_1$ can be carried out simultaneously, while incompatible measurements cannot be carried out at the same time. This perspective is studied in this subsection.

As illustrated in Figure 6.3(a), given a system in state $|\psi\rangle$ as input, a **simultaneous measurement** produces two classical outputs

$$x \sim p(|\psi\rangle|\mathcal{M}_0) \text{ and } y \sim p(|\psi\rangle|\mathcal{M}_1), \tag{6.47}$$

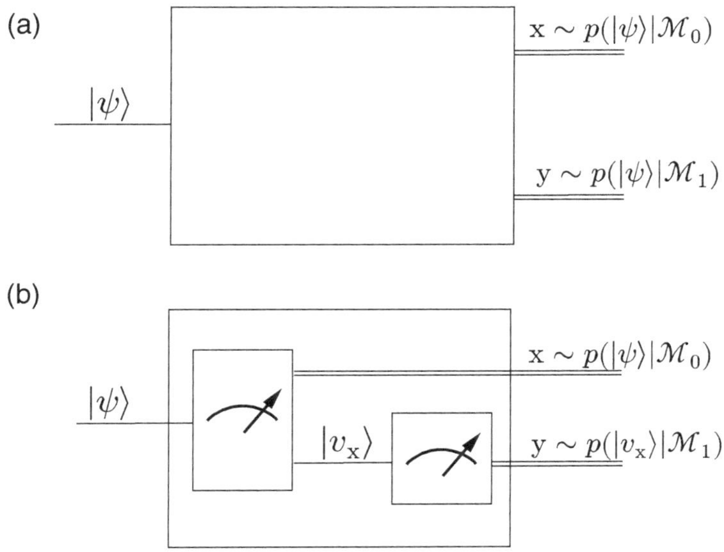

Figure 6.3 A simultaneous measurement amounts to a block that, given a system in state $|\psi\rangle$ as its input, produces two classical outputs matching the measurement distributions $p(|\psi\rangle|\mathcal{M}_0)$ and $p(|\psi\rangle|\mathcal{M}_1)$ for both settings $\mathcal{M}_0$ and $\mathcal{M}_1$.

respectively matching the measurement distributions $p(|\psi\rangle|\mathcal{M}_0)$ and $p(|\psi\rangle|\mathcal{M}_1)$ for settings $\mathcal{M}_0$ and $\mathcal{M}_1$. Note that this definition only requires the two measurement outputs to satisfy the conditions (6.47) individually, and it does not impose any requirement on the correlation, or lack thereof, between the two outputs.

For two compatible measurements, it is possible to design a simultaneous measurement block. As shown in Figure 6.3(b), this can be done by carrying out the two measurements sequentially, with the second measurement applied to the post-measurement state of the first measurement. One can readily check that this approach satisfies the condition (6.47) (see Problem 6.8).

In contrast, a simultaneous measurement block as in Figure 6.3(a) cannot be designed for incompatible measurement settings. In particular, with the sequential approach in Figure 6.3(b), the disturbance produced by the first measurement would cause the distribution of the output from the second measurement to be different from the measurement distribution $p(|\psi\rangle|\mathcal{M}_1)$ of the original state $|\psi\rangle$.

6.4.6 Generalizations of the Uncertainty Inequality

The uncertainty inequality presented in this section has been generalized in several directions.

First, there exist stronger forms of the uncertainty inequality that depend on the entire compatibility matrix, rather than just on the scalar compatibility $\mathrm{c}(\mathcal{M}_0, \mathcal{M}_1)$.

Second, there are extensions that involve other entropies, such as the **Maassen–Uffink inequality**

$$\mathrm{H}^{\mathrm{R}}_{\alpha}(|\psi\rangle|\mathcal{M}_0) + \mathrm{H}^{\mathrm{R}}_{\beta}(|\psi\rangle|\mathcal{M}_1) \geq \log_2\left(\frac{1}{\mathrm{c}(\mathcal{M}_0, \mathcal{M}_1)}\right), \tag{6.48}$$

which holds for the Rényi measured entropies with any orders $\alpha, \beta \geq 0.5$ satisfying $1/\alpha + 1/\beta = 2$. There are also related inequalities for the Tsallis measured entropy, and, more generally, for (h, ϕ)-entropies as defined in Section 3.8.

Third, one may consider more than two measurement settings at a time, and study the mutual incompatibility between any pairs of measurements. In this regard, it turns out that one can find sets of maximally incompatible measurement settings that contain at most $N + 1$ measurement settings.

Finally, there are also state-dependent uncertainty inequalities, in which the right-hand side of the inequality depends on the state $|\psi\rangle$, as will be detailed in Chapter 7. The reader is referred to the Recommended Resources for further discussions along these lines.

6.5 Projective Measurements

6.5.1 Partial Measurements

Complete measurements extract maximal information from a quantum state but disturb it to such an extent that the post-measurement state no longer depends on the pre-measurement state $|\psi\rangle$. This trade-off between information extraction and disturbance is encapsulated by the **no-information-without-disturbance** principle.

This raises an important question: Are there types of measurements that extract less information from a quantum state while being gentler, causing less disturbance to the pre-measurement state?

At one extreme, if no measurement is applied, the pre-measurement state remains unchanged. This section explores the middle ground between the extremes of no measurement and complete measurement. Specifically, it introduces **partial quantum measurements**, which provide less information than complete measurements while causing less disturbance to the original state.

A partial quantum measurement yields a classical output that takes values in a reduced set $[M]$ with $M \leq N$, where $M = N$ corresponds to a complete quantum measurement. Smaller values of M result in more **coarse-grained** measurements, extracting less information from the quantum state. In accordance with the no-information-without-disturbance principle, such coarse-grained measurements disturb the quantum state less, preserving some of its original properties.

To systematically introduce the class of partial quantum measurements discussed in this chapter, known as **projective measurements**, the presentation in this chapter will follow these steps:

1. Formalize partial classical measurements (Section 6.5.2).
2. Revisit the formalism of complete quantum measurements using the concept of projection matrices (Section 6.5.3).

3. Connect projection matrices to the span of a subset of orthonormal states (Section 6.5.4).
4. Based on this background, introduce projective measurements, providing a framework to understand partial quantum measurements and their properties (Sections 6.5.5 and 6.5.6).

6.5.2 Partial Classical Measurement

To start, consider the following examples.

Example 6.14

Partial classical measurements can be defined in terms of questions whose answers do not fully identify the state. For instance, for a classical system with $N = 4$ states taking values in set $[4] = \{0, 1, 2, 3\}$, a partial classical measurement may be defined as the response to the question "is x even?"

To elaborate, denote as $\mathrm{y} \in \{0, 1\}$ the answer, with $\mathrm{y} = 0$ representing answer yes, and $\mathrm{y} = 1$ indicating answer no. As discussed in Chapter 2, this question partitions the set [4] into $M = 2$ subsets – the subset $\mathcal{X}_0 = \{0, 2\}$ of even numbers and the subset $\mathcal{X}_1 = \{1, 3\}$ of odd numbers. Accordingly, the answer $\mathrm{y} = y$ is produced when $\mathrm{x} \in \mathcal{X}_y$, that is, when a hypothetical complete measurement – which is actually not carried out – would have returned a value $\mathrm{x} = x$ in set $\mathcal{X}_y$.

Therefore, if the complete measurement of state x is characterized by a 4×1 probability vector $p = [p_0, p_1, p_2, p_3]^T$, the partial measurement, denoted as $\mathcal{P}$, returns variable $\mathrm{y} \sim p(\mathcal{P})$ with the 2×1 probability vector

$$p(\mathcal{P}) = \begin{bmatrix} p_0(\mathcal{P}) \\ p_1(\mathcal{P}) \end{bmatrix} = \begin{bmatrix} p_0 + p_2 \\ p_1 + p_3 \end{bmatrix}, \tag{6.49}$$

where $p_0(\mathcal{P})$ is the probability of returning yes, i.e., $\mathrm{y} = 0$, and $p_1(\mathcal{P})$ is the probability of producing no as the answer, i.e., $\mathrm{y} = 1$.

Example 6.15

As another example, consider a language model producing a text $\mathrm{x} \sim p$ in response to some prompt. Different responses x may carry the same meaning y in relation to the query of interest, i.e., "The first Roman Emperor was Octavian" or "Octavian was the first Roman Emperor." Accordingly, when the only quantity of interest is the semantic content y, one can view the model's response as a partial measurement in which subsets $\mathcal{X}_y$ collect all possible output texts that carry the same meaning y. The reader is referred to Problem 2.10 for related discussions.

To generalize the two examples above, the following definition is instrumental.

Definition 6.10 Partition
A partition of the set $[N]$ consists of $M \leq N$ disjoint non-empty subsets $\mathcal{X}_y$, with $y \in [M]$, that satisfy the conditions

$$\mathcal{X}_y \cap \mathcal{X}_{y'} = \emptyset \text{ for } y \neq y' \text{ and } [N] = \bigcup_{y \in [M]} \mathcal{X}_y. \tag{6.50}$$

By (6.50), every possible value $x \in [N]$ is included in one, and only one, subset $\mathcal{X}_y$ of a partition.

A **partial classical measurement** $\mathcal{P}$ with $M \leq N$ outputs is defined by a partition $\{\mathcal{X}_y\}_{y \in [M]}$ that returns one of $M \leq N$ values in the set $[M]$. Specifically, the partial measurement $\mathcal{P}$ returns output y whenever the state of the system takes any value $x \in \mathcal{X}_y$, and hence whenever a hypothetical complete measurement would have returned output $x \in \mathcal{X}_y$.

Formally, for an $N \times 1$ probability vector p describing the underlying complete measurement, the partial measurement $\mathcal{P}$ returns a random variable $\mathrm{y} \sim p(\mathcal{P})$ with the $M \times 1$ probability vector $p(\mathcal{P}) = [p_0(\mathcal{P}), \ldots, p_{M-1}(\mathcal{P})]^T$ having entries

$$p_y(\mathcal{P}) = \sum_{x \in \mathcal{X}_y} p_x \tag{6.51}$$

for $y \in [M]$.

By discarding information about the identity of the specific output x within the corresponding subset $\mathcal{X}_y$ for the observation $y \in [M]$, a partial measurement $\mathcal{P}$ returns a less informative output than the original complete measurement $\mathcal{M}$. Mathematically, the entropy of the output $\mathrm{y} \sim p_y(\mathcal{P})$ cannot be larger than the entropy of the output for the original measurement, i.e.,

$$\mathrm{H}_\alpha(p(\mathcal{P})) \leq \mathrm{H}_\alpha(p). \tag{6.52}$$

This inequality will be proved in a more general quantum setting later in the chapter (see (6.82) below).

6.5.3 Complete Quantum Measurement via Projection Matrices

As the next step toward the definition of projective measurements, this subsection reviews complete quantum measurements in an arbitrary orthonormal basis $\mathcal{M} = \{|v_x\rangle\}_{x \in [N]}$ by using the notation of projection matrices.

Definition 6.11 Projection matrices for a complete measurement
The projection matrix associated with state $|v_x\rangle$ is given by

$$\Pi_x = |v_x\rangle\langle v_x|. \tag{6.53}$$

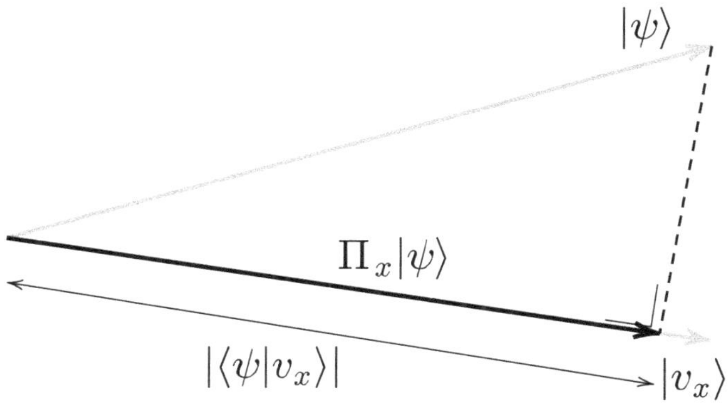

Figure 6.4 The multiplication of a state $|\psi\rangle$ by matrix Π_x returns the unnormalized vector $\Pi_x|\psi\rangle$ corresponding to the orthogonal projection of vector $|\psi\rangle$ in the direction $|v_x\rangle$.

Geometrically, as illustrated in Figure 6.4, the multiplication $\Pi_x|\psi\rangle$ of a state $|\psi\rangle$ by the projection matrix Π_x implements the orthogonal projection of vector $|\psi\rangle$ in the direction $|v_x\rangle$. Specifically, when applied to a state $|\psi\rangle$, the projection matrix Π_x returns the vector

$$\begin{aligned}\Pi_x|\psi\rangle &= (|v_x\rangle\langle v_x|)|\psi\rangle \\ &= |v_x\rangle \cdot \langle v_x|\psi\rangle \\ &= \langle v_x|\psi\rangle \cdot |v_x\rangle,\end{aligned} \tag{6.54}$$

which, as seen in the figure, is proportional to state $|v_x\rangle$, and has a 2-norm equal to the absolute value of the inner product between the two vectors, $|\langle v_x|\psi\rangle|$ (see Section 5.4.2). Note that the vector $\Pi_x|\psi\rangle$ is generally not a valid state, since it is not normalized. In particular, the inequality $|\langle v_x|\psi\rangle| < 1$ holds unless one has the equivalence condition $|v_x\rangle = |\psi\rangle$.

Using the definition (6.53), a complete quantum measurement can equivalently be described as follows.

Key Result 6.7 Complete measurements using projection matrices
Given a complete measurement in measurement basis $\mathcal{M} = \{|v_x\rangle\}_{x\in[N]}$, the measurement probabilities for a pre-measurement state $|\psi\rangle$ can be written as

$$p_x(|\psi\rangle|\mathcal{M}) = \langle\psi|\Pi_x|\psi\rangle \tag{6.55}$$

for all $x \in [N]$, and the corresponding post-measurement state can be expressed as

$$|v_x\rangle = \frac{1}{\sqrt{p_x(|\psi\rangle|\mathcal{M})}}\Pi_x|\psi\rangle. \tag{6.56}$$

By (6.55), the measurement probability $p_x(|\psi\rangle|\mathcal{M})$ is obtained by "sandwiching" the corresponding projection matrix Π_x between the bra and ket of the pre-measurement state $|\psi\rangle$. The equivalence of (6.55) with the Born rule follows from the equalities

$$\begin{aligned} p_x(|\psi\rangle|\mathcal{M}) &= |\langle v_x|\psi\rangle|^2 \\ &= (\langle v_x|\psi\rangle)^* \langle v_x|\psi\rangle \\ &= \langle\psi|v_x\rangle\langle v_x|\psi\rangle \\ &= \langle\psi|\Pi_x|\psi\rangle. \end{aligned} \tag{6.57}$$

As for the post-measurement state (6.56), it is observed that the division by $\sqrt{p_x(|\psi\rangle|\mathcal{M})}$ is necessary to ensure the normalization of the post-measurement state:

$$\left\| \frac{1}{\sqrt{p_x(|\psi\rangle|\mathcal{M})}} \Pi_x|\psi\rangle \right\|_2^2 = \frac{1}{p_x(|\psi\rangle|\mathcal{M})} \langle\psi|\Pi_x|\psi\rangle = 1. \tag{6.58}$$

In (6.58), the first equality uses the **idempotence property** of the projection matrix Π_x (see also (6.92) below for a generalization)

$$\Pi_x^\dagger \Pi_x = |v_x\rangle\langle v_x||v_x\rangle\langle v_x| = \Pi_x. \tag{6.59}$$

6.5.4 Projection Matrices and Span

The final preliminary concept to review prior to the definition of projection measurements is that of a general **projection matrix**. This is done next by first introducing the notion of the **span** of a set of orthonormal vectors.

Definition 6.11 Span of a set of orthonormal vectors
Fix $N \times 1$ orthonormal vectors $\{|v_x\rangle\}_{x\in[L]}$ with $L \leq N$. Consider the subset of the N-dimensional space that includes all $N \times 1$ vectors u that can be obtained as the superposition of the L orthonormal vectors $\{|v_x\rangle\}_{x\in[L]}$, i.e.,

$$\mathcal{S}(\{|v_x\rangle\}_{x\in[L]}) = \left\{ u = \sum_{x\in[L]} \alpha_x|v_x\rangle \text{ with complex numbers } \{\alpha_x\}_{x\in[L]} \right\}. \tag{6.60}$$

The subset $\mathcal{S}(\{|v_x\rangle\}_{x\in[L]})$ is said to be the span of the orthonormal vectors $\{|v_x\rangle\}_{x\in[L]}$, and the integer L is known as the **dimension** of the span.

The span of a set of states is a **subspace**. In linear algebra, a subspace is a subset of a vector space with the following properties.

- **Zero vector:** It contains the all-zero vector.
- **Closed under addition:** Given two vectors u and v in the subspace, it contains the sum $u + v$.
- **Closed under scalar multiplication:** Given a vector u in the subspace and a complex number α, it contains the product αu.

One can readily verify that the span $\mathcal{S}(\{|v_x\rangle\}_{x\in[L]})$ is a subspace. Furthermore, any subspace can be expressed as the span of some set of vectors. The next example illustrates that a span is a smaller vector space sitting inside a larger vector space. Furthermore, not all vectors in a span are valid quantum states, as they are not normalized in general.

Example 6.16

A span $\mathcal{S}(\{|v_0\rangle\})$ of dimension $L = 1$ is a line residing in a higher-dimensional space. In fact, it corresponds to all vectors $u = \alpha_0 v_0$ that are aligned with the given state $|v_0\rangle$.

With dimension $L = 2$, the span $\mathcal{S}(\{|v_0\rangle, |v_1\rangle\})$ is the plane with coordinate axes given by the vectors $|v_0\rangle$ and $|v_1\rangle$. As an example, for a three-dimensional space, $N = 3$, Figure 6.5 shows the span $\mathcal{S}(\{|v_0\rangle, |v_1\rangle\})$.

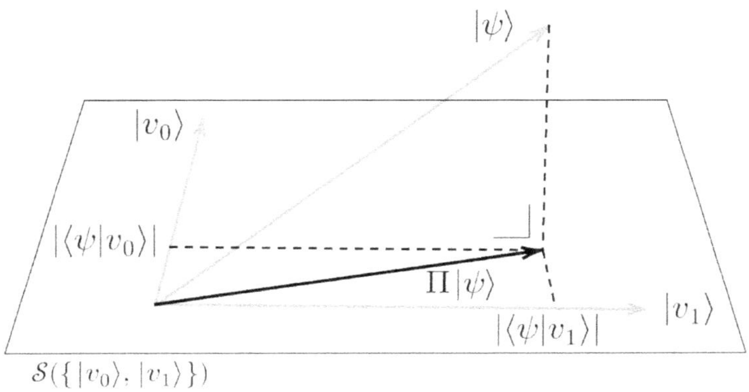

Figure 6.5 The vector $\Pi|\psi\rangle$, which is generally unnormalized, is a linear combination of the vectors $\{|v_x\rangle\}_{x\in[L]}$, and thus it lies in the span $\mathcal{S}(\{|v_x\rangle\}_{x\in[L]})$.

With $L = 3$, the span $\mathcal{S}(\{|v_0\rangle, |v_1\rangle, |v_2\rangle\})$ is a three-dimensional space with a generally larger space $N \geq L$; and so on until dimension $L = N$, for which the span $\mathcal{S}(\{|v_x\rangle\}_{x\in[N]})$ coincides with the entire N-dimensional space.

Definition 6.12 Projection matrix onto a subspace
Given $L \leq N$ orthonormal vectors $\{|v_x\rangle\}_{x\in[L]}$, the $N \times N$ matrix

$$\Pi = \sum_{x\in[L]} |v_x\rangle\langle v_x| \tag{6.61}$$

is referred to as the projection matrix onto the span $\mathcal{S}(\{|v_x\rangle\}_{x\in[L]})$. The **rank** of the projection matrix Π is equal to the dimension of the span, i.e.,

$$\text{rank}(\Pi) = L. \tag{6.62}$$

Example 6.17

The matrix (6.53) is the projection matrix onto the one-dimensional span $\mathcal{S}(|v_x\rangle)$. Thus, the rank of the matrix is given as $\text{rank}(|v_x\rangle\langle v_x|) = 1$.

More generally, the rank of an $N \times N$ projection matrix satisfies the inequalities

$$1 \leq \text{rank}(\Pi) \leq N, \tag{6.63}$$

thus ranging from 1 to the dimension of the overall space, N.

Multiplication by a projection matrix Π with any rank implements the orthogonal projection of state $|\psi\rangle$ onto the corresponding span $\mathcal{S}(\{|v_x\rangle\}_{x\in[L]})$. Mathematically, the product $\Pi|\psi\rangle$ equals

$$\begin{aligned}\Pi|\psi\rangle &= \left(\sum_{x\in[L]} |v_x\rangle\langle v_x|\right)|\psi\rangle \\ &= \sum_{x\in[L]} |v_x\rangle \cdot \langle v_x|\psi\rangle \\ &= \sum_{x\in[L]} \langle v_x|\psi\rangle \cdot |v_x\rangle.\end{aligned} \tag{6.64}$$

The resulting vector $\Pi|\psi\rangle$ is a superposition of the vectors $\{|v_x\rangle\}_{x\in[L]}$, hence lying in the span $\mathcal{S}(\{|v_x\rangle\}_{x\in[L]})$. Note that this operation generalizes the orthogonal projection (6.54) from one-dimensional spans to any dimension L. Furthermore, the vector $\Pi|\psi\rangle$ is unnormalized, unless the original state $|\psi\rangle$ is already contained in the span $\{|v_x\rangle\}_{x\in[L]}$.

Example 6.18

Figure 6.5 shows an example with an $N = 3$-dimensional space and an $L = 2$-dimensional subspace corresponding to the plane $\mathcal{S}(\{|v_0\rangle, |v_1\rangle\})$. As illustrated, the vector $\Pi|\psi\rangle$ is a superposition of the vectors $\{|v_0\rangle, |v_1\rangle\}$, hence residing on the span $\mathcal{S}(\{|v_0\rangle, |v_1\rangle\})$.

Based on the calculation (6.64), an equivalent way to think about a projection matrix Π is as the matrix that preserves a given subspace.

Key Result 6.8 Subspace preserved by a projection matrix

Fix the projection Π onto a span $\mathcal{S}(\{|v_x\rangle\}_{x\in[L]})$ as in (6.61). Multiplication by matrix Π does not change the state $|\psi\rangle$ if and only if the state $|\psi\rangle$ is in the subspace $\mathcal{S}(\{|v_x\rangle\}_{x\in[L]})$, i.e.,

$$|\psi\rangle \in \mathcal{S}(\{|v_x\rangle\}_{x\in[L]}) \iff \Pi|\psi\rangle = |\psi\rangle. \tag{6.65}$$

Given the correspondence (6.65) between projection matrix (6.61) and span $\mathcal{S}(\{|v_x\rangle\}_{x\in[L]})$, it is also useful to identify the span as a function of the projection matrix by using the notation

$$\mathcal{S}(\{|v_x\rangle\}_{x\in[L]}) = \mathcal{S}(\Pi). \tag{6.66}$$

6.5.5 Projective Measurements

Using the background built in this section, projective measurements are now introduced as a quantum version of partial classical measurements (Section 6.5.2) that use the formalism of projection matrices (Sections 6.5.3 and 6.5.4).

As shown in Figure 6.6, a projective measurement yields a random classical output $y \in [M]$, where $M \leq N$ denotes the number of possible outcomes, much as a partial classical measurement does. It also behaves like a complete quantum measurement in that it generally leaves the system in a post-measurement state that differs from the original pre-measurement state $|\psi\rangle$. However, by selecting M to be smaller than N, projective measurements introduce a form of coarse graining that reduces the disturbance to the pre-measurement state while extracting less information from it.

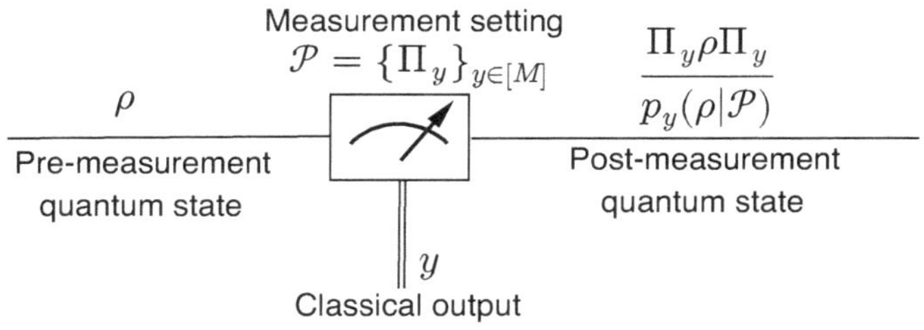

Figure 6.6 Illustration of a projective measurement $\mathcal{P}$.

A projective measurement can be defined by specifying an underlying hypothetical complete measurement with basis $\mathcal{M} = \{|v_x\rangle\}_{x\in[N]}$, along with a partition $\{\mathcal{X}_y\}_{y\in[M]}$ of the set $[N]$ (see (6.50)). This way, each possible output $y \in [M]$ of the projective measurement is associated with the projection matrix

$$\Pi_y = \sum_{x\in\mathcal{X}_y} |v_x\rangle\langle v_x|, \tag{6.67}$$

which preserves the span $\mathcal{S}(\{|v_x\rangle\}_{x\in\mathcal{X}_y})$.

Definition 6.13 Projective measurement of a pure state
Given a measurement basis $\mathcal{M} = \{|v_x\rangle\}_{x\in[N]}$ and a partition $\{\mathcal{X}_y\}_{y\in[M]}$ with $M \leq N$, a projective measurement characterized by the M projection matrices

$$\mathcal{P} = \left\{ \Pi_y = \sum_{x \in \mathcal{X}_y} |v_x\rangle\langle v_x| \right\}_{y \in [M]}, \tag{6.68}$$

when applied to a pre-measurement state $|\psi\rangle$, results in the following.

- **Classical output:** The measurement yields a classical output

$$\mathrm{y} \sim p(|\psi\rangle|\mathcal{P}) = [p_0(|\psi\rangle|\mathcal{P}), \ldots, p_{M-1}(|\psi\rangle|\mathcal{P})]^T, \tag{6.69}$$

  with probabilities

$$p_y(|\psi\rangle|\mathcal{P}) = \langle\psi|\Pi_y|\psi\rangle. \tag{6.70}$$

- **Post-measurement state:** For a given classical output $\mathrm{y} = y$, the quantum system is left in the corresponding post-measurement state

$$|\phi_y\rangle = \frac{1}{\sqrt{p_y(|\psi\rangle|\mathcal{P})}} \Pi_y|\psi\rangle. \tag{6.71}$$

Note that the classical output distribution (6.70) is a direct generalization of the Born rule (6.55) from complete to projective measurements. By extension, it is also referred to as the Born rule. In a similar way, the post-measurement state (6.71) is a direct generalization of (6.56).

Example 6.19

With $M = N$, the only possible partition of the set $[N]$ is defined by the singleton subsets $\mathcal{X}_y = \{y\}$ for $y \in [M] = [N]$. In this case, the projection matrices (6.67) coincide with the rank-1 projection matrices (6.53) of a complete quantum measurement defined by basis $\mathcal{M}$. Therefore, when the projection matrices Π_y have rank $\text{rank}(\Pi_y) = 1$ for all $y \in [M]$, projection measurements recover complete quantum measurements.

As discussed in the rest of this section, ranks larger than 1 entail a coarse graining of the underlying complete measurement setting, entailing a smaller disturbance to the pre-measurement state.

In a way similar to the classical counterpart (6.51), the measurement probability (6.70) can be interpreted in terms of the output of a hypothetical complete quantum measurement in the underlying measurement basis $\mathcal{M} = \{|v_x\rangle\}_{x \in [N]}$. Specifically, a projective measurement returns output y whenever the complete quantum measurement $\mathcal{M} = \{|v_x\rangle\}_{x \in [N]}$ would have returned a value x within the set $\mathcal{X}_y$. Indeed, the probability $p_y(|\psi\rangle|\mathcal{P})$ is the sum of the corresponding probabilities $\{p_x(|\psi\rangle|\mathcal{M})\}_{x \in \mathcal{X}_y}$ for the complete measurement $\mathcal{M}$.

To verify this point, it is sufficient to evaluate the probability $p_y(|\psi\rangle|\mathcal{P})$ of observing output $y \in [M]$ as

$$\begin{aligned}
p_y(|\psi\rangle|\mathcal{P}) &= \langle\psi|\Pi_y|\psi\rangle \\
&= \langle\psi|\left(\sum_{x\in\mathcal{X}_y} |v_x\rangle\langle v_x|\right)|\psi\rangle \\
&= \sum_{x\in\mathcal{X}_y} |\langle\psi|v_x\rangle|^2 \\
&= \sum_{x\in\mathcal{X}_y} p_x(|\psi\rangle|\mathcal{M}),
\end{aligned} \tag{6.72}$$

which is the sum of the probabilities $p_x(|\psi\rangle|\mathcal{M})$ that the complete measurement $\mathcal{M}$ returns an output x in the subset $\mathcal{X}_y$.

This coarse graining of the classical outputs $x \in \mathcal{X}_y$ into a single output y entails a loss of information compared to the hypothetical underlying complete measurement. On the flip side, a projective measurement with $M < N$ outputs generally causes a smaller disturbance on the state.

In fact, all states that belong to the span $\mathcal{S}(\Pi_y)$ of projector Π_y are left unchanged by the measurement (see (6.65)). Thus, the rank of the projection matrix Π_y, namely $\text{rank}(\Pi_y) = |\mathcal{X}_y|$, determines the level of disturbance caused by the measurement on the pre-measurement state $|\psi\rangle$. A larger rank $\text{rank}(\Pi_y)$ implies that a subspace of larger dimension, equal to $\text{rank}(\Pi_y)$, is preserved by the measurement when the classical output is y.

Larger values of $\text{rank}(\Pi_y)$ also cause the number of outputs M to be smaller, given that the equality (6.71) implies the constraint

$$N = \sum_{y\in[M]} |\mathcal{X}_y| = \sum_{y\in[M]} \text{rank}(\Pi_y). \tag{6.73}$$

This illustrates the trade-off between the amount of information extracted from a measurement, which requires larger values of M, and the disturbance caused by the measurement, which decreases with the ranks of the projection matrices.

Example 6.20

This example may be considered as the counterpart of the partial measurement for classical systems presented in Example 6.14. To describe it, take a quantum system with states of dimension $N = 4$. The underlying, hypothetical, measurement uses the computational basis $\mathcal{M} = \{|0\rangle, |1\rangle, |2\rangle, |3\rangle\}$, and the $M = 2$ subsets defining the projection matrices are given as the subset $\mathcal{X}_0 = \{0, 2\}$ of even indices and the subset $\mathcal{X}_1 = \{1, 3\}$ of odd indices. This is also known as the **parity measurement** setting.

Accordingly, the projective measurement $\mathcal{P} = \{\Pi_0, \Pi_1\}$ is defined by the two projection matrices

$$\Pi_0 = |0\rangle\langle 0| + |2\rangle\langle 2| = \begin{bmatrix} 1 & 0 & 0 & 0 \\ 0 & 0 & 0 & 0 \\ 0 & 0 & 1 & 0 \\ 0 & 0 & 0 & 0 \end{bmatrix} \tag{6.74}$$

and

$$\Pi_1 = |1\rangle\langle 1| + |3\rangle\langle 3| = \begin{bmatrix} 0 & 0 & 0 & 0 \\ 0 & 1 & 0 & 0 \\ 0 & 0 & 0 & 0 \\ 0 & 0 & 0 & 1 \end{bmatrix}, \tag{6.75}$$

with ranks $\text{rank}(\Pi_0) = \text{rank}(\Pi_1) = 2$.

Given an arbitrary pure state

$$|\psi\rangle = \psi_0|0\rangle + \psi_1|1\rangle + \psi_2|2\rangle + \psi_3|3\rangle, \tag{6.76}$$

the projective measurement $\mathcal{P} = \{\Pi_0, \Pi_1\}$ produces the classical output $\mathrm{y} = 0$ with probability (6.55) given by

$$\begin{aligned} p_0(|\psi\rangle|\mathcal{P}) &= \langle\psi|\Pi_0|\psi\rangle \\ &= \langle\psi|0\rangle\langle 0|\psi\rangle + \langle\psi|2\rangle\langle 2|\psi\rangle \\ &= |\langle\psi|0\rangle|^2 + |\langle\psi|2\rangle|^2 \\ &= |\psi_0|^2 + |\psi_2|^2; \end{aligned} \tag{6.77}$$

and, in a similar manner, the output $\mathrm{y} = 1$ is produced with probability

$$\begin{aligned} p_1(|\psi\rangle|\mathcal{P}) &= 1 - p_0(|\psi\rangle|\mathcal{P}) \\ &= |\psi_1|^2 + |\psi_3|^2. \end{aligned} \tag{6.78}$$

Moreover, when the classical output is $\mathrm{y} = 0$, the projective measurement $\mathcal{P} = \{\Pi_0, \Pi_1\}$ leaves the system in the post-measurement state

$$\begin{aligned} |\phi_0\rangle &= \frac{1}{\sqrt{p_0(|\psi\rangle|\mathcal{P})}}\Pi_0|\psi\rangle \\ &= \frac{1}{\sqrt{|\psi_0|^2 + |\psi_2|^2}}(\psi_0|0\rangle + \psi_2|2\rangle); \end{aligned} \tag{6.79}$$

while, when $\mathrm{y} = 1$, the post-measurement state is

$$|\phi_1\rangle = \frac{1}{\sqrt{|\psi_1|^2 + |\psi_3|^2}}(\psi_1|1\rangle + \psi_3|3\rangle). \tag{6.80}$$

As an example, assume that the state $|\psi\rangle$ is a superposition of even states, i.e., $\psi_1 = \psi_3 = 0$, and thus it belongs to the span $\mathcal{S}(\Pi_0)$. In this case, the classical output equals $y = 0$ with probability 1, and the state is undisturbed by the measurement, since the post-measurement state is $|\phi_0\rangle = |\psi\rangle$. In a similar manner, if the state $|\psi\rangle$ lies in the span $\mathcal{S}(\Pi_1)$, the classical output equals $y = 1$ with probability 1, and the state is also undisturbed by the measurement, yielding the post-measurement state $|\phi_1\rangle = |\psi\rangle$.

Example 6.21

The ability of a projective measurement to leave undisturbed all states within the spans of its projection matrices has several important applications. One prominent application is **quantum error correction**.

Quantum error correction works by encoding quantum information into a quantum state $|\psi\rangle$ such that the state lies within a specific subspace, known as the **coding subspace**. A projective measurement is then employed to detect and correct any "errors" that may affect the state $|\psi\rangle$.

To briefly explain the operation of quantum error correction, consider a projective measurement that includes a projection matrix Π_0, which defines the coding subspace as the span span(Π_0). The key principle is as follows: If the state $|\psi\rangle$ remains within the span $\mathcal{S}(\Pi_0)$, the measurement will leave the state undisturbed. However, if a perturbation causes the state to move outside the subspace $\mathcal{S}(\Pi_0)$, the projective measurement can detect the deviation and provide information needed to correct the error. In other words, if the state $|\psi\rangle$ remains in the coding subspace (the span $\mathcal{S}(\Pi_0)$ of matrix Π_0), the measurement yields $y = 0$ and the state is left unchanged; if $|\psi\rangle$ is perturbed out of that subspace, the measurement outcome ($y \neq 0$) flags an error and projects the state $|\psi\rangle$ onto a state that informs the error-correction procedure.

6.5.6 Direct Definition of Projective Measurements

This section has introduced projective measurements by making reference to an underlying, hypothetical, complete measurement. However, the complete measurement plays only a conceptual – and not a physical – role, as the actual measurement is a coarser-grained version of the complete measurement. In line with this observation, it is possible to introduce projective measurements directly, without first specifying a complete measurement and a corresponding partition of the complete measurement's outputs.

Instead of defining an underlying complete measurement and a partition, as done above, a projective measurement can be defined by specifying a set of M projection matrices with a specific property.

Key Result 6.9 Projective measurement setting
A projective measurement setting $\mathcal{P}$ is specified by any set of $M \leq N$ projection matrices $\{\Pi_y\}_{y\in[M]}$ that satisfy the **resolution-of-the-identity** equality

$$\sum_{y\in[M]} \Pi_y = I, \tag{6.81}$$

where I is the $N \times N$ identity matrix.

This follows because, given projection matrices $\{\Pi_y\}_{y\in[M]}$ satisfying the condition (6.81), it is always possible to identify an orthonormal basis such that the projection matrices can be expressed as in (6.67).

6.5.7 Measured Entropy for Projective Measurements

Intuitively, the uncertainty on the output of a projective measurement should be no larger than the uncertainty associated with the underlying, hypothetical, complete measurement $\mathcal{M}$. In fact, the classical output of a projective measurement $\mathcal{P}$ can be obtained by discarding information from the output of the complete measurement $\mathcal{M}$. Therefore, the classical output of the projective measurement $\mathcal{P}$ is easier to predict than the finer-grained output of the complete measurement setting $\mathcal{M}$.

To formalize this intuition, consider the measured entropy $\mathrm{H}_\alpha(|\psi\rangle|\mathcal{M})$ for the complete measurement $\mathcal{M}$ against the measured entropy $\mathrm{H}_\alpha(|\psi\rangle|\mathcal{P})$ for a projective measurement $\mathcal{P}$ derived from $\mathcal{M}$. As proved in Appendix 6.B, in a manner similar to (6.52), one has the inequality

$$\mathrm{H}_\alpha(|\psi\rangle|\mathcal{P}) \leq \mathrm{H}_\alpha(|\psi\rangle|\mathcal{M}), \tag{6.82}$$

showing that indeed the output of a projective measurement cannot be more uncertain than the output of the underlying complete measurement.

6.6 Normal Matrices and Spectral Theorem

This section introduces normal matrices, a family of matrices that play an important role in quantum physics. Normal matrices include, among others, unitary matrices and projection matrices. A central property of normal matrices is that they can always be expressed as a linear combination of rank-1 projection matrices. This result, known as the eigendecomposition, is a fundamental tool in quantum mechanics and will be utilized extensively throughout this book.

6.6.1 Normal Matrices

To start, consider the following definition.

Definition 6.14 Commuting matrices
Two $N \times N$ matrices A and B are said to commute if the product between the two matrices does not depend on the order of the terms in the multiplication, i.e.,

$$AB = BA. \tag{6.83}$$

Using this notion, normal matrices are defined as follows.

Definition 6.15 Normal matrix
An $N \times N$ matrix A is said to be normal if it satisfies the equality

$$A^\dagger A = AA^\dagger, \tag{6.84}$$

i.e., if the Hermitian transpose matrix $A^\dagger$ commutes with the matrix A.

As highlighted by the next examples, most matrices of interest in quantum mechanics are normal.

Example 6.22

Unitary matrices. An $N \times N$ unitary matrix $U = [|u_0\rangle \cdots |u_{N-1}\rangle]$, where $\{|u_x\rangle\}_{x \in [N]}$ are the N-dimensional columns, is normal since, as defined in (5.43), it satisfies the equalities

$$U^\dagger U = UU^\dagger = I. \tag{6.85}$$

Therefore, not only do matrices U and $U^\dagger$ commute, but matrix $U^\dagger$ is also the inverse of matrix U.

Example 6.23

Permutation matrices. An $N \times N$ permutation matrix S is unitary, and thus also normal. In fact, the columns of a permutation matrix are obtained by reordering the columns of the $N \times N$ identity matrix, and hence they form an orthonormal basis. Note also that, by (6.85), the inverse S^{-1} of a permutation matrix is given by the transpose $S^\dagger = S^T$.

Example 6.24

Diagonal matrices. A diagonal matrix

$$\Lambda = \text{diag}(\lambda) = \begin{bmatrix} \lambda_0 & 0 & 0 & 0 \\ 0 & \lambda_1 & 0 & 0 \\ & & & \\ 0 & 0 & \ddots & 0 \\ 0 & 0 & 0 & \lambda_{N-1} \end{bmatrix}, \tag{6.86}$$

with diagonal vector $\lambda = [\lambda_0, \ldots, \lambda_{N-1}]^T$, is normal, since it satisfies the equalities

$$\Lambda^\dagger \Lambda = \Lambda \Lambda^\dagger = \begin{bmatrix} |\lambda_0|^2 & 0 & 0 & 0 \\ 0 & |\lambda_1|^2 & 0 & 0 \\ & & & \\ 0 & 0 & \ddots & 0 \\ 0 & 0 & 0 & |\lambda_{N-1}|^2 \end{bmatrix}. \tag{6.87}$$

Example 6.25

Hermitian matrices. An $N \times N$ matrix is said to be Hermitian if it is equal to its Hermitian transpose

$$A^\dagger = A. \tag{6.88}$$

Equivalently, all the elements $A_{x,x}$ for $x \in [N]$ on the main diagonal are real-valued; and every entry $A_{x,y}$ with $x \neq y$ has the counterpart $A_{y,x}$ across the main diagonal satisfying the equality $A^*_{x,y} = A_{y,x}$. For example, matrix

$$A = \begin{bmatrix} \frac{1}{4} & \frac{1}{2}e^{i\frac{\pi}{6}} \\ \frac{1}{2}e^{-i\frac{\pi}{6}} & \frac{3}{4} \end{bmatrix} \tag{6.89}$$

is Hermitian. By the definition (6.88), the Hermitian transpose matrix $A^\dagger$ commutes with matrix A, and hence any Hermitian matrix A is normal.

Example 6.26

Pauli matrices. The Pauli matrices introduced in Chapter 5 are both unitary and Hermitian, and hence also normal, as they satisfy the conditions

$$P_x P_x^\dagger = P_x^\dagger P_x = I \text{ and } P_x^\dagger = P_x \tag{6.90}$$

for all $x \in [4]$.

Example 6.27

Projection matrices. A projection matrix Π, generally defined as in (6.61), i.e., $\Pi = \sum_{x\in[L]} |v_x\rangle\langle v_x|$, is Hermitian, and thus also normal. In fact, the Hermitian transpose of the projection matrix Π is evaluated as

$$\begin{aligned} \Pi^\dagger &= \sum_{x\in[L]} (|v_x\rangle\langle v_x|)^\dagger \\ &= \sum_{x\in[L]} |v_x\rangle\langle v_x| = \Pi, \end{aligned} \tag{6.91}$$

where the basic equality $\langle v_x|^\dagger = |v_x\rangle$ was applied.

A useful equivalent characterization of a projection matrix is that a matrix Π is a projection if and only if it is Hermitian and **idempotent**, i.e.,

$$\Pi \text{ is a projection matrix} \iff \Pi \text{ is Hermitian and } \Pi\Pi = \Pi. \tag{6.92}$$

6.6.2 Spectral Theorem

Normal matrices have the key property that, in a specific basis, they can always be represented as a diagonal matrix.

Key Result 6.10 Spectral theorem
For any $N \times N$ normal matrix A, there exist an $N \times N$ unitary matrix $U = [|u_0\rangle, \ldots, |u_{N-1}\rangle]$ and an $N\times N$ diagonal matrix $\Lambda = \text{diag}(\lambda)$ such that matrix A can be expressed as the **eigendecomposition**

$$A = U\Lambda U^\dagger \tag{6.93}$$

$$= \sum_{x\in[N]} \lambda_x |u_x\rangle\langle u_x|. \tag{6.94}$$

In the eigendecomposition (6.93) and (6.94), the diagonal elements $\{\lambda_x\}_{x\in[N]}$ in matrix Λ are known as the **eigenvalues** of matrix A; and the columns $\{|u_x\rangle\}_{x\in[N]}$ are the **eigenvectors** of matrix A.

By (6.94), any normal matrix A can be written as the linear combination of rank-1 projection matrices $|u_x\rangle\langle u_x|$, which are obtained from the eigenvectors $\{|u_x\rangle\}_{x\in[N]}$ of matrix A. The eigenvectors $\{|u_x\rangle\}_{x\in[N]}$ form an orthonormal basis of the N-dimensional space, since they are the columns of a unitary matrix U. They are accordingly also referred to collectively as an **eigenbasis** for matrix A.

Definition 6.16 Rank of a normal matrix
The rank of a normal matrix A equals the number of non-zero eigenvalues of matrix A, i.e.,

$$\mathrm{rank}(A) = |\{x \in [N]\colon \lambda_x \neq 0\}|. \tag{6.95}$$

Note that this definition of rank reduces to the definition given in (6.62) when specialized to projection matrices.

The eigendecomposition has the following properties.

- **Matrix–product multiplication as amplitude scaling:** To understand the significance of the eigendecomposition, consider the effect of a multiplication of a state $|\psi\rangle$ by a normal matrix A. Any state $|\psi\rangle$ can always be expressed as a superposition of states in the eigenbasis $\{|u_x\rangle\}_{x\in[N]}$ of matrix A as

$$|\psi\rangle = \sum_{x\in[N]} \alpha_x |u_x\rangle, \tag{6.96}$$

  where each complex amplitude α_x is obtained via the inner product $\alpha_x = \langle u_x|\psi\rangle$. Using the decomposition (6.96), the matrix–vector multiplication $A|\psi\rangle$ thus yields

$$\begin{aligned} A|\psi\rangle &= \left(\sum_{x\in[N]} \lambda_x |u_x\rangle\langle u_x|\right)\left(\sum_{x'\in[N]} \alpha_{x'}|u_{x'}\rangle\right) \\ &= \sum_{x\in[N]}\sum_{x'\in[N]} \lambda_x \alpha_{x'} |u_x\rangle\langle u_x|u_{x'}\rangle \\ &= \sum_{x\in[N]} \lambda_x \alpha_x |u_x\rangle. \end{aligned} \tag{6.97}$$

  Accordingly, as shown in Figure 6.7, multiplying a state $|\psi\rangle$ by a normal matrix A scales each amplitude α_x of the state $|\psi\rangle$ in the direction of the basis vector $|u_x\rangle$ by the corresponding eigenvalue λ_x. Note that, more broadly than in the simplified illustration in the figure, the eigenvalue λ_x is generally complex.
- **Uniqueness:** The set of eigenvalues associated with a matrix A – which is given by vector λ when disregarding the order of its entries – is unique. In contrast, there

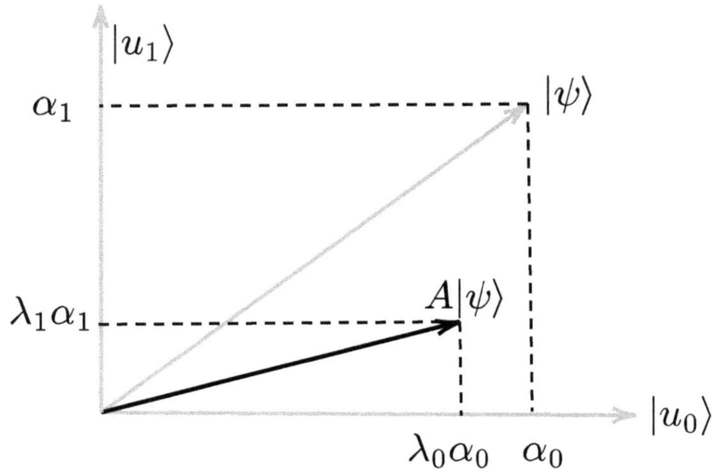

Figure 6.7 Illustration of a matrix–vector multiplication $A|\psi\rangle$ via the eigendecomposition of matrix A.

may be different eigenbases U satisfying the equality (6.93). In particular, there is a unique eigenbasis – with each eigenvector defined except for a global phase – if and only if all the eigenvalues $\{\lambda_x\}_{x\in[N]}$ are distinct. When the same eigenvalue λ_x is associated with more than one eigenvector, the eigendecomposition is not unique. In fact, in this case, the eigenvectors associated with the same eigenvalue in an eigendecomposition can be replaced by any other orthonormal basis spanning the same subspace without changing the product in (6.93).

6.6.3 Examples

The following are notable examples of eigendecompositions.

Example 6.28

Eigendecomposition of a diagonal matrix. For a diagonal matrix A, the eigendecomposition (6.93) is obtained by setting $U = I$ and $\Lambda = A$. Accordingly, the matrix–vector product $A|\psi\rangle$ simply multiplies each xth entry of the vector $|\psi\rangle$ by the xth element on the diagonal, λ_x.

Example 6.29

Eigendecomposition of a unitary matrix. For a unitary matrix U, by the spectral theorem, one has

$$UU^\dagger = U^\dagger U = \left(\sum_{x\in[N]} \lambda_x|u_x\rangle\langle u_x|\right)^\dagger \left(\sum_{x\in[N]} \lambda_x|u_x\rangle\langle u_x|\right)$$
$$= \sum_{x\in[N]} |\lambda_x|^2|u_x\rangle\langle u_x|. \tag{6.98}$$

In order to satisfy the defining property (6.85) of unitary matrices, one must thus have

$$|\lambda_x| = 1, \text{ i.e., } \lambda_x = \exp(i\gamma_x), \text{ for } x \in [N] \tag{6.99}$$

for some phases $\{\gamma_x\}_{x\in[N]}$.

That is, a matrix A is unitary if and only if it is normal and all the N eigenvalues have absolute value equal to 1. This indicates that a unitary transformation A rotates the phase of each amplitude α_x in the superposition (6.96) by an amount equal to γ_x.

Example 6.30

Eigendecomposition of a Hermitian matrix. Hermitian matrices are also normal, and hence by the spectral theorem the Hermitian transpose equals

$$\begin{aligned} A^\dagger &= \left(\sum_{x\in[N]} \lambda_x |u_x\rangle\langle u_x|\right)^\dagger \\ &= \sum_{x\in[N]} \lambda_x^* |u_x\rangle\langle u_x|. \end{aligned} \tag{6.100}$$

Therefore, in order to satisfy the defining property (6.88) of Hermitian matrices, one must have the equalities $\lambda_x^* = \lambda_x$ for all $x \in [N]$, which implies that the eigenvalues must be real:

$$\lambda_x \text{ is real for } x \in [N]. \tag{6.101}$$

Indeed, a matrix A is Hermitian if and only if it is normal and all the N eigenvalues are real-valued.

Example 6.31

Eigendecomposition of Pauli matrices. The two eigenvalues of all the Pauli matrices P_1, P_2, and P_3 are equal to $\lambda_0 = 1$ and $\lambda_1 = -1$. Since the eigenvalues are real-valued and have absolute value equal to 1, the Pauli matrices are both unitary and Hermitian.

For the Pauli X matrix, the eigendecomposition is given by

$$P_1 = |+\rangle\langle+| - |-\rangle\langle-|, \tag{6.102}$$

and thus the eigenbasis of matrix P_1 is the Hadamard basis $\{|+\rangle, |-\rangle\}$. The Hadamard basis is accordingly also known as the **X basis**.

For the Pauli Y matrix, the eigendecomposition is

$$P_2 = |+i\rangle\langle+i| - |-i\rangle\langle-i| \tag{6.103}$$

with eigenbasis

$$\left\{|+i\rangle = \frac{1}{\sqrt{2}}(|0\rangle + i|1\rangle), |-i\rangle = \frac{1}{\sqrt{2}}(|0\rangle - i|1\rangle)\right\}, \tag{6.104}$$

which is also known as the **Y basis**.

Finally, for the Pauli Z matrix, the eigendecomposition is

$$P_3 = |0\rangle\langle 0| - |1\rangle\langle 1| \tag{6.105}$$

with eigenbasis given by the computational basis $\{|0\rangle, |1\rangle\}$, which is hence also known as the **Z basis**.

Example 6.32

Eigendecomposition of a projection matrix. For a projection matrix Π, the general form (6.61) coincides with an eigendecomposition of the matrix. Accordingly, an eigenbasis is given by the set $\{|u_x\rangle\}_{x\in[N]}$ of states in the underlying measurement basis, with eigenvalues equal to

$$\lambda_x = 1 \text{ for } x \in [L] \text{ and } \lambda_x = 0 \text{ for } x \notin [L]. \tag{6.106}$$

It follows that a matrix is a projection matrix with rank r if and only if it is normal and it has r eigenvalues equal to 1 and all other eigenvalues equal to 0.

This in turn implies that, via the multiplication $\Pi|\psi\rangle$, a projection matrix Π preserves all components of the superposition (6.96) corresponding to eigenvalues equal to 1, while nulling all the other components. Geometrically, as per (6.64), this corresponds to the orthogonal projection of state $|\psi\rangle$ onto the subspace $\mathcal{S}(\Pi)$. Note that the only projection matrix that is unitary is the identity matrix.

6.6.4 Positive Semi-Definite Matrices

An important class of normal matrices is given by positive semi-definite matrices.

Definition 6.17 Positive semi-definite matrix
Hermitian matrices with all eigenvalues being non-negative, i.e., with

$$\lambda_x \geq 0 \text{ for } x \in [N], \tag{6.107}$$

are said to be positive semi-definite.

Positive semi-definite matrices can be constructed as follows (see Appendix 12.G).

Key Result 6.11 Obtaining positive semi-definite matrix
Given an arbitrary $N \times M$ matrix A, the matrices $AA^\dagger$ and $A^\dagger A$ are positive semi-definite, i.e.,

$$AA^\dagger \text{ and } A^\dagger A \text{ are positive semi-definite.} \tag{6.108}$$

Example 6.33

All rank-1 matrices are positive semi-definite, and, more generally, all projection matrices are positive semi-definite.

6.6.5 Function of a Normal Matrix

In the next chapters, it will be often useful to use functions of normal matrices, which are defined as follows.

Definition 6.18 Function of a normal matrix
Given any $N \times N$ normal matrix A with eigendecomposition (6.94), and given any scalar function $f(\cdot)$ (mapping a scalar into a scalar), the function $f(A)$ is defined as the $N \times N$ normal matrix

$$f(A) = \sum_{x \in [N]} f(\lambda_x)|u_x\rangle\langle u_x|. \tag{6.109}$$

This can equivalently be written as

$$f(A) = Uf(\Lambda)U^\dagger, \tag{6.110}$$

where $f(\Lambda)$ is a diagonal matrix with diagonal elements given by $\{f(\lambda_x)\}_{x \in [N]}$.

Accordingly, the function $f(A)$ has the same eigenvectors of matrix A, with eigenvalues obtained by passing the eigenvalues of matrix A through the function $f(\cdot)$. Note that, in case the eigendecomposition of matrix A is not unique, any eigendecomposition can be used to evaluate the matrix $f(A)$ without affecting the final result. Furthermore, it is emphasized that the matrix $f(A)$ cannot be obtained by applying function $f(\cdot)$ separately to all the entries of matrix A (except for the special case of diagonal matrices).

Example 6.34

What is matrix $\cos(\pi A)$, with

$$A = |+\rangle\langle+|? \tag{6.111}$$

First note that matrix A is normal since it is a projection matrix. Furthermore, using (6.94), the eigenbasis is given by $\{|u_0\rangle = |+\rangle, |u_1\rangle = |-\rangle\}$, and the corresponding eigenvalues are $\lambda = [\lambda_0 = 1, \lambda_1 = 0]^T$. Therefore, the scalar function $f(x) = \cos(\pi x)$ returns the 2×2 matrix

$$\begin{aligned}\cos(\pi A) &= f(1)|+\rangle\langle+| + f(0)|-\rangle\langle-| \\ &= -|+\rangle\langle+| + |-\rangle\langle-|.\end{aligned} \tag{6.112}$$

Example 6.35

Given a normal matrix A with general eigendecomposition (6.94), one has

$$A^\alpha = \sum_{x\in[N]} \lambda_x^\alpha |u_x\rangle\langle u_x| \tag{6.113}$$

for any real number α (assuming that all the quantities λ_x^α are well defined). Therefore, the definition (6.109) is consistent with the usual definition of power of a matrix when α is an integer.

As another related example, the function $|A|^\alpha$ returns the matrix

$$|A|^\alpha = \sum_{x\in[N]} |\lambda_x|^\alpha |u_x\rangle\langle u_x|. \tag{6.114}$$

This calculation yields the definition of the Schatten norm, which is given next.

6.6.6 Schatten Norm

Building on the α-norm for vectors defined in Section 3.4, the following is an important measure for matrices.

Definition 6.19 Schatten norm
For $\alpha \geq 1$, the α-Schatten norm of a normal matrix A is defined as the α-norm of the vector of eigenvalues, λ, i.e.,

$$\begin{aligned} ||A||_\alpha &= ||\lambda||_\alpha \\ &= \left(\sum_{x\in[N]} |\lambda_x|^\alpha\right)^{\frac{1}{\alpha}} = (\mathrm{tr}(|A|^\alpha))^{\frac{1}{\alpha}}, \end{aligned} \tag{6.115}$$

where the **trace operator** $\mathrm{tr}(\cdot)$ evaluates the sum of the eigenvalues (see Section 7.4.1).

The following are notable special cases of the Schatten norm.

Example 6.36

The 1-Schatten norm

$$||A||_1 = \sum_{x\in[N]} |\lambda_x| \tag{6.116}$$

is also known as the **trace norm** or **nuclear norm**.

The 2-Schatten norm is also known as the **Hilbert–Schmidt**, or **quadratic**, norm

$$||A||_2 = \sqrt{\sum_{x\in[N]} |\lambda_x|^2}. \tag{6.117}$$

The Schatten norm can be extended to general non-square and non-normal matrices as discussed in Problem 6.10.

6.7 Summary

- The measured entropy $H_\alpha(|\psi\rangle|\mathcal{M})$ is the classical entropy of the measurement probability distribution $p(|\psi\rangle|\mathcal{M})$, i.e., $H_\alpha(|\psi\rangle|\mathcal{M}) = H_\alpha(p(|\psi\rangle|\mathcal{M}))$.
- The measured entropy for a given quantum state $|\psi\rangle$ depends on the choice of the measurement setting $\mathcal{M}$, leading to measured entropy values ranging from the minimum value 0 to the maximum value $H_\alpha(u)$.
- The coherence of a quantum state $|\psi\rangle$ with respect to a measurement basis $\mathcal{M}$ quantifies the degree to which the state $|\psi\rangle$ is in a genuine superposition of states in the basis $\mathcal{M}$.
- The uncertainty associated with the output of a quantum measurement, quantified by the measured entropy $H_\alpha(|\psi\rangle|\mathcal{M})$, depends on the coherence of the pre-measurement state $|\psi\rangle$ with respect to the measurement basis $\mathcal{M}$. Minimum coherence yields a zero entropy, while maximum coherence produces maximal entropy.
- The coherence of a state $|\psi\rangle$ with respect to a measurement basis $\mathcal{M}$ also dictates the level of disturbance caused by the measurement on the quantum state. Minimum coherence yields no disturbance, while maximum coherence produces maximal disturbance.
- Coherence can be measured via the minimum distance between the pre-measurement state and the states in the measurement basis, producing the trace distance of coherence $K^{tr}(|\psi\rangle|\mathcal{M}) = \min_{x\in[N]} D^{tr}(|\psi\rangle, |v_x\rangle)$ as an important special case.
- Two quantum measurement settings $\mathcal{M}_0$ and $\mathcal{M}_1$ are incompatible if it is not possible to find an equivalent state in one basis for every state in the other basis.
- For incompatible measurement settings $\mathcal{M}_0$ and $\mathcal{M}_1$, when one measurement setting yields sharp outputs, having a low measured entropy $H_\alpha(|\psi\rangle|\mathcal{M}_0)$, the other setting must yield uncertain outputs having a sufficiently high measured entropy $H_\alpha(|\psi\rangle|\mathcal{M}_1)$. This is formalized by the uncertainty inequality.
- The level of compatibility between two measurement settings $\mathcal{M}_0 = \{|v_x\rangle\}_{x\in[N]}$ and $\mathcal{M}_1 = \{|u_x\rangle\}_{x\in[N]}$ is fully described by the compatibility matrix $C(\mathcal{M}_0, \mathcal{M}_1)$, whose (x, y)th entry is given by the fidelity $F(|v_x\rangle, |u_y\rangle)$.
- A partial description of the compatibility matrix is given by the scalar coherence measure $c(\mathcal{M}_0, \mathcal{M}_1) = \max_{x,y\in[N]} F(|v_x\rangle, |u_y\rangle)$.
- Two measurement settings $\mathcal{M}_0$ and $\mathcal{M}_1$ are maximally incompatible if and only if $c(\mathcal{M}_0, \mathcal{M}_1) = 1/N$; while $c(\mathcal{M}_0, \mathcal{M}_1) < 1$ is a sufficient condition for the two measurements settings to be incompatible.
- The uncertainty inequality $H(|\psi\rangle|\mathcal{M}_0) + H(|\psi\rangle|\mathcal{M}_1) \geq -\log_2(c(\mathcal{M}_0, \mathcal{M}_1))$ indicates that, when $c(\mathcal{M}_0, \mathcal{M}_1) < 1$ and thus $-\log_2(c(\mathcal{M}_0, \mathcal{M}_1)) > 0$, if one measured entropy is small, the other must be sufficiently large.
- Every normal matrix A, satisfying the defining equality $AA^\dagger = A^\dagger A$, admits an eigendecomposition $A = \sum_{x\in[N]} \lambda_x |v_x\rangle\langle v_x|$ with eigenvalues $\{\lambda_x\}_{x\in[N]}$ and eigenvectors $\{|v_x\rangle\}_{x\in[N]}$.

- Special cases of normal matrices include diagonal matrices, Hermitian matrices, unitary matrices, and projection matrices.
- Projective measurements $\mathcal{P} = \{\Pi_x\}_{x\in[M]}$ are specified by $M \leq N$ projection matrices $\{\Pi_x\}_{x\in[M]}$ that satisfy the resolution-of-the-identity condition $\sum_{x\in[M]} \Pi_x = I$.
- Projective measurements generalize complete quantum measurements, supporting a trade-off between information extracted by the measurement and disturbance caused to the pre-measurement state.
- The level of coarse graining applied by a projective measurement is determined by the ranks of the projection matrices $\{\Pi_y\}_{y\in[M]}$. In fact, the post-measurement state is given by $|\psi_y\rangle = (p_y(|\psi\rangle|\mathcal{P}))^{-1/2}\Pi_y|\psi\rangle$ with measurement probability $p_y(|\psi\rangle|\mathcal{P}) = \langle\psi|\Pi_y|\psi\rangle$.
- The post-measurement state $|\phi_y\rangle$ equals the pre-measurement state $|\psi\rangle$ as long as the latter is in the span $\mathcal{S}(\Pi_y)$ of the projection matrix Π_y, whose dimension equals $\text{rank}(\Pi_y)$.

6.8 Recommended Resources

A review of resource theory can be found in [1]. Extensions of the uncertainty inequality are provided in [2–5].

PROBLEMS

6.1 (a) For a general qubit state $|\psi\rangle = \psi_0|0\rangle + \psi_1|1\rangle$, find a measurement basis under which the measured entropy equals zero. (b) Then, find a measurement basis under which the Shannon and Rényi measurement entropies equal 1 bit.

6.2 Consider two measurement bases $\mathcal{M}_0$ and $\mathcal{M}_1$. Show that, if a state in $\mathcal{M}_0$ has fidelity equal to 1 with respect to a state in $\mathcal{M}_1$, it must have fidelity equal to 0 with respect to all other vectors in $\mathcal{M}_1$.

6.3 (a) For any two measurement settings $\mathcal{M}_0$ and $\mathcal{M}_1$, prove the equality $\mathrm{C}(\mathcal{M}_0, \mathcal{M}_1)^T = \mathrm{C}(\mathcal{M}_1, \mathcal{M}_0)$. (b) Show that the compatibility matrix $\mathrm{C}(\mathcal{M}_0, \mathcal{M}_1)$ is doubly stochastic.

6.4 Evaluate the compatibility matrix (6.33) for $\theta = 0$, $\theta = \pi/4$, $\theta = \pi/2$, and $\theta = 3\pi/4$. What can you conclude about the compatibility of measurement settings $\mathcal{M}_0$ and $\mathcal{M}_1$ in each case?

6.5 Given an orthonormal basis, demonstrate a general construction to obtain another basis that is complementary. *Hint:* Read Appendix 6.A.

6.6 (a) Prove the bounds (6.36) on the compatibility metric $\mathrm{c}(\mathcal{M}_0, \mathcal{M}_1)$. *Hint:* Recall that the compatibility matrix is doubly stochastic. (b) Demonstrate pairs of measurement bases $\mathcal{M}_0$ and $\mathcal{M}_1$ that attain the lower and upper bounds, respectively.

6.7 Show that, for $N = 2$, knowing the value of the metric $\mathrm{c}(\mathcal{M}_0, \mathcal{M}_1)$ allows us to fully reconstruct the overall 2×2 compatibility matrix $\mathrm{C}(\mathcal{M}_0, \mathcal{M}_1)$ apart from the arbitrary ordering of the rows and columns.

6.8 Prove that applying two compatible measurement settings sequentially produces classical outputs satisfying the condition (6.47).

6.9 Prove that the post-measurement state $|\phi_y\rangle = p_y(|\psi\rangle|\mathcal{P})^{-1/2}\Pi_y|\psi\rangle$ in (6.71) is normalized.

6.10 (a) For any $M \times N$ matrix A, show that the $N \times N$ matrix $A^\dagger A$ is Hermitian. (b) Argue that the definition of the α-Schatten norm can be extended to any $M \times N$ matrix B as

$$||A||_\alpha = \left(\mathrm{tr}\left(\left(\sqrt{A^\dagger A}\right)^\alpha\right)\right)^{\frac{1}{\alpha}}, \tag{6.118}$$

where the equality is intended as a definition. To this end, show that if A is normal, this definition coincides with (6.115).

6.11 Show that a projective measurement with $M = N$ possible classical outputs coincides with a complete measurement.

6.12 Consider the projective measurement setting $\mathcal{P} = \{\Pi_0, \Pi_1\}$ with

$$\Pi_0 = \frac{1}{2}(I + P) \text{ and } \Pi_1 = \frac{1}{2}(I - P), \tag{6.119}$$

where P is a Pauli matrix. (a) Write the projection matrices Π_0 and Π_1 for $P = P_1$, and demonstrate that $\mathcal{P}$ represents a measurement in the Hadamard basis. (b) Write the projection matrices for $P = P_3$, and demonstrate that $\mathcal{P}$ represents a measurement in the computational basis. Overall, this problem shows that the projection matrices (6.119) represent a complete measurement in the eigenbasis of the Pauli matrix P.

Appendix

6.A Constructing a Basis for Maximal Coherence

This appendix provides a general construction for a basis $\mathcal{M}_*^{\max} = \{|v_x\rangle\}_{x\in[N]}$ with respect to which a given state $|\psi\rangle$ is maximally coherent. To this end, an auxiliary orthonormal basis $\mathcal{M}^{\text{aux}} = \{|u_x\rangle\}_{x\in[N]}$ is obtained first by setting $|u_0\rangle = |\psi\rangle$ and by choosing the remaining vectors $\{|u_x\rangle\}_{x=1}^{N-1}$ to be any orthonormal basis of the $(N-1)$-dimensional subspace orthogonal to $|\psi\rangle$.

Define as $U = [|u_0\rangle, \ldots, |u_{N-1}\rangle]$ the $N \times N$ matrix collecting vectors $\{|u_x\rangle\}_{x\in[N]}$ by columns. Note that we have the equality

$$U^\dagger|\psi\rangle = |0\rangle = \begin{bmatrix} 1 \\ 0 \\ \vdots \\ 0 \end{bmatrix}, \tag{6.120}$$

since, by construction, the columns of matrix U, except the first one, are all orthogonal to vector $|\psi\rangle$. Using matrix U, the basis $\mathcal{M}_*^{\max} = \{|v_x\rangle\}_{x\in[N]}$ is obtained as

$$|v_x\rangle = U|h_x\rangle, \tag{6.121}$$

where $|h_x\rangle$ is the xth column of any Hadamard matrix $H = [|h_0\rangle, \ldots, |h_{N-1}\rangle]$, such as the Fourier matrix (see Section 5.4.3). To see that the basis $\mathcal{M}_*^{\max} = \{|v_x\rangle\}_{x\in[N]}$ has the desired property (6.16), it is sufficient to show that the amplitude $\langle v_x|\psi\rangle$ has the form $1/\sqrt{N}\exp(i\theta_x)$ for some phase θ_x.

Indeed, the amplitude at hand is given by

$$\langle v_x|\psi\rangle = \langle h_x|U^\dagger|\psi\rangle = \langle h_x|0\rangle, \tag{6.122}$$

which equals the first element of vector $\langle h_x|$. Therefore, the amplitude $\langle v_x|\psi\rangle$ corresponds to the complex conjugate of the xth element in the first (i.e., 0th) row of the Hadamard matrix H. Since each (x, y)th element of a Hadamard matrix has the form $H_{xy} = 1/\sqrt{N}\exp(i\theta_{xy})$, this completes the proof.

6.B Proof of (6.82)

This appendix aims to prove the inequality (6.82). To this end, it is sufficient to show (6.82) for the Tsallis measured entropy $\mathrm{H}_\alpha^{\mathrm{T}}(|\psi\rangle|\mathcal{M})$, i.e., to demonstrate the inequality $\mathrm{H}_\alpha^{\mathrm{T}}(|\psi\rangle|\mathcal{P}) \leq \mathrm{H}_\alpha^{\mathrm{T}}(|\psi\rangle|\mathcal{M})$. In fact, the Tsallis entropy includes the Shannon entropy as a special case, and the Rényi entropy is obtained via a one-to-one monotonic relation from the Tsallis entropy (see Appendix 3.C).

Using (6.72) and the notation $\ell_\alpha(\cdot)$ for the log-loss, the Tsallis measured entropy for the projective measurement $\mathcal{P}$ is evaluated as

$$\begin{aligned}
\mathrm{H}_\alpha^{\mathrm{T}}(|\psi\rangle|\mathcal{P}) &= \sum_{y\in[M]} p_y(|\psi\rangle|\mathcal{P})\ell_\alpha(p_y(|\psi\rangle|\mathcal{P})) \\
&= \sum_{y\in[M]} \left(\sum_{x\in\mathcal{X}_y} p_x(|\psi\rangle|\mathcal{M})\right)\ell_\alpha\left(\sum_{x\in\mathcal{X}_y} p_x(|\psi\rangle|\mathcal{M})\right) \\
&= \sum_{x\in[N]} p_x(|\psi\rangle|\mathcal{M})\ell_\alpha\left(\sum_{x'\in\mathcal{X}_y:\, x\in\mathcal{X}_y} p_{x'}(|\psi\rangle|\mathcal{M})\right) \\
&\leq \sum_{x\in[N]} p_x(|\psi\rangle|\mathcal{M})\ell_\alpha(p_x(|\psi\rangle|\mathcal{M})) \\
&= \mathrm{H}_\alpha^{\mathrm{T}}(|\psi\rangle|\mathcal{M}).
\end{aligned} \tag{6.123}$$

In the third equality, the sum $\sum_{x'\in\mathcal{X}_y:\, x\in\mathcal{X}_y}$ is over all indices x' in the set $\mathcal{X}_y$ that includes the index x. Furthermore, the inequality follows from the fact that the α-log-loss $\ell_\alpha(p)$ is decreasing in the probability p (see Figure 3.2). It is thus concluded that a projective measurement has a measurement uncertainty that is no larger than any projective measurement derived from it.

REFERENCES

[1] E. Chitambar and G. Gour, "Quantum resource theories," *Reviews of Modern Physics*, vol. 91, no. 2, art. 025001, 2019.

[2] L.-L. Sun, K. Bharti, Y.-L. Mao, X. Zhou, L.-C. Kwek, J. Fan, and S. Yu, "Disturbance enhanced uncertainty relations," *arXiv preprint arXiv:2202.07251*, 2022.

[3] S. Zozor, G. M. Bosyk, and M. Portesi, "General entropy-like uncertainty relations in finite dimensions," *Journal of Physics A: Mathematical and Theoretical*, vol. 47, no. 49, art. 495302, 2014.

[4] P. J. Coles, M. Berta, M. Tomamichel, and S. Wehner, "Entropic uncertainty relations and their applications," *Reviews of Modern Physics*, vol. 89, no. 1, art. 015002, 2017.

[5] B.-G. Englert, "Uncertainty relations revisited," *Physics Letters A*, vol. 494, art. 129278, 2024.

[6] L. Kuhn, Y. Gal, and S. Farquhar, "Semantic uncertainty: Linguistic invariances for uncertainty estimation in natural language generation," *arXiv preprint arXiv:2302.09664*, 2023.

7 Classical and Quantum Uncertainty: Mixed States and Quantum Entropy

7.1 Overview

Previous chapters have introduced the concept of quantum uncertainty, a distinctive form of randomness intrinsic to quantum measurements. Unlike classical systems, even when the quantum state is pure – and therefore fully known – the outcome of a quantum measurement is inherently uncertain. This uncertainty manifests itself as a probability distribution over possible outcomes, depending on the measurement setting.

This chapter bridges classical uncertainty, which arises from a lack of knowledge about a system's state, and quantum uncertainty, which stems from the probabilistic nature of measurements on a known quantum state. Specifically, the focus will be on quantum states that are only known to correspond to one of several possible states, each occurring with a given probability.

A collection of quantum pure states assigned with probabilities is referred to as an ensemble. When there is uncertainty about the specific quantum pure state within the ensemble, the state of the system is described as mixed. Mixed states are prevalent in practical scenarios. Even in controlled environments, such as within a quantum computer, noise and other imperfections inevitably introduce probabilistic uncertainty, rendering the state mixed.

As will be shown in this chapter, the behavior of a mixed state is fully characterized by its density matrix, which determines the probability distribution of the outcomes for any projective measurement, as well as the corresponding expected values of observables. Notably, different ensembles can result in the same density matrix, making them indistinguishable.

As seen in Chapter 6, for pure states, the measured entropy – the entropy of the outcome distribution – can be made arbitrarily small by selecting an appropriate measurement setting. However, for mixed states, the presence of classical uncertainty imposes a lower bound on this entropy. When minimized over all possible measurement settings, the measured entropy is termed quantum entropy. The quantum entropy quantifies the irreducible uncertainty associated with mixed state measurements.

When a measurement is performed and its classical output is discarded or forgotten, the resulting state becomes mixed due to the introduction of classical uncertainty about the measurement outcome. Such measurements, referred to as non-selective, are shown in this chapter to increase the quantum entropy.

Finally, the chapter extends the concept of coherence from pure states to mixed states. A mixed state is said to be incoherent with respect to a particular orthonormal basis if it can be represented as an ensemble of pure states, all of which are incoherent with respect to that basis. Alternatively, coherence can be understood as the property of states that are disturbed by a non-selective measurement in the given basis.

7.1.1 Learning Objectives and Organization

By the end of this chapter, the reader should be able to:

- understand the behavior of ensembles of pure states;
- derive the density matrix associated with an ensemble of pure states;
- understand the defining properties of density matrices;
- evaluate the measurement probability vector for a mixed state under a projective measurement;
- evaluate the expected value of observables for mixed states;
- understand and use the uncertainty inequality for observables;
- calculate the measured entropy for a mixed state under a projective measurement;
- understand the concept of quantum entropy and its relation with the measured entropy;
- understand the properties of the Shannon, Tsallis, and Rényi quantum entropies;
- evaluate the density matrix produced by a non-selective measurement;
- understand the definition of coherence for mixed states;
- and understand and apply the uncertainty inequality for mixed states.

7.2 Uncertain Quantum States

While Chapters 5 and 6 have assumed that the system is in a known, pure, state $|\psi\rangle$, this section studies the more general situation in which there is classical uncertainty on the state of the system. That is, the system is only known to be in one of a number of possible states with given probabilities. The uncertainty on the identity of the true state is classical in the sense that, like for classical systems, it refers to a lack of knowledge on a quantity that pre-exists the measurement.

Classical uncertainty about the state of a quantum system may be caused by a number of factors. Conceptually, one may think that the quantum state is prepared by some agent or phenomenon in a way that is not fully known to the observer. For instance, one may have prepared a state using a noisy quantum computer.

Overall, in the settings studied in this chapter, there are, informally, two sources of uncertainty, and hence of information.

- **Classical uncertainty:** This arises from incomplete knowledge about the preparation of the system, which is only known to be in one of several possible states, each associated with a specific probability.
- **Quantum uncertainty:** Even if the state were fully known and therefore pure, uncertainty persists due to the probabilistic nature of the outcomes in a classical measurement.

While the two sources of uncertainty are conceptually distinct, it will be seen in this chapter that, in general, their respective contributions cannot be isolated.

This section introduces the formalism of ensembles of pure states, and then describes the effect of a complete measurement for a system described by an ensemble of pure states.

7.2.1 Ensembles of Pure States

Assume that the observer is uncertain among L possible states $\{|\psi_l\rangle\}_{l\in[L]}$ of a quantum system. The number of states L is arbitrary, and it may be smaller than, equal to, or larger than the dimension N of the quantum state. The possible states $\{|\psi_l\rangle\}_{l\in[L]}$ are also arbitrary, and, in particular, they need not be orthogonal and thus perfectly distinguishable (see Section 5.6). As a point of notation, it is emphasized that $|\psi_l\rangle$ represents a quantum state and not the lth amplitude of a state.

The observer assigns a probability p_l to each state $|\psi_l\rangle$. Therefore, the knowledge of the state of the system at the observer can be described as follows.

Definition 7.1 Ensemble of pure states
An ensemble of pure states

$$\mathcal{E} = \{(|\psi_0\rangle, p_0), (|\psi_1\rangle, p_1), \ldots, (|\psi_{L-1}\rangle, p_{L-1})\} \tag{7.1}$$

is defined by

- an $L \times 1$ probability vector $p = [p_0, p_1, \ldots, p_{L-1}]^T$, and
- an ensemble of L distinct, and generally non-orthogonal, states $\{|\psi_l\rangle\}_{l\in[L]}$.

In (7.1), each pair $(|\psi_l\rangle, p_l)$ represents a possible state for the system and the corresponding probability.

7.2.2 Measuring an Ensemble of Pure States

If one of the probabilities equals one, i.e., if one has $p_l = 1$ for some $l \in [L]$, there is no uncertainty about the state of the system, which is given by the pure state $|\psi_l\rangle$. This is the situation studied in Chapters 5 and 6. In this case, for a complete measurement setting $\mathcal{M} = \{|v_x\rangle\}_{x\in[N]}$, by Born's rule, there is a probability

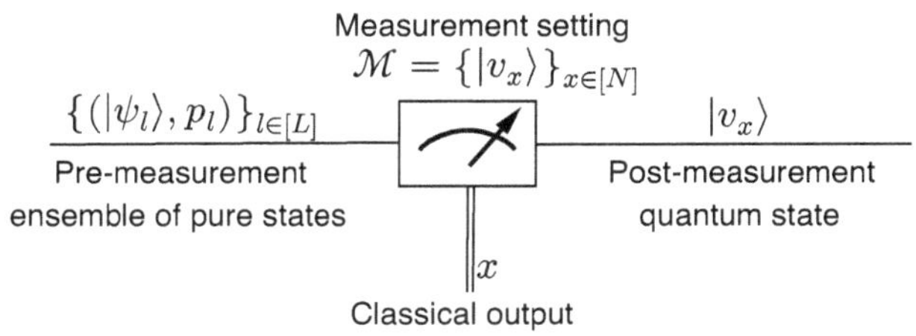

Figure 7.1 Illustration of a complete measurement in basis $\mathcal{M}$ applied to a pre-measurement ensemble of pure states $\mathcal{E} = \{(|\psi_0\rangle, p_0), (|\psi_1\rangle, p_1), \ldots, (|\psi_{L-1}\rangle, p_{L-1})\}$.

$p_x(|\psi_l\rangle|\mathcal{M}) = |\langle v_x|\psi_l\rangle|^2$ of observing the classical output x and of leaving the system in the post-measurement state $|v_x\rangle$. What is the effect of measurement $\mathcal{M}$ when applied to an ensemble of states $\mathcal{E}$?

As illustrated in Figure 7.1, prior to the measurement, the state of the system is $|\psi_l\rangle$ with probability p_l, and hence the probability that the state is $|\psi_l\rangle$ and that the measurement output is x equals the product $p_l|\langle v_x|\psi_l\rangle|^2$. Therefore, by the law of total probability, the probability of observing x as the classical output is given by the sum

$$p_x(\mathcal{E}|\mathcal{M}) = \sum_{l\in[L]} p_l|\langle v_x|\psi_l\rangle|^2 \tag{7.2}$$

over all possible states $\{|\psi_l\rangle\}_{l\in[L]}$ in the ensemble. That is, the probability of obtaining output x is the average measurement probability $|\langle v_x|\psi_l\rangle|^2$ over the probability distribution p defining the ensemble. The probability vector $p(\mathcal{E}|\mathcal{M}) = [p_0(\mathcal{E}|\mathcal{M}), \ldots, p_{N-1}(\mathcal{E}|\mathcal{M})]^T$, viewed as a function of the measurement setting $\mathcal{M}$, defines the behavior of the ensemble $\mathcal{E}$.

Furthermore, as shown in Figure 7.1, the post-measurement state is given by $|v_x\rangle$ with probability $p_x(\mathcal{E}|\mathcal{M})$.

7.3 From Ensembles of Pure States to Density Matrices

Section 7.2 has shown that the informational state of an observer that is uncertain about the pure state of a system can be described by an ensemble of pure states. This section will demonstrate that, due to the indistinguishability of non-orthogonal quantum states, there are infinitely many ensembles that have the same behavior, that is, that have the same dependence of the measurement probability $p(\mathcal{E}|\mathcal{M})$ on the measurement setting $\mathcal{M}$. In particular, all equivalent ensembles $\mathcal{E}$ – i.e., all ensembles exhibiting the same behavior $p(\mathcal{E}|\mathcal{M})$ – have the characterizing property that they yield the same correlation matrix for the random state, which is known as the **density matrix** ρ.

7.3.1 Density Matrices for Pure States

To introduce density matrices, consider first the case with no uncertainty about the quantum state, which is described by the trivial ensemble $\mathcal{E} = \{(|\psi\rangle, 1)\}$. For this ensemble, by the Born rule, the probability of observing the classical output $x \in [N]$ under a complete measurement setting $\mathcal{M} = \{|v_x\rangle\}_{x\in[N]}$ in (7.2) reduces to

$$\begin{aligned} p_x(\mathcal{E}|\mathcal{M}) &= |\langle v_x|\psi\rangle|^2 \\ &= \langle v_x|\psi\rangle(\langle v_x|\psi\rangle)^* \\ &= \langle v_x|\psi\rangle\langle\psi|v_x\rangle \\ &= \langle v_x|(|\psi\rangle\langle\psi|)|v_x\rangle. \end{aligned} \tag{7.3}$$

The expression (7.3) can be simplified by defining the $N \times N$ matrix

$$\rho = |\psi\rangle\langle\psi| \tag{7.4}$$

as

$$p_x(\mathcal{E}|\mathcal{M}) = \langle v_x|\rho|v_x\rangle. \tag{7.5}$$

Matrix ρ is referred to as the density matrix associated with the pure state $|\psi\rangle$. Note that matrix ρ in (7.4) is positive semi-definite and that it has rank 1 (see Section 6.6.4).

Example 7.1

To illustrate the definition of the rank-1 density matrix (7.4), consider the case of a qubit system ($N = 2$) in an arbitrary pure state $|\psi\rangle = [\psi_0, \psi_1]^T$. The density matrix is given by

$$\rho = \begin{bmatrix} \psi_0 \\ \psi_1 \end{bmatrix} [\psi_0^*, \psi_1^*] = \begin{bmatrix} |\psi_0|^2 & \psi_0\psi_1^* \\ \psi_0^*\psi_1 & |\psi_1|^2 \end{bmatrix}. \tag{7.6}$$

Note that the off-diagonal terms of the density matrix ρ depend on the relative phase of the amplitudes ψ_0 and ψ_1 (see Section 5.5). Moreover, they are the complex conjugate of one another, i.e., $\psi_0\psi_1^* = (\psi_0^*\psi_1)^*$, as required by the fact that the matrix is Hermitian.

Furthermore, the off-diagonal elements equal zero if the state is incoherent with respect to the computational basis, i.e., if $|\psi\rangle = |0\rangle$ or $|\psi\rangle = |1\rangle$. Thus, non-zero off-diagonal terms indicate a level of coherence of the state with respect to the computational basis.

7.3.2 Density Matrices for Ensembles of Pure States

The representation (7.5) of a measurement probability can be extended beyond the trivial ensemble $\mathcal{E} = \{(|\psi\rangle, 1)\}$ to any ensemble of pure states (7.1). To see this, in a manner similar to (7.3), rewrite the measurement probability $p_x(\mathcal{E}|\mathcal{M})$ in (7.2) as

$$\begin{aligned} p_x(\mathcal{E}|\mathcal{M}) &= \sum_{l\in[L]} p_l |\langle v_x|\psi_l\rangle|^2 \\ &= \sum_{l\in[L]} p_l \langle v_x|(|\psi_l\rangle\langle\psi_l|)|v_x\rangle \\ &= \langle v_x| \left(\sum_{l\in[L]} p_l |\psi_l\rangle\langle\psi_l| \right) |v_x\rangle \\ &= \langle v_x|\rho|v_x\rangle. \end{aligned} \tag{7.7}$$

In the last equality of (7.7), the density matrix ρ was defined as follows.

Definition 7.2 Density matrix associated with an ensemble of pure states
The density matrix associated with the ensemble of pure state $\mathcal{E} = \{(|\psi_l\rangle, p_l)\}_{l\in[L]}$ is the $N \times N$ matrix

$$\rho = \sum_{l\in[L]} p_l |\psi_l\rangle\langle\psi_l|. \tag{7.8}$$

From a probabilistic perspective, the density matrix ρ in (7.8) can be viewed as the correlation matrix of the random vector $|\psi_{\mathrm{l}}\rangle$, where l is interpreted as a random variable $\mathrm{l} \sim p$ with distribution p. As such, following Section 6.6.4, the density matrix is positive semi-definite, i.e., it is Hermitian,

$$\rho^\dagger = \rho, \tag{7.9}$$

with non-negative eigenvalues.

As seen in Section 7.3.1, for a pure state $|\psi\rangle$, the density matrix ρ in (7.4) has rank 1. In contrast, when the underlying ensemble $\mathcal{E}$ contains at least two inequivalent states with non-zero probabilities, the density matrix ρ in (7.8) has rank larger than 1.

Definition 7.3 Mixed state
When the rank of the density matrix ρ is larger than 1, the state is said to be mixed.

According to this definition, a mixed state is characterized not only by quantum uncertainty, but also by classical uncertainty.

Example 7.2

Consider the ensemble of two pure qubit states

$$\mathcal{E} = \{(|0\rangle, p_0), (|1\rangle, p_1)\} \tag{7.10}$$

with probabilities p_0 and p_1 satisfying $p_0 + p_1 = 1$. The density matrix associated with this ensemble is given by the diagonal matrix

$$\rho = p_0|0\rangle\langle 0| + p_1|1\rangle\langle 1| = \begin{bmatrix} p_0 & 0 \\ 0 & p_1 \end{bmatrix}. \tag{7.11}$$

In a manner consistent with Example 7.1, the lack of coherence of both constituent pure states in the ensemble $\mathcal{E}$ with respect to the computational basis is reflected in the zero off-diagonal elements in (7.11). Unlike the previous example, however, the rank of the density matrix equals 2, and not 1, unless the state is pure, i.e., unless $p_0 = 0$ or $p_0 = 1$. The rank is thus a measure of classical uncertainty, while the presence of off-diagonal elements indicates coherence with respect to the computational basis.

Example 7.3

As another example, take the ensemble of qubit states

$$\mathcal{E} = \{(|+\rangle, p_0), (|-\rangle, p_1)\}, \tag{7.12}$$

with probabilities p_0 and p_1 satisfying $p_0 + p_1 = 1$. The density matrix associated with this ensemble is

$$\begin{aligned} \rho &= p_0|+\rangle\langle +| + p_1|-\rangle\langle -| \\ &= \frac{1}{2}p_0\begin{bmatrix} 1 & 1 \\ 1 & 1 \end{bmatrix} + \frac{1}{2}p_1\begin{bmatrix} 1 & -1 \\ -1 & 1 \end{bmatrix} \\ &= \frac{1}{2}\begin{bmatrix} p_0 + p_1 & p_0 - p_1 \\ p_0 - p_1 & p_0 + p_1 \end{bmatrix}. \end{aligned} \tag{7.13}$$

The off-diagonal elements in (7.13) are generally non-zero, reflecting the coherence of the constituent states $\{|+\rangle, |-\rangle\}$ with respect to the computational basis. Furthermore, the rank of the matrix is 2 unless one has $p_0 = 0$ or $p_0 = 1$.

For $p_0 = p_1 = 0.5$, the density matrices (7.11) and (7.13) coincide, despite being generated based on two distinct ensembles. In this case, the two ensembles are said to be equivalent.

Since the measurement probability $p_x(\mathcal{E}|\mathcal{M})$ in (7.7) depends on the ensemble $\mathcal{E}$ only via the density matrix ρ, the Born rule for mixed states, which generalizes (7.5), can be formalized as follows.

Definition 7.4 Complete quantum measurement of a mixed state

Given a measurement basis $\mathcal{M} = \{|v_x\rangle\}_{x\in[N]}$, a complete quantum measurement applied to a pre-measurement state ρ results in the following.

- **Classical output:** The measurement yields a classical output $\mathrm{x} \sim p(\rho|\mathcal{M})$ with distribution

$$p_x(\rho|\mathcal{M}) = \langle v_x|\rho|v_x\rangle. \tag{7.14}$$

- **Post-measurement state:** For a given classical output x $= x$, the quantum system is left in the corresponding post-measurement state $|v_x\rangle$ associated with the observed outcome x.

Example 7.4

Consider a complete measurement in the computational basis $\mathcal{M} = \{|x\rangle\}_{x\in[N]}$. Given an arbitrary density matrix ρ, the measurement probabilities (7.14) are

$$p_x(\rho|\mathcal{M}) = \langle x|\rho|x\rangle = \rho_{x,x}, \tag{7.15}$$

where $\rho_{x,x}$ represents the xth element of the main diagonal of matrix ρ, i.e., the (x, x)th element of matrix ρ. For a pure state $|\psi\rangle$, by (7.4), the measurement probability (7.15) recovers the Born rule studied in Chapter 5, i.e., $p_x(\rho|\mathcal{M}) = |\psi_x|^2$.

Example 7.5

Take now the case of a measurement in the Hadamard basis $\mathcal{M} = \{|+\rangle, |-\rangle\}$ for a qubit system ($N = 2$). For any 2×2 density matrix ρ, the measurement probability (7.14) can be evaluated as

$$\begin{aligned} p_0(\rho|\mathcal{M}) &= \langle +|\rho|+\rangle \\ &= \frac{1}{2}(\rho_{00} + \rho_{11} + \rho_{01} + \rho_{10}) \\ &= \frac{1}{2}(\rho_{00} + \rho_{11} + 2\mathrm{Re}(\rho_{01})), \end{aligned} \tag{7.16}$$

with $\mathrm{Re}(\cdot)$ being the real part of a complex number. Note that the last equality in (7.16) follows from the fact that a density matrix is Hermitian, and thus the equality $\rho_{01} = \rho_{10}^*$ holds, implying $\rho_{01} + \rho_{10}^* = 2\mathrm{Re}(\rho_{01})$.

7.3.3 Equivalent Ensembles

By the Born rule (7.14), the density matrix ρ provides a complete description of the behavior of an ensemble $\mathcal{E}$. Consequently, one can refer to the quantum system as being in the state ρ, as the specific details of the ensemble are unnecessary to fully characterize the system's behavior.

Definition 7.5 Equivalent ensembles of pure states
Any two ensembles of pure states $\mathcal{E}_0$ and $\mathcal{E}_1$ that have the same density matrix are said to be equivalent, as any measurement made on the two

ensembles produces statistically equivalent outputs. This definition may be summarized as

$$\mathcal{E}_0 \text{ and } \mathcal{E}_1 \text{ equivalent } \Leftrightarrow \rho(\mathcal{E}_0) = \rho(\mathcal{E}_1), \tag{7.17}$$

where $\rho(\mathcal{E}_0)$ and $\rho(\mathcal{E}_1)$ are the density matrices (7.8) associated with the two ensembles $\mathcal{E}_0$ and $\mathcal{E}_1$, respectively.

Example 7.6

Section 5.5 showed that two pure states $|\psi\rangle$ and $|\phi\rangle$ are equivalent, in the sense that they have the same behavior, if they can be related as $|\psi\rangle = \exp(i\gamma)|\phi\rangle$ for some global phase γ. Treating the two pure states as rank-1 density matrices, one can readily verify that this definition of equivalence is consistent with (7.17). In fact, for two equivalent pure states $|\psi\rangle = \exp(i\gamma)|\phi\rangle$ and $|\phi\rangle$, the density matrices are equal:

$$|\psi\rangle\langle\psi| = \exp(i\gamma)\exp(-i\gamma)|\phi\rangle\langle\phi| = |\phi\rangle\langle\phi|. \tag{7.18}$$

Given a density matrix ρ with rank larger than 1, there are infinitely many possible ensembles $\mathcal{E}$ corresponding to the density matrix. Each such ensemble is said to **realize** the density matrix. These equivalent ensembles may have different numbers, L, of pure states.

Key Result 7.1 Rank of the density matrix and cardinality of the ensemble
Given a density matrix with rank given by rank(ρ), any ensemble $\mathcal{E}$ realizing this matrix must have

$$L \geq \text{rank}(\rho) \tag{7.19}$$

pure states. Furthermore, it is always possible to find an ensemble with exactly $L = \text{rank}(\rho)$ pure states that realizes the density matrix ρ. Therefore, the rank of the density matrix rank(ρ) represents the minimum number of pure states for any ensemble that realizes the state ρ.

The inequality (7.19) follows directly from the sum (7.8) and from the definition of rank. In fact, a density matrix of the form (7.8) can have rank not exceeding L. The second part of the claim – namely the existence of an ensemble with the minimum number of pure states – will be proved in Section 7.5 by using the eigendecomposition of the density matrix.

Example 7.7

As seen in Examples 7.2 and 7.3, both ensembles $\mathcal{E}_0 = \{(|0\rangle, 0.5), (|1\rangle, 0.5)\}$ and $\mathcal{E}_1 = \{(|+\rangle, 0.5), (|-\rangle, 0.5)\}$ yield the density matrix

$$\rho = \frac{1}{2}I, \tag{7.20}$$

where I is the 2×2 identity matrix. Therefore, the two ensembles are equivalent.

Example 7.8

Consider the following two ensembles of pure qubit states:

$$\mathcal{E}_0 = \{(|0\rangle, 1/3), (|1\rangle, 1/3), (|+\rangle, 1/3)\} \tag{7.21}$$

and

$$\mathcal{E}_1 = \{(|+\rangle, 2/3), (|-\rangle, 1/3)\}. \tag{7.22}$$

While having a different number L of pure states, the two ensembles are equivalent, as they yield the same density matrix

$$\rho = \frac{1}{2}\begin{bmatrix} 1 & 1/3 \\ 1/3 & 1 \end{bmatrix}. \tag{7.23}$$

Since different ensembles can result in identical behavior, knowledge of the underlying ensemble $\mathcal{E}$ provides additional insight into how the system was prepared – insight that cannot be obtained from the system itself through measurements. In other words, the specific ensemble used to prepare the state is not directly observable through measurement. Measurements can only reveal information about the density matrix via the probabilities of the measurement outcomes (7.7).

In this regard, an interesting theoretical question is to characterize general properties of ensembles $\mathcal{E}$ that realize a given density matrix ρ.

Key Result 7.2 Ensemble classification theorem

Consider two ensembles $\mathcal{E}_0 = \{(|\psi_l\rangle, p_l)\}_{l\in[L_0]}$ and $\mathcal{E}_1 = \{(|\phi_l\rangle, q_l)\}_{l\in[L_1]}$. Assume without loss of generality the inequality $L_0 \geq L_1$. Pad the probability distribution q with zeros, so that both vectors p and q have L_0 entries. The two ensembles are associated with the same density matrix if and only if there exists an $L_0 \times L_0$ unitary matrix U such that the equalities

$$\sqrt{p_l}|\psi_l\rangle = \sum_{l'\in[L_0]} \sqrt{q_{l'}}\, U_{l,l'} |\phi_{l'}\rangle \tag{7.24}$$

hold for all $l \in [L_0]$, where $U_{l,l'}$ represents the (l, l')th entry of matrix U.

For a proof of this result, please see the Recommended Resources.

7.4 Projective Measurements of Mixed States

This section extends the Born rule for mixed states presented in Section 7.3 (see (7.14)) to the case of projective measurements. Projective measurements were described in Section 6.5 for pure states. To this end, some useful background material concerning the trace operator is introduced first.

7.4.1 Trace

The **trace** of a square, $N \times N$, matrix A is the sum of the elements on its diagonal:

$$\mathrm{tr}(A) = \sum_{x \in [N]} A_{x,x}. \tag{7.25}$$

To write this operation in a more convenient way, note that, for any square matrix A, the (x, y)th entry can be extracted via the "sandwich" operation (see Problem 7.1)

$$A_{x,y} = \langle x|A|y\rangle. \tag{7.26}$$

Using (7.26) in (7.25), the trace can thus be written as follows.

Key Result 7.3 Trace
The trace (7.25) of an $N \times N$ matrix A can be written as

$$\mathrm{tr}(A) = \sum_{x \in [N]} \langle x|A|x\rangle. \tag{7.27}$$

This characterization has the following direct consequences.

Key Result 7.4 Linearity of the trace
By (7.27), the trace is a linear operation, in the sense that the equality

$$\mathrm{tr}(aA + bB) = a\mathrm{tr}(A) + b\mathrm{tr}(B) \tag{7.28}$$

holds for any two $N \times N$ matrices A and B and any two scalars a and b.

Key Result 7.5 The trace turns outer products to inner products
Given any two $N \times 1$ vectors a and b, the trace turns the **outer product** $ab^\dagger$, an $N \times N$ matrix, into the **inner product** $b^\dagger a$, a scalar, i.e.,

$$\mathrm{tr}(ab^\dagger) = b^\dagger a. \tag{7.29}$$

When applied to $N \times 1$ state vectors $|u\rangle$ and $|v\rangle$, the property (7.29) yields the equality

$$\mathrm{tr}(|u\rangle\langle v|) = \langle u|v\rangle, \tag{7.30}$$

where $|u\rangle\langle v|$ is the outer product of states $|u\rangle$ and $|v\rangle$. Note that the outer product $|u\rangle\langle u|$ of a state $|u\rangle$ by itself is the projection matrix in the span $\mathcal{S}(|u\rangle)$ of the state.

Key Result 7.6 Permutation invariance of the trace

For matrices A and B having suitable dimensions, the trace is invariant to permutations of the two matrices, i.e.,

$$\mathrm{tr}(AB) = \mathrm{tr}(BA). \tag{7.31}$$

The property (7.31) generalizes (7.29). In fact, the latter equality can be obtained as a special case by setting $A = a$ and $B = b^\dagger$ and observing that the trace of a scalar is the scalar itself. Furthermore, as an application of this property, the following holds.

Key Result 7.7 Cyclic shift invariance of the trace

For matrices A, B, and C having suitable dimensions, one has the equality

$$\mathrm{tr}(ABC) = \mathrm{tr}(BCA) = \mathrm{tr}(CAB). \tag{7.32}$$

Using property (7.32), the trace can be expressed as in (7.27) by substituting the computational basis with *any* orthonormal basis $\{|v_x\rangle\}_{x\in[N]}$ as (Problem 7.3)

$$\mathrm{tr}(A) = \sum_{x\in[N]} \langle v_x|A|v_x\rangle. \tag{7.33}$$

Example 7.9

Take any *normal* matrix A with eigendecomposition $A = \sum_{x\in[N]} \lambda_x |u_x\rangle\langle u_x|$ (see Section 6.6.2). Choosing the basis of eigenvectors $\{|u_x\rangle\}_{x\in[N]}$ to evaluate the trace as per (7.33) yields

$$\mathrm{tr}(A) = \sum_{x\in[N]} \langle u_x|A|u_x\rangle = \sum_{x\in[N]} \lambda_x. \tag{7.34}$$

Hence, for a normal matrix, the trace is the sum of the eigenvalues.

7.4.2 Born Rule for Projective Measurements and Mixed States

With this background in place, the Born rule for projective measurements can be formulated as follows.

First, the Born rule (7.14) for a complete measurement $\mathcal{M}$ can equivalently be expressed as

$$\begin{aligned} p_x(\rho|\mathcal{M}) &= \langle v_x|\rho|v_x\rangle \\ &= \mathrm{tr}(\langle v_x|\rho|v_x\rangle) \\ &= \mathrm{tr}(\rho|v_x\rangle\langle v_x|) \\ &= \mathrm{tr}(\rho\Pi_x), \end{aligned} \tag{7.35}$$

where the third equality uses the cyclic shift invariance of the trace (7.32); and the last equality leverages the definition of the projection matrix $\Pi_x = |v_x\rangle\langle v_x|$. Note that the probability (7.35) can be also written as $p_x(\rho|\mathcal{M}) = \mathrm{tr}(\Pi_x\rho)$, swapping density and projection matrices, by the cyclic shift invariance of the trace.

Second, using (6.70), the probability of observing classical value $y \in [M]$ for a pure state $|\psi\rangle$ under a projective measurement setting $\mathcal{P} = \{\Pi_y\}_{y\in[M]}$ is given by

$$\begin{aligned} p_y(|\psi\rangle|\mathcal{P}) &= \langle\psi|\Pi_y|\psi\rangle \\ &= \mathrm{tr}(\rho\Pi_y). \end{aligned} \tag{7.36}$$

The second equality in (7.36) follows directly from the cyclic shift invariance property of the trace by using the definition (7.4) of the density matrix $\rho = |\psi\rangle\langle\psi|$.

The expressions (7.35) and (7.36) can be generalized to projective measurements on mixed states ρ, yielding the most general form of Born rule encountered up to now. The setting is illustrated in Figure 7.2, and the result is summarized as follows.

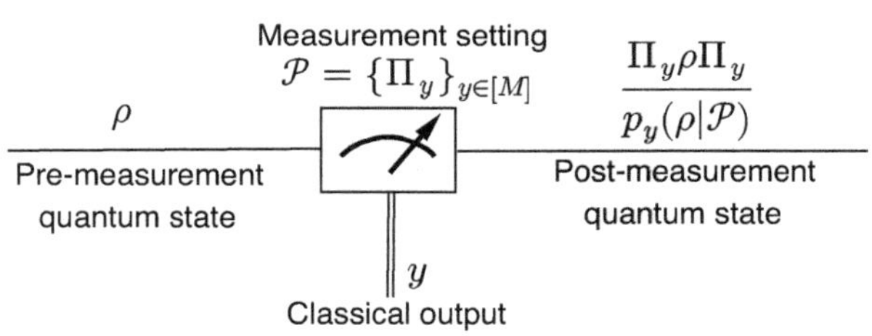

Figure 7.2 Illustration of a projective measurement in setting $\mathcal{P}$ on a mixed state defined by density matrix ρ.

Key Result 7.8 Projective measurement of a mixed state
Given a projective measurement setting $\mathcal{P} = \{\Pi_y\}_{y\in[M]}$ with $M \leq N$, a projective quantum measurement applied to a pre-measurement state ρ results in the following.

- **Classical output:** The measurement yields a classical output y $\sim p(\rho|\mathcal{P})$ with distribution

$$p_y(\rho|\mathcal{P}) = \mathrm{tr}(\rho\Pi_y) = \mathrm{tr}(\Pi_y\rho). \tag{7.37}$$

- **Post-measurement state:** For a given classical output y $= y$, the quantum system is left in the corresponding post-measurement state

$$\rho_y = \frac{1}{p_y(\rho|\mathcal{P})}\Pi_y\rho\Pi_y. \tag{7.38}$$

As detailed in Appendix 7.A, the derivation of the measurement probability (7.37) follows in a way similar to (7.2) by leveraging the law of total probability. Furthermore, the post-measurement density matrix (7.38) is a direct generalization of the post-measurement state $|\phi_y\rangle$ in (6.71) for a pre-measurement pure state $|\psi\rangle$. In fact, one can directly verify the equality $\rho_y = |\phi_y\rangle\langle\phi_y|$ for the rank-1 pre-measurement density matrix $\rho = |\psi\rangle\langle\psi|$.

As discussed in Chapter 6, the ranks of the projection matrices Π_y determine the coarse-graining level of the measurement, supporting a flexible trade-off between the informativeness of the classical output and the disturbance caused to the quantum state. To elaborate on this aspect, using the definition (7.8) of a density matrix, the post-measurement state (7.38) can be written as

$$\rho_y = \sum_{l\in[L]} p_l \frac{1}{p_y(\rho|\mathcal{P})}\Pi_y|\psi_l\rangle\langle\psi_l|\Pi_y. \tag{7.39}$$

The density matrix ρ_y is thus realized by the ensemble

$$\left\{\left(\frac{1}{\sqrt{p_y(\rho|\mathcal{P})}}\Pi_y|\psi_l\rangle, p_l\right)\right\}_{l\in[L]}, \tag{7.40}$$

where each pure state $\Pi_y|\psi_l\rangle/\sqrt{p_y(\rho|\mathcal{P})}$ lies in the span $\mathcal{S}(\Pi_y)$. It follows that all the pure states in the ensemble that are in the span $\mathcal{S}(\Pi_y)$, whose dimension is equal to the rank $\mathrm{rank}(\Pi_y)$, are preserved by the measurement when the classical output is y.

Example 7.10

For a system with $N = 4$ dimensions, consider the parity projective measurement $\mathcal{P} = \{\Pi_0, \Pi_1\}$ introduced in Chapter 6 (see Example 6.20), which is characterized by $M = 2$ measurement outputs and by the two projection matrices $\Pi_0 = |0\rangle\langle 0| + |2\rangle\langle 2|$ and $\Pi_1 = |1\rangle\langle 1| + |3\rangle\langle 3|$. Suppose that the pre-measurement state is given by the diagonal density matrix

$$\rho = p_0|0\rangle\langle 0| + p_1|1\rangle\langle 1| + p_2|2\rangle\langle 2| + p_3|3\rangle\langle 3| \tag{7.41}$$

with probability vector $p = [p_0, p_1, p_2, p_3]^T$. Note that this state is realized by the ensemble $\mathcal{E} = \{(|l\rangle, p_l)\}_{l\in[4]}$ consisting of computational-basis states.

Using (7.37), the measurement produces classical output $y = 0$ with probability

$$
\begin{aligned}
p_0(\rho|\mathcal{P}) &= \mathrm{tr}(\rho\Pi_0) \\
&= \mathrm{tr}\left(\left(\sum_{l\in[4]} p_l|l\rangle\langle l|\right)(|0\rangle\langle 0| + |2\rangle\langle 2|)\right) \\
&= \mathrm{tr}\left(\left(\sum_{l\in[4]} p_l|l\rangle\langle l|\right)|0\rangle\langle 0|\right) + \mathrm{tr}\left(\left(\sum_{l\in[4]} p_l|l\rangle\langle l|\right)|2\rangle\langle 2|\right) \\
&= \mathrm{tr}(p_0|0\rangle\langle 0|) + \mathrm{tr}(p_2|2\rangle\langle 2|) \\
&= p_0 + p_2,
\end{aligned} \tag{7.42}
$$

where the linearity of the trace, the orthonormality of the vectors $\{|l\rangle\}_{l\in[4]}$ in the computational basis, and the cyclic shift invariance property of the trace were used.

The probability (7.42) is the same obtained in Example 6.14, and it corresponds to the marginal distribution of the distribution p. Thus, the state ρ behaves in the same way as a classical system with distribution p. This is because the state can be realized by the states $\{|l\rangle\}_{l\in[4]}$, which are all incoherent with respect to the computational basis.

Furthermore, when the classical output is $y = 0$, the post-measurement state (7.38) is

$$
\begin{aligned}
\rho_0 &= \frac{1}{p_0(\rho|\mathcal{P})}\Pi_0\rho\Pi_0 \\
&= \frac{1}{p_0 + p_2}(p_0|0\rangle\langle 0| + p_2|2\rangle\langle 2|).
\end{aligned} \tag{7.43}
$$

The post-measurement state ρ_0 is thus mixed, containing some level of classical uncertainty that is inherited from the pre-measurement state ρ. This result contrasts with the case of complete measurements, for which the post-measurement state is always pure.

It also follows from (7.43) that, for any state with $p_1 = p_3 = 0$, the post-measurement state ρ_0 in (7.43) equals the pre-measurement state ρ. Hence, in this case, the measurement does not disturb the state. This is because, with $p_1 = p_3 = 0$, the state can be realized by the ensemble $\mathcal{E} = \{(|0\rangle, p_0), (|2\rangle, p_2)\}$, and both states in the ensemble $\mathcal{E}$ belong to the span $\mathcal{S}(\Pi_0)$ of the projection matrix Π_0.

Example 7.11

Consider again the parity measurement $\mathcal{P} = \{\Pi_0, \Pi_1\}$, which is applied now to the density matrix

$$
\rho = \frac{1}{2}|\psi\rangle\langle\psi| + \frac{1}{2}|\phi\rangle\langle\phi| = \frac{1}{4}\begin{bmatrix} 1 & 0 & 0 & 1 \\ 0 & 1 & 1 & 0 \\ 0 & 1 & 1 & 0 \\ 1 & 0 & 0 & 1 \end{bmatrix}, \tag{7.44}
$$

with states $|\psi\rangle = 0.5[1, 1, 1, 1]^T$ and $|\phi\rangle = 0.5[1, -1, -1, 1]^T$. An ensemble realizing the density matrix ρ is thus given by $\mathcal{E} = \{(|\psi\rangle, 0.5), (|\phi\rangle, 0.5)\}$.

The probability of obtaining output $y = 0$ can be evaluated as $p_0(\rho|\mathcal{P}) = \text{tr}(\rho\Pi_0) = 0.5$, and the corresponding post-measurement state is derived as

$$\rho_0 = \frac{1}{p_0(\rho|\mathcal{P})}\Pi_0\rho\Pi_0 = \frac{1}{2}\begin{bmatrix} 1 & 0 & 0 & 0 \\ 0 & 0 & 0 & 0 \\ 0 & 0 & 1 & 0 \\ 0 & 0 & 0 & 0 \end{bmatrix}$$
$$= \frac{1}{2}|0\rangle\langle 0| + \frac{1}{2}|2\rangle\langle 2|. \tag{7.45}$$

Finally, it is conceptually useful to interpret the Born rule (7.37) as a form of inner product between the density matrix ρ and the projection matrix Π_y.

Definition 7.6 Trace inner product
Given two matrices A and B of suitable dimensions, the operation

$$\text{tr}(A^\dagger B) \tag{7.46}$$

is defined as the **Hilbert–Schmidt inner product**, or **trace inner product**, between matrices A and B.

The trace inner product generalizes the inner product $a^\dagger b$ for $N \times 1$ vectors a and b. Like the inner product between vectors, the trace inner product can be viewed as a measure of similarity, or alignment, between matrices.

Using this definition and the fact that the density matrix is Hermitian, the Born rule (7.37) states that the measurement probability $p_y(\rho|\mathcal{P})$ equals the trace inner product $\text{tr}(\rho^\dagger\Pi_y) = \text{tr}(\rho\Pi_y)$ between matrices ρ and Π_y. The probability $p_y(\rho|\mathcal{P})$ in (7.37) is hence larger as the projection matrix Π_y becomes more aligned to the density matrix ρ.

7.5 From Density Matrices to Ensembles of Pure States

As seen earlier in this chapter, the density matrix is a mathematical description of a **class of equivalent ensembles** of pure states, each producing the same behavior. Given any of the equivalent ensembles, the corresponding density matrix is obtained as the correlation matrix (7.8) of the random state drawn from the ensemble. This section addresses the following questions.

1. **Defining properties of a density matrix:** What properties must an $N \times N$ matrix have in order to be a density matrix?
2. **Ensembles realizing a density matrix:** Given a density matrix ρ, how can one obtain an ensemble of pure states whose behavior is described by ρ?

7.5.1 Defining Properties of a Density Matrix

The answer to the first question is given as follows.

Key Result 7.9 Defining properties of a density matrix
An $N \times N$ square matrix ρ is a density matrix if and only if it satisfies the following properties.

- **Positive semi-definiteness:** Matrix ρ is positive semi-definite, i.e., it is Hermitian with non-negative eigenvalues,

$$\lambda_x \geq 0 \text{ for all } x \in [N]. \tag{7.47}$$

- **Unit trace:** The trace of matrix ρ equals 1, i.e.,

$$\text{tr}(\rho) = \sum_{x \in [N]} \lambda_x = 1. \tag{7.48}$$

Appendix 7.B shows that any density matrix $\rho = \sum_{l \in [L]} p_l |\psi_l\rangle\langle\psi_l|$ realized by a given ensemble of pure states, $\mathcal{E} = \{(|\psi_l\rangle, p_l)\}_{l \in [L]}$, satisfies the properties (7.47) and (7.48). This proves the "only if" part of the claim above.

A proof of the "if" part of the claim is provided in Section 7.5.2. As a byproduct of the proof, the discussion in Section 7.5.2 will also address the second question posed above concerning the construction of an ensemble described by a density matrix ρ.

7.5.2 Eigenensemble

Any matrix ρ satisfying the conditions (7.47) and (7.48) admits an eigendecomposition of the form

$$\rho = \sum_{x \in [N]} \lambda_x |u_x\rangle\langle u_x|, \tag{7.49}$$

with eigenvalues $\{\lambda_x \geq 0\}_{x \in [N]}$ and basis of eigenvectors $\{|u_x\rangle\}_{x \in [N]}$. Furthermore, since the eigenvalues are non-negative and satisfy the property (7.48), the vector

$$\lambda = [\lambda_0, \ldots, \lambda_{N-1}]^T \tag{7.50}$$

is a valid probability distribution.

Definition 7.7 Eigenensemble
Given a density matrix ρ with eigendecomposition (7.49), the ensemble of pure states

$$\mathcal{E}(\rho) = \{(|u_0\rangle, \lambda_0), (|u_1\rangle, \lambda_1), \ldots, (|u_{N-1}\rangle, \lambda_{N-1})\} \tag{7.51}$$

obtained from the vector of eigenvalues λ and eigenbasis $\{|u_x\rangle\}_{x\in[N]}$ realizes the density matrix ρ, and is known as an eigenensemble for matrix ρ. The eigenensemble is generally not unique, and any eigenensemble of density matrix ρ is denoted as $\mathcal{E}(\rho)$.

By definition, an eigenensemble has a number of non-zero probabilities λ_x equal to the rank of the density matrix, rank(ρ) (see (6.95)). Therefore, it represents an ensemble realizing density matrix ρ with a minimal number of pure states.

Example 7.12

The diagonal matrix

$$\rho = \mathrm{diag}(p) = \sum_{x\in[N]} p_x |x\rangle\langle x|, \tag{7.52}$$

with diagonal elements $\{p_x\}_{x\in[N]}$ forming a probability distribution, is associated with the eigenensemble

$$\mathcal{E}(\rho) = \{(|0\rangle, p_0), (|1\rangle, p_1), \ldots, (|N-1\rangle, p_{N-1})\} \tag{7.53}$$

comprising the states in the computational basis. Accordingly, the vector of eigenvalues is given by the probabilities $\lambda = [p_0, \ldots, p_{N-1}]^T$.

Example 7.13

Consider any density matrix with equal eigenvalues, i.e., with eigenvalue vector $\lambda = u = N^{-1}[1, \ldots, 1]^T$. Note that this corresponds to a situation of maximal classical uncertainty. Irrespective of the choice of the eigenbasis $\{|u_x\rangle\}_{x\in[N]}$, the density matrix is given by

$$\begin{aligned}\rho &= \frac{1}{N}\sum_{x\in[N]} |u_x\rangle\langle u_x| \\ &= \frac{1}{N} I\end{aligned} \tag{7.54}$$

by the resolution-of-the-identity equality (5.60). Quantum states described by a density matrix of the form (7.54) are said to be **maximally mixed**, since they present the maximum level of classical uncertainty.

The density matrix (7.54) of a maximally mixed state plays an important role in quantum information theory, and is reserved a special notation.

Definition 7.8 Maximally mixed state

A maximally mixed state is defined as

$$\pi = \frac{1}{N} I = \text{diag}(u), \tag{7.55}$$

where diag(u) is a diagonal matrix with elements equal to the uniform distribution.

Example 7.14

Gibbs states describe physical systems at thermal equilibrium. They are defined by the density matrix

$$\rho = \frac{1}{Z} \exp\left(-\frac{1}{T} H\right), \tag{7.56}$$

where T is a temperature parameter – more precisely, the product of the Boltzmann constant and the physical temperature; H is a Hermitian matrix that is known as the Hamiltonian of the system; and Z is a positive scalar that ensures that the trace tr(ρ) equals 1.

Being Hermitian, the Hamiltonian admits the eigendecomposition $H = \sum_{x \in [N]} h_x |v_x\rangle\langle v_x|$ with non-negative eigenvalues $\{h_x\}_{x \in [N]}$. Physically, the eigenvalues of the Hamiltonian have the significance of energy levels. Using the definition introduced in Chapter 6 of functions of normal matrices, the Gibbs density matrix (7.56) can thus be expressed in terms of the eigendecomposition of the Hamiltonian as

$$\rho = \frac{1}{Z} \sum_{x \in [N]} \exp\left(-\frac{h_x}{T}\right) |v_x\rangle\langle v_x|. \tag{7.57}$$

Accordingly, a Gibbs state ρ is realized by the eigenensemble

$$\mathcal{E}(\rho) = \{(|v_x\rangle, \lambda_x)\}_{x \in [N]} \tag{7.58}$$

with pure states given by the eigenvectors of the Hamiltonian and corresponding eigenvalues given by the Boltzmann distribution

$$\lambda_x = \frac{1}{Z} \exp\left(-\frac{h_x}{T}\right). \tag{7.59}$$

As the temperature T increases, all the non-zero eigenvalues of the density matrix ρ tend to equalize. Accordingly, if all the eigenvalues h_x of the Hamiltonian are strictly positive, when $T \to \infty$, the density ρ tends to the maximally mixed state π. In contrast, when T decreases, the Boltzmann distribution (7.59) tends to become more concentrated at the indices x corresponding to the smallest eigenvalues h_x. In the case with a single minimum eigenvalue, this reflects the physical phenomenon whereby, as the system cools down, it tends to the eigenstate $|v_x\rangle$ corresponding to minimal energy – as measured by the minimal eigenvalues h_x.

As a final note, typically, the Hamiltonian is expressed as a sum of Hermitian matrices as

$$H = \sum_x \theta_x H_x, \tag{7.60}$$

where the individual matrices H_x have the physical interpretation of describing local interactions in a composite system and the parameters θ_x are real numbers.

7.6 Uncertainty Inequality for Observables ⏭

As introduced in Section 5.5.5, observables describe numerical quantities obtained via the measurement of a quantum system. This section first shows that an observable can be specified without reference to an underlying measurement setting. Then, it formulates an uncertainty inequality that relates the uncertainty levels corresponding to two observables. It will be concluded that there is a fundamental limit to the precision at which different observables can be measured on the same state.

7.6.1 Observables as Hermitian Matrices

To review, Section 5.5.5 introduced an observable O in terms of: (i) a complete measurement setting $\mathcal{M}$; and (ii) an assignment $\{O_x\}_{x\in[N]}$ of each possible classical measurement output $x \in [N]$ to a corresponding numerical value O_x. For a state ρ, the measurement of an observable O yields the classical output O_x with probability $p_x(\rho|\mathcal{M})$ for all $x \in [N]$. Therefore, the output O_{x}, with $\mathrm{x} \sim p(\rho|\mathcal{M})$ is a random variable, whose expected value is

$$\langle O \rangle_\rho = \mathrm{E}_{p(\rho|\mathcal{M})}[O_{\mathrm{x}}]. \tag{7.61}$$

Using the linearity of the trace, the expected value (7.61) of the observable can be expressed as follows.

Definition 7.9 Observable matrix
The expected value (7.61) of observable O can be written as

$$\begin{aligned} \langle O \rangle_\rho &= \sum_{x\in[N]} O_x \mathrm{tr}(\rho|v_x\rangle\langle v_x|) \\ &= \mathrm{tr}\left(\rho \sum_{x\in[N]} O_x|v_x\rangle\langle v_x|\right) \\ &= \mathrm{tr}(\rho O), \end{aligned} \tag{7.62}$$

where the observable matrix is defined as

$$O = \sum_{x\in[N]} O_x|v_x\rangle\langle v_x|. \tag{7.63}$$

Therefore, by (7.62), the expected value $\langle O\rangle_\rho$ is given by the trace inner product $\mathrm{tr}(\rho O)$ between the density matrix ρ and the observable matrix O. Observable matrices O are also commonly referred to, for short, as *observables* in the literature.

Example 7.15

The expected value $\langle O\rangle_\rho$ recovers as a special case the Born rule (7.37) by setting as observable the projection matrix Π_y, i.e., $O = \Pi_y$. Note that a projection matrix is indeed in the form (7.63) with mapping O_x returning 1 for the states $|v_x\rangle$ within the span of the projection matrix Π_y, while returning 0 otherwise.

Example 7.16

For a pure state $\rho = |\psi\rangle\langle\psi|$, the expected value can be expressed in terms of the observable matrix O as

$$\begin{aligned}\langle O\rangle_\rho &= \mathrm{tr}(\rho O)\\ &= \mathrm{tr}(|\psi\rangle\langle\psi|O)\\ &= \langle\psi|O|\psi\rangle,\end{aligned} \tag{7.64}$$

where the third equality follows from the cyclic shift invariance of the trace.

The expression (7.63) is in the form of an eigendecomposition, with eigenbasis given by the measurement basis $\mathcal{M}$ and eigenvalues given by the set $\{O_x\}_{x\in[N]}$. Since the eigenvalues are real, the observable matrix O is Hermitian. Conversely, given any Hermitian matrix O, one can always associate with it a measurement setting $\mathcal{M}$ and assignment $\{O_x\}_{x\in[N]}$ via the eigendecomposition (7.63). In summary, one can conclude the following.

Key Result 7.10 Equivalent characterization of observable matrices
Any Hermitian matrix defines a valid observable, and any observable matrix is Hermitian.

Density matrices, which are unit-trace positive semi-definite matrices, and observables, which are Hermitian matrices, are the two most fundamental elements of quantum physics, with the former representing physical states and the latter measurable quantities.

Example 7.17

For a single-qubit system ($N = 2$), consider the observable defined by the Pauli Z matrix

$$O = P_3 = |0\rangle\langle 0| - |1\rangle\langle 1|. \tag{7.65}$$

By (7.65), measuring the observable with matrix P_3 corresponds to a standard measurement in the computational basis $\mathcal{M}$, in which the classical measurement output 0 is associated with numerical value $O_0 = 1$, while the classical measurement output 1 is associated with the numerical output value $O_1 = -1$.

Furthermore, using (7.62), for any 2×2 density matrix ρ, the expectation of the observable is given by

$$\begin{aligned}\langle P_3\rangle_\rho &= \text{tr}(\rho O)\\ &= \rho_{0,0} - \rho_{1,1},\end{aligned} \tag{7.66}$$

i.e., by the difference of the elements on the diagonal of matrix ρ.

Example 7.18

Take now the observable defined by the Pauli X matrix $O = P_1 = |+\rangle\langle +| - |-\rangle\langle -|$. This corresponds to a measurement in the Hadamard basis $\mathcal{M} = \{|+\rangle, |-\rangle\}$, in which the classical output 0 is assigned numerical value $O_0 = 1$ and the measurement output 1 is assigned output $O_1 = -1$. Using (7.62), for any 2×2 density matrix ρ, the expectation of the observable reads

$$\begin{aligned}\langle P_1\rangle_\rho &= \frac{1}{2}(\rho_{00} + \rho_{11} + 2\text{Re}(\rho_{01})) - \frac{1}{2}(\rho_{00} + \rho_{11} - 2\text{Re}(\rho_{01}))\\ &= 2\text{Re}(\rho_{01}),\end{aligned} \tag{7.67}$$

which is a function of the relative phase of the two amplitudes.

7.6.2 Compatible Observables

As defined in Section 6.4, two distinct complete measurements are compatible when they share the same measurement basis; and they are otherwise incompatible. The following definition is a direct extension to observables.

Definition 7.10 Compatible observables

Two observables O_0 and O_1 are said to be **compatible** if they share a common eigenbasis, i.e., if they admit eigendecompositions of the form

$$O_0 = \sum_{x\in[N]} O_{0,x}|v_x\rangle\langle v_x| \text{ and } O_1 = \sum_{x\in[N]} O_{1,x}|v_x\rangle\langle v_x| \tag{7.68}$$

with equal eigenvectors $\{|v_x\rangle\}_{x\in[N]}$ and generally different eigenvalues $\{O_{0,x}\}_{x\in[N]}$ and $\{O_{1,x}\}_{x\in[N]}$. Otherwise, the two observables are said to be **incompatible**.

While the characterization (7.68) requires the eigendecomposition of the two observable, it is also possible to express the compatibility of two observables in a simpler way that involves only multiplications of the two observable matrices (see Recommended Resources).

Key Result 7.11 Equivalent characterization of compatible observables
Two observables O_1 and O_2 are compatible if and only if they commute, i.e., if and only if the equality

$$O_1 O_2 = O_2 O_1 \tag{7.69}$$

holds.

The relationship (7.69) can also be expressed by introducing the **commutator**

$$[A, B] = AB - BA \tag{7.70}$$

for any two matrices A and B with suitable dimensions. The commutator of two observables O_0 and O_1, i.e., $[O_0, O_1] = O_0 O_1 - O_1 O_0$, is **skew-Hermitian**, i.e., $[O_0, O_1]^\dagger = -[O_0, O_1]$. Furthermore, two observables O_0 and O_1 are compatible if and only if the commutator equals the all-zero matrix, i.e.,

$$[O_0, O_1] = 0. \tag{7.71}$$

Therefore, any two observables with a non-zero commutator are incompatible.

Example 7.19

The Pauli matrices P_1, P_2, and P_3 define mutually incompatible pairs of observables for single-qubit systems. For Pauli X and Y matrices P_1 and P_2, whose products are $P_1 P_2 = iP_3$, and hence $P_2 P_1 = (P_1 P_2)^\dagger = -iP_3$, the commutator is

$$[P_1, P_2] = 2iP_3, \tag{7.72}$$

which is different from the all-zero matrix.

Similarly, for observables P_2 and P_3, the product is $P_2 P_3 = iP_1$, which yields the commutator

$$[P_2, P_3] = 2iP_1; \tag{7.73}$$

and for observables P_3 and P_1, one has $P_3 P_1 = iP_2$ and

$$[P_3, P_1] = 2iP_2. \tag{7.74}$$

Note the circular structure of the commutators.

7.6.3 Uncertainty Inequality

As for the underlying measurement settings, incompatible observables satisfy an uncertainty inequality. The most classical formulation, as introduced by Werner Heisenberg, is given by considering the variance of the observables. Note that this contrasts with the uncertainty inequality described in Section 6.4.4 that is based on information measures.

The variance of an observable O for state ρ is defined as

$$\mathrm{var}_\rho(O) = \langle O^2 \rangle_\rho - (\langle O \rangle_\rho)^2. \tag{7.75}$$

Denoting as $\mathcal{M}$ an eigenbasis for the observable O, the definition (7.75) can readily be seen to coincide with the classical definition of variance for the random variable O_{x} with $\mathrm{x} \sim p(\rho|\mathcal{M})$ (see Section 1.4.4), as one has the equalities

$$\begin{aligned}\mathrm{var}_\rho(O) &= \mathrm{E}_{p(\rho|\mathcal{M})}[O_{\mathrm{x}}^2] - (\mathrm{E}_{p(\rho|\mathcal{M})}[O_{\mathrm{x}}])^2 \\ &= \mathrm{E}_{p(\rho|\mathcal{M})}[(O_{\mathrm{x}} - \mathrm{E}_{p(\rho|\mathcal{M})}[O_{\mathrm{x}}])^2].\end{aligned} \tag{7.76}$$

Key Result 7.12 Uncertainty inequality for observables
Given any two $N \times N$ observables O_0 and O_1, the following inequality holds

$$\sqrt{\mathrm{var}_\rho(O_0)} \cdot \sqrt{\mathrm{var}_\rho(O_1)} \geq \frac{1}{2}|\mathrm{tr}(\rho[O_0, O_1])|. \tag{7.77}$$

By the inequality (7.77), the product of the standard deviations $\sqrt{\mathrm{var}_\rho(O_0)}$ and $\sqrt{\mathrm{var}_\rho(O_1)}$ cannot be smaller than the quantity $|\mathrm{tr}(\rho[O_0, O])|/2$. If the two observables are incompatible, this quantity can be strictly positive, $|\mathrm{tr}(\rho[O_0, O_1])|/2 > 0$. In this case, if the variance of one observable under a state ρ is small, the variance of the other observable for the same state must be large enough. For a proof of (7.77), please refer to the Recommended Resources.

Unlike the entropy-based counterpart studied in Chapter 6, the right-hand side of (7.77) depends on the state ρ. The need for this dependence is illustrated by the example below.

Example 7.20

For a system with dimension $N = 4$, consider the observables

$$O_0 = \frac{1}{2}\begin{bmatrix} 1 & 1 & 0 & 0 \\ 1 & 1 & 0 & 0 \\ 0 & 0 & 0 & 0 \\ 0 & 0 & 0 & 0 \end{bmatrix} \text{ and } O_1 = \begin{bmatrix} 1 & 0 & 0 & 0 \\ 0 & 0 & 0 & 0 \\ 0 & 0 & 0 & 0 \\ 0 & 0 & 0 & 0 \end{bmatrix}. \tag{7.78}$$

One can check that the two observables are incompatible. However, given state

$$\rho = \frac{1}{2}\begin{bmatrix} 0 & 0 & 0 & 0 \\ 0 & 0 & 0 & 0 \\ 0 & 0 & 1 & 0 \\ 0 & 0 & 0 & 1 \end{bmatrix}, \tag{7.79}$$

the variances of both observables equal zero, i.e., $\mathrm{var}_\rho(O_0) = \mathrm{var}_\rho(O_1) = 0$. This does not contradict the uncertainty inequality, since the right-hand side of (7.77) is also equal to zero for this particular state ρ.

Example 7.21

Consider the observable $O_0 = P_3$, i.e., the Pauli Z matrix, and the observable $O_1 = |\theta\rangle\langle\theta| - |\theta^\perp\rangle\langle\theta^\perp|$, where $|\theta\rangle = [\cos(\theta), \sin(\theta)]^T$ and $|\theta^\perp\rangle = [-\sin(\theta), \cos(\theta)]^T$. Note that for $\theta = 0$ the two underlying measurement bases are compatible, while for $\theta = \pi/4$ the two underlying bases are maximally incompatible. Accordingly, one can check that the commutator $[O_0, O_1]$ equals the all-zero matrix for $\theta = 0$, while it has non-zero entries for θ in the range of interest between 0 and $\pi/4$. Furthermore, the eigenvalues for both observables equal -1 and 1, and thus the maximum value for the variances $\mathrm{var}_\rho(O_0)$ and $\mathrm{var}_\rho(O_1)$ is 1.

To evaluate the effect of the choice of θ on the variance of the two observables, 1000 pure states $|\psi\rangle$ were generated at random, and the corresponding standard deviations of the observables were measured along with the right-hand side of the uncertainty inequality (7.77). Specifically, the two complex amplitudes of each state vector were generated with independent Gaussian real and imaginary parts and then normalized.

Figure 7.3(a) shows the average value of the right-hand side of the uncertainty inequality (7.77) as a function of the angle θ. The average is evaluated over the 1000 random realizations of the state vector. As the angle θ increases from 0 to $\pi/4$, this quantity is seen to grow larger, indicating observables that are increasingly incompatible.

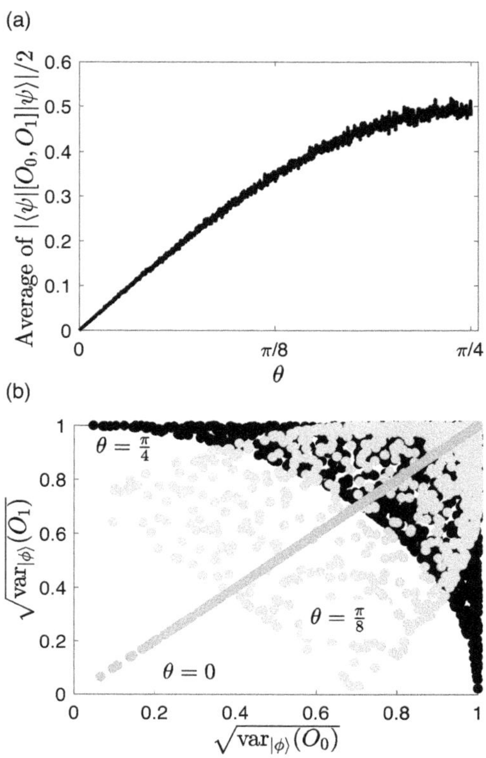

Figure 7.3 For observables $O_0 = P_3$ and $O_1 = |\theta\rangle\langle\theta| - |\theta^\perp\rangle\langle\theta^\perp|$: (a) the average value of the right-hand side of the uncertainty inequality (7.77) as a function of the angle θ; (b) a scatter plot of the pairs of standard deviations $\sqrt{\mathrm{var}_{|\psi\rangle}(O_0)}$ and $\sqrt{\mathrm{var}_{|\psi\rangle}(O_1)}$ for different values of θ. Average and scatter plots are obtained by generating 1000 random states $|\psi\rangle$.

Figure 7.3(b) shows a scatter plot of the pairs of standard deviations $\sqrt{\text{var}_{|\psi\rangle}(O_0)}$ and $\sqrt{\text{var}_{|\psi\rangle}(O_1)}$ obtained for the randomly generated states $|\psi\rangle$ for three different values of θ. For $\theta = 0$, the two observables coincide, and it is thus clearly possible for both observables to have a small variance. In contrast, as θ increases up to $\pi/4$, the two observables become increasingly incompatible. In this regime, states $|\psi\rangle$ for which the variance of one observable is large must necessarily exhibit a sufficiently large variance for the other observable. Furthermore, this trade-off becomes more pronounced as θ tends to $\pi/4$, at which point the two underlying measurement bases are maximally incompatible.

7.7 Measured Entropy for Mixed States

This section addresses the problem of quantifying the uncertainty associated with the observation of a mixed state ρ by extending the notion of measured entropy, introduced in Section 6.2.1, from pure to mixed states. Intuitively, the measured entropy for a state ρ should account for: (i) the classical uncertainty associated with the random selection of a pure state from an ensemble realizing the density matrix ρ; and (ii) the quantum uncertainty exhibited by quantum measurements on pure states. This picture is, however, complicated by the fact that, unless the state is pure, there is an infinitely large number of equivalent ensembles for a mixed state ρ.

To start, the measured entropy for mixed states is naturally defined by using the Born rule (7.37) as follows.

Definition 7.11 Shannon, Tsallis, and Rényi measured entropies
The Shannon, Tsallis, and Rényi measured entropies of quantum state ρ under measurement setting $\mathcal{P} = \{\Pi_y\}_{y\in[M]}$ are defined as

$$\text{H}(\rho|\mathcal{P}) = \text{H}(p(\rho|\mathcal{P})), \tag{7.80}$$

$$\text{H}_\alpha^{\text{T}}(\rho|\mathcal{P}) = \text{H}_\alpha^{\text{T}}(p(\rho|\mathcal{P})), \tag{7.81}$$

and

$$\text{H}_\alpha^{\text{R}}(\rho|\mathcal{P}) = \text{H}_\alpha^{\text{R}}(p(\rho|\mathcal{P})), \tag{7.82}$$

respectively, with $\alpha \geq 0$ being the order of the Tsallis and Rényi entropies.

For instance, the Shannon measured entropy (7.80) can be written more explicitly in terms of the measurement distribution $p(\rho|\mathcal{P})$ as

$$\begin{aligned}\text{H}(\rho|\mathcal{P}) &= -\sum_{y\in[M]} p_y(\rho|\mathcal{P})\log_2(p_y(\rho|\mathcal{P})) \\ &= -\sum_{y\in[M]} \text{tr}(\rho\Pi_y)\log_2(\text{tr}(\rho\Pi_y)).\end{aligned} \tag{7.83}$$

In line with the convention adopted throughout the book, the generic notation $\mathrm{H}_\alpha(\rho|\mathcal{P})$ will be used henceforth to refer to any of the measured entropies whenever the given statement applies to all entropy metrics in (7.80)–(7.82).

By definition, the measured entropy $\mathrm{H}_\alpha(\rho|\mathcal{P})$ quantifies the overall uncertainty associated with a measurement output, thus accounting for the contributions of both classical and quantum uncertainties. The next example illustrates the interplay between the two forms of uncertainty.

Example 7.22

Consider the measurement setting $\mathcal{M} = \{|+\rangle, |-\rangle\}$ in the Hadamard basis, and the density matrix

$$\rho = \frac{1}{2}\begin{bmatrix} 1 & 1/3 \\ 1/3 & 1 \end{bmatrix}. \tag{7.84}$$

The corresponding measurement probability distribution is evaluated as $p_0(\rho|\mathcal{M}) = \langle+|\rho|+\rangle = 2/3$ and $p_1(\rho|\mathcal{M}) = 1/3$. Accordingly, the Shannon measured entropy is given by

$$\mathrm{H}(\rho|\mathcal{M}) = \mathrm{h}\left(\frac{1}{3}\right) = 0.918. \tag{7.85}$$

This measure quantifies the overall uncertainty associated with the measurement output. Is it possible to determine the separate contributions of classical and quantum uncertainties?

To address this question, assume that we are told that the ensemble generating the state is the eigenensemble $\mathcal{E}(\rho) = \{(|+\rangle, 2/3), (|-\rangle, 1/3)\}$. In this case, the classical uncertainty associated with the selection of a pure state from the ensemble is described by the probability vector $p = [2/3, 1/3]^T$, and thus the corresponding Shannon entropy is

$$\mathrm{H}(p) = \mathrm{h}\left(\frac{1}{3}\right) = 0.918. \tag{7.86}$$

In contrast, the quantum uncertainty equals zero, since both states in the ensemble, namely $|+\rangle$ and $|-\rangle$, are incoherent with respect to the measurement (Hadamard) basis. Therefore, with this choice of ensemble, the overall uncertainty $\mathrm{H}(\rho|\mathcal{M})$ is explained exclusively by the classical uncertainty $\mathrm{H}(p)$.

However, the choice of the ensemble is not unique. For example, we may be told that the underlying ensemble is given by $\mathcal{E} = \{(|0\rangle, 1/3), (|1\rangle, 1/3), (|+\rangle, 1/3)\}$ instead. Recall that this ensemble also corresponds to the density matrix ρ (see Example 7.8). In this case, one may be tempted to say that the classical uncertainty is given by the Shannon entropy of the probability vector $p = [1/3, 1/3, 1/3]^T$, since this reflects the level of randomness in the selection of the quantum state. However, the Shannon entropy of this distribution is $\mathrm{H}(p) = \log_2(3) = 1.585$, which is larger than the overall uncertainty given by the measured entropy $\mathrm{H}(\rho|\mathcal{M}) = 0.918$.

The reason for this apparent contradiction is that the states in the ensemble are not orthogonal, and thus they are not perfectly distinguishable. Therefore, the effect of the random choice of the pure state in the ensemble on the overall uncertainty is mitigated by the alignment between the states in the ensemble. This yields a measured entropy that is smaller than the entropy of the ensemble probability vector p.

In conclusion, Example 7.22 has shown that the contributions of classical and quantum uncertainties to the measured entropy are generally inextricable, for the following reasons.

- **Non-uniqueness of ensembles:** For mixed states, there is an infinite number of ensembles associated with a density matrix, and the contributions to the classical and quantum uncertainties depend on the choice of the ensemble.
- **Non-orthogonality of states in an ensemble:** The states within an ensemble may not be orthogonal, causing the overall uncertainty to be possibly smaller than the classical uncertainty associated with the ensemble.

Example 7.23 shows that, even for a given ensemble consisting of orthogonal states, the contributions of classical and quantum uncertainties are not additive.

Example 7.23

Consider again the density matrix (7.84) studied in Example 7.22, and fix the eigenensemble $\mathcal{E}(\rho) = \{(|+\rangle, 2/3), (|-\rangle, 1/3)\}$, which consists of orthogonal states. For the computational basis $\mathcal{M}$, the measurement probabilities are given by $p_0(\rho|\mathcal{M}) = \rho_{0,0} = 0.5$ and $p_1(\rho|\mathcal{M}) = \rho_{1,1} = 0.5$. Accordingly, the Shannon measured entropy equals $\mathrm{H}(\rho|\mathcal{M}) = 1$.

As discussed in Example 7.22, the classical uncertainty associated with the selection of a pure state from the ensemble $\mathcal{E}(\rho)$ is evaluated as $\mathrm{H}(p) = 0.918$. Furthermore, for any of the two states $|+\rangle$ or $|-\rangle$ in the ensemble $\mathcal{E}(\rho)$, the quantum uncertainty, measured by the Shannon entropy, is given by $\mathrm{H}(p(|+\rangle|\mathcal{M})) = \mathrm{H}(p(|-\rangle|\mathcal{M})) = 1$. Thus, the sum of the classical and quantum uncertainties is larger than the Shannon measured entropy, showing that the two contributions do not add up.

7.8 Quantum Entropy

As discussed in Section 6.2.1, for any pure state, it is always possible to find measurement settings that yield a measured entropy equal to zero. This section will show that this is not true for mixed states. Specifically, when minimized over all complete measurement settings $\mathcal{M}$, the measured entropy $\mathrm{H}_\alpha(\rho|\mathcal{M})$ for mixed states is positive, and is referred to as the **quantum entropy** $\mathrm{H}_\alpha(\rho)$. The quantum entropy can be interpreted as the irreducible uncertainty level associated with a state ρ.

7.8.1 Defining the Quantum Entropy

Definition 7.12 Quantum entropy as the minimum measured entropy
Given a density matrix ρ, the quantum entropy is defined as

$$\mathrm{H}_\alpha(\rho) = \min_{\mathcal{M}} \mathrm{H}_\alpha(\rho|\mathcal{M}), \tag{7.87}$$

where the minimization is over all possible complete measurement settings.

By this definition, the quantum entropy $\mathrm{H}_\alpha(\rho)$ represents the minimum entropy associated with a complete measurement of the state.

Key Result 7.13 Entropy-minimizing measurement basis
Any eigenbasis $\{|u_x\rangle\}_{x\in[N]}$ of matrix ρ is an entropy-minimizing measurement basis, denoted as $\mathcal{M}_*^{\min}$.

The proof of this result is discussed in Section 11.2.3. With an optimized measurement setting $\mathcal{M}_*^{\min} = \{|u_x\rangle\}_{x\in[N]}$, the measurement probabilities equal

$$p_x(\rho|\mathcal{M}_*^{\min}) = \langle u_x|\rho|u_x\rangle = \lambda_x, \tag{7.88}$$

for $x \in [N]$, where $\lambda = [\lambda_0, \ldots, \lambda_{N-1}]^T$ is the vector of eigenvalues of matrix ρ. The following equivalent characterization of the quantum entropy then follows.

Key Result 7.14 Equivalent characterization of the quantum entropy
The quantum entropy of a density matrix ρ coincides with the classical entropy of the vector of eigenvalues, λ, of matrix ρ, i.e.,

$$\mathrm{H}_\alpha(\rho) = \mathrm{H}_\alpha(\lambda). \tag{7.89}$$

Accordingly, the **Shannon, Tsallis, and Rényi quantum entropies** are defined as

$$\mathrm{H}(\rho) = \mathrm{H}(\rho|\mathcal{M}_*^{\min}) = \mathrm{H}(\lambda), \tag{7.90}$$

$$\mathrm{H}_\alpha^{\mathrm{T}}(\rho) = \mathrm{H}_\alpha^{\mathrm{T}}(\rho|\mathcal{M}_*^{\min}) = \mathrm{H}_\alpha^{\mathrm{T}}(\lambda), \tag{7.91}$$

and

$$\mathrm{H}_\alpha^{\mathrm{R}}(\rho) = \mathrm{H}_\alpha^{\mathrm{R}}(\rho|\mathcal{M}_{\min}^*) = \mathrm{H}_\alpha^{\mathrm{R}}(\lambda), \tag{7.92}$$

respectively.

The following remarks on the quantum entropy are in order.

- **Quantum entropy as the minimum measurement uncertainty:** Due to the minimization over all complete measurement settings in the definition (7.87), the quantum entropy $\mathrm{H}_\alpha(\rho)$ does not depend on the measurement setting, and is only a function on the quantum state ρ.
- **Quantum entropy and measured entropy:** By its definition (7.87), the quantum entropy lower bounds the value of the measured entropy attained with any measurement setting $\mathcal{M}$, i.e.,

$$\mathrm{H}_\alpha(\rho) \le \mathrm{H}_\alpha(\rho|\mathcal{M}). \tag{7.93}$$

- **Quantum entropy as the minimum preparation randomness:** The uncertainty measured by the quantum entropy (7.89) is fully accounted for by the classical uncertainty associated with the preparation of an eigenensemble $\{(|u_x\rangle, \lambda_x)\}_{x\in[N]}$. Thus, the quantum entropy $\mathrm{H}_\alpha(\rho)$ corresponds to the minimal classical randomness required to prepare the state ρ.
- **Why is the minimization in (7.87) limited to complete measurements?** It is important to limit the optimization in (7.87) to complete measurement settings. In fact, if the optimization was extended to projective measurements, one would trivially obtain a minimum value equal to zero. In particular, a projective measurement with a single measurement outcome, i.e., $M = 1$, would produce a measured entropy equal to zero. An alternative definition of measured entropy, known as **observational entropy**, compensates for the effect of coarse graining, obtaining a metric that is no smaller than the quantum entropy for any projective measurement setting (see Appendix 7.C).
- **"Shannon" quantum entropy?** It is historically inaccurate to attribute the quantity (7.90) to Claude Shannon. This quantity was, in fact, introduced by John von Neumann in 1932, while Claude Shannon presented the classical entropy in 1948. Accordingly, it is typically referred to as the **von Neumann entropy.** Here, the name Shannon quantum entropy is preferred for consistency with the corresponding classical entropy measure.
- **Tsallis and Rényi quantum entropies vs. Shannon quantum entropy:** By the properties of their classical counterparts, the Rényi and Tsallis quantum entropies recover the Shannon quantum entropy for $\alpha = 1$, i.e.,

$$\mathrm{H}_1^{\mathrm{T}}(\rho) = \mathrm{H}_1^{\mathrm{R}}(\rho) = \mathrm{H}(\rho). \tag{7.94}$$

Example 7.24

Consider the following density matrix for a system with $N = 4$ dimensions,

$$\rho = \frac{1}{3}\begin{bmatrix} 1 & 0 & 0 & 0 \\ 0 & 0.5 & 0.5 & 0 \\ 0 & 0.5 & 0.5 & 0 \\ 0 & 0 & 0 & 1 \end{bmatrix}, \tag{7.95}$$

for which the vector of eigenvalues is $\lambda = [1/3, 1/3, 1/3, 0]^T$. Accordingly, the Shannon quantum entropy equals $\mathrm{H}(\rho) = \mathrm{H}(\lambda) = \log_2(3) = 1.585$.

In contrast, with a measurement in the computational basis $\mathcal{M}$, the measured entropy equals

$$\mathrm{H}(\rho|\mathcal{M}) = -\frac{2}{3}\log_2\left(\frac{1}{3}\right) - \frac{2}{6}\log_2\left(\frac{1}{6}\right)$$
$$= 1.918, \tag{7.96}$$

since the measurement probabilities coincide with the elements on the diagonal of matrix ρ. Therefore, an entropy-minimizing complete measurement setting $\mathcal{M}_*^{\min}$, i.e., an eigenbasis, yields a smaller entropy compared to the computational basis.

7.8.2 Properties of the Quantum Entropy

The quantum entropies inherit from their classical counterparts the following basic properties.

Key Result 7.15 Non-negativity and minimum-value property of the quantum entropy
For any density matrices ρ, the quantum entropy $\mathrm{H}_\alpha(\rho)$ of any finite order $\alpha \geq 0$ is non-negative,

$$\mathrm{H}_\alpha(\rho) \geq 0, \tag{7.97}$$

and it equals zero if and only if the state is pure, i.e.,

$$\mathrm{H}_\alpha(|\psi\rangle\langle\psi|) = 0. \tag{7.98}$$

Indeed, for a pure state $\rho = |\psi\rangle\langle\psi|$, the vector of eigenvalues of the density matrix is given by a zero-entropy distribution.

Key Result 7.16 Maximum-value property of the quantum entropy
The quantum entropy $\mathrm{H}_\alpha(\rho)$ of any order $\alpha \geq 0$ is upper bounded by the quantum entropy of the maximally mixed state, i.e.,

$$\mathrm{H}_\alpha(\rho) \leq \mathrm{H}_\alpha(\pi). \tag{7.99}$$

Thus, maximally mixed states are maximum-entropy states.

Finally, consider the property of concavity. As discussed in Section 2.5.4, concavity indicates that mixing increases uncertainty. In the case of quantum states, this refers to a situation in which there is classical uncertainty among a set of density matrices $\{\rho_x\}_{x\in[M]}$, so that each density matrix ρ_x is assigned a probability p_x for $x \in [M]$. The resulting mixed state is characterized by the density matrix

$$\rho = \sum_{x\in[M]} p_x \rho_x \tag{7.100}$$

obtained by averaging the possible density matrices $\{\rho_x\}_{x\in[M]}$. For a concave quantum entropy, the quantum entropy of the average density matrix ρ in (7.100) is no smaller than the average quantum entropy for the individual density matrices $\{\rho_x\}_{x\in[M]}$.

Key Result 7.17 Concavity of the quantum entropy
The Shannon quantum entropy, the Tsallis quantum entropy for any order $\alpha \geq 0$, and the Rényi quantum entropy for $0 \leq \alpha \leq 1$ are concave with respect to the density matrix ρ. That is, for these entropies one has the inequality

$$\mathrm{H}_\alpha\left(\sum_{x\in[M]} p_x\rho_x\right) \geq \sum_{x\in[M]} p_x \mathrm{H}_\alpha(\rho_x). \tag{7.101}$$

The concavity property (7.101) will be proved in Chapter 16 for the Shannon quantum entropy (see Problem 16.20). Proofs for the Tsallis and Rényi quantum entropies can be found in the Recommended Resources.

7.9 Shannon, Tsallis, and Rényi Quantum Entropy

In Section 7.8, the quantum entropy was defined as the classical entropy of the vectors of eigenvalues of the density matrix. As discussed in this section, the quantum entropy can also be expressed directly in terms of the density matrix ρ. Using this characterization, the Tsallis and Rényi quantum entropies will also be specialized to a number of important cases.

7.9.1 Background

To start, recall from Section 6.6.5 that, for any $N \times N$ normal matrix A with eigendecomposition $A = \sum_{x\in[N]} \lambda_x |u_x\rangle\langle u_x|$, and any real-valued scalar function $f(\cdot)$, the matrix function $f(A)$ is defined as

$$f(A) = \sum_{x\in[N]} f(\lambda_x)|u_x\rangle\langle u_x|. \tag{7.102}$$

This implies that the trace of function $f(A)$ equals the sum of the eigenvalues $\{f(\lambda_x)\}_{x\in[N]}$ of matrix (7.102) as

$$\mathrm{tr}(f(A)) = \sum_{x\in[N]} f(\lambda_x). \tag{7.103}$$

7.9.2 Shannon Quantum Entropy

Using (7.103), the Shannon quantum entropy in (7.90) can be expressed as follows.

Key Result 7.18 Equivalent definition of the Shannon quantum entropy
The Shannon quantum entropy can be written as

$$\mathrm{H}(\rho) = -\mathrm{tr}(\rho \log_2(\rho)). \tag{7.104}$$

To verify the equality between (7.90) and (7.104), it is sufficient to apply (7.103) with the function $f(x) = x\log_2(x)$ as

$$\begin{aligned} -\mathrm{tr}(\underbrace{\rho \log_2(\rho)}_{=f(\rho)}) &= -\sum_{x\in[N]} f(\lambda_x) \\ &= -\sum_{x\in[N]} \lambda_x \log_2(\lambda_x) = \mathrm{H}(\lambda). \end{aligned} \tag{7.105}$$

In comparing the expression $\mathrm{H}(p) = -\sum_{x\in[N]} p_x \log_2(p_x)$ of the Shannon entropy for a distribution p with the expression (7.104) of the Shannon quantum entropy, one may note the correspondence in Table 7.1. As discussed further in the rest of this section, this correspondence can be kept in mind as a mnemonic aid for the relationship between the classical and quantum entropies.

The inequalities (7.97) and (7.99) imply the following bounds.

Key Result 7.19 Bounds on the Shannon quantum entropy
For any $N \times N$ density matrix ρ, the Shannon quantum entropy satisfies the inequalities

$$0 \leq \mathrm{H}(\rho) \leq \mathrm{H}(\pi) = \log_2(N). \tag{7.106}$$

The lower bound attained if and only if the state is pure, and the upper bound obtained if and only if the state is maximally mixed.

The upper bound in (7.106) implies that, like the Shannon entropy, the Shannon quantum entropy can grow at most logarithmically with the dimension N.

Table 7.1 Correspondence between quantities appearing in the definitions of classical and quantum entropies

Classical	Quantum
$\sum_{x\in[N]}$	$\mathrm{tr}(\cdot)$
$f(p_x)$	$f(\rho)$

Example 7.25

Consider the density matrix

$$\rho = p_0|0\rangle\langle 0| + (1 - p_0)|\theta\rangle\langle\theta| \tag{7.107}$$

with state $|\theta\rangle = [\cos(\theta), \sin(\theta)]^T$ for some angle $0 \leq \theta \leq \pi/2$ and probability p_0. The density matrix ρ can be generated by the ensemble $\mathcal{E} = \{(|0\rangle, p_0), (|\theta\rangle, 1 - p_0)\}$. For $\theta = 0$, the two states in the ensemble are equivalent, and are thus indistinguishable. Accordingly, the state is pure, and the Shannon quantum entropy is zero.

In the other extreme case, with $\theta = \pi/2$, the two states in the ensemble are orthogonal, and thus perfectly distinguishable. As a result, the eigenvalue vector λ of the density matrix ρ coincides with the probability vector p, i.e., $\lambda = p = [p_0, 1 - p_0]^T$, and the Shannon quantum entropy equals $\mathrm{H}(\rho) = \mathrm{h}(p_0)$.

In general, the eigenvalues of the density matrix can be computed in closed form as (see the Recommended Resources)

$$\lambda_0 = \frac{1}{2}\left(1 - \sqrt{1 - 4p_0(1 - p_0)(\sin(\theta))^2}\right) \tag{7.108}$$

and $\lambda_1 = 1 - \lambda_0$. Accordingly, the Shannon quantum entropy is obtained as $\mathrm{H}(\rho) = \mathrm{h}(\lambda_0)$.

Figure 7.4 plots the Shannon quantum entropy as a function of the probability p_0 for different values of the angle θ. As illustrated by the figure, the eigenvalues λ_0 and λ_1 depend not only on the distribution p but also on the quantum properties of the ensemble $\mathcal{E}$ via the angle θ. Take, for instance, the case $p = 1/2$, which corresponds to a situation of maximal classical uncertainty in the generation of the ensemble. With $p = 1/2$, one obtains a Shannon quantum entropy equal to zero when the quantum states are indistinguishable, i.e., when $\theta = 0$; while the maximum value of 1 bit is attained when $\theta = \pi/2$ and thus the two states are orthogonal.

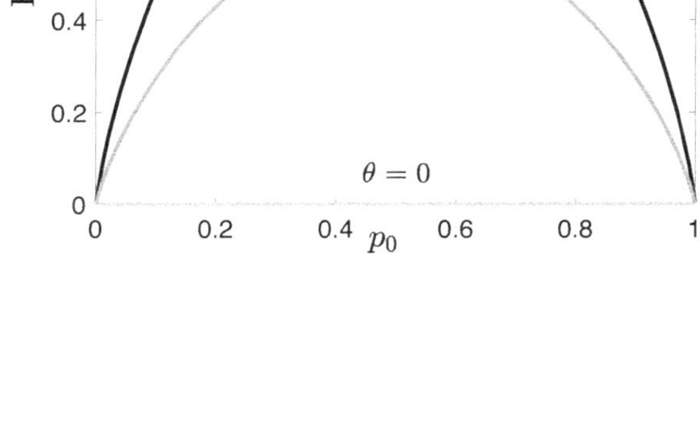

Figure 7.4 Shannon quantum entropy $\mathrm{H}(\rho)$ for the density matrix (7.107) as a function of probability p_0.

Example 7.26

Given a Hermitian matrix H and a temperature constant T, the **free energy** of a state ρ is defined as

$$\mathrm{tr}(H\rho) - T\mathrm{H}(\rho). \tag{7.109}$$

The first term in the free energy is the expected value of the observable H, while the second is the Shannon quantum entropy scaled by the temperature $T > 0$. The **Gibbs state** (7.56) with Hamiltonian H can be shown to minimize the free energy. For further discussion on this example, see Problem 7.9.

7.9.3 Thermodynamic Information and Shannon Quantum Entropy ⏭

In a manner similar to the discussion in Section 2.5.5 about classical information, the Shannon quantum entropy can be related to a dual measure that is connected to the order present in a system that can be extracted and turned into work. This measure, known as **quantum negentropy**, can be defined through the Shannon quantum entropy $\mathrm{H}(\rho)$ as

$$\mathrm{I}(\rho) = \log_2(N) - \mathrm{H}(\rho). \tag{7.110}$$

Accordingly, if the state is maximally mixed, no work can be extracted from it, i.e., $\mathrm{I}(\rho) = 0$, while the maximum amount of work can be extracted from pure states, i.e., $\mathrm{I}(\rho) = \log_2(N)$.

7.10 Tsallis Trace ⏭

As detailed in Section 3.4, the Tsallis and Rényi classical entropies of order α can be expressed as a function of the same quantity, known as **Tsallis sum** of order α. This section shows that a similar characterization can be extended to the Tsallis and Rényi quantum entropies.

7.10.1 Introducing the Tsallis Trace

The quantum counterpart of the Tsallis sum is given by the Tsallis trace, which is defined as follows.

Definition 7.13 Tsallis trace
For a given $N \times N$ density matrix ρ with $N \times 1$ eigenvalue vector λ, the Tsallis trace of order α is defined as

$$\text{Tsallis trace of order } \alpha = \mathrm{tr}(\rho^\alpha) = \sum_{x\in[N]} \lambda_x^\alpha. \tag{7.111}$$

In words, the Tsallis trace of order α is the Tsallis sum of order α for the vector of eigenvalues, λ. As summarized in Table 7.2, this implies that the Tsallis trace is a measure of the mixedness of state ρ when $0 \leq \alpha < 1$, while it is a measure of purity when $\alpha > 1$. This is in the sense that the Tsallis sum of order $0 \leq \alpha < 1$ grows as the vector λ becomes more spread out, and thus the state ρ becomes more mixed; while the Tsallis sum of order $\alpha > 1$ grows as the vector λ becomes more concentrated into a single non-zero entry, and hence the state becomes closer to a pure state (see Table 3.1).

The interpretation of the Tsallis trace as a measure of mixedness or purity is reflected by the following bounds, which follow directly from the definition (7.111).

Key Result 7.20 Bounds on the Tsallis trace

For $0 < \alpha < 1$, the Tsallis trace satisfies the inequalities

$$\underbrace{1}_{\text{pure state}} \leq \operatorname{tr}(\rho^{\alpha}) \leq \underbrace{N^{1-\alpha}}_{\text{maximally mixed state}}, \tag{7.112}$$

with the lower bound attained for a pure state and the upper bound obtained for the maximally mixed density matrix.

For $\alpha > 1$, the Tsallis trace satisfies the inequalities

$$\underbrace{\frac{1}{N^{\alpha-1}}}_{\text{maximally mixed state}} \leq \operatorname{tr}(\rho^{\alpha}) \leq \underbrace{1}_{\text{pure state}}, \tag{7.113}$$

with the lower bound obtained for the maximally mixed density matrix and the upper bound attained for a pure state.

The special case $\alpha = 0$ will be studied in Section 7.10.2, along with several other important settings.

Finally, it is noted that the same interpretations in Table 7.2 apply also to the α-Schatten norm of the density matrix ρ introduced in Section 6.6.6, which reads

$$||\rho||_{\alpha} = \operatorname{tr}(\rho^{\alpha})^{\frac{1}{\alpha}} = \left(\sum_{x\in[N]} \lambda_x^{\alpha}\right)^{\frac{1}{\alpha}}. \tag{7.114}$$

Table 7.2 Interpretation of the Tsallis trace as a function of the order α

Order	Tsallis trace/α-Schatten norm	$1/(1-\alpha)$
$0 \leq \alpha < 1$	Mixedness measure	Positive
$\alpha > 1$	Purity measure	Negative

7.10.2 Special Cases of the Tsallis Trace

Definition 7.14 Rank
When $\alpha = 0$, the Tsallis trace equals the rank of matrix ρ, i.e.,

$$\text{tr}(\rho^0) = \text{rank}(\rho) = |\text{supp}(\rho)|, \tag{7.115}$$

where it is recalled that the support of the vector of eigenvalues of density matrix ρ is defined as $\text{supp}(\lambda) = \{x \in [N]: \lambda_x > 0\}$.

The rank is indeed a measure of the mixedness of the state, since it equals the minimum number of pure states that must be included in any ensemble realizing matrix ρ. In particular, the rank satisfies the inequalities (7.112) as

$$\underbrace{1}_{\text{pure state}} \leq \text{rank}(\rho) \leq N, \tag{7.116}$$

where the upper bound is attained by any density matrix for which all N eigenvalues are positive.

Definition 7.15 Purity
When $\alpha = 2$, the Tsallis trace is referred to as the purity of the quantum state ρ, and is given by

$$\text{tr}(\rho^2) = \sum_{x \in [N]} \lambda_x^2. \tag{7.117}$$

Accordingly, the purity is obtained as the sum of the squares of the eigenvalues of the density matrix ρ.

Using (7.113), the purity satisfies the inequalities

$$\underbrace{\frac{1}{N}}_{\text{maximally mixed state}} \leq \text{tr}(\rho^2) \leq \underbrace{1}_{\text{pure state}}, \tag{7.118}$$

with the lower bound $1/N$ obtained if and only if the state is maximally mixed, and the maximum purity value of 1 attained if and only if the state is pure.

7.10.3 Tsallis Trace, Tsallis Quantum Entropy, and Rényi Quantum Entropy

Using the Tsallis trace, the Tsallis quantum entropy can be expressed as follows.

Key Result 7.21 Tsallis Quantum entropy
The Tsallis quantum entropy of order $\alpha \geq 0$, defined in (7.91), can be written as

$$\mathrm{H}_\alpha^{\mathrm{T}}(\rho) = -\mathrm{tr}(\rho \log_{2,\alpha}(\rho))$$
$$= \begin{cases} \frac{1_2}{1-\alpha}(\mathrm{tr}(\rho^\alpha) - 1) & \text{for } \alpha > 0 \text{ and } \alpha \neq 1, \\ \mathrm{H}(\rho) & \text{for } \alpha = 1. \end{cases} \tag{7.119}$$

Thus, the Tsallis quantum entropy depends through the function $1_2(1-\alpha)^{-1}(\cdot - 1)$ – first seen in (3.28) for the Tsallis entropy – on the Tsallis trace $\mathrm{tr}(\rho^\alpha)$.

Key Result 7.22 Rényi quantum entropy
The Rényi quantum entropy, defined in (7.92), can be written as

$$\mathrm{H}_\alpha^{\mathrm{R}}(\rho) = \begin{cases} \frac{1}{1-\alpha}\log_2(\mathrm{tr}(\rho^\alpha)) & \text{for } \alpha \geq 0 \text{ and } \alpha \neq 1, \\ \mathrm{H}(\rho) & \text{for } \alpha = 1. \end{cases} \tag{7.120}$$

Accordingly, the Rényi quantum entropy is obtained by applying the logarithmic function $(1-\alpha)^{-1}\log_2(\cdot)$ – first encountered in (3.58) for the Rényi entropy – to the Tsallis trace (7.111). As a result, the Rényi quantum entropy decreases with the order $\alpha \geq 0$. Furthermore, the upper bound in (7.99) is given by $\mathrm{H}_\alpha^{\mathrm{R}}(\pi) = \log_2(N)$, and thus the Rényi quantum entropy increases at most logarithmically with the dimension N.

7.10.4 Special Cases of the Tsallis Quantum Entropy

Like its classical counterpart, the Tsallis quantum entropy decreases with the order α, yielding the following important special cases.

Definition 7.16 Tsallis quantum entropy of order $\alpha = 0$, aka Tsallis quantum max-entropy
The Tsallis quantum entropy of order $\alpha = 0$ is given by

$$\mathrm{H}_0^{\mathrm{T}}(\rho) = 1_2 \cdot (\mathrm{rank}(\rho) - 1). \tag{7.121}$$

Accordingly, the Tsallis quantum entropy of order $\alpha = 0$ increases linearly with the $\mathrm{rank}(\rho)$. Specifically, it attains the upper bound in (7.99), i.e., $\mathrm{H}_0^{\mathrm{T}}(\rho) = \mathrm{H}_0^{\mathrm{T}}(\pi) = 1_2 \cdot (N-1)$ for any density matrix with maximal rank, $\mathrm{rank}(\rho) = N$. Note that this maximum value grows linearly with N.

Definition 7.17 Tsallis quantum entropy of order $\alpha = 2$
The Tsallis quantum entropy of order $\alpha = 2$ is given by

$$\mathrm{H}_2^{\mathrm{T}}(\rho) = 1_2 \cdot (1 - \mathrm{tr}(\rho^2)). \tag{7.122}$$

The Tsallis quantum entropy of order $\alpha = 2$ decreases linearly with the purity $\mathrm{tr}(\rho^2)$, attaining the upper bound

$$\mathrm{H}_2^{\mathrm{T}}(\rho) \leq \mathrm{H}_2^{\mathrm{T}}(\pi) = 1_2 \cdot \left(1 - \frac{1}{N}\right) \tag{7.123}$$

only for a maximally mixed state π. By (7.123), the Tsallis quantum entropy of order $\alpha = 2$ remains bounded as N increases.

Definition 7.18 Tsallis quantum entropy of order $\alpha = \infty$, aka Tsallis quantum min-entropy
The Tsallis quantum entropy of order $\alpha = \infty$ is given as

$$\mathrm{H}_\infty^{\mathrm{T}}(\rho) = 0. \tag{7.124}$$

Accordingly, with $\alpha \to \infty$, the Tsallis quantum entropy equals zero irrespective of the density matrix ρ.

7.10.5 Special Cases of the Rényi Quantum Entropy

Definition 7.19 Rényi quantum entropy of order $\alpha = 0$, aka Rényi quantum max-entropy
The quantum Rényi entropy of order $\alpha = 0$ is given by

$$\begin{aligned}\mathrm{H}_0^{\mathrm{R}}(\rho) &= \log_2(\mathrm{rank}(\rho)) \\ &= \log_2\left(|\mathrm{supp}(\lambda)|\right).\end{aligned} \tag{7.125}$$

Therefore, the Rényi quantum max-entropy increases logarithmically with the rank of the density matrix.

Definition 7.20 Rényi quantum entropy of order $\alpha = 2$
The quantum Rényi entropy of order $\alpha = 2$ can be expressed in terms of the purity (7.117) as

$$\mathrm{H}_2^{\mathrm{R}}(\rho) = -\log_2(\mathrm{tr}(\rho^2)). \tag{7.126}$$

Accordingly, like for the Tsallis counterpart (7.122), the quantum Rényi entropy of order $\alpha = 2$ decreases with the purity $\mathrm{tr}(\rho^2)$.

The minimal value of the Rényi quantum entropy is obtained in the limit $\alpha = \infty$, yielding the following.

Definition 7.21 Rényi quantum entropy of order $\alpha = \infty$, aka Rényi quantum min-entropy
The Rényi quantum entropy of order $\alpha = \infty$ is given by

$$H^{R}_{\infty}(\rho) = -\log_2\left(\max_{x\in[N]} \lambda_x\right). \tag{7.127}$$

The Rényi quantum min-entropy is thus a function of the maximum eigenvalue of the density matrix ρ, $\max_{x\in[N]} \lambda_x$, which is a measure of the purity of the density matrix ρ. The Rényi quantum min-entropy plays an important role in quantum cryptography, as it captures the worst-case level of uncertainty for an eavesdropper on state ρ (see Recommended Resources).

7.11 Non-Selective Measurements ⏭

What is the impact of forgetting on the uncertainty of an observer? To address this question, take first the case of a classical system and of an observer characterized by an informational state p. Assume that the observer makes a complete measurement, obtaining the state of the system $\mathrm{x} \sim p$. However, the outcome x is then forgotten or discarded. Upon forgetting the observation, the observer is just as ignorant about the state of the system as it was before making the measurement. Therefore, the informational state of the observer is still described by the same probability distribution p; and, from the observer's perspective, another measurement would return again a random observation $\mathrm{x} \sim p$. This section will show that the situation is very different for quantum systems.

Specifically, this section will study non-selective measurements, quantum measurements followed by the discarding of the classical output. The main conclusion will be that, even when the classical output of a measurement is forgotten, the measurement generally causes a disturbance to the state of the system. Non-selective measurements will also be proved to increase the quantum entropy.

7.11.1 Defining Non-Selective Measurements

As illustrated in Figure 7.5, a non-selective quantum measurement consists of a measurement followed by the discarding of the classical output. Accordingly, given a pre-measurement state ρ, a projective setting $\mathcal{P} = \{\Pi_y\}_{y\in[M]}$ is applied first. The measurement returns the classical output $y \in [M]$ with probability $p_y(\rho|\mathcal{P}) = \mathrm{tr}(\rho\Pi_\mathrm{y})$, leaving the system in the post-measurement state $\rho_y = p_y(\rho|\mathcal{P})^{-1}\Pi_y\rho\Pi_y$ (see (7.37) and (7.38)). The measurement output $\mathrm{y} = y$ is then forgotten or discarded.

Upon forgetting, from the observer's viewpoint, the classical output y is uncertain, and it is equal to $y \in [N]$ with probability $p_y(\rho|\mathcal{P})$. Thus, as far as the observer is concerned, the system is characterized by an ensemble state

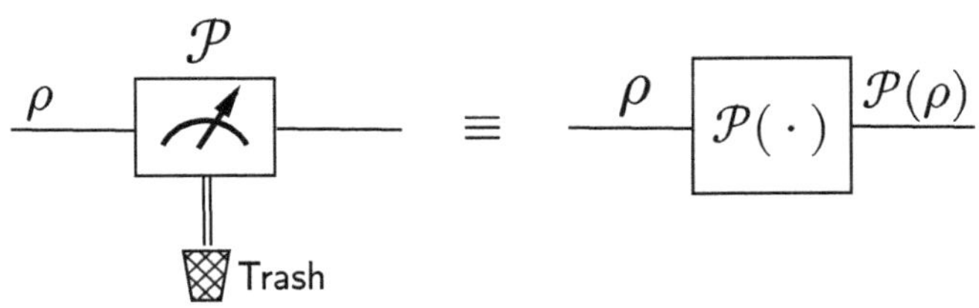

Figure 7.5 A non-selective quantum measurement consists of a measurement in setting $\mathcal{P}$ followed by the discarding of the classical output (on the left). It can equivalently be described by a transformation that modifies the pre-measurement state ρ, producing the post-measurement state $\mathcal{P}(\rho)$.

$$\mathcal{E} = \{(\rho_y, p_y(\rho|\mathcal{P}))\}_{y\in[M]}, \tag{7.128}$$

where the post-measurement state ρ_y is assigned probability $p_y(\rho|\mathcal{P})$. The density matrix corresponding to this ensemble describes the post-measurement state.

Definition 7.22 Non-selective measurement
A non-selective measurement in setting $\mathcal{P} = \{\Pi_y\}_{y\in[M]}$ with $M \leq N$ transforms the pre-measurement state ρ into the post-measurement state

$$\begin{aligned}\mathcal{P}(\rho) &= \sum_{y\in[M]} p_y(\rho|\mathcal{P})\rho_y \\ &= \sum_{y\in[M]} \Pi_y \rho \Pi_y. \end{aligned} \tag{7.129}$$

Example 7.27

In the special case of a non-selective complete measurement in setting $\mathcal{M} = \{|v_x\rangle\}_{x\in[N]}$, the post-measurement state, denoted as $\mathcal{M}(\rho)$, is obtained from (7.129) as

$$\begin{aligned}\mathcal{M}(\rho) &= \sum_{x\in[N]} |v_x\rangle\langle v_x|\rho|v_x\rangle\langle v_x| \\ &= \sum_{x\in[N]} \underbrace{\langle v_x|\rho|v_x\rangle}_{=p_x(\rho|\mathcal{M})} |v_x\rangle\langle v_x|. \end{aligned} \tag{7.130}$$

Therefore, with a non-selective complete measurement, the post-measurement state $\mathcal{M}(\rho)$ has eigenvectors equal to the measurement basis $\mathcal{M} = \{|v_x\rangle\}_{x\in[N]}$ and eigenvalues given by the measurement probability vector $p(\rho|\mathcal{M})$.

Example 7.28

Consider a non-selective measurement in the computational-basis setting $\mathcal{M} = \{|x\rangle\}_{x\in[N]}$, for which the projection matrices are given by $\{\Pi_x = |x\rangle\langle x|\}_{x\in[N]}$. The density matrix (7.130) of the state produced by this non-selective measurement is given by

$$\mathcal{M}(\rho) = \sum_{x \in [N]} \underbrace{\langle x|\rho|x\rangle}_{=\rho_{x,x}} |x\rangle\langle x| = \begin{bmatrix} \rho_{0,0} & 0 & \cdots & 0 \\ 0 & \rho_{1,1} & \ddots & 0 \\ \vdots & \ddots & \ddots & 0 \\ 0 & 0 & \cdots & \rho_{N-1,N-1} \end{bmatrix}. \tag{7.131}$$

So, a non-selective measurement in the computational basis renders the density matrix diagonal by zeroing out all the elements outside the main diagonal. Following the discussion in Examples 7.1–7.3, this implies that a non-selective measurement in the computational basis $\mathcal{M}$ destroys any coherence with respect to basis $\mathcal{M}$ that may have been present in the pre-measurement state ρ.

Example 7.29

As discussed in Section 6.5, projective measurements can be viewed as coarser-grained versions of complete measurements, trading information extracted from the measurement for disturbance to the system. To further elaborate on this perspective, consider a system with $N = 4$ dimensions. A projective measurement $\mathcal{P} = \{\Pi_0, \Pi_1\}$ is applied, with the rank-2 projection matrices $\Pi_0 = |0\rangle\langle 0| + |1\rangle\langle 1|$ and $\Pi_1 = |2\rangle\langle 2| + |3\rangle\langle 3|$. Under this non-selective measurement setting, the post-measurement state (7.130) is

$$\mathcal{P}(\rho) = \Pi_0\rho\Pi_0 + \Pi_1\rho\Pi_1 = \begin{bmatrix} \rho_{0,0} & \rho_{0,1} & 0 & 0 \\ \rho_{1,0} & \rho_{1,1} & 0 & 0 \\ 0 & 0 & \rho_{2,2} & \rho_{2,3} \\ 0 & 0 & \rho_{3,2} & \rho_{3,3} \end{bmatrix}. \tag{7.132}$$

Comparing the post-measurement state (7.131) for a complete measurement with the post-measurement state (7.132) confirms that projective measurements can be gentler than complete measurements. In fact, the post-measurement state (7.132) preserves more information about the pre-measurement state ρ than the post-measurement state (7.131). In particular, given the presence of non-zero off-diagonal terms in (7.132), the state (7.132) retains some of the coherence with respect to the computational basis.

7.11.2 Non-Selective Complete Measurements Increase Quantum Entropy

Forgetting can be thought of as a form of noise caused by an imperfect measurement. In a manner consistent with this interpretation, a non-selective measurement increases the quantum entropy.

Key Result 7.23 Non-selective measurements increase the quantum entropy
Given any non-selective measurement in setting $\mathcal{P}$ and any pre-measurement state ρ, the quantum entropy satisfies the inequality

$$\mathrm{H}_\alpha(\mathcal{P}(\rho)) \geq \mathrm{H}_\alpha(\rho). \tag{7.133}$$

This result will be proved in full generality in Section 11.7. However, for the special case of a complete measurement setting $\mathcal{M}$, the inequality in (7.133), i.e.,

$$\mathrm{H}_\alpha(\mathcal{M}(\rho)) \geq \mathrm{H}_\alpha(\rho), \tag{7.134}$$

can be shown as a direct consequence of the inequality (7.93). In fact, the following result holds for complete measurements.

Key Result 7.24 Post-measurement entropy and measured entropy for complete measurements
For a complete quantum measurement $\mathcal{M}$, the quantum entropy $\mathrm{H}_\alpha(\mathcal{M}(\rho))$ of the post-measurement state $\mathcal{M}(\rho)$ equals the measured entropy $\mathrm{H}_\alpha(\rho|\mathcal{M})$ of the pre-measurement state ρ with respect to basis $\mathcal{M}$:

$$\mathrm{H}_\alpha(\mathcal{M}(\rho)) = \mathrm{H}_\alpha(\rho|\mathcal{M}). \tag{7.135}$$

In words, the irreducible uncertainty $\mathrm{H}_\alpha(\mathcal{M}(\rho))$ associated with the post-measurement state $\mathcal{M}(\rho)$ equals the corresponding measured entropy $\mathrm{H}_\alpha(\rho|\mathcal{M})$. As detailed in Appendix 7.E, this result holds because the state $\mathcal{M}(\rho)$ has eigenbasis $\mathcal{M}$. The equality (7.135) implies (7.134) via the inequality (7.93).

It is worth noting that the equality (7.135) does not hold more generally for projective measurements. In fact, the quantum entropy $\mathrm{H}_\alpha(\mathcal{P}(\rho))$ can be larger than the measured entropy $\mathrm{H}_\alpha(\rho|\mathcal{P})$. Take, for instance, the case of a trivial projective measurement $\mathcal{P} = \{I\}$: the quantum entropy $\mathrm{H}_\alpha(\mathcal{P}(\rho)) = \mathrm{H}_\alpha(\rho)$ can be positive, while the measured entropy $\mathrm{H}_\alpha(\rho|\mathcal{P})$ equals zero.

7.12 Coherence for Mixed States ⏭

As introduced in Chapter 6, coherence is an important property of quantum systems. This section extends the definition of coherence to mixed states. Furthermore, it generalizes the uncertainty inequality, presented in Section 6.4.4 for pure states, to mixed states.

7.12.1 Defining Coherence for Mixed States

As studied in Section 6.3, the coherence of a pure state $|\psi\rangle$ with respect to basis $\mathcal{M} = \{|v_x\rangle\}_{x\in[N]}$ quantifies the level of superposition of state $|\psi\rangle$ with respect to the basis $\mathcal{M}$. As illustrated in Figure 7.6(a), a pure state is thus incoherent with respect to basis $\mathcal{M}$ if it coincides with one of the states $\{|v_x\rangle\}_{x\in[N]}$, and it is coherent otherwise.

When measured in setting $\mathcal{M}$, a pure state that is coherent with respect to basis $\mathcal{M}$ yields a random classical output and is disturbed by the measurement. This is a manifestation of quantum uncertainty. In contrast, when a measurement in setting

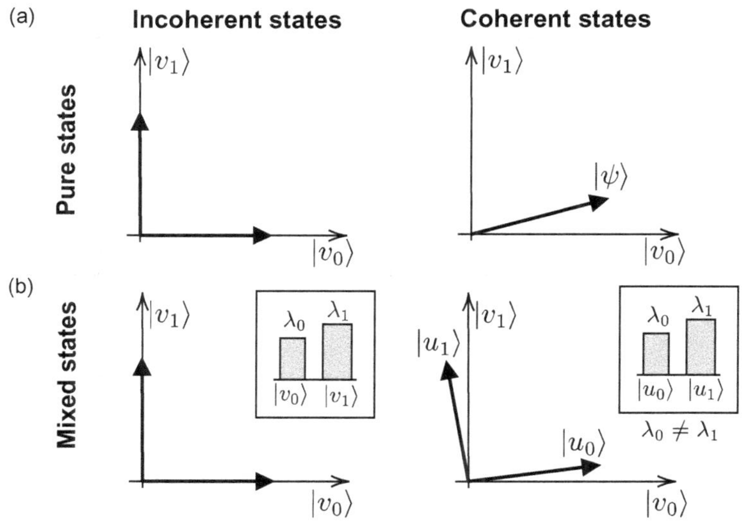

Figure 7.6 (a) A pure state $|\psi\rangle$ is incoherent with respect to a basis $\mathcal{M} = \{|v_x\rangle\}_{x\in[N]}$ if it is equivalent to one of the basis states, and it is coherent otherwise. (b) A mixed state ρ is incoherent with respect to a basis $\mathcal{M} = \{|v_x\rangle\}_{x\in[N]}$ if it has an eigenbasis equal to the measurement basis, and it is coherent otherwise.

$\mathcal{M}$ is applied to a pure state that is incoherent with respect to basis $\mathcal{M}$, the classical output is deterministic and the state is left unchanged.

As shown in Figure 7.6(b), a mixed state is incoherent if it can be generated by an ensemble $\{(|v_x\rangle, \lambda_x)\}_{x\in[N]}$ consisting only of incoherent pure states $\{|v_x\rangle\}_{x\in[N]}$. Conversely, it is coherent if every ensemble of orthogonal states realizing the state contains at least one coherent state. This definition is formalized as follows.

Definition 7.23 Coherence for mixed states
A density matrix ρ is said to be incoherent with respect to a measurement basis $\mathcal{M}$ if it has an eigenbasis equal to $\mathcal{M}$; otherwise, it is said to be coherent with respect to the basis $\mathcal{M}$.

This definition is consistent with that given above for pure states, which is recovered as a special case. However, for mixed states, it is no longer true that only coherent states yield random classical outputs and are disturbed by a measurement. Instead, owing to the presence of classical randomness on the preparation of the state, *any* mixed state yields random measurement outputs and is disturbed by a measurement.

Example 7.30

A density matrix ρ is incoherent with respect to the computational basis $\mathcal{M} = \{|x\rangle\}_{x\in[N]}$ if and only if it is diagonal. In fact, a matrix obtained from an ensemble of states of the form $\{|x\rangle\langle x|\}_{x\in[N]}$ is diagonal. Therefore, the presence of off-diagonal terms in a density matrix indicates the presence of coherence with respect to the computational basis (see also Examples 7.1–7.3 and 7.28).

Example 7.31

For a given distribution vector p and an orthonormal basis $\mathcal{M} = \{|v_x\rangle\}_{x\in[N]}$, consider the pure state $|\psi\rangle = \sum_{x\in[N]} \sqrt{p_x}|v_x\rangle$ and the density matrix $\rho = \sum_{x\in[N]} p_x|v_x\rangle\langle v_x|$. The pure state $|\psi\rangle$ is coherent with respect to the orthonormal basis $\mathcal{M}$, unless the distribution p has zero entropy. In contrast, the density state ρ is incoherent with respect to the orthonormal basis $\mathcal{M}$, irrespective of the distribution p.

Example 7.32

The single-qubit state (7.13), i.e., $\rho = p_0|+\rangle\langle+| + p_1|-\rangle\langle-|$, is coherent with respect to the computational basis, unless it is maximally mixed, i.e., unless $p_0 = p_1 = 0.5$. In contrast, it is incoherent with respect to the Hadamard basis for any p, since the Hadamard basis is an eigenbasis for density matrix ρ.

Coherence for mixed states can also be characterized from the perspective of the disturbance caused by a measurement to the given state. Indeed, while conventional measurements generally disturb a mixed state, even if it is incoherent, this is not the case for non-selective measurements.

Key Result 7.25 Coherence for mixed states (equivalent definition)
A mixed state ρ is incoherent with respect to an orthonormal basis $\mathcal{M} = \{|v_x\rangle\}_{x\in[N]}$ if and only if it is not changed by a non-selective measurement in setting $\mathcal{M}$, i.e., if and only if the equality

$$\mathcal{M}(\rho) = \sum_{x\in[N]} \langle v_x|\rho|v_x\rangle|v_x\rangle\langle v_x| = \rho \tag{7.136}$$

holds between the post-measurement state $\mathcal{M}(\rho)$ in (7.130) and the pre-measurement state ρ. Otherwise, the state is coherent with respect to a basis $\mathcal{M}$.

This is an equivalent definition of coherence, since the condition (7.136) is necessary and sufficient for the density matrix to have eigenbasis equal to $\mathcal{M} = \{|v_x\rangle\}_{x\in[N]}$.

Example 7.33

The maximally mixed state π is incoherent with respect to any measurement basis $\mathcal{M}$. In fact, the condition (7.136) is satisfied for any basis $\mathcal{M}$ as

$$\begin{aligned}\mathcal{M}(\pi) &= \sum_{x\in[N]} \langle v_x|\pi|v_x\rangle |v_x\rangle\langle v_x| \\ &= \frac{1}{N}\sum_{x\in[N]} |v_x\rangle\langle v_x| = \pi.\end{aligned} \tag{7.137}$$

7.12.2 Coherence Measures for Mixed States

As discussed in Section 7.12.1, an incoherent mixed state satisfies the equality (7.136). Thus, for a coherent state, there is a non-negligible discrepancy between the pre-measurement state ρ and the post-measurement state $\mathcal{M}(\rho)$. Based on this observation, a coherence measure for state ρ under measurement basis $\mathcal{M}$ can be defined by evaluating the extent to which the state is affected by the non-selective measurement $\mathcal{M}(\cdot)$. Measuring the change in the system caused by the measurement through the quantum entropy yields the following definition.

Definition 7.24 Coherence measure for mixed states
A coherence measure for a mixed state ρ with respect to a measurement basis $\mathcal{M}$ is defined by the difference between the quantum entropies of post-measurement and pre-measurement states for a non-selective measurement in basis $\mathcal{M}$, i.e.,

$$\begin{aligned}\mathrm{K}(\rho|\mathcal{M}) &= \mathrm{H}_\alpha(\mathcal{M}(\rho)) - \mathrm{H}_\alpha(\rho) \\ &= \mathrm{H}_\alpha(\rho|\mathcal{M}) - \mathrm{H}_\alpha(\rho),\end{aligned} \tag{7.138}$$

where the second equality follows from (7.135).

Key Result 7.26 Bounds on the coherence measure
The coherence measure $\mathrm{K}(\rho|\mathcal{M})$ satisfies the inequalities

$$\underbrace{0}_{\text{incoherent}} \leq \mathrm{K}(\rho|\mathcal{M}) \leq \underbrace{\mathrm{H}_\alpha(u)}_{\text{pure and maximally coherent}}. \tag{7.139}$$

The lower bound is attained for incoherent states, while the upper bound is obtained for pure and maximally coherent states.

To prove the lower bound in (7.139), note that, by the inequality (7.93), non-selective measurements increase the entropy and thus the quantity $\mathrm{K}(\rho|\mathcal{M})$ is non-negative. Furthermore, equality holds if and only if the basis $\mathcal{M}$ coincides with an eigenbasis of matrix ρ.

For the upper bound, the coherence measure $\mathrm{K}(\rho|\mathcal{M})$ takes the maximum value $\mathrm{K}(\rho|\mathcal{M}) = \mathrm{H}_\alpha(\pi) = \mathrm{H}_\alpha(u)$ when ρ is pure and maximally coherent with respect to basis $\mathcal{M}$ (Problem 7.12). In fact, since any mixed state has a positive quantum entropy $\mathrm{H}_\alpha(\rho)$, mixedness necessarily reduces the coherence measure (7.138). Indeed, there is generally a tension between coherence and mixedness, with the latter reducing the former (see Recommended Resources).

Example 7.34

Consider the single-qubit state

$$\rho = p_0|+\rangle\langle+| + p_1|1\rangle\langle 1| \tag{7.140}$$

and the computational basis $\mathcal{M}$. The state ρ is pure and incoherent with respect to the computational basis $\mathcal{M}$ for $p_0 = 0$, while it is pure and maximally coherent with respect to basis $\mathcal{M}$ for $p_0 = 1$. Therefore, as p_0 increases from 0 to 1, the coherence of the state ρ with respect to $\mathcal{M}$ grows larger.

Adopting the Shannon entropy, Figure 7.7 plots the coherence measure (7.138) as a function of the probability p_0. The figure shows that the coherence measure increases with p_0, achieving the maximum value $\mathrm{H}(\pi) = \log_2(2) = 1$ for $p_0 = 1$.

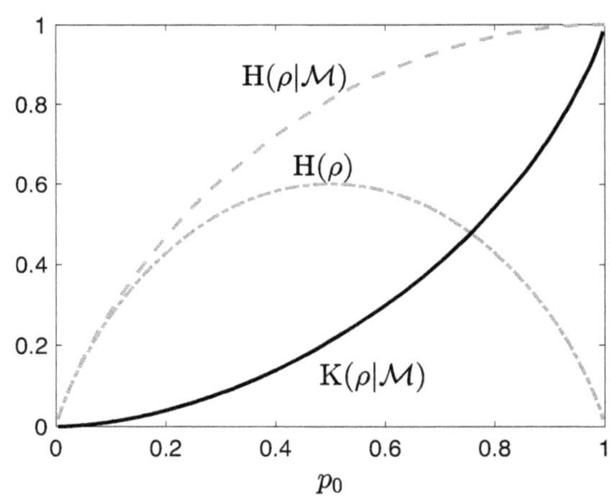

Figure 7.7 Coherence measure $\mathrm{K}(\rho|\mathcal{M})$, measured entropy $\mathrm{H}(\rho|\mathcal{M})$, and quantum entropy $\mathrm{H}(\rho)$ for the state (7.140) as a function of the probability p_0.

The figure also plots the measured entropy $\mathrm{H}(\rho|\mathcal{M})$ and quantum entropy $\mathrm{H}(\rho)$ in (7.138). The measured entropy $\mathrm{H}(\rho|\mathcal{M})$ increases with the probability p_0, since a larger coherence implies a larger uncertainty on the measurement output. In contrast, the Shannon quantum entropy $\mathrm{H}(\rho)$ attains the maximum value for $p_0 = 0.5$, while tending to zero as the state becomes pure for $p_0 = 0$ and $p_0 = 1$.

The coherence measure (7.138) can be generalized by following the resource-based approach presented in Section 6.3.4 for pure states. This perspective yields coherence measures of the form

$$\mathrm{K}(\rho|\mathcal{M}) = \min_{\{\lambda_x \geq 0\}_{x\in[N]}:\ \sum_{x\in[N]}\lambda_x=1} \mathrm{D}\left(\rho, \sum_{x\in[N]} \lambda_x |v_x\rangle\langle v_x|\right), \tag{7.141}$$

where $\mathrm{D}(\rho,\sigma)$ is a function that quantifies the difference between states ρ and σ. By minimizing over the vector of eigenvalues $\lambda = [\lambda_0, \ldots, \lambda_{N-1}]^T$, the coherence measure (7.141) evaluates the minimum distance between state ρ and the set of all possible incoherent states with respect to basis $\mathcal{M}$. Distance measures $\mathrm{D}(\rho,\sigma)$ will be introduced in Chapter 8. Furthermore, with a specific choice of such measures – namely the Shannon quantum relative entropy – it will be shown that the measure (7.138) is recovered from (7.141).

7.12.3 Uncertainty Inequality for Mixed States

The uncertainty inequality was presented in Section 6.4.4 for pure states. The following is a generalization to mixed states. A proof is provided in Appendix 11.A.

Key Result 7.27 Uncertainty inequality for mixed states
For any state ρ and for any two measurement bases $\mathcal{M}_0$ and $\mathcal{M}_1$ with compatibility $\mathrm{c}(\mathcal{M}_0, \mathcal{M}_1)$, the following inequality holds:

$$\mathrm{H}(\rho|\mathcal{M}_0) + \mathrm{H}(\rho|\mathcal{M}_1) \geq \log_2\left(\frac{1}{\mathrm{c}(\mathcal{M}_0, \mathcal{M}_1)}\right) + \mathrm{H}(\rho). \tag{7.142}$$

Unlike for the inequality (6.43) for pure states, the right-hand side in (7.142) is state dependent, being a function of the state ρ. However, the inequality (7.142) coincides with (6.43) for pure states, since pure states have zero quantum entropy, i.e., $\mathrm{H}(\rho) = 0$. As illustrated by the next example, the state-dependent term $\mathrm{H}(\rho)$ is useful to account for the inherent classical uncertainty of mixed states, making the bound tighter for mixed states.

Example 7.35

For $N = 2$, the maximally mixed density matrix $\rho = \pi$ yields the measured entropy $\mathrm{H}(\pi|\mathcal{M}) = 1$ for any measurement setting $\mathcal{M}$, and thus also the quantum entropy $\mathrm{H}(\pi) = 1$. Consider two maximally incompatible measurements $\mathcal{M}_0$ and $\mathcal{M}_1$, for which the compatibility measure is thus minimal, i.e., $\mathrm{c}(\mathcal{M}_0, \mathcal{M}_1) = 1/2$. The uncertainty inequality (7.142) yields

$$\mathrm{H}(\pi|\mathcal{M}_0) + \mathrm{H}(\pi|\mathcal{M}_1) \geq 1 + 1 = 2. \tag{7.143}$$

Hence, for a maximally mixed state π, the uncertainty inequality (7.142) is tight, confirming the important role played by the state-dependent term $\mathrm{H}(\rho)$.

More generally, the inequality (7.142) can be shown to be tight when state ρ is incoherent with respect to at least one of the measurement settings $\mathcal{M}_0$ or $\mathcal{M}_1$ (see Recommended Resources).

7.13 Decomposing Square Matrices ⏭

This last section presents some important technical results about the ways in which an $N \times N$ matrix can be decomposed as a linear combination of other, typically simpler, matrices. This is done by treating an $N \times N$ matrix as a vector in an N^2-dimensional linear space, yielding decompositions in terms of a basis set of N^2 matrices $\{B_z\}_{z\in[N^2]}$.

Given a basis $\{B_z\}_{z\in[N^2]}$ of $N \times N$ matrices, any $N \times N$ complex-valued matrix A can be expressed as a linear combination

$$A = \sum_{z\in[N^2]} \gamma_z B_z \tag{7.144}$$

for some complex numbers $\{\gamma_z\}_{z\in[N^2]}$. This section will discuss three classes of bases $\{B_z\}_{z\in[N^2]}$, namely: (i) orthonormal bases of rank-1 matrices; (ii) orthonormal bases of observables; and (iii) non-orthogonal bases of density matrices.

7.13.1 Orthonormal Basis of Matrices

To start, consider the space of all $N \times N$ complex-valued matrices, in which the trace inner product $\mathrm{tr}(A^\dagger B)$ in (7.46) measures the alignment between two $N \times N$ matrices A and B. Using the trace inner product, two $N \times N$ matrices A and B are said to be orthogonal if the equality $\mathrm{tr}(A^\dagger B) = 0$ holds, or equivalently $\mathrm{tr}(B^\dagger A) = 0$. The norm corresponding to the trace inner product is the 2-Schatten norm or quadratic norm (see Section 6.6.6)

$$||A||_2 = \sqrt{\mathrm{tr}(A^\dagger A)} = \sqrt{\mathrm{tr}(|A|^2)}. \tag{7.145}$$

Using these definitions, an orthonormal basis for the space of $N \times N$ matrices is given by N^2 matrices $\{B_z\}_{z\in[N^2]}$ that satisfy the normalization conditions

$$||B_z||_2^2 = 1 \tag{7.146}$$

for all $z \in [N^2]$, as well as the orthogonality conditions

$$\mathrm{tr}(B_z^\dagger B_{z'}) = 0 \tag{7.147}$$

for all pairs of indices $z \in [N^2]$ and $z' \in [N^2]$ with $z \neq z'$.

In a manner analogous to the case of vectors, given an orthonormal basis $\{B_z\}_{z\in[N^2]}$, any $N \times N$ matrix A can be expressed as the linear combination (7.144) with coefficients obtained via the trace inner product

$$\gamma_z = \mathrm{tr}(B_z^\dagger A) \tag{7.148}$$

for all $z \in [N^2]$.

7.13.2 Orthonormal Basis of Rank-1 Matrices

To construct a first example of an orthonormal basis for the space of $N \times N$ matrices, fix any orthonormal basis $\{|u_x\rangle\}_{x\in[N]}$ for the space of $N \times 1$ vectors. The set of N^2 rank-1 matrices

$$\{\{B_{xy} = |u_x\rangle\langle u_y|\}_{x\in[N]}\}_{y\in[N]} \tag{7.149}$$

forms an orthonormal basis for the space of $N \times N$ matrices (Problem 7.18).

Therefore, by (7.144) and (7.148), one can express any $N \times N$ matrix A as

$$A = \sum_{x\in[N]} \sum_{y\in[N]} \gamma_{xy} |u_x\rangle\langle u_y|, \tag{7.150}$$

with coefficients

$$\gamma_{xy} = \text{tr}(B_{xy}^\dagger A) = \text{tr}(|u_y\rangle\langle u_x|A) = \langle u_x|A|u_y\rangle \tag{7.151}$$

for all $x, y \in [N]$.

A particularly simple version of this equality is obtained with the computational basis $\{|u_x\rangle = |x\rangle\}_{x\in[N]}$, in which case the coefficient γ_{xy} corresponds to the (x, y)th entry $A_{x,y}$ of matrix A, i.e.,

$$A = \sum_{x\in[N]} \sum_{y\in[N]} A_{x,y}|x\rangle\langle y|. \tag{7.152}$$

Note that the outer product $|x\rangle\langle y|$ is an $N \times N$ matrix with all zeros except for position (x, y).

7.13.3 Orthonormal Basis of Observables

Another useful choice as an orthonormal basis for the set of $N \times N$ matrices is obtained by selecting all matrices $\{B_z\}_{z\in[N^2]}$ as observables, i.e., as Hermitian matrices $\{B_z = O_z\}_{z\in[N^2]}$. For a qubit system, an important example is derived from the Pauli matrices as discussed in Example 7.36.

Example 7.36

For $N = 2$, the $N^2 = 4$ Pauli matrices $\{P_0, P_1, P_2, P_3\}$ are mutually orthogonal, as one can verify the equalities $\text{tr}(P_x^\dagger P_y) = \text{tr}(P_x P_y) = 0$ for all $x \neq y$ and $x, y \in [4]$ (see Problem 7.16). However, the norm $||P_x||_2$ of each Pauli matrix equals $\sqrt{2}$, and thus the scaled Pauli matrices

$$\left\{\frac{1}{\sqrt{2}}P_0, \frac{1}{\sqrt{2}}P_1, \frac{1}{\sqrt{2}}P_2, \frac{1}{\sqrt{2}}P_3\right\} \tag{7.153}$$

form an orthonormal basis for the space of all 2×2 matrices. The set (7.153) will be referred to as the **Pauli basis**.

As a numerical example, the matrix

$$A = \begin{bmatrix} 1 & 0.1 \\ 0.3 & 2 \end{bmatrix} \tag{7.154}$$

can be expressed in terms of the Pauli basis as

$$A = \frac{3}{2}P_0 + \frac{1}{5}P_1 - \frac{1}{10}iP_2 - \frac{1}{2}P_3. \quad (7.155)$$

For example, the coefficient of the normalized Pauli X matrix ρ_1 is obtained via (7.148) as

$$\gamma_1 = \frac{1}{\sqrt{2}}\text{tr}(P_1 A) = \frac{1}{\sqrt{2}}\text{tr}\left(\begin{bmatrix} 0.3 & 2 \\ 1 & 0.1 \end{bmatrix}\right) = \frac{2}{5\sqrt{2}}. \quad (7.156)$$

Generalizing Example 7.36, for any dimension N, it is possible to design orthonormal bases of observables $\{O_z\}_{z\in[N^2]}$ that include the scaled identity, i.e.,

$$O_0 = \frac{1}{\sqrt{N}}I, \quad (7.157)$$

along with $N^2 - 1$ observables with zero trace

$$\text{tr}(O_z) = 0 \text{ for } z \in \{1, \ldots, N^2 - 1\}. \quad (7.158)$$

Pauli strings, which apply for $N = 2^S$ with integer S, provide an instance of an orthonormal basis with these properties, and they will be discussed in Chapter 8. **Gell-Mann matrices** provide another useful construction (see Recommended Resources).

The selection of an orthonormal basis made of observables $\{O_z\}_{z\in[N^2]}$ is particularly interesting when the matrix being decomposed is a density matrix ρ. In fact, in this case, each coefficient γ_z in (7.148) is given by the expectation of the observable, i.e.,

$$\gamma_z = \text{tr}(O_z\rho) = \langle O_z\rangle_\rho. \quad (7.159)$$

Therefore, the coefficients γ_z of the linear decomposition (7.144) are real-valued, and they can be obtained by evaluating the expectations $\langle O_z\rangle_\rho$ of the observables.

Given an unknown quantum state ρ and an orthonormal basis of observables, one can hence recover the state ρ by estimating all the expectations $\gamma_z = \langle O_z\rangle_\rho$ for $z \in [N^2]$ through the decomposition (7.144). This procedure, a form of quantum tomography, requires estimating the expected value of N^2 observables. This is practically problematic when the number N^2 is large as in composite systems.

Example 7.37

With $N = 2$, a density matrix ρ can be expressed as a linear combination of the observables given by the Pauli matrices. This yields the expression

$$\rho = \frac{1}{2}(P_0 + \gamma_1 P_1 + \gamma_2 P_2 + \gamma_3 P_3), \quad (7.160)$$

where $\gamma = [\gamma_1, \gamma_2, \gamma_3]^T$, known as the **Bloch vector**, has real-valued entries $\gamma_x = \text{tr}(P_x\rho)$. The decomposition (7.160) follows from the general form (7.144) since the coefficient for the basis matrix $P_0/\sqrt{2} = I/\sqrt{2}$ equals $\text{tr}(\rho)/\sqrt{2} = 1/\sqrt{2}$ for any density matrix ρ.

From (7.160), evaluating the purity of density matrix ρ yields

$$\mathrm{tr}(\rho^2) = \frac{1}{2}(1 + ||\gamma||_2^2). \tag{7.161}$$

Therefore, for pure states, the Bloch vector has 2-norm $||\gamma||_2 = 1$, while for mixed states the 2-norm is smaller than 1, i.e., $||\gamma||_2 < 1$. In the extreme case of a maximally mixed state, the Bloch vector has all entries equal to zero.

The decomposition (7.144) using an orthonormal basis of observables can also be useful when applied to an observable O. In fact, the expected value $\langle O \rangle_\rho$ of an observable $O = \sum_{z \in [N^2]} \gamma_z O_z$ can be computed by first evaluating the expected values $\langle O_z \rangle_\rho$ of the basis observables O_z for $z \in [N^2]$, and then combining the resulting expectations as

$$\langle O \rangle_\rho = \sum_{z \in [N^2]} \gamma_z \langle O_z \rangle_\rho. \tag{7.162}$$

This approach is practically relevant if calculating the expected values $\{\langle O_z \rangle_\rho\}_{z \in [N^2]}$ for the basis observables can be done more efficiently than for the observable O.

7.13.4 Basis of Density Matrices

Can one find an orthonormal basis of observables that consists of density matrices? To address this question, recall that in a space of dimension N one can define a set of at most N orthogonal, and hence perfectly distinguishable, pure states. Mixed states can be viewed as noisy pure states, and thus the consideration of a mixed state cannot increase the number of distinguishable states in a space of dimension N. Therefore, it is *not* possible to construct an orthonormal basis of N^2 density matrices. That said, one can define a set of N^2 density matrices that, while not orthogonal, form a basis for the space of $N \times N$ density matrices. This construction is illustrated by the following example.

Example 7.38

For $N = 2$, consider the density matrices

$$\begin{aligned} &\rho_0 = |0\rangle\langle 0|,\ \rho_1 = |1\rangle\langle 1|, \\ &\rho_2 = |+\rangle\langle +|, \text{ and } \rho_3 = |+i\rangle\langle +i| = \frac{1}{2}\begin{bmatrix} 1 & -i \\ i & 1 \end{bmatrix}. \end{aligned} \tag{7.163}$$

These matrices are *not* mutually orthogonal. For instance, one has the non-zero trace inner product $\mathrm{tr}(\rho_0\rho_2) = 0.5 \neq 0$. However, as shown in Appendix 7.D, any 2×2 density matrix ρ can be expressed as a linear combination of these density matrices.

For example, using the basis (7.163), the 2×2 density matrix

$$\rho = \frac{1}{2}\begin{bmatrix} 1 & a \\ a & 1 \end{bmatrix}, \tag{7.164}$$

with $0 \leq a \leq 1$, can be written as

$$\rho = \frac{1}{2}(1-a)\rho_0 + \frac{1}{2}(1-a)\rho_1 + a\rho_2. \tag{7.165}$$

Since the set (7.163) is a basis, so is the set

$$\rho_0 = \frac{1}{\sqrt{2}}I,\ \rho_1 = |0\rangle\langle 0|,$$
$$\rho_2 = |+\rangle\langle +|, \text{ and } \rho_3 = |+i\rangle\langle +i|, \tag{7.166}$$

which can also be written equivalently in terms of Pauli matrices as

$$\rho_0 = \frac{1}{\sqrt{2}}P_0,\ \rho_1 = \frac{1}{2}(I+P_1),$$
$$\rho_2 = \frac{1}{2}(I+P_2), \text{ and } \rho_3 = \frac{1}{2}(I+P_3). \tag{7.167}$$

7.14 Summary

- An ensemble of pure states $\mathcal{E} = \{(|\psi_l\rangle, p_l)\}_{l\in[L]}$ describes a situation in which there is classical uncertainty about the preparation of a quantum state.
- The behavior of an ensemble is fully described by the density matrix $\rho = \sum_{l\in[L]} p_l |\psi_l\rangle\langle\psi_l|$, with a rank larger than 1 indicating a mixed state.
- Any two ensembles that have the same density matrix are equivalent, as any measurement made on the two ensembles produces statistically equivalent outputs.
- The measurement probability for a general projective measurement $\mathcal{P} = \{\Pi_y\}_{y\in[M]}$ is given by $p_y(\rho|\mathcal{P}) = \mathrm{tr}(\rho\Pi_y)$, while the corresponding post-measurement state is $\rho_y = \Pi_y\rho\Pi_y/p_y(\rho|\mathcal{P})$, for all $y \in [M]$.
- An $N \times N$ square matrix ρ is a density matrix if and only if it is positive semi-definite and it has a trace equal to 1.
- The ensemble $\mathcal{E}(\rho)$ obtained from an eigendecomposition of matrix ρ, referred to as the eigenensemble, has the minimum number of pure states among all ensembles realizing density matrix ρ, which equals the rank of matrix ρ.
- The measured entropy $\mathrm{H}(\rho|\mathcal{P})$ of a mixed quantum state ρ under the projective measurement setting $\mathcal{P} = \{\Pi_y\}_{y\in[M]}$ is defined as the classical entropy of the corresponding measurement probability distribution $p(\rho|\mathcal{P})$, i.e., $\mathrm{H}(\rho|\mathcal{P}) = \mathrm{H}(p(\rho|\mathcal{P}))$.
- When minimized over all complete measurement settings $\mathcal{M}$, the measured entropy yields the quantum entropy $\mathrm{H}_\alpha(\rho)$, which can be interpreted as the irreducible uncertainty level associated with the preparation or measurement of state ρ.
- An optimal, entropy-minimizing, measurement setting is given by any eigenbasis $\{|u_x\rangle\}_{x\in[N]}$ of matrix ρ. This implies that the quantum entropy equals the classical entropy of the vector of eigenvalues λ, i.e., $\mathrm{H}_\alpha(\rho) = \mathrm{H}_\alpha(\lambda)$.
- The quantum entropy satisfies the bounds $0 \leq \mathrm{H}_\alpha(\rho) \leq \mathrm{H}_\alpha(\pi)$, with the lower bound attained for pure states and the upper bound obtained for maximally mixed states $\rho = \pi$.

- The Shannon quantum entropy can be expressed as $\mathrm{H}(\rho) = -\mathrm{tr}(\rho \log_2(\rho))$; the Tsallis quantum entropy as $1_2 \cdot (\mathrm{tr}(\rho^\alpha) - 1)/(1 - \alpha)$; and the Rényi quantum entropy as $\log_2(\mathrm{tr}(\rho^\alpha))/(1 - \alpha)$.
- The Tsallis and Rényi quantum entropies can be written in terms of the Tsallis trace $\mathrm{tr}(\rho^\alpha) = \sum_{x \in [N]} \lambda_x^\alpha$, which is a measure of mixedness for $0 \leq \alpha < 1$ and of purity for $\alpha > 1$.
- The post-measurement state resulting from a non-selective measurement in setting $\mathcal{P}$ of the pre-measurement state ρ is given by the density matrix $\mathcal{P}(\rho) = \sum_{y \in [M]} \Pi_y \rho \Pi_y$.
- A non-selective measurement increases the quantum entropy in the sense that the inequality $\mathrm{H}_\alpha(\mathcal{P}(\rho)) \geq \mathrm{H}_\alpha(\rho)$ holds.
- A density matrix ρ is said to be incoherent with respect to an orthonormal basis $\mathcal{M}$ if has an eigenbasis equal to $\mathcal{M}$.
- Equivalently, a mixed state ρ is incoherent with respect to an orthonormal basis $\mathcal{M}$ if and only if it is not changed by a non-selective measurement in setting $\mathcal{M}$, i.e., $\mathcal{M}(\rho) = \rho$.
- Any $N \times N$ matrix can be decomposed into a linear combination of a basis set of N^2 matrices $\{B_z\}_{z \in [N^2]}$, which may consist of: (i) an orthonormal basis of rank-1 matrices; (ii) an orthonormal basis of observables; and (iii) a non-orthogonal basis of density matrices.

7.15 Recommended Resources

Proofs of the ensemble classification theorem (7.24), the equivalence between commutativity and compatibility of observables, and the uncertainty inequality (7.142) are provided in the textbook [1]. Further insights into the uncertainty inequality can be found in [2]. Concavity and measures of classical and quantum uncertainty are discussed in detail in [3–5].

The expression (7.108) is derived in [6]. For an introduction to quantum cryptography, refer to [7]. The interplay between coherence and mixedness is explored in [8], while a description of the Gell-Mann matrices is available in [9]. For detailed information on bases for spaces of matrices, consult [10]. Lastly, the observational entropy is discussed in [11].

PROBLEMS

7.1 Prove that the (x, y)th entry of a square matrix A can be expressed as $A_{x,y} = \langle x|A|y\rangle$ for all $x, y \in [N]$.

7.2 Show the equality $\mathrm{tr}(|v\rangle\langle u|) = \langle u|v\rangle$ for any two states $|u\rangle$ and $|v\rangle$.

7.3 Using the cyclic shift invariance of the trace, prove that the trace can be written as $\mathrm{tr}(A) = \sum_{x \in [N]} \langle v_x|A|v_x\rangle$ for any orthonormal basis $\{|v_x\rangle\}_{x \in [N]}$. *Hint:* Use the fact that the trace of a scalar is the scalar itself, as well as the resolution-of-the-identity equality.

7.4 For a density matrix ρ with eigendecomposition $\rho = \sum_{l\in[N]} \lambda_l |u_l\rangle\langle u_l|$, consider a classical system that first picks an eigenvector $|u_l\rangle$ with probability λ_l and then generates output x with probability distribution $p(|u_l\rangle|\mathcal{M})$. Prove that the resulting distribution of the output x is given by the measurement distribution $p(\rho|\mathcal{M})$.

7.5 Argue that, while storing a generic density matrix ρ requires a classical memory of size that grows as N^2, if the rank of the density matrix equals r, the memory requirement scales with N as the product rN.

7.6 Prove the equality $-\mathrm{tr}(\rho \log_{2,\alpha}(\rho)) = \mathrm{H}^{\mathrm{T}}_{\alpha}(\lambda)$, where λ is the vector of eigenvalues of density matrix ρ.

7.7 Prove the bounds (7.112) and (7.113) on the Tsallis trace. *Hint:* Use the corresponding bounds derived in Chapter 3 for the Tsallis sum.

7.8 Prove the equality $(1-\alpha)^{-1}\log_2(\mathrm{tr}(\rho^{\alpha})) = \mathrm{H}^{\mathrm{R}}_{\alpha}(\lambda)$, where λ is the vector of eigenvalues of density matrix ρ.

7.9 (a) Can the Shannon quantum entropy be expressed as the expected value of an observable that does not depend on the state? (b) Can the Tsallis and Rényi quantum entropies be expressed as the expected values of observables that do not depend on the state?

7.10 Prove the non-negativity, minimal entropy, and maximal entropy properties of the Shannon, Tsallis, and Rényi quantum entropies by using the corresponding properties for the classical entropies.

7.11 Show that, for a pure state $\rho = |\psi\rangle\langle\psi|$ that is incoherent with respect to the complete measurement setting $\mathcal{M} = \{\Pi_x = |v_x\rangle\langle v_x|\}_{x\in[N]}$, the equality $\sum_{x\in[N]} \Pi_x \rho \Pi_x = \rho$ holds. Thus, a non-selective measurement in setting $\mathcal{M}$ does not disturb the pre-measurement state.

7.12 Prove that the coherence measure in (7.138) with any $\alpha > 0$ achieves the maximum value when ρ is pure and maximally coherent with respect to basis $\mathcal{M}$. *Hint:* Use the upper bound (7.99).

7.13 Prove the inequality

$$\mathrm{H}_{\alpha}(\rho) \leq \mathrm{H}_{\alpha}(\mathrm{diag}(\rho)), \tag{7.168}$$

where $\mathrm{diag}(\rho)$ is a diagonal matrix with the same elements as the density matrix ρ.

7.14 Consider the observable P given by any of the Pauli matrices P_1, P_2, or P_3, as well as the corresponding projective measurement setting $\mathcal{P} = \{\Pi_0, \Pi_1 = I - \Pi_0\}$ with $\Pi_0 = 1/2(I+P)$. Show that the expected value $\langle P\rangle_{\rho}$ for any 2×2 density matrix ρ is related to the measurement probability $p_0(\rho|\mathcal{P})$ as

$$p_0(\rho|\mathcal{P}) = \frac{1}{2}(1 + \langle P\rangle_{\rho}). \tag{7.169}$$

7.15 Show that the expected value of an observable O with eigendecomposition $O = \sum_{x\in[N]} O_x |v_x\rangle\langle v_x|$ can be expressed under a pure state $|\psi\rangle$ as

$$\langle O\rangle_{|\psi\rangle} = \sum_{x\in[N]} p_x(|\psi\rangle|\mathcal{M})a_x, \tag{7.170}$$

where $\mathcal{M} = \{|v_x\rangle\}_{x\in[N]}$ is a complete measurement setting including the eigenvectors of the observable O, and

$$a_x = \frac{\langle\psi|O|v_x\rangle}{\langle\psi|v_x\rangle}. \tag{7.171}$$

The equality (7.170) is at the core of **variational Monte Carlo**, a method for the evaluation of the eigenvector corresponding to the smallest eigenvalue of a positive semi-definite observable O.

7.16 Show that the four Pauli matrices are orthogonal with respect to the trace inner product, i.e., the equality $\mathrm{tr}(P_xP_y) = 0$ holds for all index pairs $x, y \in [4]$ with $x \neq y$.

7.17 Prove that the set of density matrices $\{\rho_0 = P_0/\sqrt{2}, \rho_1 = 0.5(I + P_1), \rho_2 = 0.5(I + P_2), \rho_3 = 0.5(I + P_3)\}$ forms a basis for the space of all 2×2 density matrices. *Hint:* A 2×2 density matrix can be expressed as a linear combination of the Pauli matrices.

7.18 Show that the set $\{\{B_{xy} = |u_x\rangle\langle u_y|\}_{x\in[N]}\}_{y\in[N]}$ forms an orthonormal basis for the space of $N \times N$ matrices.

7.19 Given any pure state $|\psi\rangle$, define as $\Pi_{|\psi\rangle} = |\psi\rangle\langle\psi|$ the projection matrix onto the line defined by vector $|\psi\rangle$. For any two states $|\psi\rangle$ and $|\phi\rangle$, show the equality $|\langle\psi|\phi\rangle|^2 = \mathrm{tr}(\Pi_{|\psi\rangle}\Pi_{|\phi\rangle})$.

7.20 Show that, when minimizing over all possible measurement settings $\mathcal{M}$, the coherence measure (7.138) equals zero for any state ρ. Therefore, any state is incoherent in some basis.

Appendix

7.A Deriving Projective Measurements for Mixed States

This appendix derives the measurement probability (7.37) and the post-measurement state (7.38) from the Born rule and the post-measurement state for pure states. To this end, consider any ensemble $\mathcal{E} = \{(|\psi_l\rangle, p_l)\}_{l\in[L]}$ of pure states realizing the density matrix ρ. When the pre-measurement state equals $|\psi_l\rangle$, by the Born rule for pure states, the post-measurement state is given by

$$|\phi_{l|y}\rangle = \frac{1}{\sqrt{p_y(|\psi_l\rangle|\mathcal{P})}}\Pi_y|\psi_l\rangle \tag{7.172}$$

with probability $p_y(|\psi_l\rangle|\mathcal{P})$. The notation $|\phi_{l|y}\rangle$ is a reminder that this is the lth possible post-measurement state when the classical measurement output is y.

Since the pre-measurement state equals $|\psi_l\rangle$ with probability p_l, the overall probability of having the system in the post-measurement state $|\phi_{l|y}\rangle$ is given by the product

$$p_{l,y} = p_l \cdot p_y(|\psi_l\rangle|\mathcal{P}). \tag{7.173}$$

Therefore, by the law of total probability, the probability of obtaining classical output $y \in [M]$ is obtained via the sum

$$\begin{aligned} p_y(\rho|\mathcal{P}) &= \sum_{l\in[L]} p_l p_y(|\psi_l\rangle|\mathcal{P}) \\ &= \sum_{l\in[L]} p_l \mathrm{tr}(|\psi_l\rangle\langle\psi_l|\Pi_y) \\ &= \mathrm{tr}(\rho\Pi_y), \end{aligned} \tag{7.174}$$

where $\rho = \sum_{l\in[L]} p_l|\psi_l\rangle\langle\psi_l|$ is the pre-measurement density matrix. This recovers the measurement probability (7.37).

When the output of the projective measurement is $y \in [M]$, by the **Bayes rule**, the probability that the pre-measurement state was $|\psi_l\rangle$ is $p_{l|y} = p_{l,y}/p_y(\rho|\mathcal{P})$. Therefore, for a given classical output y, the post-measurement state is characterized by the ensemble $\mathcal{E}_y = \{(|\phi_{l|y}\rangle, p_{l|y})\}_{l\in[L]}$, which corresponds to density matrix

$$\begin{aligned} \rho_y &= \sum_{l\in[L]} p_{l|y}|\phi_{l|y}\rangle\langle\phi_{l|y}| \\ &= \sum_{l\in[L]} \frac{p_l \cdot p_y(|\psi_l\rangle|\mathcal{P})}{p_y(|\psi_l\rangle|\mathcal{P}) \cdot p_y(\rho|\mathcal{P})} \Pi_y|\psi_l\rangle\langle\psi_l|\Pi_y \\ &= \frac{1}{p_y(\rho|\mathcal{P})}\Pi_y\rho\Pi_y, \end{aligned} \tag{7.175}$$

recovering (7.38).

7.B Defining Properties of Density Matrices

This appendix shows that any density matrix $\rho = \sum_{l\in[L]} p_l|\psi_l\rangle\langle\psi_l|$ realized by an ensemble of pure states, $\mathcal{E} = \{(|\psi_l\rangle, p_l)\}_{l\in[L]}$, is positive semi-definite and has trace equal to 1. To start, since a density matrix is Hermitian, it admits the eigendecomposition $\rho = \sum_{x\in[N]} \lambda_x|u_x\rangle\langle u_x|$. Furthermore, the xth eigenvalue λ_x can be written as

$$\begin{aligned} \langle u_x|\rho|u_x\rangle &= \sum_{l\in[L]} p_l\langle u_x|\psi_l\rangle\langle\psi_l|u_x\rangle \\ &= \sum_{l\in[L]} p_l|\langle u_x|\psi_l\rangle|^2 \geq 0, \end{aligned} \tag{7.176}$$

and is hence non-negative.

As for the trace, the following equalities hold:

$$\begin{aligned} \mathrm{tr}(\rho) &= \sum_{x\in[N]} \langle x|\rho|x\rangle \\ &= \sum_{x\in[N]} \langle x|\left(\sum_{l\in[L]} p_l|\psi_l\rangle\langle\psi_l|\right)|x\rangle \\ &= \sum_{l\in[L]} p_l \sum_{x\in[N]} |\langle x|\psi_l\rangle|^2 \\ &= \sum_{l\in[L]} p_l = 1, \end{aligned} \tag{7.177}$$

where the normalization of state vector $|\psi\rangle$ was used to justify the equality $\sum_{x\in[N]} |\langle x|\psi_l\rangle|^2 = 1$.

7.C Observational Entropy

In Section 7.11 it was argued that projective measurements $\mathcal{P}$ can reduce the measured entropy $\mathrm{H}_\alpha(\rho|\mathcal{P})$ due to the coarse graining caused by the choice of projection matrices with rank larger than 1. The level of coarse graining associated with a measurement output $x \in [M]$ is dictated by the rank $r_x = \mathrm{rank}(\Pi_x)$ of the corresponding projection matrix Π_x, with a larger rank r_x indicating a coarser-grained measurement output x.

Based on this observation, the quantum entropy was defined by limiting the minimization (7.87) of the measured entropy to complete measurements, thus disabling coarse graining. However, one may wish to define a quantum entropy measure that captures not only the intrinsic uncertainty associated with a state ρ, which is reflected by the Shannon quantum entropy $\mathrm{H}(\rho)$, but also the lack of knowledge caused by coarse graining.

With this goal in mind, given a projective measurement $\mathcal{P} = \{\Pi_x\}_{x\in[M]}$, the **Shannon observational entropy** is defined as

$$\mathrm{H}^{\mathrm{obs}}(\rho|\mathcal{P}) = \sum_{x\in[M]} p_x(\rho|\mathcal{P})(-\log_2(p_x(\rho|\mathcal{P})) + \log_2(r_x)), \tag{7.178}$$

with $r_x = \mathrm{rank}(\Pi_x)$ for all $x \in [M]$. Accordingly, the Shannon observational entropy increases the uncertainty $-\log(p_x(\rho|\mathcal{P}))$ associated with a measurement output $x \in [M]$ by an amount $\log(r_x)$ that grows with the level of coarse graining of the measurement output x.

For complete measurements, i.e., for ranks $r_x = 1$ and for $M = N$, the Shannon observational entropy coincides with the Shannon measured entropy, i.e., $\mathrm{H}^{\mathrm{obs}}(\rho|\mathcal{P}) = \mathrm{H}(\rho|\mathcal{P})$. At the other extreme, for $M = 1$, the single projection matrix $\Pi_0 = I$ has rank $r_0 = N$ and measurement probability $p_0 = 1$. Therefore, with $M = 1$, the Shannon observational entropy reduces to $\mathrm{H}^{\mathrm{obs}}(\rho|\mathcal{P}) = \log_2(N)$. This result is in stark contrast to the Shannon measured entropy $\mathrm{H}(\rho|\mathcal{P})$, which would be equal to zero when $M = 1$.

In general, one can prove the inequality

$$\mathrm{H}^{\mathrm{obs}}(\rho|\mathcal{P}) \geq \mathrm{H}(\rho), \tag{7.179}$$

for any measurement setting $\mathcal{P}$. Accordingly, the observational entropy $\mathrm{H}^{\mathrm{obs}}(\rho|\mathcal{P})$ captures not only the inherent uncertainty in state ρ, as measured by the Shannon quantum entropy $\mathrm{H}(\rho)$, but also the lack of knowledge due to coarse graining, as quantified by the difference $\mathrm{H}^{\mathrm{obs}}(\rho|\mathcal{P}) - \mathrm{H}(\rho) \geq 0$. For further details, the reader is referred to the Recommended Resources.

7.D Basis of Density Matrices

For any orthonormal basis $\{|u_x\rangle\}_{x\in[N]}$, consider the set of N matrices

$$\{|u_x\rangle\langle u_x|\}_{x\in[N]}, \tag{7.180}$$

the set of $(N^2 - N)/2$ matrices

$$\left\{\left\{\frac{1}{2}(|u_x\rangle + |u_y\rangle)(\langle u_x| + \langle u_y|)\right\}_{y\in[N]:y>x}\right\}_{x\in[N]}, \tag{7.181}$$

and the set of $(N^2 - N)/2$ matrices

$$\left\{\left\{\frac{1}{2}(|u_x\rangle + i|u_y\rangle)(\langle u_x| - i\langle u_y|)\right\}_{y\in[N]:y>x}\right\}_{x\in[N]}. \tag{7.182}$$

Note that these are rank-1 density matrices, since they are expressed as the outer product of a pure state.

Any $N \times N$ matrix A can be written as a linear combination of the N^2 orthonormal matrices $\{\{|u_x\rangle\langle u_y|\}_{x\in[N]}\}_{y\in[N]}$. Therefore, in order to prove that the N^2 matrices (7.180)–(7.182) form a basis of the space of $N \times N$ matrices, it is enough to demonstrate that any matrix $|u_x\rangle\langle u_y|$ can be written as a linear combination of such matrices. To this end, note first that matrix $|u_x\rangle\langle u_y|$ with $x = y$ is included in the set of matrices (7.180)–(7.182).

With $y > x$, one obtains

$$\begin{aligned}|u_x\rangle\langle u_y| &= \frac{1}{2}(|u_x\rangle + |u_y\rangle)(\langle u_x| + \langle u_y|)\\ &+ \frac{i}{2}(|u_x\rangle + i|u_y\rangle)(\langle u_x| - i\langle u_y|)\\ &- \frac{1}{2}(1+i)|u_x\rangle\langle u_x| - \frac{1}{2}(1+i)|u_y\rangle\langle u_y|,\end{aligned} \tag{7.183}$$

while for $y < x$ the equality

$$\begin{aligned}|u_x\rangle\langle u_y| =&\frac{1}{2}(|u_x\rangle + |u_y\rangle)(\langle u_x| + \langle u_y|)\\ &- \frac{i}{2}(|u_x\rangle + i|u_y\rangle)(\langle u_x| - i\langle u_y|)\\ &- \frac{1}{2}(1-i)|u_x\rangle\langle u_x| - \frac{1}{2}(1-i)|u_y\rangle\langle u_y|\end{aligned} \tag{7.184}$$

holds, proving the claim.

7.E Proof of Equality (7.135)

This appendix focuses on proving the equality $\mathrm{H}_\alpha(\mathcal{M}(\rho)) = \mathrm{H}_\alpha(\rho|\mathcal{M})$ in (7.135) between the quantum entropy of the post-measurement state $\mathcal{M}(\rho)$ and the measured entropy $\mathrm{H}_\alpha(\rho|\mathcal{M})$. To start, note that the optimal measurement basis for the post-measurement state $\mathcal{M}(\rho)$ is given by $\mathcal{M}$, since this corresponds to an eigenbasis of state $\mathcal{M}(\rho)$. Therefore, by the definition (7.87), the quantum entropy for the post-measurement state satisfies the equality

$$\mathrm{H}_\alpha(\mathcal{M}(\rho)) = \mathrm{H}_\alpha(\mathcal{M}(\rho)|\mathcal{M}), \tag{7.185}$$

where the right-hand side is the measured entropy of state $\mathcal{M}(\rho)$ in basis $\mathcal{M}$.

Moreover, by definition of the post-measurement state $\mathcal{M}(\rho)$, one has the equalities

$$\mathrm{H}_\alpha(\mathcal{M}(\rho)|\mathcal{M}) = \mathrm{H}_\alpha(p(\rho|\mathcal{M})) = \mathrm{H}_\alpha(\rho|\mathcal{M}). \tag{7.186}$$

In particular, the result (7.186) shows that, if we make a measurement in setting $\mathcal{M}$ and then we forget its outcome, making the same measurement again returns an output with an entropy $\mathrm{H}_\alpha(\mathcal{M}(\rho)|\mathcal{M})$ equal to the entropy $\mathrm{H}_\alpha(\rho|\mathcal{M})$ of the original measurement.

Combining (7.185) and (7.186) completes the proof.

REFERENCES

[1] M. A. Nielsen and I. L. Chuang, *Quantum Computation and Quantum Information*. Cambridge: Cambridge University Press, 2010.

[2] P. J. Coles, M. Berta, M. Tomamichel, and S. Wehner, "Entropic uncertainty relations and their applications," *Reviews of Modern Physics*, vol. 89, no. 1, art. 015002, 2017.

[3] X. Hu and Z. Ye, "Generalized quantum entropy," *Journal of Mathematical Physics*, vol. 47, no. 2, art. 023502, 2006.

[4] A. E. Rastegin, "Some general properties of unified entropies," *Journal of Statistical Physics*, vol. 143, pp. 1120–1135, 2011.

[5] S. Luo, "Quantum versus classical uncertainty," *Theoretical and Mathematical Physics*, vol. 143, pp. 681–688, 2005.

[6] T. Heinosaari and M. Ziman, *The Mathematical Language of Quantum Theory: From Uncertainty to Entanglement*. Cambridge: Cambridge University Press, 2011.

[7] T. Vidick and S. Wehner, *Introduction to Quantum Cryptography*. Cambridge: Cambridge University Press, 2023.

[8] Q.-H. Zhang and S.-M. Fei, "Coherence–mixedness trade-offs," *Journal of Physics A: Mathematical and Theoretical*, vol. 57, no. 23, art. 235301, 2024.

[9] E. W. Weisstein, "Generalized Gell-Mann matrix." [Online]. Available at: https://mathworld.wolfram.com/GeneralizedGell-MannMatrix.html

[10] S. Milz and K. Modi, "Quantum stochastic processes and quantum non-Markovian phenomena," *PRX Quantum*, vol. 2, no. 3, art. 030201, 2021.

[11] D. Šafránek, A. Aguirre, J. Schindler, and J. Deutsch, "A brief introduction to observational entropy," *Foundations of Physics*, vol. 51, pp. 1–20, 2021.

8 Quantum Relative Entropy

8.1 Overview

Generalizing the classical relative entropies discussed in Chapter 4, this chapter introduces the Shannon, Tsallis, and Rényi quantum relative entropies, which serve as distance measures between pairs of density matrices ρ and σ.

Recall from Chapter 7 that density matrices describe mixed states that are realized by ensembles of pure states. Under a mixed state, each pure state in the ensemble is prepared with some probability. Based on this, intuitively, in order to evaluate the distance between two density matrices, it is not sufficient to compare the probability distributions underlying the preparation of the corresponding mixed states. Rather, one must also consider the relationships between all possible pairs of pure states in the respective ensembles.

In particular, when density matrices ρ and σ are expressed in terms of their respective eigenensembles, the quantum relative entropies will be seen to depend on:

1. The relationship between the eigenvalues of the two density matrices, which reflect the probabilities assigned to the pure states in the ensembles.
2. The compatibility of the two eigenbases, which determines the alignment between the pure states in the ensembles.

Accordingly, a notable simplification occurs when the density matrices ρ and σ commute. In this special case, the density matrices share the same eigenbasis, and the quantum relative entropies reduce to the classical relative entropies between their corresponding vectors of eigenvalues.

Like their classical counterparts (see Chapter 4), quantum relative entropies will also be shown to be instrumental in formalizing the monotonicity property of quantum entropy, $\mathrm{H}_\alpha(\rho)$. Specifically, this property can be expressed as equalities of the form $\mathrm{H}_\alpha(\rho) = \mathrm{H}_\alpha(\pi) - \mathrm{D}_\alpha(\rho\|\pi)$, where $\mathrm{H}_\alpha(\pi)$ is the maximum entropy achieved by the maximally mixed state π, and $\mathrm{D}_\alpha(\rho\|\pi)$ is the relevant quantum relative entropy between the density matrices ρ and π. This expression underscores the relationship between a state's quantum entropy and its deviation from the state of maximal uncertainty, which is quantified by the quantum relative entropy.

As discussed in Chapter 4, classical relative entropies are closely tied to the optimal probability of detection for distinguishing between samples drawn from two probability distributions. Similarly, this chapter establishes the connection between quantum relative entropies and the informational task of detecting the state of a quantum system, when the system is known a priori to be in one of two possible states.

Through the study of this detection problem, the chapter introduces the trace distance, a fundamental measure of distinguishability between quantum states. Furthermore, it demonstrates how the trace distance is bounded by the quantum Shannon relative entropy through a quantum analog of Pinsker's inequality, providing a link between these two measures.

Finally, this chapter extends the concept of fidelity, introduced in Chapter 4 for classical distributions and in Chapter 5 for pure states, to mixed states. Unlike pure states, there are multiple approaches to defining fidelity for mixed states.

A particularly insightful method will be seen to involve interpreting mixed states as pure states of a higher-dimensional system, which includes both the system of interest and its environment. This interpretation is known as purification. From this perspective, the classical uncertainty associated with a mixed state is modeled as inaccessible information hidden from the system within the environment. This discussion will also support the introduction of the Kronecker product as a key mathematical tool for constructing purifications and for modeling composite systems.

8.1.1 Learning Objectives and Organization

By the end of this chapter, the reader should be able to:

- understand the definition and properties of the Shannon, Tsallis, and Rényi quantum relative entropies;
- formalize the property of monotonicity of the Shannon, Tsallis, and Rényi quantum entropies in terms of the corresponding quantum relative entropies;
- derive the optimal detection probability for the problem of distinguishing two mixed quantum states based on measurements of either state;
- understand definitions and properties of fidelity measures for mixed states;
- understand the definition and properties of the trace distance;
- evaluate the purification of a mixed state;
- and use the Kronecker product for vectors and matrices.

8.2 Shannon Quantum Relative Entropy

This section defines and studies the Shannon quantum relative entropy $\mathrm{D}(\rho||\sigma)$ between two density matrices ρ and σ. The Shannon quantum relative entropy will be seen to generalize the Shannon relative entropy presented in Chapter 4, which is recovered in the special case in which the two density matrices ρ and σ commute.

In a manner similar to the Shannon relative entropy, it will sometimes be useful to think of matrix ρ as the **reference**, or **true**, density matrix and of matrix σ as an **approximation** of ρ.

8.2.1 Defining the Shannon Quantum Relative Entropy Using Eigendecompositions: Commuting Density Matrices

Take any two $N \times N$ density matrices ρ and σ with eigendecompositions

$$\rho = \sum_{x\in[N]} \lambda_x |u_x\rangle\langle u_x| \tag{8.1}$$

and

$$\sigma = \sum_{x\in[N]} \mu_x |v_x\rangle\langle v_x|. \tag{8.2}$$

Note that some of the eigenvalues of the two matrices may equal zero, and that the rank of each density matrix equals the number of strictly positive eigenvalues (see Section 6.6), i.e.,

$$r_\rho = \text{rank}(\rho) = |\{x \in [N]\colon \lambda_x > 0\}| \tag{8.3}$$

and

$$r_\sigma = \text{rank}(\sigma) = |\{x \in [N]\colon \mu_x > 0\}|. \tag{8.4}$$

If more than one eigendecomposition exists for either density matrix, any eigendecomposition can be selected in what follows.

To start, consider the simpler situation in which the two density matrices commute, so that they satisfy the equality

$$\rho\sigma = \sigma\rho. \tag{8.5}$$

As seen in Section 7.6.2, in this case, the two density matrices share a basis of eigenvectors, and thus one can set $\{|v_x\rangle = |u_x\rangle\}_{x\in[N]}$ in the eigendecompositions (8.1) and (8.2), i.e.,

$$\rho = \sum_{x\in[N]} \lambda_x |u_x\rangle\langle u_x| \tag{8.6}$$

and

$$\sigma = \sum_{x\in[N]} \mu_x |u_x\rangle\langle u_x|. \tag{8.7}$$

This way, each eigenvalue λ_x of matrix ρ corresponds to the eigenvalue μ_x of matrix σ, as they both refer to the same eigenvector $|u_x\rangle$.

With commuting density matrices, the Shannon quantum relative entropy is naturally defined as the Shannon relative entropy between the vectors of eigenvalues λ and μ.

Definition 8.1 Shannon quantum relative entropy for commuting density matrices
Given commuting density matrices ρ and σ with eigendecompositions (8.6) and (8.7), respectively, the Shannon quantum relative entropy is defined as

$$\mathrm{D}(\rho||\sigma) = \mathrm{D}(\lambda||\mu). \tag{8.8}$$

Note that in evaluating (8.8) one must ensure that the eigenvalues are listed in vectors λ and μ so that the eigenvalues assigned to the same eigenvector are in the same position in the two vectors λ and μ.

8.2.2 Defining the Shannon Quantum Relative Entropy Using Eigendecompositions: General Density Matrices

More generally, for non-commuting density matrices, the two eigenbases are distinct, and thus the simple definition (8.8) is not applicable. Following Section 6.4, the alignment between the eigenbases of the two density matrices (8.1) and (8.2) is captured by the compatibility of the corresponding measurement settings

$$\mathcal{M}_\rho = \{|u_x\rangle\}_{x\in[N]} \text{ and } \mathcal{M}_\sigma = \{|v_x\rangle\}_{x\in[N]}. \tag{8.9}$$

Specifically, the compatibility between the two measurement settings is described by the $N \times N$ **compatibility matrix** $\mathrm{C}(\mathcal{M}_\rho, \mathcal{M}_\sigma)$, with (x, y)th element given by the **fidelity** $\mathrm{F}(|u_x\rangle, |v_y\rangle) = |\langle u_x|v_y\rangle|^2$:

$$[\mathrm{C}(\mathcal{M}_\rho, \mathcal{M}_\sigma)]_{x,y} = \mathrm{F}(|u_x\rangle, |v_y\rangle) = |\langle u_x|v_y\rangle|^2. \tag{8.10}$$

The two density matrices ρ and σ commute if and only if the compatibility matrix $\mathrm{C}(\mathcal{M}_\rho, \mathcal{M}_\sigma)$ is a permutation matrix, i.e., if every row and every column of matrix $\mathrm{C}(\mathcal{M}_\rho, \mathcal{M}_\sigma)$ has exactly one entry equal to 1, and all other entries equal to 0. In general, for non-commuting density matrices, matrix $\mathrm{C}(\mathcal{M}_\rho, \mathcal{M}_\sigma)$ is doubly stochastic, with entries ranging in the interval between 0 and 1.

Intuitively, in order for the two density matrices to be considered similar, each eigenvalue μ_y of matrix σ should serve as an approximation of an eigenvalue λ_x of matrix ρ to an extent that depends on the alignment of the corresponding eigenvectors $|v_y\rangle$ and $|u_x\rangle$. Using the fidelity (8.10) to measure alignment, the Shannon quantum relative entropy is defined as follows.

Definition 8.2 Shannon quantum relative entropy using eigendecompositions
For any two density matrices ρ and σ with eigendecompositions (8.1) and (8.2), the Shannon quantum relative entropy is defined as

$$\mathrm{D}(\rho||\sigma) = \sum_{x\in[N]} \sum_{y\in[N]} \mathrm{F}(|u_x\rangle, |v_y\rangle)\lambda_x \log_2\left(\frac{\lambda_x}{\mu_y}\right). \tag{8.11}$$

The definition (8.11) recovers the expression (8.8) for the case of commuting matrices (see Problem 8.1). More generally, the term $\lambda_x \log_2(\lambda_x/\mu_y)$ in (8.11) contributes to the Shannon quantum relative entropy to a degree that depends on the alignment of the corresponding eigenvectors $|u_x\rangle$ and $|v_y\rangle$ via the fidelity $\mathrm{F}(|u_x\rangle, |v_y\rangle)$. Comparing two density matrices hence requires considering *all* possible pairs of pure states the density matrices may represent – i.e., the eigenvectors.

Using the fact that the compatibility matrix is doubly stochastic, the Shannon quantum relative entropy (8.11) can also equivalently be written as

$$\mathrm{D}(\rho||\sigma) = \sum_{x\in[N]} \lambda_x \log_2(\lambda_x) - \sum_{x\in[N]} \sum_{y\in[N]} \mathrm{F}(|u_x\rangle, |v_y\rangle)\lambda_x \log_2(\mu_y). \tag{8.12}$$

Example 8.1 illustrates the interplay between eigenvalues and eigenvectors of the two density matrices in determining the value of the Shannon quantum relative entropy.

Example 8.1

Consider the 2×2 density matrices

$$\rho = \lambda_0|0\rangle\langle 0| + \lambda_1|1\rangle\langle 1| \tag{8.13}$$

and

$$\sigma = \mu_0|\theta\rangle\langle\theta| + \mu_1|\theta^\perp\rangle\langle\theta^\perp|, \tag{8.14}$$

with $|\theta\rangle = [\cos(\theta), \sin(\theta)]^T$ and $|\theta^\perp\rangle = [-\sin(\theta), \cos(\theta)]^T$. As suggested by the notation, the two density matrices are expressed in (8.13) and (8.14) directly in terms of their respective eigendecompositions (8.1) and (8.2), which are illustrated in Figure 8.1.

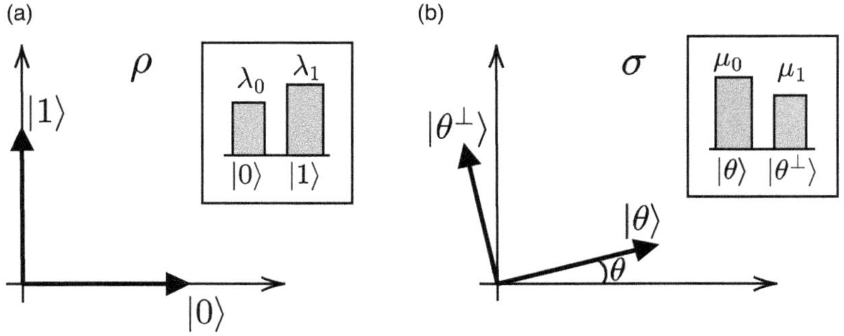

Figure 8.1 Illustration of the density matrices (a) (8.13) and (b) (8.14) via their respective eigendecompositions.

The Shannon quantum relative entropy $\mathrm{D}(\rho||\sigma)$ in (8.11) depends not only on the discrepancy between the vectors of eigenvalues $\lambda = [\lambda_0, \lambda_1]^T$ and $\mu = [\mu_0, \mu_1]^T$, but also on the compatibility of the eigenbases $\mathcal{M}_\rho = \{|0\rangle, |1\rangle\}$ and $\mathcal{M}_\sigma = \{|\theta\rangle, |\theta^\perp\rangle\}$. The compatibility matrix is given by

$$C(\mathcal{M}_\rho, \mathcal{M}_\sigma) = \begin{bmatrix} (\cos(\theta))^2 & (\sin(\theta))^2 \\ (\sin(\theta))^2 & (\cos(\theta))^2 \end{bmatrix}, \tag{8.15}$$

showing that the two matrices commute if $\theta = 0$ or $\theta = \pi/2$ in the interval $0 \leq \theta < \pi$.

Using (8.11) and (8.15), the Shannon quantum relative entropy for density matrices (8.13) and (8.14) is computed as

$$\begin{aligned} D(\rho||\sigma) =& (\cos(\theta))^2 \lambda_0 \log_2\left(\frac{\lambda_0}{\mu_0}\right) + (\sin(\theta))^2 \lambda_0 \log_2\left(\frac{\lambda_0}{\mu_1}\right) \\ &+ (\sin(\theta))^2 \lambda_1 \log_2\left(\frac{\lambda_1}{\mu_0}\right) + (\cos(\theta))^2 \lambda_1 \log_2\left(\frac{\lambda_1}{\mu_1}\right). \end{aligned} \tag{8.16}$$

Figure 8.2 shows the Shannon quantum relative entropy $D(\rho||\sigma)$ for $\mu_1 = 0.3$ for different values of θ as a function of λ_1.

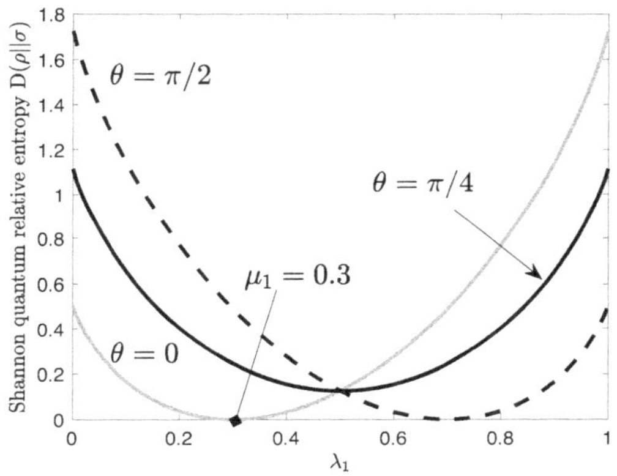

Figure 8.2 Shannon quantum relative entropy $D(\rho||\sigma)$ for density matrices (8.13) and (8.14) with $\mu_1 = 0.3$ and different values of θ as a function of λ_1.

For $\theta = 0$, the two density matrices commute. Furthermore, the eigenvalues λ_0 and μ_0 are associated with the same eigenvector $|0\rangle$, while the eigenvalues λ_1 and μ_1 are associated with the same eigenvector $|1\rangle$. Thus, the Shannon quantum relative entropy takes the simpler form (8.8), which coincides with the binary Shannon relative entropy $d(\lambda_1||\mu_1)$. As studied in Chapter 4, the relative entropy $d(\lambda_1||0.3)$ equals zero when $\lambda_1 = \mu_1 = 0.3$, and it grows as λ_1 becomes increasingly different from $\mu_1 = 0.3$.

For $\theta = \pi/2$, the density matrices ρ and σ also commute, but now eigenvalue μ_0 is associated with eigenvector $|\pi/2\rangle = |1\rangle$, while eigenvalue μ_1 is associated with eigenvector $|0\rangle$. Accordingly, the Shannon quantum relative entropy coincides with the binary Shannon relative entropy $d(\lambda_0||\mu_1)$, which attains its minimum value for $\lambda_1 = \mu_0 = 0.7$.

For other values of θ in the given interval $0 \leq \theta < \pi$, the two density matrices do not commute, and the Shannon quantum relative entropy remains strictly positive as λ_1 varies. In fact, in such cases, there does not exist a value of λ_1 that ensures a match between the two density matrices.

The impact of the compatibility of the eigenbases is further illustrated in Figure 8.3, which shows the Shannon quantum relative entropy as a function of the angle $0 \leq \theta \leq \pi$ for different pairs of values λ_1 and μ_1. For $\lambda_1 = \mu_1 = 0.5$, both density matrices coincide with the maximally mixed density, i.e., $\rho = \sigma = \pi$, and hence the Shannon quantum relative entropy equals zero for any value of angle θ. For $\lambda_1 = \mu_1 = 0.3$, the Shannon quantum relative entropy equals zero when the density matrices commute and the eigenvalues λ_x and μ_x are assigned to the same eigenvector $|x\rangle$ for $x \in [2]$. This occurs for $\theta = 0$ and $\theta = \pi$.

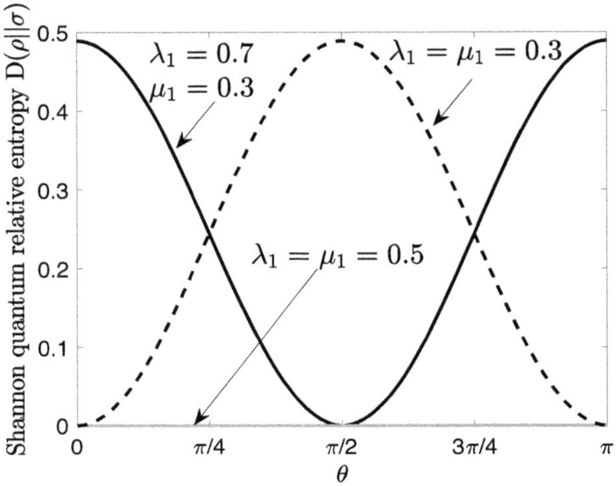

Figure 8.3 Shannon quantum relative entropy $D(\rho||\sigma)$ for density matrices (8.13) and (8.14) as a function of the angle $0 \leq \theta \leq \pi$ for different pairs of values λ_1 and μ_1.

Otherwise, the maximum of the Shannon quantum relative entropy for $\lambda_1 = \mu_1 = 0.3$ is attained at $\theta = \pi/2$, at which point the eigenbases are compatible but with swapped eigenvalues, and hence the Shannon quantum relative entropy equals $d(\lambda_1||\mu_0) = d(0.3||0.7)$. For $\lambda_1 = 0.7$ and $\mu_1 = 0.3$, conversely, the minimum of the Shannon quantum relative entropy is attained when $\theta = \pi/2$, at which point it equals $d(\lambda_1||\mu_0) = d(0.7||0.7) = 0$.

8.2.3 Impossible States, Black Swans, and Hallucinations

In order to evaluate the sum (8.11) or (8.12), it is useful to identify special states that, as in the case of the Shannon relative entropy studied in Section 4.2, have extreme contributions to the Shannon quantum relative entropy. Importantly, unlike the classical case presented in Section 4.2, states here do not refer to indices x, but rather to pure states, i.e., to N-dimensional normalized vectors.

To start, consider the notions of the support and kernel of a density matrix, which use the definition of the span of a set of orthonormal vectors introduced in Section 6.5.4.

Definition 8.3 Support of a density matrix
The support supp(ρ) of density matrix ρ with eigendecomposition (8.6) is the span of the eigenvectors of matrix ρ with strictly positive eigenvalues, i.e.,

$$\mathrm{supp}(\rho) = \mathcal{S}(\{|u_x\rangle\}_{x\in[N]:\, \lambda_x>0}). \tag{8.17}$$

Definition 8.4 Kernel of a density matrix
The kernel ker(ρ) of density matrix ρ with eigendecomposition (8.6) is the span of the eigenvectors of matrix ρ with zero eigenvalues, i.e.,

$$\ker(\rho) = \mathcal{S}(\{|u_x\rangle\}_{x\in[N]:\, \lambda_x=0}). \tag{8.18}$$

The support and kernel of a matrix contain mutually orthogonal states:

$$\langle\psi|\phi\rangle = 0 \text{ for all } |\psi\rangle \in \mathrm{supp}(\rho) \text{ and } |\phi\rangle \in \ker(\rho). \tag{8.19}$$

However, the support and kernel of a density matrix do not partition the N-dimensional space. That is, there are states that are in neither the support nor the kernel of matrix ρ. In fact, as illustrated by the following example, one can create pure states that are in a superposition of states from both support and kernel of a density matrix, thus not belonging to either subspace.

Example 8.2

For a three-dimensional space with computational basis $\{|0\rangle, |1\rangle, |2\rangle\}$, consider the density matrix $\rho = \lambda_0|0\rangle\langle 0| + \lambda_1|1\rangle\langle 1|$ with strictly positive eigenvalues λ_0 and λ_1. As illustrated in Figure 8.4, the support supp(ρ) of density matrix ρ is the plane $\mathcal{S}(|0\rangle, |1\rangle)$ spanned by the states $|0\rangle$ and $|1\rangle$. Furthermore, the kernel ker(ρ) is the line orthogonal to supp(ρ), corresponding to the span $\mathcal{S}(|2\rangle)$. The state $|\psi\rangle$ in the figure is in a superposition of states from the support and from the kernel.

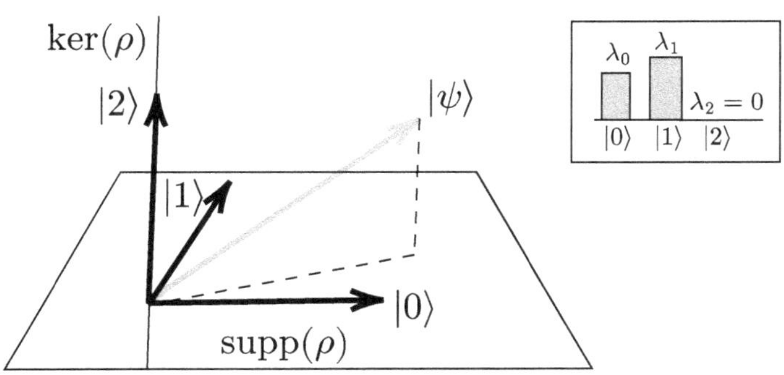

Figure 8.4 Support and kernel of density matrix $\rho = \lambda_0|0\rangle\langle 0| + \lambda_1|1\rangle\langle 1|$ with strictly positive eigenvalues λ_0 and λ_1 in an $N = 3$-dimensional space. The state $|\psi\rangle$ in the figure is in a superposition of states from the support and from the kernel.

Definition 8.5 Impossible states
Given the reference density matrix ρ with eigendecompositions (8.6), all the pure states within the subspace $\ker(\rho)$ are said to be impossible. Equivalently, a pure state $|\psi\rangle$ is impossible if it is orthogonal to all the states in the support of the matrix ρ, i.e., if $\mathrm{F}(|\psi\rangle, |u_x\rangle) = 0$ for all $x \in [N]$ with $\lambda_x > 0$.

This definition marks as impossible only the states in the span $\ker(\rho)$. This does not imply, however, that all other pure states are "possible" under the density matrix ρ. In particular, any state as in Figure 8.4, which is in a superposition of of states from both support and kernel of matrix ρ, cannot belong to any ensemble realizing matrix ρ. Based on this definition, the following is a generalization of the concept of black swans introduced in Section 4.2 from classical to quantum states.

Definition 8.6 Black swans
For reference density matrix ρ and density matrix σ with eigendecompositions (8.6) and (8.7), respectively, an eigenvector $|v_x\rangle$ of density matrix σ is a black swan if: (i) it corresponds to a zero eigenvalue, $\mu_x = 0$, of matrix σ, thus lying in the subspace $\ker(\sigma)$; and (ii) it is not impossible, i.e., it is not in the subspace $\ker(\rho)$:

$$|v_x\rangle \notin \ker(\rho) \text{ and } |v_x\rangle \in \ker(\sigma). \tag{8.20}$$

A black swan is thus an eigenvector of matrix σ that has zero probability under density matrix σ and is not in the kernel of the density matrix ρ.

Example 8.3

Figure 8.5 shows the example of two-dimensional density matrices ρ and σ with rank equal to 1. The eigenvector $|v_1\rangle$ of density matrix σ is a black swan because: (i) it corresponds to a zero eigenvalue $\mu_1 = 0$ and thus it is in the kernel of density matrix σ; and (ii) it is not contained in the kernel of density matrix ρ. In particular, the state $|u_0\rangle$, which is in the support of matrix ρ, has a positive fidelity with the state $|v_1\rangle$.

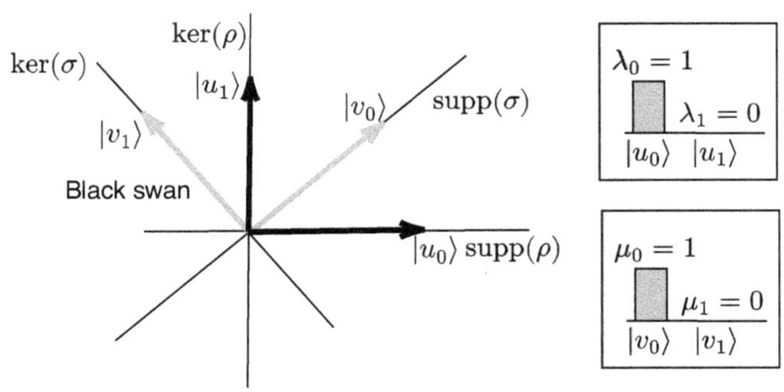

Figure 8.5 The eigenvector $|v_1\rangle$ of density matrix σ is a black swan because it is in the kernel of density matrix σ and it is not in the kernel of density matrix ρ.

In evaluating the Shannon quantum relative entropy (8.12), the following equalities are assumed.

- **Impossible states:** Impossible states do not contribute to the Shannon quantum relative entropy. In fact, as in (4.10), one sets the equality

$$\lambda \log_2(\mu) = 0 \text{ for } \lambda = 0 \text{ and } \mu \geq 0. \tag{8.21}$$

This implies that for any impossible eigenvector $|u_x\rangle$, which has a zero eigenvalue $\lambda_x = 0$, the corresponding terms $\lambda_x \log_2(\lambda_x)$ and $\lambda_x \log_2(\mu_y)$ in the sums (8.11) and (8.12) equal zero.

- **Black swans:** Black swans cause the Shannon quantum relative entropy to take an infinite value. By definition, any black-swan eigenvector $|v_y\rangle$ of density matrix σ corresponds to a zero eigenvalue, $\mu_y = 0$, and is not in the kernel of matrix ρ. As a consequence, it is not orthogonal to at least one of the eigenvectors in the support supp(ρ). Therefore, there exists at least one index $x \in [N]$ for which the fidelity $\mathrm{F}(|u_x\rangle, |v_y\rangle)$ is strictly positive and the eigenvalue λ_x is also positive. The equality

$$-\lambda \log_2(\mu) = \infty \text{ for } \lambda > 0 \text{ and } \mu = 0, \tag{8.22}$$

which follows (4.11), then implies that the corresponding terms -$\mathrm{F}(|u_x\rangle, |v_y\rangle)\lambda_x$ $\log_2(\mu_y)$ in the sums (8.11) and (8.12) equal infinity.

Example 8.4

For a system over $N = 4$ dimensions, consider the rank-1 density matrix

$$\rho = |\theta\rangle\langle\theta|, \tag{8.23}$$

with the 4×1 pure state vector $|\theta\rangle = \cos(\theta)|2\rangle + \sin(\theta)|3\rangle$, where $0 \leq \theta < \pi$ is a parameter; as well as the density matrix

$$\sigma = \begin{bmatrix} \frac{1}{3} & 0 & 0 & 0 \\ 0 & \frac{1}{3} & 0 & 0 \\ 0 & 0 & \frac{1}{6} & -\frac{1}{6} \\ 0 & 0 & -\frac{1}{6} & \frac{1}{6} \end{bmatrix}. \tag{8.24}$$

Under what conditions on parameter θ is the Shannon quantum relative entropy $\mathrm{D}(\rho||\sigma)$ finite?

The matrix σ has eigenvectors $|v_0\rangle = |0\rangle$, $|v_1\rangle = |1\rangle$, $|v_2\rangle = 1/\sqrt{2}(|2\rangle + |3\rangle)$, and $|v_3\rangle = 1/\sqrt{2}(|2\rangle - |3\rangle)$, and corresponding eigenvalues $\mu_0 = \mu_1 = \mu_3 = 1/3$, and $\mu_2 = 0$. In fact, it can be expressed as

$$\sigma = \frac{1}{3}|v_0\rangle\langle v_0| + \frac{1}{3}|v_1\rangle\langle v_1| + \frac{1}{3}|v_3\rangle\langle v_3|. \tag{8.25}$$

Therefore, the only possible black swan is state $|v_2\rangle$, since it corresponds to a zero eigenvalue, $\mu_2 = 0$, and is thus in the subspace ker(σ). To ensure that it is not a black swan, guaranteeing that the Shannon quantum relative entropy is finite, the state $|v_2\rangle$

must be impossible. This in turn requires the state $|v_2\rangle$ to be orthogonal to the support of matrix ρ.

The support of the density matrix ρ consists of the span of the state $|\theta\rangle$, i.e., $\text{supp}(\rho) = \mathcal{S}(|\theta\rangle)$. Therefore, the Shannon quantum relative entropy $D(\rho||\sigma)$ is finite as long as the state $|v_2\rangle$ is orthogonal to $|\theta\rangle$, i.e.,

$$\langle\theta|v_2\rangle = \frac{1}{\sqrt{2}}\langle\theta|(|2\rangle + |3\rangle) = 0. \tag{8.26}$$

This condition is equivalent to the equality

$$\cos(\theta) + \sin(\theta) = 0, \tag{8.27}$$

which is satisfied for $\theta = 3\pi/4$ in the interval $0 \leq \theta < \pi$, yielding $|\theta\rangle = |v_3\rangle$.

For $\theta = 3\pi/4$, the Shannon quantum relative entropy depends exclusively on the terms in the sums (8.11) and (8.12) involving the only possible state $|\theta\rangle = |v_3\rangle$, yielding

$$D(\rho||\sigma) = -\sum_{y\in[4]} F(|v_3\rangle, |v_y\rangle)\log_2(\mu_y) = \log_2(3). \tag{8.28}$$

Two density matrices ρ and σ can be considered as **maximally distinct** when the supports of the two density matrices are orthogonal. This condition may be written as

$$\text{supp}(\rho) \perp \text{supp}(\sigma), \tag{8.29}$$

and it can be considered as the counterpart of the classical setting with disjoint supports studied in Chapter 4 (see (4.7)). In this situation, every eigenvector of matrix σ in the kernel $\ker(\sigma)$ is a black swan.

Finally, for completeness, one can establish an analogous definition to the hallucinations introduced in Section 4.2 for classical states.

Definition 8.7 Hallucinations
For reference density matrix ρ and density matrix σ with eigendecompositions (8.6) and (8.7), respectively, an eigenvector $|v_x\rangle$ of density matrix σ is a hallucination if: (i) it corresponds to a positive eigenvalue, $\mu_x > 0$, of matrix σ, hence lying in the subspace $\text{supp}(\sigma)$; and (ii) it is not in the subspace $\text{supp}(\rho)$. A hallucination is thus an eigenvector of matrix σ that has positive probability under density matrix σ, and cannot belong to any ensemble realizing the density matrix ρ.

8.2.4 Defining the Shannon Quantum Relative Entropy Using Density Matrices

The definition of the Shannon quantum relative entropy (8.11) relies on the eigendecompositions of the two density matrices. As discussed next, the Shannon quantum relative entropy can also be, more conventionally, defined directly in terms of the density matrices.

To start, the condition for the Shannon quantum relative entropy to be finite – namely the absence of black swans – can be expressed through the supports of the density matrices ρ and σ. Specifically, in a way completely analogous to (4.4) for the Shannon relative entropy, the following result holds.

Key Result 8.1 Absence of black swans

For reference density matrix ρ and density matrix σ, there are no black swans if and only if the support of the density matrix σ includes the support of the reference density matrix ρ, i.e.,

$$\text{no black swans} \iff \text{supp}(\rho) \subseteq \text{supp}(\sigma). \tag{8.30}$$

This claim can readily be proved by noting that absence of black swans is equivalent to the inclusion

$$\ker(\sigma) \subseteq \ker(\rho), \tag{8.31}$$

which is equivalent to (8.30). Indeed, by the condition (8.30), any vector that is orthogonal to supp(σ) – and therefore lies in ker(σ) – must also be orthogonal to supp(ρ), placing it in ker(ρ).

Example 8.5

To illustrate the condition (8.30), take a system with dimension $N = 3$, and assume that matrix ρ has a single positive eigenvalue $\lambda_0 = 1$ with eigenvector $|u_0\rangle$, while matrix σ has two positive eigenvalues $\mu_0 > 0$ and $\mu_1 > 0$ with respective eigenvectors $|v_0\rangle$ and $|v_1\rangle$. As depicted in Figure 8.6(a), the Shannon quantum relative entropy $\mathrm{D}(\rho||\sigma)$ is finite when the support of matrix ρ, which consists of the span $\mathcal{S}(|u_0\rangle)$, lies within the span $\text{supp}(\sigma) = \mathcal{S}(\{|v_0\rangle, |v_1\rangle\})$. Otherwise, as seen in Figure 8.6(b), when the support of matrix ρ does not fall completely inside the support of matrix σ, the Shannon quantum relative entropy $\mathrm{D}(\rho||\sigma)$ is infinite.

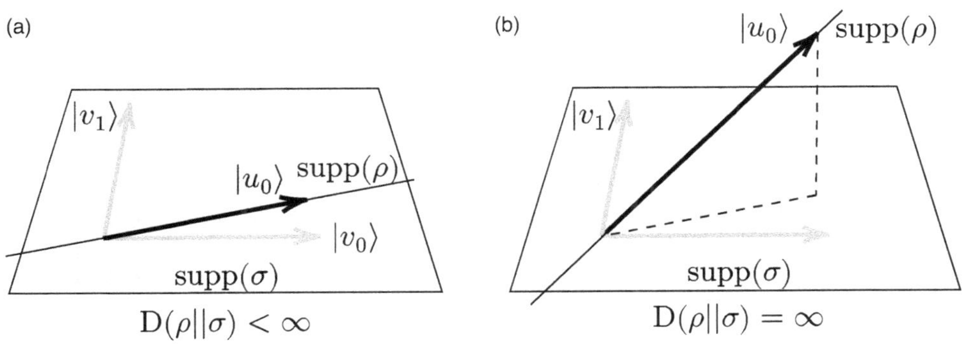

Figure 8.6 Illustration of the condition (8.30) for a system with dimension $N = 3$, in which matrix ρ has a single positive eigenvalue $\lambda_0 = 1$, while matrix σ has two positive eigenvalues $\mu_0 > 0$ and $\mu_1 > 0$. The Shannon quantum relative entropy is finite if the condition (8.30) is met (a); while a violation of this condition yields an infinite Shannon quantum relative entropy (b).

Using the condition (8.30), an equivalent characterization of the Shannon quantum relative entropy is given as follows.

Key Result 8.2 Shannon quantum relative entropy using density matrices
For any two $N \times N$ density matrices ρ and σ, the Shannon quantum relative entropy in (8.11) can equivalently be written as

$$\mathrm{D}(\rho||\sigma)=\begin{cases}\mathrm{tr}\left(\rho\left(\log_2(\rho)-\log_2(\sigma)\right)\right) & \text{if } \mathrm{supp}(\rho)\subseteq\mathrm{supp}(\sigma),\\ \infty & \text{otherwise,}\end{cases} \tag{8.32}$$

where the function $\log_2(\cdot)$ of a normal matrix is evaluated as discussed in Section 6.6.5.

The expression (8.32) can be understood by using the correspondence between trace and sum described in Table 7.1. In fact, the Shannon relative entropy for two distributions p and q is defined as $\mathrm{D}(p||q)=\sum_{x\in[N]}p_x(\log_2(p_x)-\log_2(q_x))$. Replacing the sum with the trace and the probability distributions p and q with the density matrices ρ and σ yields (8.32). The equality between (8.11) and (8.32) is proved in Appendix 8.A.

Finally, the condition (8.30) for a finite Shannon quantum relative entropy can be expressed using the projection matrices into the supports of the two density matrices (see Section 6.5.4). To elaborate on this, define as Π_ρ and Π_σ the **projection matrices** into the spans of the eigenvectors of matrices ρ and σ with strictly positive eigenvalues. Using the eigendecompositions (8.1) and (8.2), these definitions are formalized as

$$\Pi_\rho = \sum_{x\in[N]:\ \lambda_x>0} |u_x\rangle\langle u_x| \tag{8.33}$$

and

$$\Pi_\sigma = \sum_{x\in[N]:\ \mu_x>0} |v_x\rangle\langle v_x|, \tag{8.34}$$

respectively. With these definitions, the condition (8.30) is equivalent to the equality

$$\Pi_\sigma\Pi_\rho = \Pi_\rho, \tag{8.35}$$

i.e., $(I-\Pi_\sigma)\Pi_\rho = 0$. In fact, under the condition (8.30), any vector in the support $\mathrm{supp}(\rho)$ is also in the subspace $\mathrm{supp}(\sigma)$. This implies that given any vector u, one must have the equality $\Pi_\sigma\Pi_\rho u = \Pi_\rho u$, since vector $\Pi_\rho u$ is in the subspace $\mathrm{supp}(\rho)$, and thus also in $\mathrm{supp}(\sigma)$.

Example 8.6

For the two density matrices ρ and σ in (8.23) and (8.24) with dimension $N = 4$, this example identifies conditions under which the Shannon quantum relative entropy $\mathrm{D}(\rho||\sigma)$ is finite by using the equality (8.35). To this end, evaluate the projection matrices as $I - \Pi_\sigma = |v_2\rangle\langle v_2|$, with $|v_2\rangle = 1/\sqrt{2}(|2\rangle + |3\rangle)$, and $\Pi_\rho = |\theta\rangle\langle\theta|$. Then, the equality (8.35) amounts to

$$(I - \Pi_\sigma)\Pi_\rho = \langle v_2|\theta\rangle |v_2\rangle\langle\theta| = 0. \tag{8.36}$$

This equality holds if and only if condition (8.27) is verified, confirming the results in Example 8.4.

8.2.5 Non-Negativity of the Quantum Relative Entropy

The Shannon quantum relative entropy satisfies the following key property, which is proved in Appendix 8.B.

Key Result 8.3 Non-negativity and minimum-value property of the Shannon quantum relative entropy

For any two $N \times N$ density matrices ρ and σ, the Shannon quantum relative entropy is non-negative,

$$\mathrm{D}(\rho||\sigma) \geq 0, \tag{8.37}$$

with equality $\mathrm{D}(\rho||\sigma) = 0$ holding if and only if the density matrices are identical, i.e., $\rho = \sigma$.

Example 8.7

Section 7.12.2 introduced coherence measures for a density matrix ρ. Specifically, the coherence of a state ρ with respect to an orthonormal basis $\mathcal{M} = \{|v_x\rangle\}_{x\in[N]}$ was quantified via the distance between ρ and the set of coherence-free states. Using as measure of distance the Shannon quantum relative entropy, this approach yields the coherence measure

$$\mathrm{K}(\rho|\mathcal{M}) = \min_{\{\lambda_x \geq 0\}_{x\in[N]}:\ \sum_{x\in[N]} \lambda_x = 1} \mathrm{D}\left(\rho \,\middle\|\, \sum_{x\in[N]} \lambda_x |v_x\rangle\langle v_x|\right), \tag{8.38}$$

where the minimization is over all the non-negative eigenvalues $\{\lambda_x\}_{x\in[N]}$ of a density matrix $\sum_{x\in[N]} \lambda_x |v_x\rangle\langle v_x|$, which corresponds to the most general form for an incoherent state with respect to basis $\mathcal{M}$. As an example, for the computational basis $\mathcal{M}$, the problem (8.38) seeks for the diagonal matrix that is closest to state ρ in terms of the Shannon quantum relative entropy.

As shown in Appendix 8.C, the quantity (8.38), known as the **Shannon relative entropy of coherence**, can be evaluated as

$$\begin{aligned}\mathrm{K}(\rho|\mathcal{M}) &= \mathrm{D}(\rho||\mathcal{M}(\rho)) \\ &= \mathrm{H}_\alpha(\rho|\mathcal{M}) - \mathrm{H}_\alpha(\rho),\end{aligned} \tag{8.39}$$

where $\mathcal{M}(\rho) = \sum_{x\in[N]} \Pi_x \rho \Pi_x$ with $\{\Pi_x = |v_x\rangle\langle v_x|\}_{x\in[N]}$ denotes the post-measurement state for a non-selective measurement in basis $\mathcal{M}$ (see Section 7.11). Note that, for the computational basis $\mathcal{M}$, this result implies that the closest diagonal state, as per problem (8.38), has diagonal elements equal to matrix ρ.

Interestingly, the Shannon relative entropy of coherence coincides with the measure (7.138) introduced in Chapter 7. Therefore, this measure can be interpreted both as a resource-theoretic metric and as a measure of disturbance caused by a measurement in setting $\mathcal{M}$, as discussed in Section 7.12.2.

8.2.6 Monotonicity of the Shannon Quantum Entropy

The Shannon quantum relative entropy supports the formulation of a monotonicity property for the Shannon quantum entropy.

Key Result 8.4 Monotonicity property of the Shannon quantum entropy
For any density matrix ρ, the Shannon quantum entropy $\mathrm{H}(\rho)$ can be written as

$$\begin{aligned}\mathrm{H}(\rho) &= \mathrm{H}(\pi) - \mathrm{D}(\rho||\pi) \\ &= \log_2(N) - \mathrm{D}(\rho||\pi). \end{aligned} \tag{8.40}$$

Therefore, as the Shannon quantum relative entropy $\mathrm{D}(\rho||\pi)$ between the density ρ and the maximally mixed state π decreases to its minimum value of zero, the Shannon quantum entropy approaches the maximum value $\mathrm{H}(\pi) = \log_2(N)$.

The equality (8.40) can easily be verified by direct calculation as

$$\begin{aligned}\mathrm{D}(\rho||\pi) &= \mathrm{tr}(\rho\log_2(\rho)) - \mathrm{tr}(\rho\log_2(\pi)) \\ &= \mathrm{tr}(\rho\log_2(\rho)) + \log_2(N) \\ &= -\mathrm{H}(\rho) + \log_2(N). \end{aligned} \tag{8.41}$$

8.2.7 Other Definitions of the Shannon Quantum Relative Entropy ⏭

This section ends with a few notes on potential alternative definitions for the Shannon quantum relative entropy. To this end, recall that in Chapter 7 the Shannon quantum entropy was defined as the minimum over all complete measurement settings $\mathcal{M}$ of the Shannon measured entropy, i.e.,

$$\mathrm{H}(\rho) = \min_{\mathcal{M}} \mathrm{H}(\rho|\mathcal{M}) = \min_{\mathcal{M}} \mathrm{H}(p(\rho|\mathcal{M})). \tag{8.42}$$

Following a similar approach, one may attempt to define a quantum version of the Shannon relative entropy as the quantity

$$\tilde{\mathrm{D}}(\rho||\sigma) = \max_{\mathcal{M}} \mathrm{D}(p(\rho|\mathcal{M})||p(\sigma|\mathcal{M})). \tag{8.43}$$

This corresponds to the maximal Shannon relative entropy between the measurement distributions $p(\rho|\mathcal{M})$ and $p(\sigma|\mathcal{M})$ over all possible measurement settings $\mathcal{M}$. For commuting density matrices, this alternative definition indeed coincides with the Shannon quantum relative entropy (8.8). However, for general non-commuting density matrices, the quantity $\tilde{\mathrm{D}}(\rho||\sigma)$ is not equal to the Shannon quantum relative entropy. Furthermore, a closed-form expression for $\tilde{\mathrm{D}}(\rho||\sigma)$ is not known (see Recommended Resources).

As another potential quantum extension of the Shannon relative entropy, one could take

$$\tilde{\mathrm{D}}(\rho||\sigma) = \mathrm{tr}\left(\rho \log_2 \left(\sqrt{\rho}\sigma^{-1}\sqrt{\rho}\right)\right), \tag{8.44}$$

which is also known as the **Blavkin–Staszewski** definition. Note that the matrix $\sqrt{\rho}\sigma^{-1}\sqrt{\rho}$ inside the logarithm is positive semi-definite, and hence also normal, justifying the application of the logarithm to it. Observe also that this would not be the case with a matrix of the form $\rho\sigma^{-1}$, which one may expect in analogy to the classical case. Like (8.43), the definition (8.44) recovers (8.8) for commuting density matrices, but is otherwise distinct (see Problem 8.6).

8.3 Tsallis Quantum Relative Entropy ⏭

This section generalizes the Shannon quantum relative entropy to the family of Tsallis quantum relative entropies. The key element of this definition is a quantum extension of the Tsallis relative sum introduced in Chapter 4, which will also be central to the definition of the Rényi quantum relative entropy in the next section.

8.3.1 Defining the Tsallis Relative Trace

The Tsallis relative trace is defined as follows.

Definition 8.8 Tsallis relative trace using density matrices

Given two $N \times N$ density matrices ρ and σ, the Tsallis relative trace of order $\alpha \geq 0$ with $\alpha \neq 1$ is defined as

$$\alpha\text{-Tsallis relative trace} = \mathrm{tr}(\rho^{\alpha}\sigma^{1-\alpha}). \tag{8.45}$$

According to this definition, like the Shannon quantum relative entropy, the Tsallis relative trace depends on both the eigenvalues and the eigenvectors of the two density matrices. This dependence is formalized by the following equivalent definition, which uses the general eigendecompositions (8.1) and (8.2).

Table 8.1 Interpretation of the Tsallis relative trace as a function of the order α

Order	Tsallis relative trace	$1/(\alpha - 1)$
$0 \leq \alpha < 1$	Similarity measure	Negative
$\alpha > 1$	Dissimilarity measure	Positive

Key Result 8.5 Tsallis relative trace using eigendecompositions
Given two $N \times N$ density matrices ρ and σ, the Tsallis relative trace of order $\alpha \geq 0$ in (8.45) can equivalently be written as

$$\mathrm{tr}(\rho^{\alpha}\sigma^{1-\alpha}) = \sum_{x \in [N]} \sum_{y \in [N]} \mathrm{F}(|u_x\rangle, |v_y\rangle)\lambda_x^{\alpha}\mu_y^{1-\alpha}. \tag{8.46}$$

Accordingly, the Tsallis relative trace weighs each term $\lambda_x^{\alpha}\mu_y^{1-\alpha}$ in (8.46) by the fidelity $\mathrm{F}(|u_x\rangle, |v_y\rangle)$ between the corresponding eigenvectors.

The Tsallis relative trace recovers the Tsallis relative sum introduced in Section 4.4.4 for commuting density matrices. In fact, in this case, the Tsallis relative trace equals the Tsallis relative sum between the respective vectors of eigenvalues λ and μ, i.e.,

$$\mathrm{tr}(\rho^{\alpha}\sigma^{1-\alpha}) = \sum_{x \in [N]} \lambda_x^{\alpha}\mu_x^{1-\alpha}. \tag{8.47}$$

Furthermore, by the equalities (4.45) and (4.46) in Section 4.4.4, black swans cause the Tsallis relative trace to take an infinite value only for $\alpha > 1$, while they have a finite contribution for $0 \leq \alpha < 1$. In particular, for $0 \leq \alpha < 1$, the Tsallis relative trace is always finite, while for $\alpha \geq 1$, the Tsallis relative trace is only finite under the same condition (8.30) as for the Shannon quantum relative entropy.

In a manner analogous to the Tsallis relative sum studied in Section 4.4.4, as summarized in Table 8.1, the Tsallis relative trace serves as a measure of the similarity of the density matrices ρ and σ for $0 \leq \alpha < 1$; while, for $\alpha > 1$, it is a measure of dissimilarity.

Example 8.8

Consider again the 2×2 density matrices (8.13) and (8.14). Using (8.46), the Tsallis relative trace is evaluated as

$$\begin{aligned}\mathrm{tr}(\rho^{\alpha}\sigma^{1-\alpha}) =& (\cos(\theta))^2\lambda_0^{\alpha}\mu_0^{1-\alpha} + (\sin(\theta))^2\lambda_0^{\alpha}\mu_1^{1-\alpha} \\ &+ (\sin(\theta))^2\lambda_1^{\alpha}\mu_0^{1-\alpha} + (\cos(\theta))^2\lambda_1^{\alpha}\mu_1^{1-\alpha}.\end{aligned} \tag{8.48}$$

For a numerical example, fixing $\mu_1 = 0.3$, Figures 8.7 and 8.8 show the Tsallis relative trace as a function of λ_1 for two different values of α. Consider first the choice

$\alpha = 0.5$ illustrated in Figure 8.7, for which the Tsallis relative trace is a measure of similarity. If the density matrices commute, i.e., for $\theta = 0$ and $\theta = \pi/2$, the Tsallis relative trace attains its maximum value of 1 when the two density matrices are equal. This occurs for $\lambda_1 = \mu_1 = 0.3$ when $\theta = 0$ and for $\lambda_1 = \mu_0 = 0.7$ when $\theta = \pi/2$. For other values of θ, such as $\theta = \pi/4$, the eigenbases are not compatible, and the Tsallis relative trace cannot attain the maximum value of 1.

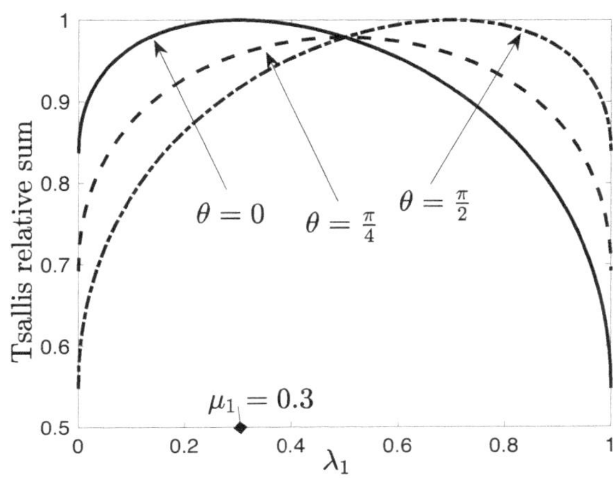

Figure 8.7 Tsallis relative trace for the density matrices (8.13) and (8.14) as a function of λ_1 with $\mu_1 = 0.3$ and order $\alpha = 0.5$.

As illustrated in Figure 8.8, take now the complementary case $\alpha = 2$, for which the Tsallis relative trace represents a measure of dissimilarity. Since both eigenvalues μ_0 and μ_1 are positive, the Tsallis relative trace is always finite in this case. Moreover, when the two density matrices commute, i.e., for $\theta = 0$ and $\theta = \pi/2$, the minimum value of the Tsallis relative trace, equal to 1, is attained when the two density matrices are equal. In contrast, for $\theta = \pi/4$, the two density matrices do not commute, and the minimum value of 1 is not attainable.

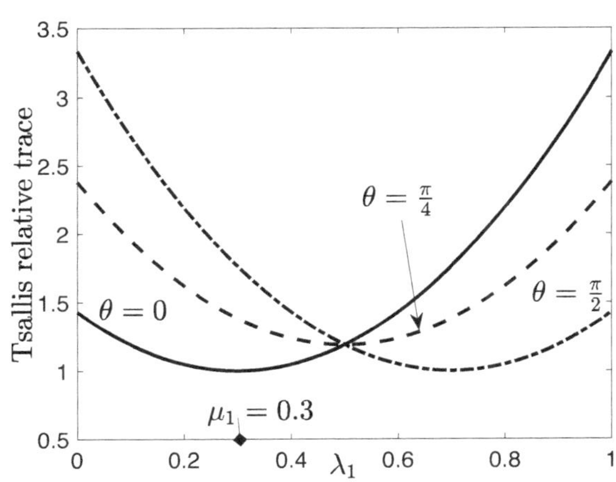

Figure 8.8 Tsallis relative trace for the density matrices (8.13) and (8.14) as a function of λ_1 with $\mu_1 = 0.3$ and order $\alpha = 2$.

What changes if, instead of varying the eigenvalues of matrix ρ, one varies the matrix σ? Figure 8.9 illustrates the Tsallis relative trace for $\alpha = 2$ when fixing $\lambda_1 = 0.3$

and varying μ_1. In this case, the general conclusions discussed for Figure 8.8 are still valid, with the important caveat that now the Tsallis relative trace can take an infinite value. In particular, this happens any time $\mu_1 = 0$ or $\mu = 1$, since, with this choice of μ_1, the support of density matrix ρ is not included in the support of matrix σ, i.e., $\text{supp}(\rho) \not\subseteq \text{supp}(\sigma)$.

Figure 8.9 Tsallis relative trace for the density matrices (8.13) and (8.14) as a function of μ_1 with $\lambda_1 = 0.3$ and order $\alpha = 2$.

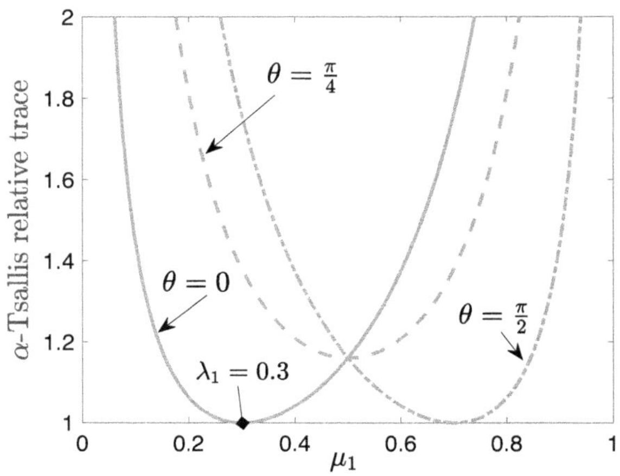

8.3.2 Bounds on the Tsallis Relative Trace

The following bounds substantiate the interpretations of the Tsallis relative trace in Table 8.1 focusing on strictly positive orders $\alpha > 0$.

Key Result 8.6 Bounds on the Tsallis relative trace
For $0 < \alpha < 1$, the Tsallis relative trace satisfies the inequalities

$$\underbrace{0}_{\text{supp}(\rho)\perp\text{supp}(\sigma)} \leq \text{tr}(\rho^\alpha \sigma^{1-\alpha}) \leq \underbrace{1}_{\rho=\sigma}, \tag{8.49}$$

with the lower bound attained for density matrices with orthogonal supports and the upper bound attained with identical density matrices $\rho = \sigma$.

For $\alpha > 1$, the Tsallis relative trace satisfies the inequalities

$$\underbrace{1}_{\rho=\sigma} \leq \text{tr}(\rho^\alpha \sigma^{1-\alpha}) \leq \underbrace{\infty}_{\text{supp}(\rho)\not\subseteq\text{supp}(\sigma)}, \tag{8.50}$$

with the lower bound attained with identical density matrices $\rho = \sigma$ and the upper bound achieved when the support of density matrix σ does not include the support of density matrix ρ.

These bounds confirm that the Tsallis relative trace with $0 < \alpha < 1$ is a measure of similarity. In fact, it attains the minimum value for density matrices that are, in a sense, maximally dissimilar, having orthogonal supports. Conversely, the maximum value is obtained when the density matrices are equal. In contrast, with $\alpha > 1$, the

Tsallis relative trace is a measure of dissimilarity, taking the minimum value for identical states, while attaining the maximum value when there are black swans.

Example 8.9

Consider the two density matrices $\rho = |0\rangle\langle 0|$ and $\sigma = 0.1|0\rangle\langle 0| + 0.9|1\rangle\langle 1|$. For any order $\alpha > 1$, the Tsallis relative trace, a measure of dissimilarity, equals

$$\mathrm{tr}(\rho^{\alpha}\sigma^{1-\alpha}) = 0.1^{1-\alpha}. \tag{8.51}$$

Reversing the roles of the two density matrices, one gets

$$\mathrm{tr}(\sigma^{\alpha}\rho^{1-\alpha}) = \infty, \tag{8.52}$$

given the condition $\mathrm{supp}(\sigma) \not\subseteq \mathrm{supp}(\rho)$. This example highlights the fact that the Tsallis relative trace is not symmetric with respect to the density matrices ρ and σ.

To address the special case of a zero order, $\alpha = 0$, note first that the density matrix ρ^0 equals the projection matrix Π_ρ (see Problem 8.11). Therefore, the Tsallis relative trace of order $\alpha = 0$ can be written as

$$\mathrm{tr}(\rho^{0}\sigma) = \mathrm{tr}(\Pi_\rho\sigma). \tag{8.53}$$

By Table 8.1, this is a measure of similarity, and it satisfies the following bounds.

Key Result 8.7 Bounds on the Tsallis relative sum for $\alpha = 0$

The Tsallis relative trace of order $\alpha = 0$ satisfies the inequalities

$$\underbrace{0}_{\mathrm{supp}(\rho)\perp\mathrm{supp}(\sigma)} \leq \mathrm{tr}(\Pi_\rho\sigma) \leq \underbrace{1}_{\mathrm{supp}(\sigma)\subseteq\mathrm{supp}(\rho)}, \tag{8.54}$$

where the lower bound is attained when the density matrices have orthogonal supports and the upper bound is obtained when the support of density matrix σ is included in the support of density matrix ρ.

It is worth emphasizing the difference between the upper bound in (8.54) for $\alpha = 0$ and the corresponding upper bound (8.49), which holds for $0 < \alpha < 1$. In the latter case, the maximum value of similarity, equal to 1, is only attained when the two density matrices are identical. In contrast, with $\alpha = 0$, the maximum value of 1 is attained as long as the support of density matrix σ is included in the support of density matrix ρ. Following the definition given in Section 8.2.3, this condition indicates that there are no hallucinations. Thus, the Tsallis relative trace of order $\alpha = 0$ provides a weaker form of similarity than the other values of order in range $0 < \alpha < 1$.

Example 8.10

For the density matrices $\rho = |0\rangle\langle 0|$ and $\sigma = 0.1|0\rangle\langle 0| + 0.9|1\rangle\langle 1|$, the measure of similarity given by the Tsallis relative trace of order $\alpha = 0.5$ is symmetric and given by

$$\mathrm{tr}(\sqrt{\rho}\sqrt{\sigma}) = \mathrm{tr}(\sqrt{\sigma}\sqrt{\rho}) = \sqrt{0.1}; \tag{8.55}$$

while with $\alpha = 0$ one gets the asymmetric similarity measures

$$\mathrm{tr}(\Pi_\rho \sigma) = 0.1 \tag{8.56}$$

and

$$\mathrm{tr}(\Pi_\sigma \rho) = \mathrm{tr}(\rho) = 1. \tag{8.57}$$

8.3.3 Tsallis Quantum Relative Entropy

In a manner completely analogous to its classical counterpart described in Chapter 4, the Tsallis quantum relative entropy of order $\alpha \geq 0$ is obtained by applying the linear function (4.53) to the Tsallis relative trace, yielding the following definition.

Definition 8.9 Tsallis quantum relative entropy
The Tsallis quantum relative entropy of order $\alpha \geq 0$ is defined as

$$\mathrm{D}_\alpha^{\mathrm{T}}(\rho||\sigma) = \begin{cases} \frac{1_2}{\alpha-1}(\mathrm{tr}(\rho^\alpha \sigma^{1-\alpha}) - 1) & \text{if } 0 \leq \alpha < 1, \\ & \text{or if } \alpha > 1 \text{ and } \mathrm{supp}(\rho) \subseteq \mathrm{supp}(\sigma)\ , \\ \infty & \text{otherwise} \end{cases} \tag{8.58}$$

and, by continuity, for $\alpha = 1$, as

$$\mathrm{D}_1^{\mathrm{T}}(\rho||\sigma) = \mathrm{D}(\rho||\sigma). \tag{8.59}$$

Using the α-log-loss, the Tsallis quantum relative entropy (8.58) can also be written as

$$\mathrm{D}_\alpha^{\mathrm{T}}(\rho||\sigma) = \sum_{x\in[N]} \sum_{y\in[N]} \mathrm{F}(|u_x\rangle, |v_y\rangle)\lambda_x \log_{2,\alpha}\left(\frac{\lambda_x}{\mu_y}\right). \tag{8.60}$$

Example 8.11

For the 2×2 density matrices (8.13) and (8.14), the Tsallis quantum relative entropy can be computed directly using the Tsallis trace (8.48) into (8.58). The result is plotted in Figure 8.10 for a fixed $\lambda_1 = 0.3$, while varying μ_1 for $\alpha = 0.5$ (a) and $\alpha = 2$ (b).

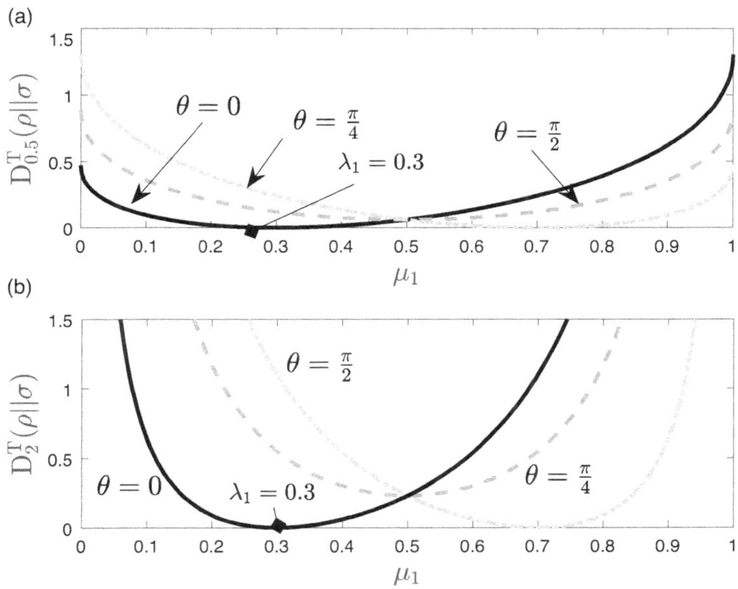

Figure 8.10 Tsallis quantum relative entropy for the density matrices (8.13) and (8.14) as a function of μ_1 with $\lambda_1 = 0.3$, as well as $\alpha = 0.5$ (a) and $\alpha = 2$ (b).

For $\theta = 0$ and $\theta = \pi/2$, the density matrices commute, and the Tsallis quantum relative entropy takes the minimum value of zero when the two densities are equal, while for $\theta = \pi/4$ the minimum value of zero cannot be attained.

For $\alpha = 0.5$, the Tsallis quantum relative entropy is always finite. In contrast, for $\alpha = 2$, it goes to infinity when $\mu_0 = 0$ or $\mu_1 = 1$, in which case condition $\text{supp}(\rho) \subseteq \text{supp}(\sigma)$ is not satisfied.

Key Result 8.8 Non-negativity and minimum-value property of the Tsallis quantum relative entropy

The Tsallis quantum relative entropy of any order $\alpha \geq 0$ is non-negative,

$$D^T_\alpha(\rho||\sigma) \geq 0. \tag{8.61}$$

Furthermore, for all positive orders, $\alpha > 0$, the Tsallis quantum relative entropy equals zero if and only if the two density matrices are equal, i.e., if and only if the equality $\rho = \sigma$ holds.

The property (8.61) can be derived from the bounds (8.49) and (8.50) on the Tsallis relative trace, and an alternative proof can be found in Appendix 8.D. The special case $\alpha = 0$ can be addressed by using the bounds (8.54), and it will be discussed in Section 8.3.4.

The following property is a direct consequence of the expression (8.60) and of the monotonicity property of the α-logarithm with respect to parameter α (see Appendix 3.A).

Key Result 8.9 The Tsallis quantum relative entropy increases with the order α
Given any fixed $N \times N$ density matrices ρ and σ, the Tsallis quantum relative entropy $\mathrm{D}_\alpha^\mathrm{T}(\rho||\sigma)$ is an increasing function of the order $\alpha \geq 0$, i.e., the inequality

$$\mathrm{D}_{\alpha'}^\mathrm{T}(\rho||\sigma) \geq \mathrm{D}_\alpha^\mathrm{T}(\rho||\sigma) \tag{8.62}$$

holds for any two orders α' and α with $\alpha' \geq \alpha$. Furthermore, for $0 \leq \alpha < 1$, the Tsallis quantum entropy is always finite, and it satisfies the upper bound

$$\mathrm{D}_\alpha^\mathrm{T}(\rho||\sigma) \leq \frac{1_2}{1-\alpha}, \tag{8.63}$$

with equality if the supports of the two density matrices are orthogonal.

8.3.4 Special Cases of the Tsallis Quantum Relative Entropy

This subsection reviews some special cases of the Tsallis quantum relative entropy.

Definition 8.10 Tsallis quantum relative entropy of order $\alpha = 0$, aka Tsallis quantum min-relative entropy
The Tsallis quantum relative entropy of order $\alpha = 0$ is obtained from the general definition (8.58) as

$$\mathrm{D}_0^\mathrm{T}(\rho||\sigma) = 1_2(1 - \mathrm{tr}(\Pi_\rho \sigma)). \tag{8.64}$$

By the bounds (8.54), the Tsallis quantum min-relative entropy obtains the minimum value of zero any time the condition $\mathrm{supp}(\sigma) \subseteq \mathrm{supp}(\rho)$ is satisfied, i.e., when there are no hallucinations:

$$\mathrm{D}_0^\mathrm{T}(\rho||\sigma) = 0 \Longleftrightarrow \mathrm{supp}(\sigma) \subseteq \mathrm{supp}(\rho). \tag{8.65}$$

Furthermore, the maximum value $\mathrm{D}_0^\mathrm{T}(\rho||\sigma) = 1_2$ is attained when the supports of the two density matrices are orthogonal.

Definition 8.11 Tsallis quantum relative entropy of order $\alpha = 0.5$, aka Tsallis quantum square Hellinger distance
The Tsallis quantum relative entropy of order $\alpha = 0.5$ is given by

$$\mathrm{D}_{1/2}^\mathrm{T}(\rho,\sigma) = 1_2 \cdot 2\left(1 - \mathrm{tr}(\sqrt{\rho}\sqrt{\sigma})\right). \tag{8.66}$$

The square Hellinger distance is symmetric with respect to the densities ρ and σ, as emphasized by the notation $\mathrm{D}^{\mathrm{T}}_{1/2}(\rho, \sigma)$, which replaces the double bar with a comma between the two densities. The quantity $\mathrm{tr}(\sqrt{\rho}\sqrt{\sigma})$ is known as the **Tsallis fidelity** and is discussed further in Section 8.4.

8.3.5 Monotonicity of the Tsallis Quantum Entropy

In a manner completely analogous to the classical case studied in Section 4.4.9, the following result holds.

> **Key Result 8.10 Monotonicity property of the Tsallis quantum entropy**
> For any density matrix ρ, the Tsallis quantum entropy $\mathrm{H}^{\mathrm{T}}_{\alpha}(\rho)$ of any order $\alpha \geq 0$ can be written as
> $$\begin{aligned}\mathrm{H}^{\mathrm{T}}_{\alpha}(\rho) &= \mathrm{H}^{\mathrm{T}}_{\alpha}(\pi) - N^{1-\alpha}\mathrm{D}^{\mathrm{T}}_{\alpha}(\rho||\pi) \\ &= -\log_{2,\alpha}\left(\frac{1}{N}\right) - N^{1-\alpha}\mathrm{D}^{\mathrm{T}}_{\alpha}(\rho||\pi). \end{aligned} \tag{8.67}$$

To prove this result, one can evaluate the Tsallis quantum relative entropy $\mathrm{D}^{\mathrm{T}}_{\alpha}(\rho||\pi)$ directly by writing the maximally mixed state as

$$\pi = \frac{1}{N}\sum_{x\in[N]} |u_x\rangle\langle u_x|, \tag{8.68}$$

where $\{|u_x\rangle\}_{x\in[N]}$ is an eigenbasis of density matrix ρ. Using the definition (8.60), this yields the equality

$$\mathrm{D}^{\mathrm{T}}_{\alpha}(\rho||\pi) = \frac{1_2}{\alpha - 1}\left(\frac{1}{N^{1-\alpha}}\sum_{x\in[N]} p_x^{\alpha} - 1\right), \tag{8.69}$$

which implies (8.67).

8.4 Fidelity between Mixed States ⏭

This section elaborates on the notion of fidelity between mixed states.

8.4.1 Fidelity: A Recap

So far, two notions of fidelity have been introduced in this book. First, in Section 4.5, the fidelity between two $N \times 1$ probability vectors p and q was defined as the Tsallis relative sum of order $\alpha = 0.5$, i.e.,

$$\mathrm{F}(p, q) = \sum_{x\in[N]} \sqrt{p_x q_x}, \tag{8.70}$$

which is a measure of similarity between probability distributions. It was also observed that, more generally, the Tsallis sum with $0 \leq \alpha < 1$ serves as a measure of similarity between distributions. Thus, one may also use alternative definitions of fidelity such as the Chernoff fidelity $\mathrm{F}^{\mathrm{C}}(p, q) = \min_{\alpha:\, 0 \leq \alpha \leq 1} \sum_{x \in [N]} p_x^{\alpha} q_x^{1-\alpha}$.

Second, in Section 5.6, the fidelity between two $N \times 1$ quantum pure states $|\psi\rangle$ and $|\phi\rangle$ was defined as

$$\mathrm{F}(|\psi\rangle, |\phi\rangle) = |\langle\psi|\phi\rangle|^2, \tag{8.71}$$

which is a measure of similarity between quantum states.

The measures (8.70) and (8.71) share the property of being bounded in the interval between 0 and 1, with the minimum value 0 indicating maximally dissimilar distributions or states and the maximum value 1 identifying identical distributions or states. Specifically, a zero fidelity $\mathrm{F}(p, q)$ is attained if the two distributions p and q have disjoint supports, while a zero fidelity $\mathrm{F}(|\psi\rangle, |\phi\rangle)$ is attained when the two states are orthogonal. This section elaborates on the problem of generalizing the notions of fidelity studied so far to mixed states ρ and σ.

8.4.2 Fidelity for Quantum Mixed States

To start, the following is a list of desirable properties that a fidelity measure $\mathrm{F}(\rho, \sigma)$ between quantum mixed states ρ and σ should ideally satisfy.

- **Symmetry:** The fidelity should not depend on the order of the two densities:

$$\mathrm{F}(\rho, \sigma) = \mathrm{F}(\sigma, \rho). \tag{8.72}$$

- **Bounds:** The fidelity should be bounded in the interval between 0 and 1:

$$0 \leq \mathrm{F}(\rho, \sigma) \leq 1. \tag{8.73}$$

- **Minimum value:** The equality $\mathrm{F}(\rho, \sigma) = 0$ should hold if the two density matrices are maximally dissimilar in the sense that they have orthogonal supports, $\mathrm{supp}(\rho) \perp \mathrm{supp}(\sigma)$.
- **Maximum value:** The equality $\mathrm{F}(\rho, \sigma) = 1$ should hold if the two density matrices are equal:

$$\mathrm{F}(\rho, \rho) = 1. \tag{8.74}$$

- **Reduction to the classical fidelity:** Given commuting density matrices ρ and σ, with eigendecompositions as in (8.6) and (8.7), the fidelity should be equal to the classical fidelity (8.70) between the corresponding vectors of eigenvalues λ and μ, i.e.,

$$\mathrm{F}(\rho, \sigma) = \mathrm{F}(\lambda, \mu) \text{ if } \rho \text{ and } \sigma \text{ commute.} \tag{8.75}$$

- **Reduction to the quantum fidelity for pure states:** Given pure states $\rho = |\psi\rangle\langle\psi|$ and $\sigma = |\phi\rangle\langle\phi|$, the fidelity should coincide with (8.71), i.e.,

$$\mathrm{F}(|\psi\rangle\langle\psi|, |\phi\rangle\langle\phi|) = \mathrm{F}(|\psi\rangle, |\phi\rangle). \tag{8.76}$$

It is noted that these conditions will be interpreted and applied somewhat flexibly in what follows. For example, a quantum fidelity may also be defined that recovers other classical fidelity measure for commuting states, such as the Chernoff fidelity, instead of satisfying the equality (8.76). As another example, later on it will be useful to allow for the equality $\mathrm{F}(\rho, \sigma) = \mathrm{F}(\lambda, \mu)^2$ for commuting matrices, deviating from (8.75).

It turns out that there are many ways to define fidelity measures that satisfy these conditions, while differing in terms of other properties, such as concavity, monotonicity under transformations (see Part III), and tensorization (see Part VI). The rest of this section focuses on the following standard definitions of fidelity:

- Tsallis fidelity $\mathrm{F}^{\mathrm{T}}(\rho, \sigma) = \mathrm{tr}(\sqrt{\rho}\sqrt{\sigma})$;
- Chernoff fidelity $\mathrm{F}^{\mathrm{C}}(\rho, \sigma) = \min_{\alpha:\, 0 \leq \alpha \leq 1} \mathrm{tr}(\rho^{\alpha}\sigma^{1-\alpha})$;
- Uhlmann–Josza fidelity $\mathrm{F}^{\mathrm{UJ}}(\rho, \sigma) = ||\sqrt{\rho}\sqrt{\sigma}||_1^2$, with $||\cdot||_1$ being the 1-Schatten norm.

Following a common convention, when not stated otherwise, the term *fidelity* will be used to refer to the Uhlmann–Josza fidelity.

8.4.3 Tsallis Fidelity

The **Tsallis fidelity** coincides with the Tsallis relative trace of order $\alpha = 0.5$, i.e.,

$$\begin{aligned} \mathrm{F}^{\mathrm{T}}(\rho, \sigma) &= \mathrm{tr}(\sqrt{\rho}\sqrt{\sigma}) \\ &= \sum_{x \in [N]} \sum_{y \in [N]} \mathrm{F}(|u_x\rangle, |v_y\rangle)\sqrt{\lambda_x \mu_y}, \end{aligned} \tag{8.77}$$

where the second equality follows from (8.46). The Tsallis fidelity is also referred to as the affinity fidelity or A-fidelity, and it is common in some references to add a square root to the definition given in (8.77). When squared, the probability (8.77) is also known as the **Holevo fidelity**. One can check directly that the Tsallis fidelity $\mathrm{F}^{\mathrm{T}}(\rho, \sigma)$ satisfies all the properties listed in Section 8.4.2 (see Problem 8.13).

8.4.4 Chernoff Fidelity

In a manner similar to the classical case, the **Chernoff fidelity** between density matrices ρ and σ is defined as

$$\mathrm{F}^{\mathrm{C}}(\rho, \sigma) = \min_{\alpha:\, 0 \leq \alpha \leq 1} \mathrm{tr}(\rho^{\alpha}\sigma^{1-\alpha}). \tag{8.78}$$

Like the Tsallis fidelity, the Chernoff fidelity also satisfies the properties listed in Section 8.4.2, with the caveat that, if the density matrices commute, the condition (8.75) is satisfied with the Chernoff fidelity in lieu of the fidelity (8.70).

By the definition of the Chernoff fidelity $\mathrm{F}^{\mathrm{C}}(\rho, \sigma)$, the inequality

$$\mathrm{F}^{\mathrm{C}}(\rho, \sigma) \leq \mathrm{F}^{\mathrm{T}}(\rho, \sigma) \tag{8.79}$$

holds, with equality if and only if $\alpha = 0.5$ is a solution of the minimization problem (8.78). The choice $\alpha = 0.5$ is indeed optimal in some important special cases (see Recommended Resources).

Example 8.12

This example highlights a potential advantage of the Chernoff fidelity over the Tsallis fidelity in terms of a specific operational interpretation. Consider the problem of distinguishing a pure state $\rho = |\psi\rangle\langle\psi|$ from a generally mixed state σ. To this end, as in Section 5.6, a complete measurement is applied with measurement basis $\mathcal{M} = \{|v_x\rangle\}_{x\in[N]}$ including the pure state $|\psi\rangle$ as $|v_0\rangle = |\psi\rangle$. Using the classical measurement output, the detector then chooses state ρ if the output is $\mathrm{x} = 0$, and selects state σ otherwise.

This detector has the useful property that, if the true state is $|\psi\rangle$, the detector always outputs the correct decision. In contrast, if the true state is σ, the probability of a correct detection is given by

$$\begin{aligned}\Pr_\sigma[\mathrm{x} \neq 0|\mathcal{M}] &= 1 - p_0(\sigma|\mathcal{M}) \\ &= 1 - \langle\psi|\sigma|\psi\rangle. \end{aligned} \tag{8.80}$$

For a pure state $\sigma = |\phi\rangle\langle\phi|$, this result recovers the probability of correct detection derived in Section 5.6, which equals

$$\Pr_{|\phi\rangle}[\mathrm{x} \neq 0|\mathcal{M}] = 1 - |\langle\psi|\phi\rangle|^2 = 1 - \mathrm{F}(|\psi\rangle, |\phi\rangle), \tag{8.81}$$

and is a function of the pure-state fidelity $\mathrm{F}(|\psi\rangle, |\phi\rangle)$ in (8.71).

In the more general case of a mixed state σ, one can show that the probability of detection (8.80) can similarly be expressed in terms of the Chernoff fidelity as (see Problem 8.14)

$$\Pr_\sigma[\mathrm{x} \neq 0|\mathcal{M}] = 1 - \mathrm{F}^{\mathrm{C}}(|\psi\rangle\langle\psi|, \sigma). \tag{8.82}$$

In contrast, the Tsallis fidelity $\mathrm{F}^{\mathrm{T}}(\rho, \sigma)$ with a pure state $\rho = |\psi\rangle\langle\psi|$ gives

$$\begin{aligned}\mathrm{F}^{\mathrm{T}}(|\psi\rangle\langle\psi|, \sigma) &= \mathrm{tr}\left(\sqrt{|\psi\rangle\langle\psi|}\sqrt{\sigma}\right) \\ &= \mathrm{tr}\left(|\psi\rangle\langle\psi|\sqrt{\sigma}\right) \\ &= \langle\psi|\sqrt{\sigma}|\psi\rangle, \end{aligned} \tag{8.83}$$

where the equality $\sqrt{|\psi\rangle\langle\psi|} = |\psi\rangle\langle\psi|$ follows by applying the square root to the rank-1, normal, matrix $|\psi\rangle\langle\psi|$. Thus, the complement $1 - \mathrm{F}^{\mathrm{T}}(|\psi\rangle\langle\psi|, \sigma)$ does not equal the probability of an incorrect detection n (8.80).

Overall, a lesson from this example is that a useful property for a fidelity measure $\mathrm{F}(\rho, \sigma)$ to have is that, when evaluated for a pure state $|\psi\rangle\langle\psi|$ and a mixed state σ, it returns

$$\mathrm{F}(|\psi\rangle\langle\psi|, \sigma) = \langle\psi|\sigma|\psi\rangle. \tag{8.84}$$

In fact, if this is the case, the complement $1 - \mathrm{F}(|\psi\rangle\langle\psi|, \sigma)$ equals the probability of incorrect detection (8.80) in the described setting.

8.4.5 Uhlmann–Josza Fidelity

In order to introduce the Uhlmann–Josza fidelity, recall the definition of the 1-Schatten norm of an arbitrary $M \times N$ matrix A (see Problem 6.10)

$$||A||_1 = \mathrm{tr}\left(\sqrt{A^\dagger A}\right), \tag{8.85}$$

which equals the sum of the square roots of the eigenvalues of matrix $A^\dagger A$.

Definition 8.12 Uhlmann–Josza fidelity
The Uhlmann–Josza fidelity between two $N \times N$ density matrices ρ and σ is defined as the square of the 1-Schatten norm of the matrix $\sqrt{\rho}\sqrt{\sigma}$, i.e.,

$$\mathrm{F}^{\mathrm{UJ}}(\rho, \sigma) = ||\sqrt{\sigma}\sqrt{\rho}||_1^2. \tag{8.86}$$

Using (8.85), the Uhlmann–Josza fidelity can equivalently be written as

$$\mathrm{F}^{\mathrm{UJ}}(\rho, \sigma) = \left(\mathrm{tr}\left(\sqrt{\sqrt{\rho}\sigma\sqrt{\rho}}\right)\right)^2. \tag{8.87}$$

Note that the expression $\sqrt{\sqrt{\rho}\sigma\sqrt{\rho}}$ is well defined, since the matrix $\sqrt{\rho}\sigma\sqrt{\rho}$ is positive semi-definite. Furthermore, the Uhlmann–Josza fidelity is symmetric, i.e., $\mathrm{F}^{\mathrm{UJ}}(\rho, \sigma) = \mathrm{F}^{\mathrm{UJ}}(\sigma, \rho)$, despite this not being apparent from the definition.

More broadly, the Uhlmann–Josza fidelity satisfies all other properties listed in Section 8.4.2 for fidelity measures, with the mentioned minor caveat that, when the two density matrices commute, one has the equality

$$\mathrm{F}^{\mathrm{UJ}}(\rho, \sigma) = \mathrm{F}(\lambda, \mu)^2, \tag{8.88}$$

in lieu of (8.75). Note also that, if the two states are pure, i.e., $\rho = |\psi\rangle\langle\psi|$ and $\sigma = |\phi\rangle\langle\phi|$, the following holds

$$\begin{aligned}\sqrt{\sqrt{\rho}\sigma\sqrt{\rho}} &= \sqrt{|\psi\rangle\langle\psi|\phi\rangle\langle\phi|\psi\rangle\langle\psi|}\\ &= |\langle\psi|\phi\rangle|\sqrt{|\psi\rangle\langle\psi|}\\ &= |\langle\psi|\phi\rangle| \cdot |\psi\rangle\langle\psi|,\end{aligned} \tag{8.89}$$

yielding the Uhlmann–Josza fidelity as

$$\begin{aligned}\mathrm{F}^{\mathrm{UJ}}(\rho, \sigma) &= |\langle\psi|\phi\rangle|^2 \cdot |||\psi\rangle\langle\psi|||_1^2\\ &= |\langle\psi|\phi\rangle|^2,\end{aligned} \tag{8.90}$$

which equals the pure-state fidelity (8.71).

The Uhlmann–Josza fidelity also meets the desirable condition (8.84) presented in Example 8.12, i.e.,

$$\begin{aligned} \mathrm{F}^{\mathrm{UJ}}(|\psi\rangle\langle\psi|,\sigma) &= ||\sqrt{|\psi\rangle\langle\psi|}\sqrt{\sigma}||_1^2 \\ &= \left(\mathrm{tr}\left(\sqrt{\sqrt{\sigma}|\psi\rangle\langle\psi|\sqrt{\sigma}}\right)\right)^2 \\ &= \langle\psi|\sigma|\psi\rangle. \end{aligned} \tag{8.91}$$

Example 8.13

When the state ρ is pure, i.e., $\rho = |\psi\rangle\langle\psi|$, using (8.91), the Tsallis quantum min-relative entropy (8.64) can be written in terms of the fidelity as

$$\begin{aligned} \mathrm{D}_0^{\mathrm{T}}(\rho||\sigma) &= 1_2(1 - \langle\psi|\sigma|\psi\rangle) \\ &= 1_2(1 - \mathrm{F}^{\mathrm{UJ}}(|\psi\rangle\langle\psi|,\sigma)). \end{aligned} \tag{8.92}$$

Thus, a larger fidelity is equivalent to a smaller Tsallis quantum min-relative entropy.

The Uhlmann–Josza fidelity can also be viewed as a special case of an alternative definition of the Tsallis relative trace (8.45), namely

$$\mathrm{tr}\left(\left(\sigma^{\frac{1-\alpha}{2\alpha}}\rho^{\alpha}\sigma^{\frac{1-\alpha}{2\alpha}}\right)^{\alpha}\right). \tag{8.93}$$

One can readily check that the two definitions (8.45) and (8.93) coincide for commuting density matrices, but they are otherwise different. With the definition (8.93), the Uhlmann–Josza fidelity (8.87) is recovered, upon squaring, as a special case for $\alpha = 0.5$. Note the parallel with the Tsallis fidelity (8.77), which coincides with the Tsallis relative trace of order $\alpha = 0.5$ defined as in (8.45).

Considering all fidelity measures introduced in this section, the following bounds hold, where the first inequality was demonstrated in (8.79) and for the other the reader is referred to the Recommended Resources.

Key Result 8.11 Relations among fidelity measures

For the fidelity measures presented in this section, the following bounds hold:

$$\mathrm{F}^{\mathrm{C}}(\rho,\sigma) \leq \mathrm{F}^{\mathrm{T}}(\rho,\sigma) \leq \sqrt{\mathrm{F}^{\mathrm{UJ}}(\rho,\sigma)}. \tag{8.94}$$

Example 8.14

For some parameter a, with $0 \leq a \leq 1$, consider the density matrix ρ with eigendecomposition

$$\rho = \frac{1}{2}(1+a)|+\rangle\langle+| + \frac{1}{2}(1-a)|-\rangle\langle-| \tag{8.95}$$

having eigenvectors $\{|+\rangle, |-\rangle\}$ and corresponding eigenvalues $\{1/2(1+a), 1/2(1-a)\}$; and the density matrix

$$\sigma = \frac{1}{2}(1+a)|0\rangle\langle 0| + \frac{1}{2}(1-a)|1\rangle\langle 1| \tag{8.96}$$

having eigenvectors $\{|0\rangle, |1\rangle\}$ and corresponding eigenvalues $\{1/2(1+a), 1/2(1-a)\}$.

Parameter a controls the purity of density matrices ρ and σ (see Section 7.10). In fact, the purity of both density matrices can be evaluated as the sum of the squared eigenvalues

$$\mathrm{tr}(\rho^2) = \mathrm{tr}(\sigma^2) = \frac{1}{4}(1+a)^2 + \frac{1}{4}(1-a)^2 = \frac{1}{2}(1+a^2). \tag{8.97}$$

This implies that a purity level $1/2 \leq P \leq 1$ is obtained with the choice $a = \sqrt{2P-1}$. Accordingly, by varying a, the two states range from maximally mixed, identical, states $\rho = \sigma = \pi$ with $a = 0$ and $P = 1/2$, to pure states $\rho = |+\rangle\langle +|$ and $\sigma = |0\rangle\langle 0|$ with $a = 1$ and $P = 1$.

The Tsallis fidelity can be evaluated using (8.77) as

$$\begin{aligned} \mathrm{F}^{\mathrm{T}}(\rho,\sigma) &= \frac{1}{4}(1+a) + \frac{1}{4}(1-a) + \frac{1}{2}\sqrt{1-a^2} \\ &= \frac{1}{2}\left(1+\sqrt{2(1-P)}\right), \end{aligned} \tag{8.98}$$

which is a decreasing function of the purity P, with maximal fidelity $\mathrm{F}(\rho,\sigma) = 1$ achieved if $P = 1/2$ and minimal fidelity $\mathrm{F}(\rho,\sigma) = |\langle +|0\rangle|^2 = 0.5$ obtained for $P = 1$. The Chernoff fidelity can be computed in a similar way, and it can be seen to equal the Tsallis fidelity in this case. The Uhlmann–Josza fidelity can be evaluated numerically by using the definition (8.86).

Figure 8.11 shows the behavior of the metrics discussed above as a function of the purity P. As the purity increases, the two states become increasingly different, and all fidelity measures decrease. For reference, the figure also reports the trace inner product $\mathrm{tr}(\rho\sigma) = 0.5$, which does not depend on the purity. Note that, for pure states, e.g., for purity $P = 1$, all the fidelity measures coincide with the pure-state fidelity $\mathrm{F}(|\psi\rangle, |\phi\rangle)$.

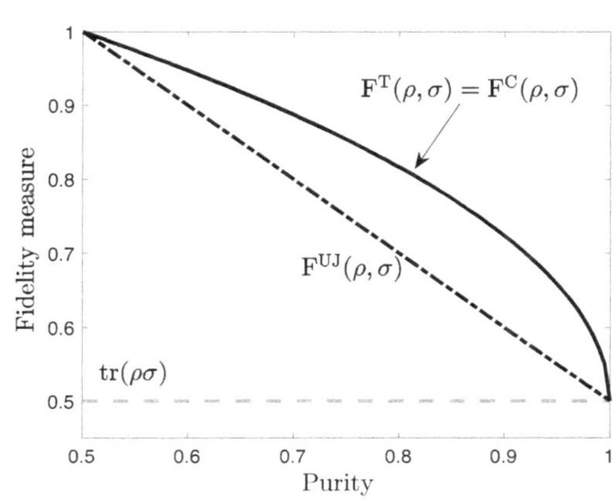

Figure 8.11 Fidelity measures for the density matrices (8.95) and (8.96) as a function of the purity of the density matrices.

8.5 Rényi Quantum Relative Entropy

8.5.1 Defining the Rényi Quantum Relative Entropy

Using arguments analogous to Section 8.4, the Rényi quantum relative entropy is defined as follows by applying the function (4.72) to the Tsallis relative trace.

Definition 8.13 Rényi quantum relative entropy, aka Petz–Rényi quantum relative entropy

The Rényi quantum relative entropy of order $\alpha \geq 0$ is defined as

$$\mathrm{D}^{\mathrm{R}}_{\alpha}(\rho||\sigma) = \begin{cases} \frac{1}{\alpha-1}\log_2(\mathrm{tr}(\rho^{\alpha}\sigma^{1-\alpha})) & \text{if } 0 \leq \alpha < 1 \text{ and } \mathrm{supp}(\rho) \not\perp \mathrm{supp}(\sigma), \\ & \text{or } \alpha > 1 \text{ and } \mathrm{supp}(\rho) \subseteq \mathrm{supp}(\sigma), \\ \infty & \text{otherwise,} \end{cases} \tag{8.99}$$

and

$$\mathrm{D}^{\mathrm{R}}_{1}(\rho||\sigma) = \mathrm{D}(\rho||\sigma). \tag{8.100}$$

As opposed to the Tsallis quantum relative entropy (8.58), the Rényi quantum relative entropy can take an infinite value even when $0 \leq \alpha < 1$, namely when the supports of the density matrices are orthogonal. In fact, in this case, the Tsallis relative trace equals zero. The following properties can be derived in a manner similar to Section 4.6.

Key Result 8.12 Non-negativity and minimum-value property of the Rényi quantum relative entropy

The Rényi quantum relative entropy is non-negative,

$$\mathrm{D}^{\mathrm{R}}_{\alpha}(\rho||\sigma) \geq 0, \tag{8.101}$$

for all orders $\alpha \geq 0$. Furthermore, for all orders $\alpha > 0$, it equals zero if and only if the two density matrices are equal, i.e., $\rho = \sigma$, while the special case $\alpha = 0$ will be detailed below.

Key Result 8.13 The Rényi quantum relative entropy increases with the inverse temperature α

For any $N \times N$ density matrices ρ and σ, the Rényi quantum relative entropy $\mathrm{D}^{\mathrm{R}}_{\alpha}(\rho||\sigma)$ is an increasing function of the order $\alpha \geq 0$, i.e., the inequality

$$\mathrm{D}^{\mathrm{R}}_{\alpha'}(\rho||\sigma) \geq \mathrm{D}^{\mathrm{R}}_{\alpha}(\rho||\sigma) \tag{8.102}$$

holds for any two orders α' and α with $\alpha' \geq \alpha$.

Key Result 8.14 Rényi quantum relative entropy vs. Tsallis quantum relative entropy
When $0 \leq \alpha < 1$, the Rényi quantum relative entropy is no smaller than the Tsallis quantum relative entropy

$$\mathrm{D}_{\alpha}^{\mathrm{R}}(\rho||\sigma) \geq \mathrm{D}_{\alpha}^{\mathrm{T}}(\rho||\sigma); \tag{8.103}$$

with $\alpha = 1$, the two relative entropies are equal and they coincide with the Shannon quantum relative entropy

$$\mathrm{D}_{\alpha}^{\mathrm{R}}(\rho||\sigma) = \mathrm{D}_{\alpha}^{\mathrm{T}}(\rho||\sigma) = \mathrm{D}(\rho||\sigma); \tag{8.104}$$

while for $\alpha > 1$, the inequality

$$\mathrm{D}_{\alpha}^{\mathrm{R}}(\rho||\sigma) \leq \mathrm{D}_{\alpha}^{\mathrm{T}}(\rho||\sigma) \tag{8.105}$$

holds.

8.5.2 Special Cases of the Rényi Quantum Relative Entropy

Some special cases of the Rényi quantum relative entropy are elaborated on next.

Definition 8.14 Rényi quantum relative entropy of order $\alpha = 0$, aka Rényi quantum min-relative entropy
The Rényi quantum relative entropy of order $\alpha = 0$ is obtained as

$$\mathrm{D}_{0}^{\mathrm{R}}(\rho||\sigma) = -\log_2(\mathrm{tr}(\Pi_{\rho}\sigma)). \tag{8.106}$$

Accordingly, as in the classical case and as for the Tsallis quantum min-relative entropy in (8.64), the Rényi quantum min-relative entropy equals zero as long as there are no hallucinations, i.e., as long as the inclusion condition $\mathrm{supp}(q) \subseteq \mathrm{supp}(p)$ holds:

$$\mathrm{D}_{0}^{\mathrm{R}}(\rho||\sigma) = 0 \Longleftrightarrow \mathrm{supp}(\sigma) \subseteq \mathrm{supp}(\rho). \tag{8.107}$$

Definition 8.15 Rényi quantum relative entropy of order $\alpha = 0.5$
For $\alpha = 0.5$, the Rényi quantum relative entropy is given by

$$\mathrm{D}_{0.5}^{\mathrm{R}}(\rho,\sigma) = -2\log_2(\mathrm{F}^{\mathrm{T}}(\rho,\sigma)), \tag{8.108}$$

which is a function of the Tsallis fidelity (8.77).

Definition 8.16 Rényi quantum relative entropy of order $\alpha = \infty$, aka Rényi quantum max-relative entropy
The Rényi quantum relative entropy of order $\alpha = \infty$ is given by

$$\mathrm{D}^{\mathrm{R}}_{\infty}(\rho||\sigma) = \log_2\left(\max_{x\in[N]} \lambda_x\left(\sigma^{-1/2}\rho\sigma^{-1/2}\right)\right), \tag{8.109}$$

where $\lambda_x(\cdot)$ represents the xth eigenvalue of the argument matrix.

For commuting matrices, this definition coincides with the classical definition (4.82), i.e., $\mathrm{D}^{\mathrm{R}}_{\infty}(\rho||\sigma) = \mathrm{D}^{\mathrm{R}}_{\infty}(\lambda||\mu) = \max_{x\in[N]} \log_2(\lambda_x/\mu_x)$, where λ and μ are the vectors of eigenvalues as per the eigendecompositions (8.6) and (8.7). The expression (8.109) can equivalently be evaluated as $\mathrm{D}^{\mathrm{R}}_{\infty}(\rho||\sigma) = \log_2(r^*)$, where r^* is the minimum value $r > 0$ such that all the eigenvalues of matrix $r \cdot \sigma - \rho$ are non-negative.

8.5.3 Monotonicity of the Rényi Quantum Entropy

Finally, the Rényi quantum relative entropy supports the following monotonicity property of the Rényi quantum entropy.

Key Result 8.15 Monotonicity property of the Rényi quantum entropy
The Rényi quantum entropy of order α satisfies the equality

$$\mathrm{H}^{\mathrm{R}}_{\alpha}(\rho) = \log_2(N) - \mathrm{D}^{\mathrm{R}}_{\alpha}(\rho||\pi), \tag{8.110}$$

which mirrors, and generalizes, the counterpart decomposition (8.40) for the Shannon entropy.

8.6 Optimal Detection, Quantum Relative Entropies, and Trace Distance

As illustrated in Figure 8.12, this section studies an informational task in which a detector is faced with a quantum system that may be in a state described by density matrix ρ or density matrix σ. The detector knows the density matrices ρ and σ, but it is not aware of which of the two density matrices describes the state of the system. The analysis of the resulting optimal detection probabilities presented in this section will offer an operational interpretation of some of the quantum relative entropies introduced in this chapter. In the process, the trace distance introduced in Chapter 4 will also be extended from probability distributions to density matrices.

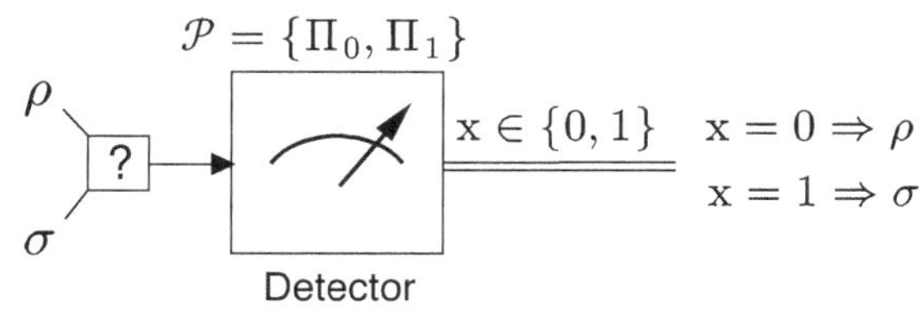

Figure 8.12 Illustration of the problem of optimal detection for mixed states.

8.6.1 Detecting Mixed States

To find out whether the state is ρ or σ, the detector in Figure 8.12 applies a projective measurement $\mathcal{P} = \{\Pi_0, \Pi_1 = I - \Pi_0\}$ to the quantum system, where I is the $N \times N$ identity matrix. When classical output $\mathrm{x} = 0$ is recorded, the detector decides that the true state is ρ, while a classical output $\mathrm{x} = 1$ yields a decision for state σ.

The design of the setting $\mathcal{P}$ must control the two detection probabilities of interest, namely the detection probability when the true density is ρ, i.e.,

$$p_0(\rho|\mathcal{P}) = \mathrm{tr}(\Pi_0 \rho); \tag{8.111}$$

and the detection probability when the true density is σ, i.e.,

$$p_1(\sigma|\mathcal{P}) = \mathrm{tr}(\Pi_1 \sigma). \tag{8.112}$$

Example 8.15

Consider the extreme case in which the two density matrices have orthogonal supports. Under this assumption, it is possible to design a detector $\mathcal{P}$ that obtains the maximum values $p_0(\rho|\mathcal{P}) = 1$ and $p_1(\sigma|\mathcal{P}) = 1$ for both detection probabilities. To this end, it is sufficient to choose the projection matrices as $\Pi_0 = \Pi_\rho$ and $\Pi_1 = I - \Pi_\rho$, where Π_ρ, defined in (8.33), is the projection matrix onto the support of density matrix ρ.

In fact, the equality $\Pi_\rho \rho = \rho$ implies the detection probability $p_0(\rho|\mathcal{P}) = \mathrm{tr}(\Pi_\rho \rho) = 1$, as well as

$$p_1(\sigma|\mathcal{P}) = \mathrm{tr}((I - \Pi_\rho)\sigma) = 1 - \mathrm{tr}(\Pi_\rho \sigma) = 1, \tag{8.113}$$

where the last equality holds by the orthogonality assumption, which implies the equality $\Pi_\rho \sigma = 0$. Hence, density matrices with orthogonal supports are maximally distinguishable.

Under more general conditions than in the example above, the optimal measurement setting $\mathcal{P}$ depends on the relative importance placed by the designer on the two detection probabilities $p_0(\rho|\mathcal{P})$ and $p_1(\sigma|\mathcal{P})$.

8.6.2 Optimal Detection with Perfect Detection for One State

To elaborate on this point, assume that the detector wishes to ensure that the detection probability when density matrix ρ is the true state of the system equals 1, i.e.,

$p_0(\rho|\mathcal{P}) = 1$. In this case, the design problem may be formulated as the constrained optimization

$$\begin{aligned} &\underset{\mathcal{P}}{\text{maximize}}\; p_1(\sigma|\mathcal{P}) \\ &\text{subject to}\; p_0(\rho|\mathcal{P}) = 1, \end{aligned} \tag{8.114}$$

whereby one tries to maximize the detection probability $p_1(\sigma|\mathcal{P})$ by picking one among all detectors that ensure probability $p_0(\rho|\mathcal{P}) = 1$.

The optimal solution to this problem is given by the measurement setting $\mathcal{P}^* = \{\Pi_\rho, I - \Pi_\rho\}$ introduced in Example 8.15. In fact, in order for the constraint in (8.114) to be satisfied, it is necessary to ensure the equality $\Pi_0\rho = \rho$, which is met by the choice $\Pi_0 = \Pi_\rho$. Furthermore, any projection matrix with a larger rank than Π_ρ cannot increase the probability $p_1(\sigma|\mathcal{P})$ beyond $\mathrm{tr}((I - \Pi_\rho)\sigma)$.

Accordingly, using the Tsallis quantum relative entropy $\mathrm{D}_0^{\mathrm{T}}(\rho||\sigma)$ in (8.64) and the Rényi quantum relative entropy $\mathrm{D}_0^{\mathrm{R}}(\rho||\sigma)$ in (8.106), the maximum value of the objective function in (8.114) is given by

$$\begin{aligned} p_1(\sigma|\mathcal{P}^*) &= \mathrm{tr}((I - \Pi_\rho)\sigma) \\ &= \frac{1}{1_2}\mathrm{D}_0^{\mathrm{T}}(\rho||\sigma) \\ &= 1 - 2^{-\mathrm{D}_0^{\mathrm{R}}(\rho||\sigma)}. \end{aligned} \tag{8.115}$$

Thus, as in the classical case studied in Section 4.7, the optimal probability of detection increases with the Tsallis or Rényi min-relative entropies. Furthermore, the dependence of the optimal detection probability is linear with respect to the Tsallis quantum min-relative entropy $\mathrm{D}_0^{\mathrm{T}}(\rho||\sigma)$, and exponential with respect to the Rényi quantum min-relative entropy $\mathrm{D}_0^{\mathrm{R}}(\rho||\sigma)$.

Example 8.16

Suppose that density matrix ρ has rank equal to N, while density matrix σ is arbitrary, In this case, in order to satisfy the constraint in problem (8.114), one must set $\Pi_0 = \Pi_\rho = I$, which yields the optimal detection probability $p_1(\sigma|\mathcal{P}^*) = 0$. In fact, in this case, one has the equalities $\mathrm{D}_0^{\mathrm{T}}(\rho||\sigma) = 0$ and $\mathrm{D}_0^{\mathrm{R}}(\rho||\sigma) = 0$, as there are no hallucinations.

8.6.3 Optimal Detection under the Average Detection Probability

In another common formulation of the design problem, the detector assigns the same a priori probability to both density matrices. Under this assumption, the detection probability of interest is the average

$$\mathrm{P_S}(\mathcal{P}) = \frac{1}{2}p_0(\rho|\mathcal{P}) + \frac{1}{2}p_1(\sigma|\mathcal{P}). \tag{8.116}$$

Therefore, the design problem is given by the maximization

$$\underset{\mathcal{P}}{\text{maximize}} \left\{ \mathrm{P_S}(\mathcal{P}) = \frac{1}{2}\left(1 + \mathrm{tr}\left(\Pi_0(\rho - \sigma)\right)\right) \right\}, \tag{8.117}$$

where the following equalities were used:

$$\begin{aligned} \mathrm{P_S}(\mathcal{P}) &= \frac{1}{2}\left(\mathrm{tr}\left(\Pi_0\rho + \Pi_1\sigma\right)\right) \\ &= \frac{1}{2}\left(\mathrm{tr}\left(\Pi_0\rho + (\mathrm{I} - \Pi_0)\sigma\right)\right) \\ &= \frac{1}{2}\left(1 + \mathrm{tr}\left(\Pi_0(\rho - \sigma)\right)\right). \end{aligned} \tag{8.118}$$

To address the problem (8.117), consider now the matrix $\rho - \sigma$ that appears in the expression of the probability of detection, $\mathrm{P_S}$. Any linear combination of Hermitian matrices is Hermitian, and hence matrix $\rho - \sigma$ is Hermitian. Therefore, it has an eigendecomposition, which may be written as

$$\rho - \sigma = \sum_{x \in [N]} \delta_x |v_x\rangle\langle v_x|. \tag{8.119}$$

Note that the eigenvectors $\{|v_x\rangle\}_{x\in[N]}$ are generally distinct from the eigenvectors of the density matrices ρ and σ, unless the two matrices ρ and σ commute. In fact, for commuting matrices, by (8.6) and (8.7), matrix $\rho - \sigma$ shares the same eigenvectors as matrices ρ and σ, i.e., $|u_x\rangle = |v_x\rangle$ for all $x \in [N]$, and each eigenvalue δ_x equals the difference of the corresponding eigenvalues λ_x and μ_x, i.e., $\delta_x = \lambda_x - \mu_x$ for $x \in [N]$.

Denote as $[N]^+ \subseteq [N]$ the subset of the indices $x \in [N]$ for which the corresponding eigenvalue δ_x of the difference matrix $\rho - \sigma$ is non-negative. With this choice, one has the equivalence

$$\delta_x \geq 0 \Longleftrightarrow x \in [N]^+. \tag{8.120}$$

As shown in Appendix 8.E, the following holds.

Key Result 8.16 Optimal measurement (Holevo–Helstrom measurement)
The optimal measurement setting $\mathcal{P}^* = \{\Pi_0^*, \Pi_1^* = I - \Pi_0^*\}$ for problem (8.117), known as the Holevo–Helstrom measurement, is given by

$$\Pi_0^* = \sum_{x \in [N]^+} |v_x\rangle\langle v_x| \tag{8.121}$$

and $\Pi_1^* = I - \Pi_0^*$. The optimal projection matrix (8.121) can equivalently be written as

$$\Pi_0^* = \mathrm{step}(\rho - \sigma), \tag{8.122}$$

where the scalar function step$(\cdot)$, extended as described in Section 6.6 to normal matrices, returns 1 if the argument is non-negative and 0 otherwise.

Intuitively, by projecting into the span $\mathcal{S}(\{|v_x\rangle\}_{x\in[N]^+})$ of the eigenvectors $|v_x\rangle$ corresponding to non-negative eigenvalues $\delta_x \geq 0$, the matrix Π_0^* selects the directions in the vector space that are more aligned with the eigenvectors of matrix ρ than they are to the eigenvectors of matrix σ.

8.6.4 Trace Distance

As also demonstrated in Appendix 8.E, the corresponding maximum probability of detection is given as follows.

Key Result 8.17 Maximum detection probability and trace distance
The maximum value of the optimal detection problem (8.117) is given by

$$\mathrm{P_S}(\mathcal{P}^*) = \frac{1}{2}\left(1 + \mathrm{D^{tr}}(\rho,\sigma)\right), \tag{8.123}$$

where the **trace distance** between density matrices ρ and σ is defined as

$$\mathrm{D^{tr}}(\rho,\sigma) = \frac{1}{2}||\rho - \sigma||_1 = \frac{1}{2}\sum_{x\in[N]}|\delta_x|. \tag{8.124}$$

This result is a direct generalization of the optimal probability of detection derived in Section 4.7 for classical probability distributions. In fact, one can check that for commuting matrices the trace distance $\mathrm{D^{tr}}(\rho,\sigma)$ is given by the trace distance between the corresponding vectors of eigenvalues λ and μ, i.e.,

$$\begin{aligned}\mathrm{D^{tr}}(\rho,\sigma) = \mathrm{D^{tr}}(\lambda,\mu) &= \frac{1}{2}||\lambda - \mu||_1 \\ &= \frac{1}{2}\sum_{x\in[N]}|\lambda_x - \mu_x|,\end{aligned} \tag{8.125}$$

where $||\cdot||_1$ is the 1-norm for vectors.

8.6.5 Properties of the Trace Distance

As reflected by the notation $\mathrm{D^{tr}}(\rho,\sigma)$, the trace distance is a symmetric function of the two densities ρ and σ. Furthermore, it satisfies the following properties.

Key Result 8.18 Reduction to the trace distance for pure states
When evaluated for pure state $\rho = |\psi\rangle\langle\psi|$ and $\sigma = |\phi\rangle\langle\phi|$, the trace distance (8.124) coincides with the trace distance (5.86) defined in Section 5.6 for pure states, i.e.,

$$\mathrm{D^{tr}}(|\psi\rangle\langle\psi|, |\psi\rangle\langle\psi|) = \mathrm{D^{tr}}(|\psi\rangle, |\phi\rangle) = \sqrt{1 - \mathrm{F}(|\psi\rangle, |\phi\rangle)}. \tag{8.126}$$

In fact, neglecting irrelevant global phases, one can always write $|\phi\rangle = \cos(\theta)|\psi\rangle + \sin(\theta)|\psi^{\perp}\rangle$, where $|\psi^{\perp}\rangle$ represents a state, dependent on $|\phi\rangle$, that is orthogonal to $|\psi\rangle$ and θ is some angle. With this notation, the trace distance is evaluated as $\mathrm{D}^{\mathrm{tr}}(|\psi\rangle\langle\psi|, |\psi\rangle\langle\psi|) = |\sin(\theta)|$ and the fidelity is $\mathrm{F}(|\psi\rangle, |\phi\rangle) = (\cos(\theta))^2 = |\langle\psi|\phi\rangle|^2$.

This said, it is important to emphasize that the relation (8.126) between fidelity and trace distance for pure states does not extend to mixed states. Rather, a useful connection between trace distance and Uhlmann–Josza fidelity is given by the following inequalities.

Key Result 8.19 Fuchs–van de Graaf inequalities
Given any two $N \times N$ density matrices ρ and σ, the following inequalities hold:

$$1 - \sqrt{\mathrm{F}^{\mathrm{UJ}}(\rho,\sigma)} \leq \mathrm{D}^{\mathrm{tr}}(\rho,\sigma) \leq \sqrt{1 - \mathrm{F}^{\mathrm{UJ}}(\rho,\sigma)}. \tag{8.127}$$

Note that, by (8.90), the upper bound in (8.127) coincides with the right-hand side of (8.126) for pure states.

Key Result 8.20 Bounds on the trace distance
The trace distance satisfies the inequalities

$$\underbrace{0}_{\rho=\sigma} \leq \mathrm{D}^{\mathrm{tr}}(\rho,\sigma) \leq \underbrace{1}_{\mathrm{supp}(\rho)\perp\mathrm{supp}(\sigma)}, \tag{8.128}$$

where the lower bound, $\mathrm{D}^{\mathrm{tr}}(\rho,\sigma) = 0$, is attained if and only if the density matrices ρ and σ are equal; while the upper bound, $\mathrm{D}^{\mathrm{tr}}(\rho,\sigma) = 1$, is obtained if and only if ρ and σ have orthogonal supports.

The bounds (8.128) can be related to the role of the trace distance in the optimal detection probability (8.123). In fact, the maximum detection probability $\mathrm{P_S}(\mathcal{P}^*) = 1$ is attained when the two density matrices have orthogonal supports, and hence they can be perfectly distinguished. In contrast, the minimum value of the detection probability, $\mathrm{P_S}(\mathcal{P}^*) = 0.5$, is obtained when the two density matrices are identical, and thus they cannot be distinguished.

Key Result 8.21 Quantum Pinsker inequality
The trace distance is related to the Shannon quantum relative entropy as

$$\mathrm{D}^{\mathrm{tr}}(\rho,\sigma) \leq \sqrt{\frac{1}{2 \cdot 1_2}\mathrm{D}(\rho||\sigma)}. \tag{8.129}$$

Therefore, by the quantum Pinsker inequality (see Recommended Resources for a proof), the maximal probability of detection (8.123) can be bounded in terms of the Shannon quantum relative entropy as

$$\mathrm{P_S}(\mathcal{P}^*) \leq \frac{1}{2}\left(1 + \sqrt{\frac{1}{2 \cdot 1_2}\mathrm{D}(\rho||\sigma)}\right). \tag{8.130}$$

Example 8.17

In this example, the optimal detection probability $\mathrm{P_S}(\mathcal{P}^*)$ is compared against the upper bound (8.130) based on the quantum Pinsker inequality. To this end, 2×2 random density matrices ρ and σ are generated by drawing 2×2 matrices G_0 and G_1 with complex Gaussian entries of zero mean and power 1; and then evaluating the density matrices as $\rho = G_0 G_0^\dagger/\mathrm{tr}(G_0 G_0^\dagger)$ and $\sigma = G_1 G_1^\dagger/\mathrm{tr}(G_1 G_1^\dagger)$ to ensure that the matrices are positive semi-definite and have trace equal to 1.

Figure 8.13 shows a scatter plot of 100 realizations of the maximal detection probabilities $\mathrm{P_S}(\mathcal{P}^*)$ on the horizontal axis, i.e., the left-hand side of (8.130), and of the corresponding upper bound on the right-hand side of (8.130) on the vertical axis. The dashed line corresponds to the case in which the upper bound matches the true probability. The figure confirms the validity of the upper bound, which, however, may be vacuous, taking values larger than 1 for some realizations of the density matrices.

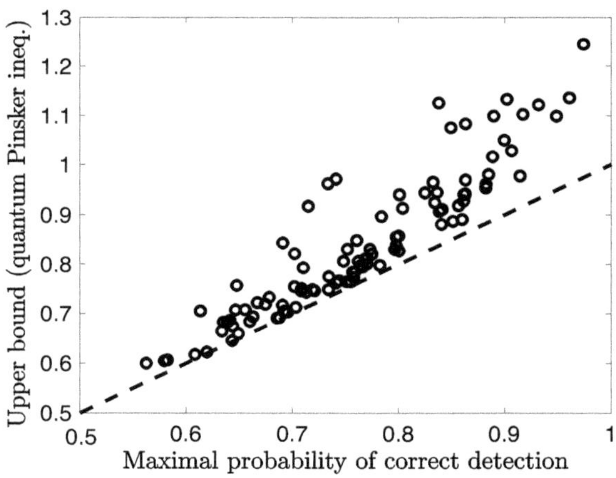

Figure 8.13 Comparison between the upper bound on the right-hand side of the quantum Pinsker inequality (8.130) (vertical axis) and the maximal detection probability $\mathrm{P_S}(\mathcal{P}^*)$ (horizontal axis) for 100 randomly generated pairs of density matrices. The dashed line corresponds to the case in which the upper bound matches the true probability.

8.6.6 Disturbance and Information

As an application of the trace distance, this section ends by revisiting the no-information-without disturbance principle. As discussed in Section 6.5, this principle states, informally, that no information can be extracted without causing disturbance to the state. Conversely, if a measurement does not extract information from a state, one may expect it not to cause the post-measurement state to differ too much from the pre-measurement state. One way to formalize this property is to relate the probability $p_y(\rho|\mathcal{P})$ of a measurement output y with the trace distance $\mathrm{D}^{\mathrm{tr}}(\rho, \rho_y)$ between pre-measurement state ρ and post-measurement state ρ_y.

In this regard, it can be proved that the following inequality holds (see Recommended Resources):

$$\mathrm{D}^{\mathrm{tr}}(\rho, \rho_y) \leq \sqrt{1 - p_y(\rho|\mathcal{P})}. \tag{8.131}$$

Accordingly, if a measurement output y has a large probability of occurring and thus it does not extract much information from the state, the corresponding post-measurement state ρ_y cannot deviate too much from the pre-measurement state ρ.

8.7 Purification and Fidelity ⏭

Density matrices represent ensembles of pure states that account for both classical and quantum uncertainties. In particular, the rank of matrix ρ,

$$r = \mathrm{rank}(\rho), \tag{8.132}$$

corresponds to the minimum number of pure states in an ensemble that realizes matrix ρ. Therefore, the rank is a measure of the mixedness of the state, determining the Tsallis quantum entropy (7.121) and the Rényi quantum entropy (7.125) of order $\alpha = 0$. This section will show that an $N \times N$ density matrix ρ can equivalently be modeled as a pure state $|\rho\rangle$ in a larger space of dimension Nr. As also discussed in this section, this process of purifying density matrices ρ and σ supports the comparison of ρ and σ in a different domain, yielding new fidelity and distance measures.

8.7.1 Introducing Purification

A mixed state with density matrix of rank r can be purified into a pure state in a larger space of dimension Nr or larger. To start, for an $N \times N$ density matrix ρ, consider any ensemble $\mathcal{E} = \{(|v_x\rangle, p_x)\}_{x\in[L]}$ of $L \geq r$ pure states that realizes it, i.e.,

$$\rho = \sum_{x\in[L]} p_x |v_x\rangle\langle v_x|. \tag{8.133}$$

Recall that, if the rank is larger than 1, i.e., $r > 1$, there are infinitely many ensembles $\mathcal{E}$ realizing the density matrix ρ as in (8.133). Furthermore, in order for the equality (8.133) to hold, it is necessary that the number L of pure states in the ensemble is no smaller than r.

Definition 8.17 Purification
Given any ensemble $\mathcal{E} = \{(|v_x\rangle, p_x)\}_{x\in[L]}$ that realizes the rank-r density matrix ρ with $L \geq r$, the corresponding purified state is defined by the $NL \times 1$ vector

$$|\rho\rangle = \begin{bmatrix} \sqrt{p_0}|v_0\rangle \\ \sqrt{p_1}|v_1\rangle \\ \vdots \\ \sqrt{p_{L-1}}|v_{L-1}\rangle \end{bmatrix}. \tag{8.134}$$

This vector stacks the L states $\{|v_x\rangle\}_{x\in[L]}$ in the ensemble, with each vector $|v_x\rangle$ multiplied by the square root $\sqrt{p_x}$ of the corresponding probability. The NL-dimensional space that includes the purified state $|\rho\rangle$ is referred to as the **purified space**.

As a first observation, vector $|\rho\rangle$ represents a valid quantum state in the NL-dimensional purified space, since its squared 2-norm is equal to 1, i.e.,

$$\langle\rho|\rho\rangle = \sum_{x\in[L]} p_x \langle v_x|v_x\rangle = \sum_{x\in[L]} p_x = 1. \tag{8.135}$$

Furthermore, the following observations are in order regarding definition (8.134).

- **Dimension of the purified space:** For rank $r = 1$, i.e., for a pure state $\rho = |\psi\rangle\langle\psi|$, the dimensions of the original space and of the purified space coincide, and the purified state $|\rho\rangle$ is equal to the state $|\psi\rangle$. More generally, for a mixed state with rank $r > 1$, the dimension of the purified state $|\rho\rangle$ grows proportionally to the rank r, and thus to the mixedness of the state. In particular, the maximum dimension of the purified state $|\rho\rangle$ is $N^2 \times 1$, which is obtained for full-rank density matrices ρ, i.e., for $r = N$.
- **Non-uniqueness:** There are as many purified states of the form (8.134) as there are ensembles realizing the density matrix ρ. Therefore, while the density matrix ρ can be reconstructed from the purified state $|\rho\rangle$, there are generally many ways to map a density matrix into an ensemble and thus into a purified state.
- **Purification via the eigenensemble:** As a notable special case of the purification (8.134), the smallest number $L = r$ of pure states is obtained with the eigenensemble $\mathcal{E}(\rho) = \{(|u_x\rangle, \lambda_x)\}_{x\in[r]}$, which corresponds to the eigendecomposition

$$\rho = \sum_{x\in[r]} \lambda_x |u_x\rangle\langle u_x|, \tag{8.136}$$

where all the r eigenvalues λ_x, with $x \in [r]$, are strictly positive. The purified state (8.134) obtained from the eigenensemble is given by the $Nr \times 1$ vector

$$|\rho\rangle = \begin{bmatrix} \sqrt{\lambda_0}|u_0\rangle \\ \sqrt{\lambda_1}|u_1\rangle \\ \vdots \\ \sqrt{\lambda_{r-1}}|u_{r-1}\rangle \end{bmatrix}. \tag{8.137}$$

- **Church of the Larger Hilbert Space:** One way to think about the extra dimensions required to purify the state is that they correspond to an external system, the **environment**, which "knows" the identity of the true pure state in the ensemble. The environment is not accessible to the N-dimensional system of interest, causing the classical uncertainty associated with the density matrix ρ. According to this view, all classical randomness arises from inaccessible information stored in a part of a larger system that includes the system under consideration. The idea that, in a sufficiently large space, all states are pure is also known as the "Church of the Larger Hilbert Space."

Example 8.18

Consider the 2×2 density matrix (8.95). The rank of this matrix equals $r = 1$ for $a = 1$, in which case the purified state coincides with the state itself, i.e., $|\rho\rangle = |+\rangle$. In contrast, for any $0 \leq a < 1$, the rank is $r = 2$, and a purified state is obtained as the 4×1 vector (8.134), which reads

$$|\rho\rangle = \begin{bmatrix} \sqrt{\frac{1}{2}(1+a)}|+\rangle \\ \sqrt{\frac{1}{2}(1-a)}|-\rangle \end{bmatrix}. \tag{8.138}$$

8.7.2 Reproducing the Behavior of a Mixed State in the Purified Space

As illustrated in Figure 8.14, any purified state $|\rho\rangle$ can reproduce the behavior of the state ρ. This is in the sense that, for any projective measurement setting $\mathcal{P}$ defined on the N-dimensional space of the density matrix ρ, there exists a measurement setting $\mathcal{P}^{\mathrm{P}}$ defined on the NL-dimensional space of the purified state $|\rho\rangle$ that produces the same probability distribution $p(\rho|\mathcal{P})$ over classical outputs, as well as consistent post-measurement states.

To elaborate, take a general projective measurement setting $\mathcal{P} = \{\Pi_x\}_{x\in[M]}$, with $M \leq N$ classical outputs, which operates in the original N-dimensional space. From Chapter 7, recall that this measurement returns the classical output $x \in [M]$ with probability $p_x(\rho|\mathcal{P}) = \mathrm{tr}(\Pi_x\rho)$, while leaving the system in the post-measurement state

$$\begin{aligned} \rho_x &= \frac{1}{p_x(\rho|\mathcal{P})}\Pi_x\rho\Pi_x \\ &= \sum_{y\in[L]} \frac{p_y}{p_x(\rho|\mathcal{P})}\Pi_x|v_y\rangle\langle v_y|\Pi_x, \end{aligned} \tag{8.139}$$

where the second equality uses the decomposition (8.133).

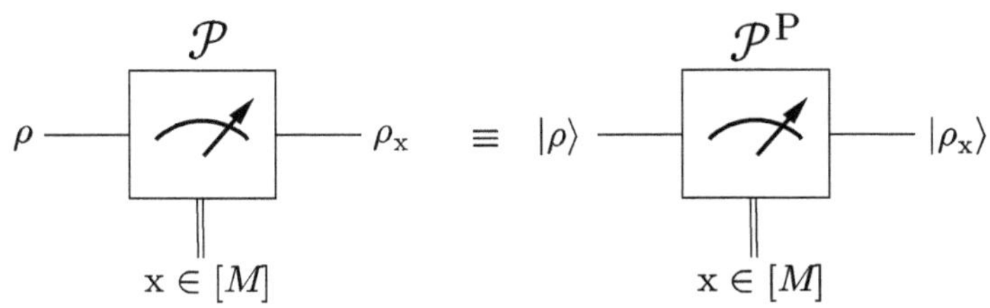

Figure 8.14 For any projective measurement setting $\mathcal{P}$ defined on the N-dimensional space of the density matrix ρ (on the left), there exists a measurement setting $\mathcal{P}^{\mathrm{P}}$ defined on the NL-dimensional space of the purified state $|\rho\rangle$ (on the right) that produces the same probability distribution $p(\rho|\mathcal{P})$ over classical outputs x, as well as consistent post-measurement states.

For each $N \times N$ projection matrix Π_x, consider the corresponding $NL \times NL$ projection matrix

$$\Pi_x^{\mathrm{P}} = \begin{bmatrix} \Pi_x & 0 & \cdots & 0 \\ 0 & \Pi_x & \ddots & \vdots \\ \vdots & \ddots & \ddots & 0 \\ 0 & \cdots & 0 & \Pi_x \end{bmatrix}, \tag{8.140}$$

where the projection matrix Π_x is repeated L times on the main diagonal. One can indeed check that this is a projection matrix, since it is idempotent (see (6.92)). Define now the projective measurement $\mathcal{P}^{\mathrm{P}} = \{\Pi_x^{\mathrm{P}}\}_{x\in[M]}$ in the purified NL-dimensional space.

By the Born rule, the setting $\mathcal{P}^{\mathrm{P}}$ applied to the purified state $|\rho\rangle$ yields the classical output $\mathrm{x} = x \in [M]$ with probability

$$\begin{aligned} p_x(|\rho\rangle|\mathcal{P}^{\mathrm{P}}) &= \langle\rho|\Pi_x^{\mathrm{P}}|\rho\rangle \\ &= \sum_{y\in[L]} p_y\langle v_y|\Pi_x|v_y\rangle \\ &= \mathrm{tr}(\Pi_x\rho) \\ &= p_x(\rho|\mathcal{P}), \end{aligned} \tag{8.141}$$

which coincides with the original measurement probability $p_x(\rho|\mathcal{P})$, as desired. Note that the second equality in (8.141) follows by direct calculation using the expression (8.134) of the purified state and the expression (8.140) of the projection matrix Π_x^{P}.

Furthermore, given the pre-measurement state $|\rho\rangle$ in (8.134), the post-measurement state for measurement setting $\mathcal{P}^{\mathrm{P}}$ is given by the $NL \times 1$ state

$$\frac{1}{\sqrt{p_x(|\rho\rangle|\mathcal{P}^{\mathrm{P}})}}\Pi_x^{\mathrm{P}}|\rho\rangle = \begin{bmatrix} \sqrt{\frac{p_0}{p_x(\rho|\mathcal{P})}}\Pi_x|v_0\rangle \\ \sqrt{\frac{p_1}{p_x(\rho|\mathcal{P})}}\Pi_x|v_1\rangle \\ \vdots \\ \sqrt{\frac{p_{L-1}}{p_x(\rho|\mathcal{P})}}\Pi_x|v_{L-1}\rangle \end{bmatrix}, \tag{8.142}$$

where the equality is again proved by a direct application of the expressions (8.134) and (8.140).

Using the definition (8.134), the state (8.142) corresponds to the purified state obtained from the ensemble

$$\left\{\left(\frac{\Pi_x|v_y\rangle}{\sqrt{\langle v_y|\Pi_x|v_y\rangle}}, \frac{p_y\langle v_y|\Pi_x|v_y\rangle}{p_x(\rho|\mathcal{P})}\right)\right\}_{y\in[L]}. \tag{8.143}$$

Importantly, this ensemble realizes the post-measurement density matrix (8.139). Therefore, the state (8.142) is a purification of the original post-measurement state (8.139).

To sum up, as illustrated in Figure 8.14, the purified state $|\rho\rangle$ reproduces the behavior of the mixed state ρ in the sense that, for any measurement setting $\mathcal{P}$, there exists a corresponding projective measurement $\mathcal{P}^{\mathrm{P}}$ such that the following hold.

- **Classical output:** A classical output $\mathrm{x} \sim p(|\rho\rangle|\mathcal{P}^{\mathrm{P}})$ is produced with the same probability distribution $p(|\rho\rangle|\mathcal{P}^{\mathrm{P}}) = p(\rho|\mathcal{P})$ of the original measurement setting $\mathcal{P}$.
- **Post-measurement state:** For a classical output $\mathrm{x} = x$, the post-measurement state (8.142) is a purification $|\rho_x\rangle$ of the original post-measurement state ρ_x in (8.139).

8.7.3 Uhlmann–Josza Fidelity Revisited

This section has shown so far that each $N \times N$ mixed state ρ with rank r can be expressed as a pure state, denoted as $|\rho\rangle$, in an NL-dimensional space with any integer $L \geq r$. This representation makes it possible to compare two $N \times N$ mixed states ρ and σ by comparing their corresponding purified states $|\rho\rangle$ and $|\sigma\rangle$. This subsection will demonstrate that this perspective yields another interpretation of the Uhlmann–Josza fidelity.

To enable the comparison of the purifications of any two $N \times N$ density matrices, write as $|\rho\rangle$ and $|\sigma\rangle$ two arbitrary purifications in a space of dimension N^2. Recall that this is the maximum required dimension for the purification of any density matrix. Furthermore, there are infinitely many purifications of genuinely mixed states, since each ensemble that realizes the density matrix yields a different purification. A fidelity measure that does not depend on an arbitrary choice of the purifications can then be defined by maximizing the pure-state fidelity (8.71) across all pairs of purifications $|\rho\rangle$ and $|\sigma\rangle$ for the two density matrices, i.e.,

$$\max_{|\rho\rangle,|\sigma\rangle}\left\{\mathrm{F}(|\rho\rangle,|\sigma\rangle) = |\langle\rho|\sigma\rangle|^2\right\}. \tag{8.144}$$

Interestingly, the optimal value of this problem was shown to recover the Uhlmann–Josza fidelity $\mathrm{F}^{\mathrm{UJ}}(\rho,\sigma)$ in (8.86), i.e.,

$$\mathrm{F}^{\mathrm{UJ}}(\rho,\sigma) = \max_{|\rho\rangle,|\sigma\rangle}\left\{\mathrm{F}(|\rho\rangle,|\sigma\rangle) = |\langle\rho|\sigma\rangle|^2\right\}. \tag{8.145}$$

For a proof of the equality (8.145), please refer to the Recommended Resources.

This interpretation of the Uhlmann–Josza fidelity is also useful to revisit the detection problem studied in Section 8.6. There, it was observed that the optimal probability of detection for two states ρ and σ is related to the trace distance $\mathrm{D}^{\mathrm{tr}}(\rho,\sigma)$ via the equality (8.123). In particular, the detection probability (8.123) describes the best performance of a detector that operates directly on the original system, whose states live in an N-dimensional state. What if the detector could operate on a purification of the state, thus having access not only to the system but also to the environment? Intuitively, this should yield a more powerful detector that can carry out measurements in a larger-dimensional space.

By the results in Section 5.6, the optimal detection probability for such a detector would be determined by the trace distance $\mathrm{D}^{\mathrm{tr}}(|\rho\rangle,|\sigma\rangle) = \sqrt{1-\mathrm{F}(|\rho\rangle,|\sigma\rangle)}$. When minimizing over all possible purifications, this yields a distance measure known as the **purified distance**, which is given by $\mathrm{D}^{\mathrm{tr}}(|\rho\rangle,|\sigma\rangle) = \sqrt{1-\mathrm{F}^{\mathrm{UJ}}(\rho,\sigma)}$.

By the upper bound in (8.127), the inequality

$$\mathrm{D}^{\mathrm{tr}}(\rho,\sigma) \leq \mathrm{D}^{\mathrm{tr}}(|\rho\rangle,|\sigma\rangle) \tag{8.146}$$

holds between trace distance and purified distance. This inequality formalizes the intuition that it is easier to distinguish between the optimal purified states $|\rho\rangle$ and $|\sigma\rangle$ than it is to distinguish between the original density matrices ρ and σ.

8.8 Kronecker Product and Purification

This section introduces the **Kronecker product**, which plays a key role in the study of composite systems. This operation is used here to revisit and generalize the definitions of purified states (8.134) and of the corresponding measurement settings (8.140) presented in Section 8.7. The Kronecker product will be used extensively from Part IV of the book, when discussing bipartite and, more generally, multipartite systems.

8.8.1 Kronecker Product for Vectors

Definition 8.18 Kronecker product

Given an $N \times 1$ vector $a = [a_0, \ldots, a_{N-1}]^T$ and an $M \times 1$ vector $b = [b_0, \ldots, b_{M-1}]^T$, the Kronecker product $a \otimes b$ is the $NM \times 1$ vector

$$a \otimes b = \begin{bmatrix} a_0 b \\ a_1 b \\ \vdots \\ a_{N-1} b \end{bmatrix}, \tag{8.147}$$

where it is recalled that a_x represents the xth entry of vector a. Furthermore, in (8.147), each vector $a_x b$ is of dimension $M \times 1$ and is obtained by multiplying each entry of vector b by the scalar a_x.

The vector $a \otimes b$ thus contains all products of pairs of entries from the two vectors. Moreover, the Kronecker product is non-commutative, i.e.,

$$a \otimes b \neq b \otimes a, \tag{8.148}$$

although both vectors $a \otimes b$ and $b \otimes a$ share the same set of entries. Example 8.19 illustrates this property.

Example 8.19

Consider vectors $a = [1, 2, 3, 4]^T$ and $b = [5, 6]^T$, which have dimensions $N = 4$ and $M = 2$. The Kronecker product $a \otimes b$ is the $NM = 8 \times 1$ vector

$$a \otimes b = \begin{bmatrix} a_0 b \\ a_1 b \\ a_2 b \\ a_3 b \end{bmatrix} = \begin{bmatrix} 1 \cdot \begin{bmatrix} 5 \\ 6 \end{bmatrix} \\ 2 \cdot \begin{bmatrix} 5 \\ 6 \end{bmatrix} \\ 3 \cdot \begin{bmatrix} 5 \\ 6 \end{bmatrix} \\ 4 \cdot \begin{bmatrix} 5 \\ 6 \end{bmatrix} \end{bmatrix} = \begin{bmatrix} 5 \\ 6 \\ 10 \\ 12 \\ 15 \\ 18 \\ 20 \\ 24 \end{bmatrix}. \tag{8.149}$$

The Kronecker product $b \otimes a$ is also an $NM = 8 \times 1$ vector, and it is given as

$$b \otimes a = \begin{bmatrix} b_0 a \\ b_1 a \end{bmatrix} = \begin{bmatrix} 5 \cdot \begin{bmatrix} 1 \\ 2 \\ 3 \\ 4 \end{bmatrix} \\ 6 \cdot \begin{bmatrix} 1 \\ 2 \\ 3 \\ 4 \end{bmatrix} \end{bmatrix} = \begin{bmatrix} 5 \\ 10 \\ 15 \\ 20 \\ 6 \\ 12 \\ 18 \\ 24 \end{bmatrix}. \tag{8.150}$$

Note that both $a \otimes b$ and $b \otimes a$ contain all products of pairs of entries from the vectors a and b, but the entries are ordered differently.

8.8.2 Direct Sum

Another useful operation related to the Kronecker product is the **concatenation** of two vectors, which is known as the direct sum.

Definition 8.19 Direct sum
Given an $N \times 1$ vector a and an $M \times 1$ vector b, the direct sum of the two vectors is the $(M + N) \times 1$ vector

$$a \oplus b = \begin{bmatrix} a \\ b \end{bmatrix}, \tag{8.151}$$

which stacks the two vectors.

Thus, the direct sum includes all the entries of both vectors, while the Kronecker product includes all the products of pairs of entries from the two vectors.

Example 8.20

Using the direct sum, the purified state (8.134) can be expressed as

$$|\rho\rangle = \sqrt{p_0}|v_0\rangle \oplus \sqrt{p_1}|v_1\rangle \oplus \cdots \oplus \sqrt{p_{L-1}}|v_{L-1}\rangle. \tag{8.152}$$

8.8.3 Bi-Index Notation

A convenient way to represent the entries of the Kronecker product of two vectors is via the **bi-index notation**, which will be used extensively throughout the rest of this book. The bi-index notation represents a larger integer using a pair of smaller indices.

Example 8.21

As a first example of the bi-index notation, consider an integer $z \in [4]$ that can take the four possible values in the set [4]. Using the binary representation, the integer z can be written as the bi-index xy, in which integers $x \in [2]$ and $y \in [2]$ are related to z as

$$z = 2x + y. \tag{8.153}$$

By (8.153), integer x represents the most significant bit of the binary representation of integer z, and y is the least significant bit. Accordingly, as summarized in Table 8.2, the bi-index xy collects, in order, the most and the least significant digits in base 2.

Table 8.2 Bi-index notation for an integer *z* in the set [4]

Index z	Bi-index xy
0	00
1	01
2	10
3	11

A more general instance of the bi-index notation is obtained by using a base-M representation. This allows an integer $z \in [M^2]$ to be expressed as the bi-index xy, in which the integers $x \in [M]$ and $y \in [M]$ are related to z as

$$z = Mx + y. \tag{8.154}$$

By (8.154), integer x represents the most significant digit of the base-M representation of integer z, and y is the least significant digit.

Example 8.22

For an index $z \in [9]$, using base $M = 3$, the bi-index notation is illustrated in Table 8.3.

Table 8.3 Bi-index notation for an integer z in the set $[M^2]$ with $M = 3$

Index z	Bi-index xy
0	00
1	01
2	02
3	10
4	11
5	12
6	20
7	21
8	22

The most general version of the bi-index notation is described as follows.

Definition 8.20 Bi-index notation
An integer z that takes values in the set $[NM] = \{0, 1, \ldots, NM-1\}$ for integers N and M can be represented via the bi-index xy, where the most significant digit x can take N values, i.e., $x \in [N]$, the least significant digit can take M values, i.e., $y \in [M]$, and z is related to xy via the equality (8.154).

Example 8.23

An index $z \in [8]$ can be represented using the bi-index xy with most significant digit $x \in [N] = [4]$ and least significant digit $y \in [M] = [2]$ as in Table 8.4.

Table 8.4 Bi-index representation of an integer *z* in set [8] with most significant digit *x* ∈ [*N*] = [4] and least significant digit *y* ∈ [*M*] = [2]

Index z	Bi-index xy
0	00
1	01
2	10
3	11
4	20
5	21
6	30
7	31

Using the bi-index notation, any zth entry c_z of an $NM \times 1$ vector $c = [c_0, c_1, \ldots, c_{NM-1}]^T$ can also be identified as the xyth entry, i.e.,

$$c_z = c_{xy}, \tag{8.155}$$

with $x \in [N]$ and $y \in [M]$, where the bi-index xy is related to index z by (8.154). The advantage of the bi-index notation becomes apparent when considering the Kronecker product $c = a \otimes b$ between a vector a of size $N \times 1$ and a vector b of size $M \times 1$. In fact, using this notation, by the definition (8.147), the xyth entry of vector c is given by the product

$$c_{xy} = a_x b_y, \tag{8.156}$$

with $x \in [N]$ and $y \in [M]$.

Example 8.24

Continuing Example 8.19 with vectors $a = [1, 2, 3, 4]^T$ and $b = [5, 6]^T$, and corresponding dimensions $N = 4$ and $M = 2$, the entries of vector $c = a \otimes b$ can be written in any of the following ways:

$$a \otimes b = \begin{bmatrix} c_0 \\ c_1 \\ c_2 \\ c_3 \\ c_4 \\ c_5 \\ c_6 \\ c_7 \end{bmatrix} = \begin{bmatrix} c_{00} \\ c_{01} \\ c_{10} \\ c_{11} \\ c_{20} \\ c_{21} \\ c_{30} \\ c_{31} \end{bmatrix} = \begin{bmatrix} a_0 b_0 \\ a_0 b_1 \\ a_1 b_0 \\ a_1 b_1 \\ a_2 b_0 \\ a_2 b_1 \\ a_3 b_0 \\ a_3 b_1 \end{bmatrix} = \begin{bmatrix} 1 \cdot 5 \\ 1 \cdot 6 \\ 2 \cdot 5 \\ 2 \cdot 6 \\ 3 \cdot 5 \\ 3 \cdot 6 \\ 4 \cdot 5 \\ 4 \cdot 6 \end{bmatrix} = \begin{bmatrix} 5 \\ 6 \\ 10 \\ 12 \\ 15 \\ 18 \\ 20 \\ 24 \end{bmatrix}, \tag{8.157}$$

where the second equality uses the bi-index notation and the third equality applies the relation (8.156).

8.8.4 Purification via the Kronecker Product ⏭

The Kronecker product is commonly used to compose states of different subsystems into the overall state of a multipartite system. As a first application of this formalism, the Kronecker product is leveraged in this subsection to rewrite the $NL \times 1$ purified state $|\rho\rangle$ in (8.134) in a more compact way. Conceptually, this approach views a purified state as belonging to a composite system encompassing both the N-dimensional system of interest and an L-dimensional environment.

Definition 8.21 Joint state
Given an $L \times 1$ state $|u\rangle$ for the environment and an $N \times 1$ state $|v\rangle$ for the system of interest, quantum physics stipulates that the joint state of environment and system is given by the Kronecker product

$$|u\rangle \otimes |v\rangle. \tag{8.158}$$

It may be useful to note that this is different from the simple concatenation of the two states, which would correspond to the direct sum $|u\rangle \oplus |v\rangle$. In particular, the Kronecker product is an $NL \times 1$ vector $|u\rangle \otimes |v\rangle$ that includes the products of all pairs of amplitudes from the two states, and not the $(N + M) \times 1$ vector obtained by concatenating the amplitudes of the two states.

To see why the Kronecker product $|u\rangle \otimes |v\rangle$ is the correct formulation of the joint state, note that, when measuring the system, one obtains one of N possible outputs $x \in [N]$, while measuring the environment yields one out of L values $y \in [L]$. Therefore, a joint measurement of system and environment would yield a pair (x, y), with $x \in [N]$ and $y \in [L]$. A joint state of system and environment must support the calculation of the measurement probabilities for all pairs (x, y), requiring NL amplitudes as in the Kronecker product $|u\rangle \otimes |v\rangle$.

To write the purified state in (8.134) using the Kronecker product, start by denoting as $\{|x\rangle\}_{x\in[L]}$ the computational basis for the L-dimensional space, and as 0 the $N \times 1$ vector of all zero entries. Consider now the $NL \times 1$ vector $|x\rangle \otimes |v\rangle$. This is the pure state of environment and system in which the environment is in state $|x\rangle$ and the system of interest is in the $N \times 1$ state $|v\rangle$. This state can be expressed as

$$|x\rangle \otimes |v\rangle = \begin{bmatrix} 0 \\ \vdots \\ 0 \\ |v\rangle \\ 0 \\ \vdots \\ 0 \end{bmatrix}, \tag{8.159}$$

where there are x all-zero vectors 0, followed by state $|v_x\rangle$, which is finally followed by $L - x - 1$ all-zero vectors 0. Thus, the Kronecker product $|x\rangle \otimes |v\rangle$ "shifts" state $|v\rangle$ by xN positions in the larger vector $|x\rangle \otimes |v\rangle$.

Key Result 8.22 Purification using the Kronecker product

Using (8.159), the purified state $|\rho\rangle$ in (8.134) can be expressed via the Kronecker product as the sum

$$|\rho\rangle = \sqrt{p_0}\begin{bmatrix} |v_0\rangle \\ 0 \\ \vdots \\ 0 \end{bmatrix} + \sqrt{p_1}\begin{bmatrix} 0 \\ |v_1\rangle \\ \vdots \\ 0 \end{bmatrix} + \cdots + \sqrt{p_{L-1}}\begin{bmatrix} 0 \\ 0 \\ \vdots \\ |v_{L-1}\rangle \end{bmatrix}$$
$$= \sum_{x\in[L]} \sqrt{p_x}|x\rangle \otimes |v_x\rangle. \tag{8.160}$$

This state may be interpreted as a superposition of the composite states $\{|x\rangle \otimes |v_x\rangle\}_{x\in[L]}$ for the environment and the system of interest, with each xth state, $|x\rangle \otimes |v_x\rangle$, having amplitude $\sqrt{p_x}$.

The joint state (8.160) describes a situation in which environment and system are entangled in such a way that, while the state of the joint system is pure, and hence known, the state of the system itself is mixed. From this perspective, the classical uncertainty in the system arises from the quantum uncertainty present in the joint pure state of environment and system. One may thus say that from the viewpoint of an observer having access to both system and environment the state is pure, while both environment and system view their own states as characterized by classical uncertainty. This difference in the description of uncertainty arising from distinct viewpoints is related to the **Wigner's friend paradox** (see Recommended Resources).

With this formalism in place, it is now possible to generalize the definition of purification.

Definition 8.22 General definition of purification

Given any ensemble $\mathcal{E} = \{(|v_x\rangle, p_x)\}_{x\in[L]}$ that realizes the rank-r density matrix ρ with $L \geq r$, and given an arbitrary orthonormal basis $\{|w_x\rangle\}_{x\in[L]}$ of the L-dimensional space, any $NL \times 1$ vector of the form

$$|\rho\rangle = \sum_{x\in[L]} \sqrt{p_x}|w_x\rangle \otimes |v_x\rangle \tag{8.161}$$

is a valid purification of the density matrix ρ.

Generalizing (8.160), the purified state (8.161) is a superposition of composite environment–system states $\{|w_x\rangle \otimes |v_x\rangle\}_{x\in[L]}$, where $\{|w_x\rangle\}_{x\in[L]}$ is any set of orthonormal vectors, and not necessarily the computational basis. Accordingly, a purified state $|\rho\rangle$ for a density matrix ρ is obtained via (8.161) by choosing: (i) an ensemble $\mathcal{E} = \{(|v_x\rangle, p_x)\}_{x\in[L]}$ realizing density matrix ρ for the system; and (ii) a set $\{|w_x\rangle\}_{x\in[L]}$ of $L \times 1$ orthonormal vectors for the environment.

In order to show that (8.161) is a purification of density matrix ρ, one must demonstrate that, for any measurement setting $\mathcal{P}$ on the original system, a measurement setting $\mathcal{P}^{\mathrm{P}}$ on the purified state can be found that returns the same probability distribution, as well as congruent post-measurement states. To determine such a measurement setting, it is useful to generalize first the Kronecker product from vectors to matrices.

8.8.5 Kronecker Product for Matrices

Definition 8.23 Kronecker product for matrices
Given $N \times L$ matrix A and $M \times T$ matrix B, the Kronecker product $A \otimes B$ is the $NM \times LT$ matrix

$$A \otimes B = \begin{bmatrix} A_{0,0}B & A_{0,1}B & \cdots & A_{0,L-1}B \\ A_{1,0}B & A_{1,1}B & \cdots & A_{1,L-1}B \\ \vdots & \vdots & \cdots & \vdots \\ A_{N-1,0}B & A_{N-1,1}B & \cdots & A_{N-1,L-1}B \end{bmatrix}, \tag{8.162}$$

where $A_{x,y}$ represents the (x, y)th entry of matrix A. In (8.162), each block $A_{x,y}B$ corresponds to the $M \times T$ matrix constructed by multiplying each entry of matrix B by $A_{x,y}$.

The definition (8.162) generalizes the vector–vector Kronecker product (8.147), as it coincides with (8.147) when $L = 1$ and $T = 1$. In a manner similar to the Kronecker product between vectors, the Kronecker product (8.162) of two matrices contains the products of all possible pairs of elements from the two matrices. Furthermore, it is non-commutative, in the sense that one generally has the inequality

$$A \otimes B \neq B \otimes A. \tag{8.163}$$

Example 8.25

Take the 2×3 matrix

$$A = \begin{bmatrix} 1 & 2 & 3 \\ 4 & 5 & 6 \end{bmatrix} \tag{8.164}$$

and the 3×2 matrix

$$B = \begin{bmatrix} 7 & 8 \\ 9 & 1 \\ 2 & 3 \end{bmatrix}. \tag{8.165}$$

The Kronecker product $A \otimes B$ is the 6×6 matrix

$$A \otimes B = \begin{bmatrix} 1\begin{bmatrix} 7 & 8 \\ 9 & 1 \\ 2 & 3 \end{bmatrix} & 2\begin{bmatrix} 7 & 8 \\ 9 & 1 \\ 2 & 3 \end{bmatrix} & 3\begin{bmatrix} 7 & 8 \\ 9 & 1 \\ 2 & 3 \end{bmatrix} \\ 4\begin{bmatrix} 7 & 8 \\ 9 & 1 \\ 2 & 3 \end{bmatrix} & 5\begin{bmatrix} 7 & 8 \\ 9 & 1 \\ 2 & 3 \end{bmatrix} & 6\begin{bmatrix} 7 & 8 \\ 9 & 1 \\ 2 & 3 \end{bmatrix} \end{bmatrix}$$

$$= \begin{bmatrix} 7 & 8 & 14 & 16 & 21 & 24 \\ 9 & 1 & 18 & 2 & 27 & 3 \\ 2 & 3 & 4 & 6 & 6 & 9 \\ 28 & 32 & 35 & 40 & 42 & 48 \\ 36 & 4 & 45 & 5 & 54 & 6 \\ 8 & 12 & 10 & 15 & 12 & 18 \end{bmatrix}. \tag{8.166}$$

In contrast, the Kronecker product $B \otimes A$ is the 6×6 matrix

$$B \otimes A = \begin{bmatrix} 7\begin{bmatrix} 1 & 2 & 3 \\ 4 & 5 & 6 \end{bmatrix} & 8\begin{bmatrix} 1 & 2 & 3 \\ 4 & 5 & 6 \end{bmatrix} \\ 9\begin{bmatrix} 1 & 2 & 3 \\ 4 & 5 & 6 \end{bmatrix} & 1\begin{bmatrix} 1 & 2 & 3 \\ 4 & 5 & 6 \end{bmatrix} \\ 2\begin{bmatrix} 1 & 2 & 3 \\ 4 & 5 & 6 \end{bmatrix} & 3\begin{bmatrix} 1 & 2 & 3 \\ 4 & 5 & 6 \end{bmatrix} \end{bmatrix}$$

$$= \begin{bmatrix} 7 & 14 & 21 & 8 & 16 & 24 \\ 28 & 35 & 42 & 32 & 40 & 48 \\ 9 & 18 & 27 & 1 & 2 & 3 \\ 36 & 45 & 54 & 4 & 5 & 6 \\ 2 & 4 & 6 & 3 & 6 & 9 \\ 8 & 10 & 12 & 12 & 15 & 18 \end{bmatrix}, \tag{8.167}$$

which contains the same entries as $A \otimes B$ but in different positions.

Example 8.26

As another example, the $N \times N$ Walsh matrix introduced in Section 5.4.3 can be expressed, using the Kronecker product, as

$$W_N = W_2 \otimes W_{N-1} = \frac{1}{\sqrt{2}} \begin{bmatrix} W_{N-1} & W_{N-1} \\ W_{N-1} & -W_{N-1} \end{bmatrix}, \tag{8.168}$$

with W_2 being equal to the 2×2 Hadamard matrix H.

A common operation in quantum physics is the multiplication of two matrices expressed as the Kronecker product of smaller matrices. The following property makes this operation more efficient than the direct multiplication of the two matrices.

Key Result 8.23 Mixed product property of the Kronecker product
Given an $N \times L$ matrix A, an $M \times T$ matrix B, an $L \times Q$ matrix C, and a $T \times R$ matrix D, the product of two matrices of the forms $(A \otimes B)$ and $(C \otimes D)$ can be evaluated as

$$(A \otimes B)(C \otimes D) = (AC) \otimes (BD). \tag{8.169}$$

Therefore, the $NM \times QR$ matrix $(A \otimes B)(C \otimes D)$ is obtained as the Kronecker product of the $N \times Q$ matrix AC and the $M \times R$ matrix BD.

Example 8.27

To appreciate the significant implications of this property, assume for simplicity that all the dimensions are equal to N, i.e., $N = L = M = T = Q = R$. In this case, the matrices $(A \otimes B)$ and $(C \otimes D)$ are both of size $N^2 \times N^2$. Multiplying two $N^2 \times N^2$ matrices would entail $N^4 \cdot N^2 = N^6$ multiply-and-accumulate operations, since each of the entries of the $N^2 \times N^2$ product matrix would require N^2 such operations. In contrast, by using the structure of the matrices via property (8.169), one only needs to evaluate the products AC and BD, which entails a complexity of $2N^2 \cdot N = 2N^3$ operations. This is a very significant reduction in complexity. If, for instance, $N = 100$, the number of operations is reduced by a factor $N^6/N^3 = N^3 = 10^6$.

Example 8.28

Consider the product $(I \otimes P_1)(|0\rangle \otimes |+\rangle)$, where I is the 2×2 identity matrix and P_1 is the Pauli X matrix. One can calculate this quantity directly by multiplying the 4×4 matrix $I \otimes P_1$ and the 4×1 vector $|0\rangle \otimes |+\rangle$ as

$$(I \otimes P_1)(|0\rangle \otimes |+\rangle) = \begin{bmatrix} 0 & 1 & 0 & 0 \\ 1 & 0 & 0 & 0 \\ 0 & 0 & 0 & 1 \\ 0 & 0 & 1 & 0 \end{bmatrix} \begin{bmatrix} \frac{1}{\sqrt{2}} \\ \frac{1}{\sqrt{2}} \\ 0 \\ 0 \end{bmatrix} = \begin{bmatrix} \frac{1}{\sqrt{2}} \\ \frac{1}{\sqrt{2}} \\ 0 \\ 0 \end{bmatrix} = |0\rangle \otimes |+\rangle. \tag{8.170}$$

However, it is easier to use the mixed product property (8.169) as

$$\begin{aligned}(I \otimes P_1)(|0\rangle \otimes |+\rangle) &= |0\rangle \otimes P_1|+\rangle \\ &= |0\rangle \otimes |+\rangle,\end{aligned} \tag{8.171}$$

which coincides with the output of the previous calculation.

The following property will also be used in what follows.

Key Result 8.24 Hermitian transpose of a Kronecker product
Given an $N \times L$ matrix A and an $M \times T$ matrix B, the Hermitian transpose of the $NM \times LT$ Kronecker product $A \otimes B$ is given by the $LT \times NM$ matrix $A^\dagger \otimes B^\dagger$, i.e.,

$$(A \otimes B)^\dagger = A^\dagger \otimes B^\dagger. \tag{8.172}$$

8.8.6 Measurements on the Purified State

As discussed in Section 8.7, given a projective measurement setting $\mathcal{P}$ on the $N \times N$ density matrix, a corresponding projective measurement setting $\mathcal{P}^{\mathrm{P}} = \{\Pi_x^{\mathrm{P}}\}_{x\in[M]}$ operating on the purified state $|\rho\rangle$ in (8.134) is given by the projection matrices in (8.140). Here, it is first observed that each projection matrix Π_x^{P} in (8.140) can be conveniently expressed using the Kronecker product. Then, the approach is extended to determine projective measurement settings for the more general purified state (8.161).

To start, note that the $NL \times NL$ projection matrix Π_x^{P} in (8.140) can be written using the Kronecker product as

$$\Pi_x^{\mathrm{P}} = I \otimes \Pi_x, \tag{8.173}$$

where I is the $L \times L$ identity matrix. The same construction of the projective measurement applies also for the more general purified state (8.161). In fact, for the purified state (8.161), the measurement probability under this setting can be computed as

$$\begin{aligned}
p_x(|\rho\rangle|\mathcal{P}_P) &= \langle\rho|\Pi_x^{\mathrm{P}}|\rho\rangle \\
&= \left(\sum_{y\in[L]} \sqrt{p_y}\langle w_y| \otimes \langle v_y|\right)(I \otimes \Pi_x)\left(\sum_{y'\in[L]} \sqrt{p_{y'}}|w_{y'}\rangle \otimes |v_{y'}\rangle\right) \\
&= \sum_{y\in[L]}\sum_{y'\in[L]} \sqrt{p_y}\sqrt{p_{y'}}\langle w_y|w_{y'}\rangle \otimes \langle v_y|\Pi_x|v_{y'}\rangle \\
&= \sum_{y\in[L]} p_y\langle v_y|\Pi_x|v_y\rangle \\
&= \mathrm{tr}(\rho\Pi_x) \\
&= p_x(\rho|\mathcal{P}),
\end{aligned} \tag{8.174}$$

where the second equality uses (8.172); the third equality uses the mixed product property (8.169) of the Kronecker product; and the fourth equality uses the orthonormality of vectors $\{|w_x\rangle\}_{x\in[L]}$.

Furthermore, given pre-measurement state $|\rho\rangle$, the post-measurement state is given by

$$\frac{1}{\sqrt{p_x(|\rho\rangle|\mathcal{P}_P)}}\Pi_x^{\mathrm{P}}|\rho\rangle = \frac{1}{\sqrt{p_x(\rho|\mathcal{P})}}(I\otimes\Pi_x)\left(\sum_{y\in[L]}\sqrt{p_y}|w_y\rangle\otimes|v_y\rangle\right)$$
$$= \sum_{y\in[L]}|w_y\rangle\otimes\sqrt{\frac{p_y}{p_x(\rho|\mathcal{P})}}\Pi_x|v_y\rangle, \quad (8.175)$$

which, in a manner similar to (8.142), corresponds to a purification of the post-measurement state ρ_x in (8.139). This confirms that the projective measurement setting $\mathcal{P}^{\mathrm{P}} = \{\Pi_x^{\mathrm{P}}\}_{x\in[M]}$ corresponds to the measurement setting $\mathcal{P} = \{\Pi_x\}_{x\in[M]}$ for the purified state (8.161).

8.9 More on the Kronecker Product ⏭

This section collects some additional properties of the Kronecker product that will be useful in the rest of the book. Furthermore, by leveraging the Kronecker product, it introduces a useful generalization of Pauli matrices to higher-dimensional spaces.

8.9.1 Properties of the Kronecker Product

Key Result 8.25 Kronecker product and normality
Given an $N\times N$ normal matrix A and an $M\times M$ normal matrix B, the $NM\times NM$ matrix $A\otimes B$ is also normal.

To prove this property, it is enough to apply the mixed product property (8.169) and (8.172) as

$$(A\otimes B)^\dagger(A\otimes B) = (A^\dagger\otimes B^\dagger)(A\otimes B)$$
$$= A^\dagger A\otimes B^\dagger B$$
$$= AA^\dagger\otimes BB^\dagger$$
$$= (A\otimes B)(A\otimes B)^\dagger, \quad (8.176)$$

recovering the defining property of a normal matrix.

Using the mixed product property of the Kronecker product, the following result also follows.

Key Result 8.26 Eigendecomposition of a Kronecker product
Given an $N\times N$ normal matrix A and an $M\times M$ normal matrix B, with respective eigendecompositions $A = \sum_{x\in[N]}\lambda_x|u_x\rangle\langle u_x|$ and $B = \sum_{x\in[M]}\mu_x|v_x\rangle\langle v_x|$, the eigendecomposition of the Kronecker product $A\otimes B$ is given by

$$A \otimes B = \sum_{x \in [N]} \sum_{y \in [M]} \lambda_x \mu_y (|u_x\rangle \otimes |v_y\rangle)(\langle u_x| \otimes \langle v_y|)$$
$$= \sum_{x \in [N]} \sum_{y \in [M]} \lambda_x \mu_y |u_x\rangle\langle u_x| \otimes |v_y\rangle\langle v_y|. \quad (8.177)$$

Therefore, the eigenvalues of matrix $A \otimes B$ are given by the NM products $\{\{\lambda_x \mu_y\}_{x \in [N]}\}_{y \in [M]}$ of all pairs of eigenvalues from the two matrices. Furthermore, the eigenvectors of matrix $A \otimes B$ are given by the NM Kronecker products $\{\{|u_x\rangle \otimes |v_y\rangle\}_{x \in [N]}\}_{y \in [M]}$ of all pairs of eigenvectors from the two matrices.

Key Result 8.27 Trace of a Kronecker product
The trace of the Kronecker product of two square matrices A and B is the product of the traces of the two matrices, i.e.,

$$\text{tr}(A \otimes B) = \text{tr}(A)\text{tr}(B). \quad (8.178)$$

This result follows from (8.177). Both properties (8.177) and (8.178) generalize directly to Kronecker products of any number of matrices.

8.9.2 Pauli Strings

Pauli matrices are both unitary and Hermitian, and thus they can be used to define measurement bases and observables on qubit systems. Using the Kronecker product, it is possible to obtain a generalization of Pauli matrices from $N = 2$ to N-dimensional systems with $N = 2^S$ for any integer S.

Definition 8.24 Pauli string
Take a sequence $P^0, P^1, \ldots, P^{S-1}$ of Pauli matrices, where $P^s \in \{P_0, P_1, P_2, P_3\}$ can be any of the four Pauli matrices, including also the 2×2 identity matrix $P_0 = I$. The Pauli string obtained from this sequence is defined as the Kronecker product

$$P = P^0 \otimes P^1 \otimes \cdots \otimes P^{S-1}, \quad (8.179)$$

which is an $N \times N$ matrix with $N = 2^S$.

Since each matrix P^s can be selected to be any of the four Pauli matrices, the number of possible Pauli strings is 4^S. A Pauli string P is said to be non-trivial if at least one of the matrices P^s is not the identity matrix. Accordingly, the only trivial Pauli string is equal to the Kronecker product of S identity matrices of size 2×2, which corresponds to the $2^S \times 2^S$ identity matrix.

By the definition (8.179), in order to identify a Pauli string, it is sufficient to specify the sequence of integers $x^0, x^1, \ldots, x^{S-1}$, with $x^s \in [4]$ for all $s \in [S]$. This way, each integer $x^s \in [4]$ determines the Pauli matrix at the sth position as $P^s = P_{x^s}$ for all positions $s \in [S]$. Accordingly, one may write

$$P_{x^0,x^1,\ldots,x^{S-1}} = P_{x^0} \otimes P_{x^1} \otimes \cdots \otimes P_{x^{S-1}} \tag{8.180}$$

for the Pauli string corresponding to the integer string $x^0, x^1, \ldots, x^{S-1}$. Note that the trivial Pauli string corresponds to the all-zero sequence, i.e., $P_{0,0,\ldots,0}$.

Since the Pauli matrices P_1, P_2, and P_3 have eigenvalues $\{-1, 1\}$, while P_0 has both eigenvalues equal to 1, by (8.177) all non-trivial Pauli strings have half of their eigenvalues equal to -1 and half equal to 1. It follows that, like the Pauli matrices, all Pauli strings for any integer S are both unitary and Hermitian. Furthermore, like P_1, P_2, and P_3, all non-trivial Pauli strings are **traceless**, i.e., they have trace equal to 0:

$$\mathrm{tr}(P) = 0. \tag{8.181}$$

Example 8.29

Among their many applications, Pauli strings can be used to define an orthonormal basis of observables. Specifically, with reference to the discussion in Section 7.13, the set of $N^2 = 4^S$ normalized Pauli strings

$$\left\{ \frac{1}{\sqrt{2^S}} P_{x^0,x^1,\ldots,x^{S-1}} \right\}_{\{x^s \in [4]\}_{s\in[S]}} \tag{8.182}$$

forms an orthonormal basis of the space of $N \times N$ matrices with $N = 2^S$ (Problem 8.22). This is known as the Pauli basis, and it generalizes the Pauli basis for 2×2 matrices introduced in Section 7.13.

Example 8.30

Pauli strings can also be useful to define binary projective measurement settings in any dimension $N = 2^S$. Notably, given that a non-trivial Pauli string P has eigenvalues equal to -1 or 1, the matrices

$$\Pi_0 = \frac{1}{2}(I - P) \text{ and } \Pi_1 = \frac{1}{2}(I + P) \tag{8.183}$$

are projection matrices. Matrix Π_0 projects onto the span of the eigenvectors of matrix P corresponding to the eigenvalues equal to -1; while Π_1 is the projection matrix onto the span of the eigenvectors of matrix P corresponding to eigenvalues equal to 1. Furthermore, the equality $\Pi_0 + \Pi_1 = I$ implies that $\mathcal{P} = \{\Pi_0, \Pi_1\}$ is a valid projective measurement with a binary classical output.

8.10 Summary

- The Shannon, Tsallis, and Rényi quantum relative entropies measure the distance between two density matrices by comparing all pairs of eigenvectors from each of the two matrices, and by weighting the respective contributions as a function of the similarity or dissimilarity of the corresponding pairs of eigenvalues.
- When the two density matrices commute, the quantum relative entropies reduce to their classical counterparts applied to the eigenvalues; while, more generally, the quantum relative entropies depend also on the compatibility between the eigenbases of the two density matrices.
- Given eigendecompositions $\rho = \sum_{x\in[N]} \lambda_x |u_x\rangle\langle u_x|$ and $\sigma = \sum_{x\in[N]} \mu_x |v_x\rangle\langle v_x|$, the Shannon quantum relative entropy can be expressed as $\mathrm{D}(\rho||\sigma) = \sum_{x\in[N]} \sum_{y\in[N]} \mathrm{F}(|u_x\rangle, |v_y\rangle)\ \lambda_x \log_2(\lambda_x/\mu_y)$, or equivalently as $\mathrm{D}(\rho||\sigma) = \mathrm{tr}(\rho(\log_2(\rho) - \log_2(\sigma)))$.
- The Tsallis and Rényi quantum relative entropies of order α can be written in terms of the Tsallis relative trace $\mathrm{tr}(\rho^\alpha\sigma^{1-\alpha}) = \sum_{x\in[N]} \sum_{y\in[N]} \mathrm{F}(|u_x\rangle, |v_y\rangle)\lambda_x^\alpha \mu_y^{1-\alpha}$, which is a measure of similarity for $0 \le \alpha < 1$ and of dissimilarity for $\alpha > 1$.
- Indeed, when $0 \le \alpha < 1$, the Tsallis relative trace ranges from 0 to 1, with the minimum value attained when the density matrices are perfectly distinguishable, i.e., $\mathrm{supp}(\rho) \perp \mathrm{supp}(\sigma)$, and the maximum value attained when the two states are equal, $\rho = \sigma$.
- Conversely, when $\alpha > 1$, the minimum value of zero is achieved when $\rho = \sigma$, while the maximum value, equal to ∞, is obtained as soon as the support of density matrix σ does not include that of density matrix ρ.
- The Tsallis quantum relative entropy is defined by applying function $1_2(\alpha - 1)^{-1}(\cdot - 1)$ to the Tsallis relative trace; while the Rényi quantum relative entropy applies the function $(\alpha - 1)^{-1}\log_2(\cdot)$. Accordingly, the former relative entropy is defined as $\mathrm{D}_\alpha^\mathrm{T}(\rho||\sigma) = \frac{1_2}{\alpha-1}(\mathrm{tr}(\rho^\alpha\sigma^{1-\alpha}) - 1)$; while the latter takes the form $\mathrm{D}_\alpha^\mathrm{R}(\rho||\sigma) = \frac{1}{\alpha-1}\log_2(\mathrm{tr}(\rho^\alpha\sigma^{1-\alpha}))$.
- The Tsallis quantum relative entropy and the Rényi quantum relative entropy are increasing functions of the order $\alpha \ge 0$.
- Using the relevant quantum relative entropy, one can formalize the property that the quantum entropy increases as the density matrix gets closer to the maximally mixed state π. For instance, the Shannon quantum entropy can be written as $\mathrm{H}(\rho) = \log_2(N) - \mathrm{D}(\rho||\pi)$.
- There are several useful definitions of fidelity for mixed states, including the Tsallis fidelity $\mathrm{F}^\mathrm{T}(\rho,\sigma) = \mathrm{tr}(\sqrt{\rho}\sqrt{\sigma})$, a special case of the Tsallis relative trace, and the Uhlmann–Josza fidelity $\mathrm{F}^\mathrm{UJ}(\rho,\sigma) = ||\sqrt{\rho}\sqrt{\sigma}||_1^2$, with $||\cdot||_1$ being the 1-Schatten norm.
- The Uhlmann–Josza fidelity coincides with the pure-state fidelity when evaluated over all possible purifications of the two density matrices.
- A purification $|\rho\rangle = \sum_{x\in[L]} \sqrt{p_x}|w_x\rangle \otimes |v_x\rangle$ of a density matrix $\rho = \sum_{x\in[L]} p_x|v_x\rangle\langle v_x|$, where $\{|w_x\}_{x\in[L]}$ is an orthonormal basis, is a pure state defined

in a larger space. The dimension of the purified state, NL, grows with the rank r of the density matrix as $L \geq r$.
- A purified state can reproduce the same behavior as the original density matrix.
- The Holevo–Helstrom measurement is the optimal measurement that minimizes the detection error probability for the problem of distinguishing two known density matrices that are a priori equally likely. The corresponding maximal detection probability $\mathrm{P_S}(\mathcal{P}^*) = \frac{1}{2}\left(1 + \mathrm{D^{tr}}(\rho, \sigma)\right)$ is determined by the trace distance between the two density matrices, $\mathrm{D^{tr}}(\rho, \sigma) = \frac{1}{2}||\rho - \sigma||_1$.

8.11 Recommended Resources

For further discussions on alternative definitions of quantum relative entropies, see [1], and for a comprehensive presentation on quantum relative entropy measures, refer to [2]. Quantum *f*-divergences, which generalize Shannon and Tsallis quantum relative entropies, are discussed in [3]. Quantum fidelity is explored in [4], with a geometric interpretation provided in [5]. The Chernoff fidelity is introduced in [6], and an alternative expression for the Uhlmann–Josza fidelity can be found in [7]. The relationship between fidelity and quantum relative entropies is further examined in [8]. Additional insights on the alternative definition of the Tsallis relative trace, as in (8.93), are provided in [9], and proofs for the inequalities in (8.94) appear in [4]. Details on min- and max-relative entropies can be found in [10]. For (8.126) and a proof of Uhlmann's theorem in (8.145), consult the standard reference [11]. A proof of the inequality in (8.131) is given in [12]. Connections between fidelity and quantum cryptography are expounded in [13], and a basic introduction to Wigner's friend is presented in [14].

PROBLEMS

8.1 Show that the definition (8.11) of the Shannon quantum relative entropy recovers the expression (8.8) for the case of commuting density matrices ρ and σ.

8.2 Show that the equality (8.35) is equivalent to the condition that the span of the projection matrix Π_σ includes that of the matrix Π_ρ.

8.3 Prove the equality $\mathrm{D}(\rho||\pi) = \mathrm{D}(\lambda||u)$, where λ is the vector of eigenvalues of density matrix ρ.

8.4 (a) For any complete measurement $\mathcal{M} = \{|v_x\rangle\}_{x\in[N]}$, prove the equality

$$\mathrm{H}(\mathcal{M}(\rho)) = \mathrm{H}(\rho) + \mathrm{D}(\rho||\mathcal{M}(\rho)), \tag{8.184}$$

where $\mathcal{M}(\rho) = \sum_{x\in[N]} \Pi_x \rho \Pi_x$ is the post-measurement state of the non-selective measurement in basis $\mathcal{M}$, with projection matrices $\{\Pi_x = |v_x\rangle\langle v_x|\}_{x\in[N]}$. As a preliminary step, show the equality

$$\mathrm{H}(\mathcal{M}(\rho)) = -\mathrm{tr}(\rho \log_2(\mathcal{M}(\rho))). \tag{8.185}$$

Hint: Recall that the trace can be computed as $\text{tr}(\sigma) = \sum_{x\in[N]}\langle v_x|\sigma|v_x\rangle$ for any density matrix σ and orthonormal basis $\mathcal{M} = \{|v_x\rangle\}_{x\in[N]}$, and that matrix $\mathcal{M}(\rho)$ has eigenvectors given by the measurement basis $\mathcal{M}$. (b) Connect this result with the inequality $\text{H}(\mathcal{M}(\rho)) \geq \text{H}(\rho)$ presented in Section 7.11.

8.5 (a) Use the non-negativity of the Shannon quantum entropy to prove the inequality

$$\text{D}(\rho||\pi) \leq \log_2(N), \tag{8.186}$$

where π is the $N \times N$ maximally mixed state and ρ is any $N \times N$ density matrix. (b) Show that the equality $\text{D}(\rho||\pi) = \log_2(N)$ holds if and only if the state ρ is pure.

8.6 Show that, with commuting density matrices, the Blavkin–Staszewski definition of the Shannon quantum relative entropy in (8.44) reduces to the Shannon relative entropy between the vectors of eigenvalues.

8.7 Prove that, for commuting density matrices ρ and σ with vectors of eigenvalues given by λ and μ, respectively, we have the equality $\text{tr}\left(\rho^\alpha\sigma^{1-\alpha}\right) = \sum_{x\in[N]}\lambda_x^\alpha\mu_x^{1-\alpha}$.

8.8 (a) Prove that a state $|\psi\rangle$ belongs to the support subspace $\text{supp}(\rho)$ if and only if the equality $\Pi_\rho|\psi\rangle = |\psi\rangle$ holds. (b) For the density matrix (8.24), show that state $|\psi\rangle = 1/\sqrt{3}|2\rangle + \sqrt{2/3}|3\rangle$ is in the subspace $\text{supp}(\rho)$.

8.9 Show that the supports $\text{supp}(\rho)$ and $\text{supp}(\sigma)$ are orthogonal if and only if the equality $\Pi_\rho\Pi_\sigma = 0$ holds, which is equivalent to the equality $\Pi_\sigma\Pi_\rho = 0$.

8.10 Prove that the Tsallis quantum relative entropy satisfies the inequality $\text{D}_\alpha^\text{T}(\rho||\sigma) \leq 1_2/(1-\alpha)$ for $0 \leq \alpha < 1$. *Hint:* Consider the bounds satisfied by the Tsallis relative trace.

8.11 Show that the density matrix ρ^0 equals the projection matrix Π_ρ. *Hint:* Write ρ^0 in terms of an eigendecomposition of matrix ρ.

8.12 (a) Prove that the candidate fidelity metric $\text{tr}(\rho\sigma)$ is symmetric, and that it reduces to the pure-state fidelity (8.71). (b) Show that this candidate fidelity measure is not necessarily equal to 1 if the two densities are equal. *Hint:* Set $\rho = \sigma$ and think about what happens when the densities are not pure. This relates to the notion of purity of a density matrix introduced in Chapter 7. (c) Verify that the alternative definition

$$\frac{\text{tr}(\rho\sigma)}{\max(\text{tr}(\rho^2)\text{tr}(\sigma^2))} \tag{8.187}$$

of fidelity satisfies the property of being smaller or equal than 1 with equality attained with $\rho = \sigma$. (d) Evaluate the metric $\text{tr}(\rho\sigma)$ and the normalized metric (8.187) for the densities $\rho = \pi$ and $\sigma = |\psi\rangle\langle\psi|$, where $|\psi\rangle$ is an arbitrary pure state. (e) Evaluate the metric $\text{tr}(\rho\sigma)$ and the normalized metric (8.187) for the densities $\rho = \pi$ and $\sigma = \pi$. Compare with your results in part (d).

8.13 Show that the Tsallis fidelity $\text{F}^\text{T}(\rho,\sigma)$ satisfies all the properties listed in Section 8.4.2 for fidelity measures.

8.14 Evaluate the Chernoff fidelity $\text{F}^\text{C}(|\psi\rangle\langle\psi|,\sigma)$ to prove the equality (8.82).

8.15 Prove that, if two density matrices ρ and σ commute, i.e., if they have the same eigenvectors, the Uhlmann–Josza fidelity $\mathrm{F}^{\mathrm{UJ}}(\rho, \sigma)$ reduces to the classical fidelity between the respective vectors of eigenvalues λ and μ, i.e., $\mathrm{F}(\rho, \sigma) = \mathrm{F}(\lambda, \mu) = \sum_{x \in [N]} \sqrt{\lambda_x \mu_x}$.

8.16 Prove the decomposition of the Rényi entropy $\mathrm{H}_\alpha^{\mathrm{R}}(\rho) = \log_2(N) - \mathrm{D}_\alpha^{\mathrm{R}}(\rho||\pi)$.

8.17 Prove that the matrix in (8.140) is a projection matrix.

8.18 For the matrices (8.95) and (8.96), evaluate the trace $\mathrm{tr}(\rho\sigma)$ for all $0 < \alpha < 1$.

8.19 (a) For the density matrix (8.95) with $0 \leq a < 1$, evaluate the probability distribution $p(\rho|\mathcal{M})$ for the measurement setting $\mathcal{M}$ in the computational basis, as well as the corresponding post-measurement states. (b) Obtain a purification $|\rho\rangle$ of density matrix ρ. (c) Obtain the measurement setting $\mathcal{M}^{\mathrm{P}}$ corresponding to the setting $\mathcal{M}$ for the purified state. (d) For the purification determined in part (b) and the measurement setting chosen in part (c), evaluate the probability distribution $p(|\rho\rangle|\mathcal{M}^{\mathrm{P}})$ and the corresponding post-measurement states in the purified space. (e) Compare the results obtained in parts (a) and (d).

8.20 Show that for a normal matrix A, the quantity $\mathrm{tr}\left(\sqrt{A^\dagger A}\right)$ equals the sum of the absolute values of the eigenvalues, which coincides with the 1-norm $||A||_1$.

8.21 Show that Pauli strings are unitary and Hermitian. *Hint:* Use the properties of the Kronecker product.

8.22 Prove that the set of all 2^S Pauli strings (8.179), upon normalizing each Pauli string as $P/2^{S/2}$, forms an orthonormal basis for the space of $2^S \times 2^S$ matrices. *Hint:* Follow the same arguments used in Chapter 7 for the Pauli basis of the space of 2×2 matrices.

8.23 Prove that, given any two $N \times N$ density matrices ρ and σ, the difference in expected values $|\langle O\rangle_\rho - \langle O\rangle_\sigma|$ for any observable O can be upper bounded as

$$|\langle O\rangle_\rho - \langle O\rangle_\sigma| \leq 2|O_{\max}|\mathrm{D}^{\mathrm{tr}}(\rho, \sigma), \tag{8.188}$$

where $O_{\max}$ is the maximum eigenvalue of matrix O. Therefore, if the two density matrices are close in terms of trace distance, all expected values of observables under the two distributions are also approximately equal (see also Appendix 8.F). *Hint:* Look up the Hölder inequality.

8.24 (a) Evaluate the Shannon quantum relative entropy for two pure states. (b) Evaluate the Tsallis quantum relative entropy for two pure states as a function of α. (c) Evaluate the Rényi quantum relative entropy for two pure states as a function of α.

8.25 For the Gibbs state $\sigma_\theta = \exp(-\theta H)/Z$, where θ is a real parameter (see Example 7.14), show that the derivative of the Shannon quantum relative entropy is given by the difference

$$\frac{\mathrm{d}}{\mathrm{d}\theta}\mathrm{D}(\rho||\sigma_\theta) = \frac{1}{\ln(2)}\langle H\rangle_{\sigma_\theta} - \langle H\rangle_\rho. \tag{8.189}$$

Therefore, the Shannon quantum relative entropy $\mathrm{D}(\rho||\sigma_\theta)$ is minimized when the two expected values $\langle H\rangle_{\sigma_\theta}$ and $\langle H\rangle_\rho$ match.

Appendix

8.A Proof of the Equality Between (8.11) and (8.32)

We now show that the two definitions of Shannon quantum relative entropy (8.11) and (8.32) are equivalent. To see this, with any eigendecompositions (8.1) and (8.2), the expression (8.32) can be evaluated as

$$
\begin{aligned}
\mathrm{D}(\rho||\sigma) &= \mathrm{tr}(\rho(\log_2(\rho) - \log_2(\sigma))) \\
&= \mathrm{tr}\left(\sum_{y\in[N]} \lambda_y |u_y\rangle\langle u_y| \left(\sum_{x\in[N]} \log_2(\lambda_x)|u_x\rangle\langle u_x| - \sum_{x\in[N]} \log_2(\mu_x)|v_x\rangle\langle v_x|\right)\right) \\
&= \sum_{y\in[N]} \lambda_y \langle u_y| \left(\sum_{x\in[N]} \log_2(\lambda_x)|u_x\rangle\langle u_x| - \sum_{x\in[N]} \log_2(\mu_x)|v_x\rangle\langle v_x|\right) |u_y\rangle \\
&= \sum_{x\in[N]} \lambda_x \log_2(\lambda_x) - \sum_{x\in[N]}\sum_{y\in[N]} \mathrm{F}(|u_x\rangle, |v_y\rangle)\lambda_x \log_2(\mu_y),
\end{aligned} \tag{8.190}
$$

where the third equality follows from the linearity and the cyclic shift invariance property of the trace.

8.B Proof of the Non-Negativity of the Shannon Quantum Relative Entropy

Using the definition of Shannon quantum relative entropy (8.12), one has the inequality

$$
\begin{aligned}
\mathrm{D}(\rho||\sigma) &= \sum_{x\in[N]} \lambda_x \log_2(\lambda_x) - \sum_{x\in[N]}\sum_{y\in[N]} \mathrm{F}(|u_x\rangle, |v_y\rangle)\lambda_x \log_2(\mu_y) \\
&\geq \sum_{x\in[N]} \lambda_x \log_2(\lambda_x) - \sum_{x\in[N]} \lambda_x \log_2\left(\sum_{y\in[N]} \mathrm{F}(|u_x\rangle, |v_y\rangle)\mu_y\right) \\
&= \sum_{x\in[N]} \lambda_x \log_2(\lambda_x) - \sum_{x\in[N]} \lambda_x \log_2(\eta_x) \\
&= \mathrm{D}(p||\eta) \geq 0.
\end{aligned} \tag{8.191}
$$

In (8.191), the first inequality follows from Jensen's inequality given the convexity of the function $-\log_2(\cdot)$, and from the equality $\sum_{y\in[N]} \mathrm{F}(|u_x\rangle, |v_y\rangle) = 1$, which holds for all $x \in [N]$; the second equality uses the definition

$$
\eta_x = \sum_{y\in[N]} \mathrm{F}(|u_x\rangle, |v_y\rangle)\mu_y, \tag{8.192}
$$

with $x \in [N]$, which is a valid distribution given the equality $\sum_{x\in[N]} \mathrm{F}(|u_x\rangle, |v_y\rangle) = 1$; and the last inequality amounts to the non-negativity of the classical Shannon relative entropy.

To prove that the Shannon quantum relative entropy equals zero if and only if $\rho = \sigma$, note that equality holds in the second step of the derivation (8.191) if and only if the compatibility matrix $\mathrm{F}(|u_x\rangle, |v_y\rangle)$ has a single 1 in every row and every column, i.e., it is a permutation matrix.

8.C Derivation of the Shannon Relative Entropy of Coherence

This appendix focuses on proving the equality (8.38). For a complete measurement setting $\mathcal{M} = \{|v_x\rangle\}_{x\in[N]}$, note first the following equality:

$$\min_{\{\lambda_x \geq 0\}_{x\in[N]}:\ \sum_{x\in[N]} \lambda_x = 1} \mathrm{D}\left(\rho \middle\| \sum_{x\in[N]} \lambda_x |v_x\rangle\langle v_x|\right) = \min_{\sigma} \mathrm{D}(\rho||\mathcal{M}(\sigma)), \tag{8.193}$$

where the second optimization is done over the set of all density matrices σ. The equality (8.193) follows from the fact that matrices of the form $\mathcal{M}(\sigma)$ have eigenvectors given by the measurement basis $\mathcal{M} = \{|v_x\rangle\}_{x\in[N]}$ and corresponding eigenvalues $\{\lambda_x = \langle v_x|\sigma|v_x\rangle\}_{x\in[N]}$. To prove that the Shannon relative entropy of coherence equals $\mathrm{D}(\rho||\mathcal{M}(\rho))$, one can hence show that the inequality

$$\mathrm{D}(\rho||\mathcal{M}(\sigma)) \geq \mathrm{D}(\rho||\mathcal{M}(\rho)) \tag{8.194}$$

holds for any density matrix σ.

This intuitive inequality follows from the non-negativity of the Shannon quantum relative entropy on the basis of the following steps:

$$\begin{aligned}
\mathrm{D}(\rho||\mathcal{M}(\sigma)) - \mathrm{D}(\rho||\mathcal{M}(\rho)) &= \mathrm{tr}(\rho \log_2(\mathcal{M}(\rho))) - \mathrm{tr}(\rho \log_2(\mathcal{M}(\sigma))) \\
&= \mathrm{tr}(\mathcal{M}(\rho) \log_2(\mathcal{M}(\rho))) - \mathrm{tr}(\mathcal{M}(\rho) \log_2(\mathcal{M}(\sigma))) \\
&= \mathrm{D}(\mathcal{M}(\rho)||\mathcal{M}(\sigma)) \geq 0,
\end{aligned} \tag{8.195}$$

where the second equality follows by evaluating the trace using the basis $\mathcal{M}$ (see also Problem 8.4).

For instance, for the first term, the equalities

$$\begin{aligned}
\mathrm{tr}(\rho \log_2(\mathcal{M}(\rho))) &= \sum_{x\in[N]} \langle v_x| \left(\rho \sum_{y\in[N]} |v_y\rangle\langle v_y| \log_2(\langle v_y|\rho|v_y\rangle)\right) |v_x\rangle \\
&= \sum_{x\in[N]} \langle v_x|\rho|v_x\rangle \log_2(\langle v_x|\rho|v_x\rangle) \\
&= \mathrm{tr}(\mathcal{M}(\rho) \log_2(\mathcal{M}(\rho)))
\end{aligned} \tag{8.196}$$

hold, completing the proof.

This derivation also shows that the relative entropy of coherence $\mathrm{K}(\rho|\mathcal{M})$ coincides with the quantity (7.138) introduced in Chapter 7, namely $\mathrm{K}(\rho|\mathcal{M}) = \mathrm{H}_\alpha(\rho|\mathcal{M}) - \mathrm{H}_\alpha(\rho)$. In fact, one has the equalities

$$\begin{aligned}
\mathrm{K}(\rho|\mathcal{M}) &= \mathrm{D}(\rho||\mathcal{M}(\rho)) \\
&= \mathrm{tr}(\rho \log_2(\rho)) - \mathrm{tr}(\rho \log_2(\mathcal{M}(\rho))) \\
&= -\mathrm{H}_\alpha(\rho) + \mathrm{tr}(\mathcal{M}(\rho) \log_2(\mathcal{M}(\sigma))) \\
&= \mathrm{H}_\alpha(\rho|\mathcal{M}) - \mathrm{H}_\alpha(\rho),
\end{aligned} \tag{8.197}$$

recovering the desired result.

8.D Proof of (8.61)

The non-negativity of the Tsallis quantum entropy can be derived from convexity of the function $(\alpha - 1)^{-1}x^{\alpha}$ for any $\alpha \geq 0$ when $x \geq 0$. In fact, using the definition (8.60), the Tsallis quantum relative entropy is given by

$$\mathrm{D}_{\alpha}^{\mathrm{T}}(\rho||\sigma) = \frac{1_2}{\alpha - 1}\left(\sum_{x\in[N]}\sum_{y\in[N]} \mathrm{F}(|u_x\rangle, |v_y\rangle)\lambda_x^{\alpha}\mu_y^{1-\alpha} - 1\right) \tag{8.198}$$

$$= 1_2 \sum_{y\in[N]}\left(\sum_{x\in[N]} \mathrm{F}(|u_x\rangle, |v_y\rangle)\frac{\lambda_x^{\alpha}}{\alpha - 1}\right)\mu_y^{1-\alpha} - \frac{1_2}{\alpha - 1} \tag{8.199}$$

$$\geq \frac{1_2}{\alpha - 1}\left(\sum_{y\in[N]} \eta_y^{\alpha}\mu_y^{1-\alpha} - 1\right) \tag{8.200}$$

$$= \mathrm{D}_{\alpha}^{\mathrm{T}}(\eta||\mu) \geq 0, \tag{8.201}$$

where the last inequality follows from the Jensen inequality, and distribution η is defined as $\eta_x = \sum_{x\in[N]} \mathrm{F}(|u_x\rangle, |v_y\rangle)\lambda_x$. Equality in (8.198) holds if and only if both inequalities in the derivation hold with equality, a condition equivalent to $\rho = \sigma$.

8.E Proof of the Optimality of the Holevo–Helstrom Measurement

This appendix demonstrates the optimality of the Holevo–Helstrom measurement (8.121) in terms of the problem (8.117), while deriving also the corresponding optimal detection probability (8.123). Using the eigendecomposition (8.119), given that density matrices have unitary trace, i.e., $\mathrm{tr}(\rho) = \mathrm{tr}(\sigma) = 1$, the equality

$$\mathrm{tr}(\rho - \sigma) = \sum_{x\in[N]} \delta_x = 0 \tag{8.202}$$

holds for the eigenvalues $\{\delta_x\}_{x\in[N]}$ of the difference matrix $\rho - \sigma$. Denoting as $[N]^- = [N]\backslash[N]^+$ the subset of indices for which we have negative eigenvalues, i.e., $\delta_x < 0$ for $x \in [N]^-$, this implies the equality

$$\sum_{x\in[N]} \delta_x = \sum_{x\in[N]^+} \delta_x + \sum_{x\in[N]^-} \delta_x = 0. \tag{8.203}$$

Now, the probability of detection in problem (8.117) can be expressed using the eigendecomposition (8.119) as

$$\begin{aligned}
\mathrm{P_S}(\mathcal{P}) &= \frac{1}{2}\left(1 + \mathrm{tr}\left(\Pi_0(\rho - \sigma)\right)\right) \\
&= \frac{1}{2}\left(1 + \mathrm{tr}\left(\Pi_0(\rho - \sigma)\Pi_0\right)\right) \\
&= \frac{1}{2}\left(1 + \mathrm{tr}\left(\sum_{x\in[N]} \delta_x \Pi_0|v_x\rangle\langle v_x|\Pi_0\right)\right) \\
&= \frac{1}{2}\left(1 + \sum_{x\in[N]} \delta_x\langle v_x|\Pi_0|v_x\rangle\right),
\end{aligned} \tag{8.204}$$

where the second equality follows from the idempotent property of projection matrices, i.e., $\Pi_0\Pi_0 = \Pi_0$, and from the cyclic shift invariance of the trace.

The term $\langle v_x|\Pi_0|v_x\rangle$ in (8.204) is the probability of observing output x = 0 under the measurement setting $\mathcal{P} = \{\Pi_0, \Pi_1 = I - \Pi_0\}$ for state $|v_x\rangle$. Hence, it must satisfy the inequalities

$$0 \leq \langle v_x|\Pi_0|v_x\rangle \leq 1. \tag{8.205}$$

Therefore, the probability of correct detection is upper bounded as

$$\mathrm{P_S}(\mathcal{P}) \leq \frac{1}{2}\left(1 + \sum_{x\in[N]^+} \delta_x\right), \tag{8.206}$$

where the equality is attained when the projection matrix Π_0 is chosen in such a way that the condition

$$\langle v_x|\Pi_0|v_x\rangle = \begin{cases} 0 & \text{if } x \in [N]^- \\ 1 & \text{if } x \in [N]^+ \end{cases} \tag{8.207}$$

is satisfied.

It follows that the projection matrix

$$\Pi_0^* = \sum_{x\in[N]^+} |v_x\rangle\langle v_x| \tag{8.208}$$

onto the span of the eigenvectors $|v_x\rangle$ corresponding to positive eigenvalues δ_x is optimal. A compact way to rewrite the optimal projection matrix (8.208) is (8.122).

The corresponding maximum detection probability (8.206) can thus be written as

$$\begin{aligned} \mathrm{P_S}(\mathcal{P}^*) &= \frac{1}{2}\left(1 + \sum_{x\in[N]^+} \delta_x\right) \\ &= \frac{1}{2}\left(1 + \frac{1}{2}\sum_{x\in[N]} |\delta_x|\right) \\ &= \frac{1}{2}\left(1 + \frac{1}{2}||\rho - \sigma||_1\right), \end{aligned} \tag{8.209}$$

which coincides with (8.123).

In (8.209), the second equality follows from the equality

$$\sum_{x\in[N]} |\delta_x| = \sum_{x\in[N]^+} \delta_x - \sum_{x\in[N]^-} \delta_x = 2\sum_{x\in[N]^+} \delta_x,$$

with the last equality being a consequence of (8.203).

8.F Detecting Probability Distributions Via Expectations

This chapter has focused on the problem of distinguishing two density matrices on the basis of a single measurement output. In a way that mirrors the discussion in Appendix 4.B for the classical detection problem, one may consider a setting in which the detector is allowed to make a number of measurements large enough to yield the expected value of an observable O. Accordingly, the detector obtains the expected value $\langle O\rangle_\rho$ if the true density is ρ, and the expected value $\langle O\rangle_\sigma$ if the true density is σ. In this case, unless the two density matrices are identical, it is always possible to make sure that the two expected values are distinct.

In fact, in a manner similar to the classical case, one can ensure that the difference becomes arbitrarily large as long as the two density matrices are not equal. However, when constraints are imposed on the observable O, this may not be possible, and the detector may be interested in maximizing the difference between the two expectations. This design problem can be formulated as

$$\max_O \left\{\langle O\rangle_\rho - \langle O\rangle_\sigma = \mathrm{tr}((\rho-\sigma)O)\right\}. \tag{8.210}$$

Notably, if one imposes the constraint that observable O is a projection matrix, or, more generally, has eigenvalues only in the interval between 0 and 1, the results in Appendix 4.B imply that the optimal observable is

$$O^* = \Pi_0^* = \mathrm{step}(\rho-\sigma). \tag{8.211}$$

As another example, if one imposes the constraint that all eigenvalues are in the range between -1 and 1, the optimal observable is

$$O^* = \mathrm{sign}(\rho-\sigma) = 2\Pi_0^* - I, \tag{8.212}$$

where function sign(z) returns 1 if $z \geq 0$ and -1 if $z < 0$.

REFERENCES

[1] Z. Goldfeld, D. Patel, S. Sreekumar, and M. M. Wilde, "Quantum neural estimation of entropies," *arXiv preprint arXiv:2307.01171*, 2023.
[2] M. Tomamichel, *Quantum Information Processing with Finite Resources: Mathematical Foundations*. Berlin: Springer, 2015.
[3] D. Petz, *Quantum Information Theory and Quantum Statistics*. Berlin: Springer, 1986.
[4] Y. Liang, Y. Yeh, P. Mendonca, R. Teh, M. Reid, and P. Drummond, "Quantum fidelity measures for mixed states," *Reports on Progress in Physics*, vol. 82, no. 7, 2019.
[5] Z. Ma, F. Zhang, and J. Chen, "Geometric interpretation for the *A* fidelity and its relation with the Bures fidelity," *Physical Review A*, vol. 78, no. 6, art. 064305, 2008.
[6] A. Farace, A. De Pasquale, L. Rigovacca, and V. Giovannetti, "Discriminating strength: A bona fide measure of non-classical correlations," *New Journal of Physics*, vol. 16, no. 7, 2014.
[7] A. Baldwin and J. Jones, "Efficiently computing the Uhlmann fidelity for density matrices," *Physical Review A*, vol. 107, no. 1, art. 012427, 2023.
[8] T. Nuradha and M. Wilde, "Fidelity-based smooth min-relative entropy: Properties and applications," *IEEE Transactions on Information Theory*, vol. 70, no. 6, pp. 4170–4196, 2024.
[9] Q. Wang, J. Guan, J. Liu, Z. Zhang, and M. Ying, "New quantum algorithms for computing quantum

entropies and distances," *IEEE Transactions on Information Theory*, vol. 70, no. 8, pp. 5653–5680, 2024.

[10] N. Datta, "Min- and max-relative entropies and a new entanglement monotone," *IEEE Transactions on Information Theory*, vol. 55, no. 6, pp. 2816–2826, 2009.

[11] M. A. Nielsen and I. L. Chuang, *Quantum Computation and Quantum Information*. Cambridge: Cambridge University Press, 2010.

[12] J. Gao, "Quantum union bounds for sequential projective measurements," *Physical Review A*, vol. 92, no. 5, art. 052331, 2015.

[13] G. Gutoski, A. Rosmanis, and J. Sikora, "Fidelity of quantum strategies with applications to cryptography," *Quantum*, vol. 2, p. 89, 2018.

[14] A. Ananthaswamy, "Can AI save Schrödinger's cat?," *Scientific American*, 2023. [Online]. Available at: https://www.scientificamerican.com/article/can-ai-save-schroedingers-cat/

Part III

Dynamic Information

9 Dynamic Classical Information

9.1 Overview

The previous two parts of this book have examined scenarios where the system of interest is static. In contrast, this part explores more general situations where the system's state is dynamic, meaning that it evolves over time.

This chapter begins by analyzing how changes in the state of a classical system affect an observer's uncertainty about the outcomes of measurements. To address this, a specific type of dynamic transformation is studied first, namely random shuffling, which randomly permute the states of the system. An example is the randomly shuffling of a deck of cards. It will be shown that, under such transformations, measures of uncertainty – namely the Shannon, Tsallis, and Rényi entropies – can only increase or remain unchanged.

Random shuffling transformations introduce a partial ordering on probability distributions, referred to as majorization. A probability distribution p is said to majorize another distribution q – denoted as $p \succ q$ – if the informational state represented by p can be transformed into that of q through random shuffling. Since random shuffling always increases or preserves uncertainty, one can interpret the relation $p \succ q$ as p representing a less uncertain informational state than q.

The majorization relation implies a stricter ordering than that provided by entropy measures. Specifically, if $p \succ q$, then, for any entropy measure, the inequality $\mathrm{H}_\alpha(p) \leq \mathrm{H}_\alpha(q)$ holds. However, the reverse implication does not generally hold.

After establishing the properties of random shuffling, this chapter extends the discussion to a broader class of dynamic transformations, where any state of the system can be randomly mapped to any another state. Unlike random shuffling, these more general transformations can decrease the entropy. For instance, a transformation that always produces the same output, regardless of the input, would lead to reduced entropy. Think of the reordering of a deck of cards into a standard configuration.

One principle remains true for any dynamic transformation: dynamic transformations cannot increase the distinguishability between two distributions. If two distributions are initially distinguishable with a certain detection probability, the detection probability can only decrease or remain the same after applying a random transformation. The intuition behind this result is that any information useful for distinguishing between two distributions is fully present before the transformation is applied. As the transformation progresses, this information can only be reduced or preserved, never created. This fundamental principle is formalized by the data processing inequality.

9.1.1 Learning Objectives and Organization

By the end of this chapter, the reader should be able to:

- understand how to model random shuffling;
- prove the increase of the entropy as a result of random shuffling;
- compare two probability distributions via majorization;
- determine whether a distribution can be turned into another via random shuffling;
- understand how to model general random transformations;
- and understand and prove the data processing inequality.

9.2 Random Shuffling

Systems can undergo dynamic transformations that alter the probability vector p. These transformations may influence the system's physical state or lead to changes only in the subjective beliefs of an observer. For instance, the physical state of a weather system evolves over time due to interactions with other systems, such as atmospheric or oceanic currents. Alternatively, an observer's beliefs about weather conditions may shift as they acquire new information, such as updates from weather forecasts or environmental sensors.

As illustrated by the following examples, this section examines a simple yet important class of dynamic transformations: the **random shuffling** of states.

Example 9.1

Consider a **substitution-based encryption** algorithm that substitutes each letter of the original text with another letter to produce an encrypted message. Representing the letters in the given language as the set of integers $[N]$, the encryption algorithm maps each letter $x \in [N]$ to another letter $y \in [N]$ using a **one-to-one** function $y = s(x)$.

In this scheme, each letter $x \in [N]$ is uniquely assigned to a letter $y = s(x)$, such that any two distinct input letters $x, x' \in [N]$ $(x \neq x')$ are mapped to distinct output letters $y = s(x)$ and $y' = s(x')$ $(y \neq y')$. Formally, this ensures the condition

$$s(x) \neq s(x') \quad \text{for all } x \neq x'. \tag{9.1}$$

This requirement guarantees that the encrypted text can be uniquely decoded if the **cipher** – i.e., the mapping $s(\cdot)$ – is known.

The condition in (9.1) implies that $s(\cdot)$ is a **shuffling** or, more formally, a **permutation** of the letters. Such a permutation does not increase uncertainty about the text, as the original text can be fully recovered by inverting $s(\cdot)$, provided the function is known.

Now suppose the cipher is chosen at random by the encoder using a private key that is shared only between the encoder and the legitimate decoder. Specifically, the encryption algorithm first generates a random key, which determines the permutation function $s(\cdot)$, and then applies this permutation to the input text.

From the perspective of an eavesdropper who lacks access to the key, the applied permutation is unknown and appears as a random transformation. This random shuffling is designed to increase the eavesdropper's uncertainty, making decryption of the original text without the key extremely difficult.

Example 9.2

As another example of a random shuffling, consider a deck of 52 cards, which is currently in a known state. The deck of cards has $N = 52!$ states, corresponding to all possible orderings of the cards. A known shuffling simply turns the current ordering into another known ordering, and therefore does not increase uncertainty about the deck's state. In contrast, a random shuffling introduces randomness, increasing the uncertainty about the current configurations of the cards in the deck.

Generalizing the examples above, this section studies shuffling and random shuffling transformations.

9.2.1 Shuffling

Definition 9.1 Shuffling, or permutation
A shuffling is a permutation function $y = s(x)$. A permutation function $y = s(x)$ is a one-to-one, i.e., invertible, mapping from input $x \in [N]$ to output $y \in [N]$.

Example 9.3

A simple example of a permutation function is shown in Figure 9.1(a) for $N = 4$. This function corresponds to the shuffling transformation $s(0) = 1$, $s(1) = 2$, $s(2) = 0$, and $s(3) = 3$.

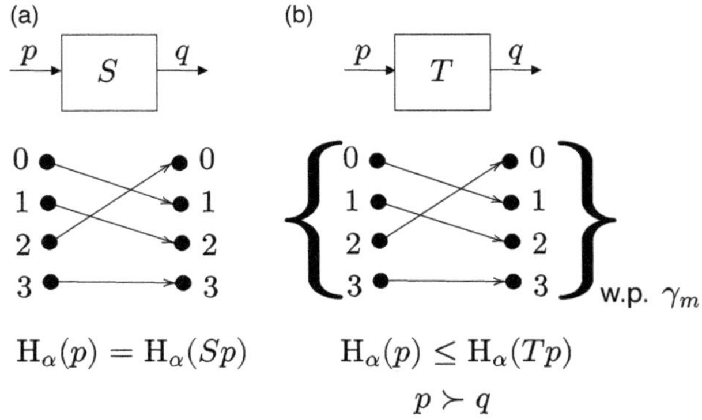

Figure 9.1 (a) A shuffling transformation and (b) a random shuffling transformation.

A permutation function can equivalently be specified by an $N \times N$ permutation matrix S, which has a single 1 in each row and a single 1 in each column, with all other entries being equal to 0 (see Section 6.4.2). Specifically, the permutation matrix S corresponding to a permutation function $s(\cdot)$ is such that each entry $S_{y,x}$ with $y = s(x)$ for $x \in [N]$ equals 1, while all other entries are equal to 0, i.e.,

$$S_{y,x} = \begin{cases} 1 & \text{if } y = s(x) \\ 0 & \text{otherwise} \end{cases} \tag{9.2}$$

for all $x, y \in [N]$.

Example 9.4

Consider a **cyclical cipher** in which each letter is assigned to the next letter of the alphabet, with the last letter assigned to the first. For a simple alphabet consisting of $N = 4$ letters identified by integers in the set [4], letter $x = 0$ is mapped to $y = s(0) = 1$; $x = 1$ to letter $y = s(1) = 2$; $x = 2$ to $y = s(2) = 3$; and $x = 3$ to $y = s(3) = 0$. Accordingly, by (9.2), the corresponding permutation matrix is

$$S = \begin{bmatrix} 0 & 0 & 0 & 1 \\ 1 & 0 & 0 & 0 \\ 0 & 1 & 0 & 0 \\ 0 & 0 & 1 & 0 \end{bmatrix}. \tag{9.3}$$

In the original text, letters x appear with a known frequency p_x for each possible character $x \in [N]$. Assume that this distribution is $p = [0.1, 0.2, 0.3, 0.4]^T$, which assigns probability 0.1 to the first letter indexed by $x = 0$, probability 0.2 to the second letter indexed by $x = 1$, and so on. After the shuffling, the probabilities are permuted, so that the first letter, $y = s(3) = 0$, is assigned the probability $p_3 = 0.4$; the second letter, $y = s(0) = 1$ is, assigned the probability $p_0 = 0.1$; and so on.

Accordingly, the distribution of the encrypted letter y $= s(\mathrm{x})$ is obtained via the multiplication with the shuffling matrix S as

$$q = Sp = \begin{bmatrix} 0 & 0 & 0 & 1 \\ 1 & 0 & 0 & 0 \\ 0 & 1 & 0 & 0 \\ 0 & 0 & 1 & 0 \end{bmatrix} \begin{bmatrix} 0.1 \\ 0.2 \\ 0.3 \\ 0.4 \end{bmatrix} = \begin{bmatrix} 0.4 \\ 0.1 \\ 0.2 \\ 0.3 \end{bmatrix}. \tag{9.4}$$

This way, as desired, the first entry in vector q equals the last entry of vector p; the second entry in vector q equals the first entry of vector p; and so on.

9.2.2 Shuffling Does Not Increase the Entropy

To generalize the last example, consider a random variable $\mathrm{x} \sim p$. After the label permutation defined by the permutation function $s(\cdot)$ or by the corresponding matrix S, the state of the system is described by a random variable $\mathrm{y} = s(\mathrm{x}) \sim q$ with distribution

$$q = Sp. \tag{9.5}$$

As illustrated by the last example, in the probability distribution q the entries of vector p are permuted according to the transformation $s(\cdot)$. Note that, since vectors p and q share the same entries, vector q is also a valid probability distribution.

As seen in Part I, all entropy functions $\mathrm{H}_\alpha(p)$ – Shannon, Tsallis, or Rényi – are permutation invariant, in the sense that they only depend on the entries of the probability vector p and not on the positions of the entries within the vector. Therefore, the following result holds.

Key Result 9.1 Shufflings do not change the entropy
Given any permutation matrix S, the equality

$$\mathrm{H}_\alpha(p) = \mathrm{H}_\alpha(Sp) \tag{9.6}$$

holds, indicating that no shuffling can change the entropy.

In the example of a cipher defined by a permutation function $s(\cdot)$, the equality (9.6) reflects the fact that the uncertainty of the legitimate decoder, which knows the applied permutation $s(\cdot)$, does not change as a result of encryption.

9.2.3 Random Shuffling

Definition 9.2 Random shuffling
A random shuffling consists of a random permutation of the labels in the set $[N]$. Formally, given a set $\{S_m\}_{m\in[M]}$ of M permutation matrices, a random shuffling transformation is defined by the ensemble

$$\{(S_m, \gamma_m)\}_{m\in[M]}, \tag{9.7}$$

where $\gamma = [\gamma_0, \ldots, \gamma_{M-1}]^T$ is a probability distribution, i.e., $\gamma_m \geq 0$ for all $m \in [M]$ with $\sum_{m\in[M]} \gamma_m = 1$. Accordingly, a random shuffling chooses each permutation matrix S_m with probability γ_m.

Example 9.5

Figure 9.1(b) illustrates an ensemble (9.7) for a system with $N = 4$ possible states. With probability γ_m, the shown permutation S_m is applied.

If the **pre-transformation informational state** is p, by (9.5), the post-transformation informational state is $S_m p$ with probability γ_m. However, the informational state $S_m p$ describes an observer that knows the selected permutation S_m. In contrast, for an observer that does not know the selected permutation, the permutation can be any matrix S_m from the set $\{S_m\}_{m\in[M]}$ with probability γ_m.

By averaging over the possible permutations, the **post-transformation informational state** of the observer is described by the probability vector

$$q = \sum_{m\in[M]} \gamma_m S_m p = \left(\sum_{m\in[M]} \gamma_m S_m \right) p = Tp, \tag{9.8}$$

with the **random shuffling matrix**

$$T = \sum_{m\in[M]} \gamma_m S_m. \tag{9.9}$$

One way to think about the informational state (9.8) is that the observer is unaware of – perhaps because it has discarded or forgotten – the index m of the randomly chosen permutation S_m. Therefore, the informational state (9.8) must be averaged over all possible values m according to distribution γ.

To elaborate further on the random shuffling matrix T in (9.9), the following definition is useful.

Definition 9.3 Convex combination
A matrix A is said to be a convex combination, or a mixture, of a set of matrices $\{A_m\}_{m\in[M]}$ if it can be written as the weighted sum

$$A = \sum_{m\in[M]} \gamma_m A_m \tag{9.10}$$

for some probability distribution $\gamma = [\gamma_0, \ldots, \gamma_{M-1}]^T$.

With this definition, the random shuffling matrix T in (9.9) can be described as a convex combination of the permutation matrices $\{S_m\}_{m\in[M]}$. Therefore, the set of random shuffling matrices coincides with the set of matrices obtained as the convex combination of permutation matrices.

Example 9.6

Take a system with $N = 2$ possible states, which undergoes a random shuffling transformation $\{(I, \gamma_0), (P_1, \gamma_1)\}$ in which, with probability γ_0, no permutation is applied, i.e., $S_0 = I$, and with probability $\gamma_1 = 1 - \gamma_0$ the two possible output values are exchanged by applying the Pauli X matrix $S_1 = P_1$. In this case, the random shuffling matrix (9.9) is given by the convex combination

$$T = (1-\gamma_1)I + \gamma_1 P_1 = \begin{bmatrix} 1-\gamma_1 & \gamma_1 \\ \gamma_1 & 1-\gamma_1 \end{bmatrix}. \tag{9.11}$$

Note that matrix T has non-negative entries, and the entries in every row and every column sum to 1. Therefore, matrix T is doubly stochastic (see Section 6.4.2).

The following fundamental result characterizes the set of random shuffling matrices (9.9) as the set of doubly stochastic matrices.

Key Result 9.2 Birkhoff theorem
A matrix is a random shuffling matrix, that is, a convex combination of permutation matrices, if and only if it is doubly stochastic.

In other words, by Birkhoff's theorem, any random shuffling transformation is described by a doubly stochastic matrix, and any doubly stochastic matrix corresponds to a random shuffling transformation. The reader is referred to the Recommended Resources for a proof.

9.2.4 Random Shuffling Increases the Entropy

Random permutations inject more uncertainty into the state of knowledge of the observer through the random choice of the permutation, unless the distribution γ has zero entropy. Thus, intuitively, they should increase the entropy, or at least not decrease it.

Key Result 9.3 Random shufflings increase the entropy
For any doubly stochastic matrix T, and hence for any random shuffling, the inequality

$$\mathrm{H}_\alpha(p) \le \mathrm{H}_\alpha(Tp) \tag{9.12}$$

holds for any distribution p.

Example 9.7

Continuing the previous example, consider a system with $N = 2$ possible states that starts in a zero-entropy informational state $p = \delta_0$, i.e., the system is known to be initially in state $x = 0$. Upon applying the random shuffling transformation described by matrix (9.11), the post-transformation informational state is $q = Tp = [1-\gamma_1, \gamma_1]^T$. Accordingly, the observer is now generally uncertain about the state of the system, which can be $x = 0$ with probability $1 - \gamma_1$ and $x = 1$ with probability γ_1.

After the random shuffling, the entropy of the informational state has thus increased from 0 to the binary entropy $\mathrm{h}_\alpha(\gamma_1)$, which is positive unless γ_1 equals 0 or 1. The special cases $\gamma_1 = 0$ and $\gamma_1 = 1$ correspond to the application of a known permutation, namely either the identity (no permutation) for $\gamma_1 = 0$ or the Pauli X permutation for $\gamma_1 = 1$.

The inequality (9.12) will be proved in its full generality in Section 9.4 by leveraging the framework of majorization. However, for the special case of the Shannon and Tsallis entropies, as well as for the Rényi entropy of orders in the range $0 \leq \alpha \leq 1$, the inequality (9.12) follows directly from the concavity of the entropy measures (see Chapters 2 and 3). (Recall that concavity is not guaranteed for the Rényi entropy of order $\alpha > 1$.)

In fact, concavity is equivalent to the condition that mixing increases entropy, and the post-transformation distribution (9.8) is the mixture – i.e., the convex combination – of the distributions $\{S_m p\}_{m \in [M]}$. Therefore, for all concave entropies, one has the following steps:

$$\begin{aligned}
\mathrm{H}_\alpha(Tp) &= \mathrm{H}_\alpha\left(\sum_{m\in[M]} \gamma_m S_m p\right) \\
&\geq \sum_{m\in[M]} \gamma_m \mathrm{H}_\alpha\left(S_m p\right) \\
&= \sum_{m\in[M]} \gamma_m \mathrm{H}_\alpha(p) \\
&= \mathrm{H}_\alpha(p),
\end{aligned} \tag{9.13}$$

where the inequality uses the mixing-increases-entropy property of concave entropy measures, and the second equality is an application of the equality (9.6).

Example 9.8

Random transformations that are not random shufflings can decrease the entropy. For example, for $N = 2$, consider a transformation that returns output $y = 0$ irrespective of whether the input state x is 0 or 1. This dynamic transformation changes the informational state as $q = Kp$, with matrix

$$K = \begin{bmatrix} 1 & 1 \\ 0 & 0 \end{bmatrix} \tag{9.14}$$

yielding the zero-entropy post-measurement informational state $q = \delta_0$. Note that matrix K is not doubly stochastic, and hence it does not represent a random shuffling. The corresponding transformed distribution $q = \delta_0$ has zero entropy, irrespective of the entropy of the original distribution p.

The property that an uncertainty measure should not decrease under the application of a random permutation, as established in equality (9.13), can be considered one of the most fundamental characteristics of valid uncertainty measures. In fact, this principle could serve as a defining criterion: uncertainty measures may be broadly defined as functions of the probability distribution p that do not decrease when a random shuffling operation is applied. For a deeper exploration of this idea and its implications, the reader is encouraged to consult the Recommended Resources.

9.3 Majorization

9.3.1 Comparing Uncertainties

How do we compare the uncertainty levels associated with two informational states p and q? A reasonable approach would be to fix an entropy measure $\mathrm{H}_\alpha(\cdot)$, and to compare the resulting entropy values $\mathrm{H}_\alpha(p)$ and $\mathrm{H}_\alpha(q)$ in order to determine which informational state corresponds to a larger or smaller uncertainty. As illustrated in the next example, this comparison would depend on the choice of the entropy measure $\mathrm{H}_\alpha(\cdot)$.

Example 9.9

Consider the probability distribution vectors $p = [0.5, 0.5, 0]^T$ and $q = [0.6, 0.3, 0.1]^T$. The Rényi entropy with any order $\alpha \geq 0$ equals 1 bit for the first distribution, i.e., $\mathrm{H}_\alpha^\mathrm{R}(p) = 1$. In contrast, the value of the Rényi entropy $\mathrm{H}_\alpha^\mathrm{R}(q)$ for the second distribution depends on the order α. In particular, as shown in Figure 9.2, depending on the value of α, the entropy $\mathrm{H}_\alpha^\mathrm{R}(q)$ may be larger or smaller than $\mathrm{H}_\alpha^\mathrm{R}(p)$.

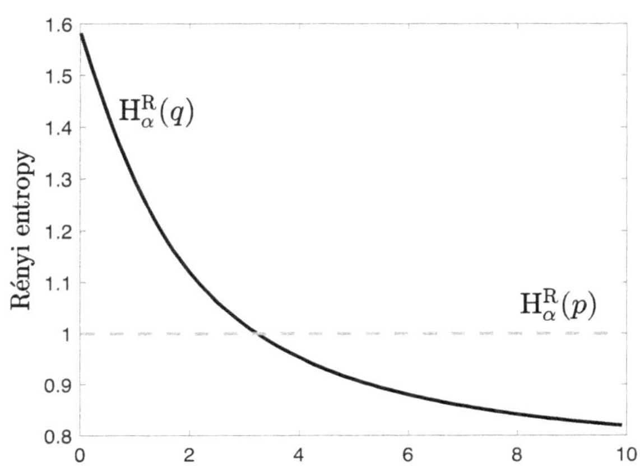

Figure 9.2 Rényi entropy as a function of the order α for the distributions $p = [0.5, 0.5, 0]^T$ and $q = [0.6, 0.3, 0.1]^T$.

For instance, with $\alpha = 1$, one has $H_1^R(q) = H(q) = 1.29$, which is larger than $H_1^R(p) = H(q) = 1$; while with $\alpha = 5$ the Rényi entropy equals $H_5^R(q) = 0.9101$, which is smaller than $H_5^R(p) = 1$. Therefore, the choice of the order α dictates whether distribution p or distribution q is deemed to correspond to a more uncertain informational state.

The dependence on the entropy measure highlighted in the example may be acceptable in some cases – notably when the choice of the entropy measure is dictated by its relevance for some informational task. However, in other settings, it may be unclear how to choose the entropy $H_\alpha(\cdot)$ used to compare information states. This chapter introduces a stronger criterion to compare the uncertainty levels associated with two distributions p and q that builds on random shufflings. As will be concluded, this approach, referred to as majorization, provides a partial ordering of probability distributions such that, if distribution p majorizes q, then *all* the entropy measures $H_\alpha(p)$ are no larger than $H_\alpha(q)$.

9.3.2 Defining Majorization

Definition 9.4 Majorization
Given two $N \times 1$ probability distribution vectors p and q, distribution p is said to **majorize** q, which is denoted as

$$p \succ q, \tag{9.15}$$

if there exists a doubly stochastic matrix T such that

$$q = Tp. \tag{9.16}$$

Equivalently, distribution q is said to be **majorized** by p, which is written as

$$q \prec p. \tag{9.17}$$

Equivalently, by Birkhoff's theorem, distribution p majorizes distribution q if there exists a random shuffling transformation $\{(S_m, \gamma_m)\}_{m\in[M]}$, for some integer M, that produces q from p. Since random shufflings inject randomness into an informational state, distribution q may be deemed to be more uncertain than distribution p. This way, the majorization relation $p \succ q$ defines an ordering between distributions p and q, according to which the majorizing distribution p is more certain than the majorized distribution q.

9.3.3 Majorization and Entropy

The definition of majorization, along with the fact, expressed by the inequality (9.12), that random shufflings increase the entropy, yields the following result.

Key Result 9.4 Majorization implies an ordering on all the entropies
Given two $N \times 1$ distributions with $p \succ q$, the inequality

$$\mathrm{H}_\alpha(p) \leq \mathrm{H}_\alpha(q) \tag{9.18}$$

holds for the Shannon entropy, as well as for the Tsallis and Rényi entropies of any order $\alpha \geq 0$. Therefore, if p majorizes q, *all* the entropy measures return a smaller, or at least no larger, value for p than for q.

This result reinforces the interpretation of majorization as a robust ordering relation between the uncertainty levels associated with two distributions. In this regard, it is important to note that the inequality $\mathrm{H}_\alpha(p) < \mathrm{H}_\alpha(q)$ for a specific entropy measure does not necessarily imply that p majorizes q. In fact, as demonstrated in the last example, for any two distributions p and q, different entropy measures can yield conflicting comparisons. It is possible to find some value of α for which $\mathrm{H}_\alpha(p) < \mathrm{H}_\alpha(q)$, while for another value of α the reverse inequality $\mathrm{H}_\alpha(p) > \mathrm{H}_\alpha(q)$ holds.

Example 9.10

For a system with $N = 2$ states, consider the random shuffling transformation (9.11) with $\gamma_1 = 0.5$, which keeps the state values unchanged or swaps them with the same probability. Given any pre-transformation distribution $p = [p_0, p_1]^T$, the post-transformation distribution is given by

$$Tp = \frac{1}{2}\begin{bmatrix} 1 & 1 \\ 1 & 1 \end{bmatrix} p = \frac{1}{2}\begin{bmatrix} p_0 + p_1 \\ p_1 + p_0 \end{bmatrix} = u. \tag{9.19}$$

Therefore, irrespective of the pre-transformation distribution p, after the transformation the informational state is uniform. This confirms the inequality (9.18), since the uniform distribution is entropy maximizing, i.e., $\mathrm{H}_\alpha(p) \leq \mathrm{H}_\alpha(u)$.

9.3.4 Comparing Distributions via Majorization

Given two $N \times 1$ probability vectors p and q, a comparison via majorization can yield one of three situations: (i) $p \succ q$; (ii) $q \succ p$; or (iii) neither $p \succ q$ nor $q \succ p$. While in the first two cases majorization orders the two distributions in terms of their respective uncertainty levels, in the third case the two distributions are not comparable in terms of majorization. Note that this cannot happen when comparing distributions using any given entropy measure, since two entropy values can always be ordered.

To ascertain which of the three situations (i), (ii) or (iii) listed above applies to a given pair of distributions p and q, one should in principle determine if there exists a random shuffling transformation T that produces one distribution from the other.

This appears to be a complicated task. However, there is a convenient equivalent definition of majorization that bypasses the explicit design of a random shuffling.

To elaborate, given any $N \times 1$ probability vector p, write as $p^{\downarrow}$ the $N \times 1$ ordered vector containing the elements of vector p listed from largest to smallest, i.e., $p^{\downarrow} = [p_0^{\downarrow}, p_1^{\downarrow}, \dots, p_{N-1}^{\downarrow}]^T$ with $p_0^{\downarrow} \geq p_1^{\downarrow} \geq \cdots \geq p_{N-1}^{\downarrow}$. For example, for vector $p = [0.1, 0.2, 0.1, 0.2, 0.4]^T$, the ordered probability vector is $p^{\downarrow} = [0.4, 0.2, 0.2, 0.1, 0.1]^T$.

If a vector p corresponds to a state of small uncertainty, the corresponding ordered vector $p^{\downarrow}$ should be concentrated at the first few entries. This can be measured via the following quantity.

Definition 9.5 Ky Fan norm

The kth Ky Fan norm of vector p is the sum of the k largest elements of vector p, i.e.,

$$||p||_{(k)} = \sum_{x \in [k]} p_x^{\downarrow}. \tag{9.20}$$

Note that the Nth Ky Fan norm $||p||_{(N)}$ is trivial for a probability vector p as one has $||p||_{(N)} = \sum_{x \in [N]} p_x^{\downarrow} = 1$. Based on this definition, as shown in Appendix 9.A, the following is an equivalent definition of majorization that makes it easier to compare two distributions.

Key Result 9.5 Equivalent definition of majorization

For two $N \times 1$ probability distributions p and q, vector p majorizes vector q, i.e., $p \succ q$, if and only if the inequalities

$$||p||_{(k)} \geq ||q||_{(k)} \tag{9.21}$$

hold for all $N-1$ integers $k \in [N-1]$.

According to this equivalent definition, the probability distribution p majorizes distribution q, i.e., $p \succ q$, if all of its Ky Fan norms are no smaller than those of vector q. Note that the inequality (9.21) is trivially satisfied for $k = N$, which is thus excluded in the definition (9.21). A way to think about the equivalent definition of majorization in (9.21) through the lens of the informational task of prediction is given in Appendix 9.B.

Example 9.11

How do the probability distributions $p = [0.3, 0.4, 0.1, 0.2]^T$ and $q = [0.2, 0.1, 0.35, 0.35]^T$ compare in terms of majorization? To address this question, first, the vectors are ordered, obtaining the ordered vectors $p^{\downarrow} = [0.4, 0.3, 0.2, 0.1]^T$ and $q^{\downarrow} = [0.35, 0.35,$

$0.2, 0.1]^T$. Then, one evaluates the $N-1 = 3$ Ky Fan norms $||p||_{(1)} = 0.4$, $||p||_{(2)} = 0.7$, $||p||_{(3)} = 0.9$, as well as $||q||_{(1)} = 0.35$, $||q||_{(2)} = 0.7$, and $||q||_{(3)} = 0.9$.

One can finally observe that all the $N - 1 = 3$ inequalities in (9.21) are satisfied, i.e.,

$$\begin{aligned} ||p||_{(1)} &= 0.4 \geq ||q||_{(1)} = 0.35 \\ ||p||_{(2)} &= 0.7 \geq ||q||_{(2)} = 0.7 \\ \text{and } ||p||_{(3)} &= 0.9 \geq ||q||_{(3)} = 0.9, \end{aligned} \tag{9.22}$$

and hence p majorizes q, i.e., $p \succ q$. It follows that all entropy measures for distribution p are no larger than those for distribution q.

Example 9.12

Consider now distributions $p = [0.5, 0.5, 0]^T$ and $q = [0.6, 0.3, 0.1]^T$. As seen in Example 9.9, some entropy measures are larger for distribution p, while others are larger for distribution q. It follows that neither of the conditions $p \succ q$ or $q \succ p$ can be satisfied. In fact, otherwise, all the entropy measures would return consistent inequalities. This can be checked via the $N - 1 = 2$ inequalities in (9.21), since one obtains $||p||_{(1)} = 0.5 > ||q||_{(1)} = 0.4$, but also $||p||_{(2)} = 0.75 < ||q||_{(2)} = 0.8$.

For any distribution p, the entropy is lower bounded by the entropy of a zero-entropy distribution δ_x and it is upper bounded by the entropy of a uniform distribution, yielding the bounds

$$0 = \mathrm{H}_\alpha(\delta_x) \leq \mathrm{H}_\alpha(p) \leq \mathrm{H}_\alpha(u). \tag{9.23}$$

These inequalities are implied by the following conditions expressed using majorization.

Key Result 9.6 Sandwich property of the majorization relation

For any probability vector p, the relationships

$$u \prec p \prec \delta_x \tag{9.24}$$

hold for any integer $x \in [N]$.

In words, the condition $p \prec \delta_x$ indicates the zero-entropy probability vector δ_x corresponds to a more certain informational state than any distribution p. Conversely, the condition $u \prec p$ states that the uniform probability vector u corresponds to a less certain informational state than any distribution p. The proof is left as an exercise in Problem 9.4.

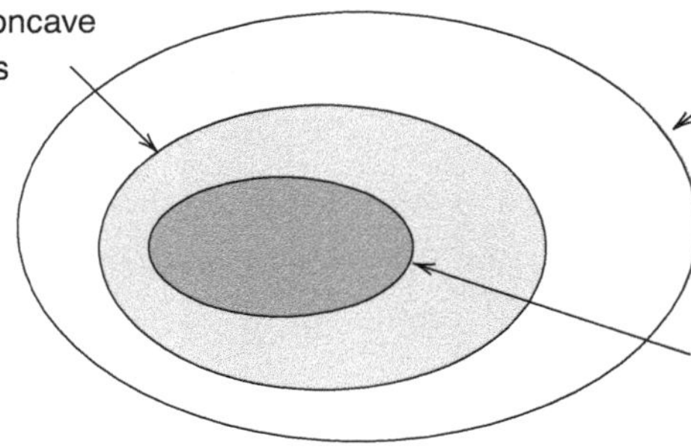

Figure 9.3 Permutation-invariant concave functions are a subset of Schur concave functions, which are in turn a subset of permutation-invariant functions.

9.4 Schur Concavity ⏭

In the previous sections, it was concluded that majorization provides a partial ordering of probability distributions that is stronger than the complete ordering obtained with any of the entropy measures. This result hinges on the inequality (9.12), which can equivalently be stated as the implication

$$p \succ q \Rightarrow \mathrm{H}_\alpha(p) \le \mathrm{H}_\alpha(q), \tag{9.25}$$

which holds for the Shannon entropy, as well as for the Tsallis and Rényi entropies of any order $\alpha \ge 0$.

The implication (9.25) was proved in Section 9.2.3 only for concave entropy measures. In this section, this result is extended to all (h, ϕ)-entropies, which, as shown in Section 3.8, encompass Shannon, Tsallis, and Rényi entropies as special cases. To this end, all (h, ϕ)-entropies are shown to be Schur concave, i.e., majorization-decreasing. For all proofs in this section, the reader is referred to the Recommended Resources.

9.4.1 Defining Schur Concavity

As illustrated in Figure 9.3, permutation-invariant concave functions, such as the Shannon entropy, the Tsallis entropy, and the Rényi entropy with $0 \le \alpha \le 1$, are a special case of Schur concave functions, which form, in turn, a subset of all permutation-invariant functions. Thus, the proof given here, which applies to all Schur concave functions, applies more broadly than the proof (9.13) based on the assumption of concavity, encompassing also the Rényi entropy of orders $\alpha > 1$.

Definition 9.6 Schur concave function
A function $f(p)$ of an $N \times 1$ probability distribution vector p is said to be Schur concave if it is majorization-decreasing. Accordingly, for any two $N \times 1$

probability vectors p and q with $p \succ q$, a Schur concave function satisfies the inequality $f(p) \leq f(q)$:

$$p \succ q \Rightarrow f(p) \leq f(q). \quad (9.26)$$

The following result provides a way to identify entropies as Schur concave functions.

Key Result 9.7 Sufficient conditions for Schur concavity
Consider a function of the form

$$f(p) = h(\varphi(p)), \quad (9.27)$$

where $\varphi(p)$ is a permutation-invariant function with a scalar output. Then, function $f(p)$ is Schur concave if one of the following two conditions is satisfied:

- function $\varphi(\cdot)$ is concave and function $h(\cdot)$ is increasing;
- or function $\varphi(\cdot)$ is convex and function $h(\cdot)$ is decreasing.

9.4.2 Shannon, Tsallis, and Rényi Entropies Are Schur Concave

As seen in Chapter 3, all the entropy measures studied in this book can be expressed as special cases of the (h, ϕ)-entropy, which is defined as

$$\mathrm{H}^{\mathrm{U}}_{h,\phi}(p) = h\left(\sum_{x\in[N]} \phi(p_x)\right), \quad (9.28)$$

where functions $\phi(\cdot)$ and $h(\cdot)$ satisfy either one of the following two conditions:

- function $\phi(\cdot)$ is concave and $h(\cdot)$ is increasing;
- or function $\phi(\cdot)$ is convex and $h(\cdot)$ is decreasing.

In order to apply the sufficient conditions described above for Schur concavity, it is enough to note that the (h, ϕ)-entropy (9.28) can be expressed in the form (9.27) by identifying the function $\varphi(p)$ in (9.27) as

$$\varphi(p) = \sum_{x\in[N]} \phi(p_x). \quad (9.29)$$

In fact, function $\varphi(p)$ is permutation invariant, since it applies the same function $\phi(\cdot)$ to all entries of vector p_x. Furthermore, the sum of concave functions is concave; and the sum of convex functions is convex (see Recommended Resources for details).

Since the (h, ϕ)-entropy is Schur concave, the Shannon, Tsallis, and Rényi entropies are majorization-decreasing and thus they satisfy the desired implication (9.25).

9.5 Channels and Data Processing Inequality

The previous sections have examined a specific class of dynamic transformations, namely random shuffling. This section broadens the discussion to encompass arbitrary random transformations, commonly referred to as channels. Unlike the case of random shuffling, an arbitrary channel does not necessarily increase the entropy. Instead, it always reduces – or at least does not improve – the capacity of a detector to differentiate between two probability distributions. This phenomenon, formalized as the data processing inequality, establishes a fundamental limit on how effectively two distributions can be distinguished after undergoing random transformations.

9.5.1 Channels

A channel is a general random transformation that maps state $x \in [N]$ into state $y \in [N]$ with some probability $p_{x \to y}$.

Definition 9.7 Channel
A channel is described by N probability distributions $\{p_{x \to y}\}_{y \in [N]}$, one for each initial state $x \in [N]$. Each probability distribution $\{p_{x \to y}\}_{y \in [N]}$ determines the probability $p_{x \to y}$ that a given input x is transformed into output y for all $y \in [N]$. Given that $\{p_{x \to y}\}_{y \in [N]}$ is a probability distribution, the conditions $p_{x \to y} \geq 0$ hold for all $x \in [N]$ and $y \in [N]$, along with the equalities

$$\sum_{y \in [N]} p_{x \to y} = 1 \text{ for all } x \in [N]. \tag{9.30}$$

The name "channel" can be motivated by viewing the transformation described by the probabilities $p_{x \to y}$ as a **noisy link** connecting a sender, holding the state x, and a recipient, holding the transformed state y. From this perspective, the transformation $p_{x \to y}$ plays the role of a random disturbance on the transmitted signal x. Note that it is possible to define channels with a number of output values y different from N, and the results in this section can readily be extended to this more general case.

Given a pre-transformation distribution p, the probability of the system being in state $y \in [N]$ after the transformation is obtained by averaging over all the ways in which state y can be produced from some state $x \in [N]$. Accordingly, by the law of total probability, the probability q_y that the system is in state $y \in [N]$ after the transformation is given by the average

$$q_y = \sum_{x \in [N]} p_x p_{x \to y} \tag{9.31}$$

over the initial state $\mathrm{x} \sim p$.

Table 9.1 Classes of transformations and corresponding matrix descriptions

Transformation	Matrix
Shuffling	Permutation matrix S
Random shuffling	Doubly stochastic matrix T
Channel	Column-stochastic matrix K

The equality (9.31) can be conveniently expressed through the $N \times N$ **channel matrix** K with each (y, x)th entry given by

$$K_{y,x} = p_{x \to y}. \tag{9.32}$$

In fact, using (9.31), the output probability vector $q = [q_0, \ldots, q_{N-1}]^T$ can be expressed as a function of the input vector p as the product

$$q = Kp. \tag{9.33}$$

The channel matrix K has non-negative entries, and, by (9.30), the sum of its entries over each column equals 1:

$$\sum_{y \in [N]} K_{y,x} = 1 \text{ for all } x \in [N]. \tag{9.34}$$

Matrices with these properties are said to be **column-stochastic**. Therefore, a channel is equivalently described by a column-stochastic matrix, and any column-stochastic matrix corresponds to a channel.

To sum up, Table 9.1 lists the transformations studied in this chapter in order of generality, from shuffling to channels. In this regard, note that the class of doubly stochastic matrices includes permutation matrices as a special case, while the class of column-stochastic matrices includes doubly stochastic matrices as a subset.

Example 9.13

Any function $f(\cdot)$ mapping integers in set $[N]$ to integers in the same set can be viewed as the channel

$$p_{x \to y} = \begin{cases} 1 & \text{if } y = f(x), \\ 0 & \text{otherwise.} \end{cases} \tag{9.35}$$

Example 9.14

For a permutation $s(\cdot)$, consider the channel that maps an input x to the output $y = s(x)$ with probability ϵ, while otherwise choosing one of the remaining $N - 1$ states $y \neq s(x)$ with equal probability. The channel distribution can be expressed as

$$p_{x \to y} = \begin{cases} \epsilon & \text{if } y = s(x), \\ \frac{1-\epsilon}{N-1} & \text{otherwise.} \end{cases} \tag{9.36}$$

Example 9.15

Given any two $N \times 1$ probability distributions p and q, it was observed in Section 9.3 that it is not always possible to obtain one of the two vectors from the other via random shuffling. In contrast, it is clearly always possible to convert any $N \times 1$ probability distribution into any other $N \times 1$ probability distribution via a channel.

For example, take the probability distribution vectors $p = [0.5, 0.5, 0]^T$ and $q = [0.6, 0.3, 0.1]^T$, for which, as seen in Example 9.12, there is no doubly stochastic matrix transforming p into q (or vice versa). However, the channel matrix

$$K = \begin{bmatrix} 0.6 & 0.6 & 0.6 \\ 0.3 & 0.3 & 0.3 \\ 0.1 & 0.1 & 0.1 \end{bmatrix}, \qquad (9.37)$$

which is column-stochastic, clearly produces q as $q = Kp$.

9.5.2 Data Processing Inequality

As a result of the application of a channel, the entropy can increase, decrease, or stay constant. This is evident from the fact, illustrated in the last example, that a channel can convert a distribution into any other distribution. In contrast, as discussed next, a channel can only decrease – or anyway not increase – the relative information between two distributions p and q, that is, the distinguishability between two distributions. The intuition behind this result, known as the **data processing inequality**, is that the random disturbance caused by a channel cannot make it easier to distinguish between two distributions.

To formalize the data processing inequality, consider again the problem studied in Section 4.7 of distinguishing two distributions p and q based on the observation of a single sample that may have been generated as $\mathrm{x} \sim p$ or as $\mathrm{x} \sim q$. As seen in Section 4.7, when both distributions are a priori equally likely, the optimal probability of detection is given by $\mathrm{P_S}(d^*) = (1 + \mathrm{D}^{\mathrm{tr}}(p, q))/2$, where $\mathrm{D}^{\mathrm{tr}}(p, q)$ is the trace distance between distributions p and q.

Thus, the larger the trace distance $\mathrm{D}^{\mathrm{tr}}(p, q)$ is, the easier it is to detect whether sample x was generated from distribution p or q. Suppose now that, as shown in Figure 9.4, the observed sample x is pre-processed via a channel K before being input to a detector. The data processing inequality states that no intermediate processing step can improve the performance of an optimal detector.

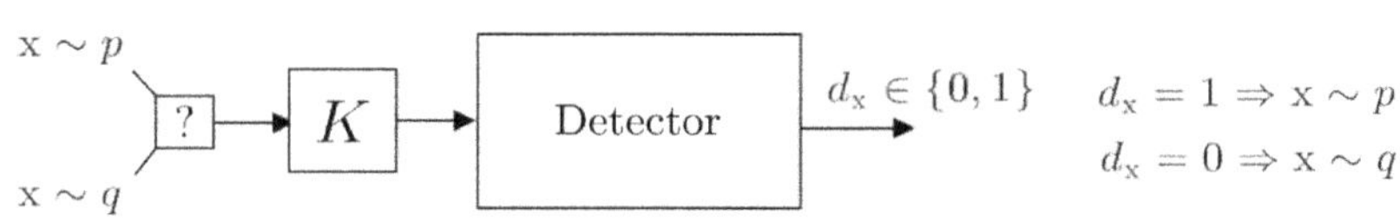

Figure 9.4 Illustration of the problem of optimal detection in which the state is pre-processed by a channel K.

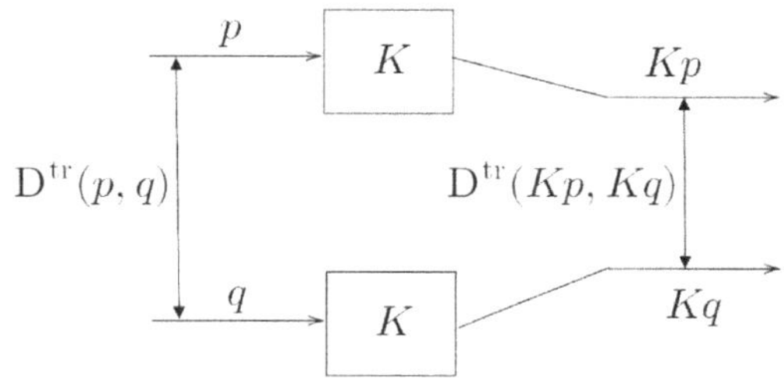

Figure 9.5 By the data processing inequality, the trace distance between the two distributions after the transformation, $D^{tr}(Kp, Kq)$, cannot be larger than the pre-transformation trace distance $D^{tr}(p, q)$.

Key Result 9.8 Data processing inequality for the trace distance
For any two $N \times 1$ distributions p and q and for any channel, the inequality

$$D^{tr}(p, q) \geq D^{tr}(Kp, Kq) \tag{9.38}$$

holds, where K is the channel matrix.

In words, as illustrated in Figure 9.5, the trace distance between the two distributions after the transformation, $D^{tr}(Kp, Kq)$, cannot be larger than the pre-transformation trace distance $D^{tr}(p, q)$. Accordingly, the optimal probability of detection $P_S(d^*)$ cannot increase as a result of the application of any channel K.

As proved in Appendix 9.C, the data processing inequalities apply also with the Shannon, Tsallis, and Rényi relative entropies, denoted generically as $D_\alpha(p||q)$, in lieu of the trace distance $D^{tr}(p, q)$.

Key Result 9.9 Data processing inequality for the relative entropies
For any two $N \times 1$ distributions p and q and for any channel, the inequality

$$D_\alpha(p||q) \geq D_\alpha(Kp||Kq) \tag{9.39}$$

holds for any order $\alpha \geq 0$, where K is the channel matrix.

Example 9.16

In the special case of a shuffling transformation S, the inequalities (9.38) and (9.39) turn into equalities, i.e.,

$$D^{tr}(p, q) = D^{tr}(Sp, Sq) \tag{9.40}$$

and

$$D_\alpha(p||q) = D_\alpha(Sp||Sq). \tag{9.41}$$

In fact, relative entropy measures are invariant to permutations of both distributions.

Example 9.17

Consider distributions $p = [0.5, 0, 0.5, 0]^T$ and $q = [0, 0.5, 0, 0.5]^T$. The supports of these distributions are orthogonal, and hence the trace distance takes the maximum value $D^{tr}(p, q) = 1$, resulting in an optimal detection probability $P_S(d^*) = 1$. Consider now the channel

$$K = \frac{1}{2}\begin{bmatrix} 1 & 0 & 0 & 1 \\ 1 & 1 & 0 & 0 \\ 0 & 1 & 1 & 0 \\ 0 & 0 & 1 & 1 \end{bmatrix}. \tag{9.42}$$

After the channel, both informational states become equal to the uniform distribution, i.e., $Kp = Kq = u$. Hence, the post-channel trace distance takes the minimum value $D^{tr}(p, q) = 0.5$, corresponding to an optimal detection probability equal to the minimum possible value $P_S(d^*) = 0.5$.

The data processing inequality states that a channel cannot improve the performance of an optimal detector. Note that the optimal detector is generally different before and after the application of the channel, and thus the data processing inequality compares the performance levels attained by two distinct detectors. However, for a fixed detector, it is possible for a channel to improve the detection probability.

Example 9.18

Consider distributions $p = [0.5, 0, 0.5, 0]^T$ and $q = u$, and adopt a (suboptimal) detector that sets $d_0 = 1$ and $d_x = 0$ otherwise, hence selecting p if $x = 0$ and q if $x \neq 0$. When both distributions are equally probable, the probability of a correct decision of this detector is

$$\begin{aligned} P_S(d) &= \frac{1}{2}\Pr_p[d_x = 1] + \frac{1}{2}\Pr_q[d_x = 0] \\ &= \frac{1}{2}\frac{1}{2} + \frac{1}{2}\frac{3}{4} = \frac{5}{8}. \end{aligned} \tag{9.43}$$

The performance of this detector can be improved if one applies first the pre-processing defined by the channel matrix

$$K = \begin{bmatrix} 1 & 0 & 1 & 0 \\ 0 & 1 & 0 & 0 \\ 0 & 0 & 0 & 0 \\ 0 & 0 & 0 & 1 \end{bmatrix}. \tag{9.44}$$

In fact, after the transformation, the informational states have changed into $Kp = [1,0,0,0]^T$ and $Kq = [0.5, 0.25, 0, 0.25]^T$. The probability of correct detection for the given detector after the transformation has thus improved to

$$\begin{aligned} \mathrm{P_S}(d) &= \frac{1}{2}\mathrm{Pr}_{Kp}[d_\mathrm{x} = 1] + \frac{1}{2}\mathrm{Pr}_{Kq}[d_\mathrm{x} = 0] \\ &= \frac{1}{2}1 + \frac{1}{2}\frac{1}{2} = \frac{3}{4} = \frac{6}{8} > \frac{5}{8}. \end{aligned} \tag{9.45}$$

9.5.3 Strong Data Processing Inequality

By how much can a channel change the relative entropy? For a given channel matrix K, this question can be addressed by studying the maximum value that the ratio $\mathrm{D}_\alpha(Kp||Kq)/\mathrm{D}_\alpha(p||q)$ can attain when maximized over all possible distributions p and q, i.e.,

$$\max_{p,q} \frac{\mathrm{D}_\alpha(Kp||Kq)}{\mathrm{D}_\alpha(p||q)}. \tag{9.46}$$

The optimized ratio in (9.46) is known as the **contraction coefficient**, and bounds on it yield **strong data processing inequalities** (see Recommended Resources for further information).

9.6 Summary

- This chapter has studied classical information in dynamic scenarios in which the information state undergoes a random transformation.
- By Birkhoff's theorem, a random shuffling can be described through the multiplication $q = Tp$ of the pre-transformation state p by a doubly stochastic matrix T.
- Random shuffling transformations cannot decrease the entropy, yielding the inequality $\mathrm{H}_\alpha(p) \leq \mathrm{H}_\alpha(Tp)$.
- A distribution p is said to majorize distribution q, which is denoted as $p \succ q$, if there exists a random shuffling producing q from p, i.e., if there exists a doubly stochastic matrix T satisfying the equality $q = Tp$.
- Majorization provides a partial ordering of probability distributions such that, if distribution p majorizes q, i.e., if $p \succ q$, then the Shannon, Tsallis, or Rényi entropies $\mathrm{H}_\alpha(p)$ are no larger than $\mathrm{H}_\alpha(q)$, i.e., $\mathrm{H}_\alpha(p) \leq \mathrm{H}_\alpha(q)$. In other words, the Shannon, Tsallis, or Rényi entropies are majorization decreasing, or Schur concave.

- The Ky Fan norm $||p||_{(k)}$ is the sum of the k largest elements of vector p.
- Equivalently, distribution p majorizes distribution q if all the Ky Fan norms $||p||_{(k)}$ of distribution p are larger than for distribution q, i.e., $||p||_{(k)} \geq ||q||_{(k)}$ for all integers $k \in [N-1]$.
- Arbitrary random transformations, known as channels, cannot increase the distinguishability between two distributions. In fact, by the data processing inequality one has the inequality $\mathrm{D}^{\mathrm{tr}}(p,q) \geq \mathrm{D}^{\mathrm{tr}}(Kp,Kq)$ for the trace distance, and $\mathrm{D}_\alpha(p||q) \geq \mathrm{D}_\alpha(Kp||Kq)$ for the Shannon, Tsallis, or Rényi relative entropies, where K is a channel matrix, i.e., a column-stochastic matrix.

9.7 Recommended Resources

For further discussion on random shuffling, please see [1]. A comprehensive introduction to majorization, including the proofs of the results presented in this chapter, is provided in [2]. Composition rules for convex functions are detailed in [3]. The data processing inequality and the strong data processing inequality are covered in [4]. In-depth discussions on the definitions of uncertainty and entropy measures, in the context of how these metrics evolve under dynamic transformations, are available in [5, 6].

PROBLEMS

9.1 Show that, for two probability vectors p and q, the conditions $\sum_{x\in[k]} p_x^{\downarrow} \leq \sum_{x\in[k]} q_x^{\downarrow}$ for all $k \in [N]$ are equivalent to the set of inequalities $\sum_{x=k}^{N-1} p_x^{\downarrow} \geq \sum_{x=k}^{N-1} q_x^{\downarrow}$ for all $k \in [N]$.

9.2 Show that the minimum value of an $N \times 1$ probability vector p cannot be larger than $1/N$. *Hint:* If, by contradiction, the smallest entry was strictly larger than $1/N$, what would happen to the sum of the probabilities in the vector?

9.3 Compare vectors $p = [2/5, 2/5, 1/10, 1/10]^T$ and $q = [1/2, 1/4, 1/4, 0]^T$ using majorization.

9.4 Prove the relations in (9.24). *Hint:* Use the equivalent definition (9.21) and insights from Problem 9.2.

9.5 Show that the uniform probability distribution u is the only vector that is majorized by all probability vectors p.

9.6 Consider a random shuffling transformation T that selects any one of the $N!$ possible permutations of the N states with equal probability. Given any pre-transformation state p, what is the post-transformation state?

9.7 Provide an example of a column-stochastic transformation K that causes the Tsallis relative entropy to strictly decrease for some pair of distributions p and q.

9.8 Show that concavity implies Schur concavity for permutation-invariant functions, while the converse implication does not hold. *Hint:* A doubly stochastic transformation can be described as a mixing operation.

9.9 An $N \times N$ matrix D is said to be **unitary-stochastic**, or **unistochastic**, if each (x, y)th entry can be expressed as $D_{x,y} = |U_{x,y}|$ for some unitary matrix U. (a) Prove that all unistochastic matrices are doubly stochastic. *Hint:* Use the definition of a unitary matrix expressed in terms of each entry of the matrix. (b) Show that the matrix

$$T = \frac{1}{2}\begin{bmatrix} 1 & 1 & 0 \\ 0 & 1 & 1 \\ 1 & 0 & 1 \end{bmatrix} \tag{9.47}$$

is doubly stochastic but not unistochastic.

9.10 Take the joint distributions $p = [0.5, 0, 0.5, 0]^T$ and $q = [0, 0.5, 0, 0.5]^T$. (a) For the permutation function $s(0) = 1$, $s(1) = 0$, $s(2) = 3$, and $s(3) = 2$, evaluate the post-permutation distributions Sp and Sq, where S is the permutation matrix for the permutation function $s(\cdot)$. (b) Evaluate the trace distance between the two distributions before and after the permutation S. (c) Assume now that the transformation S is applied with probability 0.6, while the system is left unchanged with probability 0.4. Evaluate the trace distance between the two distributions before and after the resulting transformation. (d) Comment on your results in parts (b) and (c) using the data processing inequality.

Appendix

9.A Proof of the Equivalence of the Majorization Definition (9.21)

This appendix proves that a probability vector q can be written as $q = Tp$ for some doubly stochastic matrix T if and only if the inequalities $||p||_{(k)} \geq ||q||_{(k)}$ in (9.21) hold for all $k \in [N-1]$. Throughout this proof, it will be assumed without loss of generality that the entries in vectors p and q have been rearranged in non-increasing order, so that the equalities $p^{\downarrow} = p$ and $q^{\downarrow} = q$ hold. Accordingly, the conditions (9.21) can equivalently be expressed as

$$\sum_{x \in [k]} p_x \geq \sum_{x \in [k]} q_x \tag{9.48}$$

for $k \in [N-1]$. Note also that the smallest values in the two vectors p_{N-1} and q_{N-1} must be no larger than $1/N$, i.e., $p_{N-1} \leq 1/N$ and $q_{N-1} \leq 1/N$ (see Problem 9.2).

First, for the "if" part of the claim, one must construct a doubly stochastic matrix T such that the equality $q = Tp$ holds for any two probability vectors p and q satisfying the equalities (9.48). To avoid trivial cases, assume that the two distributions are different, i.e., $p \neq q$. Define as $S_{x \leftrightarrow y}$ the $N \times N$ permutation matrix corresponding to the permutation function $s_{x \leftrightarrow y}(\cdot)$ that flips the positions of entries x and y – i.e., $s(x) = y$ and $s(y) = x$, with $s(z) = z$ for $z \notin \{x, y\}$. Also, write as

$$T^{\lambda}_{x \leftrightarrow y} = \lambda I + (1 - \lambda) S_{x \leftrightarrow y} \tag{9.49}$$

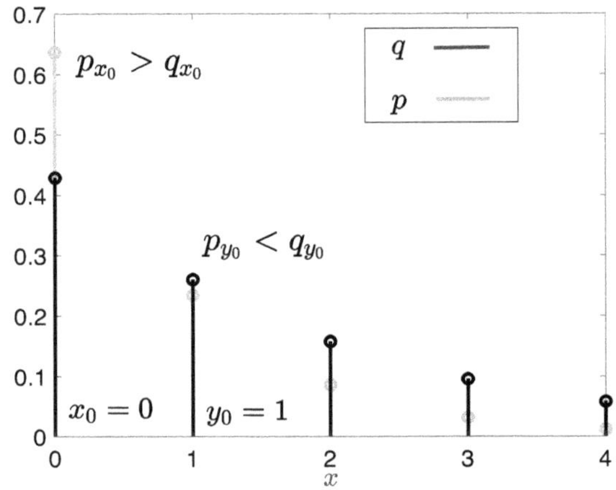

Figure 9.6 Identification of the indices x_0 and y_0 for the proof in Appendix 9.A.

the doubly stochastic matrix corresponding to a random shuffling transformation in which the states are left unchanged with probability $0 \leq \lambda \leq 1$, while the permutation $s_{x \leftrightarrow y}(\cdot)$ is applied otherwise.

The proof builds matrix T as the product of N' matrices of form $T^{\lambda}_{x \leftrightarrow y}$, i.e.,

$$T = T^{\lambda_{N'-1}}_{x_{N'-1} y_{N'-1}} \cdots T^{\lambda_1}_{x_1 y_1} T^{\lambda_0}_{x_0 y_0} \tag{9.50}$$

for some sequence of index pairs $x_t \in [N]$ and $y_t \in [N]$ with $x_t \neq y_t$ and probabilities $0 \leq \lambda_t \leq 1$ with $t \in [N']$. Note that the product of doubly stochastic matrices is doubly stochastic. The construction of matrix T is done sequentially, by correcting one difference in vectors p and q at each step t. Therefore, since there are at most N differences between the two vectors, and it is not possible for the two vectors to differ in one position (since the two vectors must both sum to 1), this requires at most $N' = N - 1$ steps and thus at most $N' = N - 1$ matrices $T^{\lambda}_{x \leftrightarrow y}$ in the product (9.50).

At the first step, $t = 0$, the indices x_0 and y_0 are identified as illustrated in Figure 9.6. Accordingly, index x_0 is the largest index $x \in [N]$ for which the inequality $p_x \geq q_x$ holds, while index y_0 is found as the smallest index y for which the inequality $p_y < q_y$ holds. Note that such indices must exist since, by assumption, the inequality $p_0 \geq q_0$ holds (as $p \succ q$), and it is not possible for inequality $p_x \geq q_x$ to be verified for all $x \in [N]$ (both vectors must sum to 1, and the case $p = q$ is excluded by assumption).

This way, one has the inequalities

$$p_{x_0} > q_{x_0} \geq q_{y_0} > p_{y_0}, \tag{9.51}$$

where the first inequality is strict by the definition of index x_0. If $|p_{x_0} - q_{x_0}| < |p_{x_1} - q_{x_1}|$, the construction sets $q = q_{x_0}$, while it chooses $q = q_{x_1}$ otherwise, with a tie broken arbitrarily. Finally, probability $0 \leq \lambda_0 \leq 1$ is selected so that mixing probabilities p_{x_0} and p_{y_0} returns q, i.e.,

$$q = \lambda_0 p_{x_0} + (1 - \lambda_0) p_{y_0}. \tag{9.52}$$

With this construction, the product $T^{\lambda_0}_{x_0y_0}p$ ensures that at least one difference between vectors p and q has been corrected.

One can now show that the majorization relation $T^{\lambda_0}_{x_0y_0}p \succ q$ holds, and hence the procedure outlined above can be repeated at most $N-1$ times to account for all differences between vectors p and q.

Consider now the reverse, "only if" part of the claim. This states that, if the equality $q = Tp$ holds for a doubly stochastic matrix T, then one must have the inequalities (9.48). To demonstrate the kth inequality in (9.48), define $T_{\leq k,y} = \sum_{x\in[k]} T_{x,y}$ as the sum of the first k entries of matrix D in the yth column.

By assumption, the equality $q_x = \sum_{y\in[N]} T_{x,y}p_y$ holds for all $x \in [N]$, which implies

$$\begin{aligned}\sum_{x\in[k]} q_x &= \sum_{x\in[k]}\sum_{y\in[N]} T_{x,y}p_y \\ &= \sum_{y\in[N]}\left(\sum_{x\in[k]} T_{x,y}\right)p_y \\ &= \sum_{y\in[N]} T_{\leq k,y}p_y. \end{aligned} \tag{9.53}$$

The target inequality (9.48) amounts to the non-negativity of the difference

$$\sum_{x\in[k]} p_x - \sum_{x\in[k]} q_x = \sum_{x\in[k]} p_x - \sum_{y\in[N]} T_{\leq k,y}p_y. \tag{9.54}$$

By the definition of a doubly stochastic matrix T, the equality $\sum_{x\in[k]}\sum_{y\in[N]} T_{x,y} = k$ is satisfied, which can also be written as $\sum_{y\in[N]} T_{\leq k,y} = k$. It follows that the difference in (9.54) can be expressed as

$$\begin{aligned}\sum_{x\in[k]} p_x - \sum_{x\in[k]} q_x &= \sum_{x\in[k]} p_x - \sum_{y\in[N]} T_{\leq k,y}p_y + \left(\sum_{y\in[N]} T_{\leq k,y} - k\right)p_{k-1} \\ &= \sum_{x\in[k]} p_x - \sum_{x\in[k]} T_{\leq k,x}p_x + \sum_{x\in[k]} D_{\leq k,x}p_{k-1} - \sum_{x\in[k]} p_{k-1} \\ &\quad - \sum_{x=k}^{n-1} T_{\leq k,x}p_x + \sum_{x=k}^{n-1} D_{\leq k,x}p_{k-1} \\ &= \sum_{x\in[k]} (1 - T_{\leq k,x})(p_x - p_{k-1}) + \sum_{x=k}^{n-1} T_{\leq k,x}(p_{k-1} - p_x), \end{aligned} \tag{9.55}$$

where in the second equality two sums were split into terms up to $k-1$ and beyond.

Now, the inequalities $1 - T_{\leq k,x} \geq 0$ hold by definition of doubly stochastic matrices for all $x \in [k]$, and one has the inequalities $p_x \geq p_{k-1}$ for all $x \in [k]$ and $p_{k-1} \geq p_x$ for all $x \in \{k, \ldots, n-1\}$ by the ordering of vector p. This implies that the difference (9.54) is indeed non-negative, concluding the proof.

9.B A Set Prediction-Based Perspective on Majorization

This appendix revisits the definition of majorization through the lens of **set prediction**. In set prediction, the goal is to select the subset of values that are most likely to be observed for a given random quantity. For a random variable $\mathrm{x} \sim p$, a **k-predictor** outputs a prediction subset $\mathcal{X} \subseteq [N]$ of cardinality $|\mathcal{X}| = k$. Given a predicted set $\mathcal{X}$, the probability of a correct detection is the probability that the set includes x, that is,

$$\Pr[\mathrm{x} \in \mathcal{X}] = \sum_{x \in \mathcal{X}} p_x. \tag{9.56}$$

In order to maximize the probability of detection, a k-predictor must thus choose the subset $\mathcal{X}^*$ of values x that have the largest probabilities p_x. Therefore, the maximal probability of prediction for a k-predictor is given by

$$\Pr[\mathrm{x} \in \mathcal{X}^*] = \sum_{x \in [k]} p_x^{\downarrow} = ||p||_{(k)}, \tag{9.57}$$

which corresponds to the kth Ky Fan norm. In this light, the majorization condition $p \succ q$ indicates that the maximal probability of prediction is larger under distribution p than it is under distribution q for all k-predictors with $k \in [N-1]$. One may thus informally say that variables following distribution p are easier to predict than variables following distribution q.

9.C Proof of the Data Processing Inequality

This appendix focuses on proving the data processing inequality (9.38) and (9.39). To this end, it is sufficient to prove the inequality

$$\mathrm{D}_f(p||q) \geq \mathrm{D}_f(Kp||Kq) \tag{9.58}$$

for any f-divergence, where, as in (9.38) and (9.39), p and q are two arbitrary distributions and K is a channel matrix. Demonstrating the inequality (9.58) would immediately prove (9.38), as well as the inequality (9.39) for the Shannon and Tsallis relative entropies. In fact, as seen in Section 4.8, the trace distance, the Shannon relative entropy, and the Tsallis relative entropy are all special cases of the f-divergence. For the Rényi relative entropy, the result then follows from the one-to-one monotonic relation between Tsallis and Rényi relative entropies (see Appendix 4.C).

To start, write $\tilde{p} = Kp$ and $\tilde{q} = Kq$ for the post-channel probabilities. These vectors have yth entries given by $\tilde{p}_y = \sum_{x \in [N]} p_x p_{x \to y}$ and $\tilde{q}_y = \sum_{x \in [N]} q_x p_{x \to y}$, respectively, for all $y \in [N]$. With this definition, the right-hand side of the desired inequality (9.58) can be written as

$$\mathrm{D}_f(Kp||Kq) = \sum_{y \in [N]} \tilde{q}_y f\left(\frac{\tilde{p}_y}{\tilde{q}_y}\right). \tag{9.59}$$

Each yth term in the sum (9.59) can be upper bounded by following the steps

$$\begin{aligned}
\tilde{q}_y f\left(\frac{\tilde{p}_y}{\tilde{q}_y}\right) &= \tilde{q}_y f\left(\sum_{x\in[N]} \frac{p_x p_{x\to y}}{\tilde{q}_y}\right) \\
&= \tilde{q}_y f\left(\sum_{x\in[N]} \frac{q_x p_{x\to y}}{\tilde{q}_y} \frac{p_x p_{x\to y}}{q_x p_{x\to y}}\right) \\
&\leq \tilde{q}_y \sum_{x\in[N]} \frac{q_x p_{x\to y}}{\tilde{q}_y} f\left(\frac{p_x}{q_x}\right) \\
&= \sum_{x\in[N]} q_x p_{x\to y} f\left(\frac{p_x}{q_x}\right),
\end{aligned} \tag{9.60}$$

where the second equality follows by multiplying and dividing by the same term; and the third step follows from Jensen's inequality. In fact, function $f(\cdot)$ is convex by the definition of f-divergence, and the quantity $q_x p_{x\to y}/\tilde{q}_y$ defines a probability distribution over index $x \in [N]$ for any fixed index $y \in [N]$.

Plugging the inequality (9.60) back into the f-divergence (9.59), one finally gets the desired inequality

$$\begin{aligned}
\mathrm{D}_f(Kp||Kq) &\leq \sum_{y\in[N]} \sum_{x\in[N]} q_x p_{x\to y} f\left(\frac{p_x}{q_x}\right) \\
&= \sum_{x\in[N]} q_x \left(\sum_{y\in[N]} p_{x\to y}\right) f\left(\frac{p_x}{q_x}\right) \\
&= \sum_{x\in[N]} q_x f\left(\frac{p_x}{q_x}\right) \\
&= \mathrm{D}_f(p||q),
\end{aligned} \tag{9.61}$$

concluding the proof.

REFERENCES

[1] M. Nielsen and G. Vidal, "Majorization and the interconversion of bipartite states," *Quantum Information & Computation*, vol. 1, no. 1, pp. 76–93, 2001.

[2] I. Olkin and A. Marshall, *Inequalities: Theory of Majorization and Its Applications*. New York: Academic Press, 2016.

[3] S. Boyd and L. Vandenberghe, *Convex Optimization*. Cambridge: Cambridge University Press, 2004.

[4] Y. Polyanskiy and Y. Wu, *Information Theory: From Coding to Learning*. Cambridge: Cambridge University Press, 2024.

[5] G. Gour and M. Tomamichel, "Entropy and relative entropy from information-theoretic principles," *IEEE Transactions on Information Theory*, vol. 67, no. 10, pp. 6313–6327, 2021.

[6] S. Brandsen, I. J. Geng, and G. Gour, "What is entropy? A perspective from games of chance," *Physical Review E*, vol. 105, no. 2, art. 024117, 2022.

10 Quantum Dynamic Information in Closed Systems

10.1 Overview

Part II explored static quantum systems, focusing on quantum systems in a specific state, whether pure or mixed. Until now, the only method discussed for changing the state of a quantum system has been coupling it with a measurement device. In fact, unlike classical systems, quantum measurements generally alter the system's state, even when the initial state is pure, i.e., fully known.

This chapter shifts the focus to closed quantum systems, aiming to understand the types of dynamic transformations a quantum state can undergo when the system is isolated from its environment and, consequently, from measurement devices.

Dynamic closed quantum systems provide an idealized model for numerous practical applications, including quantum computers. In these systems, quantum computations are represented as dynamic transformations of a quantum state, evolving in a controlled manner without external interactions.

As detailed in this chapter, the transformations of quantum states in closed systems are mathematically described by unitary matrices, which encapsulate the reversibility of changes in such systems. Consequently, these transformations preserve both the entropy and relative entropy of the system, leaving the information content unaffected.

Finally, this chapter revisits complete quantum measurements and demonstrates that any complete measurement, regardless of its basis, can be implemented using measurements in the computational basis, preceded by a pre-processing step involving a basis-specific unitary transformation.

Chapter 11 will extend the discussion to open quantum systems, exploring how their dynamics differ from the idealized behavior of closed systems.

10.1.1 Learning Objectives and Organization

By the end of this chapter, the reader should be able to:

- describe transformations in closed quantum systems via unitary matrices;
- recognize basic closed-system transformations applied to qubit systems;
- understand the concept of quantum interference caused by evolutions in a closed quantum system;

- understand the impact of unitary transformations on quantum entropies and quantum relative entropies;
- and implement a complete quantum measurement in any basis by using a unitary transformation followed by a measurement in the computational basis.

10.2 Closed-System Dynamic Transformations for Pure States

Consider a quantum system that is isolated from the rest of the world. The rest of the world is typically referred to as the *environment* in this context. Assume that the system is initially in the $N \times 1$ pure state $|\psi\rangle$. After a dynamic transformation, the system is in the $N \times 1$ state $|\phi\rangle$. What type of transformations are possible in a closed system?

10.2.1 Closed-System Transformations Are Unitary

As discussed next, dynamic transformations in a closed system must be linear, reversible, and norm-preserving.

- **Linearity:** In quantum physics all dynamic transformations are linear, and thus the post-transformation state $|\phi\rangle$ must be related to the pre-transformation state $|\psi\rangle$ via the product with an $N \times N$ matrix U as

$$|\phi\rangle = U|\psi\rangle. \tag{10.1}$$

- **Reversibility:** A key physical principle, named after the American physicist Rolf William Landauer, states that deleting information requires the expenditure of energy. Specifically, the energy required is $k_B T \ln(2)$ joules per bit, where k_B is the Boltzmann constant and T is the system's temperature in Kelvin. Since this energy must be supplied by the environment, information cannot be deleted in a closed quantum system. Consequently, closed quantum systems must conserve information. This principle implies that transformations of a quantum state in a closed system must be reversible. As a result, in the transformation (10.1), the matrix U must be invertible, i.e., it has a valid inverse U^{-1}. Recall that the inverse of a square matrix satisfies the equalities $U^{-1}U = UU^{-1} = I$, where I is the identity matrix.
- **Norm preservation:** The output vector $|\phi\rangle$ must be a valid state, requiring the normalization

$$\langle\phi|\phi\rangle = \langle\psi|U^\dagger U|\psi\rangle = 1 \tag{10.2}$$

for every possible input state $|\psi\rangle$. This condition implies the equality

$$U^\dagger U = I, \tag{10.3}$$

which in turn states that the Hermitian transpose $U^\dagger$ is the inverse of matrix U, i.e.,

$$U^{-1} = U^\dagger. \tag{10.4}$$

Note that norm preservation thus implies reversibility. As per (5.43), matrices U satisfying the condition (10.4) are said to be unitary.

Definition 10.1 Closed-system transformations are unitary
A transformation from state $|\psi\rangle$ to state $|\phi\rangle$ is physically possible if and only if it can be expressed as

$$|\phi\rangle = U|\psi\rangle, \tag{10.5}$$

where U is a unitary matrix.

The fundamental postulate of quantum mechanics expressed by the equality (10.5) is illustrated in Figure 10.1, which represents a quantum transformation U as a box, with input given by the pre-transformation state $|\psi\rangle$ and output given by the transformed state $|\phi\rangle$. Dynamic transformations described by unitary matrices are also known as **gates** in the context of quantum computing. Using this language, in Figure 10.1 a quantum gate U is applied to state $|\psi\rangle$ to produce state $|\phi\rangle$.

Definition 10.2 Equivalent closed-system transformations
Two unitary transformations U and V that differ only for a global phase, satisfying the equality

$$U = \exp(i\gamma)V \tag{10.6}$$

for some real number γ, are equivalent. This is in the sense that they produce equivalent states starting from the same pre-transformation state.

This can readily be checked since the post-transformation states

$$|\phi\rangle = V|\psi\rangle \tag{10.7}$$

and

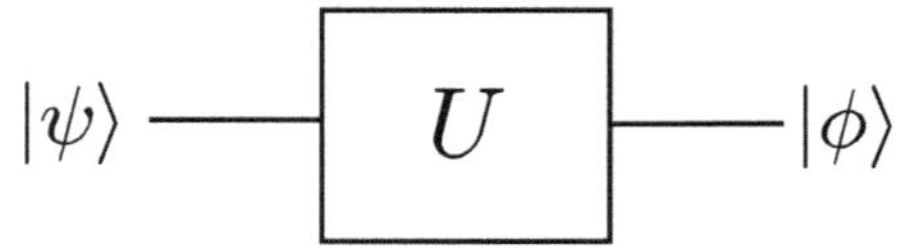

Figure 10.1 In a closed quantum system, a quantum state $|\psi\rangle$ evolves via the application of a unitary matrix U, i.e., as $|\phi\rangle = U|\psi\rangle$, which corresponds to a linear, reversible, and norm-preserving dynamic transformation.

$$\begin{aligned}|\phi'\rangle &= U|\psi\rangle \\ &= \exp(i\gamma)V|\psi\rangle \\ &= \exp(i\gamma)|\phi\rangle \end{aligned} \tag{10.8}$$

are indeed equivalent, differing only for a global phase term γ (see (5.32)).

This said, it is important to note that the phase γ in the matrix $U = \exp(i\gamma)V$ does play a role in case the unitary transformation U is applied to a subsystem of a larger system (see Section 8.7). In fact, in such situations, the phase γ contributes to the relative phase of two different subsystems.

10.2.2 Designing Unitary Transformations

In principle, any $N \times N$ unitary matrix can be physically realized in an N-dimensional quantum system. In practice, however, when implementing unitary transformations in quantum computing, one typically decomposes them into sequences of **elementary quantum gates**, which can be natively and efficiently executed on the available hardware. These elementary gates usually act on only a small subset of the larger composite system, making them easier to implement.

As illustrated in Figure 10.2, the overall transformation carried out by a quantum computer is assembled step-by-step: one first applies an elementary gate U_1, then another gate U_2, and continues this process until finally applying the last gate U_L. The overall transformation illustrated in Figure 10.2 is described by the matrix

$$U = U_L U_{L-1} \cdots U_1, \tag{10.9}$$

which can be verified to be unitary.

Note that the order of the unitary matrices in the mathematical expression (10.9) is the opposite of the order in which the unitary matrices are applied on the pre-transformation state. Indeed, the transformation (10.1) with matrix U in (10.9) yields the state

$$|\phi\rangle = U_L U_{L-1} \cdots U_1|\psi\rangle, \tag{10.10}$$

which is evaluated by computing matrix–vector multiplications from right to left, starting from $|\phi_1\rangle = U_1|\psi\rangle$, followed by $|\phi_2\rangle = U_2|\phi_1\rangle$, and so on for L steps.

In composite systems the dimension N grows exponentially in the number of subsystems S (see Section 1.3.3). With this in mind, the design of a unitary transformation U in quantum computing must ideally satisfy the following conflicting requirements.

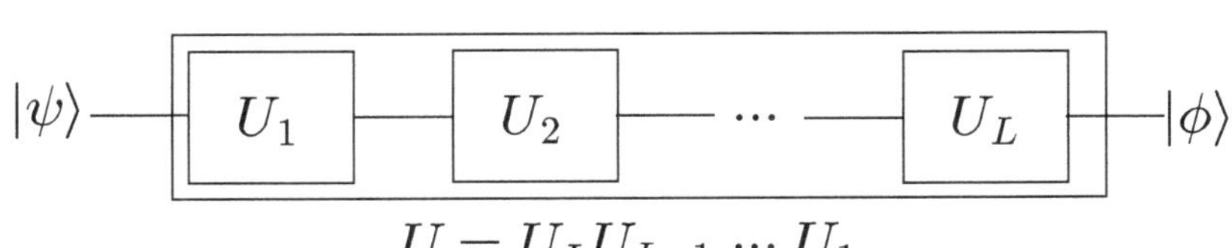

Figure 10.2 A unitary transformation U can be implemented as a sequence of elementary unitary matrices, or gates, $U_1, U_2, \ldots, U_L$.

- **Computational task:** The unitary matrix U should be part of the solution of a relevant computational problem, such as computing the Fourier transform of an input vector.
- **Efficiency:** The number of elementary gates, L, should be sufficiently small, growing slowly with the number of subsystems S. This limits the types of transformations that can be run on quantum computers, since, for an arbitrary matrix U, the number of elementary gates, L, scales exponentially with S (see Recommended Resources).
- **Quantum advantage:** In order for a quantum computing solution realized via a transformation (10.9) to be advantageous over classical computing, it is necessary that simulating the behavior of the transformed state (10.10) on a classical computer be prohibitively complex. Validating a quantum advantage turns out to be a difficult task, leading in recent years to several contentious claims.

Overall, finding applications for which quantum computing solutions are practical and advantageous over classical methods is a central challenge in the field of quantum computer science.

10.3 Interpreting Unitary Transformations ⏭

The effect of a unitary transformation (10.1) on a state $|\psi\rangle$ can be interpreted and understood in one of two ways.

- **State rotation:** A unitary matrix is a rotation of the pre-transformation state vector in the N-dimensional space.
- **Parallel phase rotations:** A unitary matrix applies N separate phase rotations of the pre-transformation state vector along the N eigenvectors of matrix U.

These two perspectives are explained in this section.

10.3.1 Unitary Transformations as State Rotations

To explain the first interpretation of a unitary transformation, take an arbitrary orthonormal basis $\{|v_x\rangle\}_{x\in[N]}$, and consider the corresponding post-transformation states

$$|w_x\rangle = U|v_x\rangle \tag{10.11}$$

for $x \in [N]$. One can readily verify that the transformed states $\{|w_x\rangle\}_{x\in[N]}$ form another orthonormal basis of the N-dimensional space. In fact, the inner products between the transformed states $\{|w_x\rangle\}_{x\in[N]}$ are given by

$$\langle w_x|w_y\rangle = \begin{cases} 0 & \text{if } x \neq y, \\ 1 & \text{if } x = y. \end{cases} \tag{10.12}$$

Therefore, as illustrated in Figure 10.3 for $N = 2$ dimensions, the transformation U effectively rotates each basis state $|v_x\rangle$ into a corresponding state $|w_x\rangle$ of another basis. The term "rotation" is used here to emphasize that the transformation maintains the mutual orthogonality between the states of the transformed basis.

Using this insight, one can now study the impact of the unitary transformation U on an arbitrary state $|\psi\rangle$ by separating the contribution of each basis state $\{|v_x\rangle\}_{x\in[N]}$. To this end, express the pre-transformation state $|\psi\rangle$ by using the orthonormal basis $\{|v_x\rangle\}_{x\in[N]}$ as the superposition

$$|\psi\rangle = \sum_{x\in[N]} \alpha_x |v_x\rangle, \tag{10.13}$$

with amplitudes $\alpha_x = \langle v_x|\psi\rangle$ for all $x \in [N]$. With this decomposition, the operation of the unitary transformation can be described as follows.

Key Result 10.1 Unitary transformations as state rotations

Given an arbitrary orthonormal basis $\{|v_x\rangle\}_{x\in[N]}$ and the pre-transformation state $|\psi\rangle$ in (10.13), the post-transformation state (10.1) is given as

$$\begin{aligned} |\phi\rangle &= U|\psi\rangle \\ &= \sum_{x\in[N]} \alpha_x U|v_x\rangle \\ &= \sum_{x\in[N]} \alpha_x |w_x\rangle. \end{aligned} \tag{10.14}$$

Therefore, as shown in Figure 10.3, the vector $|\phi\rangle$ is obtained by rotating separately each of the vectors in the underlying orthonormal basis $\{|v_x\rangle\}_{x\in[N]}$.

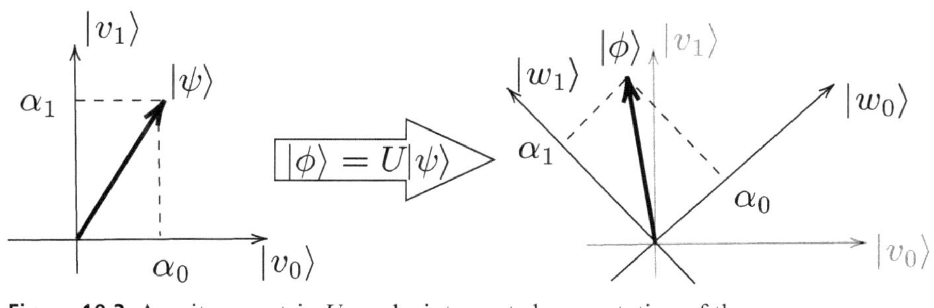

Figure 10.3 A unitary matrix U can be interpreted as a rotation of the pre-transformation state vector $|\psi\rangle$ in the N-dimensional space (here $N = 2$). In fact, the unitary transformation preserves the amplitudes, here α_0 and α_1, while mapping each basis vector $|v_x\rangle$ from an orthonormal basis $\{|v_x\rangle\}_{x\in[N]}$ into the state $|w_x\rangle = U|v_x\rangle$ from another orthonormal basis $\{|w_x\rangle\}_{x\in[N]}$.

Specifically, by the decomposition (10.14), the post-transformation state $|\phi\rangle$ has the same complex amplitudes $\{\alpha_x\}_{x\in[N]}$ as the original state $|\psi\rangle$ in (10.13), while the corresponding basis vectors have been replaced by the rotated basis $\{|w_x\rangle\}_{x\in[N]}$.

This result can be expressed by writing the unitary matrix U as the sum

$$U = \sum_{x\in[N]} |w_x\rangle\langle v_x|, \tag{10.15}$$

which emphasizes how the transformation U replaces each state vector $|v_x\rangle$ by the rotated vector $|w_x\rangle$. Indeed, using the decomposition (10.15), given input state (10.13), the output state can be written as

$$\begin{aligned}
|\phi\rangle &= \left(\sum_{x\in[N]} |w_x\rangle\langle v_x|\right)\left(\sum_{x'\in[N]} \alpha_{x'}|v_{x'}\rangle\right) \\
&= \sum_{x\in[N]}\sum_{x'\in[N]} \alpha_{x'}|w_x\rangle\langle v_x|v_{x'}\rangle \\
&= \sum_{x\in[N]} \alpha_x|w_x\rangle,
\end{aligned} \tag{10.16}$$

recovering the equality (10.14).

Example 10.1

The decomposition (10.15) of a unitary matrix applies for an arbitrary orthonormal basis $\{|v_x\rangle\}_{x\in[N]}$. In particular, choosing the basis $\{|v_x\rangle\}_{x\in[N]}$ as the computational basis $\{|x\rangle\}_{x\in[N]}$, the equality (10.15) reads

$$U = \sum_{x\in[N]} |u_x\rangle\langle x|, \tag{10.17}$$

where $|u_x\rangle$ is the xth column of matrix $U = [|u_0\rangle, \ldots, |u_{N-1}\rangle]$. This is because the post-transformation state for state $|x\rangle$ is given by

$$U|x\rangle = |u_x\rangle. \tag{10.18}$$

Example 10.2

The Pauli matrices $\{P_0, P_1, P_2, P_3\}$ are unitary. Take for example the Pauli Z matrix

$$P_3 = \begin{bmatrix} 1 & 0 \\ 0 & -1 \end{bmatrix}. \tag{10.19}$$

The Pauli Z matrix operates on the computational basis as $P_3|0\rangle = |0\rangle$ and $P_3|1\rangle = -|1\rangle$. Therefore, the decomposition (10.17) yields

$$P_3 = |0\rangle\langle 0| - |1\rangle\langle 1|. \tag{10.20}$$

Accordingly, given a general pre-transformation state $|\psi\rangle = \psi_0|0\rangle + \psi_1|1\rangle$, by (10.14), the post-transformation state is

$$|\phi\rangle = P_3|\psi\rangle = \psi_0|0\rangle - \psi_1|1\rangle = \begin{bmatrix} \psi_0 \\ -\psi_1 \end{bmatrix}. \tag{10.21}$$

Thus, the Pauli Z transformation modifies the relative phase of the amplitudes along the computational-basis states by changing the sign of the second amplitude.

To highlight the fact that the rotation implemented by a unitary matrix can be expressed in terms of any orthonormal basis, consider now the Hadamard basis $\{|v_0\rangle = |+\rangle, |v_1\rangle = |-\rangle\}$. The Pauli Z matrix acts on these basis vectors as $P_3|+\rangle = |-\rangle$ and $P_3|-\rangle = |+\rangle$, thus flipping the two basis vectors. This amounts to the decomposition (10.17) given by

$$P_3 = |-\rangle\langle+| + |+\rangle\langle-|. \tag{10.22}$$

Writing an arbitrary pre-transformation qubit state in the Hadamard basis as $|\psi\rangle = \frac{1}{\sqrt{2}}(\psi_0 + \psi_1)|+\rangle + \frac{1}{\sqrt{2}}(\psi_0 - \psi_1)|-\rangle$ (see (5.37)), the post-transformation is obtained as

$$\begin{aligned} |\phi\rangle &= P_3|\psi\rangle \\ &= \frac{1}{\sqrt{2}}(\psi_0 + \psi_1)|-\rangle + \frac{1}{\sqrt{2}}(\psi_0 - \psi_1)|+\rangle = \begin{bmatrix} \psi_0 \\ -\psi_1 \end{bmatrix}, \end{aligned} \tag{10.23}$$

which recovers the same result as (10.21).

Example 10.3

Consider now the Pauli X matrix

$$P_1 = \begin{bmatrix} 0 & 1 \\ 1 & 0 \end{bmatrix}. \tag{10.24}$$

Given the equalities $P_1|0\rangle = |1\rangle$ and $P_1|1\rangle = |0\rangle$, the Pauli X matrix rotates the vectors in the computational basis by flipping basis vector $|0\rangle$ into vector $|1\rangle$, and vice versa. Therefore, in terms of the decomposition (10.17), the Pauli X matrix can be written as

$$P_1 = |1\rangle\langle 0| + |0\rangle\langle 1|. \tag{10.25}$$

It follows that, when applied to a qubit state $|\psi\rangle$, the Pauli X matrix causes the two amplitudes to swap places, so that, by (10.14), the output state is

$$|\phi\rangle = P_1|\psi\rangle = \psi_1|0\rangle + \psi_0|1\rangle = \begin{bmatrix} \psi_1 \\ \psi_0 \end{bmatrix}. \tag{10.26}$$

Considering now the Hadamard basis, the Pauli X matrix acts on the basis vectors as $P_1|+\rangle = |+\rangle$ and $P_1|-\rangle = -|-\rangle$. This modifies the relative phase along the two basis vectors by changing the sign of the amplitude along the basis state $|-\rangle$. The corresponding decomposition (10.17) of the Pauli X matrix is

$$P_1 = |+\rangle\langle+| - |-\rangle\langle-|. \tag{10.27}$$

Example 10.4

As seen in Section 7.6.2, the transformation

$$iP_2 = \begin{bmatrix} 0 & 1 \\ -1 & 0 \end{bmatrix}, \tag{10.28}$$

which coincides with the Pauli Y matrix P_2 apart from a phase term, can be expressed as

$$iP_2 = P_3 P_1. \tag{10.29}$$

Thus, using the computational basis, this transformation first flips the two basis states and then changes the relative phase of the amplitudes. In the Hadamard basis, it first changes the relative phase of the amplitudes and then it flips the two basis states.

In fact, using the computational basis, one has the equalities

$$iP_2|0\rangle = -|1\rangle \text{ and } iP_2|1\rangle = |0\rangle, \tag{10.30}$$

while in the Hadamard basis one has

$$iP_2|+\rangle = |-\rangle \text{ and } iP_2|-\rangle = -|+\rangle. \tag{10.31}$$

Example 10.5

Another important example of a closed-system transformation for qubit states is given by the Hadamard matrix

$$H = \frac{1}{\sqrt{2}} \begin{bmatrix} 1 & 1 \\ 1 & -1 \end{bmatrix} \tag{10.32}$$

first encountered in Section 5.4.3. The Hadamard matrix acts on the computational basis as

$$H|0\rangle = |+\rangle \text{ and } H|1\rangle = |-\rangle; \tag{10.33}$$

and on the Hadamard basis as

$$H|+\rangle = |0\rangle \text{ and } H|-\rangle = |1\rangle. \tag{10.34}$$

Hence, the Hadamard matrix swaps the vectors from the Hadamard basis to the computational basis and vice versa.

Mathematically, by the decomposition (10.15), the Hadamard matrix can be expressed in one of the following ways:

$$\begin{aligned} H &= |+\rangle\langle 0| + |-\rangle\langle 1| \\ &= |0\rangle\langle +| + |1\rangle\langle -|. \end{aligned} \tag{10.35}$$

Therefore, given an arbitrary qubit state $|\psi\rangle = \psi_0|0\rangle + \psi_1|1\rangle$, the output state is given as

$$|\phi\rangle = \psi_0|+\rangle + \psi_1|-\rangle = \frac{1}{\sqrt{2}} \begin{bmatrix} \psi_0 + \psi_1 \\ \psi_0 - \psi_1 \end{bmatrix}. \tag{10.36}$$

The same result can also be obtained by using the expansion of the state in the Hadamard basis, yielding

$$|\phi\rangle = \frac{1}{\sqrt{2}}(\psi_0 + \psi_1)|0\rangle + \frac{1}{\sqrt{2}}(\psi_0 - \psi_1)|1\rangle = \frac{1}{\sqrt{2}}\begin{bmatrix} \psi_0 + \psi_1 \\ \psi_0 - \psi_1 \end{bmatrix}. \tag{10.37}$$

Example 10.6

The Hadamard transformation can be used to illustrate the important fact that the amplitudes of a quantum state can interfere with one another as they evolve under a unitary transformation. In particular, it is possible for two amplitudes to interfere destructively, making it impossible to obtain certain measurement outputs after the transformation (see Example 5.14). This is a highly non-classical behavior: as a result of a classical channel, the probabilities in a classical informational state can only add positively, with no cancellations, whereas quantum amplitudes can interfere destructively or constructively through a unitary transformation.

To elaborate, consider the setting in Figure 10.4, in which a qubit, initially in state $|+\rangle$, is modified by a Hadamard gate H, and is then measured in the computational basis. As a reference, note that if the measurement was made before the Hadamard transformation, one would obtain the classical outputs 0 or 1 with equal probability.

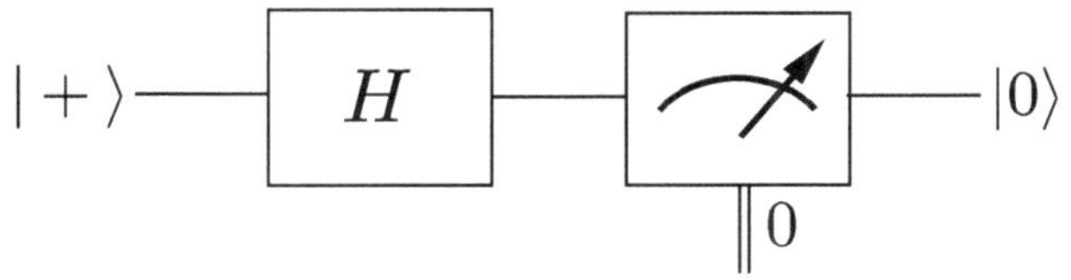

Figure 10.4 A qubit, initially in state $|+\rangle$, is modified by a Hadamard gate H, and is then measured in the computational basis. As illustrated in Figure 10.5, this causes the two amplitudes for the computational basis to combine destructively along the direction $|1\rangle$.

In contrast, after the Hadamard transformation, the state is given by $H|+\rangle = |0\rangle$. Therefore, the measurement returns the classical output $x = 0$ with probability equal to 1. Obtaining the measurement output $x = 1$ has hence become impossible. As illustrated in Figure 10.5, the two amplitudes $\psi_0 = 1/\sqrt{2}$ and $\psi_1 = 1/\sqrt{2}$, initially equal in magnitude in the pre-transformation state $|\psi\rangle = |+\rangle$, have interfered destructively with one another along the direction $|1\rangle$ when undergoing the Hadamard transformation. As a result, while the amplitude of the post-transformation state along the direction $|0\rangle$ is reinforced via the **constructive interference** of the two amplitudes in the pre-transformation state, the amplitude of the post-transformation state along the direction $|1\rangle$ is suppressed via the **destructive interference** of the two initial amplitudes.

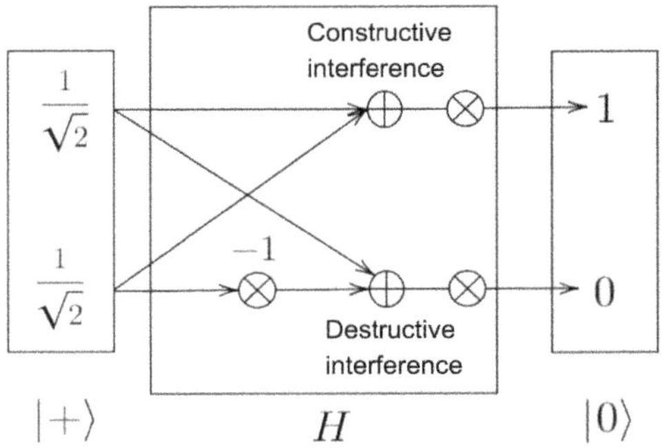

Figure 10.5 Illustration of the evolution of the two amplitudes along the computational basis for the Hadamard transformation in Figure 10.4. While the amplitude of the post-transformation state along the direction $|0\rangle$ is reinforced via constructive interference, the amplitude of the post-transformation state along the direction $|1\rangle$ is suppressed by destructive interference.

10.3.2 Qubit Quantum Gates as Rotations

In the special case of qubits, i.e., with $N = 2$ dimensions, the interpretation of a unitary transformation as a rotation can be substantiated by noting that all unitary transformations U can be realized via the cascade of three rotation matrices.

Specifically, any single-qubit unitary transformation can be expressed, apart from a global phase, in terms of three rotation angles θ_a, θ_b, and θ_c, taking values in the interval between 0 and 2π as

$$R(\theta_a, \theta_b, \theta_c) = R_x(\theta_a)R_y(\theta_b)R_x(\theta_c), \tag{10.38}$$

where x and y are two different integers in the set $\{1, 2, 3\}$. The rotation matrices $R_x(\theta)$ in (10.38) are defined next, and the reader is referred to the Recommended Resources for a proof of the characterization (10.38) and for a discussion of the geometric interpretation of the angles θ_a, θ_b, and θ_c.

The matrices in (10.38) are defined as follows.

- **Pauli X rotation matrix:** The Pauli X rotation matrix is given by

$$\begin{aligned} R_1(\theta) &= \exp\left(-i\frac{\theta}{2}P_1\right) \\ &= \cos\left(\frac{\theta}{2}\right)I - i\sin\left(\frac{\theta}{2}\right)P_1 \\ &= \begin{bmatrix} \cos\left(\frac{\theta}{2}\right) & -i\sin\left(\frac{\theta}{2}\right) \\ -i\sin\left(\frac{\theta}{2}\right) & \cos\left(\frac{\theta}{2}\right) \end{bmatrix}. \end{aligned} \tag{10.39}$$

- **Pauli Y rotation matrix:** The Pauli Y rotation matrix is defined as

$$\begin{aligned} R_2(\theta) &= \exp\left(-i\frac{\theta}{2}P_2\right) \\ &= \cos\left(\frac{\theta}{2}\right)I - i\sin\left(\frac{\theta}{2}\right)P_2 \\ &= \begin{bmatrix} \cos\left(\frac{\theta}{2}\right) & -\sin\left(\frac{\theta}{2}\right) \\ \sin\left(\frac{\theta}{2}\right) & \cos\left(\frac{\theta}{2}\right) \end{bmatrix}. \end{aligned} \tag{10.40}$$

  Thus, the Pauli Y rotation $R_2(\theta)$ has the conventional form of a rotation by an angle $\theta/2$ in a two-dimensional real space.
- **Pauli Z rotation matrix:** The Pauli Z rotation matrix is defined as

$$\begin{aligned} R_3(\theta) &= \exp\left(-i\frac{\theta}{2}P_3\right) \\ &= \cos\left(\frac{\theta}{2}\right)I - i\sin\left(\frac{\theta}{2}\right)P_3 \\ &= \begin{bmatrix} \exp\left(-i\frac{\theta}{2}\right) & 0 \\ 0 & \exp\left(i\frac{\theta}{2}\right) \end{bmatrix}. \end{aligned} \tag{10.41}$$

  Since this matrix can be expressed as

$$R_3(\theta) = \exp\left(-i\frac{\theta}{2}\right)\begin{bmatrix} 1 & 0 \\ 0 & \exp(i\theta) \end{bmatrix}, \tag{10.42}$$

  the Pauli Z rotation is equivalent to a transformation that applies a relative phase θ to the second amplitude of the state vector.

Note that, by setting $\theta = -\pi$, one finds $R_1(-\pi) = iP_1$, $R_2(-\pi) = iP_2$, and $R_3(-\pi) = iP_3$. Thus, apart from a phase term, the Pauli rotation matrices include as special cases the corresponding Pauli matrices.

10.3.3 Unitary Transformations as Phase Rotations

Consider now the second interpretation of closed-system transformations as separate phase rotations along the eigenvectors of a unitary matrix U. As discussed in Section 6.6, any unitary matrix has eigendecomposition

$$U = \sum_{x\in[N]} \lambda_x |v_x\rangle\langle v_x| \tag{10.43}$$

with eigenbasis $\{|v_x\rangle\}_{x\in[N]}$ and eigenvalues $\{\lambda_x\}_{x\in[N]}$ having absolute values equal to 1. This means that the eigenvalues can be expressed as

$$\lambda_x = \exp(i\gamma_x) \tag{10.44}$$

for some phases $\{\gamma_x\}_{x\in[N]}$.

By (10.43) and (10.44), a unitary transformation applies separate phase rotations along the directions specified by the eigenbasis $\{|v_x\rangle\}_{x\in[N]}$ of the unitary matrix U. In fact, writing the input state in the eigenbasis $\{|v_x\rangle\}_{x\in[N]}$ as $|\psi\rangle = \sum_{x\in[N]} \alpha_x |v_x\rangle$, with amplitudes $\alpha_x = \langle v_x|\psi\rangle$ for all $x \in [N]$, the post-transformation state is given by

$$\begin{aligned}|\phi\rangle &= \sum_{x\in[N]} \alpha_x \lambda_x |v_x\rangle \\ &= \sum_{x\in[N]} \alpha_x \exp(i\gamma_x)|v_x\rangle. \end{aligned} \tag{10.45}$$

Hence every amplitude α_x of the pre-transformation state is modified by the phase term $\exp(i\gamma_x)$.

Example 10.7

The **SWAP gate** is the 4×4 unitary transformation

$$U = \begin{bmatrix} 1 & 0 & 0 & 0 \\ 0 & 0 & 1 & 0 \\ 0 & 1 & 0 & 0 \\ 0 & 0 & 0 & 1 \end{bmatrix}. \tag{10.46}$$

The name comes from the fact that, in composite systems, it represents the swap of two constituent subsystems (here qubits). The SWAP gate is discussed further in Problem 15.7.

The SWAP gate is unitary with eigenvectors given by the so-called **Bell states**,

$$\begin{aligned}|v_0\rangle &= \frac{1}{\sqrt{2}}(|0\rangle + |3\rangle), \\ |v_1\rangle &= \frac{1}{\sqrt{2}}(|0\rangle - |3\rangle), \\ |v_2\rangle &= \frac{1}{\sqrt{2}}(|1\rangle + |2\rangle), \\ |v_3\rangle &= \frac{1}{\sqrt{2}}(|1\rangle - |2\rangle), \end{aligned} \tag{10.47}$$

and corresponding eigenvalues $\{\lambda_0 = 1, \lambda_1 = 1, \lambda_2 = 1, \lambda_3 = -1\}$. Note that, like the Pauli matrices, the SWAP gate is both unitary and Hermitian, since all the eigenvalues are real and they have absolute value equal to 1.

For the input state $|\psi\rangle = 1/2(|0\rangle + |1\rangle - |2\rangle - |3\rangle)$, what is the post-transformation state after a SWAP gate? The pre-transformation state can be expressed in the eigenbasis $\{|v_x\rangle\}_{x\in[4]}$ as

$$|\psi\rangle = \frac{1}{\sqrt{2}}(|v_1\rangle + |v_3\rangle). \tag{10.48}$$

Therefore, using (10.45), the post-transformation state is

$$U|\psi\rangle = \frac{1}{\sqrt{2}}(|v_1\rangle - |v_3\rangle). \tag{10.49}$$

By (10.43) and (10.44), the following characterization of a unitary matrix holds.

Key Result 10.2 Unitary matrices and Hamiltonians
Any unitary matrix U can equivalently be written as

$$U = \exp(iH), \tag{10.50}$$

with Hermitian matrix

$$H = \sum_{x \in [N]} \gamma_x |v_x\rangle\langle v_x| \tag{10.51}$$

having eigenbasis $\{|v_x\rangle\}_{x\in[N]}$ and eigenvalues equal to the phases $\{\gamma_x\}_{x\in[N]}$. Matrix H is known as the Hamiltonian of the unitary matrix U.

Example 10.8

What is the Hamiltonian of the SWAP gate (10.46)? As discussed in the last example, the phases of the eigenvalues are equal to $\gamma_0 = \gamma_1 = \gamma_2 = 0$ and $\gamma_3 = \pi$. Therefore, using the eigenvectors (10.47), the Hamiltonian of the SWAP gate is given by the matrix $H = \pi |v_3\rangle\langle v_3|$.

10.4 Dynamic Transformations for Mixed States

This section extends the description of closed-system transformations from pure to mixed states. To start, following Chapter 7, recall that a pure state $|\psi\rangle$ can be expressed as the rank-1 density matrix $\rho = |\psi\rangle\langle\psi|$. Accordingly, the post-transformation state (10.1) can be written as the density matrix

$$\begin{aligned}\sigma &= |\phi\rangle\langle\phi| \\ &= U|\psi\rangle\langle\psi|U^\dagger \\ &= U\rho U^\dagger.\end{aligned} \tag{10.52}$$

In order to extend this relation to more general density matrices, it is useful to express the pre-transformation density matrix ρ via its eigendecomposition

$$\rho = \sum_{x\in[N]} \lambda_x |u_x\rangle\langle u_x|. \tag{10.53}$$

Using the eigenensemble $\mathcal{E}(\rho) = \{(|u_0\rangle, \lambda_0), (|u_1\rangle, \lambda_1), \ldots, (|u_{N-1}\rangle, \lambda_{N-1})\}$, the mixed state ρ models a situation in which the system is in pure state $|u_x\rangle$ with probability λ_x for all $x \in [N]$.

If the system is in state $|u_x\rangle$, by (10.1), the post-transformation state is $U|u_x\rangle$. Therefore, the state after the transformation is realized by the ensemble $\{(U|u_0\rangle, \lambda_0), (U|u_1\rangle, \lambda_1), \ldots, (U|u_{N-1}\rangle, \lambda_{N-1})\}$. Note that the transformed states

$\{U|u_x\rangle\}_{x\in[N]}$ are still mutually orthogonal. Therefore, this corresponds to an eigenensemble for the post-transformation state.

Using the obtained eigenensemble, the **post-transformation density matrix** is obtained as

$$\begin{aligned}\sigma &= \sum_{x\in[N]} \lambda_x U|u_x\rangle\langle u_x|U^\dagger \\ &= U\left(\sum_{x\in[N]} \lambda_x|u_x\rangle\langle u_x|\right)U^\dagger \\ &= U\rho U^\dagger. \end{aligned} \tag{10.54}$$

Overall, the effect of a unitary matrix is to rotate the eigenvectors of the pre-transformation density matrix, yielding eigenvectors $\{U|u_x\rangle\}_{x\in[N]}$, while leaving the eigenvalues equal to those of the pre-transformation density matrix, i.e., to $\{\lambda_x\}_{x\in[N]}$. In this sense, a closed-system transformation preserves the classical uncertainty represented in the original state by the eigenvalues $\{\lambda_x\}_{x\in[N]}$.

Definition 10.3 Closed-system transformations of mixed states
In a closed system, quantum transformations of a state ρ take the form

$$\sigma = U\rho U^\dagger \tag{10.55}$$

for some unitary matrix U. Accordingly, the post-transformation state σ has the same eigenvalues as the pre-transformation state ρ, with eigenvectors rotated by the unitary matrix U.

This operation is linear, reversible, as well as positive and trace-preserving.

- **Linearity:** For any two density matrices ρ_0 and ρ_1, consider the pre-transformation state given by the mixture

$$\rho = \gamma\rho_0 + (1-\gamma)\rho_1 \tag{10.56}$$

for some probability γ with $0 \leq \gamma \leq 1$. With this input state, the post-transformation state (10.54) is given by the corresponding linear combination

$$U\rho U^\dagger = \gamma U\rho_0 U^\dagger + (1-\gamma)U\rho_1 U^\dagger. \tag{10.57}$$

Therefore, linear combinations of density matrices produce equal linear combinations of the post-transformation density matrices (10.54). This indicates that a unitary transformation is linear.
- **Reversibility:** The original state ρ can be recovered from the post-transformation state σ in (10.54) by applying the inverse transformation $U^\dagger$. In fact, using (10.54), one obtains

$$\begin{aligned} U^\dagger \sigma (U^\dagger)^\dagger &= U^\dagger \sigma U \\ &= U^\dagger U \rho U^\dagger U = \rho. \end{aligned} \tag{10.58}$$

- **Positivity and trace preservation:** The post-transformation state σ is a valid density matrix, i.e., it is a positive semi-definite matrix with trace equal to 1. This can readily be checked by recalling that the post-transformation state σ in (10.54) has the same eigenvalues as the density matrix ρ. Therefore, the eigenvalues of matrix σ are non-negative and they sum to 1.

Example 10.9

Consider the transformation of a qubit state defined by the **T gate**

$$T = \begin{bmatrix} 1 & 0 \\ 0 & \exp(i\pi/4) \end{bmatrix}, \tag{10.59}$$

which is equivalent to the Pauli Z rotation matrix $R_3(\pi/4)$ in (10.42). T gates are essential in quantum computing, and they are also known as **magic gates**. This is because the addition of a number of T gates to a unitary transformation of the form (10.10) makes it exponentially more difficult to simulate the resulting state on a classical computer.

For the pre-transformation density matrix

$$\rho = \begin{bmatrix} 0.5 & -0.3i \\ 0.3i & 0.5 \end{bmatrix}, \tag{10.60}$$

using (10.54), the post-transformation state is given by

$$\begin{aligned} \sigma &= T\rho T^\dagger \\ &= \begin{bmatrix} 1 & 0 \\ 0 & \exp(i\pi/4) \end{bmatrix} \begin{bmatrix} 0.5 & -0.3i \\ 0.3i & 0.5 \end{bmatrix} \begin{bmatrix} 1 & 0 \\ 0 & \exp(-i\pi/4) \end{bmatrix} \\ &= \begin{bmatrix} 0.5 & -0.2121 - 0.2121i \\ -0.2121 + 0.2121i & 0.5 \end{bmatrix}. \end{aligned} \tag{10.61}$$

10.5 Matrix Vectorization ⏭

As discussed in the previous sections, closed-system dynamic transformations are linear. This property can be made more apparent by adopting a matrix–vector multiplication formulation of the transformation (10.52) that more closely resembles the description used for pure states in (10.1). This section focuses on this alternative, equivalent, formulation, by introducing the operation of **matrix vectorization**. Matrix vectorization will also play an important role in Part VI in the context of tensor-based models of composite systems.

10.5.1 Defining Matrix Vectorization

Given an $N\times M$ complex-valued matrix A, matrix vectorization produces an $NM\times 1$ vector, denoted as vec(A), by stacking the rows of the matrix one on top of the other upon transposition. To detail this operation, write matrix A row-wise as

$$A = \begin{bmatrix} A_0 \\ A_1 \\ \vdots \\ A_{N-1} \end{bmatrix}, \tag{10.62}$$

where A_x represents the xth row of matrix A, which is a $1 \times M$ vector.

Definition 10.4 Vectorization
Denoting as A_x the xth row of matrix A, the vectorization of matrix A is defined as the $NM \times 1$ vector

$$\text{vec}(A) = [A_0, A_1, \ldots, A_{N-1}]^T = \begin{bmatrix} A_0^T \\ A_1^T \\ \vdots \\ A_{N-1}^T \end{bmatrix}. \tag{10.63}$$

Equivalently, using the direct sum notation introduced in Section 8.8, the vectorization of matrix A can be written as

$$\text{vec}(A) = A_0^T \oplus A_1^T \oplus \cdots \oplus A_{N-1}^T. \tag{10.64}$$

The operation of vectorization can be most conveniently expressed using the bi-index notation, which was presented in Section 8.8. Recall that, using the bi-index notation, the zth entry a_z of an $NM \times 1$ vector a, with $z \in [NM]$, can be written using two indices $x \in [N]$ and $y \in [M]$ as

$$a_z = a_{xy}, \tag{10.65}$$

where

$$z = Mx + y. \tag{10.66}$$

Key Result 10.3 Vectorization using the bi-index notation
Using the bi-index notation (10.65) and (10.66), the entries of the matrix vectorization $a = \text{vec}(A)$ for the $N \times M$ matrix A can be written as

$$a_{xy} = A_{x,y} = \langle x|A|y\rangle, \tag{10.67}$$

where $A_{x,y}$ is the (x, y)th entry of matrix A, $|x\rangle$ represents an $N \times 1$ vector from the computational basis of the N-dimensional space, and $|y\rangle$ is an $M \times 1$ vector from the computational basis of the M-dimensional space.

According to (10.67), the entries of vector $a = \text{vec}(A)$ can be indexed using either the single index $z \in [NM]$ or the two indices $x \in [N]$ and $y \in [M]$. The latter formalism is particularly useful as it facilitates the connection with matrix A thanks to the equality (10.67). It is emphasized that it is the convenience of the correspondence (10.67) between the entries $A_{x,y}$ of matrix A and the entry a_{xy} of the vectorization $a = \text{vec}(A)$ to motivate the definition of vectorization as the concatenation of rows as in (10.63). That said, it is also common in other references to adopt the definition of column-wise vectorization, and the reader is warned to verify which definition is assumed in the given text.

Example 10.10

For matrix

$$A = \begin{bmatrix} A_{0,0} & A_{0,1} & A_{0,2} \\ A_{1,0} & A_{1,1} & A_{1,2} \end{bmatrix} = \begin{bmatrix} 1+2i & 3+4i & 5+6i \\ 7+8i & 9+i & 2+3i \end{bmatrix} \tag{10.68}$$

with $N = 2$ rows and $M = 3$ columns, vectorization yields

$$\text{vec}(A) = \begin{bmatrix} a_0 \\ a_1 \\ a_2 \\ a_3 \\ a_4 \\ a_5 \end{bmatrix} = \begin{bmatrix} a_{00} \\ a_{01} \\ a_{02} \\ a_{10} \\ a_{11} \\ a_{12} \end{bmatrix} = \begin{bmatrix} A_{0,0} \\ A_{0,1} \\ A_{0,2} \\ A_{1,0} \\ A_{1,1} \\ A_{1,2} \end{bmatrix} = \begin{bmatrix} 1+2i \\ 3+4i \\ 5+6i \\ 7+8i \\ 9+i \\ 2+3i \end{bmatrix}. \tag{10.69}$$

Using the Kronecker product, the result (10.67) implies the following.

Key Result 10.4 Vectorization using the Kronecker product
The vectorization of an $N \times M$ matrix A can be expressed as

$$\text{vec}(A) = \sum_{x\in[N]} \sum_{y\in[M]} A_{x,y}(|x\rangle \otimes |y\rangle), \tag{10.70}$$

since the Kronecker product $|x\rangle \otimes |y\rangle$ is a one-hot vector with a single 1 in the position z given by (10.66).

The following property is often useful.

Key Result 10.5 Vectorization of the product of three matrices
Given any three matrices with suitable dimensions so that the product $ABC^\dagger$ is well defined, the following equality holds:

$$\text{vec}(ABC^\dagger) = (A \otimes C^*)\text{vec}(B), \tag{10.71}$$

where the matrix C^* is obtained by taking the complex conjugate of each entry of the matrix C individually.

Example 10.11

Given matrices

$$A = [\ 1+2i,\quad 3+4i\ ],$$
$$B = \begin{bmatrix} 1+i & 4+5i \\ 5+4i & 2+3i \end{bmatrix},$$
$$\text{and } C = [\ 2+i,\quad 4+3i\ ], \tag{10.72}$$

the product $ABC^\dagger$ equals $73 + 228i$, and thus one has $\text{vec}(ABC^\dagger) = 73 + 228i$.

This result can be obtained via the property (10.71) as

$$(A \otimes C^*)\text{vec}(B) = \big([\ 1+2i,\quad 3+4i\ ] \otimes [\ 2-i,\quad 4-3i\ ]\big) \begin{bmatrix} 1+i \\ 4+5i \\ 5+4i \\ 2+3i \end{bmatrix}$$
$$= [\ 4+3i,\quad 10+5i,\quad 10+5i,\quad 24+7i\ ] \begin{bmatrix} 1+i \\ 4+5i \\ 5+4i \\ 2+3i \end{bmatrix}$$
$$= 73 + 228i. \tag{10.73}$$

10.5.2 Vectorized Density Matrix

Using the definition of matrix vectorization, the density matrix can be expressed in vector form as follows.

Definition 10.5 Vectorized density
Given an $N \times N$ density matrix ρ, the vectorized density is defined as the $N^2 \times 1$ vector

$$|\rho\rangle\rangle = \text{vec}(\rho). \tag{10.74}$$

The notation $|\cdot\rangle\rangle$ is also known as a **superket**. In a manner consistent with the conventional Dirac notation, the Hermitian transpose of the vectorized density, which may be referred to as a superbra, is written as

$$\langle\langle\rho| = |\rho\rangle\rangle^\dagger. \tag{10.75}$$

Using the definition (10.74), the vectorized version of the post-transformation density matrix (10.54) can be expressed as

$$|\sigma\rangle\rangle = \mathrm{vec}(U\rho U^\dagger)$$
$$= (U \otimes U^*)|\rho\rangle\rangle, \qquad (10.76)$$

where the second equality follows from the property (10.71). Accordingly, the post-transformation vectorized density $|\sigma\rangle\rangle$ is obtained by multiplying the pre-transformation vectorized density $|\rho\rangle\rangle$ by the $N^2 \times N^2$ matrix $U \otimes U^*$. Note that, while being of size $N^2 \times N^2$, matrix $U \otimes U^*$ is fully specified by the $N \times N$ matrix U.

The matrix–vector multiplication (10.76) highlights the property of linearity of the closed-system transformation (10.1) of a density matrix.

Example 10.12

Take again the T gate (10.61) and the density matrix (10.60). Using (10.76), the vectorized density matrix for the post-transformation state is given by

$$|\sigma\rangle\rangle = (T \otimes T^*)|\rho\rangle\rangle$$
$$= \left(\begin{bmatrix} 1 & 0 \\ 0 & \exp(i\pi/4) \end{bmatrix} \otimes \begin{bmatrix} 1 & 0 \\ 0 & \exp(-i\pi/4) \end{bmatrix}\right) \begin{bmatrix} 0.5 \\ -0.3i \\ 0.3i \\ 0.5 \end{bmatrix}$$
$$= \begin{bmatrix} 0.5 \\ -0.2121 - 0.2121i \\ -0.2121 + 0.2121i \\ 0.5 \end{bmatrix}, \qquad (10.77)$$

which corresponds to the density matrix σ in (10.61).

10.6 Information Dynamics in Closed Quantum Systems

This section studies the evolution of information and relative information measures within closed quantum systems. It will be concluded that both entropy and relative entropy remain unchanged as a result of closed-system transformations. The intuition behind this result is that, due to the reversibility of unitary transformations, no information can be lost or created.

Key Result 10.6 The quantum entropy does not change in closed quantum systems

Given any $N \times N$ unitary matrix U and any density matrix ρ, the equality

$$\mathrm{H}_\alpha(U\rho U^\dagger) = \mathrm{H}_\alpha(\rho) \qquad (10.78)$$

holds for the Shannon, Tsallis, and Rényi quantum entropies.

To prove this result, recall that the quantum entropy $H_\alpha(\rho)$ equals the entropy $H_\alpha(\lambda)$ of the vector of eigenvalues, λ, of the density matrix ρ. Equality (10.78) then follows by recalling from (10.54) that the vector of eigenvalues does not change as a result of the transformation $U\rho U^\dagger$, which only rotates the eigenvectors.

It is worth noting that the conservation of quantum entropy in closed systems appears inconsistent with the thermodynamic principle stating that entropy tends to increase in isolated systems. This apparent discrepancy was well understood by John von Neumann, the pioneer who introduced quantum entropy. As von Neumann pointed out, this issue can be reconciled by reasoning in terms of observational entropy, as introduced in Appendix 7.C, rather than relying solely on quantum entropy. For further details, please refer to the Recommended Resources.

Key Result 10.7 The quantum relative entropy does not change in closed quantum systems

Given any $N \times N$ unitary matrix U and any pair of $N \times N$ density matrices ρ and σ, the equality

$$D_\alpha(U\rho U^\dagger || U\sigma U^\dagger) = D_\alpha(\rho||\sigma) \tag{10.79}$$

holds for the Shannon, Tsallis, and Rényi quantum relative entropies.

This result indicates that the relative information between the two density matrices does not change as a result of a closed-system transformation. This equality, proved in Appendix 10.A, formalizes the fact that any information that allows the density matrices to be distinguished is preserved by a unitary transformation.

10.7 Complete Measurements as Dynamic Transformations ⏭

Any orthonormal basis $\mathcal{M} = \{|v_x\rangle\}_{x\in[N]}$ defines a valid measurement setting, in the sense that it can be implemented in a quantum system. This final section demonstrates that a complete quantum measurement in setting $\mathcal{M} = \{|v_x\rangle\}_{x\in[N]}$, illustrated Figure 10.6(a), can equivalently be implemented as in Figure 10.6(b). Accordingly, the pre-measurement state is first transformed using a unitary matrix V, then a measurement in the computational basis is implemented, and finally the inverse dynamic transformation $V^\dagger$ is applied to the post-measurement state.

The upshot of this result is that all complete measurement settings can be implemented by using only measurements in the computational basis. The rest of the section will discuss the design of the transformation V as a function of the measurement basis $\mathcal{M}$ so as to ensure the equivalence illustrated in Figure 10.6.

To start, fix an arbitrary complete measurement setting $\mathcal{M} = \{|v_x\rangle\}_{x\in[N]}$. Select the unitary transformation V to have columns equal to the states in the measurement basis $\mathcal{M}$, i.e.,

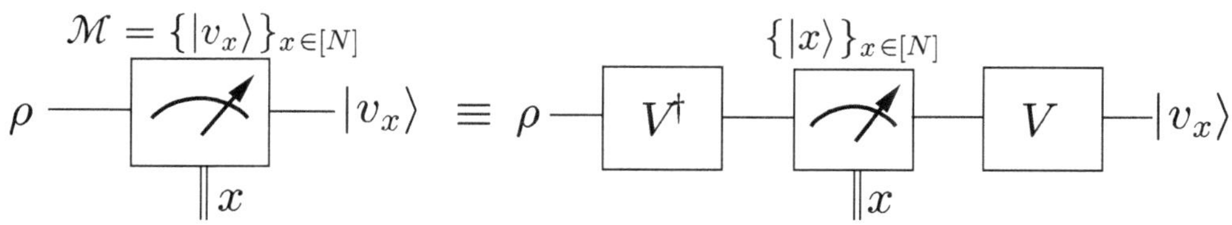

Figure 10.6 All complete measurement settings can be implemented using only measurements in the computational basis. A complete quantum measurement in setting $\mathcal{M} = \{|v_x\rangle\}_{x\in[N]}$ (on the left) can equivalently be implemented by first applying a unitary matrix $V^\dagger$, constructed from the basis $\mathcal{M}$, followed by a measurement in the computational basis and finally by the inverse dynamic transformation V (on the right).

$$
\begin{aligned}
V &= [|v_0\rangle, \ldots, |v_{N-1}\rangle] \\
&= \sum_{x\in[N]} |v_x\rangle\langle x|, \qquad (10.80)
\end{aligned}
$$

where the second equality follows as in (10.17). The unitary matrix V rotates each state $|x\rangle$ in the computational basis into the corresponding state $|v_x\rangle$ from basis $\mathcal{M}$. Furthermore, the inverse of unitary matrix (10.80) is given by

$$
V^\dagger = \sum_{x\in[N]} |x\rangle\langle v_x|. \qquad (10.81)
$$

In Figure 10.6(b), the post-transformation state after the first unitary matrix $V^\dagger$ is

$$
\sigma = V^\dagger \rho V = \sum_{x\in[N]} \sum_{y\in[N]} |x\rangle\langle v_x|\rho|v_y\rangle\langle y|. \qquad (10.82)
$$

By Born's rule, measuring this state in the computational basis $\{|x\rangle\}_{x\in[N]}$ returns classical output $\mathrm{x} = x$ with probability

$$
p_x(\sigma|\{|x\rangle\}_{x\in[N]}) = \langle x|\sigma|x\rangle = \langle v_x|\rho|v_x\rangle = p_x(\rho|\mathcal{M}). \qquad (10.83)
$$

This corresponds to the desired measurement distribution $p(\rho|\mathcal{M})$.

Furthermore, with probability $p_x(\sigma|\{|x\rangle\}_{x\in[N]}) = p_x(\rho|\mathcal{M})$, the post-measurement state is $|x\rangle$. Hence, after the last transformation, the desired post-measurement state $|v_x\rangle$ is recovered as

$$
V|x\rangle = \sum_{y\in[N]} |v_y\rangle\langle y|x\rangle = |v_x\rangle, \qquad (10.84)
$$

completing the proof of the equivalence in Figure 10.6.

10.8 Summary

- In quantum physics all dynamic transformations in closed systems are linear, reversible, and norm-preserving. Thus, they are described by a unitary matrix U. Accordingly, for a pure state $|\psi\rangle$, the post-transformation state is given by $|\phi\rangle = U|\psi\rangle$; while for a density matrix ρ the post-transformation state is $\sigma = U\rho U^\dagger$.

- Unitary transformations U rotate the states from an arbitrary orthonormal basis $\{|v_x\rangle\}_{x\in[N]}$ into the orthonormal basis $\{|w_x\rangle = U|v_x\rangle\}_{x\in[N]}$. Thus, matrix U can be written as $U = \sum_{x\in[N]} |w_x\rangle\langle v_x|$. This implies that a density matrix ρ with eigendecomposition $\rho = \sum_{x\in[N]} \lambda_x|u_x\rangle\langle u_x|$ is transformed into a density matrix with eigendecomposition $\sigma = \sum_{x\in[N]} \lambda_x|w_x\rangle\langle w_x|$ with the same eigenvalues $\{\lambda_x\}_{x\in[N]}$ and rotated eigenvectors $\{|w_x\rangle = U|v_x\rangle\}_{x\in[N]}$.
- Consider an eigendecomposition $U = \sum_{x\in[N]} \lambda_x|v_x\rangle\langle v_x|$ with eigenvalues $\{\lambda_x = \exp(i\gamma_x)\}_{x\in[N]}$ of a unitary matrix U. When applied to a pure state, the unitary matrix implements a separate phase rotation γ_x to the amplitude along the eigenvector $|v_x\rangle$.
- Unitary transformations can be written as $U = \exp(iH)$, where the Hermitian matrix H is known as the Hamiltonian of the transformation U.
- Quantum entropies and quantum relative entropies are not changed by the application of a unitary transformation.

10.9 Recommended Resources

Further discussions on the construction of unitary matrices from elementary quantum gates, as well as a proof of the result in (10.38), can be found in the standard textbook [1]. The problem of simulating the output of quantum circuits using classical computers is covered in [2]. An application of the properties of Pauli matrices as unitary matrices to quantum key distribution is presented in [3]. The foundational text on quantum mechanics by John von Neumann is [4].

PROBLEMS

10.1 Given an orthonormal basis $\{|v_x\rangle\}_{x\in[N]}$ and a unitary matrix U, show that the set of vectors $\{|w_x\rangle\}_{x\in[N]}$ with $|w_x\rangle = U|v_x\rangle$ for $x \in [N]$ also forms an orthonormal basis. *Hint:* Verify that each vector $|w_x\rangle$ has unit norm and that the inner products $\langle w_x|w_y\rangle$ are equal to zero for all $x \neq y$.

10.2 Consider the Pauli Y matrix P_2. (a) Express the dynamic transformation defined by matrix P_2 as in (10.15) by choosing as the input states $\{|v_x\rangle\}_{x\in[N]}$ the computational-basis states. (b) Repeat the same problem by choosing as input states $\{|v_x\rangle\}_{x\in[N]}$ the Hadamard-basis states.

10.3 Prove the equality

$$\exp\left(i\frac{\theta}{2}P\right) = \cos\left(\frac{\theta}{2}\right)I + i\sin\left(\frac{\theta}{2}\right)P \tag{10.85}$$

for any Pauli matrix $P \in \{P_1, P_2, P_3\}$. *Hint:* Pauli matrices are normal and they have two eigenvalues, one equal to 1 and one equal to -1.

10.4 Prove that, if matrix A and matrix B commute (i.e., if $AB = BA$), then the following equality holds:

$$(A \otimes I)\text{vec}(B) = (I \otimes A^T)\text{vec}(B). \tag{10.86}$$

10.5 (a) Show that, for a projective measurement $\mathcal{P} = \{\Pi_x\}_{x\in[N]}$, the vectorized density matrix of the post-measurement state can be written as $|\rho_x\rangle\rangle = p_x(\rho|\mathcal{P})^{-1}(\Pi_x \otimes \Pi_x^*)|\rho\rangle\rangle$, where ρ is the pre-measurement state. (b) Prove that the measurement probability can be expressed as $p_x(\rho|\mathcal{P}) = \text{tr}(\rho\Pi_x) = \text{vec}(\Pi_x^T)^T|\rho\rangle\rangle$. (c) Demonstrate that the expected value of an observable O can be expressed as $\langle O\rangle_\rho = \text{tr}(O\rho) = \text{vec}(O^T)^T|\rho\rangle\rangle$.

10.6 (a) Prove that the post-transformation density matrix $U\rho U^\dagger$ equals the original density matrix ρ if the eigenvectors of matrix U coincide with the eigenvectors of matrix ρ. (b) Find conditions under which a pure state is left unchanged by a unitary transformation.

10.7 This problem provides a comparison between channels, i.e., dynamic transformations of classical informational states, and unitary matrices, i.e., dynamic transformations of quantum states in closed systems. (a) To start, take a uniform distribution u with $N = 4$. Find the 4×4 column-stochastic matrix K of a channel that transforms the uniform distribution u into the zero-entropy distribution $\delta_0 = [1, 0, 0, 0]^T$. (b) Consider now a quantum counterpart of this setting in which the initial state $|\psi\rangle = [1, 1, 1, 1]^T/\sqrt{4}$ has all equal amplitudes, while the target final state is $|0\rangle = [1, 0, 0, 0]^T$. Would the matrix K obtained in part (a) serve as a valid closed-system quantum transformation to prepare state $|0\rangle$ from state $|\psi\rangle$? (c) Verify that the matrix

$$U = \frac{1}{2}\begin{bmatrix} 1 & 1 & 1 & 1 \\ 1 & -1 & 1 & -1 \\ 1 & 1 & -1 & -1 \\ 1 & -1 & -1 & 1 \end{bmatrix} \tag{10.87}$$

is unitary and that it satisfies the desired relation $|0\rangle = U|\psi\rangle$. (d) Describe the operation of the transformation $U|\psi\rangle$ in terms of interference between the amplitudes of the initial state $|\psi\rangle$ and compare the operation of matrix U to the classical counterpart K defined in part (a).

10.8 Express the inverse of a unitary matrix in terms of an operation on the pre-transformation superket $|\rho\rangle\rangle$.

10.9 Consider a quantum computer operating on a qutrit. The computational basis is $\{|0\rangle, |1\rangle, |2\rangle\}$. A quantum gate U is defined as

$$U = \begin{bmatrix} 1 & 0 & 0 \\ 0 & e^{i\pi/3} & 0 \\ 0 & 0 & e^{i2\pi/3} \end{bmatrix}. \tag{10.88}$$

In the computational basis, this gate applies a phase shift to the states $|1\rangle$ and $|2\rangle$, but leaves state $|0\rangle$ unchanged. (a) If the initial state of the qutrit is

$|\psi\rangle = \frac{1}{\sqrt{3}}(|0\rangle + |1\rangle + |2\rangle)$, compute the final state after applying U. (b) Verify that U is a unitary matrix.

10.10 Consider a qudit system with $N = 4$ levels, and let the computational basis states be $\{|0\rangle, |1\rangle, |2\rangle, |3\rangle\}$. A unitary gate U is defined as

$$U = \begin{bmatrix} \cos\theta & \sin\theta & 0 & 0 \\ -\sin\theta & \cos\theta & 0 & 0 \\ 0 & 0 & 1 & 0 \\ 0 & 0 & 0 & 1 \end{bmatrix}. \tag{10.89}$$

(a) Apply U to the state $|\psi\rangle = \frac{1}{\sqrt{2}}(|0\rangle + |1\rangle)$ and compute the final state. (b) For a measurement in the computational basis, calculate the measurement probabilities. (c) Discuss how matrix U acts as a rotation in the subspace spanned by $|0\rangle$ and $|1\rangle$.

Appendix

10.A Proof of (10.79)

To prove the equality (10.79), it is sufficient to focus on the Tsallis quantum relative entropy. In fact, the Shannon relative entropy is obtained as a special case, and the Rényi quantum entropy is a monotonic transformation of the Tsallis quantum relative entropy. Writing the eigendecompositions of the density matrices as $\rho = \sum_{x\in[N]} \lambda_x |u_x\rangle\langle u_x|$ and $\sigma = \sum_{x\in[N]} \mu_x |v_x\rangle\langle v_x|$, the Tsallis quantum relative entropy can be expressed as

$$\mathrm{D}_\alpha^{\mathrm{T}}(\rho||\sigma) = \frac{1_2}{\alpha - 1}\left(\sum_{x\in[N]}\sum_{y\in[N]} \mathrm{F}(|u_x\rangle, |v_y\rangle)\lambda_x^\alpha \mu_y^{1-\alpha} - 1\right). \tag{10.90}$$

Furthermore, after the transformation U, the eigendecompositions of the two matrices are changed to $U\rho U^\dagger = \sum_{x\in[N]} p_x |w_x\rangle\langle w_x|$ and $U\sigma U^\dagger = \sum_{x\in[N]} q_x |\omega_x\rangle\langle \omega_x|$, with rotated eigenvectors $|w_x\rangle = U|v_x\rangle$ and $|\omega_x\rangle = U|u_x\rangle$. The Tsallis quantum relative entropy after the transformation is therefore evaluated as

$$\begin{aligned} \mathrm{D}_\alpha^{\mathrm{T}}(U\rho U^\dagger||U\sigma U^\dagger) =& \frac{1_2}{\alpha - 1}\left(\sum_{x\in[N]}\sum_{y\in[N]} \mathrm{F}(|w_x\rangle, |\omega_y\rangle)\lambda_x^\alpha \mu_y^{1-\alpha} - 1\right) \\ =& \mathrm{D}_\alpha^{\mathrm{T}}(\rho||\sigma), \end{aligned} \tag{10.91}$$

where the second equality follows from the fact that unitary transformations do not change the fidelity, i.e.,

$$\begin{aligned} \mathrm{F}(|w_x\rangle, |\omega_y\rangle) &= |\langle w_x|\omega_y\rangle|^2 \\ &= |\langle u_x|U^\dagger U|v_y\rangle|^2 \\ &= |\langle u_x|v_y\rangle|^2 \\ &= \mathrm{F}(|u_x\rangle, |v_y\rangle). \end{aligned} \tag{10.92}$$

This concludes the proof.

REFERENCES

[1] M. A. Nielsen and I. L. Chuang, *Quantum Computation and Quantum Information*. Cambridge: Cambridge University Press, 2010.

[2] S. Aaronson and D. Gottesman, "Improved simulation of stabilizer circuits," *Physical Review A*, vol. 70, no. 5, art. 052328, 2004.

[3] S. Pirandola, U. L. Andersen, L. Banchi, M. Berta, D. Bunandar, R. Colbeck, *et al.*, "Advances in quantum cryptography," *Advances in Optics and Photonics*, vol. 12, no. 4, pp. 1012–1236, 2020.

[4] J. Von Neumann, *Mathematical Foundations of Quantum Mechanics*. Princeton, NJ: Princeton University Press, 1955.

11 Quantum Dynamic Information in Open Systems

11.1 Overview

The previous chapter explored dynamic transformations in closed quantum systems. In such systems, both information – measured by the entropy – and relative information – measured by the relative entropy – remain constant as the system evolves without interacting with its environment. This chapter shifts focus to open quantum systems, which interact with their environment.

First, a specific type of open-system transformation is studied in which the environment selects a unitary transformation at random without the system's knowledge, leaving the system uncertain about the actual transformation applied. These random unitary transformations serve as the quantum counterpart to the random shuffling transformations discussed in Chapter 9 for classical systems. Similar to their classical analogs, random unitary transformations lead to an increase – or at least no decrease – in the quantum entropy.

The study of dynamic transformations in open systems leads to the introduction of a general class of measurements known as positive operator-valued measures (POVMs). POVMs extend the concept of projective measurements, encompassing them as special cases, and allow for any number of classical outcomes. This chapter demonstrates the importance of POVMs in enabling various informational tasks, such as unambiguous state detection.

When POVMs are implemented non-selectively, they give rise to quantum channels, which generalize random unitary transformations. Similar to the classical channels discussed in Chapter 9, quantum channels can cause increases or decreases in quantum entropy. However, as in classical systems, quantum channels cannot increase the relative information, a principle formalized by the quantum data processing inequality.

Finally, this chapter introduces quantum instruments, a further generalization of both POVMs and quantum channels. Quantum instruments allow for partial selectivity in measurement outcomes and provide a more versatile framework for describing open-system dynamics.

11.1.1 Learning Objectives and Organization

By the end of this chapter, the reader should be able to:

- understand and evaluate random unitary transformations;
- describe open quantum systems within a composite system including the environment;
- describe POVMs as measurements applied to the environment;
- identify some of the informational tasks for which POVMs are required;
- describe quantum channels using Kraus matrices;
- evaluate the dynamic behavior of quantum entropies and quantum relative entropies through quantum channels;
- understand the quantum data processing inequality;
- understand and apply quantum instruments.

11.2 Random Unitary Transformations

As illustrated in Figure 11.1, this chapter studies an **open system**, i.e., a system that interacts with the rest of the world. The **environment** represents the part of the rest of the world that interacts directly with the system of interest. As will be seen, due to the interactions between system and environment, open-system transformations are generally irreversible as they may entail a loss of information from the system towards the environment.

11.2.1 Defining Random Unitary Transformations

Suppose that the environment selects a random unitary transformation U_m out of a set of M matrices $\{U_m\}_{m\in[M]}$ with probabilities $\gamma_m \geq 0$, where $\sum_{m\in[M]} \gamma_m = 1$. The environment does not inform the system of the selected transformation U_m. Therefore, from the perspective of the system, the information about the selected transformation is lost, and the post-transformation state is obtained by averaging over all possible choices of transformation U_m. This yields the following.

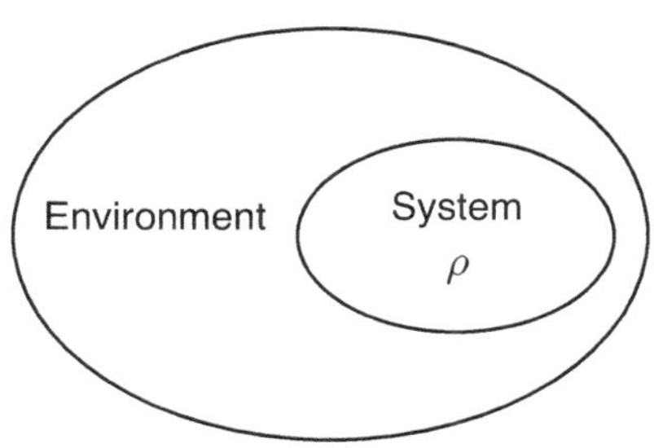

Figure 11.1 An open quantum system can be modeled as a composite system consisting of the system of interest along with a separate subsystem, referred to as the environment, that accounts for the part of the rest of the world that interacts with the system.

Definition 11.1 Random unitary transformations
Given a pre-transformation state ρ, a random unitary transformation defined by a distribution $\{\gamma_m\}_{m\in[M]}$ and by unitary matrices $\{U_m\}_{m\in[M]}$ for some integer M produces the post-transformation state

$$\sigma = \sum_{m\in[M]} \gamma_m U_m \rho U_m^\dagger. \tag{11.1}$$

Example 11.1

For an open qubit system, initially in state ρ, a Pauli X or Pauli Z transformation is selected with equal probability, yielding the post-transformation state (11.1) as

$$\sigma = \frac{1}{2}P_1\rho P_1^\dagger + \frac{1}{2}P_3\rho P_3^\dagger. \tag{11.2}$$

Suppose that the system starts in the state $|+\rangle$, which is coherent with respect to the computational basis. The corresponding density matrix is

$$\rho = \frac{1}{2}\begin{bmatrix} 1 & 1 \\ 1 & 1 \end{bmatrix}, \tag{11.3}$$

where the off-diagonal terms express the mentioned coherence of the initial state.

The post-transformation state σ is given by

$$\begin{aligned} \sigma &= \frac{1}{4}\begin{bmatrix} 1 & 1 \\ 1 & 1 \end{bmatrix} + \frac{1}{4}\begin{bmatrix} 1 & -1 \\ -1 & 1 \end{bmatrix} \\ &= \begin{bmatrix} 1/2 & 0 \\ 0 & 1/2 \end{bmatrix} = \pi, \end{aligned} \tag{11.4}$$

which corresponds to the maximally mixed density $\sigma = \pi$. This shows that, as a result of an open-system transformation, the coherence of a quantum state can be destroyed – a phenomenon known as **decoherence**. Furthermore, this result suggests that random unitary transformations tend to increase the quantum entropy, taking the state closer to the maximally mixed state, which has maximal quantum entropy.

Example 11.2

Suppose now that the initial state is maximally mixed. The post-transformation state for any random unitary transformation is also a maximally mixed state, i.e.,

$$\begin{aligned} \sigma &= \sum_{m\in[M]} \gamma_m U_m \pi U_m^\dagger \\ &= \frac{1}{N}\sum_{m\in[M]} \gamma_m U_m U_m^\dagger \end{aligned}$$

$$= \frac{1}{N}\left(\sum_{m\in[M]} \gamma_m\right) I = \pi. \tag{11.5}$$

Therefore, if the pre-transformation state has maximal quantum entropy, a random unitary transformation preserves the same level of uncertainty in the output state.

11.2.2 Random Unitary Transformations Increase the Quantum Entropy

As suggested by the last two examples, random unitary transformations inject additional uncertainty into a quantum state, generally increasing the entropy. To make progress on formalizing this observation, a relevant question is the following. Given two states ρ and σ, when can σ be obtained from ρ via a random unitary transformation? It is now shown that this question can be addressed by leveraging the framework of majorization introduced in Section 9.3.

To review the necessary background from Chapter 9, consider two arbitrary distributions p and q. Distribution p is said to majorize distribution q, which is denoted as

$$p \succ q, \tag{11.6}$$

if q can be obtained from p via random shuffling. This means that there exists a doubly stochastic matrix T that ensures the equality $q = Tp$. Checking the validity of this condition $p \succ q$ requires a simple comparison of ordered versions of the probability distributions p and q.

The Shannon, Tsallis, and Rényi entropies are majorization-decreasing, or Schur concave, in the sense that they satisfy the implication

$$p \succ q \Rightarrow \mathrm{H}_\alpha(p) \leq \mathrm{H}_\alpha(q). \tag{11.7}$$

Thus, if $p \succ q$, distribution p represents a more certain informational state than distribution q. Majorization provides a partial ordering on the space of probability distributions, as any two distributions p and q satisfy one of the following three relationships: (i) $p \succ q$; (ii) $q \succ p$; or (iii) neither $p \succ q$ nor $q \succ p$.

The definition of majorization is extended to density matrices by replacing random shufflings with random unitary transformations.

Definition 11.2 Majorization for density matrices
Given two $N \times N$ density matrices ρ and σ, the density matrix ρ is said to majorize the density matrix σ, which is denoted as

$$\rho \succ \sigma, \tag{11.8}$$

if there exists a random unitary transformation that produces state σ from state ρ as in (11.1).

That is, the majorization relation (11.8) holds if there exist unitary matrices $\{U_m\}_{m\in[M]}$ and a probability distribution $\{\gamma_m\}_{m\in[M]}$ for some integer M such that the equality (11.1) is verified. Note that, as in the classical definition of majorization, the direction of the symbol $\succ$ in (11.8) follows the transformation from input to output of the random transformation.

If $\rho \succ \sigma$, state σ can be prepared from state ρ by leveraging the additional classical randomness needed to apply a random unitary transformation. Thus, density matrix ρ is characterized by a lower level of uncertainty than σ. In this sense, one may consider state ρ as being of higher quality – of higher purity – than state σ.

Given two density matrices ρ and σ, it is generally difficult to ascertain directly if there exists a random unitary transformation producing state σ from state ρ. The following fundamental theorem makes this task simple by relating the notion of majorization between density matrices with the majorization relation between probability vectors.

Key Result 11.1 Uhlmann theorem

For any $N \times N$ density matrix ρ, write as $\lambda(\rho)$ the corresponding $N \times 1$ vector of eigenvalues. The density matrix ρ majorizes the density matrix σ, i.e., $\rho \succ \sigma$, if and only if the eigenvalue vectors $\lambda(\rho)$ and $\lambda(\sigma)$ satisfy the majorization relation

$$\lambda(\rho) \succ \lambda(\sigma). \tag{11.9}$$

This result has the following important consequence: there exists a random unitary transformation taking state ρ into state σ if and only if there exists a random shuffling transformation taking the eigenvalue vector $\lambda(\rho)$ into $\lambda(\sigma)$. As discussed, the latter condition can easily be verified based on the reordered vectors $\lambda(\rho)$ and $\lambda(\sigma)$.

Appendix 11.B proves Uhlmann's theorem by first showing how to construct matrices $\{U_m\}_{m\in[M]}$ and probability distribution $\{\gamma_m\}_{m\in[M]}$ in (11.1) for any two matrices ρ and σ satisfying the relation (11.9). Then, it demonstrates that the existence of a random unitary transformation converting state ρ into state σ implies the relation (11.9).

Example 11.3

To understand the "if" part of Uhlmann's theorem, consider the special case in which the density matrices ρ and σ are diagonal. For diagonal matrices, the elements on the diagonal coincide with the eigenvalues, and thus one can write $\rho = \text{diag}(\lambda)$ and $\sigma = \text{diag}(\mu)$, where λ and μ are the respective vectors of eigenvalues. By assumption, since $\lambda \succ \mu$, it is possible to find a random shuffling transformation $T = \sum_{m\in[M]} \gamma_m S_m$ for some probability distribution $\{\gamma_m\}_{m\in[M]}$ and permutation matrices $\{S_m\}_{m\in[M]}$ such that the equality $\mu = T\lambda$ holds.

Note now that any permutation matrix S_m is also unitary, since the Hermitian $S_m^\dagger$ is the inverse of matrix S_m, i.e., $S_m^{-1} = S_m^\dagger$. Therefore, if $\lambda \succ \mu$, the random unitary transformation with probabilities $\{\gamma_m\}_{m\in[M]}$ and unitary matrices $\{U_m = S_m\}_{m\in[M]}$ transforms state ρ into state σ as

$$\sum_{m\in[M]} \gamma_m S_m \text{diag}(\lambda) S_m^\dagger = \text{diag}\,(T\lambda)$$
$$= \text{diag}(\mu) = \sigma. \tag{11.10}$$

This example demonstrates that random unitary transformations generalize random shuffling transformations.

As mentioned, majorization provides a partial ordering on the space of probability distributions. In a similar manner, given two density matrices ρ and σ, one can have one of three cases: (i) $\rho \succ \sigma$; (ii) $\sigma \succ \rho$; or (iii) neither $\rho \succ \sigma$ nor $\sigma \succ \rho$.

Example 11.4

Consider the 3×3 density matrices

$$\rho = 0.5|0\rangle\langle 0| + 0.5|1\rangle\langle 1| \tag{11.11}$$

and

$$\sigma = 0.6|v^0\rangle\langle v^0| + 0.3|v^1\rangle\langle v^1| + 0.1|v^2\rangle\langle v^2|, \tag{11.12}$$

with $|v^0\rangle = (|0\rangle + |2\rangle)/\sqrt{2}$, $|v^1\rangle = (|0\rangle - |2\rangle)/\sqrt{2}$, and $|v^3\rangle = |1\rangle$. One can readily check that the eigenvalues are given by $\lambda(\rho) = [0.5, 0.5, 0]^T$ and $\lambda(\sigma) = [0.6, 0.3, 0.1]^T$. Therefore, following Chapter 9, neither of the majorization conditions $\rho \succ \sigma$ and $\sigma \succ \rho$ hold.

The majorization condition $\rho \succ \sigma$ provides a strict criterion to order density matrices in terms of their associated uncertainty. In particular, in a manner completely analogous to the classical counterpart result (9.18), the condition $\rho \succ \sigma$ implies that *all* the quantum entropies for state ρ are no larger than for state σ.

Key Result 11.2 Random unitary transformations increase the quantum entropy
Given any two density matrices ρ and σ such that the relation $\rho \succ \sigma$ holds, the inequality

$$\text{H}_\alpha(\rho) \leq \text{H}_\alpha(\sigma) \tag{11.13}$$

is verified for the Shannon, Tsallis, and Rényi entropies. Furthermore, the equality $\text{H}_\alpha(\rho) = \text{H}_\alpha(\sigma)$ applies in the special case in which the transformation (11.1) relating ρ and σ is *unitary*, i.e., for $M = 1$.

The inequality (11.13) follows directly from Uhlmann's theorem and from the Schur concavity of the entropy (see Section 9.4). In fact, one has

$$\mathrm{H}_\alpha(\rho) = \mathrm{H}_\alpha(\lambda(\rho)) \leq \mathrm{H}_\alpha(\lambda(\sigma)) = \mathrm{H}_\alpha(\sigma), \tag{11.14}$$

where the two equalities apply the definition of quantum entropy and the inequality is a consequence of the Schur concavity of the entropy. Furthermore, the fact that equality in (11.13) holds for $M = 1$ is a restatement of the invariance of the quantum entropy in closed systems (see (10.78)).

Further reinforcing the interpretation of the majorization between density matrices as a criterion to order the uncertainty levels associated with density matrices, the following property follows directly from the counterpart (9.24) for probability distributions.

Key Result 11.3 Sandwich property of the majorization relation for density matrices

For any $N \times N$ density matrix ρ, the majorization relations

$$\pi \prec \rho \prec |\psi\rangle\langle\psi| \tag{11.15}$$

hold, where $|\psi\rangle$ is any $N \times 1$ pure state.

The relations (11.15) are consistent with (11.13) since the maximally mixed density matrix π is characterized by maximal quantum entropy, and pure states have minimal quantum entropy.

11.2.3 Majorization and Behavior of a Quantum State ⏭

This section is concluded with a number of additional remarks on the implications of majorization for quantum systems. These properties are stated without proof, and the reader is referred to the Recommended Resources for details and discussions.

The first remark concerns constraints imposed by majorization on the behavior $p(\rho|\mathcal{M})$ of a quantum state ρ as one varies the complete measurement setting $\mathcal{M}$. Specifically, the next result addresses the following question. Given a state ρ, is it possible to find a measurement setting $\mathcal{M}$ such that the measurement distribution over the classical outputs, $p(\rho|\mathcal{M})$, equals a particular target distribution p of interest?

Key Result 11.4 Majorization and behavior of a quantum state

Given a state ρ with eigenvalues $\lambda(\rho)$ and a target $N \times 1$ probability distribution p, there exists a complete measurement setting $\mathcal{M}$ yielding the measurement probability distribution $p(\rho|\mathcal{M}) = p$ if and only if the vector of eigenvalues $\lambda(\rho)$ majorizes the probability vector p, i.e., $\lambda(\rho) \succ p$.

This result shows that any measurement distribution $p(\rho|\mathcal{M})$ must be at least as uncertain as the vector of eigenvalues $\lambda(\rho)$. In particular, this implies that, for any measurement setting $\mathcal{M}$, the following condition is satisfied:

$$\lambda(\rho) \succ p(\rho|\mathcal{M}). \tag{11.16}$$

Thus, given the Schur concavity of the entropy, the entropy associated with any measurement output, $\mathrm{H}_\alpha(\rho|\mathcal{M})$, is at least as large as the quantum entropy:

$$\mathrm{H}_\alpha(\lambda(\rho)) = \mathrm{H}_\alpha(\rho) \leq \mathrm{H}_\alpha(p(\rho|\mathcal{M})) = \mathrm{H}_\alpha(\rho|\mathcal{M}), \tag{11.17}$$

where $\mathrm{H}_\alpha(\rho|\mathcal{M})$ is the measured entropy (see Section 6.2.1).

For example, the result (11.17) implies that any measurement distribution can be produced from a pure state, while maximally mixed states can only generate a maximum-entropy measurement output.

Another important consequence of the inequality (11.17) is that the measured entropy $\mathrm{H}_\alpha(\rho|\mathcal{M})$ is minimized by choosing as measurement setting $\mathcal{M}$ an eigenbasis of matrix ρ (see Section 7.8). In fact, the eigenvalue vector $\lambda(\rho)$ corresponds to the measurement probability vector $p(\rho|\mathcal{M})$ when $\mathcal{M}$ is an eigenbasis of ρ.

A related result characterizes the amount of classical randomness required to prepare a state ρ.

Key Result 11.5 Majorization and ensemble randomness
Given a state ρ with eigenvalues $\lambda(\rho)$, there exists an ensemble $\mathcal{E} = \{(|v_x\rangle, p_x)\}_{x\in[N]}$ realizing the density matrix ρ, with a set of possibly non-orthogonal states $\{|v_x\rangle\}_{x\in[N]}$, if and only if the majorization relation $\lambda(\rho) \succ p$ holds.

Therefore, the ensemble $\mathcal{E}$ must have at least as much randomness as the density matrix ρ it realizes.

11.3 System and Environment ⏭

Section 11.2 has introduced random unitary transformations by considering a setting in which the environment selects at random a unitary transformation without informing the system. This section will formalize the notion of a **joint state** for the system of interest and for the environment. It will also show how to describe local transformations and local measurements applied to either system or environment. This formulation will support the derivation of a more general type of open-system transformation, known as quantum channels, in Section 11.4.

11.3.1 Composite Quantum States

As discussed in Section 8.7, the joint state of system and environment is defined in a larger vector space of dimension equal to the product of the dimensions of system and environment (the "Church of the Larger Hilbert Space"). Denoting as N^X the dimension of the system and as N^E the dimension of the vector space for the environment, the joint state resides in an $N^E N^X$-dimensional space. The notation N^X is adopted here in lieu of the letter N to more clearly distinguish between dimensions of system and environment.

In the simplest situation, environment and state have individual pure states. In this case, let $|\psi^X\rangle$ be the $N^X \times 1$ vector describing the pure state for the system, and $|\psi^E\rangle$ be the $N^E \times 1$ vector specifying a pure state of the environment. The following key postulate of quantum mechanics specifies how to combine the two states into a single, joint, state by using the Kronecker product (see also Section 8.7).

> **Definition 11.3 Composite quantum pure state**
> The state of a composite system encompassing a subsystem in state $|a\rangle$ and a subsystem in state $|b\rangle$ is given by the Kronecker product $|a\rangle \otimes |b\rangle$.

Using this principle, in the simple situation at hand in which environment and system have respective states $|\psi^E\rangle$ and $|\psi^X\rangle$, the joint state of system and environment is given by the $N^E N^X \times 1$ vector

$$|\psi^{EX}\rangle = |\psi^E\rangle \otimes |\psi^X\rangle, \tag{11.18}$$

where the superscript EX is used to identify the overall environment–system state. Note that the ordering of the two subsystems in the definition of the composite state is arbitrary, and hence one could also have defined the composite state as $|\psi^{XE}\rangle = |\psi^X\rangle \otimes |\psi^E\rangle$.

Example 11.5

Suppose that the environment is in state

$$|0^E\rangle = [1, 0, \ldots, 0]^T \tag{11.19}$$

for some dimension N^E, with $N^E - 1$ entries equal to 0. Then, for any state $|\psi^X\rangle$ of the system, the joint state is given by the Kronecker product

$$|0^E\rangle \otimes |\psi^X\rangle = \begin{bmatrix} |\psi^X\rangle \\ 0 \\ \vdots \\ 0 \end{bmatrix}, \tag{11.20}$$

where 0 represents the $N^E \times 1$ all-zero vector and there are $N^E - 1$ such vectors.

The composition principle based on the Kronecker product is directly extended to density matrices as follows.

> **Definition 11.4 Composite quantum mixed state**
> The state of a composite system encompassing a subsystem in state ρ and a subsystem in state σ is given by the Kronecker product $\rho \otimes \sigma$.

According to this principle, if the environment and the system have individual mixed states defined by the $N^E \times N^E$ density matrix ρ^E and the $N^X \times N^X$ density matrix ρ^X, respectively, then the state of the composite system is defined by the $N^E N^X \times N^E N^X$ density matrix

$$\rho^{EX} = \rho^E \otimes \rho^X. \tag{11.21}$$

Composite states of the form (11.18) or (11.21) are said to be uncorrelated, since each subsystem has its separate state, and thus there is no correlation between the two subsystems. Part V of this book will discuss correlated bipartite quantum states.

11.3.2 Local Unitary Transformations

If environment and system do not interact with other parts of the world, any dynamic transformation affecting the overall open system is described by an $N^E N^X \times N^E N^X$ unitary matrix U^{EX} operating on the $N^E N^X$-dimensional joint state $|\psi^{EX}\rangle$. As illustrated in Figure 11.2, a special class of unitary transformations U^{EX} apply separate unitary transformations locally on each subsystem. Such local transformations follow the usual composition rule, as detailed next.

> **Definition 11.5 Local unitary transformations**
> If an $N^E \times N^E$ unitary transformation U^E is applied locally to the environment, while an $N^X \times N^X$ unitary transformation U^X is applied locally to

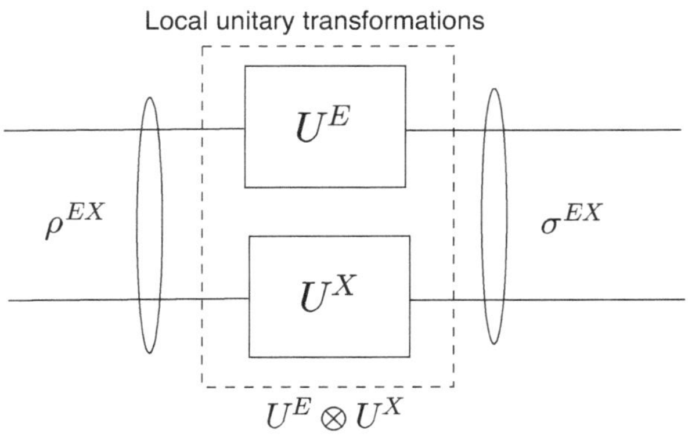

Figure 11.2 Illustration of local unitary transformations applied to environment and system of interest.

the system, then the composite unitary transformation applied to the joint environment–system state is described by the $N^E N^X \times N^E N^X$ unitary matrix

$$U^{EX} = U^E \otimes U^X. \tag{11.22}$$

Therefore, with a local unitary transformation $U^E \otimes U^X$, if the composite system is in the $N^E N^X \times 1$ pure state $|\psi^{EX}\rangle$, the post-transformation state is given by

$$(U^E \otimes U^X)|\psi^{EX}\rangle; \tag{11.23}$$

while, for an $N^E N^X \times N^E N^X$ mixed state ρ^{EX}, the post-transformation state is given by

$$\sigma^{EX} = (U^E \otimes U^X)\rho^{EX}((U^E)^\dagger \otimes (U^X)^\dagger). \tag{11.24}$$

Specializing (11.23) to uncorrelated pure states (11.18), by the mixed product property of the Kronecker product (see (8.169)), the post-transformation state is given by

$$(U^E \otimes U^X)(|\psi^E\rangle \otimes |\psi^X\rangle) = U^E|\psi^E\rangle \otimes U^X|\psi^X\rangle. \tag{11.25}$$

Thus, the post-transformation state is also uncorrelated, and the individual states environment and system are given by $U^E|\psi^E\rangle$ and $U^X|\psi^X\rangle$, respectively.

More generally, for uncorrelated mixed states (11.21), the post-transformation state is

$$\sigma^{EX} = U^E\rho^E(U^E)^\dagger \otimes U^X\rho^X(U^X)^\dagger. \tag{11.26}$$

This state is also uncorrelated, with the environment being in state $U^E\rho^E(U^E)^\dagger$ and the system being in state $U^X\rho^X(U^X)^\dagger$. Intuitively, local unitary transformations $U^E \otimes U^X$ do not create correlations between system and environment because they apply separately to the two subsystems.

As a special case of local unitary transformations, a transformation may be applied only to one of the two subsystems. Since the identity matrix describes a trivial transformation, the result of a transformation U^X applied to the system only is the post-transformation state

$$(I^E \otimes U^X)|\psi^{EX}\rangle, \tag{11.27}$$

where I^E denotes the $N^E \times N^E$ identity matrix. Note that the superscript E is added here to the identity matrix I^E in order to emphasize that the identity matrix has the dimension N^E of the environment state. Superscripts will often be removed to avoid clutter, since the dimensions are clear from the context.

Similarly, if a transformation U^E is applied to the environment only, the post-transformation state is

$$(U^E \otimes I^X)|\psi^{EX}\rangle, \tag{11.28}$$

where I^X is an $N^X \times N^X$ identity matrix. Extensions of post-measurement states (11.27) and (11.28) to mixed states ρ^{EX} follow directly from (11.24).

Example 11.6

For $N^E = N^X = 2$, consider the separable state $|0\rangle \otimes |\psi^X\rangle$ in (11.20) for an arbitrary qubit state $|\psi^X\rangle$. By (11.25), applying a Pauli X transformation P_1 only to the environment yields the post-transformation state

$$\begin{aligned}(P_1 \otimes I)(|0\rangle \otimes |\psi^X\rangle) &= P_1|0\rangle \otimes |\psi^X\rangle \\ &= |1\rangle \otimes |\psi^X\rangle = \begin{bmatrix} 0 \\ |\psi^X\rangle \end{bmatrix}. \end{aligned} \tag{11.29}$$

This example illustrates the fact that a transformation that is local to one subsystem generally has an impact on the entire post-transformation state vector.

11.3.3 Local Measurements

A general measurement of the joint state of system and environment requires access to both subsystems. However, it is often more practical to consider composite measurements that apply locally to the two subsystems.

As illustrated in Figure 11.3, consider a composite measurement setting $\mathcal{P}^E \otimes \mathcal{P}^X$, which applies a projective measurement $\mathcal{P}^E$ to the environment and a projective measurement $\mathcal{P}^X$ to the system. A composite local measurement setting is described by the following.

- **Local projective measurement setting for the environment:** A setting $\mathcal{P}^E = \{\Pi_y^E\}_{y\in[M^E]}$ corresponding to $M^E \leq N^E$ possible measurement outputs, which is characterized by $N^E \times N^E$ projection matrices $\{\Pi_y^E\}_{y\in[N^E]}$.
- **Local projective measurement setting for the system:** A setting $\mathcal{P}^X = \{\Pi_x^X\}_{x\in[M^X]}$ corresponding to $M^X \leq N^X$ outputs, which is defined by $N^X \times N^X$ projection matrices $\{\Pi_x^X\}_{x\in[M^X]}$.

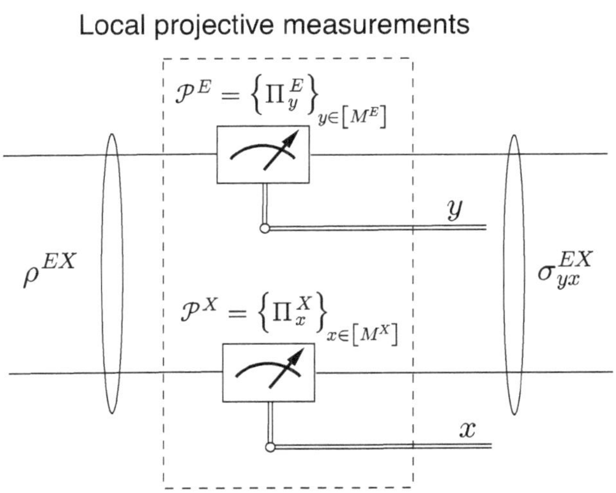

Figure 11.3 Illustration of local projective measurements applied to environment and system of interest.

Definition 11.6 Composite local measurement setting

A composite local measurement setting $\mathcal{P}^E \otimes \mathcal{P}^X$ with projection matrices $\mathcal{P}^E = \{\Pi_x^E\}_{x\in[M^E]}$ and $\mathcal{P}^X = \{\Pi_x^X\}_{x\in[M^X]}$ is a projective measurement that operates on the $N^E N^X$-dimensional space of the joint environment–system state with the following $M^E M^X$ projection matrices of dimensions $N^E N^X \times N^E N^X$:

$$\mathcal{P}^E \otimes \mathcal{P}^X = \{\{\Pi_y^E \otimes \Pi_x^X\}_{y\in[M^E]}\}_{x\in[M^X]}. \tag{11.30}$$

Accordingly, by applying the framework of projective measurements presented in Section 7.4, the operation of a composite local measurement $\mathcal{P}^E \otimes \mathcal{P}^X$ is characterized as follows.

Definition 11.7 Composite local measurements

Given a joint pre-measurement state ρ^{EX}, a composite local measurement setting $\mathcal{P}^E \otimes \mathcal{P}^X$ with projection matrices $\mathcal{P}^E = \{\Pi_x^E\}_{x\in[M^E]}$ and $\mathcal{P}^X = \{\Pi_x^X\}_{x\in[M^X]}$ results in the following.

- **Classical output:** The measurement returns two classical outputs, namely $\mathrm{y} \in [M^E]$ for the environment and $\mathrm{x} \in [M^X]$ for the system, with the probability of observing a pair $\mathrm{y} = y$ and $\mathrm{x} = x$ given as

$$p_{yx}(\rho^{EX}|\mathcal{P}^E \otimes \mathcal{P}^X) = \mathrm{tr}\left(\left(\Pi_y^E \otimes \Pi_x^X\right)\rho^{EX}\right). \tag{11.31}$$

- **Post-measurement state:** For a given classical output $(\mathrm{x} = x, \mathrm{y} = y)$, the quantum system is left in the corresponding post-measurement state

$$\sigma_{yx}^{EX} = \frac{1}{p_{yx}(\rho^{EX}|\mathcal{P}^E \otimes \mathcal{P}^X)}(\Pi_y^E \otimes \Pi_x^X)\rho^{EX}(\Pi_y^E \otimes \Pi_x^X). \tag{11.32}$$

In the special case of an uncorrelated pre-measurement state (11.21), the probability (11.31) can be written as

$$\begin{aligned} p_{yx}(\rho^{EX}|\mathcal{P}^E \otimes \mathcal{P}^X) &= \mathrm{tr}\left(\left(\Pi_y^E \otimes \Pi_x^X\right)\left(\rho^E \otimes \rho^X\right)\right) \\ &= \mathrm{tr}\left(\Pi_y^E \rho^E\right)\mathrm{tr}\left(\Pi_x^X \rho^X\right) \\ &= p_y(\rho^E|\mathcal{P}^E)p_x(\rho^X|\mathcal{P}^X). \end{aligned} \tag{11.33}$$

Therefore, the measurement probability $p_{yx}(\rho^{EX}|\mathcal{P}^E \otimes \mathcal{P}^X)$ is the product of the measurement probabilities $p_y(\rho^E|\mathcal{P}^E)$ and $p_x(\rho^X|\mathcal{P}^X)$ for environment and system. This indicates that, for uncorrelated states and local measurements, the two measurement outputs are also uncorrelated, i.e., independent (see Part IV for formal definitions).

Furthermore, the corresponding post-measurement state (11.32) is also uncorrelated, and given by

$$\sigma_{yx}^{EX} = \frac{1}{p_y(\rho^E|\mathcal{P}^E)}(\Pi_y^E \rho^E \Pi_y^E) \otimes \frac{1}{p_x(\rho^X|\mathcal{P}^X)}(\Pi_x^X \rho^X \Pi_x^X). \tag{11.34}$$

As for unitary transformations, one can also apply a measurement only to one of the two subsystems. If a projective measurement setting $\mathcal{P}^E = \{\Pi_y^E\}_{y\in[M^E]}$ is applied to the environment, using (11.31), then the probability of obtaining a classical output $y \in [M^E]$ is

$$p_y(\rho^{EX}|\mathcal{P}^E) = \mathrm{tr}\left(\left(\Pi_y^E \otimes I^X\right)\rho^{EX}\right), \tag{11.35}$$

where I^X is the $N^X \times N^X$ identity matrix. An analogous equation applies for the measurement probability when the system is measured in setting $\mathcal{P}^X = \{\Pi_x^X\}_{x\in[M^X]}$, i.e.,

$$p_x(\rho^{EX}|\mathcal{P}^X) = \mathrm{tr}\left(\left(I^E \otimes \Pi_x^X\right)\rho^{EX}\right), \tag{11.36}$$

where I^E is an $N^E \times N^E$ matrix.

The corresponding post-measurement states can be obtained in a similar way. Specifically, using (11.32), if the environment is measured in setting $\mathcal{P}^E = \{\Pi_y^E\}_{y\in[M^E]}$, the post-measurement state for classical output $y \in [M^E]$ is

$$\sigma_y^{EX} = \frac{1}{p_y(\rho^{EX}|\mathcal{P}^E)}(\Pi_y^E \otimes I^X)\rho^{EX}(\Pi_y^E \otimes I^X); \tag{11.37}$$

and, if the system is measured in setting $\mathcal{P}^X = \{\Pi_x^X\}_{x\in[M^X]}$, the post-measurement state for classical output $x \in [M^X]$ is

$$\sigma_x^{EX} = \frac{1}{p_x(\rho^{EX}|\mathcal{P}^X)}(I^E \otimes \Pi_x^X)\rho^{EX}(I^E \otimes \Pi_x^X). \tag{11.38}$$

Example 11.7

For $N^E = N^X = 2$, consider the separable state $|1\rangle \otimes |\psi^X\rangle$ in (11.29). Assume that the environment is measured in the computational basis $\mathcal{M}^E = \{|y\rangle\}_{y\in[2]}$, which corresponds to the projective measurement $\mathcal{P}^E = \{\Pi_y^E = |y\rangle\langle y|\}_{y\in[2]}$. Since the system and the environment are uncorrelated, the state of the system, $|\psi^X\rangle$, is irrelevant in determining the measurement probabilities for the measurement output y.

Indeed, by (11.33), the probability of obtaining the classical output y $\in [2]$ is

$$\begin{aligned} p_y(\rho^{EX}|\mathcal{P}^E) &= \mathrm{tr}(\underbrace{(|y\rangle\langle y| \otimes I)}_{\Pi_y^E\otimes I}\underbrace{(|1\rangle\langle 1| \otimes |\psi^X\rangle\langle\psi^X|)}_{\rho^E\otimes\rho^X}) \\ &= \mathrm{tr}(\underbrace{|y\rangle\langle y|}_{\Pi_y^E}\underbrace{|1\rangle\langle 1|}_{\rho^E}) \\ &= |\langle 1|y\rangle|^2, \end{aligned} \tag{11.39}$$

where I is a 2×2 identity matrix, which yields $p_1(\rho^{EX}|\mathcal{P}^E) = 1$. The corresponding post-measurement state σ_1^{EX} in (11.37) coincides with the pre-measurement state $|1\rangle \otimes |\psi^X\rangle$.

11.4 Positive Operator-Valued Measure (POVM) Measurements

As introduced in Section 6.5, a projective measurement in a quantum system with states in N dimensions produces classical outputs with $M \leq N$ possible values. Smaller values of M correspond to coarser-grained measurement settings, enabling a flexible trade-off between the informativeness of the classical output and the disturbance caused to the quantum state. This section extends the concept by introducing a more general class of measurements, known as **positive operator-valued measures**, or simply **POVMs**.

In a projective measurement, each orthogonal direction in the N-dimensional vector space is uniquely assigned to a single classical output value. In contrast, POVMs allow a single direction to be associated "partially" with multiple classical output values. As discussed in Section 11.5, this flexibility is particularly useful for tasks such as the unambiguous detection of two non-orthogonal quantum states. Additionally, POVMs can be designed to produce any number of classical outputs M, which may even exceed the dimension N of the quantum state space.

11.4.1 Deriving POVM Measurements ⏭

POVM measurements can be derived by considering the larger system comprising system and environment studied in Section 11.3. From this perspective, a POVM is realized by first applying a joint unitary transformation U to environment and system, and then carrying out a complete measurement on the environment state. This perspective, illustrated in Figure 11.4, is known as **Naimark dilation**. The measurement returns a classical output $y \in [M]$, where M is constrained only by the dimension N^E of the environment as $M \leq N^E$.

Assume that the system of interest is in some arbitrary pure state $\rho^X = |\psi^X\rangle\langle\psi^X|$. The extension to mixed states will be considered in Section 11.4.2. The dimension of the system is denoted as N, i.e., $N^X = N$ using the notation of the previous section.

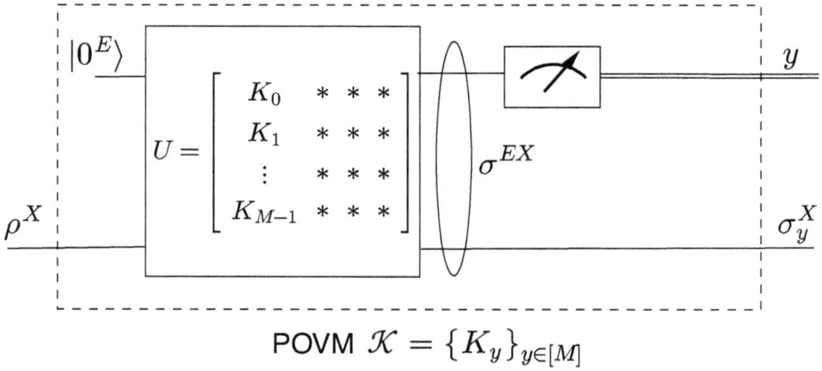

Figure 11.4 Illustration of a POVM as a unitary transformation applied jointly to system of interest and to the environment, followed by a measurement (in the computational basis) on the environment. This representation of a POVM within a larger composite system is also known as Naimark dilation.

Moreover, the dimension of the environment is denoted as M, i.e., $N^E = M$, as it corresponds to the number of possible classical outputs.

The initial state of the environment is conventionally taken to equal $|0^E\rangle$, i.e., the first vector in the computational basis for the environment space. This choice is arbitrary, and it does not affect the definition of POVMs. Therefore, the overall state of environment and system is described by the $MN \times 1$ vector $|0^E\rangle \otimes |\psi^X\rangle$.

As seen in Figure 11.4, the joint state of environment and system undergoes an $MN \times MN$ unitary transformation U. It is convenient to express this unitary transformation by dividing it into blocks $\{\{K_{x,y}\}_{x\in[M]}\}_{y\in[M]}$ of dimensions $N \times N$ as

$$U = \begin{bmatrix} K_{0,0} & K_{0,1} & \cdots & K_{0,M-1} \\ K_{1,0} & K_{1,1} & \cdots & K_{1,M-1} \\ \vdots & \vdots & & \vdots \\ K_{M-1,0} & K_{M-1,1} & \cdots & K_{M-1,M-1} \end{bmatrix}. \tag{11.40}$$

It is emphasized that here the notation $K_{x,y}$ represents an $N \times N$ matrix and not an individual entry of the matrix. Also, as discussed further later in this section, the blocks $\{\{K_{x,y}\}_{x\in[M]}\}_{y\in[M]}$ are related to each other in such a way that the overall matrix U in (11.40) is unitary.

With the notation (11.40), the post-transformation joint state is

$$|\phi^{EX}\rangle = U(|0^E\rangle \otimes |\psi^X\rangle) = U \begin{bmatrix} |\psi^X\rangle \\ 0 \\ \vdots \\ 0 \end{bmatrix} = \begin{bmatrix} K_{0,0}|\psi^X\rangle \\ K_{1,0}|\psi^X\rangle \\ \vdots \\ K_{M-1,0}|\psi^X\rangle \end{bmatrix}. \tag{11.41}$$

Therefore, as illustrated in Figure 11.4, the only part of matrix U that matters in the evaluation of the post-transformation state $|\phi^{EX}\rangle$ is the first block column, which consists of the matrices $\{K_{x,0}\}_{x\in[M]}$.

Based on this observation, one can simplify the notation by writing

$$K_{x,0} = K_x \tag{11.42}$$

for all $x \in [M]$. The unitary matrix U satisfies the equality $U^\dagger U = I$, where I is the $MN \times MN$ identity matrix, which implies that the matrices $\{K_x\}_{x\in[M]}$ guarantee the condition

$$\sum_{x\in[M]} K_x^\dagger K_x = I, \tag{11.43}$$

where I is the $N \times N$ identity matrix.

Definition 11.8 Kraus matrices

Any collection of $N \times N$ matrices $\{K_x\}_{x\in[M]}$ satisfying the condition in (11.43) is referred to as a set of Kraus matrices.

Example 11.8

The M projection matrices that define any projective measurement $\mathcal{P} = \{\Pi_x\}_{x\in[M]}$, where $M \leq N$, form a set of Kraus matrices. In fact, by definition of projective measurement, the condition (11.43) is satisfied via the equality

$$\sum_{x\in[M]} \Pi_x^\dagger \Pi_x = \sum_{x\in[M]} \Pi_x \Pi_x = \sum_{x\in[M]} \Pi_x = I, \tag{11.44}$$

where the first two equalities follow from the Hermitian and idempotence properties of projection matrices, and the last equality follows from the definition of projective measurements, i.e., from the resolution-of-the-identity relation.

Using a set $\{K_x\}_{x\in[M]}$ of Kraus matrices, the post-transformation state (11.41) of the composite system can be written as

$$\begin{aligned} |\phi^{EX}\rangle &= \begin{bmatrix} K_0|\psi^X\rangle \\ K_1|\psi^X\rangle \\ \vdots \\ K_{M-1}|\psi^X\rangle \end{bmatrix} \\ &= \sum_{x\in[M]} |x^E\rangle \otimes K_x|\psi^X\rangle. \end{aligned} \tag{11.45}$$

The second equality in (11.45) uses the fact that the term $|x^E\rangle \otimes K_x|\psi^X\rangle$ is an $MN \times 1$ vector with $K_x|\psi\rangle$ as the xth block of size $N \times 1$ and all other blocks having all zero entries.

As seen in Figure 11.4, after the unitary transformation, a measurement in the computational basis $\mathcal{M}^E = \{|x^E\rangle\}_{x\in[M]}$ is applied to the environment. Using (11.35) and (11.45), the classical output x equals $x \in [M]$ with probability

$$\begin{aligned} p_x(|\phi^{EX}\rangle|\mathcal{M}^E) &= \mathrm{tr}((|x^E\rangle\langle x^E| \otimes I^X)|\phi^{EX}\rangle\langle\phi^{EX}|) \\ &= \mathrm{tr}(K_x^\dagger K_x|\psi^X\rangle\langle\psi^X|), \end{aligned} \tag{11.46}$$

where the second equality is proved in Appendix 11.C. Note that this probability can also be written as

$$p_x(|\phi^{EX}\rangle|\mathcal{M}^E) = \mathrm{tr}(|\psi^X\rangle\langle\psi^X|K_x^\dagger K_x) = \langle\psi^X|K_x^\dagger K_x|\psi^X\rangle \tag{11.47}$$

by using the cyclic shift invariance property of the trace.

The result (11.46) generalizes the measurement probabilities for projective measurements, which take the form $\mathrm{tr}(\Pi_x|\psi^X\rangle\langle\psi^X|)$ for some projection matrix Π_x. Like projection matrices, the matrix $K_x^\dagger K_x$ in (11.46) is positive semi-definite (see Section 6.6.4). However, its eigenvalues λ are not limited to the discrete set $\{0, 1\}$ as for projection matrices. Rather, as will be seen later in this section, the condition (11.43) implies that the eigenvalues of matrix $K_x^\dagger K_x$ can range in the interval between 0 and 1. In this manner, a POVM can effectively assign a direction in the N-dimensional space, corresponding to an eigenvector of the matrix $K_x^\dagger K_x$, only partially to the classical output x, with the degree of assignment controlled by the corresponding eigenvalue.

Furthermore, as detailed in Appendix 11.C using (11.37), when the classical output is $x \in [M]$, the post-measurement state of the system is

$$\sigma_x^X = \frac{K_x|\psi^X\rangle\langle\psi^X|K_x^\dagger}{p_x(|\phi^{EX}\rangle|\mathcal{M}^E)}. \tag{11.48}$$

This is also a direct generalization of the post-measurement state for projective measurements, in which the Kraus matrix K_x is replaced by a projection matrix Π_x. In this regard, note that, given its idempotence property, a projection matrix plays the double role of the Kraus matrix K_x and of the positive semi-definite matrix $K_x^\dagger K_x$.

11.4.2 Defining POVM Measurements

Based on the derivation in Section 11.4.1, POVMs are characterized as follows.

Definition 11.9 POVM setting
For an N-dimensional quantum system, a POVM setting is defined by a set of M **Kraus matrices** $\mathcal{K} = \{K_x\}_{x\in[M]}$, i.e., by a set of matrices satisfying the condition (11.43). The number of measurement outputs, M, is arbitrary, and it can be larger than N.

Furthermore, as illustrated in Figure 11.5, a POVM operates as detailed next.

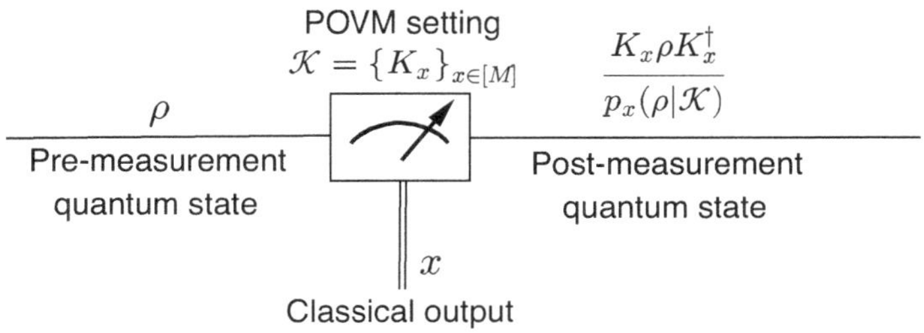

Figure 11.5 Illustration of a POVM.

Definition 11.10 POVM measurement
Given a POVM setting $\mathcal{K} = \{K_x\}_{x\in[M]}$ and a pre-measurement state ρ, the POVM measurement results in the following.

- **Classical output:** The measurement yields a classical output $\mathrm{x} = x \in [M]$ with distribution

$$p_x(\rho|\mathcal{K}) = \mathrm{tr}(K_x^\dagger K_x \rho). \tag{11.49}$$

- **Post-measurement state:** For a given classical output $\mathrm{x} = x$, the quantum system is left in the corresponding post-measurement state

$$\sigma_x = \frac{K_x \rho K_x^\dagger}{p_x(\rho|\mathcal{K})}. \tag{11.50}$$

Example 11.9

As discussed in Example 11.8, a projective setting $\mathcal{P} = \{\Pi_x\}_{x\in[M]}$, with $M \leq N$, is a special case of a POVM setting in which the Kraus matrices are given by $K_x = \Pi_x$ for $x \in [M]$. With setting $\mathcal{P}$, the measurement distribution (11.49) is given by $p_x(\rho|\mathcal{K}) = \mathrm{tr}(K_x^\dagger K_x \rho) = \mathrm{tr}(\Pi_x \rho)$, given the property $\Pi_x^\dagger \Pi_x = \Pi_x$ of projection matrices.

11.4.3 POVM Elements

While the post-measurement state (11.50) depends directly on the Kraus matrix K_x, the measurement probability (11.49) depends only on the positive semi-definite matrix

$$J_x = K_x^\dagger K_x. \tag{11.51}$$

Therefore, when one is only interested in the classical output, and not in the post-measurement state, it is common to specify a POVM setting as follows.

Definition 11.11 POVM setting (classical outputs only)
For an N-dimensional quantum system, when one is interested in modeling only the classical outputs, a POVM setting is defined by a set of M positive semi-definite matrices $\mathcal{K} = \{J_x\}_{x\in[M]}$, all with dimensions $N \times N$, known as POVM elements or effects. The POVM elements must satisfy the condition

$$\sum_{x\in[M]} J_x = I, \tag{11.52}$$

where I is the $N \times N$ identity matrix.

Definition 11.12 POVM measurement (classical outputs only)
Given a POVM setting $\mathcal{K} = \{J_x\}_{x\in[M]}$ and a pre-measurement state ρ, the measurement returns classical output $\mathrm{x} = x \in [M]$ with probability distribution

$$p_x(\rho|\mathcal{K}) = \mathrm{tr}(J_x \rho). \tag{11.53}$$

Note that the positive semi-definiteness of the POVM elements $\{J_x\}_{x\in[M]}$ gives POVMs their name. Furthermore, it is emphasized that the same notation $\mathcal{K}$ is used for both the set of Kraus matrices and the set of POVM elements, relying on the context to distinguish between the two.

Key Result 11.6 Bounds on the eigenvalues of POVM elements
Each eigenvalue λ of a POVM element J_x must satisfy the inequalities

$$0 \leq \lambda \leq 1. \tag{11.54}$$

Furthermore, given any positive semi-definite matrix J_x with eigenvalues in the range (11.54), it is always possible to construct a POVM that includes matrix J_x.

The necessity of condition (11.54) arises from **Weyl's inequality**. According to this inequality, each eigenvalue of the sum of Hermitian matrices – here, the sum $\sum_{x\in[M]} J_x$ of the POVM elements – must be no smaller than the sum of certain eigenvalues of the individual matrices $\{J_x\}_{x\in[M]}$ being summed. This leads to the conclusion that if any of the POVM elements J_x has an eigenvalue larger than 1, then the sum $\sum_{x\in[M]} J_x$ would also necessarily have an eigenvalue larger than 1. However, this is not allowed by the definition of POVMs, thereby enforcing the condition.

To construct a POVM that includes an element J_0 with all eigenvalues satisfying the condition in (11.54), consider a binary POVM defined as $\mathcal{K} = \{J_0, J_1 = I - J_0\}$. This construction forms a valid POVM because each eigenvalue of the matrix $I - J_0$ is $1 - \lambda$, where λ is an eigenvalue of J_0. Under the assumption in (11.54), all eigenvalues of $I - J_0$ are non-negative, ensuring that $I - J_0$ is positive semi-definite.

A final remark is in place regarding the relationship between Kraus matrices and POVM elements. Given a set of Kraus matrices $\{K_x\}_{x\in[M]}$, the corresponding set of POVM elements $\{J_x\}_{x\in[M]}$ is uniquely determined as $J_x = K_x^\dagger K_x$ for $x \in [M]$. Conversely, given a set of POVM elements $\{J_x\}_{x\in[M]}$, there are infinitely many Kraus matrices $\{K_x\}_{x\in[M]}$ that satisfy the relationship $J_x = K_x^\dagger K_x$ for $x \in [M]$.

A common choice for the Kraus matrix K_x corresponding to a POVM element J_x is the square root matrix

$$K_x = \sqrt{J_x}, \tag{11.55}$$

where $\sqrt{J_x}$ shares the same eigenvectors as J_x, and its eigenvalues are the square roots of the eigenvalues of J_x. However, any matrix of the form $K_x = U\sqrt{J_x}$, where U is a unitary matrix, is also a valid Kraus matrix corresponding to J_x. This is because of the equalities

$$K_x^\dagger K_x = (\sqrt{J_x}U^\dagger)(U\sqrt{J_x}) = \sqrt{J_x}U^\dagger U\sqrt{J_x} = J_x. \tag{11.56}$$

Example 11.10

For a qubit system, the only non-trivial projective measurements consist of complete measurements with two outputs, i.e., with $M = 2$. Complete measurements can potentially extract the maximum amount of information from a state, but they also potentially cause a large disturbance. POVMs allow the definitions of more general measurement settings that support a more flexible trade-off between information and disturbance.

To illustrate this point, consider the 2×2 Kraus matrices

$$\begin{aligned} K_0 &= \sqrt{1-\frac{\mu}{2}}|0\rangle\langle 0| + \sqrt{\frac{\mu}{2}}|1\rangle\langle 1| \\ \text{and } K_1 &= \sqrt{\frac{\mu}{2}}|0\rangle\langle 0| + \sqrt{1-\frac{\mu}{2}}|1\rangle\langle 1|, \end{aligned} \tag{11.57}$$

where $0 \leq \mu \leq 1$ is a parameter that determines the level of "noise" in the measurement. Matrices $\{K_0, K_1\}$ define a POVM setting $\mathcal{K}$, since they satisfy the condition (11.43), i.e., $K_0^\dagger K_0 + K_1^\dagger K_1 = I$, as can easily be checked.

When only the measurement probabilities are of interest, one can also define the POVM setting using the POVM elements $J_0 = K_0^\dagger K_0$ and $J_1 = K_1^\dagger K_1$, i.e.,

$$\begin{aligned} J_0 &= \left(1-\frac{\mu}{2}\right)|0\rangle\langle 0| + \frac{\mu}{2}|1\rangle\langle 1| \\ \text{and } J_1 &= \frac{\mu}{2}|0\rangle\langle 0| + \left(1-\frac{\mu}{2}\right)|1\rangle\langle 1|, \end{aligned} \tag{11.58}$$

which satisfy condition (11.52) as $J_0 + J_1 = I$. The POVM elements (11.58) can also be written in terms of the Pauli Z matrix P_3 as

$$\begin{aligned} J_0 &= \frac{1}{2}(I + (1-\mu)P_3) \\ \text{and } J_1 &= \frac{1}{2}(I - (1-\mu)P_3). \end{aligned} \tag{11.59}$$

With $\mu = 0$, the POVM $\mathcal{K} = \{K_0, K_1\}$ corresponds to a complete measurement in the computational basis, as the Kraus matrices equal the projection matrices $K_0 = |0\rangle\langle 0|$ and $K_1 = |1\rangle\langle 1|$. Thus, this measurement can extract maximum information, while also maximally disturbing the state.

At the other extreme, choosing $\mu = 1$ yields a completely noisy measurement with Kraus matrices equal to $K_0 = K_1 = I/\sqrt{2}$. With this maximally noisy setting, one obtains completely uninformative classical measurement outputs, which are uniformly distributed irrespective of the pre-measurement state ρ, i.e.,

$$p_x(\rho|\mathcal{K}) = \mathrm{tr}(K_x^\dagger K_x \rho) = \frac{1}{2}\mathrm{tr}(\rho) = \frac{1}{2} \tag{11.60}$$

for $x \in [2]$. On the flip side, this measurement leaves the state undisturbed, in the sense that both post-measurement states are

$$\sigma_x = \frac{K_x \rho K_x^\dagger}{p_x(\rho|\mathcal{K})} = \rho \tag{11.61}$$

for $x \in [2]$.

Intermediate values of μ in the interval between the extreme values 0 and 1 support different trade-offs between disturbance and informativeness of the measurement. To consider a numerical example, assume that the pre-measurement state is $|\psi\rangle = [\cos(\pi/8), \sin(\pi/8)]^T$. Figure 11.6 plots measures of informativeness and disturbance as a function of the parameter μ.

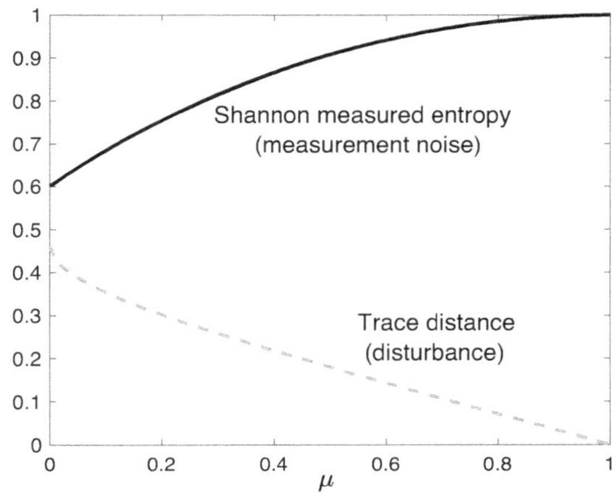

Figure 11.6 Level of noise present in the classical measurement output, accounted for by the Shannon measured entropy, and disturbance caused to the pre-measurement state, captured by the average trace distance between the pre-measurement state $|\psi\rangle$ and the post-measurement state, as a function of the parameter μ in the POVM (11.57).

To measure informativeness, or rather uninformativeness, the figure evaluates the level of noise in the classical measurement output by computing the Shannon measured entropy

$$\mathrm{H}(|\psi\rangle|\mathcal{K}) = \mathrm{H}(p(|\psi\rangle|\mathcal{K})), \tag{11.62}$$

i.e., the entropy of the classical output. A larger measured entropy indicates a noisier, or less sharp, measurement.

To quantify disturbance, the figure adopts the average trace distance between the pre-measurement state $|\psi\rangle$ and the post-measurement state, i.e.,

$$\sum_{x\in[2]} p_x(|\psi\rangle|\mathcal{K})\mathrm{D}^{\mathrm{tr}}(|\psi\rangle, |\phi_x\rangle), \tag{11.63}$$

where $|\phi_x\rangle = K_x|\psi\rangle/\sqrt{p_x(|\psi\rangle|\mathcal{K})}$ is the post-measurement state when the classical output is $x \in [2]$. A larger average trace distance indicates a more significant disturbance to the pre-measurement state.

The figure shows that, by increasing the "noise" parameter μ from 0 to 1, the POVM $\mathcal{K} = \{K_0, K_1\}$ extracts less information about the pre-measurement state $|\psi\rangle$, producing a noisier classical output, while also disturbing less the pre-measurement state $|\psi\rangle$. Therefore, an increase in parameter μ causes the POVM to become coarser grained, extracting less information from the pre-measurement state while being gentler in terms of its effect on the post-measurement state.

11.4.4 Coarse-Graining POVMs ⏭

As discussed in Example 11.10, POVM settings can control the level of coarse-graining, dictating different points on the trade-off between informativeness and disturbance, without necessarily changing the number of measurement outputs. This is in contrast to projective measurements, for which coarse graining requires decreasing the number of measurement outputs. As detailed next, a general way to obtain a coarser-grained POVM is to combine a base POVM, having any number, M, of measurement outputs, with a classical channel that adds "noise" to the measurement outputs.

To elaborate, consider a base POVM $\mathcal{K} = \{J_x\}_{x\in[M]}$ and fix a classical channel $p_{x\to y}$ with input $x \in [M]$ and output $y \in [M]$ (see Section 9.5). Then, a coarse-grained POVM can be obtained as the setting $\tilde{\mathcal{K}} = \{\tilde{J}_y\}_{y\in[M]}$ given by the POVM elements

$$\tilde{J}_y = \sum_{x\in[M]} p_{x\to y} J_x. \tag{11.64}$$

The setting $\tilde{\mathcal{K}} = \{\tilde{J}_y\}_{y\in[M]}$ defines a valid POVM, as can be proved via the equalities

$$\begin{aligned}\sum_{y\in[M]} \tilde{J}_y &= \sum_{y\in[M]} \sum_{x\in[M]} p_{x\to y} J_x \\ &= \sum_{x\in[M]} \left(\sum_{y\in[M]} p_{x\to y} \right) J_x \\ &= \sum_{x\in[M]} J_x = I. \end{aligned} \tag{11.65}$$

The channel $p_{x\to y}$ thus plays the role of noise, which may cause a decrease in the informativeness of the measurement as compared to the base POVM $\mathcal{K} = \{J_x\}_{x\in[M]}$. As a special case, the channel $p_{x\to y}$ may assign multiple values of x to the same value y, implementing the most common form of coarse graining.

If one is interested solely in the classical measurement output, the POVM $\tilde{\mathcal{K}} = \{\tilde{J}_y\}_{y\in[M]}$ can be simulated by first carrying out the base POVM, and then applying the channel $p_{x\to y}$ to the classical output as post-processing step. However, it is emphasized that this post-processing approach, while reproducing the correct measurement probability for the POVM $\tilde{\mathcal{K}} = \{\tilde{J}_y\}_{y\in[M]}$, does not leave the state in the correct post-measurement state for the POVM $\tilde{\mathcal{K}}$.

Example 11.11

The setting (11.58) is an example of a coarse-grained POVM of the form (11.64). In fact, it can be obtained from the base measurement given by the complete measurement in the computational basis along with the channel defined as

$$p_{0\to 0} = 1 - \frac{\mu}{2} \tag{11.66}$$

and

$$p_{1\to 0} = \frac{\mu}{2}. \tag{11.67}$$

11.5 Applications of POVM Measurements ⏭

The previous section has presented the general framework of POVM measurements. This section showcases the benefits of POVMs by covering three informational tasks for which projective measurements prove insufficient, while POVMs are successful. The first application involves unambiguous state detection; the second concerns the estimation of an unknown quantum state; and the third revolves around the implementation of arbitrary linear transformations of a quantum state.

11.5.1 POVM for Unambiguous Quantum State Detection

As illustrated in Figure 11.7, consider a setting in which a detector wishes to determine if a quantum system is in state $|\psi\rangle$ or in state $|\phi\rangle$. As studied in Section 5.6, when the detector is limited to a projective measurement, it is not possible to ensure correct detection irrespective of whether the true state is $|\psi\rangle$ or $|\phi\rangle$, unless the two states are orthogonal. Instead, if one wishes to detect one of the states correctly with probability 1, the probability of successful detection for the other state equals $1 - \mathrm{F}(|\psi\rangle, |\phi\rangle)$, where $\mathrm{F}(|\psi\rangle, |\phi\rangle) = |\langle\psi|\phi\rangle|^2$ is the fidelity between the two states. As detailed in this section, while non-orthogonal states cannot be perfectly distinguished even with POVMs, POVMs support the design of measurements with a rejection option such that, whenever a decision is made for either state, the decision is correct.

In the setting depicted in Figure 11.7, the detector produces $M = 3$ possible outputs such that when output $\mathrm{x} = 0$ is recorded, the detector decides that the true

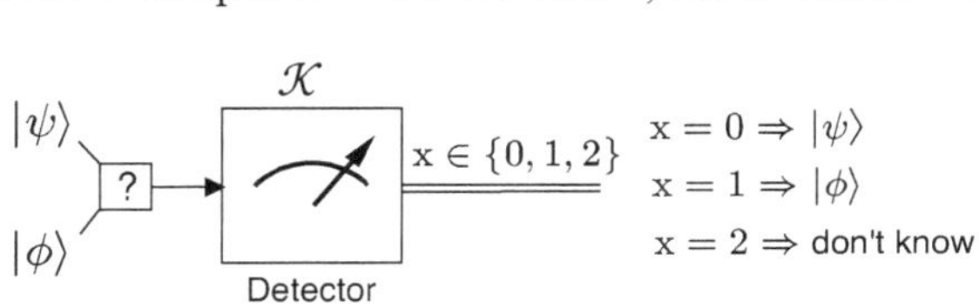

Figure 11.7 Block diagram for the problem of unambiguous quantum state detection.

state is $|\psi\rangle$; when the output is $\mathrm{x} = 1$, the decision is for state $|\phi\rangle$; and, when the output is $\mathrm{x} = 2$, the detector produces a "don't know" decision. The goal is to ensure that the probability of observing output $\mathrm{x} = 0$ when the true state is $|\phi\rangle$ is zero; and so is the probability of observing output $\mathrm{x} = 1$ when the true state is $|\psi\rangle$. Therefore, no errors are allowed, but the detector can issue a "don't know" decision when in doubt.

Mathematically, the measurement setting must be designed under the **unambiguous-detection constraint**

$$\Pr_{|\psi\rangle}[\mathrm{x} = 1] = \Pr_{|\phi\rangle}[\mathrm{x} = 0] = 0. \tag{11.68}$$

Accordingly, if the measurement returns $\mathrm{x} = 0$, the true state cannot be $|\phi\rangle$, and thus the detector can unambiguously make the decision that the true state is $|\psi\rangle$. In a similar way, when the measurement returns $\mathrm{x} = 1$, the detector can unambiguously determine the state to be $|\phi\rangle$.

The unambiguous-detection constraint (11.68) does not imply that the detector probabilities equal 1. In fact, the detection probability when the true state is $|\psi\rangle$ is given by

$$\Pr_{|\psi\rangle}[\mathrm{x} = 0] = 1 - \Pr_{|\psi\rangle}[\mathrm{x} = 2], \tag{11.69}$$

and the detection probability when the true state is $|\phi\rangle$ is

$$\Pr_{|\phi\rangle}[\mathrm{x} = 1] = 1 - \Pr_{|\phi\rangle}[\mathrm{x} = 2]. \tag{11.70}$$

The rejection probabilities $\Pr_{|\psi\rangle}[\mathrm{x} = 2]$ and $\Pr_{|\phi\rangle}[\mathrm{x} = 2]$ account for the cases in which the detector issues a "don't know" decision.

The unambiguous-detection constraint (11.68) is easily satisfied by always producing the "don't know" decision $\mathrm{x} = 2$. However, this yields zero detection probabilities in (11.69) and (11.70). As discussed next, unless the two states $|\psi\rangle$ and $|\phi\rangle$ are equivalent, a POVM can be designed that satisfies the unambiguous-detection constraint (11.68), while attaining positive detection probabilities (11.69) and (11.70).

To this end, consider the POVM $\mathcal{K} = \{J_0, J_1, J_2\}$ defined by the three POVM elements

$$\begin{aligned} J_0 &= \frac{I - |\phi\rangle\langle\phi|}{1 + |\langle\psi|\phi\rangle|}, \\ J_1 &= \frac{I - |\psi\rangle\langle\psi|}{1 + |\langle\psi|\phi\rangle|}, \\ \text{and } J_2 &= I - J_0 - J_1. \end{aligned} \tag{11.71}$$

One can verify that matrices J_0, J_1, and J_2 are positive semi-definite and that they satisfy the condition (11.52).

The POVM $\mathcal{K}$ satisfies the unambiguous-detection constraint (11.68) as

$$\begin{aligned}\Pr_{|\psi\rangle}[\mathrm{x}=1] &= p_1(|\psi\rangle|\mathcal{K}) \\ &= \mathrm{tr}(J_1|\psi\rangle\langle\psi|) \\ &= \frac{1}{1+|\langle\psi|\phi\rangle|}(1-|\langle\psi|\psi\rangle|^2)=0,\end{aligned} \tag{11.72}$$

and a similar calculation applies for the probability $\Pr_{|\phi\rangle}[\mathrm{x}=0]$, yielding $\Pr_{|\phi\rangle}[\mathrm{x}=0]=0$. Furthermore, with this POVM the rejection probabilities are given by

$$\begin{aligned}\Pr_{|\psi\rangle}[\mathrm{x}=2]=\Pr_{|\phi\rangle}[\mathrm{x}=2] &= |\langle\psi|\phi\rangle| \\ &= \sqrt{\mathrm{F}(|\psi\rangle,|\phi\rangle)}.\end{aligned} \tag{11.73}$$

It is concluded that, unlike projection measurements, the POVM $\mathcal{K}$ can meet the unambiguous-detection constraint even for non-orthogonal states $|\psi\rangle$ and $|\phi\rangle$. To this end, the POVM (11.71) refrains from making a decision with probability $\sqrt{\mathrm{F}(|\psi\rangle,|\phi\rangle)}$, irrespective of whether the true state is $|\psi\rangle$ or $|\phi\rangle$. Accordingly, the more aligned the two states are, and thus the larger the fidelity $\mathrm{F}(|\psi\rangle,|\phi\rangle)$ is, the more likely it is that the detector is not able to make a definite decision.

11.5.2 Quantum State Tomography

For the second informational task of interest, consider the setting in Figure 11.8, which depicts the problem of **quantum state tomography**. Here, the observer has access to multiple copies of a quantum system that is in an unknown, but fixed, $N \times 1$ state $|\psi\rangle$, and the goal is to construct a classical description of the state vector $|\psi\rangle$. Specifically, the observer can extract information from the copies of state $|\psi\rangle$ by applying measurements with the aim of estimating all the amplitudes of the state $|\psi\rangle$.

Given a measurement setting $\mathcal{K}$ with M outputs, if the number of copies of the system is sufficiently large, the observer can obtain a good estimate of the $M \times 1$ vector of measurement probabilities $p(|\psi\rangle|\mathcal{K})$ (see Appendix 5.A). The problem of quantum state tomography hence reduces to that of recovering the $N \times 1$ state $|\psi\rangle$ from the $M \times 1$ vector of probabilities $p(|\psi\rangle|\mathcal{K})$.

Given any projective measurement $\mathcal{P}$, this is generally not possible. In fact, the measurement probability vector $p(|\psi\rangle|\mathcal{P})$ for a projective measurement $\mathcal{P}$ consists of $M \leq N$ positive numbers, which are insufficient to determine the N complex numbers in state vector $|\psi\rangle$.

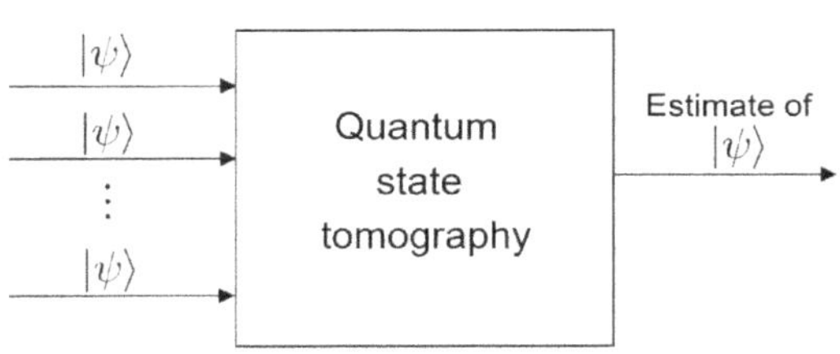

Figure 11.8 Block diagram for the problem of quantum state tomography.

Example 11.12

Consider the case of a general qubit state $|\psi\rangle = \psi_0|0\rangle + \psi_1|0\rangle$. Given that the global phase is arbitrary, one can assume without loss of generality that the amplitude ψ_0 is real and positive, i.e., $\psi_0 = |\psi_0|$, while the amplitude ψ_1 is a general complex number $\psi_1 = |\psi_1|\exp(i\gamma_1)$ for a relative phase γ_1. Therefore, quantum tomography must determine three real-valued unknowns, namely the triple $(|\psi_0|, |\psi_1|, \gamma_1)$. Any projective measurement can return at most two real numbers, making it impossible to estimate all three numbers $(|\psi_0|, |\psi_1|, \gamma_1)$.

For example, consider a measurement in the computational-basis setting $\mathcal{M} = \{|0\rangle, |1\rangle\}$. From it, one obtains the probability vector

$$p(|\psi\rangle|\mathcal{M}) = \begin{bmatrix} |\psi_0|^2 \\ |\psi_1|^2 \end{bmatrix}, \tag{11.74}$$

which brings no information about the relative phase γ_1.

In a similar way, a measurement in the Hadamard basis would also not be sufficient to determine all three parameters $(|\psi_0|, |\psi_1|, \gamma_1)$. In fact, in this case, the measurement probability vector would be given by

$$p(|\psi\rangle|\mathcal{M}) = \frac{1}{2}\begin{bmatrix} 1 + 2|\psi_0||\psi_1|\cos(\gamma_1) \\ 1 - 2|\psi_0||\psi_1|\cos(\gamma_1) \end{bmatrix}, \tag{11.75}$$

which does not provide enough information to recover the three parameters $(|\psi_0|, |\psi_1|, \gamma_1)$.

The limitation highlighted by the previous example on the use of projective measurements for quantum tomography can be alleviated by using POVMs. In fact, the probability vectors describing the output of POVMs can have any dimension M. POVMs that allow the reconstruction of any state $|\psi\rangle$ based on the corresponding probability vector $p(|\psi\rangle|\mathcal{K})$ are said to be **informationally complete**.

Example 11.13

For a qubit system, consider the measurement setting defined by the POVM elements

$$\mathcal{K} = \left\{\frac{1}{2}|0\rangle\langle 0|, \frac{1}{2}|1\rangle\langle 1|, \frac{1}{2}|+\rangle\langle +|, \frac{1}{2}|-\rangle\langle -|\right\}. \tag{11.76}$$

Note that this is indeed a valid POVM since all the POVM elements have eigenvalues $\{0, 1/2\}$, which are non-negative, and the equality

$$\frac{1}{2}(|0\rangle\langle 0| + |1\rangle\langle 1| + |+\rangle\langle +| + |-\rangle\langle -|) = \frac{1}{2}(I + I) = I \tag{11.77}$$

holds, where I is a 2×2 identity matrix.

For a general qubit state $|\psi\rangle = \psi_0|0\rangle + \psi_1|0\rangle$, with $\psi_0 = |\psi_0|$ and $\psi_1 = |\psi_1|\exp(i\gamma_1)$, the measurement probability vector is

$$p(|\psi\rangle|\mathcal{K}) = \frac{1}{2}\begin{bmatrix} |\psi_0|^2 \\ |\psi_1|^2 \\ \frac{1}{2}(1+2|\psi_0||\psi_1|\cos(\gamma_1)) \\ \frac{1}{2}(1-2|\psi_0||\psi_1|\cos(\gamma_1)) \end{bmatrix}, \tag{11.78}$$

from which one can recover the tuple $(|\psi_0|, |\psi_1|, \gamma_1)$.

Comparing (11.74) and (11.75) with (11.78) reveals that the POVM $\mathcal{K}$ can equivalently be implemented as follows.

1. **Randomized measurement selection:** Sample a uniform binary random variable $\mathrm{z} \in [2]$ and record the realized bit $\mathrm{z} = z$.
2. **Complete measurement:** If $z = 0$, apply a measurement in the computational basis $\mathcal{M}_0$, and, if $z = 1$, apply a measurement in the Hadamard basis $\mathcal{M}_1$, recording the measurement output $\mathrm{x} = x \in [2]$.
3. **Measurement output:** Return the measurement output defined by the bi-index $zx \in [4]$.

One can readily verify that the distribution of the random variable zx is indeed given by (11.78).

This example illustrates a useful approach to the construction of informationally complete POVMs, namely **randomized measurements**. The reader is referred to Problem 11.9 and the Recommended Resources for further discussion.

The definition of informationally complete POVMs can be extended from pure to mixed states. In this more general case, the goal is to obtain a classical reconstruction of a density matrix ρ by estimating all of the N^2 entries of the matrix. To this end, the POVM must have at least N^2 possible measurement outputs. There are many possible such constructions, as discussed further in the Recommended Resources.

11.5.3 Arbitrary Linear Transformations via Post-Selection

The third, and final, informational task covered in this section to showcase the benefits of POVMs consists of the application of a generic linear transformation to a quantum state. As seen in Chapter 10, closed quantum systems evolve according to a unitary transformation U, so that the post-transformation state is given by $\sigma = U\rho U^\dagger$ for a pre-transformation state ρ. Is it possible to design a transformation that produces a state of the form

$$\sigma = \frac{K\rho K^\dagger}{\mathrm{tr}(K^\dagger K\rho)} \tag{11.79}$$

for any $N \times N$ matrix K? Note that the denominator in (11.79), which is assumed to be positive, is required to ensure that σ is a valid density matrix.

Assume that the matrix K is such that the positive semi-definite matrix $K^\dagger K$ has all eigenvalues no larger than 1. Note that this condition can be satisfied for any matrix K by rescaling matrix K as γK for some scalar $0 < \gamma \leq 1$. For any such matrix K, one can define the POVM $\mathcal{K} = \{K_0 = K, K_1 = \sqrt{I - K^\dagger K}\}$.

With this measurement setting, with probability $p_0(\rho|\mathcal{K}) = \mathrm{tr}(K^\dagger K\rho)$, the post-measurement state is given by (11.79). Therefore, the state (11.79) can be produced by implementing the POVM $\mathcal{K}$ followed by post-selection, i.e., retaining the output state only if the classical measurement output is $x = 0$. Note that the probability $p_0(\rho|\mathcal{K}) = \mathrm{tr}(K^\dagger K\rho)$ of obtaining the target state can be small, making this approach potentially quite inefficient.

11.6 Quantum Channels

This section introduces quantum channels. Quantum channels describe general dynamic transformations that affect open quantum systems in the absence of (selective) measurements. Quantum channels include as special cases random unitary transformations, studied earlier in this chapter, as well as non-selective projective measurements, which were presented in Section 7.11.

11.6.1 Defining Quantum Channels

A quantum channel can be described as a non-selective version of a POVM. To explain, fix a POVM setting $\mathcal{K} = \{K_x\}_{x\in[M]}$ for some integer M. As depicted in Figure 11.9(a), following the Naimark dilation representation in Figure 11.4, a POVM $\mathcal{K}$ results from a conventional (complete) measurement on the environment that is applied after a joint unitary transformation U operating on both system and environment. The POVM $\mathcal{K}$ leaves the system, initially in state ρ, in the post-measurement state (11.50), i.e., $\sigma_x = K_x\rho K_x^\dagger/p_x(\rho|\mathcal{K})$, when the classical measurement output is x.

As also shown in Figure 11.9(a), assume now that the measurement output x $\sim p(\rho|\mathcal{K})$ is discarded or forgotten. As a result, the system does not have access to the measurement output x $\sim p(\rho|\mathcal{K})$. Therefore, the state of the system is given by the average post-measurement density matrix

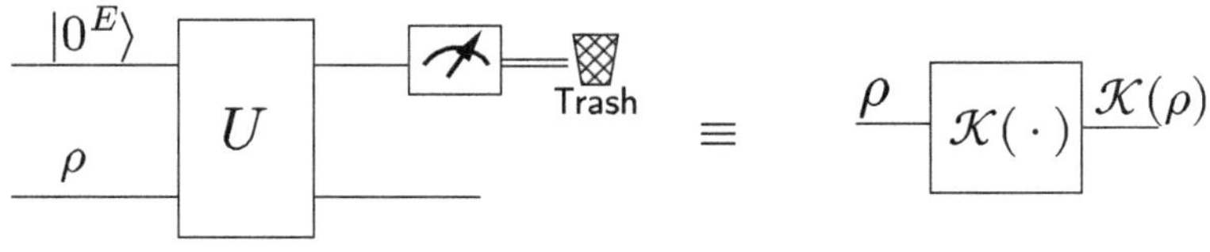

Figure 11.9 A quantum channel can be represented (on the left) via the Stinespring dilation, according to which a quantum channel is a non-selective version of a POVM, in which the classical output is discarded and (on the right) as a mapping between density matrices of the form (11.81).

$$\mathcal{K}(\rho) = \sum_{x\in[M]} p_x(\rho|\mathcal{K})\sigma_x$$
$$= \sum_{x\in[M]} K_x\rho K_x^{\dagger}. \tag{11.80}$$

Definition 11.13 Quantum channel
A quantum channel $\mathcal{K}(\cdot)$ is specified by a set of Kraus matrices $\mathcal{K} = \{K_x\}_{x\in[M]}$ (satisfying the equality (11.43)). When applied to a state ρ, the channel $\mathcal{K}(\cdot)$ produces the post-channel state

$$\mathcal{K}(\rho) = \sum_{x\in[M]} K_x\rho K_x^{\dagger}. \tag{11.81}$$

Example 11.14

Random unitary transformations. Random unitary transformations (11.1) are quantum channels obtained by choosing the Kraus matrices as $K_x = \sqrt{\gamma_x}U_x$, where U_x is a unitary matrix for all $x \in [M]$ and $\gamma = [\gamma_0, \ldots, \gamma_{M-1}]^T$ is a probability vector.

Example 11.15

Pinching channels. Non-selective projective measurements, introduced in Section 7.11, are quantum channels with Kraus matrices $K_x = \Pi_x$, where Π_x is a projection matrix for all $x \in [M]$. They are also known as pinching channels.

Example 11.16

Measure-and-prepare channels. Non-selective complete measurements $\mathcal{M}(\cdot)$ are also known as measure-and-prepare channels. To see why, take any complete measurement setting $\mathcal{M} = \{|v_x\rangle\}_{x\in[N]}$. The corresponding non-selective measurement yields the post-measurement state

$$\mathcal{M}(\rho) = \sum_{x\in[N]} \langle v_x|\rho|v_x\rangle|v_x\rangle\langle v_x|. \tag{11.82}$$

It is thus a quantum channel with Kraus matrices given by the projection matrices $K_x = |v_x\rangle\langle v_x|$ for $x \in [N]$.

This channel can be implemented by first making a complete measurement in setting $\mathcal{M}$, which returns random variable x with probabilities $\{p_x(\rho|\mathcal{M}) = \langle v_x|\rho|v_x\rangle\}_{x\in[N]}$, and then preparing a quantum system in state $|v_x\rangle$. Note that this preparation is possible because the measurement is complete, and thus the post-measurement state does not depend on the original state ρ. A more general pinching channel cannot be implemented via a measurement-and-prepare strategy.

11.6.2 Pauli Channels

Pauli channels are qubit channels whose Kraus matrices are scaled versions of the Pauli matrices, i.e.,

$$K_0 = \sqrt{\gamma_0} I, K_1 = \sqrt{\gamma_1} P_1, K_2 = \sqrt{\gamma_2} P_2, \text{ and } K_3 = \sqrt{\gamma_3} P_3, \tag{11.83}$$

for probabilities $\gamma_x \geq 0$ with $\sum_{x \in [4]} \gamma_x = 1$.

Definition 11.14 Pauli channels
A Pauli channel is defined as

$$\mathcal{K}(\rho) = \gamma_0 \rho + \gamma_1 P_1 \rho P_1 + \gamma_2 P_2 \rho P_2 + \gamma_3 P_3 \rho P_3 \tag{11.84}$$

for some probability distribution $\gamma = [\gamma_0, \gamma_1, \gamma_2, \gamma_3]^T$.

In quantum computing, the three Pauli matrices P_1, P_2, and P_3 represent different types of error that may affect a qubit. Accordingly, the probabilities γ_1, γ_2, and γ_3 determine the probability of each error type, while γ_0 is the probability that no error occurs on the qubit.

Example 11.17

Dephasing channels. When only Pauli Z-type errors are possible, one obtains the dephasing channel

$$\mathcal{K}(\rho) = (1 - \gamma)\rho + \gamma P_3 \rho P_3, \tag{11.85}$$

with $0 \leq \gamma \leq 1$. Note that for $\gamma = 0$ and $\gamma = 1$, the dephasing channel corresponds to a unitary transformation, and thus the channel is "noisy" only for values of γ strictly inside the interval between 0 and 1.

For a general pre-channel qubit density matrix ρ having entries $\rho_{x,y}$ with $x \in [2]$ and $y \in [2]$, the resulting post-channel state is given by

$$\mathcal{K}(\rho) = \begin{bmatrix} \rho_{0,0} & (1 - 2\gamma)\rho_{0,1} \\ (1 - 2\gamma)\rho_{0,1}^* & \rho_{1,1} \end{bmatrix}. \tag{11.86}$$

Therefore, the dephasing channel reduces the coherence of the state with respect to the computational basis $\mathcal{M}$, as it decreases the magnitude of the off-diagonal elements of the density matrix. In particular, with $\gamma = 0.5$, the post-transformation state becomes incoherent with respect to basis $\mathcal{M}$.

Example 11.18

Bit flip channel. When the only possible type of error is given by the Pauli X matrix P_1, the resulting channel is known as the bit flip channel, which is defined as

$$\mathcal{K}(\rho) = (1 - \gamma)\rho + \gamma P_1 \rho P_1, \tag{11.87}$$

with parameter $0 \leq \gamma \leq 1$. The name follows from the fact that, given the pre-channel state $\rho = |0\rangle\langle 0|$, the post-channel state is

$$\mathcal{K}(|0\rangle\langle 0|) = (1 - \gamma)|0\rangle\langle 0| + \gamma|1\rangle\langle 1|, \tag{11.88}$$

in which the state $|0\rangle$ is "flipped" to state $|1\rangle$ with probability γ. One also has the dual equality $\mathcal{K}(|1\rangle\langle 1|) = (1 - \gamma)|1\rangle\langle 1| + \gamma|0\rangle\langle 0|$.

Example 11.19

Depolarizing channel. Another notable example of Pauli channels is the depolarizing channel

$$\mathcal{K}(\rho) = (1 - \gamma)\rho + \gamma\pi, \tag{11.89}$$

which replaces the density matrix ρ with the maximally mixed state π with probability γ, where $0 \leq \gamma \leq 1$. Thus, for the depolarizing channel, a larger value of probability γ indicates a noisier channel. For this reason, the probability $1 - \gamma$ is also known as the **visibility** of the depolarizing measurement. The depolarizing channel is often used as an approximation of the errors incurred by a qubit in a quantum computer in order to design quantum error correction strategies.

The depolarizing channel is obtained from (11.84) with $\gamma_0 = 1 - 3\gamma/4$ and $\gamma_1 = \gamma_2 = \gamma_3 = \gamma/4$. Accordingly, as a result of a depolarizing channel, either the state is left unchanged with probability γ_0, or else an error described by the Pauli matrix P_x for some $x \in \{1, 2, 3\}$ is applied with equal probability $(1 - \gamma_0)/3$.

The post-channel density matrix is given by

$$\mathcal{K}(\rho) = \begin{bmatrix} (1-\gamma)\rho_{0,0} + \gamma/2 & (1-\gamma)\rho_{0,1} \\ (1-\gamma)\rho_{0,1}^* & (1-\gamma)\rho_{1,1} + \gamma/2 \end{bmatrix}. \tag{11.90}$$

Thus, with $\gamma = 1$, the coherence with respect to the computational basis is reduced to zero.

Example 11.20

The Kraus representation $\mathcal{K} = \{K_x\}_{x \in [M]}$ of a quantum channel (11.80) is generally not unique. To see this, consider the channel defined by Kraus matrices

$$K_0 = \begin{bmatrix} 1 & 0 \\ 0 & \sqrt{1-\mu} \end{bmatrix} \text{ and } K_1 = \begin{bmatrix} 0 & 0 \\ 0 & \sqrt{\mu} \end{bmatrix}, \tag{11.91}$$

for some $0 \leq \mu \leq 1$.

When applied to a general pre-channel qubit density matrix having entries $\rho_{x,y}$ with $x \in [2]$ and $y \in [2]$, the post-channel state is given by

$$\mathcal{K}(\rho) = \begin{bmatrix} \rho_{0,0} & \sqrt{1-\mu}\rho_{0,1} \\ \sqrt{1-\mu}\rho_{0,1}^{*} & \rho_{1,1} \end{bmatrix}. \tag{11.92}$$

Therefore, if $0 \leq \gamma \leq 0.5$, the post-channel state $\mathcal{K}(\rho)$ coincides with (11.86) by setting $\mu = 1 - (1 - 2\gamma)^2$. It is concluded that, using the mapping $\mu = 1 - (1 - 2\gamma)^2$, the quantum channel with Kraus matrices (11.91) coincides with the dephasing channel (11.85).

11.6.3 Unital Channels

All the channels discussed so far – random unitary transformations, pinching channels, and Pauli channels – are examples of unital channels.

Definition 11.15 Unital channels
A quantum channel is said to be unital if it does not modify the maximally mixed state π, i.e.,

$$\mathcal{K}(\pi) = \pi. \tag{11.93}$$

Not all channels are unital. In fact, for a general channel (11.81), the sum $\mathcal{K}(\pi) = \sum_{x \in [N]} K_x K_x^{\dagger}/N$ does not necessarily equal π. An important example of non-unital channels is presented in the following example.

Example 11.21

Amplitude-damping channel. The amplitude-damping channel is a qubit channel $\mathcal{K} = \{K_0, K_1\}$ with Kraus matrices

$$K_0 = \begin{bmatrix} 1 & 0 \\ 0 & \sqrt{1-\mu} \end{bmatrix} \text{ and } K_1 = \begin{bmatrix} 0 & \sqrt{\mu} \\ 0 & 0 \end{bmatrix} \tag{11.94}$$

for some probability $0 \leq \mu \leq 1$.

To understand the reason for the name "amplitude damping," note that a pure state $|0\rangle$ is left unchanged as $\mathcal{K}(|0\rangle\langle 0|) = |0\rangle\langle 0$, while a pure state $|1\rangle$ yields the post-channel state

$$\mathcal{K}(|1\rangle\langle 1|) = \begin{bmatrix} \mu & 0 \\ 0 & 1-\mu \end{bmatrix} = \mu|0\rangle\langle 0| + (1-\mu)|1\rangle\langle 1|. \tag{11.95}$$

Therefore, the probability associated with the "on" state $|1\rangle$ is reduced ("damped") to $1 - \mu$, while that of the "off" state $|0\rangle$ is increased to μ. Note the similarity between the Kraus matrices (11.94) and the Kraus matrices (11.91) for the dephasing channel, which is thus also known as the **phase-damping** channel.

Like the dephasing channel, the amplitude-damping channel induces decoherence, as the post-channel state for a general density matrix ρ is given by

$$\mathcal{K}(\rho) = \begin{bmatrix} \rho_{00} + \mu\rho_{1,1} & \sqrt{1-\mu}\rho_{0,1} \\ \sqrt{1-\mu}\rho_{0,1}^{*} & (1-\mu)\rho_{1,1} \end{bmatrix}, \tag{11.96}$$

which is fully incoherent, coinciding with the state $|0\rangle\langle 0|$, for $\mu = 1$.

The amplitude-damping channel is not unital. In fact, for a pre-channel maximally mixed state π, the post-channel state is

$$\mathcal{K}(\pi) = \frac{1}{2}\begin{bmatrix} 1+\mu & 0 \\ 0 & 1-\mu \end{bmatrix}, \tag{11.97}$$

which equals π if and only if the amplitude-damping probability is zero, $\mu = 0$.

11.6.4 Equivalent Characterizations of Quantum Channels ⏭

So far, two equivalent characterizations were given of a quantum channel. The first characterization views a quantum channel as a non-selective measurement applied on the environment. This representation is the counterpart of the Naimark dilation for POVMs, and is known as the **Stinespring dilation**. The second characterization is given by the Kraus representation (11.80).

Apart from the Stinespring and Kraus representations, quantum channels can also be described as a **completely positive trace-preserving (CPTP)** transformation, or via a density matrix known as the **Choi matrix**. Referring to Appendix 11.D for details on the Choi matrix representation, the CPTP characterization is described as follows.

Key Result 11.7 Quantum channels are completely positive trace-preserving (CPTP) linear maps

A mapping $\mathcal{K}(\cdot)$ between $N \times N$ matrices is a quantum channel if and only if it is as follows.

- **Linear:** Given any two density matrices ρ_0 and ρ_1, and any probability $0 \leq \gamma \leq 1$, the post-channel state for the mixture state $\rho = \gamma\rho_0 + (1-\gamma)\rho_1$ equals

$$\mathcal{K}(\rho) = \gamma\mathcal{K}(\rho_0) + (1-\gamma)\mathcal{K}(\rho_1). \tag{11.98}$$

- **Completely positive:** It transforms a positive semi-definite matrix into a positive semi-definite matrix even when applied only to a subsystem.
- **Trace preserving:** It preserves the trace in the sense that the equality

$$\mathrm{tr}(\mathcal{K}(A)) = \mathrm{tr}(A) \tag{11.99}$$

holds for any $N \times N$ matrix A.

This result implies that any CPTP transformation $\mathcal{K}(\cdot)$ can be expressed as in (11.81) for a set of M Kraus matrices. In this regard, as discussed in Appendix 11.D, the number of Kraus matrices required to describe a CPTP transformation can be shown to be no larger than N^2, i.e.,

$$M \leq N^2. \tag{11.100}$$

The minimum number M of Kraus matrices needed to describe a quantum channel can be considered a measure of the level of noise added by the channel to the post-channel state. By the Stinespring dilation, it also correspondingly quantifies the complexity of the channel. In fact, it accounts for the amount of auxiliary quantum resources – represented by the dimension of the environment – needed to implement it.

Example 11.22

The depolarizing channel (11.89) with $\gamma \neq 0$ is an example of a quantum channel with a maximum number of Kraus operators, i.e., $M = N^2 = 4$. This reflects the large level of noise associated with the application of a depolarizing channel. In contrast, the dephasing, bit flip, and amplitude damping channels require $M = 2 < N^2$ Kraus matrices.

Generalizing the special case of unitary transformations studied in Section 10.5, the linearity property (11.98) can be conveniently expressed, and verified, by using the formalism of vectorized density matrices. In particular, thanks to the vectorization property (10.71) of the product of three matrices, as well as to the Kraus representation (11.80), the vectorized version of the post-channel density matrix can be written as the matrix–vector product

$$\begin{aligned} |\mathcal{K}(\rho)\rangle\rangle &= \sum_{x \in [M]} \text{vec}\left(K_x \rho K_x^\dagger\right) \\ &= \underbrace{\sum_{x \in [M]} (K_x \otimes K_x^*)}_{K} |\rho\rangle\rangle, \end{aligned} \tag{11.101}$$

where K is an $N^2 \times N^2$ matrix that is fully described by the Kraus matrices $\mathcal{K}$.

Representing the vectorized density matrix $|\rho\rangle\rangle$ using different bases yields distinct matrix representations of a quantum channel of the form (11.101). In particular, using the Pauli basis (see Section 8.9.2), the resulting matrix K is known as the **Pauli transfer matrix** for the given quantum channel. This representation of a quantum channel is particularly convenient for Pauli channels, which can be shown to have diagonal Pauli transfer matrices.

11.6.5 Reversibility and Formal Invertibility ⏭

As seen in Chapter 10, a unitary transformation $\mathcal{K}(\cdot) = U(\cdot)U^\dagger$, where U is a unitary matrix, is reversible. In fact, as illustrated in Figure 11.10, given the output $\mathcal{K}(\rho) = U\rho U^\dagger$ of a unitary quantum channel, the **reverse quantum channel** $\mathcal{J}(\cdot) = U^\dagger(\cdot)U$ recovers any initial input state ρ as

$$\mathcal{J}(\mathcal{K}(\rho)) = U^\dagger U \rho U^\dagger U = \rho. \tag{11.102}$$

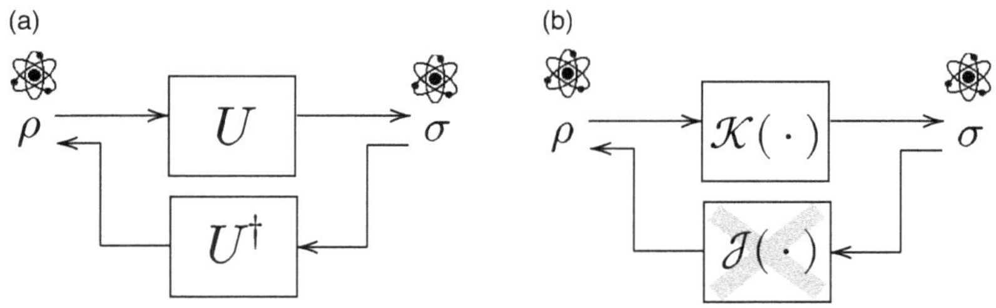

Figure 11.10 While unitary transformations are reversible (a), more general quantum channels are not (b). This means that, for more general quantum channels $\mathcal{K}(\rho)$, there is no reverse quantum channel $\mathcal{J}(\cdot)$ that satisfies the equality $\mathcal{J}(\mathcal{K}(\rho)) = \rho$ for all pre-channel states ρ.

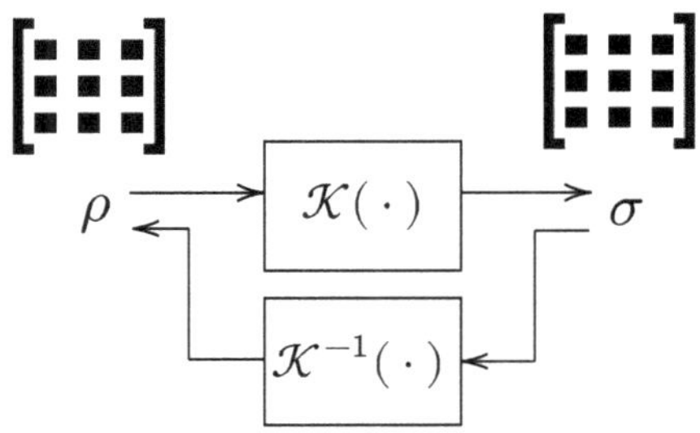

Figure 11.11 A quantum channel $\mathcal{K}(\cdot)$ can be formally inverted if the density matrix ρ can be computed from the post-channel density matrix $\sigma = \mathcal{K}(\rho)$ for all pre-channel states ρ.

In contrast, for a non-unitary quantum channel $\mathcal{K}(\cdot)$, there is no reverse quantum channel $\mathcal{J}(\cdot)$ that satisfies the equality $\mathcal{J}(\mathcal{K}(\rho)) = \rho$ for all pre-channel states ρ. In other words, non-unitary quantum channels are irreversible.

Reversibility thus refers to the possibility of implementing a physical transformation that undoes the operation of a quantum channel on a quantum system. As shown in Figure 11.10, this physical transformation applies directly on the quantum system (represented as an atom in the figure).

A less restrictive notion is that of **formal invertibility** of a channel, which refers to the feasibility, at least in principle, of computing the pre-channel state from the post-channel state. Unlike reversibility, which considers as input the post-channel quantum system, invertibility takes as the input a classical description of the actual quantum state, i.e., of the density matrix.

Accordingly, a quantum channel $\mathcal{K}(\cdot)$ is said to be formally invertible if one can evaluate, using a classical computer, the density matrix ρ from the post-channel density matrix $\sigma = \mathcal{K}(\rho)$ for all pre-channel states ρ. This is illustrated in Figure 11.11, in which the quantum systems in Figure 11.10 are replaced by classical descriptions of the quantum states.

For a classical analogy, consider the following problem. Given a channel matrix, i.e., a column-stochastic matrix K, compute the input distribution p from the output distribution $q = Kp$. This is possible if the channel matrix K is invertible, in which case one obtains

$$p = K^{-1}q. \tag{11.103}$$

Importantly, the inverse K^{-1} is generally not a channel matrix, and thus the transformation (11.103) does not correspond to a channel, that is, to a stochastic transformation of a classical system. Rather, it describes a computation that can be evaluated on a classical computer by manipulating vector q and matrix K.

Following this analogy, in a quantum setting, consider the matrix–vector representation (11.101) of the target quantum channel $\mathcal{K}(\cdot)$. Accordingly, the vectorized post-channel density matrix σ is expressed as

$$|\sigma\rangle\rangle = K|\rho\rangle\rangle, \tag{11.104}$$

where K is the $N^2 \times N^2$ matrix defined from the Kraus operators as in (11.101) and ρ is the pre-channel density matrix. When the underlying POVM $\mathcal{K}$ is informationally complete (see Section 11.5.2), the channel matrix K is formally invertible. The reader is referred to the Recommended Resources for a proof.

Therefore, in a manner analogous to the classical case (11.103), for a quantum channel with an informationally complete set of Kraus matrices, the pre-channel density matrix $|\rho\rangle\rangle$ can be evaluated from a classical description of the post-channel density matrix $|\sigma\rangle\rangle$ as

$$|\rho\rangle\rangle = K^{-1}|\sigma\rangle\rangle. \tag{11.105}$$

Rewriting (11.105) in terms of the corresponding density matrices, it is also common to use the notation

$$\rho = \mathcal{K}^{-1}(\sigma). \tag{11.106}$$

However, the transformation $\mathcal{K}^{-1}(\cdot)$ does not define a quantum channel, i.e., it cannot be expressed as in (11.81) via any set of Kraus matrices. Rather, the inverse matrix K^{-1}, and the corresponding operation $\mathcal{K}^{-1}(\cdot)$, describe mathematical operations that can be carried out on a classical computer, as long as one can store and operate on $N^2 \times N^2$ matrices.

Example 11.23

Consider the POVM setting defined by POVM elements

$$\mathcal{K} = \left\{ \frac{1}{3}|0\rangle\langle 0|, \frac{1}{3}|1\rangle\langle 1|, \frac{1}{3}|+\rangle\langle +|, \frac{1}{3}|-\rangle\langle -|, \frac{1}{3}|+i\rangle\langle +i|, \frac{1}{3}|-i\rangle\langle -i| \right\}. \tag{11.107}$$

This is an informationally complete POVM, which can be implemented, in terms of its classical output, by carrying out complete measurements in the bases $\mathcal{M}_1 = \{|0\rangle, |1\rangle\}$, $\mathcal{M}_2 = \{|+i\rangle, |-i\rangle\}$, or $\mathcal{M}_3 = \{|+\rangle, |-\rangle\}$ with equal probability (see Example 11.13).

The corresponding quantum channel $\mathcal{K}(\cdot)$ is not reversible, since it is not unitary. Thus, one cannot restore the initial state ρ from the post-channel state $\sigma = \mathcal{K}(\rho)$ by operating on the quantum state σ via a quantum channel. However, the channel can be formally inverted on a classical computer.

In particular, given a classical description of the density matrix σ, the pre-channel density matrix ρ can be computed via the formal inverse operation

$$\rho = \mathcal{K}^{-1}(\sigma) = 3\sigma - I, \tag{11.108}$$

where I is the 2×2 identity matrix. Note that the mapping $\mathcal{K}^{-1}(\sigma)$ is not a quantum channel, since it is not positive. However, it can clearly be evaluated on a classical computer.

The formal inverse mapping $\mathcal{K}^{-1}(\cdot)$ plays a central role in the framework of **classical shadows**, which aims to construct classical representations of a density matrix. Following this approach, one applies the randomized-measurement POVM (11.107) to multiple copies of the state of interest as in quantum tomography (see Section 11.5.2). Then, using the classical measurement outputs, an empirical estimate of the corresponding density matrix is obtained as in (11.82). Finally, the inverse map $\mathcal{K}^{-1}(\cdot)$ is applied to evaluate, on a classical computer, an estimate $\hat{\rho}$, known as a classical shadow, of the original density matrix ρ. This matrix $\hat{\rho}$ can finally be used to estimate the expectation of a class of observables O as $\mathrm{tr}(\hat{\rho}O)$. The reader is referred to the Recommended Resources for further discussion.

11.7 Information Dynamics through Quantum Channels

As seen so far in this chapter, random unitary transformations cannot decrease the quantum entropy, with deterministic unitary transformations leaving the quantum entropy unchanged (see Section 11.2.2). It was also shown in Section 10.6 that unitary transformations do not change the quantum relative entropy. This section extends these conclusions by demonstrating that: (i) general quantum channels can increase or decrease the quantum entropy; (ii) unital channels, including random unitary transformations as a special case, cannot decrease the quantum entropy; and (iii) the quantum relative entropy cannot be increased as a result of the application of a quantum channel, a result known as the quantum data processing inequality. For proofs of the results in this section, the reader is referred to the Recommended Resources.

11.7.1 Quantum Entropy through Quantum Channels

To start, the following example shows that, in a manner similar to the classical case studied in Section 9.5, the quantum entropy of a post-channel state can be smaller than the quantum entropy of the pre-channel state.

Example 11.24

Consider the channel described by the Kraus matrices $K_x = |0\rangle\langle x|$ for $x \in [N]$. This is a valid channel as it satisfies the condition (11.43), as

$$\sum_{x\in[N]} K_x^\dagger K_x = \sum_{x\in[N]} |x\rangle\langle x| = I, \tag{11.109}$$

where I is the $N \times N$ identity matrix. This channel replaces any pre-measurement state ρ with the pure state $|0\rangle$, i.e.,

$$\mathcal{K}(\rho) = |0\rangle\langle 0|. \tag{11.110}$$

Therefore, the quantum entropy of the post-measurement state is zero, $\mathrm{H}_\alpha(\mathcal{K}(\rho)) = 0$, irrespective of the quantum entropy $\mathrm{H}_\alpha(\rho)$ of the pre-measurement state ρ.

11.7.2 Quantum Entropy Increases through Unital Channels

While generic quantum channels may increase or decrease the quantum entropy, unital channels, which generalize random unitary transformations, cannot decrease the quantum entropy.

Key Result 11.8 Unital channels increase the quantum entropy
Given any $N \times N$ density matrix ρ and any unital channel $\mathcal{K}(\cdot)$ operating on $N \times N$ density matrices, the equality

$$\mathrm{H}_\alpha(\rho) \leq \mathrm{H}_\alpha(\mathcal{K}(\rho)) \tag{11.111}$$

is verified.

Based on this property, unital channels can be taken as a starting point for the definition of quantum uncertainty measures. Accordingly, one could define a function of a density matrix as a valid quantum uncertainty measure if it does not increase under the action of a unital channel, as per the inequality (11.111). Please consult the Recommended Resources for further discussion.

11.7.3 Quantum Data Processing Inequality

In a way analogous to classical channels, as illustrated in Figure 11.12, the relative information between two states cannot increase as the result of the application of a quantum channel.

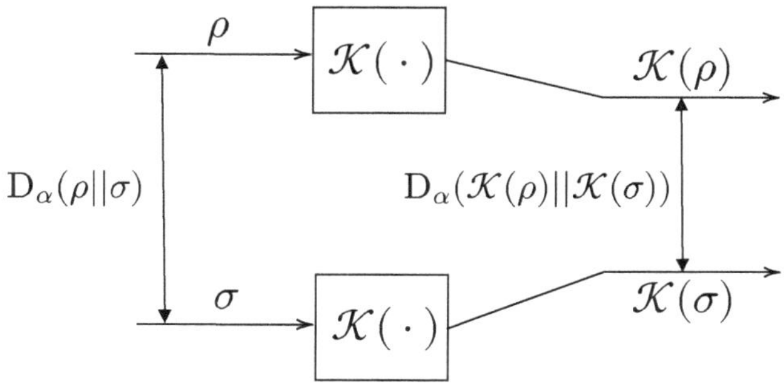

Figure 11.12 Illustration of the quantum data processing inequality.

> **Key Result 11.9 Quantum data processing inequality**
> Given any two $N \times N$ densities ρ and σ and any quantum channel $\mathcal{K}(\cdot)$ operating on $N \times N$ density matrices, the inequality
>
> $$\mathrm{D}_\alpha(\mathcal{K}(\rho)||\mathcal{K}(\sigma)) \leq \mathrm{D}_\alpha(\rho||\sigma) \tag{11.112}$$
>
> holds for the Shannon quantum relative entropy, as well as for the Tsallis and Rényi quantum relative entropies with $0 \leq \alpha \leq 1$. Furthermore, equality in (11.112) holds if the channel is a unitary transformation.

Unlike the classical case studied in Section 9.5, the quantum data processing inequality thus does not necessarily apply for all values of the order $\alpha \geq 0$.

11.8 Quantum Instruments ⏭

This chapter has made a distinction between the following two operations.

- **POVM measurements:** POVM measurements output a classical random variable x and leave the quantum system in the corresponding post-measurement state ρ_{x}.
- **Quantum channels:** Quantum channels modify the state of the system according to a deterministic transformation $\mathcal{K}(\cdot)$.

POVMs include complete and projective measurements as special cases, constituting the most general form of stochastic transformation of a quantum state investigated up to now. Conversely, quantum channels provide the most general form of deterministic transformation that may be applied to a quantum state.

This final section of the chapter describes quantum instruments, which bridge the gap between POVMs and quantum channels. They describe a more general type of stochastic transformation than POVMs, encompassing both POVMs and quantum channels as special cases. As detailed next, quantum instruments can be thought of as partially selective POVM measurements, with conventional

POVMs representing the selective extreme and quantum channels lying at the other, non-selective, end.

11.8.1 Defining Quantum Instruments

Consider a POVM defined by Kraus matrices $\mathcal{K} = \{K_x\}_{x\in[M]}$, and partition the set of possible output values $[M]$ into $M^{\mathcal{I}} \leq M$ disjoint subsets $\mathcal{X}_y \subseteq [M]$ with $y \in [M^{\mathcal{I}}]$ and

$$[M] = \bigcup_{y\in[M^{\mathcal{I}}]} \mathcal{X}_y. \tag{11.113}$$

Definition 11.16 Quantum instrument setting
A quantum instrument

$$\mathcal{I} = \{\mathcal{K}, \{\mathcal{X}_y\}_{y\in[M^{\mathcal{I}}]}\} \tag{11.114}$$

is described by an underlying POVM $\mathcal{K}$ with M Kraus matrices and by a partition $\{\mathcal{X}_y\}_{y\in[M^{\mathcal{I}}]}$ of the set $[M]$ into $M^{\mathcal{I}} \leq M$ subsets.

The quantum instrument $\mathcal{I}$ operates as follows.

1. **POVM:** The quantum instrument applies the POVM in setting $\mathcal{K}$, obtaining the corresponding classical output $\mathrm{x} = x \in [M]$.
2. **Forgetting:** The quantum instrument records only the index $y \in [M^{\mathcal{I}}]$ of the subset $\mathcal{X}_y$ that includes the classical output x, returning the classical output

$$\mathrm{y} \in [M^{\mathcal{I}}]\colon \mathrm{x} \in \mathcal{X}_{\mathrm{y}}. \tag{11.115}$$

This way, a quantum instrument $\mathcal{I}$ generalizes both POVMs and quantum channels. The POVM $\mathcal{K}$ is recovered when $M^{\mathcal{I}} = M$, and thus when each subset $\mathcal{X}_y = \{y\}$ includes a single output $y \in [N]$ and no forgetting is applied. The quantum channel $\mathcal{K}(\cdot)$ is instead obtained from the quantum instrument $\mathcal{I}$ when full forgetting of the classical output is applied, i.e., when $M^{\mathcal{I}} = 1$ and hence there is a single set $\mathcal{X}_0 = [M]$.

More generally, a quantum instrument forgets only partially the classical output x of the POVM, keeping information about the classical output x only at the resolution determined by the partition $\{\mathcal{X}_y\}_{y\in[M^{\mathcal{I}}]}$.

To describe mathematically the operation of a quantum instrument $\mathcal{I}$, suppose that the underlying POVM $\mathcal{K}$ returns output $x \in \mathcal{X}_y$. The quantum instrument forgets the specific measurement $x \in \mathcal{X}_y$, retaining only the index y of the subset $\mathcal{X}_y$. Using (11.49), the probability that output y is recorded is hence given by

$$p_y(\rho|\mathcal{I}) = \sum_{x\in\mathcal{X}_y} p_x(\rho|\mathcal{K})$$

$$= \sum_{x\in\mathcal{X}_y} \mathrm{tr}(K_x^\dagger K_x \rho)$$

$$= \mathrm{tr}\left(\left(\sum_{x\in\mathcal{X}_y} K_x^\dagger K_x\right)\rho\right). \tag{11.116}$$

Furthermore, with probability $p_y(\rho|\mathcal{I})$, the post-instrument state is obtained by averaging the post-POVM states (11.50) corresponding to the outputs in subset $\mathcal{X}_y$ as

$$\rho_y = \frac{1}{p_y(\rho|\mathcal{I})} \sum_{x\in\mathcal{X}_y} p_x(\rho|\mathcal{K}) \frac{K_x \rho K_x^\dagger}{p_x(\rho|\mathcal{K})}$$

$$= \frac{1}{p_y(\rho|\mathcal{I})} \sum_{x\in\mathcal{X}_y} K_x \rho K_x^\dagger. \tag{11.117}$$

In fact, given the output y, the post-instrument state is given by the post-POVM state $K_x\rho K_x^\dagger / p_x(\rho|\mathcal{K})$ in (11.50) with probability $p_x(\rho|\mathcal{K})/p_y(\rho|\mathcal{I})$.

These results are summarized as follows.

Definition 11.17 Quantum instrument

Given a quantum state ρ, a quantum instrument $\mathcal{I} = \{\mathcal{K}, \{\mathcal{X}_y\}_{y\in[M^{\mathcal{I}}]}\}$ results in the following.

- **Classical output:** The measurement returns output $y_{\mathrm{y}=y} \in [M^{\mathcal{I}}]$ with probability

$$p_y(\rho|\mathcal{I}) = \mathrm{tr}\left(\left(\sum_{x\in\mathcal{X}_y} K_x^\dagger K_x\right)\rho\right). \tag{11.118}$$

- **Post-measurement state:** For a classical output $\mathrm{y} = y$, the quantum system is left in the corresponding post-measurement state

$$\rho_y = \frac{1}{p_y(\rho|\mathcal{I})} \sum_{x\in\mathcal{X}_y} K_x \rho K_x^\dagger. \tag{11.119}$$

As anticipated, one can readily check that for $M^{\mathcal{I}} = M$, the quantum instrument, described by (11.118) and (11.119), recovers the POVM $\mathcal{K}$; while for $M^{\mathcal{I}} = 1$ it reduces to the quantum channel $\mathcal{K}(\cdot)$.

To conclude this section, it is useful to compare the role of the partition of the outputs $[M]$ of a POVM used in quantum instruments with the partition of the outputs $[N]$ of a complete measurement applied by projective measurements (see Section 6.5). In both cases, larger sets $\mathcal{X}_y$ in the partition entail a lower level of informativeness of the classical outputs, whose values range over smaller sets. However, the significance of the partition in the two cases is very different when viewed from the perspective of the disturbance caused to the original state.

In fact, the partition used in the definition of projective measurements describes a coarse-graining of the measurement as compared to a complete measurement setting, with larger partition sets describing gentler measurements. With projective measurements, the choice of the rank of the projective matrices, corresponding to the sizes of the subsets $\mathcal{X}_y$, determines a trade-off between information and disturbance. In contrast, the partition used in quantum instruments does not change the underlying POVM measurement, and it does not affect the level of disturbance caused to the input state. Rather, it only determines the degree to which information is lost *after* the measurement is applied.

Example 11.25

Projective measurements vs. quantum instruments. Consider a system with $N = 4$ dimensions and fix as the underlying measurement setting the complete measurement in the computational basis $\mathcal{M} = \{|0\rangle, |1\rangle, |2\rangle, |3\rangle\}$. This example compares a projective measurement and a quantum instrument that are built from the same underlying measurement setting $\mathcal{M}$ and from the same partition $\{\mathcal{X}_0 = \{0,1\}, \mathcal{X}_1 = \{2,3\}\}$ into $M^{\mathcal{I}} = 2$ subsets. Denote, as usual, the initial state as the density matrix ρ with entries $\rho_{x,y}$ with $x, y \in [N]$.

The projective measurement corresponding to this partition is characterized by the setting $\mathcal{P} = \{\Pi_0, \Pi_1\}$, with projection matrices $\Pi_0 = |0\rangle\langle 0| + |1\rangle\langle 1|$ and $\Pi_1 = |2\rangle\langle 2| + |3\rangle\langle 3|$. Accordingly, for a given density matrix ρ, this measurement returns output $\mathrm{y} = y$ with probability $p_y(\rho|\mathcal{P}) = \mathrm{tr}(\rho\Pi_y)$. Furthermore, taking as an example the output $y = 0$, the post-measurement state is

$$\frac{\Pi_0\rho\Pi_0}{\mathrm{tr}(\rho\Pi_0)} = \frac{1}{\mathrm{tr}(\rho\Pi_0)}\begin{bmatrix} \rho_{0,0} & \rho_{0,1} & 0 & 0 \\ \rho_{1,0} & \rho_{1,1} & 0 & 0 \\ 0 & 0 & 0 & 0 \\ 0 & 0 & 0 & 0 \end{bmatrix}. \tag{11.120}$$

As also discussed in Chapter 8, this measurement is gentler than a complete measurement, preserving some of the coherence with respect to the computational basis via the off-diagonal elements $\rho_{0,1}$ and $\rho_{1,0}$.

For the same partition, the quantum instrument $\mathcal{I} = \{\mathcal{M}, \{\mathcal{X}_0, \mathcal{X}_1\}\}$ applies the complete measurement $\mathcal{M}$, and then only retains the information about whether the classical output is in the subset $\{0,1\}$, in which case it outputs $\mathrm{y} = 0$, or else in the subset $\{2,3\}$, in which case it outputs $\mathrm{y} = 1$. Thus, using (11.118), the probability of producing output $\mathrm{y} = y$ coincides with that of the projective measurement, i.e.,

$$p_y(\rho|\mathcal{I}) = p_y(\rho|\mathcal{P}). \tag{11.121}$$

However, the quantum instrument is not gentle. For example, when the output is $y = 0$, using (11.119), the post-measurement state is

$$\frac{1}{p_0(\rho|\mathcal{I})}\sum_{x\in[1]} |x\rangle\langle x|\rho|x\rangle\langle x| = \begin{bmatrix} \rho_{0,0} & 0 & 0 & 0 \\ 0 & \rho_{1,1} & 0 & 0 \\ 0 & 0 & 0 & 0 \\ 0 & 0 & 0 & 0 \end{bmatrix}, \tag{11.122}$$

which is incoherent with respect to the computational basis.

11.9 Summary

- Open quantum systems support the definition of dynamic transformations that generalize unitary transformations, known as quantum channels, while also enabling POVM measurements, which generalize projective measurements.
- Quantum channels can be described as a non-selective version of POVMs, while partial selectivity yields the framework of quantum instruments, which subsume both quantum channels and POVMs.
- POVMs are described by a set $\mathcal{K} = \{K_x\}_{x\in[M]}$ of Kraus matrices that satisfy the equality $\sum_{x\in[M]} K_x^\dagger K_x = I$, where the number of outputs, M, can possibly be larger than N.
- Given a POVM setting $\mathcal{K}$ and a pre-measurement state ρ, the measurement returns the classical output $\mathrm{x} = x \in [M]$ with measurement probability $p_x(\rho|\mathcal{K}) = \mathrm{tr}(K_x^\dagger K_x \rho)$, and the corresponding post-measurement state is $\sigma_x = K_x \rho K_x^\dagger / p_x(\rho|\mathcal{K})$.
- When one is only interested in the classical output, and not in the post-measurement state, it is common to specify POVM by listing the M positive semi-definite matrices $\{J_x = K_x^\dagger K_x\}_{x\in[M]}$, which are also referred to as POVM elements.
- With this description of the POVM, the measurement probabilities are written as $p_x(\rho|\mathcal{K}) = \mathrm{tr}(J_x \rho)$ for $x \in [M]$.
- POVMs can solve informational tasks for which projective measurements are insufficient, such as unambiguous state detection and quantum state tomography.
- A quantum channel described by the set $\mathcal{K}$ of Kraus matrices produces the quantum state $\mathcal{K}(\rho) = \sum_{x\in[M]} K_x \rho K_x^\dagger$.
- Quantum channels are completely positive trace-preserving (CPTP) maps, and they are generally irreversible.
- Quantum channels can increase or decrease the quantum entropy. However, in the special case of unital channels, the quantum entropy can only increase, or at least not decrease.
- The quantum relative entropy, with order $0 \leq \alpha \leq 1$, satisfies the quantum data processing inequality $\mathrm{D}_\alpha(\mathcal{K}(\rho)||\mathcal{K}(\sigma)) \leq \mathrm{D}_\alpha(\rho||\sigma)$.

11.10 Recommended Resources

An extensive discussion on the application of majorization to the behavior of quantum systems is provided in [1]. Randomized measurements are thoroughly elaborated in [2], while further insights into informationally complete measurements are available in [3, 4]. Implications of the classical operations on POVMs of post-processing via coarse-graining and mixing via randomized measurements are elaborated on in [5, 6] and references therein. In particular, reference [7] shows that by allowing for post-selection, any POVM can be simulated using projective measurements along with randomization and post-processing, although the success probability decreases with the dimension of the vector space.

Additional information on pinching channels and random unitary transformations can be found in [8]. Quantum channels and quantum error correction are discussed in the standard textbooks [9, 10], as well as in the review article [11]. For more details on unital channels, refer to [8, Sec. 4.1.1] and [12].

The Choi matrix is introduced in [13], while mappings between channels, known as super-channels or quantum combs, are detailed in [14]. The Pauli transfer matrix is explained in [15]. A proof demonstrating that informationally complete channels can be formally inverted is presented in [16], and the concept of classical shadows is covered in [17].

For a proof that quantum entropy cannot be increased by a unital channel, see [9, Theorem 4.1.1]. The derivation of entropy measures based on the monotonicity under the application of unital channels is discussed in [18]. Extensive coverage of the quantum data processing inequality, including proofs of the relevant claims presented in this chapter, can be found in [19–21]. For more information on quantum instruments, consult [22].

PROBLEMS

11.1 Consider environment and system with dimensions $N^E = 2$ and $N^X = 2$. The state of the composite system encompassing both subsystems is described by the 4×1 vector $|\psi^{EX}\rangle = 1/\sqrt{2}[1,0,0,1]$. (a) Show that this is a valid quantum state. (b) Prove that it cannot be written as the Kronecker product of two 2×1 vector states, i.e., as $|\psi\rangle = |\psi^E\rangle \otimes |\psi^X\rangle$ for some 2×1 state vectors $|\psi^E\rangle$ and $|\psi^X\rangle$.

11.2 Given the POVM $\mathcal{K}$ in (11.107), compute the measurement probability vector $p(\rho|\mathcal{K})$ for the following pre-measurement states: (a) $\rho = 0.2|0\rangle\langle 0| + | + 0.8|1\rangle\langle 1|$ and (b) $\rho = 0.2|+\rangle\langle +| + 0.8|-\rangle\langle -|$.

11.3 For the quantum channel defined by the POVM $\mathcal{K}$ in (11.107), compute the post-channel density matrix $\mathcal{K}(\rho)$ for the pre-channel states: (a) $\rho = 0.2|0\rangle\langle 0| + | + 0.8|1\rangle\langle 1|$ and (b) $\rho = 0.2|+\rangle\langle +| + 0.8|-\rangle\langle -|$.

11.4 Consider any two complete quantum measurement settings $\mathcal{M}_0$ and $\mathcal{M}_1$ implemented one after the other on a given system in a pure state $|\psi\rangle$. The output of the first measurement is discarded. (a) Derive the probability measurement vector after the second measurement. Note that the measurement probabilities need to be averaged over the discarded output of the first measurement. Express your result by using the compatibility matrix of the two measurement settings. (b) Show that, if the two measurement settings are compatible, the entropy of the classical output of the first measurement equals the entropy of the classical output of the second measurement (when the latter is evaluated after discarding the output of the first measurement).

11.5 Prove that the quantum data processing inequality for the Rényi quantum relative entropy with any order $0 \leq \alpha \leq 1$ implies the fact that unital channels cannot decrease the Rényi quantum entropy of order α. To this end, start by writing the Rényi quantum entropy $\mathrm{H}_\alpha^{\mathrm{R}}(\rho)$ by using the decomposition discussed in Chapter 8 in terms of the Rényi quantum entropy $\mathrm{D}_\alpha^{\mathrm{R}}(\rho||\pi)$. Then, use the quantum data processing inequality and the definition of unital channel.

11.6 (a) Given L POVMs, each with M Kraus matrices, construct a single POVM with ML Kraus matrices by suitably rescaling the Kraus matrices in the original POVMs. (b) Write the ML measurement probabilities associated with the obtained POVM as a function of the measurement probabilities of the original POVMs.

11.7 This problem shows that a POVM defined by POVM elements $\mathcal{K} = \{J_x\}_{x \in [N]}$ can always be implemented via a two-step process involving a POVM with rank-1 POVM elements in the first step. Specifically, in the first step, a rank-1 POVM with rank-1 POVM elements $\{J'_z\}_{z \in [M']}$ is applied, where M' is a suitable integer, and then a coarse-graining of the first POVM is implemented, yielding the POVM elements $J_x = \sum_{z \in [M']} p_{z \to x} J'_z$, where $p_{z \to x}$ represents a channel with input z and output x. To prove this, write each rank-1 matrix J'_y as $J'_z = \lambda_z |v_z\rangle\langle v_z|$ for some positive number λ_z and state $|v_z\rangle$ by leveraging the eigendecompositions of the original POVM elements $\{J_x\}_{x \in [N]}$.

11.8 For a qubit system, consider the informationally complete POVM given by POVM elements $J_x = v_x v_x^\dagger$ with

$$\begin{aligned} v_0 &= \frac{1}{\sqrt{2}}|0\rangle, \\ v_1 &= \frac{1}{\sqrt{6}}|0\rangle + \frac{1}{\sqrt{3}}|1\rangle, \\ v_2 &= \frac{1}{\sqrt{6}}|0\rangle + \frac{1}{\sqrt{3}}\exp\left(i\frac{2\pi}{3}\right)|1\rangle, \\ v_3 &= \frac{1}{\sqrt{6}}|0\rangle + \frac{1}{\sqrt{3}}\exp\left(-i\frac{2\pi}{3}\right)|1\rangle. \end{aligned} \tag{11.123}$$

(a) Show that the setting $\mathcal{K} = \{J_x\}_{x \in [4]}$ defines a valid POVM. (b) Calculate the trace inner product $\mathrm{tr}(J_x J_y)$ for all $x = y$ and $x \in [4]$, as well as for $x = 0$ and $y = 3$.

11.9 This problem studies randomized measurements, showing how to describe them as POVMs (see Example 11.3). To this end, pre-select a set of L projective measurement settings $\mathcal{P}_z = \{\Pi_{x|z}\}_{x \in [M]}$ with $z \in [L]$. Assume that a measurement is carried out as follows. (1) A random variable $\mathrm{z} \sim p^{\mathrm{Z}}$ with $\mathrm{z} \in [L]$ is drawn to select one of the projective measurement settings $\mathcal{P}_z$ for $\mathrm{z} = z$. (2) The projective measurement $\mathcal{P}_z$ is applied to the pre-measurement state ρ, producing the measurement output $\mathrm{x} = x \in [M]$ and leaving the system in the post-measurement state $\sigma_{x|z} = \Pi_{x|z}\rho\Pi_{x|z}/p_{\mathrm{x}}(\rho|\mathcal{P}_z)$. (3) The classical output is produced as the pair (x, z) and the system is left in state $\sigma_{x|z}$. Show that this process can be described as a POVM by identifying the Kraus matrices.

Appendix

11.A Proof of the Uncertainty Inequality

This appendix proves the state-dependent uncertainty inequality (7.142) presented in Chapter 7, which implies the original uncertainty inequality (6.43) introduced in Chapter 6. To this end, consider two complete measurements $\mathcal{M}_0 = \{|v_x\rangle\}_{x\in[N]}$ and $\mathcal{M}_1 = \{|u_x\rangle\}_{x\in[N]}$. Recall from Problem 8.4 that the following relation holds between the Shannon measured entropy, the Shannon quantum entropy, and the Shannon quantum relative entropy:

$$\mathrm{H}(\rho|\mathcal{M}_0) - \mathrm{H}(\rho) = \mathrm{D}(\rho||\mathcal{M}_0(\rho)), \tag{11.124}$$

where $\mathcal{M}_0(\rho)$ indicates the post-measurement state for input state ρ under complete measurement setting $\mathcal{M}_0$. Using the quantum data processing inequality on the right-hand side of (11.124) yields

$$\begin{aligned}\mathrm{D}(\rho||\mathcal{M}_0(\rho)) \geq& \mathrm{D}(\mathcal{M}_1(\rho)||\mathcal{M}_1(\mathcal{M}_0(\rho)))\\ =& -\mathrm{H}(\mathcal{M}_1(\rho)) - \mathrm{tr}(\mathcal{M}_1(\rho)\log_2(\mathcal{M}_1(\mathcal{M}_0(\rho)))).\end{aligned} \tag{11.125}$$

Now, the density matrix in (11.125) can be expressed as

$$\begin{aligned}\mathcal{M}_1(\mathcal{M}_0(\rho)) &= \sum_{y\in[N]} |u_y\rangle\langle u_y| \left(\sum_{x\in[N]} |v_x\rangle\langle v_x| p_x(\rho|\mathcal{M}_0)\right) |u_y\rangle\langle u_y|\\ &= \sum_{y\in[N]} |u_y\rangle \left(\sum_{x\in[N]} |\langle v_x|u_y\rangle|^2 p_x(\rho|\mathcal{M}_0)\right) \langle u_y|.\end{aligned} \tag{11.126}$$

Plugging this result into (11.125) gives the inequality

$$\begin{aligned}\mathrm{D}(\rho||\mathcal{M}_0(\rho)) \geq& -\mathrm{H}(\mathcal{M}_1(\rho))\\ &- \mathrm{tr}\left(\sum_{y\in[N]} |u_y\rangle\langle u_y| \left(p_y(\rho|\mathcal{M}_1)\log_2\left(\sum_{x\in[N]} |\langle v_x|u_y\rangle|^2 p_x(\rho|\mathcal{M}_0)\right)\right)\right)\\ =& -\mathrm{H}(\mathcal{M}_1(\rho)) - \sum_{y\in[N]} p_y(\rho|\mathcal{M}_1)\log_2\left(\sum_{x\in[N]} |\langle v_x|u_y\rangle|^2 p_x(\rho|\mathcal{M}_0)\right)\\ \geq& -\mathrm{H}(\mathcal{M}_1(\rho)) - \log_2(\mathrm{c}(\mathcal{M}_0,\mathcal{M}_1)).\end{aligned} \tag{11.127}$$

The last inequality uses the definition of compatibility between measurement settings given in Chapter 6, i.e., $\mathrm{c}(\mathcal{M}_0,\mathcal{M}_1) = \max_{x,y\in[N]} |\langle v_x|u_y\rangle|^2$, as well as the fact that measurement probabilities sum to 1.

Plugging the last inequality into (11.124), along with the equality $\mathrm{H}(\mathcal{M}_1(\rho)) = \mathrm{H}(\rho|\mathcal{M}_1)$, produces the inequality

$$\mathrm{H}(\rho|\mathcal{M}_0) - \mathrm{H}(\rho) \geq -\mathrm{H}(\rho|\mathcal{M}_1) - \log_2(\mathrm{c}(\mathcal{M}_0,\mathcal{M}_1)), \tag{11.128}$$

which, upon rearranging the terms, coincides with the desired uncertainty inequality

$$\mathrm{H}(\rho|\mathcal{M}_0) + \mathrm{H}(\rho|\mathcal{M}_1) \geq \log_2\left(\frac{1}{\mathrm{c}(\mathcal{M}_0,\mathcal{M}_1)}\right) + \mathrm{H}(\rho). \tag{11.129}$$

11.B Proof of Uhlmann's Theorem

To prove Uhlmann's theorem, this appendix starts by showing how to construct matrices $\{U_m\}_{m\in[M]}$ and probability distribution $\{\gamma_m\}_{m\in[M]}$ for any two matrices ρ and σ satisfying the relation (11.9). From Section 9.3, the relation (11.9) implies that there is a doubly stochastic transformation $T = \sum_{m\in[M]} \gamma_m S_m$ with permutation matrices $\{S_m\}_{m\in[M]}$ and probability distribution $\{\gamma_m\}_{m\in[M]}$ such that the equality

$$\lambda(\sigma) = \sum_{m\in[M]} \gamma_m S_m \lambda(\rho) \tag{11.130}$$

holds. This, in turn, yields the equality

$$\Lambda(\sigma) = \sum_{m\in[M]} \gamma_m S_m \Lambda(\rho) S_m^\dagger, \tag{11.131}$$

where $\Lambda(\rho)$ and $\Lambda(\sigma)$ are $N \times N$ diagonal matrices with elements equal to the eigenvalues $\lambda(\rho)$ and $\lambda(\sigma)$, respectively.

Now, write the eigendecomposition of matrices ρ and σ, respectively, as $\rho = V\Lambda(\rho)V^\dagger$ and $\sigma = W\Lambda(\sigma)W^\dagger$ for unitary matrices V and W, collecting by columns the corresponding eigenvectors. Using (11.131), the density matrix σ can be written as

$$\begin{aligned}
\sigma &= W\Lambda(\sigma)W^\dagger \\
&= \sum_{m\in[M]} \gamma_m W S_m \Lambda(\rho) S_m^\dagger W^\dagger \\
&= \sum_{m\in[M]} \gamma_m W S_m V^\dagger \rho V S_m^\dagger W^\dagger \\
&= \sum_{m\in[M]} \gamma_m U_m \rho U_m^\dagger,
\end{aligned} \tag{11.132}$$

where the third equality follows from the equality $V^\dagger \rho V = \Lambda(\rho)$, and the last equality follows by defining the matrices $U_m = W S_m V^\dagger$. These matrices are indeed unitary, since they are products of unitary matrices.

As the last step, the proof of Uhlmann's theorem requires demonstrating that the existence of a random unitary transformation $\sigma = \sum_{m\in[M]} \gamma_m U_m \rho U_m^\dagger$, converting state ρ into state σ, implies the relation (11.9). From the eigendecompositions $\rho = V\Lambda(\rho)V^\dagger$ and $\sigma = W\Lambda(\sigma)W^\dagger$, one obtains

$$W\Lambda(\sigma)W^\dagger = \sum_{m\in[M]} \gamma_m U_m V \Lambda(\rho) V^\dagger U_m^\dagger, \tag{11.133}$$

which yields the equality

$$\Lambda(\sigma) = \sum_{m\in[M]} \gamma_m Q_m \Lambda(\rho) Q_m^\dagger, \tag{11.134}$$

with unitary matrices $Q_m = W^\dagger U_m V$.

Generalizing the equivalence between (11.130) and (11.131), the equality (11.134) can be rewritten in vector form as

$$\lambda(\sigma) = T\lambda(\rho), \tag{11.135}$$

with T being doubly stochastic. In fact, matrix T can be seen to have each entry (x, y) given by $T_{xy} = \sum_{m\in[M]} \gamma_m |Q_{xy|m}|^2$, with $Q_{xy|m}$ being the (x, y)th entry of matrix Q_m. This indicates that the desired condition $\lambda(\rho) \succ \lambda(\sigma)$ holds, concluding the proof.

11.C Proof of (11.46) and (11.48)

Using (11.35) and (11.45), the classical output y equals $y \in [M]$ with probability

$$\begin{aligned}
p_y(|\phi^{EX}\rangle|\mathcal{M}^E) &= \mathrm{tr}((|y^E\rangle\langle y^E| \otimes I^X)|\phi^{EX}\rangle\langle\phi^{EX}|) \\
&= \sum_{x\in[M]}\sum_{z\in[M]} \mathrm{tr}((|y^E\rangle\langle y^E| \otimes I^X)|x^E\rangle\langle z^E| \otimes K_x|\psi^X\rangle\langle\psi^X|K_z^\dagger) \\
&= \sum_{x\in[M]}\sum_{z\in[M]} \mathrm{tr}\left(|y^E\rangle\langle y^E|x^E\rangle\langle z^E| \otimes K_x|\psi^X\rangle\langle\psi^X|K_z^\dagger\right) \\
&= \sum_{z\in[M]} \mathrm{tr}\left(|y^E\rangle\langle z^E|\right) \mathrm{tr}\left(K_z|\psi^X\rangle\langle\psi^X|K_z^\dagger\right) \\
&= \mathrm{tr}\left(K_y|\psi^X\rangle\langle\psi^X|K_y^\dagger\right).
\end{aligned} \tag{11.136}$$

In a similar way, the post-measurement of the overall system is obtained from (11.37) and (11.45) as

$$\begin{aligned}
\sigma_y^{EX} &= \frac{1}{p_y(|\phi^{EX}\rangle|\mathcal{M}^E)}(|y^E\rangle\langle y^E| \otimes I^X)|\phi^{EX}\rangle\langle\phi^{EX}|(|y^E\rangle\langle y^E| \otimes I^X) \\
&= \frac{1}{p_y(|\phi^{EX}\rangle|\mathcal{M}^E)}|y^E\rangle\langle y^E| \otimes K_y|\psi^X\rangle\langle\psi^X|K_y^\dagger.
\end{aligned} \tag{11.137}$$

This is a separable state, and the state of the system is given by (11.48), since $|y^E\rangle\langle y^E|$ is the state of the environment.

11.D Choi Matrix Representation of Quantum Channels

Any $N \times N$ density matrix ρ can be expressed as the linear combination $\rho = \sum_{x\in[N]}\sum_{x\in[N]} \rho_{xy}|x\rangle\langle y|$, where ρ_{xy} is the (x, y)th entry of the matrix. In fact, the matrices $\{\{|x\rangle\langle y|\}_{x\in[N]}\}_{y\in[N]}$ form an orthonormal basis for the space of $N \times N$ matrices (see Section 7.13). Since quantum channels are linear, any channel $\mathcal{K}(\cdot)$ can be fully described by specifying its operation on all the N^2 matrices $|x\rangle\langle y|$, i.e., by specifying the N^2 matrices $\mathcal{K}(|x\rangle\langle y|)$. Note that, when $x \neq y$, matrix $|x\rangle\langle y|$ is not a valid density matrix, and hence the output $\mathcal{K}(|x\rangle\langle y|)$ should not be interpreted as a post-measurement density matrix.

Based on this observation, the **Choi matrix representation** of a channel $\mathcal{K}$ is given by the $N^2 \times N^2$ matrix

$$C_{\mathcal{K}} = \begin{bmatrix} \mathcal{K}(|0\rangle\langle 0|) & \mathcal{K}(|0\rangle\langle 1|) & \cdots & \mathcal{K}(|0\rangle\langle N-1|) \\ \mathcal{K}(|1\rangle\langle 0|) & \mathcal{K}(|1\rangle\langle 1|) & \cdots & \mathcal{K}(|1\rangle\langle N-1|) \\ \vdots & \vdots & \ddots & \vdots \\ \mathcal{K}(|N-1\rangle\langle 0|) & \mathcal{K}(|N-1\rangle\langle 1|) & \cdots & \mathcal{K}(|N-1\rangle\langle N-1|) \end{bmatrix}, \quad (11.138)$$

which collects the $N \times N$ responses $\mathcal{K}(|x\rangle\langle y|)$ of the channel to the matrices in the orthonormal basis $\{\{|x\rangle\langle y|\}_{x\in[N]}\}_{y\in[N]}$. The Choi matrix (11.138) can be expressed more succinctly as

$$C_{\mathcal{K}} = \sum_{x\in[N]} \sum_{y\in[N]} |x\rangle\langle y| \otimes \mathcal{K}(|x\rangle\langle y|). \quad (11.139)$$

An equivalent representation of the Choi matrix (11.138) can be proved to be

$$C_{\mathcal{K}} = \sum_{x\in[M]} \text{vec}(K_x^T)\text{vec}(K_x^T)^{\dagger}. \quad (11.140)$$

The Choi matrix is thus positive semi-definite, and it becomes a valid density matrix upon multiplication by $1/N$.

This expression shows that the Choi matrix of a unitary transformation, which corresponds to $M = 1$, has rank 1. Furthermore, it demonstrates the constraint (11.100) that the number of Kraus operators, M, can be upper bounded by N^2. In fact, any $N^2 \times N^2$ normal matrix can be expressed in the form (11.140) with at most N^2 terms in the sum.

Since the Choi matrix provides a full description of the operation of the channel, it can also be used to evaluate the post-channel state for any pre-channel state. To this end, note that the equality

$$\mathcal{K}(|x\rangle\langle y|) = (\langle x| \otimes I)C_{\mathcal{K}}(|y\rangle \otimes I), \quad (11.141)$$

where I is the $N \times N$ identity matrix. This implies that the post-transformation matrix can be written as

$$\begin{aligned} \mathcal{K}(\rho) &= \sum_{x\in[M]} \sum_{y\in[M]} \rho_{x,y}\mathcal{K}(|x\rangle\langle y|) \\ &= \sum_{x\in[M]} \sum_{y\in[M]} \rho_{x,y}(\langle x| \otimes I)C_{\mathcal{K}}(|y\rangle \otimes I). \end{aligned} \quad (11.142)$$

REFERENCES

[1] M. A. Nielsen, "An introduction to majorization and its applications to quantum mechanics," *Lecture Notes, Department of Physics, University of Queensland, Australia*, vol. 53, 2002.

[2] A. Elben, S. T. Flammia, H.-Y. Huang, R. Kueng, J. Preskill, B. Vermersch, and P. Zoller, "The randomized measurement toolbox," *Nature Reviews Physics*, vol. 5, no. 1, pp. 9–24, 2023.

[3] S. Milz and K. Modi, "Quantum stochastic processes and quantum non-Markovian phenomena," *PRX Quantum*, vol. 2, no. 3, art. 030201, 2021.
[4] J. Carrasquilla, G. Torlai, R. G. Melko, and L. Aolita, "Reconstructing quantum states with generative models," *Nature Machine Intelligence*, vol. 1, no. 3, pp. 155–161, 2019.
[5] E. Haapasalo, T. Heinosaari, and J.-P. Pellonpää, "Quantum measurements on finite dimensional systems: Relabeling and mixing," *Quantum Information Processing*, vol. 11, pp. 1751–1763, 2012.
[6] M. Kotowski and M. Oszmaniec, "Pretty-good simulation of all quantum measurements by projective measurements," *arXiv*, 2025. [Online]. Available at: https://arxiv.org/abs/2501.09339
[7] M. Oszmaniec, F. B. Maciejewski, and Z. Puchała, "Simulating all quantum measurements using only projective measurements and postselection," *Physical Review A*, vol. 100, no. 1, art. 012351, 2019.
[8] J. Watrous, *The Theory of Quantum Information*. Cambridge: Cambridge University Press, 2018.
[9] M. A. Nielsen and I. L. Chuang, *Quantum Computation and Quantum Information*. Cambridge: Cambridge University Press, 2010.
[10] N. D. Mermin, *Quantum Computer Science: An Introduction*. Cambridge: Cambridge University Press, 2007.
[11] B. M. Terhal, "Quantum error correction for quantum memories," *Reviews of Modern Physics*, vol. 87, no. 2, pp. 307–346, 2015.
[12] C. B. Mendl and M. M. Wolf, "Unital quantum channels – convex structure and revivals of Birkhoff's theorem," *Communications in Mathematical Physics*, vol. 289, pp. 1057–1086, 2009.
[13] G. Homa, A. Ortega, and M. Koniorczyk, "Choi representation of completely positive maps: A technical introduction," *arXiv*, 2024. [Online]. Available at: https://arxiv.org/abs/2402.12944
[14] G. Chiribella, G. M. D'Ariano, and P. Perinotti, "Probabilistic theories with purification," *Physical Review A*, vol. 81, no. 6, art. 062348, 2010.
[15] D. Layden, B. Mitchell, and K. Siva, "Theory of quantum error mitigation for non-Clifford gates," *arXiv*, 2024. [Online]. Available at: https://arxiv.org/abs/2403.18793
[16] T. Heinosaari and M. Ziman, *The Mathematical Language of Quantum Theory: From Uncertainty to Entanglement*. Cambridge: Cambridge University Press, 2011.
[17] H.-Y. Huang, "Learning quantum states from their classical shadows," *Nature Reviews Physics*, vol. 4, no. 2, pp. 81–81, 2022.
[18] G. Gour, M. M. Wilde, S. Brandsen, and I. J. Geng, "Inevitability of knowing less than nothing," *arXiv*, 2024. [Online]. Available at: https://arxiv.org/abs/2208.14424
[19] D. Petz, "From f-divergence to quantum quasi-entropies and their use," *Entropy*, vol. 12, no. 3, pp. 304–325, 2010.
[20] N. Datta, "Min- and max-relative entropies and a new entanglement monotone," *IEEE Transactions on Information Theory*, vol. 55, no. 6, pp. 2816–2826, 2009.
[21] C. Hirche and M. Tomamichel, "Quantum Rényi and f-divergences from integral representations," *Communications in Mathematical Physics*, vol. 405, no. 9, p. 208, 2024.
[22] E. Chitambar, D. Leung, L. Mančinska, M. Ozols, and A. Winter, "Everything you always wanted to know about LOCC (but were afraid to ask)," *Communications in Mathematical Physics*, vol. 328, pp. 303–326, 2014.

Part IV

Bipartite Classical Information

12 Bipartite Classical Information as Correlation

12.1 Overview

The previous sections of this book have explored information and uncertainty as they relate to the observation of a single, simple system. However, many systems encountered in engineering and the sciences are composite, comprising individual subsystems whose states can be observed independently. These subsystems may represent physical entities – such as cities within a country or pages in a book – or virtual quantities, like the exam scores of a group of students or the winning numbers in a lottery. In composite systems, information and uncertainty pertain not only to the individual states of the subsystems but also to their relationships. For example, knowing the weather conditions at one location might reduce uncertainty about the weather in nearby areas. This chapter focuses on bipartite systems, addressing the challenge of quantifying the correlation between two subsystems.

Like uncertainty, correlation can be defined and measured in various ways. As illustrated in Figure 12.1, this chapter examines correlation from four distinct perspectives:

- as redundant information shared between the two subsystems;
- as a resource spanning the two subsystems that cannot be created "for free," i.e., without external intervention beyond local operations;
- as the minimum amount of common randomness required to simulate the joint behavior of the subsystems in a distributed setting;
- and as the maximum amount of common information that can be extracted from the states of the subsystems in a distributed framework.

Adopting these perspectives, the chapter introduces several correlation measures, including mutual information, the relative entropy of correlation, the distributed-sampling common information, the distillable common information, and the maximal correlation. Each of these measures is symmetric, treating the two subsystems equally. Chapter 13 will study asymmetric correlation measures, which account for residual uncertainty and assign distinct roles to each subsystem.

12.1.1 Learning Objectives and Organization

By the end of this chapter, the reader should be able to:

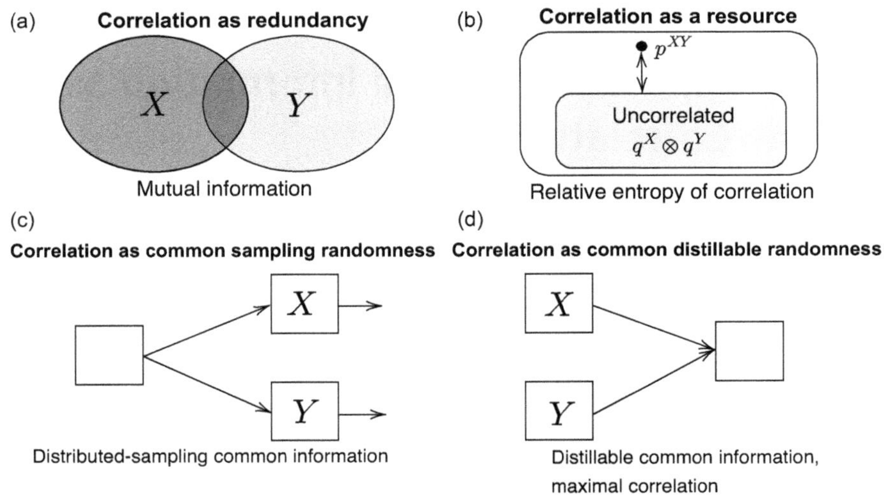

Figure 12.1 This chapter considers four perspectives on the definition of symmetric correlation measures, which yield the mutual information, the relative entropy of correlation, the distributed-sampling common information, the distillable common information, and the maximal correlation.

- evaluate joint and local entropies for a composite system consisting of two subsystems;
- understand and evaluate the Shannon mutual information as a measure of redundant information;
- understand and apply the general framework of correlation as a resource;
- understand the definition of relative entropy of correlation;
- derive the Shannon mutual information as a resource-based measure of correlation;
- understand the distributed-sampling common information as a measure of the minimum common randomness required for the distributed simulation of a bipartite system;
- understand the distillable common information as a measure of the maximum common information that can be extracted from two subsystems in a distributed way;
- understand the relation between the distillable common information and the maximal correlation;
- and evaluate the maximal correlation.

12.2 What Is Correlation?

As summarized in Figure 12.1, this chapter presents four different perspectives on the definition of correlation, along with the corresponding correlation measures. This first section presents the four approaches at a high level, providing an introduction to the content of the rest of the chapter. As in the figure, throughout this chapter, the two subsystems will be labeled as X and Y, and thus the overall composite system under study consists of subsystem X and subsystem Y.

12.2.1 Correlation as Redundancy

The first approach explored in this chapter views correlation as redundant information shared between the two subsystems. To illustrate this, consider the information-as-a-set diagram shown in Figure 12.1(a). In this diagram, the set labeled X represents the information associated with observing the state of subsystem X, quantified by the entropy of its state variable. Similarly, the set labeled Y represents the information associated with subsystem Y, as measured by the entropy of its corresponding state variable.

The entropies associated with the states of the two subsystems are referred to as local entropies, as they describe uncertainties based on local observations of each subsystem. When considering both subsystems together, the uncertainty associated with their joint observation is captured by the joint entropy. In the set diagram, the joint entropy corresponds to the union of the two sets representing the local entropies (see Table 12.1).

From this perspective, the degree of correlation between the two subsystems is determined by the size of the intersection of the sets representing their local entropies. If the intersection is empty, the subsystems contain entirely disjoint information, indicating they are uncorrelated. Conversely, a non-empty intersection signifies redundant information shared by both subsystems, which is a hallmark of correlation.

This view of correlation as redundancy leads to the definition of mutual information as a measure of correlation. Operationally, the mutual information connects to the tasks discussed in Part I for simple (i.e., non-composite) systems. For example, using the Shannon entropy to calculate local and joint entropies, the redundant information – quantified by the mutual information – represents the number of bits that can be saved by jointly compressing the states of both subsystems, as opposed to compressing them separately.

12.2.2 Correlation as a Resource ⏭

In the context of resource theory (see Section 6.3.4), correlation can alternatively be viewed as a resource shared between the two subsystems. To elaborate, consider the

Table 12.1 Mapping between information measures and elements of an information-as-a-set diagram

Information measure	Information as a set
Shannon local entropy	Individual set
Shannon joint entropy	Union
Shannon mutual information	Intersection

informational state of the system, denoted as p^{XY}, which represents the observer's knowledge about the states of the two subsystems.

As shown in Figure 12.1(b), the degree of correlation can be quantified by measuring the distance between the informational state p^{XY} and the set of all uncorrelated informational states. The rationale behind this definition is that uncorrelated states are considered free from the resource of correlation, as they can be generated solely through local operations applied to the two subsystems.

Using the relative entropy metrics introduced in Section 4.7, this approach defines the relative entropy of correlation as a correlation measure. Operationally, as discussed in Chapter 4, the relative entropy of correlation quantifies the effort required to distinguish the observed joint state of the two subsystems from a state in which the subsystems are uncorrelated.

12.2.3 Correlation as Common Sampling Randomness

The next two perspectives introduce new operational tasks involving two agents: one local to subsystem X, referred to as agent X, and the other local to subsystem Y, referred to as agent Y. The first task extends the sampling problem studied in Section 2.7 to a distributed setting, while the second addresses the complementary problem of randomness distillation.

As illustrated in Figure 12.1(c), the task of distributed sampling involves the two agents attempting to simulate observations from the two subsystems. Specifically, agent X aims to sample the state of subsystem X, while agent Y seeks to generate the state of subsystem Y. Crucially, the states sampled by the two agents must adhere to the joint distribution encoded by the given informational state p^{XY}, preserving the correlations present in the joint distribution p^{XY}.

When the states of the subsystems are correlated, distributed sampling requires the two agents to share a certain amount of common randomness. If the agents rely solely on local randomness, they can only produce uncorrelated random variables, which would fail to capture the joint distribution. Therefore, the goal of distributed sampling is for the two agents to generate the desired states while using the minimal amount of common randomness necessary.

The minimal common randomness required to successfully complete distributed sampling provides a natural measure of the correlation between the two subsystems. This perspective defines the distributed-sampling common information as a correlation measure. A possible application of this metric pertains to robotics settings in which different robots need to coordinate their (random) actions by sampling from a joint distribution.

12.2.4 Correlation as Common Distillable Randomness

As illustrated in Figure 12.1(d), a second distributed informational task is that of extracting, or distilling, common randomness at agent X and agent Y. In this task, agent X has access to the state of subsystem X, and agent Y has access to the state of subsystem Y. Through local operations on the respective states, the two agents wish to extract a common random variable.

If the two states were uncorrelated, it would be impossible to extract a non-trivial common random quantity based on local operations. In contrast, if the states are correlated, the two agents may be able to extract the same random quantity from the two respective local states. The maximum such common randomness provides a measure of correlation. Following this perspective, this chapter will present the distillable common information and the maximal correlation as correlation metrics. Applications of this metric include the distillation of a shared private cryptographic key from common randomness at two separate agents.

12.3 Joint Entropy

Throughout this part of the book, the focus is on bipartite classical systems, i.e., on systems involving two classical subsystems, which are referred to as X and Y. Subsystem X can be in one of N^X possible states, and subsystem Y in one of N^Y possible states. Accordingly, the state of subsystem X is described by a random variable $\mathrm{x} \in [N^X] = \{0, 1, \ldots, N^X - 1\}$, while the state of subsystem Y is described by a random variable $\mathrm{y} \in [N^Y] = \{0, 1, \ldots, N^Y - 1\}$. To start, this section and the next describe the joint distribution and the marginal distributions for random variables x and y, as well as the corresponding joint and local entropies.

12.3.1 Joint Distribution

The **joint state** (x, y) of a bipartite system is described by random variables

$$(\mathrm{x}, \mathrm{y}) \sim p^{XY} \tag{12.1}$$

with **joint distribution** p^{XY}. The joint distribution p^{XY} accounts for the informational state of an observer about the states of the two subsystems by specifying all the probabilities

$$p^{XY}_{xy} = \Pr[\mathrm{x} = x \text{ and } \mathrm{y} = y] \tag{12.2}$$

that the states of the two subsystems are in the configuration (x, y) with $x \in [N^X]$ and $y \in [N^Y]$. The probabilities p^{XY}_{xy} must be non-negative, i.e., $p^{XY}_{xy} \geq 0$, and they must sum to 1:

$$\sum_{x\in[N^X]} \sum_{y\in[N^Y]} p^{XY}_{xy} = 1. \tag{12.3}$$

Since there are $N^X N^Y$ possible states (x, y), the joint distribution (12.2) can be described by an $N^X N^Y \times 1$ probability vector p^{XY}, which collects the $N^X N^Y$ probabilities $\{\{p^{XY}_{xy}\}_{x\in[N^X]}\}_{y\in[N^Y]}$. In order to arrange the probabilities p^{XY}_{xy} in (12.2) into a vector, it is convenient to adopt the bi-index notation introduced in Section 8.8.3. This way, the zth element of vector p^{XY} with $z \in [N^X N^Y]$ corresponds to the probability p^{XY}_{xy}, where vector index z and measurement indices (x, y) are related as

$$z = N^Y x + y. \tag{12.4}$$

Accordingly, in the bi-index xy, the first number, x, corresponds to the most significant digit, while the second number, y, is the least significant digit.

Explicitly, the entries of the $N^X N^Y \times 1$ joint distribution vector p^{XY} can be written using the single index $z \in [N^X N^Y]$ or the bi-index xy with $x \in [N^X]$ and $y \in [N^Y]$ as

$$p^{XY} = \begin{bmatrix} p_0^{XY} \\ \vdots \\ p_{N^Y-1}^{XY} \\ p_{N^Y}^{XY} \\ \vdots \\ p_{2N^Y-1}^{XY} \\ \vdots \\ p_{(N^X-1)N^Y}^{XY} \\ \vdots \\ p_{N^X N^Y-1}^{XY} \end{bmatrix} = \begin{bmatrix} p_{00}^{XY} \\ \vdots \\ p_{0,N^Y-1}^{XY} \\ p_{10}^{XY} \\ \vdots \\ p_{1,N^Y-1}^{XY} \\ \vdots \\ p_{N^X-1,0}^{XY} \\ \vdots \\ p_{N^X-1,N^Y-1}^{XY} \end{bmatrix}. \tag{12.5}$$

Note that, in (12.5), a comma was added in the bi-index xy to separate the two constituent indices x and y when useful to avoid ambiguity.

Example 12.1

Consider a system in which both subsystems can take the same number of state values $N^X = N^Y = D$, and the joint distribution is given as

$$p_{xx}^{XY} = \frac{p^{\text{eq}}}{D} \tag{12.6}$$

for $x \in [D]$, and

$$p_{xy}^{XY} = \frac{1 - p^{\text{eq}}}{D^2 - D} \tag{12.7}$$

for $x \neq y$, where $0 \leq p^{\text{eq}} \leq 1$ is a probability.

Under this distribution, all N configurations of the form (x, x), in which the states of the two subsystems are equal, have probability p^{eq}/D. Therefore, the overall probability that the two subsystems share the same state x for any $x \in [D]$ is given by the sum

$$\sum_{x \in [D]} p_{xx}^{XY} = D\frac{p^{\text{eq}}}{D} = p^{\text{eq}}, \tag{12.8}$$

and is thus controlled by parameter p^{eq}. Intuitively, the probability p^{eq} that the two variables x and y are equal determines the level of correlation between the two subsystems under joint distribution p^{XY}.

Example 12.2

With $D = 2$, i.e., with two binary states, the distribution (12.6) and (12.7) is given by

$$p^{XY} = \begin{bmatrix} p_0^{XY} \\ p_1^{XY} \\ p_2^{XY} \\ p_3^{XY} \end{bmatrix} = \begin{bmatrix} p_{00}^{XY} \\ p_{01}^{XY} \\ p_{10}^{XY} \\ p_{11}^{XY} \end{bmatrix} = \begin{bmatrix} \frac{p^{\text{eq}}}{2} \\ \frac{1-p^{\text{eq}}}{2} \\ \frac{1-p^{\text{eq}}}{2} \\ \frac{p^{\text{eq}}}{2} \end{bmatrix}. \tag{12.9}$$

When $p^{\text{eq}} = 1$, the states of the two subsystems are always equal, and the joint distribution (12.9) equals

$$p^{XY} = \frac{1}{2}\begin{bmatrix} 1 \\ 0 \\ 0 \\ 1 \end{bmatrix}. \tag{12.10}$$

In this case, the two configurations $(x = 0, y = 0)$ and $(x = 1, y = 1)$, in which the two states are equal, are assigned equal probability 0.5, while all the other configurations have zero probability. With $p^{\text{eq}} = 1$, the correlation between the two subsystems may thus be considered to be maximal, since knowing the state of one subsystem immediately identifies also the state of the other subsystem.

When is the correlation minimal? One may be tempted to say that this is the case in the "opposite" situation in which $p^{\text{eq}} = 0$. However, in this case, the joint distribution (12.9) equals

$$p^{XY} = \frac{1}{2}\begin{bmatrix} 0 \\ 1 \\ 1 \\ 0 \end{bmatrix}, \tag{12.11}$$

which indicates that the two configurations $(x = 0, y = 1)$ and $(x = 1, y = 0)$ have equal probability, and all other configurations have zero probability. Therefore, with $p^{\text{eq}} = 0$ the two subsystems have the same maximal level of correlation as in the case $p^{\text{eq}} = 1$, since knowing the state of one subsystem unambiguously determines also the state of the other subsystem.

Take now the setting $p^{\text{eq}} = 1/D = 1/2$, for which the joint distribution (12.9) is uniform, i.e.,

$$p^{XY} = \frac{1}{4}\begin{bmatrix} 1 \\ 1 \\ 1 \\ 1 \end{bmatrix} = u. \tag{12.12}$$

Under this joint distribution, all outcomes are equally likely, and thus the state of one subsystem is not informative about the state of the other subsystem. As a result, this setting is characterized by minimal correlation. As will be discussed in this chapter, this is an example of uncorrelated subsystems.

12.3.2 Joint Entropy

The **joint entropy** $\mathrm{H}_\alpha(p^{XY})$ captures the overall uncertainty associated with the observation of the system configuration $(\mathrm{x}, \mathrm{y}) \sim p^{XY}$ under the informational state p^{XY}. Mathematically, the joint entropy $\mathrm{H}_\alpha(p^{XY})$ of a joint distribution p^{XY} is defined as the entropy $\mathrm{H}_\alpha(\cdot)$ of the $N^X N^Y \times 1$ joint distribution p^{XY}. Specializing this definition to the three entropy measures yields the following.

Definition 12.1 Shannon joint entropy
The Shannon joint entropy is defined as

$$\begin{aligned}\mathrm{H}(p^{XY}) &= -\mathrm{E}_{p^{XY}}[\log_2(p^{XY}_{\mathrm{xy}})] \\ &= -\sum_{x\in[N^X]}\sum_{y\in[N^X]} p^{XY}_{xy}\log_2(p^{XY}_{xy}).\end{aligned} \tag{12.13}$$

Definition 12.2 Tsallis joint entropy
The Tsallis joint entropy of order $\alpha \geq 0$ is defined as

$$\mathrm{H}^{\mathrm{T}}_\alpha(p^{XY}) = \begin{cases} \frac{1_2}{1-\alpha}\left(\sum_{x\in[N^X]}\sum_{y\in[N^X]}(p^{XY}_{xy})^\alpha - 1\right) & \text{for } \alpha \geq 0 \text{ with } \alpha \neq 1, \\ \mathrm{H}(p^{XY}) & \text{for } \alpha = 1. \end{cases} \tag{12.14}$$

Definition 12.3 Rényi joint entropy
The Rényi joint entropy of order $\alpha \geq 0$ is defined as

$$\mathrm{H}^{\mathrm{R}}_\alpha(p^{XY}) = \begin{cases} \frac{1}{1-\alpha}\log_2\left(\sum_{x\in[N^X]}\sum_{y\in[N^X]}(p^{XY}_{xy})^\alpha\right) & \text{for } \alpha \geq 0 \text{ with } \alpha \neq 1, \\ \mathrm{H}(p^{XY}) & \text{for } \alpha = 1. \end{cases} \tag{12.15}$$

It is common to write the Shannon, Tsallis, and Rényi joint entropies in the forms

$$\begin{aligned}\mathrm{H}(p^{XY}) &= \mathrm{H}(X, Y)_{p^{XY}}, \\ \mathrm{H}^{\mathrm{T}}_\alpha(p^{XY}) &= \mathrm{H}^{\mathrm{T}}_\alpha(X, Y)_{p^{XY}}, \\ \text{and } \mathrm{H}^{\mathrm{R}}_\alpha(p^{XY}) &= \mathrm{H}^{\mathrm{R}}_\alpha(X, Y)_{p^{XY}},\end{aligned} \tag{12.16}$$

respectively; or, when referring generically to all three joint entropies, as

$$\mathrm{H}_\alpha(p^{XY}) = \mathrm{H}_\alpha(X, Y)_{p^{XY}}. \tag{12.17}$$

The notations on the right-hand sides of (12.16) and (12.17) highlight that the joint entropy is the entropy of subsystems X and Y under the informational state defined by the joint distribution p^{XY}. When no confusion may arise, one may also drop the joint distribution in the subscript, writing $\mathrm{H}(X, Y)$ for $\mathrm{H}(X, Y)_{p^{XY}}$; $\mathrm{H}^{\mathrm{T}}_{\alpha}(X, Y)$ for $\mathrm{H}^{\mathrm{T}}_{\alpha}(X, Y)_{p^{XY}}$; $\mathrm{H}^{\mathrm{R}}_{\alpha}(X, Y)$ for $\mathrm{H}^{\mathrm{R}}_{\alpha}(X, Y)_{p^{XY}}$; and $\mathrm{H}_{\alpha}(X, Y)$ for $\mathrm{H}_{\alpha}(X, Y)_{p^{XY}}$.

Following Chapters 2 and 3, the operational significance of the joint entropy $\mathrm{H}_{\alpha}(p^{XY})$ depends on the choice of the entropy measure. For example, the Shannon joint entropy quantifies the number of bits needed to identify, compress, or sample jointly the two states $(\mathrm{x}, \mathrm{y}) \sim p^{XY}$.

Example 12.3

Returning to the joint distribution (12.9) for two binary systems with $N^X = N^Y = 2$, Figure 12.2 shows the Shannon joint entropy, as well as the Tsallis and Rényi joint entropies with $\alpha = 0.5$, as a function of the probability p^{eq}. The three entropies are minimized when the probability p^{eq} equals 0 or 1, in which cases the states of the two subsystems are respectively always opposite or equal to each other. Accordingly, with $p^{\mathrm{eq}} = 0$ or $p^{\mathrm{eq}} = 1$, the overall information returned by the observation of the state pair (x, y) is minimal, as any one of the two states carries all the information required to reconstruct the other state too. Note, however, that the joint entropies are positive even for $p^{\mathrm{eq}} = 0$ and $p^{\mathrm{eq}} = 1$ since the two states x and y are still individually uncertain.

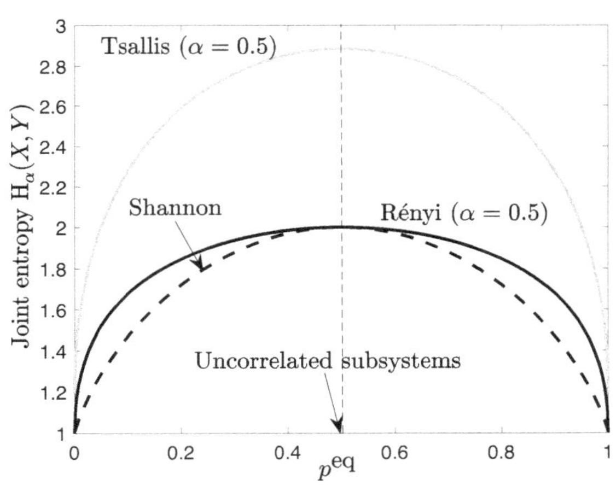

Figure 12.2 The Shannon joint entropy, as well as the Tsallis and Rényi joint entropies with $\alpha = 0.5$, as a function of the probability p^{eq} for the joint distribution (12.9).

Conversely, the three joint entropies are maximized for $p^{\mathrm{eq}} = 1/2$. As discussed in Example 12.2, under this informational state, all possible output pairs (x, y) have the same probability, and the state of one subsystem does not carry any information about the state of the other subsystem.

Overall, the joint entropies increase as the probability p^{eq} grows from $p^{\mathrm{eq}} = 0$ to $p^{\mathrm{eq}} = 1/2$, while they decrease in going from $p^{\mathrm{eq}} = 1/2$ to $p^{\mathrm{eq}} = 1$. Intuitively, when p^{eq} moves away from 1/2, the two subsystems tend to either have different states, as p^{eq} becomes smaller, or rather the same state, when p^{eq} grows larger, causing the overall uncertainty on the values of variables $(\mathrm{x}, \mathrm{y}) \sim p^{XY}$ to decrease.

Example 12.4

Consider now the joint distribution (12.6) and (12.7) with $D = 3$, for which the three joint entropies are plotted in Figure 12.3 versus the probability p^{eq}. The general conclusions here are aligned with the case $D = 2$ studied in Example 12.3, as long as one substitutes the probability $p^{\text{eq}} = 1/2$ with the probability $p^{\text{eq}} = 1/3$. Indeed, with $p^{\text{eq}} = 1/3$, the distribution (12.6) and (12.7) has all entries equal to 1/9, and the two states x and y are uncorrelated.

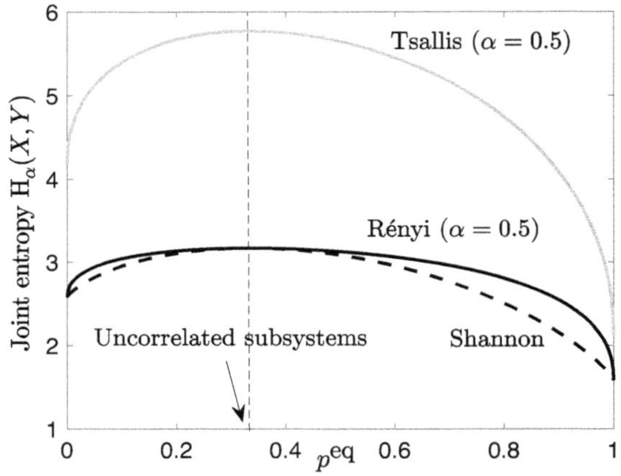

Figure 12.3 The Shannon joint entropy, as well as the Tsallis and Rényi joint entropies with $\alpha = 0.5$, as a function of the probability p^{eq} for the joint distribution (12.6) and (12.7) with $D = 3$.

That said, an interesting observation is that, unlike for the case $D = 2$, the joint entropies do not attain the minimal values for $p^{\text{eq}} = 0$. In fact, with $D = 3$, under distribution (12.6) and (12.7), a state variable does not fully determine the other when $p^{\text{eq}} = 0$. Rather, it only rules out one of the three possible values of the other state.

12.4 Local Entropies

In composite systems, one may only be interested in the state of one of the subsystems. In this case, the relevant uncertainty measure is the local entropy of the subsystem under study, which is presented in this section.

12.4.1 Marginal Distributions

To start, consider subsystem X, whose state variable $\mathrm{x} \in [N^X]$ has a distribution defined by an $N^X \times 1$ probability vector p^X. Specifically, each xth entry of vector p^X represents the probability

$$p_x^X = \Pr[\mathrm{x} = x] \tag{12.18}$$

that subsystem X is in state $\mathrm{x} = x \in [N^X]$.

The distribution p^X is known as the **marginal distribution** of subsystem X. To evaluate the marginal distribution p^X from the joint distribution p^{XY}, it is sufficient to observe that subsystem X is in state x if and only if the overall system is in any configuration of the form (x,y), where $y \in [N^Y]$ is an arbitrary state for subsystem Y. Accordingly, the marginal probability p_x^X is obtained by summing the joint distribution p_{xy}^{XY} over all possible values of y, i.e.,

$$p_x^X = \sum_{y \in [N^Y]} p_{xy}^{XY}. \tag{12.19}$$

Intuitively, one can interpret the sum over all states of subsystem Y in (12.19) as an expression of the fact that we "don't care" about the value of the state y of subsystem Y when only observing subsystem X. The sum over all values of Y is also referred to as marginalization over subsystem Y, or as marginalization over random variable y.

Local observations of subsystem Y are described in an analogous way. In particular, the marginal distribution of subsystem Y is defined by an $N^Y \times 1$ vector p^Y with entries

$$p_y^Y = \Pr[\mathrm{y} = y] = \sum_{x \in [N^X]} p_{xy}^{XY}, \tag{12.20}$$

where the sum implements the marginalization over subsystem X, i.e., over random variable x.

Example 12.5

For the distribution (12.6) and (12.7), the marginal distribution for subsystem X is given by

$$\begin{aligned} p_x^X &= \sum_{y \in [N^Y]} p_{xy}^{XY} \\ &= \frac{p^{\mathrm{eq}}}{D} + (D-1)\frac{1-p^{\mathrm{eq}}}{D^2 - D} = \frac{1}{D}, \end{aligned} \tag{12.21}$$

for all $x \in [N^X]$, and an analogous calculation applies for the marginal distribution p^Y. Therefore, the two marginal distributions – or marginals for short – are uniform, i.e., $p^X = u$ and $p^Y = u$. An interesting observation is that the marginal distributions do not depend on the probability p^{eq}, and hence they do not offer any insight into the correlation between the two subsystems.

While the marginal distributions p^X and p^Y can be obtained from the joint distribution p^{XY} via (12.19) and (12.20), the reverse is not true. In fact, it is generally impossible to reconstruct the joint distribution p^{XY} based solely on the marginals p^X and p^Y. In fact, as illustrated in Example 12.5, the marginals do not provide information about the correlation between the random variables x and y.

12.4.2 Local Entropies

The local entropies capture the levels of uncertainty associated with the states of each of the two subsystems.

Definition 12.4 Local entropy
The Shannon, Tsallis, and Rényi local entropies are the entropies of the corresponding marginal distributions. Accordingly, for subsystem X, the Shannon, Tsallis, and Rényi local entropies are defined as

$$\begin{aligned} \mathrm{H}(X)_{p^X} &= \mathrm{H}(p^X), \\ \mathrm{H}^{\mathrm{T}}_{\alpha}(X)_{p^X} &= \mathrm{H}^{\mathrm{T}}_{\alpha}(p^X), \\ \text{and } \mathrm{H}^{\mathrm{R}}_{\alpha}(X)_{p^X} &= \mathrm{H}^{\mathrm{R}}_{\alpha}(p^X), \end{aligned} \tag{12.22}$$

respectively, which can be written generically as

$$\mathrm{H}_{\alpha}(p^X) = \mathrm{H}_{\alpha}(X)_{p^X}. \tag{12.23}$$

Analogous definitions apply for the local entropies of subsystem Y.

As a point of notation, one may also write $\mathrm{H}_{\alpha}(X)_{p^{XY}}$, since the marginal distribution p^X can be recovered from the joint distribution p^{XY} via the marginalization (12.19). The relation between the joint entropy $\mathrm{H}_{\alpha}(X, Y)_{p^{XY}}$ and the marginal entropies $\mathrm{H}_{\alpha}(X)_{p^X}$ and $\mathrm{H}_{\alpha}(Y)_{p^Y}$ is the subject of the next section.

12.5 Joint Entropy vs. Local Entropies

In order to formulate correlation measures leveraging the analysis of the joint and local entropies, it is necessary to understand the relationship between the two types of entropies. To this end, this section will first formally define uncorrelated, or independent, subsystems, and then review the key properties of additivity and subadditivity of the Shannon joint entropy. It will also be observed that neither the Tsallis entropy nor the Rényi entropy satisfy the subadditivity property. Finally, the uncertainty-increases-globally property of the three entropies is presented. The material covered in this section will be instrumental to define the Shannon mutual information in Section 12.6.

12.5.1 Uncorrelated Subsystems

As discussed in Section 12.4, the joint distribution cannot be recovered from the marginal distributions, since the marginals do not provide information about the correlation between the two subsystems. However, if the two subsystems are uncorrelated, it must be possible to describe the joint distribution based solely on the marginals. In fact, in this case, there is no information about the overall system that cannot be obtained by analyzing the two subsystems separately. The following definition formalizes this observation.

Definition 12.5 Uncorrelated (or independent) subsystems
Two subsystems X and Y – as well as the corresponding state random variables $(\mathrm{x},\mathrm{y}) \sim p^{XY}$ – are said to be uncorrelated, or independent, under the joint distribution p^{XY} if the joint distribution p^{XY} equals the product of the marginals p_x^X and p_y^Y, i.e., if the equality

$$p_{xy}^{XY} = p_x^X p_y^Y \tag{12.24}$$

holds for all $x \in [N^X]$ and $y \in [N^Y]$.

Following this definition, in this book, no distinction is made between independent and uncorrelated subsystems. In other contexts, particularly when considering continuous random variables, it is useful to distinguish between uncorrelated subsystems, which cannot be related via a linear relation, and independent subsystems. In contrast, here, adopting a standard convention in quantum information theory, the qualifiers "uncorrelated" and "independent" will be used interchangeably.

Using the definition (12.24), the joint probability vector (12.5) for uncorrelated subsystems can be written as

$$p^{XY} = \begin{bmatrix} p_0^X p_0^Y \\ \vdots \\ p_0^X p_{N^Y-1}^Y \\ p_1^X p_0^Y \\ \vdots \\ p_1^X p_{N^Y-1}^Y \\ \vdots \\ p_{N^X-1}^X p_0^Y \\ \vdots \\ p_{N^X-1}^X p_{N^Y-1}^Y \end{bmatrix}. \tag{12.25}$$

Thus, using the bi-index and its relation with the Kronecker product discussed in Section 8.8.3, the definition (12.24) can equivalently be expressed as follows.

Definition 12.6 Uncorrelated (or independent) subsystems (equivalent definition)
Two subsystems X and Y – as well as the corresponding state random variables $(\mathrm{x},\mathrm{y}) \sim p^{XY}$ – are said to be uncorrelated, or independent, under the joint distribution p^{XY} if the joint distribution p^{XY} can be written as

$$p^{XY} = p^X \otimes p^Y, \tag{12.26}$$

where p^X and p^Y are the corresponding marginal distributions.

Example 12.6

For the recurring example (12.6) and (12.7), when $p^{\rm eq} = 1/D$, the joint distribution is given by $p^{XY}_{xy} = 1/D^2$ for all $x \in [D]$ and $y \in [D]$. Furthermore, as seen in Example 12.5, the marginal distributions are uniform, i.e., $p^X_x = p^Y_y = 1/D$ for all $x \in [D]$ and $y \in [D]$. Therefore, the condition (12.24) is satisfied, as the equalities

$$p^{XY}_{xy} = p^X_x p^Y_y = \frac{1}{D}\frac{1}{D} = \frac{1}{D^2} \tag{12.27}$$

hold for all $x \in [D]$ and $y \in [D]$, proving that the two variables are uncorrelated.

12.5.2 Additivity of the Shannon Joint Entropy

Intuitively, if two subsystems are uncorrelated, the total information gained from observing their states x and y should equal the sum of the uncertainties associated with observing each state separately.

This intuition is confirmed when using the Shannon surprise – introduced in Section 2.4 – as the measure of information associated with an observation (x, y) of the two random variables $(\mathrm{x}, \mathrm{y}) \sim p^{XY}$. The Shannon surprise for the joint observation is given by

$$-\log_2(p^{XY}_{xy}). \tag{12.28}$$

Similarly, the **Shannon local surprise** for an observation of the state x of subsystem X is $-\log_2(p^X_x)$, while for the state y of subsystem Y, it is $-\log_2(p^Y_y)$. Using the definition of independence (12.24) and the additivity property of logarithms, the Shannon joint surprise for uncorrelated subsystems is

$$\begin{aligned}-\log_2(p^{XY}_{xy}) &= -\log_2(p^X_x p^Y_y)\\ &= -\log_2(p^X_x) - \log_2(p^Y_y),\end{aligned} \tag{12.29}$$

which is the sum of the Shannon local surprises for the individual observations of the two subsystems. Thus, the Shannon surprise, when used as a measure of information, aligns with the intuitive property that for uncorrelated subsystems, the total uncertainty equals the sum of the local uncertainties.

By taking the expectation with respect to the joint distribution p^{XY} on both sides of the equality (12.29), one obtains

$$\begin{aligned}\underbrace{-\mathrm{E}_{p^{XY}}[\log_2(p^{XY}_{\mathrm{xy}})]}_{\mathrm{H}(X,Y)_{p^{XY}}} &= -\mathrm{E}_{p^{XY}}[\log_2(p^X_{\mathrm{x}})] - \mathrm{E}_{p^{XY}}[\log_2(p^Y_{\mathrm{y}})]\\ &= \underbrace{(-\mathrm{E}_{p^X}[\log_2(p^X_{\mathrm{x}})])}_{\mathrm{H}(X)_{p^X}} + \underbrace{(-\mathrm{E}_{p^Y}[\log_2(p^Y_{\mathrm{y}})])}_{\mathrm{H}(Y)_{p^Y}},\end{aligned} \tag{12.30}$$

which can be stated as the following key property of the Shannon joint entropy.

Key Result 12.1 Additivity of the Shannon joint entropy

For uncorrelated subsystems, i.e., for any joint distribution $p^{XY} = p^X \otimes p^Y$ with marginals p^X and p^Y, the Shannon joint entropy satisfies the equality

$$\mathrm{H}(X, Y)_{p^{XY}} = \mathrm{H}(X)_{p^X} + \mathrm{H}(Y)_{p^Y}. \tag{12.31}$$

Therefore, for uncorrelated subsystems, the Shannon joint entropy is the sum of the Shannon local entropies. In other words, uncorrelated subsystems do not have any redundant information across the two subsystems, as all the information in either subsystem contributes to the overall information about the bipartite system.

The additivity property is illustrated by using the information-as-a-set representation in Figure 12.4. In this representation, the Shannon local entropies $\mathrm{H}(X)_{p^X}$ and $\mathrm{H}(Y)_{p^Y}$ are represented by the sizes of the corresponding sets, while the joint entropy $\mathrm{H}(X, Y)_{p^{XY}}$ is represented by the size of the union of the two sets (see Table 12.1).

As seen Figure 12.4(a), for uncorrelated subsystems, by the additivity property of the Shannon joint entropy, the sets describing the two local entropies $\mathrm{H}(X)_{p^X}$ and $\mathrm{H}(Y)_{p^Y}$ are disjoint, as they do not have any informational overlap. Accordingly, the size of the union of the two sets, representing the joint entropy $\mathrm{H}(X, Y)_{p^{XY}}$, is the sum of the local entropies.

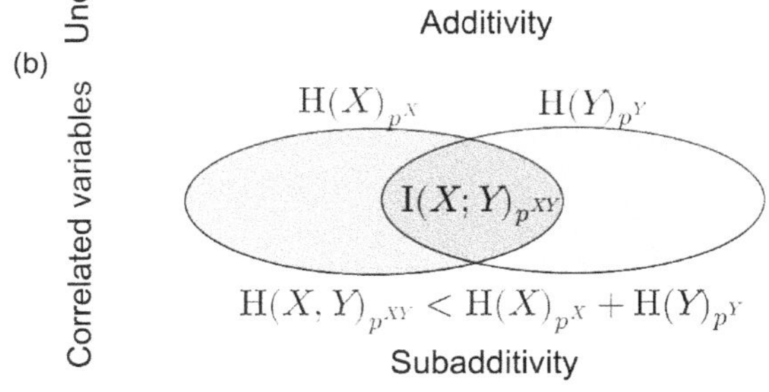

Figure 12.4 Illustration of two key properties of the Shannon joint entropy: (a) additivity; (b) subadditivity. Subadditivity includes additivity as a special case, and is central to the definition of the Shannon mutual information. The Rényi joint entropy satisfies additivity, but not subadditivity; and the Tsallis joint entropy satisfies neither property.

12.5.3 Subadditivity of the Shannon Joint Entropy

As illustrated in Figure 12.4(b), when the two subsystems are correlated, the two sets describing the local entropies overlap, and the joint entropy is smaller than the sum of the local entropies. This is formalized by the subadditivity of the Shannon joint entropy.

> **Key Result 12.2 Subadditivity of the Shannon joint entropy**
>
> For any joint distribution p^{XY} with marginals p^X and p^Y, the Shannon entropy satisfies the inequality
>
> $$\mathrm{H}(X, Y)_{p^{XY}} \leq \mathrm{H}(X)_{p^X} + \mathrm{H}(Y)_{p^Y}, \tag{12.32}$$
>
> with equality in (12.32) attained if and only if the subsystems are uncorrelated.

The subadditivity property (12.32) will be proved in Section 12.7.1. For now, note that the "if" part of the last statement is equivalent to the additivity (12.31) of the Shannon joint entropy. Thus, subadditivity subsumes additivity as a special case.

12.5.4 Additivity Beyond Shannon?

While additivity appears to be a natural property for any entropy measure, it is, in fact, not necessarily satisfied by other entropies. As summarized next, the Rényi entropy is additive, while the Tsallis entropy is not.

> **Key Result 12.3 Additivity of the Rényi joint entropy**
>
> For uncorrelated subsystems with joint distribution $p^{XY} = p^X \otimes p^Y$, the Rényi joint entropy of any order $\alpha \geq 0$ satisfies the equality
>
> $$\mathrm{H}_\alpha^{\mathrm{R}}(X, Y)_{p^X \otimes p^Y} = \mathrm{H}_\alpha^{\mathrm{R}}(X)_{p^X} + \mathrm{H}_\alpha^{\mathrm{R}}(Y)_{p^Y}. \tag{12.33}$$

This property follows directly from the definition (12.15) and the additivity of the logarithm.

> **Key Result 12.4 Pseudoadditivity of the Tsallis joint entropy**
>
> For uncorrelated subsystems with joint distribution $p^{XY} = p^X \otimes p^Y$, the Tsallis joint entropy satisfies the following equality:
>
> $$\mathrm{H}_\alpha^{\mathrm{T}}(X, Y)_{p^X \otimes p^Y} = \mathrm{H}_\alpha^{\mathrm{T}}(X)_{p^X} + \mathrm{H}_\alpha^{\mathrm{T}}(Y)_{p^Y} + \frac{(1-\alpha)}{1_2}\mathrm{H}_\alpha^{\mathrm{T}}(X)_{p^X}\mathrm{H}_\alpha^{\mathrm{T}}(Y)_{p^Y}, \tag{12.34}$$
>
> which is known as **α-additivity** or **pseudoadditivity**.

The equality (12.34) can be proved directly from the α-additivity property of the α-logarithm presented in Appendix 3.A.

The pseudoadditivity (12.34) of the Tsallis entropy confirms the non-extensivity of the Tsallis entropy discussed in Section 3.4.4. In particular, using (12.34), the joint Tsallis entropy $H_\alpha^T(X, Y)_{p^X \otimes p^Y}$ is seen to be larger than the sum of the local Tsallis entropies $H_\alpha^T(X)_{p^X}$ and $H_\alpha^T(Y)_{p^Y}$ when $0 \leq \alpha < 1$. In contrast, with $\alpha > 1$, the joint Tsallis entropy $H_\alpha^T(X, Y)_{p^X \otimes p^Y}$ is smaller than the sum of the local Tsallis entropies $H_\alpha^T(X)_{p^X}$ and $H_\alpha^T(Y)_{p^Y}$. Thus, when evaluated via the Tsallis entropy, the overall information associated with two uncorrelated subsystems can be larger or smaller than the sum of the information contents of its local parts.

12.5.5 Subadditivity Beyond Shannon?

It was just shown that the Rényi joint entropy is additive, while the Tsallis joint entropy is not. As seen for the Shannon joint entropy, subadditivity is a stronger property than additivity, and it is thus important to ask if the Rényi joint entropy is also subadditive. Note that the Tsallis joint entropy cannot be subadditive, since it is not even additive. Like the Tsallis joint entropy, the Rényi joint entropy also turns out not to be subadditive, as demonstrated by the following example. It is thus concluded that, of the three entropy measures, *only* the Shannon entropy admits the information-as-a-set representation, which requires subadditivity.

Example 12.7

Consider the following joint distribution of two random variables taking $N^X = N^Y = D = 2$ values:

$$p^{XY} = \begin{bmatrix} p_{00}^{XY} \\ p_{01}^{XY} \\ p_{10}^{XY} \\ p_{11}^{XY} \end{bmatrix} = \begin{bmatrix} 1/2 \\ 1/4 \\ 1/4 \\ 0 \end{bmatrix}. \tag{12.35}$$

The corresponding marginal distributions for subsystems X and Y are equal and given as

$$p^X = p^Y = \begin{bmatrix} 3/4 \\ 1/4 \end{bmatrix}. \tag{12.36}$$

The Rényi joint min-entropy is evaluated as

$$H_\infty^R(X, Y)_{p^{XY}} = -\log_2 \left(\max_{x,y \in [2]} p_{xy}^{XY} \right) = 1, \tag{12.37}$$

and the Rényi local min-entropies are

$$H_\infty^R(X)_{p^X} = H_\infty^R(Y)_{p^Y} = -\log_2 \left(\max_{x \in \{0,1\}} p_x^X \right) = \log_2 \left(\frac{4}{3} \right). \tag{12.38}$$

It follows that the joint entropy is larger than the sum of the marginal entropies, i.e.,

$$\begin{aligned} \mathrm{H}^{\mathrm{R}}_{\infty}(X,Y)_{p^{XY}} &= \log_2(2) = 1 \\ &> \mathrm{H}^{\mathrm{R}}_{\infty}(X)_{p^X} + \mathrm{H}^{\mathrm{R}}_{\infty}(Y)_{p^X} = \log_2\left(\frac{16}{9}\right) \approx 0.83, \end{aligned} \tag{12.39}$$

demonstrating that the Rényi entropy with infinite order, $\alpha = \infty$, does not satisfy subadditivity. This conclusion is, however, not limited to the extreme case $\alpha = \infty$.

To see this, consider Figure 12.5, which plots the Rényi joint entropy $\mathrm{H}^{\mathrm{R}}_{\alpha}(X,Y)_{p^{XY}}$ along with the Rényi local entropies $\mathrm{H}^{\mathrm{R}}_{\alpha}(X)_{p^X}$ and $\mathrm{H}^{\mathrm{R}}_{\alpha}(Y)_{p^Y}$ for the joint distribution (12.35). As can be seen, for all sufficiently large values of the order α, the joint entropy $\mathrm{H}^{\mathrm{R}}_{\alpha}(X,Y)$ is larger than the sum $\mathrm{H}^{\mathrm{R}}_{\alpha}(X)_{p^X} + \mathrm{H}^{\mathrm{R}}_{\alpha}(Y)_{p^Y}$, violating subadditivity.

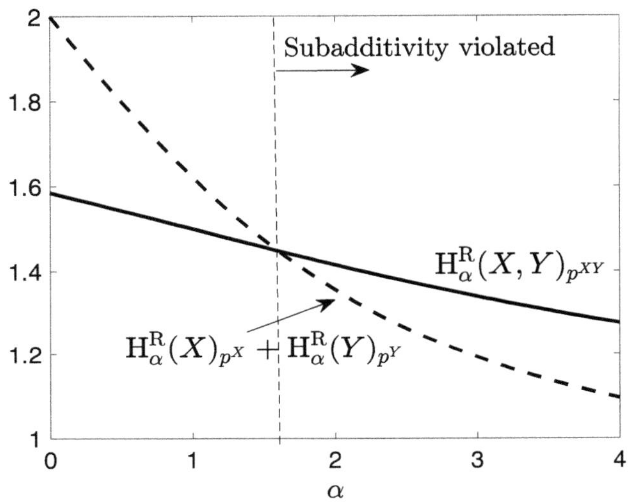

Figure 12.5 Rényi joint entropy $\mathrm{H}^{\mathrm{R}}_{\alpha}(X,Y)_{p^{XY}}$, along with the Rényi local entropies $\mathrm{H}^{\mathrm{R}}_{\alpha}(X)_{p^X}$ and $\mathrm{H}^{\mathrm{R}}_{\alpha}(Y)_{p^Y}$, as a function of the order α for the joint distribution (12.35).

12.5.6 Uncertainty Increases Globally

While subadditivity is a property satisfied only by the Shannon joint entropy, the following property applies to all entropy measures.

Key Result 12.5 Uncertainty increases globally

Given a joint distribution p^{XY} with marginal distributions p^X and p^Y, the Shannon, Tsallis, and Rényi joint entropies satisfy the inequality

$$\mathrm{H}_{\alpha}(X,Y)_{p^{XY}} \geq \max(\mathrm{H}_{\alpha}(X)_{p^X}, \mathrm{H}_{\alpha}(Y)_{p^Y}), \tag{12.40}$$

with equality if and only if at least one of the two variables x and y can be obtained as a function of the other variable.

According to the uncertainty-increases-globally property (12.40), the uncertainty on the whole subsystem – measured by the joint entropy $\mathrm{H}_\alpha(X,Y)_{p^{XY}}$ – cannot be smaller than the local uncertainty on any of the two subsystems – quantified by the maximum local entropy $\max(\mathrm{H}(X)_{p^X}, \mathrm{H}(Y)_{p^Y})$. In more physical terms, a classical bipartite state may thus be said to be more disordered, carrying more information globally than locally.

This property is particularly easy to interpret for the Shannon entropy. In fact, using the information-as-a-set representation, the joint entropy $\mathrm{H}(X,Y)_{p^{XY}}$ is represented by the size of the union of the two sets representing the local entropies, and thus it can never be smaller than either of the two set sizes $\mathrm{H}(X)_{p^X}$ or $\mathrm{H}(Y)_{p^Y}$. More formally, and more generally, a simple argument based on the Tsallis sum is sufficient to prove the inequality (12.40) also for the Tsallis and Rényi joint entropies, as shown in Appendix 12.A.

To elaborate on the conditions for equality in (12.40), take first the case in which random variable y is a function of random variable x. This means that there exists a function $f(\cdot)$ such that the equality

$$\mathrm{y} = f(\mathrm{x}) \tag{12.41}$$

holds with probability 1, that is, under the joint distribution p^{XY}, any configuration (x,y) with $y \neq f(x)$ has zero probability:

$$p^{XY}_{xy} = 0 \text{ for } y \neq f(x). \tag{12.42}$$

In this case, the relation (12.40) reduces to the equality

$$\mathrm{H}_\alpha(X,Y)_{p^{XY}} = \mathrm{H}_\alpha(X)_{p^X}. \tag{12.43}$$

From the perspective of the information-as-a-set diagram, in this situation, the set describing $\mathrm{H}_\alpha(Y)_{p^Y}$ is included in the set for $\mathrm{H}_\alpha(X)_{p^X}$, since all the information in subsystem Y can be obtained from subsystem X.

Analogous considerations apply when there exists a function $f(\cdot)$ that ensures the equality $\mathrm{x} = f(\mathrm{y})$ with probability 1, in which case the relation (12.40) reduces to the equality $\mathrm{H}_\alpha(X,Y)_{p^{XY}} = \mathrm{H}_\alpha(Y)_{p^Y}$.

Example 12.8

Subsystem X reports a high-precision rating x assigned by a user to a movie in the scale $[N^X] = [4]$. In contrast, subsystem Y only reports whether the rating is negative or positive, adopting the scale $[N^Y] = [2]$. Specifically, the state y of the second subsystem can be expressed as a function of the state x as the function $\mathrm{y} = f(\mathrm{x})$, with

$$f(x) = \begin{cases} 0 & \text{if } x \in \{0,1\} \text{ (negative rating)}, \\ 1 & \text{if } x \in \{2,3\} \text{ (positive rating)}. \end{cases} \tag{12.44}$$

Assume that the user's rating has a uniform marginal distribution p^X in the set [4]. This implies that the marginal distribution p^Y is uniform in the set [2], and thus the local entropies equal $\mathrm{H}(X)_{p^X} = \log_2(4) = 2$ and $\mathrm{H}(Y)_{p^Y} = \log_2(2) = 1$. Accordingly, the maximum entropy is given by $\mathrm{H}(X)_{p^X} = 2$, i.e., $\max(\mathrm{H}(X)_{p^X}, \mathrm{H}(Y)_{p^Y}) = \mathrm{H}(X)_{p^X} = 2$.

Furthermore, the 8×1 joint distribution vector can be evaluated as

$$p^{XY} = \frac{1}{4}[1, 0, 1, 0, 0, 1, 0, 1]^T, \tag{12.45}$$

since the configurations (x, y) given by $(0, 0)$, $(1, 0)$, $(2, 1)$, and $(3, 1)$ have probability $1/4$, with all other outputs having zero probability as in (12.42). The Shannon joint entropy is evaluated as $\mathrm{H}(X, Y)_{p^{XY}} = \log_2(4) = 2$, which satisfies the equality in (12.40) as $\mathrm{H}_\alpha(X, Y)_{p^{XY}} = \mathrm{H}_\alpha(X)_{p^X}$.

Combining the subadditivity and the uncertainty-increases-globally properties for the Shannon joint entropy, it is useful to state the following upper and lower bounds for future reference.

Key Result 12.6 Bounds on the Shannon joint entropy
The Shannon joint entropy is bounded as

$$\underbrace{\max(\mathrm{H}(X)_{p^X}, \mathrm{H}(Y)_{p^Y}) \leq}_{\text{uncertainty increases globally}} \mathrm{H}(X, Y)_{p^{XY}} \underbrace{\leq \mathrm{H}(X)_{p^X} + \mathrm{H}(Y)_{p^Y}}_{\text{subadditivity}}. \tag{12.46}$$

The first inequality says that, information-wise, the whole must be at least as large as the smallest part, while the second indicates that the whole cannot be larger than the sum of the parts. We will see in Part V of the book that, unlike subadditivity, the uncertainty-increases-globally property does not hold for quantum systems.

12.6 Correlation as Redundancy

This section focuses on the definition of a correlation measure that hinges on the evaluation of the redundant information present in both subsystems.

12.6.1 Shannon Mutual Information

The next example motivates the definition of the Shannon mutual information.

Example 12.9

As illustrated in Figure 12.6, consider the problem of compressing two successive frames in a video, which are viewed as the two subsystems X and Y of interest. A first approach would be to compress each frame separately from the other. Using the results in Chapter 2, this local compression approach would require approximately $\mathrm{H}(X)_{p^X} + \mathrm{H}(Y)_{p^Y}$ bits. In fact, the Shannon local entropies $\mathrm{H}(X)_{p^X}$ and $\mathrm{H}(Y)_{p^Y}$ represent the entropies of the frames $\mathrm{x} \sim p^X$ and $\mathrm{y} \sim p^Y$, respectively.

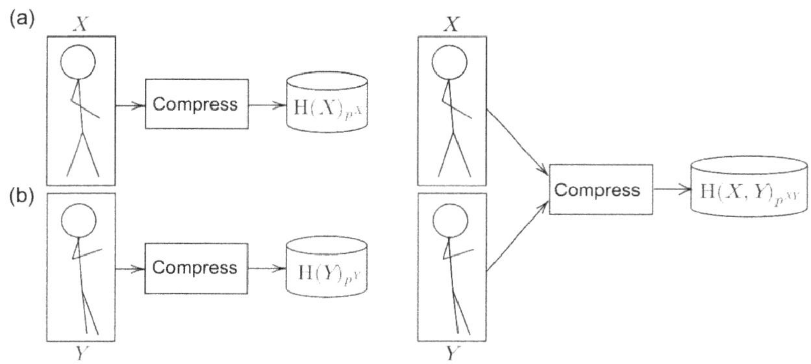

Figure 12.6 Illustration of the problem of compressing two successive frames in a video, which constitute the two subsystems X and Y of interest. Compressing the two frames separately requires approximately $\mathrm{H}(X)_{p^X} + \mathrm{H}(Y)_{p^Y}$ bits (a). In contrast, joint compression requires $\mathrm{H}(X, Y)_{p^{XY}} \leq \mathrm{H}(X)_{p^X} + \mathrm{H}(Y)_{p^Y}$ bits (b).

Local compression is not, however, how practical video compression schemes work. In fact, local compression fails to account for the fact that the two successive frames are correlated, typically having a high level of redundancy. It is thus more efficient to compress both frames jointly, requiring $\mathrm{H}(X, Y)_{p^{XY}}$ bits. Intuitively, the difference $\mathrm{H}(X)_{p^X} + \mathrm{H}(Y)_{p^Y} - \mathrm{H}(X, Y)_{p^{XY}}$ measures the level of redundancy in the two frames, capturing the increase in the length of the compressed file when the two frames are compressed separately rather than jointly.

Motivated by the last example, the Shannon mutual information is introduced next.

Definition 12.7 Shannon mutual information

For a bipartite system with joint distribution p^{XY} and corresponding marginal distributions p^X and p^Y, the Shannon mutual information is defined as

$$\mathrm{I}(X;Y)_{p^{XY}} = \mathrm{H}(X)_{p^X} + \mathrm{H}(Y)_{p^Y} - \mathrm{H}(X, Y)_{p^{XY}}. \tag{12.47}$$

This definition can be conveniently understood using the information-as-a-set representation in Figure 12.4. Recall that the local entropies are represented by the sizes of the two sets and the joint entropy by the size of the union. Thus, the definition (12.47) implies that the Shannon mutual information is represented by the size of the intersection between the two sets representing the local entropies (see Table 12.1). While the union captures the overall information present in either subsystem, the intersection captures the redundant information that is present in both subsystems.

12.6.2 Properties of the Shannon Mutual Information

To be a valid and useful measure of correlation, the Shannon mutual information must equal zero when the two subsystems are uncorrelated, while it must be positive for correlated subsystems. This property, formalized next, is satisfied by the Shannon mutual information, as it follows directly from the subadditivity property (12.32).

Key Result 12.7 Non-negativity and minimum value of the Shannon mutual information

The Shannon mutual information is non-negative

$$\mathrm{I}(X;Y)_{p^{XY}} \geq 0, \tag{12.48}$$

and it equals zero, $\mathrm{I}(X;Y)_{p^{XY}} = 0$, if and only if the two subsystems are uncorrelated, yielding the equality

$$\mathrm{I}(X;Y)_{p^X \otimes p^Y} = 0. \tag{12.49}$$

Example 12.10

For the joint distribution (12.6) and (12.7), the Shannon mutual information is plotted in Figure 12.7 as a function of the probability p^{eq} for two values of parameter D. As discussed, the subsystems are uncorrelated if $p^{\mathrm{eq}} = 1/D$, and thus with this setting the Shannon mutual information equals zero. For any other value of the probability p^{eq}, the subsystems are correlated, and the Shannon mutual information is accordingly strictly positive, increasing as p^{eq} moves away from the value $p^{\mathrm{eq}} = 1/D$.

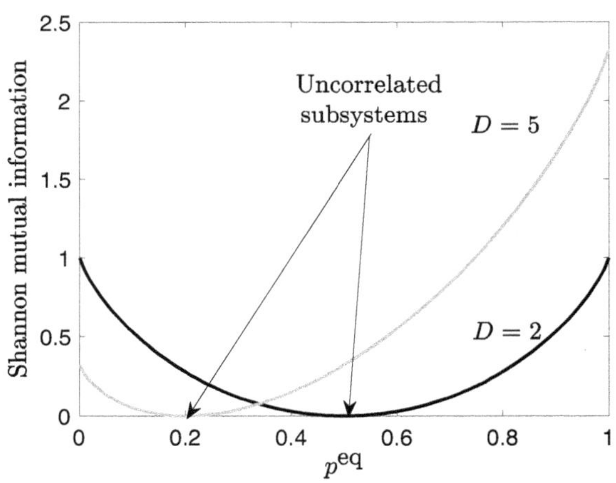

Figure 12.7 Shannon mutual information as a function of the probability p^{eq} for the joint distribution (12.6) and (12.7) for two values of parameter D.

While (12.48) provides a lower bound on the Shannon mutual information, the uncertainty-increases-globally property (12.40) directly implies the following upper bound.

> **Key Result 12.8 Maximum value of the Shannon mutual information**
> The Shannon mutual information satisfies the inequality
>
> $$\mathrm{I}(X;Y)_{p^{XY}} \leq \min(\mathrm{H}(X)_{p^X}, \mathrm{H}(Y)_{p^Y}), \tag{12.50}$$
>
> with equality if and only if at least one of the two variables x and y can be obtained as a function of the other variable.

Like the uncertainty-increases-globally property (12.40), the inequality (12.50) can be interpreted in terms of the information-as-a-set representation. In fact, with the correspondence in Table 12.1, the inequality (12.40) reflects the fact that the size of the intersection of two sets cannot exceed the size of either of the two sets.

Example 12.11

In the recurring example (12.6) and (12.7), the local entropies are equal to $\mathrm{H}(X)_{p^Y} = \mathrm{H}(Y)_{p^Y} = \log_2(D)$ irrespective of the value of the parameter p^{eq}, which determines the correlation between the subsystems. By (12.50), the maximum value of the mutual information is obtained when the equality

$$\mathrm{I}(X;Y)_{p^{XY}} = \mathrm{H}(X)_{p^Y} = \mathrm{H}(Y)_{p^Y} = \log_2(D) \tag{12.51}$$

holds. This condition is attained when one variable can be evaluated as a function of the other variables. As argued earlier, for $D = 2$, this is the case when $p^{\mathrm{eq}} = 0$ and $p^{\mathrm{eq}} = 1$, while for $D > 2$ this condition only holds for $p^{\mathrm{eq}} = 1$, in which case the two variables coincide.

12.7 Correlation as a Resource

This section presents another perspective on correlation by viewing it as a resource of the bipartite system. The approach follows resource theory, a general framework first encountered in Section 6.3.4 for the quantification of informational resources.

12.7.1 Relative Entropy of Correlation

As introduced in Section 6.3.4, a resource refers to a property of an informational state – whether classical or quantum – that distinguishes it from free states, which lack this property. For the resource of correlation, the **free informational states** consist of all joint distributions p^{XY} under which the two subsystems are uncorrelated,

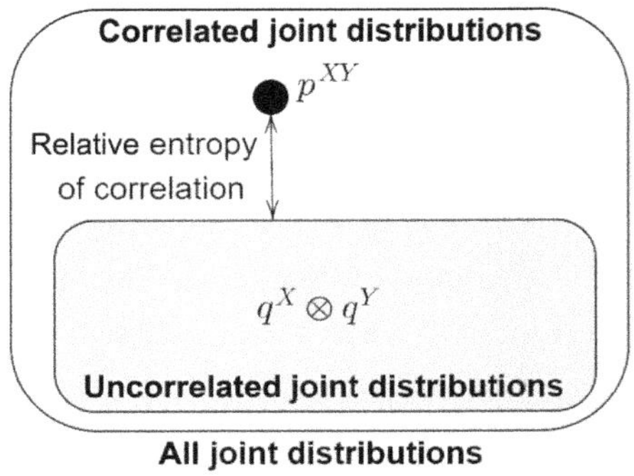

Figure 12.8 Resource-theoretic perspective on correlation. Correlation in a joint distribution p^{XY} is measured by the distance between the joint distribution p^{XY} and the set of uncorrelated distributions $q^X \otimes q^Y$.

i.e., all joint distributions of the form $p^{XY} = p^X \otimes p^Y$ for marginal distributions p^X and p^Y. According to this perspective, as illustrated in Figure 12.8, the extent to which a joint distribution p^{XY} is correlated is evaluated through a measure of the distance between the joint distribution p^{XY} and the set of uncorrelated joint distributions.

Using any relative entropy $\mathrm{D}_\alpha(\cdot||\cdot)$ introduced in Chapter 4 to measure the distance between joint distributions yields the following definition of resource-based correlation measure.

Definition 12.8 Relative entropy of correlation

Fix any relative entropy $\mathrm{D}_\alpha(\cdot||\cdot)$. For a joint distribution p^{XY}, the corresponding relative entropy of correlation is defined as

$$\mathrm{I}_{\mathrm{D}_\alpha}(X;Y)_{p^{XY}} = \min_{q^X,q^Y} \mathrm{D}_\alpha(p^{XY}||q^X \otimes q^Y), \tag{12.52}$$

where the minimization is over all possible uncorrelated joint distributions $q^{XY} = q^X \otimes q^Y$ obtained from an $N^X \times 1$ marginal probability distribution q^X and an $N^Y \times 1$ marginal probability distribution q^Y.

This definition directly implies the following key property.

Key Result 12.9 Non-negativity and minimum value of the relative entropy of correlation

For any joint distribution p^{XY}, the relative entropy of correlation is non-negative

$$\mathrm{I}_{\mathrm{D}_\alpha}(X;Y)_{p^{XY}} \geq 0. \tag{12.53}$$

Furthermore, the Shannon relative entropy of correlation, as well as the Tsallis and Rényi relative entropies of correlation for any order $\alpha > 0$, equal zero, $\mathrm{I}_{\mathrm{D}_\alpha}(X;Y)_{p^{XY}} = 0$, if and only if the two subsystems are uncorrelated, yielding the equality

$$\mathrm{I}_{\mathrm{D}_\alpha}(X;Y)_{p^X \otimes p^Y} = 0. \tag{12.54}$$

In general, evaluating the minimization in the definition (12.52) of the relative entropy of correlation is not easy. However, as shown in Appendix 12.B, in the special case of the Shannon relative entropy, the optimal distributions q^X and q^Y in (12.52) are given by the corresponding marginals p^X and p^Y of the joint distribution p^{XY}, yielding the following important result.

Key Result 12.10 The Shannon relative entropy of correlation coincides with the Shannon mutual information
The Shannon relative entropy of correlation is evaluated as

$$\mathrm{I}_{\mathrm{D}}(X;Y)_{p^{XY}} = \min_{q^X, q^Y} \mathrm{D}(p^{XY} || q^X \otimes q^Y) = \mathrm{D}(p^{XY} || p^X \otimes p^Y), \tag{12.55}$$

which equals the Shannon mutual information, i.e.,

$$\mathrm{I}_{\mathrm{D}}(X;Y)_{p^{XY}} = \mathrm{I}(X;Y)_{p^{XY}}. \tag{12.56}$$

This result reveals that the Shannon mutual information measures correlation not only as redundancy, as discussed Section 12.6, but also as a resource. To prove the equality (12.56) from (12.55), using the original definition (12.47) of the Shannon mutual information one has

$$\begin{aligned}
\mathrm{I}(X;Y)_{p^{XY}} &= \mathrm{H}(X)_{p^X} + \mathrm{H}(Y)_{p^Y} - \mathrm{H}(X, Y)_{p^{XY}} \\
&= -\mathrm{E}_{p^X}[\log_2(p^X_{\mathrm{x}})] - \mathrm{E}_{p^Y}[\log_2(p^Y_{\mathrm{y}})] + \mathrm{E}_{p^{XY}}[\log_2(p^{XY}_{\mathrm{xy}})] \\
&= -\mathrm{E}_{p^{XY}}[\log_2(p^X_{\mathrm{x}})] - \mathrm{E}_{p^{XY}}[\log_2(p^Y_{\mathrm{y}})] + \mathrm{E}_{p^{XY}}[\log_2(p^{XY}_{\mathrm{xy}})] \\
&= \mathrm{E}_{p^{XY}}\left[\log_2\left(\frac{p^{XY}_{\mathrm{xy}}}{p^X_{\mathrm{x}} p^Y_{\mathrm{y}}}\right)\right] \\
&= \mathrm{D}(p^{XY} || p^X \otimes p^Y) \\
&= \mathrm{I}_{\mathrm{D}}(X;Y)_{p^{XY}},
\end{aligned} \tag{12.57}$$

where the third equality follows from the definition of marginal distributions.

The equality (12.56) provides a simple way to prove the subadditivity property (12.32) of the Shannon joint entropy, which is equivalent to the non-negativity (12.48) of the Shannon mutual information. In fact, by the property (4.22) of the Shannon relative entropy presented in Chapter 4, the term $\mathrm{D}(p^{XY} || p^X \otimes p^Y)$ is non-negative, and it equals zero if and only if the joint distribution p^{XY} equals the Kronecker product $p^X \otimes p^Y$ and thus the two subsystems are independent.

The counterpart of the equality (12.56) does not hold for the Tsallis or Rényi relative entropies, as the solution of the problem (12.52) is generally not given by the marginals p^X and p^Y when $\alpha \neq 1$. This is illustrated with an example in Appendix 12.C.

12.7.2 Local Channels ⏭

Apart from defining free states, a central tenet of resource theory is the identification of transformations that preserve the resource level. These transformations, termed **free transformations**, cannot increase the resource level. Conversely, the resource level can be increased only via non-free transformations, also known as **resource transformations**.

For a physical analogy, consider the resource of potential kinetic energy. A ball at ground level represents a free state. Any transformation that keeps the ball at ground level is a free transformation. To transition to a resource state, i.e., to a higher potential energy, one must lift the ball to a higher level, which constitutes a resource transformation.

To analyze transformations for the resource of correlation, recall that valid transformations of probability distributions are represented by channels. Following Chapter 9, a channel acting on a joint distribution is mathematically described by a column-stochastic matrix K of size $N^X N^Y \times N^X N^Y$.

A free transformation for the resource of correlation is a channel acting on a bipartite system that preserves the uncorrelated nature of pre-channel distributions. Specifically, if the input distribution p^{XY} is uncorrelated, i.e., $p^{XY} = p^X \otimes p^Y$, the post-channel distribution q^{XY} must remain uncorrelated:

$$q^X \otimes q^Y = K(p^X \otimes p^Y) \tag{12.58}$$

for some marginals q^X and q^Y. If q^{XY} were correlated, the channel K would have introduced correlation, violating the definition of a free channel.

Intuitively, a channel satisfying the above condition for any uncorrelated input distribution must act locally. A local channel applies separate transformations to the two subsystems.

Definition 12.9 Local channels

Fix a channel that operates only on subsystem X via an $N^X \times N^X$ column-stochastic matrix, as well as a channel that operates only on subsystem Y via an $N^Y \times N^Y$ column-stochastic matrix K^Y. Given any joint distribution p^{XY}, the post-channel distribution obtained from the separate application of channel K^X to subsystem X and of channel K^Y to subsystem Y is given by

$$q^{XY} = (K^X \otimes K^Y)p^{XY}. \tag{12.59}$$

Accordingly, any channel with channel matrix of the form

$$K = K^X \otimes K^Y \tag{12.60}$$

is said to be local.

Local channels preserve the uncorrelated nature of distributions. To verify this, apply the mixed product property (8.169) of the Kronecker product to obtain

$$\begin{aligned} K(p^X \otimes p^Y) &= (K^X \otimes K^Y)(p^X \otimes p^Y) \\ &= K^X p^X \otimes K^Y p^Y. \end{aligned} \tag{12.61}$$

Hence, the post-channel distribution remains uncorrelated, with marginals $q^X = K^X p^X$ and $q^Y = K^Y p^Y$.

As illustrated in Figure 12.9, a local channel maps uncorrelated distributions to uncorrelated distributions and, therefore, cannot create correlation. To generate or increase correlation, one must apply a global channel, which cannot be expressed in the local form $K = K^X \otimes K^Y$. Intuitively, a global channel operates jointly across both subsystems, introducing interaction between them and thereby increasing the correlation level.

12.7.3 Data Processing Inequality Under Local Channels ⏭

The following result, proved in Appendix 12.D as an application of the data processing inequality (Section 9.5.2), formalizes the fact that local channels cannot increase the correlation level.

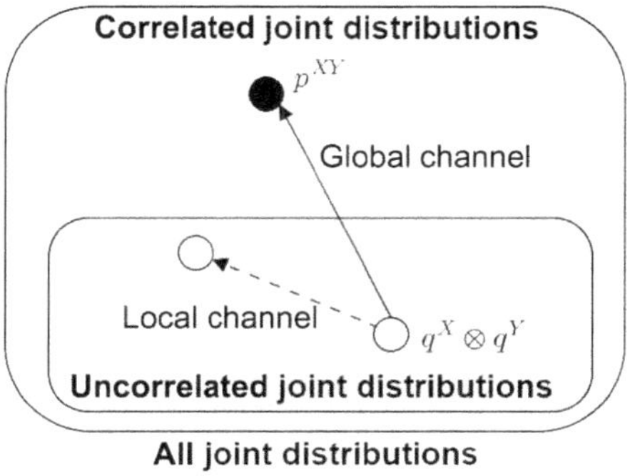

Figure 12.9 Local channels (12.60) constitute free transformations for the resource of correlation, in the sense that a local channel cannot create or increase correlation. To increase the level of correlation, one needs to implement a global channel that applies jointly on both subsystems.

Key Result 12.11 Data processing inequality for the relative entropy of correlation under local channels
For any joint distribution p^{XY} and any local channel $K^X \otimes K^Y$, the relative entropy of correlation satisfies the inequality

$$\mathrm{I}_{\mathrm{D}_\alpha}(X;Y)_{p^{XY}} \geq \mathrm{I}_{\mathrm{D}_\alpha}(X;Y)_{(K^X \otimes K^Y)p^{XY}}. \tag{12.62}$$

According to the inequality (12.62), the correlation – measured by any relative entropy of correlation – cannot increase as the result of the application of local channels. The only way to create, or increase, correlation is hence to implement a global channel.

When specialized to the Shannon relative entropy of correlation, given the equality (12.56), the inequality (12.62) yields the following important property of the Shannon mutual information.

Key Result 12.12 Data processing inequality of the Shannon mutual information under local channels
For any joint distribution p^{XY} and any local channel $K^X \otimes K^Y$, the Shannon mutual information satisfies the inequality

$$\mathrm{I}(X;Y)_{p^{XY}} \geq \mathrm{I}(X;Y)_{(K^X \otimes K^Y)p^{XY}}. \tag{12.63}$$

Example 12.12

Assume that the initial distribution p^{XY} is given by (12.9), i.e., $p^{XY} = 0.5[p^{\mathrm{eq}}, 1 - p^{\mathrm{eq}}, 1 - p^{\mathrm{eq}}, p^{\mathrm{eq}}]^T$. A local channel is applied to the bipartite system, with the two channels given by

$$K^X = K^Y = \frac{1}{2}\begin{bmatrix} 1 & 1 \\ 1 & 1 \end{bmatrix}. \tag{12.64}$$

As a result, using (12.59), the post-channel distribution is

$$q^{XY} = (K^X \otimes K^Y)p^{XY} = \frac{1}{4}\begin{bmatrix} 1 & 1 & 1 & 1 \\ 1 & 1 & 1 & 1 \\ 1 & 1 & 1 & 1 \\ 1 & 1 & 1 & 1 \end{bmatrix} p^{XY} = u. \tag{12.65}$$

Thus, irrespective of the probability p^{eq}, as a result of the free transformation $K^X \otimes K^Y$, the two subsystems become independent and the Shannon mutual information takes the minimal value $\mathrm{I}(X;Y)_{(K^X \otimes K^Y)p^{XY}} = 0$.

Example 12.13

While a local channel cannot create correlations, it is now demonstrated that a global channel K can increase the correlation level. To this end, for a system with dimensions $N^X = N^Y = 2$, take the global channel matrix

$$K = \begin{bmatrix} 1 & 1 & 0 & 0 \\ 0 & 0 & 0 & 0 \\ 0 & 0 & 0 & 0 \\ 0 & 0 & 1 & 1 \end{bmatrix}. \tag{12.66}$$

It can be verified that the channel is global, as it cannot be expressed as the Kronecker product (12.60) for some local channels K^X and K^Y.

The effect of this channel is to take any state (x, y) and transform it into the state (x, x). In other words, this channel copies the state of subsystem X into the state of subsystem Y. This can easily be checked by verifying the effect of the channel on the four zero-entropy pre-channel distributions $p^{XY} = \delta_{00}$, $p^{XY} = \delta_{01}$, $p^{XY} = \delta_{10}$, and $p^{XY} = \delta_{11}$.

Through the interactions between the two subsystems caused by the outlined copying operation, the channel can increase the correlation level. In fact, given the pre-channel distribution (12.9), the post-channel distribution is given by

$$q^{XY} = Kp^{XY} = \begin{bmatrix} \frac{1}{2} \\ 0 \\ 0 \\ \frac{1}{2} \end{bmatrix}. \tag{12.67}$$

Therefore, irrespective of the value of the probability p^{eq}, after the transformation, the two subsystems have the same state $x = y = 0$ or $x = y = 1$. Accordingly, the mutual information $\mathrm{I}(X;Y)_{(K^X \otimes K^Y)p^{XY}}$ attains the maximum value in (12.50), i.e., $\mathrm{I}(X;Y)_{(K^X \otimes K^Y)p^{XY}} = \mathrm{H}(X)_{p^X} = \mathrm{H}(Y)_{p^Y} = 1$.

Example 12.14

Following up on the last two examples, this example studies a situation in which with probability γ one applies the global channel K in (12.66), while with probability $1 - \gamma$ the local channel $K^X \otimes K^Y$ in (12.64) is applied. The resulting transformation is described by the 4×4 column-stochastic matrix

$$K_\gamma = (1 - \gamma)(K^X \otimes K^Y) + \gamma K. \tag{12.68}$$

Figure 12.10 shows the Shannon mutual information $\mathrm{I}(X;Y)_{K_\gamma p^{XY}}$ of the post-channel distribution as a function of the probability γ for $p^{\text{eq}} = 0.2$ for the pre-channel distribution (12.9).

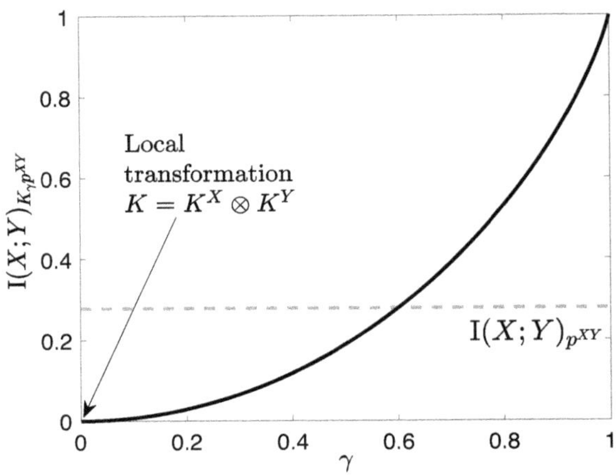

Figure 12.10 Pre-channel Shannon mutual information $I(X;Y)_{p^{XY}}$ and post-channel Shannon mutual information $I(X;Y)_{K_\gamma p^{XY}}$ for the channel K_γ in (12.68) and the pre-channel distribution (12.9) as a function of the probability γ with $p^{\text{eq}} = 0.2$.

For reference, the dashed line represents the Shannon mutual information $I(X;Y)_{p^{XY}}$ for the pre-channel distribution p^{XY}. When $\gamma = 0$, the transformation K_γ is local, and hence it cannot increase the Shannon mutual information. In this case, the mutual information is decreased to its minimum value $I(X;Y)_{K_0 p^{XY}} = 0$. In the opposite situation with $\gamma = 1$, the two subsystems become fully correlated after the transformation, and the mutual information acquires the maximum value $I(X;Y)_{K_1 p^{XY}} = 0$. As γ increases from 0 to 1, the mutual information of the post-transformation state also grows from 0 towards 1, exhibiting a larger correlation than the pre-channel distribution p^{XY} for sufficiently large values of γ.

12.7.4 Tsallis Mutual Information and Rényi Mutual Information ⏭

As discussed in this section, for the Tsallis and Rényi relative entropies of correlation, the defining problems (12.52) are not easily solved in general. For this reason, it is common to leverage the optimal solution for the Shannon relative entropy of correlation, and define alternative correlation measures by setting the local distributions q^X and q^Y to equal the marginal distributions as $(q^X = p^X,\ q^Y = p^Y)$. This approach yields the following definitions.

Definition 12.10 Tsallis mutual information and Rényi mutual information
The Tsallis mutual information of order $\alpha \geq 0$ is defined as

$$I_\alpha^{\text{T}}(X;Y)_{p^{XY}} = D_\alpha^{\text{T}}(p^{XY} || p^X \otimes p^Y), \tag{12.69}$$

while the Rényi mutual information of order $\alpha \geq 0$ is defined as

$$\mathrm{I}^{\mathrm{R}}_{\alpha}(X;Y)_{p^{XY}} = \mathrm{D}^{\mathrm{R}}_{\alpha}(p^{XY}||p^X \otimes p^Y). \tag{12.70}$$

While these metrics do not correspond to relative entropies of correlations as defined in (12.52), they still have some useful properties as measures of correlation. Indeed, they both recover the Shannon mutual information for $\alpha = 1$. Moreover, they satisfy the same non-negativity and minimum value property (12.53) and (12.54) of the relative entropies of correlation. By the properties of the relative entropy, the Tsallis mutual information and the Rényi mutual information are increasing – more precisely, non-decreasing – functions of the order α.

While the definitions (12.69) and (12.70) are common, there are other related notions of mutual information constructed from Tsallis or Rényi relative entropies. An example is the **Sibson mutual information** of order $\alpha \geq 0$, which is defined as

$$\min_{q^Y} \mathrm{D}^{\mathrm{R}}_{\alpha}(p^{XY}||p^X \otimes q^Y), \tag{12.71}$$

where the optimization is only over distribution q^Y of subsystem Y. The reader is referred to the Recommended Resources for further discussions.

12.7.5 Optimal Detection of Correlated Random Variables ⏭

The relative entropy of correlation (12.52), as well as the mutual informations (12.69)–(12.71), can be interpreted operationally through the lens of **optimal detection**. To elaborate, assume that an observation (x, y) may have been generated from either the joint distribution p^{XY} or from the joint distribution $q^{XY} = q^X \otimes q^Y$, under which the two variables are independent. Note that in the latter case the two variables x and y can be generated separately and in parallel as $\mathrm{x} \sim q^X$ and $\mathrm{y} \sim q^Y$. A detector, aware of the joint distributions p^{XY} and q^{XY}, is tasked with determining which distribution was used to generate the observed sample (x, y).

From Section 4.7, the maximal detection probability can be expressed as

$$\mathrm{P}_{\mathrm{S}}(d^*) = \frac{1}{2}(1 + \mathrm{D}^{\mathrm{tr}}(p^{XY}, q^X \otimes q^Y)), \tag{12.72}$$

where $\mathrm{D}^{\mathrm{tr}}(\cdot,\cdot)$ is the trace distance, or total variation distance. The trace distance can be related to the relative entropies as discussed in Section 4.7, thus connecting the detection probability (12.72) with the relative entropy of correlation.

Notably, using Pinsker's inequality (4.100), for the choice $q^X = p^X$ and $q^Y = p^Y$, one gets the upper bound

$$\mathrm{P}_{\mathrm{S}}(d^*) \leq \frac{1}{2}\left(1 + \sqrt{\frac{1}{2 \cdot 1_2}\mathrm{I}(X;Y)_{p^{XY}}}\right) \tag{12.73}$$

in terms of the Shannon mutual information. Hence, the more correlated the two subsystems are in terms of the Shannon mutual information, the easier it is to distinguish them from independent subsystems with the same marginal distributions.

12.7.6 Local Randomness Required to Destroy Correlations ⏭

An alternative perspective on correlation, which can be seen as dual to the resource-based approach discussed in this section, focuses on its **destruction** rather than its creation. In this view, instead of treating correlation as a global resource that requires global operations to be created, one measures the amount of local randomness required to destroy the correlation.

This perspective is illustrated next through an example.

Example 12.15

Consider the maximally correlated distribution (12.10), for which the Shannon mutual information equals $\mathrm{I}(X;Y)_{p^{XY}} = 1$. Recall that the marginal distributions p^X and p^Y are both uniform, i.e., $p^X = p^Y = [0.5, 0.5]^T$. What would it take for, say, subsystem X to delete the existing perfect correlation via a local operation, while still maintaining the marginal distributions intact?

To destroy the correlation between the two subsystems, subsystem X could generate locally a uniformly distributed bit, and then flip the local state $\mathrm{x} \in [2]$ if the coin returns heads, while leaving the state x unchanged otherwise. This local transformation is described by the 2×2 column-stochastic matrix $K^X = 0.5P_1 + 0.5I$, which coincides with (12.64), where the Pauli X matrix P_1 describes the bit flip operation.

Since no transformation is applied to subsystem Y, the overall channel is given by the 4×4 column-stochastic matrix $(K^X \otimes I)$, and thus the post-transformation distribution is given by

$$(K^X \otimes I)p^{XY} = p^X \otimes p^Y, \tag{12.74}$$

as can be checked by direct computation. By leveraging 1 bit of local randomness at subsystem X, this local transformation preserves the marginal distributions while completely erasing the correlation in distribution p^{XY}. One may hence take the 1 bit of local randomness as a measure of the correlation in the original distribution p^{XY}. For this example, this measure coincides with the Shannon mutual information.

12.8 Correlation as Common Sampling Randomness ⏭

This section analyzes correlations in bipartite systems by evaluating the complexity of simulating measurement outcomes in a distributed way at two agents.

12.8.1 Distributed Sampling

As illustrated in Figure 12.11, agent X is tasked with generating a random variable $\mathrm{x} \in [N^X]$, while agent Y must generate a random variable $\mathrm{y} \in [N^Y]$. The goal is for the pair (x, y) to be jointly distributed according to a target joint distribution p^{XY}, i.e., $(\mathrm{x}, \mathrm{y}) \sim p^{XY}$. To achieve this, each agent has unrestricted access to a

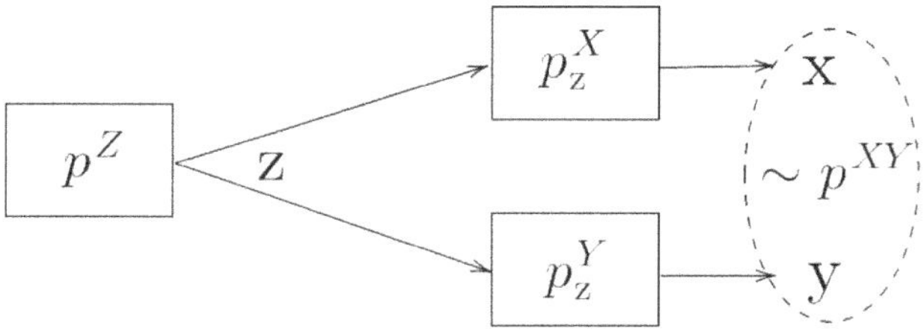

Figure 12.11 In the distributed sampling problem, agent X is tasked with generating random variable $\mathrm{x} \in [N^X]$, while agent Y must generate random variable $\mathrm{y} \in [N^Y]$, with the requirement that the two random variables (x, y) be jointly distributed according to the target joint distribution p^{XY}, i.e., $(\mathrm{x}, \mathrm{y}) \sim p^{XY}$. The minimum amount of randomness in the common random variable z represents a measure of the correlation present in the distribution p^{XY}. This measure is known as the distributed-sampling common information.

source of local randomness. However, what is limited is the availability of common randomness – randomness that is shared between both agents.

Common randomness is essential to ensure that the two random variables x and y exhibit the desired correlations. Without access to common randomness, the agents can only generate uncorrelated random variables x and y based solely on local randomness.

As explored in this section, the minimum amount of common randomness required to simulate the joint distribution $(\mathrm{x}, \mathrm{y}) \sim p^{XY}$ in a distributed manner serves as a quantitative measure of the correlations encoded in the joint distribution p^{XY}.

To formalize this setting, as illustrated in Figure 12.11, the two agents are assumed to have access to a common random variable $\mathrm{z} \sim p^Z$ taking value in the set $[N^Z] = \{0, 1, \ldots, N^Z - 1\}$. Using this resource, distributed sampling takes place as follows.

1. **Shared random variable generation:** Generate a shared realization z of the common random variable $\mathrm{z} \sim p^Z$ at both agents X and Y.
2. **Local sampling at agent X:** Given the shared realization z of the common random variable z, agent X uses a source of local randomness to generate a random variable $\mathrm{x} \sim p_z^{X|Z}$ for some $N^X \times 1$ probability vector $p_z^{X|Z}$.
3. **Local sampling at agent Y:** Given the shared realization z of the common random variable z, agent Y uses a source of local randomness to draw a random variable $\mathrm{y} \sim p_z^{Y|Z}$ for some $N^Y \times 1$ probability vector $p_z^{Y|Z}$.

It is emphasized that the bar notation in the superscripts of the distributions $p_z^{X|Z}$ and $p_z^{Y|Z}$ serves as a reminder that these distributions, referring to random variables x and y, respectively, are indexed by the integer z.

According to the procedure detailed above, a distributed sampling protocol is described by the parameters

$$N^Z, p^Z, \{p_z^{X|Z}, p_z^{Y|Z}\}_{z\in[N^Z]}, \tag{12.75}$$

where N^Z is the number of values that can be taken by the common random variable z; p^Z is the $N^Z \times 1$ distribution of the common random variable z; and the N^Z pairs of distributions $\{p_z^{X|Z}, p_z^{Y|Z}\}_{z\in[N^Z]}$ describe the local sampling operations at the agents given the realization z of the common random variable z.

Using this distributed sampling procedure, for any fixed value $\mathrm{z} = z$ of the common random variable z, the random variables x and y are uncorrelated, since they are functions solely of the respective local sources of randomness at the two agents. Correlation arises exclusively due to the dependence on the common random variable z.

Mathematically, for any fixed value $\mathrm{z} = z$ of the shared variable z, the joint distribution of the generated random variables (x, y) is given by the uncorrelated distribution

$$p_z^{XY} = p_z^{X|Z} \otimes p_z^{Y|Z}. \tag{12.76}$$

By the law of total probability, averaging over the shared randomness $\mathrm{z} \sim p^Z$, the joint distribution of the generated random variables is thus obtained as $\sum_{z\in[N^Z]} p_z^Z p_z^{X|Z} \otimes p_z^{Y|Z}$. As mentioned, the design goal is for this joint distribution to equal the target joint distribution p^{XY}, i.e.,

$$p^{XY} = \sum_{z\in[N^Z]} p_z^Z p_z^{X|Z} \otimes p_z^{Y|Z}. \tag{12.77}$$

12.8.2 Distributed-Sampling Common Information

If the two random variables are uncorrelated, i.e., if $p^{XY} = p^X \otimes p^Y$, it is clearly sufficient to use only local randomness. This can be accomplished by choosing the setting (12.75) as $N^Z = 1$, $p_0^X = p^X$, and $p_0^Y = p^Y$. This way, the shared random variable has zero entropy, $\mathrm{H}_\alpha(Z)_{p^Z} = 0$. More generally, in order to quantify the level of correlation associated with the joint distribution p^{XY}, one is interested in the minimum entropy $\mathrm{H}_\alpha(Z)_{p^Z}$ of a shared random variable $\mathrm{z} \sim p^Z$ for which one can find distributions $\{p_z^{X|Z}\}_{z\in[N^Z]}$ and $\{p_z^{Y|Z}\}_{z\in[N^Z]}$ satisfying the condition (12.77).

Definition 12.11 Distributed-sampling common information

Given a joint distribution p^{XY}, the distributed-sampling common information is defined as the result of the optimization

$$\mathrm{C}_\alpha(X;Y)_{p^{XY}} = \min \mathrm{H}_\alpha(Z)_{p^Z}, \tag{12.78}$$

where the minimization is over all the distributed sampling settings (12.75) that satisfy the requirement (12.77).

The distributed-sampling common information (12.78) requires a case-by-case non-trivial numerical evaluation in order to be computed. Approximate solutions can be found via non-negative matrix factorization algorithms (see Recommended Resources). Otherwise, in general, the distributed-sampling common information (12.78) satisfies the following properties.

Definition 12.12 Bounds on the distributed-sampling common information

For any order $\alpha \geq 0$ and any joint distribution p^{XY}, the distributed-sampling common information satisfies the inequalities

$$0 \leq \mathrm{C}_\alpha(X;Y)_{p^{XY}} \leq \min(\mathrm{H}_\alpha(X)_{p^X}, \mathrm{H}_\alpha(Y)_{p^Y}), \tag{12.79}$$

with the lower bound achieved if and only if the two subsystems are uncorrelated.

The lower bound in (12.78) follows by the non-negativity of the entropy, while the upper bound is proved in Appendix 12.D.

When specialized to the Shannon entropy, it can also be shown that the distributed-sampling common information $\mathrm{C}(X;Y)_{p^{XY}}$ is no smaller than the Shannon mutual information, i.e.,

$$\mathrm{C}(X;Y)_{p^{XY}} \geq \mathrm{I}(X;Y)_{p^{XY}}. \tag{12.80}$$

Thus, simulating correlated random variables requires no less randomness than the redundant information measured by Shannon mutual information. This will be proved in Section 12.9.

12.9 Correlation as Distillable Common Information ⏭

This section takes yet another perspective on the quantification of correlation between two subsystems by studying the problem of **distributed distillation**.

12.9.1 Distributed Distillation

As illustrated in Figure 12.12, in this problem there are two agents, X and Y, with agent X observing the state x of subsystem X and agent Y observing the state y of subsystem Y. As usual, the two observations are jointly distributed as $(\mathrm{x}, \mathrm{y}) \sim p^{XY}$. The goal is for the two agents to **distill**, i.e., extract, the same random variable z from their respective observations. The maximum entropy of a random variable produced according to this protocol represents a measure of the correlation present in the joint distribution p^{XY}.

Mathematically, agent X applies function $f^X(\cdot)$ to extract a real-valued, scalar, random variable $f^X(\mathrm{x})$, while agent Y applies a function $f^Y(\cdot)$, producing the real-valued, scalar, random variable $f^Y(\mathrm{y})$. The goal of distributed distillation is to ensure that, for all possible observations (x, y) with non-zero probability $p^{XY}_{xy} > 0$,

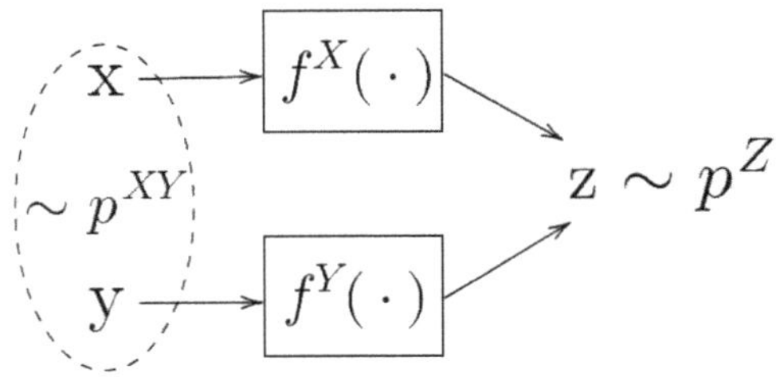

Figure 12.12 In the distributed distillation problem, also known as distributed key generation problem, agent X observes the state x of subsystem X and agent Y observing the state y of subsystem Y, with the two states having joint distribution p^{XY}. The goal is for the two agents to extract the same random variable z from the respective observations. The maximum entropy of a common random variable obtained via this type of protocols represents a measure of the correlation present in the joint distribution p^{XY}. This is known as the distillable common information.

the two functions output the same result, i.e.,

$$f^X(x) = f^Y(y) \text{ for all } (x, y) \text{ with } p^{XY}_{xy} > 0. \tag{12.81}$$

Accordingly, the random variable

$$\mathrm{z} = f^X(\mathrm{x}) = f^Y(\mathrm{y}) \tag{12.82}$$

represents common distillable information that can be obtained separately by both agents.

The problem at hand is also known as **distributed key generation**. This terminology refers to the cryptographic task of extracting a shared private key z at two agents based on locally available information. In this application, the two agents extract the common variable z in (12.82) from their respective local sources of information x and y. The variable z can then be used as a **shared private key** to enable private communication between the two agents.

12.9.2 Distillable Common Information

The condition (12.81) is easy to satisfy by choosing functions $f^X(\cdot)$ and $f^Y(\cdot)$ that return a fixed output value, irrespective of the inputs. However, the resulting shared random variable z would have zero entropy, $\mathrm{H}_\alpha(Z)_{p^Z} = 0$, yielding no insights into the correlation between the two subsystems. In order to quantify the correlation level associated with a joint distribution p^{XY}, one is interested in evaluating the maximum value of the entropy $\mathrm{H}_\alpha(Z)_{p^Z}$ for any shared random variable z that can be distilled from the two random variables (x, y) as in (12.82).

Definition 12.13 Distillable common information, aka Gaćs–Körner common information
For any order $\alpha \geq 0$ and any joint distribution p^{XY}, the distillable common information is defined as

$$\mathfrak{C}_\alpha(X;Y)_{p^{XY}} = \max_{f^X(\cdot), f^Y(\cdot):\, \mathrm{z}=f^X(\mathrm{x})=f^Y(\mathrm{y})} \mathrm{H}_\alpha(Z)_{p^Z}, \tag{12.83}$$

where the minimization is over all pairs of functions $f^X(\cdot)$ and $f^Y(\cdot)$ satisfying the condition (12.81).

Unlike the distributed-sampling common information, the distillable common information (12.83) can be evaluated explicitly, although it is often equal to zero. In particular, the optimal shared variable z, along with the corresponding functions $f^X(\cdot)$ and $f^Y(\cdot)$, can be identified from the joint probability matrix P^{XY}. This is the $N^X \times N^Y$ matrix with entries $P^{XY}_{x,y} = p^{XY}_{xy}$ for $x \in [N^X]$ and $y \in [N^Y]$. As illustrated by the following example, the procedure involves re-ordering the rows and columns until the matrix P^{XY} is in the block-diagonal form with the largest number of blocks (see Recommended Resources for more details).

Example 12.16

Consider two variables taking $N^X = N^Y = 4$ values each. The joint distribution p^{XY} has entries $p^{XY}_{00} = 1/32$, $p^{XY}_{01} = 3/32$, $p^{XY}_{10} = 3/32$, $p^{XY}_{11} = 9/32$, $p^{XY}_{22} = 1/32$, $p^{XY}_{23} = 3/32$, $p^{XY}_{32} = 3/32$, and $p^{XY}_{33} = 9/32$, while all other probabilities are equal to zero. This joint distribution yields the joint probability matrix

$$P^{XY} = \frac{1}{32}\begin{bmatrix} 1 & 3 & 0 & 0 \\ 3 & 9 & 0 & 0 \\ 0 & 0 & 1 & 3 \\ 0 & 0 & 3 & 9 \end{bmatrix}. \tag{12.84}$$

In order to identify the optimal common variable z by solving problem (12.83), it is convenient to think of the joint probability matrix (12.84) as in Table 12.2. The table highlights the values of the random variables x and y corresponding to each probability.

Table 12.2 Representation of the probability matrix (12.84) in the form of a table

$X \backslash Y$	0	1	2	3
0	1/32	3/32	0	0
1	3/32	9/32	0	0
2	0	0	1/32	3/32
3	0	0	3/32	9/32

As also made clear by Table 12.2, the probability matrix (12.84) is already in block-diagonal form. The presence of two diagonal blocks in the joint probability matrix implies that it is possible to extract a shared variable z taking $N^Z = 2$ values. In fact, one can readily check that the functions

$$f^X(x) = \begin{cases} 0 & \text{if } x = 0 \text{ or } x = 1 \\ 1 & \text{otherwise} \end{cases}$$
$$\text{and}\, f^Y(y) = \begin{cases} 0 & \text{if } y = 0 \text{ or } y = 1 \\ 1 & \text{otherwise} \end{cases} \quad (12.85)$$

return the uniformly distributed binary random variable $\mathrm{z} = f^X(\mathrm{x}) = f^Y(\mathrm{y})$ with Shannon entropy $\mathrm{H}(Z)_{p^Z} = 1$.

Furthermore, it can be verified that it is not possible to increase the number of diagonal blocks by reordering the rows and columns of the matrix. This entails that the Shannon distillable common information $\mathfrak{C}(X;Y)_{p^{XY}}$ is equal to the entropy of the random variable z, i.e., $\mathfrak{C}(X;Y)_{p^{XY}} = \mathrm{H}(Z)_{p^Z} = 1$.

In practice, the distillable common information (12.83) is often equal to zero. For instance, following the construction illustrated by Example 12.6, this happens if all the probabilities p^{XY}_{xy}, for all $x \in [N^X]$ and $y \in [N^Y]$, are strictly positive.

When using the Shannon entropy, generalizing the inequality (12.80), one can relate the distributed-sampling common information (12.78), the distillable common information (12.83), and the mutual information as follows.

Definition 12.14 Distributed-sampling common information, distillable common information, and mutual information

For any joint distribution p^{XY}, one has the inequalities

$$\mathfrak{C}(X;Y)_{p^{XY}} \leq \mathrm{I}(X;Y)_{p^{XY}} \leq \mathrm{C}(X;Y)_{p^{XY}}. \quad (12.86)$$

The inequalities in (12.86) show that the amount of shared randomness required to simulate a bipartite system in a distributed manner – quantified by the distributed-sampling common information $\mathrm{C}(X;Y)_{p^{XY}}$ – is at least as large as the mutual information $\mathrm{I}(X;Y)_{p^{XY}}$. Conversely, the shared randomness that can be extracted from the system in a distributed manner cannot exceed the mutual information $\mathrm{I}(X;Y)_{p^{XY}}$. Therefore, the shared randomness needed to simulate distributed random variables is no greater than the amount of randomness that can be extracted from them. These inequalities can be derived using the tools introduced in Chapter 19 (refer to Appendix 19.D for details).

12.9.3 Maximal Correlation

In order to ascertain whether the distillable common information equals zero or is strictly positive, it is not necessary to apply the reordering procedure outlined in Example 12.6. Instead, it is sufficient to evaluate another measure of correlation known as Hirschfeld–Gebelein–Rényi maximal correlation, or **maximal correlation** for short.

12.9.4 Defining the Maximal Correlation

The maximal correlation is defined as the maximum value of the average product $\mathrm{E}_{p^{XY}}[f^X(\mathrm{x})f^Y(\mathrm{y})]$ of two local functions $f^X(\cdot)$ and $f^Y(\cdot)$ of the random variables x and y under suitable constraints. Intuitively, the larger the product $\mathrm{E}_{p^{XY}}[f^X(\mathrm{x})f^Y(\mathrm{y})]$ can be made via the choice of functions $f^X(\cdot)$ and $f^Y(\cdot)$, the more likely it is that a shared random variable with positive entropy can be distilled by the two agents.

Definition 12.15 Maximal correlation
For any joint distribution p^{XY}, define the $N^X \times N^Y$ scaled probability matrix $\tilde{P}^{XY}$ with entries

$$\tilde{P}^{XY}_{x,y} = \frac{p^{XY}_{xy}}{\sqrt{p^X_x p^Y_y}} \tag{12.87}$$

for $x \in [N^X]$ and $y \in [N^Y]$. The maximal correlation for the joint distribution p^{XY} is defined as

$$\mathfrak{C}^{\max}(X;Y)_{p^{XY}} = \sqrt{\lambda_1\left(\tilde{P}^{XY}(\tilde{P}^{XY})^T\right)}, \tag{12.88}$$

where $\lambda_1(A)$ represents the second largest eigenvalue of matrix A.

Note that, despite the relatively complex notation, the evaluation of the maximum correlation (12.88) is computationally straightforward. Furthermore, in the special case in which at least one of the two subsystems have minimal dimension, i.e., in which N^X and/ or N^Y equal 2, one can prove that the maximal correlation can also be expressed in terms of the Tsallis mutual information (12.69) of order $\alpha = 2$ as (see Recommended Resources)

$$\begin{aligned}\mathfrak{C}^{\max}(X;Y)_{p^{XY}} &= \sqrt{\frac{1}{1_2}\mathrm{I}_2^{\mathrm{T}}(X;Y)_{p^{XY}}} \\ &= \sqrt{\sum_{x\in[N^X]}\sum_{y\in[N^Y]}\frac{(p^{XY}_{xy})^2}{p^X_x p^Y_y} - 1}.\end{aligned} \tag{12.89}$$

12.9.5 Properties of the Maximal Correlation

As follows from the derivations in Appendix 12.F, the maximal correlation satisfies the following property, which highlights its role as a correlation metric.

Key Result 12.13 Bounds on the maximal correlation
For any joint distribution p^{XY}, the maximum correlation satisfies the bounds

$$0 \leq \mathfrak{C}^{\max}(X;Y)_{p^{XY}} \leq 1, \tag{12.90}$$

with the lower bound achieved if and only if the two subsystems are uncorrelated under the joint distribution p^{XY}.

As also proved in Appendix 12.F, the maximal correlation provides conclusive evidence on whether or not it is possible to distill a non-trivial amount of common randomness from two random variables.

Key Result 12.14 Distillable common information and maximal correlation
For any joint distribution p^{XY}, the distillable common information is positive if and only if the maximal correlation takes the maximum value of 1, i.e.,

$$\mathfrak{C}(X;Y)_{p^{XY}} > 0 \Longleftrightarrow \mathfrak{C}^{\max}(X;Y)_{p^{XY}} = 1. \tag{12.91}$$

Example 12.17

For the joint distribution (12.6) and (12.7), Figure 12.13 plots the maximum correlation as a function of the probability p^{eq} for two different values of the dimension D. Note that, for $D = 2$, using (12.89) one obtains the explicit expression

$$\mathfrak{C}^{\max}(X;Y)_{p^{XY}} = 2\left|p^{\text{eq}} - \frac{1}{2}\right|. \tag{12.92}$$

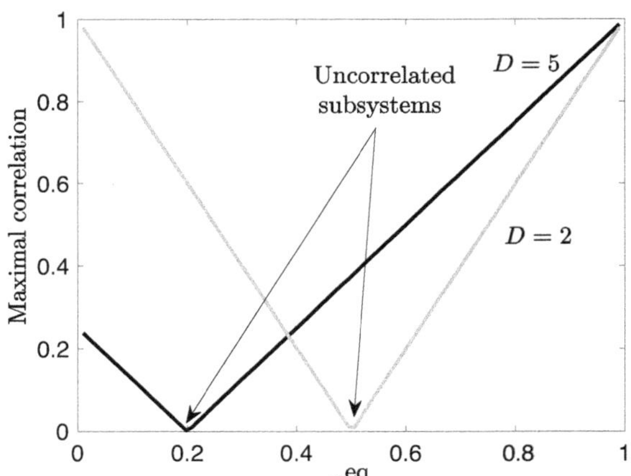

Figure 12.13 Maximum correlation as a function of the probability p^{eq} for the joint distribution (12.6) and (12.7) with two different values of the dimension D.

In a manner consistent with the bounds in (12.90), the maximal correlation equals zero when the two subsystems are uncorrelated. Furthermore, by (12.91), when the maximal correlation equals 1, it is possible to extract a positive-entropy shared random variable at the two subsystems. This happens if $p^{\mathrm{eq}} = 1$ for all values D, as well as for $p^{\mathrm{eq}} = 0$ when $D = 2$. Indeed, in all these cases, one variable can be computed as a function of the other, and thus the distillable common information is positive.

12.10 Summary

- Two subsystems X and Y are said to be uncorrelated, or independent, under the joint distribution p^{XY} if the joint distribution p^{XY} equals the product of the marginals p^X and p^Y, i.e., if $p^{XY} = p^X \otimes p^Y$.
- The Shannon entropy and Rényi entropy are additive, as they satisfy the equality $\mathrm{H}_\alpha^{\mathrm{R}}(X, Y)_{p^X \otimes p^Y} = \mathrm{H}_\alpha^{\mathrm{R}}(X)_{p^X} + \mathrm{H}_\alpha^{\mathrm{R}}(Y)_{p^Y}$, while the Tsallis entropy is not additive.
- The Shannon entropy is subadditive, satisfying the inequality $\mathrm{H}(X, Y)_{p^{XY}} \leq \mathrm{H}(X)_{p^X} + \mathrm{H}(Y)_{p^Y}$, while the Rényi and Tsallis entropies are not.
- All the entropies satisfy the inequality $\mathrm{H}_\alpha(X, Y)_{p^{XY}} \geq \max(\mathrm{H}_\alpha(X)_{p^X}, \mathrm{H}_\alpha(Y)_{p^Y})$, which indicates that uncertainty increases globally in classical systems.
- The Shannon mutual information $\mathrm{I}(X;Y)_{p^{XY}} = \mathrm{H}(X)_{p^X} + \mathrm{H}(Y)_{p^Y} - \mathrm{H}(X, Y)_{p^{XY}}$ measures the level of redundancy in the two subsystems, and is non-negative by the subadditivity of the Shannon entropy.
- The relative entropy of correlation for a joint distribution p^{XY} is defined as the minimum distance to the set of independent distributions, which are considered to be "free" states, i.e., $\min_{q^X, q^Y} \mathrm{D}_\alpha(p^{XY} || q^X \otimes q^Y)$.
- The Shannon relative entropy of correlation coincides with the Shannon mutual information, i.e., $\mathrm{I}(X;Y)_{p^{XY}} = \mathrm{D}(p^{XY} || p^X \otimes p^Y)$.
- Free operations, which do not increase correlation, are channels that act locally on each subsystem. This is reflected by the data processing inequality of the mutual information under local channels, i.e., $\mathrm{I}(X;Y)_{p^{XY}} \geq \mathrm{I}(X;Y)_{(K^X \otimes K^Y)p^{XY}}$.
- The minimum entropy $\mathrm{H}_\alpha(Z)_{p^Z}$ of a common random variable $\mathrm{z} \sim p^Z$ that enables distributed sampling is known as the distributed-sampling common information.
- The maximum entropy $\mathrm{H}_\alpha(Z)_{p^Z}$ of a common distillable random variable $\mathrm{z} \sim p^Z$ is a measure of correlation known as the distillable common information.
- The distillable common information is positive if and only if the maximal correlation takes the maximum value 1.

12.11 Recommended Resources

Extensions of Fano's inequality to Rényi entropy are explored in [1]. Sibson's mutual information is further examined in [2, 3]. Wyner's common information evaluates the solution of a specific relaxed version of problem (12.78) for the Shannon entropy in the asymptotic limit in which multiple independent copies of the variables must be generated. It is defined and elaborated upon in [4, 5]. Algorithms for non-negative matrix factorization are presented in [6]. A comprehensive overview, along with insightful details, on the concept of common information can be found in the lecture slides [7]. Additionally, further discussions on the maximal correlation are provided in [8].

PROBLEMS

12.1 For the joint distributions (12.12) and (12.35), compute joint and marginal Shannon entropies.

12.2 Prove the additivity of the Rényi entropy. *Hint:* Use the definition of the Rényi entropy to evaluate $\mathrm{H}^{\mathrm{R}}_{\alpha}(X, Y)_{p^X \otimes p^Y}$.

12.3 Prove the pseudoadditivity of the Tsallis entropy. *Hint:* Use the α-additivity of the α-logarithm, along with the definition of the joint Tsallis entropy $\mathrm{H}^{\mathrm{T}}_{\alpha}(X, Y)_{p^X \otimes p^Y}$.

12.4 Consider two random variables x and y defined on the same alphabet, i.e., $N^X = N^Y = D$ and following an arbitrary joint distribution p^{XY} with marginal distributions p^X and p^Y. Define as $p^{\mathrm{eq}} = \Pr_{p^{XY}}[\mathrm{x} = \mathrm{y}] = \sum_{x \in [D]} p^{XY}_{xx}$ the probability that the two random variables are equal. Take also a second set of random variables, $\tilde{\mathrm{x}}$ and $\tilde{\mathrm{y}}$, whose joint distribution is given by $\tilde{p}^{XY} = p^X \otimes p^Y$, and define as $\tilde{p}^{\mathrm{eq}} = \Pr_{\tilde{p}^{XY}}[\tilde{\mathrm{x}} = \tilde{\mathrm{y}}] = \sum_{x \in [D]} p^X_x p^Y_x$ the corresponding probability that the two random variables are equal. Intuitively, if the random variables x and y are correlated, the probability p^{eq} should be different from the probability $\tilde{p}^{\mathrm{eq}}$, since the latter corresponds to the case in which the two random variables are uncorrelated.
Aligned with this intuition, the **Fano inequality** relates the Shannon mutual information $\mathrm{I}(X;Y)_{p^{XY}}$ to the Shannon binary relative entropy $\mathrm{d}(p^{\mathrm{eq}}||\tilde{p}^{eq})$ in (4.18) as

$$\mathrm{I}(X;Y)_{p^{XY}} \geq \mathrm{d}(p^{\mathrm{eq}}||\tilde{p}^{eq}). \tag{12.93}$$

(A different version of the Fano inequality will be discussed in Chapter 13.)
(a) Prove Fano's inequality (12.93) by using the data processing inequality. *Hint:* Define a channel that takes as input (x, y) and outputs a random variable that equals 1 if the two variables are equal and zero otherwise. (b) For the joint distribution (12.6) and (12.7), show the equality $\mathrm{I}(X;Y)_{p^{XY}} = \mathrm{d}(p^{\mathrm{eq}}||\tilde{p}^{eq})$.

12.5 Prove that a matrix of the form $K = K^X \otimes K^Y$, where K^X is an $N^X \times N^X$ column-stochastic matrix and K^Y is an $N^Y \times N^Y$ column-stochastic matrix, is an $N^X N^Y \times N^X N^Y$ column-stochastic matrix.

12.6 (a) Show that the minimum value of the support N^Z of a shared random variable solving the problem (12.78) is lower bounded by the rank of the joint distribution matrix P^{XY}. The joint distribution matrix is the $N^X \times N^Y$ matrix with entries $P^{XY}_{x,y} = p^{XY}_{xy}$ for all $x \in [N^X]$ and $y \in [N^Y]$. (b) Show that the same quantity is upper bounded by $\min(N^X, N^Y)$.

12.7 Consider the random variable $\mathrm{x} \sim p^X = u$ with $N^X = 4$ and the random variable $\mathrm{y} = f(\mathrm{x})$ with $N^Y = 2$ where the function is defined as $f(x) = 0$ if x is even and $f(x) = 1$ if x is odd. (a) Evaluate the Shannon mutual information. (b) Compute the conditional distributions $p_x^{Y|X} = [p^{XY}_{x,0}, \ldots, p^{XY}_{x,N^Y-1}]^T / p^X_x$ for all $x \in [N^X]$. (c) Compute the conditional distributions $p_y^{X|Y} = [p^{XY}_{0,y}, \ldots, p^{XY}_{N^X-1,y}]^T / p^Y_y$ for all $y \in [N^Y]$.

12.8 Prove that the Tsallis and Rényi mutual informations in (12.69) and (12.70), respectively, are non-negative and they equal zero if and only if the two subsystems are uncorrelated.

12.9 Prove that, if random variable x and random variable y are related via a one-to-one function, then the maximal correlation $\mathfrak{C}^{\max}(X;Y)_{p^{XY}}$ equals 1.

12.10 Consider a joint distribution p^{XY} for $N^X = N^Y = 5$ with entries $p^{XY}_{xy} = 1/16$ for pairs (x, y) with $x, y \in \{0, 1\}$ and $y \in \{0, 1\}$, as well as for pairs (x, y) with $x, y \in \{3, 4\}$; and $p^{XY}_{2,2} = 1/2$. All other probabilities are zero. (a) Compute the Shannon mutual information $\mathrm{I}(X;Y)_{p^{XY}}$. (b) Evaluate the distillable common information $\mathfrak{C}(X;Y)_{p^{XY}}$. (c) Evaluate the distributed-sampling common information $\mathrm{C}(X;Y)_{p^{XY}}$.

12.11 Rare events can have significant impacts. The Shannon mutual information may not adequately capture the influence of rare events, while the Tsallis and Rényi mutual informations can provide deeper insights into the role of rare events. To elaborate on this, consider a system monitoring the occurrence of rare but impactful events, such as seismic activities, through two sensors X and Y. Each sensor can produce one of two outputs: 0, indicating no significant activity; and 1, indicating significant activity, such as an earthquake. The joint probability distribution of the states detected by sensors X and Y is given by $p^{XY}_{0,0} = 0.98$, $p^{XY}_{0,1} = p^{XY}_{1,0} = 0.05$, and $p^{XY}_{1,1} = 0.01$. (a) Compute the Shannon mutual information $\mathrm{I}(X;Y)_{p^{XY}}$. (b) Compute the Tsallis mutual information $\mathrm{I}^{\mathrm{T}}_\alpha(X;Y)$ of order $\alpha = 2$. (b) Compute the Rényi mutual information $\mathrm{I}^{\mathrm{R}}_\alpha(X;Y)$ of order $\alpha = 2$. (c) Discuss how these different mutual information measures might be applied in the context of monitoring rare events and why one might be preferred over the others in certain scenarios.

Appendix

12.A Proof of the Uncertainty-Increases-Globally Property

To prove the inequality (12.40) for all three entropies, it is sufficient to demonstrate the following inequalities for the Tsallis sum:

$$\sum_{x\in[N^X]} \sum_{y\in[N^Y]} p^\alpha_{xy} \geq \max\left(\sum_{x\in[N^X]} p^\alpha_x, \sum_{y\in[N^Y]} p^\alpha_y \right) \text{ for } 0 \leq \alpha < 1 \tag{12.94}$$

and

$$\sum_{x\in[N^X]}\sum_{y\in[N^Y]} p_{xy}^{\alpha} \le \max\left(\sum_{x\in[N^X]} p_x^{\alpha}, \sum_{y\in[N^Y]} p_y^{\alpha}\right) \text{ for } \alpha > 1. \tag{12.95}$$

In fact, once these inequalities are proved, the inequality (12.40) follows from the dependence of the entropies on the Tsallis sum discussed in Chapter 3. Note that, to simplify the notation, in this appendix, the superscripts are removed from the distributions.

At an intuitive level, to interpret these inequalities, recall from Chapter 3 that the Tsallis sum is a measure of uncertainty for $0 \le \alpha < 1$, while it is a measure of certainty for $\alpha > 1$.

To prove (12.94), write the joint distribution p_{xy} as $p_{xy} = p_x p_{y|x}$, where the conditional distribution of variable y given variable x, to be elaborated on further in Chapter 13, is defined as $p_{y|x} = p_{xy}/p_x$. With this definition, the sum in (12.94) satisfies the relations

$$\sum_{x\in[N^X]}\sum_{y\in[N^Y]} p_{xy}^{\alpha} = \sum_{x\in[N^X]} p_x^{\alpha}\left(\sum_{y\in[N^Y]} p_{y|x}^{\alpha}\right) \ge \sum_{x\in[N^X]} p_x^{\alpha}, \tag{12.96}$$

where the inequality follows from the inequalities $\sum_{y\in[N^Y]} p_{y|x}^{\alpha} \ge 1$, which hold for all $x \in [N^X]$ and for any $0 \le \alpha < 1$. The inequality (12.95) can be proved in a similar way.

12.B Derivation of Shannon Relative Entropy of Correlation

The aim of this appendix is to prove the equalities (12.55) and (12.56). To this end, it is sufficient to show that, for any pair of distributions q^X and q^Y, the quantity $\mathrm{D}(p^{XY}||q^X \otimes q^Y)$ on the left-hand side in (12.55) cannot be smaller than the Shannon mutual information. This follows as

$$\begin{aligned}\mathrm{D}(p^{XY}||q^X \otimes q^Y) &= -\mathrm{E}_{p^X}[\log_2(q_x^X)] - \mathrm{E}_{p^Y}[\log_2(q_y^Y)] - \mathrm{H}(X,Y)_{p^{XY}}\\ &\ge -\mathrm{E}_{p^X}[\log_2(p_x^X)] - \mathrm{E}_{p^Y}[\log_2(p_y^Y)] - \mathrm{H}(X,Y)_{p^{XY}}\\ &= \mathrm{I}(X;Y)_{p^{XY}},\end{aligned} \tag{12.97}$$

where the inequality is a consequence of the non-negativity of the Shannon relative entropy.

12.C Example of the Suboptimality of the Marginals for Problem (12.52)

Consider a joint distribution p^{XY} with $N^X = N^Y = D = 10$ and entries given by $p_{00}^{XY} = 0.6$ and $p_{xy}^{XY} = (1-p_{00}^{XY})/(D^2-1)$ with $(x, y) \neq (0, 0)$. Note that the marginal distributions of the joint distribution p^{XY} have entries

$$p_0^X = p_0^Y = p_{00}^{XY} + (D-1)\frac{1-p_{00}^{XY}}{D^2-1} = 0.6364, \tag{12.98}$$

$p_x^X = p_0^X/(D-1) = 0.0707$ for $x \neq 0$, and $p_y^Y = 0.0707$ for $y \neq 0$.

The relative entropy of correlation (12.52) for the Tsallis relative entropy of order $\alpha = 0.5$ reads

$$\begin{aligned} \mathrm{I}_{\mathrm{D}^{\mathrm{T}}_{0.5}}(X;Y)_{p^{XY}} &= \min_{q^X,q^Y} \mathrm{D}^{\mathrm{T}}_{0.5}(p^{XY}\|q^X \otimes q^Y) \\ &= \min_{q^X,q^Y} 2 \cdot 1_2 \left(\sum_{x\in[N^X]} \sum_{y\in[N^Y]} \sqrt{p^{XY}_{xy} q^X_x q^Y_y} - 1 \right). \end{aligned} \tag{12.99}$$

To show that the marginal distributions (12.98) do not solve problem (12.99), take the class of local distributions q^X and q^Y of the form $q^X = q^Y = [q_0, (1 - q_0)/(D-1), \ldots, (1-q_0)/(D-1)]^T$, which include the marginals by setting $q_0 = 0.6364$. Figure 12.14 plots the Tsallis relative entropy in (12.99) as a function of the parameter $q^X_0 = q^Y_0 = q_0$. It is observed that the optimal value is indeed different from the marginals $p^X_0 = p^Y_0$.

12.D Proof of the Upper Bound (12.79)

Denoting as p^X and p^Y the marginal distributions of the target joint distribution p^{XY}, a simple way to ensure the requirement (12.77) is to choose the distributed sampling parameters in (12.75) as

$$\begin{aligned} N^Z &= N^X, \\ p^Z &= p^X, \\ p^{X|Z}_z &= \delta_z \text{ for } z \in [N^X], \\ p^{Y|Z}_z &= \left[\frac{p^{XY}_{z,0}}{p^X_z}, \frac{p^{XY}_{z,1}}{p^X_z}, \ldots, \frac{p^{XY}_{z,N^Y-1}}{p^X_z} \right]^T, \end{aligned} \tag{12.100}$$

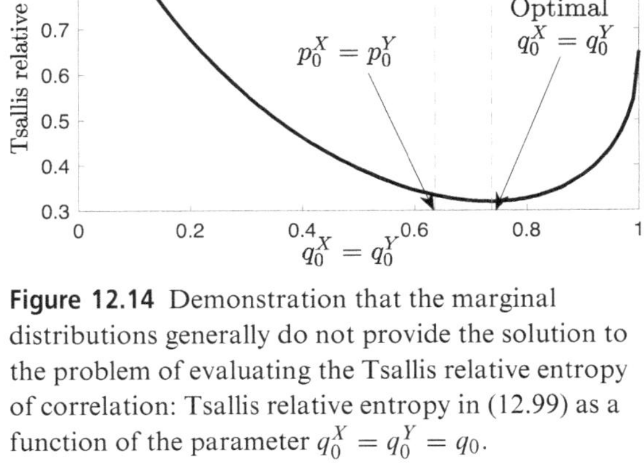

Figure 12.14 Demonstration that the marginal distributions generally do not provide the solution to the problem of evaluating the Tsallis relative entropy of correlation: Tsallis relative entropy in (12.99) as a function of the parameter $q^X_0 = q^Y_0 = q_0$.

where δ_z is the zero-entropy distribution concentrated at the value $z \in [N^X]$. The resulting distributed sampling policy starts by generating the common variable z with distribution equal to the marginal p^X. Then, agent X simply produces the common variable as its output, i.e., z = x. In contrast, agent Y produces output y $= y$ with probability $p_{y|z}^{Y|Z} = p_{z,y}^{XY}/p_z^X$, where $p_{y|z}^{Y|Z}$ is the yth entry of vector $p_z^{Y|Z}$.

The probability $p_{y|z}^{Y|Z}$ used by agent Y in this protocol is proportional to the desired probability $p_{z,y}^{XY}$ of the state (z, y). It can be verified to be a valid distribution, since it satisfies the equality

$$\sum_{y\in[N^Y]} p_{y|z}^{Y|Z} = \sum_{y\in[N^Y]} \frac{p_{z,y}^{XY}}{p_z^X} = \frac{\sum_{y\in[N^Y]} p_{z,y}^{XY}}{p_z^X} = \frac{p_z^X}{p_z^X} = 1. \tag{12.101}$$

As detailed in Chapter 13, the distribution $p_{y|z}^{Y|Z}$ is known as the conditional distribution of random variable y given that random variable x takes the value x $= z$ under the joint distribution p^{XY}.

To demonstrate that this distributed sampling strategy works as desired, it is sufficient to plug the choices (12.100) into the right-hand side of (12.77). Focusing on the (x, y)th entry of the resulting distribution, one gets

$$\begin{aligned}\left[\sum_{z\in[N^Z]} p_z^Z p_z^{X|Z} \otimes p_z^{Y|Z}\right]_{x,y} &= \sum_{z\in[N^X]} p_z^X \delta_{x|z} p_z^{Y|Z} \\ &= p_x^X \frac{p_{x,y}^{XY}}{p_x^X} \\ &= p_{x,y}^{XY},\end{aligned} \tag{12.102}$$

where $\delta_{x|z}$ is the xth entry of vector δ_z. This derivation shows that the setting (12.100) satisfies the requirement (12.77).

With the distributed sampling strategy in (12.100), the entropy of the shared random variable, $\mathrm{H}_\alpha(Z)_{p^Z}$, equals the entropy of subsystem X, i.e., $\mathrm{H}_\alpha(X)_{p^X}$. However, since an analogous strategy could also be applied by swapping the roles of the two agents, the amount of required shared randomness can potentially be reduced to

$$\mathrm{H}_\alpha(Z)_{p^Z} = \min(\mathrm{H}_\alpha(X)_{p^X}, \mathrm{H}_\alpha(Y)_{p^Y}). \tag{12.103}$$

This entropy thus yields an upper bound on the distributed-sampling common information, concluding the proof.

12.E Proof of Data Processing Inequality of the Mutual Information Under Local Channels

This section aims to prove the data processing inequality under local channels in (12.62). Since the channel $K^X \otimes K^Y$ is a transformation on the overall bipartite system, the data processing inequality yields the first inequality in

$$\min_{q^X,q^Y} \mathrm{D}_\alpha(p^{XY}||q^X \otimes q^Y) \geq \min_{q^X,q^Y} \mathrm{D}_\alpha((K^X \otimes K^Y)p^{XY}||K^X q^X \otimes K^Y q^Y) \quad (12.104)$$

$$\geq \min_{q^X,q^Y} \mathrm{D}_\alpha((K^X \otimes K^Y)p^{XY}||q^X \otimes q^Y), \quad (12.105)$$

while the second equality follows from the fact that the set of all distributions $q^X \otimes q^Y$ is no smaller than the set of distributions that can be expressed as $K^X q^X \otimes K^Y q^Y$.

12.F On the Maximal Correlation

In order to introduce the maximal correlation, it is necessary to recall first the standard definition of the Pearson correlation coefficient.

Definition 12.16 Pearson correlation coefficient
Take two real-valued random variables a and b with joint distribution p^{AB}. The two random variables $(a, b) \sim p^{AB}$ are normalized in the sense that they have zero mean and variance equal to 1, i.e., $\mathrm{E}_{p^A}[\mathrm{a}] = \mathrm{E}_{p^A}[\mathrm{b}] = 0$ and $\mathrm{E}_{p^A}[\mathrm{a}^2] = \mathrm{E}_{p^A}[\mathrm{b}^2] = 1$. The Pearson correlation coefficient is defined as the expectation of the product of the two normalized random variables, i.e., $\mathrm{E}_{p^{AB}}[\mathrm{ab}]$.

Key Result 12.15 Bounds on the Pearson correlation coefficient
Given two normalized random variables a and b, the Pearson correlation coefficient satisfies the inequalities

$$-1 \leq \mathrm{E}_{p^{AB}}[\mathrm{ab}] \leq 1, \quad (12.106)$$

where the upper bound is attained with $\mathrm{a} = \mathrm{b}$, while the lower bound is attained with $\mathrm{a} = -\mathrm{b}$.

The maximal correlation $\mathfrak{C}^{\max}(X;Y)_{p^{XY}}$ for random variables $(\mathrm{x}, \mathrm{y}) \sim p^{XY}$ is defined as the maximum Pearson correlation coefficient between all possible normalized random variables $f^X(\mathrm{x})$ and $f^Y(\mathrm{y})$ extracted from variables x and y, respectively. This definition is formalized as follows.

Definition 12.17 Hirschfeld–Gebelein–Rényi maximal correlation, aka maximal correlation
For any joint distribution p^{XY}, the maximal correlation is defined as

$$\mathfrak{C}^{\max}(X;Y)_{p^{XY}} = \max_{f^X(\cdot),f^Y(\cdot)} \mathrm{E}_{p^{XY}}[f^X(\mathrm{x})f^Y(\mathrm{y})], \quad (12.107)$$

where the maximization is over all functions $f^X(\cdot)$ and $f^Y(\cdot)$ that ensure that random variables $f^X(\mathrm{x})$ and $f^X(\mathrm{x})$ are normalized, i.e.,

$$\mathrm{E}_{p^X}[f^X(\mathrm{x})] = \mathrm{E}_{p^Y}[f^Y(\mathrm{y})] = 0 \tag{12.108}$$

and

$$\mathrm{E}_{p^X}[(f^X(\mathrm{x}))^2] = \mathrm{E}_{p^Y}[(f^Y(\mathrm{y}))^2] = 1. \tag{12.109}$$

To derive the maximal correlation expression in (12.88), it is useful to start by defining the $N^X \times 1$ vector

$$\tilde{f}^X = \left[\sqrt{p_0^X} f^X(0), \ldots, \sqrt{p_{N^X-1}^X} f^X(N^X-1)\right]^T \tag{12.110}$$

and the $N^Y \times 1$ vector

$$\tilde{f}^Y = \left[\sqrt{p_0^Y} f^Y(0), \ldots, \sqrt{p_{N^Y-1}^Y} f^X(N^Y-1)\right]^T, \tag{12.111}$$

as well as the vectors $\tilde{p}^X = \left[\sqrt{p_0^X}, \ldots, \sqrt{p_{N^X-1}^X}\right]^T$ and $\tilde{p}^Y = \left[\sqrt{p_0^Y}, \ldots, \sqrt{p_{N^Y-1}^Y}\right]^T$.

Using these definitions, the zero-mean conditions (12.108) are expressed as

$$\mathrm{E}_{p^X}[f^X(\mathrm{x})] = \sum_{x\in[N^X]} p_x^X f^X(x) = (\tilde{f}^X)^T \tilde{p}^X = 0 \tag{12.112}$$

and

$$\mathrm{E}_{p^Y}[f^Y(\mathrm{y})] = \sum_{y\in[N^Y]} p_y^Y f^Y(y) = (\tilde{f}^Y)^T \tilde{p}^Y = 0. \tag{12.113}$$

These conditions require that vectors $\tilde{f}^X$ and $\tilde{p}^X$ are orthogonal, and that vectors $\tilde{f}^Y$ and $\tilde{p}^Y$ are also orthogonal.

In a similar manner, the constraints (12.109) can be expressed as

$$\mathrm{E}_{p^X}[(f^X(\mathrm{x}))^2] = \sum_{x\in[N^X]} p_x^X (f^X(x))^2 = (\tilde{f}^X)^T \tilde{f}^X = ||\tilde{f}^X||^2 = 1 \tag{12.114}$$

and

$$\mathrm{E}_{p^Y}[(f^Y(\mathrm{y}))^2] = ||\tilde{f}^Y||^2 = 1. \tag{12.115}$$

Thus, these constraints impose that the 2-norms of vectors $\tilde{f}^X$ and $\tilde{f}^Y$ are equal to 1.

Finally, the Pearson correlation coefficient in (12.107) can be written as

$$\mathrm{E}_{p^{XY}}[f^X(\mathrm{x}) f^Y(\mathrm{y})] = \sum_{x\in[N^X]} \sum_{y\in[N^Y]} p_{xy}^{XY} f^X(x) f^Y(y) = (\tilde{f}^X)^T \tilde{P}^{XY} \tilde{f}^Y. \tag{12.116}$$

Overall, the maximum correlation (12.107) is obtained by addressing the problem

$$\mathfrak{C}^{\max}(X;Y)_{p^{XY}} = \max_{\tilde{f}^X, \tilde{f}^Y} (\tilde{f}^X)^T \tilde{P}^{XY} \tilde{f}^Y \tag{12.117}$$

under the orthogonality constraints (12.112) and (12.113) and the norm constraints (12.114) and (12.115).

To proceed, the reader is invited to review the definition and properties of the singular value decomposition in Appendix 12.G. Using the result (12.123) therein, it follows that the vectors $\tilde{p}^X$ and $\tilde{p}^Y$ are the left and right singular vectors of the matrix $\tilde{P}^{XY}$ corresponding to the largest singular value of the matrix. Furthermore, the same result (12.123) also implies that the vectors $\tilde{f}^X$ and $\tilde{f}^Y$ defined in (12.117) are the left and right singular vectors of matrix $\tilde{P}^{XY}$ corresponding to the second largest singular value $\omega_1(\tilde{P}^{XY}) = 1$ of matrix $\tilde{P}^{XY}$.

It was thus concluded that the maximal correlation in (12.117) coincides with the second largest singular value of matrix $\tilde{P}^{XY}$, i.e.,

$$\mathfrak{C}^{\max}(X;Y)_{p^{XY}} = \omega_1(\tilde{P}^{XY}). \tag{12.118}$$

By the equality (12.124) in Appendix 12.G, this quantity equals the square root of the second largest eigenvalue of matrix $\tilde{P}^{XY}(\tilde{P}^{XY})^T$, concluding the derivation of the expression (12.88).

Finally, for a proof of the equivalence in (12.91), assume that the distillable common information is positive and thus there exists a shared random variable $\mathrm{z} = f^X(\mathrm{x}) = f^Y(\mathrm{y})$. Without loss of generality, one can always ensure that the variable z is normalized. In fact, given any random variable a with mean $\mathrm{E}_{p^A}[\mathrm{a}]$ and variance $\mathrm{var}_{p^A}(\mathrm{a})$, the variable $(\mathrm{a} - \mathrm{E}_{p^A}[\mathrm{a}])/\sqrt{\mathrm{var}_{p^A}(\mathrm{a})}$ is normalized. Therefore, using the definition (12.107), a positive distillable common information implies that the maximal correlation equals 1.

Conversely, if the maximal correlation equals 1, it must be true that the equality $f^X(\mathrm{x}) = f^Y(\mathrm{y})$ holds for the optimal functions in (12.107). This, in turn, implies that there exists a shared random variable $\mathrm{z} = f^X(\mathrm{x}) = f^Y(\mathrm{y})$. Note, however, that it is possible for the maximal correlation to be positive, while the distillable common information equals zero.

12.G Singular Value Decomposition

Referring to standard textbooks on linear algebra for details, the singular value decomposition (SVD) states that, for an arbitrary, possibly complex-valued, $N \times M$ matrix A, there exist an $N \times N$ unitary matrix $U = [|u_0\rangle, \ldots, |u_N\rangle]$, an $M \times M$ unitary matrix $V = [|v_0\rangle, \ldots, |v_M\rangle]$, an integer number

$$r \leq \min(N, M), \tag{12.119}$$

and a vector $\omega = [\omega_0, \ldots, \omega_{r-1}]^T$ with positive entries, such that the matrix can be expressed as

$$A = \sum_{x \in [r]} \omega_x |u_x\rangle\langle v_x|. \tag{12.120}$$

The number r is known as the rank of matrix A. One can check that this definition is consistent with the definition of rank for normal matrices.

The r vectors $\{|u_x\rangle\}_{x\in[r]}$ and $\{|v_x\rangle\}_{x\in[r]}$ are referred to as the left and right singular vectors of matrix A, respectively; and the elements $\{\omega_x\}_{x\in[r]}$ are known as the singular values of matrix A. By the SVD (12.120), one has the equality

$$\langle u_x|A|v_x\rangle = \omega_x \tag{12.121}$$

for all $x \in [r]$.

Furthermore, ordering the singular values without loss of generality from largest to smallest, with ties broken arbitrarily, i.e.,

$$\omega_0 \geq \omega_0 \geq \cdots \geq \omega_{r-1}, \tag{12.122}$$

one has the following optimization-based characterization of singular values.

Key Result 12.16 Optimization-based characterization of singular values
Given an $N \times M$ matrix A with rank r, the zth largest singular value ω_z, with $z \in [r]$, is obtained as the result of the optimization

$$\max_{|u\rangle,|v\rangle} \langle u|A|v\rangle = \omega_z, \tag{12.123}$$

where the $N \times 1$ ket $|u\rangle$ is constrained to be orthogonal to all left singular vectors $\{|u_y\rangle\}_{y\in[z]}$, i.e., $\langle u|u_y\rangle = 0$ for all $y \in [z]$, and the $M \times 1$ ket $|v\rangle$ is constrained to be orthogonal to all right singular vectors $\{|v_y\rangle\}_{y\in[z]}$, i.e., $\langle v|v_y\rangle = 0$ for all $y \in [z]$.

Finally, the following relation holds between the singular values of the $N \times M$ matrix A and the eigenvalues of the matrices $AA^\dagger$ and $A^\dagger A$.

Key Result 12.17 Singular values and eigenvalues
The zth largest singular value of matrix A, denoted as $\omega_z(A)$, coincides with the square root of the zth largest eigenvalue of matrix $AA^\dagger$, i.e., $\lambda_z(AA^\dagger)$, and of matrix $A^\dagger A$, i.e., $\lambda_z(A^\dagger A)$; that is, the equality

$$\omega_z(A) = \sqrt{\lambda_z(AA^\dagger)} = \sqrt{\lambda_z(A^\dagger A)} \tag{12.124}$$

holds for all $z \in [r]$.

To prove the equalities (12.124), it is sufficient to use the SVD (12.120) to obtain

$$AA^\dagger = \sum_{x\in[r]} \omega_x^2 |u_x\rangle\langle u_x|,$$

which identifies the eigenvalues of matrix $AA^\dagger$ as in (12.124). The result in (12.124) also shows the fact, anticipated in Section 6.6.2, that matrices $AA^\dagger$ and $A^\dagger A$ are positive semi-definite, since all the eigenvalues are non-negative.

REFERENCES

[1] T. S. Han and S. Verdu, "Generalizing the Fano inequality," *IEEE Transactions on Information Theory*, vol. 40, no. 4, pp. 1247–1251, 1994.

[2] F. Liese and I. Vajda, "On divergences and informations in statistics and information theory," *IEEE Transactions on Information Theory*, vol. 52, no. 10, pp. 4394–4412, 2006.

[3] A. R. Esposito, M. Gastpar, and I. Issa, "Sibson's α-mutual information and its variational representations," *arXiv*, 2024. [Online]. Available at: https://arxiv.org/abs/2405.08352

[4] A. D. Wyner, "The common information of two dependent random variables," *IEEE Transactions on Information Theory*, vol. 21, no. 2, pp. 163–179, 1975.

[5] G. Xu, W. Liu, and B. Chen, "A lossy source coding interpretation of Wyner's common information," *IEEE Transactions on Information Theory*, vol. 62, no. 2, pp. 754–768, 2015.

[6] D. Lee and H. S. Seung, "Algorithms for non-negative matrix factorization," *Advances in Neural Information Processing Systems*, vol. 13, 2000.

[7] A. E. Gamal, "Viterbi lecture," 2017. Available at: https://minghsiehece.usc.edu/wp-content/uploads/2017/06/El-Gamal-Viterbi-Lecture-PDF.pdf

[8] V. Anantharam, A. Gohari, S. Kamath, and C. Nair, "On maximal correlation, hypercontractivity, and the data processing inequality studied by Erkip and Cover," *arXiv*, 2013. [Online]. Available at: https://arxiv.org/abs/1304.6133

13 Bipartite Classical Information via Residual Uncertainty

13.1 Overview

Chapter 12 has introduced correlation measures that treat the two subsystems of a bipartite system in a symmetric manner. This chapter adopts a different perspective by studying correlation from the viewpoint of one subsystem. Specifically, the chapter addresses the following question: If an observer is initially unaware of the state x of subsystem X, does acquiring this state reduce the observer's uncertainty about the state y of subsystem Y?

If the subsystems are uncorrelated, observing X provides no additional information about Y, and the observer's uncertainty about Y remains unchanged. In contrast, if correlation exists, the observer's residual uncertainty about Y after measuring X will differ from the original uncertainty about Y before the measurement.

This discussion motivates the definition of correlation measures that compare the uncertainty levels about one subsystem before and after observing the other subsystem. Importantly, these measures are inherently asymmetric, as the residual uncertainty about X given Y need not equal the residual uncertainty about Y given X.

This chapter will explore two approaches for measuring residual uncertainty – and, by extension, correlation.

- **Ensemble entropy:** The first approach quantifies the average uncertainty about the state of subsystem Y after observing the state of subsystem X. This measure, referred to as ensemble entropy, is denoted as $\mathrm{H}_\alpha(\mathcal{E}^{Y|X})_{p^{XY}}$. With this definition, the difference

$$\mathrm{H}_\alpha(Y)_{p^Y} - \mathrm{H}_\alpha(\mathcal{E}^{Y|X})_{p^{XY}}, \tag{13.1}$$

between the pre-observation uncertainty $\mathrm{H}_\alpha(Y)_{p^Y}$ and the residual uncertainty after observing X, quantified by the ensemble entropy $\mathrm{H}_\alpha(\mathcal{E}^{Y|X})_{p^{XY}}$, provides the first correlation measure to be studied in this chapter.

- **Conditional entropy:** The second approach measures residual uncertainty as the difference between the joint entropy $\mathrm{H}_\alpha(X, Y)_{p^{XY}}$ and the local entropy $\mathrm{H}_\alpha(X)_{p^X}$ of the measured subsystem X, yielding the conditional entropy $\mathrm{H}_\alpha(Y|X)_{p^{XY}}$. The rationale behind this definition is that subtracting the local entropy of X from the joint entropy leaves the residual uncertainty about subsystem Y from the vantage point of subsystem X. With this definition, the difference

$$\mathrm{H}_\alpha(Y)_{p^Y} - \mathrm{H}_\alpha(Y|X)_{p^{XY}} \tag{13.2}$$

provides an alternative correlation measure, which will also be studied in this chapter.

When the Shannon entropy (i.e., order $\alpha = 1$) is used as the uncertainty measure, both (13.1) and (13.2) will be seen to recover the Shannon mutual information, establishing a consistency with the symmetric measures from Chapter 12.

More generally, for concave entropy measures such as the Tsallis entropy and the Rényi entropy of order $0 \leq \alpha \leq 1$, the difference (13.1) defines a new, asymmetric, correlation measure called the locally accessible information or the expected information gain. This metric quantifies the information about subsystem Y that can be accessed by observing subsystem X. In contrast, the difference (13.2) does not provide a valid measure of bipartite information with entropy measures beyond the Shannon entropy (due to the fact that only the Shannon entropy is subadditive).

Finally, this chapter introduces a method for assessing residual uncertainty through the study of the sequential sampling problem, an informational task that further highlights the asymmetry inherent in the analysis of correlations.

13.1.1 Learning Objectives and Organization

By the end of this chapter, the reader should be able to:

- understand and use conditional distributions;
- understand the concept of ensembles of distributions;
- understand and calculate the ensemble entropy;
- understand and calculate the locally accessible information;
- and understand and calculate the conditional entropy.

13.2 Conditional Distributions

Throughout this chapter, as in Chapter 12, the focus is on a bipartite classical system, which consists of subsystems X and Y. The respective states of the subsystems are defined by jointly distributed random variables $(\mathrm{x}, \mathrm{y}) \sim p^{XY}$ with $\mathrm{x} \in [N^X]$ and $\mathrm{y} \in [N^Y]$. The joint distribution p^{XY} is an $N^X N^Y \times 1$ vector, and the marginal distributions for subsystems X and Y are given by the $N^X \times 1$ vector p^X and the $N^Y \times 1$ vector p^Y, respectively.

This section studies a setting in which the observer is aware of the state x of subsystem X. The goal is to characterize the informational state of the observer as it pertains to the state y of subsystem Y. Throughout this discussion, one can swap the roles of subsystems X and Y to define corresponding quantities for a setting in which the observer is informed about the state y of subsystem Y.

13.2.1 Defining Conditional Distributions

Since the observer knows the state x of subsystem X, the informational state must be described by a distribution over the possible values $y \in [N^Y]$ of the state y of subsystem Y. This distribution is generally different from the marginal p^Y, which does not take into account knowledge about the state x of subsystem X. The informational state must rather reflect the extent to which knowledge about the state x of subsystem X affects the uncertainty about the state y of subsystem Y. Thus, mathematically, it must depend on all the entries of the joint probability vector p^{XY} of the form p^{XY}_{xy}, where x is the observed state of subsystem X and index $y \in [N^Y]$ varies over all possible states of subsystem Y.

Example 13.1

Take the joint distribution $p^{XY} = [p^{XY}_{00}, p^{XY}_{01}, p^{XY}_{10}, p^{XY}_{11}]^T = [0.1, 0, 0, 0.9]^T$, for which the marginal distribution of subsystem Y is $p^Y = [0.1, 0.9]^T$. Accordingly, prior to observing the state of subsystem X, the most likely state of subsystem Y is by far $y = 1$, which has probability $p^Y_1 = 0.9$. Suppose now that the observer becomes aware that the state of subsystem X is x $= 0$. Under the joint distribution p^{XY}, the probability p^{XY}_{01} of subsystem X being in state x $= 0$ and subsystem Y being in state y $= 1$ is zero. Therefore, upon observing x $= 0$, the observer is certain that the state of subsystem Y is y $= 0$. Thus, the observer has substantially "changed its mind" about subsystem Y as a result of the observation of subsystem X.

As illustrated by the last example, when state x $= x$ of subsystem X is measured, the informational state of the observer depends on the $N^Y \times 1$ vector

$$\tilde{p}^{Y|X}_x = \begin{bmatrix} p^{XY}_{x0} \\ p^{XY}_{x1} \\ \vdots \\ p^{XY}_{x,N^Y-1} \end{bmatrix} \tag{13.3}$$

of joint probabilities $\{p^{XY}_{xy}\}_{y \in [N^Y]}$ for the given observed value x. (A comma is added to the last entry for clarity of notation.)

Since x $= x$ was observed, the marginal probability p^X_x must be positive, as, otherwise, the observation x $= x$ would have been impossible. This probability is recovered by summing the entries of vector $\tilde{p}^{Y|X}_x$:

$$\sum_{y \in [N^Y]} \tilde{p}^{Y|X}_x = \sum_{y \in [N^Y]} p^{XY}_{xy} = p^X_x. \tag{13.4}$$

Thus, the vector $\tilde{p}^{Y|X}_x$ is generally not a probability distribution, since its entries do not sum to 1, but to p^X_x.

In order to obtain a valid probability distribution over state $y \in [N^Y]$, vector $\tilde{p}^{Y|X}_x$ is divided by the probability $p^X_x > 0$ to yield

$$p_x^{Y|X} = \begin{bmatrix} p_{0|x}^{Y|X} \\ p_{1|x}^{Y|X} \\ \vdots \\ p_{N^Y-1|x}^{Y|X} \end{bmatrix} = \frac{1}{p_x^X} \begin{bmatrix} p_{x0}^{XY} \\ p_{x1}^{XY} \\ \vdots \\ p_{x,N^Y-1}^{XY} \end{bmatrix}. \tag{13.5}$$

The vector $p_x^{Y|X}$ is known as the **conditional distribution** of subsystem Y given observation $\mathrm{x} = x$ of subsystem X under the joint distribution p^{XY}.

Definition 13.1 Conditional distribution

Given a joint distribution p^{XY} and a state $x \in [N^X]$ for subsystem X with positive probability, i.e., $p_x^X > 0$, the conditional probability of subsystem Y being in state $\mathrm{y} = y$ given that the state of subsystem X is $\mathrm{x} = x$ is given by

$$\begin{aligned} p_{y|x}^{Y|X} &= \Pr[\mathrm{y} = y|\mathrm{x} = x] \\ &= \frac{\Pr[\mathrm{x} = x \text{ and } \mathrm{y} = y]}{\Pr[\mathrm{x} = x]} \\ &= \frac{p_{xy}^{XY}}{p_x^X}, \end{aligned} \tag{13.6}$$

which corresponds to the yth entry of vector $p_x^{Y|X}$ in (13.5).

In the conditional distribution $p_x^{Y|X}$, the bar in the notation $Y|X$ is read **"conditioned on."** Accordingly, $p_x^{Y|X}$ is the probability distribution of subsystem Y, on the left-hand side of the bar, conditioned on the observation $\mathrm{x} = x$ of subsystem X, on the right-hand side of the bar.

When reversing the roles of the two subsystems, one obtains the conditional distribution $p_y^{X|Y}$ of subsystem X conditioned on subsystem Y, which is well defined when $p_y^Y > 0$. The $N^X \times 1$ conditional distribution $p_y^{X|Y}$ has entries given by the ratio

$$p_{x|y}^{X|Y} = \Pr[\mathrm{x} = x|\mathrm{y} = y] = \frac{p_{xy}^{XY}}{p_y^Y} \tag{13.7}$$

for all $x \in [N^X]$.

Example 13.2

For the joint distribution $p^{XY} = [p_{00}^{XY}, p_{01}^{XY}, p_{10}^{XY}, p_{11}^{XY}]^T = [0.1, 0, 0, 0.9]^T$ studied in Example 13.1, the conditional distribution of subsystem Y given $\mathrm{x} = 0$ is

$$p_0^{Y|X} = \begin{bmatrix} 1 \\ 0 \end{bmatrix} = \delta_0. \tag{13.8}$$

Thus, as discussed, after observing $\mathrm{x} = 0$, we are certain that the subsystem Y is in state $\mathrm{y} = 0$.

Example 13.3

For the distribution (12.6) and (12.7), which has $N^X = N^Y = D$ and uniform marginal distributions for both systems, fix a state $x \in [N^X]$ for subsystem X. The conditional probability for subsystem Y given subsystem X is

$$p_{y|x}^{Y|X} = \frac{p_{xy}^{XY}}{p_x^X} = \frac{\frac{1-p^{\text{eq}}}{D^2-D}}{\frac{1}{D}} = \frac{1-p^{\text{eq}}}{D-1} \tag{13.9}$$

for all values $y \neq x$; while it equals

$$p_{x|x}^{Y|X} = \frac{p_{xx}^{XY}}{p_x^X} = \frac{\frac{p^{\text{eq}}}{D}}{\frac{1}{D}} = p^{\text{eq}} \tag{13.10}$$

for $y = x$.

Accordingly, if we have observed $\mathrm{x} = x$ for the state of subsystem X, the probability $p_{x|x}^{Y|X}$ that subsystem Y is also in state x equals the probability p^{eq} that the two states are identical; while the other $D-1$ states $y \neq x$ have the same probability $(1-p^{\text{eq}})/(D-1)$.

To take a more concrete example, set $D = 5$ and assume that subsystem X is in state $\mathrm{x} = 0$. Before observing the state of the subsystem X, the observer assigns equal marginal probabilities $p_y^Y = 1/D = 0.2$ to all possible states $y \in [5]$ of the subsystem Y. After observing the state $\mathrm{x} = 0$ for subsystem X, the observer gives the conditional probability $p_{0|0}^{Y|X} = p^{\text{eq}}$ to state $y = 0$, while assigning conditional probabilities $p_{y|0}^{Y|X} = (1-p^{\text{eq}})/4$ to all other four states $y \in \{1, 2, 3, 4\}$. Therefore, while, before the observation, all five states of subsystem Y were considered to be equally likely, after the observation $\mathrm{x} = 0$, state $y = 0$ has a generally different probability from the other states $y \neq 0$.

In particular, if $p^{\text{eq}} > 0.2$, the post-observation probability $p_{0|0}^{Y|X}$ is larger than the pre-observation probability p_0^Y; while the opposite is true for $p^{\text{eq}} < 0.2$. In the special case with $p^{\text{eq}} = 0.2$, for which the two subsystems are uncorrelated, the conditional distribution $p_x^{Y|X}$ coincides with the marginal distribution $p^Y = u$ for all $x \in [5]$. Hence, for uncorrelated systems, the observation of subsystem X does not change the informational state of the observer concerning subsystem Y.

13.2.2 Chain Rule for the Joint Distribution

Example 13.3 has illustrated the general fact that, when the two subsystems are uncorrelated, the conditional distribution and the marginal distribution coincide, and thus the residual uncertainty equals the local uncertainty prior to the observation of the other subsystem. To formalize this general result, consider first the following consequence of the definition of the conditional distribution (13.6).

Key Result 13.1 Chain rule for the joint distribution

Given a joint distribution p^{XY}, the following equality, known as the chain rule for the joint distribution,

$$p_{xy}^{XY} = p_x^X p_{y|x}^{Y|X}, \tag{13.11}$$

holds for all $x \in [N^X]$ and $y \in [N^Y]$.

The chain rule (13.11) expresses the joint distribution p_{xy}^{XY} as the product of the marginal probability for one variable, here p_x^X, and of the conditional probability for the other variable, here $p_{y|x}^{Y|X}$. It thus first accounts for the probability of the event $\mathrm{x} = x$ that subsystem X is in state x, and then for the probability that subsystem Y is in state $\mathrm{y} = y$ given that subsystem X is in state $\mathrm{x} = x$.

Note that the equality (13.11) is technically only well defined for values of x with positive marginal distribution $p_x^X > 0$, since this condition is required in the definition of the conditional distribution $p_x^{Y|X}$. However, one can write the equality (13.11) even when $p_x^X = 0$ as long as it is agreed that, when $p_x^X = 0$, the product in (13.11) is considered to equal zero. Note also that, using (13.7), one could invert the roles of the two subsystems, yielding the equality $p_{xy}^{XY} = p_y^Y p_{x|y}^{X|Y}$.

As detailed in Chapter 12, if the two subsystems are uncorrelated, or independent, the joint distribution equals the product of the marginals, i.e., the equalities

$$p_{xy}^{XY} = p_x^X p_y^Y \tag{13.12}$$

hold for all $x \in [N^X]$ and $y \in [N^Y]$. Comparing (13.12) with the chain rule (13.11) yields the following result.

Key Result 13.2 Uncorrelated system and conditional distributions

Given a joint distribution p^{XY}, two subsystems are uncorrelated if and only if the conditional distribution $p_x^{Y|X}$ equals the corresponding marginal distribution p^Y for all $x \in [N^X]$ with $p_x^X > 0$; i.e., if and only if the equalities

$$p_{y|x}^{Y|X} = p_y^Y \tag{13.13}$$

hold for all $x \in [N^X]$ and all $y \in [N^Y]$ with $p_x^X > 0$; or, equivalently, if and only if $p_{x|y}^{X|Y} = p_x^X$ for all $x \in [N^X]$ and all $y \in [N^Y]$ with $p_y^Y > 0$.

This result can also be seen by plugging (13.12) into the definition (13.6) of the conditional distribution to yield

$$p_{y|x}^{Y|X} = \frac{p_{xy}^{XY}}{p_x^X} = \frac{p_x^X p_y^Y}{p_x^X} = p_y^Y. \tag{13.14}$$

13.2.3 Chain Rule for the Shannon Surprise

From an informational standpoint, the chain rule of the joint distribution in (13.11) implies the following.

> **Key Result 13.3 Chain rule for the Shannon surprise**
> The Shannon surprise of the joint observation (x, y) is given by the sum
>
> $$-\log_2(p_{xy}^{XY}) = -\log_2(p_x^X) - \log_2(p_{y|x}^{Y|X}), \tag{13.15}$$
>
> which is known as the chain rule for the Shannon surprise.

The chain rule (13.15) expresses the Shannon joint surprise $-\log_2(p_{xy}^{XY})$ as the sum of the Shannon local surprise $-\log_2(p_x^X)$, which is accrued when observing subsystem X to be in state x, and the Shannon conditional surprise $-\log_2(p_{y|x}^{Y|X})$. The Shannon conditional surprise $-\log_2(p_{y|x}^{Y|X})$ results from the observation of subsystem Y in state y given that we have already observed that subsystem X is in state x. For independent subsystems, the equality (13.13) implies the additive property $\log_2(p_{xy}^{XY}) = -\log_2(p_x^X) - \log_2(p_y^Y)$, first encountered in (12.29), as a special case of the chain rule (13.15).

13.2.4 Conditional Distributions as Channels ⏭

The conditional distribution $p_x^{Y|X}$ was introduced in this section as the informational state of an observer that is aware of the state of subsystem X. This interpretation of the conditional distribution is static, in the sense that it does not refer to an actual change in the system, but merely a change in the state of knowledge of the observer. The distribution $p_x^{Y|X}$ has also a dual, dynamic interpretation as a channel.

As introduced in Chapter 9, a channel describes a dynamic transformation of a system. Specifically, a channel is defined by the probability $p_{x \to y}$ that the post-channel state is y when the pre-channel state is x for all possible pairs (x, y). Consider now the random variable x – the state of subsystem X – as the initial state of a given system, and the random variable y – the state of subsystem Y – as the state of the system after the channel. According to this interpretation, subsystems X and Y are thus separated in time, with subsystem X preceding subsystem Y. Note that this definition of a channel allows for a change in the number of states between pre- and post-channel systems, from N^X to N^Y, thus generalizing the definition given in Chapter 9.

Given a channel $p_{x \to y}$, the key observation is that the vector

$$\begin{bmatrix} p_{x \to 0} \\ p_{x \to 1} \\ \vdots \\ p_{x \to N^Y - 1} \end{bmatrix} \tag{13.16}$$

of transition probabilities starting from a fixed state $x \in [N^X]$ defines a distribution over variable $\mathrm{y} \in [N^Y]$. Thus, the vector (13.16) can be interpreted as the conditional distribution $p_x^{Y|X}$ of subsystem Y conditioned on subsystem X.

In particular, by the chain rule (13.11) for the joint distribution, given a channel $p_{x \to y}$ and an initial distribution p^X, the joint distribution of pre- and post-channel states can be obtained as

$$p_{xy}^{XY} = p_x^X \underbrace{p_{x \to y}}_{p_x^{Y|X}}. \tag{13.17}$$

Thus, the channel plays the role of a conditional distribution. Conversely, any conditional distribution can also be interpreted as a channel.

It is thus concluded that a joint distribution p^{XY} can be interpreted and used in one of two ways.

- **Static interpretation:** The joint distribution represents a static informational state about a bipartite system.
- **Dynamic interpretation:** The joint distribution describe the relationship between the pre- and post-channel states of a dynamic system.

As discussed in Part V of the book, this duality of interpretations does not apply to joint quantum states for bipartite systems, which are only physically meaningful as static descriptions.

13.3 Ensemble Entropy

This section focuses on the problem of quantifying the residual uncertainty as the average uncertainty about the state of one subsystem, say Y, after the state of the other subsystem, X, is observed.

13.3.1 Ensemble of Conditional Probability Distributions

As detailed in Section 13.2, the informational state of an observer that is aware of the state $\mathrm{x} = x$ of subsystem X is described by the $N^Y \times 1$ conditional probability vector $p_x^{Y|X}$. Therefore, accounting for the randomness in the state $\mathrm{x} \sim p^X$, the informational state of the observer can be described as follows.

Definition 13.2 Ensemble of conditional distributions
Given a joint distribution p^{XY}, the informational state of an observer that has access to subsystem X is characterized by the ensemble of conditional distributions

$$\mathcal{E}^{Y|X} = \{(p_x^{Y|X}, p_x^X)\}_{x \in [N^X]}, \tag{13.18}$$

since the conditional distribution $p_x^{Y|X}$ is realized with probability p_x^X for $x \in [N^X]$.

The ensemble of conditional distributions $\mathcal{E}^{Y|X}$ in (13.18) describes the perspective of an observer that is aware of state x. In a similar way, reversing the roles of the two subsystems, one can describe the perspective of an observer that has access to the state of subsystem Y through the ensemble of conditional distributions

$$\mathcal{E}^{X|Y} = \{(p_y^{X|Y}, p_y^Y)\}_{y \in [N^Y]}. \tag{13.19}$$

Henceforth, unless stated otherwise, the presentation focuses on the ensemble $\mathcal{E}^{Y|X}$, since the dual ensemble $\mathcal{E}^{X|Y}$ can be addressed in the same way.

For any ensemble $\mathcal{E}^{Y|X}$ in (13.18), the average of the distributions $\{p_x^{Y|X}\}_{x \in [N^X]}$ over the choice of variable $\mathrm{x} \sim p^X$ yields the marginal distribution p^Y, i.e.,

$$\mathrm{E}_{p^X}[p_{\mathrm{x}}^{Y|X}] = \sum_{x \in [N^X]} p_x^X p_x^{Y|X} = p^Y, \tag{13.20}$$

where the expected value is evaluated in an entry-wise fashion across the vectors $p_{\mathrm{x}}^{Y|X}$.

Furthermore, as formalized by the following result, the diversity of the distributions in the ensemble $\mathcal{E}^{Y|X}$ provides evidence about the level of correlation between the two subsystems under the joint distribution p^{XY}.

> **Key Result 13.4 Ensemble diversity as evidence of correlation**
>
> If subsystems X and Y are uncorrelated, all conditional distributions $p_x^{Y|X}$ in the ensemble $\mathcal{E}^{Y|X}$ are equal to the marginal distribution p^Y; while, if they are correlated, there are at least two different conditional distributions $p_x^{Y|X}$ in the ensemble, i.e., $p_{x'}^{Y|X} \neq p_x^{Y|X}$ for some $x' \neq x \in [N^X]$.

In other words, when X and Y are uncorrelated, the ensemble $\mathcal{E}^{Y|X}$ is minimally diverse, since all the conditional distributions $\{p_x^{Y|X}\}_{x \in [N^X]}$ are equal; while the presence of at least two different conditional distributions in the set $\{p_x^{Y|X}\}_{x \in [N^X]}$ in the ensemble implies that the two subsystems are correlated.

13.3.2 Ensemble Entropy

The description of the informational state of an observer with access to subsystem X via the ensemble $\mathcal{E}^{Y|X}$ in (13.18) makes it possible to introduce an important measure of residual uncertainty, the ensemble entropy.

Upon observing the state $\mathrm{x} = x$, the informational state of the observer is given by $p_x^{Y|X}$, and thus the uncertainty of the observer is quantified by the entropy $\mathrm{H}_\alpha(p_x^{Y|X})$. In the case of uncorrelated subsystems, the informational state does not vary as a function of x, yielding the same uncertainty level $\mathrm{H}_\alpha(p_x^{Y|X}) = \mathrm{H}_\alpha(p^Y)$ for every observation $x \in [N^X]$. In contrast, for correlated subsystems, the entropy $\mathrm{H}_\alpha(p_x^{Y|X})$ for any given state $x \in [N^X]$ can be smaller than or larger than the marginal entropy $\mathrm{H}_\alpha(p^Y)$.

Example 13.4

Consider the following joint distribution p^{XY} with $N^X = N^Y = 2$:

$$p^{XY} = \begin{bmatrix} p_{00}^{XY} \\ p_{01}^{XY} \\ p_{10}^{XY} \\ p_{11}^{XY} \end{bmatrix} = \begin{bmatrix} 0.5 \\ 0 \\ 0.25 \\ 0.25 \end{bmatrix}. \tag{13.21}$$

Since the marginal distribution for subsystem Y is $p^Y = [0.75, 0.25]^T$, a priori, before the observation of subsystem X, the uncertainty regarding the state of subsystem Y is given by the binary entropy $\mathrm{H}_\alpha(p^Y) = \mathrm{h}_\alpha(0.25)$.

Given the observation $\mathrm{x} = 0$, the conditional distribution for subsystem Y is

$$p_0^{Y|X} = \begin{bmatrix} p_{0|0}^{Y|X} \\ p_{1|0}^{Y|X} \end{bmatrix} = \frac{1}{p_0^X} \begin{bmatrix} p_{00}^{XY} \\ p_{01}^{XY} \end{bmatrix} = \begin{bmatrix} 1 \\ 0 \end{bmatrix} = \delta_0; \tag{13.22}$$

while for $\mathrm{x} = 1$, one has

$$p_1^{Y|X} = \begin{bmatrix} p_{0|1}^{Y|X} \\ p_{1|1}^{Y|X} \end{bmatrix} = \frac{1}{p_1^X} \begin{bmatrix} p_{10}^{XY} \\ p_{11}^{XY} \end{bmatrix} = \begin{bmatrix} 0.5 \\ 0.5 \end{bmatrix} = u. \tag{13.23}$$

Hence, after the observation of subsystem X, the uncertainty about subsystem Y decreases to the minimum entropy $\mathrm{H}_\alpha(p_0^{Y|X}) = \mathrm{h}_\alpha(0) = 0$ if the observation is $\mathrm{x} = 0$. In contrast, it increases to the maximum entropy $\mathrm{H}_\alpha(p_1^{Y|X}) = \mathrm{h}_\alpha(0.5)$ if the observation is $\mathrm{x} = 1$. Thus, depending on the value of the observation $\mathrm{x} = x$, the uncertainty of the observer can increase or decrease as compared to an observer that does not have access to the state of subsystem X.

The last example focused on the uncertainty at the observer for any given fixed value x of the state of subsystem X. In order to evaluate an overall measure of uncertainty, it is sensible to average over all possible values of the random variable $\mathrm{x} \sim p^X$, yielding the following definition.

Definition 13.3 Ensemble entropy

The ensemble entropy of subsystem Y given subsystem X quantifies the residual uncertainty $\mathrm{H}_\alpha(p_\mathrm{x}^{Y|X})$ averaged over the distribution of the observation $\mathrm{x} \sim p^X$ of subsystem X:

$$\mathrm{H}_\alpha(\mathcal{E}^{Y|X})_{p^{XY}} = \mathrm{E}_{p^X}\left[\mathrm{H}_\alpha(p_\mathrm{x}^{Y|X})\right]. \tag{13.24}$$

As reflected by the notation $\mathrm{H}_\alpha(\mathcal{E}^{Y|X})_{p^{XY}}$, the ensemble entropy is a function of the ensemble $\mathcal{E}^{Y|X}$ in (13.18), and the subscript p^{XY} serves as a reminder of the dependence of the ensemble on the joint distribution p^{XY}.

Definition 13.4 Shannon, Tsallis, and Rényi ensemble entropies

Specializing the definition (13.24), the Shannon ensemble entropy is given by

$$\mathrm{H}(\mathcal{E}^{Y|X})_{p^{XY}} = -\sum_{x\in[N^X]}\sum_{y\in[N^Y]} p_{xy}^{XY}\log_2(p_{y|x}^{Y|X}); \tag{13.25}$$

the Tsallis ensemble entropy of order $\alpha \geq 0$ with $\alpha \neq 1$ is

$$\mathrm{H}_\alpha^{\mathrm{T}}(\mathcal{E}^{Y|X})_{p^{XY}} = \frac{\mathrm{l}_2}{1-\alpha}\left(\sum_{x\in[N^X]}\sum_{y\in[N^Y]} p_x^X (p_{y|x}^{Y|X})^\alpha - 1\right); \tag{13.26}$$

and the Rényi ensemble entropy of order $\alpha \geq 0$ with $\alpha \neq 1$ is

$$\mathrm{H}_\alpha^{\mathrm{R}}(\mathcal{E}^{Y|X}) = \frac{1}{1-\alpha}\sum_{x\in[N^X]} p_x^X \log_2\left(\sum_{y\in[N^Y]} (p_{y|x}^{Y|X})^\alpha\right). \tag{13.27}$$

Furthermore, for $\alpha = 1$, both Tsallis and Rényi ensemble entropies coincide with the Shannon ensemble entropy, i.e.,

$$\mathrm{H}(\mathcal{E}^{Y|X})_{p^{XY}} = \mathrm{H}_1^{\mathrm{T}}(\mathcal{E}^{Y|X})_{p^{XY}} = \mathrm{H}_1^{\mathrm{R}}(\mathcal{E}^{Y|X}). \tag{13.28}$$

13.3.3 Properties of the Ensemble Entropy

The following properties of the ensemble entropy follow from the non-negativity of the classical entropy (see (2.43), (3.40), and (3.68)).

Key Result 13.5 Non-negativity of the ensemble entropy

For any joint distribution p^{XY}, the ensemble entropy is non-negative,

$$\mathrm{H}_\alpha(\mathcal{E}^{Y|X})_{p^{XY}} \geq 0, \tag{13.29}$$

with equality $\mathrm{H}_\alpha(\mathcal{E}^{Y|X})_{p^{XY}} = 0$ if and only if random variable y is a function of random variable x.

To understand the condition for equality in (13.29), recall that the entropy $\mathrm{H}_\alpha(p)$ equals zero if and only if the distribution p is concentrated at a single state value. As stated in the result above, the equality $\mathrm{H}_\alpha(\mathcal{E}^{Y|X})_{p^{XY}} = 0$ is equivalent to the existence of a function $f(\cdot)$ that relates x to y via the functional relationship y $=$ f(x). This follows because the ensemble entropy (13.24) equals zero if and only if the conditional distribution $p_x^{Y|X}$ is concentrated at a single value $y = f(x) \in [N^Y]$ for every $x \in [N^X]$ with positive probability $p_x^X > 0$.

Key Result 13.6 The ensemble entropy equals the local entropy for uncorrelated subsystems

When subsystems X and Y are uncorrelated, the ensemble entropy equals the local entropy, i.e.,

$$\mathrm{H}_\alpha(\mathcal{E}^{Y|X})_{p^X \otimes p^Y} = \mathrm{H}_\alpha(Y)_{p^Y}. \tag{13.30}$$

This property is a direct consequence of the fact that, for uncorrelated subsystems, all the conditional distributions $p_x^{Y|X}$ in the ensemble $\mathcal{E}^{Y|X}$ coincide with the marginal distribution p^Y.

Key Result 13.7 The ensemble entropy is not symmetric

Reversing the roles of the two subsystems generally changes the ensemble entropy, i.e., one has the inequality

$$\mathrm{H}_\alpha(\mathcal{E}^{Y|X})_{p^{XY}} \neq \mathrm{H}_\alpha(\mathcal{E}^{X|Y})_{p^{XY}}. \tag{13.31}$$

The next example illustrates this property of the ensemble entropy.

Example 13.5

For the joint distribution (13.21), using (13.22) and (13.23), the ensemble $\mathcal{E}^{Y|X}$ in (13.18) is given by

$$\begin{aligned}\mathcal{E}^{Y|X} &= \{(p_0^{Y|X}, p_0^X), (p_1^{Y|X}, p_1^X)\}\\ &= \{(\delta_0, 0.5), (u, 0.5)\}.\end{aligned} \tag{13.32}$$

The corresponding Shannon ensemble entropy is thus evaluated as the average

$$\mathrm{H}(\mathcal{E}^{Y|X})_{p^{XY}} = \frac{1}{2}\left(\mathrm{h}(0) + \mathrm{h}\left(\frac{1}{2}\right)\right) = \frac{1}{2}. \tag{13.33}$$

Swapping the roles of the two subsystems, the ensemble $\mathcal{E}^{X|Y}$ can be evaluated in a similar way as

$$\begin{aligned}\mathcal{E}^{X|Y} &= \{(p_0^{X|Y}, p_0^Y), (p_1^{X|Y}, p_1^Y)\}\\ &= \{([2/3, 1/3]^T, 3/4), (\delta_1, 1/4)\},\end{aligned} \tag{13.34}$$

yielding the Shannon ensemble entropy

$$\mathrm{H}(\mathcal{E}^{X|Y})_{p^{XY}} = \frac{3}{4}\mathrm{h}\left(\frac{1}{3}\right) + \frac{1}{4}\mathrm{h}(0) = 0.6887, \tag{13.35}$$

which is different from $\mathrm{H}(\mathcal{E}^{Y|X})_{p^{XY}}$.

Given the monotonicity property of the entropy as a function of the order α (see (3.21) and (3.57)), the following property holds.

Key Result 13.8 The ensemble entropy decreases with the inverse temperature α

For any joint distribution p^{XY}, the ensemble entropy is a decreasing function of the order $\alpha \geq 0$, i.e., the inequality

$$\mathrm{H}_{\alpha'}(\mathcal{E}^{Y|X})_{p^{XY}} \leq \mathrm{H}_{\alpha}(\mathcal{E}^{Y|X})_{p^{XY}} \tag{13.36}$$

holds for any two orders α' and α with $\alpha' \geq \alpha$.

13.3.4 Operational Interpretation of the Ensemble Entropy

The operational interpretation of the ensemble entropy depends on the choice of the entropy measure. Notably, with the Shannon entropy, each entropy term $\mathrm{H}(p_x^{Y|X})$ in (13.24) represents the minimum average number of bits required to compress the state y of subsystem Y assuming that both the encoder and decoder know the value of the state $\mathrm{x} = x$. In fact, when state $\mathrm{x} = x$ is known, the informational state about subsystem Y is described by the conditional distribution $p_x^{Y|X}$. Therefore, the Shannon ensemble entropy $\mathrm{H}(\mathcal{E}^{Y|X})_{p^{XY}}$ is the expected value of the minimum average number of bits $\mathrm{H}(p_{\mathrm{x}}^{Y|X})$, with average taken over the state $\mathrm{x} \sim p^X$.

As another example, with the Tsallis entropy, each entropy term $\mathrm{H}_\alpha^{\mathrm{T}}(p_x^{Y|X})$ represents the average prediction loss for the state of subsystem Y assuming that the predictor has access to the state $\mathrm{x} = x$ of subsystem X. Accordingly, the Tsallis ensemble entropy $\mathrm{H}_\alpha^{\mathrm{T}}(\mathcal{E}^{Y|X})_{p^{XY}}$ is the expected value of the prediction error over the distribution p^X of the states of subsystem X.

13.4 Locally Accessible Information, aka Expected Information Gain

As discussed in Section 13.3, the ensemble entropy $\mathrm{H}_\alpha(\mathcal{E}^{Y|X})_{p^{XY}}$ equals the local entropy $\mathrm{H}_\alpha(Y)_{p^Y}$ when the two subsystems are uncorrelated (see (13.30)). Therefore, any discrepancy between the local entropy $\mathrm{H}_\alpha(Y)_{p^Y}$ and the ensemble entropy $\mathrm{H}_\alpha(\mathcal{E}^{Y|X})_{p^{XY}}$ can be interpreted as evidence of the existence of correlation between subsystems X and Y.

13.4.1 Defining the Locally Accessible Information

Based on this observation, this subsection studies the following quantity as a correlation measure.

Definition 13.5 Locally accessible information

For any joint distribution p^{XY}, the locally accessible information from subsystem X to subsystem Y is defined as the difference

$$\mathrm{J}_\alpha(X \to Y)_{p^{XY}} = \mathrm{H}_\alpha(Y)_{p^Y} - \mathrm{H}_\alpha(\mathcal{E}^{Y|X})_{p^{XY}}. \tag{13.37}$$

An analogous definition applies for the locally accessible information from subsystem Y to subsystem X, i.e., $\mathrm{J}_\alpha(Y \to X)_{p^{XY}} = \mathrm{H}_\alpha(X)_{p^X} - \mathrm{H}_\alpha(\mathcal{E}^{X|Y})_{p^{XY}}$. Unlike the correlation metrics studied in the previous chapter, the locally accessible information is generally asymmetric, in the sense that one has the inequality

$$\mathrm{J}_\alpha(X \to Y)_{p^{XY}} \neq \mathrm{J}_\alpha(Y \to X)_{p^{XY}}. \tag{13.38}$$

To elaborate on the significance of the locally accessible information, it is useful to note that, given the trivial equality $\mathrm{E}_{p^X}[\mathrm{H}_\alpha(Y)_{p^Y}] = \mathrm{H}_\alpha(Y)_{p^Y}$, the locally accessible information (13.37) can equivalently be expressed as follows.

Key Result 13.9 Locally accessible information as expected information gain
The locally accessible information can be expressed as the average

$$\mathrm{J}_\alpha(X \to Y)_{p^{XY}} = \mathrm{E}_{p^X}[\mathrm{H}_\alpha(Y)_{p^Y} - \mathrm{H}_\alpha(p_{\mathrm{x}}^{Y|X})], \tag{13.39}$$

where each term $\mathrm{H}_\alpha(Y)_{p^Y} - \mathrm{H}_\alpha(p_x^{Y|X})$ in (13.39) quantifies the change in uncertainty about the state of subsystem Y that is caused by observing the state $\mathrm{x} = x$ of subsystem X.

By (13.39), the locally accessible information $\mathrm{J}_\alpha(X \to Y)_{p^{XY}}$ represents the average change in the uncertainty about subsystem Y caused by knowledge of the state of subsystem X.

13.4.2 Properties of the Locally Accessible Information

As illustrated in Example 13.4, the change in uncertainty $\mathrm{H}_\alpha(Y)_{p^Y} - \mathrm{H}_\alpha(p_x^{Y|X})$ corresponding to the observation of a given state x can be positive, negative, or zero. However, the average change, i.e., the locally accessible information, satisfies the following property (see Appendix 13.A for a proof).

Key Result 13.10 Non-negativity and minimum value of the locally accessible information
Given any joint distribution p^{XY}, for the Shannon entropy, the Tsallis entropy of any order $\alpha > 0$, and the Rényi entropy of order $0 < \alpha \leq 1$, the locally accessible information is non-negative,

$$\mathrm{J}_\alpha(X \to Y)_{p^{XY}} \geq 0, \tag{13.40}$$

and it equals zero if and only if the two subsystems are uncorrelated:

$$\mathrm{J}_\alpha(X \to Y)_{p^X \otimes p^Y} = 0. \tag{13.41}$$

According to this result, the locally accessible information is a meaningful measure of correlation only for some of the entropy measures. Specifically, these are the entropies that are strictly concave (see Appendix 2.D). Using these entropies, on average, the uncertainty on the state of subsystem Y is not increased as a result of the observation of subsystem X. In particular, a strictly positive locally accessible information, $\mathrm{J}_\alpha(X \to Y)_{p^{XY}} > 0$, reveals the presence of correlation under the joint distribution p^{XY}. Furthermore, a zero locally accessible information, $\mathrm{J}_\alpha(X \to Y)_{p^{XY}} = 0$, demonstrates that the random variables $(x, y) \sim p^{XY}$ are uncorrelated.

Example 13.6

For $N^X = 4$ and $N^Y = 2$, consider the probability ensemble

$$\mathcal{E}^{Y|X} = \left\{ \left(\begin{bmatrix} 0.85 \\ 0.15 \end{bmatrix}, \frac{1}{4} \right), \left(\begin{bmatrix} 0.8 \\ 0.2 \end{bmatrix}, \frac{1}{4} \right), \left(\begin{bmatrix} 0.75 \\ 0.25 \end{bmatrix}, \frac{1}{4} \right), \left(\begin{bmatrix} 0.7 \\ 0.3 \end{bmatrix}, \frac{1}{4} \right) \right\}, \quad (13.42)$$

for which the reader is invited to evaluate the corresponding joint distribution.

Adopting the Shannon entropy, Figure 13.1(a) shows: (i) the marginal entropy $\mathrm{H}(Y)_{p^Y}$ (black line); (ii) the entropies $\{\mathrm{H}(p_x^{Y|X})\}_{x\in[4]}$ in the ensemble $\mathcal{E}^{Y|X}$ (dashed lines); and (iii) the ensemble entropy $\mathrm{H}(\mathcal{E}^{Y|X})_{p^{XY}}$ as the average of the individual entropies $\mathrm{H}(p_x^{Y|X})$ over distribution $p^X = u$.

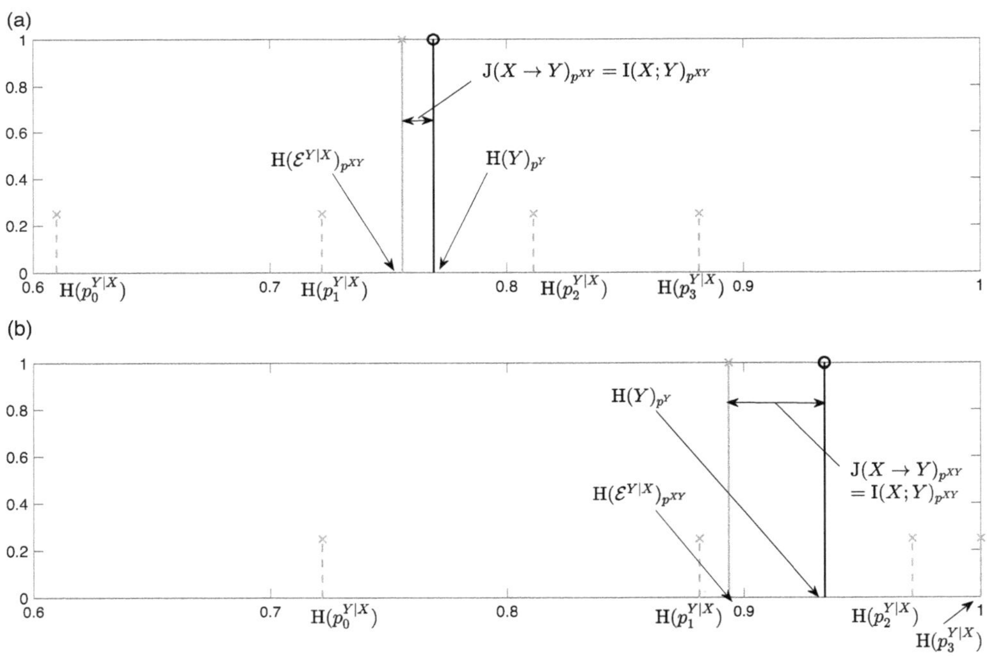

Figure 13.1 Marginal entropy $\mathrm{H}(Y)_{p^Y}$ (black line), entropies $\{\mathrm{H}(p_x^{Y|X})\}_{x\in[4]}$ for the ensemble $\mathcal{E}^{Y|X}$ (dashed lines), and ensemble entropy $\mathrm{H}(\mathcal{E}^{Y|X})_{p^{XY}}$ for the probability ensembles (13.42) (a) and (13.43) (b).

The Shannon locally accessible information $\mathrm{J}(X \to Y)_{p^{XY}}$ is marked as the difference between the marginal entropy $\mathrm{H}(Y)_{p^Y}$ and the ensemble entropy $\mathrm{H}(\mathcal{E}^{Y|X})_{p^{XY}}$. This difference is seen to be quite small for the joint distribution at hand. This is because the conditional distributions $\{p_x^{Y|X}\}_{x\in[4]}$ in the ensemble are similar, and thus the expected information gain in observing the state of subsystem X is not significant.

For comparison, take now the different ensemble

$$\mathcal{E}^{Y|X} = \left\{ \left(\begin{bmatrix} 0.8 \\ 0.2 \end{bmatrix}, \frac{1}{4} \right), \left(\begin{bmatrix} 0.7 \\ 0.3 \end{bmatrix}, \frac{1}{4} \right), \left(\begin{bmatrix} 0.6 \\ 0.4 \end{bmatrix}, \frac{1}{4} \right), \left(\begin{bmatrix} 0.5 \\ 0.5 \end{bmatrix}, \frac{1}{4} \right) \right\}, \tag{13.43}$$

in which the conditional distributions $\{p_x^{Y|X}\}_{x\in[4]}$ are more diverse. Figure 13.1(b) shows again the marginal entropy $\mathrm{H}(Y)_{p^Y}$, the entropies $\{\mathrm{H}(p_x^{Y|X})\}_{x\in[4]}$, as well as the ensemble entropy $\mathrm{H}(\mathcal{E}^{Y|X})_{p^{XY}}$ and the locally accessible information $\mathrm{J}(X \to Y)_{p^{XY}}$. Reflecting the more significant dependence of the conditional distribution $p_x^{Y|X}$ on the state $x \in [N^X]$, the locally accessible information $\mathrm{J}(X \to Y)_{p^{XY}}$ is larger for this second ensemble.

Example 13.7

Only strictly concave entropy measures yield locally accessible information that provides valid measures of correlation. To verify this with a numerical example, consider again the joint distribution (13.21). The Rényi entropy of order $\alpha > 1$ is not guaranteed to be concave, and thus the non-negativity of the difference (13.37) is generally not satisfied. For instance, with order $\alpha = \infty$, the difference in (13.37) is computed as

$$\begin{aligned} \mathrm{h}_\infty^{\mathrm{R}}\left(\frac{1}{4}\right) - \frac{1}{2}\mathrm{h}_\infty^{\mathrm{R}}\left(\frac{1}{2}\right) &= -\log_2\left(\frac{3}{4}\right) - \frac{1}{2} \\ &= -0.0850. \end{aligned} \tag{13.44}$$

13.4.3 Conditioning Reduces the Ensemble Entropy

The non-negativity and minimum-value properties of the locally accessible information (13.40) and (13.41) have the following important consequence.

Key Result 13.11 Conditioning reduces the ensemble entropy
Given any joint distribution p^{XY}, for the Shannon entropy, the Tsallis entropy of any order $\alpha > 0$, and the Rényi entropy of order $0 < \alpha \leq 1$, the ensemble entropy is upper bounded by the marginal entropy, i.e.,

$$\mathrm{H}_\alpha(\mathcal{E}^{Y|X})_{p^{XY}} \leq \mathrm{H}_\alpha(Y)_{p^Y}, \tag{13.45}$$

with equality (13.30) attained if and only if the two subsystems are uncorrelated.

In words, this property says that, when observing – "conditioning on" – one subsystem, the average uncertainty about the other subsystem, measured by the ensemble entropy, cannot increase.

13.4.4 The Shannon Locally Accessible Information Is the Shannon Mutual Information

While the locally accessible information is asymmetric, in the special case of the Shannon entropy, the locally accessible information is equal in both directions. What is more, as shown next, the two locally accessible information measures $\mathrm{J}(X \to Y)_{p^{XY}}$ and $\mathrm{J}(Y \to X)_{p^{XY}}$ coincide with the Shannon mutual information.

Key Result 13.12 The Shannon locally accessible information coincides with the Shannon mutual information

For any joint distribution p^{XY}, the following equalities hold:

$$\mathrm{J}(X \to Y)_{p^{XY}} = \mathrm{J}(Y \to X)_{p^{XY}} = \mathrm{I}(X;Y)_{p^{XY}}. \tag{13.46}$$

This result can be proved directly by using the respective definitions as

$$\begin{aligned}
\mathrm{J}(X \to Y)_{p^{XY}} &= \mathrm{H}(Y)_{p^Y} - \mathrm{H}(\mathcal{E}^{Y|X})_{p^{XY}} \\
&= \mathrm{H}(Y)_{p^Y} + \mathrm{E}_{p^{XY}}\left[\log_2\left(\frac{p^{XY}_{\mathrm{xy}}}{p^X_{\mathrm{x}}}\right)\right] \\
&= \mathrm{H}(Y)_{p^Y} - \mathrm{E}_{p^X}\left[\log_2(p^X_{\mathrm{x}})\right] - \left(-\mathrm{E}_{p^{XY}}\left[\log_2\left(p^{XY}_{\mathrm{xy}}\right)\right]\right) \\
&= \mathrm{H}(Y)_{p^Y} + \mathrm{H}(X)_{p^X} - \mathrm{H}(X,Y)_{p^{XY}} \\
&= \mathrm{I}(X;Y)_{p^{XY}}.
\end{aligned} \tag{13.47}$$

Example 13.8

For the ensemble (13.42), Figure 13.2 plots the Tsallis and Rényi locally accessible information measures $\mathrm{J}_\alpha(X \to Y)_{p^{XY}}$ and $\mathrm{J}_\alpha(Y \to X)_{p^{XY}}$ as a function of the order α in the range $0 < \alpha \leq 1$. Recall that, with this restriction on the order, all locally accessible information measures are valid metrics to quantify correlations. For reference, the figure also shows the Shannon mutual information $\mathrm{I}(X;Y)_{p^{XY}}$, which coincides with the Tsallis and Rényi locally accessible informations for $\alpha = 1$.

As the order α tends to zero, the Tsallis and Rényi locally accessible informations depend only on the support of the conditional distributions. Since, for the given joint distribution p^{XY}, the support is not changed when conditioning on either variable, the locally accessible informations $\mathrm{J}_0(X \to Y)_{p^{XY}}$ and $\mathrm{J}_0(Y \to X)_{p^{XY}}$ equal zero. At the other extreme, for $\alpha = 1$, all locally accessible informations coincide with the Shannon mutual information, irrespective of the direction.

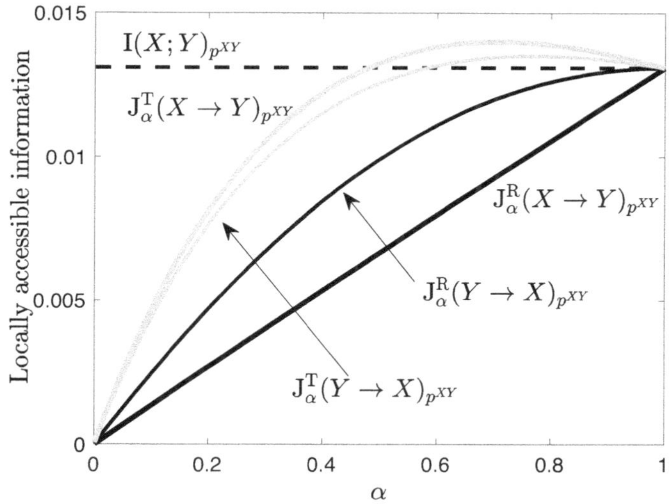

Figure 13.2 Tsallis and Rényi locally accessible informations $\mathrm{J}_\alpha(X \to Y)_{p^{XY}}$ and $\mathrm{J}_\alpha(Y \to X)_{p^{XY}}$, along with the Shannon mutual information $\mathrm{I}(X;Y)_{p^{XY}}$, as a function of the order α in the range $0 < \alpha \leq 1$ for the ensemble (13.42).

Excluding these extreme cases, the Tsallis and Rényi locally accessible informations $\mathrm{J}_0(X \to Y)_{p^{XY}}$ and $\mathrm{J}_0(Y \to X)_{p^{XY}}$ are different in the two directions. Moreover, as the order α increases, while both the marginal and ensemble entropies decrease as a consequence of the monotonicity of the entropy as a function of α (see Chapter 3), the locally accessible information is generally not monotonic as a function of the order α.

13.4.5 Shannon Mutual Information and Shannon Relative Entropy Revisited ⏭

As proved in Section 12.7, the Shannon mutual information $\mathrm{I}(X;Y)_{p^{XY}}$ is equal to the Shannon relative entropy $\mathrm{D}(p^{XY}||p^X \otimes p^Y)$ between the joint distribution p^{XY} and the product of marginals $p^X \otimes p^Y$. It is now shown that one can also express the Shannon mutual information in terms of the Shannon relative entropy between conditional and marginal distributions. This approach formalizes the intuition that the diversity of the ensemble (13.18) determines the amount of correlation present in the joint distribution p^{XY}.

To elaborate, consider the Shannon relative entropy $\mathrm{D}(p_x^{Y|X|}||p^Y)$, which captures the difference between a conditional distribution $p_x^{Y|X|}$ in the ensemble (13.18) and the marginal distribution p^Y. Averaging over the state $\mathrm{x} \sim p^X$, one obtains a measure of the diversity of the ensemble, and thus also a measure of correlation.

Interestingly, this average too coincides with the Shannon mutual information:

$$\begin{aligned} \mathrm{E}_{p^X}[\mathrm{D}(p_{\mathrm{x}}^{Y|X|}||p^Y)] &= \mathrm{E}_{p^X}\left[\mathrm{E}_{p_{\mathrm{x}}^{Y|X}}\left[\log_2\left(\frac{p_{\mathrm{y|x}}^{Y|X}}{p_{\mathrm{y}}^Y}\right)\right]\right] \\ &= \mathrm{E}_{p^{XY}}\left[\log_2\left(\frac{p_{\mathrm{xy}}^{XY}}{p_{\mathrm{x}}^X p_{\mathrm{y}}^Y}\right)\right] \\ &= \mathrm{I}(X;Y)_{p^{XY}}. \end{aligned} \tag{13.48}$$

Thus, the Shannon mutual information can also be interpreted as the average distance between conditional distributions and marginal distribution. Note that, by symmetry, one also has the equality $\mathrm{I}(X;Y)_{p^{XY}} = \mathrm{E}_{p^Y}[\mathrm{D}(p_{\mathrm{y}}^{X|Y|}||p^X)]$.

13.5 Conditional Entropy

As studied in Chapter 12, the Rényi joint entropy of any order $\alpha \geq 0$, and thus also the Shannon joint entropy, are additive. This means that, when the two subsystems are uncorrelated, the Rényi joint entropy decomposes into the sum of the local entropies as

$$\mathrm{H}_\alpha^{\mathrm{R}}(X, Y)_{p^X \otimes p^Y} = \mathrm{H}_\alpha^{\mathrm{R}}(X)_{p^X} + \mathrm{H}_\alpha^{\mathrm{R}}(Y)_{p^Y}, \tag{13.49}$$

allowing a neat separation of the contributions of the two subsystems to the overall uncertainty. This section focuses on deriving a generalization of this decomposition that applies also to correlated subsystems. The main outcome of this discussion will be the introduction of the conditional entropy as an alternative measure of residual uncertainty.

13.5.1 Defining the Shannon Conditional Entropy

Consider first the special case of the Shannon entropy, for which the information-as-a-set representation is instrumental in building intuition. According to this representation, the union of the two sets representing the local entropies $\mathrm{H}(X)_{p^X}$ and $\mathrm{H}(Y)_{p^Y}$ quantifies the joint entropy $\mathrm{H}(X, Y)_{p^{XY}}$ for a general joint distribution p^{XY}. Therefore, as seen in Figure 13.3(a), when the two sets are disjoint, representing uncorrelated subsystems, the size of the union equals the sum of the sizes of the individual sets, which recovers the additivity property (13.49).

In contrast, when the two subsystems are correlated, the information-as-a-set representation is depicted in Figure 13.3(b). In this case, the two sets representing the local entropies $\mathrm{H}(X)_{p^X}$ and $\mathrm{H}(Y)_{p^Y}$ overlap, and thus the decomposition (13.49) of the joint entropy into the sum of local entropies no longer holds. This motivates the following definition.

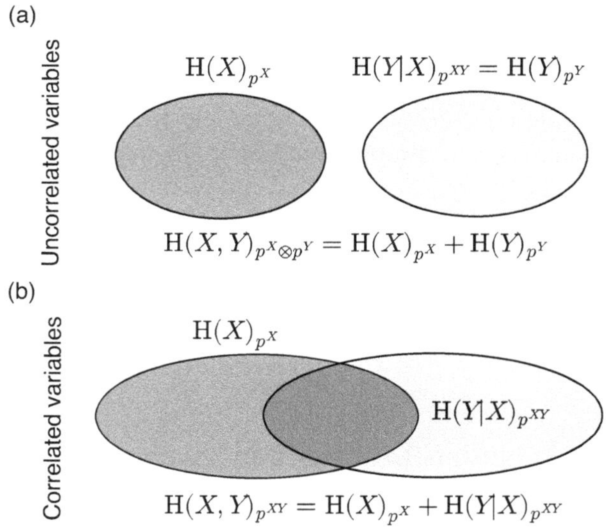

Figure 13.3 Illustration of the Shannon conditional entropy using the information-as-a-set representation.

> **Definition 13.6 Shannon conditional entropy**
> For any joint distribution p^{XY}, the Shannon conditional entropy of subsystem Y given subsystem X is defined as the difference
> $$\mathrm{H}(Y|X)_{p^{XY}} = \mathrm{H}(X, Y)_{p^{XY}} - \mathrm{H}(X)_{p^X}. \qquad (13.50)$$

The rationale for this definition is that it directly implies the following decomposition of the Shannon joint entropy.

> **Key Result 13.13 Chain rule for the Shannon joint entropy**
> For any joint distribution p^{XY}, the Shannon joint entropy can be decomposed as
> $$\mathrm{H}(X, Y)_{p^{XY}} = \mathrm{H}(X)_{p^X} + \mathrm{H}(Y|X)_{p^{XY}}, \qquad (13.51)$$
> which is known as the chain rule for the Shannon joint entropy.

In (13.51), the contribution of subsystem X is accounted for by the local entropy $\mathrm{H}(X)_{p^X}$, while the residual contribution of subsystem Y is quantified by the Shannon conditional entropy $\mathrm{H}(Y|X)_{p^{XY}}$.

The chain rule (13.51) can be understood using the information-as-a-set presentation. In particular, as seen in Figure 13.3(b), the Shannon conditional entropy $\mathrm{H}(Y|X)_{p^{XY}}$ in (13.50) is represented by the difference between the size of the union of the two sets, representing the joint entropy $\mathrm{H}(X, Y)_{p^{XY}}$, and the size of the set representing local entropy $\mathrm{H}(X)_{p^X}$. The Shannon conditional entropy $\mathrm{H}(Y|X)_{p^{XY}}$ thus provides another way to quantify the residual uncertainty about subsystem Y that is left over when the state of subsystem X is known.

Note that, as usual, one can swap the roles of the two subsystems. Specifically, defining the Shannon conditional entropy of subsystem X conditioned on subsystem Y as

$$\mathrm{H}(X|Y)_{p^{XY}} = \mathrm{H}(X, Y)_{p^{XY}} - \mathrm{H}(Y)_{p^Y}, \tag{13.52}$$

one also has the chain rule

$$\mathrm{H}(X, Y)_{p^{XY}} = \mathrm{H}(Y)_{p^Y} + \mathrm{H}(X|Y)_{p^{XY}}. \tag{13.53}$$

13.5.2 Shannon Conditional Entropy and Shannon Ensemble Entropy

So far, this chapter has introduced two measures of residual uncertainty. The first, the ensemble entropy, was derived as the average entropy (13.24) of the conditional distributions, while the second, the conditional entropy, was obtained as the difference between joint entropy and local entropy in (13.50). Remarkably, when considering the special case of the Shannon entropy, these two measures coincide.

Key Result 13.14 The Shannon conditional entropy coincides with the Shannon ensemble entropy

For any joint distribution p^{XY}, the following equality holds:

$$\mathrm{H}(Y|X)_{p^{XY}} = \mathrm{H}(\mathcal{E}^{Y|X})_{p^{XY}}. \tag{13.54}$$

A simple way to prove this result is by connecting the chain rule for the Shannon joint entropy in (13.51) with the chain rule for the Shannon surprise in (13.15). In fact, taking the expectation over the joint distribution p^{XY} on both sides of the equality (13.15) and comparing with (13.51), one obtains

$$\underbrace{\mathrm{E}_{p^{XY}}[-\log_2(p^{XY}_{\mathrm{xy}})]}_{=\mathrm{H}(X,Y)_{p^{XY}}} = \underbrace{\mathrm{E}_{p^X}[-\log_2(p^X_{\mathrm{x}})]}_{=\mathrm{H}(Y)_{p^Y}} + \underbrace{\mathrm{E}_{p^{XY}}[-\log_2(p^{Y|X}_{\mathrm{y|x}})]}_{=\mathrm{H}(Y|X)_{p^{XY}}}. \tag{13.55}$$

This implies, using the definition (13.25) of the Shannon ensemble entropy $\mathrm{H}(\mathcal{E}^{Y|X})_{p^{XY}}$, that the Shannon conditional entropy can be written as

$$\mathrm{H}(Y|X)_{p^{XY}} = \mathrm{E}_{p^{XY}}[-\log_2(p^{Y|X}_{\mathrm{y|x}})] = \mathrm{H}(\mathcal{E}^{Y|X})_{p^{XY}}, \tag{13.56}$$

proving the claim.

The equality (13.54) implies that the Shannon conditional entropy inherits all the properties of the Shannon ensemble entropy. In particular, the following bounds are direct consequences of (13.29) and (13.45).

Key Result 13.15 Bounds on the Shannon conditional entropy

For any joint distribution p^{XY}, the following bounds hold:

$$0 \leq \mathrm{H}(Y|X)_{p^{XY}} \leq \mathrm{H}(Y)_{p^Y}. \tag{13.57}$$

The lower bound $\mathrm{H}(Y|X)_{p^{XY}} = 0$ is attained if and only if random variable y is a function of random variable x, while the upper bound $\mathrm{H}(Y|X)_{p^{XY}} = \mathrm{H}(Y)_{p^Y}$ is obtained if and only if the two subsystems are uncorrelated, i.e., $p^{XY} = p^X \otimes p^Y$. The upper bound in (13.57) indicates that conditioning reduces the Shannon entropy.

Furthermore, the equality (13.54) also implies that the Shannon mutual information (13.46) can be written as follows.

Key Result 13.16 Shannon mutual information and Shannon conditional information

For any joint distribution p^{XY}, the Shannon mutual information can be expressed as

$$\begin{aligned}\mathrm{I}(X;Y)_{p^{XY}} &= \mathrm{H}(Y)_{p^Y} - \mathrm{H}(Y|X)_{p^{XY}} \\ &= \mathrm{H}(X)_{p^X} - \mathrm{H}(X|Y)_{p^{XY}}.\end{aligned} \tag{13.58}$$

Accordingly, the Shannon mutual information is the difference between the corresponding marginal and conditional entropies.

13.5.3 Shannon Conditional Entropy and Shannon Relative Entropy ⏭

As discussed in Section 4.3, the Shannon entropy $\mathrm{H}(p)$ of an $N \times 1$ distribution p satisfies the decomposition

$$\mathrm{H}(p) = \log_2(N) - \mathrm{D}(p||u). \tag{13.59}$$

This decomposition formalizes the intuition that the Shannon entropy becomes larger as the distribution p gets closer to the uniform distribution u. Does a similar decomposition apply also to the Shannon conditional entropy? If so, what is the correct reference distribution?

To address this question, apply the decomposition (13.59) to the joint Shannon entropy $\mathrm{H}(X, Y)_{p^{XY}}$ to obtain the equality

$$\mathrm{H}(X, Y)_{p^{XY}} = \log_2(N^X) + \log_2(N^Y) - \mathrm{D}(p^{XY}||u^X \otimes u^Y). \tag{13.60}$$

This follows from (13.59) given the equality $\log_2(N^X N^Y) = \log_2(N^X) + \log_2(N^Y)$, and given that the uniform joint distribution $u^{XY}_{xy} = 1/(N^X N^Y)$ for $x \in [N^X]$ and $y \in [N^Y]$ can be expressed as the Kronecker product $u^{XY} = u^X \otimes u^Y$, with $u^X_x = 1/N^X$ and $u^Y_y = 1/N^Y$ for $x \in [N^X]$ and $y \in [N^Y]$.

Now, using the definition (13.50) of the Shannon conditional entropy and the decomposition $\mathrm{H}(Y)_{p^Y} = \log_2(N^Y) - \mathrm{D}(p^Y||u^Y)$, simple algebra yields the following decomposition for the Shannon conditional entropy.

Key Result 13.17 Decomposition of the Shannon conditional entropy
For any joint distribution p^{XY}, the Shannon conditional entropy can be decomposed as

$$\mathrm{H}(Y|X)_{p^{XY}} = \log_2(N^Y) - \mathrm{D}(p^{XY}||p^X \otimes u^Y). \tag{13.61}$$

According to the equality (13.61), the Shannon conditional entropy $\mathrm{H}(Y|X)_{p^{XY}}$ grows as the distance $\mathrm{D}(p^{XY}||p^X \otimes u^Y)$ between the joint distribution and the reference distribution $p^X \otimes u^Y$ decreases, attaining the maximum value $\log_2(N^Y)$ when $p^{XY} = p^X \otimes u^Y$. Under the reference joint distribution $p^X \otimes u^Y$, the two subsystems are uncorrelated and the marginal distribution of subsystem Y is uniform. Intuitively, both conditions – independence and maximum-local entropy – are required for the conditional entropy $\mathrm{H}(Y|X)_{p^{XY}}$ to be maximized.

13.5.4 Fano's Inequality ⏭

The representation (13.61) of the conditional entropy is useful to pursue an important application of the Shannon conditional entropy to the problem of detection. To proceed, take the special case in which the two subsystems have the same number of states, i.e., $N^X = N^Y = D$. Thinking now of random variable x as a prediction of the target random variable y, the goal is to relate the probability

$$p^{\mathrm{eq}} = \Pr[\mathrm{x} = \mathrm{y}] = \sum_{x \in [D]} p^{XY}_{xx} \tag{13.62}$$

of correct prediction to the Shannon conditional entropy $\mathrm{H}(Y|X)_{p^{XY}}$.

Intuitively, if the detection probability p^{eq} is close to 1, it must be true that the Shannon conditional entropy $\mathrm{H}(Y|X)_{p^{XY}}$ is small. In fact, in this case, random variable x serves as an accurate prediction for random variable y. This intuition is formalized by the following key information-theoretic relation.

Key Result 13.18 Fano's inequality
For any joint distribution p^{XY}, the inequality

$$\mathrm{H}(Y|X)_{p^{XY}} \leq \mathrm{h}(p^{\mathrm{eq}}) + (1 - p^{\mathrm{eq}}) \log_2(D - 1) \tag{13.63}$$

holds, where p^{eq} is the probability (13.62) and $\mathrm{h}(p^{\mathrm{eq}})$ is the corresponding binary Shannon entropy.

By the Fano inequality (13.63), a large detection probability p^{eq} implies a small value for the conditional entropy $\text{H}(Y|X)_{p^{XY}}$. The Fano inequality can be proved following Problem 12.4.

13.5.5 Rényi Conditional Entropy ⏭

So far, this section has introduced the Shannon conditional entropy as the difference (13.50), which makes it possible to establish the chain rule (13.51) for the Shannon joint entropy. Can a chain rule like (13.51) be derived also for the Rényi joint entropy?

As seen in Chapter 12, the Rényi joint entropy satisfies the uncertainty-increases-globally property (12.40), which indicates that the Rényi joint entropy $\text{H}^{\text{R}}_{\alpha}(X, Y)_{p^{XY}}$ is no smaller than both of the local entropies $\text{H}^{\text{R}}_{\alpha}(X)_{p^X}$ and $\text{H}^{\text{R}}_{\alpha}(Y)_{p^Y}$. Therefore, in a manner completely analogous to the Shannon conditional entropy, one can introduce the following definition.

Definition 13.7 Rényi conditional entropy and chain rule for the Rényi joint entropy

Given any order $\alpha \geq 0$, the Rényi conditional entropy of subsystem Y given subsystem X is defined as the difference

$$\begin{aligned}\text{H}^{\text{R}}_{\alpha}(Y|X)_{p^{XY}} &= \text{H}^{\text{R}}_{\alpha}(X, Y)_{p^{XY}} - \text{H}^{\text{R}}_{\alpha}(X)_{p^X} \\ &= \frac{1}{1-\alpha}\log_2\left(\frac{\sum_{x\in[N^X]}\sum_{y\in[N^Y]}(p^{XY}_{xy})^{\alpha}}{\sum_{x\in[N^X]}(p^X_x)^{\alpha}}\right),\end{aligned} \tag{13.64}$$

yielding the chain rule for the Rényi joint entropy

$$\text{H}^{\text{R}}_{\alpha}(X, Y)_{p^{XY}} = \text{H}^{\text{R}}_{\alpha}(X)_{p^X} + \text{H}^{\text{R}}_{\alpha}(Y|X)_{p^{XY}}. \tag{13.65}$$

The uncertainty-increases-globally property (12.40) of the Rényi joint entropy implies the following property.

Key Result 13.19 Non-negativity and minimum value for the Rényi conditional entropy

For any joint distribution p^{XY}, the Rényi conditional entropy satisfies the inequality

$$\text{H}^{\text{R}}_{\alpha}(Y|X)_{p^{XY}} \geq 0, \tag{13.66}$$

with equality $\text{H}^{\text{R}}_{\alpha}(Y|X)_{p^{XY}} = 0$ if and only if random variable y is a function of random variable x.

While the definition (13.64) follows the same rationale as for the Shannon conditional entropy, the important property that conditioning decreases entropy (see

(13.57)) does *not* hold for the Rényi conditional entropy. That is, one may have the inequality $\mathrm{H}_\alpha^{\mathrm{R}}(Y|X)_{p^{XY}} > \mathrm{H}_\alpha^{\mathrm{R}}(Y)_{p^Y}$, for some joint distributions p^{XY}. This is a direct consequence of the fact that the Rényi joint entropy is not subadditive, as discussed in Section 12.5.5.

Example 13.9

For a system with dimensions $N^X = N^Y = 2$, consider the joint distribution $p^{XY} = [p_{00}^{XY}, p_{01}^{XY}, p_{10}^{XY}, p_{11}^{XY}] = [0.2, 0, 0.4, 0.4]^T$. The marginal distribution for subsystem Y is given by $p^Y = [0.6, 0.4]^T$, for which the Rényi local entropy of order $\alpha = 3$ is given by $\mathrm{H}_3^{\mathrm{R}}(Y)_{p^Y} = 0.9183$; while the marginal distribution for subsystem X is $p^X = [0.2, 0.8]^T$, which yields $\mathrm{H}_3^{\mathrm{R}}(X)_{p^X} = 0.4717$. The corresponding Rényi conditional entropy is computed as

$$\mathrm{H}_3^{\mathrm{R}}(Y|X)_{p^{XY}} = \mathrm{H}_3^{\mathrm{R}}(X, Y)_{p^{XY}} - \mathrm{H}_3^{\mathrm{R}}(X) = 0.9675, \tag{13.67}$$

which is larger than the marginal entropy $\mathrm{H}_3^{\mathrm{R}}(Y)_{p^Y} = 0.9183$, violating the conditioning-reduces-entropy property.

As another point of distinction between Shannon and Rényi conditional entropies, while the Shannon conditional entropy equals the Shannon ensemble entropy, the Rényi conditional entropy is generally different from the Rényi ensemble entropy, i.e., $\mathrm{H}_\alpha^{\mathrm{R}}(Y|X)_{p^{XY}} \neq \mathrm{H}_\alpha^{\mathrm{R}}(\mathcal{E}^{Y|X})_{p^{XY}}$. In particular, the Rényi conditional entropy cannot be obtained as the average of the entropy $\mathrm{H}_\alpha^{\mathrm{R}}(p_{\mathrm{x}}^{Y|X})$ with respect to the random variable $\mathrm{x} \sim p^X$. Instead, it corresponds to a generalized average of $\mathrm{H}_\alpha^{\mathrm{R}}(p_{\mathrm{x}}^{Y|X})$. The reader is referred to Section 3.7 for the definition of generalized average that will be adopted below.

To elaborate, recall from Section 3.7 the definition of the escort distribution of order $\alpha \geq 0$ for the marginal distribution p^X, whose xth entry is given by

$$\tilde{p}_{\alpha,x}^X = \frac{(p_x^X)^\alpha}{\sum_{x' \in \mathcal{X}} (p_{x'}^X)^\alpha}. \tag{13.68}$$

As proved in Appendix 13.B, the following result holds.

Key Result 13.20 Rényi conditional entropy as a generalized average

For any joint distribution p^{XY}, the Rényi conditional entropy can be written as

$$\mathrm{H}_\alpha^{\mathrm{R}}(Y|X)_{p^{XY}} = \frac{1}{1-\alpha} \log_2 \left(\mathrm{E}_{\tilde{p}_\alpha^X} \left[2^{(1-\alpha)\mathrm{H}_\alpha^{\mathrm{R}}(p_{\mathrm{x}}^{Y|X})} \right] \right). \tag{13.69}$$

By (13.69), the Rényi conditional entropy is thus a generalized average of the entropies $\mathrm{H}_\alpha^{\mathrm{R}}(p_x^{Y|X})$ of the conditional distributions $p_x^{Y|X}$, reducing to the conventional average $\mathrm{H}_\alpha^{\mathrm{R}}(\mathcal{E}^{Y|X})_{p^{XY}} = \mathrm{E}_{p^X}[\mathrm{H}_\alpha^{\mathrm{R}}(p_{\mathrm{x}}^{Y|X})]$, i.e., to the ensemble entropy, for $\alpha = 1$.

Therefore, both the Rényi conditional entropy $\mathrm{H}_\alpha^{\mathrm{R}}(Y|X)_{p^{XY}}$ and the Rényi ensemble entropy $\mathrm{H}_\alpha^{\mathrm{R}}(\mathcal{E}^{Y|X})_{p^{XY}}$ capture the residual uncertainty on the state subsystem Y given the state of subsystem X. However, they do so by averaging in a different manner the entropies $\mathrm{H}_\alpha^{\mathrm{R}}(p_{\mathrm{x}}^{Y|X})$ of the conditional distributions with respect to the distribution of the observed state x.

Finally, unlike the Rényi entropy and the Rényi ensemble entropy, the Rényi conditional entropy is not necessarily monotonic in the order α (see Recommended Resources). An alternative definition of Rényi conditional entropy that relates to the maximum probability of detection is provided in Appendix 13.C.

13.5.6 Tsallis Conditional Entropy ⏭

In order to define the Tsallis conditional entropy, one could proceed in a manner analogous to the Shannon and Rényi conditional entropies by leveraging the uncertainty-increases-globally property (12.40) of the Tsallis joint entropy. However, a more common approach to define the Tsallis conditional entropy reflects the pseudoadditivity (12.34) of the Tsallis entropy, as briefly discussed next.

Definition 13.8 Tsallis conditional entropy and pseudo-chain rule for the Tsallis joint entropy

Given the order $\alpha \geq 0$, the Tsallis conditional entropy of subsystem Y given subsystem X is defined as

$$\mathrm{H}_\alpha^{\mathrm{T}}(Y|X)_{p^{XY}} = \frac{\mathrm{H}_\alpha^{\mathrm{T}}(X,Y)_{p^{XY}} - \mathrm{H}_\alpha^{\mathrm{T}}(X)_{p^X}}{1 + \frac{(1-\alpha)}{\mathrm{l}_2}\mathrm{H}_\alpha^{\mathrm{T}}(X)_{p^X}}. \tag{13.70}$$

This definition satisfies the following **pseudo-chain rule** for the Tsallis joint entropy:

$$\mathrm{H}_\alpha^{\mathrm{T}}(X,Y)_{p^{XY}} = \mathrm{H}_\alpha^{\mathrm{T}}(X)_{p^X} + \mathrm{H}_\alpha^{\mathrm{T}}(Y|X)_{p^{XY}} + \frac{(1-\alpha)}{\mathrm{l}_2}\mathrm{H}_\alpha^{\mathrm{T}}(X)_{p^X}\mathrm{H}_\alpha^{\mathrm{T}}(Y|X)_{p^{XY}}, \tag{13.71}$$

which reduces to the pseudoadditivity property (12.34) in the special case of uncorrelated subsystems.

Interestingly, the Tsallis conditional entropy (13.70) can also be written as

$$\mathrm{H}_\alpha^{\mathrm{T}}(Y|X)_{p^{XY}} = \mathrm{E}_{\tilde{p}_\alpha^X}\left[\mathrm{H}_\alpha^{\mathrm{T}}(p_{\mathrm{x}}^{Y|X})\right], \tag{13.72}$$

where the expectation is taken over the escort distribution (13.68). Furthermore, the alternative expression (13.72) demonstrates the following property.

> **Key Result 13.21 Non-negativity and minimum value for the Tsallis conditional entropy**
> For any joint distribution p^{XY}, the Tsallis conditional entropy (13.70) satisfies the inequality
> $$\mathrm{H}_{\alpha}^{\mathrm{T}}(Y|X)_{p^{XY}} \geq 0, \tag{13.73}$$
> with equality $\mathrm{H}_{\alpha}^{\mathrm{T}}(Y|X)_{p^{XY}} = 0$ if and only if random variable y is a function of random variable x.

The reader is referred to the Recommended Resources for further discussion on the Tsallis conditional entropy.

13.6 Conditional Shared Randomness via Sequential Sampling ⏭

Chapter 12 introduced the informational task of distributed sampling, in which the goal was for two agents to generate random variables $(\mathrm{x}, \mathrm{y}) \sim p^{XY}$, while sharing the minimum amount of common randomness. In this final section, a sequential counterpart of the distributed sampling problem is studied in which the two agents operate not in parallel but rather one after the other.

As shown in Figure 13.4, in this problem agent X generates a random variable $\mathrm{x} \sim p^X$ from the marginal p^X of the desired joint distribution p^{XY}. The generated value x is provided to agent Y, whose goal is to draw a random variable $\mathrm{y} \sim p_{\mathrm{x}}^{Y|X}$ from the conditional distribution $p_{\mathrm{x}}^{Y|X}$ corresponding to the received value x. By the chain rule for the joint distribution in (13.11), this ensures that the joint distribution of the random variables x and y equals the desired distribution $p_{xy}^{XY} = p_x^X p_{y|x}^{Y|X}$ with $x \in [N^X]$ and $y \in [N^Y]$. This form of sampling from a joint distribution is known as **sequential**, or **ancestral**, **sampling**.

To realize this task, agent Y applies a function

$$\mathrm{y} = f(\mathrm{x}, \mathrm{z}) \tag{13.74}$$

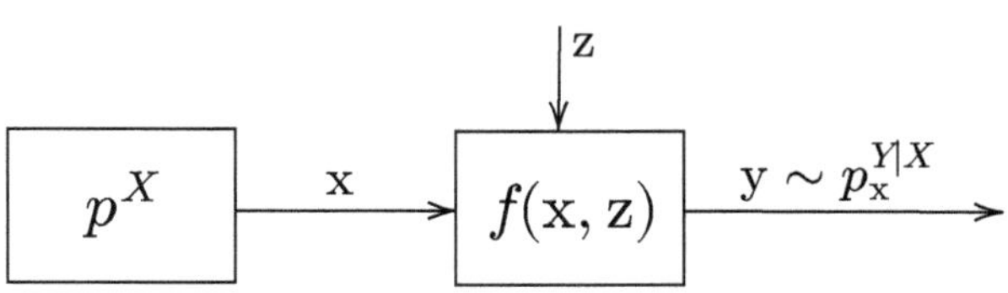

Figure 13.4 In the sequential sampling problem, agent X transmits a random variable $\mathrm{x} \sim p^X$ generated from the marginal p^X of the desired joint distribution p^{XY} to agent Y, whose goal is to draw a random variable $\mathrm{y} \sim p_{\mathrm{x}}^{Y|X}$ from the conditional distribution $p_{\mathrm{x}}^{Y|X}$ via a function $f(\mathrm{x}, \mathrm{z})$, where variable $\mathrm{z} \sim p^Z$ is uncorrelated with x.

that takes as input x from agent X, along with a random variable z $\sim$ p^Z, uncorrelated with random variable x, that plays the role of local randomness.

If the subsystems are uncorrelated under the joint distribution p^{XY}, agent Y can ignore random variable x and simply set $p^Z = p^Y$ and y = z in (13.74). In this case, the random variable y in (13.74) is a function of random variable z only, and thus the conditional entropy $\mathrm{H}(Y|Z)_{p^{YZ}}$ equals zero.

In contrast, if the two subsystems are correlated, the conditional entropy $\mathrm{H}(Y|Z)_{p^{YZ}}$ must be positive. Based on this reasoning, the conditional entropy $\mathrm{H}(Y|Z)_{p^{YZ}}$ can be used as an asymmetric measure of the correlation between random variables x and y.

Interestingly, it can be proved that, for any choice of distribution p^Z and function $f(\cdot,\cdot)$ that satisfy the equality (13.74), the Shannon conditional entropy $\mathrm{H}(Y|Z)_{p^{YZ}}$ is no larger than the Shannon mutual information (see Appendix 19.F), i.e.,

$$\mathrm{H}(Y|Z)_{p^{YZ}} \leq \mathrm{I}(X;Y)_{p^{XY}}. \tag{13.75}$$

Therefore, the measure of correlation given by the conditional entropy $\mathrm{H}(Y|Z)_{p^{YZ}}$ is more conservative than the Shannon mutual information.

13.7 Summary

- This chapter has described asymmetric correlation measures that focus on the residual uncertainty about the state of one subsystem, say Y, given knowledge of the state of the other subsystem, X. Two measures of residual uncertainty were introduced: the ensemble entropy and the conditional entropy.
- The ensemble entropy $\mathrm{H}_\alpha(\mathcal{E}^{Y|X})_{p^{XY}} = \mathrm{E}_{p^X}[\mathrm{H}_\alpha(p_{\mathrm{x}}^{Y|X})]$ is the average entropy of the conditional distribution $p_x^{Y|X}$. The conditional distribution $p_x^{Y|X}$ describes the informational state of an observer that knows the state x of subsystem X.
- The corresponding correlation measure is the locally accessible information from subsystem X to subsystem Y, which is equal to the difference $\mathrm{J}_\alpha(X \to Y)_{p^{XY}} = \mathrm{H}_\alpha(Y)_{p^Y} - \mathrm{H}_\alpha(\mathcal{E}^{Y|X})_{p^{XY}}$.
- The locally accessible information is well defined for any strictly concave entropy measure – i.e., for the Shannon entropy, for the Tsallis entropy of any order $\alpha > 0$, and for the Rényi entropy of order $0 < \alpha \leq 1$. With these choices, the locally accessible information is non-negative, $\mathrm{J}_\alpha(X \to Y)_{p^{XY}} \geq 0$, and it equals zero if and only if the two subsystems are uncorrelated, $\mathrm{J}_\alpha(X \to Y)_{p^X \otimes p^Y} = 0$.
- While, in general, the locally accessible information is asymmetric, i.e., $\mathrm{J}_\alpha(X \to Y)_{p^{XY}} \neq \mathrm{J}_\alpha(Y \to X)_{p^{XY}}$, the Shannon locally accessible information is symmetric and it coincides with the Shannon mutual information: $\mathrm{J}(X \to Y)_{p^{XY}} = \mathrm{J}(Y \to X)_{p^{XY}} = \mathrm{I}(X;Y)_{p^{XY}}$.
- The second measure of residual uncertainty is the conditional entropy $\mathrm{H}_\alpha(Y|X)_{p^{XY}} = \mathrm{H}_\alpha(X,Y)_{p^{XY}} - \mathrm{H}_\alpha(X)_{p^X}$, which removes the marginal entropy of the conditioning subsystem from the joint entropy.

- The conditional entropy $H_\alpha(Y|X)_{p^{XY}}$ is non-negative by the uncertainty-increases-globally property.
- In the special case of the Shannon entropy, the conditional entropy coincides with the ensemble entropy, i.e., $H(Y|X)_{p^{XY}} = H(\mathcal{E}^{Y|X})_{p^{XY}}$. Thus, the Shannon mutual information can be expressed as the difference $I(X;Y)_{p^{XY}} = H(Y)_{p^Y} - H(Y|X)_{p^{XY}}$.
- Conditioning reduces the conditional entropy for the Shannon entropy, but not necessarily for the Tsallis and Rényi entropies.
- A different definition of conditional entropy, accounting for pseudo-additivity, is often used for the Tsallis entropy.

13.8 Recommended Resources

A variant of the locally accessible information was introduced with the purpose of quantifying the extent to which subsystem Y can be controlled – rather than predicted – from subsystem X. This approach replaces conditioning with an intervention, a concept that is formalized within the framework of causal inference. The resulting metric, known as causal information gain, is detailed in [1].

For an example of the non-monotonicity of the conditional Rényi entropy with respect to the order α, see [2]. Further details on the Tsallis conditional entropy can be found in [3], and a comprehensive discussion on the sequential sampling problem is provided by [4].

PROBLEMS

13.1 For $N^X = 2$ and $N^Y = 3$, consider the joint distribution

$$\begin{aligned} p^{XY} &= [p_{00}^{XY}, p_{01}^{XY}, p_{02}^{XY}, p_{10}^{XY}, p_{11}^{XY}, p_{12}^{XY}]^T \\ &= [0.5, 0, 0, 0.2, 0.2, 0.1]^T. \end{aligned} \tag{13.76}$$

(a) Compute the marginal distributions p^X and p^Y. (b) Compute the conditional distributions $p^{X|Y}$ and $p^{Y|X}$. (c) Evaluate the Shannon ensemble entropies $H(\mathcal{E}^{Y|X})_{p^{XY}}$ and $H(\mathcal{E}^{X|Y})_{p^{XY}}$. (d) Evaluate the Tsallis ensemble entropies $H_\alpha^T(\mathcal{E}^{Y|X})_{p^{XY}}$ and $H_\alpha^T(\mathcal{E}^{X|Y})_{p^{XY}}$ for $\alpha = 0.5$. (e) Compute the Shannon conditional entropies $H(Y|X)_{p^{XY}}$ and $H(X|Y)_{p^{XY}}$.

13.2 Consider the joint distribution (13.76). (a) Compute the Shannon locally accessible information measures $J(X \to Y)_{p^{XY}}$ and $J(Y \to X)_{p^{XY}}$. (b) Evaluate the Tsallis locally accessible information measures $J_\alpha^T(X \to Y)_{p^{XY}}$ and $J_\alpha^T(Y \to X)_{p^{XY}}$ for $\alpha = 0.5$. (c) Compute the Shannon mutual information $I(X;Y)_{p^{XY}}$.

13.3 Use the expression (13.69) to prove that the Rényi conditional entropy is non-negative and that it equals zero if and only if random variable x is a function of random variable y.

13.4 Prove that, since the Rényi entropy is not subadditive, the Rényi conditional entropy does not satisfy the conditioning-reduces-entropy property.

13.5 Prove the equality (13.72).

13.6 Consider a binary classification task where the input X can take two values and the label Y is also binary. Suppose the joint distribution is given by $p_{00}^{XY} = 0.4, p_{01}^{XY} = 0.1, p_{10}^{XY} = 0.1$, and $p_{11}^{XY} = 0.4$. Compute the Rényi locally accessible information $\mathrm{J}_\alpha^{\mathrm{R}}(X \to Y)$ for $\alpha = 1$ and for $\alpha = \infty$. Explain the difference between the results obtained with $\alpha = 1$ and $\alpha = \infty$, illustrating how different orders emphasize distinct aspects of the correlation between the two subsystems.

13.7 Consider a simplified stock market prediction scenario with three possible outcomes for the next day's price movement: it can either go up, remain unchanged, or go down. Let X denote the true state of the stock movement and Y our predicted state. Suppose we have a prediction strategy that correctly identifies the true movement of the stock with probability $p^{\mathrm{eq}} = 0.6$ and fails otherwise. Applying the Fano inequality, derive an upper bound the conditional entropy $\mathrm{H}(Y|X)$.

13.8 In a biological experiment, let X represent two possible environmental conditions and Y represent two possible states of a microorganism. The joint distribution p^{XY} is specified as $p_{00}^{XY} = 0.48$, $p_{01}^{XY} = 0.02$, $p_{10}^{XY} = 0.12$, and $p_{11}^{XY} = 0.38$. Using the Rényi entropy of order $\alpha = 0.5$, compute the locally accessible information $\mathrm{J}_\alpha^{\mathrm{R}}(X \to Y)$ and $\mathrm{J}_\alpha^{\mathrm{R}}(Y \to X)$. Discuss the difference between these two quantities, noting that knowledge about the environmental conditions (X) may reduce uncertainty about the microorganism's state (Y) more significantly than knowledge of the microorganism's state (Y) reduces uncertainty about the environmental conditions (X).

13.9 Suppose you have two binary random variables X and Y. Let $p_0^X = 0.3$, $p_{0|0}^{Y|X} = 0.1$, and $p_{0|1}^{Y|X} = 0.9$. Sequential (ancestral) sampling of the pair (X, Y) is carried out as follows. First, agent X generates a sample $\mathrm{x} = x$ from p^X; then agent Y produces a sample y from the appropriate conditional distribution $p_x^{Y|X}$. Design a function $f(x, z)$ and a local (discrete) random variable z that takes the realized value x and z to output a realization of random variable y.

Appendix

13.A Proof of the Inequality (13.40)

The inequality (13.40), along with the conditions for the equality (13.41), follow directly from the strict concavity of the mentioned entropy measures (see Appendix 2.D). In fact, using the definition of locally accessible information, one has

$$\begin{aligned}\mathrm{J}_\alpha(X \to Y)_{p^{XY}} &= \mathrm{H}_\alpha(p^Y) - \mathrm{E}_{p^X}[\mathrm{H}_\alpha(p_{\mathrm{x}}^{Y|X})] \\ &= \mathrm{H}_\alpha(\mathrm{E}_{p^X}[p_{\mathrm{x}}^{Y|X}]) - \mathrm{E}_{p^X}[\mathrm{H}_\alpha(p_{\mathrm{x}}^{Y|X})] \geq 0,\end{aligned} \tag{13.77}$$

where the second equality follows from (13.20) and the last step is a result of the Jensen inequality (2.71). This inequality is strict unless the distributions $p_{\mathrm{x}}^{Y|X}$ are equal for all values x with $p_x^X > 0$, which require the two subsystems to be uncorrelated.

13.B Derivation of the Alternative Expression of the Rényi Conditional Entropy

In this appendix, the goal is to prove the alternative expression (13.69) for the Rényi conditional entropy. To this end, starting from (13.64) and using the definition of the escort distribution $\tilde{p}_\alpha^X$ in (13.68), the following equalities hold:

$$\begin{aligned}
\mathrm{H}_\alpha^{\mathrm{R}}(Y|X)_{p^{XY}} &= \frac{1}{1-\alpha}\log_2\left(\sum_{x\in[N^X]} \frac{(p_x^X)^\alpha}{\sum_{x'\in[N^X]}(p_{x'}^X)^\alpha} \sum_{y\in[N^Y]} \left(p_{y|x}^{Y|X}\right)^\alpha\right) \\
&= \frac{1}{1-\alpha}\log_2\left(\mathrm{E}_{\tilde{p}_\alpha^X}\left[\sum_{y\in[N^Y]} \left(p_{y|\mathrm{x}}^{Y|X}\right)^\alpha\right]\right) \\
&= \frac{1}{1-\alpha}\log_2\left(\mathrm{E}_{\tilde{p}_\alpha^X}\left[2^{(1-\alpha)\mathrm{H}_\alpha^{\mathrm{R}}(p_{\mathrm{x}}^{Y|X})}\right]\right).
\end{aligned} \tag{13.78}$$

13.C Alternative Definition of the Rényi Conditional Entropy

This appendix presents an alternative definition of the Rényi conditional entropy of order ∞, which is also known as the **Rényi conditional min-entropy**. The relevance of this definition is that it relates to the important informational task of detecting one variable given another.

Specifically, consider the binary detection problem in which the goal is estimating a binary random variable $\mathrm{y} \in [2]$ given the observation of an arbitrary variable $\mathrm{x} \in [N^X]$, with the two random variables being jointly distributed as p^{XY}. Using the results in Section 4.7, the optimal detection probability is given by

$$\mathrm{P_S}(d^*) = \mathrm{E}_{p^X}\left[\max_{y\in[2]} p_{y|\mathrm{x}}^{Y|X}\right]. \tag{13.79}$$

This probability can be expressed using the Rényi min-entropy as

$$\begin{aligned}
\mathrm{P_S}(d^*) &= \mathrm{E}_{p^X}\left[2^{-\log_2(\max_{y\in[2]} p_{y|\mathrm{x}}^{Y|X})}\right] \\
&= \mathrm{E}_{p^X}\left[2^{-\mathrm{H}_\infty^{\mathrm{R}}(p_{\mathrm{x}}^{Y|X})}\right].
\end{aligned} \tag{13.80}$$

The Rényi conditional min-entropy $\mathrm{H}_{\min}^{\mathrm{R}}(Y|X)$ is the quantity that satisfies the equality

$$\mathrm{P_S}(d^*) = 2^{-\mathrm{H}_{\min}^{\mathrm{R}}(Y|X)_{p^{XY}}}, \tag{13.81}$$

i.e.,

$$\mathrm{H}_{\min}^{\mathrm{R}}(Y|X)_{p^{XY}} = -\log_2\left(\mathrm{E}_{p^X}\left[2^{-\mathrm{H}_\infty^{\mathrm{R}}(p_{\mathrm{x}}^{Y|X})}\right]\right). \tag{13.82}$$

In a manner similar to the Rényi conditional entropy (13.69), the Rényi conditional min-entropy in (13.82) is in the form of a generalized average of the Rényi entropy of the conditional distributions.

13.D Axiomatic Definition of Entropy (Continued)

As anticipated in Appendix 3.G, the Shannon entropy is unique, among the entropy measures $\mathrm{F}(\cdot)$, in that it not only satisfies the continuity, maximum-value, and expansibility axioms, but it also guarantees the strong additivity axiom. Since the latter requires the notion of conditional distribution, it is introduced in this appendix.

Definition 13.9 Strong additivity axiom
Given two random variables x and y with joint distribution p^{XY}, the following equality holds:

$$\mathrm{F}(p^{XY}) = \mathrm{F}(p^X) + \sum_{x\in[N^X]} p_x^X \mathrm{F}(p_x^{Y|X}). \tag{13.83}$$

Strong additivity indicates that the entropy $\mathrm{F}(p^{XY})$ of a joint observation of two random variables must be the sum of the entropy of one variable – $\mathrm{F}(p^X)$ – and of the average entropy of the conditional distribution $p_x^{Y|X}$ of the other variable given the first – $\sum_{x\in[N^X]} p_x^X \mathrm{F}(p_x^{Y|X})$. Indeed, the Shannon entropy satisfies this axiom by the chain rule (13.51).

REFERENCES

[1] F. N. Simoes, M. Dastani, and T. van Ommen, "Causal entropy and information gain for measuring causal control," in *Proceedings of the European Conference on Artificial Intelligence*. Cham: Springer Nature, pp. 216–231, 2023.

[2] A. Teixeira, A. Matos, and L. Antunes, "Conditional Rényi entropies," *IEEE Transactions on Information Theory*, vol. 58, no. 7, pp. 4273–4277, 2012.

[3] S. Furuichi, "Information theoretical properties of Tsallis entropies," *Journal of Mathematical Physics*, vol. 47, no. 2, art. 023302, 2006.

[4] C. T. Li and A. E. Gamal, "Strong functional representation lemma and applications to coding theorems," *IEEE Transactions on Information Theory*, vol. 64, no. 11, pp. 6967–6978, 2018.

Part V

Bipartite Quantum Information

14 Bipartite Classical and Quantum Information for Pure States

14.1 Overview

This part of the book focuses on correlations in bipartite quantum systems. Due to the contextuality of quantum measurements, correlations in such systems can be broadly categorized as follows.

- **Classical correlations:** Classical correlations describe the relationships between the random variables obtained through local measurements at the two subsystems under a fixed measurement setting.
- **Quantum correlations:** Quantum correlations pertain to the overall behavior of the quantum joint state as the local measurement settings are varied.

This chapter and the next examine classical and quantum correlations in pure states, while Chapter 16 extends the discussion to mixed states.

For bipartite pure states, classical and quantum correlations are inherently intertwined: every correlated joint quantum state exhibits both types of correlations. A bipartite pure state is uncorrelated if it is possible to assign local pure states to both subsystems. In this case, the system possesses neither classical nor quantum correlations. Conversely, when the joint state cannot be described in terms of individual local pure states, the system is said to be entangled, and it exhibits both classical and quantum correlations.

Entanglement manifests as a profound connection between the two subsystems, with no classical counterpart. Specifically, in an entangled state, local measurements on one subsystem alter the overall joint state, producing an instantaneous effect on both subsystems – a phenomenon famously referred to as "spooky action at a distance" by Albert Einstein. However, according to the no-signaling theorem, to be discussed in this chapter, this change cannot be detected locally in the other subsystem, ensuring that faster-than-light communication remains impossible.

Mathematically, the correlation structure of a pure quantum state will be seen in this chapter to be fully characterized by the Schmidt decomposition, which represents the state as a superposition of uncorrelated states. This decomposition provides a foundation for key correlation measures, namely:

- The **quantum mutual information**, which quantifies both classical and quantum correlations.

- The **maximum measured mutual information**, which captures classical correlations.
- The **entanglement entropy**, which measures purely quantum correlations.

This chapter will also show that the uniquely quantum nature of the entanglement-induced link between the two subsystems has two significant implications.

- **Global reduction of uncertainty:** In contrast to classical systems, an entangled pure state causes global uncertainty to decrease. In fact, while it is not possible to assign pure states to the subsystems of an entangled state, each subsystem can be described locally by a mixed state known as the reduced density matrix. Entanglement thus introduces classical uncertainty at the subsystem level, even though the overall system remains in a pure state. In other words, despite the absence of uncertainty in the overall state, the local states of an entangled system are mixed and characterized by classical uncertainty.
- **Negative quantum conditional entropy:** For entangled pure states, the quantum conditional entropy, which quantifies the residual uncertainty about one subsystem given access to the other, is negative. This reflects the strength of entanglement: the information obtained about one subsystem from the other can exceed the total information contained in the subsystem itself.

These features highlight the profound and non-classical nature of quantum entanglement.

14.1.1 Learning Objectives and Organization

By the end of this chapter, the reader should be able to:

- understand the difference between uncorrelated and entangled quantum pure states;
- describe local and global measurements on a bipartite quantum pure state;
- understand the no-signaling theorem;
- compute and use the Schmidt decomposition;
- compute and use the reduced density matrices of a quantum pure state;
- evaluate the quantum mutual information, the maximum measured mutual information, the entanglement entropy, and the quantum conditional entropy for a quantum pure state;
- and understand the conditioning-reduces-quantum-entropy and uncertainty-decreases-globally properties for quantum pure states.

14.2 Bipartite Quantum Pure States

This part of the book focuses on bipartite quantum systems that consist of two subsystems X and Y, which are also known as sites in physics. Subsystems X and Y, when measured, return one out of N^X and N^Y possible values $x \in [N^X]$ and

$y \in [N^Y]$, respectively, with the usual notations $[N^X] = \{0, 1, \ldots, N^X - 1\}$ and $[N^Y] = \{0, 1, \ldots, N^Y - 1\}$. Note that, as for the bipartite classical systems studied in Part IV, the total number of possible measurement outcomes is given by the product $N^X N^Y$, which counts the number of pairs (x, y) with $x \in [N^X]$ and $y \in [N^Y]$.

14.2.1 Defining Bipartite Quantum Pure States

This chapter focuses on pure states for the bipartite quantum system. As discussed in Chapter 5, pure states describe quantum systems that are in known configurations. Furthermore, following Chapter 5, since the overall system can take $N^X N^Y$ possible values when measured, a bipartite pure state is defined as follows.

Definition 14.1 Bipartite quantum pure state
A bipartite pure state, also known as a joint state, is defined by a normalized $N^X N^Y \times 1$ complex-valued state vector

$$|\psi^{XY}\rangle = \begin{bmatrix} \psi_0^{XY} \\ \vdots \\ \psi_{N^Y-1}^{XY} \\ \psi_{N^Y}^{XY} \\ \vdots \\ \psi_{2N^Y-1}^{XY} \\ \vdots \\ \psi_{(N^X-1)N^Y}^{XY} \\ \vdots \\ \psi_{N^X N^Y-1}^{XY} \end{bmatrix} = \begin{bmatrix} \psi_{00}^{XY} \\ \vdots \\ \psi_{0,N^Y-1}^{XY} \\ \psi_{10}^{XY} \\ \vdots \\ \psi_{1,N^Y-1}^{XY} \\ \vdots \\ \psi_{N^X-1,0}^{XY} \\ \vdots \\ \psi_{N^X-1,N^Y-1}^{XY} \end{bmatrix}, \tag{14.1}$$

with the second expression using the bi-index notation. Normalization amounts to the condition

$$\begin{aligned} \langle\psi^{XY}|\psi^{XY}\rangle &= |||\psi^{XY}\rangle||_2^2 \\ &= \sum_{x\in[N^X]} \sum_{y\in[N^Y]} |\psi_{xy}^{XY}|^2 = 1. \end{aligned} \tag{14.2}$$

As a reminder, the bi-index notation represents an integer $z \in [N^X N^Y]$ via the indices xy with $x \in [N^X]$ and $y \in [N^Y]$. Index x denotes the most significant digit and y the least significant digit so that the two representations are related via the equality (see Section 8.8)

$$z = N^Y x + y. \tag{14.3}$$

As done in (14.1), the bi-index xy is also written with a comma as x, y when confusion may arise.

Example 14.1

For $N^X = 2$ and $N^Y = 3$, the 6×1 state vector $|\psi^{XY}\rangle$ can be written as

$$|\psi^{XY}\rangle = \begin{bmatrix} \psi_0^{XY} \\ \psi_1^{XY} \\ \psi_2^{XY} \\ \psi_3^{XY} \\ \psi_4^{XY} \\ \psi_5^{XY} \end{bmatrix} = \begin{bmatrix} \psi_{00}^{XY} \\ \psi_{01}^{XY} \\ \psi_{02}^{XY} \\ \psi_{10}^{XY} \\ \psi_{11}^{XY} \\ \psi_{12}^{XY} \end{bmatrix}. \tag{14.4}$$

14.2.2 Uncorrelated vs. Entangled Pure States

Assume that subsystem X is characterized by the $N^X \times 1$ state $|\psi^X\rangle$, while subsystem Y is in the $N^Y \times 1$ pure state $|\psi^Y\rangle$. As discussed in Section 11.3.1, one of the key principles of quantum physics is that the state of the resulting composite system is given by the Kronecker product $|\psi^{XY}\rangle = |\psi^X\rangle \otimes |\psi^Y\rangle$. When the joint state can be expressed as the Kronecker product of individual states, it is said to be uncorrelated.

Definition 14.2 Uncorrelated, or separable, quantum pure states
A bipartite pure state is said to be uncorrelated or separable if it can be written as the Kronecker product

$$|\psi^{XY}\rangle = |\psi^X\rangle \otimes |\psi^Y\rangle \tag{14.5}$$

of local quantum pure states $|\psi^X\rangle$ and $|\psi^Y\rangle$ for the two subsystems. Equivalently, using the bi-index notation, each xyth entry of an uncorrelated state $|\psi^{XY}\rangle$ can be expressed as the product

$$\psi_{xy}^{XY} = \psi_x^X \psi_y^Y \tag{14.6}$$

between the corresponding entries of the states $|\psi^X\rangle$ and $|\psi^Y\rangle$ for all $x \in [N^X]$ and $y \in [N^Y]$.

To verify that (14.5) is a valid state, one can evaluate its norm as

$$\begin{aligned} \langle\psi^{XY}|\psi^{XY}\rangle &= (\langle\psi^X| \otimes \langle\psi^Y|)(|\psi^X\rangle \otimes |\psi^Y\rangle) \\ &= \langle\psi^X|\psi^X\rangle \otimes \langle\psi^Y|\psi^Y\rangle \\ &= \langle\psi^X|\psi^X\rangle\langle\psi^Y|\psi^Y\rangle = 1, \end{aligned} \tag{14.7}$$

where the second equality follows from the mixed product property (8.169) of the Kronecker product.

Example 14.2

For a system consisting of a qudit and a qutrit, i.e., with $N^X = 2$ and $N^Y = 3$, given arbitrary local pure states $|\psi^X\rangle = [\psi_0^X, \psi_1^X]^T$ and $|\psi^Y\rangle = [\psi_0^Y, \psi_1^Y, \psi_2^Y]^T$, the uncorrelated state vector (14.5) is obtained as

$$|\psi^{XY}\rangle = |\psi^X\rangle \otimes |\psi^Y\rangle = \begin{bmatrix} \psi_0^X \psi_0^Y \\ \psi_0^X \psi_1^Y \\ \psi_0^X \psi_2^Y \\ \psi_1^X \psi_0^Y \\ \psi_1^X \psi_1^Y \\ \psi_1^X \psi_2^Y \end{bmatrix}. \tag{14.8}$$

Example 14.3

For a two-qubit system, consider the state

$$|\psi^{XY}\rangle = \frac{1}{2}\begin{bmatrix} 1 \\ 1 \\ 1 \\ 1 \end{bmatrix}. \tag{14.9}$$

This state is uncorrelated since it can be written as the Kronecker product

$$|\psi^{XY}\rangle = |+\rangle \otimes |+\rangle, \tag{14.10}$$

with $|+\rangle = 1/\sqrt{2}(|0\rangle + |1\rangle) = 1/\sqrt{2}[1, 1]^T$.

Example 14.4

For a two-qubit system, consider the computational basis $\mathcal{M} = \{|0\rangle, |1\rangle, |2\rangle, |3\rangle\}$ of the $N^X N^Y = 4$-dimensional space. Using the bi-index notation (14.3), the computational basis can also be expressed as $\mathcal{M} = \{|00\rangle, |01\rangle, |10\rangle, |11\rangle\}$. Note that this expression corresponds to the use of a base-2 representation of the indices of the basis vectors. The states in the computational basis $\mathcal{M}$ are all uncorrelated. For instance, one has

$$|00\rangle = \begin{bmatrix} 1 \\ 0 \\ 0 \\ 0 \end{bmatrix} = |0\rangle \otimes |0\rangle, \tag{14.11}$$

where $|0\rangle$ is one of the 2×1 vectors in the computational basis $\{|0\rangle, |1\rangle\}$ for the two-dimensional space.

In a similar way, for all the states in the computational basis $\mathcal{M}$, the equality

$$|xy\rangle = |x\rangle \otimes |y\rangle \tag{14.12}$$

holds for any $x \in [2]$ and $y \in [2]$. Thus, by (14.12), the bi-index notation xy identifying the joint state $|xy\rangle$ in the 4-dimensional space can also be interpreted as describing the Kronecker product $|x\rangle \otimes |y\rangle$ between the corresponding local states $|x\rangle$ and $|y\rangle$.

Joint pure states cannot always be expressed in the form (14.5) of uncorrelated states. That is, it is not always possible to assign local pure states $|\psi^X\rangle$ and $|\psi^Y\rangle$ to the two subsystems. This observation leads to the following definition.

Definition 14.3 Entangled quantum pure states
A bipartite pure state $|\psi^{XY}\rangle$ is said to be entangled if it is not uncorrelated, that is, if there are no local pure states $|\psi^X\rangle$ and $|\psi^Y\rangle$ for which the joint state can be expressed as the Kronecker product $|\psi^{XY}\rangle = |\psi^X\rangle \otimes |\psi^Y\rangle$.

Example 14.5

For a two-qubit system, consider the bipartite state

$$|\Phi_+^{XY}\rangle = \frac{1}{\sqrt{2}}\begin{bmatrix}1\\0\\0\\1\end{bmatrix} = \frac{1}{\sqrt{2}}(|00\rangle + |11\rangle)$$
$$= \frac{1}{\sqrt{2}}(|0\rangle \otimes |0\rangle + |1\rangle \otimes |1\rangle), \tag{14.13}$$

which is also known as a **Bell state**. Note that, given its importance, the Bell state (14.13) is reserved the special notation $|\Phi_+^{XY}\rangle$.

The Bell state (14.13) is entangled, as it cannot be written as a Kronecker product of two local qubit states. To see this, recall that, by (14.6), the general form of an uncorrelated state is

$$|\psi^X\rangle \otimes |\psi^Y\rangle = \begin{bmatrix}\psi_0^X\psi_0^Y\\ \psi_0^X\psi_1^Y\\ \psi_1^X\psi_0^Y\\ \psi_1^X\psi_1^Y\end{bmatrix}. \tag{14.14}$$

Therefore, in order to ensure the equality $|\Phi_+^{XY}\rangle = |\psi^X\rangle \otimes |\psi^Y\rangle$, one would need to impose the conditions $\psi_0^X\psi_0^Y = 1$, $\psi_0^X\psi_1^Y = 0$, $\psi_1^X\psi_0^Y = 0$, and $\psi_1^X\psi_1^Y = 1$. However, these conditions are incompatible, since the second equality requires amplitude ψ_0^X or ψ_1^Y to equal zero, making it impossible to satisfy either the first or the last equality.

Generalizing this example, the following states play an important role in quantum information theory.

Definition 14.4 Bell states
For a two-qubit system, i.e., for $N^X = N^Y = 2$, the four Bell states

$$|\Phi_+^{XY}\rangle = \frac{1}{\sqrt{2}}(|00\rangle + |11\rangle) = \frac{1}{\sqrt{2}}[1,0,0,1]^T, \tag{14.15}$$

$$|\Psi_+^{XY}\rangle = \frac{1}{\sqrt{2}}(|01\rangle + |10\rangle) = \frac{1}{\sqrt{2}}[0,1,1,0]^T, \tag{14.16}$$

$$|\Phi_-^{XY}\rangle = \frac{1}{\sqrt{2}}(|00\rangle - |11\rangle) = \frac{1}{\sqrt{2}}[1,0,0,-1]^T, \tag{14.17}$$

$$|\Psi_-^{XY}\rangle = \frac{1}{\sqrt{2}}(|01\rangle - |10\rangle) = \frac{1}{\sqrt{2}}[0,1,-1,0]^T \tag{14.18}$$

are entangled.

As an example of a practical physical application of the Bell states, consider the state $|\Psi_+^{XY}\rangle$, which is an equal superposition of the states $|01\rangle = |0\rangle \otimes |1\rangle$ and $|10\rangle = |1\rangle \otimes |0\rangle$. This state may describe the decay of a Higgs boson into an electron and a positron. In this context, the states $|0\rangle$ and $|1\rangle$ describe situations in which either particle has spin up or spin down, respectively. Since the overall angular momentum of the two particles must be zero, only the states $|01\rangle$ and $|10\rangle$ are possible. Accordingly, the state of the spins of two particles is described by the superposition $|\Psi_+^{XY}\rangle$.

14.2.3 Joint Distributions vs. Quantum States ⏭

The reader may have noted the algebraic similarities between a joint distribution p^{XY} and a pure state $|\psi^{XY}\rangle$, as they are both $N^X N^Y \times 1$ vectors. It is, however, important to emphasize the following differences.

- **Classical pure states are always uncorrelated:** A bipartite classical system in a known, i.e., pure, state is characterized by a zero-entropy joint distribution $p^{XY} = \delta_{xy}$, which is concentrated at the known state (x, y) of the two subsystems. Therefore, in the classical case, a joint pure state is such that both subsystems can be assigned individual pure states x and y. In this sense, classical pure states cannot exhibit any correlations.
- **Classical vs. quantum uncertainties:** Following the discussion in Chapter 5, a joint distribution describes an informational state with classical uncertainty, while a pure state $|\psi^{XY}\rangle$ describes a known quantum state that has only quantum uncertainty.
- **Static vs. dynamic interpretations:** As seen in Section 13.2.4, a joint distribution has dual static and dynamic interpretations. In contrast, quantum joint states can only represent static relations between two subsystems. In fact, for a quantum pure state there is no notion of a chain rule or of a "conditional state"

that can be interpreted as a channel. Accordingly, it is generally not meaningful to talk about a joint state for a system across two time instants. The reader is referred to the Recommended Resources for more discussion on this aspect.

14.3 Local Measurements

The correlations associated with a bipartite quantum state are revealed by making local measurements at the two subsystems. Section 11.3.3 presented local projective measurements, and this section extends the framework to local POVMs.

14.3.1 Local POVMs at Both Subsystems

Unlike the rest of the chapter, for generality, this section assumes a mixed state for the bipartite system. As will be explored fully in Chapter 16, a bipartite mixed state is described by an $N^X N^Y \times N^X N^Y$ density matrix ρ^{XY}. In particular, for the special case of a pure state $|\psi^{XY}\rangle$, the joint density matrix reduces to the rank-1 matrix $\rho^{XY} = |\psi^{XY}\rangle\langle\psi^{XY}|$.

Definition 14.5 Local POVM
A local POVM setting

$$\mathcal{K}^X \otimes \mathcal{K}^Y = \{\{K_x^X \otimes K_y^Y\}_{x\in[N^X]}\}_{y\in[N^Y]} \tag{14.19}$$

is characterized by the following.

- **POVM setting $\mathcal{K}^X$ for subsystem X:** The POVM setting for subsystem X consists of $N^X \times N^X$ Kraus matrices $\mathcal{K}^X = \{K_x^X\}_{x\in[M^X]}$ satisfying the equality $\sum_{x\in[M^X]}(K_x^X)^\dagger K_x^X = I$, where I is the $N^X \times N^X$ identity matrix.
- **POVM setting $\mathcal{K}^Y$ for subsystem Y:** The POVM setting for subsystem Y is described by $N^Y \times N^Y$ Kraus matrices $\mathcal{K}^Y = \{K_y^Y\}_{y\in[M^Y]}$ with $\sum_{y\in[M^Y]}(K_y^Y)^\dagger K_y^Y = I$, where I is the $N^Y \times N^Y$ identity matrix.

As special cases of local POVMs, local complete or projective measurement settings will be denoted as $\mathcal{M}^X \otimes \mathcal{M}^Y$ and $\mathcal{P}^X \otimes \mathcal{P}^Y$, respectively.

As illustrated in Figure 14.1, a local POVM in setting $\mathcal{K}^X \otimes \mathcal{K}^Y$ returns classical outputs $x \in [M^X]$ and $y \in [M^Y]$ as the outcomes of the local measurements at subsystems X and Y, respectively. Furthermore, when the classical measurement outputs are (x, y), it leaves the bipartite system in a post-measurement state σ_{xy}^{XY}. Specifically, using the general formalism of POVMs described in Section 11.4, measurement probabilities and post-measurement states are given as follows.

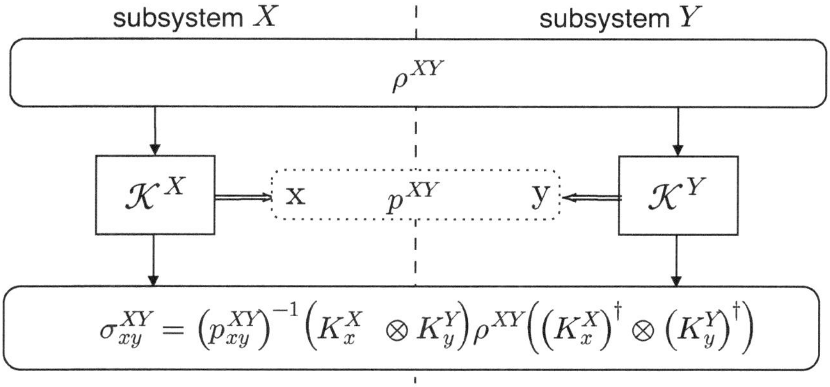

Figure 14.1 Illustration of a local POVM in setting $\mathcal{K}^X \otimes \mathcal{K}^Y$.

Definition 14.6 Measurement distribution and post-measurement state for local POVMs

Given a local POVM setting $\mathcal{K}^X \otimes \mathcal{K}^Y$ with POVM elements $\{J_x^X = (K_x^X)^\dagger K_x^X\}_{x\in[M^X]}$ and $\{J_y^Y = (K_y^Y)^\dagger K_y^Y\}_{y\in[M^Y]}$ and a pre-measurement state ρ^{XY}, the POVM measurement results in the following.

- **Classical output:** The measurement yields classical outputs x and y with the joint distribution

$$\begin{aligned} p_{xy}^{XY} &= p_{xy}^{XY}(\rho^{XY}|\mathcal{K}^X \otimes \mathcal{K}^Y) \\ &= \mathrm{tr}((K_x^X \otimes K_y^Y)\rho^{XY}((K_x^X)^\dagger \otimes (K_y^Y)^\dagger)) \\ &= \mathrm{tr}((J_x^X \otimes J_y^Y)\rho^{XY}). \end{aligned} \tag{14.20}$$

- **Post-measurement state:** For classical outputs ($\mathrm{x} = x, \mathrm{y} = y$), the quantum system is left in the corresponding post-measurement state

$$\sigma_{xy}^{XY} = \frac{1}{p_{xy}^{XY}}(K_x^X \otimes K_y^Y)\rho^{XY}((K_x^X)^\dagger \otimes (K_y^Y)^\dagger). \tag{14.21}$$

Note the use of the simplified notation p_{xy}^{XY} in (14.20) as a shorthand for the full expression $p_{xy}^{XY}(\rho^{XY}|\mathcal{K}^X \otimes \mathcal{K}^Y)$. Specializing the measurement probability (14.20) to the case of a pure state $|\psi^{XY}\rangle$, one obtains the joint probability distribution

$$p_{xy}^{XY} = \langle\psi^{XY}|(J_x^X \otimes J_y^Y)|\psi^{XY}\rangle. \tag{14.22}$$

Furthermore, in this case, the post-measurement state σ_{xy}^{XY} in (14.21) is also pure, i.e., $\sigma_{xy}^{XY} = |\phi_{xy}^{XY}\rangle\langle\phi_{xy}^{XY}|$, and given by

$$|\phi_{xy}^{XY}\rangle = \frac{1}{\sqrt{p_{xy}^{XY}}}(K_x^X \otimes K_y^Y)|\psi^{XY}\rangle. \tag{14.23}$$

Example 14.6

Consider a general local POVM setting $\mathcal{K}^X \otimes \mathcal{K}^Y$ applied to an uncorrelated state $|\psi^{XY}\rangle = |\psi^X\rangle \otimes |\psi^Y\rangle$. By (14.22), the joint probability distribution of the classical measurements $(\mathrm{x}, \mathrm{y}) \sim p^{XY}$ factorizes as the product

$$p_{xy}^{XY} = \langle \psi^X | J_x^X | \psi^X \rangle \cdot \langle \psi^Y | J_y^Y | \psi^Y \rangle = p_x^X \cdot p_y^Y, \tag{14.24}$$

where probabilities

$$p_x^X = \langle \psi^X | J_x^X | \psi^X \rangle = p_x(|\psi^X\rangle | \mathcal{K}^X) \tag{14.25}$$

and

$$p_y^Y = \langle \psi^Y | J_y^Y | \psi^Y \rangle = p_y(|\psi^Y\rangle | \mathcal{K}^Y) \tag{14.26}$$

follow the POVM measurement rule (11.49) for the individual subsystems. Therefore, the measurement outputs $\mathrm{x} \sim p^X$ and $\mathrm{y} \sim p^Y$ are uncorrelated, and both measurement probabilities p^X and p^Y and post-measurement states are described by the corresponding POVMs at either subsystem.

Furthermore, by (14.23), the post-measurement state of the overall system is uncorrelated and given by the Kronecker product of the respective post-measurement states (11.50) as

$$|\phi_{xy}^{XY}\rangle = \frac{1}{\sqrt{p_x^X}} K_x^X |\psi^X\rangle \otimes \frac{1}{\sqrt{p_y^Y}} K_y^Y |\psi^Y\rangle. \tag{14.27}$$

Overall, this example shows that, if the bipartite state is uncorrelated, the two subsystems can be treated separately when considering local measurements.

Example 14.7

Take now the two-qubit Bell state $|\Phi_+^{XY}\rangle$ and assume that local complete measurements are made at the two subsystems in the respective computational bases. In this case, the joint probability distribution (14.22) of the classical measurements is given as

$$p_{xy}^{XY} = |\langle \Phi_+^{XY} | xy \rangle|^2 = \begin{cases} 0.5 & \text{for } x = y, \\ 0 & \text{otherwise.} \end{cases} \tag{14.28}$$

Furthermore, the post-measurement state (14.23) corresponding to the output (x, x) is

$$|\phi_{xx}^{XY}\rangle = |x\rangle \otimes |x\rangle, \tag{14.29}$$

while states (x, y) with $x \neq y$ are not possible.

These results have two important implications.

- **Local measurements of entangled states return correlated classical outputs:** Unlike the case of uncorrelated states studied in Example 14.6, local measurements of entangled states return correlated classical outputs. For this example, by (14.28), the

classical outputs are, in fact, maximally correlated, obtaining the maximum value of the Shannon mutual information $\mathrm{I}(X;Y)_{p^{XY}} = \mathrm{H}(X)_{p^X} = \mathrm{H}(Y)_{p^Y} = 1$.
- **Local measurements break entanglement:** Local measurements break the entanglement of the pre-measurement state $|\Phi_+^{XY}\rangle$, yielding the uncorrelated post-measurement states (14.29).

14.3.2 Local POVMs at One Subsystem Only

The discussion above assumes that a POVM is applied to both subsystems. As a special case, a measurement may be carried out only at subsystem X or only at subsystem Y. If a non-trivial POVM $\mathcal{K}^X$ is applied only to subsystem X, the POVM setting for subsystem Y reduces to the singleton set of Kraus matrices $\mathcal{I}^Y = \{I\}$ encompassing only the $N^Y \times N^Y$ identity matrix I. Accordingly, the resulting subsystem X-only measurement setting corresponds to the local POVM setting

$$\mathcal{K}^X \otimes \mathcal{I}^Y = \{K_x^X \otimes I\}_{x\in[N^X]}. \tag{14.30}$$

Applying the general characterization (14.20) and (14.21), one thus has the following.

Definition 14.7 Local POVM at one subsystem only
Given a local POVM setting $\mathcal{K}^X \otimes \mathcal{I}^Y$ with POVM elements $\{J_x^X = (K_x^X)^\dagger K_x^X\}_{x\in[M^X]}$ and a pre-measurement state ρ^{XY}, the POVM measurement results in the following.

- **Classical output:** The classical output y is trivial, as it is always equal to $\mathrm{y} = 0$, while the distribution p^X of the classical output $\mathrm{x} \sim p^X$ is given by the probabilities $\{p_x^X = p_{x0}^{XY}\}_{x\in[M^X]}$, with
$$p_x^X = p_x^X(\rho^{XY}|\mathcal{K}^X \otimes \mathcal{I}^Y) = \mathrm{tr}((J_x^X \otimes I)\rho^{XY}). \tag{14.31}$$
- **Post-measurement state:** The corresponding post-measurement state σ_x^{XY} is
$$\sigma_x^{XY} = \frac{1}{p_x^X}(K_x^X \otimes I)\rho^{XY}((K_x^X)^\dagger \otimes I). \tag{14.32}$$

In a similar manner, if a POVM $\mathcal{K}^Y = \{K_y^Y\}_{y\in[M^Y]}$ is applied only to subsystem Y, one has the local POVM setting

$$\mathcal{I}^X \otimes \mathcal{K}^Y = \{I \otimes K_y^Y\}_{y\in[N^Y]}, \tag{14.33}$$

where I is the $N^X \times N^X$ identity matrix. The corresponding expressions for pure pre-measurement states and for the case of a measurement only at subsystem Y are evaluated in a similar way as for the local setting $\mathcal{K}^X \otimes \mathcal{I}^Y$, yielding the measurement probability

$$p_y^Y = p_y^Y(\rho^{XY}|\mathcal{I}^X \otimes \mathcal{K}^Y) = \mathrm{tr}((I \otimes J_y^Y)\rho^{XY}); \tag{14.34}$$

and the corresponding post-measurement state

$$\sigma_y^{XY} = \frac{1}{p_y^Y}(I \otimes K_y^Y)\rho^{XY}(I \otimes (K_y^Y)^\dagger). \tag{14.35}$$

Example 14.8

Consider a bipartite system in which a local complete measurement $\mathcal{M}^X = \{|v_x\rangle\}_{x\in[N^X]}$ is applied only to subsystem X. For any uncorrelated state (14.5), i.e., $|\psi^{XY}\rangle = |\psi^X\rangle \otimes |\psi^Y\rangle$, the density matrix is $\rho^{XY} = |\psi^X\rangle\langle\psi^X| \otimes |\psi^Y\rangle\langle\psi^Y|$, and, from (14.31), the probability of subsystem X obtaining output $x \in [N^X]$ is

$$\begin{aligned} p_x^X &= \langle v_x| \otimes \langle v_x||(|x\rangle\langle x| \otimes I)|v_x\rangle \otimes |v_x\rangle \\ &= |\langle v_x|\psi^X\rangle|^2. \end{aligned} \tag{14.36}$$

Furthermore, using (14.32), the post-measurement state corresponding to output x is given as

$$\begin{aligned} |\psi_x^{XY}\rangle &= \frac{1}{\sqrt{p_x^X}}(|v_x\rangle\langle v_x| \otimes I)(|\psi^X\rangle \otimes |\psi^Y\rangle) \\ &= |v_x\rangle \otimes |\psi^Y\rangle. \end{aligned} \tag{14.37}$$

Hence, given an uncorrelated state (14.5), the measurement at subsystem X behaves in the same way as the measurement in a single system characterized by the state $|\psi^X\rangle$. Indeed, the probability (14.36) and the post-measurement state (14.37) coincide with the standard results (5.65) and (5.69) covered in Chapter 5. Furthermore, the post-measurement state $|\psi_x^{XY}\rangle$ remains uncorrelated, and the corresponding state of subsystem Y is unchanged as a result of the measurement on subsystem X.

Example 14.9

For a two-qubit state, consider now the Bell state $|\Phi_+^{XY}\rangle$ in (14.15), on which a local measurement in the computational basis is carried out at subsystem X. By (14.31), the probability of obtaining $x \in [2]$ is given by

$$p_x^X = \langle\Phi_+^{XY}|(|x\rangle\langle x| \otimes I)|\Phi_+^{XY}\rangle = \frac{1}{2}, \tag{14.38}$$

and the post-measurement state (14.32) is given by

$$\begin{aligned} |\phi_x^{XY}\rangle &= \frac{1}{\sqrt{p_x^X}}(|x\rangle\langle x| \otimes I)\left(\frac{1}{\sqrt{2}}|00\rangle + \frac{1}{\sqrt{2}}|11\rangle\right) \\ &= |x\rangle \otimes |x\rangle, \end{aligned} \tag{14.39}$$

where I is a 2×2 identity matrix.

This example illustrates two important facts about local measurements on entangled states.

- **Local measurements, even when applied only at a single subsystem, break entanglement:** Even when applied only at one subsystem, local measurements break the entanglement of the pre-measurement state $|\Phi_+^{XY}\rangle$, yielding the uncorrelated post-measurement state (14.39).
- **Local measurements at one subsystem have a global effect on the overall state of the system:** Making a measurement at one subsystem changes the overall state of both subsystems. This non-local effect is what Albert Einstein famously referred to as "spooky." How can something happening here at X simultaneously also affect Y, which may be arbitrarily far away?

14.3.3 No-Signaling Theorem

The previous example demonstrated that a local measurement can instantaneously modify the overall state of a bipartite quantum system. To understand why Einstein found this phenomenon troubling, consider a scenario involving two agents: X and Y. Agent X holds subsystem X, while agent Y holds subsystem Y. The two subsystems may be arbitrarily far from one another. Agent X selects a local measurement setting $\mathcal{K}^X$ and performs a measurement on subsystem X. This measurement causes the overall state of the system to change in a manner dependent on the chosen setting $\mathcal{K}^X$.

Now imagine that agent Y could infer the measurement setting $\mathcal{K}^X$ chosen by agent X by performing a local measurement on their own subsystem Y. If this were possible, the two agents could communicate information instantaneously, regardless of the distance between them. Specifically, agent X could encode information into the choice of the measurement setting $\mathcal{K}^X$, and agent Y could recover this information without any delay.

Such faster-than-light communication would violate relativity, which prohibits information transfer at speeds exceeding the speed of light. This raises an apparent contradiction: how can the instantaneous change in the joint quantum state coexist with the prohibition of faster-than-light communication?

To answer this question, suppose that an arbitrary local measurement $\mathcal{K}^X \otimes \mathcal{K}^Y$ is made on a state ρ^{XY}. This setting may describe a situation in which agent X choose the local POVM setting $\mathcal{K}^X$ to encode information intended for agent Y, while agent Y uses the POVM setting $\mathcal{K}^Y$ to try and extract information about the setting $\mathcal{K}^X$ selected by agent X. From the perspective of subsystem Y, the probability of obtaining output y is given by the marginal distribution p^Y of the joint distribution (14.20). Therefore, in order for agent Y to be able to recover information about the setting $\mathcal{K}^X$, it is necessary for the distribution p^Y to depend on the measurement setting $\mathcal{K}^X$. It is now shown that this is not the case.

From the joint distribution p^{XY} in (14.20), the marginal distribution p^Y is evaluated as the sum

$$\begin{aligned} p_y^Y &= \sum_{x\in[N^X]} p_{xy}^{XY} \\ &= \sum_{x\in[N^X]} \operatorname{tr}((J_x^X \otimes J_y^Y)\rho^{XY}) \\ &= \operatorname{tr}\left(\left(\sum_{x\in[N^X]} J_x^X \otimes J_y^Y\right)\rho^{XY}\right) \\ &= \operatorname{tr}\left(\left(I \otimes J_y^Y\right)\rho^{XY}\right), \end{aligned} \tag{14.40}$$

where the third equality uses the linearity of the trace, while the last equality follows from the definition of POVM. Thus, the marginal distribution p^Y depends solely on the measurement setting $\mathcal{K}^Y$ for subsystem Y, and not on the measurement setting $\mathcal{K}^X$ for subsystem X. This conclusion is summarized by the following theorem.

Key Result 14.1 No-signaling theorem
Given any bipartite state ρ^{XY} and any local POVM $\mathcal{K}^X \otimes \mathcal{K}^Y$ with POVM elements $\{J_x^X = (K_x^X)^\dagger K_x^X\}_{x\in[M^X]}$ and $\{J_y^Y = (K_y^Y)^\dagger K_y^Y\}_{y\in[M^Y]}$, the marginal measurement distribution p^X depends only on the measurement setting $\mathcal{K}^X$, while the marginal measurement distribution p^Y depends only on the measurement setting $\mathcal{K}^Y$. Specifically, the marginal measurement probabilities are given by

$$p_x^X(\rho^{XY}|\mathcal{K}^X \otimes \mathcal{K}^Y) = p_x^X(\rho^{XY}|\mathcal{K}^X) = \operatorname{tr}\left(\left(J_x^X \otimes I\right)\rho^{XY}\right) \tag{14.41}$$

and

$$p_y^Y(\rho^{XY}|\mathcal{K}^X \otimes \mathcal{K}^Y) = p_y^Y(\rho^{XY}|\mathcal{K}^Y) = \operatorname{tr}\left(\left(I \otimes J_y^Y\right)\rho^{XY}\right). \tag{14.42}$$

The no-signaling theorem (14.41) and (14.42) shows that the probability distribution of the observations at one subsystem does not depend on the measurement setting at the other subsystem.

As a point of notation, given the dependence of the marginal distribution p^Y only on the measurement setting $\mathcal{K}^Y$, one may write $p^Y(\rho^{XY}|\mathcal{K}^Y)$ in lieu of $p^Y(\rho^{XY}|\mathcal{K}^X \otimes \mathcal{K}^Y)$ as done in (14.42). In fact, the probability distribution p^Y in (14.42) equals the probability distribution (14.34) of a measurement setting $\mathcal{I}^X \otimes \mathcal{K}^Y$ in which no measurement is applied to subsystem X. Therefore, from the perspective of subsystem Y, the classical measurement output y has the same distribution irrespective of whether or not any measurement is carried out at subsystem X. Furthermore, by (14.41), the same conclusions apply for subsystem X.

As a result of the no-signaling theorem, subsystem Y cannot obtain locally, based on measurements at subsystem Y, any information about the measurement setting $\mathcal{K}^X$, if any, chosen by subsystem X. This provides a resolution to the concern raised by Einstein on the implausibility of an instantaneous global effect of a local measurement. Indeed, while the measurement at one subsystem, say X, can change the global state of a bipartite system, the impact of any measurement at subsystem X cannot be detected via a local measurement at subsystem Y. Therefore, there is no way for agent X to communicate any information to agent Y by merely applying a local measurement at subsystem X. While local measurements can have a global instantaneous effect, no information can travel between the two subsystems as a result of this phenomenon.

14.4 Entanglement and Superposition

As seen earlier in this chapter, by definition, a pure state is entangled if it cannot be expressed as the Kronecker product of individual pure states for the two subsystems. That said, as shown in this section, a general entangled pure state can always be expressed as a superposition of uncorrelated states. As will be elaborated on in the rest of this chapter, this type of decomposition is extremely useful to study the correlations present in a state.

14.4.1 Entangled States as Superpositions of Uncorrelated States

An arbitrary bipartite pure state can be expressed as a superposition of any set of $N^X N^Y$ states that form an orthonormal basis for the $N^X N^Y$-dimensional space. The following result shows that one can always construct an orthonormal basis consisting exclusively of uncorrelated states (see Problem 14.2).

Key Result 14.2 Orthonormal basis of uncorrelated states

Given any orthonormal basis $\mathcal{M}^X = \{|v_x^X\rangle\}_{x \in [N^X]}$ for subsystem X and any orthonormal basis $\mathcal{M}^Y = \{|v_y^Y\rangle\}_{y \in [N^Y]}$ for subsystem Y, the set of uncorrelated states

$$\mathcal{M} = \mathcal{M}^X \otimes \mathcal{M}^Y = \{\{|v_x^X\rangle \otimes |v_y^Y\rangle\}_{x \in [N^X]}\}_{y \in [N^Y]} \tag{14.43}$$

forms an orthonormal basis for the $N^X N^Y$-dimensional space.

Thus, given two local orthonormal bases $\mathcal{M}^X$ and $\mathcal{M}^Y$, the collection of the $N^X N^Y$ Kronecker products $|v_x^X\rangle \otimes |v_y^Y\rangle$ constructed from all pairs of states $(|v_x^X\rangle, |v_y^Y\rangle)$ from the two bases forms an orthonormal basis for the $N^X N^Y$-dimensional space. As a point of notation, it is common to use the simplified expressions

$$|v_x^X\rangle \otimes |v_y^Y\rangle = |v_x^X v_y^Y\rangle \tag{14.44}$$

and

$$\langle v_x^X| \otimes \langle v_y^Y| = \langle v_x^X v_y^Y| \tag{14.45}$$

when referring to the ket and bra vectors for each of the uncorrelated basis states.

This result implies that any bipartite pure state can be expressed as a superposition of uncorrelated states.

Key Result 14.3 Quantum pure states as a superposition of uncorrelated states

Given any orthonormal basis of uncorrelated states $\mathcal{M} = \mathcal{M}^X \otimes \mathcal{M}^Y$ as in (14.43), any quantum pure state $|\psi^{XY}\rangle$ can be written as the superposition

$$\begin{aligned} |\psi^{XY}\rangle &= \sum_{x\in[N^X]} \sum_{y\in[N^Y]} \alpha_{xy}^{XY} |v_x^X\rangle \otimes |v_y^Y\rangle \\ &= \sum_{x\in[N^X]} \sum_{y\in[N^Y]} \alpha_{xy}^{XY} |v_x^X v_y^Y\rangle \end{aligned} \tag{14.46}$$

with amplitudes

$$\begin{aligned} \alpha_{xy}^{XY} &= (\langle v_x^X| \otimes \langle v_y^Y|)|\psi^{XY}\rangle \\ &= \langle v_x^X v_y^Y|\psi^{XY}\rangle \end{aligned} \tag{14.47}$$

obtained by projecting state $|\psi^{XY}\rangle$ in the direction of the vector $|v_x^X\rangle \otimes |v_y^Y\rangle = |v_x^X v_y^Y\rangle$.

Example 14.10

When choosing the local computational bases $\mathcal{M}^X = \{|x\rangle\}_{x\in[N^X]}$ and $\mathcal{M}^Y = \{|y\rangle\}_{y\in[N^Y]}$, the orthonormal basis $\mathcal{M}^X \otimes \mathcal{M}^Y$ in (14.43) consists of the states

$$\mathcal{M} = \{\{|xy\rangle = |x\rangle \otimes |y\rangle\}_{x\in[N^X]}\}_{y\in[N^Y]}. \tag{14.48}$$

In (14.48), generalizing (14.12) to any dimensions N^X and N^Y, the notation $|xy\rangle$ can be read in two equivalent ways: (i) as the computational basis vector of the $N^X N^Y$-dimensional space corresponding to the bi-index xy in (14.3); and (ii) as a shorthand for the Kronecker product $|x\rangle \otimes |y\rangle$.

Using the computational basis (14.43), a generic bipartite state $|\psi^{XY}\rangle$ in (14.1) can be written as the superposition of the corresponding $N^X N^Y$ uncorrelated states as

$$\begin{aligned} |\psi^{XY}\rangle &= \sum_{x\in[N^X]} \sum_{y\in[N^Y]} \psi_{xy}^{XY} |xy\rangle \\ &= \sum_{x\in[N^X]} \sum_{y\in[N^Y]} \psi_{xy}^{XY} |x\rangle \otimes |y\rangle. \end{aligned} \tag{14.49}$$

For example, for $N^X = 2$ and $N^Y = 2$, a generic state vector can be written as

$$\begin{aligned} |\psi^{XY}\rangle &= \begin{bmatrix} \psi_{00}^{XY} \\ \psi_{01}^{XY} \\ \psi_{10}^{XY} \\ \psi_{11}^{XY} \end{bmatrix} \\ &= \psi_{00}^{XY}|0\rangle \otimes |0\rangle + \psi_{01}^{XY}|0\rangle \otimes |1\rangle + \psi_{10}^{XY}|1\rangle \otimes |0\rangle + \psi_{11}^{XY}|1\rangle \otimes |1\rangle \\ &= \psi_{00}^{XY}|00\rangle + \psi_{01}^{XY}|01\rangle + \psi_{10}^{XY}|10\rangle + \psi_{11}^{XY}|11\rangle. \end{aligned} \tag{14.50}$$

Note that the Bell states in (14.15)–(14.18) are expressed in this form.

Example 14.11

For a two-qubit system, take the orthonormal basis obtained from the Hadamard bases of the two subsystems, i.e.,

$$\begin{aligned}\mathcal{M} &= \{|+\rangle \otimes |+\rangle, |+\rangle \otimes |-\rangle, |-\rangle \otimes |+\rangle, |-\rangle \otimes |-\rangle\} \\ &= \{|++\rangle, |+-\rangle, |-+\rangle, |--\rangle\}. \end{aligned} \tag{14.51}$$

Note that state $|++\rangle$ can be expressed as $|++\rangle = 1/2(|00\rangle + |01\rangle + |10\rangle + |11\rangle)$, and similar expressions in terms of the computational basis apply also to the other state vectors in the basis $\mathcal{M}$.

The Bell state $|\Phi_+^{XY}\rangle$ can be written in terms of this basis using (14.46) and (14.47) by evaluating the amplitudes as

$$\begin{aligned}\langle ++|\Phi_+^{XY}\rangle &= \frac{1}{2\sqrt{2}}(\langle 00| + \langle 01| + \langle 10| + \langle 11|)(|00\rangle + |11\rangle) \\ &= \frac{1}{2\sqrt{2}}(1+0+0+1) = \frac{1}{\sqrt{2}}, \end{aligned} \tag{14.52}$$

and, in a similar way, $\langle --|\Phi_+^{XY}\rangle = 1/\sqrt{2}$, which implies $\langle +-|\Phi_+^{XY}\rangle = \langle -+|\Phi_+^{XY}\rangle = 0$.

Accordingly, the Bell state $|\Phi_+^{XY}\rangle$ is given by the superposition

$$|\Phi_+^{XY}\rangle = \frac{1}{\sqrt{2}}(|++\rangle + |--\rangle). \tag{14.53}$$

The other Bell states can be evaluated analogously as (see Problem 14.3)

$$|\Psi_+^{XY}\rangle = \frac{1}{\sqrt{2}}(|++\rangle - |--\rangle), \tag{14.54}$$

$$|\Phi_-^{XY}\rangle = \frac{1}{\sqrt{2}}(|-+\rangle + |+-\rangle), \tag{14.55}$$

$$\text{and } |\Psi_-^{XY}\rangle = \frac{1}{\sqrt{2}}(|-+\rangle - |+-\rangle). \tag{14.56}$$

14.4.2 Identifying Entangled States

The results discussed above establish that *any* quantum pure state can be represented as a linear combination of uncorrelated states. Consequently, the mere ability to express a state as such a combination does not suffice to determine whether it is entangled. In fact, even uncorrelated states can be expressed as a superposition of multiple uncorrelated states in certain bases. Instead, a state is classified as entangled if and only if it *cannot* be represented as a single uncorrelated state in any basis of uncorrelated states.

It is also emphasized that a state is entangled or it is not, irrespective of the specification of any given basis. Thus, the definition of entanglement, unlike that of properties like coherence (Section 6.3), is not tied to the choice of a particular basis.

Example 14.12

The uncorrelated state $|+\rangle \otimes |-\rangle$ can be written as the superposition

$$|+\rangle \otimes |-\rangle = \frac{1}{2}(|00\rangle - |01\rangle + |10\rangle - |11\rangle) \quad (14.57)$$

of four uncorrelated states. However, the state is not entangled, since it can also be expressed as the Kronecker product $|+\rangle \otimes |-\rangle$ using the local Hadamard bases.

14.4.3 Orthonormal Bases of Entangled States

This section has proved that an orthonormal basis for the $N^X N^Y$-dimensional space can be obtained by combining any two local orthonormal bases for the two subsystems. More generally, one can design bases that contain entangled states. Measurements constructed using such bases find a number of applications, including various primitive functions for the quantum internet such as quantum teleportation and entanglement swapping.

Example 14.13

For a two-qubit system, the set of four Bell states

$$\mathcal{M} = \{|\Phi_+^{XY}\rangle, |\Psi_+^{XY}\rangle, |\Phi_-^{XY}\rangle, |\Psi_-^{XY}\rangle\} \quad (14.58)$$

forms an orthonormal basis, known as the **Bell basis**, for the $N^X N^Y = 4$-dimensional space. Indeed, one can readily check that the four Bell states are orthogonal.

Using the Bell basis, as an example, the uncorrelated state $|\psi^{XY}\rangle = |00\rangle$ can be written as the superposition

$$|\psi^{XY}\rangle = \frac{1}{\sqrt{2}}(|\Phi_+^{XY}\rangle + |\Phi_-^{XY}\rangle), \quad (14.59)$$

given the corresponding amplitudes

$$\langle\Phi_+^{XY}|\psi^{XY}\rangle = \frac{1}{\sqrt{2}} \text{ and } \langle\Phi_-^{XY}|\psi^{XY}\rangle = \frac{1}{\sqrt{2}}. \quad (14.60)$$

Thus, interestingly, while state $|\psi^{XY}\rangle = |00\rangle$ is uncorrelated, it can be expressed as the superposition of two entangled states.

14.5 Reduced Density Matrices

In Section 14.3, it was observed that, by the no-signaling theorem, the marginal measurement probabilities depend exclusively on the measurement setting at the given subsystem. Specifically, the measurement probability distributions $p^X(|\psi^{XY}\rangle|\mathcal{K}^X)$ and $p^Y(|\psi^{XY}\rangle|\mathcal{K}^Y)$ in (14.41) and (14.42) depend only on the respective measurement settings $\mathcal{K}^X$ and $\mathcal{K}^Y$. However, evaluating the distributions $p^X(|\psi^{XY}\rangle|\mathcal{K}^X)$ and $p^Y(|\psi^{XY}\rangle|\mathcal{K}^Y)$ in (14.41) and (14.42) still requires knowledge of the global state $|\psi^{XY}\rangle$, unless the state is uncorrelated.

This section will show that the measurement probability distributions $p^X(|\psi^{XY}\rangle|\mathcal{K}^X)$ and $p^Y(|\psi^{XY}\rangle|\mathcal{K}^Y)$ can, in fact, be expressed in terms of local states for the subsystems X and Y, respectively. However, unless the state $|\psi^{XY}\rangle$ is uncorrelated, the local states for the subsystems are mixed, and thus characterized by density matrices ρ^X and ρ^Y for subsystems X and Y. As a result, while, by definition, entangled systems do not admit individual pure states for the two subsystems, it is possible to describe the behavior of each subsystem of an entangled state via local mixed states.

The mixed states ρ^X and ρ^Y, referred to as the reduced density matrices, provide a local complete description of the states of the two subsystems. This is in the sense that they allow the evaluation of the respective marginal measurement probabilities $p^X(|\psi^{XY}\rangle|\mathcal{K}^X)$ and $p^Y(|\psi^{XY}\rangle|\mathcal{K}^Y)$ for any local measurement under respective settings $\mathcal{K}^X$ and $\mathcal{K}^Y$. The requirement for a local description to be mixed in the presence of entanglement suggests that classical uncertainty in a local state arises from interactions with other subsystems.

14.5.1 Amplitude Matrix

To facilitate the definition of reduced density matrices, the following definition is useful.

Definition 14.8 Amplitude matrix

For any $N^X N^Y \times 1$ quantum joint pure state $|\psi^{XY}\rangle$, the amplitude matrix Ψ^{XY} is the $N^X \times N^Y$ matrix with each (x, y)th entry $\Psi^{XY}_{x,y}$ equal to

$$\Psi^{XY}_{x,y} = \psi^{XY}_{xy}, \tag{14.61}$$

i.e., to the xyth amplitude of vector $|\psi^{XY}\rangle$, for all $x \in [N^X]$ and $y \in [N^Y]$.

By the definition (14.61), the amplitude matrix Ψ^{XY} and the state vector $|\psi^{XY}\rangle$ are related via the vectorization

$$\text{vec}(\Psi^{XY}) = |\psi^{XY}\rangle. \tag{14.62}$$

In fact, it is recalled from Section 10.5 that the vectorization vec$(\cdot)$ stacks the transpose of the rows of matrix Ψ^{XY} one on top of the other. Furthermore, by the normalization of the state $|\psi^{XY}\rangle$, the amplitude matrix Ψ^{XY} satisfies the equality

$$\sum_{x\in[N^X]} \sum_{y\in[N^Y]} |\Psi^{XY}_{xy}|^2 = 1. \tag{14.63}$$

Example 14.14

For a generic two-qubit state $|\psi^{XY}\rangle = [\psi^{XY}_{00}, \psi^{XY}_{01}, \psi^{XY}_{10}, \psi^{XY}_{11}]^T$, the amplitude matrix is given by

$$\Psi^{XY} = \begin{bmatrix} \psi^{XY}_{00} & \psi^{XY}_{01} \\ \psi^{XY}_{10} & \psi^{XY}_{11} \end{bmatrix}. \tag{14.64}$$

Note that, by (14.62), the state $|\psi^{XY}\rangle$ is recovered by stacking the transpose of the rows of the amplitude matrix Ψ^{XY}.

For any uncorrelated state $|\psi^{XY}\rangle$, the amplitude matrix takes a particularly simple form.

Key Result 14.4 Amplitude matrix and entanglement

A state $|\psi^{XY}\rangle$ is uncorrelated, taking the form $|\psi^{XY}\rangle = |\psi^X\rangle \otimes |\psi^Y\rangle$, if and only if the amplitude matrix Ψ^{XY} can be expressed as the outer product

$$\Psi^{XY} = |\psi^X\rangle\langle\psi^Y|^*, \tag{14.65}$$

where $\langle\psi^Y|^*$ denotes the element-wise complex conjugate of the bra vector $\langle\psi^Y|$.

Equivalently, a state $|\psi^{XY}\rangle$ is entangled if and only if the amplitude matrix has rank larger than 1:

$$|\psi^{XY}\rangle \text{ entangled} \iff \text{rank}(\Psi^{XY}) > 1. \tag{14.66}$$

This result can be verified by recalling the property $\text{vec}(ABC^\dagger) = (A \otimes C^*)\text{vec}(B)$ of vectorization (see (10.71)), which yields the equality

$$\text{vec}(|\psi^X\rangle\langle\psi^Y|^*) = |\psi^X\rangle \otimes |\psi^Y\rangle \tag{14.67}$$

by setting $A = |\psi^X\rangle$, $B = 1$, and $C = |\psi^Y\rangle^*$. In this regard, note that the need for the complex conjugate in (14.65) follows since the matrix $|\psi^X\rangle\langle\psi^Y|$ has (x, y)th entry equal to $\psi_x^X(\psi_y^Y)^*$. Thus, vectorizing matrix $|\psi^X\rangle\langle\psi^Y|$ would not yield the original state $|\psi^{XY}\rangle = |\psi^X\rangle \otimes |\psi^Y\rangle$.

Example 14.15

For the two-qubit state

$$|\psi^{XY}\rangle = \frac{1}{2}\begin{bmatrix} 1 \\ -i \\ i \\ 1 \end{bmatrix}, \tag{14.68}$$

the amplitude matrix is

$$\Psi^{XY} = \frac{1}{2}\begin{bmatrix} 1 & -i \\ i & 1 \end{bmatrix}. \tag{14.69}$$

This matrix can be expressed as (14.65) with

$$|\psi^X\rangle = \begin{bmatrix} \frac{1}{\sqrt{2}} \\ \frac{1}{\sqrt{2}}i \end{bmatrix} \text{ and } |\psi^Y\rangle = \begin{bmatrix} \frac{1}{\sqrt{2}} \\ -\frac{1}{\sqrt{2}}i \end{bmatrix}. \tag{14.70}$$

The rank of the amplitude matrix equals 1, and thus the state $|\psi^{XY}\rangle$ is uncorrelated and it can be written as $|\psi^{XY}\rangle = |\psi^X\rangle \otimes |\psi^Y\rangle$.

14.5.2 Defining Reduced Density Matrices

As proved in Appendix 14.A, the measurement probability for a local measurement at a subsystem can be expressed as follows.

Key Result 14.5 Marginal measurement probability and reduced density matrix (subsystem *X*)

For any quantum joint pure state $|\psi^{XY}\rangle$ with amplitude matrix Ψ^{XY} in (14.61), given any local POVM $\mathcal{K}^X \otimes \mathcal{K}^Y$ with POVM elements $\{J_x^X = (K_x^X)^\dagger K_x^X\}_{x\in[M^X]}$, the marginal measurement probabilities for subsystem X can be expressed as

$$p_x^X(|\psi^{XY}\rangle|\mathcal{K}^X) = p_x^X(\rho^X|\mathcal{K}^X) = \mathrm{tr}(J_x^X\rho^X) \tag{14.71}$$

for all $x \in [M^X]$, where the reduced density matrix for subsystem X is defined as the $N^X \times N^X$ matrix

$$\rho^X = \Psi^{XY}(\Psi^{XY})^\dagger. \tag{14.72}$$

Key Result 14.6 Marginal measurement probability and reduced density matrix (subsystem *Y*)

For any quantum joint pure state $|\psi^{XY}\rangle$ with amplitude matrix Ψ^{XY} in (14.61), given any local POVM $\mathcal{K}^X \otimes \mathcal{K}^Y$ with POVM elements $\{J_y^Y = (K_y^Y)^\dagger K_y^Y\}_{y\in[M^Y]}$, the marginal measurement probabilities for subsystem Y can be expressed as

$$p_y^Y(|\psi^{XY}\rangle|\mathcal{K}^Y) = p_y^Y(\rho^Y|\mathcal{K}^Y) = \mathrm{tr}(J_y^Y\rho^Y) \tag{14.73}$$

for all $y \in [M^Y]$, where the reduced density matrix for subsystem Y is defined as the $N^Y \times N^Y$ matrix

$$\rho^Y = ((\Psi^{XY})^\dagger\Psi^{XY})^*. \tag{14.74}$$

The characterization (14.71) of the marginal measurement probability distribution reveals that subsystem X behaves as a mixed state with density matrix ρ^X under the local POVM setting $\mathcal{K}^X$, irrespective of any possible measurement applied to subsystem Y. In a symmetric fashion, the marginal measurement probability distribution (14.73) coincides with that for a mixed state ρ^Y under the local POVM setting $\mathcal{K}^Y$, whether or not any measurement is applied to subsystem X.

Importantly, the reduced density matrices ρ^X and ρ^Y in (14.72) and (14.74) do not describe the joint distribution of the outputs of local measurements applied to both systems, but only the marginal distributions of each local measurement. Therefore, the reduced density matrices provide descriptions of the local behaviors of the two subsystems, but not of the global behavior of the overall system.

Example 14.16

Take the joint state

$$|\psi^{XY}\rangle = \frac{1}{\sqrt{2}} \begin{bmatrix} 1 \\ i \\ 0 \\ 0 \end{bmatrix}. \tag{14.75}$$

For a local measurement at subsystem X in the computational basis $\mathcal{M}^X$, this example verifies that the marginal probability distribution $p^X(|\psi^{XY}\rangle|\mathcal{M}^X)$ in (14.41) can equivalently be expressed using the reduced density matrix as $p^X(\rho^X|\mathcal{M}^X)$ in (14.71).

By (14.41), the measurement probability is given by

$$\begin{aligned} p_0^X(|\psi^{XY}\rangle|\mathcal{M}^X) &= \langle\psi^{XY}|\left(|0\rangle\langle 0| \otimes I\right)|\psi^{XY}\rangle \\ &= \langle\psi^{XY}| \begin{bmatrix} 1 & 0 & 0 & 0 \\ 0 & 1 & 0 & 0 \\ 0 & 0 & 0 & 0 \\ 0 & 0 & 0 & 0 \end{bmatrix} |\psi^{XY}\rangle = 1. \end{aligned} \tag{14.76}$$

Alternatively, one can compute the amplitude matrix (14.64) as

$$\Psi^{XY} = \frac{1}{\sqrt{2}} \begin{bmatrix} 1 & i \\ 0 & 0 \end{bmatrix}, \tag{14.77}$$

yielding the reduced density matrix (14.72) as

$$\rho^X = \frac{1}{2} \begin{bmatrix} 1 & i \\ 0 & 0 \end{bmatrix} \begin{bmatrix} 1 & 0 \\ -i & 0 \end{bmatrix} = \begin{bmatrix} 1 & 0 \\ 0 & 0 \end{bmatrix} = |0\rangle\langle 0|. \tag{14.78}$$

From this matrix, one can obtain the measurement probability as

$$p_0^X(\rho^X|\mathcal{M}^X) = \mathrm{tr}(|0\rangle\langle 0|\rho^X) = \langle 0|(|0\rangle\langle 0|)|0\rangle = 1. \tag{14.79}$$

The same calculations can also be applied for a local measurement at subsystem Y. In particular, by (14.42), a local measurement in the computational basis $\mathcal{M}^Y$ yields the measurement probability

$$\begin{aligned} p_0^Y(|\psi^{XY}\rangle|\mathcal{M}^Y) &= \langle\psi^{XY}|\left(I \otimes |0\rangle\langle 0|\right)|\psi^{XY}\rangle \\ &= \langle\psi^{XY}| \begin{bmatrix} 1 & 0 & 0 & 0 \\ 0 & 0 & 0 & 0 \\ 0 & 0 & 1 & 0 \\ 0 & 0 & 0 & 0 \end{bmatrix} |\psi^{XY}\rangle = \frac{1}{2}. \end{aligned} \tag{14.80}$$

This probability can alternatively be evaluated by computing the reduced density matrix

$$\rho^Y = \frac{1}{2}\left(\begin{bmatrix} 1 & 0 \\ -i & 0 \end{bmatrix} \begin{bmatrix} 1 & i \\ 0 & 0 \end{bmatrix} \right)^* = \frac{1}{2} \begin{bmatrix} 1 & -i \\ i & 1 \end{bmatrix}, \tag{14.81}$$

yielding

$$\begin{aligned} p_0^Y(\rho^Y|\mathcal{M}^Y) &= \mathrm{tr}(|0\rangle\langle 0|\rho^Y) \\ &= \frac{1}{2}\langle 0| \begin{bmatrix} 1 & -i \\ i & 1 \end{bmatrix} |0\rangle = \frac{1}{2}. \end{aligned} \tag{14.82}$$

14.5.3 Properties of the Reduced Density Matrices

The reduced density matrices ρ^X and ρ^Y are generally different. In fact, even their dimensions may differ, as they depend exclusively on the local respective dimensions N^X and N^Y. That said, both reduced density matrices share the same rank and eigenvalues, as summarized in the following.

Key Result 14.7 Schmidt rank and Schmidt probability vector

For any pure state $|\psi^{XY}\rangle$, the two density matrices ρ^X and ρ^Y have the same rank

$$r = \text{rank}(\rho^X) = \text{rank}(\rho^Y), \tag{14.83}$$

which is known as the **Schmidt rank**. Furthermore, the two density matrices have the same non-zero eigenvalues $\lambda = [\lambda_0, \lambda_1, \ldots, \lambda_{r-1}]^T$, which are known as the **Schmidt probabilities** and satisfy the conditions

$$\lambda_z > 0 \text{ for all } z \in [r] \text{ and } \sum_{z\in[r]} \lambda_z = 1. \tag{14.84}$$

To interpret this important result, recall first from Section 7.8 that the eigenvalues λ of a density matrix describe the classical randomness associated with the generation of the corresponding state. In particular, a state with eigenvalues λ can be prepared by first drawing a classical random variable with distribution λ, and then producing the pure state equal to the corresponding eigenvector. Thus, the number of strictly positive eigenvalues, r, is a measure of the classical uncertainty in the two subsystems. The shared eigenvalues of the reduced density matrices ρ^X and ρ^Y imply that the two subsystems have the same level of classical uncertainty.

Given that the rank of an $N \times N$ matrix cannot exceed N, the Schmidt rank r is upper bounded by the minimum of the dimensions N^X and N^Y, i.e.,

$$r \leq N_{\text{min}} = \min(N^X, N^Y). \tag{14.85}$$

A full proof of the result above will follow from the Schmidt decomposition to be described in Section 14.9. For now, it is noted that the equality (14.83) is a direct consequence of the definitions (14.72) and (14.74), which imply that the ranks of the two reduced density matrices coincide with the rank of the amplitude matrix. This observation yields the following important corollary of the equivalence (14.66).

Key Result 14.8 Schmidt rank and entanglement

A joint pure state $|\psi^{XY}\rangle$ is entangled if and only if the Schmidt rank r is larger than 1:

$$|\psi^{XY}\rangle \text{ entangled} \iff r = \text{rank}(\rho^X) = \text{rank}(\rho^Y) > 1. \tag{14.86}$$

This result formalizes the fact that, for a pure state, local classical uncertainty – manifested by a Schmidt rank r larger than 1 – is a hallmark of entanglement.

Example 14.17

Take the two-qubit state

$$|\psi^{XY}\rangle = \sqrt{p}|00\rangle + \sqrt{1-p}|11\rangle, \tag{14.87}$$

for some probability $0 \leq p \leq 1$. Note that with $p = 0.5$, the state coincides with the Bell state $|\Phi_+^{XY}\rangle$, while for $p = 0$ and $p = 1$ the state is uncorrelated.

The reduced density matrices at the two subsystems are equal to

$$\rho^X = \rho^Y = p|0\rangle\langle 0| + (1-p)|1\rangle\langle 1|, \tag{14.88}$$

since the amplitude matrix is

$$\Psi^{XY} = \begin{bmatrix} \sqrt{p} & 0 \\ 0 & \sqrt{1-p} \end{bmatrix} = \sqrt{p}|0\rangle\langle 0| + \sqrt{1-p}|1\rangle\langle 1|. \tag{14.89}$$

Thus, the Schmidt rank equals $r = 1$ if and only if $p = 0$ or $p = 1$, and in this case the state $|\psi^{XY}\rangle$ is uncorrelated. Otherwise, the Schmidt rank equals $r = 2$, and the state $|\psi^{XY}\rangle$ is entangled with Schmidt probability vector $\lambda = [p, 1-p]^T$.

14.5.4 Entanglement and Reduced Density Matrices

An important observation about the equivalence (14.86) is that entanglement in a pure state can be ascertained based exclusively on local properties, without having access to the entire system. In fact, the Schmidt probability vector λ is the vector of eigenvalues of the reduced density matrices ρ^X and ρ^Y for the local subsystems. Hence, the presence of classical uncertainty at the level of the local subsystems proves the presence of entanglement for pure states. Conversely, the absence of classical uncertainty at the subsystems shows that the system is uncorrelated.

As will be seen in Chapter 15, this simple picture breaks down for joint mixed states, for which entanglement cannot be ascertained based on the properties of the reduced density matrices. In fact, as will be discussed, a density matrix of the form $\rho^X \otimes \rho^Y$ is uncorrelated, since the overall state can be obtained from the local states ρ^X and ρ^Y, but the ranks of the reduced density matrices ρ^X and ρ^Y are not necessarily equal to 1.

14.6 Classical Correlations in Quantum Pure States

As discussed in Section 14.5, local measurements on an entangled pure state yield classically correlated observations. Furthermore, the degree of correlation between the observations generally varies based on the choice of local measurement settings.

This section first introduces a measure of the classical correlations arising from local measurements as a function of the measurement setting. The maximum level of classical correlations is then obtained by optimizing this measure over all possible local complete settings. This approach yields a classical correlation measure, known as maximum measured mutual information, that does not depend on the choice of a specific measurement setting.

Throughout this part of the book, correlations will be measured by using the Shannon mutual information, which was shown in the previous part to have useful and natural properties as a measure of classical correlations associated with a joint distribution p^{XY}. Therefore, as a point of terminology, the qualifier "Shannon" will henceforth be dropped, as this will not cause any ambiguity.

14.6.1 Maximum Measured Mutual Information

To begin, recall that, for a joint distribution p^{XY}, the Shannon mutual information is given by the difference

$$\mathrm{I}(X;Y)_{p^{XY}} = \mathrm{H}(X)_{p^X} + \mathrm{H}(Y)_{p^Y} - \mathrm{H}(X,Y)_{p^{XY}} \tag{14.90}$$

between the sum of the local entropies, $\mathrm{H}(X)_{p^X} + \mathrm{H}(Y)_{p^Y}$, and the joint entropy $\mathrm{H}(X,Y)_{p^{XY}}$.

Using this measure, given a local POVM $\mathcal{K}^X \otimes \mathcal{K}^Y$, the classical correlations associated with the measurement outputs $(\mathrm{x},\mathrm{y}) \sim p^{XY}(|\psi^{XY}\rangle|\mathcal{K}^X \otimes \mathcal{K}^Y)$ can be quantified by the mutual information

$$\mathrm{I}(X;Y)_{p^{XY}(|\psi^{XY}\rangle|\mathcal{K}^X \otimes \mathcal{K}^Y)}. \tag{14.91}$$

This quantity depends on the measurement setting $\mathcal{K}^X \otimes \mathcal{K}^Y$. This raises the question: What is the *maximum* correlation level for measurement outputs (x,y) obtained from a state $|\psi^{XY}\rangle$?

In order to address this question, one could evaluate the maximum of the mutual information (14.91) over all possible local POVMs $\mathcal{K}^X \otimes \mathcal{K}^Y$. However, in practice, it is often preferable to restrict the optimization to *complete* local measurements $\mathcal{M}^X \otimes \mathcal{M}^Y$. In fact, as discussed in Chapter 11, the use of more general POVMs implies the consideration of larger, open, systems, which may not always be physically realizable or relevant when studying the correlations present across two subsystems. Additionally, as noted in Section 6.5, projective measurements can be viewed as a coarser version of an underlying (hypothetical) complete measurement. Consequently, extending the optimization domain from complete to projective measurements would not increase the mutual information (14.91). These considerations lead to the following definition.

Definition 14.9 Maximum measured mutual information

Given a bipartite pure state $|\psi^{XY}\rangle$, the maximum measured mutual information is the maximum mutual information of the classical measurement outputs

$(x, y) \sim p^{XY}(|\psi^{XY}\rangle|\mathcal{M}^X \otimes \mathcal{M}^Y)$ over all possible local complete measurement settings $\mathcal{M}^X \otimes \mathcal{M}^Y$, i.e.,

$$J(X;Y)_{|\psi^{XY}\rangle} = \max_{\mathcal{M}^X \otimes \mathcal{M}^Y} I(X;Y)_{p^{XY}(|\psi^{XY}\rangle|\mathcal{M}^X \otimes \mathcal{M}^Y)}. \tag{14.92}$$

Example 14.18

For the Bell state $|\Phi_+^{XY}\rangle$ in (14.13), choosing local measurements in the computational bases $\mathcal{M}^X \otimes \mathcal{M}^Y$, the joint measurement distribution $p^{XY}(|\Phi_+^{XY}\rangle|\mathcal{M}^X \otimes \mathcal{M}^Y)$ is given by (14.28). Therefore, the resulting mutual information between the classical outputs takes the value $I(X;Y)_{p^{XY}(|\Phi_+^{XY}\rangle|\mathcal{M}^X \otimes \mathcal{M}^Y)} = 1$.

Since the measurement outputs are binary, the maximum measured mutual information cannot exceed 1 bit (see (12.50)). Thus, the local computational bases $\mathcal{M}^X \otimes \mathcal{M}^Y$ provide an optimal solution for problem (14.92) given the state $|\Phi_+^{XY}\rangle$, proving that the maximum measured mutual information of the Bell state is $J(X;Y)_{|\Phi_+^{XY}\rangle} = 1$. The same result can be demonstrated for all Bell states (14.15)–(14.18), yielding the equalities

$$J(X;Y)_{|\Phi_+^{XY}\rangle} = J(X;Y)_{|\Phi_-^{XY}\rangle} = J(X;Y)_{|\Psi_+^{XY}\rangle} = J(X;Y)_{|\Psi_-^{XY}\rangle} = 1. \tag{14.93}$$

Generalizing the last example, in order to evaluate the maximum measured mutual information $J(X;Y)_{|\psi^{XY}\rangle}$ in (14.92), one must find the local measurement setting $\mathcal{M}^X \otimes \mathcal{M}^Y$ that maximizes the mutual information under the joint distribution $p^{XY}(|\psi^{XY}\rangle|\mathcal{M}^X \otimes \mathcal{M}^Y)$. To proceed, consider the following definition.

Definition 14.10 Local eigenbases

Given a joint state $|\psi^{XY}\rangle$, a local eigenbasis of subsystem X is any eigenbasis

$$\mathcal{M}_*^X = \{|u_x^X\rangle\}_{x\in[N^X]} \tag{14.94}$$

of the reduced density matrix ρ^X, and a local eigenbasis of subsystem Y is any eigenbasis

$$\mathcal{M}_*^Y = \{|u_y^Y\rangle\}_{y\in[N^Y]} \tag{14.95}$$

of the reduced density matrix ρ^Y.

With these definitions, as proved in Appendix 14.B, the following result holds.

Key Result 14.19 Evaluating the maximum measured information

Given a joint state $|\psi^{XY}\rangle$, any pair of local eigenbases $\mathcal{M}_*^X$ and $\mathcal{M}_*^Y$ provides an optimal solution for the problem (14.92). Furthermore, the maximum measured mutual information is given by

$$\begin{aligned} J(X;Y)_{|\psi^{XY}\rangle} &= I(X;Y)_{p^{XY}(|\psi^{XY}\rangle|\mathcal{M}_*^X \otimes \mathcal{M}_*^Y)} \\ &= H(\lambda), \end{aligned} \tag{14.96}$$

where λ is the Schmidt probability vector for state $|\psi^{XY}\rangle$.

By the characterization (14.96), in general, the evaluation of the maximum measured mutual information requires the numerical calculation of the Schmidt probability vector λ (from the reduced density matrices). That said, as detailed in Appendix 14.C, for two-qubit states, i.e., for $N^X = N^Y = 2$, there is a closed-form formula that allows the direct evaluation of the entropy $\mathrm{H}(\lambda)$ in terms of a quantity known as **concurrence**.

Example 14.19

For the state (14.87), i.e., $|\psi^{XY}\rangle = \sqrt{p}|00\rangle + \sqrt{1-p}|11\rangle$, the local eigenbases are given by the computational bases, i.e., $\mathcal{M}_*^X = \{|0\rangle, |1\rangle\}$ and $\mathcal{M}_*^X = \{|0\rangle, |1\rangle\}$. Consider also the, generally suboptimal, measurement bases $\mathcal{M}^X = \mathcal{M}^Y = \{|\theta\rangle, |\theta^\perp\rangle\}$, where $|\theta\rangle = [\cos(\theta), \sin(\theta)]^T$ and $|\theta^\perp\rangle = [-\sin(\theta), \cos(\theta)]^T$. With this choice, for $\theta = 0$ the measurement bases $\mathcal{M}^X$ and $\mathcal{M}^Y$ coincide with the local eigenbases $\mathcal{M}_*^X$ and $\mathcal{M}_*^Y$, thus maximizing the measured mutual information; while for $\theta = \pi/4$ the measurement bases are complementary to the computational bases (see Section 6.4.2).

The mutual information $\mathrm{I}(X;Y)_{p(|\psi^{XY}\rangle|\mathcal{M}^X\otimes\mathcal{M}^Y)}$ is shown in Figure 14.2 as a function of the measurement setting parameter θ for states (14.87) characterized by different values of parameter p. For uncorrelated states, i.e., for $p = 0$, the measured mutual information $\mathrm{I}(X;Y)_{p(|\psi^{XY}\rangle|\mathcal{M}^X\otimes\mathcal{M}^Y)}$ equals zero for all measurement settings. In fact, in this case, the measurement outputs are always uncorrelated, irrespective of the adopted local measurement setting.

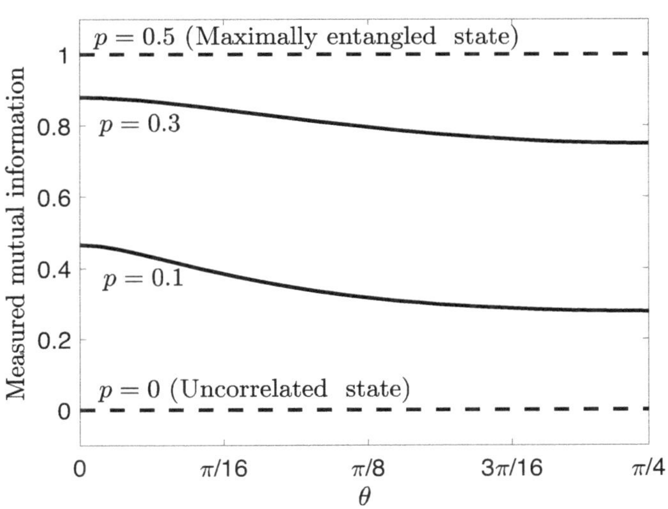

Figure 14.2 Measured mutual information $\mathrm{I}(X;Y)_{p(|\psi^{XY}\rangle|\mathcal{M}^X\otimes\mathcal{M}^Y)}$ as a function of the measurement setting parameter θ for states (14.87).

Conversely, for $p = 0.5$, i.e., for the Bell state $|\Phi_+^{XY}\rangle$, the measured mutual information equals 1 irrespective of the measurement bases. In this case, the level of correlation is always maximal, no matter which local complete measurement setting is selected.

As will be discussed in Section 14.7, states like the Bell state $|\Phi_+^{XY}\rangle$ are maximally entangled, yielding maximum correlation irrespective of the measurement bases.

For intermediate values of p, the maximum value of the measured mutual information is obtained for $\theta = 0$, i.e., for the local eigenbases. Furthermore, the resulting maximum measured mutual information $\mathrm{J}(X;Y)_{|\psi^{XY}\rangle} = \mathrm{I}(X;Y)_{p(|\psi^{XY}\rangle|\mathcal{M}_*{}^X \otimes \mathcal{M}_*{}^Y)}$ increases as the parameter p grows from $p = 0$, corresponding to an uncorrelated state, to $p = 0.5$, yielding a maximally entangled state.

14.7 Classical and Quantum Correlations in Quantum Pure States

The maximum measured mutual information (14.92) quantifies the classical correlations associated with a state $|\psi^{XY}\rangle$ by evaluating the maximum correlation that can be extracted from the state via local complete measurements. This section focuses on quantifying both classical and quantum correlations present in a state $|\psi^{XY}\rangle$ via a measure known as the **quantum mutual information**. The difference between the quantum mutual information and the maximum measured mutual information isolates the quantum correlations, and is referred to as the **entanglement entropy**.

Building on the same principles used to define the classical mutual information in Chapter 12, the definition of the quantum mutual information relies on the concepts of quantum local entropies and quantum joint entropy. As introduced next, the quantum local entropies measure the uncertainty level associated with each of the two subsystems, while the joint quantum entropy accounts for the uncertainty associated with the overall subsystem.

14.7.1 Quantum Local Entropies

Following Chapter 6, the local uncertainty on each of the subsystems is measured by the quantum entropy of the corresponding reduced density matrix. Specifically, the local uncertainty at subsystem X is quantified by the quantum entropy $\mathrm{H}(X)_{\rho^X}$, while the local uncertainty at subsystem Y is accounted for by the quantum entropy $\mathrm{H}(Y)_{\rho^Y}$. This yields the following definition.

Definition 14.11 Quantum local entropies

For any quantum joint pure state $|\psi^{XY}\rangle$ with Schmidt probability vector λ, the quantum local entropy for subsystem X is defined as

$$\mathrm{H}(X)_{\rho^X} = \mathrm{H}(\lambda), \tag{14.97}$$

and the quantum local entropy for subsystem Y is defined as

$$\mathrm{H}(Y)_{\rho^Y} = \mathrm{H}(\lambda). \tag{14.98}$$

Given that the eigenvalues of the two reduced density matrices are equal, the quantum local entropies (14.97) and (14.98) also coincide. Physically, this coincidence arises because the uncertainty in each subsystem's state is due to its entanglement with the other subsystem, a property shared by the overall system.

Key Result 14.10 Bounds on the quantum local entropies

For any pure state $|\psi^{XY}\rangle$ with Schmidt rank r, the reduced density matrices ρ^X and ρ^Y satisfy the inequalities

$$0 \leq \mathrm{H}(X)_{\rho^X} = \mathrm{H}(Y)_{\rho^Y} \leq \log_2(r), \tag{14.99}$$

where the lower bound $\mathrm{H}(X)_{\rho^X} = \mathrm{H}(Y)_{\rho^Y} = 0$ is attained if and only if the state $|\psi^{XY}\rangle$ is uncorrelated. Thus, a positive quantum local entropy is a necessary and sufficient condition for entanglement:

$$|\psi^{XY}\rangle \text{ entangled} \iff \mathrm{H}(X)_{\rho^X} = \mathrm{H}(Y)_{\rho^Y} > 0. \tag{14.100}$$

The inequality (14.99) follows directly from the fact that a density matrix of rank r has maximum entropy when all eigenvalues are equal to $1/r$, yielding the quantum entropy $\log_2(r)$. Furthermore, the equivalence (14.100) is a restatement of the result (14.86). In fact, a Schmidt rank r larger than 1 is equivalent to the positivity of the quantum local entropies $\mathrm{H}(X)_{\rho^X} = \mathrm{H}(Y)_{\rho^Y}$.

14.7.2 Quantum Joint Entropy

While the quantum local entropies are non-negative, and strictly positive for entangled states, the quantum joint entropy is equal to zero for any pure state $|\psi^{XY}\rangle$, i.e.,

$$\mathrm{H}(X, Y)_{|\psi^{XY}\rangle} = 0. \tag{14.101}$$

As discussed next, this implies that entanglement causes a seemingly natural property satisfied by classical systems to be violated by quantum systems.

14.7.3 Insubordination (Uncertainty Decreases Globally)

As detailed in Section 12.5, in classical systems uncertainty increases globally. That is, using the Shannon entropy, the inequality

$$\max(\mathrm{H}(X)_{p^X}, \mathrm{H}(Y)_{p^Y}) \leq \mathrm{H}(X, Y)_{p^{XY}} \tag{14.102}$$

relates the local entropies for the marginal distributions p^X and p^Y to the joint entropy for the joint distribution p^{XY}. In words, the uncertainty about the whole is no smaller than the uncertainty about any of its parts. This property does not hold for quantum systems.

Key Result 14.11 Uncertainty decreases globally for quantum pure states (insubordination)

For any quantum pure state $|\psi^{XY}\rangle$, the following inequality is satisfied:

$$\underbrace{\max(\mathrm{H}(X)_{\rho^X}, \mathrm{H}(Y)_{\rho^Y})}_{=\mathrm{H}(X)_{\rho^X}=\mathrm{H}(Y)_{\rho^Y}=\mathrm{H}(\lambda)} \geq \mathrm{H}(X, Y)_{|\psi^{XY}\rangle} = 0, \tag{14.103}$$

with equality holding if and only if the state is uncorrelated.

This result shows that entanglement manifests itself via the violation of the classical uncertainty-increases-globally principle. This property is also known as insubordination, and is a hallmark of entanglement. Accordingly, for an entangled pure quantum state, in a sense, the uncertainty about the whole is smaller than the uncertainty about any of the parts.

This counterintuitive conclusion reflects the fact that, while the overall state of the system is known, the individual states of the subsystems are uncertain, as they are described by mixed states. This observation was summarized by Erwin Schrödinger in 1935 as "*knowledge of the individual systems can decline to the scantiest, even to zero, while that of the combined system remains continually maximal. The best possible knowledge of a whole does not include the best possible knowledge of its parts – and this is what keeps coming back to haunt us.*"

14.7.4 Quantum Mutual Information

Like the classical mutual information (14.90), the quantum mutual information captures the amount of shared information present across the two subsystems by evaluating the difference between the sum of the local entropies and the joint entropy.

Definition 14.12 Quantum mutual information

Given a quantum joint pure state $|\psi^{XY}\rangle$ with Schmidt probability vector λ, the quantum mutual information is defined as

$$\begin{aligned}\mathrm{I}(X;Y)_{|\psi^{XY}\rangle} &= \mathrm{H}(X)_{\rho^X} + \mathrm{H}(Y)_{\rho^Y} - \mathrm{H}(X, Y)_{|\psi^{XY}\rangle} \\ &= 2\mathrm{H}(\lambda).\end{aligned} \tag{14.104}$$

Unlike the maximum measured mutual information, the quantum mutual information does not quantify the correlations arising from any specific measurement of the state. Instead, it measures the total shared information between the two subsystems, encompassing both classical and quantum correlations present in the state $|\psi^{XY}\rangle$. As summarized in Figure 14.3, for a given state $|\psi^{XY}\rangle$, the quantum mutual information $\mathrm{I}(X;Y)_{|\psi^{XY}\rangle}$ accounts for both classical and quantum

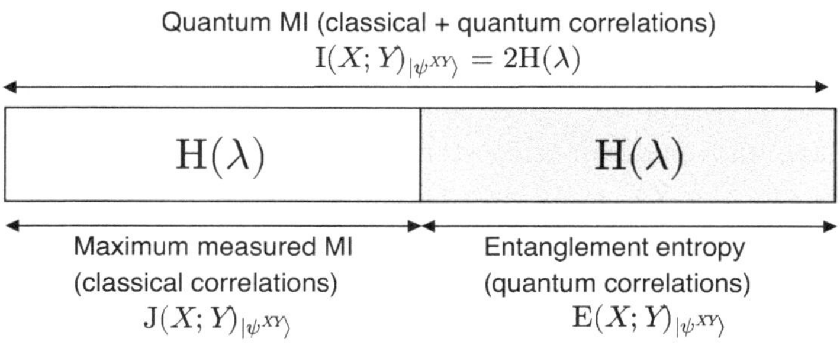

Figure 14.3 For a given state $|\psi^{XY}\rangle$, the quantum mutual information $I(X;Y)_{|\psi^{XY}\rangle}$ accounts for both classical and quantum correlations; the maximum measured mutual information $J(X;Y)_{|\psi^{XY}\rangle}$ captures only classical correlations; and the entanglement entropy $E(X;Y)_{|\psi^{XY}\rangle}$ quantifies quantum correlations (which coincide with entanglement for pure states).

correlations, while the maximum measured mutual information $J(X;Y)_{|\psi^{XY}\rangle}$ captures only classical correlations.

14.7.5 Entanglement Entropy

As highlighted in Figure 14.3, the difference between the quantum mutual information and the maximum measured mutual information provides a measure of the quantum correlations inherent in the state $|\psi^{XY}\rangle$. This yields the following definition.

Definition 14.13 Entanglement entropy

For any quantum pure state $|\psi^{XY}\rangle$ with Schmidt probability vector λ, the quantum mutual information, capturing both classical and quantum correlations, can be expressed as the sum

$$\underbrace{I(X;Y)_{|\psi^{XY}\rangle}}_{=2H(\lambda)} = \underbrace{J(X;Y)_{|\psi^{XY}\rangle}}_{=H(\lambda)} + \underbrace{E(X;Y)_{|\psi^{XY}\rangle}}_{=H(\lambda)}, \tag{14.105}$$

where the maximum measured mutual information $J(X;Y)_{|\psi^{XY}\rangle}$ accounts for the classical correlations, and the entanglement entropy

$$\begin{aligned} E(X;Y)_{|\psi^{XY}\rangle} &= I(X;Y)_{|\psi^{XY}\rangle} - J(X;Y)_{|\psi^{XY}\rangle} \\ &= H(\lambda) \end{aligned} \tag{14.106}$$

quantifies classical correlations.

Through the decomposition (14.105), any entangled quantum pure state $|\psi^{XY}\rangle$ exhibits equal amounts of classical and quantum correlations. Both types of correlations are quantified by the entropy $H(\lambda)$ of the Schmidt probability vector λ. Consequently, an uncorrelated state has zero correlations, while an entangled state possesses an equal, positive, share of classical and quantum correlations.

For pure states, quantum correlations – those not captured by the maximum measured mutual information – are synonymous with entanglement. As will be discussed in Chapter 15, these correlations enable informational tasks like superdense coding and quantum teleportation, which inherently rely on entanglement. The measure $\mathrm{E}(X;Y)_{|\psi^{XY}\rangle}$, which quantifies quantum correlations, is therefore called entanglement entropy. As explored in Chapter 16, for mixed states, quantum correlations go beyond entanglement, encompassing also weaker forms of non-classical correlations.

14.7.6 Properties of the Quantum Mutual Information

Due to the uncertainty-increases-globally property (14.102), the classical mutual information satisfies the inequality $\mathrm{I}(X;Y)_{p^{XY}} \leq \min(\mathrm{H}(X)_{p^X}, \mathrm{H}(Y)_{p^Y})$. This states that the amount of information that can be inferred from one subsystem about the other cannot exceed the information present in either subsystem. As a direct consequence of the uncertainty-decreases-globally property (14.103), this property is no longer valid for the quantum mutual information.

Key Result 14.12 Insubordination property of the quantum mutual information
For any pure state $|\psi^{XY}\rangle$, the quantum mutual information satisfies the inequality

$$\mathrm{I}(X;Y)_{|\psi^{XY}\rangle} = 2\mathrm{H}(\lambda) \geq \min(\mathrm{H}(X)_{\rho^X}, \mathrm{H}(Y)_{\rho^Y}) = \mathrm{H}(\lambda), \tag{14.107}$$

where the inequality is strict for entangled states.

In essence, this relationship suggests that having access to one subsystem allows us to infer more information about the other subsystem than is contained within the subsystem itself. This phenomenon reflects the non-classical nature of quantum correlations brought about by entanglement.

Furthermore, the correlation measures introduced in this chapter satisfy the following bounds.

Key Result 14.13 Bounds on the quantum mutual information, maximum measured mutual information, and entanglement entropy
For a state $|\psi^{XY}\rangle$ with Schmidt probability vector λ, the quantum mutual information $\mathrm{I}(X;Y)_{|\psi^{XY}\rangle} = 2\mathrm{H}(\lambda)$, the maximum measured mutual information $\mathrm{J}(X;Y)_{|\psi^{XY}\rangle} = \mathrm{H}(\lambda)$, and the entanglement entropy $\mathrm{E}(X;Y)_{|\psi^{XY}\rangle} = \mathrm{H}(\lambda)$ can be bounded by leveraging the following inequalities:

$$\underbrace{0}_{\text{uncorrelated}} \leq \mathrm{H}(\lambda) \leq \log_2(r) \leq \underbrace{\log_2(N_{\min})}_{\text{maximally entangled}}, \tag{14.108}$$

with the lower bound $\mathrm{H}(\lambda) = 0$ achieved if and only if the state $|\psi^{XY}\rangle$ is uncorrelated.

14.7.7 Maximally Entangled States

When the upper bound in (14.108) is attained, the state $|\psi^{XY}\rangle$ is said to be maximally entangled.

Definition 14.14 Maximally entangled state

A state $|\psi^{XY}\rangle$ with Schmidt probability vector λ is said to be maximally entangled if the condition $\mathrm{H}(\lambda) = \log_2(N_{\min})$ holds. This happens if and only if the Schmidt rank equals the maximum value $r = N_{\min}$ and the Schmidt probability vector is uniform, i.e., $\lambda = u$.

By this definition, a maximally entangled state can be detected by studying the local behavior of the subsystems as described by the following result.

Key Result 14.14 Maximally entangled states and reduced density matrices

A state $|\psi^{XY}\rangle$ is maximally entangled if and only if the reduced density matrices of the subsystem with the smallest dimension are maximally mixed:

$$|\psi^{XY}\rangle \text{ maximally entangled} \iff \begin{cases} \rho^X = \pi & \text{if } N_{\min} = N^X, \\ \rho^Y = \pi & \text{if } N_{\min} = N^Y. \end{cases} \tag{14.109}$$

According to this result, maximum local uncertainty – resulting in maximally mixed reduced density matrices – is associated with a maximally entangled state.

Example 14.20

For $N^X = N^Y = N_{\min} = 2$, consider again the state (14.87), i.e., $|\psi^{XY}\rangle = \sqrt{p}|00\rangle + \sqrt{1-p}|11\rangle$, for some probability p. Since the Schmidt probability vector is $\lambda = [1-p, p]^T$, the corresponding entropy is $\mathrm{H}(\lambda) = \mathrm{h}(p)$. The upper bound $\mathrm{H}(\lambda) = \log_2(N_{\min}) = 1$ in (14.108) is attained when $p = 1/2$, i.e., for the Bell state $|\psi^{XY}\rangle = |\Phi_+^{XY}\rangle$. Hence, in this case, the state is maximally entangled.

Example 14.21

Generalizing the last example, for any dimensions N^X and N^Y, consider states of the form

$$|\psi^{XY}\rangle = \sum_{x\in[N^X]} \sum_{y\in[N^Y]} \sqrt{p_{xy}^{XY}}\,|xy\rangle, \tag{14.110}$$

where p^{XY} is a joint probability distribution. Note that this is a valid quantum state, since the classical normalization condition $\sum_{x\in[N^X]} \sum_{y\in[N^Y]} p_{xy}^{XY} = 1$ implies the normalization $\langle\psi^{XY}|\psi^{XY}\rangle = 1$ for the ket (14.110).

As a first observation, when the joint distribution p^{XY} describes uncorrelated subsystems, i.e., $p_{xy}^{XY} = p_x^X p_y^Y$ for all $x \in [N^X]$ and $y \in [N^Y]$, the corresponding state $|\psi^{XY}\rangle$ in (14.110) is uncorrelated. In fact, in this case, the state can be written as

$$
\begin{aligned}
|\psi^{XY}\rangle &= \sum_{x\in[N^X]} \sum_{y\in[N^Y]} \sqrt{p_x^X}|x\rangle \otimes \sqrt{p_y^Y}|y\rangle \\
&= \left(\sum_{x\in[N^X]} \sqrt{p_x^X}|x\rangle \right) \otimes \left(\sum_{y\in[N^Y]} \sqrt{p_y^Y}|y\rangle \right).
\end{aligned} \tag{14.111}
$$

It follows that the equality $\mathrm{H}(\lambda) = 0$ holds, and the state $|\psi^{XY}\rangle$ presents zero classical and quantum correlations.

Consider now the joint distribution p^{XY} with $N^X = N^Y = N_{\min} = D$ used as a recurring example in the previous part of the book, which is defined as

$$
p_{xx}^{XY} = \frac{p^{\text{eq}}}{D} \tag{14.112}
$$

for all $x \in [D]$, and as

$$
p_{xy}^{XY} = \frac{1 - p^{\text{eq}}}{D^2 - D} \tag{14.113}
$$

for all $x, y \in [D]$ with $x \neq y$, where p^{eq} is some probability. Recall that the probability p^{eq} dictates the correlation level in the joint distribution p^{XY}. In particular, with $p^{\text{eq}} = 1/D$, the joint distribution p^{XY} describes uncorrelated random variables. Furthermore, with $p^{\text{eq}} = 1$, it yields maximally correlated random variables, which are equal with probability 1.

The corresponding quantum state (14.110) is given as

$$
|\psi^{XY}\rangle = \sqrt{\frac{p^{\text{eq}}}{D}} \sum_{x\in[D]} |xx\rangle + \sqrt{\frac{1 - p^{\text{eq}}}{D^2 - D}} \sum_{x\in[D]} \sum_{y\in[D]:\, y\neq x} |xy\rangle. \tag{14.114}
$$

If $p^{\text{eq}} = 1/D$, the joint distribution factorizes as $p_{xy}^{XY} = p_x^X p_y^Y$ with $p_x^X = p_y^Y = 1/D$ for all $x \in [N^X]$ and $y \in [N^Y]$, and hence the state $|\psi^{XY}\rangle$ is uncorrelated, i.e.,

$$
|\psi^{XY}\rangle = \left(\sum_{x\in[N^X]} \sqrt{\frac{1}{D}}|x\rangle \right) \otimes \left(\sum_{y\in[N^Y]} \sqrt{\frac{1}{D}}|y\rangle \right), \tag{14.115}
$$

yielding $\mathrm{H}(\lambda) = 0$. In this case, the state presents zero classical and quantum correlations.

For $D = 3$, Figure 14.4 shows the entropy $\mathrm{H}(\lambda)$ as a function of the probability p^{eq}. As the probability p^{eq} moves away from the value $1/D = 1/3$, the entropy $\mathrm{H}(\lambda)$ increases. Thus, both classical and quantum correlations, which are measured by the entropy $\mathrm{H}(\lambda)$, grow larger. Furthermore, the maximum entropy value $\log_2(3) = 1.5850$ is obtained for $p^{\text{eq}} = 1$, at which point the state is maximally entangled.

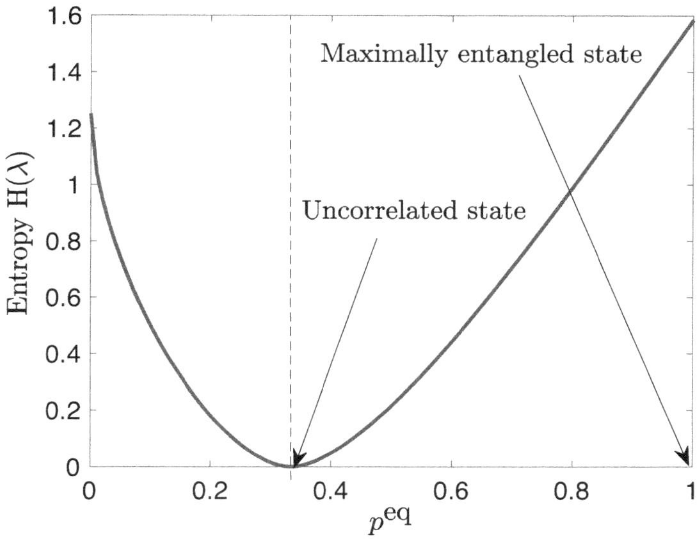

Figure 14.4 Entropy $\mathrm{H}(\lambda)$ of the Schmidt probability vector λ for the state (14.114) as a function of the probability p^{eq}.

14.7.8 Distinguishing Classical and Quantum Correlations ⏭

This section has argued that the quantum mutual information $\mathrm{I}(X;Y)_{|\psi^{XY}\rangle}$ accounts for both classical and quantum correlations, with the former being quantified by the measured mutual information $\mathrm{J}(X;Y)_{|\psi^{XY}\rangle}$. This distinction between classical and quantum correlations was predicated on the interpretation of the quantum mutual information as a measure of the overall correlations present in a state.

As discussed next, it is also possible to separate classical and quantum correlations through the lens of the informational task of erasing correlations via the introduction of local randomness. This perspective generalizes to quantum systems the approach presented in Section 12.7.6, which quantifies classical correlations via the amount of local randomness needed to destroy correlations.

To elaborate, consider two subsystems that are initially in some state $|\psi^{XY}\rangle$. Assume that each subsystem can inject "noise" in the local subsystem by applying a quantum local channel. The following questions then arise: How much local noise is needed to completely eliminate the quantum correlations? Once the quantum correlations are erased, how much additional local randomness is required to also remove the classical correlations? These two levels of randomness can be viewed as measures of the quantum and classical correlations present in the initial state $|\psi^{XY}\rangle$. This approach is illustrated with the following example.

Example 14.22

Assume that a pair of qubits is initially in the Bell state $|\Phi_+^{XY}\rangle$. To erase quantum correlations, i.e., entanglement, subsystem Y uses 1 bit of local randomness – a fair coin – to decide whether or not to apply the Pauli Z matrix P_3. That is, subsystem Y applies a dephasing quantum channel with parameter $\gamma = 0.5$ to its qubit (see Section 11.6).

If subsystem Y does not apply the Pauli Z matrix P_3, the state $|\Phi_+^{XY}\rangle$ is left unchanged, while an application of the Pauli Z matrix yields the state

$$\begin{aligned}(I \otimes P_3)|\Phi_+^{XY}\rangle &= \frac{1}{\sqrt{2}}(|0\rangle \otimes P_3|0\rangle + |1\rangle \otimes P_3|1\rangle)\\ &= \frac{1}{\sqrt{2}}(|0\rangle \otimes |0\rangle - |1\rangle \otimes |1\rangle)\\ &= |\Phi_-^{XY}\rangle. \end{aligned} \tag{14.116}$$

Therefore, the post-transformation state equals $|\Phi_+^{XY}\rangle$ or $|\Phi_-^{XY}\rangle$ with equal probability, yielding the joint density matrix

$$\sigma^{XY} = \frac{1}{2}|\Phi_+^{XY}\rangle\langle\Phi_+^{XY}| + \frac{1}{2}|\Phi_-^{XY}\rangle\langle\Phi_-^{XY}|. \tag{14.117}$$

The density matrix (14.117) can be checked by direct calculation to equal

$$\begin{aligned}\sigma^{XY} &= \frac{1}{2}(|0\rangle \otimes |0\rangle)(\langle 0| \otimes \langle 0|) + \frac{1}{2}(|1\rangle \otimes |1\rangle)(\langle 1| \otimes \langle 1|)\\ &= \frac{1}{2}|0\rangle\langle 0| \otimes |0\rangle\langle 0| + \frac{1}{2}|1\rangle\langle 1| \otimes |1\rangle\langle 1|. \end{aligned} \tag{14.118}$$

Accordingly, the state can be realized via the ensemble of uncorrelated states $\{(|0\rangle \otimes |0\rangle, 0.5), (|1\rangle \otimes |1\rangle, 0.5)\}$. So, the local operation at subsystem Y, requiring 1 bit of local randomness, has erased entanglement. Note that joint mixed states will be discussed further in Chapter 15.

While entanglement has been erased, state σ^{XY} can still lead to correlated measurement outputs. For example, under local measurements in the respective computational bases, the measurement outputs (x, y) equal (0,0) or (1,1) with probability 0.5. In fact, as mentioned, we can treat the state σ^{XY} as being equal to $|0\rangle \otimes |0\rangle$ or $|1\rangle \otimes |1\rangle$ with equal probability.

In order to erase the residual classical correlations, subsystem Y uses another bit of local randomness to decide whether to apply or not the Pauli X matrix P_1. That is, subsystem Y applies a bit flip quantum channel with $\gamma = 0.5$.

If it applies the Pauli X matrix P_1, the post-transformation state is given by

$$\begin{aligned}(I \otimes P_1)\sigma^{XY}(I \otimes P_1) &= \frac{1}{2}|0\rangle\langle 0| \otimes P_1|0\rangle\langle 0|P_1 + \frac{1}{2}|1\rangle\langle 1| \otimes P_1|1\rangle\langle 1|P_1\\ &= \frac{1}{2}|0\rangle\langle 0| \otimes |1\rangle\langle 1| + \frac{1}{2}|1\rangle\langle 1| \otimes |0\rangle\langle 0|. \end{aligned} \tag{14.119}$$

Therefore, since this transformation occurs with probability 0.5, and otherwise the state σ^{XY} is left unchanged, the overall state of the system becomes

$$\frac{1}{2}\sigma^{XY} + \frac{1}{2}(I \otimes P_1)\sigma^{XY}(I \otimes P_1) = \pi \otimes \pi, \tag{14.120}$$

where π is the maximally mixed qubit state.

Thus, both subsystems are now in the maximally mixed state π. Since they have individual states, the system is in an uncorrelated state. It is concluded that, while the first bit of local randomness has helped erase the quantum correlations in the original state $|\Phi_+^{XY}\rangle$, the second bit has removed the classical correlations left over from the first transformation.

Overall, measuring correlations by the amount of local randomness required to erase them, one may say that the state $|\Phi_+^{XY}\rangle$ has 1 bit of classical correlations and 1 bit of quantum correlations. This conclusion is in line with the decomposition (14.105). In fact, for the Bell state $|\Phi_+^{XY}\rangle$, the quantum mutual information of the state equals 2 bits, while the maximum measured mutual information and the entanglement entropy are both equal to 1 bit.

14.7.9 Entanglement Entropy Using Alternative Entropy Measures ⏭

This section concludes with a brief note about the possible use of alternative entropy measures to evaluate entanglement. Notably, using the Rényi entropy, one can define the **Rényi entanglement entropy** of order α as the Rényi entropy of the eigenvalues λ, i.e.,

$$\begin{aligned} \mathrm{E}_\alpha^\mathrm{R}(X;Y)_{|\psi^{XY}\rangle} &= \mathrm{H}_\alpha^\mathrm{R}(\lambda) \\ &= \frac{1}{1-\alpha}\log_2\left(\sum_{z\in[r]}\lambda_z^\alpha\right). \end{aligned} \tag{14.121}$$

As an important special case, the Rényi entanglement entropy of order infinity is given by

$$\mathrm{E}_\infty^\mathrm{R}(X;Y)_{|\psi^{XY}\rangle} = -\log_2\left(\max_{z\in[r]}\lambda_z\right), \tag{14.122}$$

which has the following instructive resource-theoretic interpretation.

As discussed in Chapter 12, correlations can be interpreted as a resource that can be quantified via some measure of distance to the set of uncorrelated states. Accordingly, entanglement can be measured via the minimum distance between the state $|\psi^{XY}\rangle$ and the set of uncorrelated states $|\phi^X\rangle \otimes |\phi^Y\rangle$. One way to quantify this distance is to evaluate the maximum fidelity between state $|\psi^{XY}\rangle$ and any uncorrelated state $|\phi^X\rangle \otimes |\phi^Y\rangle$, i.e.,

$$\max_{|\phi^X\rangle\otimes|\phi^Y\rangle} \mathrm{F}(|\psi^{XY}\rangle, |\phi^X\rangle \otimes |\phi^Y\rangle) = \max_{|\phi^X\rangle\otimes|\phi^Y\rangle} |\langle\psi^{XY}|(|\phi^X\rangle \otimes |\phi^Y\rangle)|^2, \tag{14.123}$$

where maximization is over all possible uncorrelated states $|\phi^X\rangle \otimes |\phi^Y\rangle$. This maximization can be shown to yield the maximum Schmidt probability $\max_{z\in[r]}\lambda_z$ in (14.122) (see Problem 14.16). Therefore, the Rényi entanglement entropy of order infinity in (14.122) provides a measure of the dissimilarity between the state $|\psi^{XY}\rangle$ and any uncorrelated state $|\phi^X\rangle \otimes |\phi^Y\rangle$.

Another common alternative entanglement metric is obtained via the Tsallis entropy of order 2, which yields

$$\begin{aligned} \mathrm{E}_2^\mathrm{T}(X;Y)_{|\psi^{XY}\rangle} &= 1_2 \cdot \left(1 - \sum_{x\in[r]}\lambda_x^2\right) \\ &= 1_2 \cdot (1 - \mathrm{tr}((\rho^X)^2)) \\ &= 1_2 \cdot (1 - \mathrm{tr}((\rho^Y)^2)). \end{aligned} \tag{14.124}$$

This measure decreases linearly with the purity $\mathrm{tr}((\rho^X)^2) = \mathrm{tr}((\rho^Y)^2)$ of the two reduced density matrices, reflecting the fact that more mixed local states indicate a larger entanglement.

14.8 Quantum Conditional Entropy

This section introduces and studies the quantum conditional entropy as the extension of the classical conditional entropy introduced in Chapter 13.

14.8.1 Classical Conditional Entropy: A Recap

As a recap, for a joint distribution p^{XY}, the classical conditional entropy

$$\mathrm{H}(Y|X)_{p^{XY}} = \mathrm{H}(X,Y)_{p^{XY}} - \mathrm{H}(X)_{p^X} \tag{14.125}$$

is a measure of the average residual uncertainty on the state of subsystem Y given the observation of the state of subsystem X. Using the definition (14.125), the mutual information (14.90) can be expressed as the difference between marginal and conditional entropies

$$\mathrm{I}(X;Y)_{p^{XY}} = \mathrm{H}(Y)_{p^Y} - \mathrm{H}(Y|X)_{p^{XY}}. \tag{14.126}$$

Complying with intuition, for classical systems, the conditional entropy $\mathrm{H}(Y|X)_{p^{XY}}$ was shown in Chapter 13 to satisfy the following properties.

- **Non-negativity:** $\mathrm{H}(Y|X)_{p^{XY}} \geq 0$.
- **Conditioning reduces entropy:** $\mathrm{H}(Y|X)_{p^{XY}} \leq \mathrm{H}(Y)_{p^Y}$.

This section studies the counterpart of the conditional entropy (14.125) for quantum pure states. It will be shown that, as a direct consequence of the insubordination of quantum states to the principle that uncertainty increases globally (Section 14.7.3), the quantum conditional entropy may be negative. In particular, the quantum conditional entropy is negative if and only if the state is entangled. Furthermore, it will also be demonstrated that the quantum conditional entropy guarantees the conditioning-reduces-entropy property.

14.8.2 Defining the Quantum Conditional Entropy

Mirroring the classical definition (14.125), given a pure state $|\psi^{XY}\rangle$, the quantum conditional entropy $\mathrm{H}(Y|X)_{|\psi^{XY}\rangle}$ is defined by enforcing the chain rule for the quantum entropy as

$$\mathrm{H}(X,Y)_{|\psi^{XY}\rangle} = \mathrm{H}(X)_{\rho^X} + \mathrm{H}(Y|X)_{|\psi^{XY}\rangle}, \tag{14.127}$$

where ρ^X is the reduced density matrix for subsystem X. Accordingly, for a quantum state $|\psi^{XY}\rangle$, the quantum conditional entropy of subsystem Y given subsystem X is defined as

$$\mathrm{H}(Y|X)_{|\psi^{XY}\rangle} = \mathrm{H}(X,Y)_{|\psi^{XY}\rangle} - \mathrm{H}(X)_{\rho^X}. \tag{14.128}$$

Loosely speaking, the quantum conditional entropy $\mathrm{H}(Y|X)_{|\psi^{XY}\rangle}$ captures the amount of information that is in the global state $|\psi^{XY}\rangle$, but is not included in the local subsystem X. Note that the analogous definition $\mathrm{H}(X|Y)_{|\psi^{XY}\rangle} = \mathrm{H}(X, Y)_{|\psi^{XY}\rangle} - \mathrm{H}(Y)_{\rho^Y}$ applies by swapping the roles of the two subsystems. Furthermore, recalling the equalities $\mathrm{H}(X, Y)_{|\psi^{XY}\rangle} = 0$ and $\mathrm{H}(X)_{\rho^X} = \mathrm{H}(Y)_{\rho^Y} = \mathrm{H}(\lambda)$, the quantum conditional entropy (14.128) can be written explicitly as follows.

Definition 14.15 Quantum conditional entropy

For any joint pure state $|\psi^{XY}\rangle$, the quantum conditional entropy of subsystem Y given subsystem X is given as

$$\mathrm{H}(Y|X)_{|\psi^{XY}\rangle} = -\mathrm{H}(\lambda), \tag{14.129}$$

and the quantum conditional entropy of subsystem X given subsystem Y is

$$\mathrm{H}(X|Y)_{|\psi^{XY}\rangle} = -\mathrm{H}(\lambda), \tag{14.130}$$

where λ is the Schmidt probability vector.

14.8.3 Properties of the Quantum Conditional Entropy

This characterization of the quantum conditional entropy implies the following properties.

Key Result 14.15 Conditioning reduces the quantum conditional entropy

For any pure state $|\psi^{XY}\rangle$, the quantum conditional entropy satisfies the inequality

$$\underbrace{\mathrm{H}(Y|X)_{|\psi^{XY}\rangle}}_{=-\mathrm{H}(\lambda)} \leq \underbrace{\mathrm{H}(Y)_{\rho^Y}}_{=\mathrm{H}(\lambda)}, \tag{14.131}$$

with equality if and only if the state is uncorrelated, in which case both quantum conditional entropy and quantum entropy equal zero, i.e., $\mathrm{H}(\lambda) = 0$.

While the conditioning-reduces-entropy property (14.131) applies also to classical systems, the next property is a defining feature of quantum systems.

Key Result 14.16 Non-positivity of the quantum conditional entropy

For any pure state $|\psi^{XY}\rangle$, the quantum conditional entropy is non-positive, i.e.,

$$\mathrm{H}(X|Y)_{|\psi^{XY}\rangle} = \mathrm{H}(Y|X)_{|\psi^{XY}\rangle} \leq 0, \tag{14.132}$$

with the equalities $\mathrm{H}(X|Y)_{|\psi^{XY}\rangle} = \mathrm{H}(Y|X)_{|\psi^{XY}\rangle} = 0$ holding if and only if the state $|\psi^{XY}\rangle$ is uncorrelated.

Accordingly, the negativity of the quantum conditional entropy is a necessary and sufficient condition for the entanglement of a pure quantum state:

$$|\psi^{XY}\rangle \text{ entangled} \iff \mathrm{H}(X|Y)_{|\psi^{XY}\rangle} = \mathrm{H}(Y|X)_{|\psi^{XY}\rangle} < 0. \tag{14.133}$$

Since a negative quantum conditional entropy $\mathrm{H}(Y|X)_{|\psi^{XY}\rangle}$, or $\mathrm{H}(X|Y)_{|\psi^{XY}\rangle}$, proves the entanglement of the state $|\psi^{XY}\rangle$, the negative quantum conditional entropy can be interpreted as a measure of entanglement. This measure is known as the **coherent information**.

14.8.4 All You Need Are the Schmidt Probabilities

This chapter has defined several measures of uncertainty and correlation for quantum pure states, which are summarized in Table 14.1. As made clear by the last column of the table, all the introduced measures are directly obtained from the entropy $\mathrm{H}(\lambda)$ of the Schmidt probability vector λ.

Furthermore, it was noted in this chapter that the presence of entanglement is reflected in, and revealed by, a violation of the principle that uncertainty increases globally, i.e., by insubordination. Mathematically, this conclusion amounts to the equivalence relationships

$$\begin{aligned} |\psi^{XY}\rangle \text{ entangled} &\iff \mathrm{H}(\lambda) > 0 \\ &\iff \mathrm{E}(X;Y)_{|\psi^{XY}\rangle} > 0 \\ &\iff \mathrm{I}(X;Y)_{|\psi^{XY}\rangle} > 0 \\ &\iff \mathrm{J}(X;Y)_{|\psi^{XY}\rangle} > 0 \\ &\iff \mathrm{H}(Y|X)_{|\psi^{XY}\rangle} = \mathrm{H}(X|Y)_{|\psi^{XY}\rangle} < 0. \end{aligned} \tag{14.134}$$

In words, either a positive correlation measure or a negative quantum conditional entropy indicate the presence of entanglement.

Table 14.1 Summary of the measures defined in this chapter

Measure	Notation	Significance	Value
Maximum measured mutual information	$\mathrm{J}(X;Y)_{\lvert\psi^{XY}\rangle}$	Classical correlations	$\mathrm{H}(\lambda)$
Entanglement entropy	$\mathrm{E}(X;Y)_{\lvert\psi^{XY}\rangle}$	Entanglement	$\mathrm{H}(\lambda)$
Quantum mutual information	$\mathrm{I}(X;Y)_{\lvert\psi^{XY}\rangle}$	Classical and quantum correlations	$2\mathrm{H}(\lambda)$
Quantum conditional entropy	$\mathrm{H}(X\lvert Y)_{\lvert\psi^{XY}\rangle}$ $\mathrm{H}(Y\lvert X)_{\lvert\psi^{XY}\rangle}$	Residual uncertainty	$-\mathrm{H}(\lambda)$

14.9 Schmidt Decomposition

A pure state $|\psi^{XY}\rangle$ can always be expressed as a superposition of uncorrelated states using (14.46). This section demonstrates that the local eigenbases $\mathcal{M}_*^X$ and $\mathcal{M}_*^Y$ – the eigenbases of the reduced density matrices ρ^X and ρ^Y, respectively – enable the decomposition of $|\psi^{XY}\rangle$ into the minimal number of uncorrelated states.

In particular, the Schmidt decomposition provides a representation of $|\psi^{XY}\rangle$ in terms of a number of uncorrelated states equal to the Schmidt rank r. Specifically, for an uncorrelated state, the Schmidt decomposition yields a single uncorrelated state, i.e., $r = 1$, while, for an entangled state, the decomposition includes up to $N_{\min} = \min(N^X, N^Y)$ uncorrelated states.

14.9.1 Defining the Schmidt Decomposition

To proceed, given a state $|\psi^{XY}\rangle$, write as $\mathcal{M}_*^X = \{|u_x^X\rangle\}_{x\in[N^X]}$ and $\mathcal{M}_*^Y = \{|u_y^Y\rangle\}_{y\in[N^Y]}$ any two local eigenbases. Define as **valid** an ordering of the eigenvectors in the local eigenbases with the property that, for all $z \in [r]$, the eigenvector $|u_z^X\rangle$ of matrix ρ^X and the eigenvector $|u_z^Y\rangle$ of matrix ρ^Y correspond to the same eigenvalue λ_z of the reduced density matrices ρ^X and ρ^Y, respectively. The remaining eigenvectors $\{|u_z^X\rangle\}_{z=r}^{N^X-1}$ and $\{|u_z^Y\rangle\}_{z=r}^{N^Y-1}$, if any, can be ordered in an arbitrary way.

It is emphasized that there is more than one valid ordering when the eigenvalues are not all distinct. Moreover, with any valid ordering, the reduced density matrices can be expressed via the eigendecompositions

$$\rho^X = \sum_{z\in[r]} \lambda_z |u_z^X\rangle\langle u_z^X| \tag{14.135}$$

and

$$\rho^Y = \sum_{z\in[r]} \lambda_z |u_z^Y\rangle\langle u_z^Y|. \tag{14.136}$$

Key Result 14.17 Schmidt decomposition

Consider any state $|\psi^{XY}\rangle$ with Schmidt rank r and Schmidt probability vector $\lambda = [\lambda_0, \lambda_1, \ldots, \lambda_{r-1}]^T$. There exist local eigenbases $\mathcal{M}_*^X = \{|u_x^X\rangle\}_{x\in[N^X]}$ and $\mathcal{M}_*^Y = \{|u_y^Y\rangle\}_{y\in[N^Y]}$ and a valid ordering that allow the state to be expressed as the superposition of r uncorrelated states

$$|\psi^{XY}\rangle = \sum_{z\in[r]} \sqrt{\lambda_z} |u_z^X\rangle \otimes |u_z^Y\rangle. \tag{14.137}$$

Furthermore, it is not possible to express state $|\psi^{XY}\rangle$ as the superposition of fewer than r uncorrelated states.

The Schmidt decomposition (14.137) contains a number of uncorrelated states equal to the Schmidt rank $r \leq N_{\min} = \min(N^X, N^Y)$, which can be much smaller than the number $N^X N^Y$ of uncorrelated states present in the generic decomposition

(14.46). Each uncorrelated state in the Schmidt decomposition is of the form $|u_z^X\rangle \otimes |u_z^Y\rangle$, and is thus formed by pairs of states $(|u_z^X\rangle, |u_z^Y\rangle)$ from the local eigenbases. The states $\{|u_z^X\rangle\}_{z\in[r]}$ are known as the **Schmidt left states** and the states $\{|u_z^Y\rangle\}_{z\in[r]}$ as the **Schmidt right states**. A derivation of the Schmidt decomposition will be presented in Section 14.9.2.

Example 14.23

For an uncorrelated state $|\psi^{XY}\rangle = |\psi^X\rangle \otimes |\psi^Y\rangle$, the Schmidt rank is $r = 1$. In this case, the Schmidt probability vector reduces to the scalar $\lambda = 1$, and the Schmidt left and right states are respectively given by

$$|u_0^X\rangle = |\psi^X\rangle \text{ and } |u_0^Y\rangle = |\psi^Y\rangle. \tag{14.138}$$

Therefore, the Schmidt decomposition coincides with the original expression $|\psi^{XY}\rangle = |\psi^X\rangle \otimes |\psi^Y\rangle$. Clearly, no decomposition of the state exists with fewer than one uncorrelated state.

Example 14.24

The state $|\psi^{XY}\rangle = \sqrt{p}|01\rangle + \sqrt{1-p}|10\rangle$ with any probability $0 \leq p \leq 1$ is also already in the form of the Schmidt decomposition (14.137). In fact, with $p = 0$ and $p = 1$, the state is uncorrelated, and thus follows the previous example. For probabilities $0 < p < 1$, using (14.72) and (14.74), the reduced density matrices are given by

$$\rho^X = \begin{bmatrix} p & 0 \\ 0 & 1-p \end{bmatrix} = p|0\rangle\langle 0| + (1-p)|1\rangle\langle 1| \tag{14.139}$$

$$\text{and } \rho^Y = \begin{bmatrix} 1-p & 0 \\ 0 & p \end{bmatrix} = (1-p)|0\rangle\langle 0| + p|1\rangle\langle 1|, \tag{14.140}$$

and thus the Schmidt rank $r = 2$ and the Schmidt probability vector is $\lambda = [p, 1-p]^T$.

The Schmidt decomposition (14.137) is obtained with the local eigenbases $\mathcal{M}_*^X = \{|0\rangle, |1\rangle\}$ and $\mathcal{M}_*^X = \{|0\rangle, |1\rangle\}$, ordered so that the Schmidt left states are given by $|u_0^X\rangle = |0\rangle$ and $|u_1^X\rangle = |1\rangle$, while the Schmidt right states are $|u_0^Y\rangle = |1\rangle$ and $|u_1^Y\rangle = |0\rangle$. Note that this is a valid ordering of the eigenbases since the eigenstates $|u_0^X\rangle = |0\rangle$ and $|u_0^Y\rangle = |1\rangle$ correspond to the eigenvalue p; while the eigenstates $|u_1^X\rangle = |1\rangle$ and $|u_1^Y\rangle = |0\rangle$ correspond to the eigenvalue $1-p$. It is also observed that for $p = 0.5$ there are two valid orderings, but only one yields the Schmidt decomposition.

Since, when $0 < p < 1$, the state is entangled, there is no decomposition with fewer than $r = 2$ uncorrelated states.

Example 14.25

As discussed in Section 14.7.7, a state $|\psi^{XY}\rangle$ is maximally entangled if and only if the Schmidt rank equals the maximum value $r = N_{\min}$ and the Schmidt probability vector is uniform, i.e., $\lambda = u = [1, \ldots, 1]/N_{\min}$. Accordingly, by (14.137), a maximally entangled state has a Schmidt decomposition of the form

$$|\psi^{XY}\rangle = \sqrt{\frac{1}{N_{\min}}} \sum_{z \in [N_{\min}]} |u_z^X\rangle \otimes |u_z^Y\rangle. \tag{14.141}$$

Example 14.26

The Schmidt decomposition may include a much smaller number of uncorrelated states as compared to the $N^X N^Y$ terms in the generic decomposition (14.46). To see this, suppose that $N^X = 2$ while $N^Y = 10^9$. While a generic decomposition (14.46) would require $N^X N^Y = 2 \cdot 10^9$ uncorrelated states, the Schmidt decomposition presents at most $N_{\min} = 2$ uncorrelated states.

14.9.2 Deriving the Schmidt Decomposition ⏭

The Schmidt decomposition can be derived directly as a consequence of the singular value decomposition (SVD) of the amplitude matrix Ψ^{XY} in (14.61). The reader is referred to Appendix 12.G for background on the SVD. When applied to the amplitude matrix, the SVD yields the following.

Key Result 14.18 Singular value decomposition of the amplitude matrix

Given the amplitude matrix Ψ^{XY} for a state $|\psi^{XY}\rangle$ with Schmidt rank r, there exist orthonormal bases $\{|v_x^X\rangle\}_{x \in [N^X]}$ and $\{|v_y^Y\rangle\}_{y \in [N^Y]}$, and a vector $\omega = [\omega_0, \omega_1, \ldots, \omega_{r-1}]^T$ of r positive real numbers, i.e., $\omega_z > 0$ for $z \in [r]$, such that the amplitude matrix can be expressed as the sum

$$\Psi^{XY} = \sum_{z \in [r]} \omega_z |v_z^X\rangle\langle v_z^Y|. \tag{14.142}$$

The vectors $\{|v_z^X\rangle\}_{z \in [r]}$ and $\{|v_z^Y\rangle\}_{z \in [r]}$ are referred to as the left and right singular vectors of the amplitude matrix Ψ^{XY}, respectively, while the numbers $\{\omega_z\}_{z \in [r]}$ are the singular values of matrix Ψ^{XY}.

Using (14.62) and the property of vectorization $\text{vec}(ABC^\dagger) = (A \otimes C^*)\text{vec}(B)$ (see (10.71)), the SVD (14.142) yields

$$\begin{aligned} |\psi^{XY}\rangle = \text{vec}(\Psi^{XY}) &= \sum_{z \in [r]} \omega_z \text{vec}(|v_z^X\rangle\langle v_z^Y|) \\ &= \sum_{z \in [r]} \omega_z |v_z^X\rangle \otimes |v_z^Y\rangle^*. \end{aligned} \tag{14.143}$$

Comparing this expression with the Schmidt decomposition (14.137), one immediately identifies Schmidt vectors and Schmidt probabilities as summarized in Table 14.2.

Table 14.2 Relationships between Schmidt decomposition and SVD of the amplitude matrix

Schmidt decomposition of $\|\psi^{XY}\rangle$	Singular value decomposition of Ψ^{XY}
Schmidt rank r	Rank r
Schmidt probabilities $\{\lambda_z\}_{z\in[r]}$	Squared singular values $\{\omega_z^2\}_{z\in[r]}$
Schmidt left states $\{\|u_z^X\rangle\}_{z\in[r]}$	Left singular vectors $\{\|v_z^X\rangle\}_{z\in[r]}$
Schmidt right states $\{\|u_z^Y\rangle\}_{z\in[r]}$	Conjugate right singular vectors $\{\|v_z^Y\rangle^*\}_{z\in[r]}$

Table 14.3 Relationships between Schmidt decomposition and reduced density matrices

Schmidt decomposition of $\|\psi^{XY}\rangle$	Reduced density matrix ρ^X	Reduced density matrix ρ^Y
Schmidt rank r	Rank r	Rank r
Schmidt probabilities $\{\lambda_z\}_{z\in[r]}$	Positive eigenvalues $\{\lambda_z\}_{z\in[r]}$	Positive eigenvalues $\{\lambda_z\}_{z\in[r]}$
Schmidt left states $\{\|u_z^X\rangle\}_{z\in[r]}$	Eigenvectors $\{\|u_z^X\rangle\}_{z\in[r]}$	N/A
Schmidt right states $\{\|u_z^Y\rangle\}_{z\in[r]}$	N/A	Eigenvectors $\{\|u_z^Y\rangle\}_{z\in[r]}$

Specifically, the left Schmidt states $\{|u_z^X\rangle\}_{z\in[r]}$ are given by the left singular vectors $\{|v_z^X\rangle\}_{z\in[r]}$, and the right Schmidt states $\{|u_z^Y\rangle\}_{z\in[r]}$ equal the entry-wise complex conjugate $\{|v_z^Y\rangle^*\}_{z\in[r]}$ of the left singular vectors $\{|v_z^Y\rangle\}_{z\in[r]}$:

$$|u_z^X\rangle = |v_z^X\rangle \text{ and } |u_z^Y\rangle = |v_z^Y\rangle^* \text{ for } z \in [r]. \tag{14.144}$$

Furthermore, the Schmidt probabilities $\{\lambda_z\}_{z\in[r]}$ equal the square of the singular values $\{\omega_z\}_{z\in[r]}$:

$$\omega_z = \sqrt{\lambda_z} \text{ for } z \in [r] \tag{14.145}$$

or equivalently

$$\omega_z^2 = \lambda_z \text{ for } z \in [r]. \tag{14.146}$$

To complete the derivation of the Schmidt decomposition (14.137), it is thus sufficient to demonstrate that the squared singular values $\{\omega_z^2\}_{z\in[r]}$ coincide with the positive eigenvalues of the reduced density matrices ρ^X and ρ^Y, as well as that the singular vectors $\{|v_z^X\rangle\}_{z\in[r]}$ and $\{|v_z^Y\rangle^*\}_{z\in[r]}$ are eigenvectors of the reduced density matrices ρ^X and ρ^Y, respectively. This is summarized in Table 14.3, and it follows directly from the definitions (14.72) and (14.74) of the reduced density matrices.

To see this, take for instance the reduced density matrix ρ^X, which, using the SVD (14.142), can be written as

$$\begin{aligned}\rho^X &= \left(\sum_{z\in[r]} \omega_z |v_z^X\rangle\langle v_z^Y|\right)\left(\sum_{z'\in[r]} \omega_{z'} |v_{z'}^X\rangle\langle v_{z'}^Y|\right)^\dagger \\ &= \sum_{z\in[r]}\sum_{z'\in[r]} \omega_z\omega_{z'} |v_z^X\rangle\langle v_z^Y|v_{z'}^Y\rangle\langle v_{z'}^X| \\ &= \sum_{z\in[r]} \omega_z^2 |v_z^X\rangle\langle v_{z'}^X|. \end{aligned} \tag{14.147}$$

This expression, which describes the eigendecomposition of matrix ρ^X, confirms the desired correspondences in Table 14.3, completing the derivation of the Schmidt decomposition.

Example 14.27

For $N^X = 2$ and $N^Y = 4$, and thus $N_{\min} = 2$, consider the state

$$|\psi^{XY}\rangle = \frac{i}{2}|00\rangle + \frac{12+5i}{26}|01\rangle - \frac{1}{2}|11\rangle + \frac{3}{10}|12\rangle - \frac{2i}{5}|13\rangle. \tag{14.148}$$

The amplitude matrix is given by

$$\Psi^{XY} = \begin{bmatrix} \frac{i}{2} & \frac{12+5i}{26} & 0 & 0 \\ 0 & -\frac{1}{2} & \frac{3}{10} & -\frac{2i}{5} \end{bmatrix}. \tag{14.149}$$

Evaluating the square of the singular values yields the Schmidt probability vector $\lambda = [0.75, 0.25]^T$, showing that the Schmidt rank is maximal $r = N_{\min} = 2$. Alternatively, one could evaluate the eigenvalues of either of the reduced density matrices, obtaining the same result.

The derivation of the Schmidt decomposition also shows that there can be no decomposition into uncorrelated states with a number of terms smaller than the Schmidt rank r. In fact, if that were the case, the reduced density matrices would have rank smaller than r, which would be inconsistent with the definition of Schmidt rank (see Problem 14.17).

14.9.3 Schmidt Decomposition and Majorization ⏭

As discussed, the Schmidt rank r determines the smallest number of uncorrelated states required in a decomposition of a given state. This result can be strengthened by using the theory of majorization. To elaborate, consider any decomposition of the state $|\psi^{XY}\rangle$ of the form

$$|\psi^{XY}\rangle = \sum_{z\in[M]} \sqrt{\beta_x}|\phi_z^X\rangle \otimes |\phi_z^Y\rangle \tag{14.150}$$

for some arbitrary (not necessarily mutually orthogonal) states $\{|\phi_z^X\rangle\}_{z\in[M]}$ and $\{|\phi_z^Y\rangle\}_{z\in[M]}$ and some probability distribution $\beta = [\beta_0, \ldots, \beta_{M-1}]^T$. The following result holds. For a proof, the reader is referred to the Recommended Resources.

Key Result 14.19 Comparing decompositions of bipartite pure states into uncorrelated states

Given a joint pure state $|\psi^{XY}\rangle$ with Schmidt probability vector λ and Schmidt rank r, any decomposition into uncorrelated states must have $M \geq r$ terms, and it must satisfy the majorization condition

$$\lambda \succ \beta, \tag{14.151}$$

with vector λ being zero padded to be of dimension $M \times 1$.

Furthermore, for any $M \times 1$ vector β satisfying the majorization relation (14.151), there exist states $\{|\phi_z^X\rangle\}_{z\in[M]}$ and $\{|\phi_z^Y\rangle\}_{z\in[M]}$ for which the decomposition (14.150) applies.

According to this result, the Schmidt probability vector λ corresponds to the least uncertain probability distribution among all the probability distributions β that permit a decomposition (14.150) of the state into a sum of uncorrelated states.

14.10 Partial Trace

The reduced density matrices in (14.72) and (14.74) are defined in terms of the amplitude matrix. This section shows that the reduced density matrices can also be obtained as the result of an operation on the joint density state $\rho^{XY} = |\psi^{XY}\rangle\langle\psi^{XY}|$ that generalizes the trace and that can be viewed as the counterpart of marginalization for classical probability distributions.

14.10.1 Defining the Partial Trace

The partial trace over subsystem X, denoted as $\text{tr}_X(\cdot)$, applies the trace only to the matrices pertaining to subsystem X, producing the reduced density matrix for subsystem Y:

$$\rho^Y = \text{tr}_X(\rho^{XY}). \tag{14.152}$$

In a symmetric way, the partial trace over subsystem Y, denoted as $\text{tr}_Y(\cdot)$, applies the trace only to the matrices relative to subsystem Y, yielding the reduced density matrix for subsystem X:

$$\rho^X = \text{tr}_Y(\rho^{XY}). \tag{14.153}$$

Throughout this section, for generality, the focus is on joint mixed states ρ^{XY}, encompassing pure states $\rho^{XY} = |\psi^{XY}\rangle\langle\psi^{XY}|$ as a special case.

To start, as shown in Appendix 14.D, any $N^X N^Y \times N^X N^Y$ matrix A, including joint density matrices, can be expressed as a sum of the form

$$A = \sum_{z \in [M]} A_z^X \otimes A_z^Y, \tag{14.154}$$

for some integer M, some $N^X \times N^X$ matrices $\{A_z^X\}_{z \in [M]}$, and some $N^Y \times N^Y$ matrices $\{A_z^Y\}_{z \in [M]}$. Therefore, in defining the operation of the partial trace, it is enough to focus on matrices of the form (14.154).

Definition 14.16 Partial trace
Given any $N^X N^Y \times N^X N^Y$ matrix A written as in (14.154), the partial trace over subsystem X returns the $N^Y \times N^Y$ matrix

$$\mathrm{tr}_X \left(\sum_{z \in [M]} A_z^X \otimes A_z^Y \right) = \sum_{z \in [M]} \mathrm{tr}(A_z^X) A_z^Y, \tag{14.155}$$

while the partial trace over subsystem Y returns the $N^X \times N^X$ matrix

$$\mathrm{tr}_Y \left(\sum_{z \in [M]} A_z^X \otimes A_z^Y \right) = \sum_{z \in [M]} A_z^X \mathrm{tr}(A_z^Y). \tag{14.156}$$

Accordingly, the partial trace over subsystem X applies the trace only to the matrices $\{A_z^X\}_{z \in [N^X]}$, which pertain to subsystem X. Moreover, the partial trace over subsystem Y applies the trace only to the matrices $\{A_z^Y\}_{z \in [N^Y]}$, which pertain to subsystem Y.

14.10.2 Computing Reduced Density Matrices via the Partial Trace

Based on this definition, it is now possible to verify the equalities (14.152) and (14.153). To this end, write the joint density matrix $\rho^{XY} = |\psi^{XY}\rangle\langle\psi^{XY}|$ using the Schmidt decomposition (14.137) as

$$\begin{aligned}\rho^{XY} &= |\psi^{XY}\rangle\langle\psi^{XY}| \\ &= \sum_{z \in [r]} \sum_{z' \in [r]} \sqrt{\lambda_z \lambda_{z'}} |u_z^X\rangle\langle u_{z'}^X| \otimes |u_z^Y\rangle\langle u_{z'}^Y|. \end{aligned} \tag{14.157}$$

Note that the density matrix is in the form (14.154).

Therefore, using the partial trace (14.155) over subsystem X, one obtains

$$\begin{aligned}\mathrm{tr}_X(|\psi^{XY}\rangle\langle\psi^{XY}|) &= \sum_{z \in [r]} \sum_{z' \in [r]} \sqrt{\lambda_z \lambda_{z'}} \mathrm{tr}(|u_z^X\rangle\langle u_{z'}^X|) \cdot |u_z^Y\rangle\langle u_{z'}^Y| \\ &= \sum_{z \in [r]} \lambda_z |u_z^Y\rangle\langle u_z^Y| = \rho^Y, \end{aligned} \tag{14.158}$$

where the last equality uses (14.136). In a similar way, using the partial trace (14.156) over subsystem Y, the reduced density matrix ρ^X in (14.136) is recovered as

$$\begin{aligned} \mathrm{tr}_Y(|\psi^{XY}\rangle\langle\psi^{XY}|) &= \sum_{z\in[r]}\sum_{z'\in[r]} \sqrt{\lambda_z\lambda_{z'}}|u_z^X\rangle\langle u_{z'}^X| \cdot \mathrm{tr}(|u_z^Y\rangle\langle u_{z'}^Y|) \\ &= \sum_{z\in[r]} \lambda_z |u_z^X\rangle\langle u_z^X| = \rho^X. \end{aligned} \tag{14.159}$$

Example 14.28

For an uncorrelated state $|\psi^{XY}\rangle = |\psi^X\rangle \otimes |\psi^Y\rangle$, the partial trace returns the reduced density matrices $\rho^X = |\psi^X\rangle\langle\psi^X|$ and $\rho^Y = |\psi^Y\rangle\langle\psi^Y|$ as

$$\mathrm{tr}_X(\rho^X \otimes \rho^Y) = \mathrm{tr}(\rho^X)\rho^Y = \rho^Y \tag{14.160}$$

and

$$\mathrm{tr}_Y(\rho^X \otimes \rho^Y) = \mathrm{tr}(\rho^Y)\rho^X = \rho^X. \tag{14.161}$$

Example 14.29

The Bell state $\rho^{XY} = |\Psi_-^{XY}\rangle\langle\Psi_-^{XY}|$ can be written as

$$\begin{aligned} \rho^{XY} =& \frac{1}{2}(|0\rangle\langle 0| \otimes |1\rangle\langle 1| + |1\rangle\langle 1| \otimes |0\rangle\langle 0| \\ & - |0\rangle\langle 1| \otimes |1\rangle\langle 0| - |1\rangle\langle 0| \otimes |0\rangle\langle 1|). \end{aligned} \tag{14.162}$$

Therefore, applying the partial trace, the reduced density matrices are obtained as

$$\begin{aligned} \rho^X =& \mathrm{tr}_Y(\rho^{XY}) \\ =& \frac{1}{2}(|0\rangle\langle 0|\mathrm{tr}(|1\rangle\langle 1|) + |1\rangle\langle 1|\mathrm{tr}(|0\rangle\langle 0|) \\ & - |0\rangle\langle 1|\mathrm{tr}(|1\rangle\langle 0|) - |1\rangle\langle 0|\mathrm{tr}(|0\rangle\langle 1|)) \\ =& \frac{1}{2}(|0\rangle\langle 0| + |1\rangle\langle 1|) = \pi, \end{aligned} \tag{14.163}$$

and, in a similar way, $\rho^Y = \mathrm{tr}_X(\rho^{XY}) = \pi$. This calculation confirms the earlier conclusion that Bell states, being maximally entangled, have maximally mixed reduced density matrices.

14.10.3 Partial Trace as a "Sandwich" Operation

As discussed in Section 7.4.1, the trace of an $N \times N$ matrix A can equivalently be expressed using the "sandwich" operation

$$\mathrm{tr}(A) = \sum_{x\in[N]} \langle v_x|A|v_x\rangle, \tag{14.164}$$

where $\{|v_x\rangle\}_{x\in[N]}$ is an arbitrary orthonormal basis for the N-dimensional space. The partial trace can be written in a similar way by applying the "sandwich" operation only on the subsystem being traced out. As detailed next, this follows by using (14.164) in the definitions of partial trace (14.155) and (14.156).

Key Result 14.20 Partial trace as a "sandwich" operation
Given any $N^X N^Y \times N^X N^Y$ matrix A, the partial trace over subsystem X can be expressed as

$$\mathrm{tr}_X(A) = \sum_{x \in [N^X]} ((\langle v_x^X| \otimes I) A (|v_x^X\rangle \otimes I), \tag{14.165}$$

where I is the $N^Y \times N^Y$ identity matrix and $\{|v_x^X\rangle\}_{x\in[N^X]}$ is an arbitrary orthonormal basis for the N^X-dimensional space. In a similar way, the partial trace over subsystem Y can be expressed as

$$\mathrm{tr}_Y(A) = \sum_{y \in [N^Y]} (I \otimes \langle v_y^Y|) A (I \otimes |v_y^Y\rangle), \tag{14.166}$$

where I is the $N^X \times N^X$ identity matrix and $\{|v_y^Y\rangle\}_{y\in[N^Y]}$ is an arbitrary orthonormal basis for the N^Y-dimensional space.

14.11 Summary

- Bipartite quantum pure states are defined by $N^X N^Y \times 1$ complex state vectors $|\psi^{XY}\rangle$ with entries ψ_{xy}^{XY} listed using the bi-index notation xy.
- States of the form $|\psi^{XY}\rangle = |\psi^X\rangle \otimes |\psi^Y\rangle$, in which the subsystems have individual pure states $|\psi^X\rangle$ and $|\psi^Y\rangle$, are said to be uncorrelated, while any other bipartite pure state is entangled.
- A local POVM returns random variables $(\mathrm{x}, \mathrm{y}) \sim p^{XY}$ with joint distribution $p_{xy}^{XY} = \mathrm{tr}((J_x^X \otimes J_y^Y)\rho^{XY})$, and the post-measurement state for observation (x, y) is given by $\sigma_{xy}^{XY} = (K_x^X \otimes K_y^Y)\rho^{XY}((K_x^X)^\dagger \otimes (K_y^Y)^\dagger)/p_{xy}^{XY}$.
- Local measurements break entanglement and have a global effect on the overall state.
- While the measurement at one subsystem, say Y, can change the global state of a bipartite system, the impact of any measurement at Y cannot be detected at subsystem X. This is the no-signaling theorem: there is no way for subsystem Y to communicate any information to subsystem X by merely applying a local measurement at subsystem Y.
- Any bipartite pure state can be expressed in the Schmidt decomposition form $|\psi^{XY}\rangle = \sum_{z\in[r]} \sqrt{\lambda_z} |u_z^X\rangle \otimes |u_z^Y\rangle$ for some integer $r \leq N_{\min} = \min(N^X, N^Y)$, known as the Schmidt rank, where $\lambda = [\lambda_0, \ldots, \lambda_{r-1}]^T$ is the Schmidt probability vector and the sets of left Schmidt states $\{|u_z^X\rangle\}_{z\in[r]}$ and right Schmidt states $\{|u_z^Y\rangle\}_{z\in[r]}$ are orthogonal.
- A state $|\psi^{XY}\rangle$ is entangled if and only if the Schmidt rank r is larger than 1.

- A state $|\psi^{XY}\rangle$ is said to be maximally entangled if it has maximal Schmidt rank $r = N_{\min}$ and all the Schmidt probabilities are equal to $\lambda_z = 1/N_{\min}$ for $r \in [N_{\min}]$.
- The Schmidt decomposition is a direct application of the SVD of the amplitude matrix derived from the joint state $|\psi^{XY}\rangle$.
- The reduced density matrix of subsystem X is obtained as $\rho^X = \text{tr}_Y(|\psi^{XY}\rangle\langle\psi^{XY}|) = \sum_{z\in[r]} \lambda_z |u_z^X\rangle\langle u_z^X|$, where $\text{tr}_Y(\cdot)$ is the partial trace with respect to subsystem Y; while the reduced density matrix of subsystem Y is obtained as $\rho^Y = \text{tr}_X(|\psi^{XY}\rangle\langle\psi^{XY}|) = \sum_{z\in[r]} \lambda_z |u_z^Y\rangle\langle u_z^Y|$, where $\text{tr}_X(\cdot)$ is the partial trace with respect to subsystem X.
- The two reduced density matrices have rank equal to the Schmidt rank, eigenvalues equal to the Schmidt probabilities λ, and eigenbases corresponding to the respective Schmidt states.
- Since the Schmidt rank is larger than 1 if and only if the state $|\psi^{XY}\rangle$ is entangled, entanglement with another subsystem causes a subsystem to become noisy, i.e., to acquire classical uncertainty.
- The marginal measurement distribution $p^X(|\psi^{XY}\rangle|\mathcal{K}^X)$ can be written directly in terms of the reduced density matrix ρ^X as $p_x^X(|\psi^{XY}\rangle|\mathcal{K}^X) = p_x^X(\rho^X|\mathcal{K}^X) = \text{tr}(J_x^X \rho^X)$; and, similarly, one has the measurement probability $p_y^Y(|\psi^{XY}\rangle|\mathcal{K}^Y) = p_y^Y(\rho^Y|\mathcal{K}^Y) = \text{tr}(J_y^Y \rho^Y)$.
- The quantum mutual information $\text{I}(X;Y)_{|\psi^{XY}\rangle}$ captures both classical and quantum correlations and is equal to $\text{I}(X;Y)_{|\psi^{XY}\rangle} = 2\text{H}(\lambda)$.
- The overall correlation level is equally split between classical and quantum correlations: the maximum measured information, quantifying classical correlations, equals $\text{J}(X;Y)_{|\psi^{XY}\rangle} = \text{H}(\lambda)$, and the entanglement entropy $\text{E}(X;Y)_{|\psi^{XY}\rangle} = \text{I}(X;Y)_{|\psi^{XY}\rangle} - \text{J}(X;Y)_{|\psi^{XY}\rangle} = \text{H}(\lambda)$ captures quantum correlations.
- For quantum pure states, uncertainty decreases globally in the sense that the inequality $\max(\text{H}(X)_{\rho^X}, \text{H}(Y)_{\rho^Y}) \geq \text{H}(X, Y)_{|\psi^{XY}\rangle} = 0$ holds, and is strict if and only if the state $|\psi^{XY}\rangle$ is entangled.
- The quantum conditional entropy of subsystem Y given subsystem X is defined as $\text{H}(Y|X)_{|\psi^{XY}\rangle} = \text{H}(X, Y)_{|\psi^{XY}\rangle} - \text{H}(X)_{\rho^X}$.
- While conditioning reduces the quantum conditional entropy, as for the classical case, the quantum conditional entropy is non-positive, i.e., $\text{H}(Y|X)_{|\psi^{XY}\rangle} \leq 0$, and the inequality is strict if and only if the state $|\psi^{XY}\rangle$ is entangled.

14.12 Recommended Resources

The relationship between static and dynamic interpretations of quantum models is thoroughly examined in [1], providing insights into quantum theory from the perspective of Bayesian inference. Reference [2] introduces the entanglement metric of concurrence, while the concept of insubordination, which highlights the unique non-local characteristics of quantum correlations, is discussed in depth by [3]. The work [4] explores the distinction between classical and quantum correlations by examining the process of erasing correlations through local randomness, thereby clarifying the nature of quantum correlations. For further details on coherent information, a foundational concept in quantum data processing and error correction, see [5, 6]. Finally, reference [7] provides a proof of the result (14.151), which addresses the comparison of bipartite pure states decomposed into uncorrelated states, leveraging majorization as a tool to analyze transformations of entangled states.

PROBLEMS

14.1 Show that all four Bell states are entangled.

14.2 Demonstrate that the $N^X N^Y$ vectors obtained by taking the Kronecker products $|v_x^X\rangle \otimes |v_y^Y\rangle$ of all pairs $|v_x^X\rangle$ and $|v_y^Y\rangle$ from any local orthonormal bases $\{|v_x^X\rangle\}_{x\in[N^X]}$ and $\{|v_y^Y\rangle\}_{y\in[N^Y]}$ for the two subsystems form an orthonormal basis for the $N^X N^Y$-dimensional space.

14.3 Write the Bell states as a superposition of vectors obtained from the local Hadamard bases.

14.4 Show that the joint distribution of measurements carried out locally at the two subsystems is the same irrespective of whether the measurements are carried out simultaneously or sequentially (i.e., one after the other). To this end, write the joint distributions in both cases and verify that they are equal.

14.5 (a) Given a system in Bell state $|\Phi_+^{XY}\rangle$, show that, if subsystem Y applies a measurement in any POVM setting $\mathcal{K}^Y = \{K_y^Y\}_{y\in[M^Y]}$, it obtains value $y \in [N^Y]$ with probability $p_y^Y = 0.5\mathrm{tr}((K_y^Y)^\dagger K_y^Y)$. (b) Show that, more generally, for a maximally entangled state $|\psi^{XY}\rangle$ with any dimensions $N^X = N^Y = D$, the measurement probability for a POVM $\mathcal{K}^Y = \{K_y^Y\}_{y\in[M^Y]}$ is given by $p_y^Y = D^{-1}\mathrm{tr}((K_y^Y)^\dagger K_y^Y)$ for all $y \in [N^Y]$.

14.6 Consider a setting in which there are two agents, one holding subsystem X and one holding subsystem Y. Assume that the two agents can exchange classical information, i.e., bits. By making local observations on their respective subsystems and exchanging classical information, they wish to determine if the joint state of the two subsystems is $|\psi_0^{XY}\rangle$ or $|\psi_1^{XY}\rangle$. (a) Find a strategy based on local measurements and classical communications between the two agents that allows the two agents to distinguish the Bell states $|\psi_0^{XY}\rangle = |\Phi_-^{XY}\rangle$ and $|\psi_1^{XY}\rangle = |\Psi_-^{XY}\rangle$. (b) Can a strategy be found that addresses the same problem for the Bell states $|\Phi_+^{XY}\rangle$ and $|\Phi_-^{XY}\rangle$?

14.7 Find the Schmidt decomposition for state $|\psi^{XY}\rangle = [0, 1, -1, 0]^T/\sqrt{2}$.

14.8 Show that for any pair of local orthonormal bases $\{|v_x^X\rangle\}_{x\in[N^X]}$ and $\{|v_y^Y\rangle\}_{x\in[N^Y]}$, any $N^X N^Y \times N^X N^Y$ matrix A can be written as

$$A = \sum_{x\in[N^X]} \sum_{y\in[N^Y]} \sum_{x'\in[N^X]} \sum_{y'\in[N^Y]} \alpha_{xy,x'y'} |v_x^X\rangle\langle v_{x'}^X| \otimes |v_y^Y\rangle\langle v_{y'}^Y|. \quad (14.167)$$

To this end, derive the expression of the coefficients $\alpha_{xy,x'y'}$ as a function of matrix A and of the orthonormal bases.

14.9 Show that for local measurements in the eigenbases of the reduced density matrices, the measurement joint probability distribution is given by $p_{xy}(|\psi^{XY}\rangle|\mathcal{M}^X \otimes \mathcal{M}^Y) = \lambda_x$ for $x = y$ and $x \in [r]$ and it is equal to zero otherwise, where λ_x is the xth Schmidt probability.

14.10 (a) Given an $N \times N$ matrix A, prove the equality

$$\mathrm{vec}(A) = (I \otimes A^T) \sum_{x\in[N]} |x\rangle \otimes |x\rangle, \tag{14.168}$$

where I is the $N \times N$ identity matrix and $\{|x\rangle\}_{x\in[N]}$ represents the computational basis for the N-dimensional vector space. (b) Use this result to show that a state $|\psi^{XY}\rangle$ with dimensions $N^X = N^Y = D$ is related to its amplitude matrix Ψ^{XY} via the equality

$$|\psi^{XY}\rangle = (I \otimes (\Psi^{XY})^T) \sum_{x\in[D]} |x\rangle \otimes |x\rangle. \tag{14.169}$$

14.11 (a) Prove that for any uncorrelated pure two-qubit state $|\psi^{XY}\rangle = |\psi^X\rangle \otimes |\psi^Y\rangle$ the maximum fidelity with any maximally entangled state equals $1/2$, i.e., the inequality

$$\mathrm{F}(|\psi^{XY}\rangle, |\phi^{XY}\rangle) \leq \frac{1}{2} \tag{14.170}$$

holds for any maximally entangled state $|\phi^{XY}\rangle$. (b) For which uncorrelated states does equality hold in (14.170)?

14.12 Which of the two states $\sqrt{0.1}|00\rangle + \sqrt{0.9}|11\rangle$ and $\sqrt{0.2}|++\rangle + \sqrt{0.8}|--\rangle$ is more entangled?

14.13 For any pure state $|\psi^{XY}\rangle$, prove the result

$$\mathrm{supp}(|\psi^{XY}\rangle\langle\psi^{XY}|) \subseteq \mathrm{supp}(\rho^X \otimes \rho^Y). \tag{14.171}$$

Hint: Use the Schmidt decomposition and recall from Chapter 8 that this condition is equivalent to the equality $\Pi_{\rho^X\otimes\rho^Y}|\psi^{XY}\rangle = |\psi^{XY}\rangle$, where $\Pi_{\rho^X\otimes\rho^Y}$ is the projection matrix into the support subspace for matrix $\rho^X \otimes \rho^Y$.

14.14 Show the equality

$$\mathrm{D}(|\psi^{XY}\rangle\langle\psi^{XY}|||\rho^X \otimes \rho^Y) = \mathrm{I}(X;Y)_{|\psi^{XY}\rangle}. \tag{14.172}$$

Thus, the quantum relative entropy between the state $|\psi^{XY}\rangle\langle\psi^{XY}|$ and the uncorrelated mixed state $\rho^X \otimes \rho^Y$ equals the quantum mutual information. *Hint:* Use the eigendecomposition of matrix $\rho^X \otimes \rho^Y$ (see Chapter 8).

14.15 Demonstrate that, if a pure state is entangled, there is no local complete measurement setting $\mathcal{M}^X \otimes \mathcal{M}^Y$ that returns outputs with zero uncertainty, i.e., the joint entropy of the joint measurement probability must be positive. *Hint:* Reason by contradiction.

14.16 Show that the maximum fidelity $\max_{|\phi^X\rangle\otimes|\phi^Y\rangle} \mathrm{F}(|\psi^{XY}\rangle, |\phi^X\rangle\otimes|\phi^Y\rangle)$ between a pure state $|\psi^{XY}\rangle$ and any uncorrelated state $|\phi^X\rangle \otimes |\phi^Y\rangle$ equals the maximum Schmidt probability $\max_{z\in[r]} \lambda_z$. *Hint:* Use the Schmidt decomposition and write the states $|\phi^X\rangle$ and $|\phi^Y\rangle$ in terms of the orthonormal bases given by the Schmidt vectors. Then optimize over the amplitudes of the states with respect to such bases. The Cauchy–Schwarz inequality may be useful.

14.17 Prove that the Schmidt decomposition expresses a state as a superposition of the minimal number of uncorrelated states. *Hint:* What would be the rank of the amplitude matrix with fewer uncorrelated states?

Appendix

14.A Proof of (14.71) and (14.73)

Writing the state $|\psi^{XY}\rangle$ as in (14.49), the measurement probability (14.41), with index $z \in [N^X]$, can be expressed as

$$\begin{aligned}
p_z^X(|\psi^{XY}\rangle|\mathcal{K}^X) &= \langle\psi^{XY}|(J_z^X \otimes I)|\psi^{XY}\rangle \\
&= \sum_{x\in[N^X]}\sum_{y\in[N^Y]}\sum_{x'\in[N^X]}\sum_{y'\in[N^Y]} (\psi_{xy}^{XY})^*\psi_{x'y'}^{XY}\langle xy|(J_z^X\otimes I)|x'y'\rangle \\
&= \sum_{x\in[N^X]}\sum_{y\in[N^Y]}\sum_{x'\in[N^X]}\sum_{y'\in[N^Y]} (\psi_{xy}^{XY})^*\psi_{x'y'}^{XY}\langle x|J_z^X|x'\rangle\langle y|y'\rangle \\
&= \sum_{x\in[N^X]}\sum_{x'\in[N^X]}\left(\underbrace{\sum_{y\in[N^Y]}\psi_{x'y}^{XY}(\psi_{xy}^{XY})^*}_{\rho_{x',x}^X}\right)[J_z^X]_{x,x'} \\
&= \underbrace{\sum_{x\in[N^X]}\sum_{x'\in[N^X]}[J_z^X]_{x,x'}\rho_{x',x}^X}_{[J_z^X\rho^X]_{x,x}} \\
&= \mathrm{tr}(J_x^X\rho^X).
\end{aligned} \tag{14.173}$$

In the derivation above, the fourth equality follows from the definition (14.72) of the reduced density matrix; the following equality from the definition of matrix product; and the last equality from the definition of trace. Similar calculations yield the equality (14.73).

14.B Derivation of the Maximum Measured Mutual Information

This appendix proves that: (i) the measurement setting $\mathcal{M}^X \otimes \mathcal{M}^Y$ that maximizes the measured mutual information of random variables $(\mathrm{x}, \mathrm{y}) \sim p^{XY}(|\psi^{XY}\rangle|\mathcal{M}^X \otimes \mathcal{M}^Y)$ is given by any pair of local eigenbases; and (ii) the maximum measured mutual information is given by (14.96). The data processing inequality under local channels, to be presented in Chapter 16, implies the inequality

$$\mathrm{I}(X;Y)_{p^{XY}(|\psi^{XY}\rangle|\mathcal{M}^X\otimes\mathcal{M}^Y)} \leq \mathrm{H}(X)_{\rho^X} = \mathrm{H}(\lambda), \tag{14.174}$$

as detailed in Appendix 16.F. To complete the proof, it is thus sufficient to note that the upper bound $\mathrm{H}(\lambda)$ can be achieved by choosing a measurement setting $\mathcal{M}_*^X \otimes \mathcal{M}_*^Y$ for any local eigenbases $\mathcal{M}_*^X$ and $\mathcal{M}_*^Y$.

By using the Schmidt decomposition (14.137), one can readily prove that, with a measurement setting of the form $\mathcal{M}_*^X \otimes \mathcal{M}_*^Y$, the joint measurement distribution is given by (Problem 14.9)

$$p_{xy}^{XY}(|\psi^{XY}\rangle|\mathcal{M}_*^X \otimes \mathcal{M}_*^Y) = \begin{cases} \lambda_x & \text{for } x = y \in [r], \\ 0 & \text{otherwise.} \end{cases} \tag{14.175}$$

By direct computation, it is thus concluded that the corresponding mutual information is given by $\mathrm{I}(X;Y)_{p^{XY}(|\psi^{XY}\rangle|\mathcal{M}_*^X\otimes\mathcal{M}_*^Y)} = \mathrm{H}(\lambda)$, completing the proof.

14.C Concurrence

A two-qubit state can always be expressed as a superposition of the vectors orthonormal basis $\{|v_0\rangle = |\Phi^+\rangle, |v_1\rangle = i|\Phi^-\rangle, |v_2\rangle = i|\Psi^+\rangle, |v_3\rangle = |\Psi^-\rangle\}$. The vectors in this basis are indeed orthogonal, as they are a scaled version of the Bell states. Accordingly, any two-qubit state can be written as

$$|\psi^{XY}\rangle = \sum_{z\in[4]} \alpha_z |v_z\rangle, \tag{14.176}$$

where each amplitude α_z is obtained via the projection $\alpha_z = \langle v_z|\psi^{XY}\rangle$.

The concurrence of state $|\psi^{XY}\rangle$ is defined as

$$\mathcal{C}(X;Y)_{|\psi^{XY}\rangle} = \left|\sum_{z\in[4]} \alpha_z^2\right|. \tag{14.177}$$

The entanglement entropy can be expressed in terms of the concurrence of the state as

$$\mathrm{E}(X;Y)_{|\psi^{XY}\rangle} = \mathrm{h}\left(\frac{1}{2} + \frac{1}{2}\sqrt{1 - (\mathcal{C}(X;Y)_{|\psi^{XY}\rangle})^2}\right). \tag{14.178}$$

Note that the entanglement entropy is minimized when the concurrence takes the minimum value $\mathcal{C}(X;Y)_{|\psi^{XY}\rangle} = 0$, and is maximized when the concurrence takes the maximum value $\mathcal{C}(X;Y)_{|\psi^{XY}\rangle} = 1$.

As an example, for Bell states, the concurrence is equal to 1 since there is a single non-zero amplitude α_z in the decomposition (14.176), which has absolute value $|\alpha_z| = 1$. It follows from (14.178) that the entanglement entropy is maximal, i.e., $\mathrm{E}(X;Y)_{|\psi^{XY}\rangle} = \mathrm{h}(1/2) = 1$. As another example, for the state (14.87), i.e., $|\psi^{XY}\rangle = \sqrt{p}|00\rangle + \sqrt{1-p}|11\rangle$, one can write (14.176) with amplitudes

$$\alpha_0 = \frac{\sqrt{p} + \sqrt{1-p}}{\sqrt{2}} \text{ and } \alpha_1 = -i\frac{\sqrt{p} - \sqrt{1-p}}{\sqrt{2}}. \tag{14.179}$$

14.D Decomposing a Matrix Into Kronecker Products

This appendix shows that a decomposition like (14.154) holds for any $N^X N^Y \times N^X N^Y$ matrix A. To see this, note that an element of matrix A can be indexed using two bi-indices, with one bi-index used for the row index and one bi-index used for the column index. That is, any element in the xyth row, i.e., the $(N^Y x + y)$th row, and $(x'y')$th column, i.e., the $(N^Y x' + y')$th column, can be expressed as $A_{xy,x'y'}$. For instance, for $N^X = N^Y = 2$, the elements of the matrix can be written as

$$A = \begin{bmatrix} A_{00,00} & A_{00,01} & A_{00,10} & A_{00,11} \\ A_{01,00} & A_{01,01} & A_{01,10} & A_{01,11} \\ A_{10,00} & A_{10,01} & A_{10,10} & A_{10,11} \\ A_{11,00} & A_{11,01} & A_{11,10} & A_{11,11} \end{bmatrix}. \quad (14.180)$$

Accordingly, a matrix A can be expressed as the sum

$$A = \sum_{x\in[N^X]} \sum_{y\in[N^Y]} \sum_{x'\in[N^X]} \sum_{y'\in[N^Y]} A_{xy,x'y'} |x\rangle\langle x'| \otimes |y\rangle\langle y'|, \quad (14.181)$$

where matrix $|x\rangle\langle x'| \otimes |y\rangle\langle y'|$ has all zero entries except for a 1 in position $(xy, x'y')$. The expression (14.154) can be recovered by lumping together all indices x, y, x', y' into a single index z. Note that, in this general formulation, the number of elements in the sum is maximal, namely $(N^X N^Y)^2$. An analogous expression applies by replacing the computational bases with any bases on the two subsystems (Problem 14.8).

REFERENCES

[1] M. S. Leifer and R. W. Spekkens, "Towards a formulation of quantum theory as a causally neutral theory of Bayesian inference," *Physical Review A*, vol. 88, no. 5, art. 052130, 2013.

[2] C. H. Bennett, D. P. DiVincenzo, J. A. Smolin, and W. K. Wootters, "Mixed-state entanglement and quantum error correction," *Physical Review A*, vol. 54, no. 5, art. 3824, 1996.

[3] R. Horodecki, P. Horodecki, M. Horodecki, and K. Horodecki, "Quantum entanglement," *Reviews of Modern Physics*, vol. 81, no. 2, art. 865, 2009.

[4] B. Groisman, S. Popescu, and A. Winter, "Quantum, classical, and total amount of correlations in a quantum state," *Physical Review A*, vol. 72, no. 3, art. 032317, 2005.

[5] B. Schumacher and M. A. Nielsen, "Quantum data processing and error correction," *Physical Review A*, vol. 54, no. 4, art. 2629, 1996.

[6] M. M. Wilde, *Quantum Information Theory*. Cambridge: Cambridge University Press, 2017.

[7] M. A. Nielsen and G. Vidal, "Majorization and the interconversion of bipartite states," *Quantum Information & Computation*, vol. 1, no. 1, pp. 76–93, 2001.

15 Bipartite Dynamic Classical and Quantum Information for Pure States

15.1 Overview

Chapter 14 has introduced entanglement as a static property of bipartite quantum systems. In contrast, this chapter investigates important dynamic aspects of entanglement, namely creating entanglement, transforming entanglement, and applying entanglement.

An entangled state cannot be created from an uncorrelated state by using transformations that act locally on the two subsystems. In the language of resource theory, local operations are considered free because they cannot generate entanglement. The creation of entanglement, therefore, requires the application of transformations that act globally on the entire system. This chapter introduces a common class of such global operations, known as controlled quantum gates, and provides specific examples for two-qubit systems.

Once an entangled state is created, it is often distributed among remote agents, each holding one of the two subsystems. In this scenario, global quantum operations are no longer feasible, but the joint state can still be manipulated through local operations applied to each subsystem. These operations can be adaptively chosen based on classical information exchanged between the agents.

This class of distributed protocols, referred to as local operations with classical communications (LOCC), cannot increase entanglement, but they can transform it in ways that can be formalized via majorization. As will be detailed in this chapter, this analysis supports the introduction of measures of entanglement, known as entanglement measures and entanglement monotones, that have the general property of being non-increasing with respect to the application of LOCC.

When entangled subsystems are distributed between two separate agents, they become a valuable resource for performing a range of informational tasks that are impossible with classical information alone. This chapter will focus on superdense coding, as well as on a distributed sampling task, while briefly mentioning also teleportation. Each of these tasks exploits the non-local properties of entanglement to achieve demonstrable improvements over their classical counterparts.

That said, the presence of entanglement also imposes significant constraints on the types of operations that can be performed on quantum systems. As this chapter will explore, no transformation can clone an unknown quantum state, nor is it possible to delete quantum information within a closed quantum system. These limitations have practical applications. For example, the no-cloning theorem serves as a foundational principle in quantum cryptography.

15.1.1 Learning Objectives and Organization

By the end of this chapter, the reader should be able to:

- understand how to prepare an entangled state using controlled quantum gates;
- assess whether a state can be converted into another state by means of LOCC protocols;
- assess whether a state can be converted into another state by means of stochastic LOCC (SLOCC) protocols;
- understand entanglement measures and entanglement monotones;
- understand superdense coding;
- understand the non-local properties of entanglement through the task of distributed sampling;
- and describe the no-cloning theorem and the no-deleting theorem.

15.2 Creating Entanglement

As illustrated in Figure 15.1(a), preparing an uncorrelated state $|\psi^{XY}\rangle = |\psi^X\rangle \otimes |\psi^Y\rangle$ can be produced by having subsystem X prepare state $|\psi^X\rangle$ and subsystem Y prepare state $|\psi^Y\rangle$. Therefore, uncorrelated states can be generated by applying local operations (LOs) independently at each subsystem, without the need for coordination or information exchange between them.

In contrast, as shown in Figure 15.1(b), preparing an entangled state requires a global operation (GO) that acts jointly on both subsystems. Entanglement, in fact, arises from the interaction between the two subsystems, which cannot be achieved

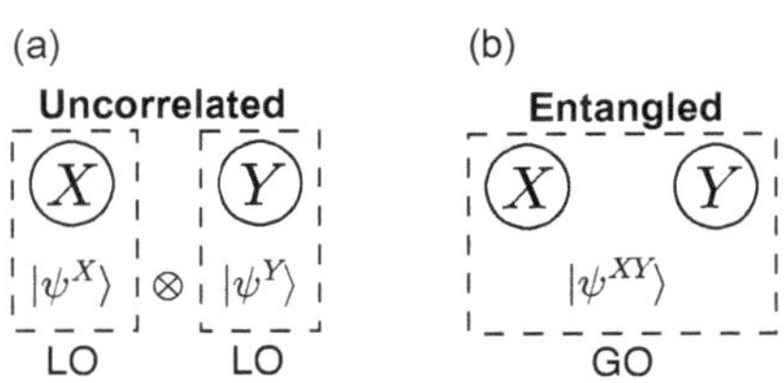

Figure 15.1 Preparing an uncorrelated state $|\psi^{XY}\rangle = |\psi^X\rangle \otimes |\psi^Y\rangle$ only requires local operations (LOs), which are applied independently at each subsystem (a); while preparing an entangled state requires implementing a global operation (GO) that acts jointly on both subsystems (b).

through local operations alone. In practice, global operations can only be performed by an agent with access to both subsystems, or by two local agents that share quantum resources.

This section presents global dynamic transformations that can create entanglement by focusing on the important class of controlled quantum gates. Specific examples will be given by concentrating on two-qubit systems, i.e., on systems with dimensions $N^X = N^Y = 2$, as this is the most common setting for the implementation of quantum computers.

15.2.1 Local Quantum Gates

As discussed in Chapter 10, the evolution of a quantum state in a closed system is governed by dynamic transformations represented by unitary matrices. Unitary matrices that act independently on each subsystem are known as local quantum gates. Accordingly, a local quantum gate for subsystem X is described by an $N^X \times N^X$ unitary matrix U^X, while a local quantum gate for subsystem Y is described by an $N^Y \times N^Y$ unitary matrix U^Y. Following Section 11.3.1, when a local transformation U^X is applied to subsystem X and a local transformation U^Y operates on subsystem Y, the overall transformation acting on a bipartite state $|\psi^{XY}\rangle$ is described by the $N^X N^Y \times N^X N^Y$ unitary matrix $U^{XY} = U^X \otimes U^Y$.

Definition 15.1 Local quantum gates

Given local quantum gates U^X and U^Y and pre-transformation state $|\psi^{XY}\rangle$, the post-transformation state is given by

$$|\psi^{XY}\rangle = (U^X \otimes U^Y)|\phi^{XY}\rangle. \tag{15.1}$$

As a special case, one of the two unitary matrices can be an identity matrix, indicating that a local gate is applied only to one of the two subsystems. For instance, when gate U^X is applied to subsystem X and no operation is applied to subsystem Y, the post-transformation state is

$$|\psi^{XY}\rangle = (U^X \otimes I)|\phi^{XY}\rangle, \tag{15.2}$$

where I is the $N^Y \times N^Y$ identity matrix.

As anticipated, local gates cannot create entanglement.

Key Result 15.1 Local quantum gates cannot create entanglement

If local gates $U^X \otimes U^Y$ are applied to a general uncorrelated state $|\psi^X\rangle \otimes |\psi^Y\rangle$, the post-transformation state

$$(U^X \otimes U^Y)(|\psi^X\rangle \otimes |\psi^Y\rangle) = U^X|\psi^X\rangle \otimes U^Y|\psi^Y\rangle \tag{15.3}$$

is uncorrelated.

15.2.2 Controlled Quantum Gates

Creating entanglement requires applying a global quantum gate that acts on the entire system. A global quantum gate is described by an $N^X N^Y \times N^X N^Y$ unitary matrix U^{XY} that cannot be expressed in the form $U^{XY} = U^X \otimes U^Y$ for any local quantum gates U^X and U^Y. An important class of global gates for bipartite systems is given by controlled quantum gates. A controlled quantum gate performs a joint operation across both subsystems, using one subsystem as a "controller" to dictate the local gate applied to the other subsystem. To elaborate, without loss of generality, consider subsystem X as the controller and subsystem Y as the controlled subsystem.

Definition 15.2 Controlled quantum gates
Fix a set $\{U_x^Y\}_{x\in[N^X]}$ of N^X possible $N^Y \times N^Y$ unitary matrices specifying local gates for subsystem Y. A controlled gate with subsystem X as the controller and Y as the controlled subsystem is then defined as

$$U^{XY} = \sum_{x\in[N^X]} |x\rangle\langle x| \otimes U_x^Y, \tag{15.4}$$

where $\{|x\rangle\}_{x\in[N^X]}$ is the computational basis for subsystem X.

The controlled gate U^{XY} in (15.4) corresponds to the $N^X N^Y \times N^X N^Y$ block-diagonal unitary matrix

$$U^{XY} = \begin{bmatrix} U_0^Y & 0 & \cdots & 0 \\ 0 & U_1^Y & \cdots & 0 \\ \vdots & \vdots & \ddots & \vdots \\ 0 & \cdots & \cdots & U_{N^X-1}^Y \end{bmatrix}, \tag{15.5}$$

where 0 denotes the $N^Y \times N^Y$ all-zero matrix.

Controlled gates from X to Y may be specified in a **quantum circuit diagram** as shown in Figure 15.2. In the diagram, each wire represents one of the two subsystems, and a dot is placed on the controlling subsystem, while the box on the controlled subsystem reports the constituent unitary transformations $\{U_x^Y\}_{x\in[N^X]}$.

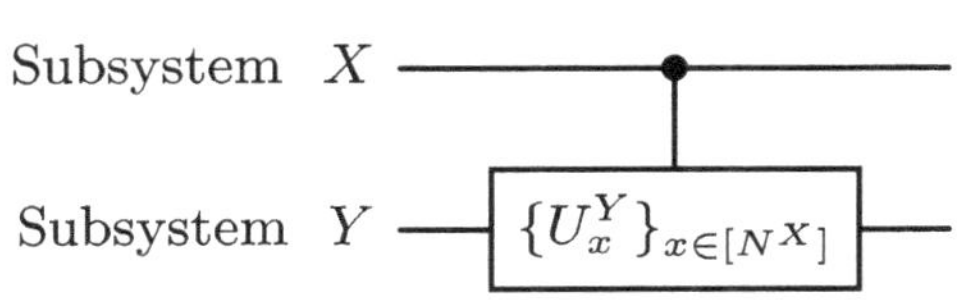

Figure 15.2 Quantum circuit diagram representing a controlled quantum gate with controlling subsystem X and controlled subsystem Y.

Intuitively, a controlled gate (15.4) applies unitary matrix U_x^Y locally to subsystem Y "when subsystem X is in state $|x\rangle$." The quotation marks here emphasize that this should not be taken as a literal description of how the unitary U^{XY} operates, since subsystem X may not actually be in any specific state $|x\rangle$ of the computational basis. In fact, subsystem X could be in a general superposition of computational basis states, and more generally, it may not even have a pure state if the pre-transformation joint state is entangled.

To understand more formally the effect of the controlled gate (15.4) on a pre-transformation bipartite state $|\psi^{XY}\rangle$, the following result, proved and generalized in Appendix 15.A, is instrumental.

Key Result 15.2 State decomposition via the partial inner product

Any state $|\psi^{XY}\rangle$ can be expressed as

$$|\psi^{XY}\rangle = \sum_{x\in[N^X]} \sqrt{\beta_x}|x\rangle \otimes |\phi_x^Y\rangle, \tag{15.6}$$

where $\{|x\rangle\}_{x\in[N^X]}$ is the computational basis for subsystem X; $\beta = [\beta_0, \ldots, \beta_{N^X-1}]^T$ is a probability distribution; and the $N^Y \times 1$ state vectors $\{|\phi_x^Y\rangle\}_{x\in[N^X]}$ are generally non-orthogonal. Furthermore, the states $\{|\phi_x^Y\rangle\}_{x\in[N^X]}$ can be obtained via the partial inner product

$$(\langle x| \otimes I)|\psi^{XY}\rangle = \sqrt{\beta_x}|\phi_x^Y\rangle \tag{15.7}$$

for all $x \in [N^X]$.

The decomposition (15.6) writes the state $|\psi^{XY}\rangle$ as a superposition of N^X uncorrelated states of the form $|x\rangle \otimes |\phi_x^Y\rangle$, where $\{|x\rangle\}_{x\in[N^X]}$ is the computational basis for subsystem X. The states $|\phi_x^Y\rangle$ are obtained via the operation (15.7), which amounts to an inner product with the computational-basis vector $|x\rangle$ that is applied only to subsystem X.

Using the decomposition (15.6), the post-transformation state produced by the controlled gate (15.4) is evaluated as

$$U^{XY}|\psi^{XY}\rangle = \sum_{x\in[N^X]} \sqrt{\beta_x}|x\rangle \otimes U_x^Y|\phi_x^Y\rangle. \tag{15.8}$$

Therefore, the output state $U^{XY}|\psi^{XY}\rangle$ is a superposition of N^X uncorrelated states in which each xth separable state, $|x\rangle \otimes U_x^Y|\phi_x^Y\rangle$, is obtained by applying the local transformation U_x^Y to subsystem Y. From this expression, one can say that the controlled gate operates across N^X distinct **superposition branches**, with each xth branch applying a different local gate U_x^Y to subsystem Y, depending on the corresponding state of subsystem X.

15.2.3 Two-Qubit Controlled Gates

Examples of local quantum gates for qubit systems, known as single-qubit gates, were given in Chapter 10. These include the Pauli matrices P_1, P_2, and P_3, as well as the Hadamard gate H. This subsection introduces two examples of two-qubit gates, showing also that they can produce entangled qubits starting from uncorrelated qubits.

A first prominent example of a controlled gate for two-qubit systems is the **controlled NOT (CNOT) gate**. Taking the qubit X as the controlling qubit, the CNOT gate is defined by the global unitary matrix (15.4) with the local unitary matrices $\{U_0^Y = I,\ U_1^Y = P_1\}$, i.e.,

$$U^{XY} = |0\rangle\langle 0| \otimes I + |1\rangle\langle 1| \otimes P_1 = \begin{bmatrix} 1 & 0 & 0 & 0 \\ 0 & 1 & 0 & 0 \\ 0 & 0 & 0 & 1 \\ 0 & 0 & 1 & 0 \end{bmatrix}. \tag{15.9}$$

Accordingly, the CNOT gate leaves the state of subsystem Y unchanged – applying the identity $U_0^Y = I$ – for the superposition branch in which qubit X is in state $|0\rangle$; while applying the Pauli X gate $U_1^Y = P_1$ for the superposition branch in which qubit X is in state $|1\rangle$. In a quantum circuit diagram, the CNOT gate is illustrated as in Figure 15.3.

The CNOT gate is not symmetric with respect to the choice of the controller. In fact, the CNOT gate with qubit Y as the controller is given by the unitary matrix

$$U^{XY} = I \otimes |0\rangle\langle 0| + P_1 \otimes |1\rangle\langle 1| = \begin{bmatrix} 1 & 0 & 0 & 0 \\ 0 & 0 & 0 & 1 \\ 0 & 0 & 1 & 0 \\ 0 & 1 & 0 & 0 \end{bmatrix}, \tag{15.10}$$

which is different from (15.9).

Another useful two-qubit gate is the **controlled Z (CZ) gate**, which is defined by the local unitary matrices $\{U_0^Y = I,\ U_1^Y = P_3\}$, yielding the global unitary transformation

$$U^{XY} = |0\rangle\langle 0| \otimes I + |1\rangle\langle 1| \otimes P_3 = \begin{bmatrix} 1 & 0 & 0 & 0 \\ 0 & 1 & 0 & 0 \\ 0 & 0 & 1 & 0 \\ 0 & 0 & 0 & -1 \end{bmatrix}. \tag{15.11}$$

Interestingly, unlike the CNOT, the CZ gate is symmetric with respect to the choice of the controlling and controlled qubits. In fact, the CNOT gate with Y as

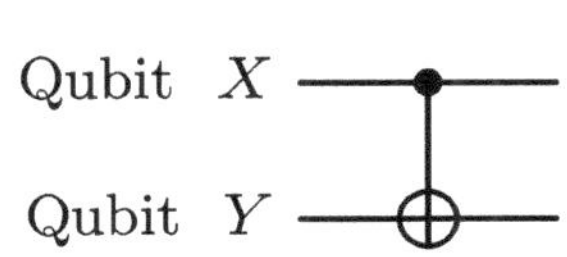

Figure 15.3 Quantum circuit diagram of a CNOT gate with controlling qubit X and controlled qubit Y.

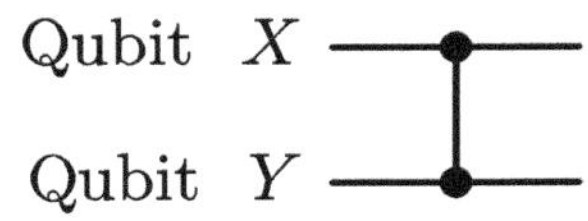

Figure 15.4 Quantum circuit diagram of a CZ gate.

the controlled qubit corresponds to the unitary

$$U^{XY} = I \otimes |0\rangle\langle 0| + P_3 \otimes |1\rangle\langle 1| = \begin{bmatrix} 1 & 0 & 0 & 0 \\ 0 & 1 & 0 & 0 \\ 0 & 0 & 1 & 0 \\ 0 & 0 & 0 & -1 \end{bmatrix}, \tag{15.12}$$

which is equal to (15.11). Given its symmetry, the CZ gate is typically denoted in quantum circuit diagrams as in Figure 15.4.

The next example demonstrates that, starting from an uncorrelated state, a two-qubit gate can create entanglement.

Example 15.1

Consider the quantum circuit in Figure 15.5, in which the bipartite system starts in the uncorrelated state $|+\rangle \otimes |0\rangle$. Then, a CNOT gate is applied with qubit X as the controller. Using (15.9), the output state is obtained as

$$\begin{aligned} U^{XY}(|+\rangle \otimes |0\rangle) &= (|0\rangle\langle 0| \otimes I + |1\rangle\langle 1| \otimes P_1)(|+\rangle \otimes |0\rangle) \\ &= |0\rangle\langle 0|+\rangle \otimes |0\rangle + |1\rangle\langle 1|+\rangle \otimes P_1|0\rangle \\ &= \frac{1}{\sqrt{2}}(|0\rangle \otimes |0\rangle + |1\rangle \otimes |1\rangle) \\ &= |\Phi_+^{XY}\rangle, \end{aligned} \tag{15.13}$$

which is a Bell state, and thus it is maximally entangled.

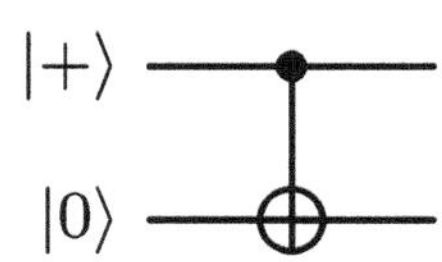

Figure 15.5 Quantum circuit generating a Bell state starting from an uncorrelated state.

In the quantum circuit in Figure 15.5, entanglement is generated by preparing the control qubit X in the superposition state $|+\rangle = \frac{1}{\sqrt{2}}(|0\rangle + |1\rangle)$. In this setup, the superposition branch corresponding to $|0\rangle$ leaves the state of subsystem Y unchanged, while the branch corresponding to $|1\rangle$ flips the state of subsystem Y to $|1\rangle$. This creates correlation between the states of the two subsystems, resulting in entanglement.

Example 15.2

One can also create entanglement via the CZ gate. To see this, consider the quantum circuit in Figure 15.6, in which the initial state $|+\rangle \otimes |+\rangle$ is uncorrelated. After the CZ gate, using (15.11), a maximally entangled state is produced as

$$U^{XY}(|+\rangle \otimes |0\rangle) = (|0\rangle\langle 0| \otimes I + |1\rangle\langle 1| \otimes P_3)(|+\rangle \otimes |+\rangle)$$
$$= |0\rangle\langle 0|+\rangle \otimes |+\rangle + |1\rangle\langle 1|+\rangle \otimes P_3|+\rangle$$
$$= \frac{1}{\sqrt{2}}(|0\rangle \otimes |+\rangle + |1\rangle \otimes |-\rangle). \quad (15.14)$$

Note that this state can be converted to the Bell state $|\Phi_+^{XY}\rangle$ by applying the Hadamard gate H to the first qubit or to the second qubit.

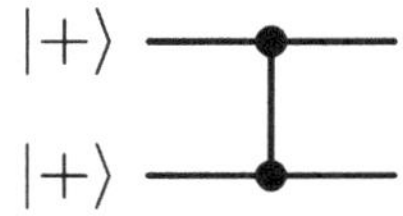

Figure 15.6 Quantum circuit generating a maximally entangled state from an uncorrelated state.

15.3 Transforming Entanglement

Suppose now that an entangled state has been created. How do the entanglement properties of the state evolve when local operations are applied to the two subsystems? This section will demonstrate that local unitary transformations do not alter the degree of entanglement, while more general local operations combined with classical communication can modify entanglement but they cannot increase it.

15.3.1 Local Unitary Transformations Do Not Change Entanglement

To begin, the following result demonstrates that local unitary transformations do not alter any of the correlation measures introduced in Chapter 14, including the entanglement entropy. This result mirrors the classical property of local channels, discussed in Section 12.7, according to which local channels cannot create or enhance correlations in classical systems. Viewing entanglement as a resource, local unitary transformations are hence free operations, as they do not increase the entanglement level.

To formalize this result, write as $\lambda(|\psi^{XY}\rangle)$ the Schmidt probability vector for state $|\psi^{XY}\rangle$.

Key Result 15.3 Local unitary transformations do not change classical and quantum correlations

For any pure state $|\psi^{XY}\rangle$, the Schmidt probability vector $\lambda(|\psi^{XY}\rangle)$ is not modified by the application of any local unitary transformation $U^X \otimes U^Y$, i.e.,

$$\lambda(|\psi^{XY}\rangle) = \lambda(U^X \otimes U^Y|\psi^{XY}\rangle). \quad (15.15)$$

This implies that the quantum mutual information, the maximum measured mutual information, the entanglement entropy, and the quantum conditional entropy are left unchanged by a local unitary transformation.

As an example of the result (15.15), considering the entanglement entropy, one has the equality

$$\begin{aligned}\mathrm{E}(X;Y)_{|\psi^{XY}\rangle} &= \mathrm{H}(\lambda(|\psi^{XY}\rangle)) \\ &= \mathrm{H}(\lambda(U^X \otimes U^Y|\psi^{XY}\rangle)) \\ &= \mathrm{E}(X;Y)_{U^X\otimes U^Y|\psi^{XY}\rangle}. \end{aligned} \tag{15.16}$$

To demonstrate the equality (15.15), write the pre-transformation state $|\psi^{XY}\rangle$ by using the Schmidt decomposition $|\psi^{XY}\rangle = \sum_{z\in[r]}\sqrt{\lambda_z}|u_z^X\rangle \otimes |u_z^Y\rangle$, with Schmidt probability vector λ, Schmidt left states $\{|u_z^X\rangle\}_{z\in[r]}$, and Schmidt right states $\{|u_z^Y\rangle\}_{z\in[r]}$. Applying a local unitary transformation $U^X \otimes U^Y$, the state $|\psi^{XY}\rangle$ is transformed into

$$U^X \otimes U^Y|\psi^{XY}\rangle = \sum_{z\in[r]} \sqrt{\lambda_z} U^X|u_z^X\rangle \otimes U^Y|u_z^Y\rangle. \tag{15.17}$$

The post-transformation local states $\{|w_z^X\rangle = U^X|u_z^X\rangle\}_{z\in[r]}$ are mutually orthogonal, and so are the states $\{|w_z^Y\rangle = U^Y|u_z^Y\rangle\}_{z\in[r]}$ (see Section 10.3). Therefore, the transformed state (15.17) can be expressed as

$$U^X \otimes U^Y|\psi^{XY}\rangle = \sum_{z\in[r]} \sqrt{\lambda_z}|w_z^X\rangle \otimes |w_z^Y\rangle. \tag{15.18}$$

This demonstrates that the transformed state has the same Schmidt probability vector λ of the original state $|\psi^{XY}\rangle$, while the Schmidt left and right states are rotated by the respective local unitary matrices as compared to the pre-transformation state $|\psi^{XY}\rangle$.

15.3.2 Local Unitary Equivalence ⏭

As formalized by the equality (15.17), local unitary transformations can turn an arbitrary state $|\psi^{XY}\rangle$ into any other state with the same dimensions N^X and N^Y as long as the transformed state has the same Schmidt probability vector. Indeed, given an arbitrary state with Schmidt decomposition $|\psi^{XY}\rangle = \sum_{z\in[r]}\sqrt{\lambda_z}|u_z^X\rangle \otimes |u_z^Y\rangle$, one can obtain any other state $|\phi^{XY}\rangle = \sum_{z\in[r]}\sqrt{\lambda_z}|w_z^X\rangle \otimes |w_z^Y\rangle$ with the same Schmidt probability vector λ by applying the local unitary transformation $U^X \otimes U^Y$ with

$$U^X = \sum_{x\in[N^X]} |w_x^X\rangle\langle u_x^X| \tag{15.19}$$

and

$$U^Y = \sum_{y\in[N^Y]} |w_y^Y\rangle\langle u_y^Y|. \tag{15.20}$$

In (15.19) and (15.20), the sets of Schmidt states have been completed in an arbitrary way to form orthonormal bases.

This observation motivates the following definition.

Definition 15.3 Local unitary equivalent states

Any two states with the same dimensions N^X and N^Y and equal Schmidt probability vectors are said to be local unitary equivalent. In fact, any one of the two states can be obtained from the other via a local unitary transformation.

This definition entails that all the $N^X N^Y \times 1$ states that share the same Schmidt probability vector λ are local unitary equivalent to the state

$$|\Phi_\lambda^{XY}\rangle = \sum_{z\in[r]} \sqrt{\lambda_z}|z^X\rangle \otimes |z^Y\rangle, \tag{15.21}$$

where $\{|z^X\rangle\}_{z\in[N^X]}$ and $\{|z^Y\rangle\}_{z\in[N^Y]}$ denote the computational bases of subsystems X and Y, respectively. In this sense, the state $|\Phi_\lambda^{XY}\rangle$ represents the entire class of states of dimensions N^X and N^Y with Schmidt probability vector λ. Indeed, any state within the class can be transformed into $|\Phi_\lambda^{XY}\rangle$ using only local unitary operations, and conversely, the state $|\Phi_\lambda^{XY}\rangle$ can be used to generate any state in the class through local unitary transformations.

When $N^X = N^Y = D$, one may write the representative state (15.21), without causing any ambiguity, via the simpler notation

$$|\Phi_\lambda^{XY}\rangle = \sum_{z\in[r]} \sqrt{\lambda_z}|z\rangle \otimes |z\rangle = \sum_{z\in[r]} \sqrt{\lambda_z}|zz\rangle, \tag{15.22}$$

where $\{|z\rangle\}_{z\in[D]}$ represents the computational basis of the D-dimensional space.

Example 15.3

For two-qubit systems, i.e., for $N^X = N^Y = D = 2$, all states with Schmidt vector $\lambda = [\lambda_0, \lambda_1]^T$ are local unitary equivalent to the state

$$|\psi^{XY}\rangle = \sqrt{\lambda_0}|00\rangle + \sqrt{\lambda_1}|11\rangle. \tag{15.23}$$

Furthermore, as a special case, all maximally entangled two-qubit states are local unitary equivalent to the Bell state $|\Phi_+^{XY}\rangle$.

Example 15.4

All maximally entangled states with dimensions N^X and N^Y are local unitary equivalent to the state

$$|\Phi_{\max}^{XY}\rangle = \frac{1}{\sqrt{N_{\min}}} \sum_{z\in[N_{\min}]} |z^X\rangle \otimes |z^Y\rangle. \tag{15.24}$$

Thus, state $|\Phi_{\max}^{XY}\rangle$ can be viewed as representative of all maximally entangled states with the same dimensions N^X and N^Y. In particular, for two-qudit states with dimensions $N^X = N^Y = D$, the representative state (15.24) can be expressed as

$$|\Phi_{\max}^{XY}\rangle = \frac{1}{\sqrt{D}} \sum_{z\in[D]} |zz\rangle, \tag{15.25}$$

which corresponds to the Bell state $|\Phi_+^{XY}\rangle$ for $D = 2$, i.e., for two-qubit states.

As discussed, any joint state with Schmidt probability vector λ can be obtained using local unitary transformations from the representative state $|\Phi_\lambda^{XY}\rangle$ in (15.21). The next result, which can be proved by direct calculation, shows that the state $|\Phi_\lambda^{XY}\rangle$ can be expressed formally by using the maximally entangled state $|\Phi_{\max}^{XY}\rangle$ in (15.24) (see Problem 15.1). It is noted that the relationships below do not correspond to physical transformations but only to convenient mathematical expressions that can be useful in reasoning about entangled states.

Key Result 15.4 Expressing representative states via maximally entangled states

Given any Schmidt probability vector λ, the state $|\Phi_\lambda^{XY}\rangle$ in (15.21) can be expressed as a function of the corresponding maximally entangled state $|\Phi_{\max}^{XY}\rangle$ as

$$|\Phi_\lambda^{XY}\rangle = (\Lambda^{1/2} \otimes I)|\Phi_{\max}^{XY}\rangle, \tag{15.26}$$

where Λ is an $N^X \times N^X$ diagonal matrix with first r diagonal elements equal to the corresponding Schmidt probabilities $\{\lambda_z\}_{z\in[r]}$. Analogously, one can also write

$$|\Phi_\lambda^{XY}\rangle = (I \otimes \Lambda^{1/2})|\Phi_{\max}^{XY}\rangle, \tag{15.27}$$

where Λ is an $N^Y \times N^Y$ diagonal matrix with first r diagonal elements equal to the corresponding Schmidt probabilities $\{\lambda_z\}_{z\in[r]}$.

15.3.3 Local Operations with Classical Communications

As discussed in this section, local unitary transformations do not change the correlations present in a pure state, as they preserve the Schmidt probability vector. However, local unitary transformations represent only one type of local transformation. They are, in fact, a special case of the larger class of local operations with classical communications (LOCC). LOCC protocols generalize local unitary transformations in the following ways.

- **Local POVMs:** Each subsystem can perform general local POVMs, with unitary transformations being a specific example of POVMs.
- **Classical communications:** The subsystems can exchange classical information, transmitting the outcomes of their respective local measurements to one another.
- **Sequential protocol:** The process unfolds in multiple steps, allowing each subsystem to adjust its measurement settings based on the classical information received from the other subsystem at each step.

Intuitively, neither local quantum measurements nor classical communication can increase entanglement, as neither operation enables global quantum interactions between the two subsystems. In fact, as demonstrated next, LOCC protocols are incapable of enhancing the level of entanglement. More generally, the subsequent analysis will establish the exact conditions under which one quantum state can be transformed into another via LOCC protocols.

In its most general form, an LOCC protocol operates along a finite number of steps, with two subsystems taking turns in carrying out a local POVM and in communicating the classical outcome of the measurement to the other subsystem. The POVM selected by a subsystem may generally depend on all the previous measurement outputs at both subsystems. More formally, an LOCC protocol can be described as follows.

Definition 15.4 LOCC protocols

As illustrated in Figure 15.7(a), an LOCC proceeds via the following steps.

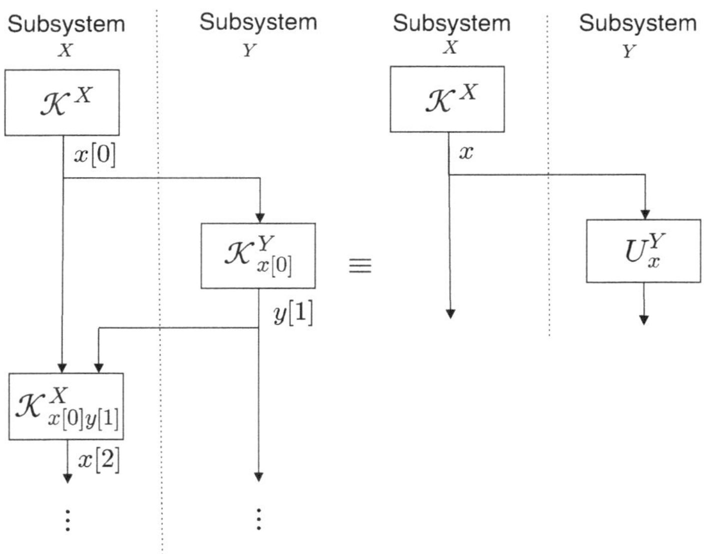

Figure 15.7 Illustration of a general LOCC protocol (on the left), along with a simplified LOCC protocol consisting of only two steps (on the right).

Step 0: At the first step, indexed as 0, subsystem X makes a local POVM $\mathcal{K}^X$, and communicates the classical outcome of the measurement $x[0]$ to the other subsystem.

Step 1: In the second step, indexed as 1, subsystem Y chooses a local POVM measurement $\mathcal{K}^X_{x[0]}$ based on the message $x[0]$ received from subsystem Y, applies it, and sends the classical outcome $y[1]$ back to subsystem X.

Steps 2, 3, . . .: In subsequent steps, the subsystems alternate, each choosing a local POVM based on all prior measurement outcomes, performing the measurement, and communicating the resulting classical outcome to the other subsystem.

Before analyzing the capabilities of an LOCC protocol, the following remarks are in place. First, the choice of the starting subsystem is irrelevant, as the first subsystem – here subsystem X – can always apply a trivial POVM incorporating only the identity matrix. Second, LOCC protocols include as a special case local unitary transformations. In fact, a local unitary transformation $U^X \otimes U^Y$ can be implemented via an LOCC in which subsystem X applies a POVM $\mathcal{K}^X = \{U^X\}$, while subsystem Y applies the POVM $\mathcal{K}_0^Y = \{U^Y\}$. Note that in this case no classical information is exchanged since the trivial POVM applied by subsystem X always returns the output $x[0] = 0$.

Having defined LOCC protocols, the following question arises: When can a state $|\psi^{XY}\rangle$ be turned into another state $|\phi^{XY}\rangle$ via an LOCC protocol? In general, the output state produced by an LOCC protocol is random, since it depends on the specific sequence of measurement outputs produced by the local POVMs. Accordingly, an LOCC transformation is defined as follows.

Definition 15.5 LOCC transformations
Given an initial state $|\psi^{XY}\rangle$, the notation

$$|\psi^{XY}\rangle \underset{\text{LOCC}}{\longrightarrow} |\phi^{XY}\rangle \tag{15.28}$$

indicates the existence of an LOCC protocol that produces state $|\phi^{XY}\rangle$ with probability one, meaning that the final state is achieved regardless of the sequence of measurement outcomes from the POVMs.

15.3.4 Majorization and Entanglement

Intuitively, a state $|\phi^{XY}\rangle$ can be obtained from a state $|\psi^{XY}\rangle$ via an LOCC protocol with probability one if and only if the state $|\phi^{XY}\rangle$ does not possess more entanglement than the state $|\psi^{XY}\rangle$. In fact, entanglement cannot be created without global quantum operations. To rigorously formalize this intuition, it is necessary to resort to tools from majorization, as detailed in this subsection. Referring to Section 9.3 for a more complete introduction to majorization, this subsection starts with a brief review of the key concepts.

Given an $r \times 1$ probability vector λ, write as $\lambda^{\downarrow}$ the $r \times 1$ vector containing the elements of vector λ listed from largest to smallest, i.e., $\lambda^{\downarrow} = [\lambda_0^{\downarrow}, \lambda_1^{\downarrow}, \ldots, \lambda_{r-1}^{\downarrow}]^T$ with $\lambda_0^{\downarrow} \geq \lambda_0^{\downarrow} \geq \cdots \geq \lambda_{r-1}^{\downarrow}$. The kth Ky Fan norm of vector λ is the sum of the k largest elements of the vector, i.e., $||\lambda||_{(k)} = \sum_{x\in[k]} \lambda_x^{\downarrow}$. A probability vector λ' is said to majorize another probability vector λ'' of the same dimension, which is denoted as

$$\lambda' \succ \lambda'', \tag{15.29}$$

if one has the inequalities

$$||\lambda'||_{(k)} \geq ||\lambda''||_{(k)} \text{ for all } k \in [N-1]. \tag{15.30}$$

Majorization defines a partial ordering on the space of probability distributions λ of the same dimension that is stronger than the entropy. This is in the sense that the majorization condition $\lambda' \succ \lambda''$ implies the inequality $H(\lambda') \leq H(\lambda'')$, i.e.,

$$\lambda' \succ \lambda'' \Longrightarrow H(\lambda') \leq H(\lambda''), \tag{15.31}$$

while there are pairs of probability vectors λ' and λ'' with $H(\lambda') < H(\lambda'')$ for which the majorization condition $\lambda' \succ \lambda''$ does not hold. Note that the two inequality signs go in different directions in the implication (15.31): if a distribution majorizes another distribution, its entropy is smaller than (or cannot exceed) the entropy of the other distribution.

With this background in place, consider now two states $|\psi^{XY}\rangle$ and $|\phi^{XY}\rangle$ with respective Schmidt probability vectors $\lambda(|\psi^{XY}\rangle)$ and $\lambda(|\phi^{XY}\rangle)$. According to the definition of majorization, the condition

$$\lambda(|\psi^{XY}\rangle) \prec \lambda(|\phi^{XY}\rangle) \tag{15.32}$$

implies that the Schmidt probability vector of $|\psi^{XY}\rangle$ is more spread out. Since a more evenly distributed Schmidt probability vector corresponds to higher entanglement, condition (15.32) indicates that state $|\psi^{XY}\rangle$ is at least as entangled as $|\phi^{XY}\rangle$. In fact, applying (15.31), this condition leads to the following result.

Key Result 15.5 Majorization implies an ordering of the entanglement entropy
Given two joint states $|\psi^{XY}\rangle$ and $|\phi^{XY}\rangle$ with the same dimensions and with respective Schmidt probability vectors $\lambda(|\psi^{XY}\rangle)$ and $\lambda(|\phi^{XY}\rangle)$, the following implication holds:

$$\lambda(|\psi^{XY}\rangle) \prec \lambda(|\phi^{XY}\rangle) \Longrightarrow E(X;Y)_{|\psi^{XY}\rangle} \geq E(X;Y)_{|\phi^{XY}\rangle}. \tag{15.33}$$

Thus, majorization provides a more refined criterion for ordering states by their entanglement as compared to the entanglement entropy.

This property is illustrated by the following example.

Example 15.5

Consider the following quantum states with $N^X = N^Y = 3$ dimensions:

$$|\psi^{XY}\rangle = \sqrt{0.6}|00\rangle + \sqrt{0.3}|11\rangle + \sqrt{0.1}|22\rangle$$
$$\text{and } |\phi^{XY}\rangle = \frac{1}{\sqrt{2}}(|00\rangle + |11\rangle). \tag{15.34}$$

The inequality $E(X;Y)_{|\psi^{XY}\rangle} = 1.2955 \geq E(X;Y)_{|\phi^{XY}\rangle} = 1$ is satisfied, and thus state $|\psi^{XY}\rangle$ is more entangled than $|\phi^{XY}\rangle$ in terms of the entanglement entropy.

However, the Schmidt probability vectors $\lambda(|\psi^{XY}\rangle) = [0.6, 0.3, 0.1]^T$ and $\lambda(|\phi^{XY}\rangle) = [0.5, 0.5, 0]^T$ do not satisfy the majorization condition $\lambda(|\psi^{XY}\rangle) \prec \lambda(|\phi^{XY}\rangle)$. Therefore, while the state $|\psi^{XY}\rangle$ is more entangled than $|\phi^{XY}\rangle$ based on the entanglement entropy, this conclusion does not hold when using majorization as the criterion for ordering entanglement levels.

15.3.5 What LOCC Can and Cannot Do

Building on the established relationship between entanglement and majorization, the following result provides a necessary and sufficient condition for LOCC transformations.

Key Result 15.6 Necessary and sufficient condition for LOCC transformations
A state $|\phi^{XY}\rangle$ can be obtained from a state $|\psi^{XY}\rangle$ via LOCC if and only if the Schmidt probability vector $\lambda(|\psi^{XY}\rangle)$ of the original state $|\psi^{XY}\rangle$ is majorized by the Schmidt probability vector $\lambda(|\phi^{XY}\rangle)$ of the transformed state, i.e.,

$$|\psi^{XY}\rangle \underset{\text{LOCC}}{\longrightarrow} |\phi^{XY}\rangle \Longleftrightarrow \lambda(|\psi^{XY}\rangle) \prec \lambda(|\phi^{XY}\rangle), \tag{15.35}$$

or equivalently, using (15.30),

$$|\psi^{XY}\rangle \underset{\text{LOCC}}{\longrightarrow} |\phi^{XY}\rangle \Longleftrightarrow ||\lambda(|\psi^{XY}\rangle)||_{(k)} \leq ||\lambda(|\phi^{XY}\rangle)||_{(k)} \text{ for all } k \in [N-1]. \tag{15.36}$$

This fundamental result demonstrates that majorization offers the appropriate method for ordering states by their entanglement levels when adopting the operational perspective that greater entanglement should enable an LOCC transformation. In this context, ranking entanglement based solely on entanglement entropy is insufficient to draw conclusions about LOCC transformability.

From a formal perspective, the feasible LOCC transformations described in (15.35) exhibit the opposite direction of majorization compared to the transformations between probability distributions discussed in Section 9.3. Specifically, as shown in Section 9.3, a distribution p can be transformed into a distribution q through random shuffling if and only if the majorization condition $p \succ q$ holds. This reflects the fact that doubly stochastic transformations increase randomness.

In contrast, an LOCC transformation cannot increase the "randomness" of the Schmidt probability vector, as this would imply an increase in the entanglement of the resulting state. In fact, LOCC transformations only operate in the reverse direction: the post-LOCC Schmidt probability vector must majorize the pre-LOCC Schmidt probability vector.

A proof of the equivalence (15.35) can be derived via the following steps.

1. **Proving the sufficiency of two-step LOCC protocols:** The first part of the proof shows that, as illustrated in Figure 15.7, any final state produced by an LOCC protocol can also be achieved through a two-step LOCC protocol. Therefore, focusing on two-step LOCC protocols is sufficient without loss of generality. A sketch of this proof is provided in Appendix 15.B.
2. **Proving that a two-step LOCC protocol $|\psi^{XY}\rangle \underset{\text{LOCC}}{\longrightarrow} |\phi^{XY}\rangle$ exists if and only if $\lambda(|\psi^{XY}\rangle) \prec \lambda(|\phi^{XY}\rangle)$:** The final part of the proof demonstrates that a two-step LOCC transformation can produce the state $|\phi^{XY}\rangle$ from $|\psi^{XY}\rangle$ with probability one if and only if the majorization condition $\lambda(|\psi^{XY}\rangle) \prec \lambda(|\phi^{XY}\rangle)$ is satisfied. The proof of this result is discussed in Appendix 15.C.

15.3.6 E-Bit

The characterization (15.35) of LOCC-based transformations has the following important consequence.

> **Key Result 15.7 Universality of maximally entangled pure states**
> Given a maximally entangled state $|\Phi_{\max}^{XY}\rangle$ one can obtain any pure state $|\phi^{XY}\rangle$ with the same dimensions via LOCC, i.e.,
>
> $$|\Phi_{\max}^{XY}\rangle \underset{\text{LOCC}}{\longrightarrow} |\phi^{XY}\rangle. \tag{15.37}$$

This result follows since a maximally entangled state has a uniform Schmidt probability vector $\lambda = u$, which is majorized by any other Schmidt probability vector λ of the same dimension, i.e., $u \prec \lambda$. This establishes maximally entangled states as a **universal entanglement currency**, meaning that if subsystems X and Y share a maximally entangled state, they can generate any other state of the same dimensions via LOCC. Because of its unique role as entanglement currency, a maximally entangled state in a qubit system – more specifically, the Bell state $|\Phi_{+}^{XY}\rangle$ – is commonly referred to as an **e-bit**.

15.3.7 What SLOCC Can and Cannot Do ⏭

LOCC transformability requires the final state to be produced with probability one. Given that general LOCC protocols may produce different states depending on the measurement outcomes, it is useful to consider the following broader definition of transformability via LOCC protocols.

> **Definition 15.6 Transformations via stochastic LOCC (SLOCC)**
> Given an initial state $|\psi^{XY}\rangle$, the notation
>
> $$|\psi^{XY}\rangle \underset{\text{SLOCC}}{\longrightarrow} \{(|\phi_z^{XY}\rangle, p_z)\}_{z\in[M]} \tag{15.38}$$
>
> indicates that there exists an LOCC that produces state $|\phi_z^{XY}\rangle$ with probability p_z for all $z \in [M]$, where M is some integer. The mapping (15.38) between state $|\psi^{XY}\rangle$ and the ensemble of pure states $\{(|\phi_z^{XY}\rangle, p_z)\}_{z\in[M]}$ is referred to as a stochastic LOCC (SLOCC) transformation.

According to this definition, if no LOCC exists that can transform state $|\psi^{XY}\rangle$ into state $|\phi^{XY}\rangle$ with probability one, one may still ask whether the transformation is possible with a probability greater than zero but less than one. A proof of the following result can be found in the Recommended Resources.

Key Result 15.8 Necessary and sufficient condition for SLOCC transformations
A state $|\phi^{XY}\rangle$ can be obtained from a state $|\psi^{XY}\rangle$ via SLOCC with positive probability if and only if the Schmidt rank $r(|\phi^{XY}\rangle)$ of the final state $|\phi^{XY}\rangle$ does not exceed the Schmidt rank $r(|\psi^{XY}\rangle)$ of the initial state $|\psi^{XY}\rangle$, i.e.,

$$|\psi^{XY}\rangle \underset{\text{SLOCC}}{\longrightarrow} |\phi_z^{XY}\rangle \text{ with } p_z > 0 \Longleftrightarrow r(|\psi^{XY}\rangle) \geq r(|\phi^{XY}\rangle). \tag{15.39}$$

The Schmidt rank is a weaker measure of entanglement as compared to majorization. Consequently, this result demonstrates that achieving a transformation with non-zero probability imposes less stringent conditions than converting a state with probability one.

Example 15.6

For qubit states, any entangled state has maximum Schmidt rank $r = 2$. Therefore, any entangled two-qubit state can be transformed into any other state with some non-zero probability. In this sense, all two-qubit entangled states form an **equivalence class** under SLOCC as they are **interconvertible**, although with generally different probabilities. However, if one requires transformations with probability one, only states that are local unitary equivalent belong to the same class.

The result (15.39) pertains to one specific post-transformation state. Alternatively, one can ask if there is an SLOCC transformation $|\psi^{XY}\rangle \underset{\text{SLOCC}}{\longrightarrow} \{(|\phi_z^{XY}\rangle, p_z)\}_{z\in[M]}$ that returns an entire target ensemble $\{(|\phi_z^{XY}\rangle, p_z)\}_{z\in[M]}$.

Key Result 15.9 Necessary and sufficient condition for SLOCC transformations
An ensemble $\{(|\phi_z^{XY}\rangle, p_z)\}_{z\in[M]}$ can be obtained from a state $|\psi^{XY}\rangle$ via SLOCC if and only if the following conditions hold:

$$|\psi^{XY}\rangle \underset{\text{SLOCC}}{\longrightarrow} \{(|\phi_z^{XY}\rangle, p_z)\}_{z\in[M]}$$
$$\Longleftrightarrow ||\lambda(|\psi^{XY}\rangle)||_{(k)} \leq \sum_{z\in[M]} p_z ||\lambda(|\phi_z^{XY}\rangle)||_{(k)} \text{ for all } k \in [N-1]. \tag{15.40}$$

The conditions (15.40) generalize (15.36), which is recovered for a zero-entropy ensemble, i.e., for $M = 1$.

By (15.40), in order for the transformation $|\psi^{XY}\rangle \underset{\text{SLOCC}}{\longrightarrow} \{(|\phi_z^{XY}\rangle, p_z)\}_{z\in[M]}$ to be possible, the output states $|\phi_z^{XY}\rangle$ must be less entangled than the initial state $|\psi^{XY}\rangle$ on average with respect to distribution p_z. The reader is referred to the Recommended Resources for proofs and discussions.

15.3.8 LOCCs with Multiple State Copies ⏭

One of the main conclusions of this section is that the entanglement entropy provides only necessary, but not sufficient, conditions for the transformability of a state into another via LOCC. The situation turns out to be different if one considers a number of copies $S_{|\psi^{XY}\rangle}$ of state $|\psi^{XY}\rangle$. The question of interest in this setting is: How many copies $S_{|\phi^{XY}\rangle}$ of state $|\phi^{XY}\rangle$ can one create via LOCC starting with $S_{|\psi^{XY}\rangle}$ copies of state $|\psi^{XY}\rangle$?

In the asymptotic regime of an arbitrarily large number of copies $S_{|\psi^{XY}\rangle}$, one can prove that the entanglement entropy provides also a sufficient condition to assess the extent to which state $|\psi^{XY}\rangle$ can be transformed into another state $|\phi^{XY}\rangle$ via LOCC. Specifically, as $S_{|\psi^{XY}\rangle}$ goes to infinity, the maximum number of copies produced for state $|\phi^{XY}\rangle$ is known to become approximately equal to

$$S_{|\phi^{XY}\rangle} = S_{|\psi^{XY}\rangle} \frac{\mathrm{E}(X;Y)_{|\psi^{XY}\rangle}}{\mathrm{E}(X;Y)_{|\phi^{XY}\rangle}}. \tag{15.41}$$

Therefore, the ratio $\mathrm{E}(X;Y)_{|\psi^{XY}\rangle}/\mathrm{E}(X;Y)_{|\phi^{XY}\rangle}$ between the entanglement entropies dictates the efficiency with which LOCC can transform state $|\psi^{XY}\rangle$ into $|\phi^{XY}\rangle$.

A key consequence of this conclusion is that, while a single copy of a state can only maintain or decrease entanglement, multiple copies allow for an increase in the entanglement entropy. This is achieved by converting a larger number of less entangled states into fewer copies of a more entangled state, a process known as **entanglement distillation**.

15.4 Entanglement Measures and Entanglement Monotones ⏭

As established in Section 15.3, if a state $|\psi^{XY}\rangle$ can be transformed into a state $|\phi^{XY}\rangle$ via LOCC, then the state $|\psi^{XY}\rangle$ must possess a higher degree of entanglement than $|\phi^{XY}\rangle$. This observation implies that any valid measure of entanglement should exhibit a monotonic behavior under LOCC protocols. Specifically, if $|\psi^{XY}\rangle$ can be converted to $|\phi^{XY}\rangle$ through LOCC, any valid measure of entanglement applied to $|\phi^{XY}\rangle$ must not exceed the value assigned to $|\psi^{XY}\rangle$. Thanks to the implication (15.33), the entanglement entropy exemplifies this monotonicity property with respect to LOCC transformations. Building on this insight, this section introduces more general classes of entanglement metrics.

15.4.1 Entanglement Measures

Generalizing the entanglement entropy, entanglement measures cannot increase as a result of an LOCC transformation.

Definition 15.7 Entanglement measures

An entanglement measure $\mathcal{E}(X;Y)_{|\psi^{XY}\rangle}$ is any function of a bipartite state $|\psi^{XY}\rangle$ that does not increase as the result of an LOCC protocol. This property is expressed by the implication

$$|\psi^{XY}\rangle \underset{\text{LOCC}}{\longrightarrow} |\phi^{XY}\rangle \Longrightarrow \mathcal{E}(X;Y)_{|\psi^{XY}\rangle} \geq \mathcal{E}(X;Y)_{|\phi^{XY}\rangle}. \tag{15.42}$$

The equivalence between LOCC transformability and the majorization of the Schmidt probability vector established in (15.35) yields the following alternative definition of entanglement measures.

Key Result 15.10 Entanglement measures and Schur concavity

A function $\mathcal{E}(X;Y)_{|\psi^{XY}\rangle}$ is an entanglement measure if and only if it is a Schur concave function of the Schmidt probability vector $\lambda(|\psi^{XY}\rangle)$ of state $|\psi^{XY}\rangle$.

In fact, as discussed in Section 9.4, a function $f(\lambda)$ is Schur concave if it is majorization-decreasing in the sense that it satisfies the inequality $f(\lambda') \leq f(\lambda'')$ for any two probability vectors λ' and λ'' with $\lambda' \succ \lambda''$.

Example 15.7

Since the Shannon, Rényi, and Tsallis entropies are all Schur concave functions of the underlying probability vector, any entropy $\mathrm{H}_\alpha(\lambda)$ of the Schmidt probability vector λ is a valid entanglement measure.

Example 15.8

Consider the function

$$\mathcal{E}_k(X;Y)_{|\psi^{XY}\rangle} = 1 - ||\lambda(|\psi^{XY}\rangle)||_{(k)}, \tag{15.43}$$

which sums the smallest $N-k$ elements of the Schmidt probability vector $\lambda(|\psi^{XY}\rangle)$. Note that this quantity can be thought of as the complement of the kth Ky Fan norm $||\lambda(|\psi^{XY}\rangle)||_{(k)}$.

As anticipated by the notation in (15.43), the quantity $\mathcal{E}_k(X;Y)_{|\psi^{XY}\rangle}$ is an entanglement measure for any $k \in [N-1]$. This follows directly from the characterization (15.36) of LOCC transformability. In fact, the inequality (15.36) implies that the function (15.43) cannot increase as a result of an LOCC transformation.

Although entanglement measures are monotonic under LOCC transformations, the possibility of an LOCC transformation cannot be determined solely by comparing the entanglement measure values of two states. Specifically, for any two states $|\psi^{XY}\rangle$ and $|\phi^{XY}\rangle$, the inequality $\mathcal{E}(X;Y)_{|\psi^{XY}\rangle} \geq \mathcal{E}(X;Y)_{|\phi^{XY}\rangle}$ does not imply that the transformation $|\psi^{XY}\rangle \overset{\text{LOCC}}{\longrightarrow} |\phi^{XY}\rangle$ is feasible. Nevertheless, while a single entanglement measure may be insufficient to determine LOCC transformability, examining multiple, carefully selected entanglement measures together might provide enough information.

Example 15.9

Consider the entanglement measures (15.43) for all $k \in [N-1]$. The inequalities

$$\mathcal{E}_k(X;Y)_{|\psi^{XY}\rangle} \geq \mathcal{E}_k(X;Y)_{|\phi^{XY}\rangle} \tag{15.44}$$

for all $k \in [N-1]$ imply – and in fact are equivalent to – the feasibility of the transformation $|\psi^{XY}\rangle \underset{\text{LOCC}}{\longrightarrow} |\phi^{XY}\rangle$. This follows directly from the equivalence relation (15.36).

15.4.2 Entanglement Monotones

Entanglement measures account only for changes that occur with probability one as a result of the application of an LOCC protocol. As detailed in Section 15.3, more generally, an SLOCC protocol transforms a state $|\psi^{XY}\rangle$ into an ensemble $\{(|\phi_z^{XY}\rangle, p_z)\}_{z\in[M]}$. SLOCC transformations motivate the following definition of a different class of entanglement metrics.

Definition 15.8 Entanglement monotone

An entanglement monotone is any function $\mathfrak{E}(X;Y)_{|\psi^{XY}\rangle}$ of state $|\psi^{XY}\rangle$ that cannot increase *on average* as a result of the application of an SLOCC protocol. This property is expressed by the implication

$$|\psi^{XY}\rangle \underset{\text{SLOCC}}{\longrightarrow} \{(|\phi_z^{XY}\rangle, p_z)\}_{z\in[M]} \Longrightarrow \mathfrak{E}(X;Y)_{|\psi^{XY}\rangle} \geq \sum_{z\in[M]} p_z \mathfrak{E}(X;Y)_{|\phi_z^{XY}\rangle}. \tag{15.45}$$

Condition (15.45) indicates that the average entanglement level following an SLOCC transformation – the right-hand side of the inequality in (15.45) – cannot exceed the initial entanglement level, represented by the left-hand side of the inequality in (15.45).

Every entanglement monotone qualifies as an entanglement measure. Indeed, the inequalities in (15.45) that define an entanglement monotone imply the conditions in (15.42) for entanglement measures in the case of a zero-entropy ensemble. However, the converse does not hold; not all entanglement measures are entanglement monotones. Consequently, an entanglement monotone offers a more stringent criterion for quantifying entanglement than a general entanglement measure.

Example 15.10

The quantity in (15.43) is an entanglement monotone for all $k \in [N-1]$, which may be written as

$$\mathfrak{E}_k(X;Y)_{|\psi^{XY}\rangle} = 1 - ||\lambda(|\psi^{XY}\rangle)||_{(k)}. \tag{15.46}$$

In fact, by the equivalence (15.40), the SLOCC transformation $|\psi^{XY}\rangle \underset{\text{SLOCC}}{\longrightarrow} \{(|\phi_z^{XY}\rangle, p_z)\}_{z\in[M]}$ implies the inequality (15.45) for the metric $\mathfrak{E}_k(X;Y)_{|\psi^{XY}\rangle}$, i.e.,

$$\mathfrak{E}_k(X;Y)_{|\psi^{XY}\rangle} \geq \sum_{z\in[M]} p_z \mathfrak{E}_k(X;Y)_{|\phi^{XY}\rangle}. \tag{15.47}$$

What is more, the equivalence (15.40) also shows the converse result that, if this inequality holds for all $k \in [N-1]$, then the SLOCC transformation $|\psi^{XY}\rangle \underset{\text{SLOCC}}{\longrightarrow} \{(|\phi_z^{XY}\rangle, p_z)\}_{z\in[M]}$ is feasible.

15.5 Applying Entanglement: Superdense Coding ⏭

Thus far, entanglement has been introduced as a formal property of a quantum state that reflects the correlations between measurement outcomes. This section and the next explore two practical perspectives on entanglement, illustrating how it enables informational tasks that are unattainable with classical resources. The first task, covered in this section, is superdense coding – a communication protocol between two agents, each holding one part of a bipartite system. The second task, discussed in the next section, is distributed sampling (see Section 12.8 for a classical version of the task).

15.5.1 Classical Communications over a Quantum Channel

This section examines a scenario in which agent X and agent Y are connected via an ideal quantum channel capable of transmitting a single qubit without distortion. A straightforward – though unremarkable! – use of this quantum channel allows agent X to send 1 bit of classical information to agent Y. To this end, agent X transmits a qubit in state $|0\rangle$ if the classical 1-bit message is 0 and a qubit in state $|1\rangle$ if the classical bit equals 1. Since the two states are orthogonal and the quantum channel is ideal, agent Y can decode the classical bit correctly with probability one (see Section 5.6).

Although using a qubit to transmit just 1 bit may seem inefficient, it is actually impossible to transmit more than 1 bit of classical information without risking decoding errors at agent Y. In fact, transmitting multiple bits would require encoding classical multi-bit messages into non-orthogonal qubit states, which agent Y could not perfectly distinguish. Therefore, through a single-qubit quantum channel, agent X can send at most 1 bit of classical information to agent Y. Furthermore, this limitation remains even if the agents share classical randomness, enabling them to randomize their encoding and decoding strategies.

Assume now that, as illustrated in Figure 15.8, the agents share a Bell state $|\Phi_+^{XY}\rangle$, with the first qubit at agent X and the second qubit at agent Y. As will be concluded, the shared Bell state can act effectively as a second parallel channel, allowing agent X to transmit 2 bits of information to agent Y. Note the stark contrast with shared classical randomness: while shared classical random bits do not increase the capacity of the channel between the two agents, a pair of entangled qubits does.

15.5.2 Superdense Coding

To support the transmission of two bits, **superdense coding** operates as follows. As detailed in Table 15.1, depending on the message consisting of two classical bits, agent X applies a local single-qubit gate to subsystem X, i.e., to the first qubit of the two-qubit Bell state. For instance, if agent X wishes to send the bits (0, 1), it applies the Pauli X gate P_1 to its qubit. A concise form to describe the encoding table is that, given the two bits (a, b) with $a, b \in \{0, 1\}$, agent X applies the gate $P_3^a P_1^b$, where P_3 is the Pauli Z gate.

After applying the relevant transformation in the table, agent X sends the first qubit of the Bell pair, i.e., subsystem X, to agent Y. This way, agent Y now has

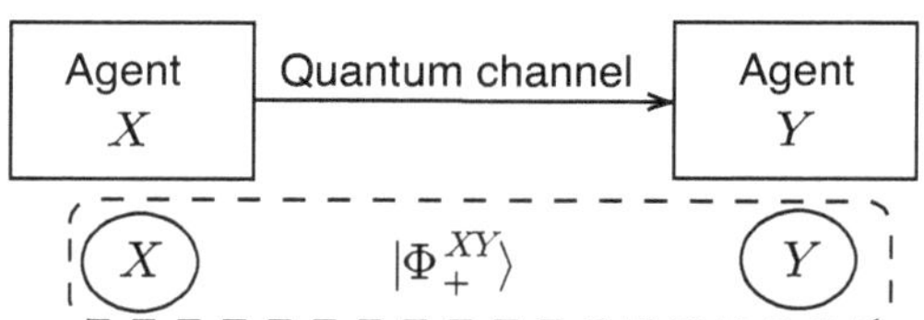

Figure 15.8 In superdense coding, the two communicating agents share a Bell state $|\Phi_+^{XY}\rangle$, with the first qubit at agent X and the second qubit at agent Y. Using this resource, along with a single-qubit noiseless channel, agent X can send 2 bits of classical information to agent Y.

Table 15.1 Operation applied by agent *X* in superdense coding

Message	Transformation
0, 0	I
0, 1	P_1
1, 0	P_3
1, 1	$P_3P_1 = \begin{bmatrix} 0 & 1 \\ -1 & 0 \end{bmatrix}$

access to the post-transformation state of the two qubits. By Table 15.1, the post-transformation states are

$$\begin{aligned}
|\phi_{0,0}^{XX}\rangle &= |\Phi_{+}^{XY}\rangle,\\
|\phi_{0,1}^{XX}\rangle &= (P_1 \otimes I)|\Phi_{+}^{XY}\rangle = \frac{1}{\sqrt{2}}(|10\rangle + |01\rangle) = |\Psi_{+}^{XY}\rangle,\\
|\phi_{1,0}^{XX}\rangle &= (P_3 \otimes I)|\Phi_{+}^{XY}\rangle = \frac{1}{\sqrt{2}}(|00\rangle - |11\rangle) = |\Phi_{-}^{XY}\rangle,\\
|\phi_{1,1}^{XX}\rangle &= (P_3 P_1 \otimes I)|\Phi_{+}^{XY}\rangle = \frac{1}{\sqrt{2}}(|01\rangle - |10\rangle) = |\Psi_{-}^{XY}\rangle,
\end{aligned} \tag{15.48}$$

where the subscript indicates the message encoded by agent X.

Since the four post-transformation states correspond to the Bell states, they are orthogonal. Therefore, they can be perfectly distinguished by agent Y via a joint measurement in the Bell basis $\{|\Phi_{+}^{XY}\rangle, |\Psi_{+}^{XY}\rangle, |\Phi_{-}^{XY}\rangle, |\Psi_{-}^{XY}\rangle\}$ (see Section 14.4.3). This confirms that the shared Bell state acts as an additional channel between the agents, enabling a single-qubit channel to transmit two classical bits instead of one.

15.5.3 Quantum Teleportation

As a brief remark, superdense coding has an important dual task known as **teleportation**. In teleportation, agent X has an unknown state $|\psi^X\rangle$ that it wishes to transmit to agent Y. However, two agents are connected only by a classical channel with a capacity of 2 bits. This channel is clearly insufficient to carry the qubit.

However, if the two agents share a Bell state as in superdense coding, it turns out that it is possible to transfer the state $|\psi^X\rangle$ to agent Y. This task is accomplished via the quantum teleportation protocol, for which the reader is referred to the Recommended Resources.

15.6 Applying Entanglement: Distributed Sampling ⏭

What makes correlations arising from entanglement different from classical correlations? As illustrated by superdense coding in Section 15.5, entanglement is a resource that can be used to carry out informational tasks that would be impossible if one substituted the entangled subsystems with classically correlated subsystems. To further elaborate on the operational differences between classical and quantum correlations, this section describes a task that helps define more precisely the non-local properties of entangled subsystems.

15.6.1 Entanglement-Aided Distributed Sampling

In the distributed sampling problem of interest, there are two agents, X and Y, which hold subsystems X and Y, respectively. The two quantum subsystems are in the joint state $|\psi^{XY}\rangle$. As illustrated in Figure 15.9(a), agent X selects at random a POVM measurement setting $\mathcal{K}^X$ from a set of possible measurement settings; and

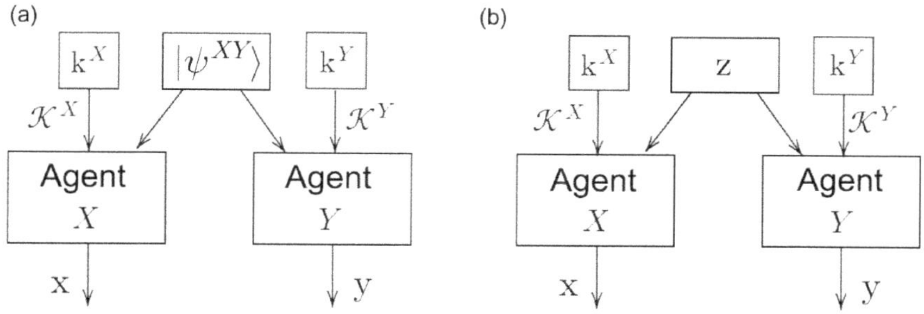

Figure 15.9 Agent X selects at random a POVM measurement setting $\mathcal{K}^X$ from a set of possible measurement settings; and subsystem Y also selects a POVM measurement setting $\mathcal{K}^Y$ from a set of possible measurement settings. Can the resulting joint distribution of measurement settings and measurement outputs for the system in part (a), in which the agents share an entangled quantum state, be simulated by the system in part (b), in which the agents only share classical randomness?

subsystem Y also selects a POVM measurement setting $\mathcal{K}^Y$ from a set of possible measurement settings. Importantly, the two measurement settings are selected independently, and no communication is allowed between the two agents.

Given the selected measurement settings $\mathcal{K}^X$ and $\mathcal{K}^Y$, agent X applies a local measurement in setting $\mathcal{K}^X$ to subsystem X, while agent Y applies a measurement in setting $\mathcal{K}^Y$ to subsystem Y. As a result of the local measurement in setting $\mathcal{K}^X \otimes \mathcal{K}^Y$, agent X obtains random variable x and agent Y random variable y. The joint distribution of random variables x and y is given by $p^{XY}(|\psi^{XY}\rangle|\mathcal{K}^X \otimes \mathcal{K}^Y)$. The task at hand is concerned with the joint distribution of the measurement settings and of the measurement outputs, which is denoted as $p^{\mathcal{K}^X\mathcal{K}^Y XY}$. Note that the measurement settings are random quantities as they are selected randomly by the two agents.

To elaborate, as in Figure 15.9, introduce random indices k^X and k^Y that run over the possible measurement settings at subsystems X and Y, respectively. Their joint distribution is denoted as

$$p_{k^X k^Y}^{\mathcal{K}^X\mathcal{K}^Y} = p_{k^X}^{\mathcal{K}^X} p_{k^Y}^{\mathcal{K}^Y}, \tag{15.49}$$

where the product on the right-hand side reflects the independence of the selections at the two agents. The measurement distribution $p^{XY}(|\psi^{XY}\rangle|\mathcal{K}^X \otimes \mathcal{K}^Y)$ is also written as $p^{XY}(|\psi^{XY}\rangle|k^X, k^Y)$ so as to emphasize the dependence of the joint distribution of the measurements on the selected indices k^X and k^Y.

Using (15.49) and the chain rule, the four random variables – the measurement setting indices k^X and k^Y, along with the measurement outputs x and y – have a joint distribution $p^{\mathcal{K}^X\mathcal{K}^Y XY}$ given by the product

$$p_{k^X k^Y xy}^{\mathcal{K}^X\mathcal{K}^Y XY} = p_{k^X}^{\mathcal{K}^X} \cdot p_{k^Y}^{\mathcal{K}^Y} \cdot p_{xy}^{XY}(|\psi^{XY}\rangle|k^X, k^Y). \tag{15.50}$$

The key question under study is whether this distribution can be simulated by the classical system in Figure 15.9(b), in which the two agents share only classical randomness and not an entangled quantum system.

15.6.2 Shared Classical Randomness-Aided Distributed Sampling

To address this question, consider now the classical system in Figure 15.9(b), in which again agent X selects at random a measurement setting $\mathcal{K}^X$ and subsystem Y draws independently a measurement setting $\mathcal{K}^Y$. The joint distribution of the corresponding measurement setting indices k^X and k^Y is thus given by (15.49).

Suppose that the two agents can communicate using a classical channel, so that they can inform each other of the respective indices k^X and k^Y. As a result, both agents know the overall setting $\mathcal{K}^X \otimes \mathcal{K}^Y$ and hence also the joint distribution $p^{XY}(|\psi^{XY}\rangle|k^X,k^Y)$.

In this case, since the joint distribution $p^{XY}(|\psi^{XY}\rangle|k^X,k^Y)$ is known by both subsystems, the outputs x and y with joint distribution $p^{XY}(|\psi^{XY}\rangle|k^X,k^Y)$ can be simulated by the classical system in the figure. Indeed, leveraging access to a shared classical random variable $\mathrm{z} \sim p^Z$ with a suitable entropy level, agents X and Y can generate locally the respective variables x and y by following the approach in Section 12.8.

Therefore, as long as the two subsystems can communicate the selected measurement settings, the outputs of a quantum measurement in a given setting $\mathcal{K}^X \otimes \mathcal{K}^Y$ can be perfectly simulated by two classical subsystems that share only classical randomness. In particular, the joint distribution of the variables k^X, k^Y, x, and y produced by the classical system coincides with the joint distribution $p^{\mathcal{K}^X\mathcal{K}^Y XY}$ in (15.50) produced by the quantum system in Figure 15.9(a).

Assume now that the two agents share only classical randomness and that they cannot communicate their respective, randomly selected, measurement settings. Accordingly, agent X does not know the measurement setting $\mathcal{K}^Y$ selected by agent Y, and, similarly, subsystem Y does not know the measurement setting $\mathcal{K}^X$ selected by subsystem X. Can the joint distribution $p^{\mathcal{K}^X\mathcal{K}^Y XY}$ in (15.50) still be simulated by leveraging only classical shared randomness?

A key result in quantum mechanics is that the answer to this question is no. In order to be able to simulate the joint distribution $p^{\mathcal{K}^X\mathcal{K}^Y XY}$ in (15.50), the two agents generally need to have access to entangled quantum subsystems. A proof of this result can be found in standard references (see Recommended Resources). This conclusion suggests again that, in some sense, the presence of entanglement between two subsystems can be thought of as a special type of communication channel.

Both superdense coding and the distributed sampling task studied in this section illustrate the non-local nature of entanglement. Although by the no-signaling principle, the "channel" created by entangled subsystems cannot be used by itself to convey information, it does act like an auxiliary channel for some information-processing tasks.

15.7 What Global Unitary Transformations Cannot Do

In Section 15.2, it was shown that global transformations U^{XY} are needed in order to create entanglement from a separable state. This last section elaborates on two limitations of the types of transformations that are possible on a bipartite quantum system, even when using a global transformation U^{XY}. Specifically, it will be demonstrated that it is not possible to design any unitary transformation U^{XY} that copies an unknown state from one subsystem to another, and that it is also not possible to delete an unknown state.

15.7.1 No-Cloning Theorem

Consider a system with dimensions $N^X = N^Y = D$. The **no-cloning theorem** concerns the setting illustrated in Figure 15.10, in which the bipartite system starts in a state $|\psi^X\rangle \otimes |0\rangle$, where $|\psi^X\rangle$ is an arbitrary unknown state and $|0\rangle$ represents the known initial state for subsystem Y. Note that the initial state $|0\rangle$ of subsystem Y does not affect the conclusions to follow, in the sense that any other known initial state could be selected for subsystem Y. The goal is to find a $D^2 \times D^2$ unitary transformation U^{XY} that transforms the state $|\psi^X\rangle \otimes |0\rangle$ into state $|\psi^X\rangle \otimes |\psi^X\rangle$, no matter what the state $|\psi^X\rangle$ for subsystem X is.

Such a unitary transformation would effectively clone the state $|\psi^X\rangle$ onto the state of subsystem Y. However, the no-cloning theorem states that no such unitary matrix U^{XY} can exist. Therefore, it is not possible to clone an arbitrary state $|\psi^X\rangle$ from subsystem X to subsystem Y.

This result follows directly from the linearity of the evolution of a closed quantum system. To see why, consider as an example a two-qubit system, i.e., $D = 2$. The same steps apply more generally. Suppose, by contradiction, that a unitary matrix U^{XY} could be found that satisfies the equalities

$$U^{XY}(|0\rangle \otimes |0\rangle) = |0\rangle \otimes |0\rangle \text{ and } U^{XY}(|1\rangle \otimes |0\rangle) = |1\rangle \otimes |1\rangle, \tag{15.51}$$

hence cloning the states in the computational basis from subsystem X to subsystem Y. Since matrix U^{XY} must apply universally to any state of subsystem X, the equality $U^{XY}(|+\rangle \otimes |0\rangle) = |+\rangle \otimes |+\rangle$ must also be verified.

However, if the conditions (15.51) were true, by linearity one would also have

$$\begin{aligned} U^{XY}(|+\rangle \otimes |0\rangle) &= U^{XY}\left(\frac{1}{\sqrt{2}}(|0\rangle \otimes |0\rangle + |1\rangle \otimes |0\rangle)\right) \\ &= \frac{1}{\sqrt{2}}(|0\rangle \otimes |0\rangle + |1\rangle \otimes |1\rangle) \\ &= |\Phi_+^{XY}\rangle, \end{aligned} \tag{15.52}$$

which is not the desired state $|+\rangle \otimes |+\rangle$. This shows that a cloning unitary U^{XY} operating as in Figure 15.10 cannot exist.

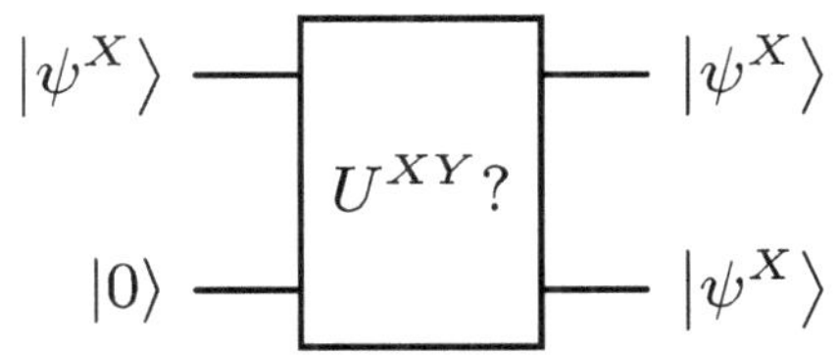

Figure 15.10 Can a global unitary transformation U^{XY} be found that can "clone" an arbitrary, unknown, state $|\psi^X\rangle$?

The no-cloning theorem has several important implications on the constraints imposed by physics on quantum information processing.

- **Quantum state accessibility:** The information that can be obtained from a single copy of an unknown state $|\psi\rangle$ is inherently limited. In fact, information can only be extracted by applying a measurement, which inevitably disturbs the state, rendering it unavailable for further examination. This limitation would be avoided if it were possible to clone the state. Indeed, in a world without the no-cloning theorem, one could produce any number of copies of an unknown quantum state and then perform quantum tomography to estimate the state with arbitrary precision.
- **No faster-than-light communications:** According to the no-signaling theorem, the choice of measurement setting on one subsystem cannot be detected by the other subsystem. This addresses the apparent paradox highlighted by Albert Einstein, which seemed to challenge the principle that information cannot travel faster than light (see Section 14.3). However, if it were possible to clone an unknown state, this principle would no longer hold, as the second subsystem could then detect the measurement basis chosen by the first (Problem 15.5).
- **Quantum cryptography:** Unconditionally secure communication schemes are possible because an eavesdropper cannot copy the quantum information exchanged between two legitimate parties.
- **No broadcasting:** The impossibility of cloning a quantum state also means that it is impossible to broadcast an unknown quantum state to multiple receivers; otherwise, each receiver would effectively receive a clone of the state.

15.7.2 No-Deleting Theorem

The no-cloning theorem can be paraphrased as stating that, in a closed system, it is impossible to increase the information available about a quantum state. Indeed, as discussed, if one could copy a quantum state, it would be possible to estimate the state with arbitrary precision through repeated measurements. Interestingly, a complementary principle holds as well: in a closed system, it is also impossible to decrease the amount of information. Together with the no-cloning theorem, this result – known as the no-deleting theorem – suggests that information is conserved within a closed system.

Specifically, as illustrated in Figure 15.11, the no-deleting theorem states that it is impossible to find a unitary matrix U^{XY} that transforms state $|\psi\rangle \otimes |\psi\rangle$ into state $|\psi\rangle \otimes |0\rangle$ for any arbitrary state $|\psi\rangle$. The reader is referred to the Recommended Resources for further discussions.

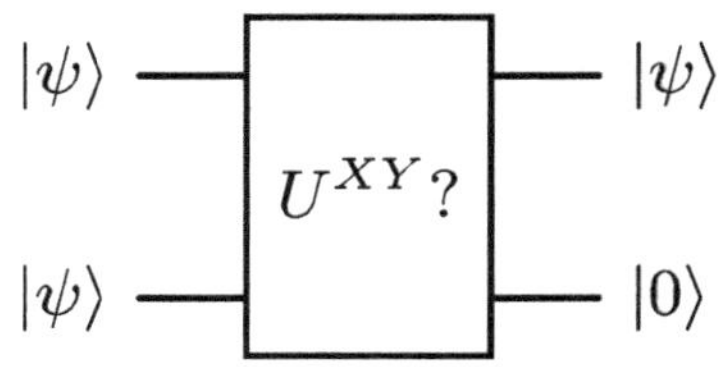

Figure 15.11 Is it possible to find a global unitary matrix U^{XY} that "deletes" an arbitrary, unknown, state $|\psi\rangle$?

15.8 Summary

- Local gates, which are of the form $U^X \otimes U^Y$, cannot generate entanglement, which requires the application of operations that access both subsystems. In fact, a local unitary transformation $U^X \otimes U^Y$ leaves the Schmidt probability vector λ of the original state unchanged, affecting only the Schmidt states.
- Any two states with the same dimensions and with the same Schmidt probability vector are said to be local unitary equivalent, since they can be interconverted by using local unitary transformations. In particular, all maximally entangled states are local unitary equivalent to the state $|\Phi^{XY}_{\max}\rangle = (N_{\min})^{-1/2} \sum_{z\in[N_{\min}]} |z^X\rangle\otimes|z^Y\rangle$.
- Notable examples of global operations include controlled gates, such as the CNOT and CZ gates in two-qubit systems.
- A controlled gate operates across a number of superposition branches, with each branch applying a different local gate to the controlled subsystem depending on the corresponding state of the controlling subsystem.
- Local operations with classical communications (LOCC) can change, but they cannot increase, entanglement, and they incorporate as a special case local unitary transformations.
- A state $|\phi^{XY}\rangle$ can be obtained from a state $|\psi^{XY}\rangle$ via LOCC if and only if the Schmidt probability vector $\lambda(|\psi^{XY}\rangle)$ of the original state $|\psi^{XY}\rangle$ is majorized by the Schmidt probability vector $\lambda(|\phi^{XY}\rangle)$ of the transformed state, i.e., $|\psi^{XY}\rangle \underset{\text{LOCC}}{\longrightarrow} |\phi^{XY}\rangle \Longleftrightarrow \lambda(|\psi^{XY}\rangle) \prec \lambda(|\phi^{XY}\rangle)$.
- A larger entanglement entropy – i.e., $\mathrm{E}(X;Y)_{|\psi^{XY}\rangle} \geq \mathrm{E}(X;Y)_{|\phi^{XY}\rangle}$ – is a necessary but not sufficient condition for a state $|\psi^{XY}\rangle$ to be convertible via LOCC into state $|\phi^{XY}\rangle$. In fact, the implication $\lambda(|\psi^{XY}\rangle) \prec \lambda(|\phi^{XY}\rangle) \Longrightarrow \mathrm{E}(X;Y)_{|\psi^{XY}\rangle} \geq \mathrm{E}(X;Y)_{|\phi^{XY}\rangle}$ does not apply in the reverse direction.
- The equivalence between LOCC transformability and majorization between the Schmidt probability vectors implies that, given a maximally entangled state $|\Phi^{XY}_{\max}\rangle$, one can obtain any pure state $|\phi^{XY}\rangle$ with the same dimensions via an LOCC protocol, i.e., $|\Phi^{XY}_{\max}\rangle \underset{\text{LOCC}}{\longrightarrow} |\phi^{XY}\rangle$.
- An entanglement measure $\mathcal{E}(X;Y)_{|\psi^{XY}\rangle}$ is any function of a bipartite state $|\psi^{XY}\rangle$ that does not increase as a result of an LOCC protocol.
- An entanglement monotone is any function $\mathfrak{E}(X;Y)_{|\psi^{XY}\rangle}$ of state $|\psi^{XY}\rangle$ that does not increase on average as a result of the application of an SLOCC protocol. An entanglement monotone is an entanglement measure, but the opposite is not necessarily true.
- Entanglement can be leveraged to carry out informational tasks, such as superdense coding, that would be impossible with classical resources alone.
- Local measurements of an entangled state can modify the global state of the system. While this property is not sufficient by itself to convey information, it can be harnessed by specific informational tasks to emulate forms of communications between the agents.

- Entanglement imposes constraints on the type of transformations that are feasible in quantum systems. In particular, it is not possible to design a unitary transformation U^{XY} that copies an unknown state from one subsystem to another, and it is also not possible to design a unitary transformation U^{XY} that deletes an unknown state.

15.9 Recommended Resources

A detailed proof of the characterization (15.39) for SLOCC transformability is provided in [1]. Proofs and discussions related to the result (15.40) are covered in [2, 3]. A standard reference for the teleportation protocol is [4]. The result that the system shown in Figure 15.9(a) cannot be classically simulated without communication between agents is demonstrated in [5]. The no-deleting theorem and its implications for the conservation of information are discussed in [6, 7]. A proof of the equivalence of multi-step and two-step LOCC protocols is provided by [8].

PROBLEMS

15.1 Prove the equalities (15.26) and (15.27).

15.2 (a) Describe in words the operation of a controlled gate with subsystem Y as the controller using the concept of superposition branches. (b) Obtain the CNOT gate for two-qubit systems in which qubit Y is the controlling subsystem by writing the corresponding 4×4 unitary matrix U^{XY}. (c) If the input state is $|\psi^{XY}\rangle = |+\rangle\otimes|+\rangle$, what is the output state $U^{XY}|+\rangle\otimes|+\rangle$ for the CNOT gate obtained at point b? (d) Is the output state $U^{XY}|+\rangle\otimes|+\rangle$ entangled?

15.3 Propose quantum circuits that can create all four Bell states starting from separable states.

15.4 Two-qubit gates can destroy entanglement. As an example, for the entangled state $|\Phi_+^{XY}\rangle$, construct a 4×4 unitary matrix that returns the separable state $|0\rangle\otimes|0\rangle$.

15.5 Prove that, in a world without the no-cloning theorem, two subsystems sharing an entangled state could communicate instantaneously, violating the constraint that information cannot travel faster than light.

15.6 (a) For a system with dimensions $N^X = N^Y = D$, prove the equality

$$(I\otimes A)|\Phi_{\max}^{XY}\rangle = (A^T\otimes I)|\Phi_{\max}^{XY}\rangle, \tag{15.53}$$

where I is the $D\times D$ identity matrix and A is any $D\times D$ matrix. (b) Use this result to show the equalities

$$\begin{aligned}(U\otimes U^*)|\Phi_{\max}^{XY}\rangle &= |\Phi_{\max}^{XY}\rangle,\\ (U^*\otimes U)|\Phi_{\max}^{XY}\rangle &= |\Phi_{\max}^{XY}\rangle,\end{aligned} \tag{15.54}$$

for any $D\times D$ unitary matrix U.

15.7 (a) Consider the two-qubit gate

$$U^{XY} = \sum_{x\in[2]}\sum_{y\in[2]} |y\rangle\langle x| \otimes |x\rangle\langle y|. \tag{15.55}$$

Prove that, given an uncorrelated input state $|\psi^{XY}\rangle = |\psi^X\rangle \otimes |\psi^Y\rangle$, this gate produces the swapped state $|\psi^Y\rangle \otimes |\psi^X\rangle$. It is therefore known as the SWAP gate (see also Section 10.3). (b) Generalize the SWAP gate to any system with dimensions $N^X = N^Y = D$. (c) Describe the effect of the SWAP gate on an arbitrary pure state for a system with dimensions $N^X = N^Y = D$ by using the Schmidt decomposition.

15.8 Consider the observable

$$O = \sum_{x\in[2]}\sum_{y\in[2]} |y\rangle\langle x| \otimes |x\rangle\langle y| \tag{15.56}$$

for a two-qubit system. Note that this matrix coincides with the SWAP gate studied in the previous problem. (a) Write the 4×4 matrix O by specifying all the entries. (b) Verify that this is a valid observable (you may use the help of a software). (c) Verify that the eigenbasis is given by vectors $\{|00\rangle, |11\rangle, |\Psi_+^{XY}\rangle, |\Psi_-^{XY}\rangle\}$ with corresponding eigenvalues $\{1, 1, 1, -1\}$. (d) Prove that the expected value of the observable O under an uncorrelated state, namely $\langle O\rangle_{|\psi^X\rangle\otimes|\psi^Y\rangle}$, equals the fidelity of the two states $|\psi^X\rangle$ and $|\psi^Y\rangle$, i.e.,

$$\langle O\rangle_{|\psi^X\rangle\otimes|\psi^Y\rangle} = \mathrm{F}(|\psi^X\rangle, |\psi^Y\rangle) = |\langle\psi^X|\psi^Y\rangle|^2. \tag{15.57}$$

Accordingly, the fidelity of two states can be evaluated by computing the expected value of the SWAP observable for a system prepared in an uncorrelated state involving the two states.

15.9 Show that any two-qubit state can be converted into a state of the form $|\psi^{XY}\rangle = \alpha|00\rangle + \beta|11\rangle$, with real non-negative amplitudes α and β with $\alpha \geq \beta \geq 0$, via the application of local unitary transformations.

Appendix

15.A Partial Inner Product

This appendix shows that a pure state $|\psi^{XY}\rangle$ can be written as a superposition of N^X separable states $\{|v_x^X\rangle \otimes |\phi_x^Y\rangle\}_{x\in[N^X]}$, where the states $\{|v_x^X\rangle\}_{x\in[N^X]}$ form an orthonormal basis for subsystem X, while the states $\{|\phi_x^Y\rangle\}_{x\in[N^X]}$ need not be orthogonal. Specifically, given any basis $\{|v_x^X\rangle\}_{x\in[N^X]}$ for subsystem X, a joint state $|\psi^{XY}\rangle$ can always be expressed as the sum of only N^X separable states as

$$|\psi^{XY}\rangle = \sum_{x\in[N^X]} \sqrt{\beta_x}|v_x^X\rangle \otimes |\phi_x^Y\rangle, \tag{15.58}$$

where the vector $\beta = [\beta_0, \ldots, \beta_{N^X-1}]^T$ forms a probability distribution.

In an analogous way, the proof below will also imply that it is possible to express a pure state $|\psi^{XY}\rangle$ as a superposition of N^Y separable states $\{|\phi_y^X\rangle \otimes |v_y^Y\rangle\}_{x\in[N^X]}$, where the states $\{|v_y^Y\rangle\}_{y\in[N^Y]}$ form an orthonormal basis for subsystem Y, while the states $\{|\phi_y^X\rangle\}_{y\in[N^Y]}$ need not be orthogonal, as

$$|\psi^{XY}\rangle = \sum_{y\in[N^Y]} \sqrt{\beta_y}|\phi_y^X\rangle \otimes |v_y^Y\rangle \tag{15.59}$$

with the vector $\beta = [\beta_0, \ldots, \beta_{N^Y-1}]^T$ forms a probability distribution. It is emphasized that the states $\{|v_x^X\rangle \otimes |\phi_x^Y\rangle\}_{x\in[N^X]}$ in (15.58) are generally not mutually orthogonal, and that the same holds for the states $\{|\phi_y^X\rangle \otimes |v_y^Y\rangle\}_{y\in[N^Y]}$ in (15.59).

To prove this result, focus, without loss of generality, on the equality (15.58). Write as $\alpha_{xy}^{XY} = \langle v_x^X v_y^Y|\psi^{XY}\rangle$ the amplitude of the state $|\psi^{XY}\rangle$ with respect to the basis vector $|v_x^X\rangle \otimes |v_y^Y\rangle$. With this definition, define state vectors $\{|\phi_x^Y\rangle\}_{x\in[N^X]}$ as

$$|\phi_x^Y\rangle = \frac{1}{\sqrt{\beta_x}} \sum_{y\in[N^X]} \alpha_{xy}^{XY}|v_y^Y\rangle, \tag{15.60}$$

where the constant β_x is chosen so as to ensure the normalization of vector $|\phi_x^Y\rangle$ as

$$\beta_x = \sum_{y\in[N^X]} |\alpha_{xy}^{XY}|^2. \tag{15.61}$$

Using (15.60), one immediately obtains the desired decomposition (15.58). In this regard, note that the vector β is indeed a probability distribution given the sum

$$\sum_{x\in[N^X]} \beta_x = \sum_{x\in[N^X]} \sum_{y\in[N^X]} |\alpha_{xy}^{XY}|^2 = 1. \tag{15.62}$$

Furthermore, given the definition (15.60), the product $\sqrt{\beta_x}|\phi_x^Y\rangle$ can be obtained from state $|\psi^{XY}\rangle$ as

$$(\langle v_x^X| \otimes I)|\psi^{XY}\rangle = \sqrt{\beta_x}|\phi_x^Y\rangle. \tag{15.63}$$

The operation on the left-hand side of (15.63) can be interpreted as a partial inner product, in which an inner product is applied to subsystem X, and not to subsystem Y. Intuitively, the vector $\sqrt{\beta_x}|\phi_x^Y\rangle$ represents the "vector amplitude" of the bipartite state $|\psi^{XY}\rangle$ when expressed in terms of the basis $\{|v_x^X\rangle\}_{x\in[N^X]}$ for subsystem X.

In a similar manner, the decomposition (15.59) is obtained by evaluating the "vector amplitude" $\sqrt{\beta_y}|\phi_y^X\rangle$ via the partial inner product over subsystem Y as

$$(I \otimes \langle v_y^Y|)|\psi^{XY}\rangle = \sqrt{\beta_y}|\phi_y^Y\rangle. \tag{15.64}$$

Example 15.11

For the state

$$|\psi^{XY}\rangle = \frac{1}{2\sqrt{2}}((1+i)|00\rangle + (1-i)|01\rangle + (-1+i)|10\rangle + (-1-i)|11\rangle), \tag{15.65}$$

consider the Hadamard basis $\{|v_0^X\rangle = |+\rangle, |v_1^X\rangle = |-\rangle\}$ for subsystem X. The decomposition (15.58) into two separable states is obtained as $|\psi^{XY}\rangle = \sqrt{\beta_0}|+\rangle \otimes |\phi_0^Y\rangle + \sqrt{\beta_1}|-\rangle \otimes |\phi_1^Y\rangle$, with

$$\begin{aligned}\sqrt{\beta_0}|\phi_0^Y\rangle &= (\langle+| \otimes I)|\psi^{XY}\rangle = \frac{1}{\sqrt{2}}((\langle 0| + \langle 1|) \otimes I)|\psi^{XY}\rangle \\ &= \frac{1}{4}((1+i)|0\rangle + (1-i)|1\rangle + (-1+i)|0\rangle + (-1-i)|1\rangle) \\ &= \frac{i}{2}(|0\rangle - |1\rangle) \end{aligned} \tag{15.66}$$

and

$$\begin{aligned}\sqrt{\beta_1}|\phi_1^Y\rangle &= (\langle-| \otimes I)|\psi^{XY}\rangle \\ &= \frac{1}{4}((1+i)|0\rangle + (1-i)|1\rangle - (-1+i)|0\rangle + (1+i)|1\rangle) \\ &= \frac{1}{2}(|0\rangle + |1\rangle), \end{aligned} \tag{15.67}$$

yielding $\beta_0 = \beta_1 = 1/2$.

15.B Proof of Equivalence of Multi-Step and Two-Step LOCC Protocols

To establish the equivalence between multi-step and two-step LOCC protocols, a key intermediate result demonstrates that measurements can be restricted to a single subsystem, as required in the two-step LOCC framework. This is achieved by showing that any measurement performed on one subsystem can equivalently be conducted on the other.

Specifically, it is shown that the bipartite states resulting from measurements on either subsystem are locally unitarily equivalent. Since locally unitarily equivalent states share the same Schmidt probability vector, this result substantiates the claim that, for analyzing the entanglement properties of states produced via LOCC, measurements can, without loss of generality, be assumed to occur entirely within subsystem X.

Having established that all measurements can be confined to one subsystem without loss of generality, the equivalence between multi-step and two-step LOCC protocols follows naturally, as outlined in the Recommended Resources.

The analysis can focus on projective measurements rather than more general POVMs. This simplification is justified under the assumption that local auxiliary systems of arbitrary dimensions are available at both subsystems. As discussed in Chapter 11, POVMs can be implemented as projective measurements on an extended system that includes an auxiliary environment. Under this assumption, one can also consider that the two subsystems have the same dimension D.

Take now an arbitrary state $|\psi^{XY}\rangle$, which is transformed using local unitary transformations into the Schmidt decomposition $|\psi^{XY}\rangle = \sum_{z\in[r]} \sqrt{\lambda_z}|z\rangle \otimes |z\rangle$

in (15.22). For any $D \times D$ projection matrix Π, write the corresponding post-measurement state for a projective measurement at subsystem X as

$$|\phi_X^{XY}\rangle = \frac{1}{\sqrt{C}}(\Pi \otimes I)|\psi^{XY}\rangle, \tag{15.68}$$

where $C = \langle\psi^{XY}|(\Pi \otimes I)|\psi^{XY}\rangle$; and as

$$|\phi_Y^{XY}\rangle = \frac{1}{\sqrt{C}}(I \otimes \Pi)|\psi^{XY}\rangle \tag{15.69}$$

the post-measurement state when the projective measurement is made at subsystem Y. Note that the normalizing constant C is the same for both states, as the equality $\langle\psi^{XY}|(\Pi \otimes I)|\psi^{XY}\rangle = \langle\psi^{XY}|(I \otimes \Pi)|\psi^{XY}\rangle$ can be proved by using the Schmidt decomposition (15.22).

The goal is to show that the two states $|\phi_X^{XY}\rangle$ and $|\phi_Y^{XY}\rangle$ are local unitary equivalent, thus demonstrating that the post-measurement state for a projective measurement on subsystem X is local unitary equivalent to the corresponding post-measurement state implemented at subsystem Y. To this end, it is sufficient to show that the two states $|\phi_X^{XY}\rangle$ and $|\phi_Y^{XY}\rangle$ have the same Schmidt probability vectors.

From Chapter 14, the Schmidt probability vector is obtained as the eigenvalue vector of any of the two reduced density matrices. The reduced density matrix for subsystem X under state $|\phi_X^{XY}\rangle$ is

$$\mathrm{tr}_Y(|\phi_X^{XY}\rangle\langle\phi_X^{XY}|) = \frac{1}{C}\sum_{z\in[r]} \lambda_z \Pi|z\rangle\langle z|\Pi, \tag{15.70}$$

which coincides with the reduced density matrix $\mathrm{tr}_X(|\phi_Y^{XY}\rangle\langle\phi_Y^{XY}|)$ for subsystem Y under state $|\phi_Y^{XY}\rangle$. This concludes the proof.

15.C Proof of the Relation Between LOCC Transformability and Majorization

Based on the result in Appendix 15.B, in order to prove (15.35), one can focus on two-step LOCC protocols. For a proof of the implication $\lambda(|\psi^{XY}\rangle) \prec \lambda(|\phi^{XY}\rangle) \Longrightarrow |\psi^{XY}\rangle \underset{\text{LOCC}}{\longrightarrow} |\phi^{XY}\rangle$, the reader is referred to the Recommended Resources, and the focus in this appendix is on demonstrating the reverse implication $|\psi^{XY}\rangle \underset{\text{LOCC}}{\longrightarrow} |\phi^{XY}\rangle \Longrightarrow \lambda(|\psi^{XY}\rangle) \prec \lambda(|\phi^{XY}\rangle)$.

As illustrated in Figure 15.7(b), a two-step LOCC protocol takes the following form.

1. **Local POVM measurement at subsystem X:** In the first step, subsystem X makes a local POVM measurement $\mathcal{K}^X = \{K_x^X\}_{x\in[M^X]}$, and sends the output of the measurement $\mathrm{x} \in [M^X]$, which is distributed as $\mathrm{x} \sim p(\rho^X|\mathcal{K}^X)$, to subsystem Y.
2. **Local unitary transformation at subsystem Y:** In the second, and last, step, subsystem Y applies a local unitary transformation U_x^Y that is generally chosen as a function of the message x received in the first step.

By construction, this protocol returns the state

$$\frac{1}{\sqrt{p_x^X(\rho^X|\mathcal{K}^X)}}(K_x^X \otimes U_x^Y)|\psi^{XY}\rangle \tag{15.71}$$

with probability $p_x^X(\rho^X|\mathcal{K}^X)$. Since, by assumption, the state $|\phi^{XY}\rangle$ is obtained with probability one, the post-LOCC state (15.71) must equal $|\phi^{XY}\rangle$ for all possible measurement outputs $x \in [M^X]$, i.e.,

$$|\phi^{XY}\rangle = \frac{1}{\sqrt{p_x^X}}(K_x^X \otimes U_x^Y)|\psi^{XY}\rangle, \tag{15.72}$$

writing p_x^X for $p_x^X(\rho^X|\mathcal{K}^X)$.

Using (15.26) and (15.27), the two states can be written without loss of generality as

$$|\psi^{XY}\rangle = (\Lambda_\psi^{1/2} \otimes I)|\Phi_{\max}^{XY}\rangle \tag{15.73}$$

and

$$|\phi^{XY}\rangle = (\Lambda_\phi^{1/2} \otimes I)|\Phi_{\max}^{XY}\rangle, \tag{15.74}$$

where $|\Phi_{\max}^{XY}\rangle$ is a maximally entangled state; I is the $D \times D$ identity matrix; and Λ_ψ and Λ_ϕ are $D \times D$ diagonal matrices with the respective Schmidt probability vectors $\lambda(|\psi^{XY}\rangle)$ and $\lambda(|\phi^{XY}\rangle)$ on the main diagonal.

Thus, plugging (15.73) and (15.74) into (15.72), one obtains the equality

$$(\Lambda_\phi^{1/2} \otimes I)|\Phi_{\max}^{XY}\rangle = \frac{1}{\sqrt{p_x^X}}(K_x^X \Lambda_\psi^{1/2} \otimes U_x^Y)|\Phi_{\max}^{XY}\rangle, \tag{15.75}$$

which must hold for all $x \in [M^X]$. To simplify the analysis, assume now that the Schmidt ranks are maximal, so that matrices Λ_ψ and Λ_ϕ are invertible. This yields the equivalent equalities

$$|\Phi_{\max}^{XY}\rangle = \left(\frac{1}{\sqrt{p_x^X}}\Lambda_\phi^{-1/2} K_x^X \Lambda_\psi^{1/2} \otimes U_x^Y\right)|\Phi_{\max}^{XY}\rangle \tag{15.76}$$

for all $x \in [M^X]$.

In order for the left-hand side of (15.76) to be a valid state, it is necessary for the matrix $\Lambda_\phi^{-1/2} K_x^X \Lambda_\psi^{1/2}/\sqrt{p_x^X}$ to be unitary. Furthermore, by (15.54), the equality (15.76) requires the condition $\Lambda_\phi^{-1/2} K_x^X \Lambda_\psi^{1/2}/\sqrt{p_x^X} = (U_x^Y)^*$, i.e.,

$$K_x^X = \sqrt{p_x^X}\Lambda_\phi^{1/2}(U_x^Y)^* \Lambda_\psi^{-1/2}. \tag{15.77}$$

Writing $V_x^Y = (U_x^Y)^T$ to simplify the notation, the construction (15.77) is valid if the matrices $\mathcal{K}^X = \{K_x^X\}_{x \in [M]}$ satisfy the defining condition of a POVM, namely

$$\sum_{x \in [M]} K_x^\dagger K_x = \sum_{x \in [M]} p_x^X \Lambda_\psi^{-1/2} V_x^Y \Lambda_\phi (V_x^Y)^\dagger \Lambda_\psi^{-1/2} = I, \tag{15.78}$$

where I is the $D \times D$ identity matrix.

The equality (15.78) can, in turn, be expressed as

$$\Lambda_\psi = \sum_{x\in[M]} p_x^X V_x^Y \Lambda_\phi (V_x^Y)^\dagger, \tag{15.79}$$

which is equivalent to the majorization relation $\Lambda_\phi \succ \Lambda_\psi$ (see Section 11.2.2). By Uhlmann's theorem from Section 11.2.2, this majorization condition holds only if the majorization condition $\lambda(\Lambda_\phi) \succ \lambda(\Lambda_\psi)$, or equivalently the condition $\lambda(|\phi^{XY}\rangle) \succ \lambda(|\psi^{XY}\rangle)$, are verified. This concludes the proof.

REFERENCES

[1] W. Dür, G. Vidal, and J. I. Cirac, "Three qubits can be entangled in two inequivalent ways," *Physical Review A*, vol. 62, no. 6, art. 062314, 2000.

[2] D. Jonathan and M. B. Plenio, "Minimal conditions for local pure-state entanglement manipulation," *Physical Review Letters*, vol. 83, no. 7, art. 1455, 1999.

[3] G. Vidal, "Entanglement monotones," *Journal of Modern Optics*, vol. 47, no. 2–3, pp. 355–376, 2000.

[4] M. A. Nielsen and I. L. Chuang, *Quantum Computation and Quantum Information*. Cambridge: Cambridge University Press, 2010.

[5] N. Gisin, *Quantum Chance: Nonlocality, Teleportation and Other Quantum Marvels*. New York: Springer, 2014.

[6] A. K. Pati and S. L. Braunstein, "Impossibility of deleting an unknown quantum state," *Nature*, vol. 404, no. 6774, pp. 164–165, 2000.

[7] M. Horodecki, R. Horodecki, A. Sen, and U. Sen, "Common origin of no-cloning and no-deleting principles conservation of information," *Foundations of Physics*, vol. 35, pp. 2041–2049, 2005.

[8] H.-K. Lo and S. Popescu, "Concentrating entanglement by local actions: Beyond mean values," *Physical Review A*, vol. 63, no. 2, art. 022301, 2001.

16 Bipartite Quantum and Classical Information for Mixed States

16.1 Overview

As discussed in this part of the book, quantum systems exhibit two types of correlations: classical correlations, which can be extracted through an optimized local measurement setting; and quantum correlations, which extend beyond what is possible to extract from any single measurement setting. Furthermore, for pure states, the following facts were established in Chapters 14 and 15.

1. **Equal presence of classical and quantum correlations:** The maximum measured mutual information, which quantifies classical correlations, and the entanglement entropy, which accounts for quantum correlations, are both equal to the entropy $H(\lambda)$ of the Schmidt probability vector λ. Therefore, for pure states, classical correlations cannot exist without quantum correlations.
2. **Equivalence of quantum correlations and entanglement:** Quantum correlations are synonymous with entanglement, a non-local resource that can be harnessed for informational tasks such as superdense coding.

This chapter generalizes these notions and conclusions to quantum mixed states, revealing a significantly more complex picture. In particular, the two properties discussed above for pure states do not hold in the case of mixed states. Instead, the following will be demonstrated.

1. **Classical correlations without quantum correlations:** There exist quantum mixed states that exhibit classical correlations without having any quantum correlations.
2. **Quantum correlations beyond entanglement:** Quantum correlations encompass forms weaker than entanglement. Specifically, some quantum mixed states exhibit non-classical correlations without being entangled. Therefore, the set of quantumly correlated mixed states strictly contains the set of entangled states.

To understand the first property, recall from Chapter 7 that mixed states are represented by density matrices, which can be seen as a generalization of probability distributions. In this context, mixed states in bipartite quantum systems can be interpreted as a quantum analog of joint probability distributions. Similar to joint distributions in classical systems, quantum mixed states can exhibit classical correlations even in the absence of quantum correlations.

The second property is more nuanced, reflecting the existence of mixed states whose preparation requires more than "classical" operations but, unlike entanglement, does not require global quantum operations. These non-classical operations involve shared classical randomness along with local quantum operations. Such states, while not as powerful as entangled states and thus unsuitable for tasks like superdense coding, can offer advantages over purely classically correlated states for some tasks.

To rigorously address these observations, this chapter will extend the definitions of classical and quantum correlations from pure to mixed states. In particular, the quantum discord will be introduced as a new correlation measure that quantifies all quantum correlations – including entanglement, and also correlations beyond entanglement.

16.1.1 Learning Objectives and Organization

By the end of this chapter, the reader should be able to:

- understand the difference between uncorrelated and entangled mixed states;
- evaluate the reduced density matrices of a joint density matrix and the corresponding marginal measurement probabilities;
- understand and evaluate the quantum mutual information as a measure of the overall correlations present in a quantum mixed state;
- partition classical and quantum correlations by evaluating the maximum measured mutual information and the quantum symmetric discord;
- understand the properties of classical–classical and classical–quantum states;
- determine if a quantum mixed state is entangled by using one of several sufficient conditions;
- understand entanglement monotones for quantum mixed states;
- and evaluate the residual uncertainty about a quantum subsystem given another subsystem via the quantum conditional entropy and via the quantum ensemble entropy.

16.2 Bipartite Mixed States

This section reviews the description of bipartite mixed states using density matrices.

16.2.1 Joint Density Matrices

As in previous chapters, the bipartite quantum system consists of subsystem X of dimension N^X and subsystem Y of dimension N^Y. In general, the state of a bipartite quantum system is described by an $N^X N^Y \times N^X N^Y$ density matrix, whose rank distinguishes between pure states, studied in prior chapters, and mixed states, which present classical uncertainty.

Definition 16.1 Joint density matrix

A bipartite mixed state is described by an $N^X N^Y \times N^X N^Y$ density matrix ρ^{XY}, which is a positive semi-definite matrix with trace equal to 1, i.e., $\mathrm{tr}(\rho^{XY}) = 1$. The rank $r^{XY} = \mathrm{rank}(\rho^{XY})$ of a joint density matrix ρ^{XY} ranges in the interval

$$1 \leq r^{XY} \leq N^X N^Y. \tag{16.1}$$

The lower bound $r^{XY} = 1$ indicates a pure state $\rho^{XY} = |\psi^{XY}\rangle\langle\psi^{XY}|$, while a strictly larger rank, $r^{XY} > 1$, corresponds to a mixed state.

When the rank r^{XY} is larger than 1, the density matrix ρ^{XY} describes a state characterized by classical uncertainty, which is realized by ensembles of at least r^{XY} pure states. A special ensemble, encompassing the minimum number of pure states, is the eigenensemble.

Key Result 16.1 Eigendecomposition of a joint density matrix

A quantum mixed state described by a joint density matrix ρ^{XY} with rank r^{XY} can be realized by an eigenensemble

$$\mathcal{E}(\rho^{XY}) = \{(|u_0^{XY}\rangle, p_0), (|u_1^{XY}\rangle, p_1), \ldots, (|u_{r^{XY}-1}^{XY}\rangle, p_{r^{XY}-1})\}, \tag{16.2}$$

where each of the orthonormal $N^X N^Y \times 1$ eigenvectors $|u_z^{XY}\rangle$ is associated with the eigenvalue p_z for $z \in [r^{XY}]$. The eigenvalues $p = [p_0, \ldots, p_{r^{XY}-1}]^T$ form a probability distribution. Using the eigenensemble (16.2), the density matrix ρ^{XY} can be expressed using the eigendecomposition

$$\rho^{XY} = \sum_{z \in [r^{XY}]} p_z |u_z^{XY}\rangle\langle u_z^{XY}|. \tag{16.3}$$

Example 16.1

The eigenvalues $p = [p_0, \ldots, p_{r^{XY}-1}]^T$ of the density matrix ρ^{XY} must not be confused with the Schmidt probability vector λ defined in Section 14.5 for pure states. In fact, the eigenvalues reflect the classical uncertainty associated with the preparation of the state, while the Schmidt probability vector determines the entanglement properties of a pure state.

To see the difference between these two notions, take for example the pure state $\rho^{XY} = |\Phi_+^{XY}\rangle\langle\Phi_+^{XY}|$. The density matrix ρ^{XY} has a single non-zero eigenvalue, $p_0 = 1$, and thus the eigenvalue vector p reduces to a scalar, but the Schmidt probability vector equals the 2×1 uniform distribution $\lambda = u$. This reflects the fact that the pure state $\rho^{XY} = |\Phi_+^{XY}\rangle\langle\Phi_+^{XY}|$ has no classical uncertainty and is maximally entangled.

A joint density matrix ρ^{XY} has $N^X N^Y$ rows and $N^X N^Y$ columns. Therefore, an entry of the matrix ρ^{XY} is identified by a pair of indices (z, z') with $z \in [N^X N^Y]$ and

$z' \in [N^X N^Y]$. As for joint pure states, it is convenient to separate the contributions of the two subsystems by adopting the bi-index notation for both row and column indices.

Recall that the bi-index notation writes an index $z \in [N^X N^Y]$ as xy, where indices $x \in [N^X]$ and $y \in [N^Y]$ are related to z as $z = N^Y x + y$. Accordingly, the (z, z') entry $\rho^{XY}_{z,z'}$ can be identified as the $(xy, x'y')$ entry $\rho^{XY}_{xy,xy'}$ with $z = N^Y x + y$ and $z' = N^Y x' + y'$.

Accordingly, any joint density matrix can be written as

$$\begin{aligned}\rho^{XY} &= \sum_{x\in[N^X]} \sum_{y\in[N^Y]} \sum_{x'\in[N^X]} \sum_{y'\in[N^Y]} \rho^{XY}_{xy,x'y'} |xy\rangle\langle x'y'| \\ &= \sum_{x\in[N^X]} \sum_{y\in[N^Y]} \sum_{x'\in[N^X]} \sum_{y'\in[N^Y]} \rho^{XY}_{xy,x'y'} |x\rangle\langle x'| \otimes |y\rangle\langle y'|,\end{aligned} \tag{16.4}$$

where the outer product

$$\begin{aligned}|xy\rangle\langle x'y'| &= (|x\rangle \otimes |y\rangle)(\langle x'| \otimes \langle y'|) \\ &= |x\rangle\langle x'| \otimes |y\rangle\langle y'|\end{aligned} \tag{16.5}$$

has all zero entries except for a single 1 in position $(xy, x'y')$. It is noted that in (16.5) the expression $|xy\rangle\langle x'y'|$ separates row bi-index, xy, from column bi-index, $x'y'$; while the expression $|x\rangle\langle x'| \otimes |y\rangle\langle y'|$ isolates the contribution pertaining to subsystem X, $|x\rangle\langle x'|$, from the contribution of subsystem Y, $|y\rangle\langle y'|$.

Example 16.2

For two-qubit systems, a generic joint density matrix ρ^{XY} can be written as

$$\begin{aligned}\rho^{XY} &= \begin{bmatrix} \rho_{0,0} & \rho_{0,1} & \rho_{0,2} & \rho_{0,3} \\ \rho_{1,0} & \rho_{1,1} & \rho_{1,2} & \rho_{1,3} \\ \rho_{2,0} & \rho_{2,1} & \rho_{2,2} & \rho_{2,3} \\ \rho_{3,0} & \rho_{3,1} & \rho_{3,2} & \rho_{3,3} \end{bmatrix} \\ &= \begin{bmatrix} \rho_{00,00} & \rho_{00,01} & \rho_{00,10} & \rho_{00,11} \\ \rho_{01,00} & \rho_{01,01} & \rho_{01,10} & \rho_{01,11} \\ \rho_{10,00} & \rho_{10,01} & \rho_{10,10} & \rho_{10,11} \\ \rho_{11,00} & \rho_{11,01} & \rho_{11,10} & \rho_{11,11} \end{bmatrix},\end{aligned} \tag{16.6}$$

with the second expression using the bi-index notation.

Example 16.3

For a system with dimensions $N^X = N^Y = D$, the density matrix corresponding to a maximally entangled pure state $|\Phi^{XY}_{\max}\rangle = (D)^{-1/2} \sum_{z\in[D]} |zz\rangle$ is given as

$$\rho^{XY}_{\max} = |\Phi^{XY}_{\max}\rangle\langle\Phi^{XY}_{\max}| = \frac{1}{D} \sum_{z\in[D]} \sum_{z'\in[D]} |zz\rangle\langle z'z'|. \tag{16.7}$$

Accordingly, the entries of matrix $\rho^{XY}_{\max}$ with coordinates $(zz, z'z')$ are equal to $1/D$ for all pairs $z, z' \in [D]$, while all the other entries are equal to zero.

As a numerical example, for two-qubit systems, the density matrix corresponding to the maximally entangled state $|\Phi^{XY}_{\max}\rangle = |\Phi^{XY}_{+}\rangle$ is given by

$$\rho^{XY}_{\max} = |\Phi^{XY}_{+}\rangle\langle\Phi^{XY}_{+}| = \frac{1}{2}(|00\rangle\langle 00| + |00\rangle\langle 11| + |11\rangle\langle 00| + |11\rangle\langle 11|)$$

$$= \begin{bmatrix} \frac{1}{2} & 0 & 0 & \frac{1}{2} \\ 0 & 0 & 0 & 0 \\ 0 & 0 & 0 & 0 \\ \frac{1}{2} & 0 & 0 & \frac{1}{2} \end{bmatrix}. \tag{16.8}$$

Example 16.4

For $N^X = N^Y = D$, a maximally mixed state is given by the density matrix $\pi^{XY} = I/D^2$, where I is the $D^2 \times D^2$ identity matrix. Accordingly, the maximally mixed state π^{XY} can be written in the form (16.4) as

$$\pi^{XY} = \frac{1}{D^2} \sum_{z\in[D]} \sum_{z'\in[D]} |zz'\rangle\langle zz'|, \tag{16.9}$$

so that the (zz', zz')th elements, i.e., all the elements on the main diagonal, are equal to $1/D$, while all other elements equal zero. Note the critical – but perhaps rather subtle at first glance – difference between (16.9), representing a maximally mixed state, and (16.7), which instead represents a maximally entangled state.

Example 16.5

Isotropic states lie midway between pure maximally entangled states and maximally mixed states. Under an isotropic state, a system is pure and maximally entangled with probability $0 \le \eta \le 1$ and is maximally mixed with probability $1-\eta$. This corresponds to the density matrix

$$\rho^{XY} = \eta|\Phi^{XY}_{\max}\rangle\langle\Phi^{XY}_{\max}| + (1-\eta)\pi^{XY}, \tag{16.10}$$

which is a convex combination of the states $|\Phi^{XY}_{\max}\rangle\langle\Phi^{XY}_{\max}|$ and π^{XY} (see Section 9.2.3 for a definition of convex combination).

Example 16.6

The **Werner state** is an isotropic state for two-qubit systems, in which the maximally entangled state is given by the Bell state $|\Psi^{XY}_{-}\rangle$. Accordingly, the Werner state is defined as

$$\begin{aligned}\rho^{XY} &= \eta|\Psi_-^{XY}\rangle\langle\Psi_-^{XY}| + (1-\eta)\pi^{XY} \\ &= \frac{\eta}{2}(|01\rangle\langle 01| + |10\rangle\langle 10| - |01\rangle\langle 10| - |10\rangle\langle 01|) + \frac{1-\eta}{4}I \\ &= \begin{bmatrix} \frac{1-\eta}{4} & 0 & 0 & 0 \\ 0 & \frac{1+\eta}{4} & -\frac{\eta}{2} & 0 \\ 0 & -\frac{\eta}{2} & \frac{1+\eta}{4} & 0 \\ 0 & 0 & 0 & \frac{1-\eta}{4} \end{bmatrix}. \end{aligned} \tag{16.11}$$

16.2.2 Uncorrelated Mixed States

As introduced in Section 14.2.2, a pure state $|\psi^{XY}\rangle$ is uncorrelated if it can be expressed as the Kronecker product $|\psi^{XY}\rangle = |\psi^X\rangle \otimes |\psi^Y\rangle$ of the individual states $|\psi^X\rangle$ and $|\psi^Y\rangle$ of the two subsystems. Uncorrelated states can be prepared by using local operations, with subsystem X preparing state $|\psi^X\rangle$ and subsystem Y preparing state $|\psi^Y\rangle$. Furthermore, they yield uncorrelated local measurements (x, y) under a local measurement setting $\mathcal{K}^X \otimes \mathcal{K}^Y$:

$$(\mathrm{x}, \mathrm{y}) \sim p(|\psi^X\rangle \otimes |\psi^Y\rangle | \mathcal{K}^X \otimes \mathcal{K}^Y) = p(|\psi^X\rangle | \mathcal{K}^X) p(|\psi^Y\rangle | \mathcal{K}^Y). \tag{16.12}$$

This definition is generalized to mixed states as follows.

Definition 16.2 Uncorrelated mixed states
A bipartite mixed state is said to be uncorrelated if the joint density matrix can be written as

$$\rho^{XY} = \rho^X \otimes \rho^Y \tag{16.13}$$

for some $N^X \times N^X$ density matrix ρ^X for subsystem X and some $N^Y \times N^Y$ density matrix ρ^Y for subsystem Y.

Uncorrelated mixed states hence can be prepared by using local operations, with agent X preparing subsystem X in state ρ^X and agent Y preparing subsystem Y in state ρ^Y.

Example 16.7

The definition (16.13) of uncorrelated mixed states generalizes the definition of uncorrelated pure states. In fact, the density matrix of an uncorrelated pure state $|\psi^X\rangle \otimes |\psi^Y\rangle$ is given by the Kronecker product

$$\begin{aligned}\rho^{XY} &= (|\psi^X\rangle \otimes |\psi^Y\rangle)(\langle\psi^X| \otimes \langle\psi^Y|) \\ &= \underbrace{|\psi^X\rangle\langle\psi^X|}_{\rho^X} \otimes \underbrace{|\psi^Y\rangle\langle\psi^Y|}_{\rho^Y}. \end{aligned} \tag{16.14}$$

The utility of the bi-index notation becomes especially evident when dealing with uncorrelated states. In fact, using (16.13), each $(xy, x'y')$th entry $\rho^{XY}_{xy,x'y'}$ of an uncorrelated mixed density matrix equals the product of the (x, x')th entry $\rho^{X}_{x,x'}$ of matrix ρ^X and the (y, y')th entry $\rho^{Y}_{y,y'}$ of matrix ρ^Y, i.e.,

$$\rho^{XY}_{xy,x'y'} = \rho^{X}_{x,x'}\rho^{Y}_{y,y'} \tag{16.15}$$

for all $x, x' \in [N^X]$ and $y, y' \in [N^Y]$.

Example 16.8

For a two-qubit system, an uncorrelated state is of the form

$$\rho^{XY} = \begin{bmatrix} \rho^X_{0,0}\rho^Y_{0,0} & \rho^X_{0,0}\rho^Y_{0,1} & \rho^X_{0,1}\rho^Y_{0,0} & \rho^X_{0,1}\rho^Y_{0,1} \\ \rho^X_{0,0}\rho^Y_{1,0} & \rho^X_{0,0}\rho^Y_{1,1} & \rho^X_{0,1}\rho^Y_{1,0} & \rho^X_{0,1}\rho^Y_{1,1} \\ \rho^X_{1,0}\rho^Y_{0,0} & \rho^X_{1,0}\rho^Y_{0,1} & \rho^X_{1,1}\rho^Y_{0,0} & \rho^X_{1,1}\rho^Y_{0,1} \\ \rho^X_{1,0}\rho^Y_{1,0} & \rho^X_{1,0}\rho^Y_{1,1} & \rho^X_{1,1}\rho^Y_{10} & \rho^X_{1,1}\rho^Y_{11} \end{bmatrix}. \tag{16.16}$$

Example 16.9

Consider the two-qubit state with density matrix

$$\rho^{XY} = \frac{1}{4}\begin{bmatrix} 1 & -1 & 1 & -1 \\ -1 & 1 & -1 & 1 \\ 1 & -1 & 1 & -1 \\ -1 & 1 & -1 & 1 \end{bmatrix}. \tag{16.17}$$

This corresponds to an uncorrelated state because the density matrix can be written as the Kronecker product $\rho^{XY} = \rho^X \otimes \rho^Y$, with $\rho^X = |+\rangle\langle+|$ and $\rho^Y = |-\rangle\langle-|$.

Example 16.10

Take the two-qubit state

$$\begin{aligned} \rho^{XY} &= \frac{1}{6}I \otimes |0\rangle\langle 0| + \frac{2}{3}|+\rangle\langle+| \otimes |1\rangle\langle 1| \\ &= \frac{1}{3}\begin{bmatrix} 0.5 & 0 & 0 & 0 \\ 0 & 1 & 0 & 1 \\ 0 & 0 & 0.5 & 0 \\ 0 & 1 & 0 & 1 \end{bmatrix}. \end{aligned} \tag{16.18}$$

This can be interpreted as a system that is in state $\rho^{XY}_0 = I/2 \otimes |0\rangle\langle 0|$ with probability 1/3 and in state $\rho^{XY}_1 = |+\rangle\langle+| \otimes |1\rangle\langle 1|$ with probability 2/3. Thus, it is a convex combination of the states ρ^{XY}_0 and ρ^{XY}_1.

While both states ρ^{XY}_0 and ρ^{XY}_1 are uncorrelated, the state ρ^{XY} is not uncorrelated. In fact, if there were matrices ρ^X and ρ^Y satisfying the equality $\rho^{XY} = \rho^X \otimes \rho^Y$, by (16.16), the entries would have to guarantee the conditions $\rho^X_{0,0}\rho^Y_{0,0} = \rho^{XY}_{00,00} = 1/6$, $\rho^X_{0,1}\rho^Y_{0,0} = \rho^{XY}_{00,10} = 0$, and $\rho^X_{0,1}\rho^Y_{1,1} = \rho^{XY}_{01,11} = 1/3$. However, these equalities are incompatible.

16.2.3 Decomposing Density Matrices Using Uncorrelated Mixed States

As shown by the expression (16.4), any density matrix can be written as a linear combination of uncorrelated states. The coefficients of this linear combination are generally complex. However, as demonstrated next, it is also always possible to write any density matrix as a linear combination with real-valued coefficients of specific uncorrelated states. As will be seen throughout this chapter, this decomposition offers useful insights into the correlation properties of quantum mixed states.

Key Result 16.2 Mixed states as real linear combination of uncorrelated states

For any quantum mixed state ρ^{XY}, there exist local density matrices $\{\rho_z^X\}_{z\in[M]}$ and $\{\rho_z^Y\}_{z\in[M]}$ for some integer M, such that the state can be expressed as the linear combination

$$\rho^{XY} = \sum_{z\in[M]} \gamma_z \rho_z^X \otimes \rho_z^Y \tag{16.19}$$

for real-valued coefficients $\{\gamma_z\}_{z\in[M]}$ satisfying the condition

$$\sum_{z\in[M]} \gamma_z = 1. \tag{16.20}$$

The proof of this result can be found in Appendix 16.A. Importantly, the vector $\gamma = [\gamma_0, \dots, \gamma_{M-1}]^T$ is generally not a probability distribution, but rather a quasi-probability distribution vector, which is defined as follows.

Definition 16.3 Quasi-probability distribution

A vector $\gamma = [\gamma_0, \dots, \gamma_{M-1}]^T$ is said to be a quasi-probability distribution if all the entries $\{\gamma_z\}_{z\in[M]}$ are real-valued and sum to 1, i.e., $\sum_{z\in[M]} \gamma_z = 1$.

When all the entries are non-negative, a quasi-distribution is a probability distribution. In this case, the linear combination (16.19) corresponds to a convex combination of uncorrelated states. The extent to which a quasi-probability γ deviates from a probability distribution can be evaluated via the 1-norm $||\gamma||_1 = \sum_{z\in[M]} |\gamma_z|$. In fact, one has the equivalence

$$\gamma \text{ is a probability distribution} \iff ||\gamma||_1 = 1, \tag{16.21}$$

and values of the norm $||\gamma||_1$ greater than 1 indicate a larger deviation from a probability distribution.

16.2.4 Reduced Density Matrices

As seen in Chapter 15, the reduced density matrices determine the marginal measurement probability distributions for local measurements. They are obtained via the partial trace operations $\rho^X = \text{tr}_Y(\rho^{XY})$ and $\rho^Y = \text{tr}_X(\rho^{XY})$. Using the bi-index notation, the partial trace, introduced in Section 14.10, can equivalently be expressed as follows.

Key Result 16.3 Partial trace via the bi-index notation

For any $N^X N^Y \times N^X N^Y$ matrix A, the (y, y')th element of matrix $\text{tr}_X(A)$ is given by

$$[\text{tr}_X(A)]_{(y,y')} = \sum_{x \in [N^X]} A_{xy,xy'} \tag{16.22}$$

with $y, y' \in [N^Y]$; and the (x, x')th element of matrix $\text{tr}_Y(A)$ is given by

$$[\text{tr}_Y(A)]_{(x,x')} = \sum_{y \in [N^Y]} A_{xy,x'y} \tag{16.23}$$

for $x, x' \in [N^X]$.

The trace sums over all equal values of the indices of rows and columns. In contrast, as shown in (16.22), the partial trace $\text{tr}_X(\cdot)$ sums over all equal values of the first part of the bi-indices of rows and columns. Conversely, by (16.23), the partial trace $\text{tr}_Y(\cdot)$ sums over all equal values of the second part of the bi-indices of rows and columns.

Example 16.11

For the joint density matrix (16.18), the reduced density matrices are given by

$$\begin{aligned}\rho^X &= \text{tr}_Y(\rho^{XY}) \\ &= \frac{1}{6}I \cdot \langle 0|0\rangle + \frac{2}{3}|+\rangle\langle +| \cdot \langle 1|1\rangle \\ &= \frac{1}{6}I + \frac{2}{3}|+\rangle\langle +| \\ &= \begin{bmatrix} 1/2 & 1/3 \\ 1/3 & 1/2 \end{bmatrix}.\end{aligned} \tag{16.24}$$

This result can also be computed from the matrix form (16.18) using (16.23). For example, for the $(0, 0)$th entry of the matrix, one has

$$\rho^X_{0,0} = \rho^{XY}_{00,00} + \rho^{XY}_{01,01} = \rho^{XY}_{0,0} + \rho^{XY}_{1,1} = 1.5/3 = 1/2. \tag{16.25}$$

In a similar way, the reduced density matrix for subsystem Y is

$$\begin{aligned}\rho^Y &= \text{tr}_X(\rho^{XY}) \\ &= \frac{1}{3}|0\rangle\langle 0| + \frac{2}{3}|1\rangle\langle 1| \\ &= \begin{bmatrix} 1/3 & 0 \\ 0 & 2/3 \end{bmatrix}.\end{aligned} \tag{16.26}$$

Alternatively, using (16.22), one can compute the entry

$$\rho^{Y}_{0,1} = \rho^{XY}_{00,01} + \rho^{XY}_{10,11} = \rho^{XY}_{0,1} + \rho^{XY}_{2,3} = 0, \tag{16.27}$$

and analogous evaluations apply to the other entries.

In the case of bipartite pure states, it was shown in Section 14.5 that the eigenvalues of the two reduced density matrices are equal, and they correspond to the Schmidt probability vector. For mixed states, there is no counterpart of the Schmidt vector, and the eigenvalues of the two reduced density matrices are generally different.

Example 16.12

Consider a joint distribution p^{XY} over two bits, and the corresponding density matrix

$$\rho^{XY} = \begin{bmatrix} p^{XY}_{00} & 0 & 0 & 0 \\ 0 & p^{XY}_{01} & 0 & 0 \\ 0 & 0 & p^{XY}_{10} & 0 \\ 0 & 0 & 0 & p^{XY}_{11} \end{bmatrix}. \tag{16.28}$$

Using the relation (16.22), one obtains the reduced density matrix

$$\begin{aligned} \rho^{Y} &= \mathrm{tr}_X(\rho^{XY}) \\ &= \begin{bmatrix} p^{XY}_{00} + p^{XY}_{10} & 0 \\ 0 & p^{XY}_{01} + p^{XY}_{11} \end{bmatrix} \\ &= \begin{bmatrix} p^{Y}_{0} & 0 \\ 0 & p^{Y}_{1} \end{bmatrix} = \mathrm{diag}(p^{Y}). \end{aligned} \tag{16.29}$$

Therefore, the reduced density matrix ρ^{Y} has diagonal elements equal to the entries of the marginal distribution p^{Y} of the joint distribution p^{XY} for subsystem Y.

In a similar way, for the reduced density matrix with respect to subsystem Y, using the relation (16.23), one obtains

$$\begin{aligned} \rho^{X} &= \mathrm{tr}_Y(\rho^{XY}) \\ &= \begin{bmatrix} p^{XY}_{00} + p^{XY}_{01} & 0 \\ 0 & p^{XY}_{10} + p^{XY}_{11} \end{bmatrix} \\ &= \begin{bmatrix} p^{X}_{0} & 0 \\ 0 & p^{X}_{1} \end{bmatrix} = \mathrm{diag}(p^{X}), \end{aligned} \tag{16.30}$$

where p^{X} is the marginal distribution of the joint distribution p^{XY} for subsystem X. Thus, in general, the two matrices ρ^{X} and ρ^{Y} have distinct eigenvalues.

These results, connecting the reduced density matrices to the marginal distributions of a joint distribution p^{XY}, generalize to a density matrix $\rho^{XY} = \mathrm{diag}(\rho^{XY})$ of any dimensions N^{X} and N^{Y} (see Problem 16.3). Accordingly, the reduced density matrices can be thought of as the quantum counterpart of marginal distributions, and the partial trace is the quantum counterpart of marginalization.

As a final note, writing a state ρ^{XY} as the linear combination (16.19), the reduced density matrices can be expressed as the sums

$$\rho^X = \sum_{z \in [M]} \gamma_z \rho_z^X \tag{16.31}$$

and

$$\rho^Y = \sum_{z \in [M]} \gamma_z \rho_z^Y . \tag{16.32}$$

16.3 Classical–Classical and Classical–Quantum States

This section introduces two key classes of mixed states for bipartite systems, which will serve as important reference points throughout the chapter. The first class consists of **classical–classical states**, which are defined as all the states that can be prepared by using only shared classical randomness and local "classical" operations. While classical–classical states lack any quantum correlations, they can exhibit classical correlations. As such, they exemplify the principle introduced in Section 16.1 that classical correlations can exist independently of quantum correlations in mixed quantum states.

The second class includes **classical–quantum states** and **quantum–classical states**, which can be generated through local quantum operations supplemented by shared classical randomness. Classical–quantum and quantum–classical states exhibit quantum correlations, though they are not entangled. Thus, they illustrate the property highlighted in Section 16.1 that quantum correlations encompass forms of correlation weaker than entanglement.

Throughout this section, agent X will refer as usual to an agent having access to subsystem X, while agent Y has access to subsystem Y.

16.3.1 Classical–Classical States

A classical–classical state can be prepared via shared randomness and local "classical" operations. To define a classical–classical state, one fixes the following quantities.

- **Joint distribution:** A joint distribution p^{XY} for variables x and y with dimensions N^X and N^Y, respectively.
- **Local orthonormal bases for both subsystems:** Local orthonormal bases $\mathcal{M}^X = \{|v_x^X\rangle\}_{x \in [N^X]}$ and $\mathcal{M}^Y = \{|v_y^Y\rangle\}_{y \in [N^Y]}$ for subsystem X and subsystem Y, respectively.

A classical–classical state can then be generated as follows.

1. **Generation of correlated random variables:** Agents X and Y generate local random variables x and y with joint distribution p^{XY} by using shared classical randomness.

2. **Local "classical" operations at the two subsystems:** Given the realized values (x, y) of the random variables (x, y), agent X produces state $|v_x^X\rangle$ and agent Y prepares state $|v_y^Y\rangle$.

The operation of generating the local states $|v_x^X\rangle$ and $|v_y^Y\rangle$ is said to be "classical" because the possible states $\{|v_x^X\rangle\}_{x\in[N^X]}$ and $\{|v_y^Y\rangle\}_{y\in[N^Y]}$ produced by the agents are mutually orthogonal, and thus are incoherent with respect to the respective local bases $\mathcal{M}^X$ and $\mathcal{M}^Y$. In other words, they are perfectly distinguishable using measurements in the respective local bases. The quotation signs are a reminder that the operations do involve quantum systems, but they have classical characteristics. With this in mind, in the following, the quotation marks will be removed when referring to operations involving the generation of orthogonal states.

By this preparation procedure, a classical–classical state is a convex combination of uncorrelated states $|v_x^X\rangle \otimes |v_y^Y\rangle$ obtained by composing pairs of states from local orthonormal bases. Specifically, the probability associated with each state $|v_x^X\rangle \otimes |v_y^Y\rangle$, i.e., with the density matrix

$$(|v_x^X\rangle \otimes |v_y^Y\rangle)(\langle v_x^X| \otimes \langle v_y^Y|) = |v_x^X\rangle\langle v_x^X| \otimes |v_y^Y\rangle\langle v_y^Y|, \tag{16.33}$$

is p_{xy}^{XY}. Note that the density matrices $\{|v_x^X\rangle\langle v_x^X|\}_{x\in[N^X]}$ are mutually orthogonal in the sense that the trace inner product (7.46) yields

$$\mathrm{tr}(|v_x^X\rangle\langle v_x^X||v_{x'}^X\rangle\langle v_{x'}^X|) = 0 \tag{16.34}$$

for all $x, x' \in [N^X]$ with $x \neq x'$. In a similar way, the density matrices $\{|v_y^Y\rangle\langle v_y^Y|\}_{y\in[N^Y]}$ are mutually orthogonal, given the equalities $\mathrm{tr}(|v_y^Y\rangle\langle v_y^Y||v_{y'}^Y\rangle\langle v_{y'}^Y|) = 0$ for all $y, y' \in [N^Y]$ with $y \neq y'$. This construction yields the following formal definition.

Definition 16.4 Classical–classical states
A quantum state ρ^{XY} is said to be classical–classical if it can be written as

$$\chi^{XY} = \sum_{x\in[N^X]} \sum_{y\in[N^Y]} p_{xy}^{XY} |v_x^X\rangle\langle v_x^X| \otimes |v_y^Y\rangle\langle v_y^Y| \tag{16.35}$$

for some joint distribution p^{XY} and for some local orthonormal bases $\mathcal{M}^X = \{|v_x^X\rangle\}_{x\in[N^X]}$ and $\mathcal{M}^Y = \{|v_y^Y\rangle\}_{y\in[N^Y]}$.

Note the use of the special notation χ^{XY} as a mnemonic that a density matrix describes a classical–classical state. By their definition (16.35), classical–classical states can equivalently be characterized as follows.

- **Eigenbasis of uncorrelated states:** Classical–classical states have an eigenbasis $\{\{|v_x^X\rangle \otimes |v_y^Y\}_{x\in[N^X]}\}_{y\in[N^Y]}$ consisting of uncorrelated states (with eigenvalues given by a joint distribution p^{XY}).
- **Incoherence with respect to a basis of uncorrelated states:** Classical–classical states are incoherent with respect to a basis $\{\{|v_x^X\rangle \otimes |v_y^Y\}_{x\in[N^X]}\}_{y\in[N^Y]}$ of uncorrelated states (see Section 7.12 for the definition of incoherence for mixed states).

Example 16.13

A density matrix of the form $\rho^{XY} = \text{diag}(p^{XY})$, which has the elements of vector p^{XY} on its main diagonal, is classical–classical. In fact, it is in the form (16.35) with bases $\mathcal{M}^X$ and $\mathcal{M}^Y$ given by the respective local computational bases. Equivalently, it is incoherent with respect to the local computational bases.

Example 16.14

Writing the identity matrix as $I = |+\rangle\langle+| + |-\rangle\langle-|$, the density matrix (16.18) can be expressed as

$$\rho^{XY} = \frac{1}{6}|+\rangle\langle+| \otimes |0\rangle\langle 0| + \frac{2}{3}|+\rangle\langle+| \otimes |1\rangle\langle 1| + \frac{1}{6}|-\rangle\langle-| \otimes |0\rangle\langle 0|, \tag{16.36}$$

which has the form (16.35) with the Hadamard basis $\mathcal{M}^X = \{|+\rangle, |-\rangle\}$ for the subsystem X and the computational basis $\mathcal{M}^Y = \{|0\rangle, |1\rangle\}$ for subsystem Y. Therefore, this is a classical–classical state with joint distribution $p^{XY} = [p_{00}^{XY} = 1/6, p_{01}^{XY} = 2/3, p_{10}^{XY} = 1/6, p_{11}^{XY} = 0]^T$.

For a classical-classical state, the bases $\mathcal{M}^X$ and $\mathcal{M}^Y$ in the definition (16.35) can readily be shown to correspond to local eigenbases $\mathcal{M}_*^X = \mathcal{M}^X$ and $\mathcal{M}_*^Y = \mathcal{M}^Y$. Recall from Section 14.6 that the local eigenbases are the eigenbases of the reduced density matrices. This can be verified by using (16.31) and (16.32), as summarized next.

Key Result 16.4 Reduced density matrices for classical–classical states
The reduced density matrices for classical–classical states (16.35) are given by

$$\text{tr}_Y(\chi^{XY}) = \sum_{x\in[N^X]} p_x^X |v_x^X\rangle\langle v_x^X| \tag{16.37}$$

and

$$\text{tr}_X(\chi^{XY}) = \sum_{y\in[N^Y]} p_y^Y |v_y^Y\rangle\langle v_y^Y|, \tag{16.38}$$

where p^X and p^Y are the marginal distributions of the joint distribution p^{XY}. Therefore, the bases $\mathcal{M}^X$ and $\mathcal{M}^Y$ are local eigenbases for state ρ^{XY}.

16.3.2 Classical–Classical States via Non-Selective Local Measurements

Intuitively, classical–classical states have only classical correlations since they can be prepared using only shared randomness and local classical operations. To further substantiate this intuition, it is now shown that classical–classical states can also be produced from any joint state ρ^{XY} as shown in Figure 16.1. Specifically, starting from any state ρ^{XY}, non-selective local complete measurements on both subsystems always result in a classical–classical state (see Section 7.11 for an introduction to non-selective measurements and Example 11.6 for a formulation as quantum

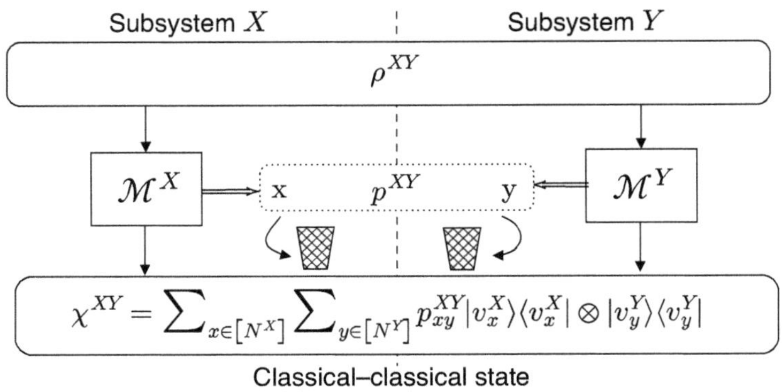

Figure 16.1 When a local complete measurement in some setting $\mathcal{M}^X = \{|v_x^X\rangle\}_{x\in[N^X]}$ and $\mathcal{M}^Y = \{|v_y^Y\rangle\}_{y\in[N^Y]}$ is applied to an arbitrary state ρ^{XY} in a non-selective way, the resulting state is the classical–classical state (16.39) with local eigenbases $\mathcal{M}^X$ and $\mathcal{M}^Y$.

channels). Therefore, local complete measurements, which remove coherence with respect to the measurement bases, followed by forgetting, which reintroduces classical randomness, yield states that are devoid of quantum correlations.

As illustrated in Figure 16.1, assume that a complete local measurement in the basis $\mathcal{M}^X \otimes \mathcal{M}^Y$, with $\mathcal{M}^X = \{|v_x^X\rangle\}_{x\in[N^X]}$ and $\mathcal{M}^Y = \{|v_y^Y\rangle\}_{y\in[N^Y]}$, is applied to an arbitrary state ρ^{XY}. After measurement, the two classical outputs $(x, y) \sim p^{XY}$ are discarded.

Following Section 14.3, the post-measurement state is

$$\mathcal{M}^X \otimes \mathcal{M}^Y(\rho^{XY}) = \sum_{x\in[N^X]} \sum_{y\in[N^Y]} p_{xy}^{XY} |v_x^X\rangle\langle v_x^X| \otimes |v_y^Y\rangle\langle v_y^Y|, \tag{16.39}$$

where

$$p_{xy}^{XY} = p_{xy}^{XY}(\rho^{XY}|\mathcal{M}^X \otimes \mathcal{M}^Y) = \langle v_x^X| \otimes \langle v_y^Y|\rho^{XY}|v_x^X\rangle \otimes |v_y^Y\rangle \tag{16.40}$$

is the probability of observing outputs (x, y) and $|v_x^X\rangle\langle v_x^X| \otimes |v_y^Y\rangle\langle v_y^Y|$ is the corresponding post-measurement state. The post-measurement state (16.39) is a classical–classical state with local eigenbases $\mathcal{M}^X$ and $\mathcal{M}^Y$.

Example 16.15

Consider any state ρ^{XY} with general dimensions N^X and N^Y to which the subsystems apply local complete measurements in the computational basis. The post-measurement state (16.39) is given by the density

$$\begin{aligned}\mathcal{M}^X \otimes \mathcal{M}^Y(\rho^{XY}) &= \sum_{x\in[N^X]} \sum_{y\in[N^Y]} p_{xy}^{XY} |x\rangle\langle x| \otimes |y\rangle\langle y| \\ &= \text{diag}(p^{XY}),\end{aligned} \tag{16.41}$$

where p^{XY} is the measurement probability vector with entries

$$p^{XY}_{xy} = \langle xy|\rho^{XY}|xy\rangle = \rho^{XY}_{xy,xy}. \tag{16.42}$$

Accordingly, the post-measurement state is diagonal with diagonal elements given by the entries of the measurement probability vector p^{XY}. Therefore, a non-selective measurement in the computational bases cancels out the elements outside the main diagonal of the density matrix, making the state incoherent with respect to the computational basis of the $N^X N^Y$-dimensional space. As discussed, this corresponds to a classical–classical state.

Example 16.16

Assume that the pre-measurement state is a generic classical–classical χ^{XY} as in (16.35), and that a local complete measurement is applied in the setting $\mathcal{M}^X \otimes \mathcal{M}^Y$ given by the local eigenbases of the state. Since the classical–classical state is incoherent with respect to the basis $\mathcal{M}^X \otimes \mathcal{M}^Y$, the local measurement $\mathcal{M}^X \otimes \mathcal{M}^Y$ does not disturb the state when implemented in a non-selective way (see Section 6.3). Specifically, using (16.39), the post-measurement state is given by the state itself, i.e.,

$$\mathcal{M}^X \otimes \mathcal{M}^Y(\chi^{XY}) = \chi^{XY}. \tag{16.43}$$

This example illustrates another equivalent perspective on classical–classical states: classical–classical states are all the states that are left undisturbed by some non-selective local complete measurement.

Example 16.17

Take the state

$$\begin{aligned}\rho^{XY} &= \eta|+\rangle\langle+|\otimes|+\rangle\langle+| + (1-\eta)|\Phi_+^{XY}\rangle\langle\Phi_+^{XY}| \\ &= \frac{\eta}{4}\begin{bmatrix} \frac{2}{\eta}-1 & 1 & 1 & \frac{2}{\eta}-2 \\ 1 & 1 & 1 & 1 \\ 1 & 1 & 1 & 1 \\ \frac{2}{\eta}-2 & 1 & 1 & \frac{2}{\eta}-1 \end{bmatrix},\end{aligned} \tag{16.44}$$

which is the convex combination of the uncorrelated state $|+\rangle\langle+|\otimes|+\rangle\langle+|$ and of the maximally entangled state $|\Phi_+^{XY}\rangle\langle\Phi_+^{XY}|$, with $0 \leq \eta \leq 1$.

The post-measurement state (16.39) for local measurements in the computational bases is given by the diagonal matrix (16.41), which yields the classical–classical state

$$\mathcal{M}^X \otimes \mathcal{M}^Y(\rho^{XY}) = \frac{\eta}{4}\begin{bmatrix} \frac{2}{\eta}-1 & 0 & 0 & 0 \\ 0 & 1 & 0 & 0 \\ 0 & 0 & 1 & 0 \\ 0 & 0 & 0 & \frac{2}{\eta}-1 \end{bmatrix}. \tag{16.45}$$

Applying another non-selective measurement using the same setting $\mathcal{M}^X \otimes \mathcal{M}^Y$ would not change the state.

16.3.3 Classical–Quantum States

While classical–classical states possess only classical correlations, classical–quantum states display both classical and quantum correlations, though they stop short of exhibiting entanglement. A classical–classical state is defined by a joint distribution p^{XY} and by local eigenbases for the two subsystems. In contrast, classical–quantum states are defined by the following.

- **Marginal distribution:** A marginal distribution p^X for a random variable x of dimension N^X.
- **Local eigenbasis for subsystem** X**:** A local orthonormal basis $\{|v_x^X\rangle\}_{x\in[N^X]}$ for subsystem X, along with the corresponding mutually orthogonal states $\{|v_x^X\rangle\langle v_x^X|\}_{x\in[N^X]}$.
- **Arbitrary states for subsystem** Y **(conditional density matrices):** A set of N^X arbitrary states $\{\rho_x^{Y|X}\}_{x\in[N^X]}$ for subsystem X, which are generally not orthogonal, i.e.,

$$\mathrm{tr}(\rho_x^{Y|X}\rho_{x'}^{Y|X}) \neq 0 \tag{16.46}$$

  for $x, x' \in [N^X]$ and $x' \neq x$. The density matrices $\{\rho_x^{Y|X}\}_{x\in[N^X]}$ are referred to as the conditional density matrices for subsystem Y given subsystem X.

A classical–quantum state can be prepared by using shared classical randomness together with local classical operations at subsystem X and local quantum operations at subsystem Y. Specifically, a state of the form (16.47) can be generated via the following steps.

1. **Generation of a shared random variable:** Agents X and Y generate a random variable x with distribution p^X using shared classical randomness.
2. **Local classical operations at subsystem** X**:** Given the realized value $\mathrm{x} = x$ of the shared random variables x, agent X produces the state $|v_x^X\rangle$.
3. **Local quantum operations at subsystem** Y**:** Given the realized value $\mathrm{x} = x$ of the shared random variables x, agent Y prepares the state $\rho_x^{Y|X}$.

The operation performed by agent X in preparing state (16.47) is considered classical in the same sense as for classical–classical states, since the states produced by agent X are mutually orthogonal. The key difference from classical–classical states is that agent Y must prepare one of the states in the non-orthogonal set $\{\rho_x^{Y|X}\}_{x\in[N^X]}$. Since the conditional density matrices $\{\rho_x^{Y|X}\}_{x\in[N^X]}$ are non-orthogonal, and thus not perfectly distinguishable, their preparation requires genuinely quantum operations by agent Y, thereby justifying the label "classical–quantum" for states prepared in this manner. This leads to the following formal definition.

Definition 16.5 Classical–quantum states

A mixed state is said to be classical–quantum if the density matrix can be written as

$$\overrightarrow{\omega}^{XY} = \sum_{x\in[N^X]} p_x^X |v_x^X\rangle\langle v_x^X| \otimes \rho_x^{Y|X} \tag{16.47}$$

for a marginal probability distribution $p^X = [p_0^X, \ldots, p_{N^X-1}^X]$, an orthonormal basis $\mathcal{M}^X = \{|v_x^X\rangle\}_{x\in[N^X]}$, and $N^Y \times N^Y$ density matrices $\{\rho_x^{Y|X}\}_{x\in[N^X]}$.

By the definition (16.47), a classical–quantum state is given by the convex combination of states of the form $|v_x^X\rangle\langle v_x^X| \otimes \rho_x^{Y|X}$ for $x \in [N^X]$, in which the states $\{|v_x^X\rangle\langle v_x^X|\}_{x\in[N^X]}$ for subsystem X are mutually orthogonal, satisfying the condition (16.34), while the states $\{\rho_x^{Y|X}\}_{x\in[N^X]}$ for subsystem Y are not necessarily orthogonal.

Every classical–classical state is also a classical–quantum state, but not vice versa. Thus, the set of classical–quantum states includes that of classical–classical states.

The following result is a consequence of the more general characterization (16.31) and (16.32).

Key Result 16.5 Reduced density matrices for classical–quantum states

The reduced density matrices for classical–quantum states (16.47) are given by

$$\mathrm{tr}_Y(\overrightarrow{\omega}^{XY}) = \sum_{x\in[N^X]} p_x^X |v_x^X\rangle\langle v_x^X| \tag{16.48}$$

and

$$\mathrm{tr}_X(\overrightarrow{\omega}^{XY}) = \sum_{x\in[N^X]} p_x^X \rho_x^{Y|X}. \tag{16.49}$$

Accordingly, for a classical–quantum state, the states $\mathcal{M}^X = \{|v_x^X\rangle\}_{x\in[N^X]}$ form the local eigenbasis for subsystem X, while the local eigenbasis of subsystem Y generally depends on all the conditional density matrices $\{\rho_x^{Y|X}\}_{x\in[N^X]}$.

16.3.4 Classical–Quantum States via Non-Selective Local Measurements at One Subsystem

As discussed in Section 16.3.2, classical–classical states are produced by non-selective local complete measurements at both subsystems. In contrast, non-selective local complete measurements at subsystem X only produce classical–quantum states.

To elaborate, consider the setting in Figure 16.2, in which, starting from an arbitrary state ρ^{XY}, a non-selective complete measurement $\mathcal{M}^X$ is applied only to subsystem X. Following Section 14.3, the classical output $\mathrm{x} = x$ is obtained with probability

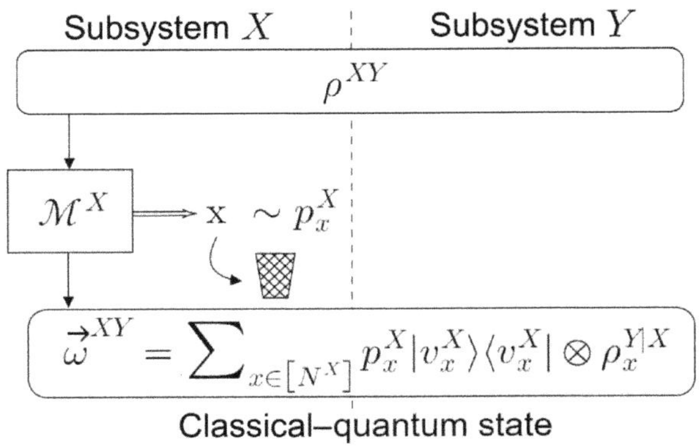

Figure 16.2 When a local complete measurement in some setting $\mathcal{M}^X$ is applied in a non-selective way to subsystem X for an arbitrary state ρ^{XY}, the resulting state is the classical–quantum state (16.47) with local eigenbasis $\mathcal{M}^X$ for subsystem X.

$$p_x^X = \text{tr}(\underbrace{(\langle v_x^X| \otimes I)\rho^{XY}(|v_x^X\rangle \otimes I)}_{\tilde{\rho}_x^{Y|X}}), \tag{16.50}$$

with the $N^Y \times N^Y$ matrix

$$\tilde{\rho}_x^{Y|X} = (\langle v_x^X| \otimes I)\rho^{XY}(|v_x^X\rangle \otimes I). \tag{16.51}$$

Note that the matrix $\tilde{\rho}_x^{Y|X}$ is not a valid density matrix, as it is generally not normalized.

Furthermore, the corresponding post-measurement for the entire system is given by

$$\begin{aligned}\frac{1}{p_x^X}(|v_x^X\rangle\langle v_x^X| \otimes I)\rho^{XY}(|v_x^X\rangle\langle v_x^X| \otimes I) &= (|v_x^X\rangle \otimes I)\rho_x^{Y|X}(\langle v_x^X| \otimes I) \\ &= |v_x^X\rangle\langle v_x^X| \otimes \rho_x^{Y|X},\end{aligned} \tag{16.52}$$

with density matrix

$$\rho_x^{Y|X} = \frac{1}{p_x^X}\tilde{\rho}_x^{Y|X}. \tag{16.53}$$

The last equality in (16.52) follows by noticing that the matrix $\rho_x^{Y|X}$ can be trivially written as $1 \otimes \rho_x^{Y|X}$.

According to (16.52), after the measurement at subsystem X, if the classical output is $x \in [N^X]$, subsystem X is left in state $|v_x^X\rangle$, while subsystem Y is in state $\rho_x^{Y|X}$. Therefore, when the measurement output is discarded, the post-measurement state is given by the average

$$\mathcal{M}^X \otimes \mathcal{I}^Y(\rho^{XY}) = \sum_{x\in[N^X]} p_x^X |v_x^X\rangle\langle v_x^X| \otimes \rho_x^{Y|X}, \tag{16.54}$$

which is a classical–quantum state. It is concluded that a non-selective local complete measurement in basis $\mathcal{M}^X$ at subsystem X causes the post-measurement state to become classical–quantum with local eigenbasis $\mathcal{M}^X$ for subsystem X.

Example 16.18

Example 16.16 illustrated the characterization of classical–classical states as the states that are not changed by some non-selective local complete measurement applied to both subsystems. Classical–quantum states can be characterized as those states for which there exists a non-selective complete measurement on subsystem X only that leaves the system undisturbed.

To see this, assume that the system is initially in the classical–quantum state $\overrightarrow{\omega}^{XY}$ in (16.47) and that a non-selective measurement is carried out only on subsystem X in the local eigenbasis $\mathcal{M}^X$. Using (16.53) and (16.51), the post-measurement state is given by the classical–quantum state (16.54) with conditional density matrices given by

$$(\langle v_x^X| \otimes I)\overrightarrow{\omega}^{XY}(|v_x^X\rangle \otimes I) = \rho_x^{Y|X}. \tag{16.55}$$

It is thus concluded that the post-measurement state coincides with the pre-measurement state, i.e.,

$$\mathcal{M}^X \otimes \mathcal{I}^Y(\overrightarrow{\omega}^{XY}) = \overrightarrow{\omega}^{XY}. \tag{16.56}$$

To summarize, for a classical–quantum system, it suffices for a complete measurement setting to exist on subsystem X such that the state remains undisturbed when the measurement is performed in a non-selective manner. If complete measurement settings can be identified for both subsystems ensuring the state remains unchanged under non-selective implementation, the state is classical–classical. This perspective confirms that all classical–classical states form a subset of classical–quantum states.

Example 16.19

For the two-qubit state (16.44), using (16.54), the post-measurement state after a non-selective local measurement in the computational basis of subsystem X is given by the classical–quantum state

$$\mathcal{M}^X \otimes \mathcal{I}^Y(\rho^{XY}) = \sum_{x\in[2]} p_x^X |x\rangle\langle x| \otimes \rho_x^{Y|X}. \tag{16.57}$$

The unnormalized conditional density matrices (16.51) are given by

$$\tilde{\rho}_0^{Y|X} = (\langle 0| \otimes I)\rho^{XY}(|0\rangle \otimes I) = \frac{\eta}{4}\begin{bmatrix} \frac{2}{\eta} - 1 & 1 \\ 1 & 1 \end{bmatrix}, \tag{16.58}$$

which corresponds to the top-left 2×2 block of matrix ρ^{XY}.

In a similar way, one gets

$$\tilde{\rho}_1^{Y|X} = (\langle 1| \otimes I)\rho^{XY}(|1\rangle \otimes I) = \frac{\eta}{4}\begin{bmatrix} 1 & 1 \\ 1 & \frac{2}{\eta} - 1 \end{bmatrix}, \tag{16.59}$$

which is the bottom-right 2 × 2 block of matrix ρ^{XY}. It follows that the measurement probabilities (16.50) are

$$p_0^X = \text{tr}(\tilde{\rho}_0^{Y|X}) = \frac{1}{2}, \tag{16.60}$$

and $p_1^X = \text{tr}(\tilde{\rho}_1^{Y|X}) = 1/2$.

Overall, the post-measurement density matrix is

$$\mathcal{M}^X \otimes \mathcal{I}^Y(\rho^{XY}) = \frac{\eta}{4}\begin{bmatrix} \frac{2}{\eta} - 1 & 1 & 0 & 0 \\ 1 & 1 & 0 & 0 \\ 0 & 0 & 1 & 1 \\ 0 & 0 & 1 & \frac{2}{\eta} - 1 \end{bmatrix}. \tag{16.61}$$

Note that this corresponds to a block-diagonal matrix, and thus it preserves some coherence with respect to the computational bases. This is in contrast to the post-measurement density matrix (16.45) for computational-basis measurements at both subsystems, which is a classical–classical state and thus incoherent with respect to the computational bases.

Example 16.20

Generalizing the previous example, consider any state ρ^{XY} with general dimensions N^X and N^Y to which subsystem X applies a non-selective local complete measurement in the computational basis. To understand the impact of this measurement on the pre-measurement state ρ^{XY}, it is useful to view the matrix ρ^{XY} as composed of $N^Y \times N^Y$ matrix blocks $\{r_{x,x'}^{XY}\}_{x,x' \in [N^X]}$ as

$$\rho^{XY} = \sum_{x \in [N^X]} \sum_{x' \in [N^X]} |x\rangle\langle x| \otimes r_{x,x'}^{XY}$$
$$= \begin{bmatrix} r_{0,0}^{XY} & r_{0,1}^{XY} & \cdots & r_{0,N^X-1}^{XY} \\ r_{1,0}^{XY} & r_{1,1}^{XY} & \cdots & r_{1,N^X-1} \\ \vdots & \vdots & \ddots & \vdots \\ r_{N^X-1,0}^{XY} & \cdots & & r_{N^X-1,N^X-1}^{XY} \end{bmatrix}. \tag{16.62}$$

The unnormalized conditional density matrix $\tilde{\rho}_x^{Y|X}$ in (16.51) can be computed as

$$\tilde{\rho}_x^{Y|X} = (\langle x| \otimes I^Y)\rho^{XY}(|x\rangle \otimes I^Y) = r_{x,x}^{XY}, \tag{16.63}$$

which corresponds to the (x, x)th $N^Y \times N^Y$ block $r_{x,x}^{XY}$ of the density matrix ρ^{XY}. It follows that the post-measurement matrix is given by the block-diagonal matrix

$$\mathcal{M}^X \otimes \mathcal{I}^Y(\rho^{XY}) = \sum_{x \in [N^X]} |x\rangle\langle x| \otimes \rho^{XY}_{x,x}$$
$$= \begin{bmatrix} r^{XY}_{0,0} & 0 & \cdots & 0 \\ 0 & r^{XY}_{1,1} & \cdots & 0 \\ \vdots & \vdots & \ddots & \vdots \\ 0 & \cdots & 0 & r^{XY}_{N^X-1,N^X-1} \end{bmatrix}, \tag{16.64}$$

where 0 represents an $N^Y \times N^Y$ all-zero matrix.

16.3.5 Quantum–Classical States

Quantum–classical states are defined by reversing the roles of the two subsystems in the definition of classical–quantum states.

Definition 16.6 Quantum–classical states
A state is said to be quantum–classical if the density matrix can be expressed as

$$\overleftarrow{\omega}^{XY} = \sum_{y \in [N^Y]} p^Y_y \rho^{X|Y}_y \otimes |v^Y_y\rangle\langle v^Y_y| \tag{16.65}$$

for some marginal distribution $p^Y = [p^Y_0, \ldots, p^Y_{N^Y-1}]^T$, orthonormal basis $\mathcal{M}^Y = \{|v^Y_y\rangle\}_{y \in [N^Y]}$, and states $\{\rho^{X|Y}_y\}_{y \in [N^Y]}$. The states $\{\rho^{X|Y}_y\}_{y \in [N^Y]}$ are referred to as the conditional density matrices for subsystem X given subsystem Y.

Example 16.21

Classical–classical states are both classical–quantum and quantum–classical.

Following the same reasoning as for classical–quantum states, quantum–classical states can be prepared by leveraging shared classical randomness, local quantum operations at subsystem X, and local classical operations at subsystem Y. Furthermore, as illustrated in Figure 16.3, a non-selective local measurement setting $\mathcal{I}^X \otimes \mathcal{M}^Y$ applied only to subsystem Y produces a quantum–classical state (16.65). Accordingly, quantum–classical states are left undisturbed by a non-selective local measurement applied only to subsystem Y in setting $\mathcal{M}^Y$.

16.3.6 Applications of Classical–Quantum States

This section is concluded by describing two practical applications of classical–quantum states, namely classical communications over quantum channels and quantum cryptography.

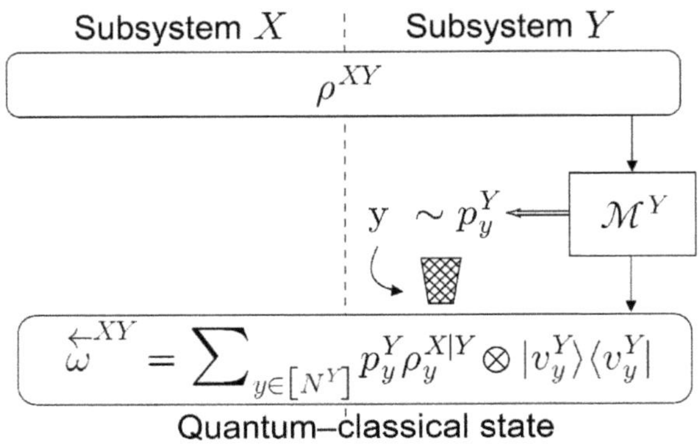

Figure 16.3 When a local complete measurement in some setting $\mathcal{M}^Y$ is applied in a non-selective way to subsystem Y for an arbitrary state ρ^{XY}, the resulting state is the quantum–classical state (16.65) with local eigenbasis $\mathcal{M}^Y$ for subsystem Y.

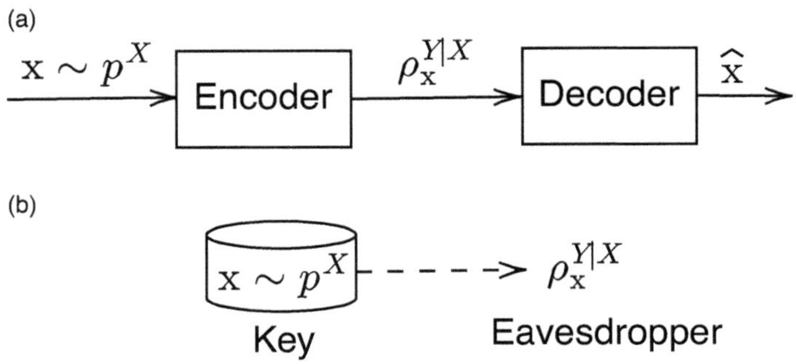

Figure 16.4 Two applications of classical–quantum states: (a) classical communication over a quantum channel and (b) quantum cryptography.

The problem of classical communication over a quantum channel is described in Figure 16.4(a). As illustrated, an encoder has a **classical message** $\mathrm{x} \sim p^X$, taking N^X possible values, to convey to a recipient, and communication takes place over a quantum channel. To this end, the encoder prepares a state $\rho_x^{Y|X}$ when the message is $\mathrm{x} = x$, and sends it to the decoder over the quantum channel. The goal of the decoder is to apply a measurement to the received quantum state to produce a good estimate $\hat{\mathrm{x}}$ of the original message x.

Since the message x is classical, its different values $[N^X]$ can be perfectly distinguished. Hence, the message x can be modeled as a quantum system that is in state $|v_x\rangle$ when the message is $\mathrm{x} = x$, where $\{|v_x\rangle\}_{x\in[N^X]}$ is an arbitrary orthonormal basis for the N^X-dimensional space. It follows that the joint state of the classical message x and of the received quantum state is $|v_x\rangle\langle v_x| \otimes \rho_x^{Y|X}$ with probability p_x^X. Therefore, averaging over the distribution of the message p^X, this coincides with the classical–quantum state (16.47).

As a second application, consider Figure 16.4(b), which depicts a setting relevant for **quantum cryptography**. In this context, the classical random variable $\mathrm{x} \sim p^X$ represents a **key**, and the state $\rho_x^{Y|X}$ is the side information available at the **eavesdropper** when the key value is $\mathrm{x} = x$. As in the case of classical communications over a quantum channel, the joint state of key and side information can thus be modeled as the classical–quantum state (16.47).

In this regard, a key is considered to be **perfect** if the distribution p^X of the key is uniform, and the two subsystems are uncorrelated. In fact, in this case, the eavesdropper is ignorant about the key, and the key has maximum entropy. This situation corresponds to the joint uncorrelated state

$$\overrightarrow{\omega}^{XY} = \pi^X \otimes \rho^Y, \tag{16.66}$$

where π^X is the maximally mixed state for subsystem X.

16.4 Separable vs. Entangled Mixed States

This section begins the study of entanglement in mixed states by defining as **separable** – or not entangled – all states that can be prepared without using global quantum operations. According to this definition, uncorrelated, classical–classical, classical–quantum, and quantum–classical states are all separable and, therefore, not entangled. However, as will be shown in the following sections, while classical–quantum and quantum–classical states are not entangled, they can still exhibit quantum correlations.

16.4.1 Defining Entangled Mixed States

A state is separable if it can be prepared via shared randomness and local quantum operations at the two subsystems.

To formalize this definition, for some integer M, fix the following.

- **Probability distribution:** A probability distribution $p = [p_0, p_1, \ldots, p_M]^T$.
- **Arbitrary local states for the two subsystems:** Arbitrary local states $\{\rho_z^X\}_{z\in[M]}$ and $\{\rho_z^Y\}_{z\in[M]}$ for subsystems X and Y. The states $\{\rho_z^X\}_{z\in[M]}$ and $\{\rho_z^Y\}_{z\in[M]}$ need not be mutually orthogonal.

A state is separable if it can be prepared via the following steps.

1. **Generation of a shared random variable:** Agents X and Y generate a shared random variable $\mathrm{z} \sim p$ by using shared classical randomness.
2. **Local quantum operations at the two subsystems:** Given the realized value $\mathrm{z} = z$ of the shared random variable, agent X produces state ρ_z^X and agent Y generates state ρ_z^Y.

Mathematically, this leads to the following definition.

Definition 16.7 Separable and entangled states
A state is said to be separable if the density matrix can be expressed as

$$\rho^{XY} = \sum_{z\in[M]} p_z \rho_z^X \otimes \rho_z^Y \tag{16.67}$$

for some integer M, probability distribution $p = [p_0, p_1, \ldots, p_M]^T$, and arbitrary local states $\{\rho_z^X\}_{z\in[M]}$ and $\{\rho_z^Y\}_{z\in[M]}$. States that are not separable are said to be **entangled**.

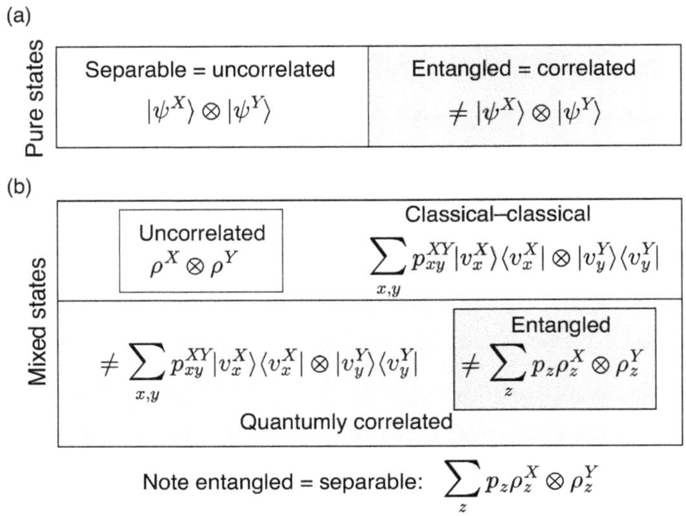

Figure 16.5 For pure states, the class of uncorrelated states coincides with the set of separable states (a). For mixed states, uncorrelated states form a strict subset of separable states, encompassing only those separable states that can be generated without the use of shared classical randomness (b).

As illustrated in Figure 16.5(a), in the special case of pure states, the class of separable states coincides with the set of uncorrelated states. In fact, for a state of the form (16.67) to be pure, the distribution p must have zero entropy, which yields an uncorrelated state $\rho^{XY} = \rho^X \otimes \rho^Y$. More broadly, as depicted in Figure 16.5(b), uncorrelated mixed states form a strict subset of separable states, encompassing only those separable states that can be generated without the use of shared classical randomness.

Example 16.22

Classical–classical states (16.39) are separable. In fact, they correspond to the states that can be generated by using shared randomness along with classical operations at the two agents. Mathematically, the definition (16.39) of a classical–classical state is obtained from that of a separable state in (16.67) by setting the distribution p to the joint distribution p^{XY} and the local states ρ_z^X and ρ_z^Y to rank-1 projection matrices obtained from local orthonormal bases.

Example 16.23

Classical–quantum states (16.47) and quantum–classical states (16.65) are also separable, as they can be prepared via shared randomness, classical operations at one agent, and quantum operations at the other agent. A classical–quantum state (16.47) is obtained from the definition (16.67) of separable states by choosing the distribution p as the marginal distribution p^X and the local states ρ_z^X as rank-1 projection matrices obtained from a local orthonormal basis, while the local states ρ_z^Y are arbitrary. Quantum–classical states can be characterized in a similar way.

Example 16.24

Consider the two-qubit density matrix

$$\rho^{XY} = \frac{1}{2}|\Phi_+^{XY}\rangle\langle\Phi_+^{XY}| + \frac{1}{2}|\Phi_-^{XY}\rangle\langle\Phi_-^{XY}| \tag{16.68}$$

which corresponds to the convex combination with equal weights of two maximally entangled states. One may thus generate this state by preparing the Bell state $|\Phi_+^{XY}\rangle$ or the Bell state $|\Phi_-^{XY}\rangle$ with equal probabilities.

Using the definitions of the Bell states $|\Phi_+^{XY}\rangle$ and $|\Phi_-^{XY}\rangle$, the state ρ^{XY} can be rewritten as

$$\begin{aligned}\rho^{XY} &= \frac{1}{2}\begin{bmatrix} 1 & 0 & 0 & 0 \\ 0 & 0 & 0 & 0 \\ 0 & 0 & 0 & 0 \\ 0 & 0 & 0 & 1 \end{bmatrix} \\ &= \frac{1}{2}|0\rangle\langle 0| \otimes |0\rangle\langle 0| + \frac{1}{2}|1\rangle\langle 1| \otimes |1\rangle\langle 1|. \end{aligned} \tag{16.69}$$

Therefore, despite being the mixture of two maximally entangled states, the state ρ^{XY} is separable.

This example highlights an important general principle: while for pure states, separability is equivalent to the reduced density matrices having rank one, this condition does not hold for mixed states. Specifically, in the case of the state ρ^{XY} in (16.68), the reduced density matrices are both maximally mixed, i.e., $\rho^X = \rho^Y = \pi$. Consequently, while maximally mixed reduced density matrices would imply a maximally entangled state for pure states, in the context of mixed states, the reduced density matrices can still be maximally mixed even when the overall state is separable. The key takeaway here is that, unlike pure states, one cannot infer whether a mixed state is separable or entangled solely based on the local properties of the reduced density matrices.

16.4.2 Two Characterizations of Entangled Mixed States

Entangled states can equivalently be characterized in the following ways. The first characterization follows by comparing the definition (16.67) of separable state with the general decomposition (16.19) into uncorrelated states.

Key Result 16.6 Entanglement and quasi-probabilities

A state ρ^{XY} is entangled if and only if any decomposition of the form (16.19) into uncorrelated states must include at least one negative number. Thus, entangled states are associated with quasi-probability distribution vectors γ, while separable state can be expressed in the form (16.19) by using a probability distribution γ.

The second characterization decomposes an entangled state into the convex combination of a separable state and a pure entangled state. For a proof of this result, please see the Recommended Resources.

Key Result 16.7 Mixed state as a convex combination of a separable state and an entangled pure state

Any bipartite density matrix ρ^{XY} can be written as

$$\rho^{XY} = \gamma \rho_S^{XY} + (1 - \gamma)|\psi^{XY}\rangle\langle\psi^{XY}|, \tag{16.70}$$

where $0 \le \gamma \le 1$ is a probability, ρ_S^{XY} is a separable state, and $|\psi^{XY}\rangle$ is a pure entangled state.

According to this decomposition, the entanglement properties of a mixed state ρ^{XY} can be fully accounted for by a pure state $|\psi^{XY}\rangle$.

16.5 Quantum Mutual Information

This section aims to quantify the total correlations in a state – including both classical and quantum correlations – by extending the concept of quantum mutual information, introduced in Section 14.7, from pure states to mixed states.

16.5.1 Defining the Quantum Mutual Information

As discussed in Section 14.7, the quantum mutual information equals the difference between the sum of the local quantum entropies and the joint quantum entropy. This yields the following generalization of the definition to mixed states.

Definition 16.8 Quantum mutual information

For any joint density matrix ρ^{XY}, the quantum mutual information is defined as the difference

$$\mathrm{I}(X;Y)_{\rho^{XY}} = \mathrm{H}(X)_{\rho^X} + \mathrm{H}(Y)_{\rho^Y} - \mathrm{H}(X,Y)_{\rho^{XY}} \tag{16.71}$$

between the sum of the local quantum entropies, $\mathrm{H}(X)_{\rho^X} + \mathrm{H}(Y)_{\rho^Y}$, and the joint quantum entropy $\mathrm{H}(X,Y)_{\rho^{XY}}$, with reduced density matrices $\rho^X = \mathrm{tr}_Y(\rho^{XY})$ and $\rho^Y = \mathrm{tr}_X(\rho^{XY})$.

There are two key differences between pure and mixed states in the evaluation of the mutual information (16.71).

- **Positive joint quantum entropy:** Unlike for pure states, the joint quantum entropy $\mathrm{H}(X,Y)_{\rho^{XY}}$, i.e.,

$$\mathrm{H}(X,Y)_{\rho^{XY}} = -\mathrm{tr}(\rho^{XY}\log_2(\rho^{XY})), \tag{16.72}$$

  is positive for mixed states, as it equals the entropy of the vector of eigenvalues of the joint density matrix ρ^{XY}.

- **Different local quantum entropies:** While for pure states the local quantum entropies are equal, the local quantum entropies

$$\mathrm{H}(X)_{\rho^X} = -\mathrm{tr}(\rho^X \log_2(\rho^X)) \tag{16.73}$$

and

$$\mathrm{H}(Y)_{\rho^Y} = -\mathrm{tr}(\rho^Y \log_2(\rho^Y)), \tag{16.74}$$

with $\rho^X = \mathrm{tr}_Y(\rho^{XY})$ and $\rho^Y = \mathrm{tr}_X(\rho^{XY})$, may be different for mixed states, since the reduced density matrices ρ^X and ρ^Y have generally distinct eigenvalues (see Example 16.12).

When evaluating the classical mutual information, the additivity of the logarithm plays an important role. In a similar way, the following property, proved in Appendix 16.B, will turn out to be useful to reason about the quantum mutual information.

> **Key Result 16.8 Additivity of the logarithm over the Kronecker product**
> For any two local density matrices σ^X and σ^Y, the following equality holds:
> $$\log_2(\sigma^X \otimes \sigma^Y) = \log_2(\sigma^X) \otimes I + I \otimes \log_2(\sigma^Y), \tag{16.75}$$
> where the first identity matrix I has dimensions $N^X \times N^X$, while the second has dimensions $N^Y \times N^Y$.

This property has the following implication.

> **Key Result 16.9 Expected value of the logarithm of a Kronecker product**
> For any joint density matrix ρ^{XY} and any local density matrices σ^X and σ^Y, the expected value $\langle \log_2(\sigma^X \otimes \sigma^Y) \rangle_{\rho^{XY}}$ can be expressed as the sum
> $$\begin{aligned} \langle \log_2(\sigma^X \otimes \sigma^Y) \rangle_{\rho^{XY}} &= \mathrm{tr}(\rho^{XY} \log_2(\sigma^X \otimes \sigma^Y)) \\ &= \mathrm{tr}(\rho^X \log_2(\sigma^X)) + \mathrm{tr}(\rho^Y \log_2(\sigma^Y)) \\ &= \langle \log_2(\sigma^X) \rangle_{\rho^X} + \langle \log_2(\sigma^Y) \rangle_{\rho^Y}. \end{aligned} \tag{16.76}$$

16.5.2 Quantum Mutual Information of Special Classes of States

This subsection evaluates the quantum mutual information for some of the special classes of states introduced in this chapter.

Uncorrelated states. For any uncorrelated state $\rho^{XY} = \rho^X \otimes \rho^Y$, using the identity (16.76), the quantum joint entropy is given by

$$\begin{aligned} \mathrm{H}(X, Y)_{\rho^X \otimes \rho^X} &= -\mathrm{tr}(\rho^X \otimes \rho^X \log_2(\rho^X \otimes \rho^X)) \\ &= \mathrm{H}(X)_{\rho^X} + \mathrm{H}(Y)_{\rho^Y}. \end{aligned} \tag{16.77}$$

Therefore, the quantum mutual information is zero,

$$\mathrm{I}(X;Y)_{\rho^X \otimes \rho^X} = 0, \tag{16.78}$$

and uncorrelated states do not present any correlations.

Classical–classical states. For a classical–classical state χ^{XY} in (16.35), the quantum mutual information equals the classical mutual information for the joint distribution p^{XY}, i.e.,

$$\mathrm{I}(X;Y)_{\chi^{XY}} = \mathrm{I}(X;Y)_{p^{XY}}. \tag{16.79}$$

As may be intuitive from this result, the correlations present in a classical–classical state χ^{XY} will be seen to only be classical.

To evaluate the quantum mutual information, note that the joint quantum entropy is given by the joint entropy of the joint distribution p^{XY}, i.e.,

$$\mathrm{H}(X, Y)_{\chi^{XY}} = \mathrm{H}(X, Y)_{p^{XY}}. \tag{16.80}$$

In fact, the eigenvalues of the density matrix χ^{XY} are given by the joint distribution p^{XY}. Moreover, in a similar way, the local quantum entropies equal the entropies of the corresponding marginal distributions, i.e.,

$$\mathrm{H}(X)_{\mathrm{tr}_Y(\chi^{XY})} = \mathrm{H}(X)_{p^X} \text{ and } \mathrm{H}(Y)_{\mathrm{tr}_X(\chi^{XY})} = \mathrm{H}(Y)_{p^Y}. \tag{16.81}$$

Classical–quantum states. Take now a classical–quantum state $\vec{\omega}^{XY}$ in (16.47). Using the reduced density matrices (16.48) and (16.49), the local quantum entropies are given by

$$\begin{aligned} \mathrm{H}(X)_{\mathrm{tr}_Y(\vec{\omega}^{XY})} &= \mathrm{H}(X)_{p^X}, \\ \mathrm{H}(Y)_{\mathrm{tr}_X(\vec{\omega}^{XY})} &= \mathrm{H}(Y)_{\sum_{x\in[N^X]} p_x^X \rho_x^{Y|X}}. \end{aligned} \tag{16.82}$$

Hence, the quantum entropy of subsystem X coincides with the classical entropy of the distribution p^X. In contrast, the quantum entropy of the subsystem Y is the quantum entropy of the average conditional density matrix $\sum_{x\in[N^X]} p_x^X \rho_x^{Y|X}$.

As proved in Appendix 16.C, the joint entropy $\mathrm{H}(X, Y)_{\vec{\omega}^{XY}}$ decomposes into the sum of the classical entropy of the marginal distribution p^X, $\mathrm{H}(X)_{p^X}$, and the average quantum entropy of the conditional density matrices $\rho_x^{Y|X}$, i.e., $\mathrm{H}(Y)_{\rho_x^{Y|X}}$, with respect to the distribution p^X.

Key Result 16.10 Joint quantum entropy of a classical–quantum state

For any classical–quantum state (16.47), the quantum joint entropy can be obtained as

$$\begin{aligned} \mathrm{H}(X, Y)_{\vec{\omega}^{XY}} &= \mathrm{H}(X)_{p^X} + \sum_{x\in[N^X]} p_x^X \mathrm{H}(Y)_{\rho_x^{Y|X}} \\ &= \mathrm{H}(X)_{p^X} + \mathrm{E}_{p^X}\left[\mathrm{H}(Y)_{\rho_x^{Y|X}}\right]. \end{aligned} \tag{16.83}$$

Thus, the overall uncertainty present in a classical–quantum state is the sum of the classical uncertainty associated with the distribution p^X and of the average residual uncertainty present in the conditional states $\{\rho_x^{Y|X}\}_{x\in[N^X]}$. The decomposition (16.83) can be interpreted as a chain rule for the joint quantum entropy of classical–quantum states, as it resembles the corresponding property for the joint classical entropy presented in Section 13.5.

Using the local and joint quantum entropies, the quantum mutual information is given as follows.

> **Key Result 16.11 Quantum mutual information of a classical–quantum state**
> The quantum mutual information for a classical–quantum state $\vec{\omega}^{XY}$ in (16.47) is given by
> $$\mathrm{I}(X;Y)_{\vec{\omega}^{XY}} = \mathrm{H}(Y)_{\sum_{x\in[N^X]} p_x^X \rho_x^{Y|X}} - \mathrm{E}_{p^X}\left[\mathrm{H}(Y)_{\rho_x^{Y|X}}\right]. \tag{16.84}$$

As indicated by expression (16.84), the total amount of correlation in a classical–quantum state depends on the degree of variation among the conditional density matrices $\{\rho_x^{Y|X}\}_{x\in[N^X]}$. If these conditional density matrices are identical, the quantum mutual information is zero, indicating that the state is uncorrelated. Conversely, if they differ, the state exhibits correlations that will be seen to encompass both classical and quantum components, though it remains separable.

The quantum mutual information (16.84) is a function of the ensemble of conditional density matrices

$$\mathcal{E}^{Y|X} = \{(\rho_x^{Y|X}, p_x^X)\}_{x\in[N^X]} \tag{16.85}$$

that underlies the definition of the classical–quantum state (16.47). It does not depend on the orthonormal basis $\mathcal{M}^X = \{|v_x^X\rangle\}_{x\in[N^X]}$. When viewed as a function of the ensemble (16.85), the quantum mutual information (16.84) is known as **Holevo information**.

16.5.3 Properties of the Quantum Mutual Information

Section 14.7 showed that the quantum mutual information for pure states satisfies the properties of non-negativity, as well as of insubordination to the classical uncertainty-increases-globally principle. It will now be observed that the same properties generalize to mixed states.

> **Key Result 16.12 Non-negativity and minimum value of the quantum mutual information**
> The quantum mutual information is non-negative,
> $$\mathrm{I}(X;Y)_{\rho^{XY}} \geq 0, \tag{16.86}$$
> and equals zero if and only if the pure state ρ^{XY} is uncorrelated.

This property is equivalent to the subadditivity of the joint quantum entropy

$$H(X, Y)_{\rho^{XY}} \leq H(X)_{\rho^X} + H(Y)_{\rho^Y}, \tag{16.87}$$

with reduced density matrices $\rho^X = \mathrm{tr}_Y(\rho^{XY})$ and $\rho^Y = \mathrm{tr}_X(\rho^{XY})$, where equality in (16.87) is attained if and only if the state ρ^{XY} is uncorrelated.

The result (16.86) follows directly from the equality (16.76), which implies that the quantum mutual information can be expressed in terms of the quantum relative entropy as

$$\begin{aligned} I(X;Y)_{\rho^{XY}} &= \mathrm{tr}(\rho^{XY}\log_2(\rho^{XY})) - \mathrm{tr}(\rho^{XY}\log_2(\rho^X \otimes \rho^Y)) \\ &= D(\rho^{XY}||\rho^X \otimes \rho^Y). \end{aligned} \tag{16.88}$$

Accordingly, the quantum mutual information measures the distance from the uncorrelated state $\rho^X \otimes \rho^Y$ obtained from the reduced density matrices ρ^X and ρ^Y.

As discussed in Section 12.5 (see (12.50)), the classical uncertainty-increases-globally principle is reflected in the inequality $I(X, Y)_{p^{XY}} \leq \min(H(X)_{p^X}, H(Y)_{p^Y})$, which holds for any joint distribution p^{XY}. This inequality indicates that the shared information at the two subsystems cannot exceed the information present at either subsystem. As formalized next, this inequality is only guaranteed for separable quantum states, and it can be violated by entangled states.

Key Result 16.13 Insubordination

The quantum mutual information satisfies the upper bound

$$I(X, Y)_{\rho^{XY}} \leq \min(H(X)_{\rho^X}, H(Y)_{\rho^Y}) \tag{16.89}$$

if the state is separable, while it satisfies the inequality

$$I(X, Y)_{\rho^{XY}} \leq 2\min(H(X)_{\rho^X}, H(Y)_{\rho^Y}) \tag{16.90}$$

for entangled states. Furthermore, there are entangled states for which the strict inequality $I(X, Y)_{\rho^{XY}} > \min(H(X)_{\rho^X}, H(Y)_{\rho^Y})$ holds.

These inequalities, proved in Appendix 16.D, demonstrate that entanglement can create correlations so strong that they surpass the information associated with the individual subsystems, thereby violating the principle that uncertainty should increase globally.

Example 16.25

As seen in Section 14.7, for an entangled pure state $|\psi^{XY}\rangle$, the quantum mutual information attains the upper bound (16.90) with equality. In fact, the quantum mutual information is given by $I(X, Y)_{|\psi^{XY}\rangle\langle\psi^{XY}|} = 2H(\lambda)$, where λ is the Schmidt probability vector, and the local quantum entropies are both equal to $H(\lambda)$.

The bounds in (16.89) and (16.90) imply the following sufficient condition for entanglement.

Key Result 16.14 Insubordination-based sufficient condition for entanglement
If the quantum mutual information of a state ρ^{XY} is larger than the minimum local quantum entropy, the state is entangled, i.e.,

$$\mathrm{I}(X;Y)_{\rho^{XY}} > \min(\mathrm{H}(X)_{\rho^X}, \mathrm{H}(Y)_{\rho^Y}) \Rightarrow \rho^{XY} \text{ is entangled.} \tag{16.91}$$

Note that this is only a sufficient condition, as there are mixed entangled states that do not satisfy this condition.

Example 16.26

Consider a generalization of the two-qubit state (16.68), in which the two Bell states are mixed with generally different probabilities η and $1-\eta$, i.e.,

$$\rho^{XY} = \eta|\Phi_+^{XY}\rangle\langle\Phi_+^{XY}| + (1-\eta)|\Phi_-^{XY}\rangle\langle\Phi_-^{XY}|. \tag{16.92}$$

Since the states $\{|\Phi_+^{XY}\rangle, |\Phi_-^{XY}\rangle\}$ are orthogonal, the eigenvalues of the density matrix ρ^{XY} are given by $[\eta, 1-\eta]^T$, and the joint entropy is $\mathrm{H}(X,Y)_{\rho^{XY}} = \mathrm{h}(\eta)$. Furthermore, the reduced density matrices ρ^X and ρ^Y are maximally mixed, and hence the local quantum entropies are $\mathrm{H}(X)_{\rho^X} = \mathrm{H}(Y)_{\rho^Y} = 1$, yielding the quantum mutual information

$$\mathrm{I}(X;Y)_{\rho^{XY}} = 2 - \mathrm{h}(\eta). \tag{16.93}$$

As discussed earlier, for $\eta = 0.5$, the state is separable. In this case, the quantum mutual information equals $\mathrm{I}(X;Y)_{\rho^{XY}} = 1$, which attains the upper bound (16.89) with equality. For a probability $\eta \neq 0.5$, the quantum mutual information is larger than 1 bit, and hence it exceeds the upper bound (16.89). By the sufficient condition (16.91), this implies that the state (16.92) is entangled for $\eta \neq 0.5$. Furthermore, with $\eta = 0$ or $\eta = 1$, the quantum mutual information equals 2 bits, achieving the upper bound (16.90) with equality. In these cases, the state is maximally entangled.

Example 16.27

For the Werner state (16.11), the quantum mutual information is given by

$$\begin{aligned}\mathrm{I}(X;Y)_{\rho^{XY}} &= 2 - \mathrm{H}(X,Y)_{\rho^{XY}} \\ &= 2 + \left(\eta + \frac{1-\eta}{4}\right)\log_2\left(\eta + \frac{1-\eta}{4}\right) + 3\left(\frac{1-\eta}{4}\right)\log_2\left(\frac{1-\eta}{4}\right),\end{aligned} \tag{16.94}$$

since the vector of eigenvalues of the joint density matrix ρ^{XY} is $[\eta+(1-\eta)/4, (1-\eta)/4, (1-\eta)/4, (1-\eta)/4]^T$. The quantum mutual information is plotted in Figure 16.6 as a function of the probability η.

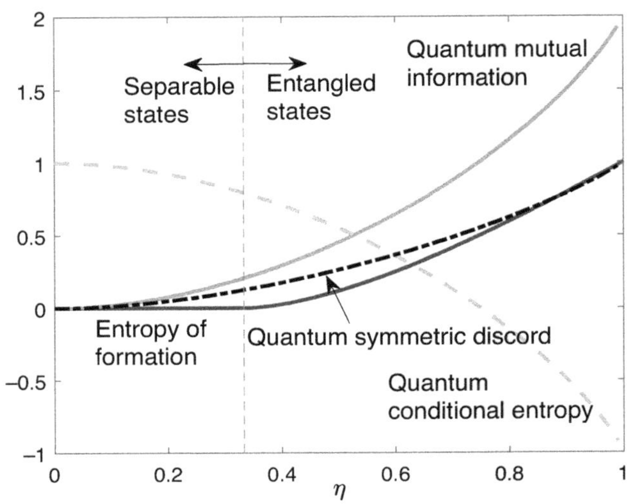

Figure 16.6 Quantum mutual information, quantum symmetric discord, entropy of formation, and quantum conditional entropy versus the probability η for the Werner state (16.11).

The quantum mutual information equals zero only for $\eta = 0$, corresponding to an uncorrelated state. In contrast, for $\eta > 0$, the quantum mutual information is positive, and thus the state exhibits some level of correlation. Moreover, the sufficient condition (16.91) implies that the Werner state is entangled for a probability η large enough to ensure that the quantum mutual information exceeds 1 bit. In fact, it will be shown later that the Werner state is entangled for $\eta > 1/3$ and separable for $0 < \eta \leq 1/3$.

Example 16.28

For dimensions $N^X = 2$ and $N^Y = 4$, consider the state

$$\rho^{XY} = \eta|\psi^{XY}\rangle\langle\psi^{XY}| + (1-\eta)|\phi^{XY}\rangle\langle\phi^{XY}| \tag{16.95}$$

with pure entangled states $|\psi^{XY}\rangle = 1/\sqrt{2}(|00\rangle + |11\rangle)$ and $|\phi^{XY}\rangle = 1/\sqrt{2}(|02\rangle + |13\rangle)$, where $0 \leq \eta \leq 1$ is a probability. Since the states $|\psi^{XY}\rangle$ and $|\phi^{XY}\rangle$ are orthogonal, the vector of eigenvalues of matrix ρ^{XY} is given by $[\eta, 1-\eta]^T$, and the quantum joint entropy is $\mathrm{H}(X,Y)_{\rho^{XY}} = \mathrm{h}(\eta)$. Furthermore, using (16.31) and (16.32), one can evaluate the local quantum entropies as $\mathrm{H}(X)_{\rho^X} = 1$ and

$$\mathrm{H}(Y)_{\mathrm{tr}_X(\rho^{XY})} = \mathrm{H}([\eta/2, \eta/2, (1-\eta)/2, (1-\eta)/2]^T). \tag{16.96}$$

For any probability $0 \leq \eta \leq 1$, this yields the quantum mutual information

$$\mathrm{I}(X;Y)_{\rho^{XY}} = 2\log_2(2) = 2. \tag{16.97}$$

The quantum mutual information thus satisfies the sufficient condition (16.91), indicating that the state ρ^{XY} is entangled.

16.5.4 Data Processing Inequality Under Local Channels ⏭

Chapter 15 highlighted another key property of quantum mutual information for pure states: its invariance under local unitary transformations (see Section 15.3). This subsection extends this essential property to mixed states by considering the effect of general local channels applied to the two subsystems.

Specifically, consider the application of channel $\mathcal{K}^X(\cdot)$ to subsystem X and of channel $\mathcal{K}^Y(\cdot)$ to subsystem Y. As formalized in the following, no local channels can increase the correlations between the two subsystems. This result provides a counterpart for the data processing inequality under local channels presented in Section 12.7.3 for classical systems.

Key Result 16.15 Data processing inequality for the quantum mutual information under local channels

For any joint state ρ^{XY}, given any two local channels $\mathcal{K}^X(\cdot)$ and $\mathcal{K}^Y(\cdot)$, the quantum mutual information satisfies the inequality

$$\mathrm{I}(X;Y)_{\mathcal{K}^X\otimes\mathcal{K}^Y(\rho^{XY})} \leq \mathrm{I}(X;Y)_{\rho^{XY}}. \tag{16.98}$$

Furthermore, in the special case of local unitary transformations, the following equality holds:

$$\mathrm{I}(X;Y)_{U^X\otimes U^Y(\rho^{XY})} = \mathrm{I}(X;Y)_{\rho^{XY}}. \tag{16.99}$$

Given the equality (16.88), the property (16.98) follows directly from the data processing inequality for the quantum relative entropy, which was introduced in Section 11.7. In fact, applying the data processing inequality to the channel $\mathcal{K}^X \otimes \mathcal{K}^Y(\cdot)$ yields the inequality

$$\begin{aligned}\mathrm{I}(X;Y)_{\rho^{XY}} =&\mathrm{D}(\rho^{XY}||\rho^X \otimes \rho^X)\\ \geq&\mathrm{D}(\mathcal{K}^X \otimes \mathcal{K}^Y(\rho^{XY})||\mathcal{K}^X(\rho^X) \otimes \mathcal{K}^Y(\rho^X))\\ =&\mathrm{I}(X;Y)_{\mathcal{K}^X\otimes\mathcal{K}^Y(\rho^{XY})}.\end{aligned} \tag{16.100}$$

The last inequality follows from the calculation of the reduced density matrices

$$\mathrm{tr}_X(\mathcal{K}^X \otimes \mathcal{K}^Y(\rho^{XY})) = \mathcal{K}^Y(\rho^Y) \tag{16.101}$$

and $\mathrm{tr}_Y(\mathcal{K}^X \otimes \mathcal{K}^Y(\rho^{XY})) = \mathcal{K}^X(\rho^X)$. These equations show that the reduced density matrices are only affected by the channels that apply on the given state.

16.6 Classical vs. Quantum Correlations

As discussed in Section 16.5, the quantum mutual information captures the total level of correlation present in a quantum state, encompassing both classical correlations, which can be extracted with a single measurement setting, and quantum correlations. This section focuses on defining a measure of classical correlations by generalizing the maximum measured mutual information, introduced in Chapter 14 for pure states, to mixed states. The difference between the quantum mutual information and the maximum measured mutual information will then be interpreted as a measure of quantum correlations, which is known as the quantum symmetric discord. Quantum correlations encompass, but are not limited to, entanglement.

16.6.1 Maximum Measured Mutual Information

As seen in Section 14.6, the maximum measured mutual information corresponds to the maximum value of the mutual information between classical measurement outputs obtained via local complete measurements.

Definition 16.9 Maximum measured mutual information
For any joint state ρ^{XY}, the maximum measured mutual information is defined as

$$\mathrm{J}(X;Y)_{\rho^{XY}} = \max_{\mathcal{M}^X \otimes \mathcal{M}^Y} \mathrm{I}(X;Y)_{p^{XY}(\rho^{XY}|\mathcal{M}^X \otimes \mathcal{M}^Y)}, \tag{16.102}$$

i.e., as the maximum mutual information between the measurement outputs $(\mathrm{x}, \mathrm{y}) \sim p^{XY}(\rho^{XY}|\mathcal{M}^X \otimes \mathcal{M}^Y)$ obtained with a local complete measurement setting $\mathcal{M}^X \otimes \mathcal{M}^Y$.

Since, by (16.39) and (16.40), the post-measurement state $\mathcal{M}^X \otimes \mathcal{M}^Y(\rho^{XY})$ has eigenvalues given by the joint measurement distribution $p^{XY}(\rho^{XY}|\mathcal{M}^X \otimes \mathcal{M}^Y)$, the maximum measured mutual information can equivalently be expressed as follows.

Key Result 16.16 Maximum measured mutual information (equivalent definition)
For any joint state ρ^{XY}, the maximum measured mutual information (16.102) can equivalently be written as

$$\mathrm{J}(X;Y)_{\rho^{XY}} = \max_{\mathcal{M}^X \otimes \mathcal{M}^Y} \mathrm{I}(X;Y)_{\mathcal{M}^X \otimes \mathcal{M}^Y(\rho^{XY})}, \tag{16.103}$$

i.e., as the maximum quantum mutual information for the post-measurement state $\mathcal{M}^X \otimes \mathcal{M}^Y(\rho^{XY})$.

By its definition (16.102), the maximum measured mutual information captures the classical correlations in a state – namely, those correlations that can be extracted via an optimized local measurement setting $\mathcal{M}^X \otimes \mathcal{M}^Y$.

Additionally, as shown by the equivalent formulation (16.103), the maximum measured mutual information can be interpreted as the highest level of correlation that remains after a local complete measurement. Indeed, the correlations quantified by the maximum measured mutual information are those retained in the post-measurement state $\mathcal{M}^X \otimes \mathcal{M}^Y(\rho^{XY})$. They are therefore not eliminated by local complete measurements.

16.6.2 Properties of the Maximum Measured Mutual Information

The maximum measured mutual information satisfies the following inequalities (see Appendix 16.F for a proof).

Key Result 16.17 Bounds on the maximum measured mutual information
The maximum measured mutual information satisfies the inequalities

$$0 \leq \mathrm{J}(X;Y)_{\rho^{XY}} \leq \min(\mathrm{H}(X)_{\rho^X}, \mathrm{H}(Y)_{\rho^Y}, \mathrm{I}(X;Y)_{\rho^{XY}}). \tag{16.104}$$

The lower bound $\mathrm{J}(X;Y)_{\rho^{XY}} = 0$ is attained if and only if the pure state ρ^{XY} is uncorrelated; and the upper bound $\mathrm{J}(X;Y)_{\rho^{XY}} = \mathrm{I}(X;Y)_{\rho^{XY}}$ is achieved if and only if the state is classical–classical, i.e.,

$$\mathrm{J}(X;Y)_{\chi^{XY}} = \mathrm{I}(X;Y)_{\chi^{XY}}. \tag{16.105}$$

The inequality $\mathrm{J}(X;Y)_{\rho^{XY}} \leq \min(\mathrm{H}(X)_{\rho^X}, \mathrm{H}(Y)_{\rho^Y})$ shows that the maximum measured mutual information cannot exceed the minimum uncertainty associated with the two subsystems, thus abiding by the classical uncertainty-increases-globally property. This reflects the interpretation of the maximum measured mutual information as a measure of classical correlations.

The equality (16.105) follows from the characterization (16.103) and from the property (16.43) of classical–classical states. In fact, classical–classical states are left undisturbed by the measurement setting $\mathcal{M}^X \otimes \mathcal{M}^Y$ constructed from the local eigenbases $\mathcal{M}^X$ and $\mathcal{M}^Y$. This equality demonstrates that classical–classical states possess only classical correlations. Specifically, as illustrated in Figure 16.7(a), for a classical–classical state χ^{XY}, all the correlations present in the state – which are quantified by the quantum mutual information $\mathrm{I}(X;Y)_{\chi^{XY}}$ – are captured by the maximum measured mutual information $\mathrm{J}(X;Y)_{\chi^{XY}}$.

More broadly, as shown in Figure 16.7(b), for states ρ^{XY} that are not classical–classical, one has the strict inequality $\mathrm{J}(X;Y)_{\rho^{XY}} < \mathrm{I}(X;Y)_{\rho^{XY}}$, and the state exhibits both classical and quantum correlations.

Furthermore, while for classical–classical states the optimal measurement setting is given by the local eigenbases, for more general states the maximization over the measurement setting in (16.102) is not straightforward. Given the difficulty in optimizing over the measurement setting, it has also been proposed to redefine the measured mutual information by fixing the measurement setting bases $\mathcal{M}^X$ and $\mathcal{M}^Y$ to the local eigenbases even for non-classical–classical states (see Recommended Resources).

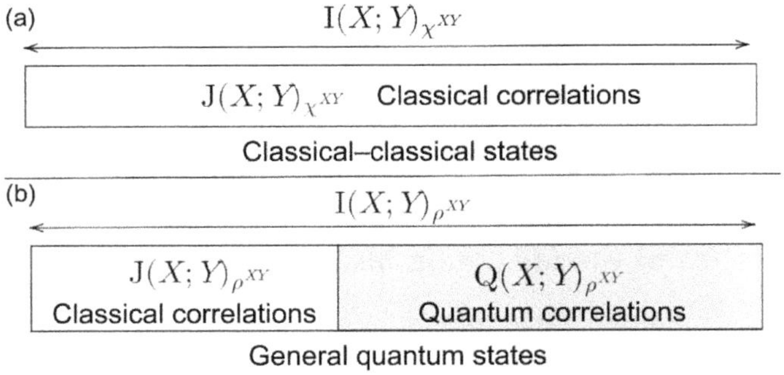

Figure 16.7 For a classical–classical state χ^{XY}, all the correlations present in the state – which are quantified by the quantum mutual information $\mathrm{I}(X;Y)_{\chi^{XY}}$ – are captured by the maximum measured mutual information $\mathrm{J}(X;Y)_{\chi^{XY}}$ (a). For states ρ^{XY} that are not classical–classical, one has the strict inequality $\mathrm{J}(X;Y)_{\rho^{XY}} < \mathrm{I}(X;Y)_{\rho^{XY}}$, and the state exhibits both classical and quantum correlations (b).

Example 16.29

Consider the convex combination of the classical–classical two-qubit state $\chi^{XY} = 0.5(|0\rangle\langle 0|\otimes|0\rangle\langle 0|+|1\rangle\langle 1|\otimes|1\rangle\langle 1|)$ and the maximally entangled pure state $|\Phi_+^{XY}\rangle\langle\Phi_+^{XY}|$,

$$\rho^{XY} = \frac{\eta}{2}(|0\rangle\langle 0| \otimes |0\rangle\langle 0| + |1\rangle\langle 1| \otimes |1\rangle\langle 1|) + (1-\eta)|\Phi_+^{XY}\rangle\langle\Phi_+^{XY}|, \qquad (16.106)$$

with some probability $0 \le \eta \le 1$. Figure 16.8 shows the quantum mutual information and the maximum measured mutual information as a function of the probability η.

Figure 16.8 Quantum mutual information and maximum measured mutual information as a function of the probability η for the state (16.106).

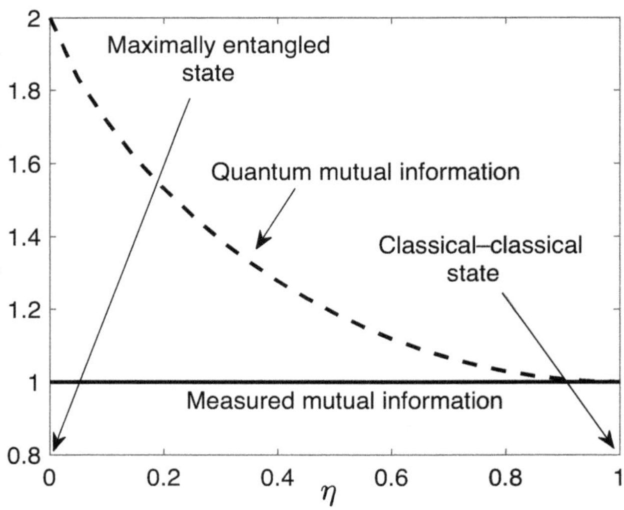

The quantum mutual information ranges from the maximum value of 2 bits – which is attained at $\eta = 0$ for the maximally entangled state $|\Phi_+^{XY}\rangle$ – to the value of 1 bit – which is attained for the classical–classical state χ^{XY}, i.e., at $\eta = 1$. In contrast, the maximum measured mutual information does not depend on the probability η, being equal to 1 bit for all values of η. This can be proved by noting that the quantum mutual information $\mathrm{I}(X;Y)_{\mathcal{M}^X \otimes \mathcal{M}^Y(\rho^{XY})}$ takes the maximum value of 1 bit for a local measurement in the computational basis.

Therefore, the quantum mutual information and the maximum measured mutual information coincide for $\eta = 1$, i.e., when the state ρ^{XY} is classical–classical. Otherwise, there is a gap between quantum mutual information and the maximum measured mutual information, indicating that the state ρ^{XY} also has some level of quantum correlation.

16.6.3 Quantum Symmetric Discord

Given that the quantum mutual information captures both classical and quantum correlations, while the maximum measured mutual information quantifies only classical correlations, their difference provides a measure for the level of quantum correlations present in a state.

Definition 16.10 Quantum symmetric discord

For any state ρ^{XY}, the quantum symmetric discord is defined as the difference between the quantum mutual information and the maximum measured mutual information:

$$\mathrm{Q}(X;Y)_{\rho^{XY}} = \mathrm{I}(X;Y)_{\rho^{XY}} - \mathrm{J}(X;Y)_{\rho^{XY}}. \tag{16.107}$$

A state is **quantumly correlated** if the quantum symmetric discord is positive. Recalling the result (16.104) and (16.105), as illustrated in Figure 16.5, this means that all states that are not classical–classical are quantumly correlated. Not all quantumly correlated states, however, are entangled.

Example 16.30

For the Werner state (16.11), Figure 16.6 reports the quantum symmetric discord, along with the quantum mutual information discussed in Example 16.27. The quantum symmetric discord is optimized numerically over the measurement bases $\mathcal{M}^X$ and $\mathcal{M}^Y$ (see Appendix 16.G). The figure shows that the quantum symmetric discord can be positive even for separable states, i.e., even when $\eta \leq 1/3$. Thus, in this regime, the state has quantum correlations that are weaker than entanglement.

An alternative measure of quantum correlations can be devised by adopting the resource-theoretic approach of measuring the distance to the set of states that are free from quantum correlations. Since the set of free states coincides with the classical–classical states, the resulting measure, known as the **relative entropy of quantumness**, is defined as

$$\mathrm{Q_R}(X;Y)_{\rho^{XY}} = \min_{\chi^{XY}} \mathrm{D}(\rho^{XY}||\chi^{XY}), \tag{16.108}$$

where minimization is over all classical–classical states χ^{XY}.

16.7 Detecting and Measuring Entanglement ⏭

Section 16.6 introduced measures of classical and quantum correlations, and highlighted the fact that for mixed states not all quantum correlations amount to entanglement. This section focuses on the problems of detecting and measuring entanglement. Specifically, a sufficient condition for the detection of entanglement is first described, and then several measures of entanglement for mixed states are introduced.

16.7.1 PPT Sufficient Condition for Entanglement

In principle, in order to determine if a state is entangled, one should verify that the state ρ^{XY} cannot be expressed in the form (16.67). However, this is generally a prohibitively complex task. A sufficient condition to determine the presence of entanglement, namely (16.91), was provided earlier in this chapter based on the insubordination property of entangled states.

An alternative condition that often proves useful is the **positive partial transpose (PPT)** criterion, which is based on the following definition.

Definition 16.11 Partial transpose

For any state ρ^{XY}, using the bi-index notation (16.4), the partial transpose $(\rho^{XY})^{T^X}$ with respect to subsystem X has entries

$$[(\rho^{XY})^{T^X}]_{xy,x'y'} = \rho^{XY}_{x'y,xy'}; \tag{16.109}$$

while the partial transpose $(\rho^{XY})^{T^Y}$ with respect to subsystem Y has entries

$$[(\rho^{XY})^{T^Y}]_{xy,x'y'} = \rho^{XY}_{xy',x'y}. \tag{16.110}$$

The partial transpose operation can equivalently be expressed by leveraging the decomposition (16.19) of an arbitrary density matrix into uncorrelated states. Specifically, using (16.19), the partial transpose $(\rho^{XY})^{T^X}$ of matrix ρ^{XY} with respect to subsystem X is given by

$$(\rho^{XY})^{T^X} = \left(\sum_{z\in[M]} \gamma_z \rho_z^X \otimes \rho_z^Y\right)^{T^X} = \sum_{z\in[M]} \gamma_z (\rho_z^X)^T \otimes \rho_z^Y, \tag{16.111}$$

thus transposing only the density matrices pertaining to subsystem X. In a similar way, the partial transpose with respect to subsystem Y, denoted as $(\rho^{XY})^{T^Y}$, is obtained as

$$(\rho^{XY})^{T^Y} = \left(\sum_{z \in [M]} \gamma_z \rho_z^X \otimes \rho_z^Y \right)^{T^Y} = \sum_{z \in [M]} \gamma_z \rho_z^X \otimes (\rho_z^Y)^T. \qquad (16.112)$$

Example 16.31

In order to evaluate the partial transpose with respect to subsystem Y using (16.112), it is particularly convenient to leverage the block decomposition (16.62), which yields

$$(\rho^{XY})^{T^Y} = \sum_{x \in [N^X]} \sum_{x' \in [N^X]} |x\rangle\langle x| \otimes (r_{x,x'}^{XY})^T$$
$$= \begin{bmatrix} (r_{0,0}^{XY})^T & (r_{0,1}^{XY})^T & \cdots & (r_{0,N^X-1}^{XY})^T \\ (r_{1,0}^{XY})^T & (r_{1,1}^{XY})^T & \cdots & (r_{1,N^X-1})^T \\ \vdots & \vdots & \ddots & \vdots \\ (r_{N^X-1,0}^{XY})^T & \cdots & & (r_{N^X-1,N^X-1}^{XY})^T \end{bmatrix}. \qquad (16.113)$$

Suppose now that the density matrix is separable, so that it admits a decomposition into uncorrelated states of the form (16.67). Using (16.111), the partial transpose with respect to subsystem X is given by

$$(\rho^{XY})^{T^X} = \left(\sum_{z \in [M]} p_z \rho_z^X \otimes \rho_z^Y \right)^{T^X} = \sum_{z \in [M]} p_z (\rho_z^X)^T \otimes \rho_z^Y. \qquad (16.114)$$

This is a valid density matrix, as it is a convex combination of the density matrices $\{(\rho_z^X)^T \otimes \rho_z^Y\}_{z \in [M]}$. Note that the matrix $(\rho_z^X)^T$ is indeed a density matrix, since it has the same eigenvalues as matrix ρ_z^X (see Problem 16.6).

This observation proves that, if the partial transpose is not a valid density matrix, then the state must be entangled. In particular, since the trace of the partial transpose is also equal to 1, the partial transpose is not a valid density matrix when it has at least one negative eigenvalue. The resulting PPT sufficient condition for entanglement, also known as the **Peres-Horodecki theorem**, is formalized as follows.

Key Result 16.18 PPT sufficient condition for entanglement
A density matrix ρ^{XY} corresponds to an entangled state if any of the eigenvalues of partial transpose matrix $(\rho^{XY})^{T^X}$ is negative, i.e.,

$$(\rho^{XY})^{T^X} \text{ is not positive semi-definite} \Rightarrow \rho^{XY} \text{ is entangled.} \qquad (16.115)$$

In the sufficient condition (16.115), one can equivalently replace the partial transpose $(\rho^{XY})^{T^X}$ with respect to subsystem X with the partial transpose $(\rho^{XY})^{T^Y}$ with respect to subsystem Y. In fact, the eigenvalues of the two partial transpose matrices are the same, given the equality

$$(\rho^{XY})^{T^X} = ((\rho^{XY})^{T^Y})^T. \tag{16.116}$$

It can also be proved that the PPT condition is both necessary and sufficient for a few special cases.

> **Key Result 16.19 PPT necessary and sufficient condition for entanglement**
> The density matrix ρ^{XY} corresponds to an entangled state if and only if any of the eigenvalues of matrix $(\rho^{XY})^{T^X}$ or $(\rho^{XY})^{T^Y}$ is negative for systems with: (i) $N^X = N^Y = 2$; (ii) $N^X = 2$ and $N^Y = 3$; or (iii) $N^X = 3$ and $N^Y = 2$.

Example 16.32

Since the Werner state (16.11) is a two-qubit state, the **PPT** criterion provides a necessary and sufficient condition for entanglement. Using (16.113), the partial transpose with respect to subsystem Y is given by

$$\begin{aligned}(\rho^{XY})^{T^Y} &= \begin{bmatrix} \begin{bmatrix} \frac{1-\eta}{4} & 0 \\ 0 & \frac{1+\eta}{4} \end{bmatrix}^T & \begin{bmatrix} 0 & 0 \\ -\frac{\eta}{2} & 0 \end{bmatrix}^T \\ \begin{bmatrix} 0 & -\frac{\eta}{2} \\ 0 & 0 \end{bmatrix}^T & \begin{bmatrix} \frac{1+\eta}{4} & 0 \\ 0 & \frac{1-\eta}{4} \end{bmatrix}^T \end{bmatrix} \\ &= \begin{bmatrix} \frac{1-\eta}{4} & 0 & 0 & -\frac{\eta}{2} \\ 0 & \frac{1+\eta}{4} & 0 & 0 \\ 0 & 0 & \frac{1+\eta}{4} & 0 \\ -\frac{\eta}{2} & 0 & 0 & \frac{1-\eta}{4} \end{bmatrix}. \end{aligned} \tag{16.117}$$

The vector of eigenvalues of the partial transpose $(\rho^{XY})^{T^Y}$ is given as $[(1-\eta)/4,$ $(1-\eta)/4,$ $(1-\eta)/4, (1-3\eta)/4]^T$, with eigenvalues listed in non-decreasing order. Therefore, as anticipated in Example 16.27, the Werner state is entangled if and only if $\eta > 1/3$.

As summarized above, the PPT condition is only sufficient, except for a number of special cases. This implies that there are entangled states that cannot be detected by using the PPT criterion. Specifically, these are the entangled states whose partial transpose is positive semi-definite. Such entangled states are known as **bound-entangled states** or **PPT states**. PPT states were shown to exist as long as N^X or N^Y is of size at least four.

While being entangled, PPT states, which are necessarily mixed, lack some of the key properties of entanglement that characterize entangled pure states. In particular, PPT states cannot be used for teleportation or dense coding. Therefore, PPT states do not exhibit some of the main informational benefits of pure entangled states, and they are said to possess **bound entanglement**. The reader is referred to the Recommended Resources for details on PPT states.

Another general framework to define classes of sufficient conditions for entanglement is based on the convexity of the set of separable states. This approach identifies observables W whose average is non-negative for all separable states ρ^{XY}, i.e., $\langle W\rangle_{\rho^{XY}} \geq 0$, while being negative, $\langle W\rangle_{\rho^{XY}} < 0$, for at least one entangled state ρ^{XY}. Any such observable is referred to as an **entanglement witness**, as it can be used to prove – but not disprove – the claim that a state is entangled (see Problems 16.10 and 16.11).

16.7.2 Entanglement Measures and Entanglement Monotones

Following the general approach introduced in Section 15.4, entanglement measures and entanglement monotones can be defined by imposing monotonicity constraints on the behavior of the metrics as a result of the application of an LOCC protocol.

As defined in Section 15.4, an entanglement measure $\mathcal{E}(X;Y)_{\rho^{XY}}$ is any function of a bipartite state ρ^{XY} that is monotonic with respect to LOCC. Accordingly, an entanglement measure satisfies the inequality

$$\mathcal{E}(X;Y)_{\rho^{XY}} \geq \mathcal{E}(X;Y)_{\sigma^{XY}} \tag{16.118}$$

for all states σ^{XY} that can be obtained from state ρ^{XY} via an LOCC protocol, i.e., $\rho^{XY} \underset{\text{LOCC}}{\longrightarrow} \sigma^{XY}$. The rationale behind this definition is that local operations and classical communications cannot increase entanglement, which requires global quantum operations to be enhanced.

Furthermore, an **entanglement monotone** is any function $\mathfrak{E}(X;Y)_{\rho^{XY}}$ of state ρ^{XY} that that cannot increase on average as a result of the application of an SLOCC. An entanglement monotone hence satisfies the inequality

$$\mathfrak{E}(X;Y)_{\rho^{XY}} \geq \sum_{z\in[M]} p_z \mathfrak{E}(X;Y)_{\sigma_z^{XY}} \tag{16.119}$$

for all ensembles $\{(\sigma_z^{XY}, p_z)\}_{z\in[M]}$ that can be obtained from state ρ^{XY} by using an SLOCC protocol, i.e., $\rho^{XY} \underset{\text{SLOCC}}{\longrightarrow} \{(\sigma_z^{XY}, p_z)\}_{z\in[M]}$. Recall that this is in the sense that state σ_z^{XY} is produced by the SLOCC protocol with probability p_z. Entanglement monotones are also entanglement measures, while the opposite is not true.

In the following, three entanglement monotones are reviewed: the relative entropy of entanglement, the entanglement of formation, and the negativity. It is emphasized that using different measures of entanglement generally yields distinct conclusions on which of two states is "more entangled" than the other. Specifically, of two states, one state may have a larger entanglement according to one metric, while a different metric may assign a larger entanglement to the other state.

16.7.3 Relative Entropy of Entanglement

The relative entropy of entanglement follows the resource-theoretic principle of evaluating the distance between state ρ^{XY} and the set of entanglement-free states.

Definition 16.12 Relative entropy of entanglement

For any state ρ^{XY}, the relative entropy of entanglement is defined as

$$\mathfrak{E}_R(X;Y)_{\rho^{XY}} = \min_{\sigma^{XY}} \mathrm{D}(\rho^{XY}||\sigma^{XY}), \tag{16.120}$$

where minimization is over all possible separable states σ^{XY}.

The reader is referred to the Recommended Resources for a proof that the relative entropy of entanglement is an entanglement monotone. The relative entropy of entanglement is non-negative, and is equal to zero if and only if the state ρ^{XY} is separable. Thus, a positive value of the relative entropy of entanglement is a necessary and sufficient condition for entanglement. A summary of all resource-based correlation measures introduced in this chapter is shown in Table 16.1.

As follows from their definitions, the three resource-theoretic measures in the table are related by the inequalities

$$\mathfrak{E}_\mathrm{R}(X;Y)_{\rho^{XY}} \leq \mathrm{Q}_\mathrm{R}(X;Y)_{\rho^{XY}} \leq \mathrm{I}(X;Y)_{\rho^{XY}}. \tag{16.121}$$

Indeed, the quantum correlations, measured by the relative entropy of quantumness $\mathrm{Q}_\mathrm{R}(X;Y)_{\rho^{XY}}$, amount to a fraction of the overall correlations captured by the quantum mutual information $\mathrm{I}(X;Y)_{\rho^{XY}}$. Furthermore, entanglement, quantified by the relative entropy of entanglement $\mathfrak{E}_\mathrm{R}(X;Y)_{\rho^{XY}}$, accounts for a portion of the quantum correlations measured by $\mathrm{Q}_\mathrm{R}(X;Y)_{\rho^{XY}}$.

16.7.4 Entanglement of Formation

The **entanglement of formation**, another entanglement monotone, serves as a natural generalization of the entanglement entropy from pure to mixed states. To extend the entanglement entropy, the entanglement of formation uses a method known as the **convex roof construction**. As explored next, the convex roof construction lifts a metric defined for pure states to mixed states by optimizing over all pure state ensembles that realize the given mixed state.

Table 16.1 Summary of resource-theoretic correlation measures for mixed states

Correlation type	Metric entropy	Definition
Classical + quantum	Quantum mutual information	$\mathrm{I}(X;Y)_{\rho^{XY}} = \mathrm{D}(\rho^{XY}\|\|\rho^X \otimes \rho^Y)$
Quantum	Relative entropy of quantumness	$\mathrm{Q}_\mathrm{R}(X;Y)_{\rho^{XY}} = \min_{\chi^{XY}} \mathrm{D}(\rho^{XY}\|\|\chi^{XY})$
Entanglement	Relative entropy of entanglement	$\mathfrak{E}_\mathrm{R}(X;Y)_{\rho^{XY}} = \min_{\sigma^{XY}} \mathrm{D}(\rho^{XY}\|\|\sigma^{XY})$

To proceed, write a density matrix ρ^{XY} in terms of any of the ensembles

$$\mathcal{E}^{XY} = \{(|\psi_0^{XY}\rangle, p_0), (|\psi_1^{XY}\rangle, p_1), \ldots, (|\psi_{M-1}^{XY}\rangle, p_{M-1})\} \tag{16.122}$$

realizing it via the average $\rho^{XY} = \sum_{m\in[M]} p_m |\psi_m^{XY}\rangle\langle\psi_m^{XY}|$. Operationally, the state ρ^{XY} can thus be prepared by first drawing a classical random variable $\mathrm{z} \sim p$ with distribution $p = [p_1, \ldots, p_{M-1}]^T$, and then generating the pure state $|\psi_z^{XY}\rangle$ for the obtained realization $\mathrm{z} = z$ of the random variable.

For each pure state $|\psi_z^{XY}\rangle$ in the ensemble, entanglement can be measured by the entanglement entropy, defined in Section 14.7, as $\mathrm{E}(X;Y)_{|\psi_z^{XY}\rangle} = \mathrm{H}(\lambda(|\psi_z^{XY}\rangle))$, where $\lambda(|\psi_z^{XY}\rangle)$ is the Schmidt probability vector for state $|\psi_z^{XY}\rangle$. Accordingly, for a given ensemble $\mathcal{E}^{XY}$ realizing the density matrix ρ^{XY}, the average amount of entanglement can be quantified by the **average entanglement entropy**

$$\mathrm{E}_p\left[\mathrm{E}(X;Y)_{|\psi_z^{XY}\rangle}\right] = \sum_{z\in[M]} p_z \mathrm{E}(X;Y)_{|\psi_z^{XY}\rangle}. \tag{16.123}$$

This measure, however, is not well defined, since there are infinitely many ensembles that realize a mixed state ρ^{XY}.

Example 16.33

Take the mixture of maximally entangled pure states $\rho^{XY} = \frac{1}{2}|\Phi_+^{XY}\rangle\langle\Phi_+^{XY}| + \frac{1}{2}|\Phi_-^{XY}\rangle\langle\Phi_-^{XY}|$, which was shown earlier to be separable (see Example 16.24). If the ensemble

$$\mathcal{E}^{XY} = \{(|\Phi_+^{XY}\rangle, 0.5), (|\Phi_-^{XY}\rangle, 0.5)\} \tag{16.124}$$

were used to evaluate the average entanglement entropy (16.123), one would obtain an average entanglement entropy

$$\frac{1}{2}\mathrm{E}(X;Y)_{|\Phi_+^{XY}\rangle} + \frac{1}{2}\mathrm{E}(X;Y)_{|\Phi_-^{XY}\rangle} = 1, \tag{16.125}$$

since both states of the ensemble are maximally entangled, i.e., $\mathrm{E}(X;Y)_{|\Phi_+^{XY}\rangle} = \mathrm{E}(X;Y)_{|\Phi_-^{XY}\rangle} = 1$. However, a valid measure of entanglement should return zero for state ρ^{XY}, given that the state is separable.

The state ρ^{XY} can also be written as $\rho^{XY} = \frac{1}{2}|00\rangle\langle 00| + \frac{1}{2}|11\rangle\langle 11|$, which corresponds to the ensemble $\mathcal{E}^{XY} = \{(|00\rangle, 0.5), (|11\rangle, 0.5)\}$. Given that both states $|00\rangle$ and $|11\rangle$ are separable, the average entanglement entropy evaluated using this ensemble equals zero, i.e.,

$$\frac{1}{2}\mathrm{E}(X;Y)_{|00\rangle} + \frac{1}{2}\mathrm{E}(X;Y)_{|11\rangle} = 0. \tag{16.126}$$

Accordingly, with this second choice of ensemble, the average entanglement entropy offers the sensible result that the state is not entangled.

The convex roof construction prescribes that, in order to define a valid measure of entanglement, one must minimize over all ensembles that can realize the given state.

Definition 16.13 Entanglement of formation
For any state ρ^{XY}, the entanglement of formation $\mathrm{E}(X;Y)_{\rho^{XY}}$ is defined as the minimum value of the average entanglement entropy over all possible ensembles $\mathcal{E}^{XY}$ that realize matrix ρ^{XY}:

$$\mathrm{E}(X;Y)_{\rho^{XY}} = \min_{\mathcal{E}^{XY}} \sum_{z\in[M]} p_z \mathrm{E}(X;Y)_{|\psi_z^{XY}\rangle}. \tag{16.127}$$

Like for the relative entropy of entanglement, a positive entanglement of formation provides a necessary and sufficient condition for entanglement.

Key Result 16.20 Bounds on the entanglement of formation
For any state ρ^{XY}, the entanglement of formation satisfies the bounds

$$0 \le \mathrm{E}(X;Y)_{\rho^{XY}} \le \log_2(N_{\min}), \tag{16.128}$$

where the lower bound is attained if and only if the state ρ^{XY} is separable.

For pure states, the entanglement of formation coincides with the entanglement entropy, which may be written, using the notation of Section 14.7, as

$$\mathrm{E}(X;Y)_{|\psi^{XY}\rangle\langle\psi^{XY}|} = \mathrm{E}(X;Y)_{|\psi^{XY}\rangle}. \tag{16.129}$$

More generally, the entanglement of formation is often hard to compute, as it requires an optimization over all possible ensembles realizing the given density matrix.

An exception is provided by the important two-qubit case, i.e., $N^X = N^Y = 2$, for which the explicit formula holds

$$\mathrm{E}(X;Y)_{\rho^{XY}} = \mathrm{h}\left(\frac{1+\sqrt{1-(\mathcal{C}(X;Y)_{\rho^{XY}})^2}}{2}\right), \tag{16.130}$$

where $\mathcal{C}(X;Y)_{\rho^{XY}}$ is the so-called **concurrence** of the 4×4 density matrix ρ^{XY}, which is defined as

$$\mathcal{C}(X;Y)_{\rho^{XY}} = \max\left\{0, \sqrt{\lambda_0} - \sqrt{\lambda_1} - \sqrt{\lambda_2} - \sqrt{\lambda_3}\right\}, \tag{16.131}$$

where $\{\lambda_0, \lambda_1, \lambda_2, \lambda_3\}$ are the eigenvalues, ordered as $\lambda_0 \ge \lambda_1 \ge \lambda_2 \ge \lambda_3$, of the Hermitian matrix

$$(\rho^{XY})^{1/2}(P_2 \otimes P_2)(\rho^{XY})^*(P_2 \otimes P_2)(\rho^{XY})^{1/2}, \tag{16.132}$$

with P_2 being the Pauli Y matrix.

The entanglement of formation $\mathrm{E}(X;Y)_{\rho^{XY}}$ is minimized when the concurrence takes the minimum value $\mathcal{C}(X;Y)_{\rho^{XY}} = 0$, and is maximized when the concurrence takes the maximum value $\mathcal{C}(X;Y)_{\rho^{XY}} = 1$. Note that this formula generalizes the expression (14.178) presented in Appendix 14.C for the case of pure states. The reader is referred to the Recommended Resources for details.

Example 16.34

Returning again to the Werner state (16.11), the entanglement of formation is plotted, using (16.130), in Figure 16.6. In the regime $\eta \leq 1/3$, by the PPT criterion, the Werner state is separable, and thus the entanglement of formation is zero, while it is positive for $\eta > 1/3$, indicating that the state is entangled. In particular, for $\eta > 1/3$, the entanglement of formation can be written explicitly using (16.130) as

$$\mathrm{E}(X;Y)_{\rho^{XY}} = \mathrm{h}\left(\frac{1-\sqrt{1-\left(\frac{1}{2}-\frac{3}{2}\eta\right)^2}}{2}\right) \quad \text{for } \frac{1}{3} < \eta \leq 1. \tag{16.133}$$

The gap between the quantum symmetric discord and the entanglement of formation highlighted in the figure can be interpreted as a measure of the weaker quantum correlations present in the state.

Example 16.35

For the two-qubit state (16.68), the entanglement of formation is equal to

$$\mathrm{E}(X;Y)_{\rho^{XY}} = 1 - \mathrm{h}(\eta). \tag{16.134}$$

Hence, apart from the case $\eta = 1/2$, this is an entangled state.

Definition 16.14 Maximally entangled states

A state ρ^{XY} with maximal entanglement of formation $\mathrm{E}(X;Y)_{\rho^{XY}} = \log_2(N_{\min})$ is said to be maximally entangled.

Maximally entangled pure states, as defined in Section 14.7.7, are maximally entangled according to this definition. More generally, mixed states are maximally entangled if and only if *all* ensembles realizing the state's density matrix consist of maximally entangled pure states.

Example 16.36

The state (16.95) studied in Example 16.28 can be shown to be maximally entangled for any probability $0 \leq \eta \leq 1$, i.e., $\mathrm{E}(X;Y)_{\rho^{XY}} = \log_2(N_{\min}) = \log_2(2)$ (see Recommended Resources).

The entanglement of formation is lower bounded by the relative entropy of entanglement, i.e.,

$$\mathfrak{E}_R(X;Y)_{\rho^{XY}} \leq \mathrm{E}(X;Y)_{\rho^{XY}}. \tag{16.135}$$

In this regard, a drawback of the adoption of the entanglement of formation as entanglement metric is that, unlike the relative entropy of entanglement (which satisfies the bounds (16.121)), it may exceed the quantum mutual information (see Recommended Resources).

16.7.5 Negativity

A third entanglement monotone, known as negativity, measures the degree to which the partial transpose $(\rho^{XY})^{T^X}$ of the joint density matrix ρ^{XY} deviates from positive definiteness.

Definition 16.15 Negativity

For any state ρ^{XY}, the negativity is defined as

$$\mathfrak{E}_N(X;Y)_{\rho^{XY}} = \frac{||(\rho^{XY})^{T^X}||_1 - 1}{2}. \tag{16.136}$$

The definition can also be stated equivalently in terms of the partial transpose $(\rho^{XY})^{T^Y}$.

The negativity measure is zero for all separable states, and it becomes positive exclusively for entangled states. However, unlike the relative entropy of entanglement and the entanglement of formation, the negativity measure also vanishes for certain entangled states, namely the PPT states. Despite this, the negativity can be practically useful, since calculating the relative entropy of entanglement and the entanglement of formation is challenging, whereas the negativity is computationally straightforward to evaluate.

16.8 Quantum Conditional Entropy ⏭

This chapter has so far examined measures of correlation in bipartite quantum systems that treat both subsystems symmetrically. However, as discussed in Chapter 13, classical systems also admit asymmetric measures of correlation, which quantify the residual uncertainty about one subsystem when the other is accessible. Notably, the conditional entropy and the ensemble entropy were introduced as such measures, which coincide when the Shannon entropy is used to quantify uncertainty.

In this section and the next, these residual uncertainty measures are extended to the quantum domain, introducing the concepts of quantum conditional entropy and quantum ensemble entropy. Unlike their classical counterparts, these quantum measures are not equal when evaluated using the quantum Shannon entropy. This fundamental difference between the two measures necessitates distinct treatments, which will be explored in detail.

Unless stated otherwise, this section and the next adopt the perspective of subsystem X in quantifying the residual uncertainty. The complementary perspective of subsystem Y can be addressed in the same way by swapping the two subsystems.

16.8.1 Defining the Quantum Conditional Entropy

The quantum conditional entropy $\mathrm{H}(Y|X)_{\rho^{XY}}$ quantifies the uncertainty about subsystem Y when the observer holds subsystem X. Accordingly, as detailed next, the quantum conditional entropy $\mathrm{H}(Y|X)_{\rho^{XY}}$ is defined as the difference between the joint entropy $\mathrm{H}(X, Y)_{\rho^{XY}}$ and the local entropy $\mathrm{H}(X)_{\rho^X}$.

Definition 16.16 Quantum conditional entropy

For any state ρ^{XY}, the quantum conditional entropy of subsystem Y given subsystem X is defined as

$$\mathrm{H}(Y|X)_{\rho^{XY}} = \mathrm{H}(X, Y)_{\rho^{XY}} - \mathrm{H}(X)_{\rho^X}, \tag{16.137}$$

or equivalently as

$$\mathrm{H}(Y|X)_{\rho^{XY}} = \mathrm{H}(Y)_{\rho^Y} - \mathrm{I}(X;Y)_{\rho^{XY}}, \tag{16.138}$$

where $\rho^X = \mathrm{tr}_Y(\rho^{XY})$ and $\rho^Y = \mathrm{tr}_X(\rho^{XY})$ are the reduced density matrices.

The quantum conditional entropy $\mathrm{H}(Y|X)_{\rho^{XY}}$, as defined in (16.137), quantifies the fraction of the irreducible uncertainty associated with the joint state ρ^{XY} that is not present in subsystem X alone. This definition ensures the **chain rule for quantum entropy**

$$\mathrm{H}(X, Y)_{\rho^{XY}} = \mathrm{H}(X)_{\rho^X} + \mathrm{H}(Y|X)_{\rho^{XY}}. \tag{16.139}$$

Furthermore, the equivalent expression (16.138) indicates that the quantum conditional entropy $\mathrm{H}(Y|X)_{\rho^{XY}}$ captures the irreducible uncertainty about system Y that cannot be explained by correlations with subsystem X.

Example 16.37

For an uncorrelated state $\rho^{XY} = \rho^X \otimes \rho^Y$, since the quantum mutual information $\mathrm{I}(X;Y)_{\rho^{XY}}$ equals zero, the quantum conditional entropy equals the quantum entropy,

$$\mathrm{H}(Y|X)_{\rho^X \otimes \rho^Y} = \mathrm{H}(Y)_{\rho^Y}. \tag{16.140}$$

Informally, in this case, there is no information stored in the correlations between the two subsystems.

Example 16.38

For a classical–classical state χ^{XY} with eigenvalues p^{XY}, the quantum conditional entropy coincides with the classical conditional entropy information of the joint distribution p^{XY}, i.e.,

$$\mathrm{H}(Y|X)_{\chi^{XY}} = \mathrm{H}(Y|X)_{p^{XY}}. \tag{16.141}$$

In fact, all quantum entropies coincide with the classical entropies of the joint distribution p^{XY}. This further reinforces the role of classical–classical states as analogous to classical joint distributions.

Example 16.39

For a classical–quantum state $\vec{\omega}^{XY}$, using the quantum mutual information (16.84), the quantum conditional entropy is given by

$$\mathrm{H}(Y|X)_{\vec{\omega}^{XY}} = \mathrm{E}_{p^X}\left[\mathrm{H}(Y)_{\rho_x^{Y|X}}\right]. \tag{16.142}$$

For a classical–quantum state, the quantum conditional entropy $\mathrm{H}(Y|X)_{\vec{\omega}^{XY}}$ is thus the average of the local entropies $\mathrm{H}(Y)_{\rho_x^{Y|X}}$ of the conditional density matrices $\{\rho_x^{Y|X}\}_{x\in[N^Y]}$. Accordingly, conditioning on the "classical" subsystem X within the classical–quantum state $\vec{\omega}^{XY}$ has an analogous effect as conditioning in classical joint distributions (see Section 13.5).

A similar expression does not hold for the quantum conditional entropy $\mathrm{H}(X|Y)_{\vec{\omega}^{XY}}$, which is conditioned on the "quantum" part of the system, namely subsystem Y. In fact, using again (16.84), one can evaluate the quantum conditional entropy as

$$\mathrm{H}(X|Y)_{\vec{\omega}^{XY}} = \mathrm{H}(X)_{\rho^X} - \mathrm{H}(Y)_{\rho^Y} + \mathrm{E}_{p^X}\left[\mathrm{H}(Y)_{\rho_x^{Y|X}}\right]. \tag{16.143}$$

As demonstrated in Section 16.8.2, this quantity is non-negative, since classical–quantum states are separable.

16.8.2 Properties of the Quantum Conditional Entropy

As formalized next, the quantum conditional entropy satisfies the conditioning-reduces-entropy property. Furthermore, using (16.91), it provides a sufficient condition for the detection of entanglement via insubordination.

Key Result 16.21 Conditioning reduces the quantum conditional entropy
The quantum conditional entropy satisfies the inequality

$$\mathrm{H}(Y|X)_{\rho^{XY}} \leq \mathrm{H}(Y)_{\rho^Y}, \tag{16.144}$$

with equality if and only if the state ρ^{XY} is uncorrelated.

Using (16.138), this property follows directly from the non-negativity (16.86) of the quantum mutual information.

Key Result 16.22 Insubordination-based sufficient condition for entanglement
If the quantum conditional entropy of a state ρ^{XY} is negative, the state is entangled:

$$\mathrm{H}(Y|X)_{\rho^{XY}} < 0 \text{ and/or } \mathrm{H}(X|Y)_{\rho^{XY}} < 0 \Rightarrow \rho^{XY} \text{ is entangled.} \quad (16.145)$$

Equivalently, a separable state must have a non-negative quantum conditional entropy. Note that, unlike the case of pure states, this condition is only sufficient, but not necessary. Therefore, there are entangled states with a positive quantum conditional entropy.

Example 16.40

For the Werner state (16.11), using (16.94), the quantum conditional entropy is given by

$$\begin{aligned}\mathrm{H}(Y|X)_{\rho^{XY}} = &\left(\eta + \frac{1-\eta}{4}\right)\log_2\left(\eta + \frac{1-\eta}{4}\right)\\ &- 3\left(\frac{1-\eta}{4}\right)\log_2\left(\frac{1-\eta}{4}\right) - 1,\end{aligned} \quad (16.146)$$

and one has the equality $\mathrm{H}(Y|X)_{\rho^{XY}} = \mathrm{H}(X|Y)_{\rho^{XY}}$. As shown in Figure 16.6, the quantum conditional entropy can be positive for entangled states, i.e., for $\eta > 1/3$. Furthermore, when it is negative, by the sufficient condition (16.145), the state is guaranteed to be entangled.

In Section 13.5, the classical conditional entropy was shown to satisfy the equality $\mathrm{H}(Y|X)_{p^{XY}} = \log_2(N^Y) - \mathrm{D}(p^{XY}||p^X \otimes u^Y)$. It can be verified that an analogous characterization applies for the quantum conditional entropy, i.e.,

$$\mathrm{H}(Y|X)_{\rho^{XY}} = \log_2(N^Y) - \mathrm{D}(\rho^{XY}||\pi^X \otimes \rho^Y). \quad (16.147)$$

Therefore, the quantum conditional entropy $\mathrm{H}(Y|X)_{\rho^{XY}}$ grows as the distance $\mathrm{D}(\rho^{XY}||\pi^X \otimes \rho^Y)$ to the joint density matrix $\pi^X \otimes \rho^Y$ decreases. This reference state is uncorrelated, and the reduced density matrix of subsystem X is maximally mixed.

Using (16.147), it is possible to generalize the definition of quantum conditional entropy to Rényi variants by replacing the Shannon quantum relative entropy with the Rényi quantum relative entropy. The reader is referred to the Recommended Resources for a review.

16.9 Quantum Ensemble Entropy and Quantum Discord ⏭

This section introduces a second measure of residual uncertainty, known as the quantum ensemble entropy, which captures the average uncertainty about one subsystem when a complete measurement is made on the other. While quantum conditional entropy pertains to a scenario in which an observer has direct access to one subsystem, the quantum ensemble entropy emphasizes the information obtainable through a measurement on the other subsystem. Using the quantum ensemble entropy, an asymmetric measure of quantum correlations, known as quantum discord, is finally defined.

16.9.1 Quantum Ensemble Entropy

Consider a state ρ^{XY} and a measurement setting $\mathcal{M}^X = \{|v_x^X\rangle\}_{x\in[N^X]}$ on subsystem X. As discussed earlier in this chapter, the post-measurement state for subsystem Y can be described by the ensemble

$$\mathcal{E}^{Y|X} = \{(\rho_x^{Y|X}, p_x^X)\}_{x\in[N^X]}, \tag{16.148}$$

with measurement probabilities p_x^X in (16.50) and conditional density matrices $\{\rho_x^{Y|X}\}_{x\in[N^X]}$ in (16.53).

The quantum ensemble entropy $\mathrm{H}(\mathcal{E}^{Y|X})_{\rho^{XY}}$ is defined as the average quantum entropy of the conditional density matrices resulting from a local measurement at subsystem X. Accordingly, the quantum ensemble entropy is a measure of the uncertainty at subsystem X about subsystem Y as a result of a local measurement on subsystem X. Specifically, the average quantum entropy is minimized over the measurement setting $\mathcal{M}^X$ in order to remove any dependence on the measurement setting and to obtain an irreducible measure of uncertainty.

Definition 16.17 Quantum ensemble entropy

For a joint state ρ^{XY}, the quantum ensemble entropy $\mathrm{H}(\mathcal{E}^{Y|X})_{\rho^{XY}}$ is defined as

$$\mathrm{H}(\mathcal{E}^{Y|X})_{\rho^{XY}} = \min_{\mathcal{M}^X} \mathrm{E}_{p^X}\left[\mathrm{H}(Y)_{\rho_x^{Y|X}}\right], \tag{16.149}$$

where $\mathcal{E}^{Y|X} = \{(\rho_x^{Y|X}, p_x^X)\}_{x\in[N^X]}$ is the ensemble corresponding to the measurement setting $\mathcal{M}^X$ and the minimization is over all complete measurement settings $\mathcal{M}^X$.

16.9.2 Quantum Locally Accessible Information

Since the uncertainty about subsystem Y for an observer that does not have access to subsystem X is quantified by the quantum entropy $\mathrm{H}(Y)_{\rho^Y}$, the difference

$\mathrm{H}(Y)_{\rho^Y} - \mathrm{H}(\mathcal{E}^{Y|X})_{\rho^{XY}}$ may be interpreted as the expected information gain resulting from a local measurement on subsystem X. This expected information gain yields the following quantity.

Key Result 16.23 Quantum locally accessible information

For any state ρ^{XY}, the quantum locally accessible information from subsystem X to subsystem Y is defined as

$$\mathrm{J}(X \to Y)_{\rho^{XY}} = \mathrm{H}(Y)_{\rho^Y} - \mathrm{H}(\mathcal{E}^{Y|X})_{\rho^{XY}}. \tag{16.150}$$

As in the classical case studied in Section 13.5, the notation $\mathrm{J}(X \to Y)_{\rho^{XY}}$ indicates that the quantum locally accessible information measures the information that is accessible about subsystem Y by making a local measurement at subsystem X. The quantum locally accessible information $\mathrm{J}(Y \to X)_{\rho^{XY}}$ from subsystem Y to subsystem X is defined in an analogous way. In general, the quantum locally accessible information is asymmetric, i.e.,

$$\mathrm{J}(X \to Y)_{\rho^{XY}} \neq \mathrm{J}(Y \to X)_{\rho^{XY}}. \tag{16.151}$$

16.9.3 Quantum Discord

By subtracting the locally accessible information $\mathrm{J}(X \to Y)_{\rho^{XY}}$ from the quantum mutual information $\mathrm{I}(X;Y)_{\rho^{XY}}$, one obtains a measure of the correlations in the system that cannot be accessed by making a local measurement at subsystem X. Such correlations may thus be considered to be of quantum nature.

Definition 16.18 Quantum discord

For any state ρ^{XY}, the quantum discord about subsystem Y from subsystem X is defined as the difference

$$\mathrm{Q}(X \to Y)_{\rho^{XY}} = \mathrm{I}(X;Y)_{\rho^{XY}} - \mathrm{J}(X \to Y)_{\rho^{XY}}. \tag{16.152}$$

Using the respective definitions of quantum mutual information and quantum locally accessible information, the quantum discord can equivalently be expressed as the following difference.

Key Result 16.24 Equivalent characterization of the quantum discord

For any state ρ^{XY}, the quantum discord can be expressed as

$$\mathrm{Q}(X \to Y)_{\rho^{XY}} = \mathrm{H}(\mathcal{E}^{Y|X})_{\rho^{XY}} - \mathrm{H}(Y|X)_{\rho^{XY}}. \tag{16.153}$$

Accordingly, the quantum discord is the difference between two measures of residual uncertainty, namely the quantum ensemble entropy $\mathrm{H}(\mathcal{E}^{Y|X})_{\rho^{XY}}$ studied in this section and the quantum conditional entropy $\mathrm{H}(Y|X)_{\rho^{XY}}$ introduced in Section 16.8. The ensemble entropy $\mathrm{H}(\mathcal{E}^{Y|X})_{\rho^{XY}}$ measures the residual uncertainty resulting from a classical measurement of subsystem X, while the quantum conditional entropy $\mathrm{H}(Y|X)_{\rho^{XY}}$ accounts for residual uncertainty when subsystem X is directly accessible. For classical systems, as seen in Section 13.5, these two residual uncertainty measures are equal. Therefore, the quantum discord (16.153) can be interpreted as a measure of the quantum information available at subsystem X about subsystem Y.

As a summary, Table 16.2 highlights how quantum mutual information, quantum locally accessible information, and quantum discord can be all expressed as the difference of suitable entropy measures.

The quantum discord satisfies the following property.

Key Result 16.25 Non-negativity and minimum value of the quantum discord
The quantum discord is non-negative,

$$\mathrm{Q}(X \to Y)_{\rho^{XY}} \geq 0, \tag{16.154}$$

with the lower bound $\mathrm{Q}(X \to Y)_{\rho^{XY}} = 0$ attained if and only if the pure state ρ^{XY} is classical–quantum, i.e.,

$$\mathrm{Q}(X \to Y)_{\vec{\omega}^{XY}} = 0. \tag{16.155}$$

The inequality (16.154) can be proved by leveraging the data processing inequality (16.98). In fact, the quantum locally accessible information $\mathrm{J}(X \to Y)_{\rho^{XY}}$ can be verified to equal the quantum mutual information of the state obtained after the application of an optimized non-selective measurement at subsystem X. Furthermore, the equality (16.155) follows directly from (16.142).

The quantum discord vanishes for all classical–quantum states, while the quantum symmetric discord (16.107) equals zero only for classical–classical states. Therefore, states with a positive quantum discord pass a stricter test for the presence of quantum correlations.

Table 16.2 Quantum mutual information, quantum locally accessible information, and quantum discord as differences of suitable entropy measures

Measure	Expression
$\mathrm{I}(X;Y)_{\rho^{XY}}$	$\mathrm{H}(Y)_{\rho^X} - \mathrm{H}(Y\|X)_{\rho^{XY}}$
$\mathrm{J}(X \to Y)_{\rho^{XY}}$	$\mathrm{H}(Y)_{\rho^X} - \mathrm{H}(\mathcal{E}^{Y\|X})_{\rho^{XY}}$
$\mathrm{Q}(X \to Y)_{\rho^{XY}}$	$\mathrm{H}(\mathcal{E}^{Y\|X})_{\rho^{XY}} - \mathrm{H}(Y\|X)_{\rho^{XY}}$

From an operational perspective, a positive quantum discord is equivalent to the possibility for distributed agents to create entanglement with auxiliary subsystems. The quantum discord is generally not easy to compute owing to the need to optimize over the measurement setting. However, there are known expression in some cases. The reader is referred to the Recommended Resources for details and further information, including on the connections to **quantum steerability**.

16.10 Summary

- Uncorrelated mixed states can be expressed as the Kronecker product $\rho^{XY} = \rho^X \otimes \rho^Y$ of local density matrices ρ^X for subsystem X and ρ^Y for subsystem Y.
- Any state ρ^{XY} can be written as the linear combination $\rho^{XY} = \sum_{z\in[M]} \gamma_z \rho_z^X \otimes \rho_z^Y$ of M uncorrelated states for a quasi-probability distribution $\{\gamma_z\}_{z\in[M]}$, i.e., for real numbers $\{\gamma_z\}_{z\in[M]}$ satisfying the condition $\sum_{z\in[M]} \gamma_z = 1$.
- Using this decomposition and the partial trace, the reduced density matrices can be expressed as $\rho^X = \sum_{z\in[M]} \gamma_z \rho_z^X$ and $\rho^Y = \sum_{z\in[M]} \gamma_z \rho_z^Y$.
- The reduced density matrix ρ^X determines the marginal measurement probability distribution $p^X(\rho^X|\mathcal{K}^X)$ for a local measurement POVM $\mathcal{K}^X$ on subsystem X, while the reduced density matrix ρ^Y determines the marginal measurement probability distribution $p^Y(\rho^Y|\mathcal{K}^Y)$ for a local measurement POVM $\mathcal{K}^Y$ on subsystem Y.
- The quantum mutual information, which captures both classical and quantum correlations, is defined as the difference $\mathrm{I}(X;Y)_{\rho^{XY}} = \mathrm{H}(X)_{\rho^X} + \mathrm{H}(Y)_{\rho^Y} - \mathrm{H}(X,Y)_{\rho^{XY}}$ between the sum of the local entropies, $\mathrm{H}(X)_{\rho^X} + \mathrm{H}(Y)_{\rho^Y}$, and the joint entropy $\mathrm{H}(X,Y)_{\rho^{XY}}$. It is equal to zero if and only if the state ρ^{XY} is uncorrelated.
- The quantum mutual information satisfies the data processing inequality under local channels. Given any two local channels $\mathcal{K}^X$ and $\mathcal{K}^Y$, the quantum mutual information satisfies the inequality $\mathrm{I}(X;Y)_{\mathcal{K}^X\otimes\mathcal{K}^Y(\rho^{XY})} \leq \mathrm{I}(X;Y)_{\rho^{XY}}$.
- The overall level of correlation, quantified by the quantum mutual information, can be partitioned into classical correlations and quantum correlations. To this end, classical correlations are measured by the maximum measured mutual information $\mathrm{J}(X;Y)_{\rho^{XY}}$. Accordingly, the quantum correlations are quantified by the difference between quantum mutual information and maximum measured mutual information $\mathrm{Q}(X;Y)_{\rho^{XY}} = \mathrm{I}(X;Y)_{\rho^{XY}} - \mathrm{J}(X;Y)_{\rho^{XY}}$, which is known as the quantum symmetric discord.
- Classical–classical states, which are of the form $\chi^{XY} = \sum_{x\in[N^X]} \sum_{y\in[N^Y]} p_{xy}^{XY} |v_x^X\rangle\langle v_x^X| \otimes |v_y^Y\rangle\langle v_y^Y|$, have only classical correlations in the sense that one has the equalities $\mathrm{I}(X;Y)_{\chi^{XY}} = \mathrm{J}(X;Y)_{\chi^{XY}}$ and $\mathrm{Q}(X;Y)_{\chi^{XY}} = 0$.
- A mixed state is entangled if it is not separable, i.e., if it cannot be written as a convex combination $\rho^{XY} = \sum_{z\in[M]} p_z \rho_z^X \otimes \rho_z^Y$ of uncorrelated states.
- While for pure states quantum correlations coincide with entanglement, for mixed states, there are states with positive quantum symmetric discord that are not entangled.

- For entangled states, the quantum mutual information can exceed the quantum entropy of either subsystem. This suggests that quantum correlations can be so powerful that one can infer more about a subsystem by accessing the other subsystem than from the information contained within the subsystem itself.
- There are several sufficient conditions to detect entanglement for mixed states, such as the PPT criterion.
- Furthermore, one can define entanglement measures and entanglement monotones by imposing constraints on the behavior of the metric as a result of the application of an LOCC protocol.
- The quantum conditional entropy $\mathrm{H}(Y|X)_{\rho^{XY}}$ quantifies the uncertainty about subsystem Y when the observer holds subsystem X.
- A key property of the quantum conditional entropy is that, if the quantum conditional entropy of a state ρ^{XY} is negative, the state is entangled.
- A different way to assess residual uncertainty is by assuming that a measurement in some complete basis $\mathcal{M}^X$ is made on subsystem X. In this case, the post-measurement state of subsystem Y is described by an ensemble $\{(\rho_x^{Y|X}, p_x^X)\}_{x\in[N^X]}$, and the residual uncertainty can be evaluated by the quantum ensemble entropy $\mathrm{H}(\mathcal{E}^{Y|X})_{\rho^{XY}} = \min_{\mathcal{M}^X} \mathrm{E}_{p^X}[\mathrm{H}(Y)_{\rho_x^{Y|X}}]$.
- The quantum discord is defined as the difference $\mathrm{Q}(X \to Y)_{\rho^{XY}} = \mathrm{H}(\mathcal{E}^{Y|X})_{\rho^{XY}} - \mathrm{H}(Y|X)_{\rho^{XY}}$ between the quantum ensemble entropy $\mathrm{H}(\mathcal{E}^{Y|X})_{\rho^{XY}}$ and the quantum conditional entropy $\mathrm{H}(Y|X)_{\rho^{XY}}$, Accordingly, the quantum discord is a measure of the quantum information accessible from subsystem X about subsystem Y.
- The quantum discord is equal to zero if and only if the state is classical–quantum, i.e., if and only if the state can be expressed as $\vec{\varpi}^{XY} = \sum_{x\in[N^X]} p_x^X |v_x^X\rangle\langle v_x^X| \otimes \rho_x^{Y|X}$.

16.11 Recommended Resources

Quantum cryptography, a foundational aspect of secure quantum communication, is introduced in [1]. The decomposition (16.70) of a mixed quantum state into a convex combination of a separable state and an entangled pure state is detailed in [2]. The implications of limiting the optimization in the definition (16.102) of the maximum measured mutual information to complete measurements are discussed in [3–7]. Furthermore, algorithmic considerations for calculating the maximum measured mutual information are presented in [8], with alternative definitions involving fixed local eigenbases discussed in [9].

For insights into distance-based measures of correlation, the reader is directed to [10] for the relative entropy of quantumness and to [11] for the relative entropy of entanglement. Additionally, the negativity is explored in [12], and the entanglement of formation is examined in [4, 13]. For those interested in the information-theoretic aspects of the Werner state, the reference [14] provides a detailed discussion.

In terms of conditions for entanglement, a stronger sufficient condition based on insubordination and majorization is available in [15]. Further information on

PPT states, including their properties and distillability, is discussed in [16–18]. For an exploration of maximal entanglement in mixed states, see [19]. Entanglement witnesses, an important tool for detecting entanglement, are reviewed in [6]. Additionally, their relationship to the PPT criterion through Lagrange duality is explored in [20].

Conditional quantum entropy and various generalizations are presented in [10, 11]. The metric of concurrence, a measure of entanglement, is initially introduced in [21] and later extended to systems of arbitrary dimensions in [22]. Quantum discord is explored in [23, 24], providing discussions on steerability and other properties. For computational approaches to quantum discord, see [5, 25].

Finally, the proof of the strong subadditivity property (16.182), a fundamental inequality in quantum information theory, is presented in [26].

PROBLEMS

16.1 Show that if A and B are positive semi-definite matrices with the same dimensions, then the Kronecker product $A \otimes B$ is also positive semi-definite. *Hint:* The eigenvalues of the Kronecker product are equal to the products of all pairs of eigenvalues of the individual matrices.

16.2 Show that if A and B are positive semi-definite matrices with the same dimensions, then the linear combination $aA+bB$ with coefficients $a \geq 0$ and $b \geq 0$ is also positive semi-definite. *Hint:* An $N \times N$ matrix C is positive semi-definite if and only if for any $N \times 1$ vector x we have the inequality $x^T Cx \geq 0$.

16.3 Prove that, given a joint distribution p^{XY} and corresponding density matrix $\rho^{XY} = \text{diag}(p^{XY})$, for any dimensions N^X and N^Y, the reduced density matrices are given by $\text{tr}_X(\rho^{XY}) = \text{diag}(p^Y)$ and $\text{tr}_Y(\rho^{XY}) = \text{diag}(p^Y)$, where p^X and p^Y are the marginal distributions.

16.4 As in (16.62), denote as $r^{XY}_{x,x'}$ the (x, x')th $N^Y \times N^Y$ block of a density matrix ρ^{XY}. Prove that blocks of the form $r^{XY}_{x,x}$, which are on the main block diagonal of the matrix, are positive semi-definite but their trace may be smaller than 1, while blocks $\rho^{XY}_{x,x'}$ are not even Hermitian.

16.5 Show that for any state of the form $\rho^{XY} = \sum_{z\in[M]} \gamma_z \rho^X_z \otimes \rho^Y_z$, with real-valued coefficients $\{\gamma_z\}_{z\in[M]}$, the reduced density matrices can be obtained as $\rho^X = \sum_{z\in[M]} \gamma_z \rho^X_z$ and $\rho^Y = \sum_{z\in[M]} \gamma_z \rho^Y_z$.

16.6 Show that for any density matrix ρ, the transpose ρ^T is a density matrix. *Hint:* Use the equality $(|v\rangle\langle v|)^T = |v\rangle^* \langle v|^*$, and the fact that the entry-wise conjugate $|v\rangle^*$ of a state $|v\rangle$ is also a valid state vector.

16.7 Give an example of a mixed state for which the reduced density matrices have distinct eigenvalues. Can a similar example be given for pure states?

16.8 Apply the PPT criterion to the state $\rho^{XY} = \eta|\Psi^{XY}_+\rangle\langle\Psi^{XY}_+| + (1 - \eta)|00\rangle\langle 00|$, and show that the state is entangled for any value of probability η not equal to 0.

16.9 Consider any two separable states ρ^{XY} and σ^{XY}, and show that the state $\eta\rho^{XY} + (1-\eta)\sigma^{XY}$ is also separable for any probability $\eta \in [0, 1]$.

16.10 A set is said to be convex if it contains all convex combinations of its elements. Thus, a set of density matrices is convex if, for any two states ρ^{XY} and σ^{XY} in the set, every convex combination $\eta\rho^{XY} + (1-\eta)\sigma^{XY}$ for any probability $0 \leq \eta \leq 1$ is also in the set. (a) Show that the set of separable states is convex. (b) Prove that the sets of uncorrelated states is not convex. (c) Show that the set of entangled states is not convex. (d) Prove that the sets of classical–classical states, classical–quantum states, and quantum–classical states are not convex.

16.11 For two-qubit systems, consider the joint observable $W = \frac{1}{2}I - |\Phi_+^{XY}\rangle\langle\Phi_+^{XY}|$. (a) Show that for any pure uncorrelated state $|\psi^X\rangle \otimes |\psi^Y\rangle = |\psi^X\psi^Y\rangle$ the inequality $\langle W\rangle_{|\psi^X\psi^Y\rangle\langle\psi^X\psi^Y|} \geq 0$ holds. (b) Use this result to demonstrate that for any mixed separable state ρ^{XY} one also has the inequality $\langle W\rangle_{\rho^{XY}} \geq 0$. *Hint:* Any mixed separable state can be expressed as a convex combination of pure uncorrelated states.

16.12 Prove that a classical–quantum state is locally unitary equivalent to a state of the form $\overrightarrow{\omega}^{XY} = \sum_{x\in[N^X]} p_x^X |x\rangle\langle x| \otimes \rho_x^{Y|X}$.

16.13 Evaluate the quantum mutual information for state $\rho^{XY} = \frac{\eta}{2}(|0\rangle\langle 0| \otimes |0\rangle\langle 0| + |1\rangle\langle 1| \otimes |1\rangle\langle 1|) + (1-\eta)|\Phi_+^{XY}\rangle\langle\Phi_+^{XY}|$ for any probability η.

16.14 Classical–quantum states can be used to describe bipartite systems in which subsystem X is a measuring device and subsystem Y is the quantum system being measured. To elaborate on this point, consider the general case of a quantum instrument $\mathcal{I} = \{\mathcal{K}_0(\cdot), \ldots, \mathcal{K}_{M-1}(\cdot)\}$, which, as seen in Section 11.8, unifies quantum channel and POVMs. Argue that after the quantum instrument is applied to subsystem Y, the overall system consisting of measuring instrument and measured subsystem has the joint state

$$\rho^{XY} = \sum_{x\in[M]} |x\rangle\langle x| \otimes \mathcal{K}_x(\rho), \tag{16.156}$$

where $\rho_x = p_x(\rho|\mathcal{I})^{-1}\mathcal{K}_x(\rho)$ is the post-instrument state for subsystem Y when outcome x is obtained and $p_x(\rho|\mathcal{I}) = \text{tr}(\mathcal{K}_x(\rho))$ is the corresponding measurement probability.

16.15 Prove the non-negativity and minimum value of the quantum locally accessible information.

16.16 Prove that the quantum locally accessible information is no larger than the maximum measured mutual information by using the data processing inequality.

16.17 As discussed in Appendix 11.D, a quantum channel $\mathcal{K}(\rho)$ can be expressed in terms of the Choi matrix $C_{\mathcal{K}} = \sum_{x\in[N]}\sum_{y\in[N]} |x\rangle\langle y| \otimes \mathcal{K}(|x\rangle\langle y|)$, which is positive semi-definite. (a) For the case of a qubit channel $\mathcal{K}(\cdot)$, show that the Choi matrix can equivalently be written as

$$C_{\mathcal{K}} = 2(\mathcal{I}^X \otimes \mathcal{K})(|\Phi_+^{XY}\rangle\langle\Phi_+^{XY}|). \tag{16.157}$$

Accordingly, the matrix $C_{\mathcal{K}}$ is obtained by applying the channel $\mathcal{K}(\cdot)$ to subsystem Y of the Bell state $|\Phi_+^{XY}\rangle$. *Hint:* Note the equality $(\mathcal{I}^X \otimes \mathcal{K})(|x\rangle\langle y| \otimes |x\rangle\langle y|) = |x\rangle\langle y| \otimes \mathcal{K}(|x\rangle\langle y|)]$. (b) Generalize this result to any dimension $N^X = N^Y = D$ by using the maximally entangled state $|\psi^{XY}\rangle = 1/\sqrt{D}\sum_{x\in[D]}|xx\rangle$ in place of the Bell state $|\Phi_+^{XY}\rangle$ and by showing the equality

$$C_{\mathcal{K}} = N(\mathcal{I}^X \otimes \mathcal{K})(|\psi^{XY}\rangle\langle\psi^{XY}|). \tag{16.158}$$

(c) Recalling from Section 11.6 that the post-channel state can be expressed as

$$\mathcal{K}(\rho) = \sum_{x\in[N]}\sum_{x\in[N]}\langle x|\rho|y\rangle(\langle x| \otimes I)C_{\mathcal{K}}(|y\rangle \otimes I), \tag{16.159}$$

show the equality

$$\mathcal{K}(\rho) = \mathrm{tr}_X((\rho^T \otimes I)C_{\mathcal{K}}). \tag{16.160}$$

16.18 Given two density matrices with eigendecompositions $\rho = \sum_{x\in[N]} p_x|v_x\rangle\langle v_x|$ and $\sigma = \sum_{x\in[N]} q_x|u_x\rangle\langle u_x|$, prove that the Tsallis and Rényi quantum relative entropies between the two densities are equal to their classical counterparts between the joint distributions $p_{xy}^{XY} = p_x\mathrm{F}(|v_x\rangle, |u_y\rangle)$ and $q_{xy}^{XY} = q_x\mathrm{F}(|v_x\rangle, |u_y\rangle)$. *Hint:* Recall the expression of the relative entropy as a function of the eigenvalues from Chapter 8.

16.19 Show that, if the supports of the density matrices $\{\rho_z\}_{z\in[M]}$ are orthogonal, then the quantum entropy of any convex combination $\sum_{z\in[M]} p_z\rho_z$ can be evaluated as

$$\mathrm{H}\left(\sum_{z\in[M]} p_z\rho_z\right) = \mathrm{H}(p) + \sum_{z\in[M]} p_z\mathrm{H}(\rho_z). \tag{16.161}$$

Therefore, the quantum entropy equals the sum of the classical entropy of the mixing distribution $p = [p_0, \ldots, p_{M-1}]^T$ and the average quantum entropy of the density matrices $\{\rho_z\}_{z\in[M]}$.

16.20 (a) Show that discarding a subsystem cannot increase the quantum relative entropy, i.e.,

$$\mathrm{D}(\rho^{XY}||\sigma^{XY}) \geq \max\left(\mathrm{D}(\rho^X||\sigma^X), \mathrm{D}(\rho^Y||\sigma^Y)\right). \tag{16.162}$$

Hint: The partial trace is a quantum channel. (b) Prove that given two classical–quantum states $\rho^{XY} = \sum_{x\in[N^X]} p_x^X|v_x^X\rangle\langle v_x^X| \otimes \rho_x^{Y|X}$ and $\sigma^{XY} = \sum_{x\in[N^X]} q_x^X|v_x^X\rangle\langle v_x^X| \otimes \sigma_x^{Y|X}$ with the same local basis $\mathcal{M}^X = \{|v_x^X\rangle\}_{x\in[N^X]}$, the quantum relative entropy is given by

$$\mathrm{D}(\rho^{XY}||\sigma^{XY}) = \mathrm{D}(p^X||q^X) + \sum_{x\in[N^X]} p_x^X\mathrm{D}(\rho_x^{Y|X}||\sigma_x^{Y|X}). \tag{16.163}$$

Thus, the quantum relative entropy equals the sum of the classical relative entropy between the mixing distributions p^X and q^X, and the average quantum relative entropy between the conditional density matrices at subsystem

Y. (c) Using the results in parts (a) and (b), show the **joint convexity property** of the quantum relative entropy

$$\mathrm{D}\left(\sum_{z\in[M]} p_z\rho_z \middle\| \sum_{z\in[M]} p_z\sigma_z\right) \leq \sum_{z\in[M]} p_z \mathrm{D}(\rho_z||\sigma_z). \tag{16.164}$$

(d) Use the last result to show that the quantum entropy is a concave function of the input density matrix.

16.21 Show that the trace distance between two classical–quantum states with the same local basis $\mathcal{M}^X$, i.e., $\rho^{XY} = \sum_{x\in[N^X]} p_x^X |v_x^X\rangle\langle v_x^X| \otimes \rho_x^{Y|X}$ and $\sigma^{XY} = \sum_{x\in[N^X]} q_x^X |v_x^X\rangle\langle v_x^X| \otimes \sigma_x^{Y|X}$, is given by

$$\mathrm{D}^{\mathrm{tr}}(\rho^{XY}||\sigma^{XY}) = \frac{1}{2}\sum_{x\in[N^X]} ||p_x^X\rho_x^{Y|X} - q_x^X\sigma_x^{Y|X}||_1. \tag{16.165}$$

16.22 Prove that the maximum measured mutual information, the quantum discord, the quantum locally accessible information, and the quantum discord are invariant to local unitary transformations.

16.23 (a) Calculate the quantum discord $\mathrm{Q}(X \to Y)_{\rho^{XY}}$ for the state $\rho^{XY} = (|00\rangle\langle 00| + |1+\rangle\langle 1+|)/2$. (b) To estimate the quantum discord $\mathrm{Q}(Y \to X)_{\rho^{XY}}$ for the same state ρ^{XY}, use an approximation of the quantum ensemble entropy $\mathrm{H}(\mathcal{E}^{X|Y})_{\rho^{XY}}$ obtained by choosing as measurement setting $\mathcal{M}^Y$ the Hadamard basis.

16.24 Show that any two-qubit state can be expressed as

$$\rho^{XY} = \frac{1}{4}\left(I\otimes I + \sum_{x=1}^{3} a_x(P_x\otimes I) + \sum_{y=1}^{3} b_y(I\otimes P_y) + \sum_{x=1}^{3}\sum_{y=1}^{3} c_{x,y}(P_x\otimes P_y)\right), \tag{16.166}$$

where $\{P_0 = I, P_1, P_3, P_3\}$ are the Pauli matrices and the linear-combination scalars are obtained as $a_x = \mathrm{tr}(P_x\rho^X)$, $b_y = \mathrm{tr}(P_y\rho^Y)$, and $c_{x,y} = \mathrm{tr}((P_x \otimes P_y)\rho^{XY})$.

16.25 Prove that the convex combination of entanglement monotones is an entanglement monotone.

16.26 Show that the probability $p_{\rho^{XY}\to\sigma^{XY}}$ of producing a state σ^{XY} from a state ρ^{XY} via SLOCC is upper bounded as

$$p_{\rho^{XY}\to\sigma^{XY}} \leq \frac{\mathfrak{E}(X;Y)_{\rho^{XY}}}{\mathfrak{E}(X;Y)_{\sigma^{XY}}} \tag{16.167}$$

for any entanglement monotone $\mathfrak{E}(X;Y)_{\rho^{XY}}$. *Hint:* Use (16.119). Thus, the more entangled state σ^{XY} is compared to state ρ^{XY}, the less probable it is that state σ^{XY} can be obtained from ρ^{XY} via SLOCC.

Appendix

16.A Decomposing Density Matrices using Uncorrelated Mixed States

This appendix presents two ways to construct the local density matrices in the decomposition (16.19). The first approach is to select a basis of $(N^X)^2$ density matrices $\{\rho_x^X\}_{x\in[(N^X)^2]}$ for subsystem X and a basis of $(N^Y)^2$ density matrices $\{\rho_y^Y\}_{y\in[(N^Y)^2]}$ for subsystem Y. The joint states $\{\{\rho_x^X\otimes\rho_y^Y\}_{x\in[(N^X)^2]}\}_{y\in[(N^Y)^2]}$ provide a basis for the space of $N^X N^Y \times N^X N^Y$ density matrices. Following the discussion in Section 7.13, while this basis is not orthonormal, one can always express a quantum mixed state ρ^{XY} as

$$\rho^{XY} = \sum_{x\in[(N^X)^2]}\sum_{y\in[(N^Y)^2]} \gamma_{xy}\rho_x^X \otimes \rho_y^Y \tag{16.168}$$

with real-valued coefficients γ_{xy} for all $x \in [(N^X)^2]$ and $y \in [(N^Y)^2]$ that sum to 1.

Another approach, also discussed in Section 7.13, is to use local observables. Accordingly, one fixes an orthonormal basis of $(N^X)^2$ observables $\{O_x^X\}_{x\in[(N^X)^2]}$ for subsystem X and an orthonormal basis of $(N^Y)^2$ observables $\{O_y^Y\}_{y\in[(N^Y)^2]}$ for subsystem Y. Examples of such bases include Pauli strings, which apply to settings with dimensions $N^X = 2^{D^X}$ and $N^Y = 2^{D^Y}$ for some integers D^X and D^Y (see Section 8.9.2).

Using such local bases of observables, one can check that the set of all pairs of Kronecker products $\{\{O_x^X \otimes O_y^Y\}_{x\in[(N^X)^2]}\}_{y\in[(N^Y)^2]}$ forms an orthonormal basis for the space of $N^X N^Y \times N^X N^Y$ density matrices. It follows that any $N^X N^Y \times N^X N^Y$ density matrix ρ^{XY} can be written as

$$\rho^{XY} = \sum_{x\in[(N^X)^2]}\sum_{y\in[(N^Y)^2]} \gamma_{xy} O_x^X \otimes O_y^Y \tag{16.169}$$

for real-valued coefficients

$$\gamma_{xy} = \mathrm{tr}((O_x^X \otimes O_y^Y)\rho^{XY}) = \langle O_x^X \otimes O_y^Y\rangle_{\rho^{XY}}. \tag{16.170}$$

16.B Proof of (16.75) and (16.76)

This appendix proves the equalities (16.75) and (16.76).

For the first equality, use the eigendecompositions $\sigma^X = \sum_{x\in[N^X]} p_x^X |v_x^X\rangle\langle v_x^X|$ and $\sigma^Y = \sum_{y\in[N^Y]} p_y^Y |v_y^Y\rangle\langle v_y^Y|$ to obtain the eigendecomposition $\sigma^X \otimes \sigma^Y = \sum_{x\in[N^X]}\sum_{y\in[N^Y]} p_x^X p_y^Y |v_x^X v_y^Y\rangle\langle v_x^X v_y^Y|$, which implies the equalities

$$\begin{aligned}
\log_2(\sigma^X \otimes \sigma^Y) &= \sum_{x\in[N^X]}\sum_{y\in[N^Y]} \log_2(p_x^X p_y^Y)|v_x^X v_y^Y\rangle\langle v_x^X v_y^Y| \\
&= \sum_{x\in[N^X]} \log_2(p_x^X)|v_x^X\rangle\langle v_x^X| \otimes \sum_{y\in[N^Y]} |v_y^Y\rangle\langle v_y^Y| \\
&\quad + \sum_{x\in[N^X]} |v_x^X\rangle\langle v_x^X| \otimes \sum_{y\in[N^Y]} \log_2(p_y^Y)|v_y^Y\rangle\langle v_y^Y| \qquad (16.171)\\
&= \log_2(\sigma^X)\otimes I + I \otimes \log_2(\sigma^Y), \qquad (16.172)
\end{aligned}$$

where the first equality follows from the definition of a function of a normal matrix.

For (16.76), using the general decomposition (16.19) into uncorrelated states for matrix ρ^{XY}, one obtains

$$\begin{aligned}
&\mathrm{tr}(\rho^{XY}\log_2(\sigma^X\otimes\sigma^Y))\\
&\quad=\sum_{z\in[M]}\gamma_z\mathrm{tr}\left(\rho_z^X\otimes\rho_z^Y\log_2(\sigma^X\otimes\sigma^Y)\right)\\
&\quad=\sum_{z\in[M]}\gamma_z\mathrm{tr}(\rho_z^X\log_2(\sigma^X)\otimes\rho_z^Y)+\sum_{z\in[M]}\gamma_z\mathrm{tr}(\rho_z^X\otimes\rho_z^Y\log_2(\sigma^Y))\\
&\quad=\sum_{z\in[M]}\gamma_z\mathrm{tr}(\rho_z^X\log_2(\sigma^X))+\sum_{z\in[M]}\gamma_z\mathrm{tr}(\rho_z^Y\log_2(\sigma^Y))\\
&\quad=\mathrm{tr}(\rho^X\log_2(\sigma^X))+\mathrm{tr}(\rho^Y\log_2(\sigma^Y)),
\end{aligned}\tag{16.173}$$

where the second equality follows from (16.75); the third equality from the relation $\mathrm{tr}(A^X\otimes A^Y)=\mathrm{tr}(A^X)\mathrm{tr}(A^Y)$; and the fourth equality from the expressions (16.31) and (16.32) for the reduced density matrices.

16.C Derivation of the Quantum Joint Entropy of Classical–Quantum States

This appendix derives the expression (16.83) for the joint entropy of a classical–quantum state $\overrightarrow{\omega}^{XY}=\sum_{x\in[N^X]}p_x^X|v_x^X\rangle\langle v_x^X|\otimes\rho_x^{Y|X}$. Writing as $\rho_x^{Y|X}=\sum_{y\in[N^Y]}p_{y|x}^{Y|X}|v_{y|x}^{Y|X}\rangle\langle v_{y|x}^{Y|X}|$ the eigendecomposition of the conditional density matrix $\rho_x^{Y|X}$, one has the equality

$$\log_2(\overrightarrow{\omega}^{XY})=\sum_{x\in[N^X]}\sum_{y\in[N^Y]}(\log_2(p_x^X)+\log_2(p_{y|x}^{Y|X}))|v_x^X\rangle\langle v_x^X|\otimes|v_{y|x}^{Y|X}\rangle\langle v_{y|x}^{Y|X}|.\tag{16.174}$$

Therefore, the quantum joint entropy can be computed as

$$\begin{aligned}
&\mathrm{H}(X,Y)_{\overrightarrow{\omega}^{XY}}\\
&\quad=-\mathrm{tr}\left(\overrightarrow{\omega}^{XY}\log_2\left(\overrightarrow{\omega}^{XY}\right)\right)\\
&\quad=-\mathrm{tr}\left(\sum_{x\in[N^X]}\sum_{y\in[N^Y]}(p_x^X\log_2(p_x^X)+p_x^X\log_2(p_{y|x}^{Y|X}))|v_x^X\rangle\langle v_x^X|\otimes\rho_x^{Y|X}|v_{y|x}^{Y|X}\rangle\langle v_{y|x}^{Y|X}|\right)\\
&\quad=\mathrm{H}(X)_{p^X}+\sum_{x\in[N^X]}p_x^X\mathrm{H}(Y)_{\rho_x^{Y|X}},
\end{aligned}\tag{16.175}$$

completing the proof.

16.D Proof of the Insubordination Property of the Quantum Mutual Information

The inequalities (16.89) and (16.90) are equivalent, respectively, to the inequalities

$$\mathrm{H}(X,Y)_{\rho^{XY}}\geq\max(\mathrm{H}(X)_{\rho^X},\mathrm{H}(Y)_{\rho^Y})\tag{16.176}$$

and

$$\begin{aligned}\mathrm{H}(X,Y)_{\rho^{XY}} &\geq |\mathrm{H}(X)_{\rho^X} - \mathrm{H}(Y)_{\rho^Y}| \\ &= \max(\mathrm{H}(X)_{\rho^X}, \mathrm{H}(Y)_{\rho^Y}) - \min(\mathrm{H}(X)_{\rho^X}, \mathrm{H}(Y)_{\rho^Y}).\end{aligned} \tag{16.177}$$

The latter inequality is also known as the Araki–Lieb inequality. Given the equivalence to (16.89) and (16.90), this appendix is devoted to proving the inequalities (16.176) and (16.177)

To demonstrate the Araki–Lieb inequality (16.177), it is sufficient to prove the inequality $\mathrm{H}(X,Y)_{\rho^{XY}} \geq \mathrm{H}(X)_{\rho^X} - \mathrm{H}(Y)_{\rho^Y}$, as the inequality $\mathrm{H}(X,Y)_{\rho^{XY}} \geq \mathrm{H}(Y)_{\rho^Y} - \mathrm{H}(X)_{\rho^X}$ follows in the same way by symmetry. The proof applies the idea of purification introduced in Section 8.7. To begin, consider a larger system including not only subsystems X and Y, but also the environment E. The environment is selected in such a way that the state of the overall system encompassing subsystem X, Y, and E is pure.

Denoting the resulting pure state as $|\psi^{XYE}\rangle$, the joint entropy is given as $\mathrm{H}(X,Y,E)_{|\psi^{XYE}\rangle} = 0$. Furthermore, by considering the bipartite system with subsystem Y on one side and subsystems X and E on the other, the equality

$$\mathrm{H}(X,E)_{\rho^{XE}} = \mathrm{H}(Y)_{\rho^Y} \tag{16.178}$$

holds, where ρ^{XE} and ρ^Y are the respective reduced density matrices. In fact, as seen in Section 14.5, the reduced density matrices for a pure state have the same eigenvalues. In a similar way, one has the equality

$$\mathrm{H}(E)_{\rho^E} = \mathrm{H}(X,Y)_{\rho^{XY}}. \tag{16.179}$$

The subadditivity of the quantum entropy implies the inequality

$$\mathrm{H}(X,E)_{\rho^{XE}} \leq \mathrm{H}(X)_{\rho^X} + \mathrm{H}(E)_{\rho^{XE}}, \tag{16.180}$$

which is equivalent, using (16.178) and (16.179), to the inequality

$$\mathrm{H}(Y)_{\rho^Y} \leq \mathrm{H}(X)_{\rho^X} + \mathrm{H}(X,Y)_{\rho^{XE}}, \tag{16.181}$$

concluding the proof of the Araki–Lieb inequality.

The inequality (16.176) for a separable state $\rho^{XY} = \sum_{z\in[M]} p_z \rho_z^X \otimes \rho_z^Y$ can be proved by leveraging the strong subadditivity property of the quantum entropy. This property states that, for a quantum system consisting of subsystems X, Y, and Z, the following inequality holds:

$$\mathrm{H}(X|Y,Z)_{\rho^{XYZ}} \leq \mathrm{H}(X|Y)_{\rho^{XY}}, \tag{16.182}$$

where $\mathrm{H}(X|Y,Z)_{\rho^{XYZ}} = \mathrm{H}(X,Y,Z)_{\rho^{XYZ}} - \mathrm{H}(Y,Z)_{\rho^{YZ}}$, $\rho^{YZ} = \mathrm{tr}_X(\rho^{XYZ})$, and $\rho^{XY} = \mathrm{tr}_Z(\rho^{XYZ})$. Accordingly, the strong subadditivity property is a generalization of the property that conditioning reduces entropy. The reader is referred to the Recommended Resources for a proof.

In order to prove (16.176), it is sufficient to show the inequality $\mathrm{H}(X,Y)_{\rho^{XY}} \geq \mathrm{H}(X)_{\rho^X}$, or equivalently $\mathrm{H}(X|Y)_{\rho^{XY}} \geq 0$. The symmetric inequalities $\mathrm{H}(X,Y)_{\rho^{XY}} \geq \mathrm{H}(Y)_{\rho^X}$ and $\mathrm{H}(Y|X)_{\rho^{XY}} \geq 0$ would then follow in a similar way.

Using the strong subadditivity property (16.182), the inequality $\mathrm{H}(X|Y)_{\rho^{XY}} \geq 0$ is implied by the inequality $\mathrm{H}(X|Y,Z)_{\rho^{XYZ}} \geq 0$ for a state ρ^{XYZ} with reduced density matrix $\rho^{XY} = \mathrm{tr}_Z(\rho^{XYZ})$.

To leverage this insight, consider the state

$$\rho^{XYZ} = \sum_{z\in[M]} p_z \rho_z^X \otimes \rho_z^Y \otimes |z\rangle\langle z|, \tag{16.183}$$

in which the added subsystem Z is classical. Note that this state satisfies the desired equality $\rho^{XY} = \mathrm{tr}_Z(\rho^{XYZ})$.

Since the state ρ^{XYZ} is quantum–classical, the conditional entropy can be evaluated as

$$\begin{aligned} \mathrm{H}(X|Y,Z)_{\rho^{XYZ}} &= \mathrm{E}_p[\mathrm{H}(X|Y)_{\rho_z^X \otimes \rho_z^Y}] \\ &= \mathrm{E}_p[\mathrm{H}(X)_{\rho_z^X}] \geq 0, \end{aligned} \tag{16.184}$$

concluding the proof.

16.E Classical and Quantum Correlations as Resources

Taking the perspective of quantifying quantum and classical correlations as a resource (see Section 12.7), this appendix shows that the quantum mutual information measures the minimum distance of the state ρ^{XY} to the set of states that are free from correlations, i.e., from the set of uncorrelated states, which are of the form $\sigma^X \otimes \sigma^Y$. That is, one has the equality

$$\mathrm{I}(X;Y)_{\rho^{XY}} = \min_{\sigma^X,\sigma^Y} \mathrm{D}(\rho^{XY}||\sigma^X \otimes \sigma^Y). \tag{16.185}$$

To prove (16.185), the uncorrelated state $\rho^X \otimes \rho^Y$ will be shown here to minimize the quantum relative entropy $\mathrm{D}(\rho^{XY}||\sigma^X \otimes \sigma^Y)$ among all the uncorrelated states $\sigma^X \otimes \sigma^Y$. To this end, it is sufficient to show the inequality $\mathrm{D}(\rho^{XY}||\sigma^X \otimes \sigma^Y) - \mathrm{D}(\rho^{XY}||\rho^X \otimes \rho^Y) \geq 0$ for any uncorrelated state $\sigma^X \otimes \sigma^Y$. This inequality follows as

$$\begin{aligned} &\mathrm{D}(\rho^{XY}||\sigma^X \otimes \sigma^Y) - \mathrm{D}(\rho^{XY}||\rho^X \otimes \rho^Y) \\ &= \mathrm{tr}(\rho^{XY}\log_2(\rho^X \otimes \rho^Y)) - \mathrm{tr}(\rho^{XY}\log_2(\sigma^X \otimes \sigma^Y)) \\ &= \mathrm{tr}(\rho^X\log_2(\rho^X)) + \mathrm{tr}(\rho^Y\log_2(\rho^Y)) \qquad (16.186) \\ &- \mathrm{tr}(\rho^X\log_2(\sigma^X)) - \mathrm{tr}(\rho^Y\log_2(\sigma^Y)) \\ &= \mathrm{D}(\rho^X||\sigma^X) + \mathrm{D}(\rho^Y||\sigma^Y) \geq 0, \qquad (16.187) \end{aligned}$$

where the second equality is from (16.76). The inequality is strict if and only if the equalities $\rho^X = \sigma^X$ and $\rho^Y = \sigma^Y$ hold, concluding the proof.

16.F Proof of the Bounds (16.104)

The lower bound in (16.104) follows directly from the definition (16.102), indicating that only uncorrelated states have zero classical correlations, i.e., $\mathrm{J}(X;Y)_{\rho^X\otimes\rho^Y} = 0$.

The upper bounds in (16.104) can be derived from the data processing inequality under local channels (16.98). In particular, given any optimized measurements $\mathcal{M}_*^X$ and $\mathcal{M}_*^Y$, one has

$$\begin{aligned} \mathrm{J}(X;Y)_{\rho^{XY}} &= \mathrm{I}(X;Y)_{\mathcal{M}_*^X \otimes \mathcal{M}_*^Y(\rho^{XY})} \\ &\leq \mathrm{I}(X;Y)_{\mathcal{I}^X \otimes \mathcal{M}_*^Y(\rho^{XY})} \\ &= \mathrm{H}(X)_{\rho^X} - \sum_{y \in [N^Y]} p_y^Y \mathrm{H}(\rho_y^{X|Y}) \\ &\leq \mathrm{H}(X)_{\rho^X}, \end{aligned} \tag{16.188}$$

where the first inequality follows from data processing inequality under local channels, while the second inequality follows from the non-negativity of the entropy.

The upper bound $\mathrm{J}(X;Y)_{\rho^{XY}} \leq \mathrm{I}(X;Y)_{\rho^{XY}}$ follows in a similar way by noting that the data processing inequality under local channels also implies the inequality

$$\begin{aligned} \mathrm{J}(X;Y)_{\rho^{XY}} =& \mathrm{I}(X;Y)_{\mathcal{M}_*^X \otimes \mathcal{M}_*^Y(\rho^{XY})} \\ \leq& \mathrm{I}(X;Y)_{\rho^{XY}}. \end{aligned} \tag{16.189}$$

16.G Numerical Evaluation of the Quantum Symmetric Discord

For two-qubit states, the local measurement setting at subsystem X can be described as $\mathcal{M}^X = \{\Pi_0^X, \Pi_1^X\}$, with projection matrices expressed without loss of generality by using a decomposition into scaled Pauli matrices as

$$\begin{aligned} \Pi_0^X &= \frac{1}{2}(I + (v_1^X P_1 + v_2^X P_1 + v_3^X P_3)), \\ \Pi_1^X &= \frac{1}{2}(I - (v_1^X P_1 + v_2^X P_1 + v_3^X P_3)), \end{aligned} \tag{16.190}$$

where $v^X = [v_1^X, v_2^X, v_3^X]^T$ is a real-valued vector with norm equal to 1, i.e., $||v^X|| = 1$. The measurement at subsystem Y can be parameterized in a similar way, and thus optimization over the local measurement settings $\mathcal{M}^X$ and $\mathcal{M}^Y$ can be carried out over the unitary-norm vectors $v^X = [v_1^X, v_2^X, v_3^X]^T$ and $v^Y = [v_1^Y, v_2^Y, v_3^Y]^T$.

REFERENCES

[1] T. Vidick and S. Wehner, *Introduction to Quantum Cryptography*. Cambridge: Cambridge University Press, 2023.

[2] M. Lewenstein and A. Sanpera, "Separability and entanglement of composite quantum systems," *Physical Review Letters*, vol. 80, no. 11, art. 2261, 1998.

[3] S. Wu, U. Poulsen, and K. Mølmer, "Correlations in local measurements on a quantum state, and complementarity as an explanation of nonclassicality," *Physical Review A*, vol. 80, no. 3, art. 032319, 2009.

[4] L. Henderson and V. Vedral, "Classical, quantum and total correlations," *Journal of Physics A: Mathematical and General*, vol. 34, no. 35, art. 6899, 2001.

[5] M. Ali, A. Rau, and G. Alber, "Quantum discord for two-qubit X

states," *Physical Review A*, vol. 81, no. 4, art. 069902, 2010.
[6] A. Bera, T. Das, D. Sadhukhan, S. Roy, A. De, and U. Sen, "Quantum discord and its allies: A review of recent progress," *Reports on Progress in Physics*, vol. 81, no. 2, art. 024001, 2017.
[7] J. Oppenheim, M. Horodecki, P. Horodecki, and R. Horodecki, "Thermodynamical approach to quantifying quantum correlations," *Physical Review Letters*, vol. 89, no. 18, art. 180402, 2002.
[8] D. Girolami, *Quantum Correlations in Information Theory*. PhD dissertation, University of Nottingham, 2012.
[9] J. Maziero, L. Celeri, and R. Serra, "Symmetry aspects of quantum discord," *arXiv*, 2010. [Online]. Available at: https://arxiv.org/abs/1004.2082
[10] K. Modi, T. Paterek, W. Son, V. Vedral, and M. Williamson, "Unified view of quantum and classical correlations," *Physical Review Letters*, vol. 104, no. 8, art. 080501, 2010.
[11] V. Vedral, M. Plenio, M. Rippin, and P. Knight, "Quantifying entanglement," *Physical Review Letters*, vol. 78, no. 12, art. 2275, 1997.
[12] G. Vidal and R. F. Werner, "Computable measure of entanglement," *Physical Review A*, vol. 65, no. 3, art. 032314, 2002.
[13] M. Horodecki, P. Horodecki, and R. Horodecki, "Limits for entanglement measures," *Physical Review Letters*, vol. 84, no. 9, art. 2014, 2000.
[14] N. Li and S. Luo, "Information-theoretic aspects of Werner states," *Annals of Physics*, vol. 424, art. 168371, 2021.
[15] M. Nielsen and J. Kempe, "Separable states are more disordered globally than locally," *Physical Review Letters*, vol. 86, no. 22, art. 5184, 2001.
[16] W. Dür, J. Cirac, M. Lewenstein, and D. Bruß, "Distillability and partial transposition in bipartite systems," *Physical Review A*, vol. 61, no. 6, art. 062313, 2000.
[17] C. Bennett, H. Bernstein, S. Popescu, and B. Schumacher, "Concentrating partial entanglement by local operations," *Physical Review A*, vol. 53, no. 4, art. 2046, 1996.
[18] B. C. Hiesmayr, C. Popp, and T. C. Sutter, "Bipartite bound entanglement," *arXiv*, 2024. [Online]. Available at: https://arxiv.org/abs/2406.13491
[19] Z. Li, M. Zhao, S. Fei, H. Fan, and W. Liu, "Mixed maximally entangled states," *Quantum Information & Computation*, vol. 12, no. 1–2, pp. 63–73, 2012.
[20] A. Tavakoli, A. Pozas-Kerstjens, P. Brown, and M. Araújo, "Semidefinite programming relaxations for quantum correlations," *arXiv*, 2024. [Online]. Available at: https://arxiv.org/abs/2307.02551
[21] W. Wootters, "Entanglement of formation of an arbitrary state of two qubits," *Physical Review Letters*, vol. 80, no. 10, art. 2245, 1998.
[22] P. Rungta, V. Bužek, C. Caves, M. Hillery, and G. Milburn, "Universal state inversion and concurrence in arbitrary dimensions," *Physical Review A*, vol. 64, no. 4, art. 042315, 2001.
[23] D. Cavalcanti and P. Skrzypczyk, "Quantum steering: A review with focus on semidefinite programming," *Reports on Progress in Physics*, vol. 80, no. 2, art. 024001, 2016.
[24] F. Fanchini, L. Castelano, M. Cornelio, and M. De Oliveira, "Locally inaccessible information as a fundamental ingredient to quantum information," *New Journal of Physics*, vol. 14, no. 1, art. 013027, 2012.

[25] X. Lu, J. Ma, Z. Xi, and X. Wang, "Optimal measurements to access classical correlations of two-qubit states," *Physical Review A*, vol. 83, no. 1, art. 012327, 2011.

[26] M. Nielsen and D. Petz, "A simple proof of the strong subadditivity inequality," *arXiv*, 2004. [Online]. Available at: https://arxiv.org/abs/quant-ph/0408130

Part VI

Multipartite Classical and Quantum Information

17 A Brief Introduction to Tensor Networks

17.1 Overview

This last part of the book explores multipartite systems consisting of an arbitrary number of subsystems. To initiate the study of correlations across multiple subsystems, this chapter introduces the formalism of tensor networks. Tensor networks offer an efficient framework to describe both multipartite probability distributions and multipartite quantum states, encoding the structure of the correlations among subsystems.

For D-dimensional subsystems, the state of a multipartite system is described by vectors with dimension D^S, which grows exponentially with the number of subsystems, S. When S is large, directly analyzing the correlation properties of such high-dimensional states becomes impractical. In these cases, establishing an efficient, tensor network-based description for joint distributions and joint quantum states becomes essential for studying correlations in composite systems.

This chapter covers foundational aspects of tensors and tensor networks, focusing on topics that will be crucial in the following chapters. It begins by introducing the general mathematical definition of a tensor, along with its corresponding graphical representation. Following this, it reviews key operations such as tensor contraction and the tensor product, as well as their representations within tensor networks. The chapter also demonstrates how tensor network representations simplify the evaluation of operations involving high-dimensional states, such as inner products and matrix–vector multiplications. Lastly, it discusses the concept of tensor rank, which serves as a valuable measure of the correlations present in a system.

17.1.1 Learning Objectives and Organization

By the end of this chapter, the reader should be able to:

- understand and use tensor representations of high-dimensional vectors;
- construct and interpret tensor networks;
- understand, use, and represent the operations of contraction and tensor product;
- apply the fixed-value and copy tensors;
- apply the operations of vectorization and matrization to an arbitrary tensor;

- and understand the canonic polyadic decomposition and the notion of tensor rank.

17.2 Tensors

In this section, we introduce tensors by emphasizing their mathematical formalism and graphical representation. As discussed in previous chapters, vectors and matrices can be viewed as indexed arrays of numbers. Their structures support essential algebraic operations, such as inner products and matrix multiplication. Tensors extend the concepts of vectors and matrices to higher dimensions.

17.2.1 Mathematical Definition of a Tensor

A $D \times 1$ vector a is an array arranged along a single index $x \in [D]$ with dimension D. Specifying the index x retrieves the xth entry a_x.

More generally, a $D^0 \times D^1$ matrix A is an array arranged along two indices, $x^0 \in [D^0]$ and $x^1 \in [D^1]$, with dimensions D^0 and D^1, respectively. By specifying both indices x^0 and x^1, one retrieves the (x^0, x^1)th entry A_{x^0,x^1} of the matrix.

A tensor is a **multi-index array**, in which an entry is identified by multiple indices. The order of a tensor refers to the number of indices needed to specify an entry. For example, a vector is an order-1 tensor and a matrix is an order-2 tensor.

For an order-S tensor, each index $x^s \in [D^s]$ has dimension D^s, in the sense that it can take one out of D^s possible values, for all $s \in [S]$. This structure yields the following definition.

Definition 17.1 Tensor
An order-S tensor T with dimensions $(D^0, D^1, \cdots, D^{S-1})$ defines a function

$$x^0, x^1, \ldots, x^{S-1} \mapsto T_{x^0,x^1,\ldots,x^{S-1}}, \tag{17.1}$$

that assigns a real or complex number $T_{x^0,x^1,\ldots,x^{S-1}}$ to each choice $x^0, x^1, \ldots, x^{S-1}$ of the S indices. The number $T_{x^0,x^1,\ldots,x^{S-1}}$ is referred to as the $(x^0, \ldots, x^{S-1})$th entry, or element, of the tensor.

As will be discussed, tensors with real-valued elements are useful to represent probability distributions, while complex-valued tensors are useful to represent quantum states.

The indices $x^0, x^1, \ldots, x^{S-1}$ of an order-S tensor T can take

$$D^0 D^1 \cdots D^{S-1} = \prod_{s\in[S]} D^s \tag{17.2}$$

possible values, and hence the tensor T has $\prod_{s\in[S]} D^s$ entries. For instance, if each index has the same number of dimensions, D, the number of elements in an order-S tensor is D^S, which grows exponentially with the order S.

In order to streamline the notation, henceforth, the indices $x^0, x^1, \ldots, x^{S-1}$ will also be denoted as $x^{[S]}$, with the superscript $[S]$ denoting as usual the set of integers $[S] = \{0, 1, \ldots, S-1\}$, i.e.,

$$x^{[S]} = x^0, x^1, \ldots, x^{S-1}. \tag{17.3}$$

Accordingly, the $(x^0, \ldots, x^{S-1})$th entry of tensor T is also denoted as

$$T_{x^{[S]}} = T_{x^0, x^1, \ldots, x^{S-1}}, \tag{17.4}$$

and referred to as the $x^{[S]}$th entry of the tensor.

Using a similar logic, the dimensions of a tensor may also be written with the notation

$$D^{[S]} = (D^0, D^1, \ldots, D^{S-1}) = (D^s)_{s \in [S]}. \tag{17.5}$$

Finally, the set of all possible values for the indices $x^0, x^1, \ldots, x^{S-1}$ is denoted as $[D^{[S]}]$. Thus, a tensor T contains the $\prod_{s \in [S]} D^s$ entries

$$\{T_{x^0, x^1, \ldots, x^{S-1}}\}_{x^0 \in [D^0], x^1 \in [D^1], \ldots, x^{S-1} \in [D^{S-1}]} = \{T_{x^{[S]}}\}_{x^{[S]} \in [D^{[S]}]}. \tag{17.6}$$

Example 17.1

A tensor T with order $S = 3$ and dimensions $D^0 = D^1 = D^2 = 2$ is determined by $2^3 = 8$ numbers. These correspond to the entries T_{x^0, x^1, x^2} for all indices $x^0 \in [2]$, $x^1 \in [2]$, and $x^2 \in [2]$, i.e., $T_{0,0,0}$, $T_{0,0,1}$, $T_{0,1,0}$, $T_{0,1,1}$, $T_{1,0,0}$, $T_{1,0,1}$, $T_{1,1,0}$, and $T_{1,1,1}$.

The tensor T can equivalently be specified via two 2×2 matrices. For instance, one can describe the matrices T_{0,x^1,x^2} and T_{1,x^1,x^2} that are obtained by varying the indices $x^1 \in [2]$ and $x^2 \in [2]$. Alternatively, the tensor T can be described via the two matrices obtained by fixing any of the other two indices, i.e., via the pair of matrices $T_{x^0,0,x^2}$ and $T_{x^0,1,x^2}$ or the pair $T_{x^0,x^1,0}$ and $T_{x^0,x^1,1}$.

Indices x^s of dimension $D^s = 1$ can be removed from the description of a tensor without loss of information. In fact, an index x^s of order $D^s = 1$ takes values in the trivial set $[1] = \{0\}$, and it is hence fixed to the value $x^s = 0$. Therefore, an index x_s with dimension $D_s = 1$ does not contribute to the value returned by the tensor and it can be dropped from the tensor's description. For example, an order-3 tensor T_{x^0,x^1,x^2} with dimensions $(1, 2, 2)$ can equivalently be viewed as an order-2 tensor with dimensions $(2, 2)$ in which the first index is removed. Furthermore, an order-0 tensor corresponds to a tensor of order 0, i.e., to a scalar.

17.2.2 Graphical Representation of a Tensor

Useful tensors are typically **structured**. This means that the entries of the tensor are interrelated, allowing it to be fully specified with fewer independent degrees of freedom than the total number of entries, $\prod_{s \in [S]} D^s$. Furthermore, in order to encode and reason about the structure of a tensor, it is useful to adopt a graphical

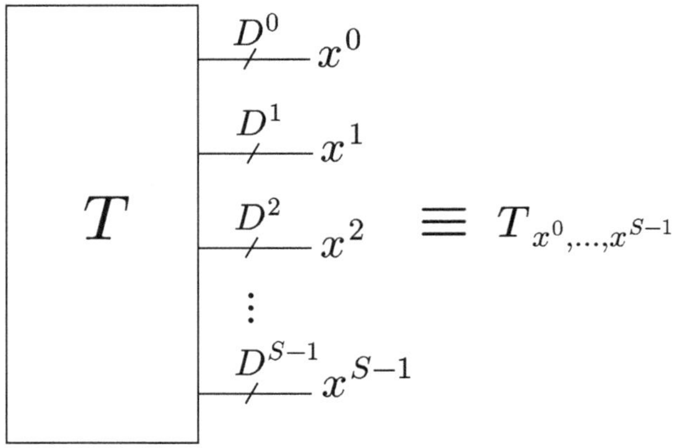

Figure 17.1 An order-S tensor with dimensions $(D^0, \ldots, D^{S-1})$ is represented by a block with S legs, one for each sth index $x^s \in [D^s]$. Once a value x^s is specified for each sth leg of a tensor, with $s \in [S]$, the tensor returns the $x^{[S]}$th entry $T_{x^{[S]}}$.

representation. This graphical approach also proves convenient for describing and implementing various operations on tensors.

As illustrated in Figure 17.1, an order-S tensor is represented by a block with S **legs**, one for each sth index x^s. Each sth leg corresponds to a specific index $x^s \in [D^s]$, and is instantiated by the value of the index $x^s \in [D^s]$. Once a value x^s is specified for each sth leg of a tensor, with $s \in [S]$, the tensor returns the value $T_{x^{[S]}}$.

The dimension D^s of each index s – that is, the number of possible values it can assume – is indicated in the diagram. However, this information may be omitted in more complex tensor diagrams to reduce visual clutter. Moreover, since each mathematical index x^s has a direct counterpart in the graphical representation as the sth leg, the terms legs and indices will be used interchangeably from this point forward.

As will be discussed, legs from different tensors can be connected to create bonds, supporting the graphical representation of operations such as inner products and the trace. Indices corresponding to legs that are not connected to any other tensor, as is the case for the tensor in Figure 17.1, are said to be **open**.

17.2.3 Vectors and Matrices as Tensors

As discussed, an order-1 tensor corresponds to a vector, while an order-2 tensor corresponds to a matrix with one tensor index running over the rows and the other over the columns of the matrix. While the legs of a tensor can be placed anywhere along the boundary of the corresponding block representation, for vectors and matrices it will be convenient to follow an explicit convention.

Specifically, as illustrated in Figure 17.2(a), an order-1 tensor with leg on the right is interpreted as a column vector, while order-1 tensors with a single index pointing to the left are considered as row vectors. In a manner consistent with this

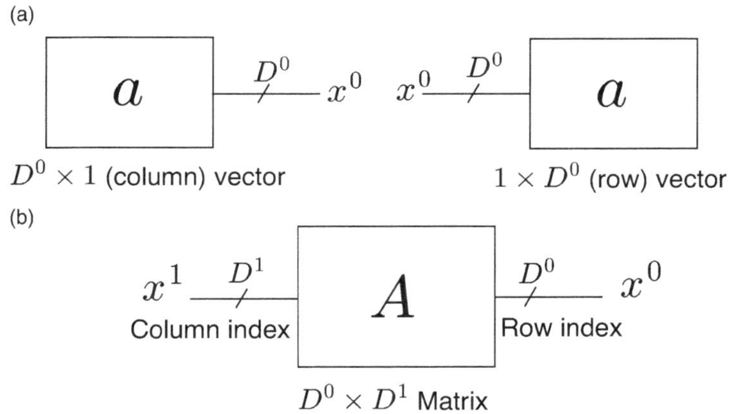

Figure 17.2 (a) An order-1 tensor with leg on the right is interpreted as a column vector, while order-1 tensors with a single index pointing to the left are considered as row vectors. (b) A matrix is represented as an order-2 tensor with the left-pointing index as the column index and the right-pointing index representing the row index.

convention, as shown in Figure 17.2(b), a matrix is represented as an order-2 tensor with the left-pointing index as the column index and the right-pointing index representing the row index.

17.3 Tensor Product

Two or more tensors can be combined to form a tensor that inherits some, or all, of the indices of the constituent tensors. The description of a tensor via the combination of, typically simpler, tensors is known as a **tensor network**. The most straightforward way to combine tensors is via the **tensor product**, which is described in this section.

The tensor product of two tensors T^0 and T^1 produces a larger-dimensional tensor T that inherits all of the indices of the constituent tensors. Each entry in tensor T is given by the product of the corresponding entries of the tensors T^0 and T^1.

Graphically, the tensor product is represented by placing the two tensors side by side in an arbitrary arrangement, illustrating their separate identities within the product. The next section will discuss a more general operation in which the indices of different tensors can be connected via a bond. Such connections can encode more complex structures in the resulting tensor.

Formally, consider any order-S^0 tensor T^0 and any order-S^1 tensor T^1, with indices having arbitrary dimensions. Denote as $x^{[S^0]}$ the indices for tensor T^0, with corresponding entries $T^0_{x^{[S^0]}}$; and as $y^{[S^1]}$ the indices for tensor T^1, with corresponding entries $T^1_{y^{[S^1]}}$. The tensor product, denoted as $T = T^0 \otimes T^1$, is a tensor T with $S^0 + S^1$ indices, which correspond to the collection of the indices $x^{[S^0]}$ of tensor T^0 and the indices $y^{[S^1]}$ of tensor T^1. Therefore, each entry $T_{x^{[S^0]},y^{[S^1]}}$ of tensor T is identified by the indices $(x^{[S^0]}, y^{[S^1]})$.

Definition 17.2 Tensor product
For any order-S^0 tensor T^0 and any order-S^1 tensor T^1, with indices having arbitrary dimensions, the tensor product

$$T = T^0 \otimes T^1 \tag{17.7}$$

is the order-$(S^0 + S^1)$ tensor with entries given by the products

$$T_{x^{[S^0]}, y^{[S^1]}} = T^0_{x^{[S^0]}} T^1_{y^{[S^1]}}, \tag{17.8}$$

with $T^0_{x^{[S^0]}}$ and $T^1_{y^{[S^1]}}$ denoting the $x^{[S^0]}$th and $y^{[S^1]}$th entries of tensors T^0 and T^1, respectively.

As will be detailed in Section 17.4, the tensor product is a generalization of the Kronecker product for vectors and matrices introduced in Section 8.8. Furthermore, as illustrated by the next example, a tensor product is represented diagrammatically by simply placing the two tensors side by side without adding any bond between the tensors.

Example 17.2

Consider the top part of Figure 17.3, which shows two tensors T^0 and T^1 with orders $S^0 = 5$ and $S^1 = 4$, respectively. The tensor product $T^0 \otimes T^1$ can be represented diagrammatically as in the figure, by placing the two tensors side by side. This is an example of a tensor network.

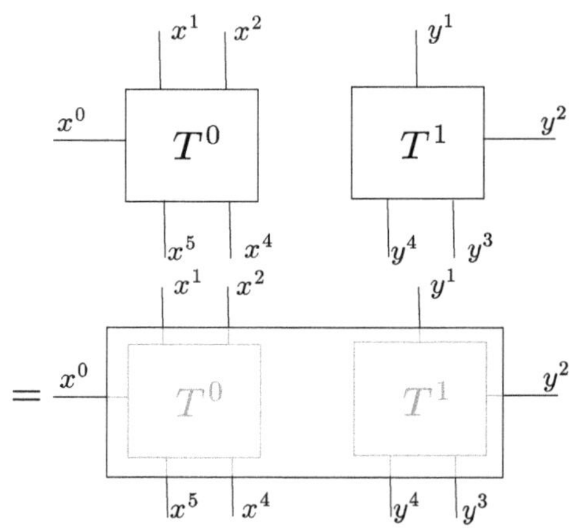

Figure 17.3 For tensors T^0 and T^1 with orders $S^0 = 5$ and $S^1 = 4$, the resulting tensor $T^0 \otimes T^1$ has order $S^0 + S^1 = 9$, and its indices include all the indices of the two tensors T^0 and T^1. The top part of the figure shows the tensor network corresponding to the tensor product $T^0 \otimes T^1$, while the bottom part of the figure highlights the resulting higher-dimensional tensor.

As illustrated in the bottom part of Figure 17.3, the resulting tensor $T^0 \otimes T^1$ has order $S^0 + S^1 = 9$, since its indices include all the indices of the two tensors T^0 and T^1.

Example 17.3

For a numerical example, assume that all the indices have the same dimension D. Furthermore, the entries of tensor T^0 and tensor T^1 equal 1 when all the indices take the same value, while all the other entries are equal to 0. That is, the entries are given by $T^0_{x,x,x,x,x} = 1$ for all $x \in [D]$, and $T^0_{x[5]} = 0$ otherwise; as well as by $T^1_{x,x,x,x} = 1$ for all $x \in [D]$, and $T^1_{x[4]} = 0$ otherwise.

The tensor $T = T^0 \otimes T^1$ has entries

$$T_{x,x,x,x,x,x,x,x,x} = T^0_{x,x,x,x,x} T^1_{x,x,x,x} = 1 \tag{17.9}$$

for all $x \in [D]$ and $T_{x[9]} = 0$ otherwise. In words, the tensor T has D entries equal to 1 – namely all the entries corresponding to equal indices – with all other entries being equal to 0.

Example 17.4

Take S tensors $T^0, T^1, \ldots, T^{S-1}$, where each tensor T^s has order 1 and dimension D^s for $s \in [S]$. The tensor

$$T = T^0 \otimes T^1 \otimes \cdots \otimes T^{S-1} \tag{17.10}$$

obtained via the tensor product of all the order-1 tensors $\{T^s\}_{s \in [S]}$ has order S and dimensions $(D^0, D^1, \ldots, D^{S-1})$. The corresponding tensor network representation is shown in Figure 17.4.

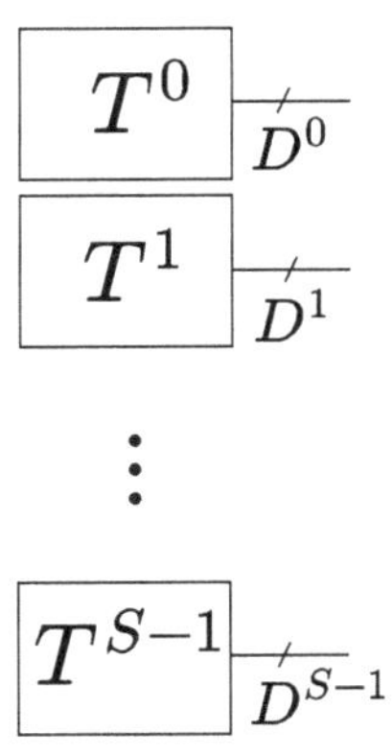

Figure 17.4 Tensor network representing the tensor product $T^0 \otimes T^1 \otimes \cdots \otimes T^{S-1}$ for order-1 tensors T^s with dimensions D^s for all $s \in [S]$.

An important observation about this example is that the tensor T is highly structured. Graphically, while a general order-S tensor with the same dimensions would be represented by a single tensor with S indices, tensor T is broken down into S separate smaller tensors, each with a separate index. Furthermore, an arbitrary order-S tensor with the same dimensions would require $\prod_{s\in[S]} D^s$ entries to be fully specified. In contrast, the entries of tensor T depend only on $\sum_{s\in[S]} D^s$ numbers, namely all the entries of the individual tensors $\{T^s\}_{s\in[S]}$.

17.4 Tensor Contraction

As discussed in Section 17.3, tensor products produce larger tensors that inherit all the indices of the original tensors. The corresponding tensor network representation is obtained by juxtaposing the constituent tensors. Beyond tensor products, tensors can be combined through an algebraic operation called **tensor contraction**, which generalizes inner products, matrix–vector multiplications, and matrix–matrix multiplications. Tensor contractions are represented graphically by connecting the indices of two tensors. The connected indices are subsequently eliminated from the resulting tensor, leaving it with only a subset of the indices from the original tensors.

To introduce the concept of tensor contraction, this section first presents inner products, matrix–vector multiplications, and matrix–matrix multiplications through the graphical formalism described in Section 17.3. Then, tensor contractions are presented in full generality by considering arbitrary tensors.

17.4.1 Inner Product as a Contraction

To start, consider the inner product

$$a^T b = \sum_{z\in[D]} a_z b_z \tag{17.11}$$

between two $D \times 1$ real-valued vectors a and b. The inner product (17.11) sums over all the values of the common index $z \in [D]$ of the two vectors, thus contracting the two vectors into a scalar.

The operation (17.11) is represented by the tensor network in Figure 17.5, which consists of two order-1 tensors, one for vector a and one for vector b. Following the convention described earlier in this chapter, the tensor b is represented as a column vector, since its leg points to the right. In contrast, having a left-pointing leg, the tensor a is represented as a row vector, corresponding to the transpose a^T. In this regard, note that one could also label the tensor as the transpose a^T. However, this

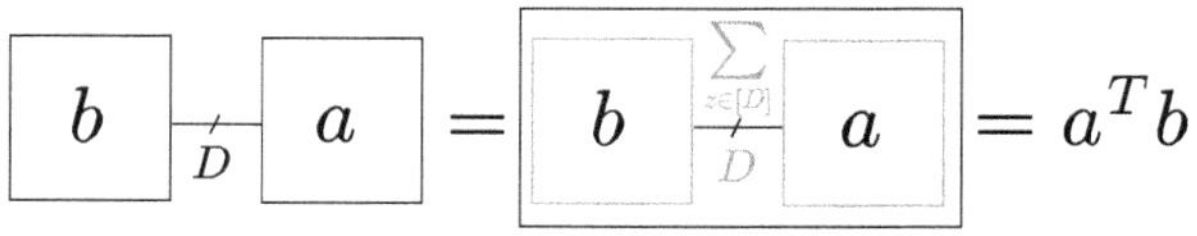

Figure 17.5 Tensor network representation of the inner product $a^T b = \sum_{z\in[D]} a_z b_z$.

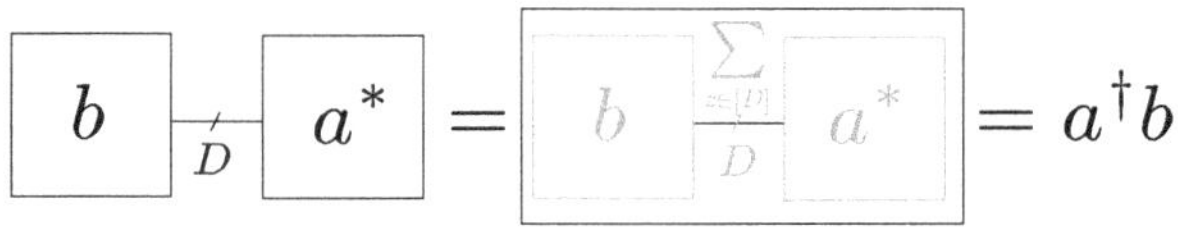

Figure 17.6 Tensor network representation of the inner product $a^\dagger b = \sum_{z \in [D]} a_z^* b_z$.

would not make any difference, since the tensor merely specifies a mapping a_z from index z to entry a_z. The two tensors are connected along their respective legs to form a **bond** with dimension equal to their shared dimension, D. Note that, due to the mentioned conventions about the positions of the legs, the order in which the terms appear in the mathematical expression $a^T b$ is the reverse of their arrangement in the tensor network.

As shown in Figure 17.5, the bond between the tensors represents the operation of summing over all the shared values of the connected indices. This operation is known as the **contraction** of the two indices, as it eliminates the two indices from the resulting tensor. In particular, the tensor formed by connecting tensors a and b as in the figure does not have any open index, and hence it represents a scalar.

For complex-valued tensors, the inner product is defined as

$$a^\dagger b = \sum_{z \in [D]} a_z^* b_z. \tag{17.12}$$

Following the same arguments as for (17.11), this operation can be represented by the tensor network in Figure 17.6, in which the entries of tensor a are complex conjugate to yield the tensor a^*.

17.4.2 Matrix–Vector Multiplication as a Contraction

Consider now the matrix–vector multiplication Ab between the $D^0 \times D^1$ matrix A and the $D^1 \times 1$ vector b. The xth entry of the $D^0 \times 1$ vector Ab, for $x \in [D^0]$, is given by the sum

$$[Ab]_x = \sum_{z \in [D^1]} A_{x,z} b_z \tag{17.13}$$

along the common index of dimension D^1. This operation is depicted as in the tensor network shown in Figure 17.7.

In it, there is a tensor for vector b and a tensor for matrix A, with the two tensors being connected by a bond along the common index of dimension D^1. The bond represents the sum in (17.13), which extends across all the values of the row index of matrix A and of the single index of vector b. The tensor resulting from the contraction has a single open index, and is thus a vector. Moreover, the open index is of dimension D^0 and it points to the right, describing the $D^0 \times 1$ (column) vector Ab.

As for the inner product, due to the convention of placing row indices on the right and column indices on the left, the arrangement of the tensors A and b in the tensor

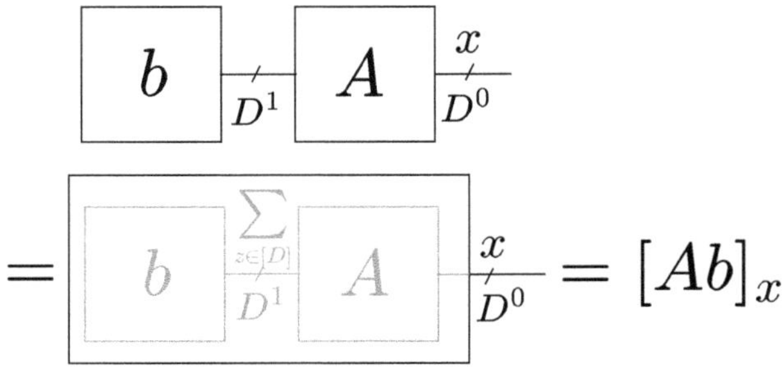

Figure 17.7 Tensor network representation of the matrix–vector multiplication Ab.

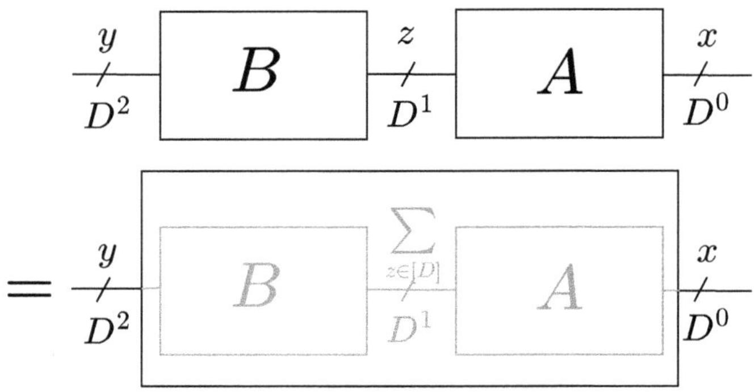

Figure 17.8 Tensor network representation of the matrix–matrix multiplication AB.

network follows the reverse order with respect to the mathematical expression Ab. This property is useful for settings in which the vector b plays the role of the input, while the matrix A represents some linear operation applied to the input vector. In this case, the tensor network in the figure is consistent with a typical block diagram in which operations are read from left to right, from the inputs to the outputs.

17.4.3 Matrix–Matrix Multiplication as a Contraction

Finally, fix a $D^0 \times D^1$ matrix A and a $D^1 \times D^2$ matrix B, and consider the product AB, which is a $D^0 \times D^2$ matrix. Each (x, y)th element of this matrix is given by the sum

$$[AB]_{x,y} = \sum_{z \in [D^1]} A_{x,z} B_{z,y} \tag{17.14}$$

along the common index of dimension D^1.

The corresponding tensor network, shown in Figure 17.8, includes one tensor for matrix A and one for matrix B, with a bond along the index being summed over in (17.14). Specifically, the bond index corresponds to the column (left) index for matrix A and to the row (right) index for matrix B. Once the two matrices have been contracted, one obtains a tensor representing the resulting matrix AB of dimension $D^0 \times D^2$, with row index x on the right and column y index on the left.

17.4.4 Contractions and Tensor Networks

Contractions can be combined to create tensor networks with any number of tensors. The following example illustrates this approach, which will be thoroughly discussed in the next sections.

Example 17.5

The tensor network in Figure 17.9 contains a $D^0 \times 1$ vector a, a $D^0 \times D^1$ matrix B, and a $D^1 \times 1$ vector c. The tensors c and B are connected via a bond of dimension D^0, while the tensors B and a are connected via a bond of dimension D^1. The two bonds correspond to two contractions, one along the index of vector a and one along the index of vector c. The tensor network thus represents the operation

$$\sum_{x\in[D^1]} \sum_{y\in[D^0]} a_x B_{x,y} c_y = a^T Bc. \quad (17.15)$$

Note again the reverse orders used in the mathematical expression and in the tensor network.

Figure 17.9 Tensor network representation of the scalar $a^T Bc$.

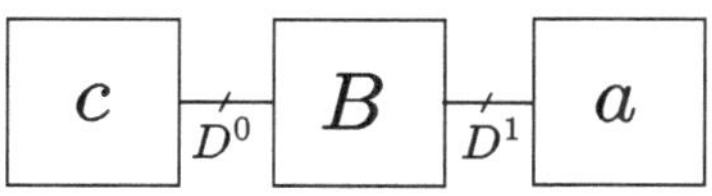

The operation (17.15) can be evaluated by applying the two contractions in any order. Specifically, one could evaluate first the right contraction, obtaining the vector $d^T = a^T B$, followed by the left contraction $d^T c$. Alternatively, the left contraction could be evaluated first, yielding the vector $f = Bc$, followed by the right contraction $a^T f$. The freedom to choose an ordering in the evaluation of contractions is common to all tensor networks, and it follows from the **associative property**

$$a^T Bc = (a^T B)c = a^T (Bc). \quad (17.16)$$

Importantly, the complexity of evaluating the contractions in a tensor network may vary dramatically depending on the order in which the contractions are evaluated. As an example, consider a special case of the tensor network in Figure 17.9 corresponding to the expression $ab^T c$, where a, b, and c are $D \times 1$ vectors. If one contracts first the right bond, computing the $D \times D$ matrix ab^T requires D^2 multiplications, as well as D^2 memory elements. Then, contracting the left bond requires computing the matrix–product vector $(ab^T)c$, which entails D^2 multiply-and-accumulate operations. Thus, the memory and computational complexities obtained with this ordering scale with the dimension D like D^2.

In contrast, contracting first the left bond and then the right bond yields a linear computation and memory complexity with respect to the dimension D. In fact, the inner product $b^T c$ requires D multiply-and-accumulate operations to obtain a scalar, which is finally multiplied element-by-element with vector a. Therefore, in this example, the order of the contractions determines whether the complexity is linear or quadratic in the problem size D.

17.4.5 Defining Tensor Contractions

To provide a general definition of tensor contraction, consider now two arbitrary tensors T^0 and T^1. The two tensors may have different orders S^0 and S^1, but they must share at least one index with the same dimension D. Denote as $x^{[S^0]}$ the indices for tensor T^0, with corresponding entries $T^0_{x^{[S^0]}}$; and as $y^{[S^1]}$ the indices for tensor T^1, with corresponding entries $T^1_{y^{[S^1]}}$. Assume that, for some integers $s^0 \in [S^0]$ and $s^1 \in [S^1]$, index x^{s^0} for tensor T^0 and index y^{s^1} for tensor T^1 have the same dimension D.

A contraction of the indices x^{s^0} and y^{s^1} yields a tensor that inherits all the remaining indices of the two tensors, namely $x^{[S^0]\setminus s^0}$ and $y^{[S^1]\setminus s^1}$, where $[S^0]\setminus s^0$ represents set $[S^0]$ with the exclusion of index s^0 and $[S^1]\setminus s^1$ represents set $[S^1]$ with the exclusion of index s^1. Furthermore, the entries $T_{x^{[S^0]\setminus s^0}, y^{[S^1]\setminus s^1}}$ of the resulting tensor are obtained by summing the products $T^0_{x^{[S^0]}} T^1_{y^{[S^1]}}$ of the corresponding entries of the two tensors T^0 and T^1 along the common value of the indices x^{s^0} and y^{s^1}. The following examples illustrates this operation.

Example 17.6

Figure 17.10 shows a tensor network that represents a tensor contraction between tensors of orders $S^0 = 6$ and $S^1 = 5$. The two tensors are contracted along the indices x^6

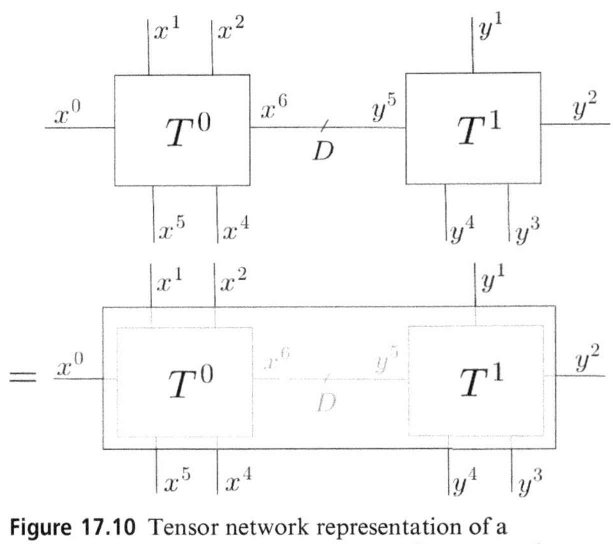

Figure 17.10 Tensor network representation of a contraction between tensors of orders $S^0 = 6$ and $S^1 = 5$, where the contraction is applied along a bond formed by connecting the indices x^6 of tensor T^0 and index y^5 of tensor T^1.

of tensor T^0 and index y^5 of tensor T^1, which are assumed to have the same dimension D. The resulting tensor has $S^0 + S^1 - 2 = 9$ legs, retaining all indices of the two original tensors that have not been contracted.

Example 17.7

Consider the tensors in Figure 17.11, in which all the indices have dimensions $D = 2$, and thus they take values 0 or 1. Tensor T^0 has entries $T^0_{x^0,z^0} = 1$ if $x^0 = z^0$ and $T^0_{x^0,z^0} = 0$ otherwise; and, in a similar way, tensor T^2 has entries $T^2_{x^1,z^1} = 1$ if $x^1 = z^1$ and $T^2_{x^1,z^1} = 0$ otherwise. Furthermore, tensor T^1 has entries $T^1_{z^0,z^1} = 0.5$ if $z^0 = z^1$ and $T^1_{z^0,z^1} = 0$ otherwise. Accordingly, tensors T^0 and T^2 return 1 if the two indices are equal, while producing 0 otherwise; and tensor T^1 yields 0.5 if the two indices are equal, and 0 otherwise.

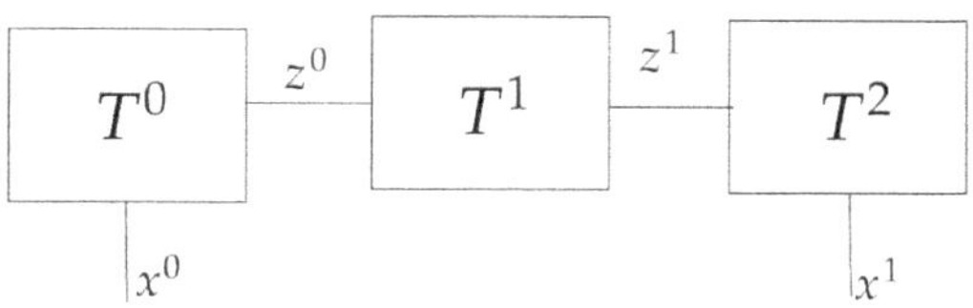

Figure 17.11 Example of a tensor network.

The contraction shown in Figure 17.11 returns a tensor with open indices x^0 and x^1 and corresponding dimensions $(2, 2)$. In particular, the contracted tensor returns the values

$$T_{x^0,x^1} = \sum_{z^0 \in [2]} \sum_{z^1 \in [2]} T^0_{x^0,z^0} T^1_{z^0,z^1} T^2_{x^1,z^1} \tag{17.17}$$

for $x^0, x^1 \in [2]$. In fact, the two bonds correspond to the two sums in (17.17) over indices z^0 and z^1.

Using (17.17), one can readily check that the only non-zero entries of tensor T are

$$\begin{aligned} T_{0,0} &= T^0_{0,0} T^1_{0,0} T^2_{0,0} = \frac{1}{2}, \\ T_{1,1} &= T^0_{1,1} T^1_{1,1} T^2_{1,1} = \frac{1}{2}, \end{aligned} \tag{17.18}$$

with the other entries $T_{0,1}$ and $T_{1,0}$ being equal to 0.

Example 17.8

For the same tensor network in Figure 17.11, consider now adding a bond between the free indices x^0 and x^1. Note that this operation is legitimate, given that the two indices have the common dimension $D = 2$. Since there are no more free indices, the result is a scalar, and it equals the sum

$$T_{0,0} + T_{1,1} = 1. \tag{17.19}$$

As illustrated in the examples above, the graphical representation of a contraction is straightforward, entailing the connection of indices with the same order for a pair of tensors. Mathematically, a contraction amounts to a sum over the connected indices, which effectively removes these indices from the resulting tensor. This is formalized in the next definition.

Definition 17.3 Tensor contraction

Given order-S^0 tensor T^0 and order-S^1 tensor T^1 with indices x^{s^0} and y^{s^1} having a common dimension D, a contraction of the two indices x^{s^0} and y^{s^1} returns the order-$(S^0 + S^1 - 2)$ tensor

$$T_{x^{[S^0]\setminus s^0}, y^{[S^1]\setminus s^1}} = \sum_{x^{s^0}=y^{s^1}\in[D]} T^0_{x^{[S^0]}} T^1_{y^{[S^1]}}, \tag{17.20}$$

where $[S^0]\setminus s^0$ represents set $[S^0]$ with the exclusion of index s^0 and $[S^1]\setminus s^1$ represents set $[S^1]$ with the exclusion of index s^1.

Bonds can also be created between indices of the same tensor; and one may even create a **self-bond** on a single index of a tensor. As demonstrated in the next examples, this type of contraction is useful to represent operations such as the trace, the partial trace, and sums over rows or columns.

Example 17.9

Trace as a contraction. Take a $D \times D$ matrix A, which is represented as an order-2 tensor with dimensions (D, D). As illustrated in Figure 17.12(a), one can add a bond between the row and column indices, since they have the same dimension. The resulting tensor has no legs, and hence it corresponds to a scalar. Specifically, the contraction of a matrix along the row and column indices yields the trace

$$\mathrm{tr}(A) = \sum_{z\in[D]} A_{z,z}. \tag{17.21}$$

Example 17.10

Self-bond. Take now Figure 17.12(b), in which a self-bond is applied to the column index of a $D^0 \times D^1$ matrix A. The resulting tensor has a single open index, and hence it corresponds to a $D^0 \times 1$ vector. Specifically, by contracting over the column index y, one obtains the sum $\sum_{y\in[D^1]} A_{x,y}$, which corresponds to a column vector obtained by summing up the columns of matrix A. In a similar manner, by contracting the row index x, one obtains a row vector $\sum_{x\in[D^0]} A_{x,y}$ that sums over the rows of the matrix.

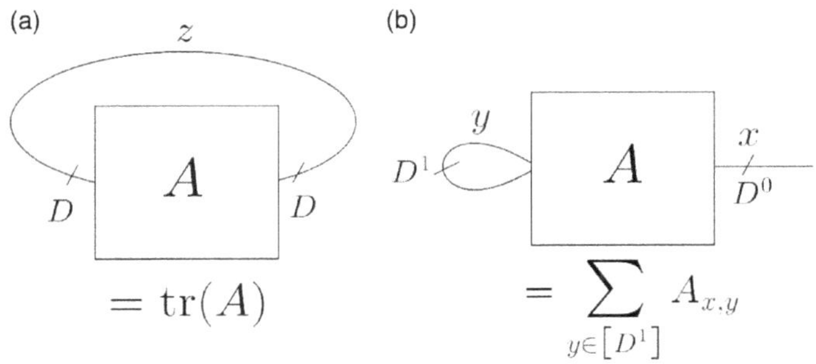

Figure 17.12 Graphical representation of (a) the trace as a contraction and (b) a self-bond.

Example 17.11

Partial trace as a contraction. Consider an $N^X N^Y \times N^X N^Y$ density matrix ρ^{XY}. Using the bi-index notation, described in Section 16.2, the density matrix can be represented via the tensor network in Figure 17.13(a), in which xy is the row bi-index and $x'y'$ is the column bi-index.

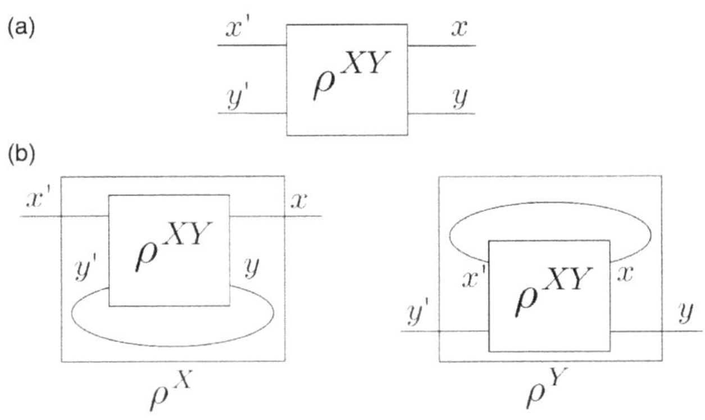

Figure 17.13 (a) Density matrix ρ^{XY} as a tensor using the bi-index notation xy for the row bi-index and $x'y'$ for the column bi-index and (b) tensor network representation of the reduced density matrices ρ^X and ρ^Y.

As seen in Section 14.10, each (x, x')th entry of the reduced density matrix ρ^X for $x, x' \in [N^X]$ is given by the partial trace

$$[\rho^X]_{x,x'} = [\mathrm{tr}_Y(\rho^{XY})]_{x,x'} = \sum_{y\in[N^Y]} \rho^{XY}_{xy,x'y}. \tag{17.22}$$

As shown in Figure 17.13(b), this amounts to contracting the indices x and x' of the tensor ρ^{XY}.

In a similar way, each (y, y')th entry of the reduced density matrix ρ^Y for $x, x' \in [N^Y]$ is evaluated as

$$[\rho^Y]_{y,y'} = [\text{tr}_X(\rho^{XY})]_{y,y'} = \sum_{x \in [N^X]} \rho^{XY}_{xy,xy'}, \tag{17.23}$$

which corresponds to contracting the indices y and y' as also illustrated in Figure 17.13.

17.5 Fixed-Value Tensor and Copy Tensor

This section discusses two special tensors that will be useful in the next chapters. The first is the fixed-value tensor, which can be used to specify the value of the index of another tensor, thus effectively "closing" an open index. The second is the copy tensor, which can be used to "copy" the value of an index into another index.

17.5.1 Fixed-Value Tensor

A fixed-value tensor δ is an order-1 tensor that returns 1 for a specific value $\bar{x}$ of its index $x \in [D]$, while producing output 0 otherwise.

Definition 17.4 Fixed-value tensor
Given an integer D and an index value $\bar{x} \in [D]$, the fixed-value tensor δ is the order-1 tensor with entries

$$\delta_x = \begin{cases} 1 & \text{if } x = \bar{x}, \\ 0 & \text{otherwise.} \end{cases} \tag{17.24}$$

Note that, when viewed as a vector, the fixed-value tensor corresponds to a zero-entropy distribution, justifying the use of the same notation δ used in Part I. In this regard, it is emphasized that the notation δ in (17.24) does not report the fixed value $\bar{x}$, using the subscript x in δ_x to index the entries of the tensor instead.

As shown in Figure 17.14, when connected to the index of another tensor, a fixed-value tensor "closes" the open index of the connected tensor, fixing the value of the open index to value $\bar{x}$. This effectively corresponds to reducing the order of index x to 1, making it possible to remove the index from the resulting tensor $\tilde{T}$.

To elaborate, consider a tensor $T_{x,y}$ with an index $x \in [D]$, as well as an arbitrary number of other indices y of any dimension. As in Figure 17.14, assume that the tensor T is connected via the index x to the fixed-value tensor δ in (17.24).

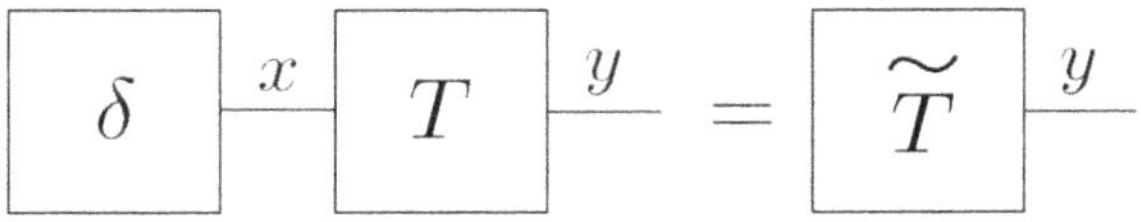

Figure 17.14 When connected to the index of another tensor, a fixed-value tensor "closes" the open index of the connected tensor, fixing the value of the index to value $\bar{x}$. The closed index can then be removed from the tensor network.

The contracted tensor, denoted as $\tilde{T}$, has entries given by

$$\tilde{T}_y = \sum_{x \in [D]} T_{x,y} \delta_x = T_{\bar{x},y}, \tag{17.25}$$

for all values of indices y. Hence, the contraction has effectively removed index x from the tensor, fixing it to value $\bar{x}$ and leaving only the indices y.

17.5.2 Copy Tensor

The copy tensor uses the contraction operation to effectively copy the value of one index onto all the other indices of the tensor. It is defined as follows.

Definition 17.5 Copy tensor
For integers D and $S \geq 2$, the copy tensor δ is the order-S tensor with all dimensions equal to D, whose entries are given by

$$\delta_{x^0,\ldots,x^{S-1}} = 1 \tag{17.26}$$

if all indices take the same value $x^0 = x^1 = \cdots = x^{S-1}$, and

$$\delta_{x^0,\ldots,x^{S-1}} = 0 \tag{17.27}$$

otherwise.

Note that sharing the same notation δ for both fixed-value tensors and copy tensors does not cause any ambiguity, since the fixed-value tensor is an order-1 tensor, while copy tensors must have order S no smaller than 2. Some useful properties of the copy tensor are summarized in Appendix 17.A.

Example 17.12

Consider order-2 tensors T^0 and T^1 with dimensions (D^0, D^1) and (D^0, D^2), respectively. The first dimension is common to both tensors, and the entries of the tensors can hence be written as $T^0_{x^0,x^1}$ and $T^1_{x^0,x^2}$ with indices $x^0 \in [D]$, $x^1 \in [D^1]$, and $x^2 \in [D^2]$. Using these two tensors, an order-3 tensor T with dimensions (D^0, D^1, D^2) is constructed that has entries

$$T_{x^0,x^1,x^2} = T^0_{x^0,x^1} T^1_{x^0,x^2}. \tag{17.28}$$

How to represent this tensor graphically using the tensors T^0 and T^1?

As a first observation, the tensor in (17.28) cannot be expressed as the tensor product of the two tensors T^0 and T^1. In fact, this would return an order-4 tensor, which inherits all the indices of the two tensors T^0 and T^1. As shown in Figure 17.15, a graphical representation of the tensor (17.28) can be obtained by using a copy tensor of order 3 to copy the index x^0 at the input of the two tensors T^0 and T^1.

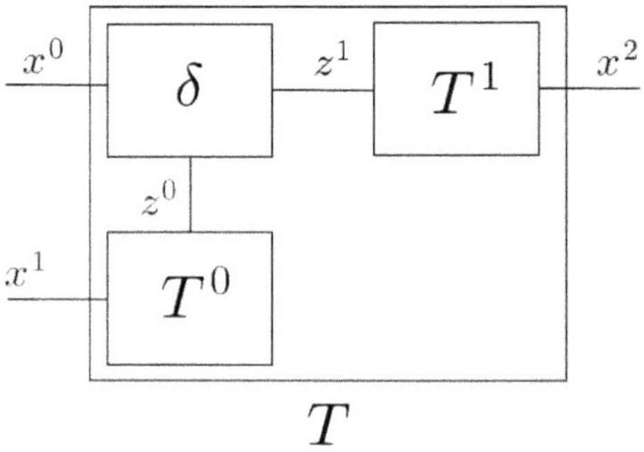

Figure 17.15 Tensor network representation of a tensor with entries $T_{x^0,x^1,x^2} = T^0_{x^0,x^1} T^1_{x^0,x^2}$.

To verify the validity of this representation, write the copy tensor as δ_{x^0,z^0,z^1}, where the first index corresponds to the index $x^0 \in [D]$ of tensor T, and the other two indices $z^0 \in [D]$ and $z^1 \in [D]$ are used to create bonds to the tensors T^0 and T^1, respectively. By using the definition (17.26) and (17.27) of a copy tensor and the properties of the tensor contraction, the resulting tensor has entries

$$\begin{aligned} T_{x^0,x^1,x^2} &= \sum_{z^0 \in [D]} \sum_{z^1 \in [D]} \delta_{x^0,z^0,z^1} T^0_{z^0,x^1} T^1_{z^1,x^2} \\ &= T^0_{x^0,x^1} T^1_{x^0,x^2} \end{aligned} \tag{17.29}$$

as desired.

17.6 Vectorization and Matrization

Tensors can be formatted into vectors and matrices to extend algebraic operations such as inner products, matrix–vector multiplications, and eigendecompositions to tensors. Conversely, interpreting large-scale vectors and matrices as structured tensors can significantly simplify the evaluation of algebraic operations involving vectors and matrices. This section will address these topics by discussing the vectorization and matrization of tensors.

17.6.1 Vectorization

The vectorization of a tensor generalizes the vectorization of matrices introduced in Section 10.5. To start, recall that the vectorization $a = \text{vec}(A)$ of a $D^0 \times D^1$ matrix

A is the $D^0D^1 \times 1$ vector a obtained by stacking the transpose of all the rows of A into one column vector.

This operation can conveniently be expressed using the bi-index notation x^0x^1, which represent the integer

$$x^0x^1 = D^1x^0 + x^1, \tag{17.30}$$

with $x^0 \in [D^0]$ being the most significant digit and $x^1 \in [D^1]$ being the least significant digit. Indeed, with this notation, each x^0x^1th entry of the vector $a = \text{vec}(A)$ for the $D^0 \times D^1$ matrix A can be written as

$$a_{x^0x^1} = A_{x^0,x^1} = \langle x^0|A|x^1\rangle, \tag{17.31}$$

where A_{x^0,x^1} is the (x^0, x^1)th entry of matrix A; $|x^0\rangle$ represents a $D^0 \times 1$ vector from the computational basis of the D^0-dimensional space; and $|x^1\rangle$ is a $D^1 \times 1$ vector from the computational basis of the D^1-dimensional space.

Noting that the Kronecker product $|x^0\rangle \otimes |x^1\rangle$ is a one-hot vector with a single 1 in the position x^0x^1 given by (17.30), an equivalent way to express the vectorization (17.31) is as follows.

Definition 17.6 Vectorization of a matrix
Given any $D^0 \times D^1$ matrix A, the vectorization operation returns the $D^0D^1 \times 1$ vector

$$\text{vec}(A) = \sum_{x^0\in[D^0]} \sum_{x^1\in[D^1]} A_{x^0,x^1}|x^0\rangle \otimes |x^1\rangle. \tag{17.32}$$

Given any order-S tensor with dimensions $(D^0, \ldots, D^{S-1})$, the vectorization operation returns a $\prod_{s\in[S]} D^s \times 1$ vector, which collects all the entries (17.6) of the tensor. This is done via a direct generalization of the sum (17.32). To introduce it, define as $\{|x^s\rangle\}_{x^s\in[D^s]}$ the computational basis for a space of dimension D^s. Moreover, for any value $x^{[S]} = x^0, \ldots, x^{S-1} \in [D^{[S]}]$ of the tensor indices, write the Kronecker product of the corresponding computational-basis vectors $|x^0\rangle, \ldots, |x^{S-1}\rangle$ as

$$|x^{[S]}\rangle = |x^0, \ldots, x^{S-1}\rangle = |x^0\rangle \otimes \cdots \otimes |x^{S-1}\rangle. \tag{17.33}$$

Definition 17.7 Vectorization of a tensor
Given any order-S tensor T with dimensions $(D^0, \ldots, D^{S-1})$, the vectorization operation returns the $\prod_{s\in[S]} D^s \times 1$ vector

$$\begin{aligned}\text{vec}(T) &= \sum_{x^0\in[D^0]} \cdots \sum_{x^{S-1}\in[D^{S-1}]} T_{x^0,\ldots,x^{S-1}}|x^0\rangle \otimes \cdots \otimes |x^{S-1}\rangle \\ &= \sum_{x^{[S]}\in[D^{[S]}]} T_{x^{[S]}}|x^{[S]}\rangle.\end{aligned} \tag{17.34}$$

Example 17.13

Consider the order-2 copy tensor δ with dimensions ($D^0 = 3, D^1 = 3$), which, by (17.26) and (17.27), has entries $\delta_{x^0,x^1} = 1$ if $x^0 = x^1 \in [3]$ and $\delta_{x^0,x^1} = 0$ otherwise. Using (17.34), the vectorized version is given by the 9×1 vector

$$\text{vec}(\delta) = |00\rangle + |11\rangle + |22\rangle = \begin{bmatrix} 1 \\ 0 \\ 0 \\ 0 \\ 1 \\ 0 \\ 0 \\ 0 \\ 1 \end{bmatrix}. \tag{17.35}$$

Example 17.14

Consider the order-3 copy tensor with dimensions ($D^0 = 2, D^1 = 2, D^2 = 2$), which has entries $\delta_{x^0,x^1,x^2} = 1$ for $x^0 = x^1 = x^2 \in [2]$ and $\delta_{x^0,x^1,x^2} = 0$ otherwise. Using (17.34), the vectorized tensor is given by the 8×1 vector

$$\text{vec}(\delta) = |000\rangle + |111\rangle = \begin{bmatrix} 1 \\ 0 \\ 0 \\ 0 \\ 0 \\ 0 \\ 0 \\ 1 \end{bmatrix}. \tag{17.36}$$

Given the one-to-one mapping (17.34) between a tensor T and the vectorized version vec(T), the two notions of tensor T and vector vec(T) can be treated as effectively being equivalent. Reflecting this equivalence, the notation (17.34) will also be used for the tensor T, writing the tensor as

$$T = \sum_{x^{[S]}} T_{x^{[S]}} |x^{[S]}\rangle \tag{17.37}$$

when no confusion may arise. Note that, to lighten the notation, whenever sums are extended to all values of a set of indices, as in (17.37), from now on the domains of the sum will not be specified.

17.6.2 Entry-Wise Characterization of Vectorization

The vectorized tensor (17.34) places the entry $T_{x^{[S]}}$ of the tensor in the position of the vector that is determined by the single entry equal to 1 of the vector $|x^{[S]}\rangle$. As elaborated next, this observation yields a generalization of the entry-wise characterization (17.31) of the vectorized matrix through the introduction of the multi-index notation as an extension of the bi-index (17.30). To describe it, take first the case in which all dimensions are equal to the same value D, i.e., $D^s = D$ for $s \in [S]$.

Definition 17.8 Multi-index notation with equal-dimension indices
Given indices with $x^s \in [D]$ for all $s \in [S]$, the multi-index $x^0x^1 \dots x^{S-1}$ corresponds to the integer in the set $[D^S]$ given by

$$\begin{aligned} x^0x^1 \dots x^{S-1} &= \sum_{s \in [S]} D^{S-1-s}x^s \\ &= D^{S-1}x^0 + D^{S-2}x^1 + \cdots + x^{S-1}. \end{aligned} \tag{17.38}$$

Accordingly, the multi-index $x^0x^1 \dots x^{S-1}$ is the base-D integer representation of an integer, with digits listed from the most significant, x^0, corresponding to the power D^{S-1}, to the least significant, x^{S-1}, corresponding to the power $D^0 = 1$.

As an important special case, for $D = 2$, the multi-index $x^0x^1 \dots x^{S-1}$ corresponds to the binary integer representation

$$x^0x^1 \dots x^{S-1} = 2^{S-1}x^0 + 2^{S-2}x^1 + \cdots + x^{S-1}, \tag{17.39}$$

which ranges in the set $[2^S]$. In general, given indices $x^0, x^1, \dots, x^{S-1}$ with different dimensions $D^{[S]}$, the concatenated label $x^0x^1 \dots x^{S-1}$ can be interpreted as a single index with digits listed in lexicographic order with varying bases. This definition is formalized as follows.

Definition 17.9 Multi-index notation
Given indices with $x^s \in [D^s]$ for all $s \in [S]$, the multi-index $x^0x^1 \dots x^{S-1}$ represents the integer in the set $[\prod_{s \in [S]} D^s]$ given by

$$\begin{aligned} x^0x^1 \dots x^{S-1} &= \left(D^1D^2 \cdots D^{S-1}\right)x^0 + \left(D^2 \cdots D^{S-1}\right)x^1 + \cdots + x^{S-1} \\ &= \sum_{s \in [S]} \left(\prod_{s'=s+1}^{S-1} D^{s'} \right) x^s, \end{aligned} \tag{17.40}$$

with the convention $\prod_{s'=S}^{S-1} D^{s'} = 1$.

With the definition of the multi-index in place, one can readily recognize that the vector $|x^{[S]}\rangle = |x^0, \dots, x^{S-1}\rangle$ is one-hot, with all zero entries except for a 1 in position given by the multi-index $x^0x^1 \dots x^{S-1}$. Therefore, the definition (17.34) of a vectorized tensor equivalently states that the $x^0x^1 \dots x^{S-1}$th entry of the vectorized tensor vec(T) is given by the tensor entry $T_{x^0,x^1,\dots,x^{S-1}}$, i.e.,

$$[\text{vec}(T)]_{x^0x^1\dots x^{S-1}} = T_{x^0,x^1,\dots,x^{S-1}}. \tag{17.41}$$

Altogether, thanks to the multi-index definition (17.41), a straightforward mapping exists between a tensor and its vectorized form. Following this one-to-one correspondence, the notation $x^{[S]}$ will be used to represent either the collection of indices $x^0, x^1, \dots, x^{S-1}$ or the concatenated multi-index $x^0x^1 \dots x^{S-1}$; the intended meaning will be evident from the context.

17.6.3 Kronecker Product and Tensor Product

The vectorization of the tensor product $T^0 \otimes T^1$ in (17.7) and (17.8) corresponds to the Kronecker product between the corresponding vectorizations vec(T^0) and vec(T^1). This equivalence justifies the use of the same symbol $\otimes$ to denote both operations.

To formalize this point, consider any two tensors T^0 and T^1 with orders S^0 and S^1 and arbitrary dimensions. By the definition (17.8) of a tensor product, the tensor $T = T^0 \otimes T^1$ has entries $T_{x^{[S^0]},y^{[S^1]}} = T^0_{x^{[S^0]}} T^1_{y^{[S^1]}}$ with indices $x^{[S^0]}$ and $y^{[S^1]}$ running over the corresponding dimensions. Using (17.34), the two vectorized tensors can be written as $\text{vec}(T^0) = \sum_{x^{[S^0]}} T^0_{x^{[S^0]}} |x^{[S^0]}\rangle$ and $\text{vec}(T^1) = \sum_{y^{[S^1]}} T^1_{y^{[S^1]}} |y^{[S^1]}\rangle$. The Kronecker product between the two vectors is thus

$$\begin{aligned} \text{vec}(T^0) \otimes \text{vec}(T^1) &= \sum_{x^{[S^0]}} \sum_{y^{[S^1]}} T^0_{x^{[S^0]}} T^1_{y^{[S^1]}} |x^{[S^0]}\rangle \otimes |y^{[S^1]}\rangle \\ &= \sum_{x^{[S^0]}} \sum_{y^{[S^1]}} T^0_{x^{[S^0]}} T^1_{y^{[S^1]}} |x^{[S^0]} y^{[S^1]}\rangle, \end{aligned} \tag{17.42}$$

which is the vectorization of the tensor product (17.8). Note that the entry of vector (17.42) corresponding to the product $T^0_{x^{[S^0]}} T^1_{y^{[S^1]}}$ is at the position given by the multi-index $x^{[S^0]}y^{[S^1]}$, in which the most significant digits $x^{[S^0]}$ come from tensor T^0 and the least significant digits are from tensor T^1.

The upshot of the equality (17.42) is that one can talk interchangeably of tensor product and Kronecker product, with the understanding that in the latter case one is technically referring to the corresponding operation on the vectorized tensors. By the same reasoning, the equivalence between Kronecker product and tensor product extends to any number of tensors, as illustrated by the next example.

Example 17.15

One of the key advantages of tensor vectorization is that it enables the application of vector operations on tensors. To elaborate, consider two order-S tensors a and b of the form

$$a = a^0 \otimes a^1 \otimes \cdots \otimes a^{S-1} \tag{17.43}$$

and

$$b = b^0 \otimes b^1 \otimes \cdots \otimes b^{S-1}, \tag{17.44}$$

where $\{a^s\}_{s\in[S]}$ and $\{b^s\}_{s\in[S]}$ are order-1 tensors with dimension D.

Using the mixed product property of the Kronecker product (8.169), the inner product between the two vectors can be evaluated as

$$\begin{aligned} a^T b &= (a^0 \otimes a^1 \otimes \cdots \otimes a^{S-1})^T (b^0 \otimes b^1 \otimes \cdots \otimes b^{S-1}) \\ &= ((a^0)^T \otimes (a^1)^T \otimes \cdots \otimes (a^{S-1})^T)(b^0 \otimes b^1 \otimes \cdots \otimes b^{S-1}) \\ &= \prod_{s\in[S]} (a^s)^T b^s. \end{aligned} \tag{17.45}$$

This operation can be represented by the tensor network in Figure 17.16. In it, the tensors a and b are represented via tensor networks consisting of separate order-1 tensors, while the inner products $\{(a^s)^T b^s\}_{s\in[S]}$ are represented by local bonds between the respective order-1 tensors as illustrated in Figure 17.5.

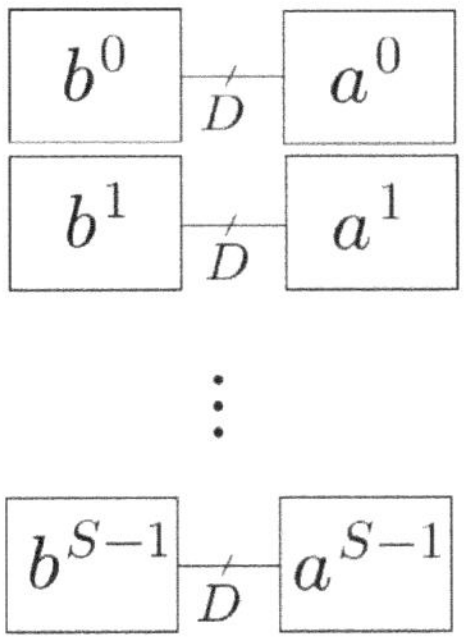

Figure 17.16 Tensor network representation of the inner product $a^T b$ between vectors a and b with structure described by (17.43) and (17.44), respectively.

The expression (17.45) demonstrates that the structure of the two tensors a and b can be leveraged to significantly reduce the complexity of evaluating the inner product $a^T b$. In fact, for arbitrary $D^S \times 1$ vectors a and b, the inner product $a^T b$ would require D^S multiply-and-accumulate operations. In stark contrast, for the structured vectors in this example, the complexity of evaluating (17.45) is of the order SD, since each of the S inner products $(a^s)^T b^s$ requires D operations.

This entails a dramatic complexity reduction, from exponential, D^S, to linear, SD, in the order S. For example, if $S = 100$ and $D = 2$, the complexity for generic vectors is of the order $D^S = 2^{100} \approx 10^{33}$, while the structured vectors in this example require a complexity of the order $SD = 100 \cdot 2 = 200$.

Example 17.16

Generalizing the previous example, consider any order-S tensor b with dimensions all equal to D, as well as a tensor a of the form (17.43). Note that the tensor b is arbitrary and thus generally unstructured. Given that the $x^{[S]}$-entry of vector a is given by the product $\prod_{s\in[S]} a^s_{x^s}$, the inner product is given by the sum

$$a^T b = \sum_{x^{[S]}} b_{x^{[S]}} \prod_{s\in[S]} a^s_{x^s}. \quad (17.46)$$

Graphically, as illustrated in Figure 17.17, this corresponds to a contraction along all indices $x^{[S]}$ of the two tensors a and b.

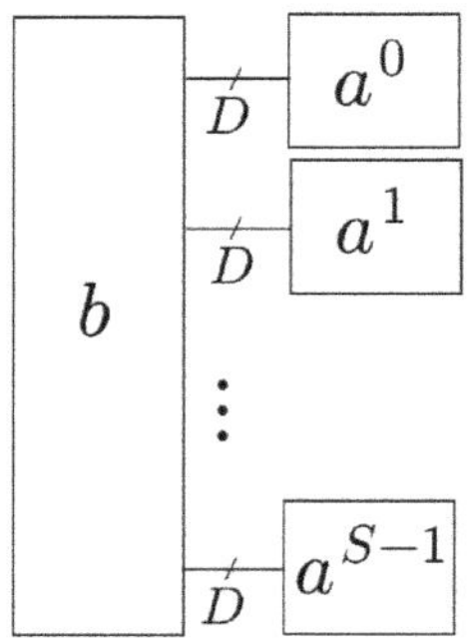

Figure 17.17 Tensor network representation of the inner product $a^T b$ between vectors a with structure described by (17.43) and an unstructured vector b.

Example 17.17

Another important operation that is unlocked by treating tensors as vectors is matrix–vector multiplication. For example, take an arbitrary order-S tensor b with all dimensions equal to D, as well as a $D^S \times D^S$ matrix expressed as the Kronecker product

$$A = A^0 \otimes A^1 \otimes \cdots \otimes A^{S-1} \quad (17.47)$$

of smaller $D \times D$ matrices $\{A^s\}_{s\in[S]}$.

Using the same reasoning as in Example 17.16, the matrix–vector product Ab, which corresponds to an order-S tensor with all dimensions equal to D, is represented by the tensor network in Figure 17.18(a). Accordingly, each matrix A^s is contracted with the sth index of tensor b along the column index.

As a special case, if the tensor b factorizes as in (17.44), the matrix–vector multiplication can be expressed as the Kronecker product

$$Ab = A^0 b^0 \otimes A^1 b^1 \otimes \cdots \otimes A^{S-1} b^{S-1}. \quad (17.48)$$

This operation is represented as a tensor network as shown in Figure 17.18(b).

In this regard, it is important to stress again the importance of structure in simplifying the evaluation of algebraic operations involving tensors. In fact, for a generic $D^S \times D^S$ matrix A and $D^S \times 1$ vector b, evaluating the product Ab requires a number of operations proportional to D^{2S}, since each of the D^S output values requires D^S

multiply-and-accumulate operations. In contrast, evaluating (17.48) only entails the evaluation of the S products $\{A^s b^s\}_{s\in[S]}$, each requiring a complexity of the order D^2. Therefore, the computational complexity has decreased from an exponential scaling, D^{2S}, to a linear scaling, SD^2, in the order S.

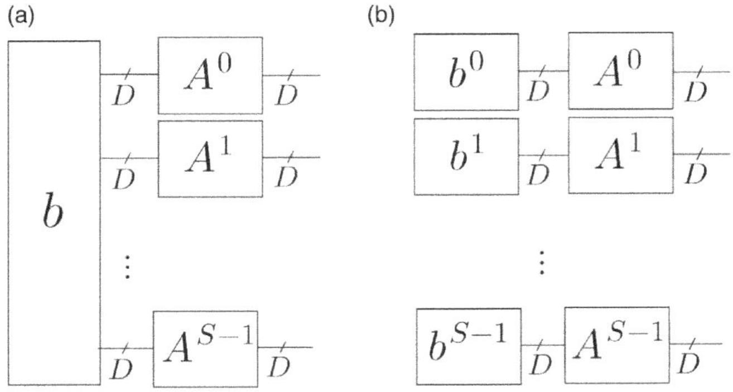

Figure 17.18 Tensor network representation of the matrix–vector product Ab with matrix A structured as in (17.47) and (a) an unstructured vector b or (b) a vector b with structure described by (17.44).

17.6.4 Matrization

A tensor can also be formatted into a matrix in order to unlock matrix-based operations such as the eigendecomposition or the SVD.

Unlike vectorization, matrization requires the choice of a partition of the set of S indices into two non-empty subsets. The first subset of indices is treated as a multi-index that identifies the row of the matrix, while the second subset of indices is a multi-index that identifies the column of the matrix. As a result, there are as many possible matrizations of a tensor as there are partitions of S indices into two non-empty subsets.

To start, it is important to set the notation for subsets of indices that will be adopted throughout this part of the book.

Definition 17.10 Subset of indices
Fix a sequence of indices $x^{[S]} = (x^0, \ldots, x^{S-1})$ with any dimensions $D^{[S]}$. For a subset of integers $\mathcal{S} = \{s_0, \ldots, s_{|\mathcal{S}|-1}\} \subseteq [S]$ with cardinality $|\mathcal{S}| \leq S$, the notation $x^{\mathcal{S}}$ represents the **ordered sequence** of the corresponding indices

$$x^{\mathcal{S}} = (x^{s_0}, \ldots, x^{s_{|\mathcal{S}|-1}}), \tag{17.49}$$

where the natural order

$$s_0 \leq s_1 \leq \cdots \leq s_{|\mathcal{S}|-1} \tag{17.50}$$

is assumed. The sequence of indices $x^{\mathcal{S}}$ can also be interpreted as the multi-index $x^{\mathcal{S}_0} \dots x^{\mathcal{S}_{|\mathcal{S}|-1}}$.

Example 17.18

For a tensor with order $S = 6$ and all dimensions equal to $D = 2$, consider the sequence of indices

$$\begin{aligned} x^{[6]} &= (x^0, x^1, x^2, x^3, x^4, x^5) \\ &= (0, 1, 1, 1, 0, 0). \end{aligned} \tag{17.51}$$

For the subset $\mathcal{S} = \{3, 5\}$, the corresponding indices are thus given by

$$\begin{aligned} x^{3,5} &= (x^3, x^5) \\ &= (1, 0). \end{aligned} \tag{17.52}$$

Note that, to avoid clutter, the notation $x^{3,5}$ is used in lieu of $x^{\{3,5\}}$ – a convention that will be adopted henceforth.

With these definitions in place, the matrization of a tensor is introduced as follows. Fix a subset $\mathcal{S} \subset [S]$, as well as the complement subset $\bar{\mathcal{S}} = [S] \backslash \mathcal{S}$. As formalized next, subset $\mathcal{S}$ is used to identify the indices $x^{\mathcal{S}}$ of the original tensor that specify the row of the matrix. These are referred to as the **row indices**. The complement subset $\bar{\mathcal{S}}$ corresponds to the indices $x^{\bar{\mathcal{S}}}$ that specify the column of the matrix, i.e., to the **column indices**.

Definition 17.11 Matrization of a tensor
Given an arbitrary order-S tensor T and a subset $\mathcal{S} \subset [S]$ of row indices, the matrization $\text{mat}_{\mathcal{S}}(T)$ is the $\prod_{s \in \mathcal{S}} D^s \times \prod_{s \in \bar{\mathcal{S}}} D^s$ matrix

$$\text{mat}_{\mathcal{S}}(T) = \sum_{x^{\mathcal{S}}} \sum_{x^{\bar{\mathcal{S}}}} T_{x^{[S]}} |x^{\mathcal{S}}\rangle \langle x^{\bar{\mathcal{S}}}|. \tag{17.53}$$

Equivalently, the $(x^{\mathcal{S}}, x^{\bar{\mathcal{S}}})$th entry of matrix $\text{mat}_{\mathcal{S}}(T)$ is given by the tensor entry $T_{x^{[S]}}$.

Example 17.19

Consider an order-3 tensor T with all dimensions equal to $D = 2$. The tensor has entries $T_{0,1,1} = 1$, $T_{1,0,0} = 1$, while all the other entries are equal to zero. Choosing the set of row indices as $\mathcal{S} = \{0, 1\}$, and thus as the set of column indices the complement $\bar{\mathcal{S}} = \{2\}$, the matrization (17.53) yields the $D^2 \times D$, i.e., 4×2, matrix

$$\begin{aligned}\text{mat}_{\{0,1\}}(T) &= T_{0,1,1}|01\rangle\langle 1| + T_{1,0,0}|10\rangle\langle 0| \\ &= |01\rangle\langle 1| + |10\rangle\langle 0| \\ &= \begin{bmatrix} 0 & 0 \\ 0 & 1 \\ 1 & 0 \\ 0 & 0 \end{bmatrix}. \end{aligned} \tag{17.54}$$

For another example, with the choice $\mathcal{S} = \{1\}$ and $\bar{\mathcal{S}} = \{0, 2\}$, the resulting 2×4 matrix is

$$\begin{aligned}\text{mat}_{\{1\}}(T) &= T_{0,1,1}|1\rangle\langle 01| + T_{1,0,0}|0\rangle\langle 10| \\ &= |1\rangle\langle 01| + |0\rangle\langle 10| \\ &= \begin{bmatrix} 0 & 0 & 1 & 0 \\ 0 & 1 & 0 & 0 \end{bmatrix}. \end{aligned} \tag{17.55}$$

The mathematical definition (17.53) is best understood via the corresponding tensor network representation. In fact, as illustrated in Figure 17.19, a tensor representation of matrix $\text{mat}_{\mathcal{S}}(T)$ is simply obtained by moving the row indices $x^{\mathcal{S}}$ on the right and the column indices $x^{\bar{\mathcal{S}}}$ on the left of the tensor (see Figure 17.2), while following the natural integer ordering specified in (17.49) and (17.50).

17.7 Canonic Polyadic Decomposition

As discussed in Section 17.6, order-S tensors obtained from order-1 tensors $\{T^s\}_{s\in[S]}$ via the tensor product $T^0 \otimes T^1 \otimes \cdots \otimes T^{S-1}$ have a highly structured form, which facilitates the evaluation of important operations such as inner products and matrix–vector multiplications. While not all tensors can be expressed in this way, it is often useful to write a tensor T as a sum of tensors of this form. This yields the following type of tensor decomposition.

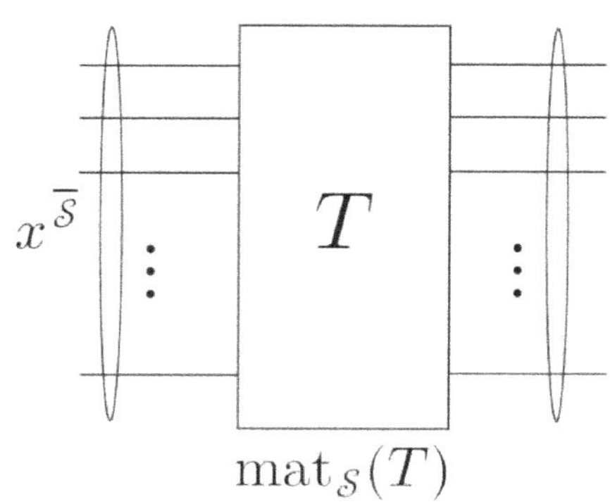

Figure 17.19 Tensor representation of the tensor matrization $\text{mat}_{\mathcal{S}}(T)$ with row indices $x^{\mathcal{S}}$.

Definition 17.12 Canonic polyadic decomposition
Given an order-S tensor T with arbitrary dimensions, a canonic polyadic decomposition (CPD) expresses the tensor in the form

$$T = \sum_{z\in[r]} \gamma_z T_z^0 \otimes T_z^1 \otimes \cdots \otimes T_z^{S-1} \tag{17.56}$$

for some integer r, r scalars $\{\gamma_z\}_{z\in[r]}$, and Sr order-1 tensors $\{\{T_z^s\}_{z\in[r]}\}_{s\in[S]}$.

By the equality (17.34), for any tensor, a CPD exists with $r = \prod_{\in[S]} D^s$ terms, but structured tensors may have CPDs with significantly lower values of the integer r. Of significant interest for the characterization of the properties of a tensor is the minimum value of integer r such that a CPD exists for the given tensor.

Definition 17.13 Tensor rank
The tensor rank r_{min} of an order-S tensor T with arbitrary dimensions is the minimum value of the integer

$$r_{\text{min}} \leq \prod_{s\in[S]} D^s \tag{17.57}$$

for which a CPD (17.56) exists for tensor T.

A tensor network representation of the CPD (17.56) is shown in Figure 17.20. The tensor network contains an order-1 tensor γ, with entries $\{\gamma_z\}_{z\in[r]}$ and dimension r; a copy tensor δ of order $S+1$, with dimensions all equal to r; and S tensors T^s, with order equal to 2 and dimensions (D^s, r) for all $s \in [S]$. Each tensor T^s has each entry $T^s_{x^s,z}$ given by the x^sth entry of tensor T_z^s. The copy tensor δ is used to "broadcast" the index z to all tensors $\{T^s\}_{s\in[S]}$, thus describing the sum (17.56) in the form of contractions involving the tensor γ and the tensors $\{T^s\}_{s\in[S]}$.

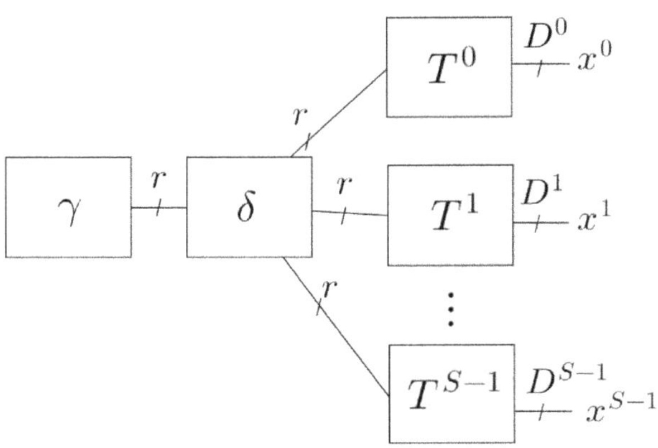

Figure 17.20 Tensor network representation of a CPD (17.56).

For order-2 tensors, CPDs can be constructed from the eigendecomposition or the SVD of a matrization of the tensor (Problem 17.6). Furthermore, for order-2 tensors, the tensor rank $r_{\min}$ coincides with the rank of any matrization of the tensor.

While for a matrix the rank can be computed efficiently, this is not the case for more general tensors, and the problem of identifying the tensor rank $r_{\min}$ is known to be computationally intractable. Furthermore, while for an $M \times N$ matrix the rank is known to be upper bounded by $\min(M, N)$, bounds stricter than (17.57) for the tensor rank $r_{\min}$ are not well understood.

17.8 Summary

- Joint probability distributions and joint quantum states in multipartite systems can conveniently be represented and manipulated by treating them as tensors.
- An order-S tensor T with dimensions $(D^0, D^1, \ldots, D^{S-1})$ is a multi-index array, in which an entry is identified by S indices $(x^0, \ldots, x^{S-1})$ with each index x^s taking D^s possible values. Accordingly, each $(x^0, \ldots, x^{S-1})$th entry of the tensor is a number $T_{x^0, x^1, \ldots, x^{S-1}}$.
- A tensor T can also be viewed as a $\prod_{s \in [S]} D^s \times 1$ vector in which the entries are ordered by mapping the indices $x^0, x^1, \ldots, x^{S-1}$ into the multi-index $x^0 x^1 \ldots x^{S-1}$.
- When all dimensions are equal to the same value D, i.e., $D^s = D$ for $s \in [S]$, the multi-index $x^{[S]}$ corresponds to a base-D representation of an integer with x^0 corresponding to the most significant digit and x^{S-1} to the least significant digit.
- The vector corresponding to tensor T can be written as $T = \sum_{x^{[S]}} T_{x^{[S]}} |x^{[S]}\rangle$, where the notation $x^{[S]}$ represents both the indices $x^0, x^1, \ldots, x^{S-1}$ and the multi-index $x^0 x^1 \ldots x^{S-1}$, and the notation $|x^{[S]}\rangle = |x^0\rangle \otimes \cdots \otimes |x^{S-1}\rangle$ is used.
- Useful tensors are typically structured, in the sense that the entries of the tensor are related to one another, and thus the number of independent degrees of freedom required to specify the tensor is smaller than the overall number of entries $\prod_{s \in [S]} D^s$.
- A tensor network provides a graphical description of the structure of a tensor.
- The two main operations that can be represented via a tensor network are tensor contractions and tensor products.
- Tensor contractions amount to sums along all the values of two common indices with the same dimension, and can be used to describe operations such as inner product, trace, partial trace, and matrix–vector multiplication.
- When viewing tensors as vectors, the tensor product corresponds to the Kronecker product.
- When the tensors are structured into the tensor product of multiple smaller tensors, operations such as inner product and matrix–product multiplication can be carried out at a much reduced complexity by operating separately on each tensor.
- A tensor can also be formatted into a matrix in order to unlock matrix-based operations such as the eigendecomposition.

- Unlike vectorization, matrization requires the choice of a partition of the set of S indices into two non-empty subsets.
- Graphically, matrization is obtained by moving the row indices on the right and the column indices on the left of the tensor.
- A canonic polyadic decomposition (CPD) of an order-S tensor T expresses the tensor as a decomposition into the weighted sum of r tensors factorized into order-1 tensors. Mathematically, a CPD is thus of the form $T = \sum_{z\in[r]} \gamma_z T_z^0 \otimes T_z^1 \otimes \cdots \otimes T_z^{S-1}$, for some scalars $\{\gamma_z\}_{z\in[r]}$ and order-1 tensors $\{\{T_z^s\}_{z\in[r]}\}_{s\in[S]}$.
- CPDs generalize matrix decompositions such as the eigendecomposition and the SVD.
- The minimum value of the integer r for which a CPD exists is known as the tensor rank $r_{\min}$ of tensor T.

17.9 Recommended Resources

A more extensive introduction to tensors can be found in [1]. Among the more advanced topics covered there is a generalization of the CPD, known as Tucker decomposition, which allows for a generic "core" tensor connected to all the tensors $\{T^s\}_{s\in[S]}$. For implications of the tensor network formalism to physics, the reader is referred to [2].

PROBLEMS

17.1 What is the result of contracting a vector along its only index via a self-bond?

17.2 Consider an order-3 tensor with all dimensions equal to 2 that returns 1 if all indices equal 0 or if all indices are equal to 1, while returning 1 otherwise. (a) Write the tensor in vectorized form. (b) Write the matrization of the tensor with $\mathcal{S} = \{0\}$. (c). Evaluate the entries of the tensor $T \otimes T'$, where T' is an order-2 tensor with all dimensions equal to 2 that returns 1 if all indices are equal to 1, while returning 0 otherwise.

17.3 Compute $(P_1 \otimes P_2 \otimes P_3)(|+\rangle \otimes T)$, where $\{P_s\}$ are the Pauli matrices and the 4×1 vector T is given by $T = |00\rangle + |11\rangle$.

17.4 For any two tensors T^0 and T^1 with orders S^0 and S^1, respectively, prove the equality $\mathrm{mat}_{\mathcal{S}}(T^0 \otimes T^1) = \mathrm{mat}_{\mathcal{S}}(T^0) \otimes \mathrm{mat}_{\{0\}}(T^1)$ for any subset $\mathcal{S} \subseteq [S^0]$. This equality shows that the tensor product $T^0 \otimes T^1$ is also consistent with the Kronecker product between matrices. *Hint:* Note the equality $|x\rangle\langle y| \otimes \langle z| = |x\rangle\langle yz|$, where $|x\rangle$, $|y\rangle$, and $|z\rangle$ are computational-basis vectors in spaces of generally different dimensions.

17.5 The **Khatri–Rao product** between two matrices $A = [a_0 \cdots a_{N-1}]$ and $B = [b_0 \cdots b_{N-1}]$, where $\{a_n\}_{n\in[N]}$ and $\{b_n\}_{n\in[N]}$ are the columns of the respective matrices, is defined as the column-wise Kronecker product $[a_0 \otimes b_0 \cdots a_{N-1} \otimes$

b_{N-1}]. Represent this operation using a tensor network encompassing one tensor for matrix A and one tensor for matrix B.

17.6 (a) Write the eigenvalue decomposition of a normal matrix in the form of a CPD. (b) Write the SVD of any matrix in the form of a CPD.

17.7 Let $A = \begin{bmatrix} 1 & 2 \\ 2 & 1 \end{bmatrix}$, which is a symmetric matrix. (a) Find the eigenvalue decomposition of A and represent it as a CPD. (b) Compute the SVD of matrix A and represent it as a CPD.

17.8 Consider an order-3 tensor T of dimensions $(3, 3, 3)$, where the entries are defined as $T_{x,y,z} = x + y + z$ for $x, y, z \in [3]$. (a) Contract the tensor T by adding a self-bond on the first index x, obtaining an order-2 tensor M with dimensions $(2, 2)$. (b) Further contract M along index y with vector $v = [1, -2, 1]^T$, obtaining a vector w. (c) Compute the contraction of vector w with a vector $u = [1, 0, -1]^T$, resulting in a scalar.

Appendix

17.A Properties of the Copy Tensor

In evaluating contractions involving copy tensors, the following properties, illustrated in Figure 17.21, are often useful.

> **Property 1. Copy tensor with two indices**
> If a copy tensor has only two indices it can be replaced with a plain edge in a tensor network.

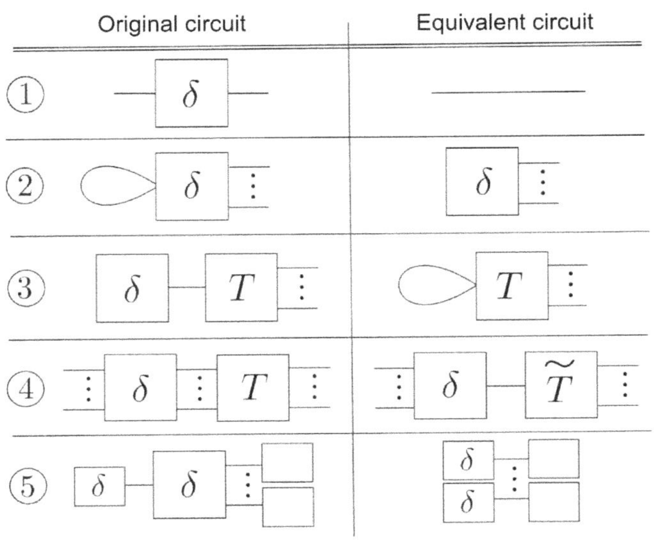

Figure 17.21 Properties of the copy tensor.

This property follows from the fact that the role of a copy tensor is merely to ensure that the two indices report the same numerical value, and the same operation can be encoded by a simple edge.

Property 2. Copy tensor with self-bonds
Self-bonds on indices of a copy tensor can be removed.

To see this, consider a copy tensor with any number S of indices $x^{[S]}$ in which a self-bond is added to the index $x^s \in [D]$. The resulting tensor is given by

$$\sum_{x^s \in [D]} \delta_{x^{[S]}} = \delta_{x^{[S] \setminus s}}, \tag{17.58}$$

which is a copy tensor of order $S - 1$.

Property 3. Copy tensor with a single bond and no free indices
If a copy tensor has a single bond and no other indices, it can be replaced by a self-bond.

This property follows because contracting the bond is equivalent to summing over all the values of the given index.

Property 4. Copy tensor with multiple bonds to a tensor
If a copy tensor has multiple bonds connecting it to the same tensor, with each bond having the same dimension D, the multiple bonds can be replaced with a single bond of dimension D.

To formalize this property, consider a tensor $T_{x^{[M]},y}$, where the indices $x^{[M]} = x^0, \ldots, x^{M-1}$ have the same dimension D. Assume that the indices $x^{[M]}$ form separate bonds with a copy tensor $\delta_{x^{[M]},z}$. Both tensor T and copy tensor δ generally have other indices y and z, which may be open or connected in a bond to another tensor.

Denoting as $[D]^M$ the set of all possible vales for the indices $x^{[M]}$, the contracted tensor is given by

$$\begin{aligned} \sum_{x^{[M]} \in [D]^M} \delta_{x^{[M]},z} T_{x^{[M]},y} &= \sum_{x \in [D]} \delta_{x,\ldots,x,z} T_{x,\ldots,x,y} \\ &= \sum_{x \in [D]} \delta_{x,z} \tilde{T}_{x,y}, \end{aligned} \tag{17.59}$$

where in the second equality the notation $x, \ldots, x$ represents M copies of index x; while the last equality is based on the definition of tensor $\tilde{T}_{x,y} = T_{x,\ldots,x,y}$.

Property 5. Copy tensor with a fixed-value tensor
Consider a fixed-value tensor closing one of the open edges of a copy tensor. The copy tensor is generally connected to other tensors. In this case, the copy tensor can be eliminated, and the fixed value can be applied directly to the tensors originally connected to the copy tensor.

REFERENCES

[1] T. G. Kolda, "Multilinear operators for higher-order decompositions," Technical Report, Sandia National Laboratories, April 2006.

[2] R. Penrose, *The Road to Reality: A Complete Guide to the Laws of the Universe*. London: Vintage, 2006.

18 Multipartite Classical Information: Fragmentation, Scale, and Strength

18.1 Overview

As explored in Part IV, correlations in bipartite classical systems can be measured by using metrics like mutual information. This chapter and the next reveal that studying correlations in multipartite classical systems introduces additional layers of complexity, involving aspects such as fragmentation, scale, strength, and structural characteristics.

Fragmentation describes the extent to which a system can be divided into groups of subsystems that exhibit no correlations with one another. For instance, in a social network, distinct cliques of individuals sharing no information across groups would indicate a high level of fragmentation.

Scale pertains to the size of the largest correlated subsets within the system. In a social network example, this may be represented by large cliques, such as fans of popular artists, where information flow within these groups is substantial.

Strength measures the degree to which subsystems deviate from mutual independence. For example, indicators of correlation strength may quantify how much information can flow to an individual within a social network, or assess the extent to which a group may be influenced by messages from individuals outside the group.

This chapter introduces tools to analyze the fragmentation, scale, and strength of multipartite correlations, while the following chapter addresses the structural aspects of these correlations. Additionally, this chapter demonstrates that multipartite correlations can vary in nature, encompassing both shared information – present across multiple subsystems – and synergistic information, which only emerges when multiple subsystems are considered together. For instance, the pages of a book share some basic information, such as language and genre. However, understanding other elements, like the plot of a novel, requires a synergistic examination of multiple pages.

Similar to Part IV, this part of the book consistently employs Shannon information measures due to their distinctive effectiveness in quantifying correlations.

18.1.1 Learning Objectives and Organization

By the end of this chapter, the reader should be able to:

- represent a joint distribution by using a tensor network that accounts for the fragmentation and scale of the available correlations;
- represent marginal distributions by using a tensor network;
- evaluate entropies and mutual informations for groups of subsystems;
- understand the definitions of globally correlated and globally uncorrelated systems;
- understand the concepts of shared, unique, and synergistic information;
- and understand and evaluate multipartite information measures based on cuts or on distance to correlation-free informational states.

18.2 Multipartite Classical Systems

As illustrated in Figure 18.1, this chapter studies a scenario involving S subsystems, thus generalizing the bipartite setting, with $S = 2$, covered by Part IV of the book. Each subsystem $s \in [S]$ is assumed to be in one of D possible states, and the state of subsystem s is denoted as $x^s \in [D]$. It is emphasized that the framework could be generalized to allow subsystems to have distinct numbers of possible states, but this would come at the cost of a more cumbersome notation.

Following Chapter 17, the notation $x^{[S]}$ is used as a shorthand for the joint state

$$x^{[S]} = x^{0,1,\ldots,S-1} = (x^0, \ldots, x^{S-1}), \tag{18.1}$$

which specifies the state $x^s \in [D]$ of each subsystem $s \in [S]$. Note that, since each state x^s can take one of D values, the total number of joint states $x^{[S]}$ is D^S, which grows exponentially in the number S of subsystems. Accordingly, the joint state $x^{[S]}$ can also be interpreted as a multi-index identifying an integer in the set $[D^S]$.

As in Chapter 17, the notation (18.1) is extended to any subset $\mathcal{S} \subseteq [S]$ of indices by following the natural order of the integers in set $\mathcal{S}$. Specifically, the states of all subsystems in the subset $\mathcal{S} = \{s_0, \ldots, s_{|\mathcal{S}|-1}\} \subseteq [S]$, with ordering $s_0 \leq s_1 \leq \cdots \leq$

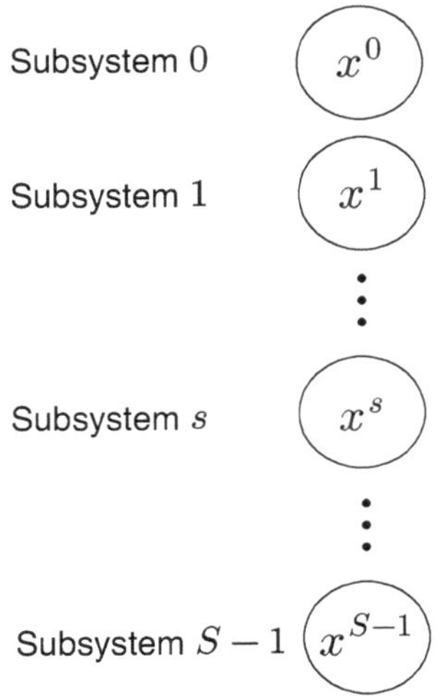

Figure 18.1 A multipartite system consisting of S subsystems.

$s_{|\mathcal{S}|-1}$, are represented by the ordered list of integers

$$x^{\mathcal{S}} = (x^{s_0}, \ldots, x^{s_{|\mathcal{S}|-1}}). \tag{18.2}$$

The joint state $x^{\mathcal{S}}$ can take $D^{|\mathcal{S}|}$ values, and, accordingly, the list $x^{\mathcal{S}}$ can also be interpreted as a multi-index identifying an integer in the set $[D^{|\mathcal{S}|}]$.

Example 18.1

Consider a system with $S = 10$ bits, i.e., with dimensions all equal to $D = 2$, that is in a joint state

$$\begin{aligned} x^{[10]} &= (x^0, x^1, x^2, x^3, x^4, x^5, x^6, x^7, x^8, x^9) \\ &= (0, 1, 1, 1, 0, 0, 1, 0, 1, 0). \end{aligned} \tag{18.3}$$

Note that there are $2^{10} = 1024$ possible joint states for this system. The joint state of the subsystems identified by the subset $\mathcal{S} = \{3, 5, 9\}$ is thus

$$\begin{aligned} x^{3,5,9} &= (x^3, x^5, x^9) \\ &= (1, 0, 0), \end{aligned} \tag{18.4}$$

and there are $2^3 = 8$ possible states for the subset $\mathcal{S}$ of subsystems. The joint state $x^{3,5,9} = (1, 0, 0)$ can also be interpreted as the multi-index $100 = 4$.

18.2.1 Joint and Partial Measurements

Measuring a subsystem $s \in [S]$ returns the state of the subsystem, which is described by a random variable $\mathrm{x}^s \in [D]$. Accordingly, when making a **joint measurement** across all subsystems, the collection of all random variables

$$\mathrm{x}^{[S]} = (\mathrm{x}^0, \ldots, \mathrm{x}^{S-1}) \tag{18.5}$$

is recorded, with $\mathrm{x}^{[S]}$ representing the overall state of the system.

More generally, it is possible to carry out a **partial measurement** that returns the states of a group of subsystems. Specifically, a measurement of the subsystems in the subset $\mathcal{S} \subseteq [S]$ produces the $|\mathcal{S}|$ random variables $\mathrm{x}^{\mathcal{S}}$ encompassing the states of all subsystems s identified by the subset $\mathcal{S}$.

18.2.2 Joint Distribution

An informational state is described by a joint probability distribution $p^{[S]}$ that assigns a probability

$$\begin{aligned} p^{[S]}_{x^{[S]}} &= p^{[S]}_{x^0, \ldots, x^{S-1}} \\ &= \Pr[\mathrm{x}^0 = x^0 \text{ and } \mathrm{x}^1 = x^1 \ldots \text{and } \mathrm{x}^{S-1} = x^{S-1}] \end{aligned} \tag{18.6}$$

to each possible configuration $x^{[S]}$ of the S subsystems. The probabilities $p^{[S]}_{x^{[S]}}$ are non-negative, i.e., $p^{[S]}_{x^{[S]}} \geq 0$ for all $x^{[S]} \in [D]^S$, and they satisfy the constraint

$$\sum_{x^{[S]}} p^{[S]}_{x^{[S]}} = 1, \tag{18.7}$$

ensuring that $p^{[S]}$ is a valid probability distribution.

As a point of notation, the superscript $[S]$ identifies $p^{[S]}$ as the joint distribution for all S subsystems. However, when considering the individual probabilities $p^{[S]}_{x^{[S]}}$ in (18.6), the superscript $[S]$ in the expression $p^{[S]}_{x^{[S]}}$ is redundant. Thus, from now on, the superscript $[S]$ will be removed from the expression $p^{[S]}_{x^{[S]}}$ when this does not create any ambiguity.

18.2.3 Joint Distribution as a Tensor

As illustrated in Figure 18.2, the joint distribution $p^{[S]}$ can be treated as an order-S tensor with all S dimensions equal to D. In fact, the joint distribution $p^{[S]}$ contains D^S entries, and each entry $p_{x^{[S]}}$ is indexed by the joint state $x^{[S]}$, which consists of S indices. The tensor representation of the joint distribution $p^{[S]}$ has S legs, with each sth leg corresponding to index x^s.

Using the tensor-to-vector mapping and the multi-index notations presented in Chapter 17, the joint distribution can also be treated as a $D^S \times 1$ vector as follows.

Key Result 18.1 Joint probability distribution as a vector

The joint probability tensor $p^{[S]}$ can be represented as the $D^S \times 1$ vector

$$p^{[S]} = \sum_{x^{[S]}} p_{x^{[S]}} |x^{[S]}\rangle \tag{18.8}$$

with $|x^{[S]}\rangle = |x^0 \cdots x^{S-1}\rangle = |x^0\rangle \otimes \cdots \otimes |x^{S-1}\rangle$. Equivalently, using the multi-index notation, the $(x^{[S]} = \sum_{s\in[S]} D^{S-s-1} x^s)$th entry of vector $p^{[S]}$ is the probability $p_{x^{[S]}}$.

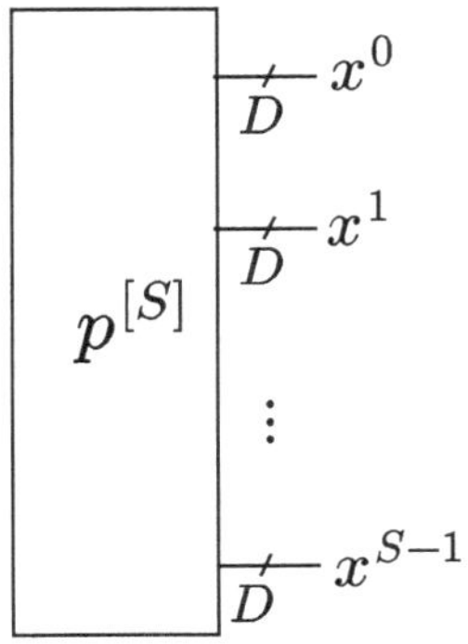

Figure 18.2 Tensor network representation of a joint distribution.

Example 18.2

For a system of three bits, consider the joint distribution with probabilities

$$p_{0,0,0} = p_{1,1,1} = 0.5 \tag{18.9}$$

and $p_{x^0,x^1,x^2} = 0$ for all other indices x^0, x^1, x^2. This corresponds to a situation in which the three bits are all equal either to 0 or to 1 with equal probability. The tensor $p^{[3]}$ can be represented by the 8×1 vector

$$\begin{aligned} p^{[3]} &= 0.5|000\rangle + 0.5|111\rangle \\ &= [0.5, 0, 0, 0, 0, 0, 0, 0.5]^T, \end{aligned} \tag{18.10}$$

since the multi-indices corresponding to the positive entries in (18.9) are 000 = 0 and 111 = 7. For future reference, this system will be referred to as encompassing **three equal uniform bits**.

Example 18.3

For another system of three bits, take the joint distribution

$$\begin{aligned} p^{[3]} &= \frac{1}{3}|001\rangle + \frac{1}{3}|010\rangle + \frac{1}{3}|100\rangle \\ &= \frac{1}{3}[0, 1, 1, 0, 1, 0, 0, 0]^T. \end{aligned} \tag{18.11}$$

This informational state assigns the same probabilities

$$p_{0,0,1} = p_{0,1,0} = p_{1,0,0} = \frac{1}{3} \tag{18.12}$$

to all three configurations in which one of the three bits is equal to 1 and the other two bits are equal to zero. This will be referred to as a system with **three uniform one-hot configurations**, since the three possible, equally likely, configurations have a single 1.

18.2.4 Marginal Distributions

Given a joint distribution $p^{[S]}$, the marginal distribution $p^{\mathcal{S}}$ of the states $\mathrm{x}^{\mathcal{S}}$ for a subset $\mathcal{S} \subseteq [S]$ of subsystems is obtained by summing over all possible values of the complement subset $\bar{\mathcal{S}} = [S] \backslash \mathcal{S}$ of subsystems.

Definition 18.1 Marginal distribution

Given a joint distribution $p^{[S]}$ and a subset $\mathcal{S} \subseteq [S]$, the marginal distribution $p^{\mathcal{S}}$ of the states $\mathrm{x}^{\mathcal{S}}$ assigns probability

$$\begin{aligned} p^{\mathcal{S}}_{x^{\mathcal{S}}} &= \Pr[\mathrm{x}^s = x^s \text{ for all } s \in \mathcal{S}] \\ &= \sum_{x^{\bar{\mathcal{S}}}} p_{x^{[S]}} \end{aligned} \tag{18.13}$$

to each configuration $x^{\mathcal{S}} \in [D]^{|\mathcal{S}|}$. In the sum (18.13), the indices in $x^{[S]}$ corresponding to the subset $\mathcal{S}$ are fixed to $x^{\mathcal{S}}$.

The sum operation in (18.13) is referred to as the marginalization over the group $\bar{\mathcal{S}}$ of subsystems, and is evaluated by summing over the $D^{|\bar{\mathcal{S}}|}$ possible values of the states $x^{\bar{\mathcal{S}}}$. The marginal distribution is a valid distribution in its own right, and hence it satisfies the conditions $p^{\mathcal{S}}_{x^{\mathcal{S}}} \geq 0$ for all entries $x^{\mathcal{S}} \in [D]^{|\mathcal{S}|}$, as well as the equality $\sum_{x^{\mathcal{S}}} p^{\mathcal{S}}_{x^{\mathcal{S}}} = 1$.

Generalizing (18.8), the marginal distribution $p^{\mathcal{S}}$ for a subset $\mathcal{S} = \{s_0, \ldots, s_{|\mathcal{S}|-1}\} \subseteq [S]$ with ordering $s_0 \leq s_1 \leq \cdots \leq s_{|\mathcal{S}|-1}$ can be expressed as the $D^{|\mathcal{S}|} \times 1$ vector

$$p^{\mathcal{S}} = \sum_{x^{\mathcal{S}}} p^{\mathcal{S}}_{x^{\mathcal{S}}} |x^{\mathcal{S}}\rangle, \tag{18.14}$$

where

$$|x^{\mathcal{S}}\rangle = |x^{s_0}\rangle \otimes |x^{s_1}\rangle \cdots \otimes |x^{s_{|\mathcal{S}|-1}}\rangle. \tag{18.15}$$

Example 18.4

Consider again the joint distribution (18.10) corresponding to three equal uniform bits. Given that the three bits take the same value 0 or 1 with probability 0.5, the marginal distribution for each subsystem is uniform. For instance, for subsystem 0, the marginal distribution is obtained via the sum

$$p^0_0 = \sum_{x^1 \in [2]} \sum_{x^2 \in [2]} p_{0,x^1,x^2} = p_{0,0,0} = 0.5, \tag{18.16}$$

and a similar calculation applies to the other two subsystems, yielding $p^1_0 = p^2_0 = 0.5$. Note that in the expressions p^0_0, p^1_0, and p^2_0, the superscripts are important to specify the identity of the subsystem.

The marginal distribution for any subset of two subsystems assigns equal probability to the two states $(0, 0)$ and $(1, 1)$, while assigning zero probability to the other two states $(0, 1)$ and $(1, 0)$. For instance, for the subset $\mathcal{S} = \{0, 1\}$, one has the marginal probabilities

$$p^{0,1}_{0,0} = \sum_{x^2 \in [2]} p_{0,0,x^2} = p_{0,0,0} = 0.5 \tag{18.17}$$

and

$$p^{0,1}_{1,1} = \sum_{x^2 \in [2]} p_{1,1,x^2} = p_{1,1,1} = 0.5. \tag{18.18}$$

Again, in the expressions $p^{0,1}_{0,0}$ and $p^{0,1}_{1,1}$, the superscripts are essential to identify the group of subsystems to which the probabilities are assigned.

Diagrammatically, given the expression (18.13), the tensor corresponding to the marginal distribution $p^{\mathcal{S}}$ is obtained by adding a self-bond at each one of the $|\bar{\mathcal{S}}|$

indices $x^{\bar{\mathcal{S}}}$ that need to be marginalized over. In fact, as seen in Chapter 17, a self-bond indicates a contraction along the given index, which in turns amounts to a sum over all possible index values. This is illustrated in Figure 18.3, which highlights the fact that the marginalized tensor $p^{\mathcal{S}}$ has indices $x^{\mathcal{S}}$, while the indices $x^{\bar{\mathcal{S}}}$ are removed via the self-bonds.

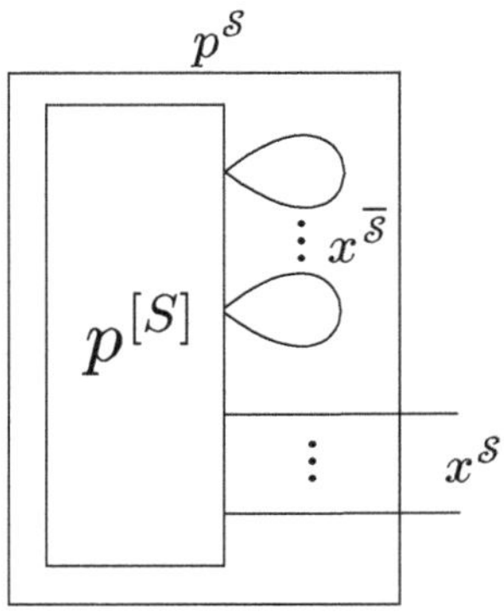

Figure 18.3 Tensor network representation of the marginal distribution $p^{\mathcal{S}}$ for subset $\mathcal{S}$.

Example 18.5

Consider the joint distribution (18.11) corresponding to three uniform one-hot configurations. The marginal distribution for each bit assigns probability 2/3 to state 0 and 1/3 to state 1. For instance, for subsystem 0, one obtains

$$p_0^0 = \sum_{x^1\in[2]} \sum_{x^2\in[2]} p_{0,x^1,x^2} = p_{0,0,1} + p_{0,1,0} = \frac{2}{3}. \tag{18.19}$$

Furthermore, the marginal distribution for any subset of two subsystems assigns probability 1/3 to the state (0, 0), probability 1/3 to the state (0, 1), and probability 1/3 to the state (1, 0). For instance, for the subset $\mathcal{S} = \{0, 1\}$, the marginal distribution is obtained as

$$p_{0,0}^{0,1} = \sum_{x^2\in[2]} p_{0,0,x^2} = p_{0,0,1} = \frac{1}{3}, \tag{18.20}$$

$$p_{0,1}^{0,1} = \sum_{x^2\in[2]} p_{0,1,x^2} = p_{0,1,0} = \frac{1}{3}, \tag{18.21}$$

and

$$p_{1,0}^{0,1} = \sum_{x^2\in[2]} p_{1,0,x^2} = p_{1,0,0} = \frac{1}{3}. \tag{18.22}$$

Example 18.6

Take the joint distribution over three bits given by

$$p^{[3]} = \frac{1}{4}|000\rangle + \frac{1}{4}|001\rangle + \frac{1}{4}|110\rangle + \frac{1}{4}|111\rangle, \tag{18.23}$$

which corresponds to the vector

$$p^{[3]} = \frac{1}{4}[1, 1, 0, 0, 0, 0, 1, 1]^T. \tag{18.24}$$

The marginal distribution for subsystems 0 and 1 is given by

$$p^{0,1} = \frac{1}{2}|00\rangle + \frac{1}{2}|11\rangle = \frac{1}{2}[1, 0, 0, 1]^T, \tag{18.25}$$

as it can readily be checked by evaluating each probability separately, as done in the previous examples. Furthermore, the marginal distribution of subsystem 2 is

$$p^2 = \frac{1}{2}|0\rangle + \frac{1}{2}|1\rangle = \frac{1}{2}[1, 1]^T. \tag{18.26}$$

For this example, one can now check that the overall tensor $p^{[3]}$ can be written as the Kronecker product

$$p^{[3]} = p^{0,1} \otimes p^2 \tag{18.27}$$

of the two marginal distributions derived above. Equivalently, in terms of the individual entries of tensor $p^{[3]}$, one has the equalities

$$p_{x^0,x^1,x^2} = p^{0,1}_{x^0,x^1} \cdot p^2_{x^2} \tag{18.28}$$

for all indices $x^0, x^1, x^2 \in [2]^3$. Using the definition of uncorrelated subsystems introduced in Section 12.5, the joint distribution $p^{[3]}$ in (18.27) describes a system in which the group consisting of subsystems 0 and 1 is uncorrelated with subsystem 2.

18.2.5 Globally Uncorrelated Systems

The main focus of this chapter, as well as the next, is the study of the correlations present within a group of subsystems. To begin, it is useful to define conditions under which a group of subsystems exhibit no correlations. Such a group is said to be **globally uncorrelated** and is defined formally as follows.

Definition 18.2 Globally uncorrelated vs. correlated systems

Under a joint distribution $p^{[S]}$, a subset $\mathcal{S} = \{s_0, \ldots, s_{|\mathcal{S}|-1}\} \subseteq [S]$ of subsystems with ordering $s_0 \leq s_1 \leq \cdots \leq s_{|\mathcal{S}|-1}$ is said to be globally uncorrelated if the corresponding marginal distribution $p^{\mathcal{S}}$ can be written as the Kronecker product

$$p^{\mathcal{S}} = p^{s_0} \otimes p^{s_1} \otimes \cdots \otimes p^{s_{|\mathcal{S}|-1}} \tag{18.29}$$

of the marginal distributions p^s for the individual subsystems $s \in \mathcal{S}$. If this condition is not satisfied, the subset $\mathcal{S}$ of subsystems is said to be correlated.

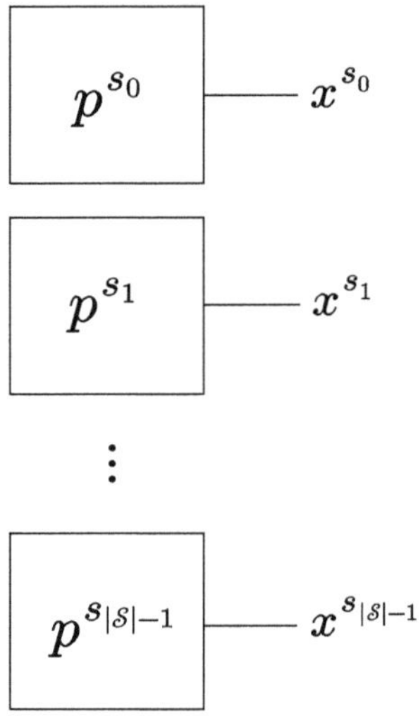

Figure 18.4 Tensor network representation of a globally uncorrelated subset $\mathcal{S} = \{s_0, \ldots, s_{|\mathcal{S}|-1}\} \subseteq [S]$ of subsystems with ordering $s_0 \leq s_1 \leq \cdots \leq s_{|\mathcal{S}|-1}$.

According to this definition, a subset $\mathcal{S}$ of subsystems exhibits a non-negligible level of correlation if its joint distribution $p^{\mathcal{S}}$ cannot be factorized into marginal distributions as shown in (18.29). An important consequence of the definition (18.29) is that the joint distribution of globally uncorrelated subsystems can be evaluated based exclusively on the marginal distributions $\{p^s\}_{s\in\mathcal{S}}$.

As depicted in Figure 18.4, the condition in (18.29) can be illustrated using the diagrammatic representation of the tensor product introduced in Chapter 17. In the tensor network, the marginal distributions $\{p^s\}_{s\in\mathcal{S}}$ are represented as order-1 tensors that remain disconnected from one another. In fact, the absence of bonds between these tensors visually encodes the lack of correlations implied by the factorization in (18.29).

The graphical representation in Figure 18.4 is also useful to determine that any two disjoint groups of subsystems within the subset $\mathcal{S}$ are uncorrelated under the joint distribution (18.29).

Example 18.7

For a system with three bits, consider the joint distribution $p^{[3]}$ with all entries equal to $1/8$. It is readily verified that all the marginal distributions are also uniform. This corresponds to a globally uncorrelated system, since each entry of the joint distribution can be expressed in terms of the marginal distributions as

$$p_{x^0,x^1,x^2} = p^0_{x^0} p^1_{x^1} p^2_{x^2} = \frac{1}{2}\frac{1}{2}\frac{1}{2} = \frac{1}{8} \tag{18.30}$$

for all $x^0 \in [2]$, $x^1 \in [2]$, and $x^2 \in [2]$.

18.3 Bipartite Correlations in Multipartite Systems

A common approach to studying correlations in a multipartite system is to evaluate the bipartite correlations that flow across a cut partitioning the system into two groups of subsystems. These bipartite correlations are typically quantified by the mutual information between the two groups. Before discussing the notion of a cut and the study of correlations between partitions of a system, this section introduces the mutual information between two groups of subsystems.

18.3.1 Joint Entropy

In order to introduce the mutual information between groups of subsystems, it is important to first extend the definition of joint entropy to any subset $\mathcal{S} \subseteq [S]$ of subsystems as follows.

Definition 18.3 Joint entropy

Given a joint distribution $p^{[S]}$, the joint entropy of a subset $\mathcal{S} \subseteq [S]$ of subsystems is defined as

$$\mathrm{H}(X^{\mathcal{S}})_{p^{\mathcal{S}}} = -\sum_{x^{\mathcal{S}}} p^{\mathcal{S}}_{x^{\mathcal{S}}} \log_2(p^{\mathcal{S}}_{x^{\mathcal{S}}}), \tag{18.31}$$

where $p^{\mathcal{S}}$ is the marginal distribution in (18.13).

According to the definition (18.31), the joint entropy $\mathrm{H}(X^{\mathcal{S}})_{p^{\mathcal{S}}}$ depends solely on the subset $\mathcal{S}$, and is independent of the ordering of the subsystems. In other words, the joint entropy is permutation invariant with respect to the indices of the subsystems. This property reflects the interpretation of the joint entropy $\mathrm{H}(X^{\mathcal{S}})_{p^{\mathcal{S}}}$ as a measure of uncertainty associated with the collective state of the subset $\mathcal{S}$ of subsystems.

As a special case, for the subset $\mathcal{S} = \{s\}$ encompassing a single subsystem $s \in [S]$, the definition (18.31) yields the local entropy for subsystem s as

$$\mathrm{H}(X^{s})_{p^{s}} = -\sum_{x^{s} \in [D]} p^{s}_{x^{s}} \log_2(p^{s}_{x^{s}}), \tag{18.32}$$

which is the entropy of the corresponding marginal distribution p^{s}.

Generalizing the result (12.40) from bipartite to multipartite systems, the joint entropy is non-negative, and it satisfies the classical property that uncertainty increases globally.

Key Result 18.2 Non-negativity and uncertainty increases globally

For any joint distribution $p^{[S]}$ and for any subset $\mathcal{S} \subseteq [S]$, the joint entropy satisfies the inequalities

$$\mathrm{H}(X^{\mathcal{S}})_{p^{\mathcal{S}}} \geq \max_{s \in \mathcal{S}} \mathrm{H}(X^{s})_{p^{s}} \geq 0. \tag{18.33}$$

The equality $\mathrm{H}(X^{\mathcal{S}})_{p^{\mathcal{S}}} = 0$ is satisfied if and only if the distribution $p^{\mathcal{S}}$ assigns probability 1 to a given value of the joint state $x^{\mathcal{S}}$.

The first inequality (18.33) indicates that the joint entropy is no smaller than any of the local entropies $\mathrm{H}(X^s)_{p^s}$ of the constituent subsystems. This inequality is tight, and thus one has the equality $\mathrm{H}(X^{\mathcal{S}})_{p^{\mathcal{S}}} = \max_{s\in[S]} \mathrm{H}(X^s)_{p^s}$, in the special case in which all random variables $\{\mathrm{x}^s\}_{s\in\mathcal{S}}$ are a function of a single random variable x^s.

The first inequality (18.33) can be generalized by considering any subset $\mathcal{S}' \subseteq \mathcal{S}$ as

$$\mathrm{H}(X^{\mathcal{S}})_{p^{\mathcal{S}}} \geq \mathrm{H}(X^{\mathcal{S}'})_{p^{\mathcal{S}'}}, \tag{18.34}$$

further substantiating the property that, in classical systems, uncertainty increases globally.

Example 18.8

The joint distribution (18.10) of a system of three equal uniform bits assigns equal probability to the configurations $(0,0,0)$ and $(1,1,1)$. Therefore, the joint entropy $\mathrm{H}(X^{[3]})_{p^{[3]}}$ for the entire system equals 1 bit. The local entropies $\mathrm{H}(X^s)$ are also equal to 1 bit for all subsystems $s \in [3]$, since all marginal distributions $\{p^s\}_{s\in[3]}$ are uniform. In this case, the first inequality (18.33) is satisfied with equality, since all the information about the system is present in any one of the subsystems.

The joint entropy also satisfies the following upper bounds.

Key Result 18.3 Subadditivity and additivity of the joint entropy

For any joint distribution $p^{[S]}$ and for any subset $\mathcal{S} \subseteq [S]$, the joint entropy satisfies the inequalities

$$\mathrm{H}(X^{\mathcal{S}})_{p^{\mathcal{S}}} \leq \sum_{s\in\mathcal{S}} \mathrm{H}(X^s)_{p^s} \leq |\mathcal{S}| \log_2(D), \tag{18.35}$$

where the first inequality indicates that the joint entropy is subadditive. Furthermore, the first inequality is attained if and only if the subset $\mathcal{S} = \{s_0, \ldots, s_{|\mathcal{S}|-1}\}$ is globally uncorrelated as in (18.29), yielding the additivity property

$$\mathrm{H}(X^{\mathcal{S}})_{p^{s_0}\otimes p^{s_1}\otimes\cdots\otimes p^{s_{|\mathcal{S}|-1}}} = \sum_{s\in\mathcal{S}} \mathrm{H}(X^s)_{p^s}. \tag{18.36}$$

In words, by the subadditivity property, the joint entropy cannot exceed the sum of the local entropies. Furthermore, for globally uncorrelated subsystems, all the information in any of the subsystems contributes in full to the overall information about the multipartite system. Thus, by the upper bound (18.35), the joint entropy can grow at most linearly with the number of subsystems, attaining the maximum value $\mathrm{H}(X^{\mathcal{S}})_{p^{\mathcal{S}}} = |\mathcal{S}| \log_2(D)$ if and only if the subsystems are globally uncorrelated and the marginal distributions are uniform.

Example 18.9

For the system of three equal uniform bits with joint distribution (18.10), the joint entropy $\mathrm{H}(X^{[3]})_{p^{[3]}}$ for the entire system equals 1 bit, and the local entropies $\mathrm{H}(X^{s})_{p^{s}}$ are also equal to 1 bit for all subsystems $s \in [3]$. The joint entropy $\mathrm{H}(X^{[3]})_{p^{[3]}}$ is thus smaller than the sum of the local entropies $\sum_{s \in [3]} \mathrm{H}(X^{s})_{p^{s}}$. This result confirms that the three subsystems are not globally uncorrelated, as the correlations among the three subsystems decrease the collective uncertainty from the maximum value of 3 bits to 1 bit.

In a similar way, for any group $\mathcal{S}$ of any two subsystems, the joint entropy equals $\mathrm{H}(X^{\mathcal{S}})_{p^{\mathcal{S}}} = 1$ bit, which is smaller than the sum of the local entropies. Thus, correlation is present between any two subsystems.

Example 18.10

Take now the three-bit joint distribution (18.11) corresponding to three uniform one-hot configurations. For individual subsystems, the entropy is given by

$$\mathrm{H}(X^{s})_{p^{s}} = \mathrm{h}\left(\frac{1}{3}\right) = 0.9183; \tag{18.37}$$

and for subsets $\mathcal{S}$ of cardinality $|\mathcal{S}| = 2$, one has

$$\mathrm{H}(X^{\mathcal{S}})_{p^{\mathcal{S}}} = \log_2(3) = 1.5850. \tag{18.38}$$

Since this is smaller than the sum of the local entropies – $\log_2(3) < 2\mathrm{h}(1/3)$ – any two subsystems are correlated.

The presence of correlations across pairs of subsystems implies that the system is not globally uncorrelated. This can readily be checked by comparing the joint entropy $\mathrm{H}(X^{[3]})_{p^{[3]}} = \log_2(3)$ with the sum of the three local entropies, as the inequality $\log_2(3) < 3\mathrm{h}(1/3)$ holds.

18.3.2 Mutual Information

The mutual information between any two groups of subsystems follows directly from the definition given in Section 12.6. To elaborate, take any two disjoint subsets $\mathcal{U} \subset [S]$ and $\mathcal{V} \subset [S]$ with $\mathcal{U} \cap \mathcal{V} = \emptyset$. The goal is to quantify the correlations that exist between the two groups of subsystems $\mathcal{U}$ and $\mathcal{V}$. Write as $p^{\mathcal{U},\mathcal{V}}$ the joint distribution of the two groups of subsystems $\mathcal{U}$ and $\mathcal{V}$, with indices ordered by concatenating the integers in subset $\mathcal{U}$ with those in subset $\mathcal{V}$.

Definition 18.4 Mutual information in multipartite systems

Given a joint distribution $p^{[S]}$, the mutual information between two disjoint groups of subsystems $\mathcal{U} \subset [S]$ and $\mathcal{V} \subset [S]$, with $\mathcal{U} \cap \mathcal{V} = \emptyset$, is defined as

$$\mathrm{I}(X^{\mathcal{U}};X^{\mathcal{V}})_{p^{\mathcal{U},\mathcal{V}}} = \mathrm{H}(X^{\mathcal{U}})_{p^{\mathcal{U}}} + \mathrm{H}(X^{\mathcal{V}})_{p^{\mathcal{V}}} - \mathrm{H}(X^{\mathcal{U},\mathcal{V}})_{p^{\mathcal{U},\mathcal{V}}}. \tag{18.39}$$

The mutual information $\mathrm{I}(X^{\mathcal{U}};X^{\mathcal{V}})_{p^{\mathcal{U},\mathcal{V}}}$ in (18.39) coincides with the mutual information between the joint states $\mathrm{x}^{\mathcal{U}}$ and $\mathrm{x}^{\mathcal{V}}$. These can in fact be viewed as random variables with joint distribution $p^{\mathcal{U},\mathcal{V}}$. Accordingly, the mutual information $\mathrm{I}(X^{\mathcal{U}};X^{\mathcal{V}})_{p^{\mathcal{U},\mathcal{V}}}$ is defined as the difference between the sum of the local entropies for subsets $\mathcal{U}$ and $\mathcal{V}$, $\mathrm{H}(X^{\mathcal{U}})_{p^{\mathcal{U}}} + \mathrm{H}(X^{\mathcal{V}})_{p^{\mathcal{V}}}$, and the joint entropy $\mathrm{H}(X^{\mathcal{U},\mathcal{V}})_{p^{\mathcal{U},\mathcal{V}}}$. As such, the mutual information $\mathrm{I}(X^{\mathcal{U}};X^{\mathcal{V}})_{p^{\mathcal{U},\mathcal{V}}}$ captures the redundant information present across the two groups of subsystems $\mathcal{U}$ and $\mathcal{V}$.

Note that the mutual information $\mathrm{I}(X^{\mathcal{U}};X^{\mathcal{V}})_{p^{\mathcal{U},\mathcal{V}}}$ is symmetric under a swap of the two groups $\mathcal{U}$ and $\mathcal{V}$, and is invariant to permutations of the subsystems within each group.

Definition 18.5 Uncorrelated groups of subsystems

Given a joint distribution $p^{[S]}$, two disjoint groups of subsystems $\mathcal{U} \subset [S]$ and $\mathcal{V} \subset [S]$, with $\mathcal{U} \cap \mathcal{V} = \emptyset$, are said to be uncorrelated if the joint distribution $p^{\mathcal{U},\mathcal{V}}$ can be written as the Kronecker product

$$p^{\mathcal{U},\mathcal{V}} = p^{\mathcal{U}} \otimes p^{\mathcal{V}} \tag{18.40}$$

of the marginal distributions $p^{\mathcal{U}}$ and $p^{\mathcal{V}}$ for the two groups.

The properties of the mutual information studied in Section 12.6 imply the following bounds.

Key Result 18.4 Bounds on the mutual information in multipartite systems

Given any two disjoint subsets $\mathcal{U} \subset [S]$ and $\mathcal{V} \subset [S]$ with $\mathcal{U} \cap \mathcal{V} = \emptyset$, the mutual information $\mathrm{I}(X^{\mathcal{U}};X^{\mathcal{V}})_{p^{\mathcal{U},\mathcal{V}}}$ satisfies the bounds

$$0 \leq \mathrm{I}(X^{\mathcal{U}};X^{\mathcal{V}})_{p^{\mathcal{U},\mathcal{V}}} \leq \min(\mathrm{H}(X^{\mathcal{U}})_{p^{\mathcal{U}}}, \mathrm{H}(X^{\mathcal{V}})_{p^{\mathcal{V}}}). \tag{18.41}$$

Furthermore, the lower bound $\mathrm{I}(X^{\mathcal{U}};X^{\mathcal{V}})_{p^{\mathcal{U},\mathcal{V}}} = 0$ is attained if and only if the subsets $\mathcal{U}$ and $\mathcal{V}$ are uncorrelated, i.e.,

$$\mathrm{I}(X^{\mathcal{U}};X^{\mathcal{V}})_{p^{\mathcal{U}} \otimes p^{\mathcal{V}}} = 0; \tag{18.42}$$

and the upper bound $I(X^{\mathcal{U}};X^{\mathcal{V}})_{p^{\mathcal{U},\mathcal{V}}} = \min(H(X^{\mathcal{U}})_{p^{\mathcal{U}}}, H(X^{\mathcal{V}})_{p^{\mathcal{V}}})$ is attained if and only if either variable $\mathrm{x}^{\mathcal{U}}$ or $\mathrm{x}^{\mathcal{V}}$ is a function of the other variable under the joint distribution $p^{\mathcal{U},\mathcal{V}}$.

Note that the upper bound in (18.41) is equivalent to the uncertainty-increases-globally property (18.34) of the joint entropy.

Example 18.11

For a system of four bits, consider the joint distribution

$$p^{[4]} = \left(\frac{1}{2}|00\rangle + \frac{1}{2}|11\rangle\right) \otimes \left(\frac{1}{2}|00\rangle + \frac{1}{2}|11\rangle\right)$$
$$= \frac{1}{4}\left(|0000\rangle + |0011\rangle + |1100\rangle + |1111\rangle\right). \tag{18.43}$$

It can readily be verified that, under this joint distribution, any two groups $\mathcal{U} \subseteq \{0,1\}$ and $\mathcal{V} \subseteq \{2,3\}$ of subsystems are uncorrelated, yielding a zero mutual information $I(X^{\mathcal{U}};X^{\mathcal{V}})_{p^{\mathcal{U},\mathcal{V}}} = 0$.

For example, using the definition (18.39), the mutual information between subsystems 0 and 2 is equal to

$$I(X^0;X^2)_{p^{0,2}} = H(X^0)_{p^0} + H(X^2)_{p^2} - H(X^{0,2})_{p^{0,2}}$$
$$= 1 + 1 - 2 = 0. \tag{18.44}$$

In fact, the marginal distribution $p^{0,2}$ is obtained from (18.43) as

$$p^{0,2}_{0,0} = \sum_{x\in[2]} \sum_{y\in[2]} p_{0,x,0,y} = \frac{1}{4}, \tag{18.45}$$

and one can evaluate the other probabilities in the same way as $p^{0,2}_{x,y} = 1/4$ for all $x \in [2]$ and $y \in [2]$.

18.3.3 Cuts

Given a composite system, the most basic way to quantify correlations is by considering a cut that partitions the system into two disjoint groups of subsystems. The mutual information between the states of the subsystems on either side of the cut quantifies the correlations flowing across the cut.

Example 18.12

Consider a time series of S samples $\mathrm{x}^0, \mathrm{x}^1, \ldots, \mathrm{x}^{S-1}$ describing the evolution of the price of a stock over a discrete time index $s \in [S]$ representing successive days. The subsystems thus correspond to the different states of the price of the stock over time.

Figure 18.5 shows a cut $\mathcal{S}$ that partitions the samples into two groups $\mathcal{S}$ and $\bar{\mathcal{S}}$. Analyzing the correlations across this cut quantifies how well past prices can predict future prices.

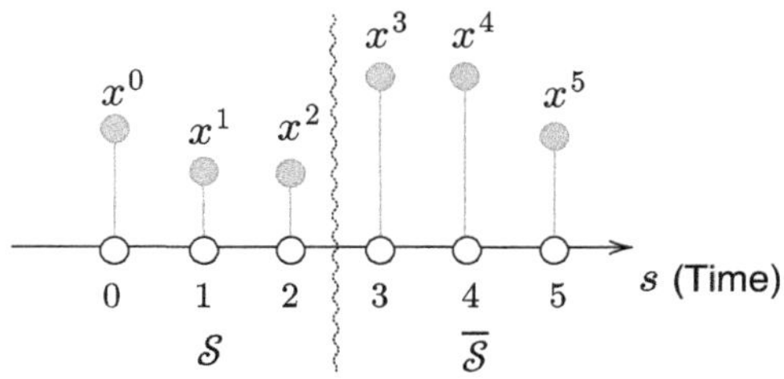

Figure 18.5 Example of a cut partitioning a time series into past and future samples. Analyzing the correlations across this cut quantifies how well information about the past can predict the future (and vice versa).

Example 18.13

For an even integer S, consider now two time series x^0, x^1,..., $\mathrm{x}^{S/2-1}$ and y^0, y^1,..., $\mathrm{y}^{S/2-1}$ that are recorded over the same discrete time instants $s \in [S/2]$. As illustrated in Figure 18.6 for $S = 12$, this set of variables forms a system consisting of S subsystems with states $\mathrm{x}^0, \mathrm{x}^1, \ldots, \mathrm{x}^{S/2-1}$, $\mathrm{x}^{S/2} = \mathrm{y}^0$, $\mathrm{x}^{S/2+1} = \mathrm{y}^1, \ldots$, $\mathrm{x}^{S-1} = \mathrm{y}^{S/2-1}$. A cut $\mathcal{S}$ such as the one in the figure divides the variables into two groups $\mathcal{S}$ and $\bar{\mathcal{S}}$, with each group corresponding to one of the two time series. Accordingly, correlations across this cut quantify the extent to which one of the time series can be estimated from the other.

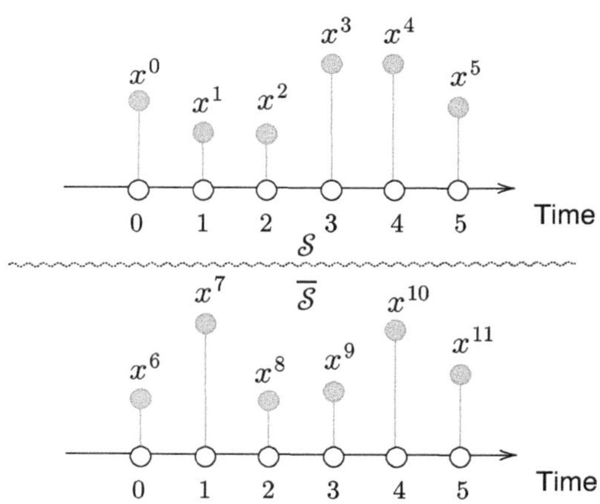

Figure 18.6 Example of a cut partitioning two time series. Analyzing the correlations across this partition quantifies how well information about one sequence can estimate the other time series.

Example 18.14

Take now a social network of S individuals, which are connected on a **graph**. An **edge** on the graph connects two individuals that have an established connection. Define as x^s the state of individual $s \in [S]$, which may represent, for instance, the expected vote in an upcoming election, or the rating given to some item. Only states that are connected on the graph, possibly indirectly via a number of intermediate individuals, can be correlated.

In Figure 18.7, cut 1 separates groups of individuals with no edge crossing the cut. In this case, there is no correlation between the variables on either side of the cut. In other words, knowing the states of individuals on one side of the cut carries no information about the states of the individuals on the other side of the cut. The same is true for cut 3.

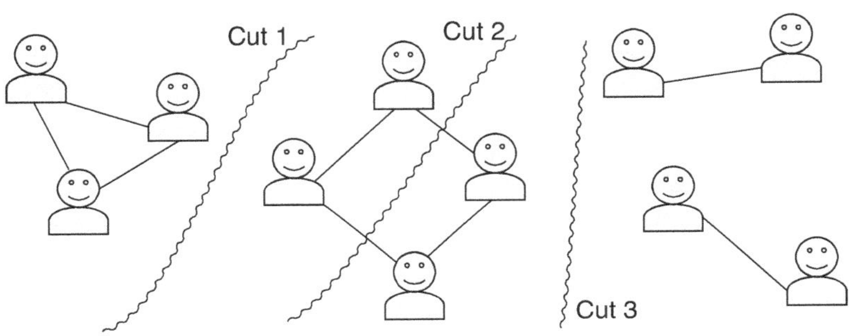

Figure 18.7 Example of cuts partitioning groups of individuals in a social network graph.

In contrast, for cut 2, there are two edges crossing the cut, and thus the mutual information across the cut is generally positive. A positive mutual information indicates that knowing the states of the individuals on one side of the cut provides information about the states of the individuals on the other side of the cut.

A cut can be formally defined as follows.

Definition 18.6 Cuts
A C-cut is a partition $(\mathcal{S}, \bar{\mathcal{S}})$ of the set $[S]$ of subsystems into a non-empty subset $\mathcal{S} \subset [S]$ of

$$|\mathcal{S}| = C \tag{18.46}$$

subsystems and its complement subset $\bar{\mathcal{S}} = [S] \backslash \mathcal{S}$ of $|\bar{\mathcal{S}}| = S - C$ subsystems. Without loss of generality, one can constrain the cut size C to satisfy the inequality

$$C \leq \left\lfloor \frac{S}{2} \right\rfloor. \tag{18.47}$$

With this convention, one has the inequality $C \leq S - C$, and thus $\mathcal{S}$ has cardinality no larger than its complement $\bar{\mathcal{S}}$. The integer C is referred to as the **cardinality** of the cut, and the cut is identified via the smaller subset $\mathcal{S}$.

The constraint (18.47) comes with no loss of generality in the sense that considering all C-cuts with C satisfying the inequality (18.47) exhausts all the cuts in the system.

Example 18.15

With $S = 2$ or $S = 3$ subsystems, the inequality (18.47) indicates that the maximum cardinality of a cut is $\lfloor S/2 \rfloor = 1$. Thus, all cuts are 1-cuts, i.e., cuts between one subsystem and the rest. For instance, in the case of $S = 3$ subsystems, the cuts are defined by the subsets $\mathcal{S} = \{0\}$, $\mathcal{S} = \{1\}$, and $\mathcal{S} = \{2\}$ including a single subsystem.

Given a composite system with S subsystems, the number of C-cuts is equal to the number of subsets of cardinality $\mathcal{S}$, i.e., to $\binom{S}{C}$ for $C < \left\lfloor \frac{S}{2} \right\rfloor$; while it is equal to $\frac{1}{2}\binom{S}{C}$ for $C = S/2$. In fact, in the latter case, which is applicable only if S is even, two subsets $\mathcal{S}$ and $\mathcal{S}'$ correspond to the same cut if $\mathcal{S}' = \bar{\mathcal{S}}$ (see Problem 18.2).

Example 18.16

With $S = 4$ subsystems, given the equality $\lfloor S/2 \rfloor = 2$, both 1-cuts and 2-cuts must be considered to account for all possible partitions of the system. These include the $\binom{4}{1} = 3$ 1-cuts $\mathcal{S} = \{0\}$, $\mathcal{S} = \{1\}$, $\mathcal{S} = \{2\}$, and $\mathcal{S} = \{3\}$; along with the $\frac{1}{2}\binom{4}{1} = 3$ 2-cuts $\mathcal{S} = \{0, 1\}$, $\mathcal{S} = \{0, 2\}$, and $\mathcal{S} = \{0, 3\}$. Note that the remaining $\binom{4}{1} - 3 = 3$ 2-cuts defined by subsets $\mathcal{S} = \{2, 3\}$, $\mathcal{S} = \{1, 3\}$, and $\mathcal{S} = \{1, 2\}$ are equivalent to the three 2-cuts listed above.

18.3.4 Bipartite Correlations

For any cut $\mathcal{S}$, consider now the states $\mathrm{x}^{\mathcal{S}}$ of the subsystems on one side of the cut and the states $\mathrm{x}^{\bar{\mathcal{S}}}$ for the subsystems on the other side of the cut. Using the definition (18.40), the system is **uncorrelated across the cut** $\mathcal{S}$ if the joint distribution factorizes as the product of the marginal distributions $p^{\mathcal{S}}$ and $p^{\bar{\mathcal{S}}}$, i.e., $p^{\mathcal{S},\bar{\mathcal{S}}} = p^{\mathcal{S}} \otimes p^{\bar{\mathcal{S}}}$. More generally, as illustrated in Figure 18.8, correlations across a cut can be quantified as follows.

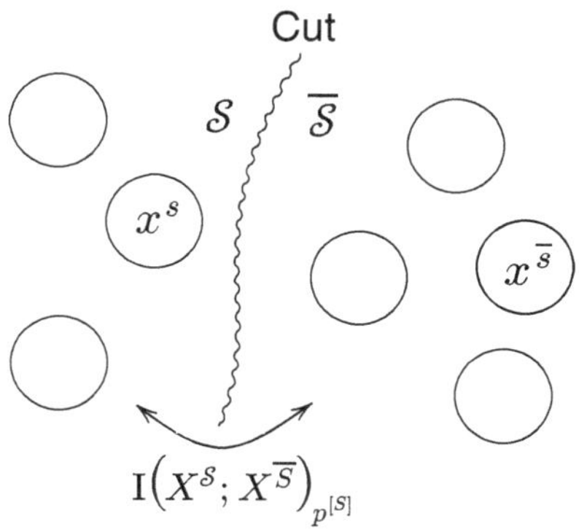

Figure 18.8 Mutual information across a cut $\mathcal{S}$.

Definition 18.7 Mutual information across a cut
Given a joint distribution $p^{[S]}$, the correlations across a C-cut $\mathcal{S} \subset [N]$ are quantified by the mutual information

$$\mathrm{I}(X^{\mathcal{S}};X^{\bar{\mathcal{S}}})_{p^{[S]}} = \mathrm{H}(X^{\mathcal{S}})_{p^{\mathcal{S}}} + \mathrm{H}(X^{\bar{\mathcal{S}}})_{p^{\bar{\mathcal{S}}}} - \mathrm{H}(X^{[S]})_{p^{[S]}} \tag{18.48}$$

between the states $\mathrm{x}^{\mathcal{S}}$ and $\mathrm{x}^{\bar{\mathcal{S}}}$ on the two sides of the cut.

Using the bounds (18.41), the following inequalities are satisfied by the mutual information (18.48).

Key Result 18.5 Bounds on the mutual information across a cut
For any C-cut $\mathcal{S}$, the mutual information (18.41) satisfies the inequalities

$$0 \le \mathrm{I}(X^{\mathcal{S}};X^{\bar{\mathcal{S}}})_{p^{[S]}} \le \min(\mathrm{H}(X^{\mathcal{S}})_{p^{\mathcal{S}}}, \mathrm{H}(X^{\bar{\mathcal{S}}})_{p^{\bar{\mathcal{S}}}}) \le C\log_2(D). \tag{18.49}$$

The lower bound $\mathrm{I}(X^{\mathcal{S}};X^{\bar{\mathcal{S}}})_{p^{[S]}} = 0$ is attained if and only if the system is uncorrelated across cut $\mathcal{S}$, i.e.,

$$\mathrm{I}(X^{\mathcal{S}};X^{\bar{\mathcal{S}}})_{p^{\mathcal{S}} \otimes p^{\bar{\mathcal{S}}}} = 0. \tag{18.50}$$

The upper bound $\mathrm{I}(X^{\mathcal{S}};X^{\bar{\mathcal{S}}})_{p^{[S]}} \le C\log_2(D)$ in (18.49) shows that the mutual information across a cut can grow at most linearly with the cut size C and at most logarithmically with the per-subsystem dimension D.

Definition 18.8 Maximum correlation across a cut
When the equality $\mathrm{I}(X^{\mathcal{S}};X^{\bar{\mathcal{S}}})_{p^{[S]}} = C\log_2(D)$ holds in the upper bound (18.49), the system is said to be maximally correlated across cut $\mathcal{S}$ under distribution $p^{[S]}$.

Example 18.17

For the system of three equal uniform bits with joint distribution (18.10), the mutual information is equal to 1 bit across every cut, attaining the upper bound that the maximum value of the mutual information is given by $C\log_2(D) = \log_2(2) = 1$. Therefore, the system is maximally correlated across all cuts. This occurs because, once the state of any one subsystem is known, the states of the other two subsystems are also determined, and, additionally, each subsystem has maximum entropy.

To see this, it is sufficient to consider only 1-cuts, i.e., with $C = 1$, which separate one subsystem from the rest of the system. For example, for the cut with $\mathcal{S} = \{0\}$, the mutual information is given by

$$\begin{aligned} \mathrm{I}(X^0;X^{1,2})_{p^{[3]}} &= \mathrm{H}(X^0)_{p^0} + \mathrm{H}(X^{1,2})_{p^{1,2}} - \mathrm{H}(X^{[3]})_{p^{[3]}} \\ &= 1 + 1 - 1 = 1, \end{aligned} \tag{18.51}$$

since all entropies are equal to 1 bit.

Example 18.18

For the system with three uniform one-hot configurations, i.e., for the joint distribution (18.11), the situation is intuitively different from the previous example, since knowing the state of one subsystem does not uniquely determine the states of the other two. To corroborate this intuition, consider again cut $\mathcal{S} = \{0\}$, across which the mutual information is given by

$$\begin{aligned} \mathrm{I}(X^0;X^{1,2})_{p^{[3]}} &= \mathrm{H}(X^0)_{p^0} + \mathrm{H}(X^{1,2})_{p^{1,2}} - \mathrm{H}(X^{[3]})_{p^{[3]}} \\ &= \mathrm{h}\left(\frac{1}{3}\right) + \log_2(3) - \log_2(3) \\ &= \mathrm{h}\left(\frac{1}{3}\right), \end{aligned} \tag{18.52}$$

and the same result is obtained for the other two possible cuts. Therefore, the system is correlated across each cut, but not maximally so, as shown by the inequality $\mathrm{h}(1/3) < 1$.

Example 18.19

Under the joint distribution (18.43), the system of four bits is maximally correlated across all 1-cuts, as the equality $\mathrm{I}(X^s;X^{[S]\setminus s})_{p^{[S]}} = 1$ holds for all subsystems $s \in [S]$. This is because, given access to the state of the other three bits, it is always possible to identify the state of the remaining bit, which is uniformly distributed. For 2-cuts, the bipartite correlation depends on the choice of the cut.

In particular, for the cut $\mathcal{S} = \{0, 1\}$, the mutual information is zero, since the joint distribution factorizes as $p^{[4]} = p^{0,1} \otimes p^{2,3}$, and thus the groups of subsystems

$\mathcal{S} = \{0,1\}$ and $\bar{\mathcal{S}} = \{2,3\}$ are uncorrelated. In contrast, for any other 2-cut, the mutual information is positive.

For instance, for the cut $\mathcal{S} = \{0,3\}$, the mutual information is

$$\begin{aligned} \mathrm{I}(X^{0,3};X^{1,2})_{p^{[4]}} &= \mathrm{H}(X^{0,3})_{p^{0,3}} + \mathrm{H}(X^{1,2})_{p^{1,2}} - \mathrm{H}(X^{[4]})_{p^{[4]}} \\ &= \sum_{s\in[4]} \mathrm{H}(X^s)_{p^s} - \mathrm{H}(X^{0,1})_{p^{1,2}} - \mathrm{H}(X^{2,3})_{p^{1,2}} \\ &= 4 - 1 - 1 = 2, \end{aligned} \tag{18.53}$$

where the second equality follows because random variables x^0 and x^3 are uncorrelated; random variables x^1 and x^2 are also uncorrelated; and so are the subsets $\mathrm{x}^{0,1}$ and $\mathrm{x}^{2,3}$. Therefore, the system is maximally correlated across cut $\mathcal{S} = \{0,3\}$, and the same conclusion applies to cut $\mathcal{S} = \{0,2\}$.

18.4 Decomposing Bipartite Correlations ⏭

Section 18.3 introduced the mutual information across a cut, which quantifies the correlations present between two groups of subsystems. A positive mutual information across a cut suggests that some information flows across the cut. When there are multiple states on at least one of the two sides of a cut, the nature of this information flow can be of different types, namely shared, unique, synergistic, or redundant.

This section describes how the mutual information across a cut can be decomposed into shared, unique, synergistic, and redundant contributions. To this end, the section focuses on a setting with three subsystems, which are referred to as X, Y, and Z, rather than as X^0, X^1, and X^2, to simplify the notation. The goal is to study the information flowing across the cut between subsystems X and Y on one side and subsystem Z on the other side. As will become apparent, the discussion becomes significantly more complex for cuts involving more than two subsystems on either side.

Example 18.20

Consider a medical diagnosis scenario in which one wishes to predict a patient's disease status Z based on their genetic markers X and based on a blood test result Y. What type of information can be carried by X and Y about Z?

First, some information about the disease status Z may be obtained *either* from the genetic markers X or from the blood result Y. This is the case when the condition being tested, Z, is determined by one's genetics, X, but it can also be detected via a blood test, Y. In this situation, information about Z is shared across subsystems X and Y.

Second, some other information about Z may be *exclusively* retrieved from X or from Y. For example, some genetic variants may predispose individuals to certain diseases, which cannot be detected by a blood test. This type of information is unique to either subsystem X or Y.

Finally, some information about Z may only be available when *both* features X and Y are considered together, beyond what each feature provides individually or redundantly. This type of information may capture complex interactions between genetics and other factors in determining a health condition, and is said to be synergistic.

As discussed in Section 18.3, the bipartite correlation between subsystems X and Y on one side of the cut and subsystem Z on the other side is quantified via the mutual information $\mathrm{I}(X, Y;Z)_{p^{XYZ}}$, where p^{XYZ} denotes the joint distribution for the three subsystems. Using the framework of information as a set introduced in Section 12.2, Figure 18.9 illustrates the decomposition of the mutual information $\mathrm{I}(X, Y;Z)_{p^{XYZ}}$ into the different contributions that were identified in the previous example. Note that in the figure and in the rest of this section, the subscripts denoting distributions are dropped from mutual informations in order to further streamline the notation.

As illustrated in Figure 18.9, the mutual information $\mathrm{I}(X, Y;Z)$ can be decomposed as the sum of four separate contributions.

- **Shared information** $\mathrm{Sh}(X, Y;Z)$, which represents the information about Z that can be obtained from *either X or Y*.
- **Unique information** $\mathrm{Un}(X;Z|Y)$, which represents the information about Z that can be obtained *only* from X and not from Y.
- **Unique information** $\mathrm{Un}(Y;Z|X)$, which represents the information about Z that can be obtained *only* from Y and not from X.
- **Synergistic information** $\mathrm{Sy}(X, Y;Z)$, which represents the information about Z that can be obtained only through a joint consideration of *both* X and Y.

As can be inferred from Figure 18.9, the overall amount of information about Z that can be extracted from X, i.e., the mutual information $\mathrm{I}(X;Z)$, is given by the sum

$$\mathrm{I}(X;Z) = \mathrm{Un}(X;Z|Y) + \mathrm{Sh}(X, Y;Z) \tag{18.54}$$

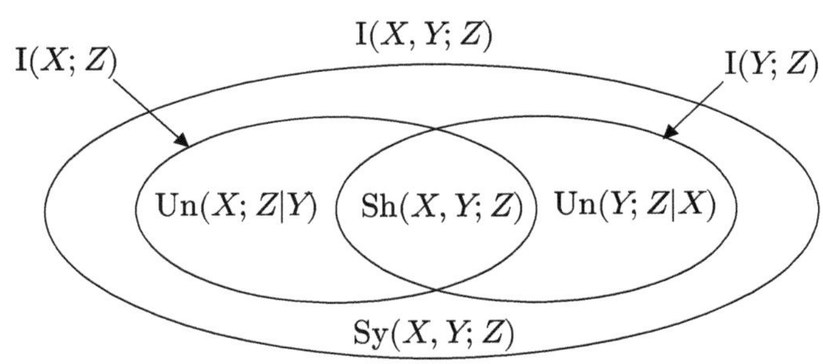

Figure 18.9 Illustration of the information decomposition of the mutual information $\mathrm{I}(X, Y;Z)_{p^{XYZ}}$ into shared information $\mathrm{Sh}(X, Y;Z)$, unique information $\mathrm{Un}(X;Z|Y)$ and $\mathrm{Un}(Y;Z|X)$, and synergistic information $\mathrm{Sy}(X, Y;Z)$.

of the contributions given by unique and shared information. In an analogous way, the mutual information $\mathrm{I}(Y;Z)$ is given by the sum

$$\mathrm{I}(Y;Z) = \mathrm{Un}(Y;Z|X) + \mathrm{Sh}(X, Y;Z). \tag{18.55}$$

Finally, the synergistic information $\mathrm{Sy}(X, Y;Z)$ can be written as the difference

$$\mathrm{Sy}(X, Y;Z) = \mathrm{I}(X, Y;Z) - \mathrm{Un}(Y;Z|X) - \mathrm{Un}(X;Z|Y) - \mathrm{Sh}(X, Y;Z). \tag{18.56}$$

The next example illustrates a situation in which the evaluation of the quantities in the information decomposition (18.54)–(18.56) is straightforward. More generally, as will be discussed, there are many possible ways to define decompositions of the form illustrated in Figure 18.9.

Example 18.21

Consider four uncorrelated binary variables $\mathrm{b}^{[4]}$, with each bit b^t being uniformly distributed for $t \in [4]$. The state of subsystem X encompasses the bits $\mathrm{x} = (\mathrm{b}^0, \mathrm{b}^1, \mathrm{b}^2 \oplus \mathrm{b}^3)$, where $\oplus$ is the exclusive OR (XOR); subsystem Y holds bits $\mathrm{y} = (\mathrm{b}^1, \mathrm{b}^2)$; while the state of subsystem Z is $\mathrm{z} = (\mathrm{b}^0, \mathrm{b}^1, \mathrm{b}^3)$. Recall that the XOR of two bits returns 1 if the two bits are different, while yielding 0 otherwise. It is also useful to note that, for any bit b, one has the equalities $b \oplus b = 0$ and $b \oplus 0 = b$.

The mutual information $\mathrm{I}(X, Y;Z)$ can be evaluated as

$$\begin{aligned} \mathrm{I}(X, Y;Z) &= \mathrm{H}(X, Y) + \mathrm{H}(Z) - \mathrm{H}(X, Y, Z) \\ &= 4 + 3 - 4 = 3, \end{aligned} \tag{18.57}$$

since both the pair of variables x, y and the three variables $\mathrm{x}, \mathrm{y}, \mathrm{z}$ are in a one-to-one correspondence with the four uncorrelated bits $\mathrm{b}^{[4]}$. The three bits of mutual information $\mathrm{I}(X, Y;Z)$ confirm that the variable z, which has entropy $\mathrm{H}(Z) = 3$, is a function of the variables x and y.

In this example, the individual terms in the information decomposition illustrated in the figure can easily be identified. Subsystem X carries the unique information given by bit b^0, yielding the unique information contribution $\mathrm{Un}(X;Z|Y) = 1$. The unique information carried by subsystem Y equals zero, i.e., $\mathrm{Un}(Y;Z|X) = 0$. The bit b^1 amounts to shared information across subsystems X and Y about Z, and thus one can set $\mathrm{Sh}(X, Y;Z) = 1$. Finally, the last of the three bits in the mutual information $\mathrm{I}(X, Y;Z)$ is accounted for by the synergistic information $\mathrm{Sy}(X, Y;Z) = 1$. This corresponds to bit b^2, which is recovered by using both variables x and y.

Example 18.22

The information decomposition outlined in this section makes it possible to distinguish joint distributions that could not be differentiated on the basis of standard information measures. As an example, consider the joint distributions over $S = 3$ subsystems with

$D = 4$ given by

$$p^{XYZ} = \frac{1}{8}(|000\rangle + |022\rangle + |202\rangle + |222\rangle + |111\rangle + |133\rangle + |331\rangle + |313\rangle) \quad (18.58)$$

and

$$q^{XYZ} = \frac{1}{8}(|000\rangle + |120\rangle + |201\rangle + |321\rangle + |012\rangle + |132\rangle + |213\rangle + |333\rangle). \quad (18.59)$$

Under both distributions, the variable z can be recovered exactly from x and y, and the random variable z is uniformly distributed. Therefore, for both distributions, the mutual information $I(X, Y;Z)$ takes the maximum value $\log_2(D) = 2$ bits. However, the two distributions differ in the way the two bits of information are decomposed into separate contributions.

In particular, under the first distribution, knowing either x or y determines whether z is even or odd, thus providing 1 bit of shared information. Furthermore, knowing both x and y pins down the exact value of z via one additional bit of synergistic information.

In contrast, for the second distribution, variables x and y each provide a separate bit of information about z, and thus there is 1 bit of unique information for either variable. For instance, when x = 0, variable z can be either 0 or 2, and knowing y resolves the remaining bit of uncertainty.

While, in the examples above, one can unambiguously determine the separate contributions of the information decomposition, this is not possible for general joint distributions. However, by the equalities (18.54)–(18.56), once one defines either the shared information or the unique information, all other information measures are also well defined. There are different ways in which one may, for instance, define the shared information $\mathrm{Sh}(X, Y;Z)$, and one is presented in Appendix 18.A.

When considering cuts involving more than two subsystems on one side of the cut, the information decomposition outlined in this section becomes exponentially more complex. In fact, with more than two subsystems, the unique, shared, and synergistic contributions to the overall mutual information across a cut can be apportioned in many distinct ways. For instance, one could divide the subsystems on one side of the cut into any two subsets, and then carry out the analysis of contributions as presented in this section.

18.5 Fragmentation and Scale of Correlations

So far, this chapter has studied correlations across cuts, thus describing flows of information between two different parts of a system. As detailed in this section, one can combine insights obtained from multiple cuts in order to produce a more comprehensive description of the correlations in the system. In particular, this section introduce ways to characterize the level of fragmentation, as well as the scale, of the correlations present in a system.

18.5.1 Fragmentation of Correlations

The fragmentation of the correlations present in a system can be evaluated by assessing the maximum number of mutually uncorrelated groups in which a system can be partitioned under the given joint distribution. To formalize this definition, recall that subsets $\{\mathcal{S}^i\}_{i\in[I]}$, with $\mathcal{S}^i \subseteq [S]$ for all $i \in [I]$, form a partition of the set $[S]$ of all subsystems if they satisfy the conditions that their union recovers the entire system and that they are disjoint, i.e.,

$$\bigcup_{i\in[I]} \mathcal{S}^i = [S] \text{ and } \mathcal{S}^i \cap \mathcal{S}^{i'} = \emptyset \text{ for all } i \neq i'. \tag{18.60}$$

Given a partition $\{\mathcal{S}^i\}_{i\in[I]}$, the notation $p^{\mathcal{S}^0,\mathcal{S}^1,\ldots,\mathcal{S}^{I-1}}$ is used to represent the joint distribution $p^{[S]}$ with indices reordered according to the list $\mathcal{S}^0, \mathcal{S}^1, \ldots, \mathcal{S}^{I-1}$ obtained by concatenating the indices of the ordered subsets $\{\mathcal{S}^i\}_{i\in[I]}$. The following definition generalizes (18.40) to any number of groups of subsystems.

Definition 18.9 Uncorrelated groups of subsystems

Given a joint distribution $p^{[S]}$, the groups of subsystems specified by the partition $\{\mathcal{S}^i\}_{i\in[I]}$ are uncorrelated if, upon reordering of the indices, the joint distribution $p^{[S]}$ can be written as the Kronecker product

$$p^{\mathcal{S}^0,\mathcal{S}^1,\ldots,\mathcal{S}^{I-1}} = p^{\mathcal{S}^0} \otimes \cdots \otimes p^{\mathcal{S}^{I-1}} \tag{18.61}$$

of the marginal distributions $\{p^{\mathcal{S}^i}\}_{i\in[I]}$ for the groups $\{\mathcal{S}^i\}_{i\in[I]}$.

The fragmentation of correlations in a system describes the extent to which the system can be divided into uncorrelated groups of subsystems. The following definition introduces a single integer parameter, $I^{\max}$, to characterize the level of fragmentation of a joint distribution.

Definition 18.10 Fragmentation of correlations

Given its joint distribution $p^{[S]}$, a system is said to be $I^{\max}$-uncorrelated if it can be partitioned into $1 \leq I^{\max} \leq S$ mutually uncorrelated subsets of subsystems, while it cannot be partitioned into $I^{\max}+1$ mutually uncorrelated subsets.

A way to conceptualize an $I^{\max}$-uncorrelated system is that its joint state $\mathrm{x}^{[S]}$ can be generated by sharing randomness only among the subsystems within the same group $\mathcal{S}^i$ for $i \in [I^{\max}]$. In other words, no communications or randomness sharing is required for subsystems belonging to distinct subsets $\mathcal{S}^i$ and $\mathcal{S}^{i'}$ with $i' \neq i$ in order to simulate the state variables $\mathrm{x}^{[S]}$. So, an $I^{\max}$-uncorrelated system can effectively be viewed as $I^{\max}$ separate systems, each corresponding to a subset $\mathcal{S}^i$.

This perspective is made particularly clear by the tensor network representation of an $I^{\max}$-uncorrelated joint distribution. As shown in Figure 18.10, the tensor network encodes the tensor product of the $I^{\max}$ order-$|\mathcal{S}^i|$ tensors $p^{\mathcal{S}^i}$ for $i \in [I^{\max}]$. It should be noted that the correlations present within each subset $\mathcal{S}^i$ following the distribution $p^{\mathcal{S}^i}$ may be small. What matters is that each group of variables $\mathrm{x}^{\mathcal{S}^i}$ cannot be further partitioned into two uncorrelated groups of variables.

According to the definition of an $I^{\max}$-uncorrelated system, a larger value of the integer $I^{\max}$ indicates a more significant fragmentation of the system into isolated groups that present no mutual correlations. In particular, with $I^{\max} = S$, the system is globally uncorrelated, satisfying the equality (18.29). Conversely, with $I^{\max} = 1$, the system cannot be partitioned even into two groups of uncorrelated subsystems.

Definition 18.11 Globally correlated systems

For a multipartite system consisting of S subsystems, the integer $I^{\max}$ satisfies the bounds

$$\underbrace{1}_{\text{globally correlated}} \leq I^{\max} \leq \underbrace{S}_{\text{globally uncorrelated}} . \tag{18.62}$$

Accordingly, while with $I^{\max} = S$ the system is globally uncorrelated, meeting the definition (18.29), with $I^{\max} = 1$ the system is said to be globally correlated. For any $I^{\max} \in \{2, \ldots, S-1\}$ the system is correlated, while not being globally correlated.

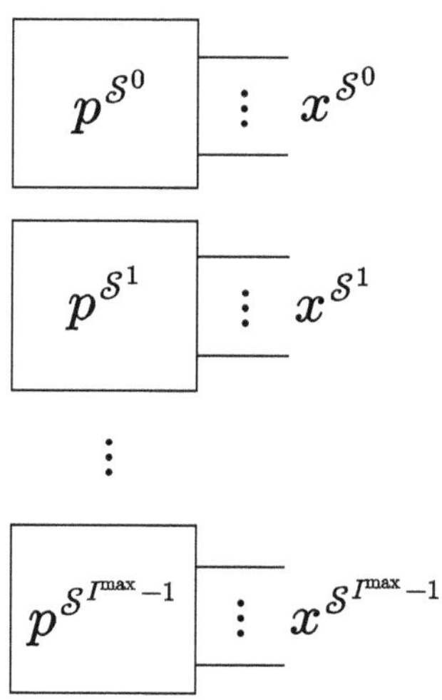

Figure 18.10 The tensor network representation of a joint distribution of the form (18.61).

Example 18.23

Take a system of three bits with joint distribution

$$p^{[3]} = \frac{1}{4}[\epsilon, \epsilon, 1-\epsilon, 1-\epsilon, 1-\epsilon, 1-\epsilon, \epsilon, \epsilon] \tag{18.63}$$

for some probability ϵ. When $\epsilon = 0.5$, the joint distribution is globally uncorrelated, corresponding to $I^{\max} = 3$. In fact, in this case, the joint distribution $p^{[3]}$ is uniform, and thus it can be expressed as the Kronecker product of $I^{\max} = 3$ marginal uniform distributions as $p^{[3]} = u \otimes u \otimes u$, with $u = [0.5, 0.5]^T$. Otherwise, the joint distribution is correlated, and thus the parameter $I^{\max}$ is smaller than 3.

To determine the value of $I^{\max}$ for probabilities $\epsilon \neq 0.5$, it is convenient to rewrite the distribution $p^{[3]}$ in tensor form as

$$\begin{aligned} p^{[3]} =& \frac{1}{4}\epsilon|000\rangle + \frac{1}{4}\epsilon|001\rangle + \frac{1}{4}\epsilon|110\rangle + \frac{1}{4}\epsilon|111\rangle \\ &+ \frac{1}{4}(1-\epsilon)|010\rangle + \frac{1}{4}(1-\epsilon)|011\rangle + \frac{1}{4}(1-\epsilon)|100\rangle + \frac{1}{4}(1-\epsilon)|101\rangle. \end{aligned} \tag{18.64}$$

This expression makes it easier to see that the distribution can be factorized as $p^{[3]} = p^{0,1} \otimes p^2$, with marginal distributions

$$p^{0,1} = \frac{1}{2}\epsilon|00\rangle + \frac{1}{2}\epsilon|11\rangle + \frac{1}{2}(1-\epsilon)|01\rangle + \frac{1}{2}(1-\epsilon)|10\rangle \tag{18.65}$$

for subsystems 0 and 1, and marginal distribution p^2 being the uniform distribution $p^2 = (|0\rangle + |1\rangle)/2$.

Accordingly, the groups of subsystems $\mathcal{S} = \{0, 1\}$ and $\bar{\mathcal{S}} = \{2\}$ are mutually uncorrelated, indicating that parameter $I^{\max}$ equals 2. Indeed, under the marginal distribution $p^{0,1}$ in (18.65), the variables x^0 and x^1 are correlated, being equal with probability $\epsilon = \Pr[\mathrm{x}^0 = \mathrm{x}^1]$. Thus, unless $\epsilon = 0.5$, the distribution $p^{0,1}$ cannot be factorized further.

In conclusion, when $\epsilon = 0.5$, the system is 3-uncorrelated, i.e., $I^{\max} = 3$; or, equivalently, it is globally uncorrelated. In contrast, when $\epsilon \neq 0.5$, the system is correlated. Specifically, it is 2-uncorrelated, i.e., $I^{\max} = 2$, since it can be broken up into two, and no more than two, uncorrelated groups of subsystems.

A factorization of the form (18.61) provides important information about the structure of the joint distribution $p^{[S]}$. This structure translates into a smaller number of parameters required to specify the joint distribution $p^{[S]}$.

An arbitrary tensor $p^{[S]}$ contains D^S probabilities, which need to sum to 1, requiring the specification of only $D^S - 1$ probabilities in the absence of any structural information. In contrast, to write down a joint distribution in the form (18.61), one needs to specify only the $I^{\max}$ tensors $\{p^{\mathcal{S}^i}\}_{i\in[I^{\max}]}$. Therefore, the overall number of parameters is $\sum_{i\in[I^{\max}]}(D^{|\mathcal{S}^i|} - 1)$, since each tensor $p^{\mathcal{S}^i}$ is described by $D^{|\mathcal{S}^i|} - 1$ probabilities. For example, if $S = 100$ and $D = 2$, the number of parameters for a generic joint distribution is $D^S - 1 = 2^{100} - 1 \approx 10^{33}$, while a globally uncorrelated distribution, i.e., with $I^{\max} = S$, requires $S(D-1) = 100$ parameters.

18.5.2 Mutual Information and Fragmentation of Correlations

A system is globally uncorrelated ($I^{\max} = S$) if and only if each subsystem is uncorrelated with the rest of the system. In contrast, the system is globally correlated ($I^{\max} = 1$) as long as it cannot be divided into any two uncorrelated groups. It follows that, by definition, globally uncorrelated systems present a zero mutual information across every 1-cut, and hence also across every other type of cut. In contrast, correlated systems have a positive mutual information across all cuts. This yields the following formal result.

> **Key Result 18.6 Globally uncorrelated and globally correlated subsystems via mutual information across cuts**
> Given its joint distribution $p^{[S]}$, a system is globally uncorrelated if and only if the mutual information equals zero across all 1-cuts:
>
> $$\text{globally uncorrelated} \Longleftrightarrow \mathrm{I}(X^s;X^{[S]\setminus s})_{p^{[S]}} = 0 \text{ for all } s \in [S]; \tag{18.66}$$
>
> or, equivalently, one has the equivalence
>
> $$\text{correlated} \Longleftrightarrow \mathrm{I}(X^s;X^{[S]\setminus s})_{p^{[S]}} > 0 \text{ for at least one } s \in [S]. \tag{18.67}$$
>
> Conversely, the system is globally correlated if and only if the mutual information is positive across all cuts:
>
> $$\text{globally correlated} \Longleftrightarrow \mathrm{I}(X^{\mathcal{S}};X^{\bar{\mathcal{S}}})_{p^{[S]}} > 0 \text{ for all cuts } \mathcal{S}. \tag{18.68}$$

Example 18.24

Under both the three-equal-uniform-bits distribution (18.10) and the three-uniform-one-hot-configurations distribution (18.11), the system of three bits is globally correlated, yielding $I^{\max} = 1$. In fact, as seen earlier, the two distributions present a positive mutual information across each cut, and thus, by (18.68), they cannot be separated into two subsets of uncorrelated subsystems.

Example 18.25

Consider five uncorrelated bits $\mathrm{b}^{[5]}$, with each bit b^t being uniformly distributed for $t \in [5]$. As illustrated in Figure 18.11, assume that subsystem 0 holds the bits $\mathrm{x}^0 = (\mathrm{b}^0, \mathrm{b}^1)$; subsystem 1 holds the bits $\mathrm{x}^1 = (\mathrm{b}^2, \mathrm{b}^3)$; subsystem 2 has the bits $\mathrm{x}^2 = (\mathrm{b}^3, \mathrm{b}^4)$; and subsystem 3 has the bits $\mathrm{x}^3 = (\mathrm{b}^0, \mathrm{b}^4)$. Note that the figure includes an edge between subsystems that include the same variable, and are thus correlated. Subsystem 0 and subsystem 1 are mutually uncorrelated, i.e., $\mathrm{I}(X^0;X^1)_{p^{0,1}} = 0$, since they do not share any bit b^t. However, there are no cuts across which the mutual information is zero, and thus the system is globally correlated.

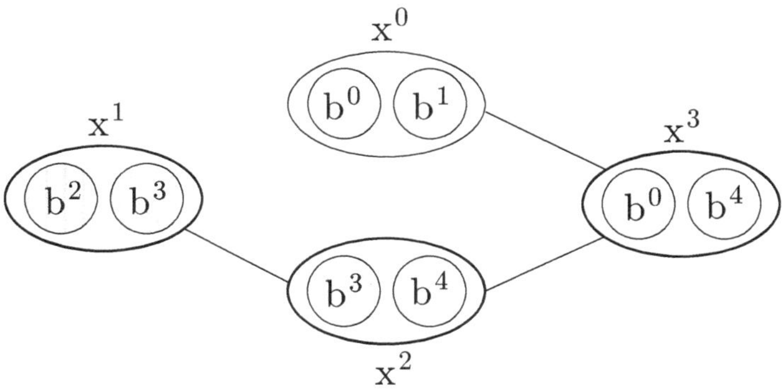

Figure 18.11 Illustration of a system of $S = 4$ subsystems constructed based on five uncorrelated bits $\mathrm{b}^{[5]}$.

18.5.3 Fragmentation of Correlations in a Group of Subsystems

While this section has focused on the entire system of S subsystems, the definitions given here apply naturally even when restricted to a subset of the subsystems. Indeed, given an initial system of S subsystems, one may consider a subset $\mathcal{S} \subseteq [S]$ of $|\mathcal{S}|$ subsystems, and study the fragmentation of the correlations associated with the marginal distribution $p^{\mathcal{S}}$. In particular, a subset $\mathcal{S}$ is said to be $I^{\max}$-uncorrelated if the subset $\mathcal{S}$ can be partitioned into $I^{\max}$ smaller subsets of mutually uncorrelated groups, and cannot be partitioned into more than $I^{\max}$ such subsets. Accordingly, a group $\mathcal{S}$ of subsystems is globally correlated if it is 1-uncorrelated, and thus it cannot be partitioned into any two mutually uncorrelated groups.

18.5.4 Scale of Correlations

As discussed, the parameter $I^{\max}$ quantifies the fragmentation of the correlations present in the system. A related, yet distinct, aspect is the scale of the system's correlations. Specifically, one can quantify how extensive the correlations are by determining the size of the largest group of subsystems that cannot be partitioned into two mutually uncorrelated subsets.

Definition 18.12 Scale of correlations
Given its joint distribution $p^{[S]}$, a system is said to be $S^{\max}$-partite correlated if (i) there exists at least one subset of $S^{\max} \leq S$ subsystems that are globally correlated, indicating that they cannot be partitioned into two mutually uncorrelated subsets; and (ii) no larger subset with this property exists, so there is no group of more than $S^{\max}$ subsystems that is globally correlated. The parameter $S^{\max}$ is also known as the **correlation depth**.

An informational state $p^{[S]}$ that is $S^{\max}$-partite correlated must be generated by a mechanism that enables the communication or the sharing of randomness across at least $S^{\max}$ subsystems. In this regard, the two parameters $I^{\max}$ and $S^{\max}$ are related, with a larger $S^{\max}$ limiting the possible values of $I^{\max}$. In particular, in a manner similar to (18.62), one has the bounds

$$\underbrace{1}_{\text{globally uncorrelated}} \leq S^{\max} \leq \underbrace{S}_{\text{globally correlated}}, \tag{18.69}$$

so that, if $S^{\max} = 1$ and thus $I^{\max} = S$, the system is globally uncorrelated, while, if $S^{\max} = S$ and hence $I^{\max} = 1$, the system is globally correlated.

Example 18.26

For a system with $S = 10$ subsystems, consider the two joint distributions

$$p^{[10]} = p^{0,1,2,3,4} \otimes p^{5,6,7,8,9} \tag{18.70}$$

and

$$p^{[10]} = p^{0} \otimes p^{1,2,3,4,5,6,7,8,9}, \tag{18.71}$$

where the distributions $p^{0,1,2,3,4}$, $p^{5,6,7,8,9}$, and $p^{1,2,3,4,5,6,7,8,9}$ cannot be further divided into the Kronecker product of any two marginal distributions. It follows that the system is 2-uncorrelated, i.e., $I^{\max} = 2$, under both distributions.

However, the first distribution is 5-partite correlated, i.e., $S^{\max} = 5$, while the second is 9-partite correlated, i.e., $S^{\max} = 1$. Therefore, both distributions have correlations with the same fragmentation level, but the correlations in the second distribution have a larger scale.

Example 18.27

Consider now a system consisting of any number, S, of bits, whose joint distribution assigns an equal probability to all configurations of bits $x^{[S]} \in [2]^S$ with even parity and a zero probability to all other configurations. Recall that a sequence of bits $x^{[S]}$ has an even parity when the number of bits equal to 1, which is computed as $\sum_{s\in[S]} \mathrm{x}^s$, is even. In formulas, the joint distribution is thus given by

$$p_{x^{[S]}} = \begin{cases} \frac{1}{2^{S-1}} & \text{if } \sum_{s\in[S]} \mathrm{x}^s \text{ is even,} \\ 0 & \text{otherwise.} \end{cases} \tag{18.72}$$

Under this joint distribution, the marginal distribution of any subset $\mathcal{S}$ of $|\mathcal{S}| \leq S - 1$ variables is globally uncorrelated. Moreover, each bit is marginally uniformly distributed. Specifically, considering any subset of $|\mathcal{S}| = S - k$ variables, with $k \in \{1, 2, \ldots, S - 1\}$, the marginal distribution is given by

$$p^{\mathcal{S}}_{x^{\mathcal{S}}} = \frac{1}{2^{S-1}}(2^{k-1}) = \frac{1}{2^{S-k}} = \frac{1}{2^{|\mathcal{S}|}}. \tag{18.73}$$

This is because there are $2^{k-1} = 2^k/2$ even-parity configurations $x^{[S]}$ that are consistent with the configuration $x^{\mathcal{S}}$ of the subset $\mathcal{S}$. In fact, depending on the parity of the bits $x^{\mathcal{S}}$, the remaining k bits in the sequence $x^{[S]}$ that are not in subset $\mathcal{S}$ must have an even or odd parity in order for the sum $\sum_{s\in[S]} x^s$ to be even.

However, the joint distribution is not globally uncorrelated, as knowing any $S-1$ bits completely specifies the remaining bit. If the $S-1$ variables have an even parity, then the remaining variable must be zero; while, if the $S-1$ variables have an odd parity, the remaining variable must be equal to 1. What is more, the joint distribution $p^{[S]}$ is in fact globally correlated. This is because the mutual information across each cut $\mathcal{S}$, $\mathrm{I}(X^{\mathcal{S}};X^{\bar{\mathcal{S}}})_{p^{[S]}}$, is positive:

$$\begin{aligned}\mathrm{I}(X^{\mathcal{S}};X^{\bar{\mathcal{S}}})_{p^{[S]}} &= \mathrm{H}(X^{\mathcal{S}})_{p^{\mathcal{S}}} + \mathrm{H}(X^{\bar{\mathcal{S}}})_{p^{\bar{\mathcal{S}}}} - \mathrm{H}(X^{[S]})_{p^{[S]}} \\ &= |\mathcal{S}| + |\bar{\mathcal{S}}| - (S-1) = 1. \end{aligned} \tag{18.74}$$

Beside being globally correlated, the system is also maximally correlated across 1-cuts (but not across larger cuts). Furthermore, for 1-cuts, the correlations captured by the mutual information $\mathrm{I}(X^s;X^{[S]\setminus s})_{p^{[S]}} = 1$ are of synergistic nature, since only by observing all the bits $\mathrm{x}^{[S]\setminus s}$ at once can one extract information about the remaining bit x^s.

18.6 Quantifying Multipartite Correlations

As discussed in Section 18.5, the degree of fragmentation and the scale of the correlations within a system can be quantified by the parameters $I^{\max}$ and $S^{\max}$, respectively. These parameters, though, do not capture the strength of the correlations present. To fully analyze the strength of correlations represented by the joint distribution $p^{[S]}$, one would need to evaluate the mutual information $\mathrm{I}(X^{\mathcal{S}};X^{\bar{\mathcal{S}}})_{p^{[S]}}$ across all possible cuts $\mathcal{S}$. Yet interpreting and utilizing this information is challenging, as the number of cuts grows exponentially with the number of subsystems, S. This section introduces scalar measures for assessing the strength of correlations in the system, termed multipartite information metrics.

18.6.1 Multipartite Information

A **multipartite information metric** $\mathrm{I}(X^0;X^1;\ldots;X^{S-1})_{p^{[S]}}$ is a function of the joint distribution $p^{[S]}$ that is designed to reveal specific characteristics of the correlations present in the system. Typical desirable properties for a multipartite information metric include the following.

- **Multipartite information for quantifying arbitrary correlations:** To measure the strength of any form of correlation, it is useful for the metric $\mathrm{I}(X^0;X^1;\ldots;X^{S-1})_{p^{[S]}}$ to be positive if and only if the system is correlated, i.e.,

$$\text{correlated} \Longleftrightarrow \mathrm{I}(X^0;X^1;\ldots;X^{S-1})_{p^{[S]}} > 0 \tag{18.75}$$

so that a value of zero, $\mathrm{I}(X^0;X^1;\ldots;X^{S-1})_{p[S]} = 0$, indicates a globally uncorrelated system.

- **Multipartite information for quantifying global correlations:** If the goal is to measure global correlations, the metric $\mathrm{I}(X^0;X^1;\ldots;X^{S-1})_{p[S]}$ should be positive exclusively when the system is globally correlated:

$$\text{globally correlated} \Longleftrightarrow \mathrm{I}(X^0;X^1;\ldots;X^{S-1})_{p[S]} > 0. \tag{18.76}$$

  Metrics meeting this criterion are sensitive exclusively to global correlations. In contrast, metrics that fulfill the condition (18.75) respond to any correlation, even if only local.

Metrics with these properties can be constructed in several ways. Two primary construction methods, covered in this section, are the following.

- **Cut-based multipartite information metrics:** These metrics aggregate mutual information values across various cuts to capture the strength of multipartite correlations.
- **Distance-based multipartite information metrics:** Based on a resource-theoretic framework, these metrics measure the distance to correlation-free states.

18.6.2 Cut-Based Multipartite Information

A cut-based multipartite information metric is a function that combines mutual information terms $\mathrm{I}(X^{\mathcal{S}};X^{\bar{\mathcal{S}}})_{p[S]}$ evaluated across different cuts $\mathcal{S}$ with the aim of satisfying a list of desirable properties, including (18.75) or (18.76). According to this definition, one can define a variety of cut-based multipartite information metrics by the following.

1. **Selection of cuts:** Selecting different subsets of cuts along which to evaluate the mutual informations.
2. **Combination across cuts:** Combining the per-cut mutual informations in distinct ways to capture different aspects of multipartite correlation.

To start, the upper bound (18.49) implies that, for any C-cut $\mathcal{S}$ with $|\mathcal{S}| = C$, the mutual information $\mathrm{I}(X^{\mathcal{S}};X^{\bar{\mathcal{S}}})_{p[S]}$ can be normalized to satisfy the inequalities

$$\underbrace{0}_{\text{uncorrelated across cut } \mathcal{S}} \leq \frac{\mathrm{I}(X^{\mathcal{S}};X^{\bar{\mathcal{S}}})_{p[S]}}{|\mathcal{S}|\log_2(D)} \leq \underbrace{1}_{\text{maximum correlation across cut } \mathcal{S}} . \tag{18.77}$$

Thus, the ratio $\mathrm{I}(X^{\mathcal{S}};X^{\bar{\mathcal{S}}})_{p[S]}/(|\mathcal{S}|\log_2(D))$ is a normalized measure of correlations across cut $\mathcal{S}$.

These ratios can be combined in different ways to meet the properties (18.75) or (18.76). Though not comprehensive, the following definitions exemplify this approach to define cut-based multipartite information metrics.

Definition 18.13 Cut-based multipartite information

For a joint distribution $p^{[S]}$, examples of cut-based multipartite information metrics include the following.

- **Maximum cut-based multipartite information:** The maximum cut-based multipartite information is defined as

$$\mathrm{I}_{\max}(X^0;X^1;\ldots;X^{S-1})_{p^{[S]}} = \max_{\mathcal{S}} \frac{\mathrm{I}(X^{\mathcal{S}};X^{\bar{\mathcal{S}}})_{p^{[S]}}}{|\mathcal{S}|\log_2(D)}. \tag{18.78}$$

- **Minimum cut-based multipartite information:** The minimum cut-based multipartite information is defined as

$$\mathrm{I}_{\min}(X^0;X^1;\ldots;X^{S-1})_{p^{[S]}} = \min_{\mathcal{S}} \frac{\mathrm{I}(X^{\mathcal{S}};X^{\bar{\mathcal{S}}})_{p^{[S]}}}{|\mathcal{S}|\log_2(D)}. \tag{18.79}$$

In (18.78) and (18.79), the optimizations are over all C-cuts $\mathcal{S}$ with $C \leq \left\lfloor \frac{S}{2} \right\rfloor$ (see (18.47)). Note that the metrics (18.78) and (18.79) are unit-free, since they are obtained from ratios of information measures in bits.

It can readily be verified that the metric $\mathrm{I}_{\max}(X^0;X^1;\ldots;X^{S-1})_{p^{[S]}}$ satisfies the condition (18.75), since it is positive if and only if there is correlation across *any* cut. In contrast, the metric $\mathrm{I}_{\min}(X^0;X^1;\ldots;X^{S-1})_{p^{[S]}}$ meets the requirement (18.76), given that it is positive if and only if the mutual information is non-zero across *all* cuts.

By the upper bound in (18.77), both metrics cannot exceed the maximum value 1. Furthermore, for the metric $\mathrm{I}_{\max}(X^0;X^1;\ldots;X^{S-1})_{p^{[S]}}$, this maximum value is achieved when maximal correlation is present at any one of the cuts. In contrast, the metric $\mathrm{I}_{\min}(X^0;X^1;\ldots;X^{S-1})_{p^{[S]}}$ assumes the maximum value of 1 when maximum correlation is exhibited across all cuts. In this case, the joint distribution is said to be **globally maximally correlated**.

A summary of the definitions of multipartite correlation strength classes based on the analysis of the mutual information across cuts can be found in Table 18.1. The accompanying illustration in Figure 18.12 summarizes the relationships between the different classes. As emphasized by the figure, globally maximally correlated distributions are a subset of globally correlated distributions, which are in turn included in the set of correlated distributions.

Table 18.1 Summary of multipartite correlation strength classes based on the analysis of the mutual information across cuts

Multipartite correlation type	Definition
Globally uncorrelated	$\mathrm{I}(X^s;X^{[S]\setminus s})_{p^{[S]}} = 0$ for all $s \in [S]$
Correlated	$\mathrm{I}(X^s;X^{[S]\setminus s})_{p^{[S]}} > 0$ for at least one $s \in [S]$
Globally correlated	$\mathrm{I}(X^{\mathcal{S}};X^{\bar{\mathcal{S}}})_{p^{[S]}} > 0$ for all cuts $\mathcal{S}$
Globally maximally correlated	$\mathrm{I}(X^{\mathcal{S}};X^{\bar{\mathcal{S}}})_{p^{[S]}} = |\mathcal{S}|\log_2(D)$ for all cuts $\mathcal{S}$

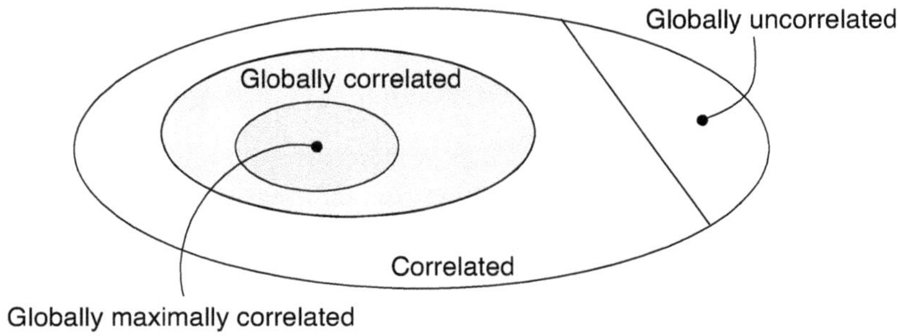

Figure 18.12 Relationships between different types of multipartite correlations (see Table 18.1 for definitions).

Example 18.28

As seen in Examples 18.17 and 18.18, both systems with three equal uniform bits and three uniform one-hot configurations – i.e., (18.10) and (18.11), respectively – correspond to globally correlated systems of three bits.

However, for the three equal uniform bits, the mutual information across each cut equals 1 bit, and hence both metrics $\mathrm{I}_{\max}(X^0;X^1;X^2)_{p[3]}$ and $\mathrm{I}_{\min}(X^0;X^1;X^2)_{p[3]}$ take the maximum value of 1 bit. The system is globally maximally correlated.

In contrast, for the three uniform one-hot configurations, the mutual information across each cut equals $\mathrm{h}(1/3)$. It follows that the system is not globally maximally correlated, and both metrics $\mathrm{I}_{\max}(X^0;X^1;X^2)_{p[3]}$ and $\mathrm{I}_{\min}(X^0;X^1;X^2)_{p[3]}$ equal $\mathrm{h}(1/3) < 1$.

Example 18.29

The joint distribution (18.43) is correlated but not globally correlated. Therefore, the maximum cut-based multipartite information is positive, while the minimum cut-based multipartite information equals zero. Specifically, the maximum cut-based multipartite information equals 1, since the mutual information equals 1 bit for all 1-cuts, and it is equal to 2 for the 2-cuts $\mathcal{S} = \{0, 2\}$ or $\mathcal{S} = \{1, 3\}$.

As a closing, non-technical, remark, examining the correlations that flow across various cuts of a complex system has been proposed as a potential indicator of the system's level of **consciousness**. This hypothesis rests on the idea that significant interactive information flows between distinct parts of a system could signal the presence of closely interlinked feedback loops – potentially indicative of conscious processing. Known as **integrated information theory**, this hypothesis has been explored extensively in the literature (see Recommended Resources).

18.6.3 Distance-Based Multipartite Information and Total Correlation

Distance-based multipartite information metrics are grounded in the resource-theoretic approach, which quantifies the strength of correlations within a joint distribution $p^{[S]}$ by measuring its distance from the set of correlation-free distributions. This subsection considers a distance-based measure designed to capture the total correlation strength across the entire system, thereby fulfilling the condition (18.75). In this context, the reference set of resource-free distributions comprises all globally uncorrelated distributions, $q^{[S]} = \otimes_{s\in[S]} q^s$, where each term q^s represents the marginal distribution of the state variable x^s.

Using the Shannon relative entropy to measure the distance between distributions, this perspective yields the following definition.

Definition 18.14 Distance-based multipartite information
Given a joint distribution $p^{[S]}$, the distance-based multipartite information is defined via the minimization

$$\mathrm{I_D}(X^0;X^1;\cdots;X^{S-1})_{p^{[S]}} = \min_{q^0,q^1,\ldots,q^{S-1}} \mathrm{D}(p^{[S]}||q^0 \otimes \cdots \otimes q^{S-1}) \quad (18.80)$$

over all possible marginal distributions $\{q^s\}_{s\in[S]}$ for all the subsystems $s \in [S]$.

By construction, the distance-based multipartite information satisfies the condition (18.75), and is thus sensitive to any correlation present in the system. Furthermore, in a manner similar to the case $S = 2$ studied in Chapter 12 (see (12.55)), one can prove the equality

$$\mathrm{I_D}(X^0;X^1;\ldots;X^{S-1})_{p^{[S]}} = \mathrm{D}(p^{[S]}||p^0 \otimes \cdots \otimes p^{S-1}), \quad (18.81)$$

which shows that the closest globally uncorrelated distribution has the same marginal distributions p^s of the original joint distribution $p^{[S]}$. This expression also makes it clear that, for $S = 2$, the distance-based multipartite information reduces to the mutual information.

Using (18.81), the distance-based multipartite information can be expressed in the following equivalent form, which is also known as total correlation.

Key Result 18.7 Total correlation
The distance-based multipartite information can equivalently be written in terms of entropy measures as the total correlation

$$\mathrm{I_D}(X^0;X^1;\ldots;X^{S-1})_{p^{[S]}} = \sum_{s\in[S]} \mathrm{H}(X^s)_{p^s} - \mathrm{H}(X^{[S]})_{p^{[S]}}. \quad (18.82)$$

Example 18.30

For the distribution (18.10) corresponding to three equal uniform bits, the total correlation equals

$$\begin{aligned} \mathrm{I_D}(X^0;X^1;X^2)_{p[3]} &= \sum_{s\in[3]} \mathrm{H}(X^s)_{p^s} - \mathrm{H}(X^{[3]})_{p[3]} \\ &= 3 - 1 = 2. \end{aligned} \tag{18.83}$$

For three uniform one-hot configurations, i.e., for the joint distribution (18.11), the total correlation is evaluated as

$$\mathrm{I_D}(X^0;X^1;X^2)_{p[3]} = 3\mathrm{h}\left(\frac{1}{3}\right) - \log_2(3) = 1.169 < 2. \tag{18.84}$$

Thus, the second distribution is less correlated than the first when using the total correlation as the correlation measure.

Example 18.31

Compare now the following two distributions over S bits via the total correlation. The first distribution assigns equal probability $1/2$ to the two configurations in which all bits are equal; while the second distribution is given by (18.72), thus assigning equal probability to all configurations having an equal parity. Both distributions are globally correlated. However, using (18.82), the total correlation for the first distribution is given by

$$\mathrm{I_D}(X^0;X^1;\ldots;X^{S-1})_{p[S]} = S - 1, \tag{18.85}$$

while for the second distribution, one gets

$$\mathrm{I_D}(X^0;X^1;\ldots;X^{S-1})_{p[S]} = S - (S-1) = 1. \tag{18.86}$$

Thus, as is intuitively the case, the first distribution, under which all bits are equal with probability 1, corresponds to a significantly more correlated system.

An equivalent perspective on the total correlation is obtained by using the notion of negentropy presented in Section 2.5.5. As a reminder, given a random variable $\mathrm{x} \sim p$ that can take N values, the negentropy $\mathrm{I}(p) = \log_2(N) - \mathrm{H}(p)$ can be interpreted as the amount of order, or structure, present in the system. The distance-based multipartite information (18.82) can be rewritten using the negentropy as the difference

$$\mathrm{I_D}(X^0;X^1;\ldots;X^{S-1})_{p[S]} = \mathrm{I}(X^{[S]})_{p[S]} - \sum_{s\in[S]} \mathrm{I}(X^s)_{p^s}. \tag{18.87}$$

This expression indicates that the distance-based multipartite information can be viewed as the overall amount of order present in the system, quantified by the negentropy $\mathrm{I}(X^{[S]})_{p[S]}$, discounted by the local order present separately in each subsystem, quantified by the sum $\sum_{s\in[S]} \mathrm{I}(X^s)_{p^s}$. Intuitively, this difference thus captures the overall correlations across the subsystems, which manifest themselves in inter-subsystem structure. Using this perspective, it is also possible to decompose the total correlation $\mathrm{I_D}(X^0;X^1;\ldots;X^{S-1})_{p^{[S]}}$ into contributions that can be ascribed to redundant information across subsystems or to synergistic information distributed across multiple subsystems (see Problem 18.7).

18.7 Summary

- The joint distribution $p^{[S]}$ of a composite system of S subsystems, with D-dimensional states, is described by an order-S tensor with D^S entries $p_{x^{[S]}}$ for $x^{[S]} \in [D]^S$.
- The tensor can be represented in vectorized form as the $D^S \times 1$ vector $p^{[S]} = \sum_{x^{[S]}} p_{x^{[S]}} |x^{[S]}\rangle$.
- Tensor representations for marginal distributions are obtained from the tensor network representing the joint distribution by adding self-bonds on the indices of the variables being marginalized.
- For any subset $\mathcal{S} \subseteq [S]$, the joint entropy $\mathrm{H}(X^{\mathcal{S}})_{p^{\mathcal{S}}}$ measures the uncertainty associated with the group of subsystems identified by $\mathcal{S}$.
- By the subadditivity and additivity properties of the Shannon entropy, the joint entropy $\mathrm{H}(X^{\mathcal{S}})_{p^{\mathcal{S}}}$ is upper bounded by the sum of the marginal entropies $\sum_{s\in\mathcal{S}} \mathrm{H}(X^s)_{p^s}$, with equality attained if the subsystems in group $\mathcal{S}$ are uncorrelated. As a result, the joint entropy of a group of subsystems grows at most linearly with the number of subsystems in group $\mathcal{S}$, i.e., as $|\mathcal{S}|\log_2(D)$.
- Given a composite system, the most basic way to quantify correlations is by considering a cut that partitions the system into two disjoint groups of subsystems, namely $\mathcal{S} \subset [S]$ and its complement $\bar{\mathcal{S}} = [S]\setminus\mathcal{S}$.
- A C-cut is a cut in which the smallest cardinality of the two subsets $\mathcal{S}$ and $\bar{\mathcal{S}}$ equals C.
- The bipartite correlations across cut $\mathcal{S}$ under a joint distribution $p^{[S]}$ can be quantified by the mutual information $\mathrm{I}(X^{\mathcal{S}};X^{\bar{\mathcal{S}}})_{p^{[S]}}$.
- The mutual information $\mathrm{I}(X^{\mathcal{S}};X^{\bar{\mathcal{S}}})_{p^{[S]}}$ equals zero if and only if random variables $\mathrm{x}^{\mathcal{S}}$ and $\mathrm{x}^{\bar{\mathcal{S}}}$ are uncorrelated, i.e., if and only if the joint distribution can be written as $p^{\mathcal{S},\bar{\mathcal{S}}} = p^{\mathcal{S}} \otimes p^{\bar{\mathcal{S}}}$.
- The mutual information $\mathrm{I}(X^{\mathcal{S}};X^{\bar{\mathcal{S}}})_{p^{[S]}}$ can grow at most linearly with the cut size C and at most logarithmically with the per-subsystem dimension D, i.e., as $C\log_2(D)$.
- When the equality $\mathrm{I}(X^{\mathcal{S}};X^{\bar{\mathcal{S}}})_{p^{[S]}} = C\log_2(D)$ holds, the system is said to be maximally correlated across cut $\mathcal{S}$ under distribution $p^{[S]}$.
- The mutual information $\mathrm{I}(X, Y;Z)$ can be decomposed as the sum of four separate contributions: the shared information $\mathrm{Sh}(X, Y;Z)$ representing the information about Z that can be obtained from either X or Y; the unique informations $\mathrm{Un}(X;Z|Y)$ and $\mathrm{Un}(Y;Z|X)$ representing the information about Z that can be

obtained only from X and from Y, respectively; and the synergistic information $\mathrm{Sy}(X, Y;Z)$ representing the information about Z that can only be obtained through a joint consideration of both X and Y.

- A system is said to be $I^{\max}$-uncorrelated under distribution $p^{[S]}$ if it can be partitioned into $1 \leq I^{\max} \leq S$ mutually uncorrelated groups of subsystems, but not into $I^{\max}+1$ mutually uncorrelated subsets.
- A system is said to be $S^{\max}$-partite correlated under the joint distribution $p^{[S]}$ if (i) there exists at least one subset of $S^{\max} \leq S$ subsystems that is globally correlated, and (ii) no larger globally correlated subset exists.
- To quantify the strength of correlations, there are two main types of multipartite information metrics: (i) cut-based multipartite information measures, which combine bipartite informations corresponding to different cuts; and (ii) distance-based multipartite information measures, which quantify the distance from correlation-free joint distributions.
- An important distance-based multipartite information known as total correlation is defined as $\mathrm{I_D}(X^0;X^1;\ldots;X^{S-1})_{p^{[S]}} = \sum_{s\in[S]} \mathrm{H}(X^s)_{p^s} - \mathrm{H}(X^{[S]})_{p^{[S]}}$.

18.8 Recommended Resources

The reader can consult [1] for a detailed discussion on the information decomposition. Integrated information theory is comprehensively explored in [2, 3], providing both foundational insights and advanced frameworks. The concept of total correlation is examined in [4, 5], while broader multivariate measures of correlations are elaborated in [6, 7]. The definition of shared information, as discussed in Appendix 18.A, is based on the framework introduced in [8]. An analysis of cut-based mutual information measures for large language models can be found in [9].

PROBLEMS

18.1 For $S = 4$ and $D = 2$, i.e., for a system of 4 bits, consider the joint distribution $p^{[4]}$ under which one has the equalities $\mathrm{x}^0 = \mathrm{x}^2$ and $\mathrm{x}^1 = \mathrm{x}^3$ with random variables x^0 and x^1 being uncorrelated and uniformly distributed. (a) Write the joint probability distribution $p^{[4]}$ as a vector. (b) Express the joint distribution using the computational basis vectors. (c) Evaluate the marginal distribution $p^{0,1}$ over subsystems $\{0, 1\}$. (d) Draw the tensor network representation of the marginal distribution $p^{0,1}$. (e) For what value of $I^{\max}$ is the joint distribution $p^{[4]}$ $I^{\max}$-uncorrelated? (f) For what value of $S^{\max}$ is the joint distribution $p^{[4]}$ $S^{\max}$-partite correlated?

18.2 (a) Prove that for $C < \lfloor N/2 \rfloor$, the number of C-cuts for a system with S subsystems is given by $\binom{S}{C}$. (b) For $C = N/2$, which is applicable if N is even, show that the number of C-cuts is $\binom{S}{C}/2$. (c) What is the total number of cuts when considering all possible values of C?

18.3 Consider the joint distribution over 4 bits given by

$$p^{[4]} = \frac{3}{8}|0000\rangle + \frac{1}{8}|0111\rangle + \frac{3}{8}|1011\rangle + \frac{1}{8}|1100\rangle. \qquad (18.88)$$

(a) Evaluate the marginal distribution of subsystems 0 and 1. (b) Calculate the marginal distribution of subsystems 2 and 3. (c) Is the system globally uncorrelated? (d) Evaluate the joint entropy $\mathrm{H}(X^{[4]})_{p^{[4]}}$. (e) Compute the mutual information between 0 and subsystems 3. (f) Calculate the conditional entropy $\mathrm{H}(X^3|X^0)_{p^{0,3}}$. (g) Is the joint distribution $p^{[4]}$ maximally correlated across at least one of the cuts?

18.4 Consider three uncorrelated and uniformly distributed bits $\mathrm{b}^{[3]}$. The three subsystems of interest have states $\mathrm{x} = (\mathrm{b}^0 \oplus \mathrm{b}^1, \mathrm{b}^2)$, $\mathrm{y} = \mathrm{b}^1$, and $\mathrm{z} = \mathrm{b}^0$. (a) Evaluate the mutual information $\mathrm{I}(X, Y;Z)_{p^{XYZ}}$. (b) Do X and Y have any synergistic information about Z? (c) Is the system globally correlated?

18.5 Given five uncorrelated and uniformly distributed bits $\mathrm{b}^{[5]}$, take a system with $S = 4$ subsystems in which subsystem 0 holds the bits $\mathrm{x}^0 = (\mathrm{b}^0, \mathrm{b}^1)$; subsystem 1 holds the bits $\mathrm{x}^1 = (\mathrm{b}^1, \mathrm{b}^2)$; subsystem 2 has the bits $\mathrm{x}^3 = (\mathrm{b}^3, \mathrm{b}^4)$; and subsystem 3 has the bits $\mathrm{x}^3 = \mathrm{b}^4$. (a) For what value of $I^{\max}$ is the system $I^{\max}$-uncorrelated? (b) For what value of $S^{\max}$ is the system $S^{\max}$-partite correlated? (c) What are the values of $I^{\max}$ and $S^{\max}$ when x^3 is changed to encompass the variables $\mathrm{x}^3 = (\mathrm{b}^0, \mathrm{b}^4)$?

18.6 (a) Prove the equality

$$\sum_{s\in[S]} \mathrm{H}(X^s)_{p^s} - \mathrm{H}(X^{[S]})_{p^{[S]}} = \sum_{s\in[S]} \mathrm{I}(X^s;X^{[s]})_{p^{[s+1]}}. \qquad (18.89)$$

(b) Show that this equality applies for any reordering of the subsystems in the right-hand side.

18.7 (a) Express the alternative cut-based multipartite information measure

$$\tilde{\mathrm{I}}(X^0;X^1;\ldots;X^{S-1})_{p^{[S]}} = \sum_{s\in[S]} \mathrm{I}(X^s;X^{[S]\setminus s})_{p^{[S]}} \qquad (18.90)$$

as

$$\sum_{s\in[S]} \mathrm{I}(X^s;X^{[S]\setminus s})_{p^{[S]}} = \sum_{s\in[S]} \mathrm{I}(X^s;X^{\{s+1,\ldots,S-1\}})_{p^{s,s+1,\ldots,S-1}} + \sum_{s\in[S]} \mathrm{I}(X^s;X^{[s-1]}|X^{\{s+1,\ldots,S-1\}})_{p^{[S]}}. \qquad (18.91)$$

(b) Interpret the first term on the right-hand side as redundant information and the second term as synergistic information when treating the subsystem index s as a discrete time index.

18.8 Consider a system consisting of any number, S, of bits ($D = 2$), for which the bits $\mathrm{x}^1, \ldots, \mathrm{x}^{S-1}$ are uncorrelated and uniformly distributed, while random variable x^0 is the logical OR of $\mathrm{x}^1, \ldots, \mathrm{x}^{S-1}$. The logical OR returns 1 as long as any of the input variables equals 1, while returning 0 otherwise. (a) Find the values of $I^{\max}$ and $S^{\max}$ for the resulting joint distribution. (b) Calculate the mutual information $\mathrm{I}(X^0;X^{1,\ldots,S-1})_{p^{[S]}}$ across the cut $\mathcal{S} = \{0\}$.

18.9 For the joint distribution described in Problem 18.8 with $S = 3$, calculate the total correlation.

18.10 In a neuroscience study, the firing rates of two neurons X and Y are measured in response to a common stimulus Z. The following entropies are estimated from data: $\mathrm{H}(X|Z)_{p^{XZ}} = 1.5$, $\mathrm{H}(Y|Z)_{p^{YZ}} = 2$, and $\mathrm{H}(X, Y|Z)_{p^{XYZ}} = 3$. (a) Calculate the conditional mutual information $\mathrm{I}(X;Y|Z)_{p^{XYZ}}$ and interpret its significance in terms of shared information between neurons X and Y given the stimulus Z. (b) If the entropy of the stimulus $\mathrm{H}(Z)_{p^Z} = 1.8$, determine the mutual information $\mathrm{I}(X, Y;Z)_{p^{XYZ}}$ between the neurons and the stimulus.

18.11 Consider the joint distribution

$$p^{[3]} = \frac{1}{6}\left(|012\rangle + |021\rangle + |102\rangle + |120\rangle + |201\rangle + |210\rangle\right), \tag{18.92}$$

for a system with $S = 3$ subsystems, each having dimension equal to $D = 3$. This joint distribution assigns equal probability to all permutations of configurations in which each subsystem occupies a distinct state in the set [3]. (a) Evaluate the total correlation $\mathrm{I_D}(X^0;X^1;X^2)_{p^{[3]}}$. (b) Is the total correlation maximal? *Hint:* Evaluate the total correlation for a joint distribution in which the three subsystems are marginally uniform and assume the same state with probability 1. (c) Is the system globally correlated? (d) Is it globally maximally correlated? Conclude that a distribution can be globally maximally correlated, while not exhibiting the maximum value for the total correlation.

Appendix

18.A A Definition of Shared Information

Define

$$\mathrm{I}(X;z) = -\log_2(p_z^Z) + \mathrm{E}_{p_z^{X|Z}}[\log_2(p_{\mathrm{x}}^{Z|X})] \tag{18.93}$$

as a measure of the information about the specific value z carried by the subsystem X. The intuition behind this definition is that the average yields $\mathrm{E}_{p^Z}[\mathrm{I}(X;z)] = \mathrm{H}(Z) - \mathrm{H}(Z|X) = \mathrm{I}(X;Z)$. Note that the subscripts are dropped from the information measures to streamline the notation. Define in a similar way

$$\mathrm{I}(Y;z) = -\log_2(p_z^Z) + \mathrm{E}_{p_z^{Y|Z}}[\log_2(p_{\mathrm{y}}^{Z|Y})] \tag{18.94}$$

with $\mathrm{E}_{p^Z}[\mathrm{I}(Y;z)] = \mathrm{I}(Y;Z)$. Then, a possible definition of the shared information is

$$\mathrm{Sh}(XY;Z) = \mathrm{E}_{p^Z}[\min(\mathrm{I}(X;z), \mathrm{I}(X;z))]. \tag{18.95}$$

Intuitively, for each value z, this measure accounts for the minimum value of the information brought by either variable x or y.

REFERENCES

[1] R. G. James and J. P. Crutchfield, "Multivariate dependence beyond Shannon information," *Entropy*, vol. 19, no. 10, art. 531, 2017.

[2] G. Tononi, "Consciousness as integrated information: A provisional manifesto," *The Biological Bulletin*, vol. 215, no. 3, pp. 216–242, 2008.

[3] M. Oizumi, N. Tsuchiya, and S.-i. Amari, "Unified framework for information integration based on information geometry," *Proceedings of the National Academy of Sciences*, vol. 113, no. 51, pp. 14 817–14 822, 2016.

[4] S. Watanabe, "Information theoretical analysis of multivariate correlation," *IBM Journal of Research and Development*, vol. 4, no. 1, pp. 66–82, 1960.

[5] M. Studeny and J. Vejnarova, "The multiinformation function as a tool for measuring stochastic dependence," in M. I. Jordan (Ed.), *Learning in Graphical Models*. New York: Springer, 1998, pp. 261–297.

[6] F. E. Rosas, P. A. M. Mediano, M. Gastpar, and H. J. Jensen, "Quantifying high-order interdependencies via multivariate extensions of the mutual information," *Physical Review E*, vol. 100, no. 3, art. 032305, 2019.

[7] H. J. Jensen, *Complexity Science: The Study of Emergence*. Cambridge: Cambridge University Press, 2022.

[8] P. L. Williams and R. D. Beer, "Nonnegative decomposition of multivariate information," *arXiv*, 2010. [Online]. Available at: https://arXiv:1004.2515

[9] Z. Chen, O. M. i Comas, Z. Jin, D. Luo, and M. Soljacic, "L^2M: Mutual information scaling law for long-context language modeling," *arXiv*, 2025. [Online]. Available: https://arxiv.org/abs/2503.04725

19 Multipartite Classical Information: Structure

19.1 Overview

Chapter 18 analyzed the fragmentation, scale, and strength of correlations in multipartite systems. This chapter leverages the tensor formalism to elaborate on the structure of multipartite correlations.

As will be discussed, the structure of multipartite correlations can be specified by introducing additional variables alongside the system variables to mediate correlations across different parts of the system. The joint distribution is then derived by marginalizing over these added variables, which are known as latent random variables.

To provide an example of the operation of latent variable models, consider the problem of modeling the correlations between weather conditions in two neighboring geographical areas. These correlations can be accounted for via the introduction of latent variables describing physical quantities such as atmospheric pressure and humidity levels.

At a technical level, the relations between the random variables within the system and the latent variables are described by conditional independence properties. Accordingly, this chapter will discuss first how to measure conditional correlations, that is, correlations between subsystems that arise once the information about the state of other subsystems has been accounted for. Then, it will cover tensor network representations of joint distributions encompassing global and local latent variables, and present ways to evaluate the correlations present in such systems. This discussion will introduce important correlation models such as canonical polyadic decompositions and matrix product distributions.

19.1.1 Learning Objectives and Organization

By the end of this chapter, the reader should be able to:

- understand and use the definitions of conditional mutual information and conditional independence;
- build a tensor network representation for an arbitrary joint distribution that factorizes into smaller tensors, possibly including latent variables;
- understand models with global and local latent variables;

- and establish bounds on the mutual information in systems with global and local latent random variables.

19.2 Conditional Correlations

Chapter 18 examined correlations by using the mutual information between two groups of subsystems. However, when multiple subsystems are involved, a new scenario emerges: correlations between two groups of subsystems can be evaluated *after* observing the states of a third group of subsystems.

19.2.1 Reviewing Conditional Distributions and Conditional Entropy

Measures of information evaluated after observing a subsystem rely on the concepts of conditional distribution and conditional entropy. Thus, to set the notation, this subsection recalls the definition of the conditional entropy from Section 13.5, as well as the definition of conditional distribution from Section 13.2.

To this end, consider any number of jointly distributed random variables divided into two disjoint groups A and B. As described in Section 13.5, the joint entropy $\mathrm{H}(A,B)_{p^{AB}}$ satisfies the chain rule

$$\mathrm{H}(A,B)_{p^{AB}} = \mathrm{H}(A)_{p^{A}} + \mathrm{H}(B|A)_{p^{AB}}, \tag{19.1}$$

where the conditional entropy $\mathrm{H}(B|A)_{p^{AB}} = \mathrm{H}(A,B)_{p^{AB}} - \mathrm{H}(A)_{p^{A}}$ quantifies the residual uncertainty about group B once the states of group A are known. The conditional entropy $\mathrm{H}(B|A)_{p^{AB}}$ equals the entropy $\mathrm{H}(B)_{p^{B}}$ if and only if the groups of random variables A and B are independent, i.e., the joint distribution factorizes as $p^{AB} = p^{A} \otimes p^{B}$.

Denote as $\mathcal{A}$ and $\mathcal{B}$ the set of possible values that can be assumed by the random variables in groups A and B, respectively. As defined in Section 13.2, the conditional distribution of group B given group A is evaluated as

$$p^{B|A}_{b|a} = \frac{p^{AB}_{ab}}{p^{A}_{a}} \tag{19.2}$$

for all $a \in \mathcal{A}$ and $b \in \mathcal{B}$. When fixing any value $a \in \mathcal{A}$, the tensor $p^{B|A}_{a}$ with entries $\{p^{B|A}_{b|a}\}_{b\in\mathcal{B}}$ is a valid probability distribution, i.e., $\sum_{b\in\mathcal{B}} p^{B|A}_{b|a} = 1$ for all $a \in \mathcal{A}$.

As presented in Section 13.3, the conditional entropy $\mathrm{H}(B|A)_{p^{AB}}$ can be expressed as the expectation

$$\begin{aligned}\mathrm{H}(B|A)_{p^{AB}} &= \mathrm{E}_{p^{A}}[\mathrm{H}(B)_{p^{B|A}_{\mathrm{a}}}] \\ &= -\mathrm{E}_{p^{AB}}[\log_2(p^{B|A}_{\mathrm{b|a}})],\end{aligned} \tag{19.3}$$

i.e., as the average of the entropy $\mathrm{H}(B)_{p^{B|A}_{a}}$ of the conditional distribution $p^{B|A}_{a}$.

19.2.2 Defining the Conditional Mutual Information

In order to introduce the key measure of conditional correlations – the conditional mutual information – it is useful to begin with a few examples.

Example 19.1

Imagine you are an old-fashioned pollster going door to door to determine a person X's likelihood of voting for a specific party. When you knock on X's door, a family member, Y, answers, and you record her political preference. With this information, you might now predict X's political leaning with some accuracy, even before speaking to X directly. For instance, you might assume – correctly or not – that X and Y's opinions are more likely to align than to differ, as described by some joint distribution p^{XY}. The correlation between their voting tendencies can be quantified by the mutual information $\mathrm{I}(X;Y)_{p^{XY}}$, based on the assumed joint distribution p^{XY}.

Now, suppose that before knocking on the door, you already have access to some information Z. For example, you might know that X lives in a neighborhood with a longstanding preference for a particular party. In this case, you may expect less additional insight from Y's preference when predicting X's stance. Consequently, the mutual information between X and Y conditioned on Z – denoted by $\mathrm{I}(X;Y|Z)_{p^{XYZ}}$ – is lower

The **conditional mutual information** $\mathrm{I}(X;Y|Z)_{p^{XYZ}}$, which is based on the joint distribution p^{XYZ} of all three variables, reflects the correlation between X and Y after accounting for knowledge of Z. While in this example the conditional mutual information $\mathrm{I}(X;Y|Z)_{p^{XYZ}}$ is smaller than $\mathrm{I}(X;Y)_{p^{XY}}$, representing a **residual correlation**, this is not always the case. In fact, the conditional mutual information $\mathrm{I}(X;Y|Z)_{p^{XYZ}}$ can be either larger or smaller than $\mathrm{I}(X;Y)_{p^{XY}}$, depending on the relation between Z and the pair (X, Y).

For instance, suppose you acquire, before knocking, the information Z that the two household members are split across different political parties. With this information, once you know Y's preference, you can deduce X's stance with certainty. As a result, the conditional mutual information $\mathrm{I}(X;Y|Z)_{p^{XYZ}}$ exceeds the mutual information $\mathrm{I}(X;Y)_{p^{XY}}$.

Example 19.2

For a numerical example, consider two random bits x and y that are independent and uniformly distributed, along with a third bit, z, which equals the XOR of x and y, i.e.,

$$\mathrm{z} = \mathrm{x} \oplus \mathrm{y}. \tag{19.4}$$

Since bits x and y are independent, their mutual information is zero, i.e., $\mathrm{I}(X;Y)_{p^{XY}} = 0$.

However, if we know the value of the bit z, the bits x and y become maximally correlated. In fact, if z is known, having x allows one to recover y as $y = x \oplus z$. Therefore, the conditional mutual information $I(X;Y|Z)_{p^{XYZ}}$ takes the maximum value of 1 bit, since y becomes a function of x when z is known.

Example 19.3

As another numerical example, consider a situation in which the random bit z is uniformly distributed, while bits x and y are equal to z:

$$x = y = z. \tag{19.5}$$

Variables x and y are maximally correlated, since knowing one bit identifies the other bit with certainty, attaining the maximal mutual information $I(X;Y)_{p^{XY}} = 1$. However, when z is known, there is no longer any uncertainty left in either variable x or y, and thus the expected information gain conditioned on the state variable z is zero, i.e., $I(X;Y|Z)_{p^{XYZ}} = 0$.

Note that, while in the previous example conditioning increased the mutual information, here the conditional mutual information is smaller than the mutual information.

The previous examples used an informal, intuitive, definition of conditional mutual information. Consider now three generic subsystems X, Y, and Z whose respective states $x \in [N^X]$, $y \in [N^Y]$, and $z \in [N^Z]$ have joint distribution p^{XYZ}. A formal definition of the conditional mutual information follows.

Definition 19.1 Conditional mutual information

Given a joint distribution p^{XYZ}, the conditional mutual information between subsystem X and Y given subsystem Z is defined as

$$I(X;Y|Z)_{p^{XYZ}} = H(X|Z)_{p^{XZ}} + H(Y|Z)_{p^{YZ}} - H(X, Y|Z)_{p^{XYZ}}, \tag{19.6}$$

where p^{XZ} and p^{YZ} represent the respective marginal distributions obtained from the joint distribution p^{XYZ}.

By its definition (19.6), the conditional mutual information $I(X;Y|Z)_{p^{XYZ}}$ captures the discrepancy between the sum of the residual local uncertainties, $H(X|Z)_{p^{XZ}} + H(Y|Z)_{p^{YZ}}$, and the global residual uncertainty $H(X, Y|Z)_{p^{XYZ}}$, when the state of subsystem Z is known. Informally, the conditional mutual information is thus positive if subsystems X and Y have non-negligible redundant information when Z is known.

Using the chain rule (19.1), one can equivalently express (19.6) by using conventional entropies, rather than conditional entropies, as

$$\mathrm{I}(X;Y|Z)_{p^{XYZ}} = \mathrm{H}(X,Z)_{p^{XZ}} + \mathrm{H}(Y,Z)_{p^{YZ}} - \mathrm{H}(X,Y,Z)_{p^{XYZ}} - \mathrm{H}(Z)_{p^{Z}}. \tag{19.7}$$

This expression, while less intuitive, can be useful when evaluating the conditional mutual information, as demonstrated in the following examples.

Example 19.4

Consider now the two distributions over three bits studied informally in Examples 19.2 and 19.3. In the first case, bits X and Y are uniformly distributed, while bit Z is related to X and Y as in (19.4), yielding the joint distribution

$$p^{XYZ} = \frac{1}{4}|000\rangle + \frac{1}{4}|110\rangle + \frac{1}{4}|011\rangle + \frac{1}{4}|101\rangle, \tag{19.8}$$

which assigns equal probability to all even-parity configurations. In contrast, the second setting assumes that the three bits are uniformly distributed and equal with probability 1 as in (19.5), corresponding to the joint distribution

$$p^{XYZ} = \frac{1}{2}|000\rangle + \frac{1}{2}|111\rangle. \tag{19.9}$$

Using the definition (19.7), the conditional mutual information for the even-parity distribution is evaluated as

$$\mathrm{I}(X;Y|Z)_{p^{XYZ}} = 2 + 2 - 2 - 1 = 1, \tag{19.10}$$

since the joint distribution p^{XYZ} has uniform marginal distributions p^{XZ}, p^{YZ}, and p^{Z}, yielding the entropies $\mathrm{H}(X,Z)_{p^{XZ}} = \mathrm{H}(Y,Z)_{p^{YZ}} = 2$ and $\mathrm{H}(Z)_{p^{Z}} = 1$, and the joint distribution p^{XYZ} has joint entropy $\mathrm{H}(X,Y,Z)_{p^{XYZ}} = 2$. Thus, although bits x and y are uncorrelated, knowing z makes x and y maximally correlated.

For the equal-bit distribution p^{XYZ}, the conditional mutual information is given as

$$\mathrm{I}(X;Y|Z)_{p^{XYZ}} = 1 + 1 - 1 - 1 = 0, \tag{19.11}$$

since all the entropies in the definition (19.7) equal 1 bit. Thus, in this case, although bits x and y are maximally correlated, knowing z renders x and y uncorrelated.

As seen in Section 13.5, the mutual information not only captures redundant information present across two subsystems, but also expected information gains. In particular, the mutual information can be written as $\mathrm{I}(X;Y)_{p^{XY}} = \mathrm{H}(Y)_{p^{Y}} - \mathrm{H}(Y|X)_{p^{XY}}$, which measures the expected information gain about the state of subsystem Y given the observation of subsystem X. By symmetry of the mutual information, one also has the reverse equality $\mathrm{I}(X;Y)_{p^{XY}} = \mathrm{H}(X)_{p^{X}} - \mathrm{H}(X|Y)_{p^{XY}}$, reflecting the interpretation of the mutual information as the expected information gain about the state of subsystem X given the observation of subsystem Y.

By using the chain rule of the joint entropy in (19.7) on the definition (19.6), a similar interpretation can be extended to the conditional mutual information (see Problem 19.1).

Key Result 19.1 Conditional mutual information via conditional entropies
The conditional mutual information (19.7) can be equivalently expressed as

$$\mathrm{I}(X;Y|Z)_{p_{XYZ}} = \mathrm{H}(Y|Z)_{p_{YZ}} - \mathrm{H}(Y|X,Z)_{p_{XYZ}}, \tag{19.12}$$

and as

$$\mathrm{I}(X;Y|Z)_{p_{XYZ}} = \mathrm{H}(X|Z)_{p_{XZ}} - \mathrm{H}(X|Y,Z)_{p_{XYZ}}. \tag{19.13}$$

According to the equivalent formulation (19.12), the conditional mutual information $\mathrm{I}(X;Y|Z)_{p_{XYZ}}$ measures the expected information gain about the state of subsystem Y obtained by observing the state of subsystem X, given that the state of subsystem Z is already known. In fact, the conditional entropy $\mathrm{H}(Y|Z)_{p_{YZ}}$ represents the residual uncertainty about Y when Z is known, while the conditional entropy $\mathrm{H}(Y|X,Z)_{p_{XYZ}}$ quantifies the residual uncertainty about Y upon learning about X when Z is known. A symmetric interpretation applies to (19.13) by swapping the roles of the two subsystems.

Example 19.5

For the joint distribution (19.8), which assigns an equal probability to all bit sequences with an even parity, the bits Y and Z are independent, yielding the equality $\mathrm{H}(Y|Z)_{p_{YZ}} = \mathrm{H}(Y)_{p_Y} = 1$, while the state y is a function of the states (x, z), which implies the equality $\mathrm{H}(Y|X,Z)_{p_{XYZ}} = 0$. Overall, using (19.12), the mutual information is confirmed to be equal to 1 bit:

$$\mathrm{I}(X;Y|Z)_{p_{XYZ}} = 1 - 0 = 1. \tag{19.14}$$

19.2.3 Non-Negativity of the Conditional Mutual Information

As proved in Appendix 19.A, the conditional mutual information is non-negative.

Key Result 19.2 Non-negativity of the conditional mutual information
The conditional mutual information is non-negative

$$\mathrm{I}(X;Y|Z)_{p_{XYZ}} \geq 0. \tag{19.15}$$

Using the equivalent definitions (19.7) and (19.13) of the conditional mutual information, the inequality (19.15) is equivalent to the following two important properties.

- **Strong subadditivity of the entropy:** The joint entropy satisfies the inequality

$$\mathrm{H}(X,Y,Z)_{p_{XYZ}} \leq \mathrm{H}(X,Z)_{p_{XZ}} + \mathrm{H}(Y,Z)_{p_{YZ}} - \mathrm{H}(Z)_{p_Z}. \tag{19.16}$$

This inequality generalizes the subadditivity property (12.32) presented in Section 12.5, i.e., $\mathrm{H}(X, Y)_{p^{XY}} \leq \mathrm{H}(X)_{p^X} + \mathrm{H}(Y)_{p^Y}$, which is recovered when subsystem Z is uncorrelated with X and Y.

- **Further conditioning reduces entropy:** The conditional entropy satisfies the inequality

$$\mathrm{H}(X|Y,Z)_{p^{XYZ}} \leq \mathrm{H}(X|Z)_{p^{XZ}}. \tag{19.17}$$

The condition (19.17) is a generalization of the conditioning-reduces-entropy property (13.57) introduced in Section 13.5. It states that adding the conditioning on Y once one has already conditioned on Z cannot increase the resulting residual uncertainty.

19.2.4 Conditionally Uncorrelated Subsystems

When is the lower bound in (19.15) achieved with equality, i.e., $\mathrm{I}(X;Y|Z)_{p^{XYZ}} = 0$? This is intuitively true when all the redundant information present across subsystems X and Y is accounted for by subsystem Z. In other words, this occurs when learning about X carries no information about Y when one already knows the state of subsystem Z. In this case, subsystems X and Y are said to be **conditionally uncorrelated** given Z.

To formalize this definition, write as

$$p^{XY|Z}_{xy|z} = \frac{p^{XYZ}_{xyz}}{p^Z_z}, \tag{19.18}$$

with $x \in [N^X]$ and $y \in [N^Y]$, the conditional distribution of X and Y given Z for a given value $\mathrm{z} = z \in [N^Z]$, where p^Z is the marginal distribution of subsystem Z. Note that, for a fixed value $z \in [N^Z]$, the entries $p^{XY|Z}_{xy|z}$ form an order-2 tensor $p^{XY|Z}_z$ with indices $x \in [N^X]$ and $y \in [N^Y]$.

Definition 19.2 Conditionally uncorrelated, or conditionally independent, subsystems

For a given joint distribution p^{XYZ}, subsystems X and Y are said to be conditionally uncorrelated given Z if the conditional distribution $p^{XY|Z}_{xy|z}$ of subsystems X and Y given Z factorizes as

$$p^{XY|Z}_{xy|z} = p^{X|Z}_{x|z} \cdot p^{Y|Z}_{y|z} \tag{19.19}$$

for all $x \in [N^X]$, $y \in [N^Y]$, and $z \in [N^Z]$. Equivalently, this happens if and only if, for any fixed $z \in [N^Z]$, the tensor $p^{XY|Z}_z$ can be expressed as the tensor product

$$p^{XY|Z}_z = p^{X|Z}_z \otimes p^{Y|Z}_z. \tag{19.20}$$

Key Result 19.3 Minimum value of the conditional mutual information
The conditional mutual information equals zero, $\mathrm{I}(X;Y|Z)_{p^{XYZ}} = 0$, if and only if the subsystems X and Y are conditionally uncorrelated given subsystem Z.

This result can readily be shown using the original definition (19.6) of the conditional mutual information by noting that the equality (19.19) is equivalent to the condition

$$\mathrm{H}(X, Y|Z)_{p^{XYZ}} = -\mathrm{E}_{p^{XYZ}}[\log_2(p^{XY|Z}_{\mathrm{xy|z}})] \\ = \mathrm{H}(X|Z)_{p^{XZ}} + \mathrm{H}(Y|Z)_{p^{YZ}}. \tag{19.21}$$

An equivalent definition of conditionally uncorrelated subsystems can be given in terms of the conditional distribution of one subsystem, say X, given the other two, Y and Z.

Key Result 19.4 Conditionally uncorrelated subsystems (equivalent definition)
Subsystems X and Y are conditionally uncorrelated given Z if and only if the conditional distribution $p^{X|YZ}_{x|yz} = p^{XYZ}_{xyz}/p^{YZ}_{yz}$ of subsystems X given Y and Z depends only on the state of subsystem Z, i.e.,

$$p^{X|YZ}_{x|yz} = p^{X|Z}_{x|z} \tag{19.22}$$

for all $x \in [N^X]$, $y \in [N^Y]$, and $z \in [N^Z]$.

This equivalence can be checked by using the chain rule $p^{YZ}_{yz} = p^Z_z \cdot p^{Y|Z}_{y|z}$ to obtain (19.22).

19.2.5 Chain Rule of the Mutual Information

As demonstrated in Appendix 19.B, the conditional mutual information supports a chain rule for the mutual information.

Key Result 19.5 Chain rule for the mutual information
Given a joint distribution p^{XYZ}, the mutual information $\mathrm{I}(X;YZ)_{p^{XYZ}}$ can be expressed as the sum

$$\mathrm{I}(X;Y,Z)_{p^{XYZ}} = \mathrm{I}(X;Y)_{p^{XY}} + \mathrm{I}(X;Z|Y)_{p^{XYZ}}. \tag{19.23}$$

Accordingly, the mutual information $\mathrm{I}(X;YZ)_{p^{XYZ}}$ between subsystem X and the pair of subsystems Y and Z equals the sum of the mutual information between X and Y, i.e., $\mathrm{I}(X;Y)_{p^{XY}}$, and of the conditional mutual information between X and

Z given Y, $\mathrm{I}(X;Z|Y)_{p^{XYZ}}$. Alternatively, one can also exchange the roles of Y and Z to obtain the equality

$$\mathrm{I}(X;Y,Z)_{p^{XYZ}} = \mathrm{I}(X;Z)_{p^{XZ}} + \mathrm{I}(X;Y|Z)_{p^{XYZ}}. \tag{19.24}$$

19.3 Structured Joint Probability Distributions as Tensor Networks

Chapter 18 examined the fragmentation of systems into groups of uncorrelated subsystems. Joint distributions that cannot be divided into uncorrelated groups are referred to as globally correlated. However, identifying a joint distribution as globally correlated offers little insight into the structure of these global correlations.

Consider, for example, a scenario where the subsystems represent distinct geographical regions, with state variables corresponding to weather conditions. While weather patterns are often globally correlated – since no region is entirely isolated – the structure of these correlations can vary. For instance, the weather in the eastern part of a country might be influenced by conditions in the west, but only through the mediation of meteorological factors in central regions. Such dependencies suggest that some groups of subsystems might be uncorrelated with others when conditioned on a third subset of subsystems.

Conditional independence properties like these reveal structural relationships within a system's correlations. This section introduces a tensor network-based representation of joint distributions that is designed to capture and express conditional independence properties. This representation hinges on the factorization of a joint distribution according to a Markov network model. Throughout the rest of this chapter, for notational simplicity, subscripts alone will henceforth denote relevant subsystems. For instance, the entries $p_{x^{\mathcal{S}}}$ refer to the tensor $p^{\mathcal{S}}$, and the entries $p_{x^s|x^{0,\ldots,s-1}}$ refer to the conditional distribution tensor $p^{s|0,\ldots,s-1}_{x^s|x^{0,\ldots,s-1}}$.

19.3.1 Markov Network Model

A Markov network model represents a joint distribution as a product of smaller tensors, also referred to as factors. By structuring distributions this way, Markov network models extend the concept of $I^{\max}$-uncorrelated distributions, where a joint distribution is represented as a tensor product of tensors with separate indices. In a more general Markov network model, constituent tensors can share indices, allowing the model to capture a broader range of structured correlations.

To formalize the definition of Markov network models, fix a collection of I groups of subsystems, $\mathcal{S}^i \subseteq [S]$ with $i \in [I]$, under the only requirement that the union of the subsets encompasses the entire system $[S]$, i.e., $\bigcup_{i\in[I]} \mathcal{S}^i = [S]$. The subsets $\{\mathcal{S}^i\}_{i\in[I]}$ need not form a partition of the set $[S]$, and thus they can have non-empty intersections. A Markov network model introduces one tensor, $\phi^{\mathcal{S}^i}$, for each subset $\mathcal{S}^i$, which is meant to describe the correlations that are internal to the variables $\mathrm{x}^{\mathcal{S}^i}$ within each subset $\mathcal{S}^i$. The subset $\mathcal{S}^i$ is known as the **scope** of tensor $\phi^{\mathcal{S}^i}$, which is referred to as a **factor**.

Definition 19.3 Markov network models
Fix an arbitrary collection of I subsets $\mathcal{S}^i \subseteq [S]$ of the subsystems, with $i \in [I]$, such that the union of the subsets spans the entire set $[S]$, i.e., $\bigcup_{i\in[I]} \mathcal{S}^i = [S]$. A Markov network model with memory $M = 1$ describes joint distributions of the form

$$p_{x^{[S]}} = \prod_{i\in[I]} \phi_{x^{\mathcal{S}^i}}, \tag{19.25}$$

for order-$|\mathcal{S}^i|$ tensors $\phi^{\mathcal{S}^i}$ with non-negative entries $\phi_{x^{\mathcal{S}^i}}$.

Example 19.6

As defined in Chapter 18, consider an $I^{\max}$-uncorrelated distribution

$$\begin{aligned} p_{x^{[S]}} &= \prod_{i\in[I^{\max}]} p_{x^{\mathcal{S}^i}} \\ &= \prod_{i\in[I^{\max}]} \phi_{x^{\mathcal{S}^i}}, \end{aligned} \tag{19.26}$$

where the tensor $\phi^{\mathcal{S}^i}$ coincides with the marginal distribution $p^{\mathcal{S}^i}$. Comparing (19.26) with (19.25) reveals that an $I^{\max}$-uncorrelated distribution is a Markov network model with $I^{\max}$ factors, in which the scopes $\{\mathcal{S}^i\}_{i\in[I]}$ form a partition of the integers $[S]$. Furthermore, in this special case, the tensors $\phi^{\mathcal{S}^i}$ are valid distributions, satisfying the equality $\sum_{x^{\mathcal{S}^i}} p_{x^{\mathcal{S}^i}} = 1$.

Any joint distribution can be written in the form of a Markov network model. In fact, generalizing the chain rule for the joint distribution of two random variables (see (13.11)), the following important result applies to any joint distribution $p_{x^{[S]}}$ (see Appendix 19.C for a proof).

Key Result 19.6 Chain rule for joint distributions
Any joint distribution $p_{x^{[S]}}$ over a system of S subsystems can be expressed as

$$\begin{aligned} p_{x^{[S]}} &= p_{x^0} \cdot p_{x^1|x^0} \cdot p_{x^2|x^{0,1}} \cdots p_{x^{S-1}|x^{0,\dots,S-2}} \\ &= \underbrace{p_{x^0}}_{\phi_{x^0}} \prod_{s\in[S]} \underbrace{p_{x^s|x^{0,1,\dots,s-1}}}_{\phi_{x^{[s]}}}. \end{aligned} \tag{19.27}$$

Accordingly, the joint distribution $p_{x^{[S]}}$ represents a Markov network model with order-$(s+1)$ tensors $\phi_{x^{[s]}}$ for all $s \in [S]$. Furthermore, the decomposition (19.27) applies for any ordering of the S subsystems.

Unlike the case of an $I^{\max}$-uncorrelated distribution (19.26), the general decomposition (19.27) contains tensors with overlapping scopes. Furthermore, the tensors need not be probability distributions, since the sum of the elements of tensor $\phi_{x^{[s]}} = p_{x^s|x^{0,1,\ldots,s-1}}$ does not equal 1 in general.

The decomposition (19.27) implied by the chain rule applies to any joint distribution, and hence it does not impose any special structure on the correlations present in the system. An important situation in which the chain rule can be used as the starting point for the definition of a structured joint distribution is discussed next.

19.3.2 Linear Markov Network Models

Consider a text consisting of S words, with each word x^s selected from a dictionary of D possible terms. Typically, each word x^s depends directly only on the most recent words, maintaining indirect correlations with earlier words in the sequence.

More formally, given a **memory** length parameter M, each word x^s may be modeled as depending only on the previous M words, $x^{s-1}, x^{s-2}, \ldots, x^{s-M}$. Thus, when producing word x^s, the system must only retain the last M words, $x^{s-1}, x^{s-2}, \ldots, x^{s-M}$. This sliding-window memory mechanism reflects the operation of large language models that power chatbots at the time of writing.

This memory constraint can be stated formally as the condition that the random variable x^s is uncorrelated with the words preceding x^{s-M} when conditioning on words $x^{s-1}, x^{s-2}, \ldots, x^{s-M}$. This property can be expressed using the conditional mutual information as

$$\mathrm{I}(X^s; X^{0,\ldots,s-M-1} | X^{s-M,\ldots,s-1})_{p^{[s]}} = 0. \tag{19.28}$$

Note that for time indices $s < 0$, one can set the variable x^s to be a fixed, known, value when evaluating (19.28). Using the equivalent definition (19.22), the equality (19.28) amounts to the conditions

$$p_{x^s|x^{0,\ldots,s-1}} = p_{x^s|x^{s-M,\ldots,s-1}} \tag{19.29}$$

for all indices $s \in [S]$. Joint distributions that satisfy the condition (19.29) for a fixed memory M are referred to as **linear Markov models.**

In the special case with memory $M = 1$, the condition (19.29) yields

$$p_{x^s|x^{0,\ldots,s-1}} = p_{x^s|x^{s-1}}. \tag{19.30}$$

Using (19.30) in the chain rule (19.27), one obtains that joint distributions satisfying the conditional independence property (19.30) factorize as follows.

Definition 19.4 Linear Markov network models (with memory $M = 1$)
A linear Markov network model with memory $M = 1$ describes joint distributions of the form

$$\begin{aligned} p_{x^{[S]}} &= p_{x^0} \cdot p_{x^1|x^0} \cdot p_{x^2|x^1} \cdots p_{x^{S-1}|x^{S-2}} \\ &= \underbrace{p_{x^0}}_{\phi_{x^0}} \prod_{s\in[S]} \underbrace{p_{x^s|x^{s-1}}}_{\phi_{x^{s-1,s}}}. \end{aligned} \tag{19.31}$$

Unlike the general chain rule (19.27), the model (19.31) contains only order-2 tensors $p_{x^s|x^{s-1}}$, along with one order-1 tensor p_{x^0}. Thus, the order-S tensor (19.31) for a linear Markov network model is fully specified by the D entries of the order-1 tensor, together with the D^2 entries of the $S-1$ order-2 tensors, for a total of $D + D^2(S-1)$ numbers. This amounts to an exponential decrease compared to the D^S entries of an arbitrary order-S probability tensor $p^{[S]}$.

A linear Markov network model (19.31) is, generally, globally correlated, i.e., it cannot be partitioned into two uncorrelated groups of subsystems. However, the correlations are highly structured, with strong localization properties governed by the conditional independence relationships (19.29). Consequently, linear Markov network models effectively describe systems with a **linear correlation structure**, like text.

More broadly, linear Markov network models exemplify a common structure for joint distributions characterized by localized correlations. Other examples of such localized correlations include those between neighboring pixels in an image, consecutive frames in a video, stock prices in a financial time series, weather conditions across different regions, and consecutive lines of code.

19.3.3 Tensor Network Representation of Linear Markov Network Models

How can the linear Markov network model (19.31) be represented via a tensor network? Since the scopes of the tensors overlap, the tensor network cannot be described using tensor products. Rather, as discussed in Section 17.5, it is necessary to include a copy tensor δ for each of the variables $\mathrm{x}^0, \mathrm{x}^1, \ldots, \mathrm{x}^{S-2}$ that appear in two tensors. Note that variable x^{S-1} is the only one that appears in one tensor, namely $p_{x^{S-1}|x^{S-2}}$.

To illustrate this construction, consider Figure 19.1, which considers the case of $S = 4$ variables. The topmost copy tensor, corresponding to $s = 0$, has open index x^0 and it is connected to tensors p^0 and $p^{1|0}$. Every other sth copy tensor with $s \in \{1, \ldots, S-2\}$ has an open leg corresponding to the index x^s, as well as two bonds of dimension D with the two tensors $p^{s|s-1}$ and $p^{s+1|s}$ that have variable x^s in their scope. Finally, the open index x^{S-1} is directly linked to the tensor $p_{x^{S-1}|x^{S-2}}$.

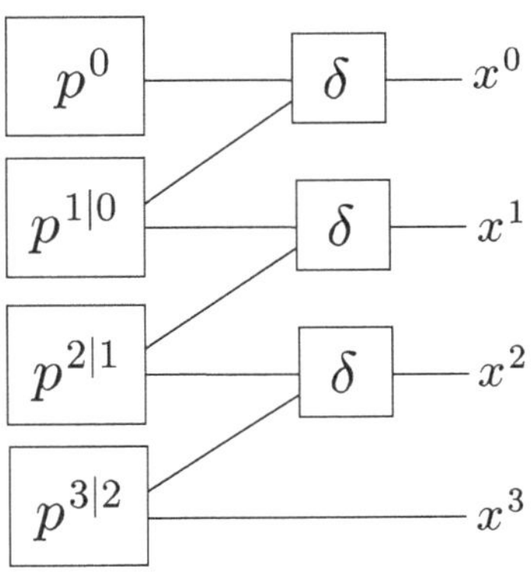

Figure 19.1 Tensor network representation of a linear Markov model with $S = 4$ subsystems.

Table 19.1 Algorithm that produces a tensor network representation of a Markov network model

Tensor network representation of Markov network models

Input: A Markov network model $p_{x^{[S]}} = \prod_{i\in[I]} \phi_{x^{\mathcal{S}^i}}$, as in (19.25)
Output: A tensor network representation of the joint distribution $p^{[S]}$
Include the order-$|\mathcal{S}^i|$ tensor $\phi^{\mathcal{S}^i}$ for all $i \in [I]$
For every variable x^s that appears in multiple, say M^s, factors, with $s \in [S]$, do:
- Include an order-$(M^s + 1)$ copy tensor
- Create M^s bonds, with each bond connecting one index from the copy tensor to the index x^s of each tensor $\phi^{\mathcal{S}^i}$ with $s \in \mathcal{S}^i$
- Mark the remaining open index of the copy tensor as x^s

For every variable x^s that appears in a single factor, mark the corresponding open index of the factor as x^s

19.3.4 Tensor Network Representation of Markov Network Models

The tensor network representation of linear Markov network models discussed above can be extended to arbitrary Markov network models (19.25) by including, along with the tensors $\{\phi^{\mathcal{S}^i}\}_{i\in[I]}$, one copy tensor for each variable x^s that appears in the scope of more than one of the tensors $\{\phi^{\mathcal{S}^i}\}_{i\in[I]}$.

In the tensor network, the tensor $\phi^{\mathcal{S}^i}$ has one open index for each variable x^s that appears exclusively in the scope of that factor (if any). Furthermore, each copy tensor has one open index, corresponding to the given assigned variable x^s. Copy tensors and tensors $\{\phi^{\mathcal{S}^i}\}_{i\in[I]}$ are connected via bonds by following the algorithm detailed in Table 19.1. The next examples illustrate the procedure.

Example 19.7

Consider a multipartite system with joint distribution

$$p_{x^{[S]}} = \underbrace{p_{x^0}}_{\phi_{x^0}} \prod_{s=1}^{S-1} \underbrace{p_{x^s|x^0}}_{\phi_{x^0,x^s}}, \tag{19.32}$$

which is characterized by the S tensors $\phi_{x^0} = p_{x^0}$ and $\phi^{0,s} = p^{s|0}\}_{s=1}^{S-1}$. The model (19.32) describes a globally correlated system in which the random variables $\mathrm{x}^1, \ldots, \mathrm{x}^{S-1}$ are conditionally independent given x^0. Thus, the variable x^0 mediates all the correlations between the other random variables.

To apply the procedure in Table 19.1, it is first observed that the variable x^0 appears in the scope of all tensors, while other variables x^s with $s \in \{1, 2, \ldots, S-1\}$ appear in a single tensor, namely $p^{s|0}$. Accordingly, a single copy tensor is required. The copy tensor has open index x^0, and one bond to each of the other tensors p_{x^0} and $\{p^{s|0}\}_{s=1}^{S-1}$. The resulting tensor network is shown in Figure 19.2.

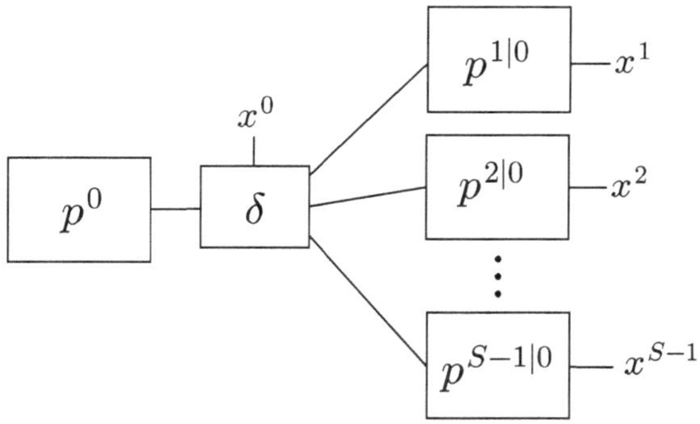

Figure 19.2 Tensor network representation for the joint distribution (19.32).

19.3.5 Gibbs Distributions ⏭

In both classical physics and machine learning, it is often useful to express the factors ϕ^{S^i} in the Markov network model (19.25) in terms of a tensor H^{S^i} known as the energy, or the Hamiltonian. The entries $H_{x^{S^i}}$ of the Hamiltonian are non-negative, and they define the entries of the tensor ϕ^{S^i} via the function

$$\phi_{x^{S^i}} \propto \exp\left(-H_{x^{S^i}}\right). \tag{19.33}$$

The proportionality factor in (19.33) accounts for the need for the factors to be multiplied by a normalization constant in order to ensure that the product (19.25) is a valid distribution. Distributions (19.25) with factors of the form (19.33) are known as **Gibbs distributions**.

By the definition (19.33), a large value of the Hamiltonian $H_{x^{S^i}}$ indicates that the tensor ϕ^{S^i} contributes to reducing the probability of joint states $x^{[S]}$ for which the states of the subsystems in group S^i are given by x^{S^i}. Conversely, a small value of the energy $H_{x^{S^i}}$ ensures that the tensor ϕ^{S^i} contributes to increasing the probability of joint states $x^{[S]}$ for which the group S^i is in the state x^{S^i}. The reader is referred to Problem 19.8 for further discussion on this aspect.

19.3.6 Tensor Networks and Factor Graphs ⏭

The reader familiar with probabilistic graphical models may have noted the relationship between the tensor network representation of joint distributions discussed in this section and the representation of joint distributions based on factor graphs. A **factor graph** contains two types of nodes, **variable nodes** and **factor nodes**. One variable node, typically drawn as a circle, is included for each random variable $\{\mathrm{x}^s\}_{s\in[S]}$. Furthermore, given a Markov network model (19.25), one factor, drawn as a square, is introduced for each tensor ϕ^{S^i}. Finally, in order to describe the factorization (19.25), each variable node is connected to all factor nodes whose tensors contain the variable in their scope.

Given that both the tensor representation and the factor graph representation provide a description of the same factorization (19.25), there is a one-to-one mapping between the two representations.

Example 19.8

For the joint distribution (19.32), the factor graph and the tensor representations are compared in Figure 19.3, with the tensor representation shown in Figure 19.3(a) and the tensor network shown in Figure 19.3(b). As illustrated in the figure, the factor graph includes one variable node (circle) for each of the S variables $\mathrm{x}^{[S]}$, along with one factor node (square) for each of the tensors p^0 and $\{p^{s|0}\}_{s=1}^{S-1}$.

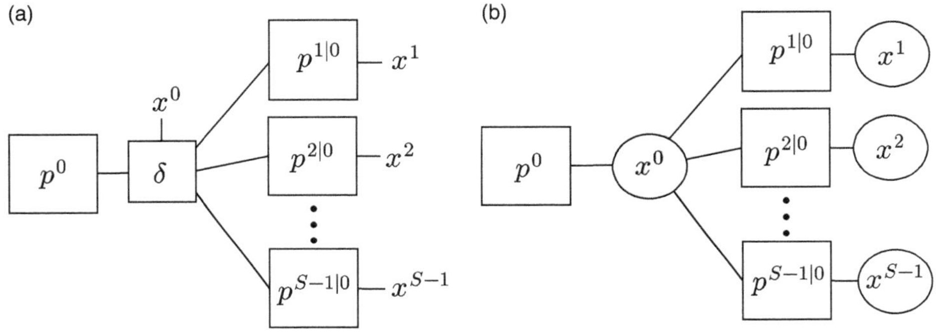

Figure 19.3 Illustration of the tensor network representation (a) and the factor graph (b) for the joint distribution (19.32).

19.4 Latent-Variable Models

A versatile approach to modeling correlations in multipartite systems involves introducing additional random variables, known as latent variables. Latent variables serve to mediate the correlations between the system variables $\mathrm{x}^{[S]}$, capturing the mechanisms underlying relationships among different parts of the system. This section begins by presenting general models of joint distributions that incorporate latent variables, along with their tensor representations. It then discusses two important examples that illustrate model classes with global and local latent variables.

19.4.1 Defining Latent-Variable Models

Latent-variable models introduce L latent variables, denoted as $\mathrm{z}^{[L]}$, in addition to the random variables $\mathrm{x}^{[S]}$ that describe the state of the system. Each latent variable z^l takes values in the set $[\chi] = \{0, 1, \ldots, \chi - 1\}$. Broadly, each latent variable

z^l is introduced to model correlations among specific subsets of state variables. Depending on the system, latent variables may represent physical quantities that are unmeasured or not of direct interest, or they may be abstract quantities introduced solely to describe correlations within the system. This section will focus on two extreme cases: global latent variables, which connect all variables in the system, and local latent variables, which are specific to pairs of subsystems.

A latent-variable model is specified as a joint distribution

$$p_{z^{[L]},x^{[S]}} = \Pr[\mathrm{z}^l = z^l \text{ for all } l \in [L] \text{ and } \mathrm{x}^s{=}x^s \text{ for all } s \in [S]] \tag{19.34}$$

for both latent and system variables. From the joint distribution $p_{z^{[L]},x^{[S]}}$, the joint distribution $p_{x^{[S]}}$ for the variables of interest is obtained via marginalization as

$$p_{x^{[S]}} = \sum_{z^{[L]}} p_{z^{[L]},x^{[S]}}, \tag{19.35}$$

where the sum is over all the χ^L values of the latent variables $z^{[L]}$.

19.4.2 Global Latent-Variable Models

Consider now a latent-variable model in which there is a single latent variable $\mathrm{z}^0 \in [\chi]$, along with the S system variables $\mathrm{x}^{[S]}$. The latent variable z^0 is introduced to capture the correlations among *all* the state variables $\mathrm{x}^{[S]}$, and it is hence referred to as a **global latent variable**.

The joint distribution of latent and system variables is accordingly defined as

$$p_{z^0,x^{[S]}} = p_{z^0} \prod_{s\in[S]} p_{x^s|z^0}. \tag{19.36}$$

This ensures that the random variables $\mathrm{x}^{[S]}$ are globally uncorrelated when conditioned on the latent variable z^0, i.e.,

$$p_{x^{[S]}|z^0} = \prod_{s\in[S]} p_{x^s|z^0}. \tag{19.37}$$

Therefore, all the correlations among the variables $\mathrm{x}^{[S]}$ are mediated through the latent variable z^0.

The joint distribution $p_{x^{[S]}}$ for the system states is obtained via the marginalization (19.35), as detailed next.

Definition 19.5 Naive Bayes model

Naive Bayes models are joint distributions of the form

$$\begin{aligned} p_{x^{[S]}} &= \sum_{z^0} p_{z^0,x^{[S]}} \\ &= \sum_{z^0} p_{z^0} \prod_{s\in[S]} p_{x^s|z^0}. \end{aligned} \tag{19.38}$$

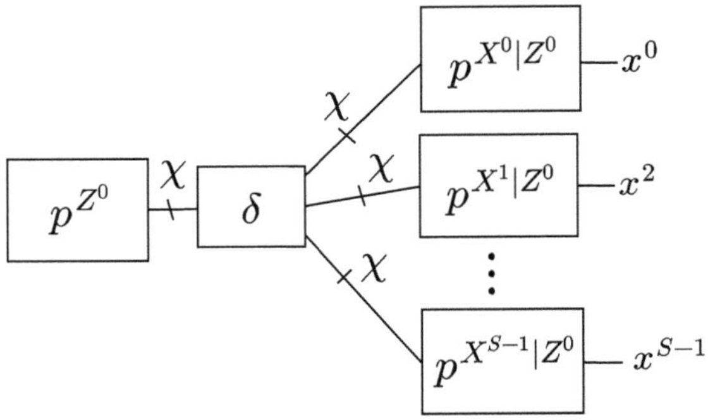

Figure 19.4 Tensor network representation of a naive Bayes model, an example of a global latent-variable model.

Thanks to the shared dependence of all subsystems on the latent variable z^0, naive Bayes models are globally correlated.

A tensor network representation for the naive Bayes model $p_{x[S]}$ in (19.38) is obtained as in Figure 19.4. In the tensor network, p^{Z^0} is the order-1 tensor with entries p_{z^0}, and $p^{X^s|Z^0}$ is the tensor with entries $p_{x^s|z^0}$ for all $s \in [S]$. The figure emphasizes that the bonds have all dimensions equal to the number χ of possible values for the latent variable z^0.

To derive the tensor network representation in Figure 19.4, note first that the joint distribution $p_{x[S],z^0}$ is represented by the tensor network in Figure 19.2, with the latent variable z^0 playing the role of the random variable x^0. Following Section 18.2.4, the marginalization in (19.38) over the latent variable z^0 is represented by adding a self-bond on the leg corresponding to z^0. Finally, this leg can be removed by using the properties of the copy tensor in Appendix 17.A (namely Property 2 in Figure 17.21).

Example 19.9

With $\chi = 1$, a naive Bayes model (19.38) is globally uncorrelated. In fact, as discussed in Chapter 17, indices of order 1 can be removed. Therefore, as illustrated in Figure 19.5, the joint distribution reduces to a globally uncorrelated distribution. In the resulting distribution, each tensor p^{X^s}, representing the marginal distribution of subsystem s, has entries $p^{X^s}_{x^s} = p^{X^s|Z^0}_{x^s|0}$ for all $x^s \in [D]$ and $s \in [S]$.

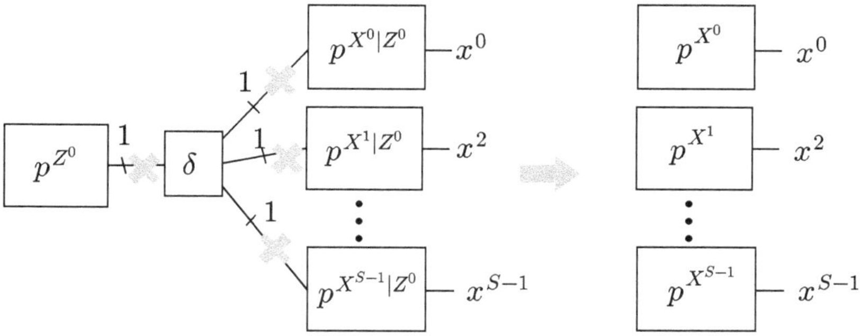

Figure 19.5 Graphical proof that a naive Bayes model with latent-variable dimension equal to $\chi = 1$ is globally uncorrelated.

Since for naive Bayes models (19.38), the latent random variable z^0 mediates all correlations among the random variables $\mathrm{x}^{[S]}$, a larger value of the cardinality χ generally indicates a larger correlation. Formally, as shown in Appendix 19.D, the mutual information $\mathrm{I}(X^s;X^{s'})_{p^{s,s'}}$ between any two variables x^s and $\mathrm{x}^{s'}$ with $s' \neq s$ can be bounded as follows.

> **Key Result 19.7 Upper bound on the mutual information for naive Bayes models**
> Given random variables $\mathrm{x}^{[S]}$ with joint distribution given by the naive Bayes model (19.40), the mutual information between any two random variables x^s and $\mathrm{x}^{s'}$ with $s, s' \in [S]$ and $s' \neq s$ is upper bounded as
>
> $$\mathrm{I}(X^s;X^{s'})_{p^{s,s'}} \leq \mathrm{H}(Z^0)_{p^{Z^0}} \leq \log_2(\chi). \tag{19.39}$$

Thus, the entropy $\mathrm{H}(Z^0)_{p^{Z^0}}$ of the latent variable z^0 upper bounds the mutual information $\mathrm{I}(X^s;X^{s'})_{p^{s,s'}}$. The intuition behind this result is that the level of correlation between any two random variables, measured by $\mathrm{I}(X^s;X^{s'})_{p^{s,s'}}$, cannot exceed the information $\mathrm{H}(Z^0)_{p^{Z^0}}$ present in the global latent variable z^0. This also entails that the mutual information $\mathrm{I}(X^s;X^{s'})_{p^{s,s'}}$ cannot exceed the logarithm of the number of possible values, χ, that can be assumed by the latent variable.

As a final note, following Chapter 17, the naive Bayes model (19.38) corresponds to a canonic polyadic decomposition (CPD) for the tensor $p^{[S]}$. In fact, the entries (19.38) form the tensor

$$p^{[S]} = \sum_{z^0 \in [\chi]} p_{z^0} \left(p_{z^0}^{X^0|Z^0} \otimes p_{z^0}^{X^0|Z^0} \otimes \cdots \otimes p_{z^0}^{X^0|Z^0} \right), \tag{19.40}$$

where the order-1 tensors $p_{z^0}^{X^s|Z^0}$, all with dimension D, have entries $p_{x^s|z^0}$ for all $z^0 \in [\chi]$. In (19.40), each term $p_{z^0}^{X^0|Z^0} \otimes p_{z^0}^{X^0|Z^0} \otimes \cdots \otimes p_{z^0}^{X^0|Z^0}$ corresponds to a globally uncorrelated distribution, reflecting the conditional independence (19.37) of the random variables $\mathrm{x}^{[S]}$ given the latent random variable z^0. Note that the minimum value of the integer χ for which the joint distribution $p^{[S]}$ can be expressed as the naive Bayes model (19.40) is related, but distinct, from the tensor rank, which does not assume any specific constraint for the terms in the CPD.

19.4.3 Local Latent-Variable Models

An important example of joint distributions with correlations described by local latent variables is given by the hidden Markov model. A hidden Markov model introduces S latent variables $\mathrm{z}^{[S]}$. Each latent variable $\mathrm{z}^s \in [\chi]$ is associated with a state variable x^s for all subsystems $s \in [S]$. As will be detailed in this subsection, the latent variable z^s serves to describe the correlation between successive variables x^s and x^{s+1}. Furthermore, the domain size χ of the latent variables dictates the strength of such pairwise correlations.

In a hidden Markov model, the latent variables $\mathrm{z}^{[S]}$ follow a linear Markov network model (19.31), i.e.,

$$p_{z^{[S]}} = p_{z^0} \prod_{s=1}^{S-1} p_{z^s|z^{s-1}}. \tag{19.41}$$

Furthermore, each variable x^s is modeled as a, generally noisy, function of the corresponding latent variable z^s. This assumption is encoded by the conditional distribution

$$p_{x^{[S]}|z^{[S]}} = \prod_{s\in[S]} p_{x^s|z^s}, \tag{19.42}$$

yielding the following definition.

Definition 19.6 Hidden Markov model
A joint distribution $p^{[S]}$ is said to follow a hidden Markov model with bond dimension χ if it can be expressed as the marginal distribution

$$p_{x^{[S]}} = \sum_{z^{[S]}} p_{z^{[S]},x^{[S]}} \tag{19.43}$$

of a joint distribution of the form

$$p_{z^{[S]},x^{[S]}} = p_{z^0} p_{x^0|z^0} \prod_{s=1}^{S-1} p_{z^s|z^{s-1}} p_{x^s|z^s} \tag{19.44}$$

for latent variables $\mathrm{z}^{[L]}$ with $\mathrm{z}^l \in [\chi]$ for all $l \in [L]$.

Parsing an expression such as (19.44) can be quite intricate. However, the tensor network representation provides valuable insights into the structure of the correlations inherent in a hidden Markov model. As shown in Figure 19.6, the tensor network representation of a hidden Markov model includes order-1 tensor p^{Z^0}, along with order-2 tensors $\{p^{X^s|Z^s}\}_{s\in[S]}$ and $\{p^{Z^s|Z^{s-1}}\}_{s=1}^{S-1}$. The tensor network

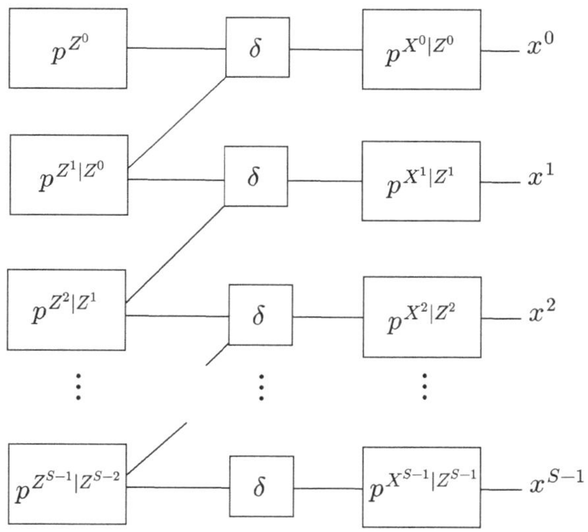

Figure 19.6 Tensor network representation for a hidden Markov model (19.43) and (19.44).

makes it visually clear that the correlations among the random variables $\mathrm{x}^{[S]}$ are mediated by the Markov process given by the local latent variables $\mathrm{z}^{[S]}$. This is in the sense that variable x^s is only connected to variable $\mathrm{x}^{s'}$ with $s' \neq s$ through the corresponding bonds of the latent Markov model $\mathrm{z}^{[S]}$.

The bond dimension χ constrains the correlation level between any two random variables x^s and $\mathrm{x}^{s'}$ in the system. The following result, proved in Appendix 19.E, formalizes this observation.

Key Result 19.8 Upper bound on the mutual information for hidden Markov models

Given random variables $\mathrm{x}^{[S]}$ with joint distribution given by the hidden Markov model (19.44), the mutual information between any two random variables x^s and $\mathrm{x}^{s'}$ with $s, s' \in [S]$ and $s' \neq s$ is upper bounded as

$$\mathrm{I}(X^s;X^{s'})_{p^{s,s'}} \leq \log_2(\chi). \qquad (19.45)$$

19.5 Summary

- The structure of the correlations in a multipartite system can be studied by analyzing conditional correlations in larger systems that include also latent variables.
- Conditional correlations are quantified by the conditional mutual information $\mathrm{I}(X;Y|Z)_{p^{XYZ}}$, which captures the correlations between subsystems X and Y once knowledge about the state of subsystems Z is accounted for.

- The conditional mutual information can be defined in several equivalent ways, including the basic form $\mathrm{I}(X;Y|Z)_{p^{XYZ}} = \mathrm{H}(X|Z)_{p^{XZ}} + \mathrm{H}(Y|Z)_{p^{YZ}} - \mathrm{H}(X,Y|Z)_{p^{XYZ}}$.
- The conditional mutual information can be larger than, smaller than, or equal to the mutual information: $\mathrm{I}(X;Y|Z)_{p^{XYZ}} \lesseqgtr \mathrm{I}(X;Y)_{p^{XY}}$.
- The conditional mutual information is non-negative, and this property implies: (i) the strong subadditivity of the entropy $\mathrm{H}(X,Y,Z)_{p^{XYZ}} \leq \mathrm{H}(X,Z)_{p^{XZ}} + \mathrm{H}(Y,Z)_{p^{YZ}} - \mathrm{H}(Z)_{p^Z}$; and (ii) the further-conditioning-reduces-entropy property $\mathrm{H}(X|Y,Z)_{p^{XYZ}} \leq \mathrm{H}(X|Z)_{p^{XZ}}$.
- The conditional mutual information $\mathrm{I}(X;Y|Z)_{p^{XYZ}}$ equals zero when the subsystems X and Y are conditionally independent given Z, i.e., when $p_z^{XY|Z} = p_z^{X|Z} \otimes p_z^{Y|Z}$ for all values of index z.
- In a manner similar to the conditional entropy, the conditional mutual information supports a chain rule for the mutual information, i.e., $\mathrm{I}(X;Y,Z)_{p^{XYZ}} = \mathrm{I}(X;Y)_{p^{XY}} + \mathrm{I}(X;Z|Y)_{p^{XYZ}}$.
- In the absence of latent variables, structural properties described in terms of conditional independence conditions imply a factorization of the joint distribution of the form $p_{x^{[S]}} = \prod_{i\in[I]} \phi_{x^{S^i}}$, where each tensor $\phi_{x^{S^i}}$ depends on a subset $\mathcal{S}^i$, known as the scope, of the variables in set $[S]$.
- A linear Markov network model (with memory $M = 1$) factorizes as $p_{x^{[S]}} = p_{x^0}\prod_{s=1}^{S-1} p_{x^s|x^{s-1}}$, implying the conditional independence of present and past beyond one sample given the previous sample, i.e., $\mathrm{I}(X^s;X^{0,\dots,s-2}|X^{s-1})_{p^{[s]}} = 0$.
- An important example of latent-variable models is given by the naive Bayes model $p^{[S]} = \sum_{z^0\in[\chi]} p_{z^0}(p_{z^0}^{X^0|Z^0} \otimes p_{z^0}^{X^1|Z^0} \otimes \cdots \otimes p_{z^0}^{X^{S-1}|Z^0})$, which introduces a global random variable $\mathrm{z}^0 \in [\chi]$.
- Another example is given by the hidden Markov model, which introduces local latent variables $\mathrm{z}^{[S]}$, with each latent variable $\mathrm{z}^s \in [\chi]$ mediating the correlation between two successive system variables x^s and x^{s+1}.
- For both global and local latent-variable models, the dimension χ dictates the maximum mutual information between two subsystems, which is given by $\log_2(\chi)$.

19.6 Recommended Resources

Factor graphs provide a powerful framework for representing probabilistic models, as discussed extensively in [1]. Furthermore, the connection between tensor networks and factor graphs has been explored in [2].

For applications of the conditional mutual information to fairness models, the reader is referred to [3]. The conditional mutual information also plays a pivotal role in quantifying correlations in causal probabilistic models, and it is central to defining concepts like transfer entropy [4, 5] and directed information. The directed information from a time series $\mathrm{a}^{[S]}$ to another time series $\mathrm{b}^{[S]}$ quantifies causal correlations via a sum of conditional mutual information terms [6–8]

$$\mathrm{I}(A \to B) = \sum_{s\in[S]} \mathrm{I}(A^{[s]};B^s \mid B^{[s-1]}). \tag{19.46}$$

By the definition (19.46), the directed information quantifies the correlation between the last s samples $\mathrm{a}^{[s]}$ of the time series $\mathrm{a}^{[S]}$ and the current sample b^s of time series $\mathrm{b}^{[S]}$, when the past $s-1$ samples $\mathrm{b}^{[s-1]}$ are already known. Accordingly, the directed information quantifies the level of predictability of sequence $\mathrm{b}^{[S]}$ based on causal access to the samples of sequence $\mathrm{a}^{[S]}$ and on knowledge of the past samples of sequence $\mathrm{b}^{[S]}$.

PROBLEMS

19.1 Prove the equality (19.12) starting from (19.7).

19.2 Consider the joint distribution over $S = 3$ subsystems with $D = 2$ given by

$$p^{XYZ} = \frac{1}{8}|000\rangle + \frac{1}{8}|110\rangle + \frac{3}{8}|011\rangle + \frac{3}{8}|101\rangle. \tag{19.47}$$

(a) Evaluate the mutual information $\mathrm{I}(X;Y)_{p^{XY}}$. (b) Evaluate the conditional mutual information $\mathrm{I}(X;Y|Z)_{p^{XYZ}}$ by choosing one of the four equivalent definitions (19.6), (19.7), (19.12), and (19.13).

19.3 Prove the equality

$$\mathrm{I}(X;Y|Z)_{p^{XYZ}} = \mathrm{E}_{p^Z}\left[\mathrm{D}(p_{\mathrm{z}}^{XY|Z}||p_{\mathrm{z}}^{X|Z} \otimes p_{\mathrm{z}}^{Y|Z})\right]. \tag{19.48}$$

Hint: Note that the conditional mutual information can be written as the average $\mathrm{I}(X;Y|Z)_{p^{XYZ}} = \mathrm{E}_{p^Z}[\mathrm{I}(X;Y)_{p_{\mathrm{z}}^{XY|Z}}]$.

19.4 Evaluate the conditional mutual information $\mathrm{I}(X;Y|Z)_{p^{XYZ}}$ for the joint distribution $p^{XYZ} = \frac{1}{3}|001\rangle + \frac{1}{3}|010\rangle + \frac{1}{3}|100\rangle$ using the equality derived in Problem 19.3.

19.5 (a) For any joint distribution p^{XYZ}, prove the equalities

$$\mathrm{H}(X|Y,Z)_{p^{XYZ}} = \log_2(N^X) - \mathrm{D}(p^{XYZ}||u^X \otimes p^{YZ}) \tag{19.49}$$

and

$$\mathrm{H}(X|Z)_{p^{XZ}} = \log_2(N^X) - \mathrm{D}(p^{XZ}||u^X \otimes p^Z), \tag{19.50}$$

where u^X is a uniform distribution for subsystem X and N^X is the number of values that can be assumed by variable x. (b) Interpret these equalities in a manner similar to the entropy decomposition $\mathrm{H}(X)_{p^X} = \log_2(N^X) - \mathrm{D}(p^{XZ}||u^X \otimes p^Z)$ discussed in Chapter 4.

19.6 Assume that the joint distribution of three random variables x, y, and z factorizes as $p_{x,y,z} = p_z p_{y|z} p_{x|y}$. (a) Prove the equality $\mathrm{I}(X;Z|Y)_{p^{XYZ}} = 0$, which indicates that X and Z are independent conditioned on Y. (b) Show that the inequality $\mathrm{I}(X;Y|Z)_{p^{XYZ}} \leq \mathrm{I}(X;Y)_{p^{XY}}$ holds, and thus, for the joint distribution p^{XYZ}, the conditional mutual information has the interpretation of residual correlations. *Hint:* Use the chain rule of mutual information twice.

19.7 For a joint distribution over $S = 5$ subsystems that factorizes as $p_{x^{[5]}} = \phi_{x^{0,1}}\phi_{x^{1,2}}\phi_{x^{0,3}}\phi_{x^4}$, obtain the tensor network representation.

19.8 Consider a Gibbs distribution over $S = 3$ bits given by $p_{x^{[3]}} = \phi_{x^{0,1}}\phi_{x^{1,2}}$ with factors characterized by the Hamiltonians $H_{x^0,x^1} = \gamma(2x^0 - 1)(2x^1 - 1)$ and $H_{x^1,x^2} = \gamma(2x^1 - 1)(2x^2 - 1)$, where γ is some real number. (a) Writing the joint distribution as $p_{x^0,x^1,x^2} = \eta \exp(-H_{x^0,x^1}) \exp(-H_{x^1,x^2})$, find the value of the scalar η that ensures the validity of the joint distribution $p_{x^{[3]}}$. Note that the constant η is a function of γ. (b) For the obtained value of η, evaluate the mutual information $\mathrm{I}(X^0;X^2)_{p^{0,2}}$ as a function of γ. (c) Evaluate the mutual information $\mathrm{I}(X^0;X^2)_{p^{0,2}}$ in the limit $\gamma \to \infty$. (d) Evaluate the mutual information $\mathrm{I}(X^0;X^2)_{p^{0,2}}$ for $\gamma = 0$. (e) Evaluate the mutual information $\mathrm{I}(X^0;X^2)_{p^{0,2}}$ in the limit $\gamma \to -\infty$.

19.9 For a joint distribution $p^{[3]}$ obtained from a latent-variable model $p_{z^0,x^{[3]}} = \phi_{z^0,x^{0,1}}\phi_{z^0,x^{1,2}}$, evaluate the tensor network representation.

19.10 Given a joint distribution of the form $p_{x,y,z} = \sum_{z \in [2]} p_z p_{x|z} p_{y|z}$, can the mutual information $\mathrm{I}(X;Y)_{p_{XY}}$ be equal to 1.5 bits?

19.11 Consider a system with $S = 3$ bits that is initially in a zero-entropy state in which all bits are equal to 0. The system undergoes a dynamic transformation consisting of two successive layers of operations. In the first layer, channels are applied in parallel to the three bits that are described by the channel matrix

$$K = \frac{1}{2}\begin{bmatrix} 1 & 1 \\ 1 & 1 \end{bmatrix}. \tag{19.51}$$

In the second layer, the channel

$$K' = \begin{bmatrix} 1 & 1 & 0 & 0 \\ 0 & 0 & 0 & 0 \\ 0 & 0 & 0 & 0 \\ 0 & 0 & 1 & 1 \end{bmatrix} \tag{19.52}$$

is applied jointly to the first bit and to the third bit. (a) Derive the channel matrix for the first layer considering its operation on the three bits. *Hint:* Channels operating on different subsystems can be composed using the Kronecker product. (b) Derive the channel matrix for the second layer considering its operation on the three bits. (c) Evaluate the joint distribution of the three bits at the output of the second layer. (d) Were the three bits correlated initially? Are they correlated after the transformation applied by the two layers?

Appendix

19.A Proof of the Strong Subadditivity of the Entropy

The non-negativity of the conditional mutual information, which is equivalent to the strong subadditivity of the entropy, follows directly from the equality (see Problem 19.2)

$$\mathrm{I}(X;Y|Z)_{p^{XYZ}} = \mathrm{E}_{p^Z}[\mathrm{D}(p_z^{XY|Z}||p_z^{X|Z} \otimes p_z^{Y|Z})], \tag{19.53}$$

and from the non-negativity of the Shannon relative entropy.

19.B Proof of the Chain Rule for the Mutual Information

The chain rule for the mutual information (19.23) follows from the definition of conditional mutual information and from the chain rule for the joint entropy via the equalities

$$\begin{aligned}\mathrm{I}(X;Y,Z)_{p^{XYZ}} &= \mathrm{H}(X)_{p^X} + \mathrm{H}(Y,Z)_{p^Y} - \mathrm{H}(X,Y,Z)_{p^{XYZ}} \\ &= \mathrm{H}(X)_{p^X} + \mathrm{H}(Y)_{p^Y} + \mathrm{H}(Y|Z)_{p^{YZ}} \\ &\quad - \mathrm{H}(X,Z)_{p^{XYZ}} - \mathrm{H}(Y|X,Z)_{p^{XYZ}} \\ &= \mathrm{I}(X;Z)_{p^{XY}} + \mathrm{I}(X;Y|Z)_{p^{XYZ}},\end{aligned} \tag{19.54}$$

where the second equality follows from the chain rule for the entropy and the third from the definitions of mutual information and conditional mutual information.

19.C Proof of the Chain Rule for the Joint Probability Distribution

As introduced in Chapter 13, given a joint distribution p_{ab} for two arbitrary random variables A and B, the chain rule amounts to the equality $p_{ab} = p_a p_{b|a}$ for all values of the indices. In the case of a joint distribution $p^{[S]}$, one can obtain (19.27) by applying the relation $p_{ab} = p_a p_{b|a}$ iteratively. At the first step, set as random variables A the first $S-1$ states $(\mathrm{x}^0, \ldots, \mathrm{x}^{S-2})$ and as random variable B the state x^{S-1}. Then, the chain rule for two variables implies the equality

$$p_{x^{[S]}} = p_{x^{0,\ldots,S-2}} \cdot p_{x^{S-1}|x^{0,\ldots,S-2}}. \tag{19.55}$$

At the second step, one applies the chain rule by setting A to $(\mathrm{x}^0, \ldots, \mathrm{x}^{S-3})$ and B to x^{S-2}; and the procedure repeats in a similar way until the expression (19.27) is recovered.

19.D Proof of (19.39) and (12.86)

This appendix starts by proving the inequality (19.39) and then addresses the related inequalities (12.86) for the common information measures introduced in Chapter 12. To prove (19.39), it is sufficient to consider three subsystems X, Y, and Z, which stand in for subsystems X^s, $X^{s'}$, and the latent variable Z^0, respectively. In particular, to reflect the naive Bayes model, the joint distribution p^{XYZ} must satisfy the equality $p_{xyz} = p_z p_{x|z} p_{y|z}$, yielding the marginal distribution $p^{XY} = \sum_{z \in [N^Z]} p_z^Z p_z^X \otimes p_z^Y$. This implies the conditional independence property

$$\mathrm{I}(X;Y|Z)_{p^{XYZ}} = 0. \tag{19.56}$$

With this notation, the desired inequality (19.39) is expressed as $\mathrm{I}(X;Y)_{p^{XY}} \leq \mathrm{H}(Z)_{p^Z}$. Using the chain rule for the conditional mutual information (19.23), the mutual information $\mathrm{I}(X;Y,Z)_{p^{XYZ}}$ can be written in the following two ways:

$$\begin{aligned}\mathrm{I}(X;Y,Z)_{p^{XYZ}} &= \mathrm{I}(X;Y)_{p^{XY}} + \mathrm{I}(X;Z|Y)_{p^{XYZ}} \\ &= \mathrm{I}(X;Z)_{p^{XZ}} + \mathrm{I}(X;Y|Z)_{p^{XYZ}}.\end{aligned} \tag{19.57}$$

Thus, using the assumption (19.56) and the non-negativity of the conditional mutual information, one obtains the inequalities

$$\mathrm{I}(X;Y)_{p^{XY}} \leq \mathrm{I}(X;Z)_{p^{XZ}} \leq \mathrm{H}(Z)_{p^Z}, \tag{19.58}$$

where the last inequality follows by the non-negativity of the conditional entropy. This completes the proof of the inequality (19.56).

Consider now the inequalities (12.86). The inequality (19.58) directly implies the inequality $\mathrm{I}(X;Y)_{p^{XY}} \leq \mathrm{C}(X;Y)_{p^{XY}}$, where $\mathrm{C}(X;Y)_{p^{XY}}$ is the distributed-sampling common information introduced in Chapter 12. In fact, the latter coincides with the minimum entropy $\mathrm{H}(Z)_{p^Z}$ of a random variable z that ensures the condition (19.56).

Consider now the inequality $\mathfrak{C}(X;Y)_{p^{XY}} \leq \mathrm{I}(X;Y)_{p^{XY}}$ for the distillable common information $\mathfrak{C}(X;Y)_{p^{XY}}$. Following the definition given in Chapter 12, in this case, the joint distribution p^{XYZ} is such that the equalities $\mathrm{H}(Z|X)_{p^{XZ}} = \mathrm{H}(Z|Y)_{p^{YZ}} = 0$ are satisfied, since random variable z is a function of x, and also a function of y. This implies the equalities $\mathrm{I}(X;Z|Y)_{p^{XYZ}} = 0$ and $\mathrm{I}(X;Z)_{p^{XZ}} = \mathrm{H}(Z)_{p^Z}$.

Thus, starting from (19.57), one obtains

$$\begin{aligned}\mathrm{I}(X;Y,Z)_{p^{XYZ}} &= \mathrm{I}(X;Y)_{p^{XY}} \\ &= \mathrm{H}(Z)_{p^Z} + \mathrm{I}(X;Y|Z)_{p^{XYZ}},\end{aligned} \tag{19.59}$$

which entails the inequality $\mathrm{H}(Z)_{p^Z} \leq \mathrm{I}(X;Y)_{p^{XY}}$. This, in turn, implies the desired bound $\mathfrak{C}(X;Y)_{p^{XY}} \leq \mathrm{I}(X;Y)_{p^{XY}}$, since the distillable common information $\mathfrak{C}(X;Y)_{p^{XY}}$ is defined as the minimum entropy $\mathrm{H}(Z)_{p^Z}$ of a distillable random variable.

19.E Proof of the Inequality (19.45)

To prove the inequality (19.45), it is sufficient to argue that the joint distribution of variables X^s and $X^{s'}$ can be expressed in the form $p_{x^s,x^{s'}} = \sum_{z\in[\chi]} p_z (p_{x^s|z} \otimes p_{x^{s'}|z})$. In fact, in this case, the inequality (19.45) follows directly from the inequality (19.39) proved in Appendix 19.D.

To this end, it is sufficient to observe, as illustrated in Figure 19.7, that by marginalizing over all other variables x^t with $t \notin \{s, s'\}$, one can write the resulting distribution as the contraction

$$p_{x^s,x^{s'}} = \sum_{z\in[\chi]} A_{x^s,z} B_{x^{s'},z} \tag{19.60}$$

for order-2 tensors A and B.

Then, defining the conditional distributions

$$p_{x^s|z} = \frac{A_{x^s,z}}{\sum_{x\in[D]} A_{x,z}}, \quad p_{x^{s'}|z} = \frac{B_{x^{s'},z}}{\sum_{x\in[D]} B_{x,z}}, \tag{19.61}$$

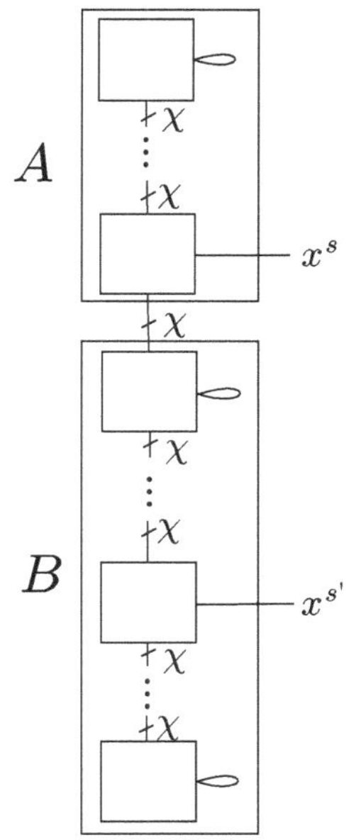

Figure 19.7 Illustration of the proof of the inequality (19.45).

as well as the distribution

$$p_z = \left(\sum_{x\in[D]} A_{x,z}\right)\left(\sum_{x\in[D]} B_{x,z}\right), \tag{19.62}$$

yields the desired distribution. This concludes the proof.

19.F Generation-Based Shared Information via Sequential Sampling

Referring to Section 13.6 for context, consider a joint distribution $p^{XYZ} = p^{XY} \otimes p^Z$ such that variable y can be obtained via a function $\mathrm{y} = f(\mathrm{x}, \mathrm{z})$. This appendix focuses on proving the inequality $\mathrm{H}(Y|Z)_{p^{YZ}} \leq \mathrm{I}(X;Y)_{p^{XY}}$.

Removing the subscripts from the information measures to simplify the notation, the following sequence of steps holds:

$$\begin{aligned}
\mathrm{H}(Y|Z) &= \mathrm{I}(X;Y|Z) + \mathrm{H}(Y|X,Z) \\
&= \mathrm{I}(X;Y|Z) \\
&= \mathrm{H}(X|Z) - \mathrm{H}(X|Y,Z) \\
&= \mathrm{H}(X) - \mathrm{H}(X|Y,Z) \\
&\leq \mathrm{H}(X) - \mathrm{H}(X|Y) \\
&= \mathrm{I}(X;Y),
\end{aligned} \tag{19.63}$$

where the first equality follows from the definition of conditional mutual information; the second equality follows because y is a function of random variables x and z, implying the equality $\mathrm{H}(Y|X,Z) = 0$; the fourth equality follows from the independence of x and z; and the inequality follows by the further-conditioning-reduces-entropy property.

REFERENCES

[1] D. Koller and N. Friedman, *Probabilistic Graphical Models: Principles and Techniques*. Cambridge, MA: MIT Press, 2009.

[2] I. Glasser, N. Pancotti, and J. I. Cirac, "From probabilistic graphical models to generalized tensor networks for supervised learning," *IEEE Access*, vol. 8, pp. 68 169–68 182, 2020.

[3] M. Hardt, E. Price, and N. Srebro, "Equality of opportunity in supervised learning," *Advances in Neural Information Processing Systems*, vol. 29, 2016.

[4] R. Vicente, M. Wibral, M. Lindner, and G. Pipa, "Transfer entropy—a model-free measure of effective connectivity for the neurosciences," *Journal of Computational Neuroscience*, vol. 30, no. 1, pp. 45–67, 2011.

[5] M. Oizumi, N. Tsuchiya, and S.-I. Amari, "Unified framework for information integration based on information geometry," *Proceedings of the National Academy of Sciences*, vol. 113, no. 51, pp. 14 817–14 822, 2016.

[6] G. Kramer, "Capacity results for the discrete memoryless network," *IEEE Transactions on Information Theory*, vol. 49, no. 1, pp. 4–21, 2003.

[7] M. Raginsky, "Directed information and Pearl's causal calculus," in *Annual Allerton Conference on Communication, Control, and Computing*, 2011, pp. 958–965.

[8] J. Jiao, H. H. Permuter, L. Zhao, Y.-H. Kim, and T. Weissman, "Universal estimation of directed information," *IEEE Transactions on Information Theory*, vol. 59, no. 10, pp. 6220–6242, 2013.

20 Multipartite Quantum Information: Fragmentation, Scale, and Strength

20.1 Overview

As seen in Part V of the book, entanglement between two subsystems is a uniquely quantum form of correlation that can be leveraged to support non-classical applications such as superdense coding and teleportation. This chapter and the next generalize the definition and analysis of entanglement from bipartite to multipartite quantum systems. Specifically, this chapter introduces multipartite quantum states as tensors, examines local measurements, and focuses on defining and analyzing entanglement measures across various system partitions. This enables the study of the fragmentation, scale, and strength of entanglement in multipartite quantum systems.

For systems with three or more qudits, the chapter demonstrates the existence of distinct entanglement classes, where states belonging to different classes cannot be transformed into one another using only local operations and classical communication (LOCC). The chapter further examines the robustness and extendibility of entanglement. In particular, entanglement is shown to be fragile under the loss of subsystems, a phenomenon commonly referred to as decoherence. Furthermore, entanglement is demonstrated to be monogamous, meaning that the presence of entanglement between two subsystems restricts the entanglement they can share with a third subsystem.

Building on this foundation, Chapter 21 will discuss the structure of entanglement in quantum states. To manage the complexity of the subject matter, both chapters will primarily focus on pure states. Throughout, comparisons with multipartite classical systems, as introduced previously, will be highlighted to underscore both parallels and fundamental differences.

20.1.1 Learning Objectives and Organization

By the end of this chapter, the reader should be able to:

- understand the formalism of multipartite quantum systems;
- represent quantum states using tensor networks;
- understand the canonic polyadic decomposition (CPD) representation of multipartite quantum systems and the notion of tensor rank;

- study and apply local measurements in a multipartite quantum system;
- calculate the bipartite entanglement entropy across any cut of the system;
- calculate the reduced density matrix for any subset of qudits;
- understand the fragmentation and scale of entanglement;
- appreciate the concept of entanglement classes for systems with at least three qudits;
- and understand the robustness and monogamy properties of entanglement.

20.2 Multipartite Quantum Systems

Consider a quantum system composed of S subsystems. When measured using a complete measurement setting, each subsystem $s \in [S]$ returns a classical output in the set $[D]$. Thus, using the terminology introduced in Chapter 5, with $D = 2$, the system is a collection of S qubits; with $D = 3$ it contains S qutrits; while for the general case $D > 3$ the system encompasses S qudits. For generality, throughout this chapter, the term qudit will be used to refer to a subsystem when the dimension D is arbitrary.

As illustrated in Figure 20.1, in analogy with joint probability vectors $p^{[S]}$, the joint pure state of a multipartite quantum system with S subsystems is defined by an order-S tensor with dimensions equal to D.

Definition 20.1 Joint pure states

The joint pure state of a multipartite quantum system with S qudits is defined by an order-S tensor $|\psi^{[S]}\rangle$, with all dimensions equal to D, such that each entry

$$\psi_{x^0,\ldots,x^{S-1}} = \psi_{x^{[S]}} \tag{20.1}$$

of the tensor, with $x^s \in [D]$ for all $s \in [S]$, is generally complex, and the tensor is normalized, i.e.,

$$\sum_{x^{[S]}} |\psi_{x^{[S]}}|^2 = 1. \tag{20.2}$$

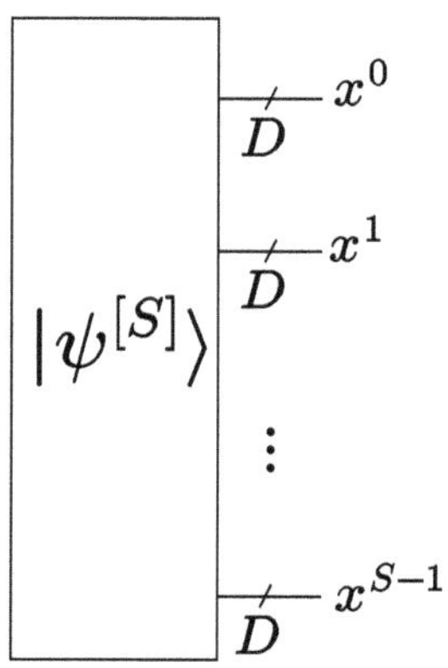

Figure 20.1 Tensor network representation of a generic joint pure state.

As a note, from a quantum physics perspective, the entire universe must be described as a joint state, which is often referred to as the **wave function of the universe**.

20.2.1 Globally Uncorrelated Quantum States

By the given definition, describing general joint quantum states $|\psi^{[S]}\rangle$ requires the specification of around D^S complex numbers. In practice, useful states are often structured, and the structure of the state can conveniently be described via a tensor network. The simplest example of a structured joint quantum state is one in which each qudit is assigned its own local pure state. As discussed in Section 14.2.2, the composition rule for separable states dictates that the joint state of such a system is represented as the Kronecker product of these local pure states.

Definition 20.2 Globally uncorrelated, or globally separable, quantum states
A quantum pure state $|\psi^{[S]}\rangle$ for a multipartite system with S qudits is said to be globally uncorrelated, or globally separable, if it can be expressed as the Kronecker product of local pure states $\{|\psi^s\rangle\}_{s\in[S]}$ for the S qudits, i.e.,

$$|\psi^{[S]}\rangle = |\psi^0\rangle \otimes |\psi^1\rangle \otimes \cdots \otimes |\psi^{S-1}\rangle. \tag{20.3}$$

State of the form (20.3) can be thought of as the counterpart of globally uncorrelated joint distributions $p^{[S]}$. In a manner analogous to globally uncorrelated distributions, the tensor network representing a globally uncorrelated quantum state contains S separate order-1 tensors, $\{|\psi^s\rangle\}_{s\in[S]}$, each of dimension D, is as shown in Figure 20.2. Accordingly, a globally uncorrelated state is fully described by the SD entries of the S local tensors $\{|\psi^s\rangle\}_{s\in[S]}$, yielding an exponential reduction in number of parameters as compared to a generic joint state $|\psi^{[S]}\rangle$.

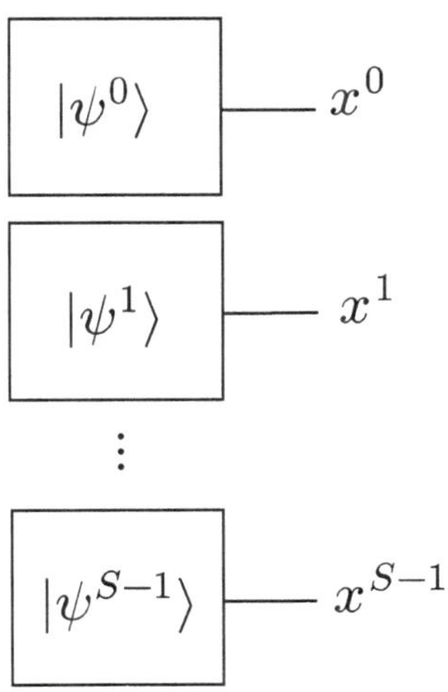

Figure 20.2 Tensor network representation of a globally uncorrelated quantum state.

The tensor product (20.3) can equivalently be written in terms of the entries of the individual states $\{|\psi^s\rangle\}_{s\in[S]}$ by using the multi-index notation

$$x^{[S]} = x^0 \dots x^{S-1} = \sum_{s\in[S]} D^{S-1-s} x^s. \tag{20.4}$$

To elaborate on this equivalent formulation, write as ψ_{x^s} the x^sth complex amplitude, with $x^s \in [D]$, of the state $|\psi^s\rangle$ of the sth qudit. Note that the superscript s is removed from the entry $\psi^s_{x^s}$ since the identity of the qudit is clear from the notation used for the index x^s. Each $x^{[S]}$th entry of the globally separable state $|\psi^{[S]}\rangle$ in (20.3) can then be written as the product of per-qudit amplitudes as

$$\psi_{x^{[S]}} = \psi_{x^0}\psi_{x^1}\cdots\psi_{x^{S-1}} = \prod_{s\in[S]} \psi_{x^s}. \tag{20.5}$$

20.2.2 Entangled Quantum States

States that are not globally uncorrelated are said to be entangled. In other words, for an entangled state, it is not possible to assign a pure state $|\psi^s\rangle$ to all the qudits $s \in [S]$.

Definition 20.3 Entangled quantum states

Any multipartite quantum state $|\psi^{[S]}\rangle$ that is not globally uncorrelated is said to be entangled.

Using the representation of a tensor as a vector introduced in Chapter 17, any joint pure state $|\psi^{[S]}\rangle$ can be expressed as the superposition

$$|\psi^{[S]}\rangle = \sum_{x^{[S]}} \psi_{x^{[S]}} |x^{[S]}\rangle, \tag{20.6}$$

where

$$|x^{[S]}\rangle = |x^0 \cdots x^{S-1}\rangle = |x^0\rangle \otimes \cdots \otimes |x^{S-1}\rangle \tag{20.7}$$

represents a globally uncorrelated state consisting of local computational-basis vectors. Thus, any joint state can be written as a superposition of globally uncorrelated states.

Example 20.1

Three-qubit GHZ and W states. Two important three-qubit states are the Greenberger–Horne–Zeilinger (GHZ) state, known for short as the **GHZ state,**

$$\begin{aligned} |\mathrm{GHZ}^{[3]}\rangle &= \frac{1}{\sqrt{2}}|000\rangle + \frac{1}{\sqrt{2}}|111\rangle \\ &= \frac{1}{\sqrt{2}}[1,0,0,0,0,0,0,1]^T; \end{aligned} \tag{20.8}$$

and the **W state**, named after the Austrian physicist Wolfgang Dür,

$$|\mathrm{W}^{[3]}\rangle = \frac{1}{\sqrt{3}}|001\rangle + \frac{1}{\sqrt{3}}|010\rangle + \frac{1}{\sqrt{3}}|100\rangle$$
$$= \frac{1}{\sqrt{3}}[0, 1, 1, 0, 1, 0, 0, 0]^T. \tag{20.9}$$

Formally, the GHZ state is the quantum counterpart of the distribution (18.9), corresponding to three equal uniform bits. Moreover, the W state is the counterpart of the distribution (18.11), which corresponds to three uniform one-hot configurations (in which each of the three bit-strings has exactly one 1 and two 0s).

The GHZ and W states can also be viewed as generalizations to three qubits of the two-qubit Bell states introduced in Chapter 14. In particular, the GHZ state is a natural generalization of the Bell state $|\Phi_+^{[2]}\rangle = \frac{1}{\sqrt{2}}|00\rangle + \frac{1}{\sqrt{2}}|11\rangle$, while the W state is a natural generalization of the Bell state $|\Psi_+^{[2]}\rangle = \frac{1}{\sqrt{2}}|01\rangle + \frac{1}{\sqrt{2}}|10\rangle$.

Example 20.2

General GHZ state. The GHZ state (20.8) can be extended to any number, S, of qudits. This is done by considering the state resulting from an equal superposition of globally uncorrelated states of the form

$$|z^{\otimes S}\rangle = |\underbrace{z \cdots z}_{S \text{ times}}\rangle$$
$$= \underbrace{|z\rangle \otimes \cdots \otimes |z\rangle}_{S \text{ times}}$$
$$= |z\rangle^{\otimes S} \tag{20.10}$$

for all $z \in [D]$. In the joint state (20.10), each subsystem is in the computational-basis state $|z\rangle$. GHZ states are thus expressed as the superposition

$$|\mathrm{GHZ}^{[S]}\rangle = \frac{1}{\sqrt{D}} \sum_{z \in [D]} |z^{\otimes S}\rangle. \tag{20.11}$$

The tensor representation of the GHZ state (20.11) is shown in Figure 20.3. Since only the entries with equal values of the indices are non-zero, the tensor coincides with the copy tensor δ with dimension D introduced in Chapter 17, apart from a scaling by $1/\sqrt{D}$.

$$|\psi^{[S]}\rangle = \frac{1}{\sqrt{D}} \sum_{z \in [D]} |z^{\otimes S}\rangle \equiv \frac{1}{\sqrt{D}} \cdot$$

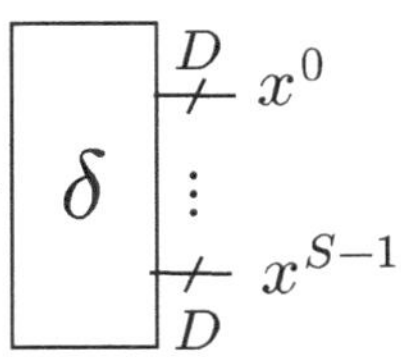

Figure 20.3 Tensor network representation of a GHZ state.

Example 20.3

Two-qudit maximally entangled state. GHZ states with $S = 2$ qudits are maximally entangled as they can be expressed as

$$|\Phi^{[2]}_{\max}\rangle = \frac{1}{\sqrt{D}}(|00\rangle + |11\rangle + \cdots + |D-1, D-1\rangle)$$
$$= \frac{1}{\sqrt{D}} \sum_{z\in[D]} |zz\rangle. \tag{20.12}$$

Thus, they have a uniform Schmidt probability vector.

The tensor representation of a two-qudit maximally entangled state (20.12) is shown in Figure 20.4. The last equivalence in the figure follows from Property 1 of the copy tensor in Appendix 17.A, which shows that the tensor representation of the maximally entangled state $|\Phi^{[2]}_{\max}\rangle$ is simply a wire connecting the two indices.

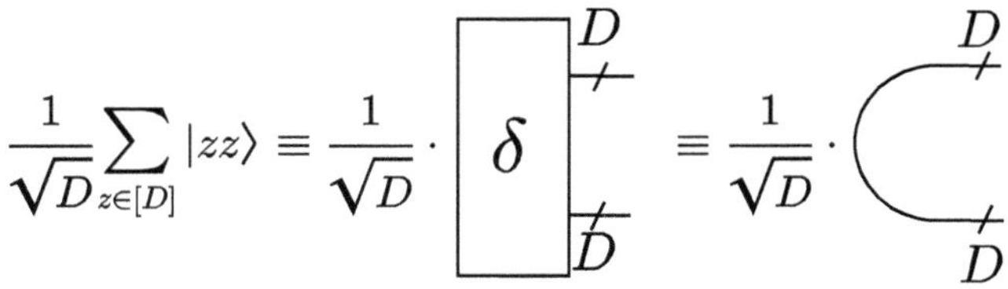

Figure 20.4 Tensor network representation of the two-qudit maximally entangled state (20.12).

The superposition (20.6) can be generalized beyond the computational basis by using the following result (see Problem 20.2).

Key Result 20.1 Globally uncorrelated orthonormal basis
Fix an orthonormal basis $\{|v_{x^s}\rangle\}_{x^s\in[D]}$ for each sth qudit with $s \in [S]$. Then, the collection of the D^S globally uncorrelated states $\{|v_{x^{[S]}}\rangle\}_{x^{[S]}\in[D]^S}$, with

$$|v_{x^{[S]}}\rangle = |v_{x^0}\rangle \otimes \cdots \otimes |v_{x^{S-1}}\rangle, \tag{20.13}$$

is an orthonormal basis for the D^S-dimensional vector space.

This implies that, given any local orthonormal bases $\{|v_{x^{[S]}}\rangle\}_{x^{[S]}\in[D]^S}$, an arbitrary state $|\psi^{[S]}\rangle$ can be expressed as the superposition

$$|\psi^{[S]}\rangle = \sum_{x^{[S]}\in[D]^S} \phi_{x^{[S]}} |v_{x^{[S]}}\rangle \tag{20.14}$$

with complex amplitudes given by the inner products

$$\phi_{x^{[S]}} = \langle v_{x^{[S]}} | \psi^{[S]}\rangle. \tag{20.15}$$

20.2.3 Canonic Polyadic Decomposition

As discussed in this section, any state $|\psi^{[S]}\rangle$ can be expressed as the superposition of globally uncorrelated states. Any such superposition must have at least two terms in order to describe an entangled state, since a single term would indicate a globally uncorrelated state. The minimum number of globally uncorrelated states required in any such superposition of a state $|\psi^{[S]}\rangle$ provides a measure of the level of entanglement present in $|\psi^{[S]}\rangle$.

For a two-qudit state, this minimum number is given by the Schmidt rank. For systems with more than two qudits, however, there is no direct extension of the Schmidt decomposition, and finding the minimum number of globally uncorrelated states for a state $|\psi^{[S]}\rangle$ is an involved problem.

This problem, in fact, coincides with that of identifying the tensor rank $r_{\min}$ of the state $|\psi^{[S]}\rangle$. When specialized to a quantum state $|\psi^{[S]}\rangle$, the CPD introduced in Section 17.7 is a decomposition of the form

$$|\psi^{[S]}\rangle = \sum_{z\in[r]} \sqrt{p_z}|\psi_z^0\rangle \otimes |\psi_z^1\rangle \otimes \cdots \otimes |\psi_z^{S-1}\rangle, \tag{20.16}$$

for some local state $\{|\psi_z^s\rangle\}_{z\in[r]}$ of qudits $s \in [S]$ and for some probability vector $p = [p_0, \ldots, p_{r-1}]^T$. Furthermore, when minimized over all possible CPDs, integer r is referred to as the tensor rank $r_{\min}$ of the state.

A tensor network representation of the CPD (20.16) is shown in Figure 20.5. Following Section 17.7, the tensor network encompasses the order-1 tensor p, the order-2 tensors Ψ^s with (x^s, z)th entry given by the x^sth entry of state $|\psi_z^s\rangle$, as well as a copy tensor δ to describe the sum in (20.16).

As mentioned, while for $S = 2$ the tensor rank can readily be computed as the Schmidt rank, for $S > 2$ qudits the evaluation of the tensor rank $r_{\min}$ is often prohibitively complex. The next examples illustrate some special cases in which the tensor rank $r_{\min}$ can be determined even for $S > 2$ subsystems.

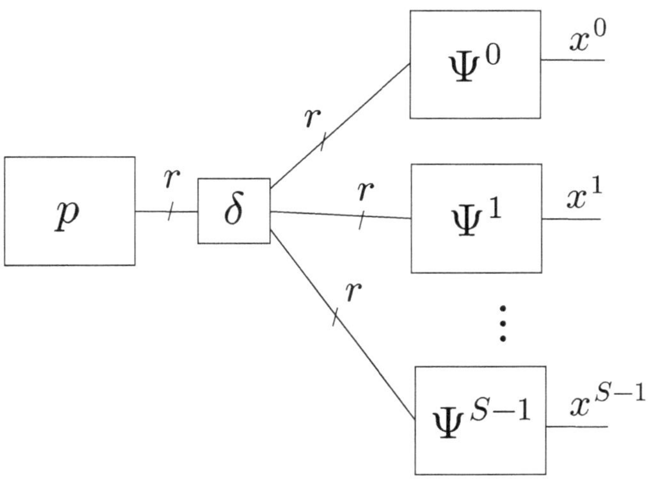

Figure 20.5 Tensor network representation of a joint state whose structure is described by the CPD (20.16).

Example 20.4

The three-qubit state

$$|\psi^{[3]}\rangle = \frac{1}{2\sqrt{2}}[1,1,1,1,1,1,1,1]^T$$
$$= \frac{1}{2\sqrt{2}}(|000\rangle + |001\rangle + |010\rangle + |011\rangle + |100\rangle + |101\rangle + |110\rangle + |111\rangle) \quad (20.17)$$

is an equal superposition of all states in the computational basis. The state (20.17) is thus in the form of a CPD with $r = 8$ globally uncorrelated states.

However, this is not the CPD of state (20.17) with the smallest possible number r of globally uncorrelated states. In particular, the state (20.17) can also be written as

$$|\psi^{[3]}\rangle = |+\rangle \otimes |+\rangle \otimes |+\rangle, \quad (20.18)$$

showing that the state is globally uncorrelated. This decomposition clearly yields the minimum number of globally uncorrelated states, and thus the tensor rank is $r_{\min} = 1$.

Example 20.5

The W state (20.9) is in the CPD form with $r = 3$ globally uncorrelated states, and with probability vector equal to the uniform distribution. It can be proved that the tensor rank of the W state equals $r_{\min} = 3$ (see Recommended Resources).

Example 20.6

The GHZ state (20.11) is in the CPD form (20.16) with $r = D$ globally separable states, probability vector p given by the uniform distribution, and local states $|\psi_z^0\rangle = |z\rangle$ for all $z \in [D]$. As will be shown in Section 20.5, the tensor rank equals $r_{\min} = D$.

20.3 Local Measurements

This section elaborates on the definition and implementation of local complete measurements in multipartite systems. Local POVM settings may be defined in a similar way, although this will not be detailed here.

20.3.1 Defining Local Measurements

A local complete measurement setting

$$\mathcal{M}^{[S]} = \mathcal{M}^0 \otimes \mathcal{M}^1 \otimes \cdots \otimes \mathcal{M}^{S-1} \quad (20.19)$$

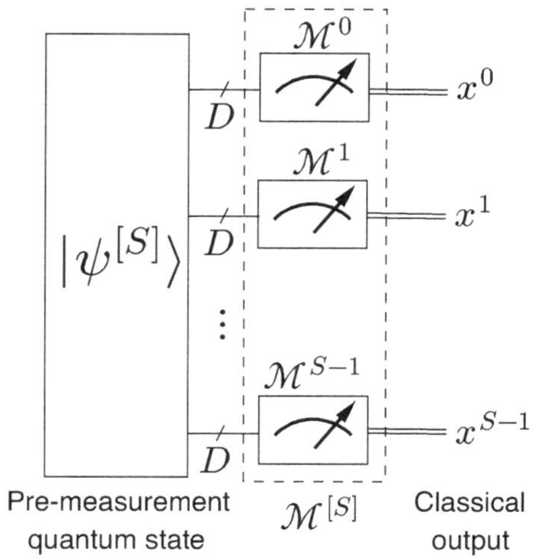

Figure 20.6 Illustration of local complete measurements in setting $\mathcal{M}^{[S]}$ in (20.19).

is specified by S orthonormal bases $\mathcal{M}^s = \{|v_x^s\rangle\}_{x\in[D]}$, one for each qudit $s \in [S]$. As illustrated in Figure 20.6, under setting $\mathcal{M}^{[S]}$, each qudit s is measured in basis $\mathcal{M}^s$, returning as classical output the realization $x^s \in [D]$ of a random variable x^s. As a result, a local measurement in setting $\mathcal{M}^{[S]}$ yields as classical outputs a collection of S jointly random variables $\mathrm{x}^{[S]} = (\mathrm{x}^0, \ldots, \mathrm{x}^{S-1})$. Note that the figure only shows the classical outputs, without depicting the post-measurement states.

The local complete measurement setting $\mathcal{M}^{[S]}$ can be viewed as a complete measurement in the globally uncorrelated basis $\mathcal{M}^{[S]}$ given by (20.13), i.e.,

$$\mathcal{M}^{[S]} = \{|v_{x^{[S]}}\rangle\}_{x^{[S]}\in[D]^S}. \tag{20.20}$$

Thus, applying the Born rule one obtains the following characterization of the joint distribution of the classical outputs $\mathrm{x}^{[S]}$.

Key Result 20.2 Local measurements of a multipartite quantum system

Given a local measurement setting $\mathcal{M}^{[S]}$ as in (20.19), for a joint pure state $|\psi^{[S]}\rangle$, the measurement results in the following.

- **Classical output:** The measurement yields classical measurement outputs $\mathrm{x}^{[S]} = x^{[S]}$ with probability

$$\begin{aligned} p_{x^{[S]}} &= p_{x^{[S]}}(|\psi^{[S]}\rangle|\mathcal{M}^{[S]}) \\ &= |\langle v_{x^{[S]}}|\psi^{[S]}\rangle|^2. \end{aligned} \tag{20.21}$$

- **Post-measurement state:** For a given classical output $\mathrm{x}^{[S]} = x^{[S]}$, the quantum system is left in the corresponding post-measurement state $|v_{x^{[S]}}\rangle$, which is globally uncorrelated.

This result implies that local complete measurements destroy entanglement, causing the post-measurement state to be globally uncorrelated.

Example 20.7

For the three-qubit GHZ state (20.8), local measurements in the computational basis produce the three bits $\mathrm{x}^{[3]} = (\mathrm{x}^0, \mathrm{x}^1, \mathrm{x}^2)$ with joint probability distribution given by

$$p_{0,0,0} = p_{1,1,1} = |\mathrm{GHZ}_{0,0,0}|^2 = |\mathrm{GHZ}_{1,1,1}|^2 = 0.5, \tag{20.22}$$

and $p_{x^0,x^1,x^2} = 0$ for any other choice of the indices $x^{[3]} \in [2]^3$. Therefore, the three measurement outputs are globally maximally correlated (see Example 18.28).

Example 20.8

For a GHZ state (20.11) with any number S of qudits, the classical measurement outputs $\mathrm{x}^{[S]} = (\mathrm{x}^0, \mathrm{x}^1, \ldots, \mathrm{x}^{S-1})$ produced by local measurements in the computational basis are also marginally uniform and equal with probability 1. Hence, the classical outputs are globally correlated, but not globally maximally correlated. In fact, using the definitions in Section 18.3, they are maximally correlated across 1-cuts, attaining the mutual information $\mathrm{I}(X^s;X^{[S]\setminus s})_{p^{[S]}} = \log_2(D)$ for all $s \in [S]$, but they are not maximally correlated across C-cuts with $C > 1$.

Example 20.9

For the three-qubit W state (20.9), the measurement probability distribution under the computational basis is obtained from the Born rule as

$$p_{0,0,1} = p_{0,1,0} = p_{1,0,0} = \frac{1}{3} \tag{20.23}$$

and $p_{x^0 x^1 x^2} = 0$ for any other index $x^{[3]} \in [2]^3$. Therefore, the measurement outputs $\mathrm{x}^{[3]} = (\mathrm{x}^0, \mathrm{x}^1, \mathrm{x}^2)$ are globally correlated, but not maximally globally correlated (see Examples 18.28 and 18.30).

20.3.2 Local Measurements via Local Unitary Transformations

As detailed in Section 10.7, a complete measurement in any basis $\mathcal{M}$ can be implemented by carrying out a measurement in the computational basis. To this end, the computational-basis measurement is preceded by a unitary transformation with a unitary matrix that collects by column the states of the basis $\mathcal{M}$. As discussed next, this implementation can be generalized to local measurement settings in a multipartite system.

To elaborate, consider any local complete measurement setting $\mathcal{M}^{[S]}$ with local measurement bases $\mathcal{M}^s = \{|v_x^s\rangle\}_{x \in [D]}$ for all qudits $s \in [S]$. As illustrated in Figure 20.7, this measurement can be implemented by first applying a local unitary transformation $(V^s)^\dagger$ to each qudit $s \in [S]$, and then carrying out a measurement in the

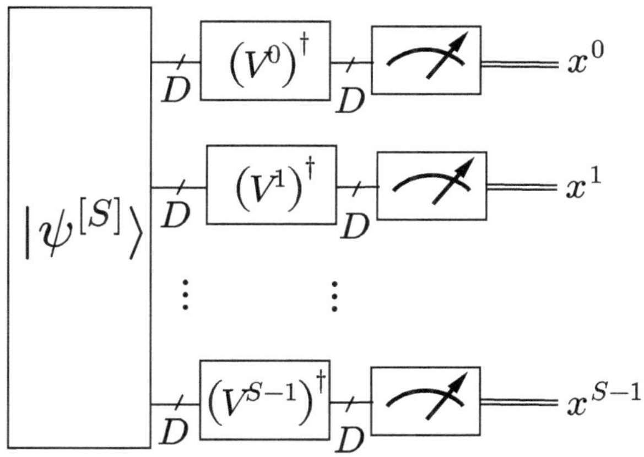

Figure 20.7 A local complete measurement setting $\mathcal{M}^{[S]}$ with local measurement bases $\mathcal{M}^s = \{|v_x^s\rangle\}_{x\in[D]}$ for all qudits $s \in [S]$ can be implemented by first applying a local unitary transformation $(V^s)^\dagger$ to each qudit $s \in [S]$ and then applying a measurement in the computational basis.

computational basis. The unitary matrix for qudit s collects as columns the vectors of the measurement basis $\mathcal{M}^s$, i.e.,

$$V^s = [|v_0^s\rangle \cdots |v_{D-1}^s\rangle]. \tag{20.24}$$

The operation of the S parallel unitary transformations is described by the $D^S \times D^S$ unitary matrix

$$V^{[S]} = V^0 \otimes V^1 \otimes \cdots \otimes V^{S-1} \tag{20.25}$$

obtained as the Kronecker product of all the local unitary matrices V^s, one for each qudit $s \in [S]$. Furthermore, the post-transformation state of the overall system is given by

$$|\phi^{[S]}\rangle = V^{[S]}|\psi^{[S]}\rangle. \tag{20.26}$$

To show that the system in Figure 20.7 implements the desired measurement in basis $\mathcal{M}^{[S]}$, one must demonstrate the equality between the measurement probability distribution $p(|\psi^{[S]}\rangle|\mathcal{M}^{[S]})$ and $p(|\phi^{[S]}\rangle|\mathcal{M}_c^{[S]})$, where $\mathcal{M}_c^{[S]}$ denotes the computational basis. This condition amounts to the equalities

$$|\langle v_{x^{[S]}}|\psi^{[S]}\rangle|^2 = |\langle x^{[S]}|\phi^{[S]}\rangle|^2 \tag{20.27}$$

for all indices $x^{[S]} \in [D]^S$.

Starting with the left-hand side of the equality (20.27) and using (20.26), one gets

$$\begin{aligned} |\langle x^{[S]}|\phi^{[S]}\rangle|^2 &= |\langle x^{[S]}|(V^{[S]}) \dagger |\psi^{[S]}\rangle|^2 \\ &= |\langle v_{x^{[S]}}|\psi^{[S]}\rangle|^2, \end{aligned} \tag{20.28}$$

where the last step uses the equality

$$\begin{aligned} V^{[S]}|x^{[S]}\rangle &= V^0|x^0\rangle \otimes \cdots \otimes V^{S-1}|x^{S-1}\rangle \\ &= |v^0_{x^0}\rangle \otimes \cdots \otimes |v^{S-1}_{x^{S-1}}\rangle \\ &= |v_{x^{[S]}}\rangle. \end{aligned} \tag{20.29}$$

Note, in fact, that the operation $V^s|x^s\rangle$ selects the x^sth column of matrix V^s. This proves the equivalence between the two measurements.

20.4 Bipartite Entanglement

As in the classical case explored in Chapter 18, the correlations in a composite quantum system can be analyzed by examining bipartite correlations across different partitions of its subsystems. Building on the discussion in Chapter 14, any partition of the system divides it into two groups of subsystems, which exhibit equal amounts of classical and quantum correlations across the two groups. In fact, both the maximum mutual information, representing classical correlations, and the entanglement entropy, representing quantum correlations, are equal to the entropy of the Schmidt probability vector for the given bipartite system.

Based on this result, this section focuses on the entanglement entropy across partitions of a multipartite quantum system, keeping in mind that an equivalent amount of classical correlations exists across each partition.

20.4.1 Cuts

As illustrated in Figure 20.8, consider a subset $\mathcal{S} \subset [S]$ of qudits and its complement subset $\bar{\mathcal{S}} = [S] \backslash \mathcal{S}$. As in Section 18.3, a subset $\mathcal{S}$ defines a C-cut that splits the system into the two disjoint subsets $\mathcal{S}$ and $\bar{\mathcal{S}}$, where C is the cardinality of the smaller subset

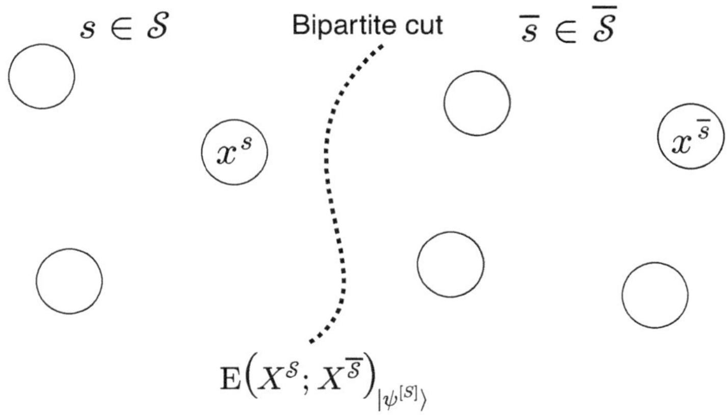

Figure 20.8 Illustration of a C-cut that splits the system into the two disjoint subsets $\mathcal{S}$ and $\bar{\mathcal{S}}$, where C is the cardinality of subset $\mathcal{S}$, i.e., $C = |\mathcal{S}|$.

$\mathcal{S}$, i.e., $C = |\mathcal{S}|$. Furthermore, the size of the smaller subset $\mathcal{S}$ is bounded as

$$C \leq \left\lfloor \frac{S}{2} \right\rfloor. \tag{20.30}$$

20.4.2 Entanglement Across a Cut

Given a cut $\mathcal{S}$, the goal in this section is to study the entanglement present between the groups of qudits $\mathcal{S}$ and $\bar{\mathcal{S}}$ for any given pure state $|\psi^{[S]}\rangle$. At an operational level, entanglement across a cut indicates that the two subsets of qudits share an informational resource that can be leveraged to carry out tasks such as superdense coding and teleportation. Conversely, when no entanglement is present across a cut, the two groups of qudits are limited by classical constraints on the information tasks that they can carry out jointly.

To study the entanglement across a cut $\mathcal{S}$ for a state $|\psi^{[S]}\rangle$, it is notationally convenient to reorder the indices of the tensor $|\psi^{[S]}\rangle$ by listing first the C indices of the qudits in subset $\mathcal{S}$ and then the $S - C$ indices for the qudits in complementary subset $\bar{\mathcal{S}}$. The resulting tensor with reordered entries is denoted as $|\psi^{\mathcal{S},\bar{\mathcal{S}}}\rangle$ and its entries are written as $\psi_{x^{\mathcal{S}}x^{\bar{\mathcal{S}}}}$, where $x^{\mathcal{S}}$ collects the indices for the qudits in subset $\mathcal{S}$ and $x^{\bar{\mathcal{S}}}$ collects the indices for the qudits in subset $\bar{\mathcal{S}}$. As in previous chapters (see Section 17.6), the indices $x^{\mathcal{S}}$ and $x^{\bar{\mathcal{S}}}$ are listed according to the natural ordering of the integers in subsets $\mathcal{S}$ and $\bar{\mathcal{S}}$, respectively.

Example 20.10

For the three-qubit state $|\psi^{[3]}\rangle = 1/\sqrt{3}[1,0,1,0,0,0,0,1]^T$, consider the cut $\mathcal{S} = \{1\}$. The state vector $|\psi^{[3]}\rangle$ with indices reordered according to this cut has entries ψ_{x^1,x^0,x^2}, which can be written as the vector

$$|\psi^{1,0,2}\rangle = \frac{1}{\sqrt{3}}[1,0,0,0,1,0,0,1]^T. \tag{20.31}$$

For example, the entry with multi-index $x^0x^1x^2 = 010 = 2$ in the original tensor $|\psi^{[3]}\rangle$, which equals $\psi^{[3]}_{0,1,0} = 1$, is moved to the entry $x^1x^0x^2 = 100 = 4$ in the tensor $|\psi^{1,0,2}\rangle$, i.e., $\psi^{1,0,2}_{1,0,0} = 1$.

Definition 20.4 Uncorrelated and entangled states across a cut

A state $|\psi^{[S]}\rangle$ is said to be uncorrelated, or separable, across a cut $\mathcal{S}$ if it can be written, upon reordering of the indices, as the Kronecker product

$$|\psi^{\mathcal{S},\bar{\mathcal{S}}}\rangle = |\psi^{\mathcal{S}}\rangle \otimes |\psi^{\bar{\mathcal{S}}}\rangle \tag{20.32}$$

between a $D^C \times 1$ state vector $|\psi^{\mathcal{S}}\rangle$ for the group $\mathcal{S}$ of qudits and a $D^{S-C} \times 1$ state vector $|\psi^{\bar{\mathcal{S}}}\rangle$ for the group $\bar{\mathcal{S}}$ of qudits. When state $|\psi^{[S]}\rangle$ cannot be factorized as in (20.32), it is said to be entangled across the cut $\mathcal{S}$.

In other words, state $|\psi^{[S]}\rangle$ is uncorrelated across cut $\mathcal{S}$ if the two groups of qudits $\mathcal{S}$ and $\bar{\mathcal{S}}$ have individual states $|\psi^{\mathcal{S}}\rangle$ and $|\psi^{\bar{\mathcal{S}}}\rangle$, and it is entangled across the cut otherwise. Note that, for a state that is uncorrelated across a cut $\mathcal{S}$, each state $|\psi^{\mathcal{S}}\rangle$ and $|\psi^{\bar{\mathcal{S}}}\rangle$ may or may not be factorizable into separate states across internal partitions for the respective subsets $\mathcal{S}$ and $\bar{\mathcal{S}}$.

Example 20.11

For a system of $S = 3$ qubits, consider the state

$$|\psi^{[3]}\rangle = \frac{1}{\sqrt{2}}[1, 0, 0, 1, 0, 0, 0, 0]^T. \tag{20.33}$$

As can be verified directly, this state can be written as

$$|\psi^{[3]}\rangle = |0\rangle \otimes |\Phi_+\rangle, \tag{20.34}$$

where $|\Phi_+\rangle = 1/\sqrt{2}(|00\rangle + |11\rangle) = [1, 0, 0, 1]^T$ is the Bell state, which is maximally entangled. Therefore, this state is uncorrelated across the cut $\mathcal{S} = \{0\}$, but it is entangled across the other two cuts $\mathcal{S} = \{1\}$ and $\mathcal{S} = \{2\}$.

20.4.3 Reduced Density Matrices

Following Chapter 14, in order to quantify the entanglement across a cut $\mathcal{S}$, one needs to evaluate the Schmidt probability vector $\lambda^{\mathcal{S}}$ for the bipartite system consisting of the groups of qudits $\mathcal{S}$ and $\bar{\mathcal{S}}$. As detailed in Section 14.5, the Schmidt probability vector $\lambda^{\mathcal{S}}$ can be computed either from the singular values of the amplitude matrix corresponding to cut $\mathcal{S}$, or, equivalently, as the eigenvalues of the reduced density matrices for groups $\mathcal{S}$ and $\bar{\mathcal{S}}$. The following definition uses the operation of matrization introduced in Section 17.6.

Definition 20.5 Amplitude matrix for a cut

Given a pure state $|\psi^{[S]}\rangle$, the amplitude matrix corresponding to a C-cut $\mathcal{S}$ is given by the $D^C \times D^{S-C}$ matrix $\Psi^{\mathcal{S}}$ obtained as the matrization of tensor $|\psi^{[S]}\rangle$ with row indices $\mathcal{S}$, i.e.,

$$\begin{aligned}\Psi^{\mathcal{S}} &= \text{mat}_{\mathcal{S}}(|\psi^{[S]}\rangle) \\ &= \sum_{x^{\mathcal{S}}} \sum_{x^{\bar{\mathcal{S}}}} \psi_{x^{\mathcal{S}}, x^{\bar{\mathcal{S}}}} |x^{\mathcal{S}}\rangle\langle x^{\bar{\mathcal{S}}}|,\end{aligned} \tag{20.35}$$

where $\psi_{x^{\mathcal{S}}, x^{\bar{\mathcal{S}}}}$ represents the the amplitude of the reordered tensor $|\psi^{\mathcal{S}, \bar{\mathcal{S}}}\rangle$ corresponding to indices $(x^{\mathcal{S}}, x^{\bar{\mathcal{S}}})$.

The squared singular values of the amplitude matrix $\Psi^{\mathcal{S}}$ yield the Schmidt probability vector $\lambda^{\mathcal{S}}$ for cut $\mathcal{S}$. The Schmidt probability vector $\lambda^{\mathcal{S}}$ can also be obtained as the eigenvalues of the reduced density matrices for group $\mathcal{S}$ or group $\bar{\mathcal{S}}$. The reduced density matrices can be computed from the amplitude matrix, as detailed in Section 14.5, or via the partial trace operation introduced in Section 14.10.

In particular, from Section 16.2, the reduced density matrix for group $\mathcal{S}$ can be written as

$$\rho^{\mathcal{S}} = \mathrm{tr}_{\bar{\mathcal{S}}}(|\psi^{[S]}\rangle\langle\psi^{[S]}|), \tag{20.36}$$

where $\mathrm{tr}_{\bar{\mathcal{S}}}(\cdot)$ represents the partial trace with respect to subsystems in group $\bar{\mathcal{S}}$. Accordingly, the $(x^{\mathcal{S}}, y^{\mathcal{S}})$th entry of the reduced density matrix $\rho^{\mathcal{S}}$ is given by the sum

$$\begin{aligned}\rho^{\mathcal{S}}_{x^{\mathcal{S}},y^{\mathcal{S}}} &= \sum_{x^{\bar{\mathcal{S}}}} [|\psi^{\mathcal{S},\bar{\mathcal{S}}}\rangle\langle\psi^{\mathcal{S},\bar{\mathcal{S}}}|]_{x^{\mathcal{S}}x^{\bar{\mathcal{S}}},y^{\mathcal{S}}x^{\bar{\mathcal{S}}}} \\ &= \sum_{x^{\bar{\mathcal{S}}}} \psi_{x^{\mathcal{S}}x^{\bar{\mathcal{S}}}}\psi^*_{y^{\mathcal{S}}x^{\bar{\mathcal{S}}}}\end{aligned} \tag{20.37}$$

over the equal values of the row and column indices for group $\bar{\mathcal{S}}$, where $[\cdot]_{a,b}$ represents the (a, b)th entry of the argument matrix.

As shown in Figure 20.9, the reduced density matrix $\rho^{\mathcal{S}}$ can readily be understood via the tensor matrix representation of the density matrix $|\psi^{[S]}\rangle\langle\psi^{[S]}|$. In particular, the figure depicts the density matrix $|\psi^{[S]}\rangle\langle\psi^{[S]}|$ as a tensor formatted as a matrix with separate row indices $x^{\mathcal{S}}, x^{\bar{\mathcal{S}}}$ and column indices $y^{\mathcal{S}}, y^{\bar{\mathcal{S}}}$. Following Section 14.10, the partial trace $\mathrm{tr}_{\bar{\mathcal{S}}}(\cdot)$ is represented by creating a bond between the row and column indices corresponding to the subset $\bar{\mathcal{S}}$, so that each row index x^s for qudit s is connected to the corresponding column index y^s, for all $s \in \bar{\mathcal{S}}$. The remaining row and column open indices, pointing towards the right and left, respectively, determine the entries $\rho^{\mathcal{S}}_{x^{\mathcal{S}},y^{\mathcal{S}}}$ of the reduced density matrix $\rho^{\mathcal{S}}$ in (20.37).

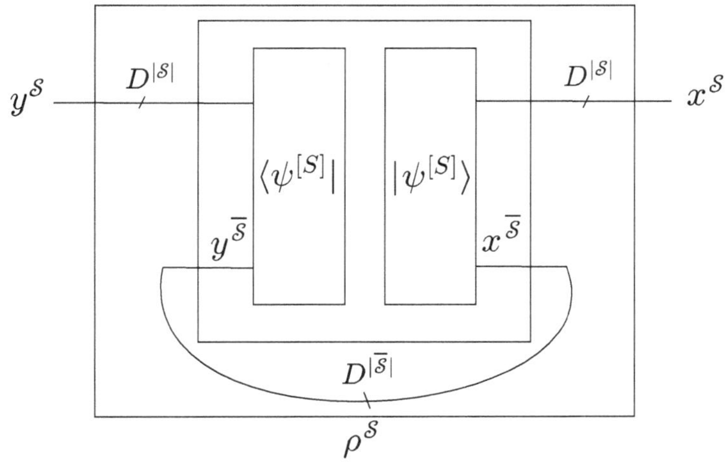

Figure 20.9 Tensor network representation of the reduced density matrix $\rho^{\mathcal{S}}$ via the tensor matrix representation of the density matrix $|\psi^{[S]}\rangle\langle\psi^{[S]}|$.

The reduced density matrix for subset $\bar{\mathcal{S}}$ can similarly be written as the result of the partial trace over subsystem $\mathcal{S}$, i.e.,

$$\rho^{\bar{\mathcal{S}}} = \mathrm{tr}_{\mathcal{S}}(|\psi^{[S]}\rangle\langle\psi^{[S]}|). \tag{20.38}$$

Following Section 14.10, the partial trace is linear. Furthermore, it satisfies the equalities

$$\mathrm{tr}_{\bar{\mathcal{S}}}(\rho^{\mathcal{S}} \otimes \rho^{\bar{\mathcal{S}}}) = \rho^{\mathcal{S}} \cdot \mathrm{tr}(\rho^{\bar{\mathcal{S}}}) = \rho^{\mathcal{S}} \tag{20.39}$$

and

$$\mathrm{tr}_{\mathcal{S}}(\rho^{\mathcal{S}} \otimes \rho^{\bar{\mathcal{S}}}) = \mathrm{tr}(\rho^{\mathcal{S}}) \cdot \rho^{\bar{\mathcal{S}}} = \rho^{\bar{\mathcal{S}}} \tag{20.40}$$

for any mixed state of the form $\rho^{\mathcal{S}} \otimes \rho^{\bar{\mathcal{S}}}$, with $\rho^{\mathcal{S}}$ and $\rho^{\bar{\mathcal{S}}}$ being the mixed states of groups $\mathcal{S}$ and $\bar{\mathcal{S}}$.

Example 20.12

Consider a diagonal density matrix

$$\rho^{[S]} = \mathrm{diag}(p^{[S]}), \tag{20.41}$$

with diagonal elements given by the entries of the joint probability vector $p^{[S]}$. Using (20.37), the reduced density matrix for any subset $\mathcal{S}$ is also diagonal, with diagonal element in position $(x^{\mathcal{S}}, x^{\mathcal{S}})$ given as

$$\rho^{\mathcal{S}}_{x^{\mathcal{S}},x^{\mathcal{S}}} = \sum_{x^{\bar{\mathcal{S}}}} p_{x^{\mathcal{S}}x^{\bar{\mathcal{S}}}} = p_{x^{\mathcal{S}}}, \tag{20.42}$$

where $p_{x^{\mathcal{S}}}$ represents an entry of the marginal distribution $p^{\mathcal{S}}$ for subset $\mathcal{S}$. Therefore, the reduced density matrix is given by

$$\rho^{\mathcal{S}} = \mathrm{diag}(p^{\mathcal{S}}). \tag{20.43}$$

This example confirms that the partial trace corresponds to the operation of marginalization for probability distributions.

Example 20.13

The GHZ state (20.11) has a maximally mixed reduced density matrix ρ^s for every qudit $s \in [S]$, i.e.,

$$\rho^s = \mathrm{tr}_{[S]\setminus s}\left(|\mathrm{GHZ}^{[S]}\rangle\langle\mathrm{GHZ}^{[S]}|\right) = \pi = \frac{1}{D}I, \tag{20.44}$$

where I is the $D \times D$ identity matrix. This result can be derived as follows.

The reordered GHZ state can be expressed as $|\mathrm{GHZ}^{s,[S]\setminus s}\rangle = \frac{1}{\sqrt{D}}\sum_{z\in[D]}|z\rangle \otimes |z^{\otimes S-1}\rangle$. Thus, the reduced density matrix for any qudit s is calculated using the linearity and the property (20.40) of the partial trace as

$$\frac{1}{D}\mathrm{tr}_{[\mathcal{S}]\setminus s}\left(\sum_{z\in[D]}\sum_{z'\in[D]}|z\rangle\langle z'|\otimes(|z\rangle\langle z'|)^{\otimes S-1}\right)$$

$$=\frac{1}{D}\sum_{z\in[D]}\sum_{z'\in[D]}|z\rangle\langle z'|\cdot\mathrm{tr}\left((|z\rangle\langle z'|)^{\otimes S-1}\right)$$

$$=\frac{1}{D}\sum_{z\in[D]}|z\rangle\langle z|=\pi, \tag{20.45}$$

where

$$A^{\otimes M}=\underbrace{A\otimes\cdots\otimes A}_{M} \tag{20.46}$$

indicates the Kronecker product of M copies of matrix A.

More generally, considering any C-cut $\mathcal{S}$, the reduced density matrix $\rho^{\mathcal{S}}$ can be obtained as

$$\rho^{\mathcal{S}}=\frac{1}{D}\mathrm{tr}_{\bar{\mathcal{S}}}\left(\sum_{z\in[D]}\sum_{z'\in[D]}(|z\rangle\langle z'|)^{\otimes C}\otimes(|z\rangle\langle z'|)^{\otimes S-C}\right)$$

$$=\frac{1}{D}\sum_{z\in[D]}\sum_{z'\in[D]}(|z\rangle\langle z'|)^{\otimes C}\cdot\mathrm{tr}\left((|z\rangle\langle z'|)^{\otimes S-C}\right)$$

$$=\frac{1}{D}\sum_{z\in[D]}(|z\rangle\langle z|)^{\otimes C}$$

$$=\frac{1}{D}\sum_{z\in[D]}|z^{\otimes C}\rangle\langle z^{\otimes C}|. \tag{20.47}$$

This is a diagonal matrix with D non-zero elements at positions characterized by the multi-index

$$z^{\otimes C}=z\sum_{s=0}^{C-1}D^{C-1-s}=z\frac{D^{C}-1}{D-1} \tag{20.48}$$

for $z\in[D]$. Thus, the reduced density matrix $\rho^{\mathcal{S}}$ is not maximally mixed unless $|\mathcal{S}|=C=1$.

Example 20.14

For the W state (20.9), the reduced density matrices for the individual qubits are not maximally mixed. In fact, considering without loss of generality qudit $\mathcal{S}=\{0\}$, one can write the W state as $|\mathrm{W}\rangle=\sqrt{\frac{2}{3}}|0\rangle\otimes|\Psi_{+}\rangle+\sqrt{\frac{1}{3}}|1\rangle\otimes|00\rangle$, with the Bell state $|\Psi_{+}\rangle=(|01\rangle+|10\rangle)/\sqrt{2}$. This yields the partial trace

$$\rho^0 = \mathrm{tr}_{1,2}\left(\frac{2}{3}|0\rangle\langle 0| \otimes |\Psi_+\rangle\langle\Psi_+| + \frac{1}{3}|1\rangle\langle 1| \otimes |00\rangle\langle 00| \right.$$
$$\left. + \frac{\sqrt{2}}{3}|0\rangle\langle 1| \otimes |\Psi_+\rangle\langle 00| + \frac{\sqrt{2}}{3}|1\rangle\langle 0| \otimes |00\rangle\langle\Psi_+|\right)$$
$$= \frac{2}{3}|0\rangle\langle 0| + \frac{1}{3}|1\rangle\langle 1|, \tag{20.49}$$

which is diagonal, but not maximally mixed.

20.4.4 Schmidt Probability Vector Across a Cut

The Schmidt rank $r^{\mathcal{S}}$ – i.e., the rank of the reduced density matrices $\rho^{\mathcal{S}}$ and $\rho^{\bar{\mathcal{S}}}$ – determines whether or not the state $|\psi^{[S]}\rangle$ is entangled across cut $\mathcal{S}$.

Definition 20.6 Schmidt rank across a cut

For any pure state $|\psi^{[S]}\rangle$ and a C-cut $\mathcal{S}$, the rank of the reduced density matrices $\rho^{\mathcal{S}}$ and $\rho^{\bar{\mathcal{S}}}$,

$$r^{\mathcal{S}} = \mathrm{rank}(\rho^{\mathcal{S}}) = \mathrm{rank}(\rho^{\bar{\mathcal{S}}}) \leq D^C, \tag{20.50}$$

is known as the Schmidt rank for cut $\mathcal{S}$.

Key Result 20.3 Schmidt rank and entanglement across a cut

A joint pure state $|\psi^{XY}\rangle$ is entangled across cut $\mathcal{S}$ if and only if the corresponding Schmidt rank is larger than 1:

$$|\psi^{XY}\rangle \text{ entangled across cut } \mathcal{S} \Longleftrightarrow r^{\mathcal{S}} > 1. \tag{20.51}$$

20.4.5 Entanglement Entropy Across a Cut

Following Section 14.7, the level of entanglement across a cut can be quantified via the entanglement entropy, which is a function of the corresponding Schmidt probability vector.

Definition 20.7 Schmidt probability vector across a cut

For any pure state $|\psi^{[S]}\rangle$ and a C-cut $\mathcal{S}$, the $r^{\mathcal{S}}$ non-zero eigenvalues $\lambda^{\mathcal{S}} = [\lambda_0^{\mathcal{S}}, \lambda_1^{\mathcal{S}}, \ldots, \lambda_{r^{\mathcal{S}}-1}^{\mathcal{S}}]^T$ of the reduced density matrices $\rho^{\mathcal{S}}$ and $\rho^{\bar{\mathcal{S}}}$ form the Schmidt probability vector for cut $\mathcal{S}$, which satisfies the conditions

$$\lambda_z^{\mathcal{S}} > 0 \text{ for all } z \in [r] \text{ and } \sum_{z\in[r]} \lambda_z^{\mathcal{S}} = 1. \tag{20.52}$$

Each Schmidt probability $\lambda_z^{\mathcal{S}}$ coincides with the square root of the corresponding singular value of the amplitude matrix $\Psi^{\mathcal{S}}$.

Definition 20.8 Entanglement entropy across a cut

For a state $|\psi^{[S]}\rangle$, the entanglement entropy across a cut $\mathcal{S}$ is defined as

$$\begin{aligned} \mathrm{E}(X^{\mathcal{S}};X^{\bar{\mathcal{S}}})_{|\psi^{[S]}\rangle} &= \mathrm{H}(\lambda^{\mathcal{S}}) \\ &= \mathrm{H}(X^{\mathcal{S}})_{\rho^{\mathcal{S}}} \\ &= \mathrm{H}(X^{\bar{\mathcal{S}}})_{\rho^{\bar{\mathcal{S}}}}, \end{aligned} \tag{20.53}$$

where $\lambda^{\mathcal{S}}$ is the Schmidt probability vector for cut $\mathcal{S}$ for state $|\psi^{[S]}\rangle$.

Note that the last two equalities in (20.53) follow from the definition of the quantum Shannon entropy.

By the properties of the entanglement entropy reviewed in Chapter 14 (see (14.108)), using the inequality (20.50) on the Schmidt rank, the entanglement entropy satisfies the following properties.

Key Result 20.4 Bounds on the entanglement entropy across a cut

For any C-cut $\mathcal{S}$, the entanglement entropy across the cut satisfies the inequalities

$$\begin{aligned} 0 \le \mathrm{E}(X^{\mathcal{S}};X^{\bar{\mathcal{S}}})_{|\psi^{[S]}\rangle} &\le \log_2(r^{\mathcal{S}}) \\ &\le C\log_2(D), \end{aligned} \tag{20.54}$$

where $r^{\mathcal{S}}$ is the Schmidt rank for cut $\mathcal{S}$. The lower bound $\mathrm{E}(X^{\mathcal{S}};X^{\bar{\mathcal{S}}})_{|\psi^{[S]}\rangle} = 0$ is obtained if and only if the Schmidt rank $r^{\mathcal{S}}$ equals 1 and thus the system is uncorrelated across cut $\mathcal{S}$.

The entanglement entropy across a cut $\mathcal{S}$ thus grows at most linearly with the size of the cut C, and logarithmically with the dimension D of each qudit. When the upper bound in (20.54) is attained, the following definition applies.

Definition 20.9 Maximal entanglement across a cut

The state $|\psi^{[S]}\rangle$ is said to be maximally entangled across a C-cut $\mathcal{S}$ if the corresponding entanglement entropy takes the maximal value $\mathrm{E}(X^{\mathcal{S}};X^{\bar{\mathcal{S}}})_{|\psi^{[S]}\rangle} = C\log_2(D)$.

20.4.6 Alternative Definitions of the Entanglement Entropy Across a Cut ⏭

As discussed in Section 14.7.9, the entanglement entropy can be evaluated using entropy measures beyond the Shannon entropy. In practice, a particularly convenient choice is the Tsallis entropy of order 2, which yields the per-cut entanglement entropy

$$\mathrm{E}_2^{\mathrm{T}}(X^{\mathcal{S}};X^{[S]\setminus\mathcal{S}})_{|\psi^{[S]}\rangle} = 1_2 \cdot (1 - \mathrm{tr}((\rho^{\mathcal{S}})^2)) \tag{20.55}$$

with reduced density matrix $\rho^{\mathcal{S}} = \mathrm{tr}_{[S]\setminus\mathcal{S}}(|\psi^{[S]}\rangle\langle\psi^{[S]}|)$. This measure decreases with the purity $\mathrm{tr}((\rho^{\mathcal{S}})^2)$ of the reduced density matrix $\rho^{\mathcal{S}}$, reflecting the local classical uncertainty on group $\mathcal{S}$ caused by the entanglement with other subsystems. See the Recommended Resources for further discussion.

20.5 Fragmentation, Scale, and Strength of Entanglement

Building on the analysis of entanglement across cuts, fragmentation, scale, and strength of entanglement can be studied in a way analogous to classical correlations in multipartite systems (see Sections 18.5 and 18.6). This section elaborates on these aspects.

20.5.1 Fragmentation of Entanglement

The fragmentation of entanglement for a state $|\psi^{[S]}\rangle$ can be described in terms of the maximum number $I^{\max}$ of subsets of qudits that have individual states under $|\psi^{[S]}\rangle$.

Definition 20.10 $I^{\max}$-uncorrelated systems

A state $|\psi^{[S]}\rangle$ is said to be $I^{\max}$-uncorrelated if there exists a partition $\{\mathcal{S}^i\}_{i\in[I^{\max}]}$ of the qudits such that the state can be written as the Kronecker product

$$|\psi^{[S]}\rangle = |\psi^{\mathcal{S}^0}\rangle \otimes \cdots \otimes |\psi^{\mathcal{S}^{I^{\max}-1}}\rangle \tag{20.56}$$

of the corresponding individual states $\{|\psi^{\mathcal{S}^i}\rangle\}_{i\in[I^{\max}]}$, and no such partition exists with more than $I^{\max}$ groups.

The number $I^{\max}$ ranges in the interval

$$\underbrace{1}_{\text{globally entangled}} \leq I^{\max} \leq \underbrace{S}_{\text{globally uncorrelated}}, \tag{20.57}$$

yielding the following definition.

Definition 20.11 Globally entangled states
States with positive entanglement entropy across *every* cut, i.e., 1-uncorrelated states, are said to be globally entangled.

Globally entangled states are also said to be **genuinely multipartite entangled**. A special class of globally entangled states has maximal entanglement across every cut.

Definition 20.12 Globally maximally entangled states
States with maximum entanglement across every cut are said to be globally maximally entangled. Equivalently, a globally maximally entangled state has maximally mixed reduced density matrices $\rho^{\mathcal{S}}$ for all cuts $\mathcal{S}$.

Globally maximally entangled states are also referred to as **absolutely maximally entangled** or, when viewed as tensors, as **perfect tensors**. It is generally not known for which values of S and D globally maximally entangled states exist (see Recommended Resources).

Overall, globally uncorrelated, entangled, globally entangled, and globally maximally entangled states are defined in a manner analogous to classical correlations (see Table 18.1 and Figure 18.12) by replacing the entanglement entropy for the mutual information.

Example 20.15

The three-qudit GHZ state, i.e., (20.11) with $S = 3$, is an example of a globally maximally entangled state, since it is maximally entangled across every cut. In fact, with $S = 3$ subsystems, it is sufficient to consider 1-cuts, and the reduced density matrices ρ^s in (20.44) are maximally mixed for all qudits $s \in [3]$. Equivalently, the entanglement entropy achieves the maximum value $\mathrm{E}(X^s;X^{[S]\setminus s})_{|\psi^{[S]}\rangle} = \log_2(D)$ across every 1-cut $\mathcal{S} = \{s\}$ for $s \in [3]$.

Example 20.16

The GHZ state (20.11) for a larger number $S > 3$ of qudits is globally entangled. It is also maximally entangled across 1-cuts, given that the reduced density matrices (20.47) are maximally mixed for qudits $s \in [S]$. However, the state is not globally maximally entangled. In fact, for any cut $\mathcal{S}$, by (20.47), the Schmidt probability vector $\lambda^{\mathcal{S}}$ is given by the $D \times 1$ vector $\lambda^{\mathcal{S}} = [1, 1, \ldots, 1]^T/D$. Thus, the Schmidt rank is $r^{\mathcal{S}} = 1$ and the entanglement entropy equals $\mathrm{E}(X^{\mathcal{S}};X^{\bar{\mathcal{S}}})_{|\mathrm{GHZ}^{[S]}\rangle} = \log_2(D)$, falling short of the upper bound $C\log_2(D)$ in (20.54).

Example 20.17

The three-qubit W state (20.9) has reduced density matrices ρ^s given by (20.49), and thus the Schmidt probability vector for 1-cuts is $\lambda^s = [2/3, 1/3]^T$ for all qudits $s \in [S]$. The corresponding entanglement entropy across the cut is

$$\mathrm{E}(X^s;X^{[S]\setminus s})_{|\psi^{[S]}\rangle} = -\frac{2}{3}\log_2\left(\frac{2}{3}\right) - \frac{1}{3}\log_2\left(\frac{1}{3}\right) = 0.9183 < 1. \tag{20.58}$$

It follows that the W state is globally entangled, although it is not globally maximally entangled.

Example 20.18

For $S = 4$ and $D = 3$, i.e., for four qutrits, the state

$$\begin{aligned}|\psi^{[4]}\rangle = \frac{1}{3}(&|0000\rangle + |0112\rangle + |0221\rangle + |1011\rangle \\ &+ |1120\rangle + |1202\rangle + |2022\rangle + |2101\rangle + |2210\rangle)\end{aligned} \tag{20.59}$$

can be verified to be globally maximally entangled. In fact, the reduced density matrices are maximally mixed for all 1-cuts as well as for all 2-cuts.

Example 20.19

Consider now any state $|\psi^{[S]}\rangle$ with tensor rank $r_{\min}$, so that the state admits a CPD (20.16) with integer $r = r_{\min}$. Given any cut $\mathcal{S}$, the reordered state can be expressed as

$$|\psi^{\mathcal{S},\bar{\mathcal{S}}}\rangle = \sum_{z\in[r_{\min}]} \sqrt{p_z}|\psi_z^{\mathcal{S}}\rangle \otimes |\psi_z^{\bar{\mathcal{S}}}\rangle, \tag{20.60}$$

with $|\psi_z^{\mathcal{S}}\rangle$ and $|\psi_z^{\bar{\mathcal{S}}}\rangle$ being the Kronecker products of the states for groups $\mathcal{S}$ and $\bar{\mathcal{S}}$, respectively. These states may be written as $|\psi_z^{\mathcal{S}}\rangle = \otimes_{z\in\mathcal{S}}|\psi_z^s\rangle$ and $|\psi_z^{\bar{\mathcal{S}}}\rangle = \otimes_{z\in\bar{\mathcal{S}}}|\psi_z^s\rangle$.

The corresponding reduced density matrix for group $\mathcal{S}$ can be written as

$$\begin{aligned}\rho^{\mathcal{S}} &= \mathrm{tr}_{\bar{\mathcal{S}}}\left(\sum_{z\in[r_{\min}]}\sum_{z'\in[r_{\min}]} \sqrt{p_z p_{z'}}|\psi_z^{\mathcal{S}}\rangle\langle\psi_{z'}^{\mathcal{S}}| \otimes |\psi_z^{\bar{\mathcal{S}}}\rangle\langle\psi_{z'}^{\bar{\mathcal{S}}}|\right) \\ &= \sum_{z\in[r_{\min}]}\sum_{z'\in[r_{\min}]} \left(\sqrt{p_z p_{z'}}\langle\psi_z^{\bar{\mathcal{S}}}|\psi_{z'}^{\bar{\mathcal{S}}}\rangle\right)|\psi_z^{\mathcal{S}}\rangle\langle\psi_{z'}^{\mathcal{S}}|.\end{aligned} \tag{20.61}$$

The rank of this matrix, i.e., the Schmidt rank $r^{\mathcal{S}}$, is no larger than the tensor rank $r_{\min}$ (see Problem 20.13). This yields the inequality

$$r^{\mathcal{S}} \leq r_{\min}. \tag{20.62}$$

Given the upper bound (20.54), the inequality (20.62) implies that the entanglement entropy across any cut $\mathcal{S}$ is upper bounded by the logarithm of the tensor rank:

$$\mathrm{E}(X^{\mathcal{S}};X^{\bar{\mathcal{S}}})_{|\psi^{[S]}\rangle} \leq \log_2(r_{\min}). \tag{20.63}$$

Thus, the tensor rank $r_{\min}$ limits the entanglement level of a state across any cut $\mathcal{S}$. Accordingly, the tensor rank is a global property of the state, which influences the entanglement properties across the entire system.

Example 20.20

As mentioned in Example 20.6, the tensor rank of the GHZ state (20.11) equals $r_{\min} = D$. To see this, it will be shown in this example that the tensor rank satisfies both inequalities $r_{\min} \leq D$ and $r_{\min} \geq D$. To prove the first inequality, note that the form (20.11) of the state corresponds to a CPD with D terms, implying that the tensor rank cannot exceed D, i.e., $r_{\min} \leq D$. As for the second inequality, observe that the entanglement entropy across 1-cuts equals $\mathrm{E}(X^s;X^{[S]\setminus s})_{|\psi^{[S]}\rangle} = \log_2(D)$ for all $s \in [S]$. Thus, using (20.63), one also has the lower bound $r_{\min} \geq D$ on the tensor rank, yielding the desired equality $r_{\min} = D$.

20.5.2 Scale of Entanglement

The scale of entanglement can be assessed by adopting the following definitions.

Definition 20.13 $S^{\max}$-partite entangled states

A state $|\psi^{[S]}\rangle$ is said to be $S^{\max}$-partite entangled if (i) there exists at least one subset of $S^{\max} \leq S$ subsystems that are globally entangled; and (ii) no larger subset with this property exists, so there is no group of more than $S^{\max}$ subsystems that is globally entangled. The parameter $S^{\max}$ is also known as the **entanglement depth**.

Example 20.21

This example examines three different four-qubit states that have distinct properties in terms of fragmentation and scale of entanglement. The first state

$$|\psi^{[4]}\rangle = |0\rangle \otimes |+\rangle \otimes |1\rangle \otimes |-\rangle, \tag{20.64}$$

in which the first qubit is in state $|0\rangle$, the second qubit is in state $|+\rangle$, and so on, is globally uncorrelated, i.e., is characterized by fragmentation and scale parameters $I^{\max} = 4$ and $S^{\max} = 1$.

The second state

$$|\psi^{[4]}\rangle = |\Phi_+\rangle \otimes |0\rangle \otimes |0\rangle, \tag{20.65}$$

with Bell state $|\Phi_+\rangle = (|0\rangle + |1\rangle)/\sqrt{2}$, is 3-uncorrelated, i.e., $I^{\max} = 3$, and 2-partite entangled, i.e., $S^{\max} = 2$. In fact, it can be partitioned into exactly three groups having individual states, and the largest entangled subset, with Bell state $|\Phi_+\rangle$, contains $S^{\max} = 2$ qubits.

Finally, the state

$$|\psi^{[4]}\rangle = |\Phi_+\rangle \otimes |\Phi_+\rangle \tag{20.66}$$

is 2-separable and 2-partite entangled, i.e., $I^{\max} = 2$ and $S^{\max} = 2$.

Example 20.22

Take now the following two states with $S = 10$ qubits:

$$|\psi^{[10]}\rangle = |\text{GHZ}\rangle \otimes |\text{GHZ}\rangle \otimes |0\rangle \otimes |0\rangle \otimes |0\rangle \otimes |0\rangle \tag{20.67}$$

and

$$|\psi^{[10]}\rangle = |\Phi_+\rangle \otimes |\Phi_+\rangle \otimes |\Phi_+\rangle \otimes |\Phi_+\rangle \otimes |0\rangle \otimes |0\rangle. \tag{20.68}$$

Both states are 6-separable ($I^{\max} = 6$), but the first is 3-partite entangled ($S^{\max} = 3$), while the second is 2-partite entangled ($S^{\max} = 2$).

20.5.3 Area Law of the Bipartite Entanglement Entropy

By the bound (20.54), the bipartite entanglement entropy across a cut $\mathcal{S}$ grows at most linearly with the size of the cut C, that is, with the number of qudits in the subset $\mathcal{S}$. When this upper bound on the scaling of bipartite entanglement is attained, the system is said to obey the **volume law of entanglement**. Generic isolated quantum systems tend to exhibit this behavior.

However, in many physical systems at equilibrium, the growth of entanglement entropy is typically slower as more qudits are encompassed within the subset $\mathcal{S}$.

In fact, many such systems obey the **area law**, which posits that the bipartite entanglement entropy scales with the size of the boundary separating $\mathcal{S}$ from its complement $\bar{\mathcal{S}}$. The area law is closely connected to the **holographic principle**, which asserts that the information contained within a volume of space can be encoded on its lower-dimensional boundary.

These concepts are central to significant problems in physics, such as the study of **black hole entropy**. Systems characterized by the area law of entanglement, as will be explored in the next chapter, can effectively be described using specialized tensor networks.

20.5.4 Strength of Multipartite Entanglement

As in the case of classical correlations discussed in Section 18.6, the strength of entanglement can be quantified by using either cut-based or distance-based measures of multipartite entanglement.

Cut-based measures of entanglement are defined by combining the entanglement entropy $\mathrm{E}(X^{\mathcal{S}};X^{\bar{\mathcal{S}}})_{|\psi^{[S]}\rangle}$ across several cuts $\mathcal{S}$. This follows the same approach adopted in Section 18.6 to quantify multipartite classical correlations. For more details, the reader is referred to the Recommended Resources.

Distance-based measures of entanglement are defined by evaluating the distance between the state of interest, $|\psi^{[S]}\rangle$, and the set of globally uncorrelated states. Such measures are strictly positive if and only if the quantum state exhibits any degree of entanglement. In other words, a zero value indicates a globally uncorrelated state, while a positive value confirms the presence of entanglement.

A notable example of a distance-based measure involves the trace distance introduced in Section 8.6. The trace distance is defined in terms of the fidelity $\mathrm{F}(|\psi\rangle,|\phi\rangle) = |\langle\psi|\phi\rangle|^2$ as $\mathrm{D}^{\mathrm{tr}}(|\psi\rangle,|\phi\rangle) = \sqrt{1-\mathrm{F}(|\psi\rangle,|\phi\rangle)}$ for any two states $|\psi\rangle$ and $|\phi\rangle$.

Definition 20.14 Geometric measure of entanglement

The geometric measure of entanglement of a state $|\psi^{[S]}\rangle$ is defined as

$$\begin{aligned}\mathrm{E}_{\mathrm{D}}(X^0;X^1;\ldots;X^{S-1})_{|\psi^{[S]}\rangle} &= \min \mathrm{D}^{\mathrm{tr}}(|\psi^{[S]}\rangle, |\phi^0\rangle\otimes|\phi^1\rangle\otimes\cdots\otimes|\phi^{S-1}\rangle) \\ &= \sqrt{1-\max \mathrm{F}(|\psi^{[S]}\rangle, |\phi^0\rangle\otimes|\phi^1\rangle\otimes\cdots\otimes|\phi^{S-1}\rangle)},\end{aligned} \tag{20.69}$$

where the minimization and maximization are over all possible globally uncorrelated states $|\phi^0\rangle\otimes|\phi^1\rangle\otimes\cdots\otimes|\phi^{S-1}\rangle$.

The geometric measure of entanglement satisfies the property

$$\text{entangled} \Longleftrightarrow \mathrm{E}_{\mathrm{D}}(X^0;X^1;\ldots;X^{S-1})_{|\psi^{[S]}\rangle} > 0. \tag{20.70}$$

Thus, this measure is sensitive to any level of entanglement, even if not present globally across the entire system. The advantage of using the trace distance over other distance measures is the relative ease of calculating the expression (20.69).

As demonstrated by the next two examples, states can be globally maximally entangled according to the definition given in Section 20.4 – having maximum entanglement entropy across all cuts – while not attaining the maximum value of the geometric measure of entanglement. This highlights the critical importance of selecting appropriate metrics for quantifying entanglement in multipartite systems. The choice of metric should align with the specific requirements and goals of the application, as different measures capture distinct aspects of entanglement.

Example 20.23

The three-qudit GHZ state with any dimension D is globally maximally entangled. Its geometric measure of entanglement can readily be seen to equal $\sqrt{1-1/D}$, which is attained by choosing in (20.69) the state $|\phi^0\rangle\otimes|\phi^1\rangle\otimes\cdots\otimes|\phi^{S-1}\rangle = |z\rangle^{\otimes S}$ for any $z\in[D]$.

Example 20.24

For $D = 3$ consider now the state

$$|\psi^{[3]}\rangle = \frac{1}{\sqrt{6}}(|012\rangle + |021\rangle + |102\rangle + |120\rangle + |201\rangle + |210\rangle), \tag{20.71}$$

which is a symmetric superposition of all permutations of the basis states in which each qutrit occupies a distinct state. The geometric measure of entanglement for this state can be obtained via the direct optimization of problem (20.69), which yields the solution $|\phi\rangle \otimes |\phi\rangle \otimes |\phi\rangle$ with $|\phi\rangle = (|0\rangle + |1\rangle + |2\rangle)/\sqrt{3}$. Accordingly, the fidelity is given by $\mathrm{F}(|\psi^{[3]}\rangle, |\phi\rangle \otimes |\phi\rangle \otimes |\phi\rangle) = 2/9$ and the geometric measure of entanglement is given by

$$\mathrm{E_D}(X^0;X^1;X^2)_{|\psi^{[3]}\rangle} = \sqrt{1 - \frac{2}{9}} \approx 0.8819. \tag{20.72}$$

This is larger than the geometric measure of entanglement for the GHZ state obtained in the previous example, which is evaluated as $\mathrm{E_D}(X^0;X^1;X^2)_{|\mathrm{GHZ}^{[3]}\rangle} = \sqrt{1-1/3} \approx 0.8164$.

20.6 Decoherence and Entanglement

Given a system of S qudits, assume that a group of qudits is made unavailable, is lost, or is discarded. If the original system of S qudits is globally entangled, is the remaining group of qudits still globally entangled?

This question is particularly important for quantum computers, whose states, in practice, must comprise not only the register of the quantum computer, but also qudits describing the environment. In this context, the question posed in the previous paragraph can be rephrased as: Does entanglement with the environment affect the entanglement created by the quantum computer within the qudits in its register?

To put this question into context, take first the case of a classical multipartite system. The following examples show that, if the system is globally correlated, i.e., correlated across every cut, it is possible for a group of qudits not to be globally correlated. That is, losing access or discarding a number of subsystems could affect the correlations present across cuts in the remaining group of qudits. However, in many practical settings, global correlation is preserved even when discarding some subsystems.

Example 20.25

For two independent and uniformly distributed bits b^0 and b^1, consider the three random variables $\mathrm{x}^0 = \mathrm{b}^0$, $\mathrm{x}^1 = \mathrm{b}^1$, and $\mathrm{x}^2 = \mathrm{b}^0 \oplus \mathrm{b}^1$, where $\oplus$ is the XOR operation. These variables are globally correlated, achieving a positive, and in fact maximal, mutual information across each cut. However, discarding any one of the three bits makes the remaining two bits uncorrelated – e.g., random variables x^0 and x^1 are clearly independent. Thus, in this example, while the three bits together are globally correlated, any group of two bits is globally uncorrelated.

Example 20.26

Take now three equal uniform bits with joint distribution

$$p^{[3]} = \frac{1}{2}(|000\rangle + |111\rangle). \tag{20.73}$$

This joint distribution may be thought of as the counterpart of the GHZ state (20.8). The system is globally correlated, attaining a positive, and maximal, mutual information across each cut. Furthermore, unlike the previous example, in this case, any group of two bits is (globally) correlated. For instance, the marginal distribution of bits $\mathcal{S} = \{1, 2\}$ is given by

$$p^{1,2} = \frac{1}{2}(|00\rangle + |11\rangle), \tag{20.74}$$

yielding the maximum mutual information $\mathrm{I}(X^1;X^2)_{p^{1,2}} = 1$.

Example 20.27

Consider now the joint distribution

$$p^{[3]} = \frac{1}{3}(|001\rangle + |010\rangle + |100\rangle) \tag{20.75}$$

over three bits, which corresponds to three uniform one-hot configurations. This joint distribution may be viewed as the counterpart of the W state (20.9). The three random bits are globally correlated under distribution $p^{[3]}$. Furthermore, as in the joint distribution (20.73), correlation is still preserved even after discarding one bit.

For instance, discarding the third bit produces the marginal distribution

$$p^{0,1} = \frac{1}{3}(|00\rangle + |01\rangle + |10\rangle), \tag{20.76}$$

which is (globally) correlated. In fact the mutual information is evaluated as

$$\mathrm{I}(X^0;X^1)_{p^{0,1}} = 2\mathrm{h}\left(\frac{1}{3}\right) - \log_2(3) = 0.251. \tag{20.77}$$

As illustrated by the next examples, entanglement tends to be less robust to losses of subsystems than classical correlations. The possible destruction of entanglement caused by the loss of a part of an entangled system accounts for the phenomenon of decoherence in quantum computers, which arises from interactions of the quantum computer's register with the environment.

Example 20.28

For the three-qubit GHZ state (20.8), take as an example the subset $\mathcal{S} = \{1,2\}$ obtained by discarding the first subsystem $s = 0$. Writing the GHZ state as $|\mathrm{GHZ}^{[3]}\rangle = 1/\sqrt{2}\sum_{z\in[2]} |z\rangle \otimes |zz\rangle$, the corresponding reduced density matrix is given by

$$\begin{aligned}\rho^{1,2} &= \frac{1}{2}\mathrm{tr}_0\left(\sum_{z\in[2]}\sum_{z'\in[2]} |z\rangle\langle z'| \otimes (|zz\rangle\langle z'z'|)^{\otimes 2}\right)\\ &= \frac{1}{2}\sum_{z\in[2]} |zz\rangle\langle zz|\\ &= \frac{1}{2}(|00\rangle\langle 00| + |11\rangle\langle 11|).\end{aligned} \tag{20.78}$$

Thus, the reduced density matrix $\rho^{1,2}$ represents a mixed state that assigns equal probability to the two uncorrelated states $|0\rangle \otimes |0\rangle$ and $|1\rangle \otimes |1\rangle$. As defined in Chapter 16, the mixture of uncorrelated states is separable (but generally not uncorrelated), and thus it does not present any entanglement. What is more, the reduced density matrix $\rho^{1,2}$ is a classical–classical state, and thus it presents only classical correlations.

It is hence concluded that the GHZ state presents a fragile form of entanglement that is erased by any loss of an individual qubit. While the GHZ state is globally entangled, discarding one qubit renders the state of the other two qubits separable, and, specifically, classical–classical.

It is emphasized that the state $\rho^{1,2}$ still presents correlations, but they are purely of classical nature. In fact, a local measurement in the computational basis of state $\rho^{1,2}$ returns classically correlated outputs with joint distribution equal to the marginal distribution (20.74) obtained for the "classical GHZ state" (20.73). Accordingly, the classical measurement outputs are maximally correlated, but the underlying state $\rho^{1,2}$ is separable.

Example 20.29

For the W state (20.9), which may be written as $|\mathrm{W}\rangle = \sqrt{\frac{2}{3}}|0\rangle \otimes |\Psi_+\rangle + \sqrt{\frac{1}{3}}|1\rangle \otimes |00\rangle$, the reduced matrix over qubits $\{1,2\}$ is given by

$$\rho^{1,2} = \frac{2}{3}|\Psi_+\rangle\langle\Psi_+| + \frac{1}{3}|00\rangle\langle 00|. \tag{20.79}$$

Therefore, with probability 2/3, qudits 1 and 2 are in the maximally entangled Bell state $|\Psi_+\rangle$, while with probability 1/3 they are in the uncorrelated state $|00\rangle$.

This state can be checked to be entangled by using the PPT sufficient condition introduced in Section 16.7. Therefore, for the W state, discarding one of the qubits does not completely destroy the entanglement between the other two qudits.

Overall, for the GHZ state, entanglement is completely lost as soon as any of the three qubits is discarded. In contrast, the "connectivity" among the three qubits under the W state is less strong, and the loss of one qubit preserves some degree of entanglement between the other two qubits.

20.7 Classes of Multipartite Entanglement ⏭

As seen in Section 15.3, all two-qubit entangled states can be converted into one another via stochastic LOCC (SLOCC) protocols with some non-zero probability. From this perspective, all two-qubit entangled states belong to the same equivalence class, as they can be generated from one another using only classical resources. Moreover, maximally entangled two-qubit states, such as Bell states, can be transformed into any other entangled two-qubit state with probability one, i.e., deterministically via LOCC. A maximally entangled state thus acts as a **universal currency** for bipartite entanglement, enabling the generation of any other two-qubit entangled state through LOCC.

As discussed in this section, for systems with more than two qubits, the situation becomes significantly more complex, as not all entangled states can be interconverted via SLOCC. Specifically, there exist distinct **entanglement classes** such that no SLOCC protocol can transform a state from one class into a state from another. This implies the absence of a universal currency for entanglement in systems with three or more qubits.

20.7.1 SLOCC Protocols in Multipartite Systems

To explore this result in greater depth, it is important to first generalize the presentation in Section 15.3 to characterize the operation of SLOCC protocols in multipartite systems. At each round of an SLOCC protocol, a POVM is applied to one qudit, and the resulting classical output is broadcast to all other qudits. The choice of the POVM setting implemented at each round generally depends on all the classical outputs of the previous measurements. The following result characterizes the relationship between input and output states for general multipartite qudit states.

Key Result 20.5 SLOCC transformations

An S-qudit state $|\phi^{[S]}\rangle$ can be obtained from another S-qudit state $|\psi^{[S]}\rangle$ with positive probability through an SLOCC protocol if and only if there exist $D \times D$ matrices $\{A^s\}_{s\in[S]}$ such that the equality

$$|\phi^{[S]}\rangle = (A^0 \otimes \cdots \otimes A^{S-1})|\psi^{[S]}\rangle \tag{20.80}$$

holds.

This result follows from the fact that, after post-selection, the output of a POVM is described via the multiplication of the original state by the Kraus matrix corresponding to the given measurement output (see Section 11.5.3). Accordingly, the output of a branch of an SLOCC protocol in a multipartite system is given by the multiplication of the initial state $|\psi^{[S]}\rangle$ by the tensor product of the Kraus operators

$\{K^s\}_{s\in[S]}$ corresponding to the local measurement outputs. Upon normalization, this yields the post-transformation state (20.80).

20.7.2 Inequivalent Classes of Multipartite Entanglement

For two-qubit states, i.e., $S = 2$ and $D = 2$, the result (20.80) confirms that all entangled states are interconvertible via SLOCC with some positive probabilities. However, for $S \geq 3$ qubits, this is not the case, and there are multiple **inequivalent classes** of multipartite entanglement. To elaborate, this subsection considers the case of $S = 3$ qubits, focusing on two basic globally entangled states: the GHZ state and the W state. As shown next, the multipartite entanglement present in these two states is not equivalent, in the sense that the two states cannot be converted into one another via SLOCC.

Starting with the GHZ state, using the characterization (20.80), after an SLOCC one obtains a state of the form

$$\begin{aligned}|\phi^{[3]}\rangle &= (A^0 \otimes A^1 \otimes A^2)|\text{GHZ}^{[3]}\rangle \\ &= (A^0|0\rangle \otimes A^1|0\rangle \otimes A^2|0\rangle + A^0|1\rangle \otimes A^1|1\rangle \otimes A^2|1\rangle) \qquad (20.81)\end{aligned}$$

for some 2×2 matrices $\{A^s\}_{s\in[3]}$. However, as discussed in Example 20.5, the W state has tensor rank equal to $r_{\min} = 3$. Thus, a state of the form (20.81), which is the superposition of two globally uncorrelated states, cannot represent a W state.

This derivation proves that it is not possible to convert GHZ states into W states via SLOCC. Furthermore, it can be demonstrated that it is also not possible to transform the W state into the GHZ state via SLOCC (see Recommended Resources). The general conclusion from this discussion is that the two globally entangled three-qubit states $|\text{GHZ}^{[3]}\rangle$ and $|\text{W}^{[3]}\rangle$ are inequivalent in the sense that they cannot be converted into one another via SLOCC.

In fact, it turns out that, for three qubits, globally entangled states can be divided into exactly two classes of states, such that states within one class can be obtained via SLOCC from the GHZ state, and states within the other class can be prepared via SLOCC from the W state. What is more, starting from $S = 4$, it can be proved that there are infinitely many different types of entangled states that are not mutually convertible via SLOCC. Refer to the Recommended Resources for details.

20.8 Monogamy of Entanglement ⏭

In Section 20.6, it was observed that entanglement is a fragile form of correlation that can be reduced, or even completely destroyed, when a subset of the system is inaccessible or lost. This section covers a property that may be considered as dual to the outlined fragility to losses, namely the **monogamy** of quantum correlations in the face of the introduction of new qudits. The question of interest here is: Can we add a qudit to an existing entangled system so that the overall system is globally entangled?

As will be discussed, entanglement is monogamous in the sense that the presence of entanglement between two qudits precludes the establishment of strong correlations with a third qudit. To elaborate on this property, throughout this section, the focus is on a system with $S = 3$ qubits, referred to as X, Y, and Z.

To put the topic of this section into context, consider first an analogous setting for classical systems. For classical systems, if subsystem X is correlated with subsystem Y, there is no constraint on the level of correlation that X can have with a third subsystem Z. More formally, given a joint distribution p^{XY}, one can create a joint distribution $p^{XYZ}_{xyz} = p^{XY}_{xy} p^{Z|XY}_{z|xy}$ such that Z is arbitrarily correlated with subsystems X and Y.

Example 20.30

Consider two maximally correlated bits X and Y with joint distribution $p^{XY} = (|00\rangle + |11\rangle)/2$. Despite the fact that X is maximally correlated with Y, one can add a third bit Z that is also maximally correlated with X. This result is obtained with the conditional distribution $p^{Z|XY}_{z|xy}$ that assigns probability 1 to the bit value $z = x = y$. In fact, with this choice, the joint distribution p^{XYZ} is given by $p^{XYZ} = (|000\rangle + |111\rangle)/2$. Under this joint distribution, both mutual informations $\mathrm{I}(X;Y)_{p^{XY}}$ and $\mathrm{I}(X;Z)_{p^{XZ}}$ attain the maximum value of 1 bit.

Consider now a three-qubit quantum system consisting of qubits X, Y, and Z. The next example shows that, if qubits X and Y are maximally entangled, then qubit X cannot be maximally entangled with a third qubit Z.

Example 20.31

Assume that qubits X and Y are maximally entangled, being in the Bell state $|\Phi^{XY}_{+}\rangle = (|00\rangle + |11\rangle)/\sqrt{2}$. This implies that qubit X cannot also be entangled with another qubit Z. In fact, if X and Z were entangled, then X and Y together would also be entangled with Z, causing the reduced density matrix ρ^{XY} to be mixed. However, by assumption, the state $|\Phi^{XY}_{+}\rangle$ of subsystems X and Y is pure, yielding a contradiction.

To generalize the conclusions in the example above, recall from Section 16.7 that the entanglement between two qubits A and B can be measured by a quantity called concurrence, which is denoted as $\mathcal{C}(A;B)_{\rho^{AB}}$ (see (16.130)). In particular, the entanglement of formation $\mathrm{E}(A;B)_{\rho^{AB}} = \mathrm{h}\left(\left(1 + \sqrt{1 - (\mathcal{C}(X;Y)_{\rho^{XY}})^2}\right)/2\right)$ is minimized when the concurrence takes the minimum value $\mathcal{C}(X;Y)_{\rho^{XY}} = 0$, and is maximized when the concurrence takes the maximum value $\mathcal{C}(X;Y)_{\rho^{XY}} = 1$.

For a pure state $|\psi^{XYZ}\rangle$, it is possible to prove the following inequality (see Recommended Resources)

$$(\mathcal{C}(X;Z)_{\rho^{XZ}})^2 \leq (\mathcal{C}(X;YZ)_{|\psi^{XYZ}\rangle})^2 - (\mathcal{C}(X;Y)_{\rho^{XY}})^2, \tag{20.82}$$

where $\rho^{XZ} = \mathrm{tr}_Y(|\psi^{XYZ}\rangle\langle\psi^{XYZ}|)$ and $\rho^{XY} = \mathrm{tr}_Z(|\psi^{XYZ}\rangle\langle\psi^{XYZ}|)$. This inequality states that the entanglement between qubits X and Z, measured by the concurrence $\mathcal{C}(X;Z)_{\rho^{XZ}}$, must decrease as the entanglement between X and Y, quantified by the concurrence $\mathcal{C}(X;Y)_{\rho^{XY}}$, increases. For instance, if qubit X is maximally entangled with Y, i.e., $\mathcal{C}(X;Y)_{\rho^{XY}} = 1$, then qubits X and Z must be uncorrelated, i.e., $\mathcal{C}(X;Z)_{\rho^{XZ}} = 0$ since the concurrence is upper bounded by 1.

20.9 Summary

- Globally uncorrelated states are of the form $|\psi^{[S]}\rangle = |\psi^0\rangle \otimes |\psi^1\rangle \otimes \cdots \otimes |\psi^{S-1}\rangle$, and thus, under a globally uncorrelated state, all qudits have an individual state. All other states are said to be entangled.
- A state $|\psi^{[S]}\rangle$ is said to be uncorrelated across a cut $\mathcal{S}$ if it can be written, upon reordering of the indices, as the Kronecker product $|\psi^{\mathcal{S},\bar{\mathcal{S}}}\rangle = |\psi^{\mathcal{S}}\rangle \otimes |\psi^{\bar{\mathcal{S}}}\rangle$, so that the two subsets $\mathcal{S}$ and $\bar{\mathcal{S}}$ of qudits have individual states $|\psi^{\mathcal{S}}\rangle$ and $|\psi^{\bar{\mathcal{S}}}\rangle$.
- States that are entangled across every cut are said to be globally entangled or genuinely multipartite entangled.
- Examples of three-qubit globally entangled states include the GHZ state $|\mathrm{GHZ}^{[3]}\rangle = \frac{1}{\sqrt{2}}|000\rangle + \frac{1}{\sqrt{2}}|111\rangle$ and the W state $|\mathrm{W}^{[3]}\rangle = \frac{1}{\sqrt{3}}|001\rangle + \frac{1}{\sqrt{3}}|010\rangle + \frac{1}{\sqrt{3}}|100\rangle$.
- A CPD of a state $|\psi^{[S]}\rangle$ expresses the state as the superposition $|\psi^{[S]}\rangle = \sum_{z\in[r]} \sqrt{p_z}|\psi_z^0\rangle \otimes |\psi_z^1\rangle \otimes \cdots \otimes |\psi_z^{S-1}\rangle$ of globally separable states, with each zth state being weighted by a positive coefficient $\sqrt{p_z}$. When minimized over all possible CPDs, the integer r is referred to as the tensor rank, $r_{\min}$, of the state $|\psi^{[S]}\rangle$.
- The Schmidt rank $r^{\mathcal{S}}$ of a state $|\psi^{[S]}\rangle$ across a cut $\mathcal{S}$ is the rank of the reduced density matrix $\rho^{\mathcal{S}}$ (or of $\rho^{\bar{\mathcal{S}}}$), while the non-zero eigenvalues of matrix $\rho^{\mathcal{S}}$ (or of $\rho^{\bar{\mathcal{S}}}$) form the Schmidt probability vector $\lambda^{\mathcal{S}}$.
- The state $|\psi^{[S]}\rangle$ is entangled across cut $\mathcal{S}$ if and only if the Schmidt rank $r^{\mathcal{S}}$ is larger than 1.
- The entanglement entropy across a cut $\mathcal{S}$ for state $|\psi^{[S]}\rangle$ is the entropy of the Schmidt probability vector $\lambda^{\mathcal{S}}$, i.e., $\mathrm{E}(X^{\mathcal{S}};X^{\bar{\mathcal{S}}})_{|\psi^{[S]}\rangle} = \mathrm{H}(\lambda^{\mathcal{S}})$.
- The state $|\psi^{[S]}\rangle$ is said to be maximally entangled across cut $\mathcal{S}$ if the upper bound $\mathrm{E}(X^{\mathcal{S}};X^{\bar{\mathcal{S}}})_{|\psi^{[S]}\rangle} = |\mathcal{S}|\log_2(D)$ is attained.
- The fragmentation of entanglement for a state $|\psi^{[S]}\rangle$ can be described in terms of the maximum number $I^{\max}$ of subsets of qudits that have individual, per-subset, pure states, as well as in terms of the maximum size $S^{\max}$ of a globally entangled subset of qudits. In other words, the parameter $I^{\max}$ is the largest number of subsets you can break the state into such that each subset of qudits has no entanglement with the other subsets, while parameter $S^{\max}$ is the size of the biggest subset of qudits that are mutually entangled with each other within the state.

- States with maximum entanglement across every cut, like the GHZ state for three qudits, are known as globally maximally entangled.
- The strength of entanglement can be measured via the cut-based or distance-based multipartite entanglement measures, such as the geometric measure of entanglement.
- While for two-qubit systems there is a universal currency for entanglement, namely maximally entangled states, for systems with more than two qubits, there are different inequivalent classes of entangled states, which cannot be mutually converted via SLOCC.
- Notably, for three qubits, globally entangled states can be divided into two classes, such that states within one class can be obtained via SLOCC from the GHZ state, and states within the other class can be obtained via SLOCC from the W state.
- Starting from a globally entangled state, entanglement can be destroyed or reduced due to losses of a subset of the qudits.
- Entanglement is monogamous in the sense that the existence of a certain amount of entanglement between X and Y entails that the entanglement between X and Z must be limited.

20.10 Recommended Resources

The study of the tensor rank, including for the W state, is addressed in [1], which explores its implications for quantum entanglement. For a detailed discussion of multipartite entanglement measures, the reader is referred to [2–6].

The classification of three-qubit entanglement into two inequivalent forms is discussed in [7], while transformations among pure multipartite entangled states under local operations are studied extensively in [8, 9]. Connections between LOCC protocols in multipartite systems and thermodynamic measures of quantum correlations can be found in [10].

The monogamy inequality is proved in [11], and a related inequality is introduced in [12].

PROBLEMS

20.1 Check that the vectors $\{|x^{[S]}\rangle\}_{x^{[S]}\in[D]^S}$ obtained from the local computational basis as in (20.7) are orthonormal. *Hint:* Use the mixed property of the Kronecker product.

20.2 Show that, given S orthonormal bases $\{|v_x^s\rangle\}_{x\in[D]}$, one for each qudit $s \in [S]$, the collection of D^S vectors $\{|v_{x^{[S]}}\rangle\}_{x^{[S]}\in[D]^S}$, with $|v_{x^{[S]}}\rangle = |v_{x^0}\rangle \otimes |v_{x^1}\rangle \otimes \cdots \otimes |v_{x^{S-1}}\rangle$, forms an orthonormal basis for the space of $D^S \times 1$ vectors.

20.3 Verify that the parallel local operation $V^{[S]} = V^0 \otimes V^1 \otimes \cdots \otimes V^{S-1}$, in which each matrix V^s for $s \in [S]$ is unitary, corresponds to a unitary transformation.

20.4 Evaluate the measurement probability distribution for the W state under a local measurement setting that uses the local Hadamard bases for all three qubits.

20.5 For the three-qubit state

$$\psi^{[3]} = \frac{1}{2\sqrt{2}}[1, -1, 1, -1, 1, -1, 1, -1]^T, \tag{20.83}$$

find the tensor rank.

20.6 (a) For the three-qubit state $|\psi^{[3]}\rangle = \frac{1}{\sqrt{2}}[1, 0, 0, 0, 0, 0, -1, 0]^T$, determine a cut across which the state is uncorrelated. (b) Repeat for state $|\psi^{[3]}\rangle = [0, 0, 0, 1, 0, 0, 1, 0]^T$. *Hint:* Try reordering the qubits so that the second and third qubits are swapped.

20.7 Derive the Schmidt probabilities for the W state (20.9) across cut $\mathcal{S} = \{2\}$ by evaluating the SVD of the corresponding amplitude matrix.

20.8 (a) Draw the tensor diagram for the measurement probability $p_{x^{[S]}} = \langle v_{x^{[S]}}|\rho^{[S]}|v_{x^{[S]}}\rangle$ obtained from a multipartite density matrix $\rho^{[S]}$. Include in the diagram one tensor for each local unitary matrix $V^s = [|v_0^s\rangle \cdots |v_{D-1}^s\rangle]$ for $s \in [S]$. (b) Show that the expression $\langle v_{x^{[S]}}|\rho^{[S]}|v_{x^{[S]}}\rangle$ is equivalent to $\mathrm{tr}(\rho^{[S]}|v_{x^{[S]}}\rangle\langle v_{x^{[S]}}|)$ with projection matrix $\Pi_{x^{[S]}} = |v_{x^{[S]}}\rangle\langle v_{x^{[S]}}|$ by using diagrammatic operations on the tensor network. *Hint:* Recall that the trace operation corresponds to a contraction between corresponding row and column indices.

20.9 (a) For an arbitrary quantum channel $\mathcal{K} = \{K_z\}_{z\in[M]}$, draw the tensor network representation for the post-channel density matrix by incorporating tensors for the pre-channel density ρ as well as for the Kraus operators $\{K_z\}_{z\in[M]}$. *Hint:* Represent each Kraus operator K_z and its Hermitian $K_z^\dagger$ with a tensor having three indices, one for the rows, one for the columns, and one for z. (b) Consider now a channel applied only to one qudit in a bipartite system in state $\rho^{[2]}$. For instance, if the channel $\mathcal{K}$ is applied only to the second qudit, the post-channel output density matrix is written as

$$I \otimes \mathcal{K}(\rho) = \sum_{z\in[M]} (I \otimes K_z)\rho^{[2]}(I \otimes K_z^\dagger). \tag{20.84}$$

Draw the tensor network representation for the post-channel state (20.84). *Hint:* Use the bi-index notation for the row and column indices of the density matrix ρ.

20.10 Prove the inequality $\mathrm{H}(X^{\mathcal{S}})_{\rho^{\mathcal{S}}} \leq |\mathcal{S}| \log_2(D)$, which indicates that the entropy $\mathrm{H}(X^{\mathcal{S}})_{\rho^{\mathcal{S}}}$ of a subset $\mathcal{S}$ of subsystems cannot grow more quickly than a linear function of the number of subsystems $|\mathcal{S}|$.

20.11 For the three-qubit state $|\psi^{[3]}\rangle = \sqrt{1/3}|000\rangle + \sqrt{2/3}|111\rangle$: (a) Evaluate the reduced density matrix ρ^s for any qubit $s \in [3]$. (b) Compute the reduced density matrix $\rho^{\mathcal{S}}$ for subset $\mathcal{S} = \{0, 1\}$. What is the reduced density matrix for the other 2-cuts? (c) Is the state $\rho^{\mathcal{S}}$ for any subset $\mathcal{S}$ of two qubits entangled? *Hint:* As will be detailed in Chapter 21, a mixed state that is the mixture of uncorrelated states is uncorrelated.

20.12 For the three-qubit state $|\psi^{[3]}\rangle = \sqrt{1/4}|001\rangle + \sqrt{1/4}|010\rangle + \sqrt{1/2}|100\rangle$: (a) Evaluate the reduced density matrix ρ^s for any qubit $s \in [3]$. (b) Compute the reduced density matrix $\rho^{\mathcal{S}}$ for subset $\mathcal{S} = \{0, 1\}$.

20.13 Take any $N \times N$ matrix B. Show that the rank of an $N \times N$ matrix C of the form $C = \sum_{z\in[M]} \sum_{z'\in[M]} B_{z,z'} a_z a_{z'}^\dagger$, where the $M \leq N$ vectors $\{a_z\}_{z\in[M]}$ have size $N \times 1$, is upper bounded by M. *Hint:* Write C as a function of matrix $A = [a_0, a_1, \ldots, a_{M-1}]$.

20.14 For a multipartite quantum system, assume that the rank of the reduced density matrix for at least one qudit s equals 1. What can be concluded about the fragmentation of entanglement in the system? In particular, provide a lower bound on the possible values for parameter $I^{\max}$, as well as an upper bound for the parameter $S^{\max}$.

20.15 For the states $|\psi^{[8]}\rangle = |\mathrm{W}\rangle \otimes |\mathrm{W}\rangle \otimes |1\rangle \otimes |-\rangle$ and $|\phi^{[8]}\rangle = |0\rangle \otimes |\Phi_+\rangle \otimes |1\rangle \otimes |-\rangle \otimes |\Phi_+\rangle \otimes |0\rangle$, where $|\mathrm{W}\rangle$ represents the W state: (a) Find the values of the parameters $I^{\max}$ and $S^{\max}$. (b) Calculate the entanglement entropy across all 1-cuts, i.e., for all $s \in [8]$.

20.16 For the extension of the W state to any number, S, of qubits, which is defined as

$$|\mathrm{W}^{[S]}\rangle = \frac{1}{\sqrt{S}}(|00\cdots 1\rangle + |00\cdots 10\rangle + \cdots |10\cdots 00\rangle): \tag{20.85}$$

(a) Calculate the reduced density matrix for any subsystem. (b) Evaluate the entanglement entropy $\mathrm{E}(X^s;X^{[S]\setminus s})_{|\mathrm{W}^{[S]}\rangle}$ for all 1-cuts $s \in [S]$. (c) What happens to the entanglement entropy $\mathrm{E}(X^s;X^{[S]\setminus s})_{|\mathrm{W}^{[S]}\rangle}$ in the limit $S \to \infty$?

REFERENCES

[1] W. Bruzda, S. Friedland, and K. Życzkowski, "Rank of a tensor and quantum entanglement," *Linear and Multilinear Algebra*, vol. 72, no. 11, pp. 1796–1859, 2023.

[2] G. De Chiara and A. Sanpera, "Genuine quantum correlations in quantum many-body systems: A review of recent progress," *Reports on Progress in Physics*, vol. 81, no. 7, art. 074002, 2018.

[3] G. Rigolin, T. R. de Oliveira, and M. C. de Oliveira, "Operational classification and quantification of multipartite entangled states," *Physical Review A*, vol. 74, no. 2, art. 022314, 2006.

[4] K. Modi, T. Paterek, W. Son, V. Vedral, and M. Williamson, "Unified view of quantum and classical correlations," *Physical Review Letters*, vol. 104, no. 8, art. 080501, 2010.

[5] J. L. Beckey, N. Gigena, P. J. Coles, and M. Cerezo, "Computable and operationally meaningful multipartite entanglement measures," *Physical Review Letters*, vol. 127, no. 14, art. 140501, 2021.

[6] M. Ma, Y. Li, and J. Shang, "Multipartite entanglement measures:

A review," *arXiv*, 2023. [Online]. Available at: https://arxiv.org/abs/2309.09459

[7] W. Dür, G. Vidal, and I. Cirac, "Three qubits can be entangled in two inequivalent ways," *Physical Review A*, vol. 62, no. 6, art. 062314, 2000.

[8] D. Sauerwein, N. R. Wallach, G. Gour, and B. Kraus, "Transformations among pure multipartite entangled states via local operations are almost never possible," *Physical Review X*, vol. 8, no. 3, art. 031020, 2018.

[9] E. Chitambar, D. Leung, L. Mančinska, M. Ozols, and A. Winter, "Everything you always wanted to know about LOCC (but were afraid to ask)," *Communications in Mathematical Physics*, vol. 328, pp. 303–326, 2014.

[10] T. Yada, N. Yoshioka, and T. Sagawa, "Measuring multipartite quantum correlations by thermodynamic work extraction," *arXiv*, 2024. [Online]. Available at: https://arxiv.org/abs/2407.04058

[11] V. Coffman, J. Kundu, and W. K. Wootters, "Distributed entanglement," *Physical Review A*, vol. 61, no. 5, art. 052306, 2000.

[12] M. Koashi and A. Winter, "Monogamy of quantum entanglement and other correlations," *Physical Review A*, vol. 69, no. 2, art. 022309, 2004.

21 Multipartite Quantum Information: Structure

21.1 Overview

This chapter elaborates on the structure of entanglement in multipartite quantum pure states by focusing on two important frameworks: matrix product states and quantum circuits. Matrix product states (MPSs) are widely used to model quantum states where entanglement adheres to an area law. In MPSs, the strength of entanglement across a cut is constrained by the size of the boundary of the partition, rather than by the size of the groups of subsystems on the two sides of the cut. Diagrammatically, MPSs are characterized by linear tensor networks where a single parameter, the bond dimension, constrains entanglement strength across the system.

Quantum circuits provide a framework to describe multipartite quantum states by specifying the dynamic transformations required to prepare a state from an initially globally uncorrelated system. These transformations involve sequences of unitary operations applied to groups of subsystems. Quantum circuits play a crucial role in the design and implementation of quantum algorithms on gate-based quantum computers. They can often be simplified through the use of diagrammatic operations, enabling the study of state properties without relying on purely analytical or numerical methods. This chapter will specifically focus on ZX calculus as a framework for the diagrammatic analysis of quantum circuits.

21.1.1 Learning Objectives and Organization

By the end of this chapter, the reader should be able to:

- comprehend and apply matrix product states to describe the structure of quantum pure states with localized entanglement properties;
- understand and use quantum circuits to model the structure of quantum pure states through dynamic transformations;
- and leverage the ZX calculus to simplify tensor network representations derived from quantum circuits.

21.2 Matrix Product States

As in Chapter 20, throughout this chapter the focus will be on pure states consisting of S qudits with equal dimensions D. MPSs are pure states characterized by tensor networks that consist of chains – or trains – of tensors, one for each qudit. They are thus useful to represent systems with a linear entanglement structure, satisfying a one-dimensional version of the area law of entanglement (see Section 20.5.3).

21.2.1 Defining Matrix Product States

Formally, MPSs are defined as follows.

Definition 21.1 Matrix product states (MPSs)
A quantum pure state $|\psi^{[S]}\rangle$ is said to be an MPS with bond dimension χ if each $x^{[S]}$th entry of the order-S tensor $|\psi^{[S]}\rangle$ can be expressed as the product

$$\psi_{x^{[S]}} = T^0_{x^0} T^1_{x^1} \cdots T^{S-1}_{x^{S-1}}, \tag{21.1}$$

where $T^0_{x^0}$ is a $1 \times \chi$ vector for every index $x^0 \in [D]$; $T^s_{x^s}$ is an $\chi \times \chi$ matrix for all $x^s \in [D]$ and for all $s = 1, \ldots, S-2$; and $T^{S-1}_{x^{S-1}}$ is an $\chi \times 1$ vector for all $x^{S-1} \in [D]$.

By the definition (21.1), an MPS is represented graphically by the **tensor train** in Figure 21.1. In it, with the exception of the last qudit, each tensor T^s for qudit s is connected to the next tensor T^{s+1} for qudit $s+1$ via a bond, describing a tensor contraction. As will be seen, the bond dimension, denoted as χ, provides a bound on the level of entanglement for the cut separating the two subsets of qudits above and below the bond.

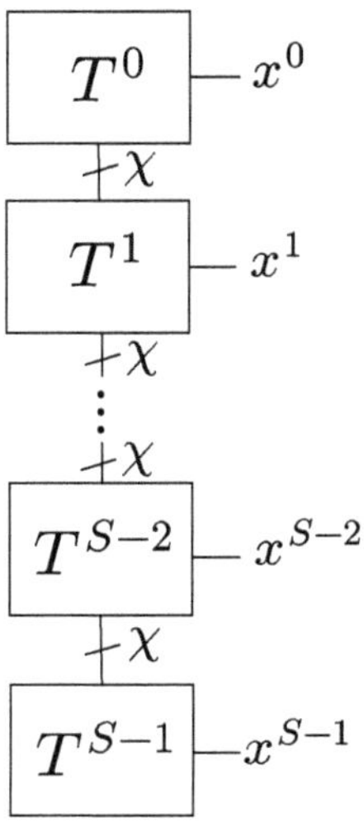

Figure 21.1 Tensor network representation of an MPS.

Example 21.1

A globally uncorrelated state $|\psi^{[S]}\rangle = |\psi^0\rangle \otimes |\psi^1\rangle \otimes \cdots |\psi^{S-1}\rangle$ is an MPS that contains no bonds, i.e., for which the bond dimension is $\chi = 1$. Recall from Chapter 17 that indices with dimension equal to 1 can be eliminated from a tensor network. In the resulting tensor network, each tensor T^s corresponds to the order-1 individual qudit state $|\psi^s\rangle$ for all $s \in [S]$.

An MPS is specified by the SD matrices $T^s_{x^s}$ for all indices $x^s \in [D]$ and subsystems $s \in [S]$. Apart from the vectors $T^0_{x^0}$ and $T^{S-1}_{x^{S-1}}$ used to bookend the product (21.1), each matrix $T^s_{x^s}$ is of dimension $\chi \times \chi$. Therefore, the overall number of entries required to specify an MPS is around $SD\chi^2$. Accordingly, for a fixed bond dimension χ, the number of parameters of an MPS scales linearly with the number of systems S. This is notably different from the exponential scaling in S of the number of entries of a generic state vector $|\psi^{[S]}\rangle$.

Example 21.2

For an arbitrary two-qudit state $|\psi^{[2]}\rangle$, a tensor train representation of the state of the form (21.1) can be obtained directly from the Schmidt decomposition. As detailed in Section 14.9, the Schmidt decomposition expresses the state as the sum $|\psi^{[2]}\rangle = \sum_{z\in[r]} \sqrt{\lambda_z}|u^0_z\rangle \otimes |u^1_z\rangle$, where $r \leq D$ is the Schmidt rank, the Schmidt left states $\{|u^0_z\rangle\}_{z\in[r]}$ are mutually orthogonal, and so are the Schmidt right states $\{|u^1_z\rangle\}_{z\in[r]}$. Thus, each (x^0, x^1)th entry of the tensor $|\psi^{[2]}\rangle$ can be written as

$$\psi_{x^0,x^1} = \sum_{z\in[r]} \sqrt{\lambda_z} u^0_{x^0|z} u^1_{x^1|z}, \tag{21.2}$$

where $u^s_{x^s|z}$ is the x^sth entry of state $|u^s_z\rangle$ for $x^s \in [D]$ and $s \in [2]$.

Then, defining the $1 \times r$ vector $T^0_{x^0} = [u^0_{x^0|0}, \ldots, u^0_{x^0|r-1}]$ and the $r \times 1$ vector

$$T^1_{x^1} = \left[\sqrt{\lambda_0} u^1_{x^1|0}, \ldots, \sqrt{\lambda_{r-1}} u^1_{x^1|r-1}\right]^T, \tag{21.3}$$

the state $|\psi^{[2]}\rangle$ can be represented as the MPS

$$\psi_{x^0,x^1} = T^0_{x^0} T^1_{x^1} \tag{21.4}$$

with bond dimension $\chi = r$ equal to the Schmidt rank.

In closing this example, it is useful to note that there is some freedom in the definition of the two order-1 tensors T^0 and T^1 in (21.4). For example, one can assign the square roots $\{\sqrt{\lambda_z}\}_{z\in[r]}$ either to tensor T^0 or to tensor T^1.

Any state can always be represented in the form of an MPS, and thus of a tensor train. However, as demonstrated in the next example, for a generic state, the bond dimension grows exponentially with the number of qudits, S. Consequently, when modeling a state with an MPS, it is typically assumed that the bond dimension χ is either constant or scales slowly with S. In practice, this assumption may entail some approximation, whose accuracy improves as the bond dimension grows. As will be detailed later in this section, MPS states with a constant bond dimension χ encode structural constraints on the entanglement present in the system.

Example 21.3

Consider the left column of Figure 21.2, in which a generic state $|\psi^{[S]}\rangle$ with $S = 5$ subsystems is represented in the form of a tensor with S open indices $x^{[S]}$. The discussion in this example generalizes directly to the case of any odd number of subsystems, S, and the main conclusion applies also to an even number of subsystems. As shown in the figure for $S = 5$, with an odd number of qudits, one has an equal number $S' = (S-1)/2$ of qudits below and above the half-point qudit with index $s = S'$. In particular, for this example, the half point is at qudit $S' = 2$.

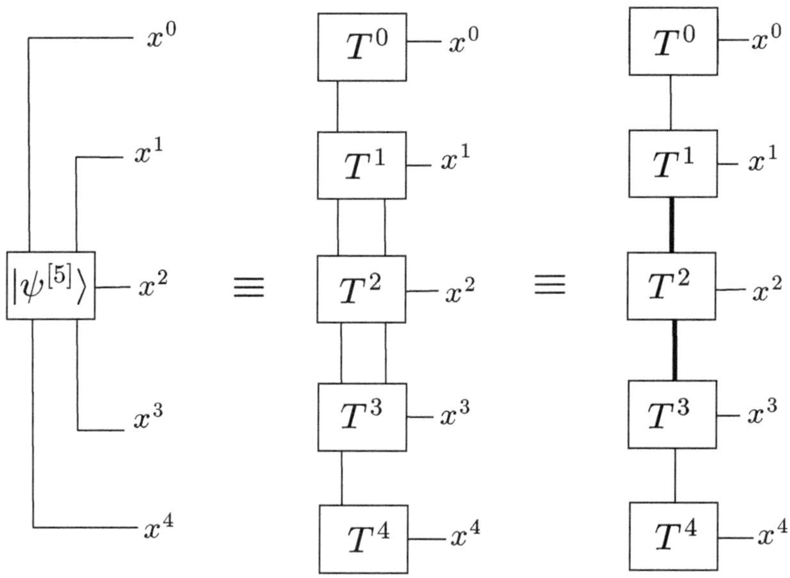

Figure 21.2 Illustration of the process of representing a generic pure state in the form of an MPS. A generic pure state requires a bond dimension that grows exponentially with the number of subsystems, S. The figure illustrates the case with $S = 5$ and $S' = 2$. Bolder lines illustrate bonds with a higher dimension.

As shown in the central and right columns of Figure 21.2, an MPS representation of the state can be obtained as follows. First, define the central tensor $T^{S'} = |\psi^{[S]}\rangle$ with open index $x^{S'} \in [D]$ and $S-1$ additional indices used to create bonds. Specifically, half of the bonds are linked to tensor $T^{S'-1}$, while the remaining half are connected to the tensor $T^{S'+1}$. Accordingly, there are $S' = (S-1)/2$ bonds from tensor $T^{S'}$ to each tensor $T^{S'-1}$ and $T^{S'+1}$. The S' bonds to tensor $T^{S'-1}$ can equivalently be viewed as a single bond of dimension $D^{S'} = D^{\frac{S-1}{2}}$, and similarly the S' bonds to tensor $T^{S'+1}$ can be viewed as a single bond of dimension $D^{S'} = D^{\frac{S-1}{2}}$.

All other tensors T^s with $s \neq S'$ have an even number of legs, including the only open index x^s. The legs are paired in such a way that the tensor returns 1 if and only if the indices in each pair take the same value, returning 0 otherwise. For example, the tensor T^1 in the figure has open index x^1 and three other indices. The rightmost bottom bond is paired with the open index x^1, while the other two indices are paired with each other. One can check that this construction yields a tensor network that represents the original tensor $|\psi^{[S]}\rangle$.

As illustrated in the right column of Figure 21.2, this construction yields bonds with exponentially increasing dimension as one moves towards the central tensor $T^{S'}$. In particular, the maximum dimension $D^{S'}$ grows exponentially with the number of subsystems, S.

Overall, this example suggests that, in order to express a general state $|\psi^{[S]}\rangle$ in the form of an MPS, which requires the same bond dimension for all tensors, the bond dimension χ must scale exponentially with S.

21.2.2 Circular Matrix Product States

The definition (21.1) of an MPS has the potential drawback that qudits 0 and 1 are treated differently compared to the other qudits in the chain. Specifically, these qudits are connected to only one neighboring qudit, whereas all other qudits are connected to two. To achieve a symmetric parameterization across all qudit indices, it is common to adopt an alternative definition of an MPS, which is given next.

Definition 21.2 Circular matrix product states (MPSs)

A quantum pure state $|\psi^{[S]}\rangle$ is said to be a circular MPS with bond dimension χ if each $x^{[S]}$th entry of the order-S tensor $|\psi^{[S]}\rangle$ can be expressed as

$$\psi_{x^{[S]}} = \mathrm{tr}(T^0_{x^0} T^1_{x^1} \cdots T^{S-1}_{x^{S-1}}), \tag{21.5}$$

in which all matrices $\{T^s_{x^s}\}$ for all $x^s \in [D]$ and for all qudits $s \in [S]$ have dimensions $\chi \times \chi$.

As shown in Figure 21.3, a circular MPS is described by a tensor train network with an additional bond connecting first and last tensors. In fact, the trace operation (21.5) indicates that the additional indices of dimension χ in tensors T^0 and T^{S-1} are connected to form a bond (see Section 17.4).

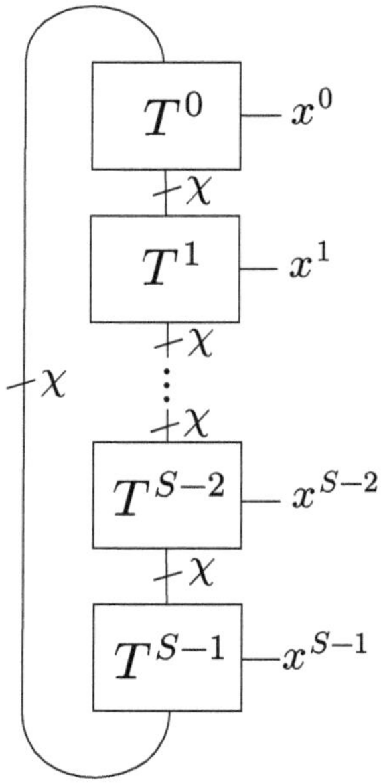

Figure 21.3 Tensor network representation of a circular MPS.

Example 21.4

The three-qubit GHZ state can be expressed as a circular MPS of the form (21.5) with 2×2 matrices

$$T_0^0 = T_0^1 = T_0^2 = |0\rangle\langle 0| \tag{21.6}$$

and

$$T_1^0 = T_1^1 = T_1^2 = |1\rangle\langle 1|. \tag{21.7}$$

Accordingly, the bond index is $\chi = 2$.

To verify that the expression (21.5) returns the GHZ state, it is enough to note: (i) the equalities $\psi_{000} = \mathrm{tr}(T_0^0 T_0^1 T_0^2) = \mathrm{tr}(|0\rangle\langle 0|) = 1$ and $\psi_{111} = \mathrm{tr}(T_1^0 T_1^1 T_1^2) = \mathrm{tr}(|1\rangle\langle 1|) = 1$; as well as (ii) the equalities $\mathrm{tr}(T^0_{x^0} T^1_{x^1} T^2_{x^2}) = 0$ for any triple (x^0, x^1, x^2) with at least two different indices, which follow given the orthogonality of the states $|0\rangle$ and $|1\rangle$.

21.2.3 Area Law of Entanglement for Matrix Product States

The bond dimension χ provides a bound on the entanglement entropy between any group of adjacent qudits and the rest of the system. Consequently, MPSs with a fixed bond dimension χ satisfy the one-dimensional area law of entanglement in the following sense.

Take a circular MPS (21.5), and define a cut $\mathcal{S}$ as **adjacent** if the subset $\mathcal{S}$ forms an "interval" along the circular tensor train. Formally, for an adjacent cut, the subset $\mathcal{S}$ is of the form $\mathcal{S} = \{s, \overline{s+1}, \dots, \overline{s+|\mathcal{S}|-1}\}$, where integer $s \in [S]$ represents the start of the interval and the notation $\overline{S+s} = s$ for $s \in [S]$ is used to capture the circular structure of the state (21.5).

Example 21.5

In Figure 21.4 for $S = 6$ qudits, the subset $\mathcal{S} = \{5, 0, 1\}$ defines an adjacent cut, starting at $s = 5$, which separates group $\mathcal{S}$ from the complementary group $\bar{\mathcal{S}} = \{2, 3, 4\}$. In the figure, the tensors corresponding to group $\mathcal{S}$ are shaded.

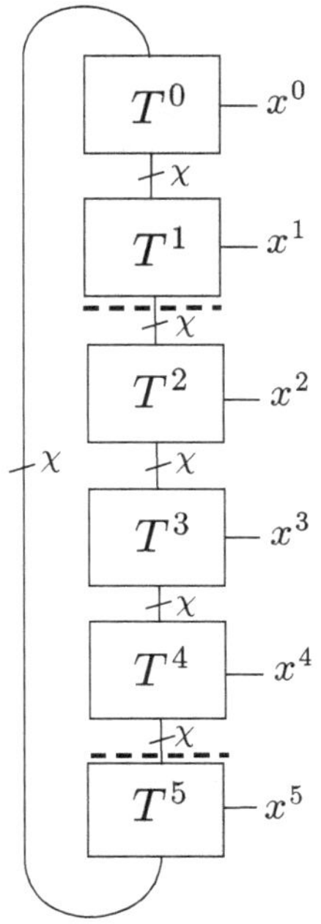

Figure 21.4 Example of an adjacent cut, which is given by $\mathcal{S} = \{5, 0, 1\}$ (shaded tensors). The dashed lines show the boundary between the subset $\mathcal{S}$ and the rest of the system.

The following result is proved in Appendix 21.A.

Key Result 21.1 Upper bound on the entanglement entropy for MPSs (one-dimensional area law of entanglement)

Given any circular MPS $|\psi^{[S]}\rangle$ with bond dimension χ, the entanglement entropy of state $|\psi^{[S]}\rangle$ across a cut $\mathcal{S}$ consisting of adjacent qudits satisfies the upper bound

$$\mathrm{E}(X^{\mathcal{S}};X^{\bar{\mathcal{S}}})_{|\psi^{[S]}\rangle} \leq 2\log_2(\chi). \tag{21.8}$$

The inequality (21.8) shows that the entanglement between a group of adjacent qudits and the rest of the system depends solely on the size of the boundary between the two groups, and not on the number of qudits in each group. In fact, for any adjacent cut $\mathcal{S}$, there are two bonds separating the group $\mathcal{S}$ from the rest of the system, and each bond has dimension χ. According to the bound (21.8), each bond crossing the cut contributes a logarithmic factor $\log_2(\chi)$ to the upper bound

(21.8). Furthermore, by the definition of entanglement entropy, $\mathrm{E}(X^{\mathcal{S}};X^{\bar{\mathcal{S}}})_{|\psi[S]\rangle} = \mathrm{H}(X)_{\rho^{\mathcal{S}}}$, the upper bound (21.8) implies that the entropy of any group $\mathcal{S}$ of adjacent qudits in an MPS cannot exceed $2\log_2(\chi)$.

Example 21.6

For the cut in Figure 21.4, the two groups $\mathcal{S} = \{5,0,1\}$ and $\bar{\mathcal{S}} = \{2,3,4\}$ contain three subsystems each. Thus, the entanglement entropy across this 3-cut could be as high as $3\log_2(D)$, as the quantum entropy of both groups of subsystems could take values up to $3\log_2(D)$ bits in a generic state. However, the upper bound (21.8) indicates that, if the state is a circular MPS, the entanglement entropy is bounded by $2\log_2(D)$ irrespective of the dimensions C and $S-C$ of the groups on either side of the cut.

21.2.4 Projected Entangled-Pair State ⏭

A general way to generate an MPS (21.5) with bond dimension χ is to "project" a system of maximally entangled pairs of qudits of dimension χ into a system of S qudits of dimension D. This construction is an example of a **projected entangled-pair state (PEPS)**. Focusing for simplicity on MPSs of the form (21.1), a high-level sketch of the construction is shown in Figure 21.5(a). Here, given a number of pairs of maximally entangled qudits with dimension χ, local operations are applied across qudits pertaining to different entangled pairs. Each such local operation produces one of the S qudits of the final state.

This approach can be generalized to grids in two dimensions as shown in Figure 21.5(b). In this system, the maximally entangled pairs of qudits are arranged on a

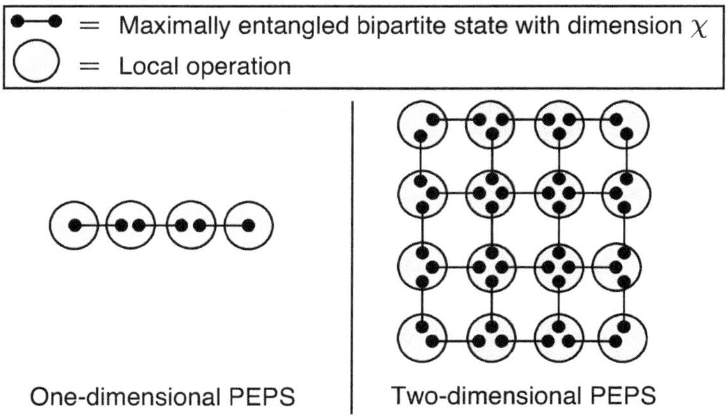

Figure 21.5 Illustration of a linear projected entangled-pair state (PEPS), corresponding to (a) an MPS and (b) a two-dimensional PEPS.

grid, and linear operations are applied as indicated to produce a number of qudits, S, equal to the product of the number of rows and columns. Two-dimensional states obtained using this approach satisfy a two-dimensional area law of entanglement. Accordingly, the entanglement entropy across any cut separating an adjacent group of qudits from the rest of the system cannot exceed $B\log_2(\chi)$, where B is the number of bonds crossing the cut. The reader is referred to the Recommended Resources for further discussion.

21.3 Quantum Circuits as Tensor Networks

A **quantum circuit** specifies a sequence of quantum transformations, or **gates**, that prepare a state $|\psi^{[S]}\rangle$ starting from a globally uncorrelated state. This section introduces quantum circuits as a structured tensor network representation for a multipartite state $|\psi^{[S]}\rangle$, offering a systematic framework to describe preparation and entanglement properties of the state.

21.3.1 Defining Quantum Circuits

A quantum circuit prepares a joint state $|\psi^{[S]}\rangle$ starting from a globally uncorrelated state, typically initialized as

$$|0\rangle^{\otimes S} = |0\rangle \otimes |0\rangle \otimes \cdots \otimes |0\rangle. \tag{21.9}$$

To achieve this, unitary transformations are applied sequentially to subsets of qudits across multiple layers. The initial state $|0\rangle$ for each qudit is chosen following a widely accepted convention. However, any other fixed initialization could also be used depending on the specific requirements of the system.

Diagrammatically, a quantum circuit is represented by a tensor network as in Figure 21.6, in which one wire represents one qudit, and time flows from left to right. The quantum circuit applies gates sequentially, with each gate represented as a box in the figure. A gate can be applied to a single qudit, or to multiple qudits. Each gate corresponds to a unitary matrix, which describes the evolution of the qudits

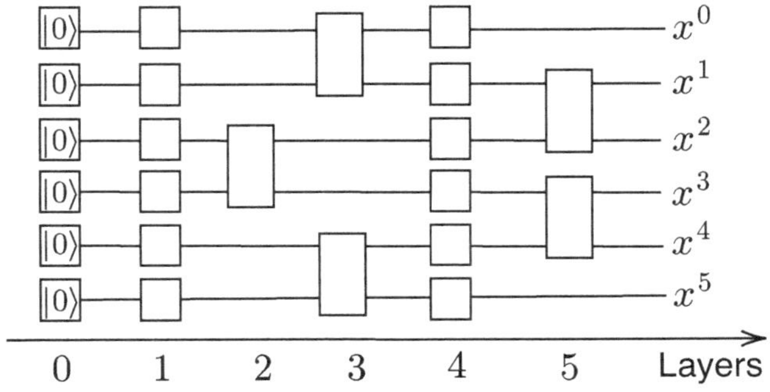

Figure 21.6 Tensor network representation of a quantum circuit.

it operates on. The presence of a gate operating on a subset of qudits describes an interaction between the qudits, which may create or modify entanglement.

Example 21.7

In the quantum circuit shown in Figure 21.6, the leftmost tensors represent the initial, globally uncorrelated, state (21.9), while all the other tensors to the right of the leftmost tensors represent quantum gates, i.e., unitary matrices. The operation of the tensor proceeds sequentially as follows. First, at layer 1, separate gates are applied to each qudit. Then, at layer 2, a two-qudit gate is applied jointly to the qudits 2 and 3. At layer 3, two two-qudit gates are applied jointly to the pairs of qudits 0 and 1 and qudits 4 and 5. Layers 4 and 5 apply separate gates to each qudit, and two two-qudit gates to qudits 1 and 2 and qudits 3 and 4, respectively.

Generalizing the last example, a quantum circuit operates across a number of layers, L, with each layer incorporating a number of local quantum gates acting on different subsets of qudits. If the quantum gates act on groups of at most B qudits, they are said to be **B-local**. Accordingly, the joint states prepared by a quantum circuit can be described as follows.

Definition 21.3 Quantum circuit
A quantum circuit L layers and B-local unitary matrices produces a joint state

$$|\psi^{[S]}\rangle = U_{L-1} \cdots U_1 U_0 |0\rangle^{\otimes S}, \tag{21.10}$$

for a sequence of $D^S \times D^S$ unitary matrices $\{U_l\}_{l \in [L]}$, one for each layer $l \in [L]$. Each unitary matrix U_l describes the parallel application of unitary matrices that act on disjoint groups of at most B qudits.

Quantum computers typically operate on registers of qubits, which sets $D = 2$. Moreover, most quantum computers are restricted to using either single-qubit or two-qubit gates, corresponding to 2-local gates ($B = 2$). While this limitation is technological in nature, it can be mathematically proven that single- and two-qubit gates are universal. Using only 2-local gates, the transformation described in (21.10) can implement *any* unitary transformation across the entire qubit register. Furthermore, it suffices to use only the two-qubit CNOT gate, combined with the single-qubit Hadamard gate and the Pauli Z rotation matrices with angles $\theta = \pi$ and $\theta = \pi/4$ (i.e., the T gate, see Section 10.4), to approximate any unitary transformation to arbitrary precision.

However, it is important to note that the number of layers, L, required for such implementations generally grows exponentially with the number of qubits. The reader is referred to the Recommended Resources for details.

21.3.2 Representing Quantum Circuits as Tensor Networks

A quantum circuit can be interpreted not only as an algorithmic description of the sequence of unitary transformations applied to the qudits to prepare a state $|\psi^{[S]}\rangle$, but also as a tensor network representation of the state $|\psi^{[S]}\rangle$. In this tensor network, the S open wires on the right side of the circuit directly correspond to the S open indices $x^{[S]}$ of the S-dimensional tensor $|\psi^{[S]}\rangle$. Through this representation, a quantum circuit provides insights into the structure of entanglement in the state $|\psi^{[S]}\rangle$.

Example 21.8

The three-qubit GHZ state $|\text{GHZ}^{[3]}\rangle$ can be generated via the quantum circuit in Figure 21.7. This quantum circuit encompasses two-qubit CNOT gates and a single-qubit Hadamard gate. The first layer of the circuit applies a Hadamard gate H to qubit 1; the second layer applies a CNOT gate on qubits 0 and 1, with qubit 1 serving as the controlling qubit; and the third layer applies a CNOT gate on qubits 1 and 2, with qubit 1 as the controlling qubit.

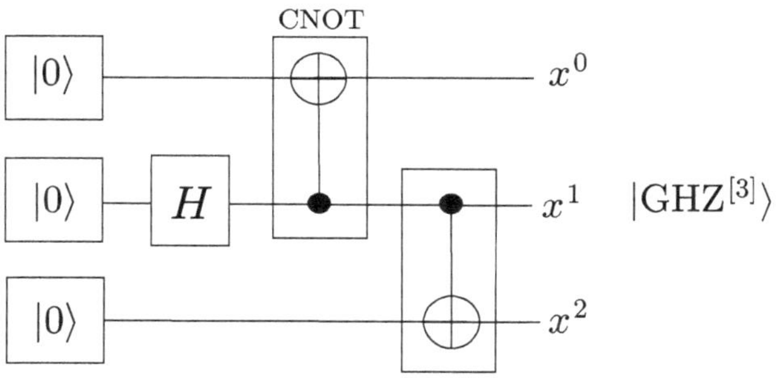

Figure 21.7 Quantum circuit preparing a three-qubit GHZ state.

Mathematically, the first layer produces the separable state

$$(I \otimes H \otimes I)(|0\rangle \otimes |0\rangle \otimes |0\rangle) = |0\rangle \otimes |+\rangle \otimes |0\rangle, \tag{21.11}$$

where I is the 2×2 identity matrix. Then, the second layer entangles the first two qubits, producing the state $|\Phi_+\rangle \otimes |0\rangle$ (see Section 15.2).

Finally, to elaborate on the impact of the last layer, recall from Section 15.2 that the CNOT gate with first qubit controlling the second can be expressed as the 4×4 unitary matrix $U = |0\rangle\langle 0| \otimes I + |1\rangle\langle 1| \otimes P_1$, where P_1 is the Pauli X gate. Therefore, the last layer, which entangles the last two qubits, yields the GHZ state as

$$\begin{aligned}
&(I \otimes U)(|\Phi_+\rangle \otimes |0\rangle) \\
&= (I \otimes |0\rangle\langle 0| \otimes I + I \otimes |1\rangle\langle 1| \otimes \sigma_1)\left(\frac{1}{\sqrt{2}}(|0\rangle \otimes |0\rangle \otimes |0\rangle + |1\rangle \otimes |1\rangle \otimes |0\rangle)\right) \\
&= \frac{1}{\sqrt{2}}(|0\rangle \otimes |0\rangle \otimes |0\rangle + |1\rangle \otimes |1\rangle \otimes |1\rangle) \\
&= |\text{GHZ}^{[3]}\rangle.
\end{aligned} \tag{21.12}$$

Apart from providing insights into the entanglement present in a joint state, the correspondence between quantum circuits and tensor networks paves the way for possible efficient calculations of the state $|\psi^{[S]}\rangle$ on a classical computer. In fact, for some quantum circuits, the contractions described by the tensor network may be carried out efficiently on a classical computer by following specific orders. In this regard, the choice of the contraction order is generally essential in order to ensure an efficient calculation on a classical computer. Section 21.4 describes a graphical methodology to evaluate the contractions in a quantum circuit.

21.4 ZX Calculus ⏭

Section 21.3 established that a quantum circuit preparing a state $|\psi^{[S]}\rangle$ can be directly interpreted as a tensor network representation for $|\psi^{[S]}\rangle$. Consequently, the quantum state $|\psi^{[S]}\rangle$ prepared by a quantum circuit can be evaluated by performing the tensor contractions corresponding to the bonds in the tensor network. This section introduces a diagrammatic approach, known as **ZX calculus**, which simplifies the computation of these tensor contractions and provides valuable insights into the properties of the prepared state $|\psi^{[S]}\rangle$.

The ZX calculus is a framework that expresses quantum gates using a small set of fundamental building blocks, each associated with simple contraction rules. This approach eliminates the need for explicit – symbolic or numerical – implementations of matrix contractions. Instead, all computations are carried out through diagrammatic transformations, making the analysis of quantum circuits potentially more intuitive and efficient.

In the ZX calculus, quantum states are represented using **unnormalized** tensor network representations, meaning that the 2-norm of the represented tensor does not necessarily equal 1. Note that this does not cause any ambiguity, since the final state can always be obtained – conceptually, if not computationally – by dividing the tensor by its norm. These tensor networks are constructed using tensors chosen from a predefined library, which is introduced next. Throughout this section, the focus is on qubit systems, and thus the dimension is set to $D = 2$.

21.4.1 Tensor Types in the ZX Calculus

The ZX calculus employs two primary types of tensors to represent a wide range of quantum states and gates. These are known as **Z-spiders** and **X-spiders**, and they are represented by white and gray circles, respectively. Regarding the terminology adopted by the ZX calculus, the "legs" of the spiders correspond to the indices of the tensors, representing the connections to other tensors or open indices.

In addition to Z-spiders and X-spiders, which will be detailed below, the ZX calculus also uses the Hadamard gate, which is expressed in unnormalized form as (see Example 10.5)

$$H = \begin{bmatrix} 1 & 1 \\ 1 & -1 \end{bmatrix}. \tag{21.13}$$

The Hadamard gate is represented in the ZX calculus by a box with two open two-dimensional indices corresponding to the row and column indices of matrix H. A summary of the tensors used in the ZX calculus and of their graphical representations can be found in Figure 21.8.

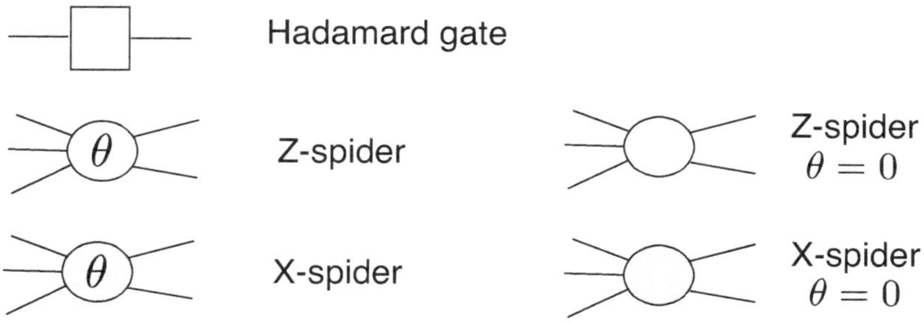

Figure 21.8 Library of tensors used in the ZX calculus.

21.4.2 Z-Spiders

The first main type of tensor used in ZX calculus is given by Z-spiders.

> **Definition 21.4 Z-spider**
> A Z-spider with phase θ, denoted as $Z(\theta)$, is an order-L tensor with dimensions all equal to 2 that is defined as
>
> $$Z(\theta) = |0\rangle^{\otimes L} + \exp(i\theta)|1\rangle^{\otimes L}. \tag{21.14}$$

Accordingly, the Z-spider tensor $Z(\theta)$ returns 1 if all the L indices are set to 0; it returns $\exp(i\theta)$ if all the L indices are set to 1; and it returns 0 otherwise. As shown in Figure 21.8, a Z-spider is represented as a white circle, and the phase θ is reported in the middle of the circle. If no phase is indicated, the phase is implicitly assumed to equal zero, i.e., $\theta = 0$.

As illustrated in Figure 21.9, Z-spiders recover a number of tensors as special cases.

- **Copy tensor:** For $\theta = 0$, the Z-spider corresponds to the copy tensor δ, as it returns 1 if all the indices are set to the same value – 0 or 1 – and it returns 0 otherwise.
- **Hadamard-basis states:** With a single index, i.e., with $L = 1$, the Z-spider recovers the $|+\rangle$ state when $\theta = 0$ and the $|-\rangle$ state when $\theta = \pi$.
- **Bell states:** For $L = 2$, the Z-spider yields the Bell state $|\Phi_+\rangle$ for $\theta = 0$ and the Bell state $|\Phi_-\rangle$ for $\theta = \pi$. As illustrated in Figure 21.9, this implies that with $\theta = 0$ and $L = 2$, the Z-spider is equivalent to a plain wire.

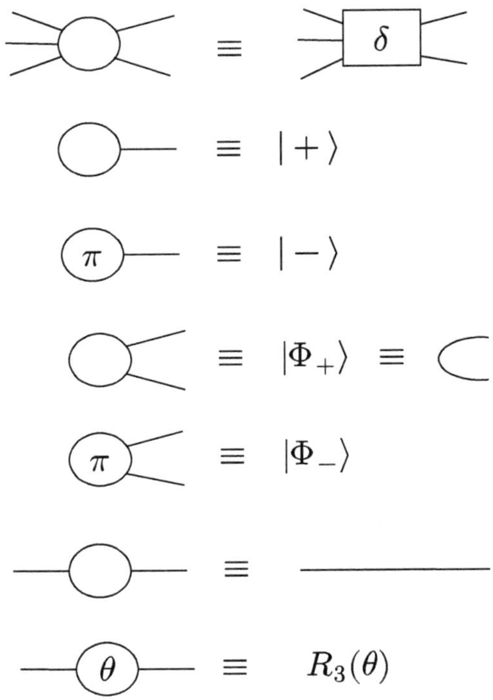

Figure 21.9 Z-spiders recover a number of tensors as special cases.

- **Pauli Z rotation:** With $L = 2$ legs, if formatted into a 2×2 matrix, the Z-spider is closely related to the Pauli Z rotation $R_3(\theta)$ presented in Section 10.3. More precisely, via matrization (see Section 17.6), when $L = 2$, one obtains

$$\begin{aligned} \text{mat}_0(Z(\theta)) &= |0\rangle\langle 0| + \exp(i\theta)|1\rangle\langle 1| \\ &= \begin{bmatrix} 1 & 0 \\ 0 & \exp(i\theta) \end{bmatrix} \\ &= \exp\left(i\frac{\theta}{2}\right) R_3(\theta), \end{aligned} \tag{21.15}$$

which reduces to the identity matrix, i.e., to a plain wire, for $\theta = 0$.

21.4.3 X-Spiders

The second main class of tensors used in the ZX calculus is given by X-spiders.

Definition 21.5 X-spider

An X-spider with phase θ is an order-L tensor $X(\theta)$ with all dimensions equal to 2 that is defined as

$$X(\theta) = |+\rangle^{\otimes L} + \exp(i\theta)|-\rangle^{\otimes L}. \tag{21.16}$$

Accordingly, the X-spider tensor $X(\theta)$ has the same form as a Z-spider, but with the computational-basis vectors replaced by the Hadamard-basis vectors. As a result, it can be checked that the tensor returns the value $1 + \exp(i\theta)$ if the parity of the indices is even, i.e., if there is an even number of L indices set to 1; while it returns $1 - \exp(i\theta)$ otherwise. Note that, unlike the Z-spider, the X-spider is dense rather than sparse, in the sense that, in general, all of its entries are non-zero.

As shown in Figure 21.8, an X-spider is represented as a gray circle, in which the phase θ is reported in the middle of the circle. As for the Z-spider, if no phase is indicated, the phase is implicitly assumed to equal zero, i.e., $\theta = 0$. In a manner similar to Z-spiders, as shown in Figure 21.10, X-spiders recover a number of important tensors as special cases.

- **Computational-basis states:** For $L = 1$, the X-spider corresponds to the $|0\rangle$ state when $\theta = 0$, and to the $|1\rangle$ state when $\theta = \pi$.
- **Bell states:** When $L = 2$, the X-spider recovers the Bell state $|\Phi_+\rangle$ for $\theta = 0$, like the Z-spider, while yielding the Bell state $|\Psi_+\rangle$ for $\theta = \pi$. As for the Z-spider, this implies that with $\theta = 0$ and $L = 2$, the X-spider is equivalent to a plain wire.
- **Pauli X rotation:** For $L = 2$, when formatted into a 2×2 matrix, the X-spider is closely related to the Pauli X rotation $R_1(\theta)$, as one has the equalities

$$\begin{aligned} \mathrm{mat}_0(X(\theta)) &= |+\rangle\langle+| + \exp(i\theta)|-\rangle\langle-| \\ &= \exp\left(-i\frac{\theta}{2}\right) R_1(\theta), \end{aligned} \tag{21.17}$$

and thus it corresponds to the identity, i.e., to a plain wire, when $\theta = 0$.

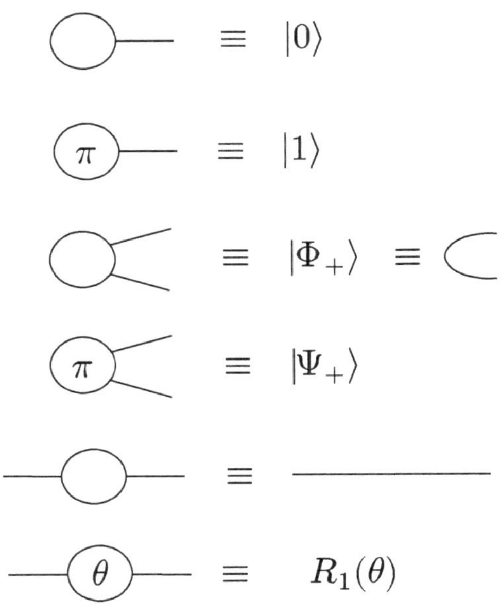

Figure 21.10 X-spiders recover a number of tensors as special cases.

21.4.4 Completeness of the ZX Calculus

X-spiders, Z-spiders, and Hadamard gates are sufficient to express *any* pure state. For this reason, the ZX calculus is said to be **complete**. To prove the completeness of the ZX calculus, recall from the previous section that any unitary matrix can be expressed using quantum circuits constructed from single-qubit unitaries and from the CNOT gate. Accordingly, proving completeness requires demonstrating that the tensors in the ZX library are sufficient to represent both arbitrary single-qubit unitaries and the CNOT gate.

Start from single-qubit gates. As detailed in Section 10.2, any single-qubit gate can be written, apart from phase terms, as a rotation $R_3(\vartheta_1)R_1(\vartheta_2)R_3(\vartheta_3)$ for some angles ϑ_1, ϑ_2, and ϑ_3. Setting $\vartheta_1 = \theta_1$, $\vartheta_2 = \theta_2$, and $\vartheta_3 = \theta_3$, the corresponding tensor network using X-spiders and Z-spiders is shown in Figure 21.11(a). Note that, being a single-qubit gate, the Hadamard gate can also be expressed in this way, but it is convenient to use a separate symbol for the Hadamard gate due to its frequent use in ZX-calculus diagrams.

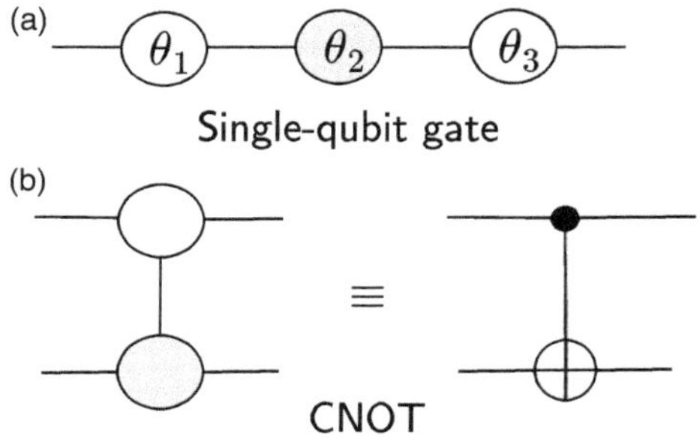

Figure 21.11 Tensor network representations using X-spiders and Z-spiders of (a) a single-qubit gate and (b) the CNOT gate.

Finally, the CNOT gate can also be expressed using X-spiders and Z-spiders, as shown in Figure 21.11(b). To check this, recall that the CNOT gate is described by the matrix

$$U = |0\rangle\langle 0| \otimes I + |1\rangle\langle 1| \otimes P_1 = \begin{bmatrix} 1 & 0 & 0 & 0 \\ 0 & 1 & 0 & 0 \\ 0 & 0 & 0 & 1 \\ 0 & 0 & 1 & 0 \end{bmatrix}, \tag{21.18}$$

which has the only non-zero entries $U_{00,00} = U_{01,01} = U_{10,11} = U_{11,10} = 1$. The tensor network made of X-spiders and Z-spiders in Figure 21.11 yields, for each entry (x^0x^1, y^0y^1), the output

$$\sum_{b\in[2]} Z_{x^0y^0b}(0)X_{x^1y^1b}(0) = \sum_{b\in[2]} \delta_{x^0y^0b}X_{x^1y^1b}(0), \tag{21.19}$$

where $b \in [2]$ represents the bond between the spiders, and the equality follows by replacing the Z-spider with the equivalent order-3 copy tensor δ. One can readily check that this coincides with the matrix representation of a CNOT in (21.18).

Example 21.9

The CZ gate (see Section 15.2) can be expressed using the ZX calculus as in Figure 21.12 by using Z-spiders and a Hadamard gate.

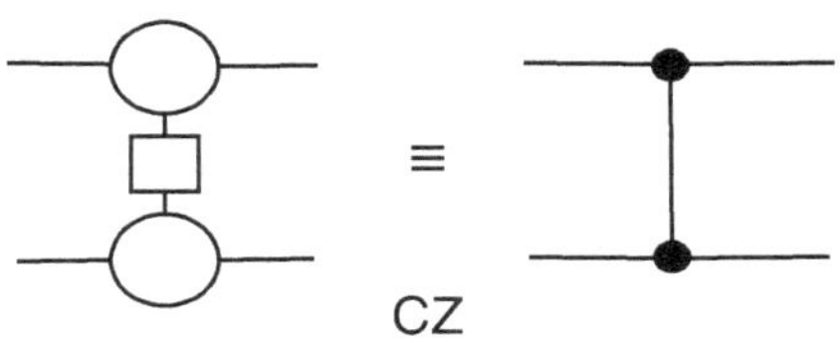

Figure 21.12 Tensor network representation using X-spiders and Z-spiders of the CZ gate.

21.4.5 Contraction Rules

The ZX calculus provides a set of diagrammatic rules that can be used to simplify a tensor network involving Hadamard gates, X-spiders, and Z-spiders. These rules describe the effect of contractions purely in terms of graphical modifications to the tensor network. The following are two important examples of rules applied by the ZX calculus. The reader is referred to the Recommended Resources for proofs and details.

Key Result 21.2 Two diagrammatic contraction rules in the ZX calculus

The following diagrammatic contraction rules can be applied to transform the tensor network representing a quantum circuit into an equivalent tensor network.

Rule 1 – Spider fusion: Replace connected spiders of the same type – X or Z – with a spider of the same type having the sum of the phases of the original spiders.

Rule 2 – Color change: Remove the Hadamard gate connected to a leg of a spider, and replace it with Hadamard gates connected to all the other legs. Then, change the color of the spider.

The following examples illustrate the use of these rules.

Example 21.10

This example recovers the result, proved in Section 15.2, that a Bell state can be created via a Hadamard gate followed by a CNOT gate. Importantly, unlike the analytical proof in Section 15.2, the proof here only uses diagrammatic steps by leveraging the ZX calculus via the two rules listed above.

As illustrated in Figure 21.13, first the quantum circuit is represented using the ZX calculus diagrammatic notation. Observe that there are two X-spiders with zero phase (gray circle) for the two states $|0\rangle$, a Hadamard gate (rectangle), as well as an X-spider connected to a Z-spider. The two rules above are then applied in the sequence illustrated in the figure to reduce the circuit to a bent wire, which corresponds to a Bell state.

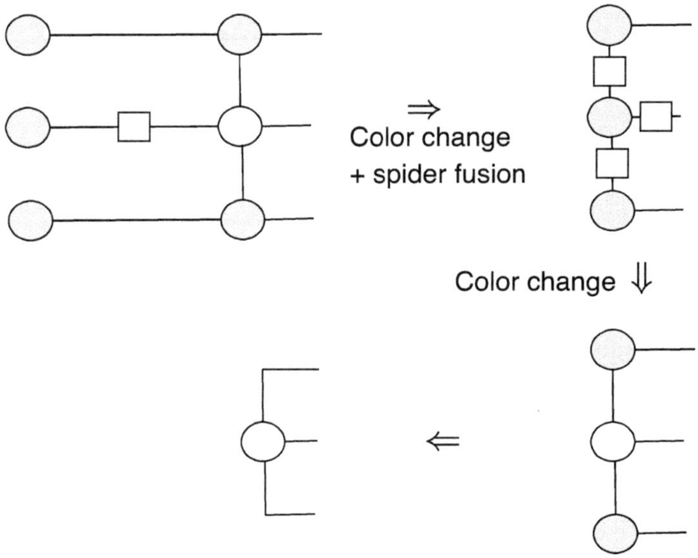

Figure 21.13 ZX calculus-based diagrammatic proof that a Bell state can be created via a Hadamard gate followed by a CNOT gate.

Note that, after the second spider fusion step, the proof uses the fact that two successive Hadamard gates cancel each other out, since matrix H is unitary and Hermitian. Furthermore, the last step uses the fact that X-spiders and Z-spiders with zero phase and two legs correspond to plain wires.

21.5 Summary

- An MPS is a quantum pure state $|\psi^{[S]}\rangle$ with each entry given by the matrix product $\psi_{x[S]} = T^0_{x^0} T^1_{x^1} \cdots T^{S-1}_{x^{S-1}}$ or $\psi_{x[S]} = \mathrm{tr}(T^0_{x^0} T^1_{x^1} \cdots T^{S-1}_{x^{S-1}})$, with bonds between matrices of dimension χ.
- MPSs satisfy a one-dimensional area law for the entanglement, in the sense that the entanglement between a group of adjacent qudits and the rest of the system depends solely on the size of the boundary between the two groups, and not on the number of qudits in each group.
- Quantum circuits provide another way to express a quantum state via a tensor network, while also representing a quantum algorithm preparing the state.

- A quantum circuit applies quantum gates sequentially, with single- and two-qubit gates being sufficient to describe any unitary transformation.
- The ZX calculus supports diagrammatic transformations in tensor networks, making it possible to evaluate the effect of contractions without the need for analytical or numerical tools.

21.6 Recommended Resources

Reference [1] explores how MPSs can be utilized to simulate and approximate quantum systems efficiently, highlighting the relationship between entropy scaling and the feasibility of such simulations. The work [2] provides a comprehensive introduction to tensor network methods, including MPS and PEPS.

References [3–6] cover quantum circuit complexity and the resources required for quantum state preparation. For an illustration of the role of the order of contractions in the efficiency of the simulation of a quantum circuit on a classical computer, the reader is referred to [7].

In-depth introductions to the ZX calculus are provided by [8, 9].

PROBLEMS

21.1 Draw the tensor network for the density matrix of an MPS.

21.2 Consider a circular MPS (21.5) for $S = 4$ qubits, i.e., the entries of the tensor are given by $\psi_{x^{[4]}} = \mathrm{tr}(T^0_{x^0} T^1_{x^1} T^2_{x^2} T^3_{x^3})$. The 2×2 matrices in the MPS are $T^s_0 = (0.2)^\eta |0\rangle\langle 0|$ and $T^s_1 = (0.8)^\eta |1\rangle\langle 1|$ for $s \in [4]$. (a) Find the exponent η that ensures the vector $|\psi^{[4]}\rangle$ is a valid quantum state. (b) Evaluate the entries of the resulting state vector. (c) Calculate an upper bound on the entanglement entropy $\mathrm{E}(X^{0,1};X^{2,3})_{|\psi^{[4]}\rangle}$ using the bond dimension. (d) Can the state $|\psi^{[4]}\rangle$ be globally maximally entangled?

21.3 This problem aims to show that for an MPS (21.1) with bond dimension χ, the entanglement entropy across a cut $\mathcal{S} = [s]$ is upper bounded by $\log_2(\chi)$. Note that the cut separates the lower qudits indexed by $[s] = 0, 1, \ldots, s-1$ from the upper qudits $s, s+1, \ldots, S-1$. (a) To this end, consider the bipartite state $|\psi^{\mathcal{S},\bar{\mathcal{S}}}\rangle$, and show that its tensor rank cannot exceed the bond dimension χ. *Hint:* Use the tensor network representation to show that the state can be represented as the linear combination of at most χ states that are uncorrelated across the cut $\mathcal{S}$. (b) As a result, argue that the Schmidt rank for the bipartite state $|\psi^{\mathcal{S},\bar{\mathcal{S}}}\rangle$ cannot be larger than χ, and complete the proof.

21.4 Prove that the quantum conditional mutual information

$$\begin{aligned}\mathrm{I}(X;Y|Z)_{\rho^{XYZ}} &= \mathrm{I}(X;YZ)_{\rho^{XYZ}} - \mathrm{I}(X;Z)_{\rho^{XZ}} \\ &= \mathrm{H}(YZ)_{\rho^{YZ}} + \mathrm{H}(XZ)_{\rho^{XZ}} - \mathrm{H}(Z)_{\rho^{Y}} - \mathrm{H}(XYZ)_{\rho^{XYZ}} \quad (21.20)\end{aligned}$$

is non-negative. To this end, prove the steps

$$\begin{aligned} \mathrm{I}(X;Y|Z)_{p^{XYZ}} &= \mathrm{D}(\rho^{XYZ}||\rho^X \otimes \rho^{YZ}) - \mathrm{D}(\rho^{XZ}||\rho^X \otimes \rho^Z) \\ &= \mathrm{D}(\rho^{XYZ}||\rho^X \otimes \rho^{YZ}) - \mathrm{D}(\mathrm{tr}_Z(\rho^{XYZ})||\mathrm{tr}_Z(\rho^X \otimes \rho^{YZ})) \\ &\geq 0. \end{aligned} \tag{21.21}$$

Hint: The partial trace is a quantum channel.

21.5 Given two circular MPSs (21.5) with bond dimensions χ and χ', show that any state obtained by linearly combining the two states can have a bond dimension no larger than the sum $\chi + \chi'$.

21.6 Consider a system with four qubits in a state $|\Phi_+\rangle \otimes |\Phi_+\rangle$. Assume that the first qubit is at Alice, the second and third qubits are at Bob, and the last qubit is at Charlie. We wish to fuse the two qubits at Bob, so that Alice, Bob, and Charlie share a three-qubit GHZ state. To this end, Bob applies a CNOT gate in which the third qubit acts as the control qubit for the second qubit. Then, Bob measures the second qubit in the computational basis. Show that by applying a suitable unitary matrix to the first qubit as a function of the classical measurement output, the first, third, and fourth qubits are left in the GHZ state.

21.7 Using the ZX calculus, show that a CNOT gate with Hadamard gates acting on the controlled qubit before and after the CNOT gate is equivalent to a CZ gate.

21.8 Using the ZX calculus, prove that the quantum circuit in Figure 21.7 prepares the GHZ state. *Hint:* Use the fact that a Z-spider with zero phase and $L = 3$ legs corresponds to a GHZ state.

Appendix

21.A Proof of the Bound (21.8)

The bound (21.8) can be proved as illustrated in Figure 21.14. To this end, as shown in Figure 21.14(a), given a group $\mathcal{S}$ of adjacent qudits, one combines all tensors into two tensors, one with indices $x^{\mathcal{S}}$ and the other with indices $x^{\bar{\mathcal{S}}}$. Then, indices and tensors are rearranged so as to obtain the tensor representation in Figure 21.14(b). This tensor network corresponds to the amplitude matrix $\Psi^{\mathcal{S}}$ defined in Section 20.4.

The two indices, each of dimension χ, that connect the two tensors can be combined into a single bond of dimension χ^2. The resulting tensor network representation in Figure 21.14(c) shows that the amplitude matrix can be written as the contraction $\Psi^{\mathcal{S}} = \sum_{z\in[\chi^2]} A_{x^{\mathcal{S}}|z} B_{x^{\bar{\mathcal{S}}}|z}$ for order-2 tensors A and B with entries $A_{x^{\mathcal{S}}|z}$ and $B_{x^{\bar{\mathcal{S}}}|z}$.

This implies that the rank of the amplitude matrix, which corresponds to the Schmidt rank $r^{\mathcal{S}}$, is upper bounded by χ^2. Recalling that the entanglement entropy $\mathrm{E}(X^{\mathcal{S}};X^{\bar{\mathcal{S}}})_{|\psi[\mathcal{S}]\rangle}$ is upper bounded by $\log_2(r^{\mathcal{S}})$ (see (20.54)) concludes the proof.

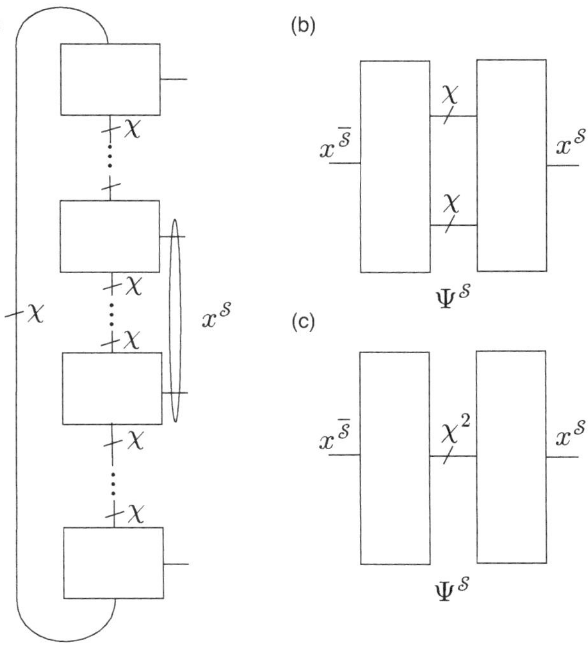

Figure 21.14 Illustration of the proof of the bound (21.8).

REFERENCES

[1] N. Schuch, M. M. Wolf, F. Verstraete, and J. I. Cirac, "Entropy scaling and simulability by matrix product states," *Physical Review Letters*, vol. 100, no. 3, art. 030504, 2008.

[2] R. Orús, "A practical introduction to tensor networks: Matrix product states and projected entangled pair states," *Annals of Physics*, vol. 349, pp. 117–158, 2014.

[3] M. A. Nielsen and I. L. Chuang, *Quantum Computation and Quantum Information*. Cambridge: Cambridge University Press, 2000.

[4] D. N. Page, "Average entropy of a subsystem," *Physical Review Letters*, vol. 71, no. 9, pp. 1291–1294, 1993.

[5] J. Haferkamp, P. Faist, N. B. Kothakonda, J. Eisert, and N. Yunger Halpern, "Linear growth of quantum circuit complexity," *Nature Physics*, vol. 18, no. 5, pp. 528–532, 2022.

[6] M. Plesch and Č. Brukner, "Quantum-state preparation with universal gate decompositions," *Physical Review A*, vol. 83, no. 3, art. 032302, 2011.

[7] J. Tindall, M. Fishman, M. Stoudenmire, and D. Sels, "Efficient tensor network simulation of IBM's eagle kicked Ising experiment," *PRX Quantum*, vol. 5, 2024.

[8] J. van de Wetering, "ZX-calculus for the working quantum computer scientist," *arXiv*, 2020. [Online]. Available at: https://arxiv.org/abs/2012.13966

[9] B. Coecke and S. Gogioso, *Quantum in Pictures: A New Way to Understand the Quantum World*. Cambridge: Quantinuum, 2023.

Postface

As of this writing, "Claude" is mostly recognized as the name of an AI chatbot. When asked about the origin of its name, Claude responded:

It's possible I was named after Claude Shannon, given his significant contributions to information theory and early AI, but that's just speculation on my part.

With this in mind, it feels appropriate to conclude this book by asking Claude the following question:

What would Claude Shannon work on today?

This is what Claude had to say:

If Claude Shannon were alive and working today, he would likely be at the forefront of several cutting-edge fields that combine his interests in mathematics, information theory, and technology. Here are some areas he might focus on.

- ***Quantum Information Theory:*** *Given his foundational work in classical information theory, Shannon would likely be deeply interested in its quantum counterpart. He might work on quantum error correction, quantum cryptography, or developing new mathematical frameworks for quantum communication.*
- ***Machine Learning and AI:*** *Shannon's work on learning machines and his interest in artificial intelligence would probably lead him to explore modern deep learning techniques. He might focus on:*
  - *Developing more efficient neural network architectures*
  - *Improving the theoretical understanding of why deep learning works so well*
  - *Creating new learning algorithms based on information-theoretic principles*
- ***Information-Theoretic Approaches to Biology:*** *Shannon was interested in the application of information theory to biology. Today, he might work on:*
  - *Analyzing genetic codes using information theory*
  - *Studying neural information processing in the brain*
  - *Developing models for complex biological systems*
- ***Advanced Cryptography:*** *Given his work during World War II and his interest in secure communication, Shannon might be working on post-quantum cryptography or developing new encryption methods based on chaos theory or other advanced mathematical concepts.*
- ***Network Information Theory:*** *Shannon might be interested in extending his work to modern network scenarios, potentially developing new theories for wireless networks, the Internet of Things, or decentralized systems like blockchain.*
- ***Algorithmic Information Theory:*** *He might explore the connections between information theory and computational complexity, perhaps working on problems related to Kolmogorov complexity or algorithmic randomness.*
- ***Information-Based Physics:*** *Shannon might be intrigued by attempts to reformulate physics in terms of information, such as in quantum information theory or theories that treat information as fundamental to the nature of reality.*

- ***Advanced Data Compression:*** *Given the explosion of data in the modern world, Shannon might work on developing new, more efficient data compression techniques, perhaps based on deep learning or other advanced algorithms.*
- ***Artificial General Intelligence (AGI):*** *Shannon's broad interests and his early work on thinking machines might lead him to tackle the challenge of creating more general, human-like artificial intelligence.*
- ***Interdisciplinary Applications:*** *Known for his ability to connect disparate fields, Shannon might work on applying information theory to areas like economics, climate science, or social network analysis. Given Shannon's track record of groundbreaking work and his penchant for tackling fundamental problems, he would likely be working on something that bridges multiple disciplines and pushes the boundaries of our current understanding of information and computation. His approach would probably involve developing new mathematical frameworks to address these complex, modern challenges.*

Index

For EU product safety concerns, contact us at Calle de José Abascal, 56–1°,
28003 Madrid, Spain or eugpsr@cambridge.org.

www.ingramcontent.com/pod-product-compliance
Ingram Content Group UK Ltd.
Pitfield, Milton Keynes, MK11 3LW, UK
UKHW050323250226
468380UK00008B/88
9781009579520